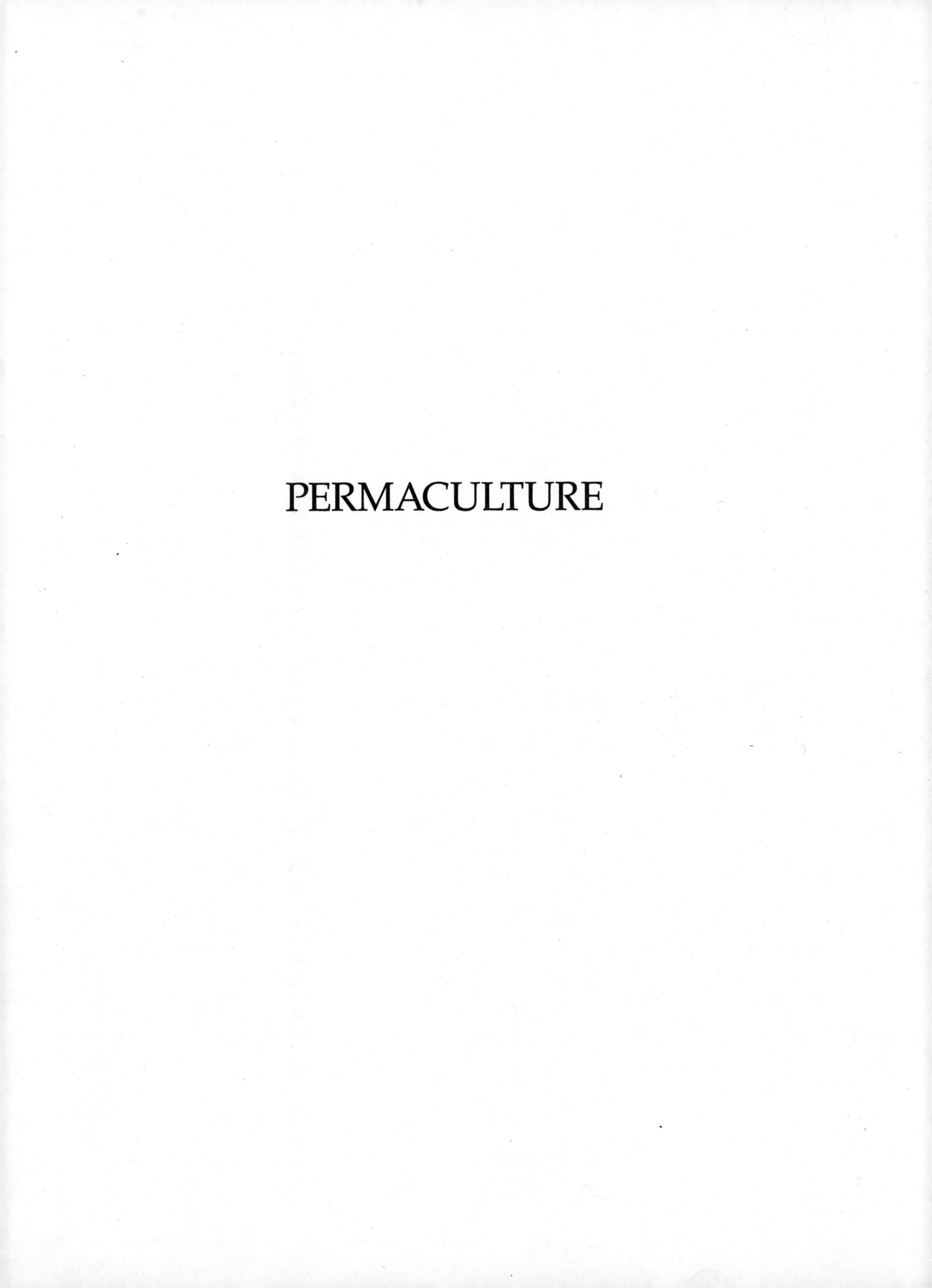

PERMACULTURE

PERMACULTURE

A Practical Guide for a Sustainable Future

Bill Mollison

ISLAND PRESS
WASHINGTON, D.C. ❑ COVELO, CALIFORNIA

ABOUT ISLAND PRESS

Island Press, a nonprofit organization, publishes, markets, and distributes the most advanced thinking on the conservation of our natural resources—books about soil, land, water, forests, wildlife, and hazardous and toxic wastes. These books are practical tools used by public officials, business and industry leaders, natural resource managers, and concerned citizens working to solve both local and global resource problems.

Founded in 1978, Island Press reorganized in 1984 to meet the increasing demand for substantive books on all resource-related issues. Island Press publishes and distributes under its own imprint and offers these services to other nonprofit organizations.

Support for Island Press is provided by Apple Computers Inc., Mary Reynolds Babcock Foundation, Geraldine R. Dodge Foundation, The Educational Foundation of America, The Charles Engelhard Foundation, The Ford Foundation, Glen Eagles Foundation, The George Gund Foundation, William and Flora Hewlett Foundation, The Joyce Foundation, The J. M. Kaplan Fund, The John D. and Catherine T. MacArthur Foundation, The Andrew W. Mellon Foundation, The Joyce Mertz-Gilmore Foundation, The New-Land Foundation, The Jessie Smith Noyes Foundation, The J.N. Pew, Jr. Charitable Trust, Alida Rockefeller, The Rockefeller Brothers Fund, The Florence and John Schumann Foundation, The Tides Foundation, and individual donors.

Island Press, Suite 300, 1718 Connecticut Avenue NW, Washington, D.C. 20009.

Library of Congress Cataloging-in-Publication Data

Mollison, B. C.
Permaculture : a practical guide for a sustainable future / by Bill Mollison ; illustrated by Andrew Jeeves.
p. cm.
Includes bibliographical references.
ISBN 1-55963-048-5
1. Permaculture—Handbooks, manuals, etc. I. Title.
S494.5.P47M64 1990
631.5'8—dc20 90-32109
CIP

Printed on recycled, acid-free paper

Manufactured in the United States of America

10 9 8 7 6 5 4 3 2 1

CONTENTS

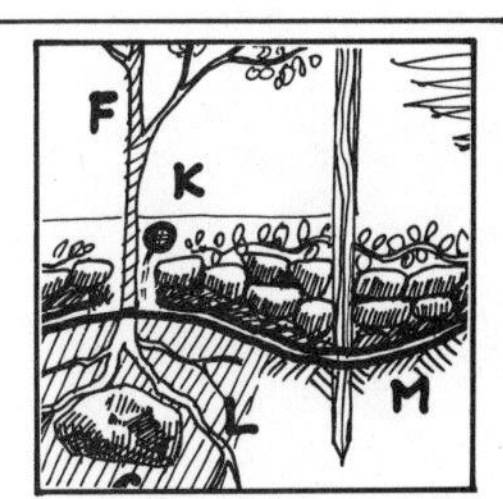
F
K
M

EARTH BANK
EARTH BANK

PREFACE

To many of us who experienced the ferment of the late 1960's, there seemed to be no positive direction forward, although almost everybody could define those aspects of the global society that they rejected, and these include military adventurism, the bomb, ruthless land exploitation, the arrogance of polluters, and a general insensitivity to human and environmental needs.

From 1972-1974, I spent some time (latterly with David Holmgren) in developing an interdisciplinary earth science (permaculture) with a potential for positivistic, integrated, and global outreach. It was January 1981 before the concept of permaculture seemed to have matured sufficiently to be taught as an applied design system, when the first 26 students graduated from an intensive 140–hour lecture series. Today, we can count thousands of people who have attended permaculture design courses, workshops, lectures, and seminars. Graduates now form a loose global network, and are effectively acting in many countries. The permaculture movement has no central structure, but rather a strong sense of shared work. Everybody is free to act as an individual, to form a small group, or to work within any other organisation. We cooperate with many other groups with diverse beliefs and practices; our system includes good practices from many disciplines and systems, and offers them as an integrated whole.

Great changes are taking place. These are not as a result of any one group or teaching, but as a result of millions of people defining one or more ways in which they can conserve energy, aid local self-reliance, or provide for themselves. All of us would acknowledge our own work as modest; it is the totality of such modest work that is impressive. There is so much to do, and there will never be enough people to do it. We must all try to increase our skills, to model trials, and to pass on the results. If a job is not being done, we can form a small group and do it (when we criticise others, we usually point the finger at ourselves!) It doesn't matter if the work we do carries the "permaculture" label, just that we do it.

By 1984, it had become clear that many of the systems we had proposed a decade earlier did, in fact, constitute a sustainable earth care system. Almost all that we had proposed was tested and tried, and where the skills and capital existed, people could make a living from products derived from stable landscapes, although this is not a primary aim of permaculture, which seeks first to stabilise and care for land, then to serve household regional and local needs, and only thereafter to produce a surplus for sale or exchange.

In 1984, we held our first international permaculture conference, and awarded about 50 applied diplomas to those who had served two years of applied work since their design course. Those of us who belong to the permaculture family have cause to be proud, but not complacent. Work has scarcely begun, but we have a great team of people which increases in numbers daily. To empower the powerless and create "a million villages" to replace nation-states is the only safe future for the preservation of the biosphere. Let interdependence and personal responsibility be our aims.

AUTHOR'S NOTE

This volume was written for teachers, students, and designers; it follows on and greatly englarges on the initial introductory texts*Permaculture One* (1978) and *Permaculture Two* (1979), both of which are still in demand a decade after publication. Very little of the material in this book is reproduced from the foundation texts.

Each volume of this work carries a surcharge of 50¢ which will be paid by Tagari Publications to the Permaculture Institute. The Institute (a public trust) holds the funds so generated in trust for tree-planting, and from time to time releases monies to selected groups who are active in permanent reafforestation. In this way, both publisher and readers can have a clear conscience about the use of the paper in this volume, or in any book published by Tagari Publications. Our trust funds are open to receive any such levy from other ethical publishers.

PERMACULTURE DEFINED AND ITS USE

Permaculture is a word coined by the author. Its copyright is vested in the Permaculture Institutes and their College of Graduates, and is guarded by them for the purposes of consistent education.

Permaculture (**perma**nent agri**culture**) is the conscious design and maintenance of agriculturally productive ecosystems which have the diversity, stability, and resilience of natural ecosystems. It is the harmonious integration of landscape and people providing their food, energy, shelter, and other material and non–material needs in a sustainable way. Without permanent agriculture there is no possibility of a stable social order.

Permaculture design is a system of assembling conceptual, material, and strategic components in a pattern which functions to benefit life in all its forms.

The philosophy behind permaculture is one of working with, rather than against, nature; of protracted and thoughtful observation rather than protracted and thoughtless action; of looking at systems in all their functions, rather than asking only one yield of them; and of allowing systems to demonstrate their own evolutions.

The word "permaculture" can be used by anybody adhering to the ethics and principles expressed herein. The only restriction on use is that of teaching; only graduates of a Permaculture Institute can teach "permaculture", and they adhere to agreed–on curriculae developed by the College of Graduates of the Institutes of Permaculture.

CONVENTIONS USED

References and Abbreviations: Minor references are given in the text, and those useful to chapter contents only are located at the close of those chapters. Key references are assembled at the close of the book, and are superscripted as numbers in the text.

Seasons and Directions: So that the text and figures are useful and readable in both hemispheres, I have used the words "sun-side" or "sunwards", and "shade-side" or "polewards" rather than south and north, and converted months to seasons as below:

Used Here		Northern Hemisphere	Southern Hemisphere
Summer	early	June	December
	mid	July	January
	late	August	February
Autumn	early	September	March
	mid	October	April
	late	November	May
Winter	early	December	June
	mid	January	July
	late	February	August
Spring	early	March	September
	mid	April	October
	late	May	November

These may be further refined by the use of "first week of...". For the same reason, the symbol below is used in figures to indicate the sun direction rather than the north or south symbol:

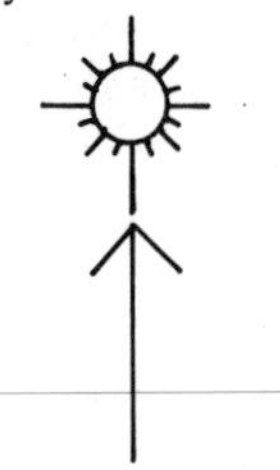

One hopes this prevents the problem of all those good North Americans wandering on the north face of their hills, looking for the sun, poised dangerously upside-down on the Earth as they are.

ACCESS TO INFORMATION

Material in this work can be fairly easily accessed in these ways: chapter and section contents are listed in the Table of Contents. Main subjects are listed in the Index. There is a list of the common and Latin names of plants used in the text located in the Appendix. Also located in the Appendix is a glossary of terms used; some few words are recently coined or (like "permaculture") are the conceit of the author.

To forestall needless correspondence, subscription to the *International Permaculture Journal* (113 Enmore Rd, Enmore, NSW 2042, Australia) gives information on permaculture themes, reviews recent publications, gives news of events, publishes a directory of permaculture centres, and has a host of other useful data. Further resources are also listed in the Appendix.

FOREWORD

We have been born into a time in human history during which a smaller percentage of our species is observing and responding to the immediate natural environment than ever before. Ecologist David Orr has called us the generation of "landscape illiteracy"; fewer and fewer people can read the landscape about them for indications of its health and hidden wealth. Sustainable agriculture advocate Wes Jackson has lamented the consequences of this syndrome as it affects the ecology of food production. More tractors, but fewer eyes and feet upon the land, mean that we fail to notice the impoverishment of the biological community and the loss of soil: an erosion of both genetic and environmental resources that have evolved in place over millennia. The ecological crisis, in a very real sense, is a failure of vision.

Permaculture: A Practical Guide for a Sustainable Future alleviates this failure by demonstrating a way of opening our eyes to the world again. By recommitting ourselves to the protracted and thoughtful observation of patterns in nature, we will discover ways to work with nature to which we might otherwise be blind. We will notice which birds are dispersing seeds of useful forest trees, so it may dawn on us that this process can be encouraged to promote reforestation. We may observe where storm runoff accumulates in the desert, so we can concentrate our initial plantings there, rather than having to drill boreholes and artificially irrigate plantings placed where they cannot easily sustain themselves. In short, this designer's manual does not give you cookbook recipes to fix quickly problems related to food production or natural resource conservation. Instead, it offers a constellation of principles and exercises so that each of us may see the earth freshly, and more fully understand the ecological context of our own actions.

In that sense, all permaculture texts have been profoundly revolutionary, for they empower people at the grass-roots level to take responsibility for the care of resources that our society has relegated to the professional elite. This book is not simply the distillation of knowledge from one more specialist. To the contrary, it is intended to stimulate your own thinking and your own experiments, so that you can become more knowledgeable about where you live than any outside authority could ever be. It is meant to make you work, and to work *with* the land. Complete answers are not in this book; they will emerge from your own dialectic with the earth.

Although Bill Mollison has defined a new interdisciplinary field of applied ecology and has become its acknowledged global expert, he seems better suited to be an earth-watcher, a backyard tinkerer, a revolutionary, a student of indigenous peoples and of natural landscapes than the head of a movement or an ecological guru. He is more prone to contemplation or working with the poor than administrating a new order. He is apt to change his mind as new information becomes available; he is prone to learn as much from mistakes made in early experiments as he is from his successes. He is also likely to laugh heartily when the world does not respond to plan, whether it is his plan or someone else's.

I have had the good fortune to host Bill twice on his travels through the Sonoran Desert, a bioregion that he finds fascinating for its diversity of drought-hardy life-forms and for its persistent indigenous cultures. One time, Bill came to look at the Desert Botanical Garden's agroecology project, where his permaculture students had helped me put into place several polycultures of native plants fed by desert storm runoff. He carefully looked over our attempts to use a plant guild that we had noticed in nature—wild chile peppers growing under the protective canopies of hackberry trees—as a model component of a desert agroforestry system. But his mind did not dwell for long on our reductionistic experiments; it raced on to imagine how these species might fit into a larger whole. Then he turned his attention to the cobblestone gabions we had built to divert floodwater from a usually dry (arroyo or wadi) watercourse into a shelterbelt of desert legumes.

"Bloody gabions!" he exclaimed. "They change the whole dynamics here, don't they? You can't just expect to put a few into place, and that's that! They've healed this gully so quickly that you're going to have to learn how to manage a much more shallow but wider flow the next time a hard rain falls. . . ."

His message was clear: One permacultural application does not necessarily make a long-term solution; it generates new challenges in an ever-evolving setting. Our role is to participate in ecological succession, not to freeze the landscape.

On another visit from Bill, we traveled into a desert landscape that was hot enough to thaw out any of our unconscious attempts to freeze it. As we drove out to O'odham Indian villages, remote from Arizona's artificial urban oases, Bill delighted us with his observations on similarities and differences between Australian and American deserts. He was so well versed in reading dryland vegetation that he could anticipate much of the historical ecology of overgrazing and arroyo-cutting that I could recall to him. At the same time, our conversations rambled through other topics: the O'odham floodwater fields where crops have been produced on one runoff event alone; living fences of cacti and ocotillo; hunter-gatherer traditions of periodically robbing the seed caches of small mammals, rather than killing the animals themselves.

Yet the moment I remember most vividly was when Bill learned that an elderly O'odham friend of mine had stopped farming due to the pain of arthritis affecting his shoulders and upper arms. Bill asked him politely if the man would mind a little temporary discomfort if Bill could help him through his arthritis. My friend replied that he would not, because he was eager to be physically active again. Then Bill leapt upon the task, first surveying the Indian man's shoulders for "the pattern," then massaging out the stiffness and choked circulation. Both Bill and the O'odham man appeared exhausted after the work-over. Following some subdued discussion, we left the village; I remember being unsure whether Bill's massage had actually helped the farmer. It was still too early to tell.

The next time I visited that particular desert village, I noticed my friend working out in his garden, his movements freer than I had seen them in some time. I greeted him.

"Where did you say that medicine man came from? Somewhere around Australia? If you ever talk to him, tell him I'm feeling better, and am trying to grow some native crops again. . . ."

Let us all learn from Bill Mollison's inspiration, to see new ways of healing ourselves, of healing the earth, and of wisely growing and gathering the earth's bounty.

Gary Paul Nabhan
Cofounder of Native Seeds/SEARCH

Chapter 1

INTRODUCTION

1.1

PERMACULTURE DESIGN PHILOSOPHY

Although this book is about design, it is also about values and ethics, and above all about a sense of peronal responsibility for earth care. I have written at times in the first person, to indicate that it is not a detached, impersonal, or even unbiased document. Every book or publication has an author, and what that author chooses to write about is subjective, for that person alone determines the subject, content, and the values expressed or omitted. I am not detached from, but have been passionately involved with this earth, and so herein give a brief vision of what I think can be achieved by anyone.

The sad reality is that we are in danger of perishing from our own stupidity and lack of personal responsibility to life. If we become extinct because of factors beyond our control, then we can at least die with pride in ourselves, but to create a mess in which we perish by our own inaction makes nonsense of our claims to consciousness and morality.

There is too much contemporary evidence of ecological disaster which appals me, and it should frighten you, too. Our consumptive lifestyle has led us to the very brink of annihilation. We have expanded our right to live on the earth to an entitlement to conquer the earth, yet "conquerors" of nature always lose. To accumulate wealth, power, or land beyond one's needs in a limited world is to be truly immoral, be it as an individual, an institution, or a nation–state.

What we have done, we can undo. There is no longer time to waste nor any need to accumulate more evidence of disasters; the time for action is here. I deeply believe that people are the only critical resource needed by people. We ourselves, if we organise our talents, are sufficient to each other. What is more, we will either survive together, or none of us will survive. To fight between ourselves is as stupid and wasteful as it is to fight during times of natural disasters, when everyone's cooperation is vital.

A person of courage today is a person of peace. The courage we need is to refuse authority and to accept only personally responsible decisions. Like war, growth at any cost is an outmoded and discredited concept. It is *our* lives which are being laid to waste. What is worse, it is our children's world which is being destroyed. It is therefore our only possible decision to withhold all support for destructive systems, and to cease to invest our lives in our own annihilation.

<u>The Prime Directive of Permaculture.</u>
The only ethical decision is to take responsibility for our own existence and that of our children.
Make it now.

Most thinking people would agree that we have arrived at final and irrevocable decisions that will abolish or sustain life on this earth. We can either ignore the madness of uncontrolled industrial growth and defence spending that is in small bites, or large catastrophes, eroding life forms every day, or take the path to life and survival.

Information and humanity, science and understanding, are in transition. Long ago, we began by wondering mainly about what is most distant; astronomy and astrology were our ancient pre-occupations. We progressed, millenia by millenia, to enumerating the wonders of earth. First by naming things, then by categorising them, and more recently by deciding how they function and what work they do within and without themselves. This analysis has resulted in the development of different sciences, disciplines and technologies; a welter of names and the sundering of parts; a proliferation of specialists; and a consequent inability to foresee results or to design integrated systems.

The present great shift in emphasis is on how the parts interact, how they work together with each other, how dissonance or harmony in life systems or society is achieved. Life *is* cooperative rather than competitive, and life forms of very different qualities may interact beneficially with one another and with their physical environment. Even "the bacteria... live by collaboration, accommodation, exchange, and barter" (Lewis Thomas, 1974).

Principle of Cooperation.
Cooperation, not competition, is the very basis of existing life systems and of future survival.

There are many opportunities to *create* systems that work from the elements and technologies that exist. Perhaps we should do nothing else for the next century but apply our knowledge. We already know how to build, maintain, and inhabit sustainable systems. Every essential problem is solved, but in the everyday life of people this is hardly apparent. The wage–slave, peasant, landlord, and industrialist alike are deprived of the leisure and the life spirit that is possible in a cooperative society which applies its knowledge. Both warders and prisoners are equally captive in the society in which we live.

If we question why we are here and what life is, then we lead ourselves into both science and mysticism which are coming closer together as science itself approaches its conceptual limits. As for life, it is the most open of open systems, able to take from the energy resources in time and to re–express itself not only as a lifetime but as a descent and an evolution.

Lovelock (1979) has perhaps best expressed a philosophy, or insight, which links science and tribal beliefs: he sees the earth, and the universe, as a thought process, or as a self–regulating, self–constructed and reactive system, creating and preserving the conditions that make life possible, and actively adjusting to regulate disturbances. Humanity however, in its present mindlessness, may be the one disturbance that the earth cannot tolerate.

> The Gaia hypothesis is for those who like to walk or simply stand and stare, to wonder about the earth and the life it bears, and to speculate about the consequences of our own presence here. It is an alternative to that pessimistic view which sees nature as a primitive force to be subdued and conquered. It is also an alternative to that equally depressing picture of our planet as a demented spaceship, forever travelling, driverless and purposeless, around an inner circle of the sun.
>
> (J.E. Lovelock, 1979).

For every scientific statement articulated on energy, the Aboriginal tribespeople of Australia have an equivalent statement on life. Life, they say, is a totality neither created nor destroyed. It can be imagined as an egg from which all tribes (life forms) issue and to which all return. The ideal way in which to spend one's time is in the perfection of the expression of life, to lead the most evolved life possible, and to assist in and celebrate the existence of life forms other than humans, for all come from the same egg.

The totality of this outlook leads to a meaningful daily existence, in which one sees each quantum of life eternally trying to perfect an expression towards a future, and possibly transcendental, perfection. It is all the more horrific, therefore, that tribal peoples, whose aim was to develop a conceptual and spiritual existence, have encountered a crude scientific and material culture whose life aim is not only unstated, but which relies on pseudo–economic and technological systems for its existence.

The experience of the natural world and its laws has almost been abandoned for closed, artificial, and meaningless lives, perhaps best typified by the dreams of those who would live in space satellites and abandon a dying earth.

I believe that unless we adopt sophisticated aboriginal belief systems and learn respect for all life, then we lose our own, not only as lifetime but also as any future opportunity to evolve our potential. Whether we continue, without an ethic or a philosophy, like abandoned and orphaned children, or whether we create opportunities to achieve maturity, balance, and harmony is the only real question that faces the present generation. This is the debate that must never stop.

A young woman once came to me after a lecture in which I wondered at the various concepts of afterlife; the plethora of "heavens" offered by various groups. Her view was,"This is heaven, right here. This is it. *Give it all you've got.*"

I couldn't better that advice. The heaven, or hell, we live in is of our own making. An afterlife, if such exists, can be no different for each of us.

1.2
ETHICS

In earlier days, several of us researched community ethics, as adopted by older religious and cooperative groups, seeking for universal principles to guide our own actions. Although many of these guidelines contained as many as 18 principles, most of these can be included in the three below (and even the second and third arise from the first):

The Ethical Basis of Permaculture
1. CARE OF THE EARTH: Provision for all life systems to continue and multiply.
2. CARE OF PEOPLE: Provision for people to access those resources necessary to their existence.
3. SETTING LIMITS TO POPULATION AND CONSUMPTION: By governing our own needs, we can set resources aside to further the above principles.

This ethic is a very simple statement of guidance, and serves well to illuminate everyday endeavours. It can be coupled to a determination to make our own way: to be neither employers nor employees, landlords nor tenants, but to be self–reliant as individuals and to cooperate as groups.

For the sake of the earth itself, I evolved a philosophy close to Taoism from my experiences with natural systems. As it was stated in *Permaculture Two*, it is a philosophy of working with rather than against nature; of protracted and thoughtful observation rather than protracted and thoughtless action; of looking at systems and people in all their functions, rather than asking only one yield of them; and of allowing systems to demonstrate their own evolutions. A basic question that can be asked in two ways is:

"What can I get from this land, or person?" or

"What does this person, or land, have to give if I cooperate with them?"

Of these two approaches, the former leads to war and waste, the latter to peace and plenty.

Most conflicts, I find, lay in how such questions are asked, and not in the answers to any question. Or, to put it another way, we are clearly looking for the right questions rather than for answers. We should be alert to rephrase or refuse the "wrong" question.

It has become evident that unity in people comes from a common adherence to a set of ethical principles, each of us perhaps going our own way, at our own pace, and within the limits of our resources, yet all leading to the same goals, which in our own case is that of a living, complex, and sustainable earth. Those who agree on such ethics, philosophies, and goals form a global nation.

How do a people evolve an ethic, and why should we bother to do so?

Humans are thinking beings, with long memories, oral and written records, and the ability to investigate the distant past by applying a variety of techniques from dendrochronology to archaeology, pollen analysis to the geological sciences. It is therefore evident that behaviours in the natural world which we thought appropriate at one time later prove to be damaging to our own society in the long–term (e.g. the effects of biocidal pest controls on soils and water).

Thus, we are led by information, reflection, and careful investigation to moderate, abandon, or forbid certain behaviours and substances that in the long–term threaten our own survival; *we act to survive.* Conservative and cautious rules of behaviour are evolved. This is a rational and sensible process, responsible for many taboos in tribal societies.

From a great many case histories we can list some rules of use, for example the RULE OF NECESSITOUS USE—that we leave any natural system alone until we are, of strict necessity, forced to use it. We may then follow up with RULES OF CONSERVATIVE USE—having found it necessary to use a natural resource, we may insist on every attempt to:

- Reduce waste, hence pollution;
- Thoroughly replace lost minerals;
- Do a careful energy accounting; and
- Make an assessment of the long–term, negative, biosocial effects on society, and act to buffer or eliminate these.

In practice, we evolve over time to various forms of *accounting for our actions*. Such accounts are fiscal, social, environmental, aesthetic, or energetic in nature, and all are appropriate to our own survival.

Consideration of these rules of necessitous and conservative use may lead us, step by step, to the basic realisation of our interconnectedness with nature; that we depend on good health in all systems for our survival. Thus, we widen the self–interested idea of human survival (on the basis of past famine and environmental disaster) to include the idea of "the survival of natural systems", and can see, for example, that when we lose plant and animal species due to our actions, we lose many survival opportunities. Our fates are intertwined. This process, or something like it, is common to every group of people who evolve a general earthcare ethic.

Having developed an earthcare ethic by assessing our best course for survival, we then turn to our relationships with others. Here, we observe a general rule of nature: that cooperative species and associations of self–supporting species (like mycorrhiza on tree roots) make healthy communities. Such lessons lead us to a sensible resolve to cooperate and take support roles in society, to foster an interdependence which values the individual's contributions rather than forms of opposition or competition.

Although initially we can see how helping our family and friends assists us in our own survival, we may evolve the mature ethic that sees all humankind as family, and all life as allied associations. Thus, we expand *people care* to *species care*, for all life has common origins. All are "our family".

We see how enlightened self–interest leads us to evolve ethics of sustainable and sensible behaviour. These then, are the ethics expressed in permaculture. Having evolved *ethics*, we can then devise *ways to apply them* to our lives, economies, gardens, land, and nature. This is what this book is about: the mechanisms of mature ethical behaviour, or how to act to sustain the earth.

There is more than one way to achieve permanence and stability in land or society. The peasant approach is well described by King[6] for old China. Here people hauled nutrients from canals, cesspits, pathways and forests to an annual grain culture. We could describe this as "feudal permanence" for its methods, period and politics. People were bound to the landscape by unremitting toil, and in service to a state or landlord. This leads eventually to famine and revolution.

A second approach is on permanent pasture of prairie, pampas, and modern western farms, where large holdings and few people create vast grazing leases, usually for a single species of animal. This is best described as "baronial permanence" with

(Continued on page 6...)

A. CONTEMPORARY/WESTERN AGRICULTURE **YEAR 1**

B. TRANSITIONAL AND CONSERVATION FARMING **YEAR 4**

C. PERMACULTURE; 70% cropland devoted to forage farming **YEAR 8**

FIGURE 1.1.
EVOLUTION FROM CONTEMPORY AGRICULTURE TO A PERMACULTURE.
I have attempted to cost contempory agriculture against a changeover to permaculture over a period of 3 8 years *(the transition period)*. Basic changes involve replacing animal forage grains with tree crop, increasing forest cover, adopting low to no tillage on remaining croplands, retrofitting the house for energy conservation, and producing some (if not all) fuel on the farm.

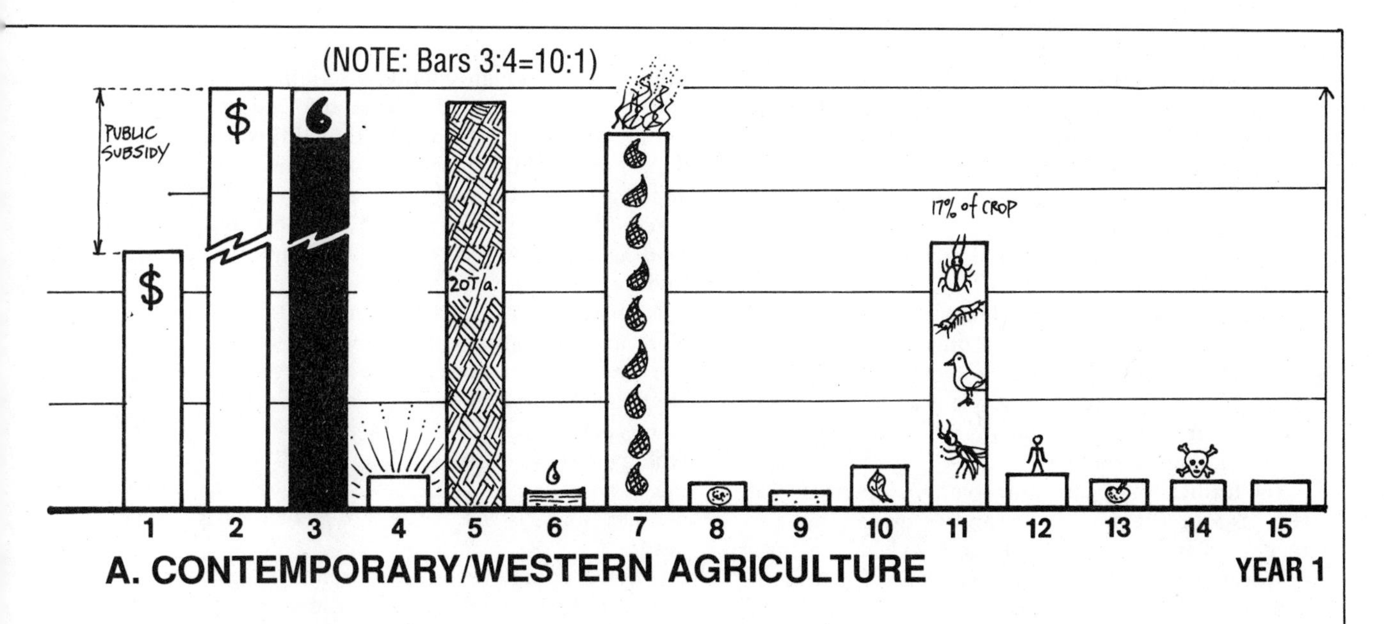

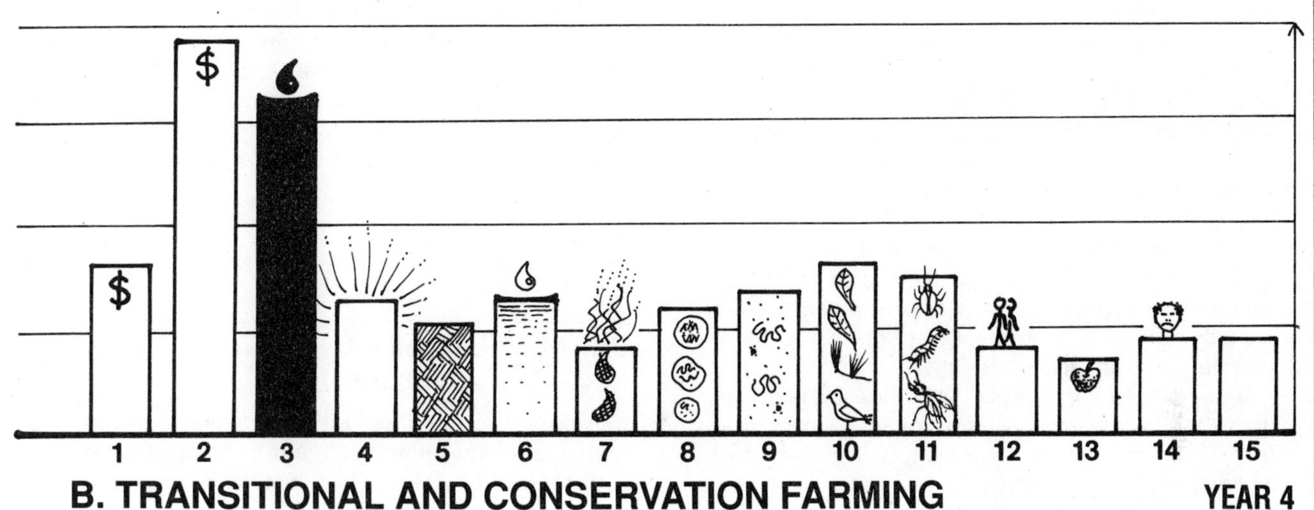

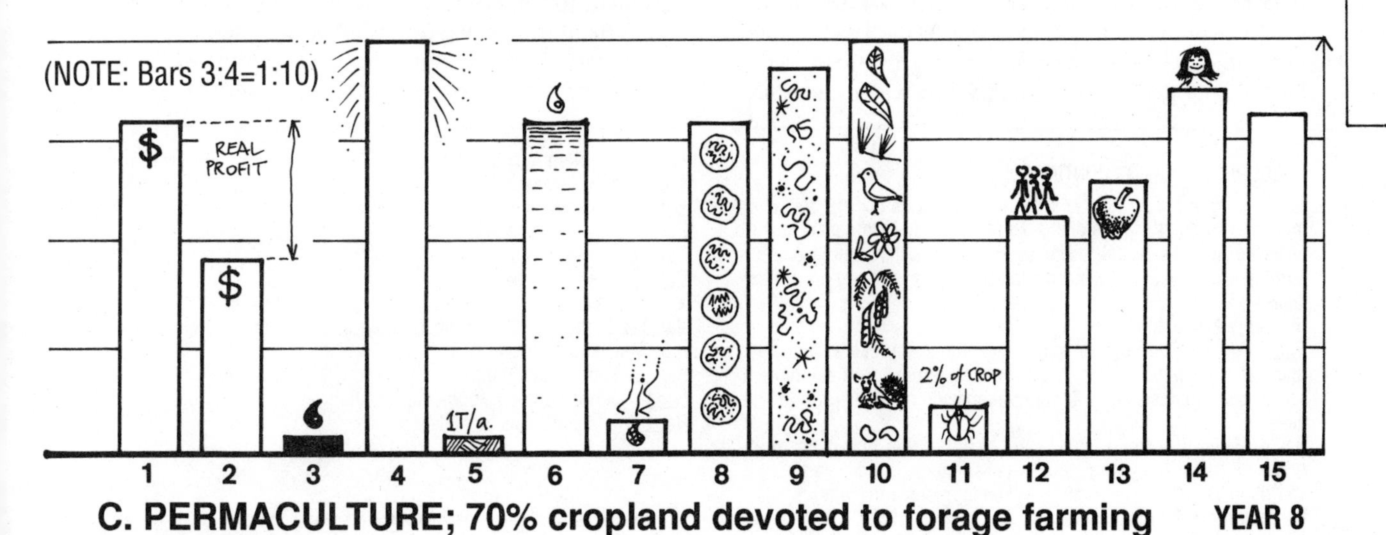

FIGURE 1.1 (Continued)
ANNOTATIONS TO THE BAR DIAGRAM; ACCOUNTING THE COSTS OF FARMING.
The accounting is in sections as follows:

I Cash (Dollar) Accounting.
Bar 1: Income from total product on the farm.

{continued next page....}

{...from page 3}

near–regal properties of immense extent, working at the lowest possible level of land use (pasture or cropland is the least productive use of land we can devise). Such systems, once mechanised, destroy whole landscapes and soil complexes. They can then best be typified as agricultural deserts.

Forests, not seen by industrial man as anything but wood, are another permanent agriculture. But they need generations of care and knowledge, and hence a tribal or communal reverence only found in stable communities. This then, is the communal permanence many of us seek: to be able to plant a pecan or citrus when we are old, and to know it will not be cut down by our children's children.

The further we depart from communal permanence, the greater the risk of tyranny, feudalism, and revolution and the more work for less yield. Any error or disturbance can then bring disaster, as can a drought year in a desert grain crop or a distant political decision on tariffs.

The real risk is that the needs of those people working "on the ground", the inhabitants, are overthrown by the needs (or greeds) of commerce and centralised power; that the forest is cut for warships or newspaper and we are reduced to serfs in a barren landscape. This has been the fate of peasant Europe, Ireland, and much of the third world.

The characteristic that typifies all permanent agricultures is that the needs of the system for energy are provided by that system. Modern crop agriculture is totally dependent on external energies—hence the oil problem and its associated pollution.

Figure 1.1 is a very simple but sufficient illustration of the case I am making. Selected forests not only yield more than annual crops, but provide a diverse nutrient and fuel resource for such crops.

Without permanent agriculture there is no possibility of a stable social order. Thus, the move from productive permanent systems (where the land is held in common), to annual, commercial agricultures where land is regarded as a commodity, involves a departure from a low– to a high–energy society, the use of land in an exploitative way, and a demand for external energy resources, mainly provided by the third world. People think I am slightly crazy when I tell them to go home and garden, or *not* to involve themselves in broadscale mechanised agriculture; but a little thought and reading will convince them that this is, in fact, the solution to many world problems.

What is now possible is a totally new synthesis of plant and animal systems, using a post–industrial or even computerised approach to system design, applying the principles of whole–system energy flows as devised by Odum (1971), and the principles of ecology as enunciated by Watt[13] and others. It is, in the vernacular, a whole new ball game to devise permaculture systems for local, regional, and personal needs.

Had we taught this approach from the beginning, we would all be in a stable and functional landscape, but our grandparents failed us, and (perhaps for lack of time or information) set up the present, and continuing, mis–designed households, towns, and cities. The concept of "free" energy put the final nail in the coffin of commonsense community, and enabled materialistic societies to rob distant peoples, oblivious of the inevitable accounting to come.

1.3

PERMACULTURE IN LANDSCAPE AND SOCIETY

As the basis of permaculture is beneficial design, it can be added to all other ethical training and skills, and has the potential of taking a place in all human endeavours. In the broad landscape, however, permaculture concentrates on already–settled areas and agricultural lands. Almost all of these need drastic rehabilitation and re–thinking. One certain result of using our skills to integrate food supply and

{...FIGURE 1.1 CAPTION CONTINUED}

Bar 2: Cost of producing that income in real terms (excess cost over income represents subsidies. Note that any farm "profits" are achieved by subsidy; the dollar costs do not balance until organic farming is achieved. Farm income is achieved by reducing production costs).

II Energy Accounting.

Bar 3: Oil (or calories) as machinery, fuels, fertilisers, biocides. Starts at 10:1 *against* (loss) in conventional farming, and can reach a 1:120 *gain* in conservation farming/permaculture with firewood and fuels.

Bar 4: Energy produced on farm; includes fuel oils from crop, firewood, calories in food produced (solar energy is a constant, but it contributes most energy in conservation farming/permaculture).

III Environmental Accounting.

Bar 5: Soil loss; includes humus loss and mineral nutrient loss.

Bar 6: Efficiency of water use and soil water storage.

Bar 7: Pollution produced (poisoning of atmosphere, soils, water) by fuels, biocides, and fertilisers. Soils are created in conservation farming/permaculture, water conserved, and pollutants removed.

IV Conservation Accounting; Life Form Richness.

Bar 8: Genetic richness in crops and livestock.

Bar 9: Soil life (biomass).

Bar 10: Forest biomass and wildlife richness.

Bar 11: Loss to pests.

V Social Accounting

Bar 12: Employment on farm (human design and/or skills replace most machine systems).

Bar 13: Food quality produced.

Bar 14: Human and environmental health.

Bar 15: Life quality, as "right livelihood".

Thus, it can be seen that a transition from contempory western agriculture to conservation farming and permaculture has most benefits for people and to other life forms; farming can become energy productive; and farms can produce real income without public subsidy, in particular if farm products are already matched to local or regional demand.

settlement, to catch water from our roof areas, and to place nearby a zone of fuel forest which receives wastes and supplies energy, will be to free most of the area of the globe for the rehabilitation of natural systems. These need never be looked upon as "of use to people", except in the very broad sense of global health.

The real difference between a cultivated (designed) ecosystem, and a natural system is that the great majority of species (and biomass) in the *cultivated* ecology is intended for the use of humans or their livestock. We are only a small part of the total primeval or natural species assembly, and only a small part of its yields are directly available to us. But in our own gardens, almost every plant is selected to provide or support some direct yield for people. Household design relates principally to the needs of people; it is thus human–centred (anthropocentric).

This is a valid aim for *settlement design*, but we also need a nature–centred ethic for wilderness conservation. We cannot, however, do much for nature if we do not govern our greed, and if we do not supply our needs from our existing settlements. If we can achieve this aim, we can withdraw from much of the agricultural landscape, and allow natural systems to flourish.

Recycling of nutrients and energy in nature is a function of many species. In our gardens, it is our own responsibility to return wastes (via compost or mulch) to the soil and plants. We actively create soil in our gardens, whereas in nature many other species carry out that function. Around our homes we can catch water for garden use, but we rely on natural forested landscapes to provide the condenser leaves and clouds to keep rivers running with clean water, to maintain the global atmosphere, and to lock up our gaseous pollutants. Thus, even anthropocentric people would be well–advised to pay close attention to, and to assist in, the conservation of existing forests and the rehabilitation of degraded lands. Our own survival demands that we preserve all existing species, and allow them a place to live.

We have abused the land and laid waste to systems we need never have disturbed had we attended to our home gardens and settlements. If we need to state a set of ethics on natural systems, then let it be thus:

- Implacable and uncompromising opposition to further disturbance of any remaining natural forests, where most species are still in balance;
- Vigorous rehabilitation of degraded and damaged natural systems to stable states;
- Establishment of plant systems for our own use on the *least* amount of land we can use for our existence; and
- Establishment of plant and animal refuges for rare or threatened species.

Permaculture as a design system deals primarily with the third statement above, but all people who act responsibly in fact subscribe to the first and second statements. That said, I believe we should use all the species we need or can find to use in our own settlement designs, *providing they are not locally rampant and invasive.*

Whether we approve of it or not, the world about us continually changes. Some would want to keep everything the same, but history, palaeontology, and commonsense tells us that all has changed, is changing, will change. In a world where we are losing forests, species, and whole ecosystems, there are three concurrent and parallel responses to the environment:

1. CARE FOR SURVIVING NATURAL ASSEMBLIES, to leave the wilderness to heal itself.
2. REHABILITATE DEGRADED OR ERODED LAND using complex pioneer species and long–term plant assemblies (trees, shrubs, ground covers).
3. CREATE OUR OWN COMPLEX LIVING ENVIRONMENT with as many species as we can save, or have need for, from wherever on earth they come.

We are fast approaching the point where we need refuges for *all* global life forms, as well as regional, national, or state parks for indigenous forms of plants and animals. While we see our local flora and fauna as "native", we may also logically see all life as "native to earth". While we try to preserve systems that are still local and diverse, we should also build new or recombinant ecologies from global resources, especially in order to stabilise degraded lands.

In your own garden, there are likely to be plants, animals, and soil organisms from every major landmass and many islands. Jet travel has merely accelerated a process already well–established by continental drift, bird migration, wind transport, and the rafting of debris by water. Everything will, in time, either become extinct, spread more widely, or evolve to new forms. Each of these processes is happening at once, but the rate of extinction and exchange is accelerating. Rather than new species, adapted hybrids are arising for example as palms, sea grasses, and snails, and micro–organisms from many continents meet, mix, and produce new accommodations to their "new" environments.

The very chemistry of the air, soil, and water is in flux. Metals, chemicals, isotopes, gases, and plastics are loose on earth that have never before been present, or never present in such form and quantity before we made it so.

It is my belief that we have two responsibilities to pursue:

- Primarily, it is to get our house and garden, our place of living, in order, so that it supports us.
- Secondarily, it is to limit our population on earth, or we ourselves become the final plague.

Both these duties are intimately connected, as stable regions create stable populations. If we do not get our cities, homes, and gardens in order, so that they feed and shelter us, we must lay waste to all other natural systems. Thus, truly responsible conservationists have gardens which support their food needs, and are working to reduce their own energy needs to a modest consumption, or to that which can be supplied by local

TABLE 1.1
PERMACULTURE DESIGN
The result of a unique assembly of constructs, species, and social systems into a unique pattern suited to a specific site and set of occupants.

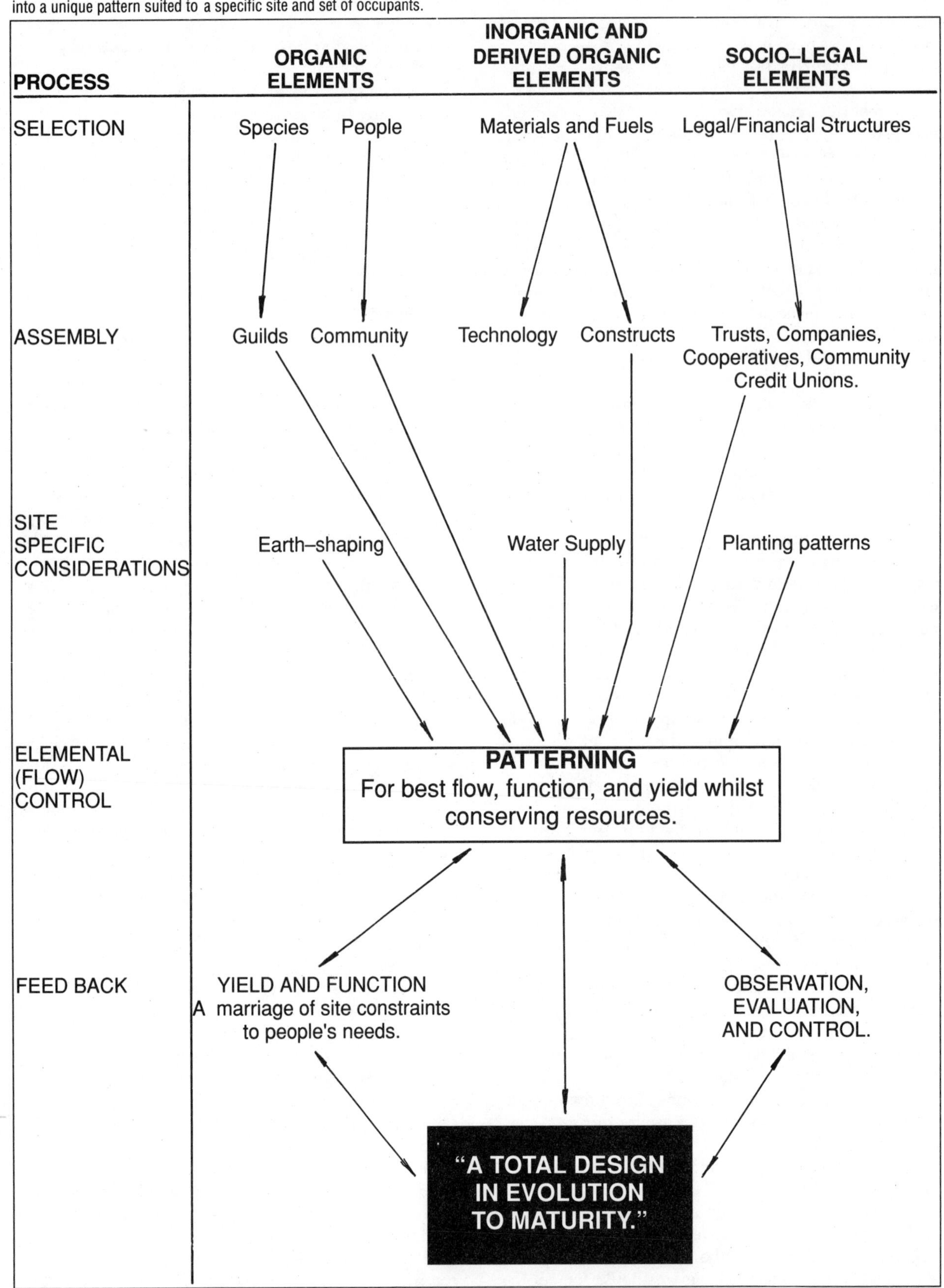

wind, water, forest, or solar power resources. We can work on providing biomass for our essential energy needs on a household and regional scale.

It is hypocrisy to pretend to save forests, yet to buy daily newspapers and packaged food; to preserve native plants, yet rely on agrochemical production for food; and to adopt a diet which calls for broadscale food production.

Philosopher–gardeners, or farmer–poets, are distinguished by their sense of wonder and real feeling for the environment. When religions cease to obliterate trees in order to build temples or human artefacts, and instead generalise love and respect to all living systems as a witness to the potential of creation, they too will join the many of us now deeply appreciating the complexity and self–sustaining properties of natural systems, from whole universes to simple molecules. Gardener, scientist, philosopher, poet, and adherent of religions all can conspire in admiration of, and reverence for, this earth. We create our own life conditions, now and for the future.

In permaculture, this means that all of us have some part in identifying, supporting, recommending, investing in, or creating wilderness habitats and species refuges. the practical way to proceed (outside the home garden) is to form or subscribe to institutes or organisations whose aims under their legal charter are to carry out conservation activities. While the costs are low, in sum total the effects are profound. Even the smallest garden can reserve off a few square metres of insect, lizard, frog, or butterfly habitat, while larger gardens and farms can fence off forest and wetland areas of critical value to local species. Such areas should be *only* for the conservation of local species.

Permaculture as a design system contains nothing new. It arranges what was always there in a different way, so that it works to conserve energy or to generate more energy than it consumes. What is novel, and often overlooked, is that *any* system of total common-sense design for human communities is revolutionary!

Design is the keyword of this book: design in landscape, social, and conceptual systems; and design in space and time. I have attempted a treatment on the difficult subject of patterning, and have tried to order some complex subjects so as to make them accessible. The text is positivistic, without either the pretended innocence or the belief that everything will turn out right. Only if we make it so will this happen.

As will be clear in other chapters of this book, the end result of the adoption of permaculture strategies in *any* country or region will be to dramatically reduce the area of the agricultural environment needed by the households and the settlements of people, and to release much of the landscape for the sole use of wildlife and for re–occupation by endemic flora. Respect for all life forms is a basic, and in fact essential, ethic for all people.

1.4

REFERENCES

Lovelock, J. E., *Gaia: A New Look at Life on Earth*, Oxford University Press, 1979.

Odum, Eugene, *Fundamentals of Ecology*, W. B. Saunders, Toronto, 1971.

Thomas, Lewis, *The Lives of a Cell*, The Viking Press, Inc., 1974.

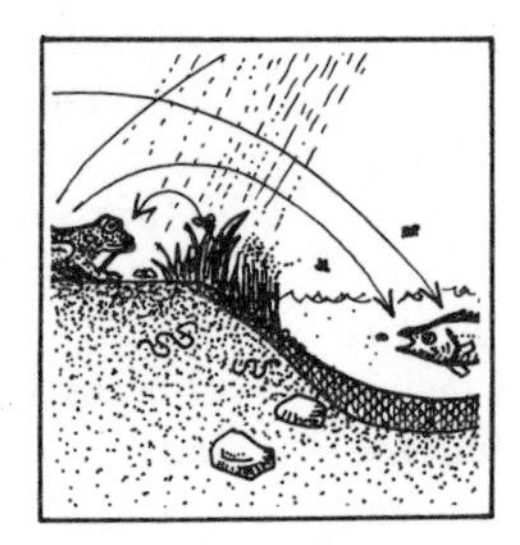

Chapter 2

CONCEPTS AND THEMES IN DESIGN

The world teeters on the threshold of revolution. If it is a bloody revolution it is all over. The alternative is a design science revolution... Design science produces so much performance per unit of resource invested as to take care of all human needs.

(Buckminster Fuller)

All living organisms... are 'open systems'; that is to say, they maintain their complex forms and functions through continuous exchanges of energies and materials with their environment. Instead of 'running down' like a mechanical clock that dissipates its energy through friction, the living organism is constantly 'building up' more complex substances from the substance it feeds on, more complex forms of energies from the energies it absorbs, and more complex patterns of information... perceptions, feelings, thoughts... form the input of its receptor organs.

(Arthur Koestler, 1967, *The Ghost in the Machine*)

Most **thermodynamic** problems concern 'closed' systems, where the reactions take place in confinement, and can be reversed; an example is the expansion and compression of gas in a cylinder. But in an open system, energy is gained or lost irreversibly, and the system, its environment, or both are changed by the interaction... the second law of thermodynamics [states that] energy tends to dissipate and organized systems drift inevitably towards **entropy**, or chaos. In seeming violation of that law, biological systems tend to become increasingly complex and efficient.

(*Newsweek*, October 24, 1977, on the Nobel Prize awarded to Ilya Prigogine.)

Lovelock shows that the biosphere, or Gaia as he calls it, actually created those conditions that are required for its support... and systematically builds up the stock of materials that it requires to move... towards increasng complexity, diversity, and stability.

(Edward Goldsmith, 1981, "Thermodynamics or Ecodynamics", *The Ecologist*.)

Man did not weave the web of life, he is merely a strand in it. Whatever he does to the web, he does to himself... to harm the earth is to heap contempt upon the creator... contaminate your bed, and you will one night suffocate in your own waste.

(Chief Seattle, 1854, responding to a U. S. government offer to buy Indian land.)

2.1
INTRODUCTION

It is alarming that in western society no popular body of directives has arisen to replace the injunctions of tribal taboo and myth. When we left tribal life we left with it all guides to sensible behaviour in the natural world, of which we are part and in which we live and die. More to the point, by never having the time or commonsense to evolve new or current guiding directives, we have forgotten how to evolve self–regulating systems. Hence, the call for a society in which we are all designers, based on an ethical and applied education, with a clear concept of life ethics.

The Gaia hypothesis, as formulated by James Lovelock, is that the earth less and less appears to behave like a material assembly, and more and more appears to act as a thought process. Even in the inanimate world we are dealing with a life force, and our acts are of great effect. The reaction of the earth is to restore equilibrium and balance. If we maltreat, overload, deform, or deflect natural systems and processes, then we will get a reaction, and this reaction may have long–term consequences. Don't do anything unless you've thought out all its consequences and advantages.

Aboriginal cultures used *myth* to show how unnecessary acts and unthinking destruction of elements brings about catastrophe and suffering. The usual structure of myth has these sequences:

1. A willful act of an individual or group.
2. A transmutation (animate to inanimate or the reverse, e.g. Lot's wife turns into a pillar of salt). This is by way of a warning.
3. Invocation of an elemental force (fire, storm, earthquake, flood, tidal wave, plague) as a result of any set of willful acts.
4. Necessary atonement by suffering, isolation, migration, or death.

So the act of a child or individual is given a meaning which relates to the whole of nature, and rebounds on the society. Reared on such myths, we go carefully in the world, aware that every unthinking act can have awful consequences.

Because we have replaced nature–based myth with a set of fixed prohibitions relating only to other people, and unrelated to nature, we have developed destructive and people–centred civilisations and religions.

In life and in design, we must accept that *immutable rules will not apply*, and instead be prepared to be guided on our continuing exploration by *flexible principles and directives*.

Thus, this book emphasises self–reliance, responsibility, and the functions of living things. Within a self–regulated system on earth, energy from the sun can be trapped and stored in any number of ways. While the sun burns, we are in an open system. If we don't destroy the earth, open–system energy saving will see us evolve as conscious beings in a conscious universe.

A Policy of Responsibility (to relinquish power)

The role of beneficial authority is to return function and responsibility to life and to people; if successful, no further authority is needed. The role of successful design is to create a self–managed system.

2.2

SCIENCE AND THE THOUSAND NAMES OF GOD

Although we can observe nature, living systems do not lend themselves to strict scientific definition for two reasons. Firstly, life is always in process of change, and secondly, life systems *react* to investigation or experiments. We must always accept, therefore, that there will never be "laws" in the area of biology.

"Hard" science, such as we apply to material systems (physics, mathematics, inorganic chemistry), studiously avoids life systems, regarding as not quite respectable those sciences (botany, zoology, psychology) which try to deal with life. Rigorous scientific method deals with the necessity of rigorous control of variables, and in a life system (or indeed any system), this presumes two things that are impossible:

1. That you know all variables (in order to control some of them and measure others) before you start; or
2. That you can in fact control all or indeed any variables without creating disorder in the life system.

Every experiment is carried out by people, and the results are imparted to people. Thus living things conduct and impart knowledge. To ignore life in the system studied, one has to ignore oneself. Life exists in conditions of flux, not imposed control, and responds to any form of control in a new fashion. Living things respond to strict control (either by removal of stimuli or by constant input of stimuli) by becoming *uncontrolled*, or (in the case of people and rats at least) by dysfunction, or by going mad.

Experiments, therefore, are not decisive, rigid, or true findings but an eternal search for the variables that have *not* been accounted for previously. This is the equivalent of true believers, in their empirical approach to the knowledge of God's name. They simply keep chanting variables of all possible names until (perhaps) they hit on the right one. Thus does science proceed in biological experiments.

Scientists who "know" and observe, don't usually apply their knowledge in the world. Those who "act", often don't know or observe. This has resulted in several tragic conditions, where productive natural ecosystems have been destroyed to create unproductive cultivated systems, breaking every sane environmental principle to do so. Energy–efficient animals (deer, kangaroo, fish) have been displaced by inefficient animal systems (sheep, cattle). Every widespread modern agricultural system needs great energy inputs; most agriculture destroys basic resources and denies future yields.

As Edward Goldsmith makes clear ("Thermodynamics or Ecodynamics", *The Ecologist*, 1981), many scientists refuse to consider the function of life in such systems. Natural systems disintegrate and decay, producing more and more helpless plants, animals, and people, and the State or the farmer takes over the function of natural processes. (The State becomes the father of the orphaned child, the farmer the father of the orphaned chicken.) It is only by returning self–regulating function and responsibility to living things (such as people) that a stable life system can evolve.

Scientific method is one of the ways to know about the real world, the world we are part of and live in. Observation and contemplative understanding is another. We can find out about many things, both living and inorganic, by timing, measuring, and observing them; enough to make calendars, computers, clocks, meters, and rulers, but not ever enough to understand the complex actions in even a simple living system. You can hit a nail on the head, or cause a machine to do so, and get a fairly predictable result. Hit a dog on the head, and it will either dodge, bite back, or die, but it will never again react in the same way. We can predict only those things we set up to be predictable, not what we encounter in the real world of living and reactive processes.

Ecologists and "whole systems" people struggle to understand open and complex systems, even though they realise that they too are a part of the system they study. In fact, given enough limnologists (those who study freshwater lakes and lake organisms), these become the most important factor in the spread of lake organisms via their boats, boots, and nets! (It is also time, I feel, for students of communities to form a *community* of students of communities and keep out of everybody else's hair!)

Overseas aid is perilously close to being a very good reason for overseas aid to be necessary, as spies need counterspies. I shudder to think that if we train more brain surgeons, they must cut open more brains in order to support themselves... imagine! I think it fair to say that if you submit to poverty, you equip yourself to know about poverty, and the same goes for lobotomy.

There are several ways not to face life: by taking drugs, watching television, becoming a fakir in a cave, or reading in pure science. All are an abdication of personal responsibility for life on earth (including, of course, one's own life). Value- and ethic-free lifestyles are as aberrant in science as in society.

> It is the quantifiability of many... scientific concepts that have led to their adoption by scientists often regardless of the fact that, as they are defined, they correspond to nothing whatsoever in the world of living things.
>
> (E. Goldsmith, 1981 "Thermodynamics or Ecodynamics", *The Ecologist.)*

Perverse planning is everywhere obvious: houses face not the sun, but rather the road, lawns replace gardens, and trees are planted to be pruned and tended. Make-work is the rule, and I suspect that most theoretical scientists inhabit demented domestic environments, just as many psychiatrists are inhabitants of mental institutions.

Scientific (and non-scientific) groups or individuals can make progress in finding solutions to specific problems. The following approaches do very well (designers please note):

1. IMPROVING TOOLS, or inventing new tools for specific jobs.
2. COLLECTING A LARGE SET OF OBSERVATIONS on occurrences, or samples of a set of phenomena, and sorting them on the basis of likeness–unlikeness (by establishing systems and system boundaries, categories, and keys to systems). This process often reveals common characteristics of diverse elements, and leads to an understanding of common traits, suggesting (by analogy) strategies in design.
3. INSIGHT : the "Aha!" or "Eureka!" response to observation. This, as is well recorded, comes to the individual as though by special gift or providence. In fact, it is quite probably the end point of 2.
4. TRIALS: "give it a try and see if it works". This empirical approach simply eliminates those things that don't work. It does not necessarily establish how or why something works, or even if it works in the long run.
5. GUESSING: the best guesses are based on trials that are already known to work.
6. OBSERVING UNIQUE EVENTS and taking note of them (the "discovery" of penicillin).
7. ACCIDENT: trials set up for one reason work in a way not predicted or foreseen; compounds made for one purpose are applied to another.
8. IMITATION: by testing already-known effects (discovered by others).
9. PATTERNING: by seeing a pattern to events of often very different natures, and thus producing insights into underlying effects. Often preceded by 2 above, but rare in science.
10. COMMONSENSE: often called "management" in business and natural systems control. This consists of staying with and steering a system or enterprise through constant adjustment to a successful conclusion or result. It also suits evolving systems, and is the basis of continuous change and adjustment.

2.3
APPLYING LAWS AND PRINCIPLES TO DESIGN

Principles differ from dogmas in that there are no penalties for error, but only learning from error, which leads to a new evolution. Dogmas are rules which are intended to force centralised control (often by guilt), and it is obvious that every such rule or law represents a failure of the social system. It is too late to fail, but never too late to adopt sensible principles for our guidance, and to throw away the rule book.

Life Intervention Principle
In chaos lies unparalleled opportunity for imposing creative order.

Just join with one or two friends to make your way in the confusion. Others will follow and learn.

There is only one law that is offered to us by such education as we derive from nature, and that is the law of return, which can be stated in many ways:

Law of Return
"Whatever we take, we must return", or

"Nature demands a return for every gift received," or "The user must pay."

We should examine, and act on, the forms of this law. It is the reason why this book carries a tree tax: that we may be able to continue in the use of books. It is why we must never buy books or newspapers that do *not* tax, nor goods where the manufacturer does not recycle or replant the materials of the manufacture. It is why we must carefully study how to use our wastes, and this includes our body wastes. Put in the form of a directive or policy statement, this law would read:

Every object must responsibly provide for its replacement; society must, *as a condition of use*, replace an equal or greater resource than that used.

Inherent in such a law are the concepts of replanting, recycling, durability, and the correct or beneficial disposal of wastes. Nature has extreme penalties for those who break such laws, and for their descendants and neighbours.

Nor can we deny immanence; if a landscape delights us, we should not insult it with castles on peaks, roadways, and clear–cuts. We should return the pleasure we get from natural prospects, and maintain their integrity. It would be pleasant indeed were the land around us always to appear welcoming or non–threatening. This effect, too, can be created or destroyed. There is no reason not to bury our necessary constructs in earth, or clothe them with vegetation. If we want pleasure in life, then we should preserve the life around us.

Energies enter a system, and either remain or escape. Our work as permaculture designers is to prevent energy leaving before the basic needs of the whole system are satisfied, so that growth, reproduction, and maintenance continue in our living components.

All permaculture designers should be aware of the fundamental principles that govern natural systems. These are not immutable rules, but can be used as a set of directives, taking each case as unique but gaining confidence and inspiration from a set of findings and solutions in other places and other times. We can use the guiding principles and laws of natural systems, as formulated by such people as Watt, Odum, and Birch, and apply some of them to our consciously–designed ecologies.

One such law is the basic law of thermodynamics, as restated by Watt[13]:

> All energy entering an organism, population or ecosystem can be accounted for as energy which is stored or leaves. Energy can be transferred from one form to another, but it cannot disappear, or be destroyed, or created. No energy conversion system is ever completely efficient.

As stated by Asimov (1970):

> The total energy of the universe is constant and the total entropy is increasing.

Entropy is bound or dissipated energy; it becomes unavailable for work, or not useful to the system. It is the waters of a mountain stream that have reached the sea. It is the heat, noise, and exhaust smoke that an automobile emits while travelling. It is the energy of food used to keep an animal warm, alive, and mobile. Thus, ambient and useful energy storages are degraded into less useful forms until they are no longer of any use to our system.

The question for the designer becomes, "How can I best use energy before it passes from my site, or system?" Our strategy is to set up an interception net from "source to sink". This net is a compound web of life and technologies, and is designed to catch and store as much energy as possible on its way to increasing entropy (as in **Figure 2.1**).

Therefore, we design to catch and store as much water as possible from the hills before it ends up at its "sink" in the quiet valley lake. If we made no attempt to store or use it as it passes through our system, we would suffer drought, have to import it from outside our system, or use energy to pump it back uphill.

Although the material world can perhaps be predictably measured (at least over a wide range of phenomena), by applications of the laws of thermodynamics, these relate mainly to non–living or experimentally "closed systems". The concept of entropy is not necessarily applicable to those living, open earth systems with which we are involved and in which we are immersed. Such laws are more useful in finding an effective path through material technologies than through a life–complexed world. The key word in open systems is "exchange". For example, on the local level, cities *appear* to be "open", but as they return little energy to the systems that supply them, and pass on their wastes as pollutants to the sea, they are *not* in exchange but in a localised one–way trade with respect to their food resource. All cities break the basic "law of return".

Life systems constantly organise and create complex storages from diffuse energy and materials, accumulating, decomposing, building, and transforming them for further use. We can use these effects in the design process by finding pathways or routes by which life systems convert diffuse materials into those of most use. For example, if we have a "waste" such as manure, we can leave it on a field. Although this is of productive use, we have only achieved one function. Alternatively, we can route it through a series of transformations that give us a variety of resources.

First we can ferment it, and distill it to alcohol, and secondly route the waste through a biogas digester, where anaerobic organisms convert it to methane, of use as a cooking or heating gas, or as fuel for vehicles. Thirdly, the liquid effluent can be sent to fields, and the

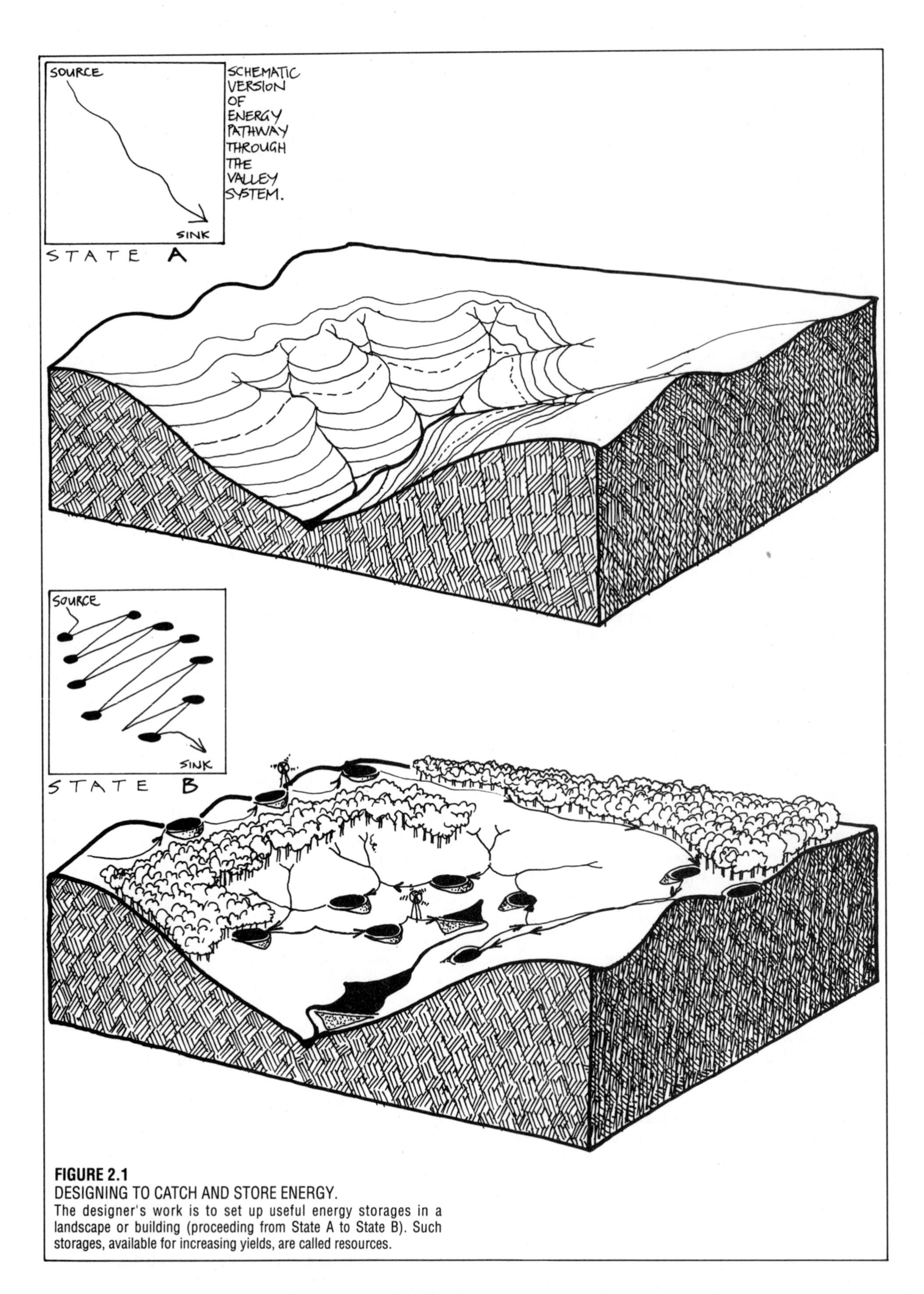

FIGURE 2.1
DESIGNING TO CATCH AND STORE ENERGY.
The designer's work is to set up useful energy storages in a landscape or building (proceeding from State A to State B). Such storages, available for increasing yields, are called resources.

solid sludge fed to worms, which convert it to rich horticultural soil. Fourthly, the worms themselves can be used to feed fish or poultry.

Birch states six principles of natural systems:

1. "Nothing in nature grows forever."

   (There is a constant cycle of decay and rebirth.)

2. "Continuation of life depends on the maintenance of the global bio–geochemical cycles of essential elements, in particular carbon, oxygen, nitrogen, sulphur, and phosphorus."

   (Thus, we need to cycle these and other minor nutrients to stimulate growth, and to keep the atmosphere and waters of earth unpolluted.)

3. "The probability of extinction of populations or a species is greatest when the density is very high or very low."

   (Both crowding and too few individuals of a species may result in reaching thresholds of extinction.)

4. "The chance that species have to survive and reproduce is dependent primarily upon one or two key factors in the complex web of relations of the organism to its environment."

   (If we can determine what these critical factors are, we can exclude, by design, some limiting factors, e.g. frost, and increase others, e.g. shelter, nest sites).

5. "Our ability to change the face of the earth increases at a faster rate than our ability to foresee the consequence of such change."

   (Hence the folly of destroying life systems for short–term profit.)

6. "Living organisms are not only means but ends. In addition to their instrumental value to humans and other living organisms, they have an intrinsic worth."

   (This is the *life ethic* thesis so often missing from otherwise ethical systems.)

Although these principles are basic and inescapable, what we as designers have to deal with is survival on a particular site, here and now. Thus, we must study whether the resources and energy consumed can be derived from renewable or non–renewable resources, and how non–renewable resources can best be used to conserve and generate energy in living (renewable) systems. Fortunately for us, the long–term energy derived from the sun is available on earth, and can be used to renew our resources if life systems are carefully constructed and preserved.

There are thus several practical design considerations to observe:

- The systems we construct should last as long as possible, and take least maintenance.
- These systems, fueled by the sun, should produce not only their own needs, but the needs of the people creating or controlling them. Thus, they are sustainable, as they sustain both themselves and those who construct them.
- We can use energy to construct these systems, providing that in their lifetime, they store or conserve more energy than we use to construct them or to maintain them.

The following are some design principles that have been distilled for use in permaculture:

1. WORK WITH NATURE, RATHER THAN AGAINST IT. We can assist rather than impede natural elements, forces, pressures, processes, agencies, and evolutions. In natural successions, grasses slowly give way to shrubs, which eventually give way to trees. We can actively assist this natural succession not by slashing out weeds and pioneers, but by using them to provide microclimate, nutrients, and wind protection for the exotic or native species we want to establish.

"If we throw nature out the window, she comes back in the door with a pitchfork" (Masanobu Fukuoka). For example, if we spray for pest infestations, we end up destroying both pests and the predators that feed on them, so the following year we get an explosion of pests because there are no predators to control them. Consequently, we spray more heavily, putting things further out of balance. Unfortunately, all the pests are never killed, and the survivors breed more resistent progeny (nature's pitchfork!)

2. THE PROBLEM IS THE SOLUTION. Everything works both ways. It is only how we see things that makes them advantageous or not. If the wind blows cold, let us use both its strength and its coolness to advantage (for example, funneling wind to a wind generator, or directing cold winter wind to a cool cupboard in a heated house). A corollary of this principle is that everything is a positive resource; it is up to us to work out *how* we may use it as such. A designer may recognise a specific site characteristic as either a problem or as a unique feature capable of several uses, e.g. jagged rock outcrops. Such features can only become "problems" when we have already decided on imposing a specific site pattern that the rock outcrop interferes with. It is not a problem, and may be an asset if we accept it for the many values it possesses. "The problem is the solution" is a Mollisonism implying that only our fixed attitudes are problems when dealing with things like rock outcrops! A friend has included several natural boulders in her home, with excellent physical, aesthetic, and economic benefit; the builder would have removed them as "problems", at great expense.

3. MAKE THE LEAST CHANGE FOR THE GREATEST POSSIBLE EFFECT. For example, when choosing a dam site, select the area where you get the most water for the least amount of earth moved.

4. THE YIELD OF A SYSTEM IS THEORETICALLY UNLIMITED. The only limit on the number of uses of a resource possible within a system is in the limit of the information and the imagination of the designer. If you think you have fully planted an area, almost any other innovative designer can see ways to add a vine, a fungus, a beneficial insect, or can see a yield potential that has been ignored. Gahan Gilfedder at the Garden of Eden in Australia found an unsuspected market for cherimoya seed, required by nurseries as seed stock for

grafting. This made a resource from a "waste" product derived from damaged fruit.

5. EVERYTHING GARDENS. A Mollisonian principle is that "everything makes it own garden", or everything has an effect on its environment. Rabbits make burrows and defecation mounds, scratch out roots, create short swards or lawns, and also creates the conditions favourable for weeds such as thistles. People build houses, dispose of sewage, dig up soils for gardens, and maintain annual vegetable patches. We can "use" the rabbit directly as food, to help in fire control, to prepare soil for "thistles" (cardoons and globe artichokes), and to shelter many native animal species in their abandoned burrows. Rabbits maintain species–rich moorland swards suited to many orchids and other small plants. It is a matter of careful consideration as to where this rabbit, and ourselves, belong in any system, and if we should control or manage their effects or tolerate them. When we examine how plants and animals change ecosystems, we may find many allies in our efforts to sustain ourselves and other species. (See **Figure 2.2**).

2.4 RESOURCES

The energies coming into our system are such natural forces as sun, wind, and rain. Living components and some technological or non–living units built into the system translate the incoming energies into useful reserves, which we can call *resources*. Some of these resources have to be used by the system for its own purposes (stocks of fish must be maintained to produce more fish). An ideal technology should at the very least fuel itself.

The surplus, over and above these system needs, is our *yield*. Yield, then, is any useful resource surplus to the needs of the local system and thus available for use, export or trade. The way to obtain yield is to be conservative in resource use, for energy, like money, is much more easily saved than generated. Resource saving involves recycling waste, insulating against heat loss, etc. Then, we can work out paths or routes to send resources on to their next "use point".

If the aim of functional design is to obtain yields, or to provide a surplus of resources, it is as well to be clear about just what it is that we call a resource, and what categories of resource there are, as these latter may affect our strategies of use. In short, we cannot use all resources in the same way and to the same ends. Ethics of resource use are evolved by knowing about the results of resource exploitation. Forests, soils, air, water, sunlight, and seeds are resources that we all regard as part of a common heritage.

A second category of resource is that which belong to us as group, family, or person: those fabricated, ordered, or otherwise developed resources that people create by their work, and of which a presence or absence does not apparently affect the common resource. What we create, however, is *always made from the common resource*, so that it is impossible to draw a line between these categories.

What other ways can we look at resources? Let us try a use–and–results approach. What happens if we use some resources, if we look upon them as a yield? We then find that a response or result follows. Resources are:

1. THOSE WHICH INCREASE BY MODEST USE. Green browse is an example: if deer do not browse shrubs, the latter may become woody and unpalatable. Also, a browsed biennial, unable to flower, may tiller out and become perennial (e.g. the fireweed *Erechthites* nibbled by wallaby in Tasmania). Seedling trees can be maintained at browse height, but if ungrazed, "escape" to unbrowsable height and shade out other palatable plants. *Overgrazing* may (by damage) cause extinction of palatable selected browse and browsers, but underbrowsing may cause similar effects. *Information* is another resource that can increase with use. It withers or is outdated if not used. Too little impoverishes a system, but when freely used and exchanged, it flourishes and increases.

2. THOSE UNAFFECTED BY USE. In impalpable terms, a view or a good climate is unaffected by use. In palpable terms the diversion of a part of a river to hydroelectric generation or irrigation (the water returned to the stream after use), is also unaffected, as is a stone pile as mulch, heat store, or water run–off collector. A well–managed ecosystem is an example of resources unaffected by use.

3. THOSE WHICH DISAPPEAR OR DEGRADE IF NOT USED. For example an unharvested crop of an annual, or a grass which could be stored for the winter, irruptions of oceanic fish, swarms of bees or grasshoppers, ripe fruit, and water run–off during rains.

4. THOSE REDUCED BY USE. For example a fish or game stock unwisely used, clay deposits, mature forests, and coal and oil.

5. THOSE WHICH POLLUTE OR DESTROY OTHER RESOURCES IF USED. Such as residual poisons in an ecosystem, radioactives, super-highways, large buildings or areas of concrete, and sewers running pollutants to the sea.

Categories 1 to 3 are those most commonly produced in natural systems and rural living situations, and are the only sustainable basis of society. Categories 4 and 5 are as a result of urban and industrial development, and if not used to produce permanent beneficial changes to the ecosystem, become pollutants (some are permanent pollutants in terms of the lifetimes of people).

It follows that a sane society manages resources categories 1 to 4 wisely, bans the use of resource category 5, and regulates all uses to produce sustainable yield. This is called resource management, and has been successfully applied to some fish and animal populations, but seldom to our own lives. Investment priorities can be decided on the same criteria, at both the national and household level.

Policy of Resource Management

A responsible human society bans the use of resources which permanently reduce yields of sustainable resources, e.g. pollutants, persistent poisons, radioactives, large areas of concrete and highways, sewers from city to sea.

Failure to do this will cause the society itself to fail, so that programmes of highway building and city expansion, the release of persistent biocides, and loss of soil will bring any society down more surely and permanently than war itself. Immoral governments tolerate desertification and land salting, concreted highways and city sprawl, which take more good land permanently out of life production than the loss of territory to a conqueror. Immorality of this nature is termed "progress" and "growth" to confuse the ignorant and to supplant local self-reliance for the temporary ends of centralised power.

The key principle to wise resource use is the principle of "enough". This is basic to understanding

FIGURE 2.2
EVERYTHING GARDENS.
A – Pruning, B – Digging, C – Mowing, D – Typical plant assembly for species. Some species (*Oryctolagus, Cuniculus, Macropus, Gallus, Cairina,* and *Homo sapiens*) at work in their fields. Plants developed by each species are maintained in similar deflection states as lawns, pruned trees, flat weeds, and characteristic herbage around dwellings.

societies in chaos or systems in disorder. Today superhighways and overpasses in Massachusetts alone need some 400 billion dollars to repair, and the collapsing sewer systems of London and New York some 80 billions. Neither Massachusetts, London, nor New York can raise this money, which shows that an unthinking historical development strategy can cripple a future society. Today's luxuries are tomorrow's disasters.

Principle of Disorder

Any system or organism can accept only that quantity of a resource which can be used productively. Any resource input beyond that point throws the system or organism into disorder; oversupply of a resource is a form of chronic pollution.

Both an over– and undersupply of resources have much the same effect, except that oversupply has more grotesque results in life systems than undersupply. To a degree, undersupply can be coped with by reduced growth and a wider spacing or dispersal of organisms, but oversupply of a resource can cause inflated growth, crowding, and sociopathy in social organisms. In people, both gross over– and under–nutrition are common. Ethical resource management is needed to balance out the pathologies of famine and obesity.

2.5 YIELDS

Yields can be thought of in immediate, palpable, and material ways, and are fairly easily measured as:

1. PRODUCT YIELD: The sum of primary and derived products available from, or surplus to, the system. Some of these are intrinsic (or precede design), others are created by design.
2. ENERGY YIELD: The sum of conserved, stored, and generated energy surplus to the system, again both intrinsic and those created by design.

Impalpable yields are those related to health and nutrition, security, and a satisfactory social context and lifestyle. Not surprisingly, it is the search for these invisible yields that most often drives people to seek good design or to take up life on the land, for "what does it benefit a man if he gains the whole world and loses his soul?" Thus, we see the invisible yields in terms of values and ethics. This governs our concept of needs and sets the limits of "enough". Here, we see an ethical basis as a vital component of yield.

Although all systems have a natural or base yield depending on their productivity, our concern in permaculture is that this essential base yield is sustainable. Several factors now operate to reduce the yield of natural systems. In the simplest form, this is the overuse of energy in degenerative systems due to the unwise application of fossil fuel energy. "Poisoning by unproductive use" is observable and widespread. Thus we must concentrate on productive use, which implies that the energy used is turned into biological growth and held as basic living material in the global ecosystem. Unused, wasted, or frivolously used resources are energies running wild, which creates chaos, destroys basic resources, and eventually abolishes all yield or surplus.

In design terms, we can find yields from those living populations or resources which are the stocks of the biologist (the so–called standing crop) or from non–living systems such as the climatic elements, chemical energy, and machine technology. There is energy stored by extinct life as coal, oil, and gas; energy left over from the formation of the earth as geothermal energy; tides; and electromagnetic and gravitational forces. Cosmic and solar energies impinge on the earth, and life intercepts these flows to make them available for life forms.

In our small part of the system (the design site) our work is to store, direct, conserve, and convert to useful forms those energies that exist on, or pass through, the site. The total sum of our strategy, in terms of surplus energy usefully stored, is the *system yield* of design.

Definition of System Yield

System yield is the sum total of surplus energy produced by, stored, conserved, reused, or converted by the design. Energy is in surplus once the system itself has available all its needs for growth, reproduction, and maintenance.

Some biologists may define yield or production in more narrow terms, accepting that a forest, lake, or crop has a finite upper limit of surplus due to substrate conditions and available energy. We do not have to accept this, as it is a passive approach, inapplicable to active and conscious design or active management using, for example, fertilisers, windbreaks, or selected species.

Even more narrowly defined is the yield of agricultural economists, who regard a single product (peaches/ha) as the yield. It may be this approach itself itself which is the true limit to yield!

A true accounting of yield takes into consideration both upstream costs (energy) and downstream costs (health). The "product yield" may create problems of pollution and soil mineral loss, and cost more than it can replace.

The very concept of surplus yield supposes either flow through or growth within our system. Coal and rock do not have yield in this sense; they have a finite or limited product. Only life and flow can yield continually, or as long as they persist. Thus the energy stocks of any system are the flows and lives within it. The flow may exist without life (as on the moon), where only technology can intervene to obtain a yield, but on earth at least, life is the intervening strategy for capturing flow and producing yield. And technology depends on the continuation of life, not the opposite.

The Role of Life in Yield

Living things, including people, are the only effective intervening systems to capture resources on this planet, and to produce a yield. Thus, it is the sum and capacity of life forms which decide total system yield and surplus.

We have long been devising houses, farms, and cities which are energy–demanding, despite a known set of strategies and techniques (all well tried) which could make these systems energy–producing. It has long been apparent that this condition is deliberately and artificially maintained by utilities, bureaucracies, and governments who are composed of those so dependent on the consumption and sale of energy resources that without this continuing exploitation they themselves would perish.

In permaculture, we have abundant strategies under the following broad categories which can create yields instead of incurring costly inputs or energy supply.

STRATEGIES THAT CREATE YIELDS.

Physical–Environmental:

- The creation of a niche in space; the provision of a critical resource.
- The rehabilitation and creation of soils.
- The diversion of water, and water recycling.
- The integration of structures and landscape.

Biological:

- The selection of low–maintenance cultivars and species for a particular site.
- Investigation of other species for usable yields.
- Supplying key nutrients; biological waste recycling (mulch, manure).
- The assembly of beneficial and cooperative **guilds** of plants and animals.

Spatial and Configurational:

- **Annidation** of units, functions, and species (annidation is a design or pattern strategy of "nesting" or stacking one thing within another, like a bowl in a bowl, or a vine in a tree).
- **Tessellation** of units, functions, and species (tessellation is the forming or arranging of a mosaic of parts).
- Innovative spatial geometry of designs as **edge** and **harmonics**.
- Routing of materials or energy to next best use.
- Zone, sector, slope, orientation, and site strategies (Chapter 3).
- Use of special patterns to suit irrigation, crop systems, or energy conservation.

Temporal:

- Sequential annidation (interplant, intercrop).
- Increasing cyclic frequency.
- Tessellation of cycles and successions, as in browsing sequences.

Technical:

- Use of appropriate and rehabilitative technology.
- Design of energy–efficient structures.

Conservation:

- Routing of resources to next best use
- Recycling at the highest level.
- Safe storage of food product.
- **No–tillage** or low–tillage cropping.
- Creation of very durable systems and objects.
- Storage of run–off water for extended use.

Cultural:

- Removing cultural barriers to resource use.
- Making unusual resources acceptable.
- Expanding choices in a culture.

Legal/Administrative:

- Removing socio–legal impediments to resource use.
- Creating effective structures to aid resource management.
- Costing and adjusting systems for *all* energy inputs and outputs.

Social:

- Cooperative endeavours, pooling of resources, sharing.
- Financial recycling within the community.
- Positive action to remove and replace impeding systems.

Design:

- Making harmonious connections between components and sub–systems.
- Making choices as to where we place things or how we live.
- Observing, managing, and directing systems.
- Applying information.

This approach to potential production is beyond that of product yield alone. It is theoretically unlimited in its potential, for system yield results from the number of strategies applied, what connections are made, and what information is applied to a particular design.

Now we see that yield in design is not some external, fixed, immutable quantity limited by circumstances that previously existed, but results from our behaviour, knowledge, and the application of our intellect, skill, and comprehension. These can either limit or liberate the concept of yield. Thus, the profound difference between permaculture design and nature, is that in permaculture we actively intervene to supply missing elements and to guide system evolution.

Limits to Yield

Yield is not a fixed sum in any design system. It is the measure of the comprehension, understanding, and ability of the designers and managers of that design.

Defined in this way, yield has no known limits, as we cannot know all ways to conserve, store, and save energy, nor can we fail to improve any system we build and observe. There is always room for another plant, another cycle, another route, another arrangement, another technique or structure. We can thus continually shrink the area we need to survive. The critical yield strategy is in governing our own appetites!

Just as we can increase yield, so we can decrease it. The perverse aims of some politicians, developers, and even religious dogmatists limit yield by disallowing certain products as a yield. Just as one's neighbours may refuse the snail and eat the lettuce, refuse the blackbird and eat the strawberry, so we may only "allow" certain types of toilets, or certain plants in gardens or parks. And thus people are the main impediment to using their potential yields.

FARM STRATEGIES:
CATEGORIES AND EXAMPLES.

If we take as a condition the "fencepost–to–fencepost" grasslands or crops now developing in the western world, and apply the strategies given, then yields will increase. How these systems interact raises yield even more, but on their own they are sufficiently impressive.

Water Storage:
(12–20% of landscape).

1. Product increase, e.g. animal protein production (water is more productive per unit area than land; fish more efficient at food conversion than cattle).

2. Product increase on land remaining due to:
  - irrigation; and
  - water nutrient quality from, e.g. fish manure.

3. Interaction, e.g. ducks on water to increase yields in and around ponds (e.g. pest and weed control, manure).

4. **Microclimatic** buffering due to water bodies (see Chapter 5, Climatic Factors).

Land Forming:

1. Product increase due to even irrigation (no dry areas or waterlogging).

2. Land stability due to reduction of soil loss from water run–off or salting.

3. Gravity flow replaces pumped water (depends on site).

4. Recycling of water possible.

Soil Reconditioning:

1. Product increase due to deeper root penetration.

2. Water infiltration (zero run–off) due to absorption.

3. Buffering of soil microclimate (see Chapter 8, Soils).

4. Supply of essential nutrients.

Establishing of Windbreak and Forage Forest:
(20–30% of landscape)

1. Shelter effects, e.g. increase in plant yields, animal protein, and microclimate buffering both above and below ground.

2. Increase in carrying capacity due to shrub and tree forage.

3. Savings on nutrients recycled via legumes and trees.

4. Intrinsic products of the forest, e.g. nectar for honey, seeds, firewood from fallen timber).

5. Insect and bird escapement, and **pest predator** habitat.

6. **Wildlife corridors.**

Selective Farm Reafforestation:
(*not* industrial forestry)

1. Increase precipitation due to night condensation, water penetration (see Chapter 6, Trees and Their Energy Transactions).

2. Product increase due to superiority of perennials over annuals in bulk, energy savings, and length of yield (**Figure: 1.1**).

3. Increase in rainfall due to trees cross–wind (see Chapter 6).

4. Reduced cost and increased capacity due to selected self–forage browse, e.g. drought–proof stockfeed, medicinal qualities of some perennial plants.

5. Reduced cost due to on–farm durable timber, e.g. fence posts, construction material.

6. Reduced carcass loss due to shivering, sweating, exposure.

7. Increased crop production in sheltered areas.

8. Increased carcass weight due to increased food intake in sheltered conditions (not the same as 6. above), i.e. on hot days cattle will graze all day when they are on shaded pasture, instead of sheltering from the sun.

9. Reduced evaporation from ponds due to less wind over water surfaces (see Chapter 5).

Market and Process Strategies:

1. Selected crop for specialty market for price/ha increase, e.g. fresh herbs near a concentration of restaurants.

2. Marketing by self–pick, mail order, direct dispatch, way–side sale.

3. Processing to a higher order of product (e.g. seed to oil).

4. Processing to refined order (e.g. crude eucalyptus oil to fractions).

5. Money saved by processing fuels on farm; plus sale of surplus fuel.

Social/Financial:

1. Market stability gained by farm–link strategy, where an urban group contracts to buy specific produce from the farmer.

2. Income from field days and educational courses.

3. Rental or income from urban visitors e.g. a guest house or holiday farm.

4. Direct investment by city people in a particular farm.

5. Formation of a local credit union and bank for the district, thus recycling money locally.

6. Vehicle and implement pool with neighbours; schedules of sowing and reaping worked out (capital saved 90%).

7. Labour exchange with neighbours.

8. Produce and marketing cooperatives.

Crop techniques:

1. Low or no–tillage farming saves:
  - energy in reduced tillage;
  - soil;
  - water and reduces evaporation; and
  - time between crops.

To put these into practical terms, I have culled from an interview with a farmer (Norm Sims, *Weekly Times*, 5 Jan. 1983) statements on savings due to some site strategies applied. On land–forming: "We expect to double production over the next few years, using half the irrigation waters" (4 times benefit); "Salinity is reduced". In severe drought: "Pasture production has never looked better and water is available". "It took us six days to irrigate what we now do in two..." and, "Rather than restricting watering intervals we are restricting the area" (aiming to milk 185 cows on 24 ha. On grazing rotation and electric fencing: "26 paddocks are grazed in a 21 day rotation" (average field of 1.6 ha each with a trough water–point for cattle).

Here, there are these specific strategies in use:

- laser levelling of fields for even irrigation;
- water reticulation;
- water storage and recycling;
- grazing rotation of 21 days;
- central access road;
- crop for concentrated rations grown; and
- pasture area reduced to give best watering regime.

It seems obvious from the foregoing that the *primary* and *certain* increases in crop yield do not just come from varietal selections (a fiction promulgated by agricultural companies, seed patent holders, agricultural researchers, or extension officers), but from attention to site design and development, followed by wise enterprise selection to suit the (modified) site, concurrently with a marketing and processing strategy.

As these are often permanent or durable strategies, it is not in the commercial interest to encourage them, as the continuous benefit is to the farmer alone, and the role of middlemen and traders is reduced. But, in the western world, the 4–6% of us in essential production are in fact enslaved, while the remaining 96% are deriving secondary or tertiary benefits without adequate return to the primary producers. This can only result in a weak economy, waste, and irresponsibility for life existence based on the expectation that the world owes politicians, students, and middlemen a living.

Benefits, like wastes, must be *returned* or recycled to keep any system going. Accumulations of unused benefits are predictive of a collapse at production level; thence, throughout all tiers of the system.

EXTENDING YIELDS

The concentration of yields into one short period is a fiscal, not an environmental or subsistence strategy, and has resulted in a "feast and famine" regime in markets and fields, and consequent high storage costs. Our aim should be to disperse food yield over time, so that many products are available at any season. This aim is achieved, in permaculture, in a variety of ways:

- By selection of early, mid and late season varieties.
- By planting the same variety in early or late–ripening situations.
- By selection of varieties that yield over a long season.
- By a general increase in diversity in the system, so that:
- Leaf, fruit, seed and root are all product yields.
- By using self–storing species such as tubers, hard seeds, fuelwood, or rhizomes which can be cropped on demand.
- By techniques such as preserving, drying, pitting, and cool storage.
- By regional trade between communities, or by the utilisation of land at different altitudes or latitudes.

YIELDS AND STORAGE

How yields endure is important, for there are unlimited opportunities to use durable yields in terms of season or lifetime.

By a series of preservation strategies, food can be stored for days, weeks, or years. Water not open to evaporation and pollution, or with natural cleansing organisms, will keep indefinitely. Shelters may outlast the forests that build them, or can be made of living or durable materials such as ivy, concrete, or stone. Energy alone (like the food which is part of energy) is difficult to store. Batteries leak or decay, heat escapes, and insulation breaks down. Only living things, like forests, increase their energy store.

Because of seasonal or diurnal cycles, we should pay close attention to storage strategies. Very little famine would occur could grains, fish, and fruit available in good times be stored for lean times. The strategies of food storage are critical. I believe that people should therefore mulch their recipe books, which often specify out–of–season or not–in–garden foods, and replace them with books that stress either low–energy methods of food preservation, or how to live easily from your garden in season.

CULTURAL IMPEDIMENTS TO YIELD.

I confess to a rare problem—gynekinetophobia, or the fear of women falling on me—but this is a rather mild illness compared with many affluent suburbanites, who have developed an almost total zoophobia, or fear of anything that moves. It is, as any traveller can confirm, a complaint best developed in the affluent North American, and seems to be part of blue toilet dyes, air fresheners, lots of paper tissues, and two showers a day.

It is very difficult, almost taboo, to talk of using rabbits, quail, pigs, poultry, or cows in city farms or urban gardens in the United States. They are commonplace city farm animals in England, and are ordinary village animals in Asia. Australians feel no repulsion towards them, and the edible guinea–pig lives comfortably in the homes of South Americans. But in the USA, no!

Useful animals are effectively abolished from American cities, leaving the field wide open for a host of others: pigeons forage the streets, thousands of gulls defecate in New York City reservoirs (fresh from the garbage piles); gigantic garbage bins are tipped over by large, flea-ridden dogs in Los Angeles; rats half the size of dogs (and also flea-ridden) are waiting for the garbage left by the dogs, and have tunnelled under the bus stops in their millions in Washington, D.C. (not far from the White House). They in turn are stalked by mangy cats, who also keep a desultory eye on the billions of cockroaches crawling in most houses. Not to mention the flies.

We will omit the legendary albino alligators of the sewers, and the rejected boa constrictors that pop up in the blue-rinse toilets. So much wasted food breeds its own population of pests. A sensible re-routing of edible garbage through a herd of pigs or a legion of guinea-pigs would abolish much of this nuisance, and a few good Asian restaurants could deal with the cats and dogs. The gulls would starve if chickens were fed on household wastes, and the besieged American might add a very large range of foods to those now available in cities. I mention this only to show that cultural prejudices can grossly reduce the available food resources, and that if we refuse to take sensible actions, some gross results can follow, with the biomass of useful foragers such as domesticated animals replaced by an equivalent biomass of pests.

MAXIMUM PRODUCT YIELD CONCEPT: THE "BIG PUMPKIN" FALLACY.

In a fluctuating climatic and market environment, the concept of forcing a *maximum product yield* is courting disaster. This is, however, the whole impetus of selling (e.g., the "big pumpkin" and "giant new variety" advertisements in seed catalogues), or in prizes awarded at agricultural shows. Better by far are more crop mixes and fail-safe systems that can produce in most conditions (wet or dry, cold or hot), or that hold constant value as subsistence (potato, taro, arrowroot) or have special value (vanilla, quinine, bamboo), or high food value per volume (fish, chicken).

The factors which can increase product yield are these:

- Genetic selection;
- Increased fertiliser (to a limited extent);
- Increased water (to a limited extent);
- Decreased competition from other non-beneficial species; and
- Better management in utilisation of yield and of harvest, timing, and integration.

They are the same factors which cause imbalance, as the selection of types for a particular yield need not be the factor that enables it to produce consistently in field field conditions (whether it be feathering to a "standard" in a chicken, redness in a rose, or weight in a fish). High-producing hens need biennial replacement (thus a constant breeding program) and may not even set their own eggs, thus needing artificial aids. A water and fertiliser-dependent crop is liable to collapse when it is water or nutrient stressed, or becomes too expensive to maintain in any market downturn. To go for one such crop, and so decrease diversity, is to decrease insurance for yield if one species or variety fails or is susceptible to change. Peasant farmers rightly reject advice based on maximum yield fallacies, and even more so if they share crops with a landlord, for they also value their spare time.

In the case of livestock, forced production is eventually limited by insoluble or intractable illnesses, so that in high-producing New Zealand herds, veterinary costs reach $120 per stock unit (for chronic illnesses such as facial eczema and white muscle disease). On less stressed pastures and farms, veterinary costs drop away to $20 or so per unit, top-dressing of pasture is reduced, and healthier herds give healthier yields. In the end, the forcing of product yields creates unique and inflexible health problems in plants, soils, and animals. Such yields become economically and ecologically unsustainable, and a danger to public health. 93% of chickens in battery cages develop cancers. If we eat cancer, we must risk cancer, for "we are what we eat" in a very real sense.

Insurance of some yield on a sustainable basis is better than expensive "feast and famine" regimes. The home garden is one such secure approach, where it is rare for all crops to fail, because of the innate diversity of such a mixed system. In fact, it is commonplace for gardeners to find a garden plant or some varieties fail in any one season, but no great harm results, as many other crops or varieties are available. Thus, species and and variety diversity are what people really need. Plant Variety Rights legislation, plant patenting, and multinational seed resource ownership has had a disastrous effect on the availability of hardy, adapted local varieties of plants, especially in Europe, where some 85% of locally-adapted seed crops have become "illegal", or have disappeared from seed company catalogues.

There are several paths open to us in design, and the least energy path is the one we seek, or evolve towards. There are two ways of producing an egg: the first has become the normal way in the western world (**Figure 2.3**), and the second is the way proposed by permaculture systems (**Figure 2.4**).

Some ridiculous systems have been evolved in which people, machines, time, and energy are expended in vast quantities on the chicken, perhaps with the aim of maximum product yield, regardless of costs. We can short-cut these systems with great gains in personal and planetary health, and with a far greater variety of yields available for local ecologies. These illustrations also bring home the commonsense nature of self-regulated systems.

2.6

CYCLES: A NICHE IN TIME

Cycles are any recurring events or phenomena. They have another implication, which is one of diversion. A cycle is, if you like, an interruption or eddy in the straight–line progression towards entropy. It is the special provenance of life to cycle materials. So efficiently does this happen that in a tropical forest almost all material nutrients are in cycle in life forms. It is this very complex cycling in the tropics which opened up so many opportunities for yield that thousands of species have evolved to take advantage of these.

If NICHES are opportunities in space, CYCLES are opportunities in time (a time–slot) and both together give harbour to many events and species. Geese eat grass, digest it, moult, produce waste products, add parasites, digestive enzymes, acids and alkalis, and defecate. The ground receives the rejecta, the sun shines, and rain may fall. Fungi, bacteria, grass roots and foliage work on feathers and faeces, and re–metabolise them into life. If we reorganise and encourage such cycles, our opportunities to obtain yields multiply. Every peasant farmer who keeps pigeons (as they still do in the Mediterranean borders) knows this truth. Here, every thinking farmer builds his own phosphate factory, as a pigeon loft.

Each such cycle is a *unique event*; diet, choice, selection, season, weather, digestion, decomposition, and regeneration differ each time it happens. Thus, it is the number of such cycles, great and small, that decide the potential for diversity. We should feel ourselves privileged to be part of such eternal renewal. Just by living we have achieved immortality—as grass, grasshoppers, gulls, geese, and other people. We are *of* the diversity we experience in every real sense.

If, as physical scientists assure us, we all contain a few molecules of Einstein, and if the atomic particles of our physical body reach to the outermost bounds of the universe, then we are all *de facto* components of all things. There is nowhere left for us to go if we are already everywhere, and this is, in truth, all we will ever have or need. If we love ourselves at all, we should respect all things equally, and not claim any superiority over what are, in effect, our other parts. Is the hand superior to the eye? The bishop to the goose? The son to the mother?

Principle of Cyclic Opportunity

Every cyclic event increases the opportunity for yield. To increase cycling is to increase yield.

People are built up molecule by molecule, cycling through themselves the materials of their environment: its air, soils, foods, minerals, and pathogens. Over time, people create their own local ecology (as do wombats and all sedentary animals); their wastes, exudae, and rejecta eventually create the very soils in which they garden. "Garbage in, garbage out" applies equally to computers and people. We gardeners are constantly cycling ourselves, and by a generational pattern of adjustment become "eco–compatible" with our landscape and climate. We are not the end point of evolution but a step on the way, and part of a whole sequence of cycles.

It is the number of such degenerative–regenerative cycles, unknowable to us, which determine the number of *opportunities* in the system, and its potential to change, mutate, diversify, and reintegrate. Not only can we never cross the same river twice, we can never see the same view twice, nor know the same system twice. Every cycle is a new opportunity. In nature, it is our right to die and make way for our successors, who are ourselves re–expressed in different forms.

It is our tolerance of the proliferation of life which permits such cycles. Deprived systems, like those blasted by biocides, lose most or all opportunity to transcend their prior state, and the egg of life is broken, degrades, and assumes a lower potential.

Tribal peoples are very much aware of, and tied to, their soil and landscapes, so that their mental and physical health depend on these ties being maintained. The rest of us have suffered forcible, historic dislocations from home sites, and many no longer know where home is, although there are new and conscious moves to reinhabit the earth and to identify with a **bioregion** as "home."

Travel itself causes stress and morbidity. Travellers both carry and acquire pathogens and spread them to other cultures. New settlers bring new species, new timetables, and new concepts. Local systems have to readjust, or fail. These processes are analagous to the disturbance of old ecosystems by new ecological or climatic forces. The post–invasion evolution contains part of the old and part of the new system, so is itself a new assembly with new potentials. Too often, however, we have destroyed very productive local ecologies, only to replace them with energy–consuming "improvements" of our own making. We have assumed the role of the creator, and destroyed the creation to do so.

Cycling of nutrients is continuous in the tropics, but is interrupted wherever drought, cold, or low nutrient status reduces the "base opportunity", just as the killing of fish stocks reduces the yield. Such cycles are slowed or even stopped by climatic factors or by our interference.

Cycles in nature are diversion routes away from entropic ends—life itself cycles nutrients—giving opportunities for yield, and thus opportunities for species to occupy time niches.

Cycles, like comets, have schedules or times to occur. Some are frequent and obvious like day and night, others long–term like sunspot cycles. Both short and long cycles are used in phenomenological reckoning by aborigines, who use cycle–indicators as time maps.

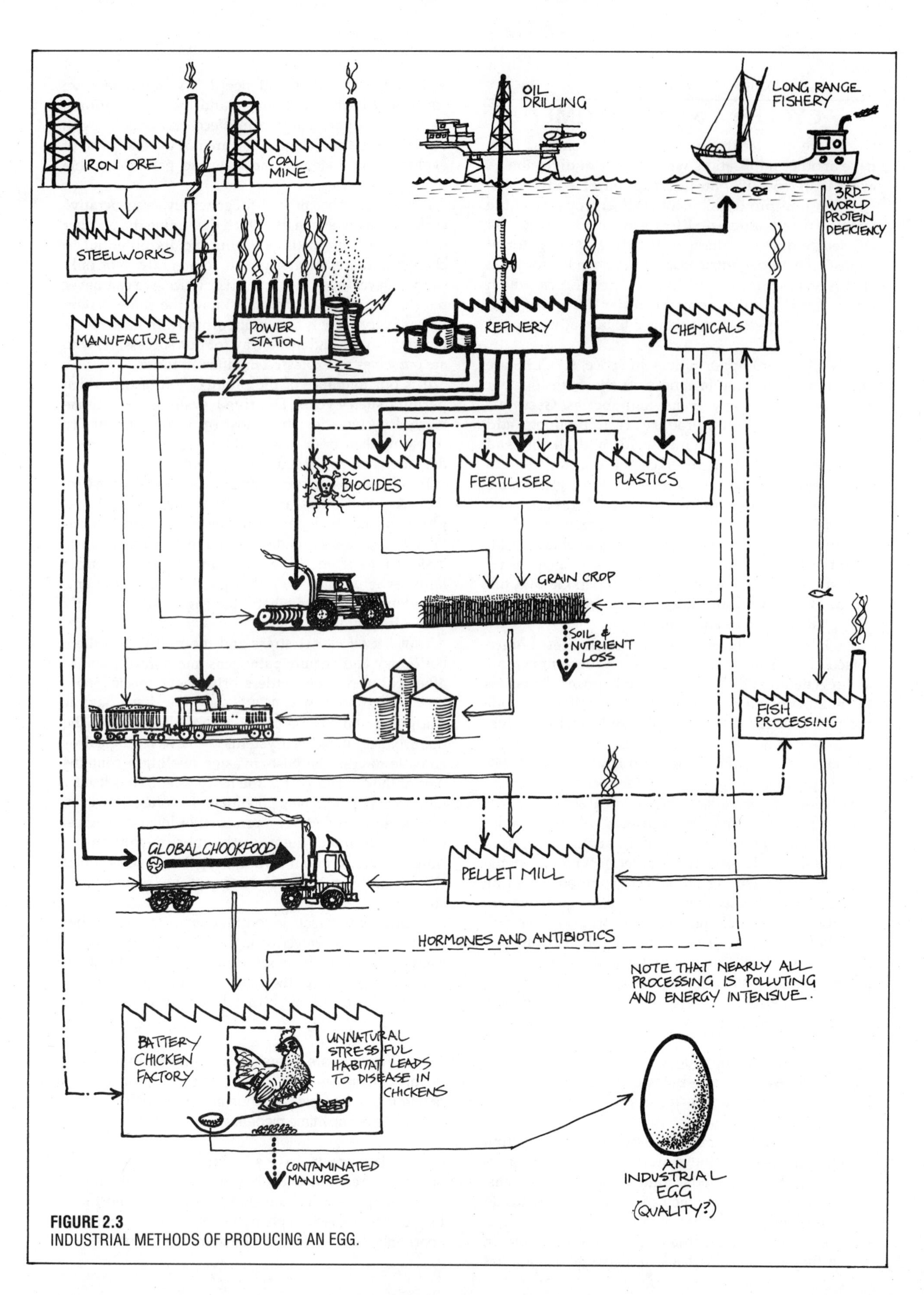

FIGURE 2.3
INDUSTRIAL METHODS OF PRODUCING AN EGG.

FIGURE 2.4
PERMACULTURE METHODS OF PRODUCING AN EGG.

ASPECTS OF THE TIME RESOURCE.

Time is a resource which can accumulate in ecosystems. It can be "lost" to an evolving or evolved system by setback (adverse disturbance), just about the same way as we can set back the hands of a clock. Such setbacks are termed *deflection states* by ecologists.

Ecosystems, especially those we are in process of constructing or destroying, are always proceding to some other state of evolution. Left alone, they may evolve at their own pace to some unknowable (or imaginary) endpoint, which we once called a climax state. However, forest climax states are temporary events in the long span of geological time.

Australian studies show that old dune forests lose the battle to mobilise nutrients, and begin to show a net nutrient loss, aided by rainfall and occasional fire, until they begin to recede to a less vigorous shrubbery system. Most other (disturbed) forests appear to be building, but (if disturbed too often) never reach the previous vigour, height, or yield. This is obvious to many of us who have seen original, regrowth, and second regrowth tree stands. These show signs of decay at progressively lower heights, and no doubt these too are losing vitality with age. I can sympathise.

Time can work as a rehabilitative resource, for active intervention in such successions enables us to analyse and to supply key nutrients and soil treatments, if needed, to assist maximum forest rejuvenation.

A second time concept is that of life–time, or the "quality time" that we have to enjoy, examine, and understand our world. To the interested observer, it would seem that life–time is very short indeed for those mobile, power–using, bombarded, employed, make–work, and busy humans who make up non–tribal societies, while many tribal peoples still manage to preserve a high quota of the celebrations, discussions, contemplations, mutual preening, and creative artwork on which many of us "wish we had time to spend..."

This erosion of the lifetimes of people, exacerbated by the media and messages of the consumer society, is perhaps the most serious effect of that society.

> Life is too much with us, late and soon
> Getting and gaining, we lay waste our years...
> (W. B. Yeats)

People so harried that they have "no time for anything else", may find that time has run out to save themselves, their lives, or those of their children.

A NICHE IN TIME AND SPACE.

Niche is a place to be, to fit in and find food, shelter, and room to operate. Many such niches are unfilled due to chance factors. Many are wiped out by agriculture or urban sprawl. Many can be created. But in pursuit of a simple food product, most farmers give no place for wildlife, no nesting sites or unbrowsed grass for quail or pheasant (both industrious insect eaters), and often no time for any intelligent assessment of the potential benefits of other species.

Existence is not only a matter of product yield, but a question of appreciating variety in landscape. Evolving plant systems and existing animals provide niches for new species: the cattle egret follows cattle; the burrows of rabbits are occupied by possum, bandicoot, snakes, frogs, and feral cats; and the growing tree becomes a trellis, shade spot, and a host to fungus and epiphytes.

Every large tree is a universe in itself. A tree offers many specialty–forage niches to bird, mammal, and invertebrate species. For instance, yellow–throated honeyeaters (in Tasmania) search the knot–holes for insects, treecreepers the bark fissures, strongbilled honeyeaters the rolls of branch bark and hanging strips of bark, and blackheaded honeyeaters the foliage, where pardalotes specify the scale insects as *their* field. As for time–sharing, the yellow–throats are permanent and territory–holding residents, the treecreepers migrants, the strongbills and blackheads roving flock species, and all of them scatter as breeding pairs in the spring and summer, so that it is rare to find any one tree fully occupied at any one time. There is also a pronounced post–breeding tendency for several bird species to form **consociations** for foraging and travelling in autumn and winter. Five to eight species travel together, some (e.g. fly-catchers) gathering insects disturbed by the others, with all species reacting to the alarm calls of any one species, but some species (mynahs for example) acting as sentinels for the whole mixed company.

Here, we see time, space, and functions all used in a complex and non–competitive way, and glimpse something of the potential for designers to enrich human societies *providing that no individual or group claims a right to sole use at all times* for an area. The failure of a monoculture to produce, sustain, or persist is thus easily explained, as many species are invading or trying to use more efficiently the complex resources of time and space.

A combined space–time factor is called a **schedule**: a time to be in that place. Any observer of public park use sees the usage change hour by hour. Morning joggers give way to lunch–time office workers, who are succeeded by older, retired people playing draughts, later displaced by evening entertainment crowds, and late at night, the people on the edge of time: the semi–legal, the unemployed, and the lonely. Towards dawn, only the lame and isolated strollers, often with dogs for companions, remain on the streets.

Many mammals, forced to develop tracks and resting places, do not control "areas", but rather time–slots in space. My own studies of wild wallaby, urban people, and possum show this to be the case. Fighting occurs when one is *out of schedule*, and ceases when that place is vacated for use.

Schedules may run on long cycles, tuned to the level of browse or succession of vegetation, e.g. a sequence of grazing has been observed for African herds, so that

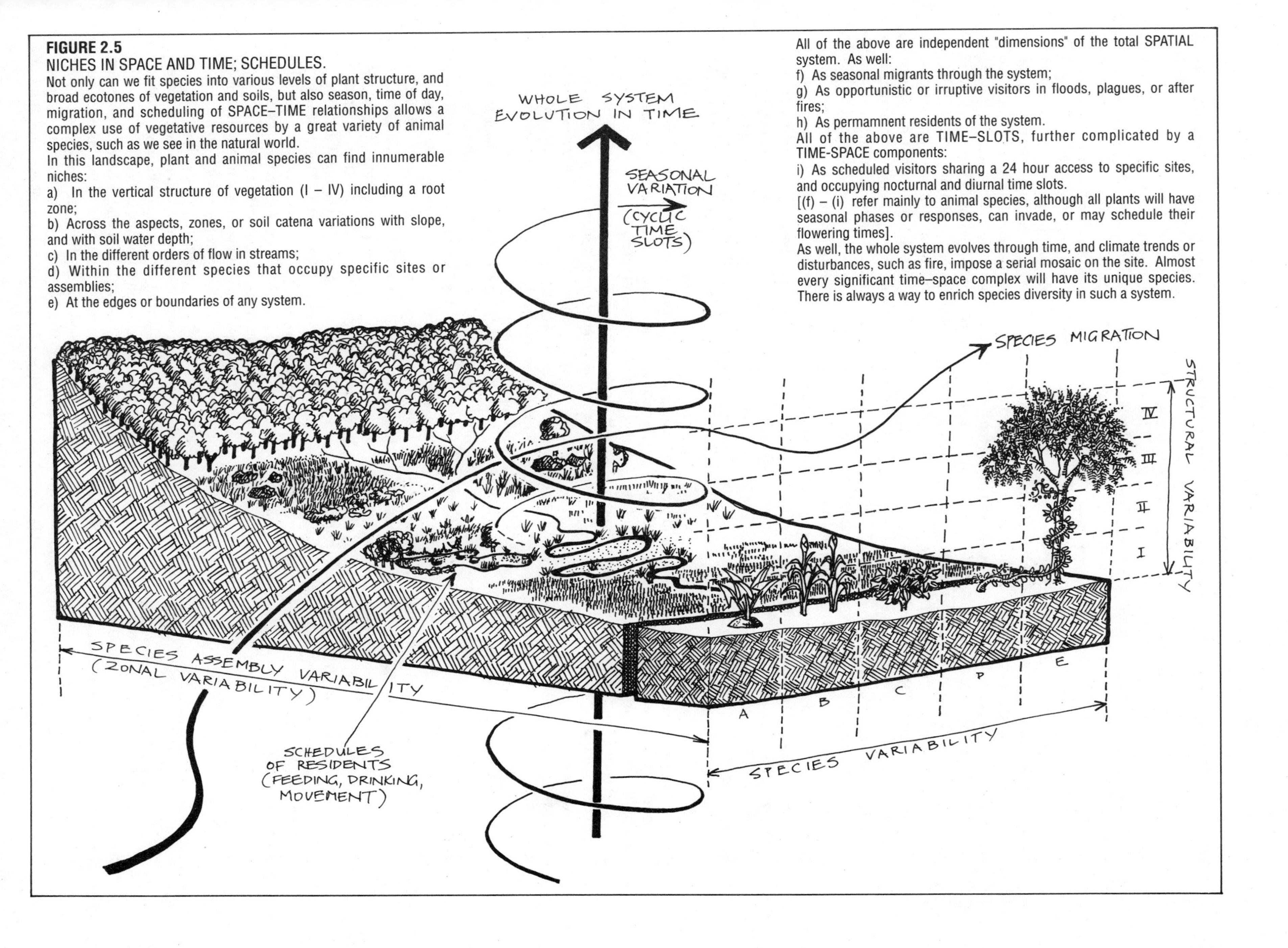

FIGURE 2.5

NICHES IN SPACE AND TIME; SCHEDULES.

Not only can we fit species into various levels of plant structure, and broad ecotones of vegetation and soils, but also season, time of day, migration, and scheduling of SPACE–TIME relationships allows a complex use of vegetative resources by a great variety of animal species, such as we see in the natural world.

In this landscape, plant and animal species can find innumerable niches:

a) In the vertical structure of vegetation (I – IV) including a root zone;

b) Across the aspects, zones, or soil catena variations with slope, and with soil water depth;

c) In the different orders of flow in streams;

d) Within the different species that occupy specific sites or assemblies;

e) At the edges or boundaries of any system.

All of the above are independent "dimensions" of the total SPATIAL system. As well:

f) As seasonal migrants through the system;

g) As opportunistic or irruptive visitors in floods, plagues, or after fires;

h) As permamnent residents of the system.

All of the above are TIME–SLOTS, further complicated by a TIME-SPACE components:

i) As scheduled visitors sharing a 24 hour access to specific sites, and occupying nocturnal and diurnal time slots.

[(f) – (i) refer mainly to animal species, although all plants will have seasonal phases or responses, can invade, or may schedule their flowering times].

As well, the whole system evolves through time, and climate trends or disturbances, such as fire, impose a serial mosaic on the site. Almost every significant time–space complex will have its unique species. There is always a way to enrich species diversity in such a system.

antelope follow wildebeest follow elephant (or some such sequence) for many herd species.

This suggests that informed graziers, knowing the preferences of different species (sheep follow cattle follow horses follow goats) can make much better use of the basic browse resource by scheduling rotation (not to keep one level of browse constant, but to dynamically balance levels by species succession).

Scheduling (the "right" to use a particular space at a specific time) occurs within species, where dominant animals use prime grazing land at prime time, and sub–dominants are pushed to the edge of time and space, or between species, so that sequences of different species use the same area of vegetation at different seasons or stages of growth. No individual "owns" the area, just a time–space slot (like a chair in a family kitchen at dinnertime). In Tasmania, there are two prime time activity peaks for wallaby over 24 hours, both at night: the main one is crepuscular (just after sundown), and the secondary one is auroral (just before dawn). This permits digestive and recuperative rest periods, denied to weaker animals who cannot compete for preferred periods. Within this framework, any possum can, by aggression, displace a wallaby at a feeding–place. Any individual holds a place only for a short time, moving on to contest another area until satiated. Thus, the sharing of resources is a complex dynamic, but no species or individual has sole rights. A human analogy would be that of a sports–ground used by different sports groups at times, by gulls or rodents whenever sports are not being played, and by worms at all times.

To summarise, we have:

- Niche in space, or "territory" (nest and forage sites);
- Niche in time (cycles of opportunity); and
- Niche in space–time (schedules).

Between these, there is always space or time available to increase turnover. Niches enable better utilisation and greater diversity, hence more yield. Of all of these niches, schedules are the best strategy for fitting in new species of mammals, providing these are not territorial species (which try to hold their own space at all times), but are chosen from cooperative species which yield space when the time is right (see **Figure 2.5**). There are lessons here for people: those who try to hold on to all things at all times prevent their use by others.

2.7

PYRAMIDS, FOOD WEBS, GROWTH, AND VEGETARIANISM

A figure often used to explain how much of a food or forage is needed to grow another animal is the *trophic pyramid*. While the pyramid is a useful concept, it is very simplistic, and in all but laboratory conditions or feed–lot situations, it is unrelated to field reality, and may only apply where we actually provide simple food to captive species. The field condition is very different (**Figure 2.6**).

The pyramid is often used to support claims that we should all become vegetarians, or herbivores. This is perhaps not so far from the truth, but there are real–world factors to consider. I have shown the pyramid and also a direct path (herbage to human) to illustrate how we would support more people if we ate vegetation. But we need to re–examine this concept for people who return their wastes to gardens. There are the following factors to consider:

1. NATURE IS MUCH MORE COMPLEX than is shown in a pyramid. Instead of simple "trophic levels", we have a complex interaction of the same species, largely governed not by food habits, but by pasture management practices. Such a complex diagram is called a *food web*, and is the normality in field conditions.

2. PYRAMIDS IGNORE FEEDBACK. In a very real sense vegetation eventually "eats" grasshoppers, frogs, fish, and people. Not only that, but as an animal grows, it returns nutrient to the soil via excreted, moulted, or discarded body wastes, and even if the frog eats 10 kg of grasshoppers to make one kilo of frog, it doesn't (obviously) keep the 10 kg in a bag, but excretes 9 kg or more back to earth as manures. This causes more vegetation to grow, thus producing more grasshoppers. The manure from insect "pests" may be the basis of a regenerative future evolution.

With these obvious feedbacks, the web itself becomes much more complex, and it starts to resemble less of a one–way staircase (the pyramid) than a series of cyclic events; less of a ziggurat and more like a spider's web. So that the real position is that waste recycling to herbage is the main producer of that herbage.

3. WHAT OF MATURITY? If our fish (level 4) was a carp, and that carp was more than a year or two old, then it would probably have reached full size, although it may then continue to live for another 80–100 years.

So now, the carp (at 80 years old and 10 kg weight) has eaten 100 x 10 kg = 1000 kg of frogs and insects, and has returned 990 kg of digested material per year to the pond, to grow more herbage. Thus, in order to keep the system in growth, we must be able to efficiently crop *any* level just before maturity is reached. We can see that old or mature systems no longer use food for growth, but for *maintenance*. So it is with mature fish, frogs, forests, and people.

Old organisms thus become constant recyclers (food in, waste out) and cease to grow, or they even begin to lose weight. This is why we try to use only young and growing plants and animals for food, if food is scarce. An exception is a fruit or nut tree, where we consume seed or fruit (seed is an immature tree).

4. ARE FOOD CHAINS SO SIMPLE? We know that people normally eat vegetation, and that many people eat grasshoppers, frogs, fish, and (at times) other people. Even a cow eats grasshoppers as it eats grass, and of course every eater ingests large quantities of

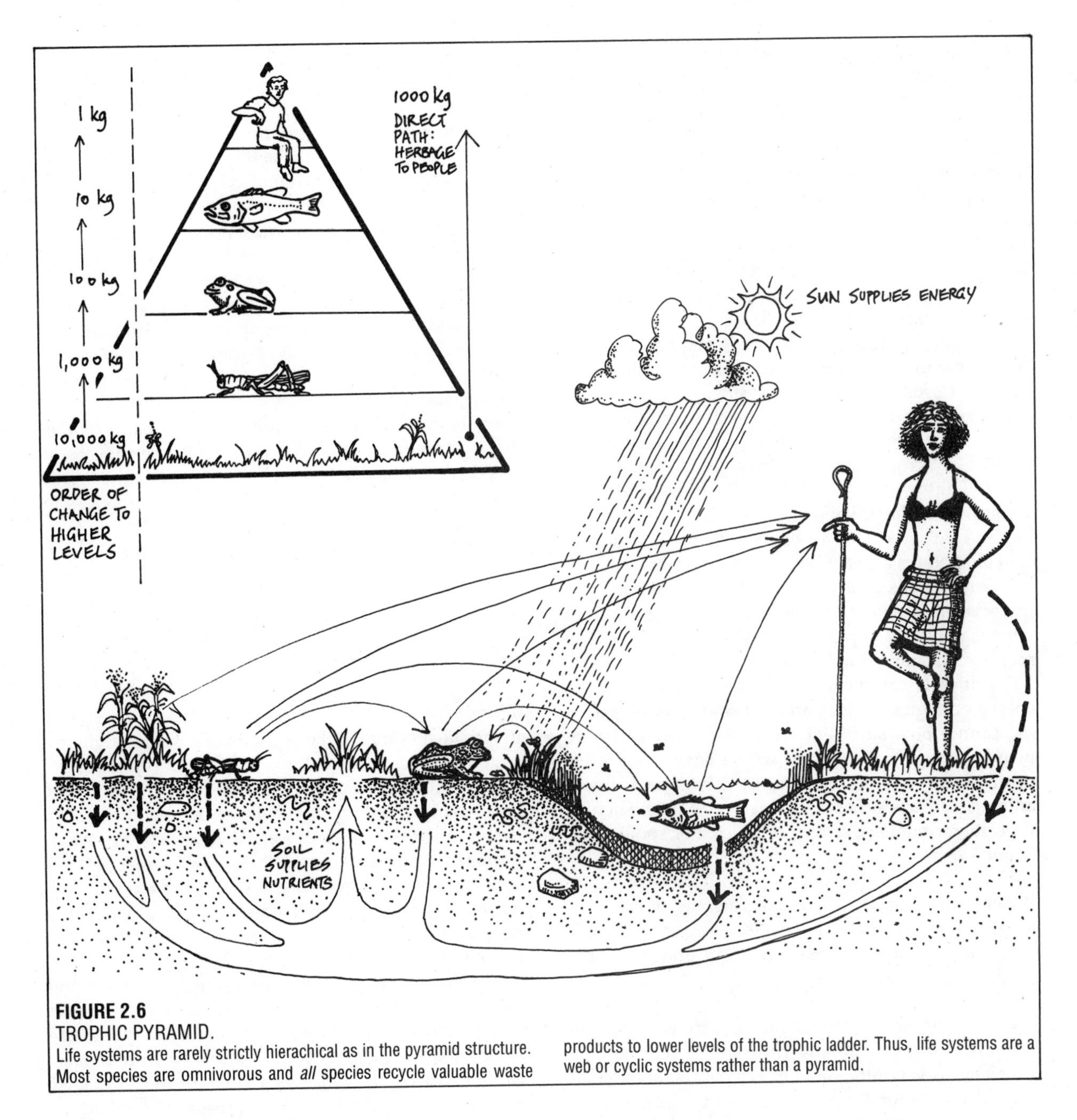

FIGURE 2.6
TROPHIC PYRAMID.
Life systems are rarely strictly hierachical as in the pyramid structure. Most species are omnivorous and *all* species recycle valuable waste products to lower levels of the trophic ladder. Thus, life systems are a web or cyclic systems rather than a pyramid.

bacteria and small animals living on vegetation. "Man cannot live by bread alone", unless in a sterile laboratory condition!

As our one–way pyramid is very suspect, so is the argument that we should become vegetarians to ameliorate the world food shortage problem. Only in home gardens is most of the vegetation edible for people; much of the earth is occupied by inedible vegetation. Deer, rabbits, sheep, and herbivorous fish are very useful to us, in that they convert this otherwise unusable herbage to acceptable human food. Animals represent a valid method of storing inedible vegetation as food. If we convert all vegetation to edible species, we assume a human priority that is unsustainable, and must destroy other plants and animals to do so.

In the urban western world, vegetarianism relies heavily on grains and grain legumes (e.g. the soya bean). Even to cook these foods, we need to use up very large quantities of wood and fossil fuels. Worse, soya beans are one of the foods owned (100% of patent rights) by a few multinationals. They are grown on rich bottomland soils, in large monocultural operations, and in 1980–82 caused more deforestation in the USA and Brazil than any other crop. Worse still, about 70% of the beans were either fed to pigs, or used in industry as a base for paint used on motor vehicles!

Much worse again, grains and grain legumes account for most of the erosion of soils in every agricultural region, and moreover, very few home gardeners in the developed world ever grow grains or grain legumes, so that much of what is eaten in the West is grown in areas where real famine threatens (mung beans from India, chick peas from Ethiopia, soya beans from Africa and India).

Old farmers and my own great–grandfather had a saying that bears some consideration. This was: "We will sell nothing from our farm that will not walk or fly off." In effect, the farmer was concerned to sell *only* animals, never crops or vegetation, because if the farm was to survive without massive energy inputs, animals were the only traditional recycling strategy for a sustainable export market.

What does all this mean to concerned and responsible people in terms of their diet and food habits, with respect to a sustainable natural system?

1. Vegetarian diets *are* very efficient, providing:
   - They are based on easily cooked or easily processed crop grown in *home gardens;*
   - That wastes, especially body wastes, are returned to the soil of that garden; and
   - That we eat from where we live, and do not exploit others or incur large transport costs.
2. Omnivorous diets (any sort of food) make the best use of *complex natural systems;* that we should eat from what is edible, at any level (except for other people in most circumstances, and under most laws!)
3. Primarily carnivorous diets have a valid place in special ecologies, such as areas of cold, where gardening cannot be a sufficient food base; in areas where people gather from the sea; where harsh conditions mean reliance on animals as gatherers; and where animals can use otherwise–waste products, such as vegetable trimmings, scraps, or rejected or spoilt vegetation.
4. We should always do our energy budgets. Whatever we eat, if we do not grow any of our own food, and over–use a flush toilet (sending our wastes to sea) we have lost the essential soil and nutrients needed for a sustainable life cycle.

While a tropical gardener can be very efficient and responsible by developing fruit and vegetable crop which needs very little cooking, sensible omnivorism is a good choice for those with access to semi–natural systems. City people using sewers would be better advised to adopt a free–range meat diet than to eat grain and grain legumes. Better still, all city waste should be returned to the soils of their supply farms.

Even in our garden, we need to concentrate on cycles and routes rather than think in pyramids. Simplistic analysis of trophic levels fails to note that some food resources are unusable by people, either because the energy needed in processing is too much (since some products of nature are poisonous or unpalatable foods), or because the resources are too scattered to repay our collection. Such resources are often harvested into useful packages by other species. Herons, themselves edible, eat poisonous toadfish, and goats will browse thorny and bitter shrubs. Thus we can specify these useful conversions before blindly eliminating a life element *of any type* from our diets.

While it is manifestly immoral to feed edible Peruvian fish to hogs in the USA, it may be of great value to convert forest acorns (unharvested in the USA) to hogs, and let the pilchards and anchovies feed the hungry of Peru (which includes the pelicans!)

The trophic pyramid is valid enough as a conceptual model, for we can see that poisons at the base concentrate at the top. In fact, the highest level of some radioactives and DDT measured are found in mothers' milk. We can see that the generalised or omnivorous (non–selective) feeder is buffered from catastrophic famine by a complex web of trophic connections, so that some losses and some gains accrue to people, being generally omnivorous.

In short, people need to discard fixed ideas, examine their kitchen cupboards, and try to reduce food imports, waste, and energy loss. A responsible diet is not easy to achieve, but the solutions lie very close to home. Viva the home gardener!

2.8 COMPLEXITY AND CONNECTIONS

There are ecologies on very flat and somewhat invariant sites that in the end simplify, or are originally simple because that condition itself is not typical of the earth's crust, just as a field, levelled, drained, and fertilised for a specific crop will not support the species that it once did when it varied in micro–elevation, drainage, and plant complexities. Other simple natural ecologies occur where rapid change can occur (sea coasts), or where we deliberately fire or plough on a regular basis, so that there is never enough time for a diverse system to establish.

Marshes, swamps, tidal flats, salt–pans, and level deserts support less diversity than adjoining hill and valley systems, but nevertheless in sum (if species are assembled from global environments or from similar climatic areas) are still very rich, and, in the case of mangrove and tidal marsh, extremely productive ecologies.

It is not that a single stand of one mangrove species is itself so diverse, it is the mobile species working at different stages of decomposition of the mangrove leaf. Each of these species in turn feeds others.

Thus, very simple plant associations may support very productive and complex animal associations. Mobile species are capable of occupying a great variety of niches in one mangrove tree or swamp stand, from underground to canopy, and of schedules from low to high tide. Time and space are needed for tree species to evolve a complex stand in such situations, and as they are often obliterated and re–established by a world–wide change due to a sea level fluctuation, relatively little time can be allowed for mangrove species to themselves develop and colonise the new, and potentially short–term, shoreline.

Old deserts, like that of Central Australia, may exhibit some 3000 species of woody plants, while recently desertified areas, like those of southwest Asia, may have as few as 150 plants surviving the recent changes from forest. We can, in these cases, act as the agents of constructive change, bringing species to assist local re–colonisation from the world's arid lands. Such species will assist in pioneering natural reafforestation. This has not generally been our aim, and we annually destroy such invaluable species complexes to grow a single crop such as wheat, thus laying waste to the future.

The number of elements in an aggregate or system certainly affects its potential complexity, if complexity is taken to be the number of functional connections between elements. In fact, as Waddington (1977) points out, in the case of a single interaction (a conversation) between elements, complexity goes up roughly as the square of the number of elements: "Two's company, three's a crowd"...and five or six is getting to be a shambles!

This is bad enough, but if we consider the number of possible connections to and from an element such as a chicken (**Figure 3.1**), we can see that these potential connections depend on the information we have about the chicken, so that the complexity of a system depends on the information we have about its components, always providing that such information is used in design. As we cannot know everything, or even know more than the approximate categories and quantity of things which are (for example) eaten by chickens, thus in permaculture we always suppose that the chicken is busy *making connections itself*, about which we could not know and, of course, for which we could not design. We must simply trust the chicken.

Thus, in commonsense, we can design for what we believe to be essentials, and let the chicken attend to all the details, checking at later stages to see that yields (our ultimate products) are satisfactory, the chickens healthy and happy, and the system holding up fairly well.

It is important to concentrate on the nature or value of *connections between elements*. In nature, we can rarely connect components as easily as a wire or piece of pipe can be fixed into place. We do not "connect" the legume to the orange tree, the chicken to the seed, or the hedgerow to the wind; we have to understand how they function, and then place them where we trust they will work. They then proceed to do additional tasks and to provide other connections themselves. They do not confine their functions to our design concepts!

Evolving complex species assemblies in isolated sites, like the Galapagos Islands, may depend more on a species–swarm arising from pioneer or survivor species than on invaders adapting from borderlands. Only when many niches are empty is a species able to differentiate and survive without competition; so the dodo and Darwin's finches arose. Having arisen, they may then well prove to be very useful to other systems. Unique island species often have functions not easily found in continental and crowded ecologies; frequently, hardy travelers like reptiles and crustaceans take up those niches that, on continents, are occupied by species of mammals and birds.

It is not enough to merely specify the *number* of connections, and not note their value in the system as a whole; it may be possible that complex social situations and cultivated or chance complexity may occur in natural systems by introductions or migrations. These new events, although increasing complexity, may reduce stability with respect to a desirable local yield. Thus, where the *benign* complexity of cooperative organisms is useful, competitive or inharmonious complexity is potentially destructive. Again, it is a question of matching needs with products, and of the values given to connections.

2.9
ORDER OR CHAOS

It follows that order and disorder arise not from some remote and abstract energy theory but from actual ground conditions or contexts, both in natural and designed systems. Entropy is the result of the framework, not the complexity. A jumble of diverse elements is disordered. An element running wild or in an active destructive mode (bull in a china shop) is disordered, and too few or too many forced connections lead to disorder.

Order is found in things *working beneficially together*. It is not the forced condition of neatness, tidiness, and straightness all of which are, in design or energy terms, disordered. True order may lie in apparent confusion; it is the acid test of entropic order to test the system for yield. If it consumes energy beyond product, it is in disorder. If it produces energy to or beyond consumption, it is ordered.

Thus the seemingly–wild and naturally–functioning garden of a New Guinea villager is beautifully ordered and in harmony, while the clipped lawns and pruned roses of the pseudo–aristocrat are nature in wild disarray.

Principle of Disorder

Order and harmony produce energy for other uses. Disorder consumes energy to no useful end.

Neatness, tidiness, uniformity, and straightness signify an energy–maintained disorder in natural systems.

2.10
PERMITTED AND FORCED FUNCTIONS

All key living elements may supply many functions in a system, but if we try to force too many work functions on an element, it collapses. One cannot reasonably expect a cow to give milk, raise a calf, forage its own food, plough, haul water, and tread a corn mill. *Forcing* an element to function, however, is a very different proposition from putting it in position where its natural or everyday behaviours permit benefits to other parts of the system.

Placed correctly, a tree or chicken experiences no stress not common to all trees and chickens about their daily business. Further, if we place any of the other elements needed close by, the tree or chicken has less stress than normal. It is the design approach itself that permits components to provide many functions without forcing functions (that are not in any case inherent) upon that element. The chicken may be busy, but not overworked.

People, too, like to be where their very different and complementary capabilities are used rather than being forced to either a single function (like a 300–egg–a–year chicken or a typist confined to a computer operation in an office), or so many functions that they suffer deprivation or overload (like our cow above).

Principle of Stress and Harmony

Stress here may be defined as either prevention of natural function, or of forced function. Harmony may be defined as the integration of chosen and natural functions, and the supply of essential needs.

2.11 DIVERSITY

Diversity is the number of different components or constructs in the system; an enumeration of elements and of parts. It has no relationship to connections between components, and little to the function or the self–regulating capacity of any real system (within the boundaries of too few or too many components). Thus diversity either of components or assemblies does not of itself guarantee either stability or yield. Where we *maintain* such diversity, as in our gardens, then this may guarantee yield, but if we leave our gardens, they will simplify, or simply be obliterated by non–maintained and hardy species adapted to that site (as is evident in any abandoned garden).

Thus, our own efforts are an integral part of maintaining diversity in a permaculture system. Few species grown by people persist beyond the lifetime of those species if we leave the situation alone. Australia is a country where towns may arise and be abandoned to serve a mining or port operation. Where these were built in forested areas, they are obliterated by forest in 30–80 years, with perhaps a few trees such as dates, mulberries, and figs persisting in savannah or isolated dryland locations. These "survivor" trees are important to note in planning longer–term stability for that region.

Great diversity may create chaos or confusion, whereas multiple function brings order and develops resources. I believe that a happy medium is to include as much diversity in a cultivated ecosystem as it can maintain itself, and to let it simplify or complicate further if that is its nature.

Very diverse things, especially such abstract systems as competing beliefs, are difficult to make compatible with any natural system, or knowledge, so that some sorts of dogmatic diversity are as incompatible as a chicken and a fox. Although true incompatibility may be rare, one should be prepared for it to exist, and an intervening neutral component can be introduced, as is the case when growing those "bad neighbours" apples and walnuts, where it is necessary to intervene with a mulberry, which gets along with them both.

Principle of Stability

It is not the number of diverse things in a design that leads to stabilty, it is the number of beneficial connections between these components.

It follows that adding in a technology or living species "just to have it there" has no sense to it. Adding it in to supply a need or consume an otherwise wasted resource—*to do something useful*—makes a great deal of sense. Often, however, we lack functional information on components and may therefore leave out technologies or species in designs which would have been useful had we known. Thus,

Information is the critical potential resource. It becomes a resource only when obtained and acted upon.

In the real world, resources are energy storages; in the abstract world, useful information or time. Watt[13], in his categories of resources, includes time and diversity. Diversity *of itself* is now not seen as a resource, but a diversity of *beneficial functional connections* certainly is a resource. *Complexity*, in the sense of some powerful interconnections between species, is what we are really seeking in food systems. Such complexity has its own rules, and we are slowly evolving those rules as recommendations for polycultures (dealt with elsewhere in this book under their climatic characteristics), or as "guilds" of plants and animals that assist each other.

Peter Moon (*New Sientist*, 28 Feb. '85) differentiates between *richness* (the number of species per unit area), *diversity* (the relative abundance of species), and *evenness* (how species contribute to the biomass total). He notes that richness may decrease in plants as systems age, when shade and competition reduce annuals or weaker species, but that richness may then increase in animals such as decomposers, due to the development of a greater range of niches and microclimate (more animals live in ungrazed or uncut grasslands, but less plant species survive).

Richness of tree species has very recently been correlated to the energy use of that plant community, as measured by evapotranspiration (*New Scientist*, 22

Oct '87) . Thus species–rich regions are not so much correlated to latitude, allied to richness in birds and mammals, or as result of prior events such as glaciation or fire, but are essentially linked to the basic productivity of the region. Within this broader framework, local niches or a range of altitudes can create more diversity; such measures refer to present, not past, climate.

Some disturbance or "moderate stress" such as we achieve in gardens provides the richest environment. We can actively design to allow some undisturbed (low stress) islands of vegetation, while mowing or digging in other areas (high stress), thus getting the best of both worlds in terms of a stress **mosaic**. We can also be active in plant and animal maintenance, increasing or decreasing grazing pressures, thus managing species abundance locally.

2.12 STABILITY

The short meaning of stability in an ecosystem is *self–regulation* rather than a climax (end–point) stability. Nothing in nature remains forever, not soil or hills or forests. For our foreseeable future we can have dynamic life–support systems, as tribal people have demonstrated to us all over the world, sometimes for thousands of years of constructive regulation.

Thus, stability in ecosystems or gardens is not the stability of a concrete pylon; it is the process of constant feedback and response that characterises such endeavours as riding a bike. We are also in an area of uncertainty about the concept of end states or climax in systems—the state to which they tend to evolve. It is doubtful if any such state ever existed, as inexorable climatic change, fire, nutrient leaching, and invasion deflect systems from their apparent endpoints.

Moreover, it is probable that very old systems are also fragile, having been long in a state of maintenance, and we may see sudden or slow collapse in such evolved states. John Seymour (*Ecos*, Summer '81–82), notes the slow loss of nutrients in an old stable dune system at Cooloola in Australia. Here, climax is a passing phase as the virgin dunes lose nutrient status to fire and water filtration to great depths, where nutrients become unavailable to trees. Thus, the study of very old systems shows a retreat from the "most evolved" (greatest biomas) condition unless some new factor is introduced (ash from a volcano, fertiliser applied by people).

Daniel Goodman [*Quartertly Review Biology*, 50(3)] notes that "wild fluctuations" may occur in tropical forests, or in savannah grasslands. Epidemics of pathogens may affect a plant or animal species and sadly decrease its numbers. Although these natural fluctuations pale beside our own effects on ecosystems, such disturbances, providing they affect only a few species, are not as severe as persistent nutrient loss (or acid rain).

All these effects are under some human control in a developed ecosystem. Protection from fire, positive nutrient supply to plants, and long–term evolutions are possible *in terms of human occupancy*. In the longer term, however, we too will be gone, and other species will arise to replace us (unless we take the earth with us, as megalomaniacs would do if we give them that chance: "If I can't take it with me, I'm not going....!") Just as it was the habit of kings to be buried with their riches, horses, and slaves, so modern warlords threaten to bury all humanity as they depart.

2.13 TIME AND YIELD

Old systems store up their energy in bulky unproductive forms, e.g. an old forest has large trunks, roots and limbs, and old fish are "on maintenance". Such ancient systems composed of large individuals (trees or animals) need energy just to maintain their health, and thus they can use less of the available sun energy, so that flow of energy *through* the system is less. Therefore the yield, or turnover of matter, is less. This too is a function of time (ageing). Matter is used up in system maintenance, and is not available as yield, or as increasing size or weight in life components.

Against this factor, species diversity (richness) works to make the most of incoming energy.

> Carlander has shown that the standing crop of fish in different reservoirs is an increasing function of the number of species present.
>
> (Watt[13])

This is also true of studies in most "wild" systems, where the complexity and standing crop are both much more than the simple cultivated ecology which replaces them. Thus, the clearing of an African veld or an Australian savannah of their web of species, and their replacement with a few perennial pasture plants and beef cattle, or with a single–species pine forest not only takes enormous energy but also grossly decreases total yields.

We would do better to try to understand how to manage natural yields, and modify such systems by management than to replace them with "economic" (here economic means monetary rather than energy return) systems which impoverish the yield and encourage disaster via pests and soil loss. Economics in future will inevitably be tied to yield judged on energy rather than on monetary return. In the present economy, we waste energy to make money. But in the very near future, any system which wastes energy must fail.

Pond and hedgerows both slowly gain species as they age, probably as a function of natural dispersal plus new niche evolution created by other species. This continues until the system begins to be overshadowed by a few large dominants or hyper-predators whose biomass represents an end storage of energy, and a decreasing yield in the total system.

Only local disturbance (fire, flood, death) renews the flow of energy through old systems. The time of cycling of natural systems may be a very long period, but in annual cropping it may be reduced to just one season or less. Permaculture thus uses the time resource much better than does annual gardening alone, and so uses sun energy to better effect. The mixed ecology of annuals and perennials maximises not only product yield, but also the resourcefulness of the men and women who establish, control, and harvest, it. It is only in a thoughtless, monetary, and doomed economy that we can evolve the concept of unemployed and unwanted human beings.

Death in over–mature systems is thus seen as the essential renewal of life, not in the negativistic sense of the fatalist, but in a positivistic and natural way. It is better that elements die, and are renewed by other species, than the system simplifies to extinction. It is better for the tribe if its components change than if it turns in on itself, ages, and decays as a whole. Life is then seen as a preparation for succession and renewal, rather than a journey to extinction.

Time as Watt notes is a resource. Like all resources, too much of it becomes counterproductive, and a system in which too much time is accumulated becomes chronically polluted, as a system in which not enough time has accumulated is below peak yield. A strawberry seedling and an old strawberry bush are equally unproductive, as are the very young and the very old in society. As there are age–specific diseases in people (whooping cough, prostate hypertrophy) so there are age–specific diseases in whole systems, and a mixed–age stand is the best insurance against complete failure or epidemic disease of this nature. As individuals, we have a right to live a responsible life, and a right to die. If our efforts to prevent ageing succeed, we may produce a crowded, unstable, and unproductive society subject to gerontocratic peevishness!

2.14 PRINCIPLE SUMMARY

The Prime Directive of Permaculture: The only ethical decision is to take responsibility for our own existence and that of our children's.

Principle of Cooperation: Cooperation, not competition, is the very basis of future survival and of existing life systems.

The Ethical Basis of Permaculture:

1. CARE OF THE EARTH: Provision for all life systems to continue and increase.

2. CARE OF PEOPLE: Provision for people to access those resources necessary to their existence.

3. SETTING LIMITS TO POPULATION AND CONSUMPTION: By governing our own needs, we can set resources aside to further the above principles.

Rules of Use of Natural Resources:

- Reduce waste, hence pollution;
- Thoroughly replace lost minerals;
- Do a careful energy accounting; and
- Make a biosocial impact assessment for long term effects on society, and act to buffer or eliminate any negative impacts.

Life Intervention Principle: In chaos lies unparalleled opportunity for imposing creative order.

Law of Return: Whatever we take, we must return, or
Nature demands a return for every gift received, or
The user must pay.

Directive of Return: Every object must responsibly provide for its replacement. Society must, *as a conditions of use*, replace an equal or greater resource than that used.

Set of Ethics on Natural Systems:

- Implacable and uncompromising opposition to further disturbance of any remaining natural forests;
- Vigorous rehabilitation of degraded and damaged natural systems to a stable state;
- Establishment of plant systems for our own use on the *least* amount of land we can use for our existence; and
- Establishment of plant and animal refuges for rare or threatened species.

The Basic Law of Thermodynamics [as restated by Watt[13]]:

"All energy entering an organism, population or ecosystem can be accounted for as energy which is stored or leaves. Energy can be transferred from one form to another, but it cannot disappear, or be destroyed, or created. No energy conversion system is ever completely efficient."

[As stated by Asimov (1970)]: "The total energy of the universe is constant and the total entropy is increasing."

Birch's Six Principles of Natural Systems:

1. Nothing in nature grows forever. There is a constant cycle of decay and rebirth.

2. Continuation of life depends on the maintenance of the global bio–geochemical cycles of essential elements, in particular carbon, oxygen, nitrogen, sulphur, and phosphorus.

3. The probability of extinction of populations or a species is greatest when the density is very high or very low. Both crowding and too few individuals of a species may reach thresholds of extinction.

4. The chance that a species has to survive and reproduce is dependent primarily upon one or two key factors in the complex web of relations of the organism to its environment.

5. Our ability to change the face of the earth increases at a faster rate than our ability to foresee the consequence of change.

6. Living organisms are not only means but ends. In addition to their instrumental value to humans and other living organisms, they have an intrinsic worth.

Practical Design Considerations:

• The systems we construct should last as long as possible, and take least maintenance.

• These systems, fueled by the sun, should produce not only their own needs, but the needs of the people creating or controlling them. Thus, they are sustainable, as they sustain both themselves and those who construct them.

• We can use energy to construct these systems, providing that in their lifetime, they store or conserve more energy than we use to construct them or to maintain them.

Mollisonian Permaculture Principles:

1. Work with nature, rather than against the natural elements, forces, pressures, processes, agencies, and evolutions, so that we assist rather than impede natural developments.

2. The problem is the solution; everything works both ways. It is only how we see things that makes them advantageous or not (if the wind blows cold, let us use both its strength and its coolness to advantage). A corollary of this principle is that everything is a positive resource; it is just up to us to work out *how* we may use it as such.

3. Make the least change for the greatest possible effect.

4. The yield of a system is theoretically unlimited. The only limit on the number of uses of a resource possible within a system is in the limit of the information and the imagination of the designer.

5. Everything gardens, or has an effect on its environment.

A Policy of Responsibility (to relinquish power): The role of beneficial authority is to return function and responsibility to life and to people; if successful, no further authority is needed. The role of successful design is to create a self–managed system.

Categories of Resources:

1. Those which increase by modest use.
2. Those unaffected by use.
3. Those which disappear or degrade if not used.
4. Those reduced by use.
5. Those which pollute or destroy other resources if used.

Policy of Resource Management: A responsible human society bans the use of resources which permanently reduce yields of sustainable resources, e.g. pollutants, persistent poisons, radioactives, large areas of concrete and highways, sewers from city to sea.

Principle of Disorder: Any system or organism can accept only that quantity of a resource which can be used productively. Any resource input beyond that point throws the system or organism into disorder; oversupply of a resource is a form of chronic pollution.

Definition of System Yield: System yield is the sum total of surplus energy produced by, stored, conserved, reused, or converted by the design. Energy is in surplus once the system itself has available all its needs for growth, reproduction, and maintenance.

The Role of Life in Yield: Living things, including people, are the only effective intervening systems to capture resources on this planet, and to produce a yield. Thus, it is the sum and capacity of life forms which decide total system yield and surplus.

Limits to Yield: Yield is not a fixed sum in any design system. It is the measure of the comprehension, understanding, and ability of the designers and managers of that design.

Dispersal of Food Yield Over Time:

• By selection of early, mid and late season varieties.

• By planting the same variety in early or late-ripening situations.

• By selection of long–yielding varieties.

• By a general increase in diversity in the system, so that:

• Leaf, fruit, seed and root are all product yields.

• By using self–storing species such as tubers, hard seeds, fuelwood, or rhizomes which can be "cropped on demand".

• By techniques such as preserving, drying, pitting, and cool storage.

• By regional trade between communities, or by the utilisation of land at different altitudes or latitudes.

Principle of Cyclic Opportunity: Every cyclic event increases the opportunity for yield. To increase cycling is to increase yield.

Cycles in nature are diversion routes away from entropic ends—life itself cycles nutrients—giving opportunities for yield, and thus opportunities for species to occupy time niches.

Types of Niches:

• Niche in space, or "territory" (nest and forage sites).

• Niche in time (cycles of opportunity).

• Niche in space–time (schedules)

Principle of Disorder: Order and harmony produce energy for other uses. Disorder consumes energy to no useful end.

Neatness, tidiness, uniformity, and straightness signify an energy–maintained disorder in natural systems.

Principle of Stress and Harmony

Stress may be defined as either prevention of natural function, or of forced function; and (conversely) harmony as the permission of chosen and natural functions and the supply of essential needs.

Principle of Stability: It is not the number of diverse things in a design that leads to stability, it is the number of beneficial connections between these components.

Information as a Resource: Information is *the* critical potential resource. It *becomes* a resource only when obtained and acted upon.

2.15 REFERENCES

Waddington, C. H., *Tools for Thought*, Paladin, UK, 1977.

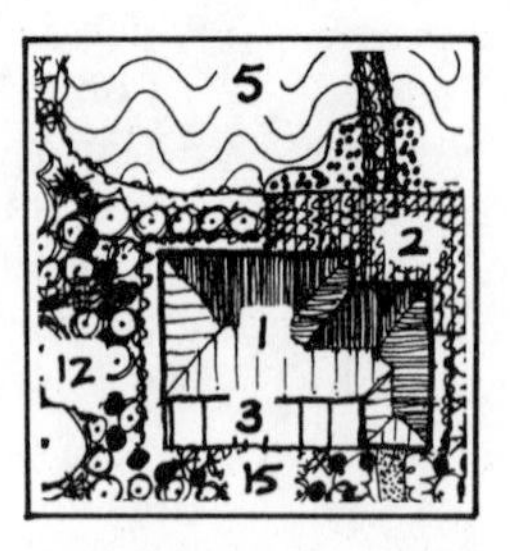

Chapter 3

METHODS OF DESIGN

3.1
INTRODUCTION

Any design is composed of concepts, materials, techniques, and strategies, as our bodies are composed of brain, bone, blood, muscles, and organs, and when completed functions as a whole assembly, with a unified purpose. As in the body, the parts function *in relation to each other*. Permaculture, as a design system, attempts to integrate fabricated, natural, spatial, temporal, social, and ethical parts (components) to achieve a whole. To do so, it concentrates not on the components themselves, but on the *relationships between them*, and on how they function to assist each other. For example, we can arrange any set of parts and design a system which may be self–destructive or which needs energy support. But by using the same parts in a different way, we can equally well create an harmonious system which nourishes life. It is in the *arrangement of parts* that design has its being and function, and it is the adoption of a purpose which decides the direction of the design.

Definition of Permaculture Design
Permaculture design is a system of assembling conceptual, material, and strategic components in a pattern which functions to benefit life in all its forms. It seeks to provide a sustainable and secure place for living things on this earth.

Functional design sets out to achieve specific ends, and the prime directive for function is:

Every component of a design should function in many ways. Every essential function should be supported by many components.

A flexible and conceptual design can accept progressive contributions from any direction, and be modified in the light of experience. Design is a continuous process, guided in its evolution by information and skills derived from earlier observations of that process. All designs that contain or involve life forms undergo a long–term process of change.

To understand design, we must differentiate it from its component parts, which are techniques, strategies, materials and assemblies:

- TECHNIQUE is "one–dimensional" in concept; a technique is *how* we do something. Almost all gardening and farming books (until 1950) were books on technique alone; design was largely overlooked.
- STRATEGIES, on the other hand, add the dimension of *time* to technique, thus expanding the conceptual dimensions. Any planting calendar is a "strategic" guide. Strategy is the use of technique to achieve a future goal, and is therefore more directly value–oriented.
- MATERIALS are those of, for instance, glass, mud, and wood. ASSEMBLIES are the putting together of technologies, buildings, and plants and animals.

There are many ways to develop a design on a particular site, some of them relying on observation, some on traditional skills usually learned in universities. I have outlined some methods as follows:

ANALYSIS: Design by listing the characteristics of components (**3.2**).

OBSERVATION: Design by expanding on direct observation of a site (**3.3**).

DEDUCTION FROM NATURE: Design by adopting the lessons learnt from nature (**3.4**).

OPTIONS AND DECISIONS: Design as a selection of options or pathways based on decisions (**3.5**).

DATA OVERLAY: Design by map overlays (**3.6**).

RANDOM ASSEMBLY: Design by assessing the results of random assemblies (**3.7**).

FLOW DIAGRAMS: Design for workplaces (**3.8**).

ZONE AND SECTOR ANALYSIS: Design by the application of a master pattern (**3.9**).

All these methods can be used to start on sensible and realistic design, with innovative characteristics. Each method is described below.

3.2
ANALYSIS
DESIGN BY LISTING THE CHARACTERISTICS OF COMPONENTS

The components of a total design for a site may range from simple technological elements to more complex economic and legal systems. How are we to make decisions about the patterning and placement of our components (systems, elements, or assemblies)? We can list what we know about the characteristics of any one component, and see where this leads us in terms of beneficial connections.

Principle of Self–Regulation
The purpose of a functional and self–regulating design is to place elements or components in such a way that each serves the needs, and accepts the products, of other elements.

To illustrate, we could select a homely and universally–known component, a chicken. What do we know about this hen? We can list its PRODUCTS (materials, behaviours, derived products), NEEDS (what the

TABLE 3.1
ELEMENTS OF A TOTAL DESIGN

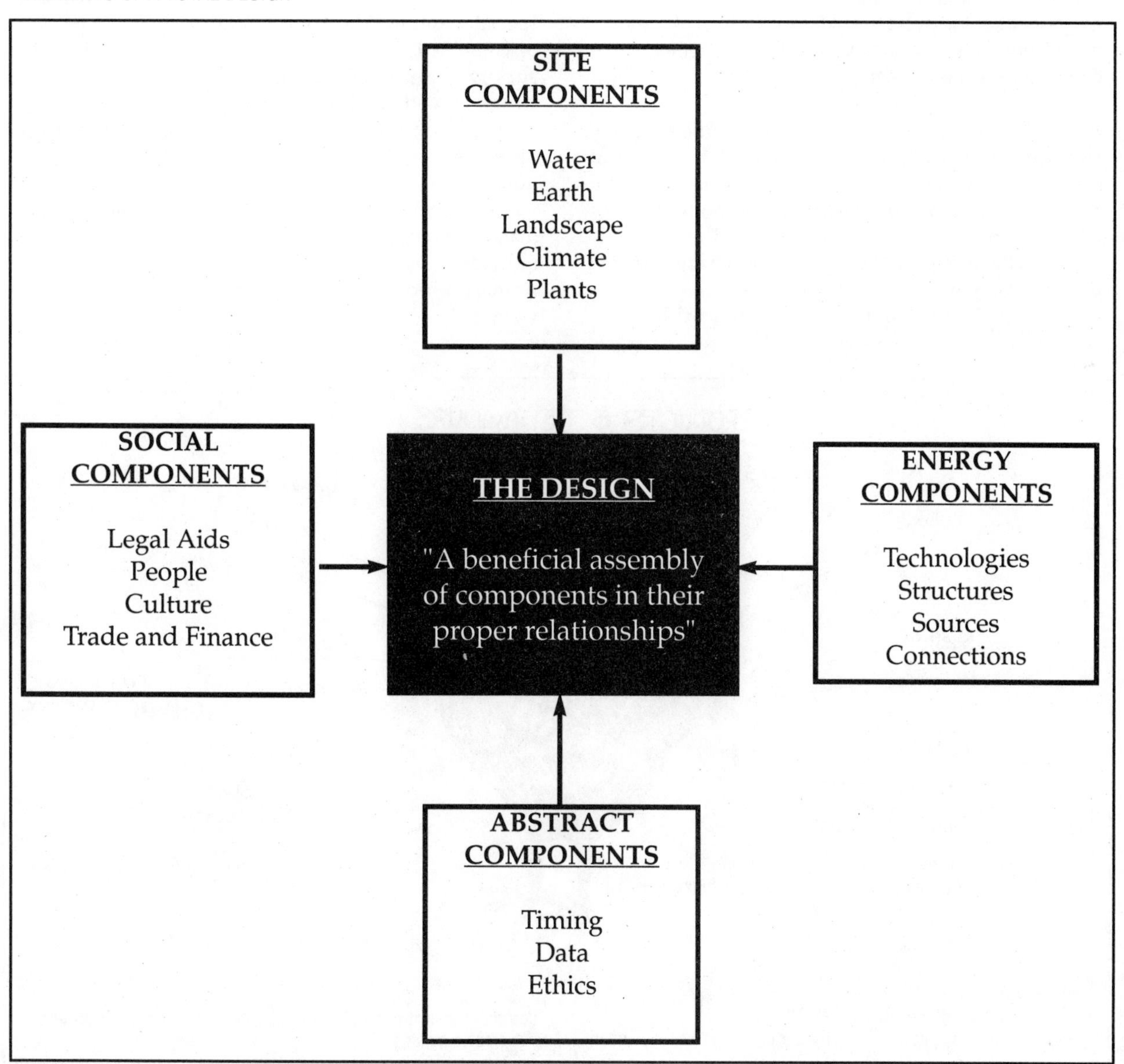

chicken requires to lead a full life), and BREED CHARACTERISTICS (the characteristics of this special kind of chicken, whether it be a Rhode Island Red, Leghorn, Hamburg, etc). See **Figure 3.1**.

A broader classification would have only two categories: "outputs" and "inputs". Outputs are the yields of a chicken, inputs are its requirements in order to give those yields. Before we list either, we should reflect on these latter categories:

OUTPUTS, YIELDS or PRODUCTS are RESOURCES if they are used productively, or can become POLLUTANTS if not used in a constructive way by some other part of the system.

INPUTS, NEEDS, or DEMANDS have to be supplied, and if not supplied by other parts of the system, then EXTERNAL ENERGY or EXTRA WORK must be found to satisfy these demands. Thus:

A POLLUTANT is an output of any system component that is not being used productively by any other component of the system. EXTRA WORK is the result of an input not automatically provided by another component of the system.

As pollution and extra work are both unneccessary results of an incompletely designed or unnatural system, we must be able to connect our component, in this case the chicken, to other components. The essentials are:

- That the inputs needed by the chicken are supplied by other components in the system; and
- That the outputs of the chicken are used by other components (including people).

We can now list the characteristics of the chicken, as we know them. Later, we can see how these need to be linked to other components to achieve our self–regulated system, by a ground strategy of *relative placement* (putting components where they can serve each other).

1. Inputs (Needs) of the Chicken

Primary needs are food, warmth, shelter, water, grit, calcium, dust baths, and other chickens.

Secondary needs are for a tolerable social and physical environment, giving a healthy life of moderate stress.

2. Outputs (Products and Behaviours) of the Chicken

Primary products are, for instance: eggs, feathers, feather dust, manure, various exhaled or excreted gases, sound, and heat.

Derived products are many. From eggs we can make a variety of foods, and derive albumen. From feathers we can make dusters, insulation, bedding, rope, and special manures. Manure is used directly in the garden or combined with leaf and stem materials (carbon) to supply compost heat. Composted anaerobically, it supplies methane for a house. Heat and gases both have a use in enclosed glasshouses, and so on. Our list of derived products is limited only by lack of specific information and by local needs for the products.

Behaviours: chickens walk, fly, perch, scratch, preen, mate, hatch eggs, care for young, form flocks of 20–30 individuals, and forage. They also process food to form primary products and to maintain growth and body weight.

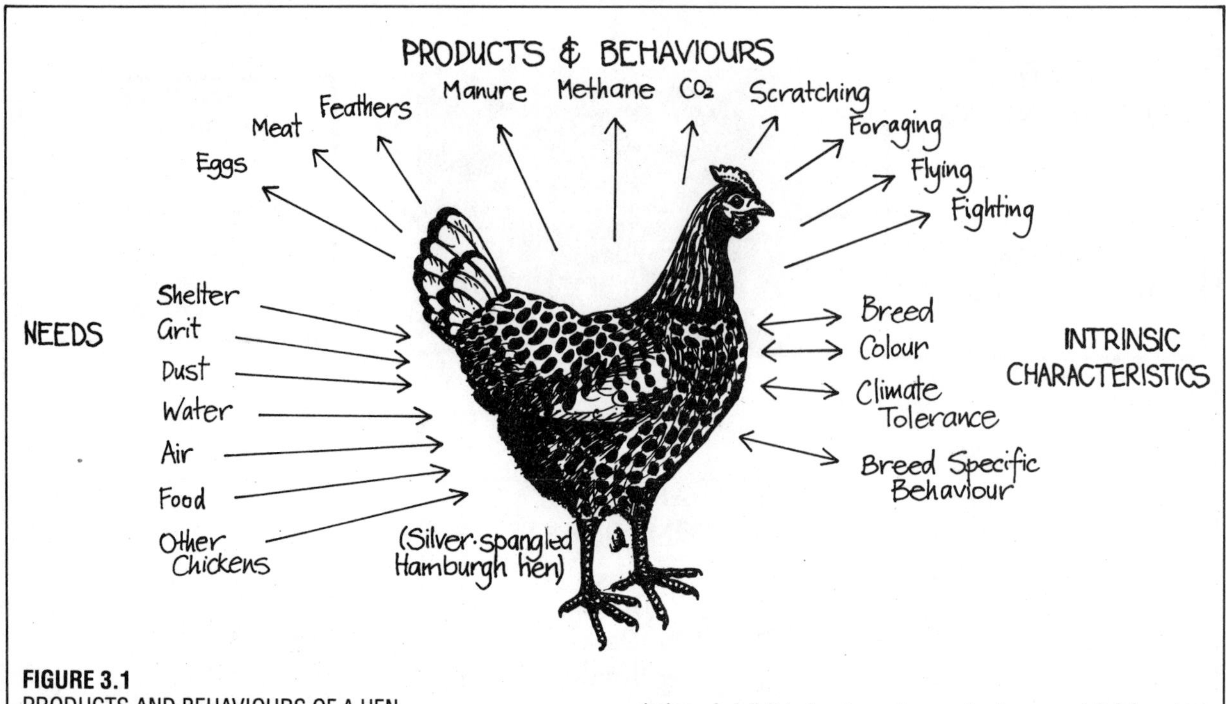

FIGURE 3.1
PRODUCTS AND BEHAVIOURS OF A HEN.
Analysis of these inputs and outputs are critical to self–governing design. A deficit in inputs creates *work*, whereas a deficit in output use creates *pollution*.

3. Intrinsics

Instrinsics are often defined as "breed characteristics". They are such factors as colour, form, weight, and how these affect behaviour, space needed, and metabolism; how climate and soil affect that chicken; or what its tolerances or limits are in relation to heat, cold, predation, and so on. For instance, white chickens survive extreme heat, while thickly–feathered large dark chickens survive extreme cold. Heavy breeds (Australorps) cannot fly over a 1.2 m fence, while lighter breeds (Leghorns) will clear it easily.

We can add much more to the above lists, but that will do to start with (you can add data to any component list as information comes in).

MAKING CONNECTIONS BETWEEN COMPONENTS.

To enable a design component to function, we must *put it in the right place.* This may be enough for a living component, e.g. ducks placed in a swamp may take care of themselves, producing eggs and meat and recycling seeds and frogs. For other components, we must also *arrange some connections*, especially for non–living components, e.g. a solar collector linked by pipes to a hot water storage. And we should *observe and regulate* what we have done. Regulation may involve confining or insulating the component or guiding it by fencing, hedging, or the use of one–way valves. Once all this is achieved, we can relax and let the system, or this part of the system, self–regulate.

Having listed all the information we have on our component, we can proceed to placement and linking strategies which may be posed as questions:

- Of what use are the products of this particular component (e.g. the chicken) to the needs of other components?
- What needs of this component are supplied by other components?
- Where is this component incompatible with other components?
- Where does this component benefit other parts of the system?

The answers will provide a plan of *relative placement* or assist the access of one component to the others.

We can choose our other components from some common elements of a small family farm where the family has stated their needs as a measure of self–reliance, not too much work, a lot of interest, and a product for trade (no millionaire could ask for more!) The components we can bring to the typical small farm are:

- Structures: House, barn, glasshouse, chicken–house.
- Constructs: Pond, hedgerow, trellis, fences.
- Domestic Animals: Chickens, cows, pigs, sheep, fish.
- Land Use: Orchard, pasture, crop, garden, woodlot.
- Context: Market, labour, finance, skills, people, land available, and cultural limits.
- Assemblies: Most technologies, machines, roads and water systems.

We will not list the characteristics of all of these elements here, but will proceed in more general terms.

In the light of linking strategies, we know where we *can't* put the chicken (in a pond, in the house of most societies, in the bank, and so on), but we *can* put the chicken in the barn, chicken–house, orchard, or with other components that either supply its needs or require its life products. Our criteria for placement is that, if possible, such placement enables the chicken to function naturally, in a place where its functions are beneficial to the whole system. If we want the chicken to work for us, we must list the energy and material needs of the other elements, and see if the chicken can help supply those needs. Thus:

THE HOUSE needs food, cooking fuel, heat in cold weather, hot water, lights, bedding, etc. It gives shelter and warmth for people. Even if the chicken is not allowed to enter, it can supply some of these needs (food, feathers, methane). It also consumes most food wastes coming from the house.

THE GLASSHOUSE needs carbon dioxide for plants, methane for germination, manure, heat, and water. It gives heat by day, and food for people, with some wastes for chickens. The chicken can obviously supply many of these needs, and utilise most of the wastes. It can also supply night heat to the glasshouse in the form of body heat.

THE ORCHARD needs weeding, pest control, manure, and some pruning. It gives food (as fruit and nuts), and provides insects for chicken forage. Thus, the orchard and the chickens seem to need each other, and to be in a beneficial and mutual exchange. They need only to be placed together.

THE WOODLOT needs management, fire control, perhaps pest control, some manure. It gives solid fuel, berries, seeds, insects, shelter, and some warmth. A beneficial interaction of chickens and woodlot is indicated.

THE CROPLAND needs ploughing, manuring, seeding, harvesting, and storage of crop. It gives food for chickens and people. Chickens obviously have a part to play in this area as manure providers and cultivators (a large number of chickens on a small area will effectively clear all vegetation and turn the soil over by scratching).

THEPASTURE needs cropping, manuring, and storage of hay or silage. It gives food for animals (worms and insects included).

THE POND needs some manure. It yields fish, water plants as food, and can reflect light and absorb heat.

In such a listing, it becomes clear that many components provide the needs and accept the products of others. However, there is a problem. On the traditional small farm the main characteristic is that *nothing is connected to anything else,* thus no component supplies

FIGURE 3.2
A TYPICAL SMALL FARM.
Villages and farms may contain all the components for self–governance but unless these components are placed in harmonious relationships to each other time, energy, and resources are wasted. In this figure unplanned and segregated systems all demand inputs.

FIGURE 3.3
A RE–DESIGNED SMALL FARM.
In this figure many elements supply the energy inputs for others, and the system can be largely self–regulating.

the needs of others. In short, the average farm does not enjoy the multiple benefits of correct relative placement, or needful access of one system or component to another. This is why most farms are rightly regarded as places of hard work, and are energy–inefficient See **Figures 3.2** and **3.3**.

Now, *without inventing anything new,* we can *redesign* the existing components to make it possible for each to serve others. See **Figure 3.4– 3.5**.

Just by moving the same components into a beneficial design assembly, we can ensure that the chicken, glasshouse or orchard is working for us, not us working for it. If we place essential components carefully, in relation to each other, not only is our maintenance work minimised, but the need to import energies is greatly reduced, and we might expect a modest surplus for sale, trade, or export. Such surplus results from the conversion of "wastes" into products by appropriate use.

The chicken–house heats (and is heated by) the glasshouse, and both are heated by the chimney. The chickens range in the orchard, providing manure and getting a large part of their food from orchard wastes and pests, and from interplants of woodlot or forest components. A glasshouse also heats the house, and part of the woodlot is a forage system and a shelter–belt. Thus, sensible placements, minimising work, have been made. Market and investment control have been placed in the house, together with an information service using a computer, which can link us to the world.

Each part of this sort of design will be dealt with in greater detail in this book, but a simple transformation such as we made from **Figure 3.2** to **Figure 3.3** is enough to show what is meant by functional design.

A great part of this design can be achieved, as it was here, by analytical methods *unrelated to any real site conditions*. Note that before we actually implement anything, before we even leave our desk, we have developed a lot of good ideas about patterns and self–regulatory systems for a family farm. It only remains to see if these are feasible on the ground, and if the family can manage to achieve them. This is the benefit of the analytical design approach: it can operate without site experience! This is also its weakness. Until the chicken is actually heating the greenhouse, manuring the orchard, or helping to produce methane for the house, our system is just information, or potential. Until that chicken is actually in function, we have produced no real resources, nor have we solved any real problems on our family farm.

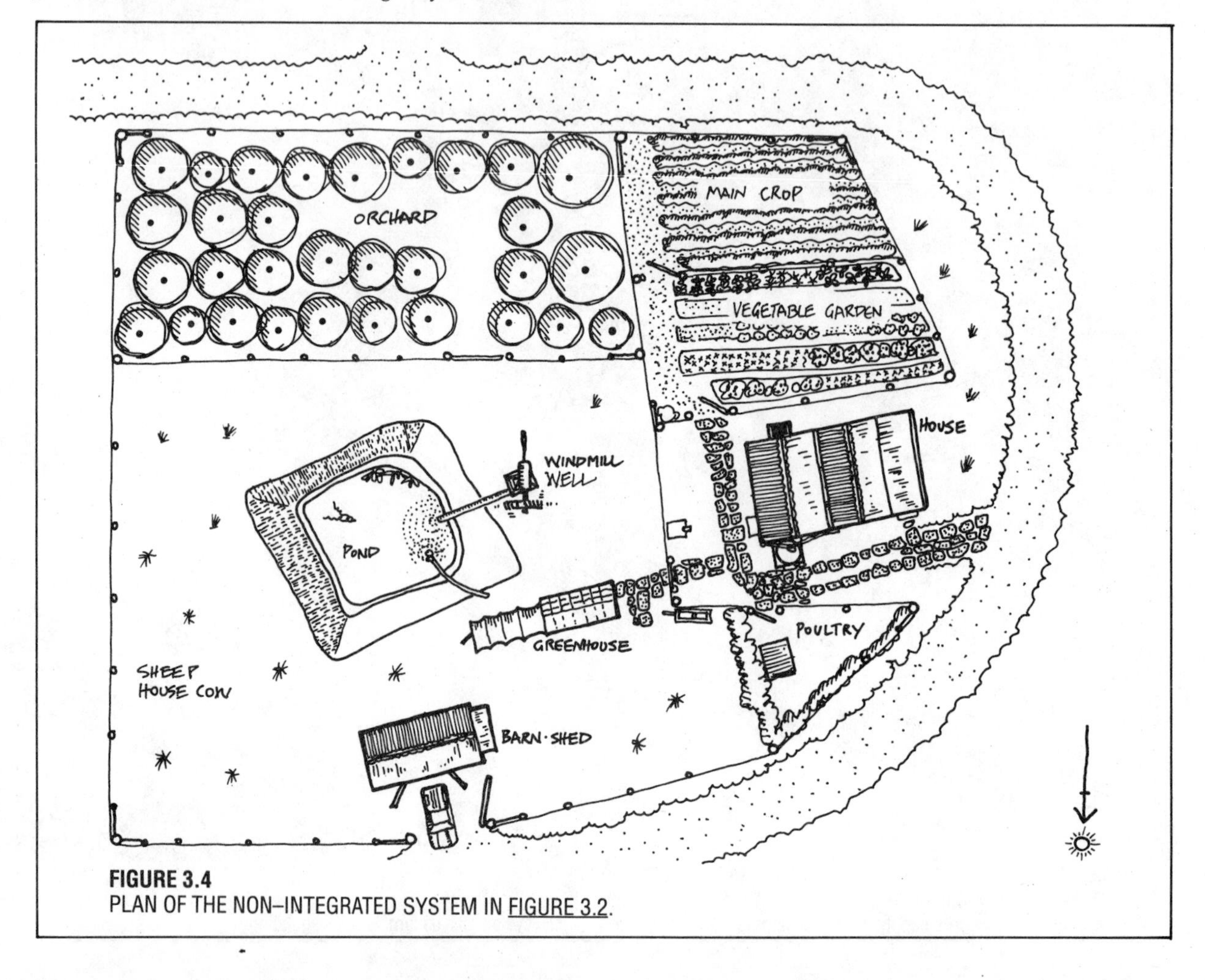

FIGURE 3.4
PLAN OF THE NON–INTEGRATED SYSTEM IN FIGURE 3.2.

Information as a Resource
RESOURCES are practical and useful energy storages, while INFORMATION is only a potential resource, until it is put to use.

We must never confuse the assembling of information with making a real resource difference. This is the academic fallacy: "I think, therefore I have acted."

Note also that we have arrived, analytically, at the need for *cooperation* within the system, and that any *competition* absorbs energy, hence consumes part of our slender resources. Our ideal is to allow the free expression of all the beneficial characteristics of the chicken, so that we avoid conflict and further regulate the system we have designed in light of real–life experience on the site.

3.3

OBSERVATION

DESIGN BY EXPANDING ON DIRECT OBSERVATION OF A SITE

Unlike the preceding analytic method, this way of arriving at design strategies starts on and around the site. Short practice at refining field observation as a design tool will convince you that no complex of map overlays, library, computer data, or remote analysis will ever supplant field observation for dependability and relevance.

Observation is not easily directed, and it is therefore regarded as largely unscientific and individualistic. Process and events, as we encounter them on a real site, are never revealed just by maps or other fixed data. Yet it is from the observation of processes and events (such as heavy rain and subsequent run–off) that we can devise strategies of "least change", and so save energy and time. No static method can reveal processes or dynamic interactions.

A camera and a notebook are great aids to observation, allowing a re–examination of information if necessary. A good memory for events helps. Video recorders are very useful to review processes.

How do we proceed? As we approach the problem, we can adopt any or all of these attitudes:

- A CHILD–LIKE AND NON–SELECTIVE APPROACH, in which "I wonder why..." may pre-face our actual observation.
- A THEMATIC APPROACH, where we try to observe a theme such as water, potential energy sources, or the conditions for natural regeneration.

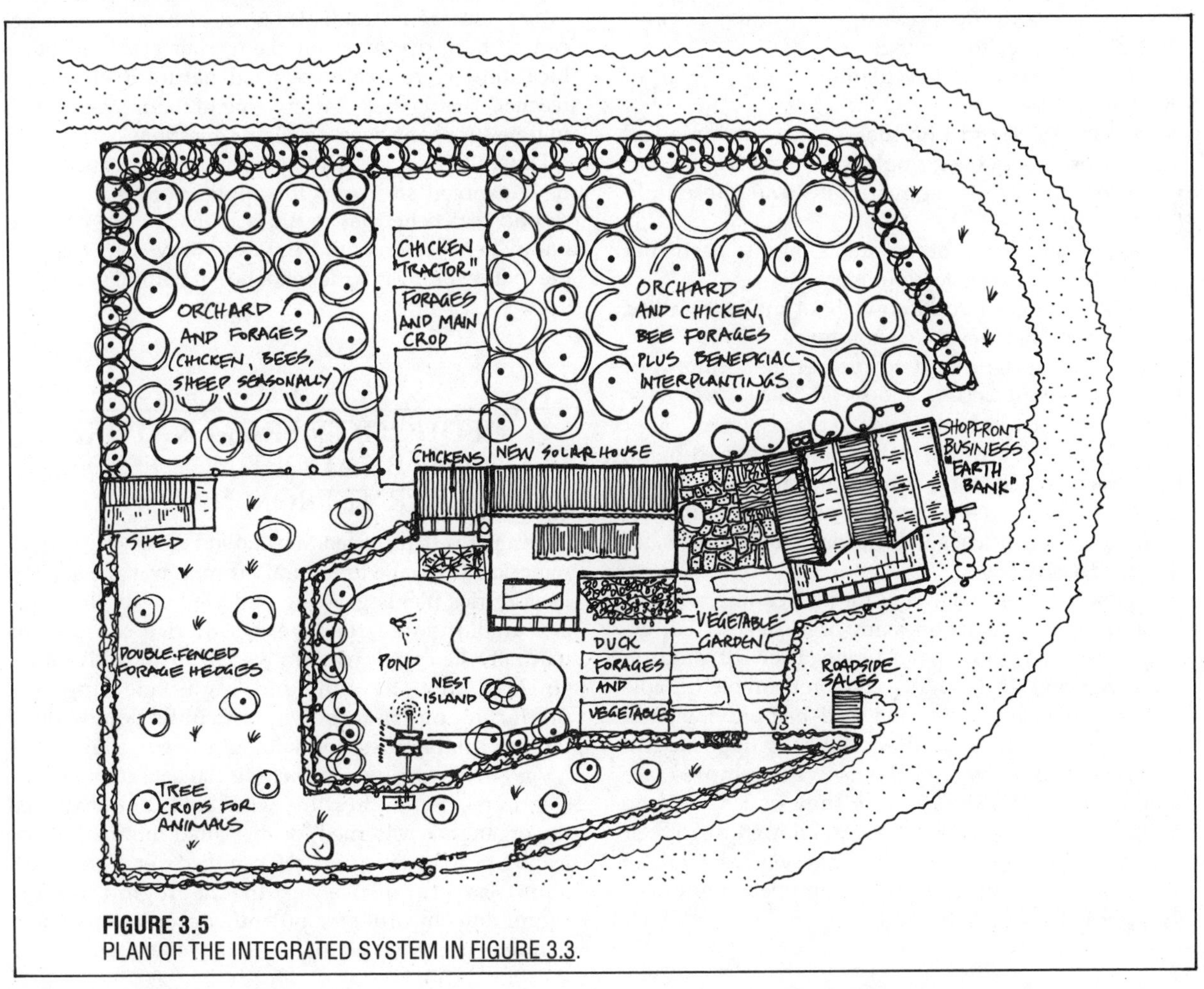

FIGURE 3.5
PLAN OF THE INTEGRATED SYSTEM IN FIGURE 3.3.

• AN INSTRUMENTAL APPROACH, where we measure, perhaps using equipment, a factor such as temperature gradients, wind, or reflection from trees.

• AN EXPERIENTIAL APPROACH, using all our senses as our instruments, trying to be fully conscious both of specific details, sensations, and the total ambience of the site.

In order to develop a design strategy, possible procedural stages are as follows:

1. Make value–free and non–interpretative notes about what is seen, measured, or experienced, e.g. that "moles have thrown up earth mounds on the field." Make no guesses or judgements at this stage (this takes some discipline but gets a lot of primary data listed).

2. Later, select some observations which interest you, and proceed to list under each of them a set of SPECULATIONS as to possible meanings, e.g. (on the moles):

- That molehills are only conspicuous on fields, and may actually occur elsewhere.
- Or that they occur only on fields.
- That fields are particularly attractive to moles.

And so on. Many speculations can arise from one observation! Speculations are a species of hypothesis, a guess about which you can obtain more information. To further examine these speculations, several strategies are open to the observer:

3. Confirm or deny speculations by any or all of these methods:

- Library research on moles, and even on allied burrowing species (e.g. gophers).
- Asking others about moles and their field behaviour.
- Devising more observations on one particular THEME just to test out your ideas.
- Recalling all you know about moles or allied species in other areas or circumstances.

This process will start to further elaborate your knowledge of an existing and *specific site characteristic*, and may already be leading you to the next step:

4. Examination of all the evidence now to hand. Have we evolved any patterns, any mode of operating? What *other* creatures burrow in fields and are predators, prey, or just good friends of moles? Now, for the last decisive step:

5. How can we find a USE for all this information? What design strategies does any of it suggest? For example, we may now have found a *lot* of data on burrowers and fields, and look upon the mole (if mole it is) as a fine soil aerator and seed–bed provider, and therefore to be encouraged—*or* the very opposite. We may have discovered places where moles are beneficial, and places where they could well be excluded. Or possibly they are best allowed to go their way as natural components in the system. Methods of mole–control or data on how to prepare moles for eating may have surfaced, and so on.

As the research and observation phase (plus others' observations) goes on, the mole will gradually be seen to be *already connected* in one or other way to worms, upturned soils, fields, lawns, gardens, pastures, water percolation, and even perhaps soil production. Dozens of useful strategies may have evolved from your first simple observations, and the site *begins to design itself*. You may begin sensible trials to test some of your hypotheses.

Some cautious trials and further observation will, in time, confirm the benefits (or otherwise) of moles in the total system or in specific parts of it. A great deal of practical information will be gathered, which will carry over to other sites and to allied observations. A study of earthworms may have co–evolved, and the interconnectedness of natural systems has become evident.

No analytic method can involve one in the world as much as observation, but observation and its methods need to be practised and developed, whereas analysis needs no prior practice and requires less field research or first–hand knowledge. As an observer, however, you are very likely to stumble on *unique and effective strategies*, and thus become an innovator!

The uses and strategies derived from observation, experience or experiments on site are the basic tools of aware, long–term residents. A set of reliable strategies can be built up, many of them transferable to other locations. Here, we have used nature itself as our teacher. That is the greatest value of nature, and it will in time supply answers to all our questions.

Thus, the end result of systematic observation is to have evolved strategies for application in design. A second and beneficial result is that we have come to know, in a personal and involved way, something of the totality of the interdependence of natural systems.

3.4

DEDUCTION FROM NATURE
DESIGN BY ADOPTING LESSONS LEARNT FROM NATURE

The impetus that started Masanobu Fukuoka[3,4] on his remarkable voyage to natural farming was the sight of healthy rice plants growing and yielding in untended and uncultivated road verges. If rice can do this naturally, he asked, why do we labour to cultivate the soil? In time he achieved high– yielding rice production on his farm without cultivation, without fertilisers or biocides, and without using machinery.

Via our senses (which include the sensations of the skin in relation to pressure, wind chill, and heat), and the organised, patterned, or measured information we extract from observation, we can discover a great deal about natural processes in the region we are examining. In order to put our observations about nature to use, we need to look at the following:

STRUCTURE

We can imitate the structure of natural systems. If we have palms, vines, large evergreen trees, an "edge" of herbaceous perennials, a groundcover of bulbs or tubers, and a rich bird fauna in the *natural* system of the region, then we can reconstruct or imitate such a system structure on our site, using some native species for pioneers, bird forage, or vine supports. We can add to this the palms, vines, trees, tubers, and poultry that are of great use to our settlement (over that broad range of uses that covers food, crafts, medicines, and fuels).

After studying the natural placement of woody legumes or windbreak in natural systems, we can imitate these in designed systems. We can improve on local species by finding out–of–region or exotic species even better suited to those roles than those of an impoverished or degraded native flora and fauna. Certainly, we can carefully select species of a wider range of use to settlements than the natural assembly.

PROCESS

Apart from the structure of natural systems, we need most of all to study process. Where does water run? How does it absorb? Why do trees grow in some special sites in deserts? Can we construct or use such processes to suit ourselves? Some of the processes we observe are processes "energised" by animals, wind, water, pioneer trees or forbs, and fire. How does a tree or herb propagate itself in this region? As every design is a continuous process, we should most of all try to create useful self–generating systems. Some examples would be:

- On Lake Chelan (Washington state, USA), walnuts self–generate from seed rolling downhill in the valleys of intermittent streams. Similar self– propagation systems work for palms in the tropics, *Aleurites* (candle–nut) in Hawaii, and asparagus along sandy irrigation channels. Thus, we save ourselves a lot of work by setting up headwater plantations and allowing these to self–propagate downstream (as for willows, Russian olive, and hundreds of water–plant species, including taro in unstable flood–water lowlands), as long as these are not a problem locally.
- Birds spread useful bird forages such as elderberries, *Coprosma*, *Lycium*, autumn olive, pioneer trees or herbs, and preferred grains such as *Chenopodium* species. If we place a few of these plants, and allow in free–ranging pigeons or pheasants, they will plant more. The same applies to dogs or foxes in the matter of loquats, bears for small fruits, and cattle for hard seeds such as honey locusts. Burrowers and hoarders such as gophers will carry bulbs and root cuttings into prairie, and jays and squirrels, choughs, or currawongs spread oaks when they bury acorns.

If, in grasslands or old pastures, we see that a "pioneer" such as tobacco bush, a pine, or an *Acacia* provides a site for birds to roost, initiating a soil change so that clumps or coppices of forest form there, we can use the same techniques and allied species to pioneer our food forests, but selecting species of more direct use to us. Many native peoples do just this, evolving scattered forest nucleii based on a set of pioneer trees, termite mounds, compost heaps, and so on. We can provide perches for birds to drop pioneer seeds, and so set up plant nucleii in degraded lands around simple perches placed on disturbed sites.

- We can provide nest holes so that owls may then move in to control rodents, purple martins to reduce mosquitoes, or woodpeckers to control codling moth. Many nurse plants allow insect predators to overwinter, feed, or shelter within our gardens, as do small ponds for frogs and rock piles for lizards. If we want these aids to pest control, we need to provide a place for them. Some of these natural workers are very effective (woodpeckers alone reduce codling moth by 40–60%).
- To limit a rampant plant, or to defeat invasive grasses, we need only to look to nature. Nature imposes successions and limits on every species, and once we know the rules, we can use this succession to limit or exclude our problem species. Many soft vines will smother prickly shrubs. Browsed or cut out, they allow trees to permanently shade out the shrub, or rot its seeds in mulch. Kikuyu grass is blocked from spreading by low hedges of comfrey, lemongrass, arrowroot (*Canna spp.*), or nasturtiums. We can use some or all of these species at tropical garden borders, or around young fruit trees. We can smother rampageous species such as *Lantana* by vines such as chayote (*Sechium edulis*) and succeed them with palm/legume forests, by cutting or rolling tracks and then planting legumes, palms, and vines of our choice. Where rampageous grasses smother the trees, we set our trees out in a protecting zone of "soft" barrier plants such as comfrey, nasturtium, or indeed any plant we locally observe to "beat the grass", and we surround our mulched gardens with belts of such plants.

There are hundreds of such botanical lessons about us. Look long enough, and the methodologies of nature become clear. This is design by analogy: we select analagous or botanically–allied species for trials. If thistles grow around a rabbit warren, then perhaps if we disturb the soil, supply urine and manure, and sow seed, we will get globe artichokes (and so I have!) Or we can pen goats or sheep on a place, then shut them out and plant it. It was by such thinking that the idea of chicken or pig "tractors" evoled to remove such stubborn weeds as nut–grass, *Convolvulus*, onion–weed, and twitch before planting a new succession of useful plants. Or we can provide fences or pits to trap wind–blown debris (dried leaves, rabbit and sheep manures, seagrasses), which can be gathered for garden use. And so on...

All these strategies can be derived from observing natural processes, and used consciously in design to achieve a great reduction in work, hence energy inputs.

LANDSCAPE

Gullies, ridgetops, natural shade, the sides of multi–storey buildings, and exposed sunny sites all demonstrate different opportunities, just as various velocities and grades of streams or rock–falls present specific niches. We can find a use for each and every such special site, whether as an aid to food storage, food dehydration, as an energy source in itself, or as a site for a special animal or plant. We also create such opportunities over time as we grow groves of trees, raise earthbanks, build houses, or excavate caves. It is in the creation of microclimates that we find a natural diversity and richness increasing. Every clump of trees invites new species to establish, every shaded area provides a refuge from heat, and every stone pile a moist and shaded soil site. We can plan such evolutions, and plant to take advantage of them, using data derived from a close observation of natural systems.

PHILOSOPHY

Life is not all survival in a stable ecosystem. First by designing well, and then observing system evolution, we gain contemplative and celebratory time. In celebration we can incorporate the myths and skills that are important to future generations. In contemplation we find more refined, profound, or subtle insights into good procedures (Fukuoka[3,4]). To implement and manage a constructed or natural system inevitably leads to a more revelationary lifestyle, a more satisfied and contented life, and a sense of one's place in nature.

To become a philosopher is not necessarily to be of benefit to the natural world, but to become a designer or gardener is to directly benefit nature or society, and one will inevitably generate natural ethics and philosophies. To become a good designer is to be in search of an understanding of nature, and to be content with the search itself. It is to design by natural example, becoming aware, taking notes, sitting a long time in one place, watching the wind behave and the trees respond, thrusting your hand into the soil to feel it for moisture (it is always more moist on the shade side of tussock grasses, for example), and becoming sensitive to the processes and sights about you.

In microcosm and macrocosm, we can learn from the world, and these are the very best lessons to adopt. There are a thousand lessons to learn, some so obvious that we could pinch ourselves for failing to notice them. Such an experiential system of design, in broad and in detail, is almost obliterated by the classroom, the sterile playground, toys, and didactic education. The huge information store that is nature is a primary reason for its preservation. We can never afford such a fine teacher or an equivalent education system that operates without cost or bureaucratic involvement.

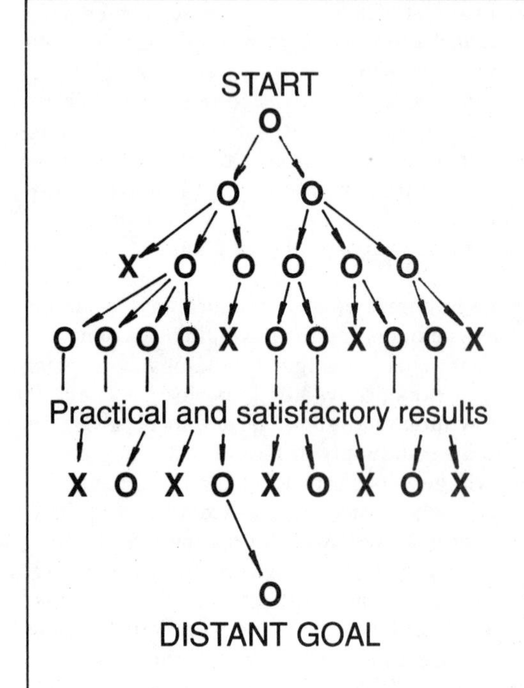

PRIORITIES Decided by ethics of use.

STAGES of procedure by urgency, finance, skills, resorces available, energy......

X A deferred, unnecessary, impractical, or unethical path.
O Possible choices.

And so on to evolutions decided by experience, returns, benefits.

FIGURE 3.6
OPTIONS AND DECISIONS.

3.5

OPTIONS AND DECISIONS
DESIGN AS A SELECTION OF OPTIONS OR PATHWAYS BASED ON DECISIONS

For a specific site and specific occupants (or clients), a design is a sequence of options based on such things as:

- Product or crop options.
- Social investment options (capital available or created).
- Skills and occupations (education available).
- Processing opportunities on or off site.
- Market availability, or specific market options.
- Management skills.

That is, any design has many potential outcomes, and it is above all the stated aims, lifestyle, and resources of the client(s) that decide their options. Any sensible design gives a *place to start*. The evolution of the design is a matter for trial, following observation, and then acting on that information.

I sometimes think that the only real purpose of an initial design is to evolve *some sort of plan* to get one started in an otherwise confusing and complex situation. If so, a design has a value for this reason alone, for as soon as we decide to start doing, we learn how to proceed.

The sort of options open to people start with a general decision (a distant goal), which is often set by ethical considerations (e.g. "care of the earth"). This may lead directly to a second set of possible options, of which erosion control, minimal tillage, and perhaps revegetation of steep slopes are firmly indicated for a specific site in the light of this ethic.

Thus, an option, *once decided on*, also indicates other options, priorities, and management decisions. In practical terms, we may also have to consider costs, and perhaps decide to generate some short– or long–term income. This, in turn, may depend on whether we maintain a part–time, non–farm income, or (taking the leap) gather up our retirement allowance and go to it.

All of this can be plotted, rather like the decision pattern a tree makes as it branches upwards. Some options are impractical, or in conflict with other decisions and ethics, and are therefore unavailable. (See **Figure 3.6**).

Following through the options that arise from either our decisions, or the constraints of site and resources, we can see an apparently endless series of pathways. The process itself is inevitable, in that it leads to a series of innovative and practical procedural pathways, some of which may be very promising, and all of which agree with the ethical, financial, cultural, and ground constraints decided by the site and/or its occupants.

As a bonus, not one or two, but several dozen options may remain open, and this is always a secure position in which to be. In an uncertain world we need all possible doors open!

Options open up or close down on readily available evidence or as decision–points are reached. All will affect the number and direction of future actions, hence the overall design. To a great extent, this approach covers the economic and legal constraints not dealt with by either of the preceding analytic or observational approaches. It is wise, however, to implement a limited range of options for trial, or we may incur stress and work as a result of taking too much on.

3.6

DATA OVERLAY
DESIGN BY MAP OVERLAYS

In design courses at modern colleges, students are taught to labour assiduously over maps, overlays on those maps, and overlays on the overlays. This approach should also be considered. However, as a methodology, it is at once more expensive, possibly more time–consuming, and potentially the most confusing of all approaches. Like the system of options, it leads to certain inevitable ground placements, and perhaps to uneasy compromises not necessarily inherent in the preceding methods. The danger here is that the map overlays omit minutiae, and can never reveal evolutionary processes.

Where a mapping and hard data approach is weakest, however, is that some factors are not able to be mapped (ethical, financial, and cultural constraints), and that it is very difficult to include those site–relevant details revealed by observation, or indicated at once by our analytic method of component inputs and outputs. Despite this, a good site map makes any *landscape* design (and this is only part of the total design) much easier, and far more visual. A good map indicates a lot of sensible options and hypotheses (dam sites, soil/crop suitability) which can later be checked with actual site conditions, available clay for dams, existing useful vegetation, threatened habitat and so on.

The danger of the purely analytic and overlay approaches is that the very remoteness of such systems makes flexibility difficult, occasioning unforeseen work and expense, which are not incurred by the more empirical and flexible "observation" and "option" systems. The latter both allow a flexible response to fresh conditions.

3.7

RANDOM ASSEMBLY
DESIGN BY ASSESSING THE RESULTS OF RANDOM ASSEMBLIES

This is another analytic method, removed from the site

itself. It is of value in assessing energy flows in the system, and is also a generator of creativity. Because it is based on a set of essentially random selections, it may reveal some very innovative designs.

The process is as follows: we select and list a set of design components, and with them a set of placement or connective strategies. If our components are arranged in a circle around these "connections", we can join them up at random, make a sketch of the results, and see what it is that we have achieved. This frees us from "rational" decisions, and forces us to consider unusual connections for their value; connections that would be inhibited from proposing by our limited education, by cultural restraints, or by normal usage. (See **Table 3.2**).

	House	
Windmill		Storage box
Glasshouse	**ATTACHED TO**	Yard or compound
Animal shelter	**BESIDE**	Caves
Trellis	**AROUND**	Trenches
Mounds	**OVER**	Swales
Compost heaps	**IN**	Ponds
Plants	**ON**	Chickens
Ducks	**UNDER**	Fish
Windbreak	**CONTAINING**	Barn
	Fence	

TABLE 3.2
RANDOM ASSEMBLY SELECTION

Having laid out a simple diagram, we can select any one component and connect it to others, creating images for further examination as to their particular uses and functions. Some simple examples are:

- Glasshouse OVER house
- Storage box IN glasshouse
- Raft ON pond
- Glasshouse ON raft
- House BESIDE pond

And, using more connections: glasshouse CONTAINING compost heap ATTACHED TO house BESIDE pond with cave UNDER, containing storages boxes with plants IN these.

We can sketch these, and see just what it is we have achieved in terms of energy savings, unique assemblies, special effects for climate, increased yield, compact design, or easier accessibility. As we do not usually think of these units *with respect to their connections*, this simple design strategy frees us to do so, and to achieve innovative results.

Having illustrated (by way of a diagram) random assemblies, we can then think out what would happen if we did in fact build them or model them. Rafts can, of course, be oriented quickly to suit seasons. Caves are cool and ponds in them almost immune from evaporation. Ducks are safe from predators on rafts. Glasshouses on rafts will warm contained water and create thermal storages and currents. Solar cells will light caves, and caves below houses supply storage and cool or warm air. Trees shade houses, and so on.

Thus, immune from ridicule and criticism, we can try various unlikely combinations and links of components (all of which probably exist somewhere), and try to assess what we have done *in terms of function*. This is, if you like, working backwards from assembly to function to benefits and system characteristics. The value of this approach is that it frees us to create novel assemblies and to assess them before trials.

Creative solutions may also be arrived at by constantly re–examining a problem, and by considering every form of solution, including that important strategy of doing nothing! (Fukuoka[3,4])

CREATIVE PROBLEM SOLVING
Restate a problem many ways, reverse the traditional approaches, and allow every solution to be considered. Simple solutions may be found by this process.

The art of thinking backwards, or in opposites, is often very effective in problem–solving. It is easier to drive an axle out of a wheel than to knock a wheel off an axle, easier to lower a potted vine down a dark shaft over a period of months than to grow it up from the bottom. So, if we worry away at problems in terms of restatements, turning things on their head and stating the opposite, we may find that real solutions lie in areas free from acquired knowledge and values.

3.8
FLOW DIAGRAMS
DESIGN FOR WORK PLACES

For designing any special work place, from a kitchen to a plant nursery, the preceding methods have limited uses. Here, we call in a different method—the "flow chart". We imagine how the process flows. In the kitchen we take from storage, prepare, cook, serve, and gather in the plates and food for waste disposal and return to storage.

Thus the processes follow a certain path. The best kitchens are U–shaped or compact, so that least movement is necessary. Storages are near the place where food, plates, or pots and pans are needed. Frequently–used items are to hand on benches, or in special niches. Strong blocks, bench tops, or tables are built to take the heavy work of chopping and the clamping on of grinders and flour mills. We can mark such designs out on the ground, and walk around these, preparing an imaginary meal, measuring the space taken up by trays, pots, and potato storages, and so creating an efficient work place. It should also involve the placement of traditional items, and agree with cultural uses.

It is advisable to involve an experienced worker in any such design, and to research prior designs or new aids to design, such as we find in office furniture which can be adjusted to the person. I have seen some excellent farm buidings such as shearing sheds and their associated yards built by worker–designers after years of observation and experience. Some people specialise in such design for schools, wineries, and golf courses. In general, it is mainly work–places which need such careful attention. Most other areas in buildings are of flexible use, and have the potential for multiple function.

The technique of flow charts is also applicable to traffic–ways and transport lines serving settlements, where loads or cargoes are to be received and sorted, and where schedules or time–place movements are integral to the activity.

3.9
ZONE AND SECTOR ANALYSIS
DESIGN BY THE APPLICATION OF A MASTER PATTERN

Zone and sector analysis is a primary energy–conserving placement pattern for the whole site. When we come to an actual site design, we must pay close attention to locating components relative to the two energy sources of the site:

First, energy available *on site*: the people, machines, wastes, and fuels of the family or society. For these, we establish ZONES of use, of access, and of time available.

Second, energy entering or flowing *through the site*: wind, water, sunlight and fire may enter the site. To govern these energies we place intervening components in the SECTORS from which such energies arise, or can be expected to enter. We also define sectors for views, for wildlife, and for temperature (as air flow). To proceed to a discussion of the pattern in its parts:

ZONES

We can visualise zones as series of concentric circles, the innermost circle being the area we visit most frequently and which we manage most intensively. Zones of use are basic to conservation of energy and resources *on site*. We do not have endless time or energy, and the things we use most, or which need us often, must be close to hand. We plan our kitchens in this way, and we can plan our living sites with equal benefit to suit our natural movements.

We should not pretend that any real site will neatly accept this essentially conceptual conformation of pattern, which will usually be modified by access, site characteristics such as slope and soils, local wind patterns, and the technical problems of, for example, constructing curved fences in societies where title boundaries, materials, and even the education available is "straight".

In *zonation*, the village or dwelling itself is Zone 0, or the origin from which we work. The available energy in Zone 0 is human, animal, piped–in, or created on site. Whatever the sources, these energies can be thought of as *available* or *on–site energies*. In order to conserve them, and those other essential re-sources of work and time, we need to place components as follows:

Zone 0 (the house or the village).
In this zone belongs good house design, attached glasshouse or shadehouse, and the integration of living components as sod roof, vines, trellis, potplants, roof gardens, and companion animals. In some climates, many of these structures are formed of the natural environment, and will in time return to it (bamboo and rattan, wattle and daub, thatch, and earth–covered or sheltered structures).

Zone 1
Those components needing continual observation, frequent visits, work input, complex techniques (fully–mulched and pruned gardens, chicken laying boxes, parsley and culinary herbs) should be placed *very close to hand*, or we waste a great deal of time and energy visiting them. Within 6 m (20 feet) or so of a home, householders can produce most of the food necessary to existence, with some modest trade requirements. In this home garden are the seedlings, young trees for outer zone placement, perhaps "mother plants" for cuttings, rare and delicate species, the small domestic and quiet animals such as fish, rabbits, pigeons, guinea pigs, and the culinary herbs used in food preparation. Rainwater catchment tanks are also placed here. Techniques include complete mulching, intensive pruning of trees, annuals with fast replacement of crop, full land use, and nutrient recycling of household wastes. In this zone, we arrange nature to serve our needs.

Zone 2
This zone is less intensively managed with spot–mulched orchards, main–crop beds, and ranging domestic animals, whose shelters or sheds may nevertheless adjoin Zone 1 or, as in some cultures, be integrated with the house. Structures such as terraces, small ponds, hedges, and trellis are placed in this zone. Where winter forces all people and animals indoors, joint accommodation units are the normality, but in milder climates, forage ranges for such domestic stock as milk cows, goats, or poultry can be placed in Zone 2. Home orchards are established here, and less intensive pruning or care arranged. Water may be piped from Zone 3, or conserved by species selection.

Zone 3

This area is the "farm" zone of commercial crop and animals for sale or barter. It is managed by green manuring, spreading manure from Zone 2, and soil conditioning. It contains natural or little–pruned trees, broadscale farming systems, large water storages, soil absorption of water, feed–store or barns, and field shelters as hedgerow or windbreak.

Zone 4

This zone is an area bordering on forest or wilderness, but still managed for wild gathering, forest and fuel needs of the household, pasture or range, and is planted to hardy, unpruned, or volunteer trees. Where water is stored, it may be as dams only, with piped input to other zones. Wind energy may be used to lift water to other areas, or other dependable technology used.

Zone 5

We characterise this zone as the natural, unmanaged environment used for occasional foraging, recreation, or just let be. This is where we learn the rules that we try to apply elsewhere.

Now, any one component can be placed in its right zone, at the best distance from our camp, house, or village. As our very perfect "target" model does not fit on real sites, we need to deform it to fit the landscape, and we can in fact bring "wedges" of a wilderness zone right to our front door: a corridor for wildlife, birds, and nature (**Figure 3.7**). Or we can extend a more regularly used zone along a frequently used path (even make a loop track to place its components on).

Zoning (distance from centre) is decided on two factors:

1. The number of times you need to visit the plant, animal or structure; and
2. The number of times the plant, animal or structure needs you to visit it.

For example, on a yearly basis, we might visit the poultry shed:

- for eggs, 365 times;
- for manure, 20 times;
- for watering, 50 times;
- for culling, 5 times; and
- other, 20 times.

Total = 460 visits; whereas one might visit an oak tree twice only, to collect acorns. Thus the zones are "frequency zones for visits", or "time" zones, however

TABLE 3.3:
SOME FACTORS WHICH CHANGE IN ZONE PLANNING AS DISTANCE INCREASES.

Factor or Strategy	ZONE I	ZONE II	ZONE III	ZONE IV
Main design for:	House climate, domestic sufficiency.	Small domestic stock & orchard.	Main crop forage, stored.	Gathering, forage, forestry, pasture.
Establishment of plants	Complete sheet mulch.	Spot mulch and tree guards.	Soil conditioning and green mulch.	Soil conditioning only.
Pruning and trees	Intensive cup or espallier trellis.	Pyramid and built trellis.	Unpruned and natural trellis.	Seedlings, thinned to selected varieites.
Selection of trees	Selected dwarf or multi-graft.	Grafted varieties and plants managed.	Selected seedlings for later grafts. by browse.	Thinned to selected varieties, or
Water provision	Rainwater tanks, bores wind pumps. reticulation.	Earth tank and wells, bores,	Water storage fire control.	Dams, rivers, in soils, dams.
Structures	House/green-house, storage integration.	Greenhouse and barns, poultry sheds.	Feed store, field shelter.	Field shelter grown as hedgerow and woodlot
Information	Stored or generated by people.	In part affected by other species.	As for II.	Arising from natural processes.

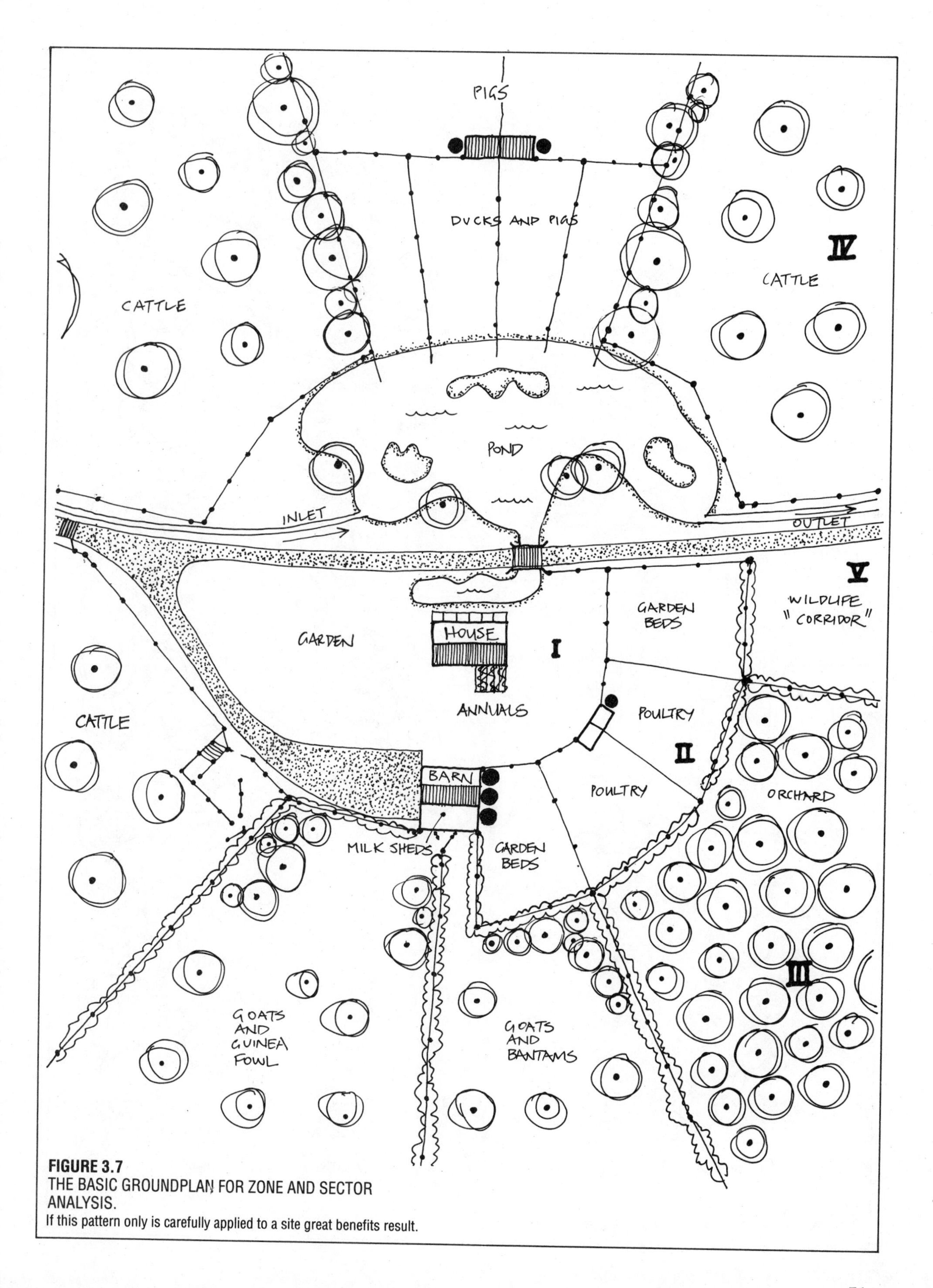

FIGURE 3.7
THE BASIC GROUNDPLAN FOR ZONE AND SECTOR ANALYSIS.
If this pattern only is carefully applied to a site great benefits result.

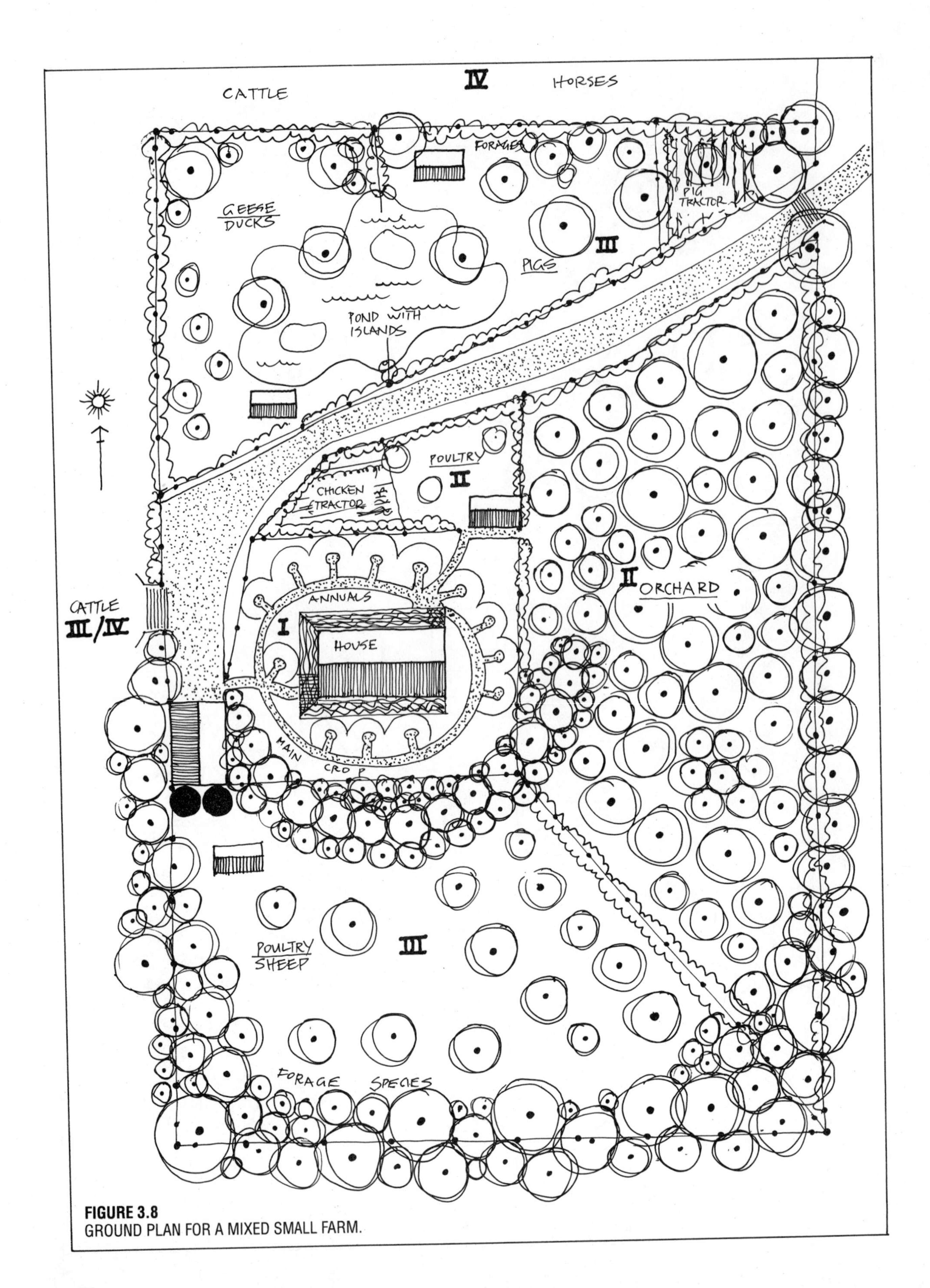

FIGURE 3.8
GROUND PLAN FOR A MIXED SMALL FARM.

you like to define them. The more visits needed, the closer the objects need to be. As another example, you need a fresh lemon 60–100 times a year, but the tree needs you only 6–12 times a year, a total of 66 to 112 times. For an apple tree, where gathering is less, the total may be 15 times visited. Thus, the components or species space themselves in zones according to the number of visits we make to them annually.

The golden rule is to develop the nearest area first, get it under control, and then expand the perimeter. A single perimeter will then enclose all your needs.

Too often, the novice selects a garden away from the house, and neither reaps the plants efficiently, nor cares for them well enough. Any soil, with effort and the compost from the recycling of wastes, will grow a good garden, so stay close to the home.

Let us think of our zones in a less ordered way, as was well described by Edgar Anderson for Central Honduras (Anderson, E., 1976):

> Close to the house and frequently more or less surrounding it is a compact garden–orchard several hundred square feet in extent. No two of these are exactly alike. There are neat plantations more or less grouped together. There are various fruit trees (nance, citrus, melias, a mango here and there, a thicket of coffee bushes in the shade of the larger trees)... There are tapioca plants of one or two varieties, grown more or less in rows at the edge of the trees. Frequently there are patches of taro; these are the framework of the garden– orchards. Here and there in rows or patches are corn and beans. Climbing and scrambling over all are vines of various squashes and their relatives; the chayote (choko) grown for the squashes, as well as its big starchy root. The luffa gourd, its skeleton used for dishrags and sponges. The cucurbits clamber over the eaves of the house and run along the ridgepole, climb high in the trees, or festoon the fence. Setting off the whole garden are flowers and various useful weeds (dahlias, gladioli, climbing roses, asparagus fern, cannas). Grain amaranth is a 'sort of encouraged weed that sows itself.'

Around the "dooryard gardens" described above, Anderson notes the fields (in Mexico) "dotted here and there with volunteer guavas and guamuchiele trees, whose fruit was carefully gathered. Were they orchards or pastures? What words are there in English to describe their groupings?"

Anderson is contrasting the strict, ordered, linear, segmented thinking of Europeans with the productive, more natural polyculture of the dry tropics. The order he describes is a semi–natural order of plants, in their right relationship to each other, but not rigorously separated into various artificial groups. More than that, the house and fence form essential trellis for the garden, so that it is no longer clear where orchards, field, house and garden have their boundaries, where annuals and perennials belong, or indeed where cultivation gives way to naturally–evolved systems.

Monoculture man (a pompous figure I often imagine to exist, sometimes fat and white like a consumer, sometimes stern and straight like a row–crop farmer) cannot abide this complexity in his garden or his life. His is the world of order and simplicity, and therefore chaos.

When thinking of placing components into zones, remember that intrinsic properties and species–specific yields are available from a component wherever it is placed (all trees give shade), so that we don't include these "intrinsics" in assessing function in design.

JUDGING ZONAL PLACEMENT

Place a component in relation to other components or functions, and for more efficient use of space or nutrient. Look for products that serve special needs not otherwise locally available.

The amount of management we must always provide in a cultivated ecosystem is characterised by conscious placement, establishment, guidance, and control energies, akin to the adjustments we normally make to our environment as we traverse it on our daily tasks.

FIGURE 3.9
Design for a 1,000 square metre block. See text.

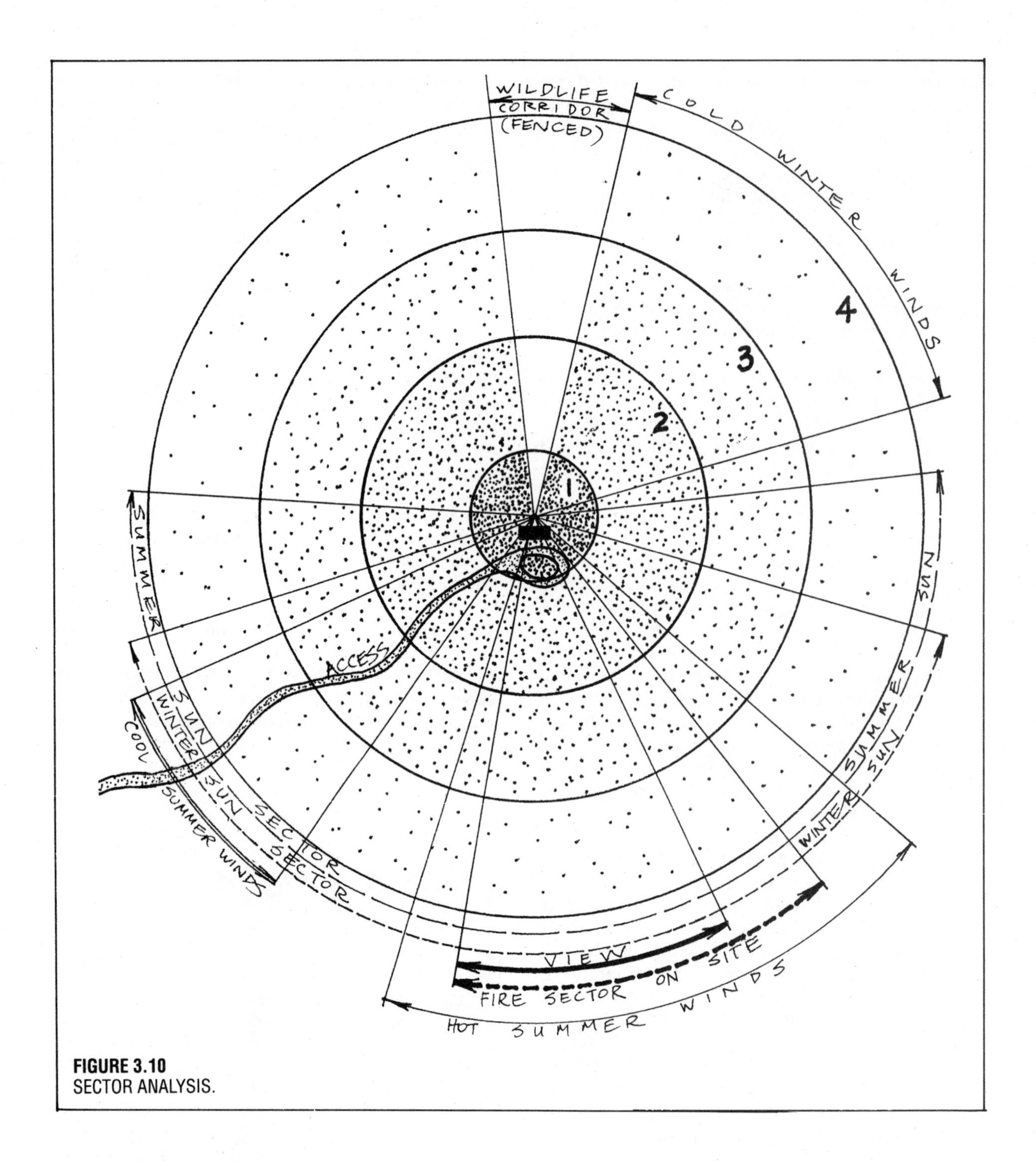

FIGURE 3.10
SECTOR ANALYSIS.

PLACEMENT IN SECTORS

Next in a permaculture design, we consider the wild energies, the "elements" of sun, light, wind, rain, wildfire, and water flow. These all come from *outside* our system and pass through it: a flow of energies generated elsewhere. For these, we plan a "sector" diagram based on the real site.

Our sectors are more site–specific than are the conceptualised zones. They outline the compass directions from which we can expect energy or other factors. Some factors we may invite in to our homes; we need sunlight for technologies and plant growth. Some we may exclude (such as an unpleasant view of a junkyard). More commonly, we plan to regulate such factors to our advantage, placing our zonal components to do so.

Whereas settlement or house is the ground zero for zones, it is the through point for sectors. Energies from outside can be thought of as so many arrows winging their way towards the home, carrying both destruc-tive and beneficial energies; we need to erect shields, deflectors, or collectors. Our choice in each and every sector is to block or screen out the incoming energy or distant view, to channel it for special uses, or to open out the sector to allow, for example, maximum sunlight. We guard against catastrophic fire, wind, or flood by protective embankments, dense trees, ponds, roads, fences, or stone walls, and we likewise invite in or exclude free–ranging or undomesticated wildlife by placements of forage systems, fences, nest boxes, and so on. Thus we place hedges, ponds, banks, walls, screens, trellises, hedgerows or any other component of design to *manage incoming energy*.

If you like, we place our components in each zone as though zones could be rotated about Zone 0. For any one component, it stops rotating when it is working to govern energies in the sector diagram. Thus, by "revolving" our zones, we find a place where our selected component (a tree, fence, pond, wall, or animal shed) works to govern sector factors. Given that we have both zone and sector energies controlled, then our component is well placed. Then we combine the two diagrams to make a spiderweb of placements, putting every main system in its right place in terms of energy analysed on the site (**Figure 3.10**).

To sum up, there should be no tree, plant, structure, or activity that is not placed according to these criteria and the ground plan. For instance, if we have a pine tree, it goes in Zone 4 (infrequent visits) AWAY from the fire danger sector (it accumulates fuel and burns like a tar barrel), TOWARDS the cold wind sector (pines are hardy windbreaks), and it should also bear edible nuts as forage.

Again, if we want to place a small structure such as a poultry shed, it should BORDER Zone 1 (for frequent visits), be AWAY from the fire sector, BORDER the annual garden (for easy manure collection), BACK ONTO the forage system, possibly ATTACH to a greenhouse, and form part of a windbreak system.

There is no mystery nor any great problem in such commonsense design systems. It is a matter of bringing to consciousness the essential factors of active planning. To restate:

The Basic Energy–Conserving Rules

Every element (plant, animal or structure) must be placed so that it serves at least two or more functions.
Every function (e.g. water collection, fire protection) is served in two or more ways.

With the foregoing rules, strategies, and criteria in mind, you can't go far wrong in design.

Placement Principle

If broad initial patterning is well analysed, and good placements made, many more advantages than we would have designed for become obvious.
Or, if we start well, other good things naturally follow on as an unplanned result.

This is the broad pattern approach. Given that the scene has been set, *observation* comes into play to evolve other pattern strategies. If we watch just how our animals move, how winds vary, or how water flows, we can evolve guiding patterns that achieve other desirable ends, e.g. making animals easy to muster, bringing them to sites where their manure is needed, steering cool winds to ameliorate excess heat input or to direct them to wind turbines, and directing water to where it is needed in our system.

SLOPE, ASPECT, ELEVATION, AND ORIENTATION.

No site is quite flat, and many have irregular configurations; thus our neat spider webs of zone and sector overlays are distorted by a more realistic landscape. To use these irregularities to our advantage, we need to further consider these factors:

With zones and sectors sketched in on the ground plan, slope analysis may proceed. High and low access roads, the former for heavy cargo or mulch, the latter for fire control, can now be placed. Provision for attached glasshouse, hot air collectors, reflection pond, solar pond, and shadehouse should be made at all homestead sites where climatic variation is experienced.

Slope determines the unpowered flow of water from source to use point, and slope and elevation will permit the placement of hot air or hot water collectors *below* their storages, where the **thermosiphon** effect can operate without external energy inputs. The simple physics of flow and thermal movements can be applied to the placement of technological equipment e.g. solar hot water panels, taking advantage of slope. Where no slope exists, towers for water tanks and hollows for heat collectors (or solar ponds) can be raised or excavated for the same effect.

However, in the normal humid landscape (where precipitation exceeds evaporation), hill profiles develop a flattened "S" curve that presents opportunities for placement analysis of components and systems, as per **Figure 3.11**.

The ancient occupied ridgeways of England testify to the commonsense of the megalithic peoples in landscape planning, but their present abandonment for industrial suburbs in flatlands does little credit to the palaeolithic planning of modern designers. The difference may be that the former planned for themselves, while the latter design for "other people".

Slope gives immense planning advantages. There is hardly a viable traditional human settlement that is not sited on those critical junctions of two natural ecologies, whether on the area between foothill forests and plains, or on the edge of plain and marsh, land and estuary, or some combination of all of these. Planners who place a housing settlement on a plain, or on a plateau, may have the "advantage" of plain planning, but abandon the inhabitants to failure if transport fuels dry up. They then have to depend on the natural environment for their varied needs but have only a monoultural landscape on which to do so. Successful and permanent settlements have always been able to draw from the resources of at least two environments.

Similarly, any settlement which fails to *preserve* natural benefits, and, for example, clears all forests, is bent on eventual extinction.

The descending slopes allow a variety of aspects, exposures, insolation, and shelter for people to manage. Midslope is our easiest environment, the shelter of forests at our back, the view over lake and plain, and the sun striking in on the tiers of productive trees above and below. **Figure 3.11** shows a broad landscape profile, typical of many humid tropical to cool climates, which we can use as a model to demonstrate some of the principles of landscape analysis.

On the high plateaus (A) or upper erosion surface, snow is stored, and trees and shrubs prevent quick water run–off. The headwaters of streams seek to make sense of a sometimes indefinite slope pattern, giving way to the steep upper slopes (B), rarely (or catastrophically) of use to agriculture, but unfortunately often cleared of protecting forest and subject to erosion because of this.

The lower slopes (C) are potentially very productive mixed agricultural areas, and well suited to the structures of people and their domestic animals and implements. Below this are the gently– descending foothills and plains (D) where cheap water storage is available as large shallow dams, and where extensive cropping can take place.

This simplified landscape should dictate several strategies for permanent use, and demands of us a careful analysis of techniques to be used on each area.

The main concern is water, as it is both the chief agent of erosion and the source of life for plants and animals. Thus the high plateau is a vast roof where rain and snow gather and winds carry saturated cloud to great heights. At night the saturated air deposits droplets on the myriad leaves of the ridge forests.

The gentle foothill country of area (C), brilliantly analysed for water conservation by Yeomans[5], supports the most viable agricultures, if the forest above is left uncut. Here, the high run–off can be led to midslope storage dams at the "Keypoint" indicated in **Figure 3.11** (examined in much more detail in Yeomans' books). Using the high slopes as a watershed, and a series of diversion catchment drains and dams, water is conserved at the keypoints for later frugal use in fields and buildings. The water is passed with its nutrients to low dams, and released as clean water from the site. (This is the ideal: the reality often falls far short of it.) The lower slopes—those safe to use tractors on at least—can be converted to immense soil–water storage systems in a very short time (a single summer often suffices). This is a matter of soil conditioning, afforestation, water interception, or a combination of these.

The plains of area (D) are the most open to wind erosion and the most resistant to water erosion. However, it is here that great damage can occur by salting. Red and dusty rains and plagues of locusts are a result of the delinquent use of the plough, heavy machinery, and clean tillage of these flattish lands, together with the removal of trees and hedgerows, and the conversion of the plains to monocultures of extensive grazing and grain cropping.

It is on the plains areas that water is most cheaply stored, in soil and in large surface dams, where no–tillage crops, **copses** and **hedgerows** are desperately needed. This is where broadscale revolutions in technique can be implemented to improve soil health, reduce wind and water losses, and produce healthy foods.

The forests on the high slopes, coupled with the thermal belt [Geiger[12]] of the house site make a

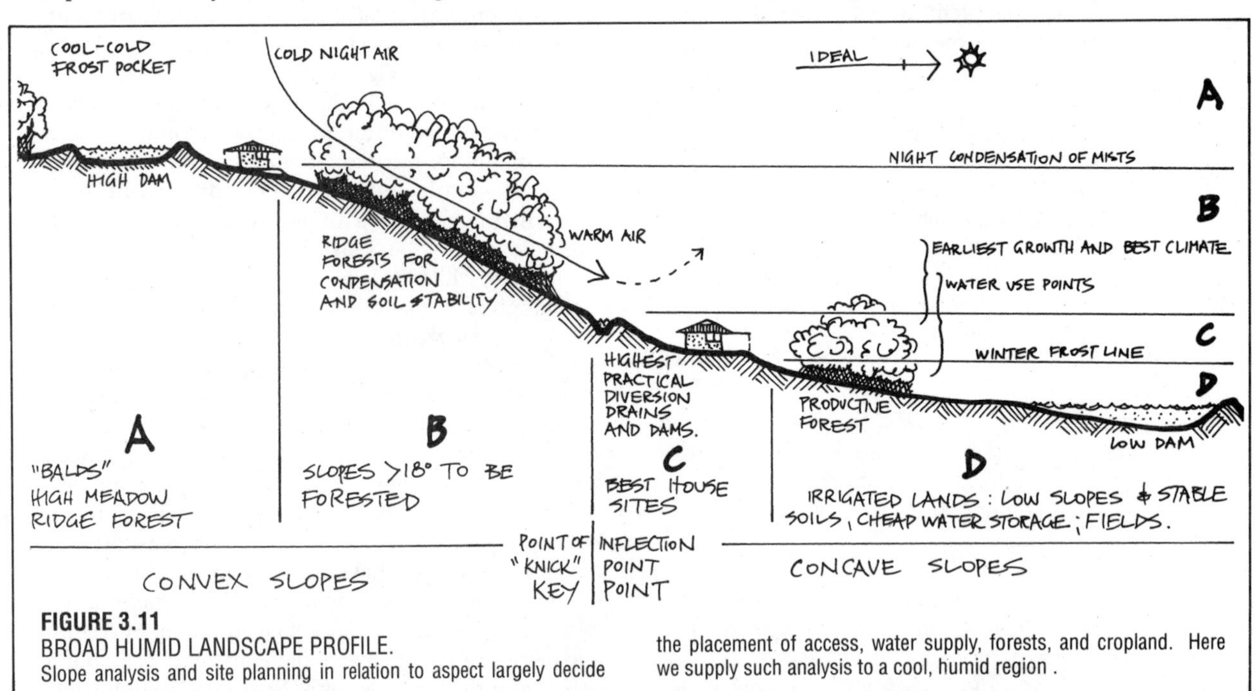

FIGURE 3.11
BROAD HUMID LANDSCAPE PROFILE.
Slope analysis and site planning in relation to aspect largely decide the placement of access, water supply, forests, and cropland. Here we supply such analysis to a cool, humid region .

remarkable difference to midslope climate and soil temperatures. Anyone who doubts this should walk towards an uphill forest on a frosty night, and measure or experience the warm down–draught from high forests. If these are above Zones 1 and 2, they present little or no fire danger. Their other functions of erosion control and water retention are well attested.

Downslope, reflection from dams adds to the warmth. Solar collectors placed here transmit heat, as air or water circulating by thermosiphoning alone, and assist house, glasshouse or garden to function more efficiently. Even very slight slopes of 1:150 function to collect water and heat if well used in the design.

The easy, rounded ridges of non–eroded lower slopes and their foothill pediments are a prime site for settlement. They allow filtration of wastes, inseparable from large populations, through lowland forest and lake, and the conversion of these wastes into useful timber, trees, fruits, and aquatic life.

If zone and sector are imposed in plan, sun angle and landscape slope are assessed in elevation. These determine the following placements:

SUN ANGLE describes the arc of the sun during summer and winter months, and so decides eave and sill placement of windows, areas of shade, and reflection or absorption angles of surfaces. Also, in *every* situation (even hot deserts), some part of the system should be left open to the sun for its energy potential.

ASPECT describes the orientation of the slope. A slope facing the sun will receive considerably more direct solar radiation than a slope facing the shade side (south in the southern hemisphere, north in the northern). The shaded side of hills may delay thaw and thus moderate frost effects in vegetation. In mid–latitudes, we seek the sunny aspect of slopes to achieve maximum sunlight absorption for our settlements and gardens.

The final act in site planning is to orient all buldings and structures or constructs correctly, to face **mid– sky**, the sun, or the wind systems they refer to, or to shelter them from detrimental factors e.g. cold winter winds or late afternoon sun.

In summary, if the elements of the design are carefully zoned, the sectors well analysed, the sun angle and slope benefits maximised for use, and the constructed environment oriented to function, then a better ground design results than most that now exist. As I reassure all would–be permaculture designers, *you can do no worse* than those prior designs you see about you, and by following the essentially simple outline above, you may well do much better. Incredible as it seems, these essential factors are the most frequently overlooked or ignored by designers to the present day, and **retrofit** is then the only remedy for ineffective design.

3.10

ZONING OF INFORMATION AND ETHICS

In this book, I am concentrating on people and their place in nature. Not to do so is to ignore the most destructive influence on all ecologies: the unthinking appetite of people—appetite for energy, newspaper, wrappings, "art", and "recreation". We can think of our zones in other than product terms and management, as a gradation between an ecosystem (the home garden) managed primarily for people, and the wilderness, where all things have their right to exist, and we are only supplicants or visitors. Only excessive energy (human or fuel) enables us to assert dominance over distant resources. When we speak of dominance, we really mean destruction.

What is proposed herein is that we have no right, nor any ethical justification, for clearing land or using wilderness while we tread over lawns, create erosion, and use land inefficiently. Our responsibility is *to put our house in order*. Should we do so, there will never be any need to destroy wilderness. Indeed, most farmers can become stewards of forest and wildlife, as they will have to become in any downturn in the energy economy. Unethical energy use is what is destroying distant resources for short–term use.

Our zones, then, represent zones of destruction, information, available energy, and human dependency. The "ecologist" with large lawns, or no food garden, is as hypocritical as the "environmentalist" drinking from from an aluminium beer can and buying newspapers to read of destructive exploits. Both occupations exploit wilderness and people.

In Zone 1 we are *information developers*; we tend species selected by, and dependent on, mankind. All animal species tend their "home gardens", and an interdependency arises that is not greatly different from the parasite–prey dependency.

In Zone 2, already nature is making our situation more complex, and we start to *learn* from species other than our people–dependent selections. As we progress outwards, we can lose our person–orientation and gain real understanding of the necessity for all life forms, as we do not "need" to exploit most species. We in fact need and use only a few species of the hundreds of thousands that exist.

In wilderness, we are visitors or strangers. We have neither need nor right to interfere or dominate. We should not settle there, and thus leave wastelands at our back. In wilderness we may learn lessons basic to good design, but we cannot improve on the information already available there. In wilderness, we learn of our little part in the scheme of all things.

Understandings

1. Everything is of use. It is not necessarily needed by people, but it is needed by the life complex of which we are a dependent part.

2. We cannot order complex functions. They must

evolve of themselves.
3. We cannot know a fraction of what exists. We will always be a minor part of the total information system.

Thus, we are *teachers* only in our home gardens, and *learners* elsewhere. Nowhere do we create. Everything we depend on we have evolved from *what is already created*, and that includes ourselves. Thoughtful people (those who get recreation from trying to understand) need wilderness as schools need teachers. Should we lose the wilderness, or suffer it to be destroyed, we will be recycled for more appropriate life in any number of ways, some very painful and protracted. We can also state our first "error" thesis here; such errors, once made, lead us into increasing problems.

Type 1 Error
When we settle into wilderness, we are in conflict with so many life forms that we have to destroy them to exist. Keep out of the bush. It is already in good order.

3.11
INCREMENTAL DESIGN

Almost all engineering design is based on small changes to existing designs, until some ultimate limit in efficiency or performance is reached. The whole process can take centuries, and the end result can be mass-produced if necessary. Kevin Lynch (1982) in his book *Site Planning* writes of site designing by incremental adaptation of already-existing designs: design by following physical systems that have been shown to work. He believes the best site planning of the past to be a result of this process, and that it in fact works very well unless some external and important condition (e.g. market or land ownership) changes. He maintains that this fine-tuning of *successful* design for a specific climate and purpose can be totally inappropriate if transferred out of culture, climate, or if applied to a different purpose.

It is, however, the most successful way to proceed *after* selective placement and energy conservation is paid sufficient attention. Known effective design units and specifications, whether of roads, culverts, houses, garden beds, or technologies have been subjected to long tests, and have evolved from trials (or prototypes) to working and reliable standards. Even if "old-fashioned" (like overshot water wheels), they may yet represent a simplicity and an efficiency hard to beat without a considerable increase in expense and complexity.

Such continual adaptation is the basis of feedback in systems undergoing establishment, where we make additions or changes to houses or plant systems. It is not the way, however, to satisfy the demands of a complex system which (like a private home and garden) has to satisfy a complex set of priorities. Nor does it cope with changing futures, new information and sets of values, or simply self-reliance and self-governance.

3.12
SUMMARY OF DESIGN METHODS

To sum up, in whole farm planning and in report writing, outlining areas of like soil, slope, or drainage will suggest sensible crops, treatments, fencelines and land use generally. If we accept what is there, ethical land use dictates conservative and appropriate usage. But (as may happen) if someone is determined to raise wheat on all that land regardless of variations, they can probably do so only if they command enough energy or resources. I am sure we can grow bananas in Antarctica if we are prepared to spend enough money, or can persuade penguins to heat a glasshouse!

Insofar as we enter into village design, we may have financial and space constraints on upper or lower sizes: a "break-even" point and an "optimum" number of houses per unit area. However, if we neglect a foray into the social effects of settlement size and into the needful local functions related to settlement size, we may be designing for human misery, the under-servicing of needs, and even for such sociopathologies as riot and crime. Such designs may be economic (in cash terms), efficient for one use only, and totally inhuman. But they are built every day. For example, it was found that rats subjected to breathing in the same airstream of their fellows experienced severe instability, physiological stress, and consequent pathologies akin to crowding stress. This is called the "Bruce effect" after the experimenter who discovered it, and this effect may apply equally well to people. Yet in almost all cities, one can see people crammed into 16–40 storey office buildings with no opening windows and only a single airstream!

Site designing needs not a specialist approach, but rather a multi-disciplinary and bio-social approach that takes into account the effects the environment has on its intended occupants.

Perhaps if we assembled all our considerable, diverse, and effective knowledge of both the parts and the whole order of design into a type of computer search or game-playing programme, we might advance the whole design process as a realm of continuing and additive human knowledge, available to everyone. Such programmes could deal with a great deal of the fussy detail that now slows design—from plant list specifications to home construction details—leaving the designer with those imponderables about the processes observed on the land, the likely trends of future societies, future needs, and a measure of human satisfaction.

In elaborating just some of the basic approaches to design, without including specialist solutions, I want to stress that all the approaches outlined are not only useful, but necessary. Only by some sensible combination of all the methods given can one select and assess all the elements that enter into a total design assembly, and so evolve a design that includes a large degree of self–management, takes account of details on site, suits the ethics and resources of people, locates ground features in an integrated way, and provides for natural systems and access routes to be properly located.

3.13

THE CONCEPT OF GUILDS IN NATURE AND DESIGN

The methodologies of polyculture design rely more on species interaction than on configuration, although both are necessary inputs to a design. Thus, in designing for best (or most beneficial) species assemblies, we need to know about, and use, the concepts of species guilds and the co–actions of species.

In the natural world, we may often notice assemblies of plants or animals of different species that never-

TABLE 3.4
IMPETUS TO DESIGN

	PERMACULTURE	MOST PRESENT–DAY DESIGN
STIMULATED BY:	A perceived social problem.	A drive to erect monuments or make money.
SUBJECT TO:	Values of energy conservation, self–reliance, and harmonious human occupancy.	Economic (as cash) considerations and the desire for profit.
MEDIATED BY:	Consideration of long–term biosocial factors.	Purely functional values for short–term cost and material factors.
ACHIEVED BY:	Research and consultation with clients, or assisting people to gain an education in design.	External funding, little consultation, little client/people education.
REFINED BY:	Allowing space, finance, and feedback to adjust activity, and allowing for new or overlooked needs as they occur.	Selling off and not taking responsibility for the results.
LEADING TO:	A dynamic and healthy area inhabited by people with the power and understanding to make necessary changes.	A dislocated populatio, powerless to effect change easily. Hence, dependency and anxiety.
RESULTS	THE STABILITY OF DYNAMIC LOCAL ADJUSTMENT.	THE INSTABILITY OF PERCEIVED INDIVIDUAL POWERLESSNESS.

theless occur together over their range. Closer examination of such mixed assemblies often reveals a set of mutual benefits that arise from such convivial togetherness. These benefits offer help or protection to the whole assembly (as when one bird species acts as "lookout" for another, or defends others from hawks). When we *design* plant guilds, as we always try to do in a polyculture, we try to maximise the benefits of each species to the others. We can also add factors of convenience to ourselves, or which save us inputs of fertiliser or pesticides, as in the "apple–centred" guild described below .

A **guild**, then, is an harmonious assembly of species clustered around a central element (plant or animal). This assembly acts in relation to the element to assist its health, aid our work in management, or buffer adverse environmental effects (see **Figure 3.12**). Let us list some of the reasons to place species in association:

<u>To benefit as selected species by:</u>

- Reducing root competition from (e.g.) invasive grasses. Almost all our cultivated food trees thrive in herbal ground covers, not grasses.
- Assisting pest control in various ways:
  - — by providing **anti–feedants** (bitter or unpalatable browse or chemical deterrents), e.g. nasturtium roots provide root chemicals to tomatoes or gooseberries which deter whitefly. Many plants, fermented or in aqueous extraction, deter pests or act as anti–feedants when sprayed on leaves of the species we wish to protect.
  - — by killing root parasites or predators, e.g. *Crotalaria* captures nematodes that damage citrus and solanaceous roots; *Tagetes* marigolds "fumigate" soils against grasses and nematodes.
  - — by hosting predators, as almost all small–flowered plants [especially*Quillaja*, many *Acacia* species, tamarisk, *Compositae* (the daisy family) and *Umbelliferae* such as dill, fennel, carrot, and coriander] host robber–flies and predatory wasps.
- Creating open soil surface conditions, or providing mulch. For example, comfrey and globe artichokes allow tree roots to feed at the surface (unlike grasses, which competes with tree roots), while spring bulbs (daffodils) or winter–grown wild *Allium* species, whose tops die down in mid–spring do not compete with deciduous tree roots in summer dry periods, nor do they intercept light rains.
- Providing free nutrients: woody or herbaceous legumes fix nitrogen or other essential nutrients via root associates, stimulate soil bacteria or fungi, and benefit associated trees. Clovers; trees such as *Acacia, Casuarina,* and *Pultenaea*; sugar–providing grasses (sugar cane); and high humus producers (bananas) all assist orchard species. Many can be slashed or trimmed to give rich mulch below trees or between crop rows.
- Providing physical shelter from frost, sunburn, or the drying effects of wind. Many hardy windbreak species of equal or greater height, both as edge windbreak or in–crop crown cover exclude frost, nullify salty or hot winds, provide mulch, and moderate the environment towards protecting our selected species. Examples are borders of bamboo, cane grasses, *Casuarina*, hardy palms, and tamarisks. In–crop shade shelter of legumes are needed by such crops as avocado, citrus, and cocoa or coffee (or any crops needing partial shade). In–crop trees can eliminate frost effects in marginal frost areas.

<u>To assist us in gathering:</u>

- Culinary associates: it is of some small benefit in detailed planning to keep common culinary associates together (tomatoes with parsley and basil; potatoes with a tub of mint) so that we also gather them together for cooking, salads, or processing (dill with cucumbers). Thus we reduce work. Dill and apples also go well together, raw or cooked, and dill is one of the *Umbelliferae* that host predatory wasps below apple trees.

<u>Specific animal associates of a guild:</u>
We have made reference, in pest control, to host plants. These can be best specified by observing, researching, or selecing plants to host quite specific predatory wasps, lacewings, or ladybirds. Vertebrates that assist our selected crop species are:

- Ground foragers, e.g. pigs or poultry specifically used to clear up the fallen fruit that host fruit fly or larval forms of pests. Foragers can be run in orchards for that relatively short period of the year when fruit is falling and rotting, or they can be used to eat reject fruit and deposit manures.
- Insectivores: birds, in particular, that search bark crevices (woodpeckers, honey–eaters) for resting larvae and egg masses. To encourage these, plant a very few scattered flowering shrubs and herbaceous plants such as *Kniphofia, Banksia, Salvia, Buddleia,* and *Fuschia.* All of these provide insect and nectar foods for insectivorous birds.
- Mollusc control: snails and slugs are almost totally controlled by a duck flock on range, and several large lizards (*Tiliqua* spp.) also feed primarily on snails. Ducks can be ranged seasonally (autumn to spring) in plant systems, and in summer on marshlands. Ducks will eat seedlings, so that appropriate scheduling is essential.
- Guard dogs: for deer, rabbits, and other vertebrate pests. A small number of guard dogs, fed and kennelled in orchards, are sufficient control for fox predation on orchard poultry foragers. Such dogs, reared *with* domestic poultry, do not attack the flocks themselves.
- Hawk kites suspended over a berry crop, or flown as light model planes over an extensive grain crop deter all flock–bird predators of the crop, and are more dependable than natural hawks. They need to be removed when not needed, so that birds do not get

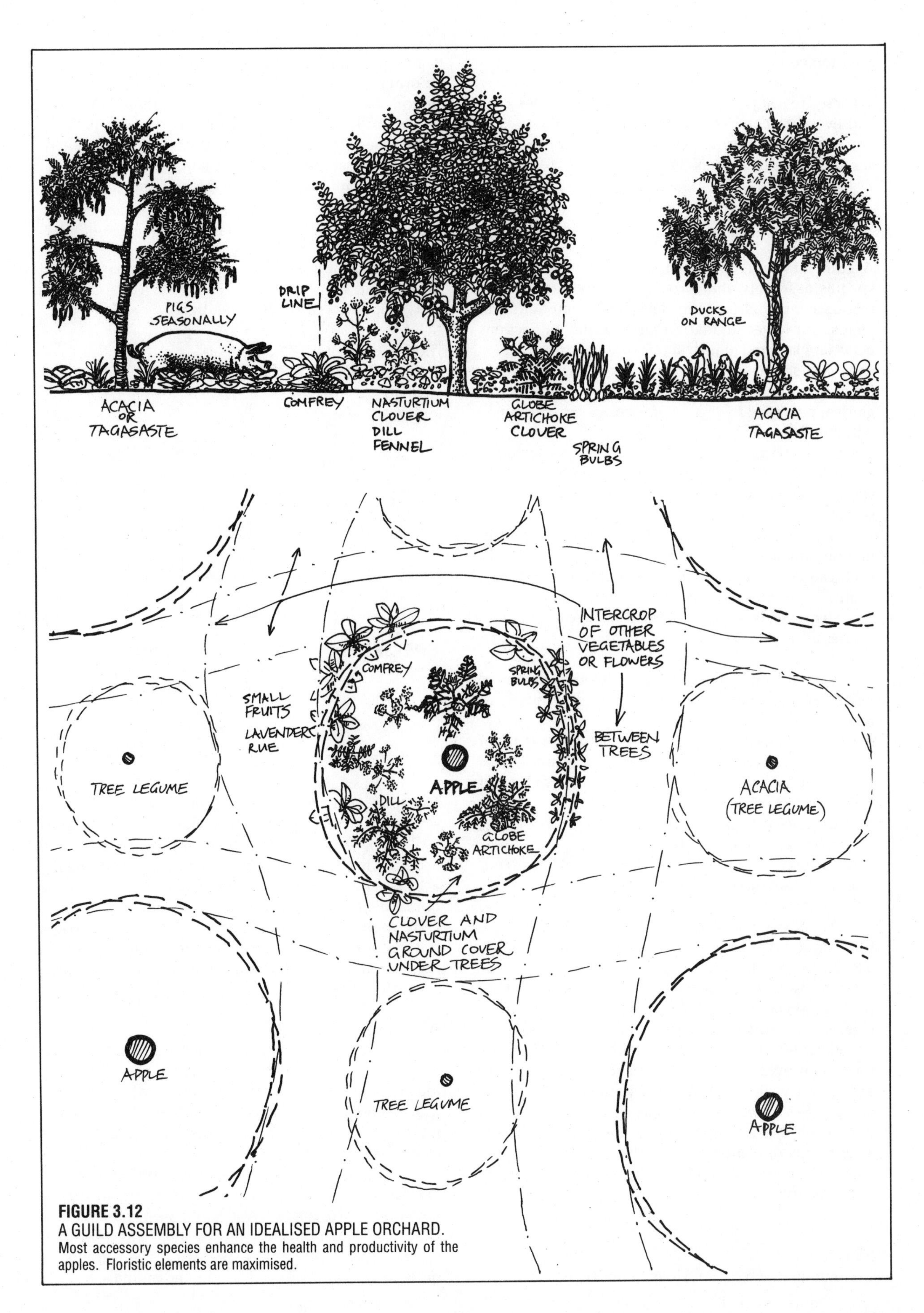

FIGURE 3.12
A GUILD ASSEMBLY FOR AN IDEALISED APPLE ORCHARD.
Most accessory species enhance the health and productivity of the apples. Floristic elements are maximised.

accustomed to them.

These are just part of the total guilds. Every designer, and every gardener, can plan such guilds for specific target species, specific pests and weed control, and specific garden beds or orchards.

ANALYTIC APPROACH TOWARDS SELECTING A GUILD

A guild of plants and animals is defined here as a species assembly that provides many benefits for resource production and self–management (more yields, but lower inputs). In general, the interactions between plant and animal species are thus:

- Most species get along fine; this is obvious from a study of any complex home garden or botanical garden; perhaps 80% of all plant species can co– mingle without ill effect.
- Some species greatly assist others in one or other of many ways. Positive benefits arise from placing such species together where they can interact (10–15% of all species).
- A minority of species show antagonistic behaviour towards one or more other species. This in itself can be a benefit (as in the case of biological pest control) or a nuisance (as in the case of rampancy or persistent weeds or pests). Perhaps as few as 5% of all species act in this way.

Now, to give the above classes of interaction a more useful analytic structure, we will allot symbols, as follows:

+ : this is used to indicate a *beneficial* result of interaction, with a yield above that of some base level (judged from a monoculture or control crop of the species).

o : this is used to indicate *"no change"* as a result of interaction, on the same basis.

–: this is used to indicate a *reduction* in yield or vigour as a result of interaction with another species.

Thus, for two useful species (each selected for a useful product), we have the simple tabulation of **Table 3.5**, which gives us all possible interactions.

The array is such that only three interactions benefit us, three are neutral, and three are antagonistic in effect effect. By grouping scores, we can analyse for beneficial effects in our interaction table, and act on these. However, because of the vagaries of weather in any given year, many times a peasant farmer may accept a (– +) effect just to ensure that he at least gets a crop, even if it is of the "losing" species. It is *always* safer to mix or complicate crop than to pin hopes on a single main crop. In fact, be guided by analyses but study reality!

STATES OF ACTION

In common usage, COACTION implies a force at work: one that restrains, impedes, compels, or even coerces another object. INTERACTION implies reciprocal action: two things acting on each other. This is an important distinction. A final category is INACTION, or an absence of any detectable action.

We cannot at this point guess which state applies, but when we put two species together, there are these possibilities:

- One acts on the other (co–action or unilateral action);
- Both act on each other (interaction or mutual action); and
- Neither act (inaction or neutrality).

TABLE 3.5
INTERACTION MATRIX OF TWO SPECIES (plant or animal).

		SPECIES A		
		+	o	–
	+	++	+o	+–
SPECIES B	o	o+	oo	o–
	–	–+	–o	––

It would seem probable that in the case of (++) and (– –) we have mutual action or interaction. In the case of (–o), (o–), (+o), (o+), (+–), (–+) *one only* needs to be acting, a form of co–action. In the case of (**oo**) neither acts, no effects appear, and both are inactive insofar as our measures can detect.

We need to observe and perhaps analyse each case, but it does seem probable that such states of action apply. Some such states can be named and examples given, for instance:

A. Mutual Action States

++ This is called symbiosis, and is common both in nature and in society. It is a "win–win" situation ideally suited to guild development. An example is the mycorrhizal associates of higher plants, where mutualism or fair trade occurs between a plant and its root associate.

-- Haskell (1970) has coined the word synnecrosis (*The Science Teacher 37(9)* Supplement), and it is obviously uncommon. War is our best example of a "lose–lose" situation, but there are also battles between plants for light, nutrients, and space. There are forms of chemical warfare in both plants and animals.

B. Single Action States

–o Haskell calls this amensalism. It hurts the *actor*, not the other. A butterfly attacking a rhinocerous would fit, or a wasp parasite "glued" to a tree it attacks, as is the case with some pine trees and Sirex wasps.

o– Called allolimy by Haskell, it leaves the actor unaffected but hurts the other, e.g. a walnut tree beside an apple tree yields well, but the juglones secreted by its roots act to kill or weaken the apple. In the same way grasses act to weaken most deciduous fruit trees.

+o Termed commensalism. Even though the actor benefits, the other remains unaffected, e.g. an epiphyte attached to a sturdy tree, such as vanilla on a coconut trunk.

o+ Called allotrophy by Haskell. The actor is unaffected, the other benefits. Examples are a teacher and student relationship, or a charity where one hands on surplus goods to another person less fortunate.

+– Called parasitism, the actor benefits, the other loses *if* the actor is the parasite. All pathogens and parasites tend to weaken or take from the host.

–+ Self-sacrifice. The actor loses. This is the reverse of parasitism, and a better word might be self-deprivation to help others. This is often seen in nature, mostly as individuals helping members of the same family or species. Medals are awarded for this in human society, and we call it selflessness or even heroics.

oo Neither one acts. No one is hurt, no one wins. Neutrality pacts may achieve this result in society, or we observe it commonly in nature. There are critical areas in nature (water holes, salt licks, grooming stations) where antagonistic species agree on neutrality. In fact, many plant species appear to be basically neutral in behaviour.

Such analyses suit two-species interactions, but where we depart (in the designed system) from nature is that we may value (in the sense of obtaining a yield from) only one of these species. Let this be species A in **Table 3.6**. The other can be a weed or a species such as *Lantana*, which we might wish to eliminate. In this case, we can set up a matrix as diagrammed in **Table 3.6.**

This is a very necessary type of analysis for selecting useful plants that will eliminate or weaken an unwanted weed species. All such analyses can be made using plant/plant, animal/animal, or plant/ animal pairs.

How do we *observe* co-action? This is quite simple in the field, *providing* there are plenty of examples to score, and we have set some criteria to score by. For example, take a town or area with a great many trees planted in the backyards. Select any one of these species for criteria, say an apple, then decide on how to score, e.g. (in compounds with apples and other species of plants growing):

+: apple tree healthy, bearing very well, not stunted or over-vigorous.
o: apple tree healthy, in fair order, bearing.
–: apple tree bearing poorly, sick or dying.
x: no apple tree in this yard.

TABLE 3.6
INTERACTION MATRIX OF TWO DIFFERENT SPECIES.

		SPECIES A (a palm)		
		+	**o**	**–**
	+'	**+'+**	**+'o**	**+'–**
SPECIES B (lantana)	**o'**	**o'+**	**o'o**	**o'–**
	–'	**–'+**	**–'o**	**–'–**

In order of benefit (increase in palms, less increase or decrease in lantana):
(–'+)>(o'+)>(–'o)>(+'+)>(o'o)>(–'–)>('–)>(o'–)>(+'o)
Best result..................>Neutral...............>Worst result.

TABLE 3.7
CO-ACTION MATRIX

	APPLE SCORE			
Other trees near, or in, yard.	**+**	**o**	**–**	**x**
WALNUTS	.	.	7	15
MULBERRY	5	5	1	3
ACACIA	7	5	.	3

Scoring can be of specific pairings:

		APPLE SCORE			
		+	**o**	**–**	**x**
	+	.	.	.	5
WALNUT SCORE	**o**	.	.	5	4
	-	.	.	.	.
	x	.	.	.	.

Then, we draw up a co–action matrix on a piece of paper, with the "apple" score at the top and "other trees" down the left side (**Table 3.7**). Tally the scores by walking from yard to yard.

We quickly see that where there are walnuts, apples are sick or absent (**o–**). However, healthy apple trees coexist with both mulberries and *Acacias* (**+o**) and (**oo**). Ideally, we use a similar score for *each species* of other tree, so that our co–action results score the same criteria for walnut, mulberry, and *Acacia* that we score for apple.

Additional field notes are useful. Healthy, untended apple trees often have quite a specific understory of spring bulbs, comfrey, clover, iris, nasturtium, etc. This too should be noted as we go. I have, in fact, carried out such analyses, and some of the results will be used as a real example in the next section.

BUILDING UP GUILDS FROM CO–ACTION ANALYSIS.

If we wish to construct a guild, then we need to bring two or more species into close proximity where we can judge the effects of one on the other. If we have a () result *anywhere*, we might be able to intervene with a third or fourth party which we can call an arbitrator, a buffer, or an intervenor.

- Apple next to walnut produces (**–o**): *not* desirable; the apple sickens or dies.
- Apple next to mulberry produces (**+o**): a good result.
- Mulberry next to walnut produces (**oo**): mutual inaction.

Thus, apple *then* mulberry *then* walnut gives us (**+oo**). By this *intervention strategy*, we have, in effect, cancelled out the (––) and have a *nett benefit* in a three–three–species array. That is, we can use several two–species results to achieve a better result with three species, which goes beyond accepting (fatalistically) the primary conflict. Here, a mulberry is the *intervenor* or critical species or element in conflict resolution. We can take this further again by examining yet other co–actions:

- *Acacia* next to walnut gives (**o+**)
- *Acacia* next to mulberry gives (**o+**)

Now, apple–mulberry–*Acacia*–walnut gives us (**++o+**), which is much better again. So we proceed to isolating and *arranging* guilds to maximise benefits and eliminate conflicts. This is part of the skill of planning strip or zone placements of mixed species.

THE ROLE OF CONFIGURATIONS IN GUILDS

Here, we have to consider *placement* of interactive elements. Obviously, there is a commonsense close spacing for many plants and machine components, but as the distance between *living* components widens, we can never be quite sure that chemical or behavioural interaction ceases. Consider the case of two territorial birds, displaying or calling a mile or more apart. To us, they appear as individuals; to each bird, the other is in clear interaction. There is distant interaction, too, via pollen or spores in plants, and perhaps even by gaseous or chemical "messengers". This is certainly true of some mammals, so that effects of one on the other can be passed on by a sense of smell, even though they are not nearby at that time, e.g. urine marking territory. The great whales may well be communicating by sound around the whole globe.

Configuration in planning a guild with *intervening* species between hostile () species, comes in assessing the distance across the interaction boundary that the effect takes place, and in then arranging the guild species to obtain a maximum of (++), (o+), or (+o) effects. For example, we find a (++) condition with legume/grain or fruit–tree/tree legume interplants. We know that the effect, for grains, extends from 1.5–2 m into the crop; thus for a configurational design, we can spiral or strip–plant these two species for a *total positive edge interaction effect in crop*. Such careful guild analyses and configurations are the basis of species planning in permaculture.

For more critical geometric analyses, see such texts as Rolfe A. Leary's *Interaction Geometry: An Ecological Perspective* , USDA Forest Service, General Technical Report NC–22, 1976. This text has a useful reference list and is issued by the North Central Forest Experimental Station, Fulwell Ave, St. Paul, MN, 55108, USA.

3.14
SUCCESSION: EVOLUTION OF A SYSTEM

Nature shows us that a *sequence of processes* arise in the establishment of "new" systems on such devastated landscapes as basalt flows and ice–planed or flood–swept sites. The first living components are hardy *pioneer species*, which establish on these damaged or impoverished environments. Thus we see "weeds" (thistles, *Lantana*) occupying overgrazed, eroded, or fired areas. These pioneer species assist the area by stabilising water flow in the landscape, and later they give shelter, provide mulch, or improve soil quality for their successors (the longer–term forest or tree crop species).

To enable a cultivated system to evolve towards a long–term stable state, we can construct a system of mixed tree, shrub, and vegetable crop, utilising livestock to act as foragers, and carefully planning the succession of plants and animals so that we receive short–, medium–, and long–term benefits. For example, a forest will yield first coppice, then pole timbers, and eventually honey, fruit, nuts, bark, and plank timber as it evolves from a pioneer and young, or crowded, plantation to a well–spaced mature stand over a period of 15–50 years.

Unlike the processes of nature, however, we can

place most of the elements of such a succession *in one planting*, so that the pioneers, ground covers, under–story species, tree legumes, herbage crop, mulch species, the long–term windbreak and the tree crop are all set out at once. So many species and individuals of each species are needed to do this that it is usually necessary to first create a small plant nursery to supply the 4,000–8,000 plants that can be placed on a hectare. While these are growing in their pots, we can fence and prepare the soil, and then plant them out to a carefully–designed long–term plan.

Where this approach is used, as it has been by many permaculturalists on their properties, quite remarkable changes occur over two to three years. Mulch is produced on site for the long–term crop, while weed competition, wind, and frost effects are nullified or moderated. Cropping can be continuous as the annuals or herbaceous perennials effectively control unwanted grasses and weed species. For instance, radish or turnip planted with tree seedlings control grasses until the small tree provides its own grass control by shading. **Figure 3.13** gives an indication of how a system can accept different species of plants and of animal browsers as it evolves.

3.15 THE ESTABLISHMENT AND MAINTENANCE OF SYSTEMS

Every design is an assembly of components. The first priority is to locate and cost those components. Where our resources are few, we look closely at the site itself, thinking of everything as a potential resource (clay, rock, weeds, animals, insects). We can think of labour, skill, time, cash, and site resources as our interchangeable energies: what we lack in one we can make up for by exchange for another (e.g. clothes–making in exchange for roof tiles). The best source of seed and plants is always neighbours, public nurseries, or forestry departments. From the early planning stages, it pays to collect seed, pots, and hardy cuttings for the site, just as it pays to forage for second–hand bricks, wood, and roofing.

The planning stage is critical. As we draw up plans, we need to take the evolution in *stages*, to break up the job into easily–achieved parts, and to place components in these parts that will be needed *early in development* (access ways, shelter, plant nursery, water supply, perhaps an energy source). Thus, we *design*, assess resources, locate components, decide priorities, and place critical systems. Because impulsive sidetracks are usually expensive, it is best to fully plan the site and its development, changing plans and designs only if the site and subsequent information forces us to do so.

On a rural (and sometimes urban) site, FENCING or hedgerow, SOIL REHABILITATION by mulch (or loosening by tools), EROSION CONTROL, and WATER SUPPLY are *the* essential precursors to successful plant establishment, for we can waste time and money putting out scattered plants in compacted, impractical, and dry sites. Any soil shaping for roads, dams, swales, terraces, or paths needs to be finalised before planting commences.

For priority in *location*, we need to first attend to Zone 1 and Zone 2; these support the household and save the most expense. What is perhaps of greatest importance, and cannot be too highly stressed, is the need to develop *very compact systems*. In the Philippines, people are encouraged to plant $4m^2$ of vegetables—a tiny plot—and from this garden they get 40–60% of their food! We can all make a very good four metres square garden, where we may fail to do so in 40 square metres.

Similarly, we plant and care for ten critical trees (for oils, citrus, nuts, and storable fruit). We can take good care of these, whereas if we plant one hundred or one thousand, we can lose up to 60% of the trees from lack of site preparation and care. Thus, ten trees and four metres square , well protected, manured, and watered, will start the Zone 1 system.

Starting with a *nucleus* and expanding outwards is the most successful, morale–building, and easily–achieved way to proceed. Broadscale systems have broadscale losses and inefficiencies. As I have made every possible mistake in my long life, the advice above is based on real–life experience. To sum up:

- Design the site thoroughly on paper.
- Set priorities based on economic reality.
- Locate and trade for components locally or cheaply.
- Develop a *nucleus* completely.
- Expand on information and area using species proved to be suited to site.

Precisely the same sort of planning (nucleus development) applies to any system of erosion control, rehabilitation of wildlife or plants, writing books, and creating nations. Break up the job into small, easily achieved, basic stages and complete these one at a time. Never draw up long lists of tasks, just the next stage. It is only in the design phase that we plan the system as a whole, so that our smaller nucleus plans are always in relation to a larger plan.

Instead of leaping towards some imaginary end point, we need to prepare the groundwork, to make modest trials, and to evolve from small beginnings. A process of constant transition from the present to the future state is an inevitable process, modest in its local effect and impressive only if widespread. Thus, we seek first to gain a foothold, next to stabilise a small area, then to develop self–reliance, and only after this is achieved to look for exportable yields or commercial gain.

Even in a commercial planting it is wise to restrict the total commercial species to 3–10 reliable plants and trees, so that easier harvesting and marketing is achievable, although the home garden and orchard can maintain far greater diversity of from 25–75 species or

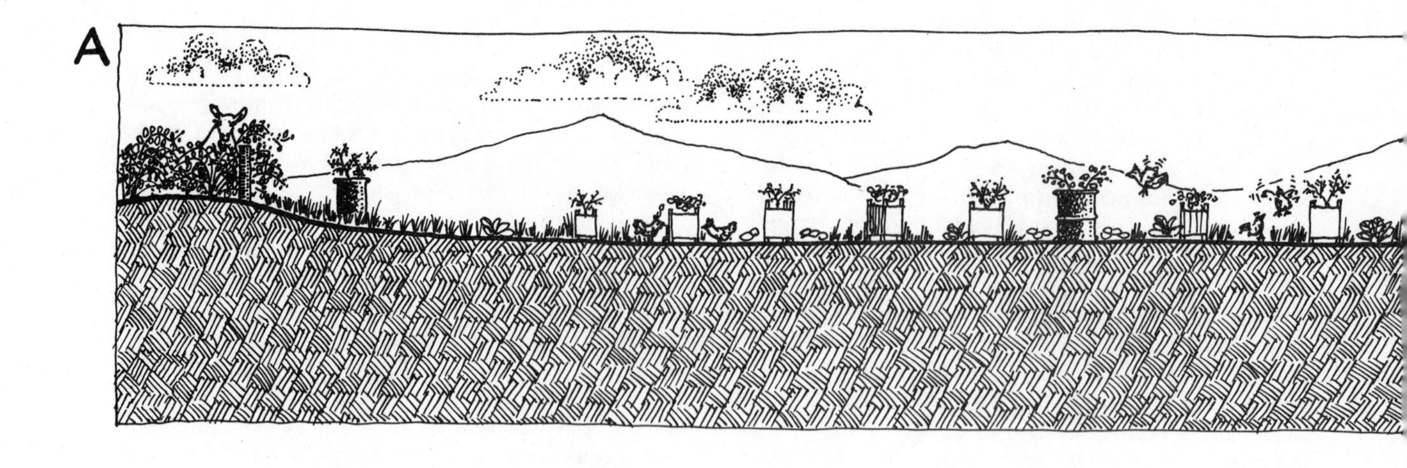

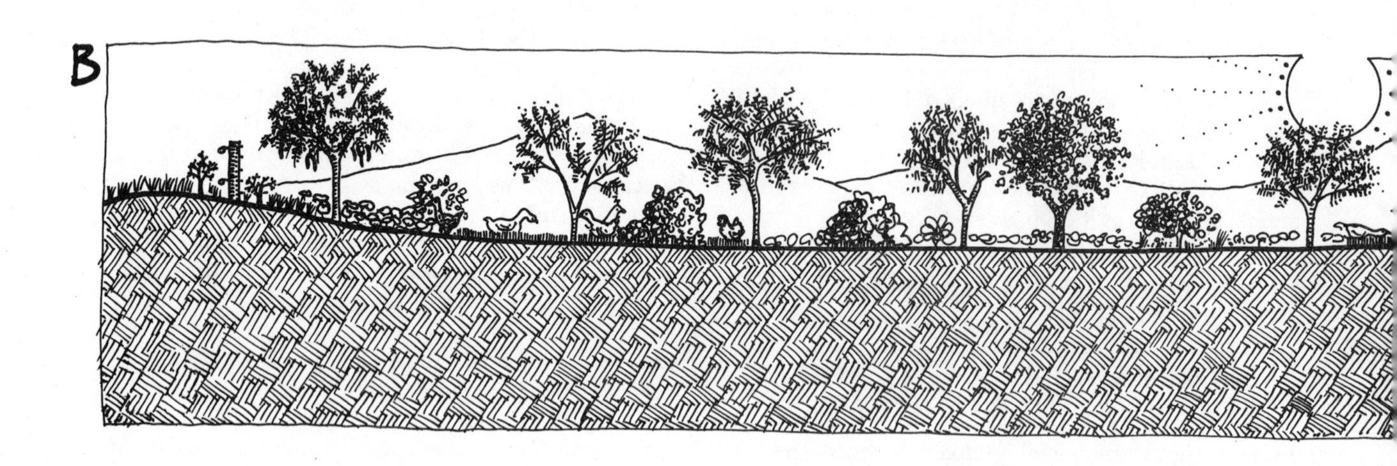

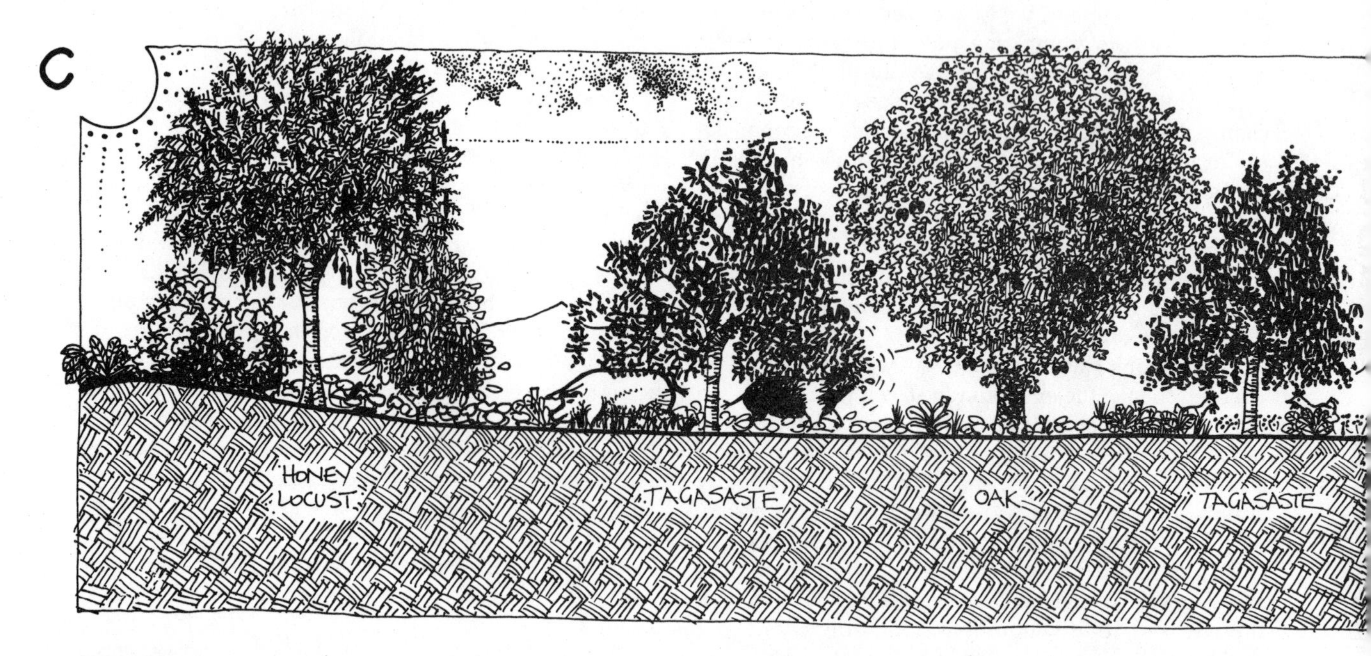

FIGURE 3.13
EVOLUTION IN A DESIGNED SYSTEM.
Pioneer species prepare for long–term evolution to a stable and productive system over a period of from 5 – 15 years.

A. System establishment; an area is fenced and a complex of species planted and protected from grazers by fencing and tree guards. Ponds are established. Only small livestock (chickens) and some annual crops can be harvested.

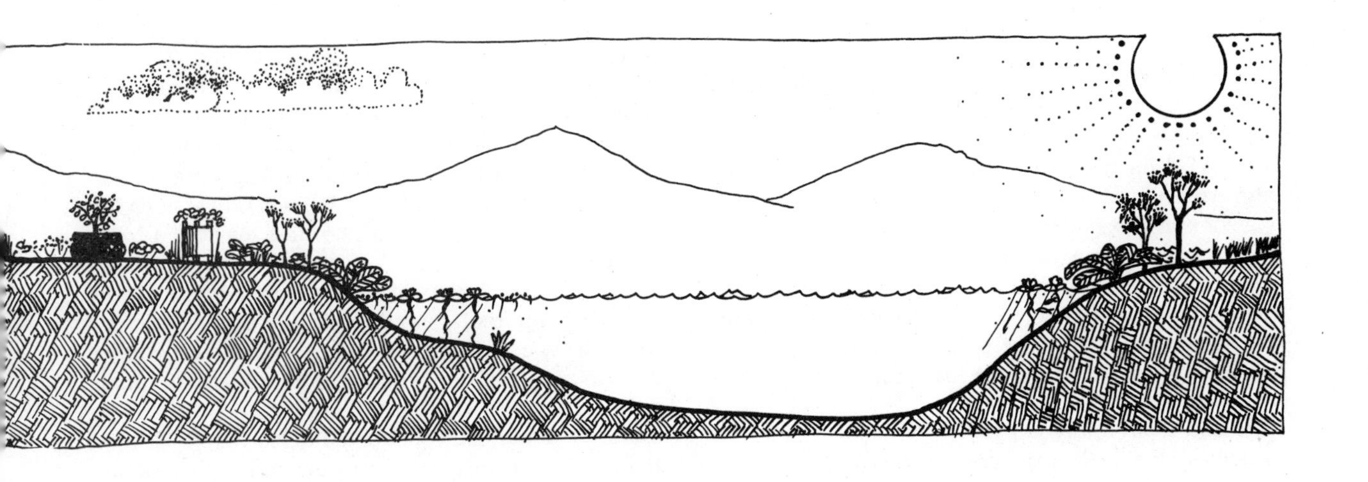

B. The system evolves to a semi-hardy stage. Geese, fish, and shellfish are introduced, and crops include some aquatic plant species.

C. An evolved system provides forage, firewood, aquatic and animal products. Larger foragers (sheep, pigs) can be grown seasonally. The system provides its own mulch and fertilizers. The mature system requires management rather than energy input, and has a variety of marketable yields (including information).

more.

Thus, our design methodologies seek to take into account all known intervening factors. But in the end it comes down to flexibility in management, to steering a path based on the results of trials, to acting on new information, and to continuing to observe and to be open or non–discriminatory in our techniques.

The success of any design comes down to how it is accepted and implemented by the people on the ground, and this factor alone explains why grand centralised schemes more often result in ruins and monuments than in stable, occupied, and well–maintained ecologies.

We can design any expensive, uncomfortable, or ruinous system as long as we do not have to live in it, or fund it ourselves. Responsible design arises from recommending to others the way you have found it possible to work or to live in a similar situation. It is much more effective to educate people to plan for themselves than to pay for a permanent and expensive corps of "planners" who lead lives unrelated to those conditions or people for whom they are employed to design.

3.16 GENERAL PRACTICAL PROCEDURES IN PROPERTY DESIGN

Except for the complex subject of village design, a property design from one–fourth to 50 ha needs firstly a clear assessment of "client or occupier needs", and stated aims or ideas from *all* potential occupiers (including children). A clear idea of the financial and skill resources of occupants is necessary so that the plan can be financially viable.

With a base map, aerial photograph, or a person as a guide, the designer can proceed to observe the site, making notes and selecting places for:

- Access ways and other earthworks;
- Housing and buildings;
- Water supply and purification, irrigation;
- Energy systems; and
- Specific forest, crop, and animal system placement.

All the above are in relation to slope, soil suitability and existing landforms. By inspection, some priorities may be obvious (fire control, access, erosion prevention). Other factors need to be tackled in stages as time, money, and species permit. At the end of each stage, trial, or project, both past performance and future stage evolution should be assessed, so that a guide to future adjustments, additions, or extension is assembled as a process. In all of this, design methodologies *plus* management is involved, and it is therefore far better to train an owner–designer who can apply long–term residential management than to evolve a roving designer, except as an aide to initial placements, procedures, and resource listings.

The restrictions on site use must first be ascertained before a plan is prepared or approved. In the matter of buildings, easements, health and sewage requirements, permits, and access there will probably be a local authority to consult. If water (stream) diversions are foreseen, state or federal authorities may need to be consulted.

The homely, but probably essential process of building up real friendship between residents, designers, officials, and neighbours should be a conscious part of new initiatives. Small local seminars help a lot, as district skills and resources can be assessed. There is no better guide to plant selection than to note district successes, or native species and exotics that usually accompany a recommended plant. Nearby towns, in gardens and parks, often reveal a rich plant resource.

As every situation is *unique*, the skill of design (and often of market success) is to select a few unique aspects for every design. These can vary from unique combinations of energy systems, sometimes with surplus for sale, to social income from recreational or accommodation uses of the property. This unique aspect may lie in special conditions of existing buildings, vegetation, soil type, or in the social and market contact of the region. Wherever occupants have special skills, a good design can use these to good effect, e.g. a good chemist can process plant oils easily.

A design is a marriage of landscape, people, and skills in the context of a regional society. If a design ended at the physical and human aspects, it would be still incomplete. Careful financial and legal advice, plus an introduction to resources in these areas, and a clear idea for marketing or income from services and products (with an eye to future trends) is also essential.

Over a relatively short evolution of three to six years, a sound design might well achieve:

- Reduction in the need to earn (conservation of food and energy costs).
- Repair and conservation of degraded landscapes, buildings, soils, and species at risk.
- Sustainable product in short–, medium–, and long–terms.
- A unique, preferably essential, service or product for the region.
- Right livelihood (good work) for occupants in services or goods.
- Sound and safe legal status for the occupiers.
- An harmonious and productive landscape without wastes or poisons.
- A cooperative and information–rich part of a regional society.

These then, or factors allied to them, are the test of good design over the long term. For many regions, a designer or occupant can provide species (as nursery), resources (as education), services (as food processing or lease), or simply an example of sustainable future occupations. Pioneer designers in a region should seek to capitalise on that pioneer aspect, and provide resources for newcomers to the region.

3.17
PRINCIPLE SUMMARY

Definition of Permaculture Design: Permaculture design is a system of assembling conceptual, material, and strategic components in a pattern which functions to benefit life in all its forms. It seeks to provide a sustainable and secure place for living things on this earth.

Functional Design: Every component of a design should function in many ways. Every essential function should be supported by many components.

Principle of Self–Regulation: The purpose of a functional and self–regulating design is to place elements or components in such a way that each serves the needs, and accepts the products, of other elements.

Chapter 4

PATTERN UNDERSTANDING

The curve described by the earth as it turns is a spiral, and the pattern of its moving about the sun... The solar system itself being part of a spiral galaxy also describes a spiral in its movement... Even for the case of circular movement, when one adds in the passage of time, the total path is a spiral... The myriad things are constantly moving in a spiral pattern... and we live within that spiral movement.

(Hiroshi Nakamura, from *Spirulina: Food for a Hungry World,* University of the Trees Press, P.O. Box 66, Boulder, California 95006, USA.)

The patterns and forms of a tree are found in many natural and evolved structures; an explosion, event, erosion sequence, idea, germination, or rupture at an edge or interface of two systems or media (here, earth and atmosphere) may generate the tree form in time and space. Many threads spiral together at the point of deformation of the surface and again disperse. The tree form may be used as a general teaching model for geography, ecology, and evolution; it portrays the movement of energy and particles in time and space. Foetus and placenta; vertebrae and bones; vortices; mushrooms and trees; the internal organs of man; the phenomena of volcanic and atom bomb explosions; erosion patterns of waves, rivers, and glaciers; communication nets; industrial location nets; migration; genealogy; and perhaps the universe itself are of the general tree form portrayed.

Simple or multiple pathways describe yin–yang, swastika, infinity, and mandala symbols. A torus of contained forces evolves with the energies of the pattern, like the doughnut of smoke that encircles the pillar of the atomic explosion.

(Bill Mollison, *Permaculture One,* 1978.)

Everything the Power of the World does is done in a circle... The wind, in its greatest power, whirls... The life of a man is a circle from childhood to childhood, and so is everything where power moves. Our teepees were round like the nests of birds, and these were always set in a circle, the nation's hoop, a nest of many nests....

(Black Elk.)

4.1 INTRODUCTION

It is with some trepidation that I attempt a treatise on patterns. Nevertheless, it must be attempted, for in patterning lies much of the ground skill and the future of design. Patterns are forms most people understand and remember. They are as memorable and repeatable as song, and of the same nature. Patterns are all about us: waves, sand dunes, volcanic landscapes, trees, blocks of buildings, even animal behaviour. If we are to reach an understanding of the basic, underlying patterns of natural phenomena, we will have evolved a powerful tool for design, and found a linking science applicable to many disciplines. For the final act of the designer, once components have been assembled, is to make a sensible pattern assembly of the whole. Appropriate patterning in the design process can assist the achievement of a sustainable yield from flows, growth forms, and timing or information flux.

Patterning is the way we frame our designs, the template into which we fit the information, entities, and objects assembled from observation, map overlays, the analytic divination of connections, and the selection of specific materials and technologies. It is this patterning that permits our elements to flow and function in beneficial relationships. The pattern *is* design, and design is the subject of permaculture.

Beyond the rigour of the simple Euclidean regularities beloved of technologists and architects, there remains most (or all) of nature. Nature imperfectly round, never flat or square, linear only for infinitesimal distances, and stubbornly abnormal. Nature flowing, crawling, flying, weeping, and in apparent disarray. Nature beyond precise measurement, and comprehensible only as sensation and system.

Nothing we can observe is regular, partly because we ourselves are imperfect observers. We tell fortunes (or lose them) on the writhing of entrails or cathode ray graphics, on the scatters of dice or bones, or on arrays of measures. Are the readings of tea leaves any less reliable than the projections of pollsters? Regular things are those few that are mechanical or shaped (temporarily) by our own restricted world view; they soon become irregular as time erodes them. Truth, like the world, changes in response to information.

There are at least these worthwhile tasks to attempt:

1. A MORE GENERAL PATTERN UNDERSTANDING, both as attempts at forming more general pattern models, and as examples of natural phenomena that demonstrate such models.

2. A LINKING DISCIPLINE that equally applies to geography, geology, music, art, astronomy, particle physics, economics, physiology, and technology. This linking discipline would apply to conscious design itself and to the information flow and transfer processes that underlie all our disciplines. Such a unifying concept has great relevance to education, at every level from primary to post-graduate disciplines.

3. GUIDES TO PATTERN APPLICATION: some examples of how applied patterning achieves our desired ends in everyday life, where rote learning, linear thinking, or Euclidean geometry have all failed to aid us in formulating sustainable settle-ments. It is in the application of harmonic patterns that we demonstrate our comprehension of the meaning of nature and life.

There have been many books on the subject of symbols, patterns, growth, form, deformation, and symmetry. The authors often abandon the exercise short of devising general models, or just as a satisfactory mathematical solution is evolved for one or more patterns, and almost always before attempting to create applied illustrations of how their efforts assist us in practical life affairs. Some are merely content to list examples, or to make catalogues of phenomena. Others pretend that meanings lie in pattern or number alone—that patterns are symbols of arcane knowledge, and they assert that only an unquestioned belief unlocks their powers.

The simple pattern models figured herein are intended to be a useful adjunct to designers and educators. They also illustrate how we can portray our thinking about life, landscapes, and the communality that is nature. Learning a master pattern is very like learning a principle; it may be applicable over a wide range of phenomena, some complex and some simple. As an abstraction, it assists us to gain meaning from life and landscape and to comprehend allied phenomena.

One can spend endless hours seeking further scientific, mystical, or topological insights into pattern. The process is addictive, and I am as unwilling to abandon this chapter as I was to start it, but I trust that others, better equipped, will expand and further explain the basic concepts. I believe that it is in sophisticated pattern application that the future of design lies, and where many solutions to intractable problems may be found.

We have a good grasp on the behaviours of pattern in natural phenomena if we can explain the SHAPES of things (in terms of their general pattern outlines); the networks and BRANCHING of tributaries (gathering flows) and distributaries (dispersal flows); the PULSING and flow regulation within organisms or the elements of wind, water, and magma; and illuminate how SCATTERED PHENOMENA arise.

Further, if WAVE phenomena and STREAMLINES are contained within our pattern analysis, as real waves or as time pulses, these and their refraction and interference patterns form another set of pattern generators, responsible for coasts, clouds, winds, and turbulent or streamlined flow. And, if we can show how the pattern outlines of landscapes, skeletal parts, or flow phenomena fit together as MATRICES (interlocking sets), or arise from such matrices (e.g. whirlwinds from thermal cells), then we can generate whole landscape systems or complete organisms from a mosaic of such patterns.

In nature, events are ordered or spaced in discrete units. There are smaller and larger orders of events, and if we arrange like forms in their *orders*, we will find clusters of measures at certain sizes, volumes, lengths, or other dimensions. This is true for river branches, social castes, settlement size, marsupials of the same form, and arrays of dunes, planets, or galaxies.

In the following pages, I will try to include all this and to derive it from the basis of a single "simple" model (**Figure 4.1**), which, understood in all its parts, has each of these phenomena, and a great many more subtle inferences, within it. Not all, or even many, of these shapes, symbols, symmetries, scatters, or forms will be individually described or figured here, but the basic pattern parts will be briefly described and related to each other. The basic model itself is derived from a stylised tree form.

We should not confuse the comprehension of FORM with the knowledge of SUBSTANCE—"the map is not the territory"—but an understanding of form gives us a better comprehension of function, and suggests appropriate strategies for design.

4.2
A GENERAL PATTERN MODEL OF EVENTS

When we look about us in the world, we see the hills,

rivers, trees, clouds, animals, and landforms generally as a set of shapes, apparently unrelated to each other, at least as far as a common underlying pattern is concerned. What *do* we see? We can list some of the visible forms as follows:

- WAVES on water and "frozen" as ripples in dunes and sandstones, or fossilised quartzites and slates.
- STREAMLINES, as foam strips on water, and in streams themselves.
- CLOUD FORMS in travertine (porous calcite from hot springs), tree crowns, and "puffy" clouds or as cloud streams.
- SPIRALS in galaxies, sunflowers, the global circulation of air, whirlpools, and chains of islands in arcs.
- LOBES, as at the edge of reefs, in lichens, and fringing the borders of salt pans.
- BRANCHES, in trees and streams converging or diverging; explosive shatter zones.
- SCATTERS of algae, tree clumps in swamps, islands, and lichens on rocks.
- NETS as cracks in mud, honeycomb, inside bird bones, in the columns of basalt (as viewed from above), and cells of rising and falling air on deserts.

The NETS or cracks in mud and cooling lava are shrinkage patterns caused not by flow or growth, but by the *lateral tension* of drying or cooling, as are many patterns in iceflows and the cracked pattern of pottery, or the cracks in bark on trees. Thermal wind cells arise at the confluence of large heat cells on desert floors, forming a net pattern if viewed from above or below.

In all of these categories, I hope to show that one master pattern is applicable, and that even the bodies of animals are made up of bones, organs, and muscles of one or more of the forms above. I will link these phenomena—generated by growth and flow— into a single model form. That form is a stylised tree (**Figure 4.1**). Around the central tree form of this model are arranged various cross-sections, plans, longitudinal sections, and streamline paths, all derived from real sections, paths, or projections of the tree.

The evolution of such a form from an initial point in space–time, I call an EVENT. Such events can be abstract or palpable. They have in common an origin (O), a phase of growth (T1–T6: an expression of their energy potential), decay, and dissolution into other events of a like or unlike form. The event of a tree is at least three–dimensional, and must be thought of as extending into and out of a plane (P). However, many similar events such as migration patterns or glaciation can be as well portrayed (as they are seen in aerial photographs) as two–dimensional.

The curvilinear STREAMLINES (S1–S9), are seen to curve or spiral through the Origin, just as (in fact) the phloem (storage cells) and xylem (sapbearing cells) spiral through the X–X' axis, or earth surface plane (P), of a real tree. Not so easy to portray is the fact that the xylem is *external* to the stems and *internal* to the roots, and the phloem the reverse. At a zone in the plane (P), therefore, these cells INTERWEAVE or cross over as they spiral out of or into the media.

This deceptively simple "apple core" or tree shape, spiralling out of the plane (P) is a slow–moving vortex such as we see in tornadoes and whirlpools. Traffic through the streamlines is *in both directions*. In trees, sugars and photosynthetic products travel from crown to root margin, and water and minerals from roots to crown. Thus, each margin of our pattern is both collecting and distributing materials from different media. The tree trades both ways with elements of the media, and there is an active water and gaseous exchange with the media (M1, M2). Two–way trade is the normality of plants, organs, and natural forms.

As we know, a crosscut of a tree stem, the basis of the study of dendrochronology, reveals a target pattern of expanding growth (by which the tree adds bulk annually) and from which we can discover much about past occurrences of drought, seasonal changes, atmospheric compositon, fire, and wind (**Figure 4.1–F**).

Screw palms (*Pandanus* spp.) of the tropics develop *ascending* stem spirals, very reminiscent of fan turbine blades, and sunflowers create open seed spirals (in two directions), so common in many whorled plants. The stem itself forces open an ever–expanding flow through the X–Y plane between the media, allowing more material to pass through as time accumulates. The event expands the initial rupture of the surface between the media, allowing greater flow to take place, and this too is recorded in the target pattern of the stem, at the point of germination of the event (O).

4.3 MATRICES AND THE STRATEGIES OF COMPACTING AND COMPLEXING COMPONENTS

A set of intersecting sine waves developed over a regular square or hexagonal matrix will set up a surface composed of our core model shapes. It doesn't matter if we see the sine waves as static or flowing, the core model will still maintain its shape, and flow in the system does not necessarily deform the pattern. Such a pattern matrix (**Figure 4.2**) shows that our models **tessellate** (from the Latin *tesserae*, meaning tiles) to create whole surfaces. If landscapes are, in fact, a set of such models, they must be able to tessellate.

Convection cells on deserts arise from a roughly hexagonal matrix of air cells 1–5 km across, and matrices also underlie the spacing of trees in forests.

Glacial landscapes show whole series of such patterns, as do regular river headwaters. We could equally well have created a matrix by adding in samples of our core pattern as we add tiles to a floor. Thus we see the Euclidean concept of points and lines underlies our curvilinear forms. Even irregular models (**Figure 4.3**) tessellate. Such tessellae are centred on nets or regular grids.

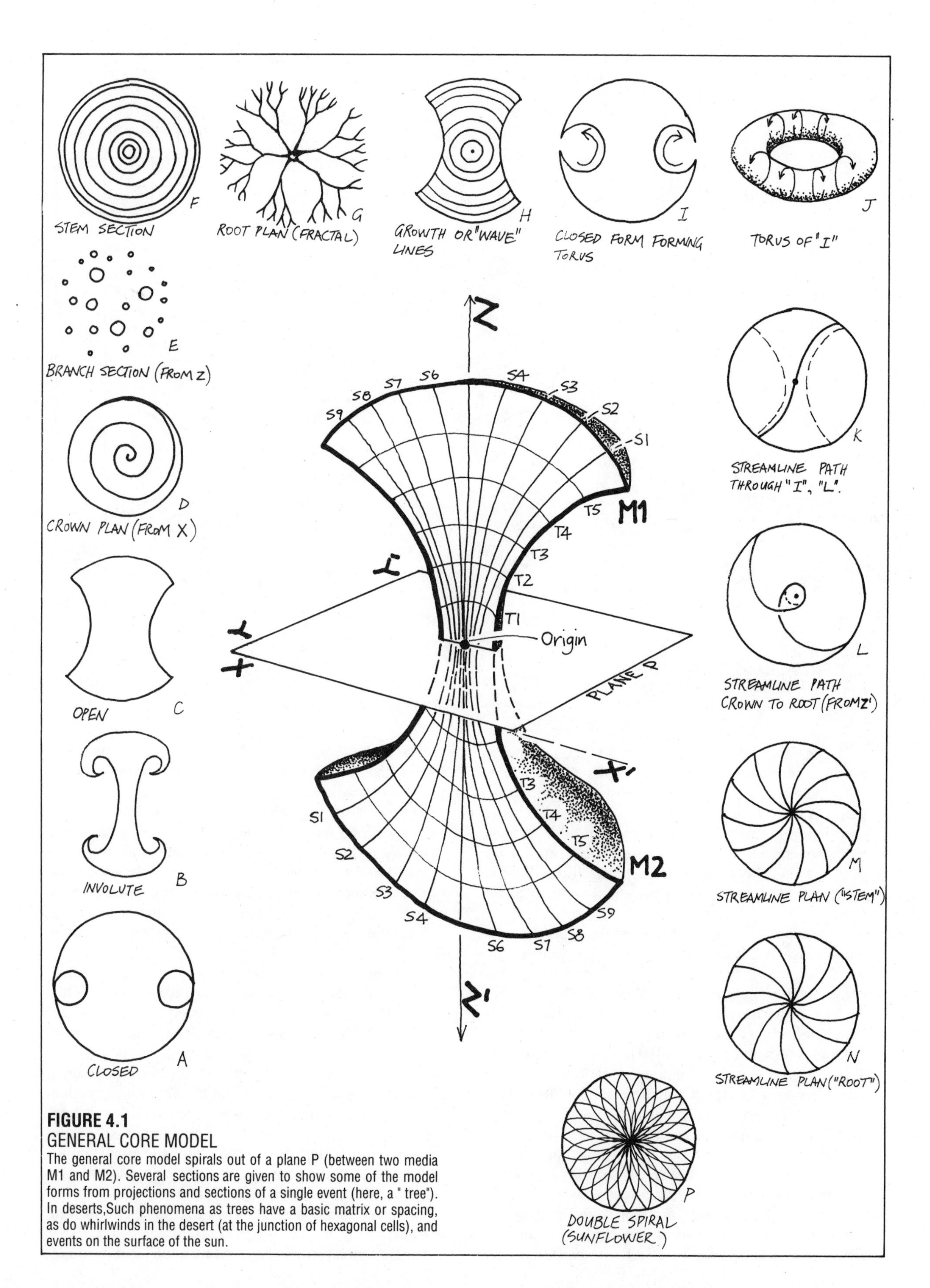

FIGURE 4.1
GENERAL CORE MODEL
The general core model spirals out of a plane P (between two media M1 and M2). Several sections are given to show some of the model forms from projections and sections of a single event (here, a " tree"). In deserts,Such phenomena as trees have a basic matrix or spacing, as do whirlwinds in the desert (at the junction of hexagonal cells), and events on the surface of the sun.

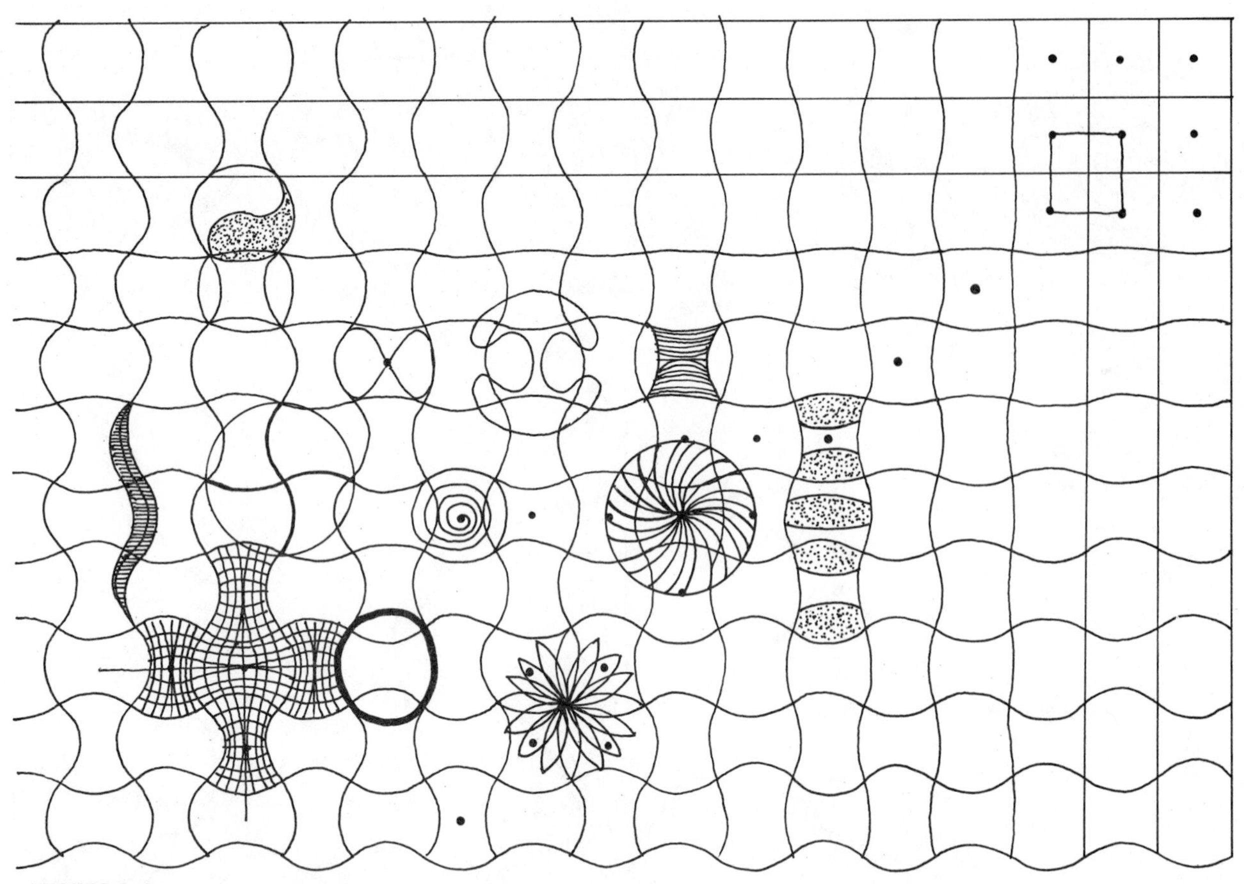

FIGURE 4. 2
PATTERN MATRIX OF TESSELATED PATTERNS.
Underlying many natural distributions (e.g. trees in a desert, heat or convection cells) and forming many patterns (such as honeycomb and cracks in mud) are matrices or grids based on approximate squares, hexagons, or intersecting sine waves.

The "growth lines" (T–series) of our models are, in effect, a series of smaller and smaller forms nested within the larger boundaries, as is the case with target patterns or tree cross–sections. The process is termed **annidation** (Latin *nidus*, a nest) and is used in practice to compactly store bowls or glasses, one within the other; it then becomes a strategy for fitting–in like components of the same or different size in a compact way.

If we *superimpose* two spirals of the opposite sense (spirals twisting in the opposite way), we develop the petal patterns of flowers and the whorls of leaves so common in vegetation, well illustrated by the seed patterns of sunflowers. The effect is also reproduced by simple reflection of such curves.

Thus we see that tessellation, annidation, or superimposition gives us a strategy set for developing complex and compact entities, or for analysing complex landscapes. As Yeomans[5] points out, ridges and valleys in landscape are identical reflections. If we model a landscape and pour plaster of Paris on the model, we reproduce the landscape in a reversed plaster model, but now the ridges are valleys.

Further, a set of our models invading into or generating from a portion of the landscape produce EXPANSION and CONTRACTION forms (**Figure 4.4**) typical of the edges of inland dunes and salt pans. This crenellated (wavy) edge produces **edge harmonics** of great relevance to design.

The study of matrices reveals that the T (time) lines are **ogives** of a tessellated model and develop from the "S" (stream) lines of the next model adjoining. We then come to understand something of the co–definitions in our core model, and its inter-dependent properties. Sets of such models and their marginal crenellations provide a complex interface in natural systems, often rich in production potential.

The earth itself is "a great tennis ball" (*New Scientist*, 21 April '77) formed of two core model forms. This earth pattern (**Figure 4.5**) of two nested core models can be re–assembled into a single continent and one sea if the present globe is shrunk to 80% of its present diameter. My old geology professor, Warren ("Sam") Carey may have been justified in his 1956 assertion that the earth was originally that much smaller. When re–assembled in this way, the globe shows an origin ("O" of our model) over each pole; the north polar origin is that of the seas, the antarctic origin that of the continents. At that time in earth history, all life forms were native to a single continent and all fish swam in one sea.

The pattern has been shattered by a total expansion of

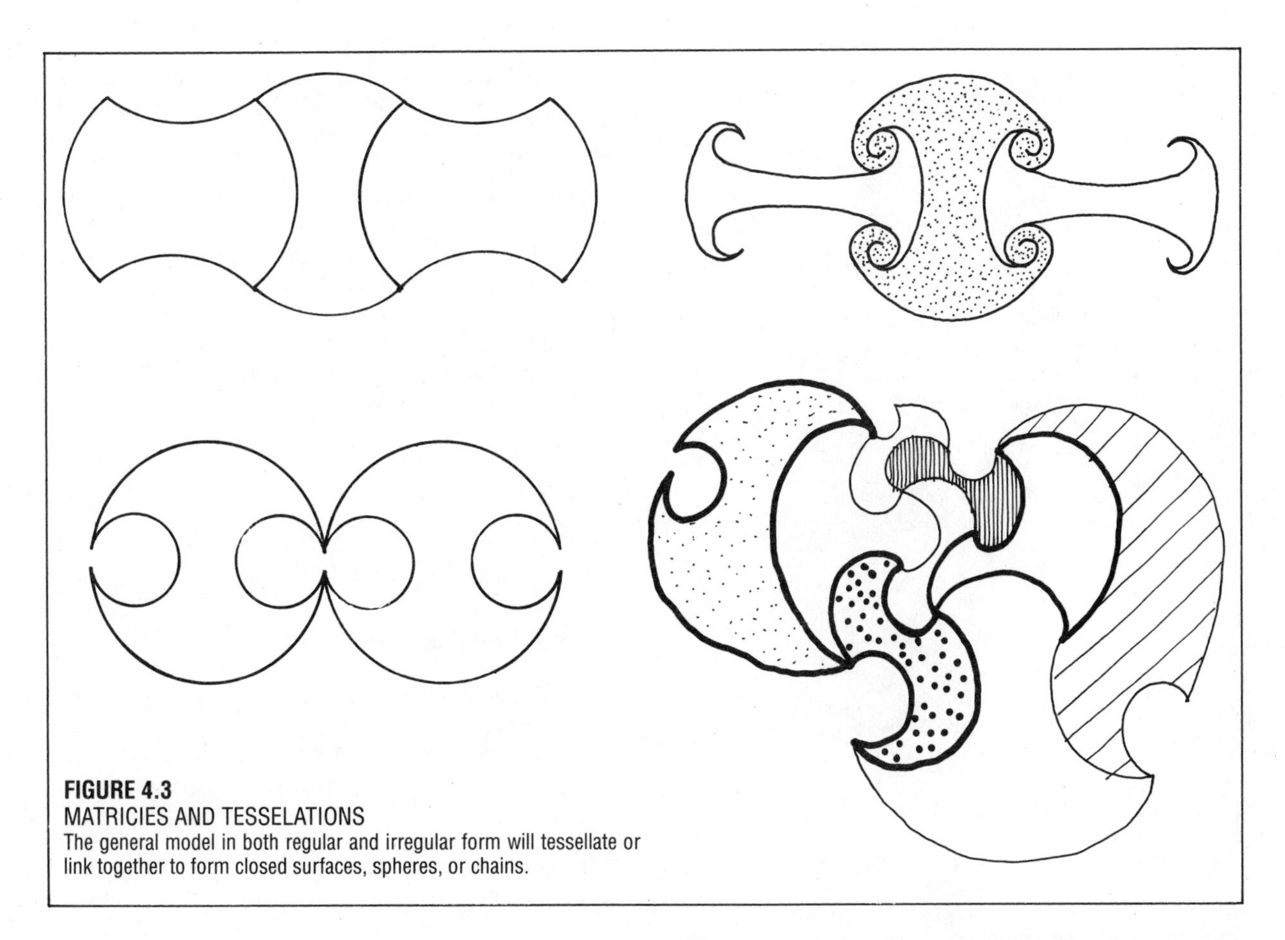

FIGURE 4.3
MATRICIES AND TESSELATIONS
The general model in both regular and irregular form will tessellate or link together to form closed surfaces, spheres, or chains.

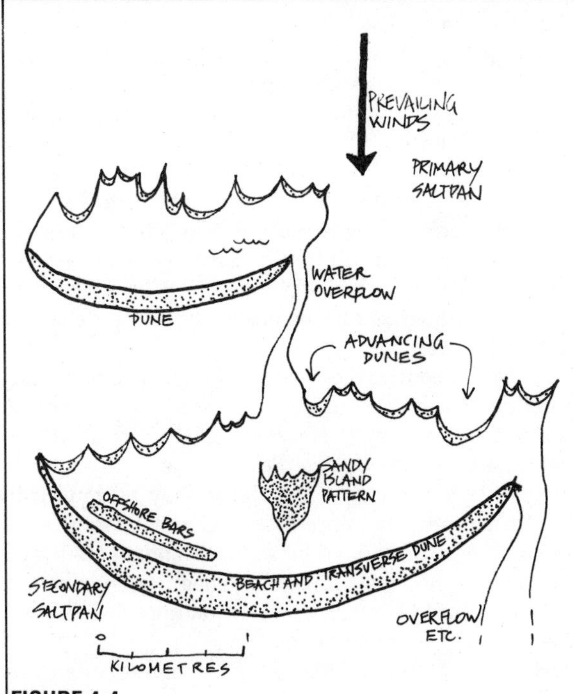

FIGURE 4.4
PATTERNS IN DESERTS
Repetitive patterning due to wind, sand, and rare heavy rains, is typical of deserts; this series of lunulate lagoons occurs in South Australia. Longitudinal dunes become crescentic transverse dunes due to water action.

the globe or by the spreading of oceanic plates cracking the continents apart rather like the net patterns on a mud patch, and isolating species for their present endemic development. The whole story is being slowly assembled by generations of biologists (Wallace, Darwin), geologists, and technicians analysing data from satellite surveys of the globe.

The original pattern shattered, continents now drift, collide, and form their own life pattern by isolation, recombination, and the slow migration of natural processes. The process also illustrates how irregularities may arise on an expansion of a previously regular matrix of forms; tension caused by expanding phenomena shatters the smooth flow of primary global events. At the end of a certain energy sequence, old patterns shatter or erode to make way for new patterns and succeeding forms of energy, as a decaying tree gives life to fungi and to other trees.

4.4
PROPERTIES OF MEDIA

Media, as a result of their chemistry, physical properties, or abstract characteristics, can be identified by us because they *differ* from each other. We distinguish not only air, water, earth, and stone but also hot, cold, salty, acid, and even some areas of knowledge as having different properties or validity. Every such

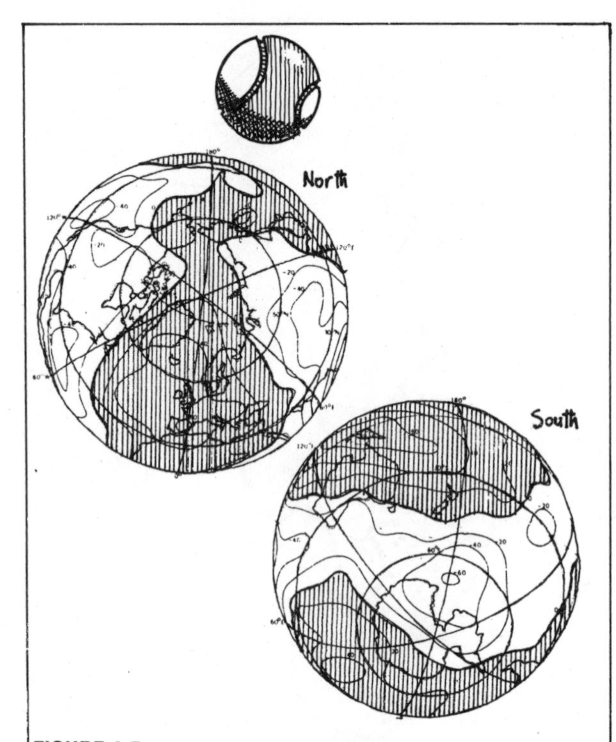

FIGURE 4.5
"THE GREAT TENNIS BALL OF EARTH"
Only two core models make up the geoid. Each is detectably higher than, or lower than, the other. Origins lie near the poles, and when the continents are re–assembled we have one great continent and one great sea.

difference has a more or less well defined BOUNDARY CONDITION, surface, or interface to other media or systems. Permaculture itself acts as a translator between many disciplines, and brings together information from several areas. It can be described as a framework or pattern into which many forms of knowledge are fitted in relation to each other. Permaculture is a synthesis of different disciplines.

Any such boundary is at times between, at times within, media, and (as in the case of the earth/air surface) these boundaries, surfaces, or perceptible differences present a place for things to happen, for *events to locate*. Thus, boundaries present an opportunity for us to place a translatory element in a design, or to deform the surface for specific flow or translation to occur.

If the media are in gaseous or liquid form, or composed of mobile particles like a crowd of people, swarm of flies, water, or dust clouds, then the media are themselves capable of flow and deformation.

In nature, many such media and boundaries can be distinguished. As one example, a pond (with part of its margin) is shown in **Figure 4.6**.

Although differently named (or not named at all), all these surfaces, edges, and boundaries separate different media, ecological assemblies, physical states, or flow conditions. Every boundary has a unique behaviour and a translation potential. Living translators (trees, fish, molluscs, water striders) live at each and every boundary. We can see that the establishment of complex boundary conditions is another primary strategy for generating complex life assemblies and energy translators.

"Most biologists," (says Vogel, 1981) "seem to have heard of the boundary layer, but they have a fuzzy notion that it is a discrete region, rather than the discrete notion that it is a fuzzy region."

Boundary/Edge Design Strategy
The creation of complex boundary conditions is a basic design strategy for creating spatial and temporal niches.

4.5
BOUNDARY CONDITIONS

Boundaries are commonplace in nature. Media are variously liquid, gaseous, or solid, in various states of flow or movement. They have very different inherent characteristics, such as relatively hotter, more acid, rough, harder, more absorbent, less perforated, darker, and so on. Even in abstract terms, society divides itself in terms of sex, age, culture, language, belief, disciplines, and colour (just to enumerate a few perceived differences).

In this confusion of definitions, social and physical, we can make one statement with certainty. People discriminate (in its true meaning, of *detecting a difference*) between a great many media or systems, and therefore recognise boundary conditions or "sorts", enabling them to define like and unlike materials or groups in terms of a large number of specific criteria.

Differences, whether in nature or society, set up a potential STRESS CONDITION. This may demonstrate itself as media boundary disturbances, friction, shear, or turbulence caused by movement, sometimes violent chemical reactions, powerful diffusion forces, or social disruption. Seldom do two different systems come in contact without a boundary reaction of one sort or another, as quiet as rust, as noisy as political debate, or as lethal as war.

If we concentrate our attention on the boundary condition, there are, crudely, two common or possible motions or particle flows—ALONG or ACROSS boundaries. In *longitudinal* flows (shear lines) between media, deflections and turbulence may be caused by local friction or the more cosmic Coriolus (spin) force. In *crossing* a boundary between media, the surfaces themselves may resist invaders (chemical or social); or various nets, sieves, or criteria may have to be by–passed by a potential invader.

However, these boundaries are, in nature, often very rich places for organisms to locate, for at least these reasons:

- Particles may naturally accumulate or deposit there (the boundary itself acts as a net or blockade).
- Special or unique niches are available in space or time within the boundary area itself.

• The resources of the two (or more) media systems are available at the boundary or nearby.

Special physical, social, or chemical conditions exist on the boundary, because of the reaction between the adjacent media. As all boundary conditions have some fuzzy depth, they constitute a third media (the media of the boundary zone itself).

This last statement is especially true of diffusive or flowing media, and of turbulent effects. Turbulence, in effect, creates a mix of the two or more media which may itself form another recognisable medium (e.g. foam on water, an emulsion of oil and water).

In our world of constant events, especially in the living world, more events occur at boundaries than occur elsewhere, because of these special conditions or differences. It is common to find that there are more different types of living species at any such boundary or edge than there are within the adjoining system or medium. Boundaries tend to be species–rich.

This "edge effect" is an important factor in permaculture. It is recognised by ecologists that the interface between two ecosystems represents a third, more complex, system which combines both. At interfaces, species from both systems can exist, and in many cases the boundary also supports its own species.

Gross photosynthetic production is higher at inter-faces. For example, the complex systems of land/ocean interface—such as estuaries and coral reefs—show the highest production per unit area of any of the major ecosystems (Kormondy, E.J., 1959, *Concepts of Ecology*, Prentice Hall, NJ, USA).

Forest/pasture interfaces show greater complexity than either system in both producers (plants) and consumers (animals). It seems that the Tasmanian Aborigines burnt forest to maintain a large interface of forest/plain, since these transitional areas provided a great variety and amount of food. Animals are found in greater numbers on edges, for example, and a fire mosaic landscape is rich in species. Such mosaics were the basis of Australian Aboriginal landscape management.

In view of the edge effect, it seems worthwhile to increase interface between particular habitats to a maximum. A landscape with a complex edge mosaic is interesting and beautiful; it can be considered the basis of the art of productive landscape design. And most certainly, increased edge makes for a more stimulating landscape. As designers we can also *create* harmonic edge with plants, water, or buildings.

There are aspects of boundaries that deserve con-

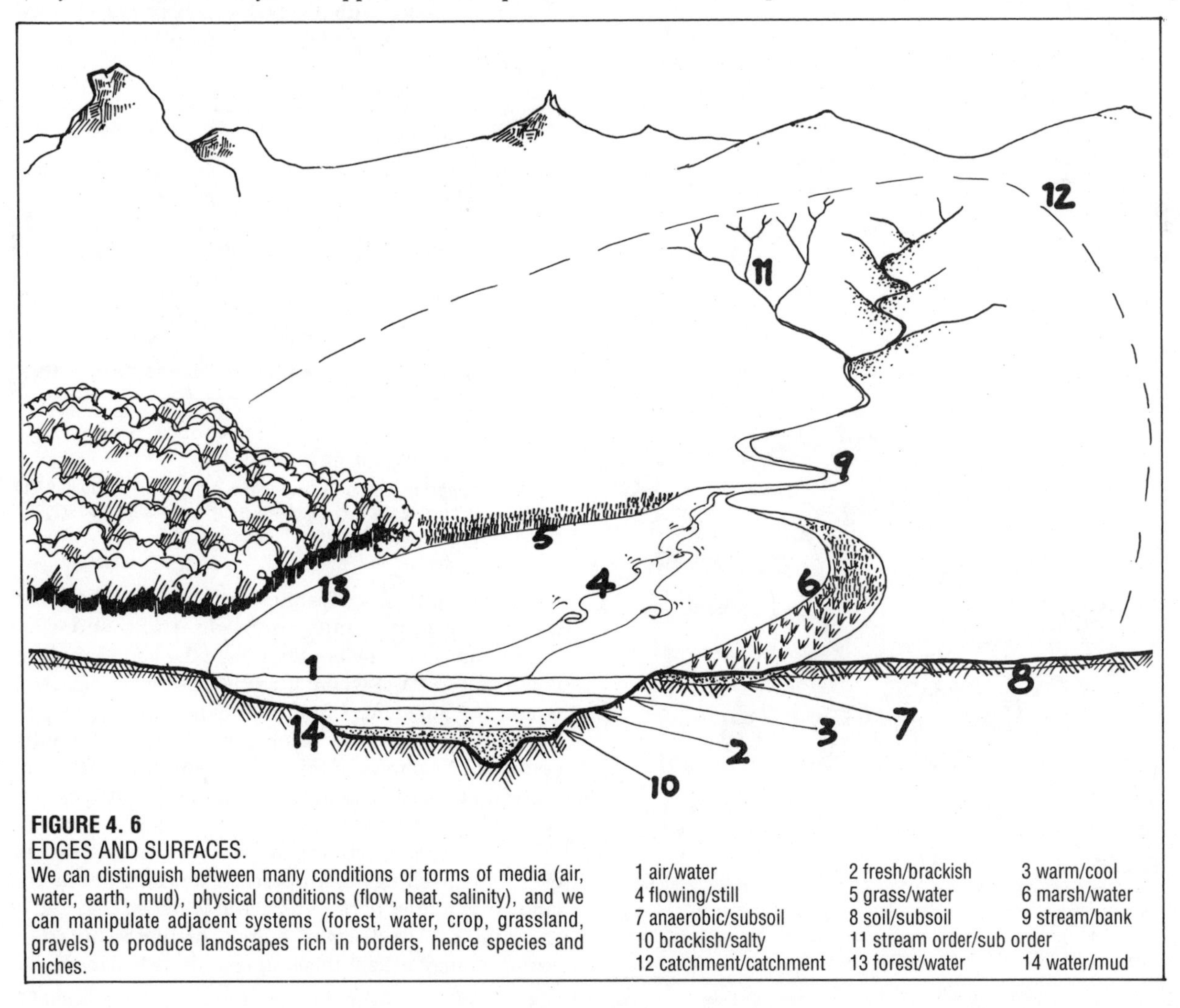

FIGURE 4. 6
EDGES AND SURFACES.
We can distinguish between many conditions or forms of media (air, water, earth, mud), physical conditions (flow, heat, salinity), and we can manipulate adjacent systems (forest, water, crop, grassland, gravels) to produce landscapes rich in borders, hence species and niches.

1 air/water 2 fresh/brackish 3 warm/cool
4 flowing/still 5 grass/water 6 marsh/water
7 anaerobic/subsoil 8 soil/subsoil 9 stream/bank
10 brackish/salty 11 stream order/sub order
12 catchment/catchment 13 forest/water 14 water/mud

siderable design intervention:

- The geometry or harmonies of any particular edge; how we **crenellate** the edge.
- Diffusion of the media across boundaries (this may make either a third system or a broader area in which to operate—few boundaries are very strictly defined).
- Effects which actively convey material to or across boundaries; in nature, these are often living organisms or flow (bees, for example).
- The compatibility (or **allelopathy**) of species or elements brought into proximity by edge design.
- Boundaries as accumulators on which we can collect mulch or nutrients.

4.6
THE HARMONICS AND GEOMETRY OF BOUNDARIES

The amplitude, configuration, and periodicity of an edge, surface, or boundary may be varied by design.

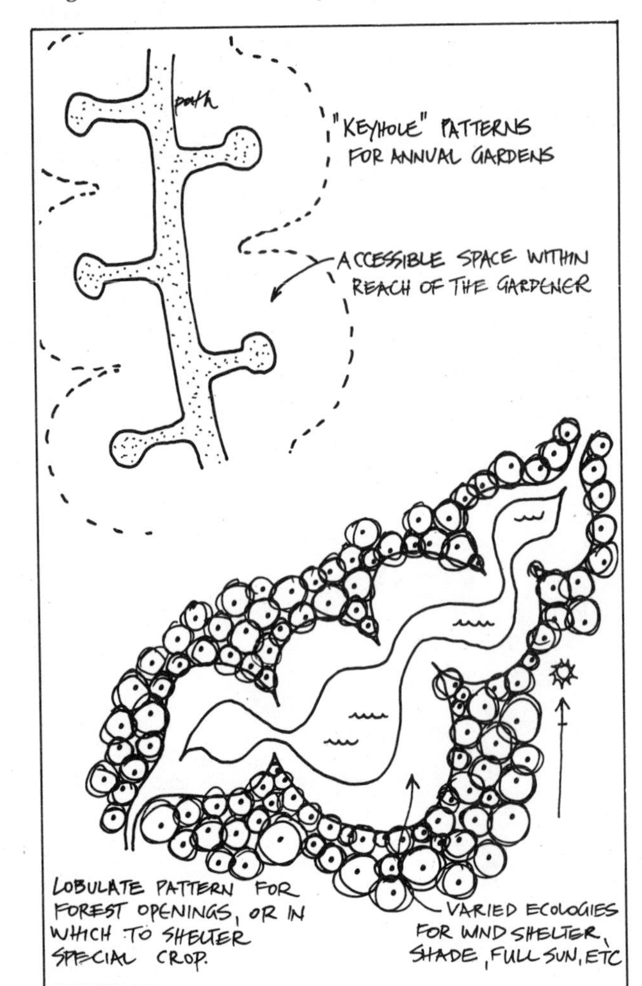

FIGURE 4.7
"Least path" design for home gardens, a keyhole pattern (common in nature) allows us to access garden beds most efficiently. Parallel paths take up to 50% of the area; keyhole beds <30% of the ground. (See also FIGURE 10.26 Gangamma's Mandala).

Edges and surfaces may be sinuous, lobular, serrate, notched, or deliberately smoothed for more efficient flow. While we may deliberately induce turbulence in salmon streams by using weirs, we are painstaking in using smooth and even conduits for energy generation in wind or hydraulic systems. We can deepen areas of shallow streams to make pools, or to prevent stream bank erosion, or to reflect sun energy to buildings; all these are manipulated to achieve specific effects on their boundaries or surfaces.

Notched or lobular edges, such as we achieve in plan by following hill contours, afford sheltered, wetter, drier, hotter, or more exposed micro–habitats for a variety of species. Serrate or zig–zag fences not only stand on their own, but resist wind–throw much better than straight barriers. Lobular embayments, like the keyhole beds of **Figure 4.7**, are obviously sheltered, spacious habitats for gardens and settlements.

As for surface and flow phenomena, we can partition water surfaces to reduce wind effect, or design to deliberately create turbulence and wind overturn. Islands, quoins, and rafts of many shapes have as many uses, and deflect flow to increase condensation or to encourage sand and snow deposition or removal. Surfaces can be pitted, ridged, spiralled, mounded, tessellated, tassled with plants or brush, paved, sprayed to stabilise mulch, mulched, or smoothed for water run–off.

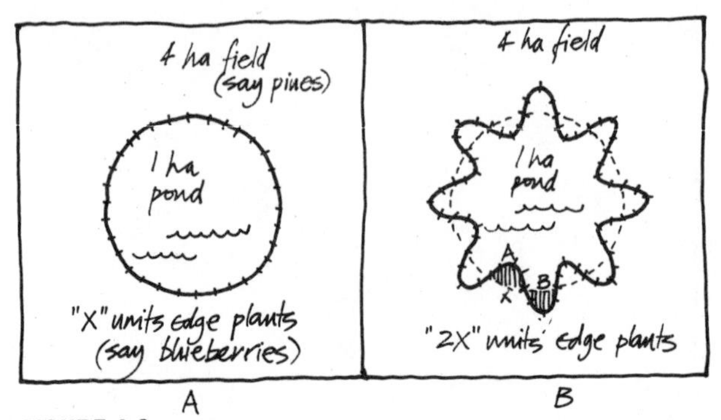

FIGURE 4.8
CRENELLATED POND EDGE
Without altering the area of a field and a pond, we can double the plants on the pond–edge (e.g. blueberries) by crenellating this edge to increase the earth/water interface.

When a boundary separates two things which differ, there is an opportunity for trade, transactions, or translation across the border. Where the boundary itself is difficult to pass, where it represents a trap or net, or where the substances and objects attempting to pass have no ability to do so, accumulations may occur at the boundary. Examples of this lie all about us, as stranded shells on the beach, people lined up at visa offices, and cars at stop lights or kerbsides.

People are, at heart, strandlopers and beach-combers; even our dwellings pile up at the junction of sea and land, on estuaries (80% of us live at water edges), and at the edge of forest, river, marsh, or plain. Invariant ecologies may attract the simple– minded planner, but they will not attract people as inhabitants or explorers.

In design, we can arrange our edges to net, stop, or sieve–through animals, plants, money, and influence. However, we face the danger of accumulating so much trash that we smother ourselves in it. Translators keep flow on the move, thereby changing the world and relieving it of its stresses. The sensible translator *passes on* resources and information to build a new life system.

There are innumerable resources in flow. Our work as designers is to make this flow function in our local system before allowing it to go to other systems. Each function carried out by information flow builds a local resource and a yield.

If you now carefully observe *every natural accumulation* of particles, you will find they lie on edges, or surfaces, or scattered nearby, like brush piled up against a fence (**Figure 4.9**). We can use these processes to gather a great variety of yields.

It follows that edges, boundaries, and interfaces have rich pickings, from trade both ways or from constant accumulations. Our dwellings and activities benefit from placement at edges, so that designing differences into a system is a resource–building strategy, whereas smoothing out differences or landscapes a deprivation of potential resources.

Objects in transit can be stopped by filters and nets. The edges of forests collect the aerial plankton that pass in the winds. Boundaries may accumulate a special richness of resource, as a coral reef collects the oxygen and energy of the sea, and the canopy of the trees the energy of the wind. We rely on translators, such as trees and coral, to store such impalpable resources, to process them into useful products, and to store them for use in their own system, with some surplus for our essential use.

Transactions at boundaries are a great part of trade and energy changes in life and nature. It seems that differences *make* trade; that every medium seeks to gather in those things it lacks, and which occur in the other medium. However, we should also look at the translator, which is often of neither medium but *a thing in itself*, the "connection or path between", created from the media, but with its own unique characteristics.

Plants, people, and pipes are translators. Nets, sieves, passes, and perforations are openings for translators to use, and (as traders know) there is no border so tight that a way does not exist for trade. Go–betweens or traders, like many plants and animals, are creatures of the edge. They seek to relieve the stresses caused by too much or too little in one place or another; or to accumulate resources (make differences) if they operate as storages. We can use naturally–occurring turbulence, trade, and accumulations to work for us, and by carefully observing, find the nets and go–betweens of use. We can use naturally–occuring turbulence, trade, and accumulations to work for us, and by carefully observing, find the nets and go–betweens of use.

FIGURE 4.9
At powerlines and fences, perched thrushes and wood pigeons defecate, so that each post gains seed and manure, and each may generate a plant from nearby forests. Perches plus disturbed soil produce this result. Fences also act as mulch accumulators across wind.

4.7
COMPATIBLE AND INCOMPATIBLE BORDERS AND COMPONENTS

There are only limited interactions possible between two abstract or real systems brought into boundary contact. The sum of possible effects available are these:

- No difference in yields, stability, or growth (o,o)
- One benefits, at the expense of the other (+,–) (–,+)
- Both benefit (+,+)
- Both are decreased in yield or vitality (–,–)
- One benefits, the other is unaffected (+,o) (o,+)
- One is decreased, the other is unaffected (–,o) (o,–)

Almost all organisms or systems get along fine. A great many derive mutual benefit, and a very few decrease the yield of others or wipe each other out. It simply doesn't pay to attack others. In the long run one destroys oneself by accumulated injury or, more certainly, by pathogens in an animal or conflicts within a society that await a monocultural crop or repressive society. For our domestic plant groups, a powerful design strategy for yield and system stability is to select compatible components for complex edge and surface phenomena.

Many crops, like wheat and pulse grains, trees which bear on the crown, and mass–planted vegetable species, yield much better on the crop edge than they do within the crop. Taking examples where edge yield is marked (e.g. in wheat, lucerne); where there is a (+,+)

relationship, as is the case of crops such as wheat and lucerne (alfalfa); and presuming a two–fold yield increase on edges (it can be more for such trees as *Acacias* with hazelnuts), we can proceed as follows.

First, we need to measure just how far into each crop the edge effect extends, so that we can estimate a finite width of higher yield. We will assume 1 m for wheat and the same for lucerne, giving a 2 m width as a double edge. It is now quite feasible to sow a field in 2 m wide alternate strips of each crop, giving us (in effect) *nothing but edge*, and obtaining from this field about the same yield as we would have had we sown twice the area to single crop stands (**Figure 4.10**).

Two crops are a simple example, but if we extend the principle to many and varied crops on an even broader scale, we approach a new concept of growing, which we can call ZONE or EDGE CROPPING. These would produce a matrix of hedgerows or edge–rows, each suited in width to a particular crop. Such zonal strips are seen naturally occurring on coasts and around saltpans or waterholes.

This sort of setup might be a nightmare for the bulk–cropper (or it may not), but has immense potential for small shareholders in a single land trust, each of whom tend one or more crop strips. It is very like the older patterns of French–intensive agriculture and the farmed strips of modern Quebec, which produce a very productive crop mosaic. Polycultures can be composed of such mosaics or zonal strips.

For cases of (–,–) interactions, both crops suffer, but active intervention with a component acceptable to both systems may work:

Place an intervening, multually–compatible component between two incompatible systems.

Compatible components may simply differ in sex, colour, chemistry, belief, or political conviction from the warring parties. However, in time a beneficial mosaic will impose itself on all expansionist systems, arising from the potential for differences carried *within* all life systems. Natural interveners arise, often as hybrids between apparently antagonistic systems. Our design intention in landscape systems is to build *interdependence into mosaics..*

Select and place components so that incompatibity is nullified, interdependence maximised.

After all, in the absence of tigers, Hindus need Muslims to eat cows; they may also need a Christian businessman between them to effect the transaction. The interdependence of mosaics of belief are called for as much as mosaics of plants.

The stupidity principle may here be stated in a different way:

Stupidity is an attempt to iron out all differences, and not to use or value them creatively.

It is our skill in organising spatial or functional distribution that may create beneficial interdependence in incompatible components. When we know enough to be able to select mutually–beneficial assemblies of plant and animal species

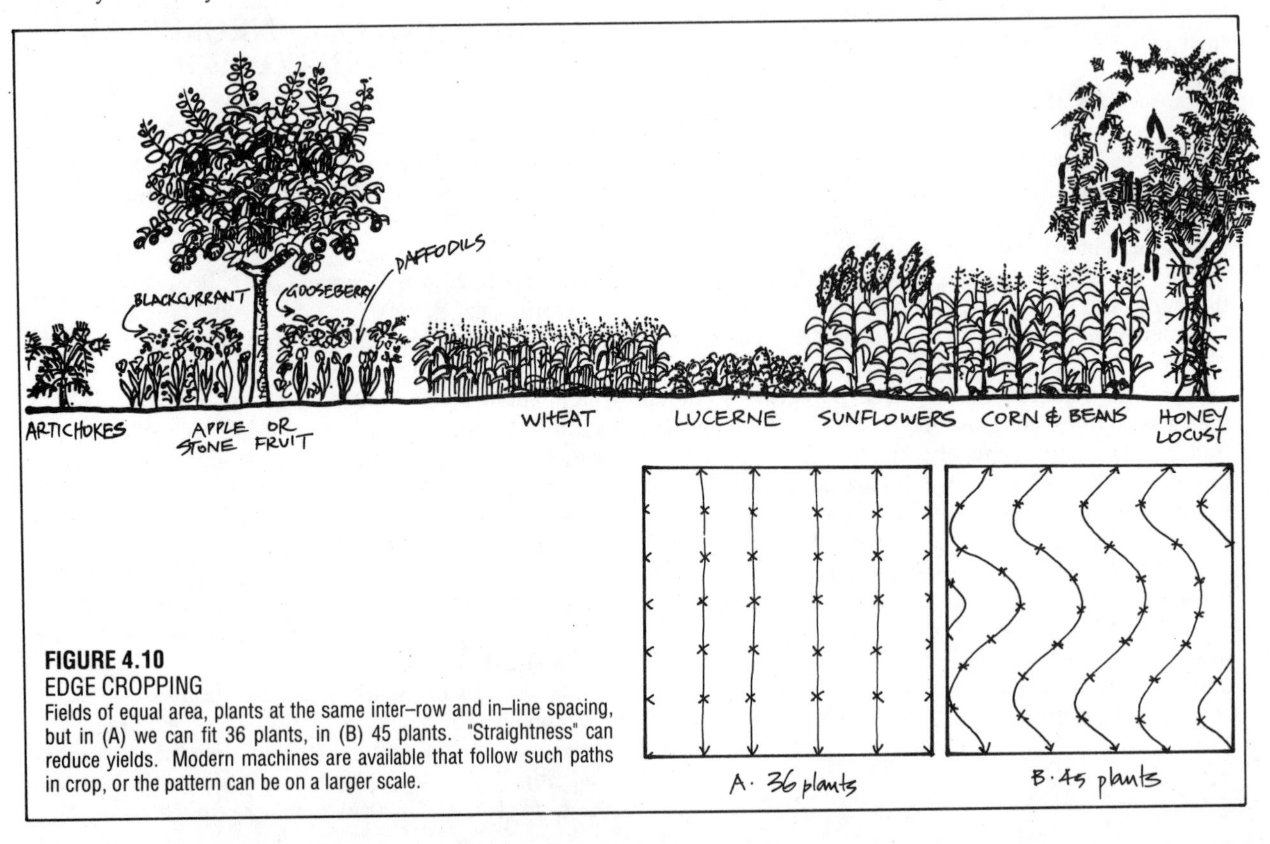

FIGURE 4.10
EDGE CROPPING
Fields of equal area, plants at the same inter–row and in–line spacing, but in (A) we can fit 36 plants, in (B) 45 plants. "Straightness" can reduce yields. Modern machines are available that follow such paths in crop, or the pattern can be on a larger scale.

(guilds) then we have two powerful interactive strategies (edge harmonics and species compatibility) for design applications.

Mosaic design (the opposite of monoculture) means the creation of many small areas of differences. A few mistakes will occur, but good average benefit will result. This was the tribal strategy.

A Golden Rule of Design
Keep it small, and keep it varied.

Our tree model is not only *different from* its supporting media, but exists *because of* them. Stress builds *because of* impermeable boundaries. If a fence allows mice through but restricts rabbits, it is the rabbit plague that will break it down. If too much money accumulates on one side of a door it will either force the door open of itself, or those deprived of it will break in. The terrible pressures that gases and molecules can exert are harmless only when that pressure is free to disperse, or where potentially destructive energies are quietly released where there are no boundaries, multiple translators, or stress–relief mechanisms.

Because the event itself creates a third medium, it again sets up stress between itself and the media (M1 and M2). It can be seen, therefore, that *once any one difference of any sort, even an idea, exists anywhere, then it demands or creates conditions for the evolution of subsequent events*. That first event itself became *yet another difference*, which in turn needed translation, and so on. The process is self–complicating, continually creating of itself all that follows, and all that continues. All is stress, or the relief of stress, and that stress and relief is located between existing differences. One difference in the beginning was enough to generate the total range of subsequent events. There are no "new" events, just a continual expression of all possible events, each arising from some recombination of preceding differences. There are no miracles, just a realisation of infinite possibilities. Any event has the potential to spawn all possible events.

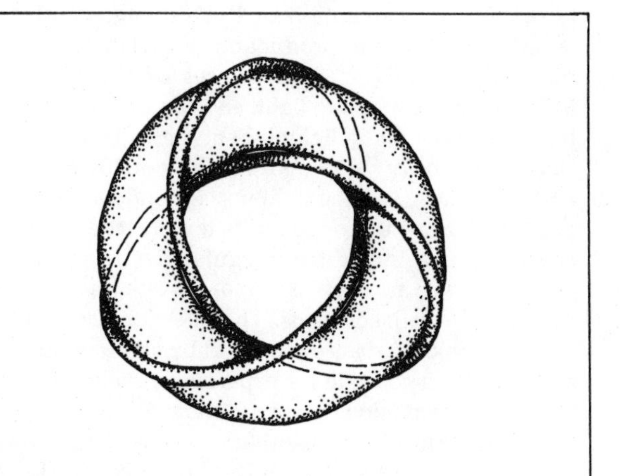

FIGURE 4.11
DNA, coiled around a plus–torus like a single path in the Robinson Congruence.

There are no new orders of events, just a discovery of existing events.
Every event we can detect is a result of a preceding event, and gives rise to subsequent events.

Between all media, some DIFFUSION can take place. This is greatly enhanced by such phenomena as surface turbulence, wave overturn, temperature differences, and pressure differentials. Boundaries between diffusing media are blurred, often seasonally different or sporadic in occurrence, and always in flux. Plants give pollens and chemicals to air, and actively intervene in radiative, gaseous, liquid, and general energy transactions with the atmosphere. Between plant groups, leaf, root and mulch exudates diffuse as chemical messengers. Water is the "universal solvent" of substances diffusing through the earth's crust, in plant systems, and in the atmosphere.

Diffusion is a quiet process operating on a broad front or over the entire surface of some media. It is analagous to, but differs from, the active transport systems that we have called events or translators. However, once an event has occurred, it *also* uses diffusive processes to gather or distribute materials, and thus *events merely enlarge the total diffusive area available*. A tree may have many acres of leaf, and evapo–transpiration will then exceed evaporation at that place by a factor of forty or more. We can grow many such trees on one acre, and thus increase the diffusion effect by factors of 1000 or more, so that gaseous exchange from leaves, and sugars in soils (or soil life) are both assisted by the trees.

4.8 THE TIMING AND SHAPING OF EVENTS

We can see how an event takes place, but how is it shaped? Our bodies arise from the origin (O) of a zygote (a fertilised egg) on the surface of the uterus; the placenta is our root, the foetus the tree of ourselves. Animals are thus events broken free from the coiling connective cord or umbilical stem of their origins. Their eventual shape is a pattern laid down or encoded by the DNA of their cells, coiled as it is around a plus–torus like a ribbon around a doughnut (**Figure 4.11**).

When my son Bill was four, we were in the bath together, and he pointed to his toes. "Why are these toes?" he asked.

"What do you mean?" I hedged.

"Well, why don't they get bigger and bigger or longer and longer? Why do they stop at being toes?"

What limits size and growth? All flows pulse, whether they are blood, wind, water, lava, or traffic.

The pulsing may be organised by PULSERS (e.g. traffic lights), and results in WAVES, or time–fronts, or particles on fixed schedules. Such pulsers (**Figure 4.12**) are located in our bodies as chemical or physio–chemical spirals in sheets of cells that swirl in sequence to create a pulsing movement in our heart, organs, and viscera. Pulsers can start, run for a preset time, and stop. This is how we grow, and why we eventually die. All mammals have an allotted number of heartbeats in relation to their body size, and when these run out, we die. (A friend also theorises that we have a set quota of words; and when they are said, we die.)

FIGURE 4.12
PULSERS
Pulsing, here from Winfree's "doped" chemicals, may take the form of spirals revolving about a locus, in this case centripetal in action like a low pressure wind cell. Pulsing is regular in these chemical systems, as in muscle or flow phenomena.

Pulsers *plus* patterns account for shapes. They determine that a toe will stop at being a toe, and not grow into a monstrous appendage or stay as a midget toe. Thus, all living events carry their characteristic time–shape memories, and (it would appear) so do rivers, volcanoes, and the sun itself. The sun "pulses" every 11 years or so, affecting our ozone and climatic factors such as rainfall. Our own pulses have characteristic or normal resting rates, as do our peristaltic or visceral movements.

Pulsers act in concert to create peristaltic or heart contractions, but if they get the wrong signals, can move out of phase and send the organ into seizure. This spasm may cause damage or death (a heart attack). The pulses drive fluids or particles through vessels or arteries in cities and in bodies, and those then branch to serve specific cells, organs, or regions.

Figure 4.12 is a quite extraordinary spiral pattern which arises from the pulsing reactions of organic acids seeded with ionic (iron, cerium) catalysts. The pulses are quite regular, "at intervals of about a minute, but these may vary up to 5 minutes in living systems such as nerve tissue and a single layer of a social amoeba" (Winfree, 1978, "Chemical Clocks: A Clue to Biological Rhythm" *New Scientist*, 5 Oct '78). The system is one of spirals rotating about a pivot point which is not a source but an invariable locus around which a spiral wave is generated. It is sequences of such phenomena that create a peristaltic system. Spirals of this nature can revolve in two senses: either organising material *to* the pivot, or (revolving in the opposite sense) *dispersing* material to the periphery. We can envisage counter–rotating spirals doing both as they do in the circulation of the atmosphere as high or low pressure cells.

The phenomena is shared by nerve, heart, and brain tissue; organic and inorganic oxidation on two–dimensional surfaces; and in thin tissue subject to exciting stimuli. Ventricular fibrilation (a potentially fatal quivering of the heart) may derive from the spasm effect distributed over heart or nerve tissue, causing an "ineffectual churning" (Winfree, *ibid*). It may also account for involuntary spasm in muscle. Spasms can damage the cells of blood vessels, and cause a build–up of scar tissue or cholesterol at the injury site, or in muscle tissue—an area of hard waste products. The social amoeba *Dictyostelium* uses the pattern to move *towards* the pivot point where "they construct a multi–cellular organism which then crawls away to complete the life cycle" (Winfree, *ibid.*), a process resembling the precursor of hormonal control in the nervous system. Some such process may assemble more complex multi–species organisms like ourselves.

The cycling spirals can be found in biological clocks, such as those which govern the 24–hour metabolism of flowers and fruit–flies, stimulated by oxygen or light pulses.

Within a specific organism, specific pulsers exist; the 24–hour rhythm (CIRCADIAN) of birds is controlled by the pineal gland (*New Scientist*, 11 Oct '84) which secretes a regular nocturnal pulse of the hormone MELATONIN (the changing levels of melatonin trigger the annual cycles of breeding and nest–building in birds). Visual perception of light changes and day lengths regulate the production of melatonin in the pineal gland. Even small pieces of the gland in isolation will respond to light, and can be disrupted by flashes of light (as in lightning) at night. Thus, we see that not only expansion, but DISCHARGE PHENOMENA such as lightning (or sudden shock in people) disrupt or trigger initiatory reactions in life rhythms, and introduce irregularities in cycles or pulsers, just as expansion or shock introduces irregularity in fixed forms. The question arises as to whether the disturbance produced by shock or sudden stimulus is responsible for expansions, cyclic changes, or shape deformations on a more general scale.

Species and individual organisms need both SHAPERS (DNA) and TIMERS (biological clocks) to achieve a specific size and shape. The two must work synchronously to achieve the correct proportions of parts such as fingers and toes, but both are critical to the organism.

Branching patterns in bodies must have (already encoded) the correct angles and placements for their main branches, leaving room for sideshoots and forks, but not for interweaves or cross–points which damage the function of the organism. In order to generate the surface or boundary of a person, and their reticulation systems, patterns of incredible complexity and strict limits must be "known" by the cells or the cell organisers.

We ourselves are part of a guild of species that lie within and without our bodies. Aboriginal peoples and the Ayurvedic practitioners of ancient India have names for such guilds, or beings made up (as we are) of two or more species forming one organism. Most of nature is composed of groups of species working interdependently, and this complexity too must have its synchronistic regulators.

4.9 SPIRALS

Implicit in many of the phenomena discussed are the forms of spirals. These may be revolving (dynamic) or fixed (static), and arise as a consequence of deformations in flow, or are rather an intrinsic property of a specific velocity of flow over surfaces. Other spiral paths are traced out by orbiting bodies over time, or are shapes developed by organisms developing a compact form (e.g. molluscs) that is analagous to annidation. Spiral forms are made visible by plants as whorls of leaves and branches.

D'arcy Thompson (1952) in his book *On Growth and Form,* discusses some of the quantitative or geometric qualities of spiral phenomena, which are hidden or revealed in many natural forms. A long spiral in section is the "S" form of humid landscape slopes (and the yin–yang symbol). Three–dimensional spirals form long ribbons of complex shape. Even within the molecular forms of matter, DNA reveals a double–helix form. Spirals are, in effect, single streamlines of vortices, tori, or sap flows.

Spirals arise from the interaction of streaming and its subsequent deflection of flow around vortices. Storl (1978, *Culture and Horticulture,* Biodynamic Literature Rhode Island) points out the spiral arrangement of leaves in many plants, where leaves are from one–half way, one–third way, and so on around the stem from the preceding leaf, or to the next leaf. Such placements may progress in a regular (Fibonacci) series, each following on from the sum of the two preceding ratios: 1:2 + 1:3 = 2:5; 1:3 + 2:5 = 3:8, etc., so we get 5:13, 8:21, 13:34 and onwards. These sequences are found in plants and in planetary orbits, so that "Venus forms five loops (retrogressions) below the ecliptic in eight years." Storl sees a relation between the forms of plants and of planets in these progressions, as we can see in the orders of size.

Like so many real–life phenomena, natural spirals are not "perfect", but "show slight progression" and gradually lose phase over long periods (Storl, *ibid*). We can often use the spiral form in design, both to create compact forms of otherwise spread–out placements and to guide water and wind flows to serve our purposes in landscape. We can see the application of spiral forms to technology in everyday life as screws, propellers, impellers, turbines, and some gears. Some species of sharks and invertebrates develop spiral gut lining to increase absorption, or spiral cilia to convey mucus and food or particles in or out of the organism. Plants such as *Convolvulus* use spiral anchors in earth, as do some parasites in animal flesh.

Thus, spirals are found where harmonic flow, compact form, efficient array, increased exchange, transport, or anchoring is needed. We can make use of such forms at appropriate places in our designs.

4.10 FLOW OVER LANDSCAPES AND OBJECTS

The simple involuted mushroom, called an "Overbeck

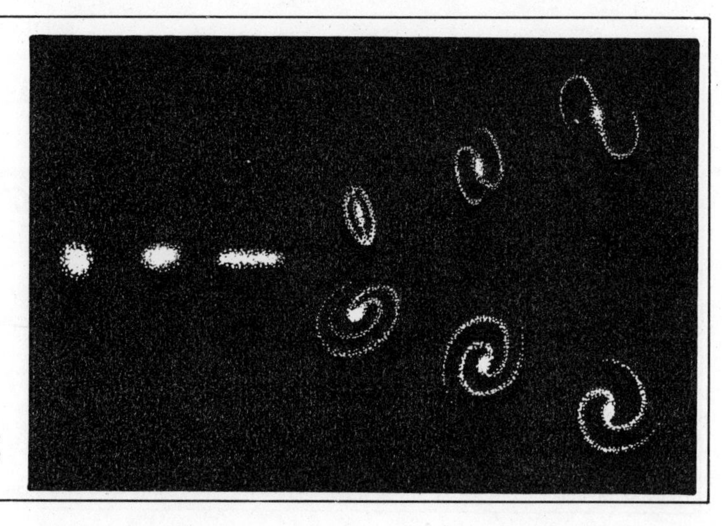

HUBBARD'S GALACTIC CLASSIFICATION
Regular and usual galaxies have limited forms (eliptical, spiral, or barred) and slow rotations. Even galaxies can be ordered in terms of form.

jet" by D'Arcy Thompson (1942), is also shown in its "apple core" model form in **Figure 4.13**. While we can produce these patterns by jetting smoke, fluid, gases or oils into other media, they occur as a part of the natural streaming of fluids and gases past fixed objects such as bluff bodies (e.g. posts) in streams, islands in tides, and trees in wind. Jet streams at altitude can generate such vortices by pushing into different air masses, as can muddy water entering the sea.

FIGURE 4.13
OVERBECK JET MODELS
The Overbeck jet is a simple half–form of the basic model, and occurs commonly in nature e.g. mushrooms, rivers flooding into the sea, and jellyfish.

Whirlpools or VORTICES are shed alternately from a fixed bluff body located in flow, each side generating its own vortex, each with a different rotation. Beautiful and complex forms are thus generated (**Figure 4.14**) and these are the basis of the work at the Virbela Institute on flowforms. The sets of vortices shed or generated downstream from fixed bodies in flow are called Von Karman trails.

The trails are stable at the 1:3.6 ratio shown in **Figure 4.14**. In many streams, and on foreshores, the clay–beds, silts, and underlying rock may develop such patterns, and posts fixed in streams commonly produce them in water. Trees and windbreaks produce similar effects in wind, as do waves at sea. In wind, they are called EKMAN SPIRALS, and in air the spiral lift effect compresses air streamlines to a height 20–40 times the height of the tree or fence fixed in the air flow.

It is obvious that the stable spirals of the Von Karman trails will produce successive pulses downstream, and this is in fact how we observe most flow phenomena to behave. Thus the pulsing of wind, water, and flow in general may rely on the elastic or deformation properties of the medium itself rather than on electro–chemical "timers" as found in organisms. In nature, there are many fixed impediments to perfect streamlined flow.

It is typical of Von Karman trails generated from a fixed body that the effect persists as 4–5 repeats (**Figure 4.14. C**), and then the stream of water gradually resumes streamlined flow. At higher velocities of stream flow, chaotic turbulence occurs, and at slower velocities, simple streaming persists around objects. Thus we see that the Von Karman trail is just one form

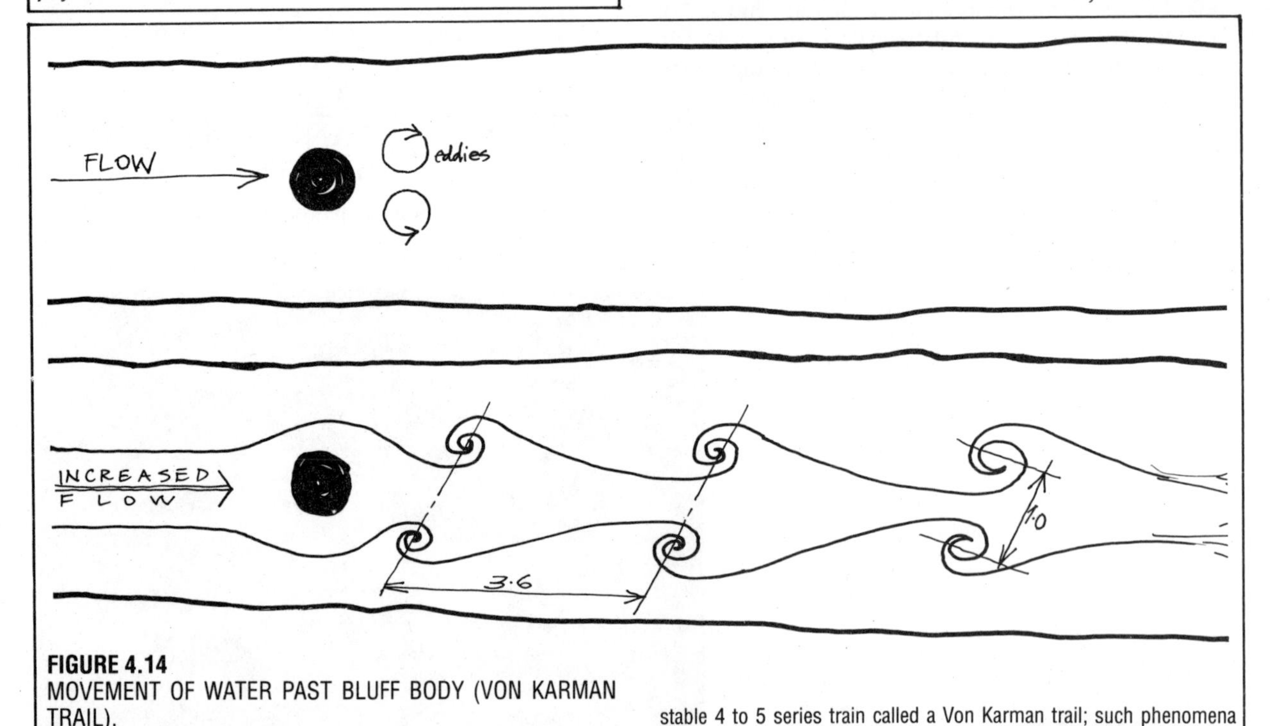

FIGURE 4.14
MOVEMENT OF WATER PAST BLUFF BODY (VON KARMAN TRAIL).
At certain modest flow speeds, water, wind, and even clouds form a stable 4 to 5 series train called a Von Karman trail; such phenomena were vital guides to Polynesian navigators (see Figure 4.26).

of pattern generated by specific flow conditions. It is nevertheless a common form in nature.

The spiralling of wind over tree lines produces a secondary effect, analagous to the streaming of tides around atolls; the wind *changes direction* past the obstacle (about 15°). Such effects may occur within media of different densities (temperatures) as when warm high-pressure wind cells ride over colder low-pressure fronts. The temperature, pressure, and velocity of wind or gas systems are often related:

- Low pressure – high velocity – cool temperature – (expansion).
- High pressure – low velocity – warmer temperature – (contraction).

Velocity in gases and fluids is strictly governed by contact with stationary surfaces, so that the velocity is effectively nil very close to static surfaces, increasing as a series of (imaginary) laminar sheet flows above that surface (**Figure 4.17**). This is the effect that is observed in viscous flow in small canals or vessels, and that governs the shapes and strategies of organisms such as limpets and starfish.

Thus, we see that media in flow can produce pulsers, vortices, and spirals as a result of irregular or obstructing objects or resistant media, and that these phenomena are interconnected.

The relationships between fluid flow, boundary conditions, and the form these impose on organisms is clear from our pattern models and their deflection states. Life as evolved by its internal and external patterns and flows is very well discussed by Vogel (1981) in a lively and scholarly book entitled *Life in*

FIGURE 4.15
MOVEMENT OF BLUFF BODY THROUGH WATER.
A bluff body drawn through still water can make a theoretically endless trail of Overbeck jet forms, superficially resembling a Von Karman trail.

FIGURE 4.16
DIFFERENT TYPES OF TRAILS AND SPIRALS.
The effects of fixed bodies on wind, waves, tides, and cloud streets; various phenomena arise from flow forms.

Moving Fluids.

Carried in the flow of media, as a thistle–down in air, are many events looking for a place to happen. A "net", resting surface, or detonator is needed for these potential events to express themselves. We can provide many such receptors or triggers in our design systems, catching nutrients in flow and ensuring events for future growth in our system. Some such nets form starting–places for events, while others are resting or death places for those entities dependent on flow, or stranded out of their nutrient media.

Just as a series of corks floating on the sea have a predictable path to shore perpendicular to the wave fronts, so does matter flow in a wave–tank model. Similarly, drift–lines form at sea (STREAMLINES in the core model), and as these end on shorelines, they deposit or remove material.

It is along these streamlines that energy acts, by medium of the waves of growth or surface waves in motion. This is how the event and its material expands: streamlines diverge as wave fronts and *disseminate* into open media, but are strong, concentrated, and visible at constrictions, near origins, or in powerful or refracted flows. A small restricted orifice in the time–front or wave–front acts as a secondary origin. Just as a grub encircling a tree causes it to branch out at that point, so constrictions in the flow of

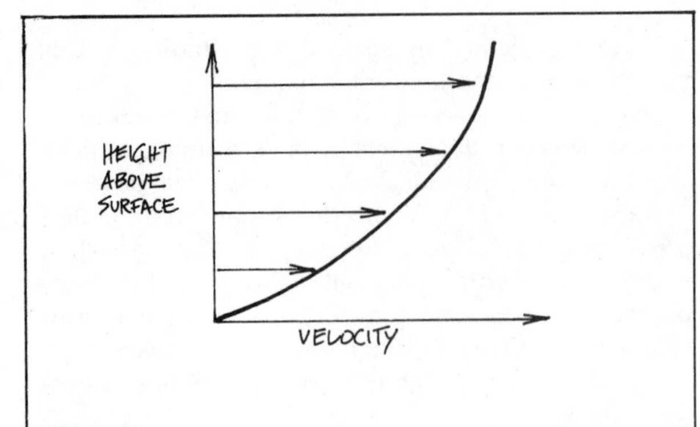

FIGURE 4.17
DIAGRAM OF LAMINAR SHEET FLOW.
Surface drag greatly affects stream flow in water or air. Velocities near a surface are close to zero, giving viscous flow, and only slowly increase as distance from the surface increases.

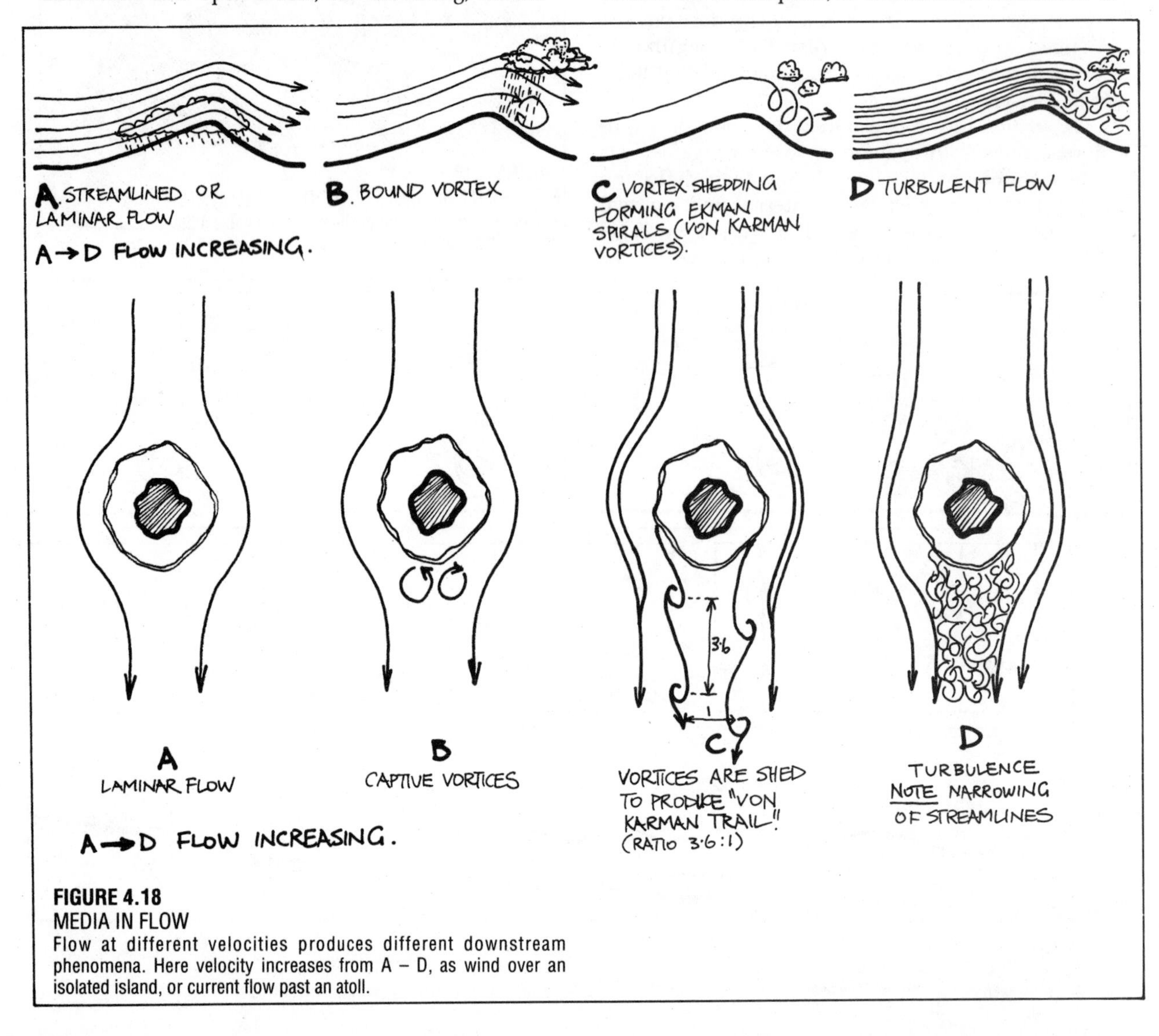

FIGURE 4.18
MEDIA IN FLOW
Flow at different velocities produces different downstream phenomena. Here velocity increases from A – D, as wind over an isolated island, or current flow past an atoll.

events generate secondary origins. I once watched this happen with a eucalypt that I planted in 1953, and saw it twin and twin again as swift-moth grubs encircled its stem. It is now a mighty tree, much branched, but the evidence of these minute constrictive events are still buried in its stem. As designers, we can impose small constricting events or place fixed objects in flow to produce such specific results. We can then be the external shapers of patterned events.

4.11
OPEN FLOW AND FLOW PATTERNS

Creatures that live in open flow conditions are specially shaped and adapted to surface or low-flow (high pressure) phenomena, and may erect or develop "chimneys" to draw fluids or gases through their burrows or bodies. Some life forms combine chimneys, spirals, and crenellations to effect an exchange between them and their fluid surrounds (Vogel, 1981).

All of these effects of flow are of great relevance to designers, engineers, and biologists, and their effects can be increased, nullified, or decreased by design. Natural effects can be used in a variety of ways, and the effects of orders may impose limits on design. Further data is given under Chapter 6: "Trees and Energy Transactions".

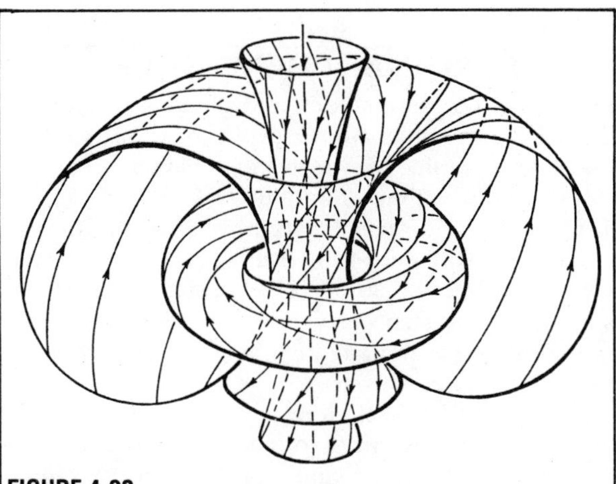

FIGURE 4.20
THE ROBINSON CONGRUENCE
Pattern for an electron, a massless particle, travelling up this page at the speed of light; energy as form, congruent also with the general model.

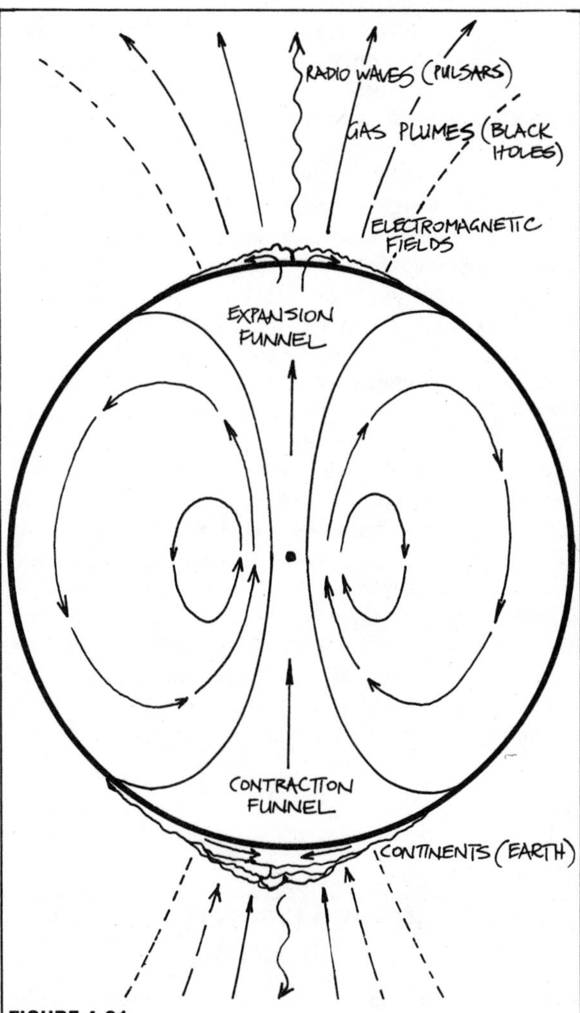

FIGURE 4.21
MODEL OF SPHERICAL BODIES IN SPACE.
A simplified model of a dense planetry body (c.f. the general model). Electro-magnetic fields and thermal convection create special conditions along the axis of spin. Many such bodies emit material at the poles before cooling.

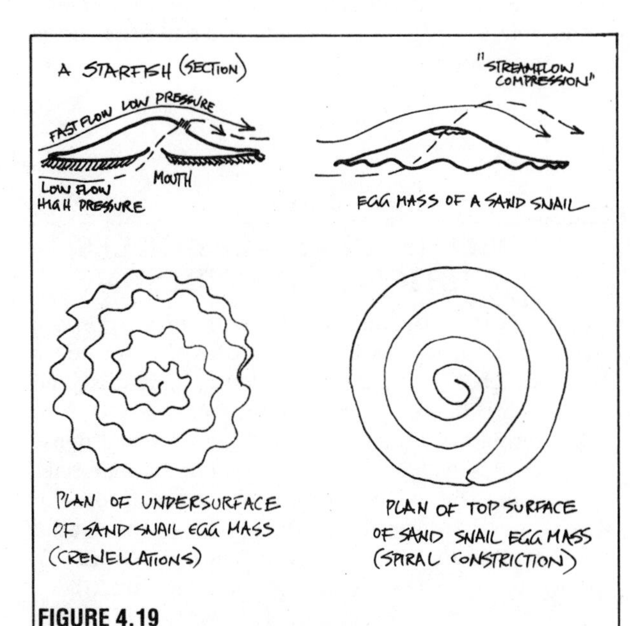

FIGURE 4.19
LAMINAR FLOW NEAR A SURFACE
(Arrows represent wind or water speed). Pressure is high near the surface, lower at a distance, thus a starfish (or a chimney) experiences a flow "under and out the top" in flow conditions. The sand snail egg mass is built to be easily irrigated, the starfish gets food in the water stream by their configuaration in flow.

4.12
TOROIDAL PHENOMENA

Implicit in our core model, and obvious in violent detonations such as atomic explosions, puffs from diesel exhaust pipes, or deliberately blown as smoke rings, are the rolling doughnut-shapes of TORI. A TORUS is a widespread natural phenomenon. The closed models (**Figure 4.13**) enclose such a torus, and we can imagine a slow-cycling torus of nutrients surrounding the stem of any tree as crown-drip carries nutrients to the ground, and the roots again return them via the stem of the tree.

Photographic stills of atomic explosions may reveal violently rotating tori around and crowning the ascending column of smoke and debris. Violent up-draughts caused by local heat create such tori in atmospheric thermals, much appreciated by soaring birds and glider pilots, who ride the inner (ascending) circle of the doughnuts of hot air that are generated, for example, over deserts on hot afternoons.

A complex toroidal form is the "Robinson Congruence" (**Figure 4.20**), portraying the space–time form of a mass–less particle such as a proton, representing, in effect, annidated tori.

DNA is also portrayed as encircling such an imaginary plus–torus in **Figure 4.11**. A Möbius strip—a one–sided twisted toroid (often portrayed by M.C. Escher in his art)—enables us to cross an edge without lifting our pencil. Many life forms produce tori (e.g. some sea snail egg masses). We rely on a torus of rubber to inflate our tyres and to seal circular hatches as O–rings.

A torus is, in effect, a special or truncated case of the Overbeck jet (**Figure 4.13**), as a foetus is a truncated "tree", and can be generated by *discontinuous* or explosive flow, or pulses in flow. A torus is a closed three–dimensional vortex. One such closed toroidal form, as found in black holes, is shown in **Figure 4.21**. Here, accretion of matter causes gaseous ejection at the poles, as the earth may have "ejected" seas and continents at the magnetic poles, or gathers in the violent energy of ionised particles that form the auroras, visible as polar tori in satellite images. Even the long curtains of the auroras seen from the ground contain vertical spiral columns (J. Reid *pers. comm.*).

4.13
DIMENSIONS AND POTENTIALS; GENERATORS

Our patterned systems may exist in two or more dimensions. We can tessellate two–dimensionally but need to envisage three dimensions for a tree form or glacier. The tree–forms of rivers flow down along S–shaped gradients; the generator of such a pattern is gravity. Sand dunes form on near–flat platforms of the desert, and have wind as their generator, as do waves on the sea. Neither gravity nor wind may much affect the creeping tree patterns of mosses, dendrites in shales, or the tree–like forms of mycorrhiza in plant cells. It is here that we see our tree form as the best way to grow or to gather nutrients in the absence of violent kinetic processes. The generator here is life or growth itself.

When kinetic forces do not act strongly, as in flat and essentially sheltered desert environments, **lobulation** and **latticing** still occur as freeze–thaw or swell–shrink patterns, as they do in ice floes on quiet ponds or in the hexagonal patterns of stones on tundra. The slow growth of crystals into rock cavities or ice is still related to our general model; the generator of pattern here being at the level of molecular forces, as in many purely chemical processes, and the forms generated are fractals.

In hill country, energies are usually a combination of stream flow and gravity. On plains, icepacks or flat snowfields, it is freeze–thaw or the swelling of clay in rain that produces lobulations or networks of earth patterns. Lobulation, the production of such shapes as in **Figure 4.7**, differs in origin and mode of expression from the kinetic–energy (flow) systems we have been discussing. I sometimes think of the lobulated forms as a response of nature, or life, to a world that threatens "no difference".

If the hills wear down, then the antepenultimate surfaces will produce their lateral, two–dimensional life patterns, as does the lichen on a rock. Kinetic erosion processes are then exchanged for physical and chemical process at the molecular level, but even this creates a sufficient difference in media for life forms to express themselves, and for differences to arise in the patterns of surfaces.

4.14
CLOSED (SPHERICAL) MODELS; ACCRETION AND EXPULSION

Although trees (including tree roots) may approach spherical form, the best examples are found in spherical bodies in space. These deflect light, dust, and gas towards them, and may capture materials. In their early formation, they themselves may have had dense cores that assembled their share of galactic materials, and around these cores a torus of matter of low– or high–speed rotation can form. This is the model presented for most bodies (*New Scinetist*, 4 April '85, pp. 12–16). A general model is given in **Figure 4.21.**

As matter accumulates in this way, bodies can respond by:

- Becoming more dense – to a limit of 10^{14} g/cm^{3};
- Swelling or expanding (producing shatter effects); and
- Ejecting material at the poles.

Or any combination of these depending on the state

of the matter attached or attracted to the core.

For pulsars, the ejection is radio waves, and for black holes high–speed gas plumes. For trees, of course, we find expansion and transpiration, not localised to the axis of growth.

However, along the Z–Z (ejection) axis of **Figure 4.1**, rotating tori speed up ejection at north poles, and slow it down at south poles, so that less viscous materials are likely to be emitted at north polar emitters. This general effect may be portrayed in one model, but each case needs study. Weak gravitational waves permeate the astronomical system as pulsers permeate or orchestrate biological systems, aiding both dispersal and accumulation depending on the sense of rotation of the accreting system, or the electromagnetic fields interacting with incoming particles. It seems probable that weak fields within the sun creates its pulsers, which proceed from pole to equator as a roll or torus of turbulence over an 11–year period.

4.15 BRANCHING AND ITS EFFECTS; CONDUITS

Various sections, plans, and views of our one tree model reveal very different sectional PATTERNS, all of which are inherent and most of which recur in many other natural forms. Benoit Mandelbrot assembled his own insights, and the speculations of others, to found a mathematics of **fractals** (his term, from the Latin *fractus*, or shattered), which is evolving to make sense of irregular phenomena, as Euclid did for more regular and measurable forms (*New Scientist*, 26 Apr '84, p.17 and 4 Apr '85, p.31–35).

Fractals are as common in nature as in abstractions, and examples are as diverse as impact shatter–zones, clouds, forked lightning, neurone nets and their signals, computer searching procedures, plant identification keys, snowflakes, and tree branches or roots. Some typical fractal forms are illustrated (**Figure 4.22**). Others make up the complex lengths of coastlines and the intricacies of turbulence.

In our tree form (**Figure 4.1**), these fractal patterns (as branches and roots) are contained within a form that would be comprehensible to Euclid, having straight axes, a plane, and regular curved lines, which can be drawn as arcs of perfect circles. Thus the apparent chaos of fractals can be seen to underlie quite regular (but never perfect) shapes in nature as branches underlie the crown canopy of a tree. As Mandelbrot has demonstrated, fractals have their own regular generators and evolutions.

FIGURE 4.22
SOME FRACTAL FORMS
Fractal forms mat be generated by repetition of relatively simple form generators, as are some crystalline formations and such phenomena as tree roots, tree branches, coasts, lightning strikes, shatter zones, information nets, and so on.

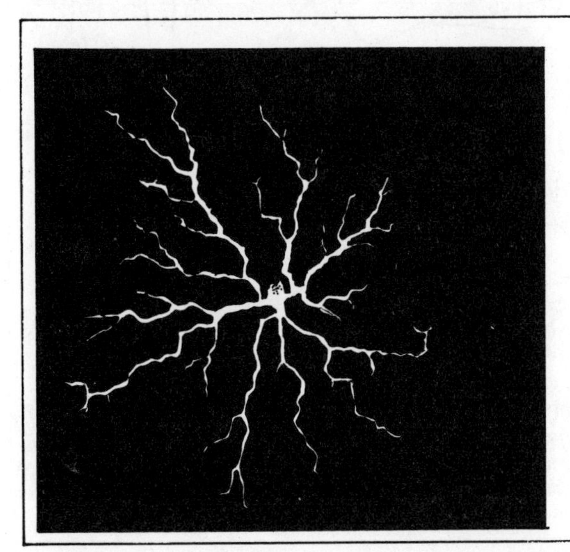

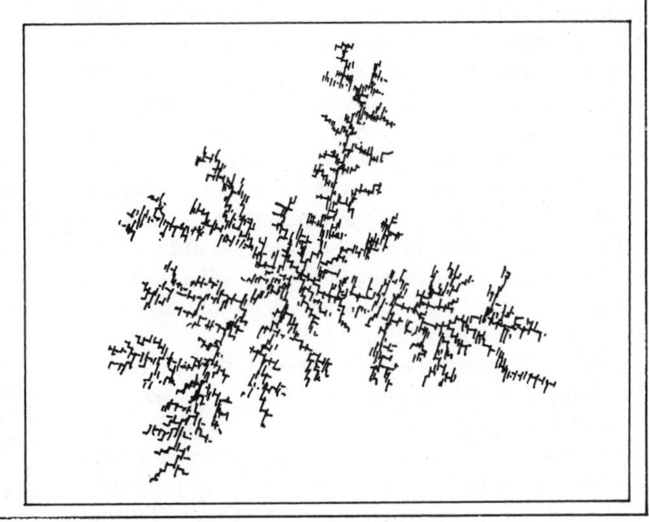

Looking down on a bare winter–deciduous tree, we see a typical fractal, which we can also find in the fulgurites (sand fused by lightning) in sand dunes, and in the shatter zones of explosions. Tree roots are, in fact, a slow shatter or explosion underground. One way to plant an apple tree in very hard ground is to detonate a small plug of gelignite a foot or two below the surface; the roots will follow the shatter pattern, and further elaborate it.

Scatters of objects may at first seem to present a class of events unrelated to either flow models or frac-tals, but fractals are being used to describe the scatters of tree clumps in grassland, or lichen on a stone. In a sense, the surface of spheroids created by branched phenomena (like the plan view of a tree crown) may show such apparently random scatters as growth points; or a curved section through the cut or pruned branches below the crown of a tree would also appear to be a scatter of points (**Figure 4.1.E**). These can be measured by fractal analysis.

Fractal theory may give us a way to measure, compute, and design for branched or scattered phenomena, but we also need to understand the physical advantages of developing ever smaller conduits. Vogel (1981) gives many insights into this process and its effects. Large conduits are of use in mass transport, but both the laminar flow patterns within them and the fact that they have a small surface area relative to their volume makes them inefficient for the *diffusion* of materials or the conduction of heat across their walls.

Ever smaller conduits have different qualities: flow is slow, almost viscous in very small tubes or branches; direction changes in small branches are therefore possible without incurring turbulence or energy losses. Walls can be permeable, and efficient collection, exchange, and transfer is effected (whether of materials or physical properties such as heat and light). Many small conduits efficiently interpenetrate the exchange media.

Wherever there is a need to collect or distribute materials, or to trade both ways with media, branching is an effective response. In design, therefore, we need to use "many paths" in such situations as home gardens, where we are always trading nutrients as our main activity. There is little advantage in forming these paths as straight lines (speed is not of the essence), but rather in developing a set of *cul de sacs* or keyhole–shaped beds (this is also the shape of sacs in lungs). Convoluted paths in gardens have the same effect. They either bring the gardener into better contact with the garden, enabling collection and servicing to occur, or create better mutual exchange between the species in the garden.

The high-pressure/low–flow nature of minor branches demands a very large total cross–sectional area of these in relation to the main supply arteries. Such small conduits may develop areas which in sum are 300–1,600 times that of the supply artery (our main roads are therefore much less in area than the foot tracks that lead off them). As an applied strategy, multiple small paths enhance our access to food systems, or in fact any system where we both take and give materials.

In organisms, the multiple branches give the being a chance to recover from injury, preserve information, and permit regrowth in the event of minor damage. It is a fool–proof system of interchange. Another way to effect interchange is to elaborate on the walls of larger conduits by involutions, attached fins, irregular surfaces, or to create spirals in fluids or gases by bending or spiralling the conduits themselves, and in general inducing a larger surface of contact between the material transported and the media with which we wish to exchange nutrients, heat, or gases.

Branching in trees is as often a result of external forces (wind and salt pruning, secateurs, or insect attack) as it is a result of internal cell patterns; it is as much forced upon things as it is the "best thing to do." We must therefore see the branched form as an interaction between an organism or process, the purpose it serves, and the external forces of the media in which the organism is immersed (the forces acting on it externally to deform the perfect pattern).

Along the streamlines (S1–S9 of our model **Figure 4.1**), fluids and gases may pass in conduits or along "transmission cords", food and signals are relayed to cells, and gases exchanged. Organs served by or serving these systems are half–models of our tree (kidneys, lungs) or branching fractals (mesenteries).

No matter how long or complex conduits are, in the end their contents diverge, escape, and disperse, and at the intake materials are gathered from dispersed sources. It is this gathering and dispersal from both

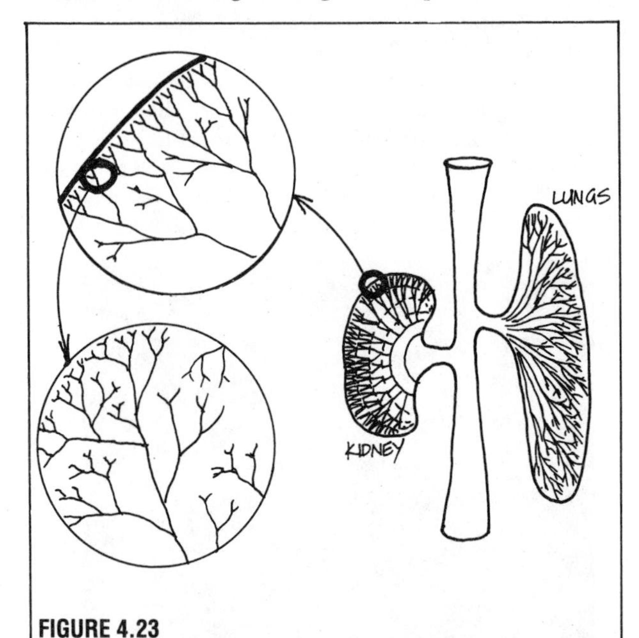

FIGURE 4.23
Any form specialised for diffusion or infusion (lungs, kidneys, mushrooms, palm trees) is usually subject to branching, and develops a "half–model" in one medium. The alvioli of the lungs further resemble the "keyhole beds" of a well–designed garden.

ends or margins of events that is a basic function of the tree–like forms that pervade living natural systems and such phenomena as rivers or lava flows.

4.16
ORDERS OF MAGNITUDE IN BRANCHES

Streams take up many ground patterns depending on the processes that have formed the underlying landscape (block faulting, folding, volcanism) and the erosion and permeability characteristics of the underlying rock itself (limestone, mudstone, sandstone, clay). That is, the ultimate pattern of a stream network in landscape depends on *process* and *substrate;* or we could call these process and *media* in terms of our model.

We can easily see that stream patterns are the sum of preceding events that gave rise to the geological processes and rock types, so that streams have a lot to tell us about such processes, a skill learnt in photo interpretation. **Figure 4.24** demonstrates some of the information so clearly told by stream patterns alone.

However, if we abstract a fairly normal dendritic (tree–like) stream branching pattern as in **Figure 4.25**, we can find out these things from the pattern alone:

- The ORDER of channels; the volume, or SIZE, of branches.
- The NUMBER OF BRANCHES in each order.
- The TOTAL CHANNEL LENGTH in each order.
- The MEANDER FREQUENCES in each order, or the behaviour of flow in the orders of branches.

Streams usually have from one to seven orders, depending on their age, size, or gradient (fall over distance). An easy gradient develops as streams cut back their headwaters and fill in (aggrade) their lower reaches; meanders increase, and the velocity of flow decreases. These older streams, like a mature tree, have developed all their branches (as has an old company or an old army). Unless stream conditions themselves change (by a process of stream capture, an increase in rainfall, or a change in landscape), streams (and businesses) maintain an equilibrium of order. Looking at our dendritic, peaceful stream, we may find something as can be seen in **Table 4.1**.

As the branches join up to make ever larger orders of channels, then about 3 times as many smaller branches join up to make each larger group, and so on. However, the individual lengths (of any one branch in each order) increase by 2 times as the order increases from 1–6. This is a very general rule of stream branching, even in non–dendritic patterns, and holds true for many streams. Similarly, meanders or bends also occur in a predictable way depending on the volume and gradient (flow). Regular meanders depend on certain velocities and stream width (as do stability of Von Karman vortex trails; **Figure 4.14**). The ratio for meanders or trails is about 1:3.6 (Vogel, 1981).

Such regularity in branching may remind us of PULSERS (wave fronts), and indeed as each size order changes, so does the behaviour not only of the water flow, but of its associated flora and fauna and their shapes. In the rills and runnels, streamlines and turbulent flow is observed. High in the stream gradients (the flattened S–curve of the stream bed in profile) we find insects and fish with suctorial parts able to stick on rocks, flattened fins to press them into the stream bed, flattened bodies and very streamlined profiles. In the middle orders, we get less turbulent water flow, more spiralling, less oxygenation, and more

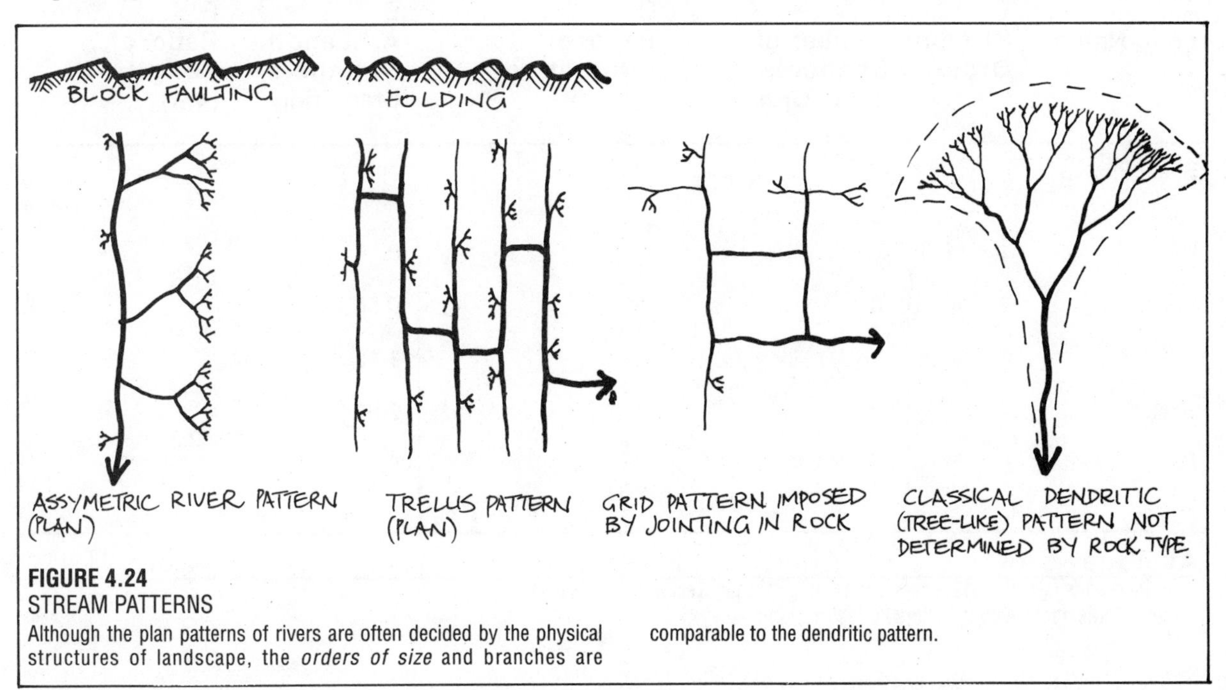

FIGURE 4.24
STREAM PATTERNS
Although the plan patterns of rivers are often decided by the physical structures of landscape, the *orders of size* and branches are comparable to the dendritic pattern.

free–swimming but very active fish of high oxygen demand; these may not live in the still water of higher order streams and low oxygen levels. Thus, we see that gaseous exchange is affected by turbulent flow, and that this in turn determines the life forms in these areas (Vogel, 1981).

In the lower stream or estuaries, we get weak swimmers, less streamlined shapes, flat fish such as flounders, bulky molluscs, jellyfish in quiet areas and lower oxygen levels. We can list many of these life changes which are coincidental with changes in stream order, so we see that the order of streams is very much connected to the behaviour of the water, the landscape, and the shape of life forms in the watershed. Branching of pathways therefore changes species, behaviour, flow, and rates of exchange of nutrients or materials carried by the stream. When we examine a tree, we find that birds and insects are also confined to, or modified to suit, the orders of branching.

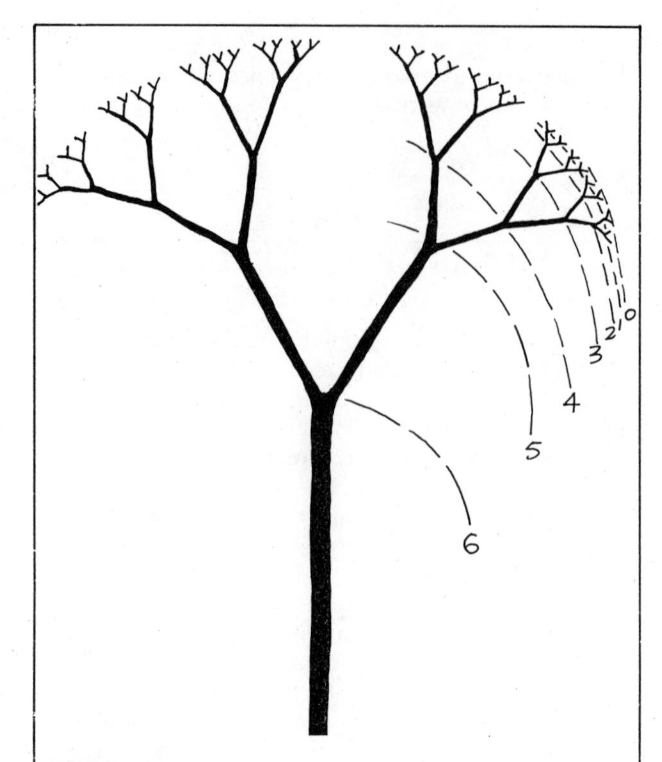

FIGURE 4.25
DENDRITIC BRANCHING.
A regular "tree" based on the proportion of real rivers. The ogives, or curved lines, can be viewd as pulses of growth, waves approaching the viscous "shoreline" of the leaves or (in the case of rivers) the slow seepage of upland rills. Here, seven orders of branches exist; more orders become difficult to develop towards the diffusion surface, where viscous flow slows the movement of fluids.

4.17
ORDERS AND DIMENSIONS

It is in the order of branching (as in our river) that we can gain insight into the order of orders, and the functions of orders. At each point of branching (or size and volume change) everything else changes, from pressures, flows, velocities, and gaseous exchange, to the life forms that associate with the specific size of

TABLE 4.1:
STREAM ORDERS AND SOME RATIOS.

A **Folk Name**	B **Stream Order**	C **Number of Channels in the Order**	D **Ratio of bifurcation/**	E **A. Length Channels branching**	F **Ratio of Length (km)**
Sheet Flow	0	-	-	-	-
Rill	1	308		0.28	
			x 3.5		x 2.0
Runnel	2	87		0.56	
			x 3.3		x 2.0
Creek	3	26		1.12	
			x 3.3		x 2.3
Stream	4	8		2.56	
			x 2.7		x 2.2
River	5	3		5.76	
			x 3.0		-
Estuary	6	1		-	
Average			(≈3.0)		(≈2.0)

(Modified after Tweadie, *Water and the World*, Thomas Nelson, Australia, 1975.)
(Arrows indicate ascending orders by factor of increase)

branches. This is how we make sense of the fish species in streams, and the bird species in a tree. Each has its place in a set or order of branches, on the bark of stems, or the leaf laminae of the tree.

This order, or size–function change, produces physical, social, and organic series. Not only species change with order, but so does behaviour. We cannot get a riot of one person, and fewer than 15 rarely clap to applaud as an audience. When we therefore come to *construct* hierarchies, the rules of order should guide us.

An array of orders is observed in a wide range of phenomena such as human settlement size, numbers in social hierarchies, trophic levels (food pyramids), and the size of animals in allied zoological families. The size of the factor itself (times 3 for river branches) changes with the dimensions of the system (times 10 for trophic pyramids). Physical entities from protons to universes display such order, with a consequent increase in the ratio, dimensions, and behaviours associated with size change (giant and dwarf stars behave very differently).

As designers, we need to study and apply branching patterns to roads or trails, and to be aware of the stable orders of such things as human settlements, or we may be in conflict both with orderly flow (can we increase the size of a highway and not alter *all* roads?), with settlement size (villages are conservative at about the 1000–people order, and unstable much below or above that number), with sequences of dam spillways, and even with the numbers of people admitted to functional hierarchies where information is passed in both directions. We can build appropriate or inappropriate systems by choosing particular orders, but we do better to study and apply *appropriate* and stable size and factor classes for specific constructs. For example, in designing a village, we should study the orders of size of settlements, and choose one of these. This decides the number and types of services and the occupations needed, which in turn decides the space and types of shops and offices for a village of that size, and the access network needed.

In human systems, we have confused the order of hierarchical function with status and power, as though a tree stem were less important than the leaves in total. We have made "higher" mean desirable, as though the fingers were less to be desired than the palm of the hand. What we should recognise is that each part needs the other, and that none functions without the others. When we remove a dominant animal from a behavioural hierarchy, another is created from lower orders. When we remove subordinates, others are created from within the dominants. So it is with streams.

Thus, we can see how rivers change their whole regime if we alter one aspect. We should also see that water is *of the whole*, not to be thought of in terms of its parts. Thus we refute the concept of *status* and assert that of *function*. It is not what you *are*; it is what you *do* in relation to the society you choose to live in. We need each other, and it is a reciprocal need wherever we have a function in relation to each other.

4.18

CLASSIFICATION OF EVENTS

All events are susceptible to classification over a variety of characteristics, and as, for example, clouds and galaxies have their pattern–names, so do many other phenomena. Some basic ways to classify events in a unified system are:

A. NATURE.

A1. Explosive, disintegration, erosion, impact, percussion.

A2. Growth, integration, construction, translation.

A3. Conceptual, idea, creative thought, insight.

B. STAGE.

B1. Potential only (ungerminated seed, unexplored idea, unexploded bomb).

B2. In process of evolution.

B3. Completed (growth and expansion ceased), articulated.

B4. Decaying (disintegrative, replaced or invaded by new events) disarticulated.

C. DIMENSION.

C1. One (linear phenomena), curves.

C2. Two (surface phenomena), tessellae, dendrites.

C3. Three (solid phenomena), trees.

C4. Four (moving solid phenomena), includes the time dimension.

C5. More (conceptual phenomena) models of particles or forces, states of energy.

D. LOCATION

D1. Generating across equi–potential surfaces (storms at sea).

D2. Within media (weather "frontal" systems).

D3. Through surfaces at 90° or so (trees).

D4. Englobements (some explosions and organisms).

D5. An idea, located out of normal dimensions of space–time.

By extending and applying these categories, all events can be given short annotations, e.g.:

- A sapling is: A2, B2, C3, D3.
- A falling bomb is: A1, B1, C4, D2.

4.19

TIME AND RELATIVITY IN THE MODEL

As we see the seed as the origin of the tree, so we can broaden our view, and our dimensions, and view the tree as the current time–focus of its own genealogy.

Before it in time lie its ancestors, and after it its progeny. It lies on the plane between past and future, and (like the seed) determines by its expression the forms of both, and is in turn determined by them. Just as the stem of the tree now encapsulates its history as smaller and smaller growth rings, so universal time encapsulates the tree in its own evolutionary history. This is difficult to portray, and has more dimensions that we can illustrate on a page. It is the basis of the Buddhist belief that all time is enfolded or *implicate* in the present, and that current events are part of a total sequence, all of which are enfolded in the present tree as ancestors, or siblings.

As we read this, we stand in the plane of the present; we are the sum of all our ancestors, and the origin of all our descendants. In terms of our model, we are at an ever–changing origin, located on the boundary of past and future. As well, we are spinning with the earth, spiralling with the galaxy, and expanding or contracting with the universe. As origins, we are on the move in time and space, and all these movements have a characteristic pulse rate.

Our bodies contain the potential for future generations, awaiting the events of pairing to create their own future events. Like a seed origin buried in the tree stem, we are buried in the stem of our siblings in a genealogy, whose branches thrive, die, and put forth new shoots and roots over time.

This is the case with all origins; they can all, even if ancient, be located in this matrix. If we know how to reconstruct the tree, we can find the place of the seed and vice versa. All rivers, erosion cells, and all glaciers originated, therefore, at the *central* stem of their courses, and built their pattern both ways along the kinetic gradient of their flow. Thus, in terms of the time dimension we see the present as the ORIGIN of both the past and the future (located as it is in the centre of our pattern).

Designers can move sideways in the waves of time (as a surfer on a wave–front), transporting seed from continent to continent, permitting natural or induced hybrid palms and legumes to weave an alternative future. Mankind is an active translator of life, and, of course, of death.

In all core models, including our own genealogy, the point where all the important action takes place is through the point of origin, which is always in the present. How we behave *now* may determine not only the future, but the past (and all time). Think of that, and realise that you are really where it's at, no matter when you are! I find great personal meaning in the Australian aboriginal life ethic, and little enough comfort in any pie–in–the–sky. If it *is* my actions which determine the sky, I want it to be full of life, and I choose to believe that I am part of all that action, with my own job to do in this life form, and other jobs to do in other phases.

4.20
THE WORLD WE LIVE IN AS A TESSELLATION OF EVENTS

I live in the crater of an ancient volcano, the caldera of which is in part eroded by the sea. Trees rise from the soils, and birds nest in them. From the seeds and eggs in the trees arise new life forms. Great wind spirals sweep in from the west with almost weekly regularity, bearing the fractal forms of turbulent clouds and causing, in autumn and mid–summer, lightning and thunder.

On this peninsula, the terminal volcanic core stands fast, refracting waves to either side, and creating a pinched neck of sand which joins us to the mainland. The hills are stepped by successive sea–level changes, and record the pulses of long–term cycles and successions. Day follows night, and life follows death follows life.

All of these phenomena are a unity of patterns long repeated and based on one master pattern, each one preparing for new evolutions and dissolutions. It is the number and complexity of such cycles that give us life opportunities, and life is the only integrative force in this part of the universe. Let us respect and preserve it.

An understanding (even a partial understanding) of the underlying patterns that link all phenomena creates a powerful abstract tool for designers. At any point in the design process, appropriate patterning can assist the achievement of a sustainable yield from flows, growth forms, or information flux. Patterns imposed on constructs in domestic or village assemblies can result in energy savings, and satisfactory aesthetics and function, while sustaining those organisms inhabiting the designed habitat.

Patterning is the way we frame our designs, the template into which we fit the information, entities, and objects assembled from observation, map overlays, the analytic divination of connections, and the selection of specific materials and technologies. It is this patterning that permits our elements to flow and function in beneficial relationships. The pattern *is* design, and design is the subject of permaculture.

Bohm (1980) urges us to go beyond regarding ourselves as interactive with each other and the environment, and to see all things as "projections of a single totality". As we experience this totality, incorporate new information, and develop our consciousness, we ourselves are fundamentally changed. "To fail to take this into account must inevitably lead one to a serious and sustained confusion in all that one does." The word "implicate" in the title of Bohm's work comes from the Latin "enfolded", and when we separate individuals, effects, or disciplines from this enfolded order, we must recognise only that we have part of the unknowable totality, not the truth itself. There are no opposites, just phases of the one phenomena.

For myself, and possibly for you if you take up the study of patterns, the contemplation of the forms of life and flow has enabled me to bring to consciousness the unity of all things, and the enfolded nature of Nature. In the matter of genealogy we can become conscious of ourselves in the time and pattern stream, and it is startling to realise that (as origin) we "determine", or rather define and are defined by, our ancestry as much as we define and are defined by our descent. We do not doubt our physical connection to either ancestry or descent. It is the sense of "all are present here" that is revealed by pattern: to be encapsulated in, and a pervading part of, a personal genealogical pattern which is itself a result of a pattern of innumerable variables.

Patterns tell us that all is streams, all particles, all waves. Each defines the other. It tells us that all is one plan. Although we find it difficult to see pattern in all the plan, it is there. We are the universe attempting to define its processes. A Kalahari bushman would say we are the dreams of a dreamer. What I feel we can never define is substance (except *as* process; this is all it may be). We can only know a few local patterns, and thus have some weak predictive capacity. It is the pattern that our local patterns cannot know that will surprise us, the strike of cosmic lightning from an unguessed source or stress.

Finally, pattern understanding can only contribute to the current and continuing evolution of new world views based on the essential one-ness of all phenomena. Lovelock (1979) has perhaps best expressed that combination of scientific insights and older tribal beliefs which assert the interdependence of animate and inanimate events. The universe, and this earth, behave as self-regulating and self- generated constructs, very much akin to a single organism or a thought process.

The conditions which make life possible are balanced about such fine tolerances that it seems close to certainty that many processes exist just to preserve this equilibrium in its dynamic stability.

From the point of view of biologists, Birch and Cobb's *The Liberation of Life* [1984; see a review by Warwick Fox in *The Ecologist* 14(4)] denies the validity of the existence of *individual* organisms or *separate* events; all exist in a *field of such events* or as an expression of one life force. Organisms such as ourselves exist only as an inseparable part of our event environments, and are in continual process of exchange with the animate and inanimate entities that surround us. We are acted upon and acting, created and creating, shaped and shaping. Fox asserts, as I have here, that "we must view the cosmos as an infinite complex of interrelated events"; all things "are in actuality enduring societies of events."

Theoretical physicists (Capra, 1976) contribute to such world views, all of which are in conflict with the current ethics that govern political, educational, and economic systems, but all of which are contributing to an increasing effort to unify and cooperate in a common ethic of earth-care, without which we have no meaning to the universe.

4.21
INTRODUCTION TO PATTERN APPLICATIONS

There are two aspects to patterning: the perception of the patterns that already exist and how these function, and the imposition of pattern on sites in order to achieve specific ends. Both are skills of sophisticated design, and may result in specific strategies, the harmonious resolution of problems, or work to produce a local resource. Given that we have absorbed some of the information inherent in the general pattern model, we need some examples of how such patterning has been applied in real-life situations.

A bird's-eye view of centralised and disempowered societies will reveal a strictly rectilinear network of streets, farms, and property boundaries. It is as though we have patterned the earth to suit our survey instruments rather than to serve human or environmental needs. We cannot perhaps blame Euclid for this, but we can blame his followers. The straight-line patterns that result prevent most sensible landscape planning strategies and result in neither an aesthetically nor functionally satisfactory landscape or streetscape. Once established, then entered into a body of law, such inane (or insane) patterning is stubbornly defended. But it is created by, and can be dismantled

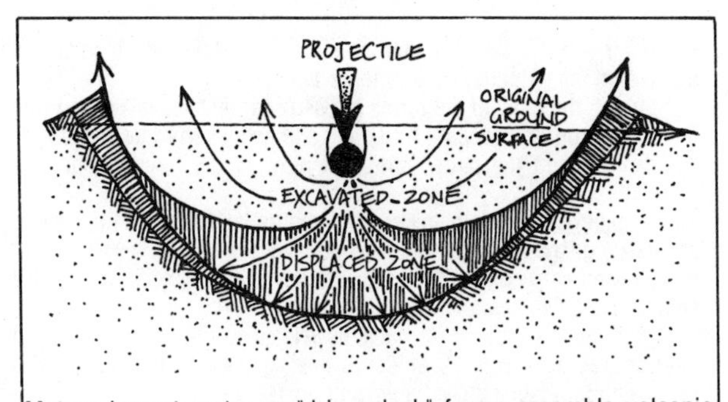

Meteor impact crater or "drip splash" forms resemble volcanic calderas.

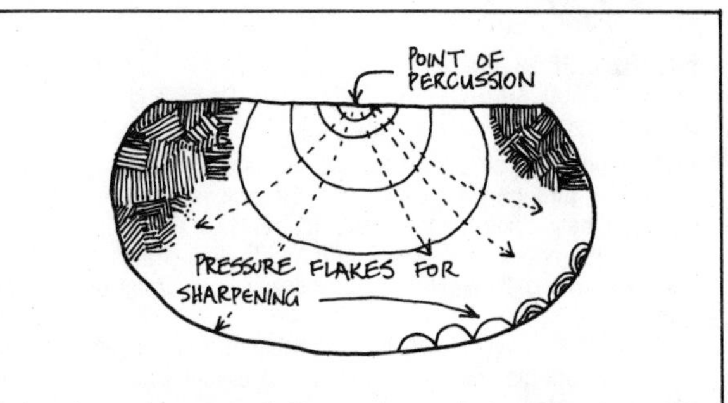

Percussion and pressure flaking produce typical patterns in obsidian or chert tools shaped by stone workers.

by, people.

A far more sensible approach was developed by Hawaiian villagers, who took natural ridgelines as their boundaries. As the area was contained in one water catchment, they thus achieved very stable and resource–rich landscapes reaching from dense cloud–forests to the outer reefs of their islands. The nature of conic and radial volcanic landscapes with their radial water lines suits such a method of land division. It is also possible for a whole valley of people to maintain a clean catchment, store and divert mid–slope water resources for their needs, catch any lost nutrients in shallow ocean enclosures (converting first to algae, then to crabs and fish), and thus to preserve the offshore reef area and the marine environment. Zulus and American Indians adopt the circular or zonal modes in their plains settlements.

Such models can be studied and adopted by future (**bioregional**) societies as sane and caring people become the majority in their region, and set about the task of landscape rehabilitation. Sensible land division is a long–delayed but essential precursor to a stable society.

4.22 THE TRIBAL USES OF PATTERNING

As I travel about the world, I find tribal peoples using an enormous variety of traditional patterns. These decorate weapons, houses, skin, and woven textiles or

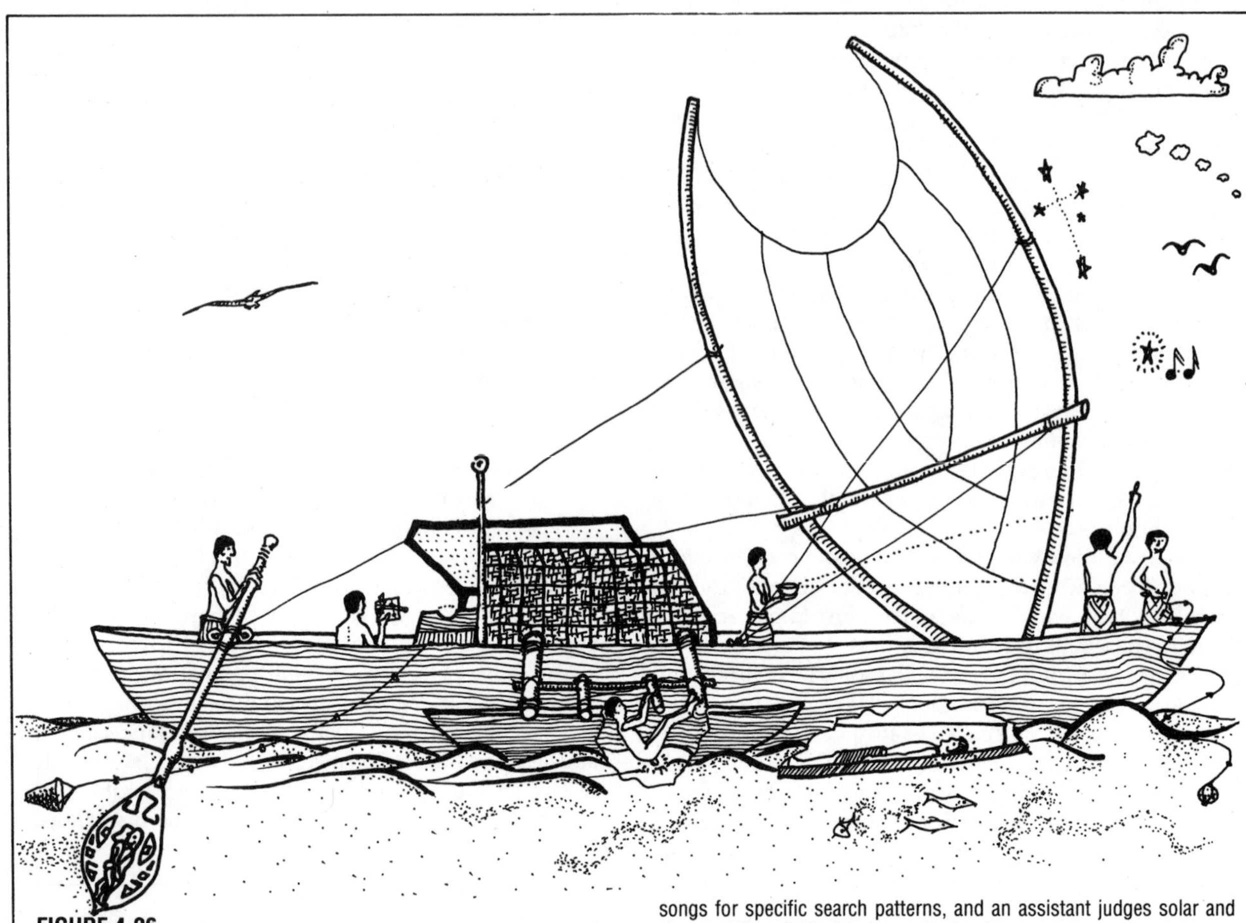

FIGURE 4.26
TRADITIONAL NAVIGATION RELIES ON PATTERN RECOGNITION.

- The helmsman sings to time the log line, and feels for current deflections through his steering oar. He or she sings to record the stages of the journey and the "rivers" (currents) of the sea.
- Lookouts note star patterns, bird flight, cloud trains, or light refracted from coastal lagoons. Water speed is recorded by a knotted float line.
- A skilled listener in a "black box" near the keel listens to wave refraction from the hull, predicts storms or islands from wave periods timed by chants.
- A navigator consults a voyage "chart" of sticks and cowries representing the interaction of phenomena and sings the navigation songs for specific search patterns, and an assistant judges solar and star elevations (latitude) via a water level.
- At night, the glow of life forms in deep sea trenches is awaited (the "lightning" from the deeps"), bird song and echoes from headlands are listened for, and the phosphorescent glow of the forests ashore is awaited.
- An experienced man lowers his testicular sac into the sea to accurately guage water mass temperature.
- The wind carries a variety of scents from forests, rock lichen, bird colonies, and fish schools; all are recorded.
- A net catches indicator organisms related to the water mass, and near shores a "lead line", or depth line, is used to find banks and sample bottom fauna picked up in soft grease.
- A crystal of calcite may be used to predict rain or to find the sun on overcast days by polarisation of light—"the lodestone" of navigators.

baskets. Many patterns have sophisticated meaning, and almost all have a series of songs or chants associated with them. Tribal art, including the forms of Celtic and ancient engraving have a pattern complexity that may have had important meanings to their peoples. We may call such people illiterate only if we ignore their patterns, songs and dances as a valid literature and as an accurate recording system.

Having evolved number and alphabetical symbols, we have abandoned pattern learning and recording in our education. I believe this to be a gross error, because simple patterns link so many phenomena that the learning of even one significant pattern, such as the model elaborated on in this chapter, is very like learning an underlying principle, which is always applicable to specific data and situations.

The Maori of New Zealand use tattoo and carved patterns to record and recall genealogical and saga information. Polynesians used pattern maps, which lacked scale, cartographic details, and trigonometric measures, but nevertheless sufficed to find 200–2,000 island specks in the vastness of the Pacific! Such maps are linked to star sets and ocean currents, and indicate wave interference patterns; they are made of sticks, flexed strips, cowries, and song cycles (**Figure 4.26**).

Pitjantjatjara people of Australia sing over sand patterns (**Figure 4.27**), and are able to "sing" strangers to a single stone in an apparently featureless desert. Many of their designs accurately reflect the lobular shapes and elaborate micro–elevations of the desert, which are nevertheless richly embroidered by changes in vegetation, and are richly portrayed in what (to Westerners) appears as abstract art. Some pattern mosaics are that of fire, pollen, or the flowering stages of a single plant, others are of rain tracks and cloud streets, and yet others involve hunting, saga, or climatic data.

Children of many tribes are taught hundreds of simple chants, the words of which hide deeper, secondary meanings about medicinal, sacred, or navigational knowledge. All this becomes meaningful when the initiate is given the decoding system, or finds it by personal revelation (intuition). A pattern map may have little meaning without its song keys to unlock that meaning. Initiation can also unlock mnemonic patterns for those who have a first clue as to meaning.

Dances, involving muscular learning and memory, coupled with chants, can carry accurate long–term messages, saga details, and planting knowledge. Many dances and chants are in fact evolved from work and travel movements. Even more interesting are the dance–imitations of other animal species, which in fact interpret for people the postural meanings of these species, although in a non–verbal and universally–transmittable way. We may scarcely be aware that many of our formal attitudes of prayer and submission are basic imitations of primate postures, for the most part taken from other species. Even the chair enables us (as it did the Egyptians) to maintain the postures of baboons, and baboons were revered as gods and embalmed by the Egyptian chair–makers. We can remember hundreds of songs, postures, and chants, but little of prose and even less of tabulated data.

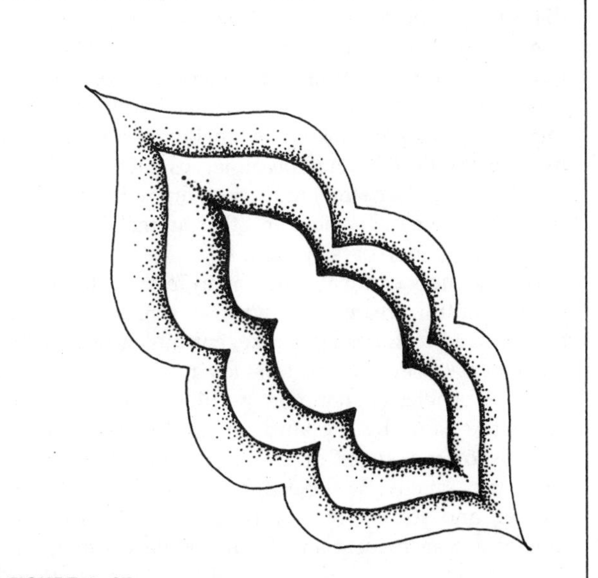

FIGURE 4. 27
SAND PATTERN MAPS (PITJANTJATJARA).
A song map of the Pitjitjantjara women. Such forms closely resemble desert claypans if the long axis is regarded as a flow axis, and the zones as lobular vegetation.

Anne Cameron (*Daughters of Copper Woman*, Vancouver Press, 1981) writes of song navigation in the Nootka Indians of British Columbia: "There was a song for goin' to China, and a song for goin' to Japan, a song for the big island and a song for the smaller one. All she {the navigator} had to know was the song and she knew where she was..."

The navigation songs of the women on canoe voyages record "the streams and creeks of the sea"—the ocean currents, headlands and bays, star constellations, and "ceremonies of ecstatic revelation". From California to the Aleutians, the sea currents were fairly constant in both speed and direction, and assisted the canoes. The steerswomen used the (very accurate) rhythm of the song duration to time both the current speed and the boat speed through the water. Current speed would be (I presume) timed between headlands, and boat speed against a log or float in the water. The song duration was, in fact, an accurate timing mechanism, as it can be for any of us today.

Song stanzas are highly accurate *timers*, accurate over quite long periods of time, and of course reproducible at any time. The song *content* was a record of the observations from prior voyages, and no doubt was open to receive new data.

People who can call the deer (Paiute wise men), the dolphin (Gilbert Islanders), the kangaroo (tribal Tasmanians) and other species to come and present themselves for death had profound behavioural, interspecific, "pulser" pattern–understanding. Just as the Eskimo navigated, in fog, by listening to the quail dialects specific to certain headlands, we can achieve similar insights if our ear for bird dialect is trained, so that song and postural signals from other species make a rich encyclopaedia of a world that is unnoticed by those who lack pattern knowledge. People who can kill by inducing fibrillation in heart nerves have a practical insight into pulser stress induction; many tribespeople can induce such behaviour in other animal species, or in people (voodoo or "singing").

The attempts of tribal shamans to foresee the future and to control dreams or visions by sensory deprivation, to read fortunes by smoke, entrails, water, or the movement of serpents, or to study random scatters of bones or pebbles are not more peculiar than our efforts to do the same by the study of the distribution of groups of measures or the writhing of lines on computer screens. By subjecting ourselves to isolation, danger, and stress, we may pass across the folds of time and scan present and future while we maintain these "absent" states, as described by Dunn (1921) in his *Experiment with Time*, and as related by participants in the sun dances of the Shoshone nation. As we drown, or fall from cliffs, our lives "pass before our eyes" (we can see the past and future).

We need to think more on these older ways of imparting useful or traditional information, and of keeping account of phenomena so that *they are available to all people*. Number and alphabet alone will not do this. Pattern, song, and dance may be of great assistance to our education, and of great relevance to our life; they are the easiest of things to accurately reproduce.

Apparently simple patterns may encode complex information. There may be no better example than that of the Anasazi spiral, with 19 intercepts on its "horizon" "horizon" line (**Figure 4.28**). This apparently simple spiral form is inscribed on a rock surface near the top of a mesa in desert country in the southwest USA. Three rock slabs have been carefully balanced and shaped, as gnomons which cast moon–shadows or (by their curvature) direct *vertical* daggers of sunlight to the points of the spiral. The 19 points at which the spiral intersects the horizontal axis are those at which the shadow of the moon is cast by a gnomon on the spiral, and indicate the moon elevation or 19–year (actually 18.6) cycle caused by the sway of the earth's axis.

Thus, one simple spiral records lunar and solar cycles for the regulation of planting, the timing of ceremonies, and (as modern science has just realised), the prediction of the 19–year (18.6 year) cycle of drought and flood. A very simple pattern encodement thus represents a practical long–term calendar for all people who live nearby. The Anasazi culture is extinct, and only a persistent investigation by Anna Soaer (an artist with *intuitive* observational skills) has revealed the significance of this arrangement. Scientists have often doubted the capacity of tribal peoples to pattern such long–term and complex events, which in terms of our clumsy alphabetical and numerical symbols are not only forgettable, but would take a small library to encode. The knowledge so presented is available to very few (ABC TV Science programme, Australia, 20th January '84). However, wherever tribes remain intact, there are many such sophisticated pattern–meanings still intact, all as complex and information–dense.

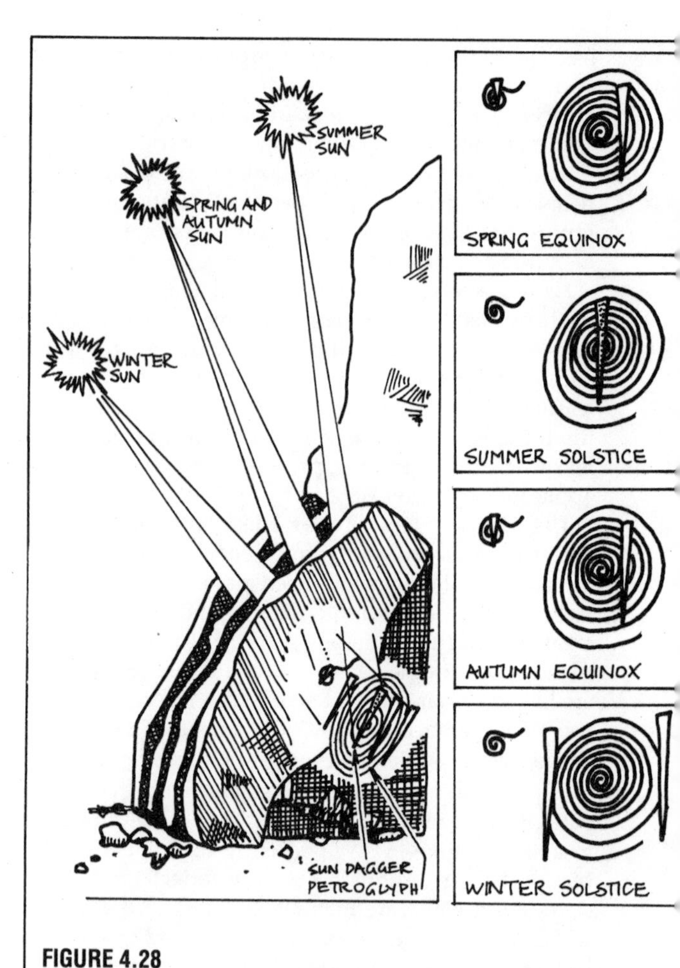

FIGURE 4.28
A petroglyph (rock carving) of the Anastasi Indians (North America) forms a long–term calendar of sun and moon cycles.

In the complex of time–concepts evolved by Australian Aborigines, only one (and the least important) is the linear concept that we use to govern our life and time. Of far greater everyday use was phenomenological (or phenological) time; the time as given not by clocks, but by the life–phenomena of flowers, birds, and weather. An example from real life is that of an old Pitjatjantjara woman who pointed out a small desert flower coming into bloom. She told me that the dingoes, in the ranges of hills far to the north, were now rearing pups, and that it was time for their group to leave for the hills to collect these pups. Thousands of such relationships are known to tribal peoples. Some such signals may not occur in 100 or 500 years (like the flowering of a bamboo), but when it does occur, special actions and ceremonies are indicated, and linked phenomena are known.

Finally, in tribal society, one is not wise by years, but by degree of revelation. Those who understand and

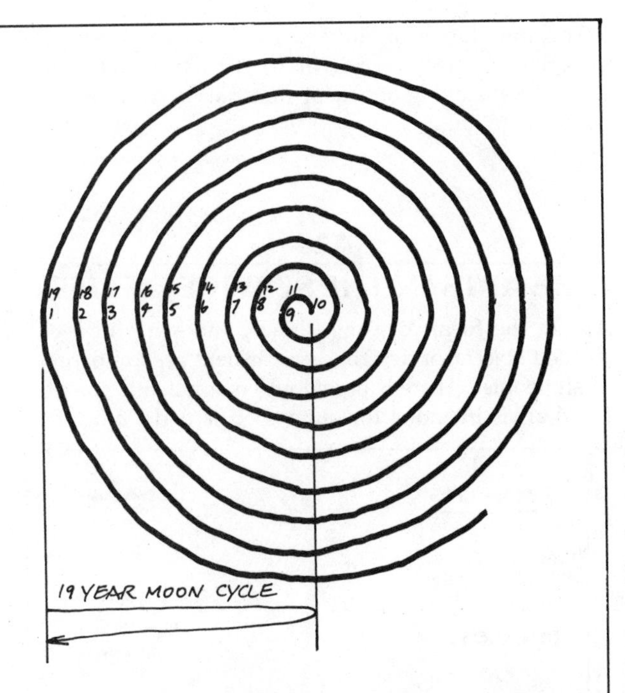

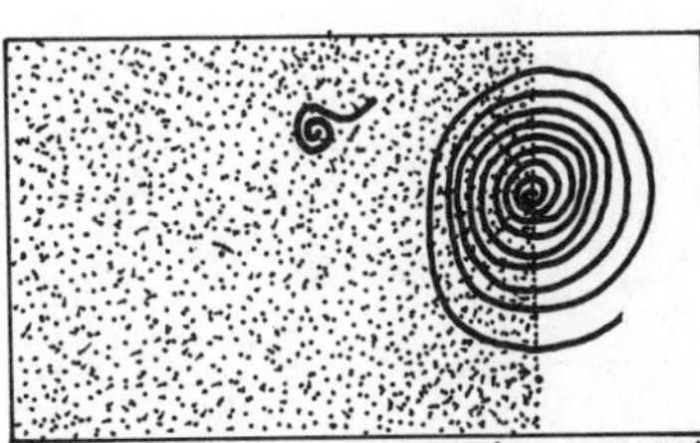

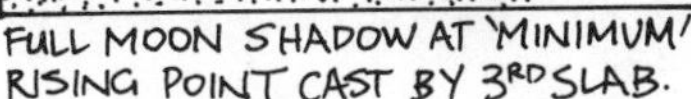

FULL MOON SHADOW AT 'MINIMUM' RISING POINT CAST BY 3RD SLAB.

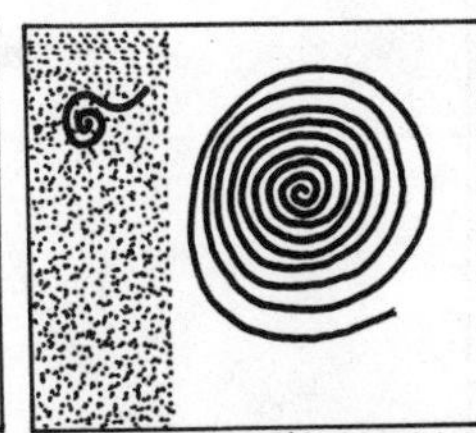

FULL MOON SHADOW AT 'MAXIMUM' RISING POINT CAST BY THIRD SLAB.

embody advanced knowledge are the most intuitive, and therefore most entitled to special veneration. Such knowledge is almost invariably based on pattern understanding, and is independent of sex or even age, so that one is "aged" by degree of revelation, not time spent in living (there are some very unrevealed "elders" in the world!).

4.23
THE MNEMONICS OF MEANING

Buddhists remind themselves of the pattern of **events** with their oft–repeated chant "Om mani padme hum"; pronounced "Aum ma–ni pay–may hung" by Tibetans and Nepalese, and meaning:

Om: the jewel in the lotus : hum

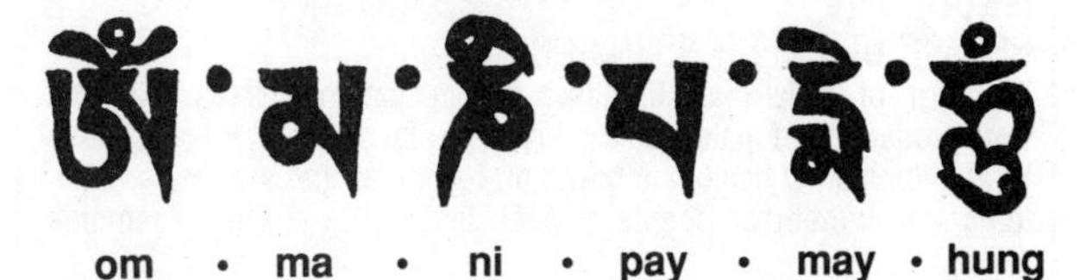

om • ma • ni • pay • may • hung

As Peter Matthiessen explains it (*The Snow Leopard*, Picador, 1980):

Aum (signing on) is the awakening or beginning harmonic, the sound of all stillness and the sounds of all time; it is the fundamental harmonic that recalls to us the universe itself.

Ma–ni: The unchanging essence or diamantine core of all phenomena; the truth, represented as a diamond, jewel, or thunderbolt. It is sometimes represented in paintings as a blue orb or a radiant jewel, and sometimes as a source of lightning or fire.

Pay–may: "Enfolded in the heart of the lotus" (mani enfolded). The visible and everyday unfolding of events, petals, or patterns thus revealing the essential unchanged core (mani) to our understanding. The core itself, or the realisation of it, is *nirvana* (the ideal state of Buddhism). The lotus represents the implicate order of tessellated and annidated events, and the process of unfolding the passage of time to successive revelations. At the core is unchanging understanding.

Hung (signing off): "It is here, *now*." A declamation of belief of the chanter in the words. It also prefaces the "Om" or beginning of the new chant cycle, although in a long sequence of such short chants, all words follow their predecessors. This is the reminder mnemonic of implicate time; all events are present *now*, and forever repeated in their form.

DORJE, or Dorje–chang, is the Tibetan Buddha–figure who holds the *dorje* (thunderbolt), represented as a radiant stone which symbolises cosmic energy. Dorje is "the primordial Buddha of Tibet", who began the great succession of current and past reincarnations. His colour is blue, for eternity, and he may carry a bell to signify the voiceless wisdom of the inanimate, or the sound of the void.

Dorje is an alter ego of Thor of the Norsemen, Durga of the Hindu, and of thunderbolts and "thunderers" of other tribal peoples. The *mani* or stone of Thor was Mjollnir, his hammer, from which derives Mjollnirstaun, and (eventually) Mollison (by way of invasions into Scotland, and migrations). Thus, even our own names may remind us of the essential oneness of the events and beliefs around us.

We can choose from tribal chants, arts, and folk decoration many such mnemonic patterns, which in their evolution over the ages express very much the same world concept as does modern physics and biology. Such thoughtful and vivid beliefs come close to realising the actual nature of the observed events around us, and are derived from a contemplation of such events, indicating a way of life and a philo-sophy rather than a dogma or set of measures.

Beliefs so evolved precede, and transcend, the emphasis on the individual, or the division of life into disciplines and categories. When we search for the roots of belief, or more specifically *meaning*, we come again and again to the one–ness underlying science, word, song, art, and pattern: "The jewel in the heart of the lotus".

Thus we see that many world beliefs share an

essential core, but we also see the drift from such nature–based and essentially universal systems towards personalised or humanoid gods, dogma, and fanaticism, and to symbols without meaning or use in our lives, or to our understanding of life. Many other world–concepts based on the analogies of rainbows, serpents, and song cycles relate to aspects of the integrated world view, and are found in Amerindian and Australian tribal cultures.

4.24

PATTERNS OF SOCIETY

We can pattern the behavior of human and other social animals to represent aspects of their society. A set of such patterns, derived from studies I and my students made in Tasmania from 1969 to 1974 are illustrated in **Figure 4.29**. The central pattern represents the orders or castes of occupational level (status) in its long axis. There are seldom more than seven major occupational levels even in such rigorously–stratified hierarchies as the army. The width of the **Figure 4.29** represents the numbers of people at each level, and for this configuration we summed the numbers of people in several organisations (to sample some 35,000 people), including the local army, a multinational company, some churches and many small businesses.

Within the general "boat" pattern form so evolved, I have marked some arrows to represent genetic streaming (by marriage or sexual congress); important classes of occupation are:

1. LOW OCCUPATIONAL (RESOURCE) AREA—MANUAL AND UNSKILLED URBAN:
Characteristics are a general dearth of material resources, low status, part–time occupations, and a remarkable preponderance of male births and survived male children (about 140 males per 100 females). Large families. Serial polyandry is common or acceptable.

2. THE CENTRAL OR MOST POPULATED LEVEL; THE "MIDDLE CLASS":
Adequate resources, nine–to–five jobs, some job tenure, and a "normal" birth ratio of 104 males per 100 females. Mixed white collar and skilled technical workers, average family sizes. Monogamy is an ideal, but is often expressed as serial monogamy.

3. THE UPPER LEVELS:
Few people, extensive resources, flexible and often self–set times, and a high proportion of female children (about 100 females to 70 males or less); urban professionals or managers would typify the group. Small families, effective polygyny via concubines or mistresses.

4. VERY HIGH LEVELS:
Executive directors and landed nobility. Variable family sizes but a preponderance of female children (as per 3. above), and a habit of lateral intermarriage for economic alliances, facilitated by exclusive schools and resorts.

The imbalance of the sex ratios in these strata ensures a genetic turnover or diffusion between classes; a streaming of genetic materials between levels over generations.

4.25

THE ARTS IN THE SERVICE OF LIFE

Art, in the forms of song, dance, and sculptural or painted objects, or designs, is an ancient preoccupation of all peoples. There is little doubt that most (if not all) tribal art is intended for quite specific ends; much of

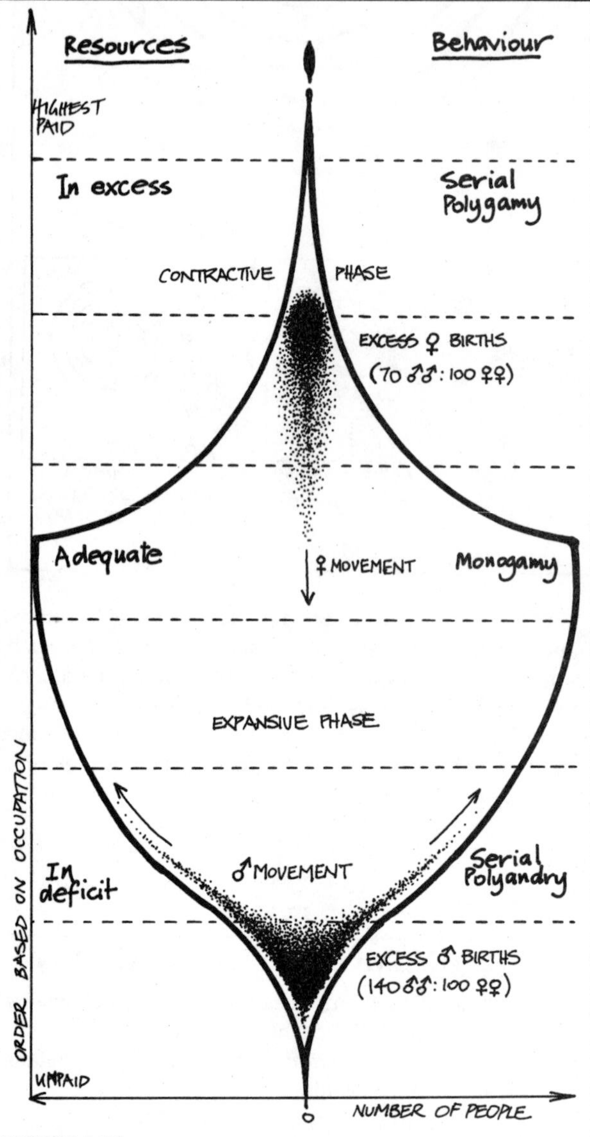

FIGURE 4.29
BEHAVIOUR PATTERNS IN SOCIETY.
The form of social hierachy based on occupational status. Note the over production of males (as a primary sex ratio) at the lower resource level, thus marital habits are based on resources and sex ratios. Width indicates number of people at each level. (n = 35,000 Tasmania, 1975).

tribal art is a public and ever–renewed mnemonic, or memory–aid. Apparently simple spiral or linear designs can combine thousands of bits of information in a single, deceptively simple pattern. The decorative function is incidental to the educational and therefore sacred information function in such patterns. "Decoration" is the trivial aspect of such art.

Much of modern art is individualistic and decorative; some "motif" art is plagiarised from ancient origins, but no longer has an educational or sacred function. Entertainment and decoration *is* a valid and important function of the arts, but it is a minor or incidental function. Social comment is a common art form in theatre and song, and spirited dances and songs are cheering or uplifting. But I know of no meaningful songs or patterns in my own "monoculture", based as it is on the jingles of advertisements and purely decorative and trivial patterns of art, and on education divorced from relevant long–term observations of the natural world.

The induction of moods and the record of ephemera are not the primary purposes of sacred or tribal art, which is carefully assembled to assist the folk records of the function and history of their society. Some modern sculptural forms, such as the "Flowform" systems of the Virbella Institute, Emerson College, Kent, UK (**Figure 4.33**) are modelled on older Roman water cascades, and serve both an aesthetic and a water–oxygenation function, assisting water purification. This is a small step towards applied art as patterning in everyday use, as are some engineering designs. We could well reintroduce or evolve pattern education, which gives *every* member of society access to profound concepts or specific knowledge.

Art belongs to, and relates to, people. It is not a way to waste energy on resources for the few. Sacred calendars melted down to bullion or objets d'art are a degradation of generations of human effort and knowledge, and the sacred art of tribal peoples hidden in museum storerooms are a form of cultural genocide, removing knowledge from its context, and trivialising objects to decorations or loot.

Human information, as a tribal art form, is most frequently debased and destroyed less for monetary gain than for the replacement of *public* information by an exotic, secretive, irrelevant and basically uninformed centralised belief system. The fanatic cares not what is destroyed if it empowers the repressive hierarchy that is then imposed. Most tribal art has been burnt, looted, destroyed, and broken by invading belief systems, destroyed by those seeking secret power rather than open knowledge, or by those who are merely destructive. Book–burning and image–breaking is the reaction of the alienated or intellectually–deprived to the accumulated wisdom of its revelationary ancestors. We most damage ourselves when we destroy information and aids to understanding.

It is a challenge to artists to study and portray knowledge in a compact, memorable, and transmissible form, to research and recreate for common use those surviving art forms which still retain their meaning, and to re–integrate such art with science and with society and its functions and needs. It is a challenge to educators to revive the meaningful geometries, songs, and dances that gave us, and our work, meaning.

4.26 ADDITIONAL PATTERN APPLICATIONS

A sophisticated application of pattern is found in the herb spiral (**Figure 4.30**) which I evolved in 1978 as a kitchen–door design. All the basic culinary herbs can be planted in an ascending spiral of earth on a 2 m wide base, ascending to 1 or 1.3 m high. All herbs planted on the spiral ramp are accessible. The construct itself gives variable aspects and drainage, with sunny

FIGURE 4. 30
HERB SPIRAL.
Pattern applied. A modest 2 m diameter by 1 m high earth spiral accommodates all necessary culinary herbs close to the kitchen door and can be watered with one 2 m sprinkler—a considerable saving in space and water as the ramp and walls exceed 9 m of plant space.

dry sites for oil–rich herbs such as thyme, sage, and rosemary, and moist or shaded sites for green foliage herbs such as mint, parsley, chives, and coriander.

This is a rare three–dimensional earth construct on a small scale, and compactly coils up a linear path or bed of herbs into one mound at the kitchen door, thus making the herbs accessible and convenient to the kitchen itself. If kitchens are not at ground level, roof or balcony gardens can carry pot–herbs in stepped walls, on wall shelves, in window boxes, or as stacks of pots in earth mounds.

Pattern analyses can also be applied to water conservation. For example, a mulch–pit (60 cm wide and deep), surrounded by a planting shelf and spill bank totalling 1.2 m (4 feet) across has a 3.8 m (12 foot) perimeter, but can be efficiently watered with *one* low–pressure sprinkler, whereas a 3.8 m straight row takes *three* such sprinklers.

Another advantage is the central (one–drop) mulch pit, so that the plants eventually overshade the centre to prevent evaporation. Such circle–mulch– grow pits are made 1.8 m (6 feet) across for bananas, and 1.8–3 m (6–10 feet) across for coconuts; all out– produce row crop for about one–third of the water use. A series or set of such gardens greatly reduce the path space and land area needed for home gardens, or orchards of banana and coconut (**Figures 4.31** and **10.26**).

A field application of patterned ground designed to direct flow, and capture materials in flow, is that of flood–plain embankments or tree lines (poplar, willow, tamarack), or both combined. These are very effective pattern impositions on landscape (although all occur naturally as rock dykes or resistant rock strata in the field) that can have several beneficial effects for a household or settlement nearby (**Figure 4.32**).

A more conscious and portable applied pattern set is that of the "Flowform" models being developed at the Virbella Institute by a small group of artist–technicians. Such turbulence basins are apparent in nature as shaped basins in streams flowing over massive sandstones or mudstone rocks. They are even in antiquity modelled in pozzelanic cement by Roman hydrologists. Flowforms are artificial replicates of the rock forms carved by turbulent streams, cast in concrete or fibreglass (**Figure 4.33**).

Stacked in sets below sewage pipe outfalls or above fish ponds at pipe inlets, they efficiently mix air and water by inducing turbulence in flow. Three distinct mixing effects are noticeable; the first a *plunge* or vertical overturn as fluid drops from one basin to another; the second as a figure–8 or lateral flow around the basins themselves; and the third (a fascinating process) as an interaction between these two, as water coursing around the basins deflects the vertical drop flow and switches it from side to side in a regular rhythm.

Within these major turbulence patterns [so clearly portrayed by da Vinci (Popham, A.E., *The Drawings of Leonardo da Vinci*, Jonathan Cape, London, 1946) and further analysed in terms of computer models and catastrophic theory by Chappell (in: *Landform Evolution in Australia*, ANU Press, Canberra 1978) for coastal uprush and backwash turbulence] are distinct vortices and counterflow, overfolds and cusps that further mix air and water at the edges of the basins and in the main flow stream.

Thus, artificial Flowform basins induce aeration, oxidise pollutants, and are themselves aesthetically pleasing and instructive hydrological pattern–models of naturally–occuring constructs. They have practical use in the primary treatment of sewage and organically polluted waters, and in the oxygenation of ponds for aquatic species production. Models of this type are the result of a long evolution beginning with wonder, sketches, analysis, observations, and then proceeding via constructed hydrological basins to practical applications over a wide variety of sites. In nature and in the Flowform system, the basins can be elongate, truncate, symmetrical, asymmetical, stepped in line, stacked like ladders, or spiralled to conserve space.

4.27
REFERENCES AND FURTHER READING

Alexander, Christopher *et al*, *A Pattern Language*, Oxford University Press, 1977.
(Instances successful design strategies for towns, buildings)

Bascom, Willard, *Waves and Beaches*, Anchor Books, New York, 1980.

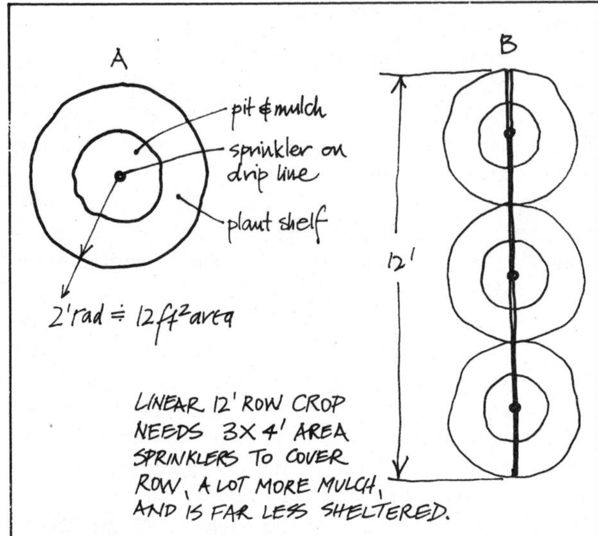

FIGURE 4.31
ROW CROP OR CIRCLE CROP
Here, a 4 m row of crop needs three 1.2 m sprinklers, while a circle of radius 0.6 m (4 m circumference) needs only one 1.2 m sprinkler, a saving of 60% in water use. Such systems apply only on the small scale, where plants can shade the inner circle.

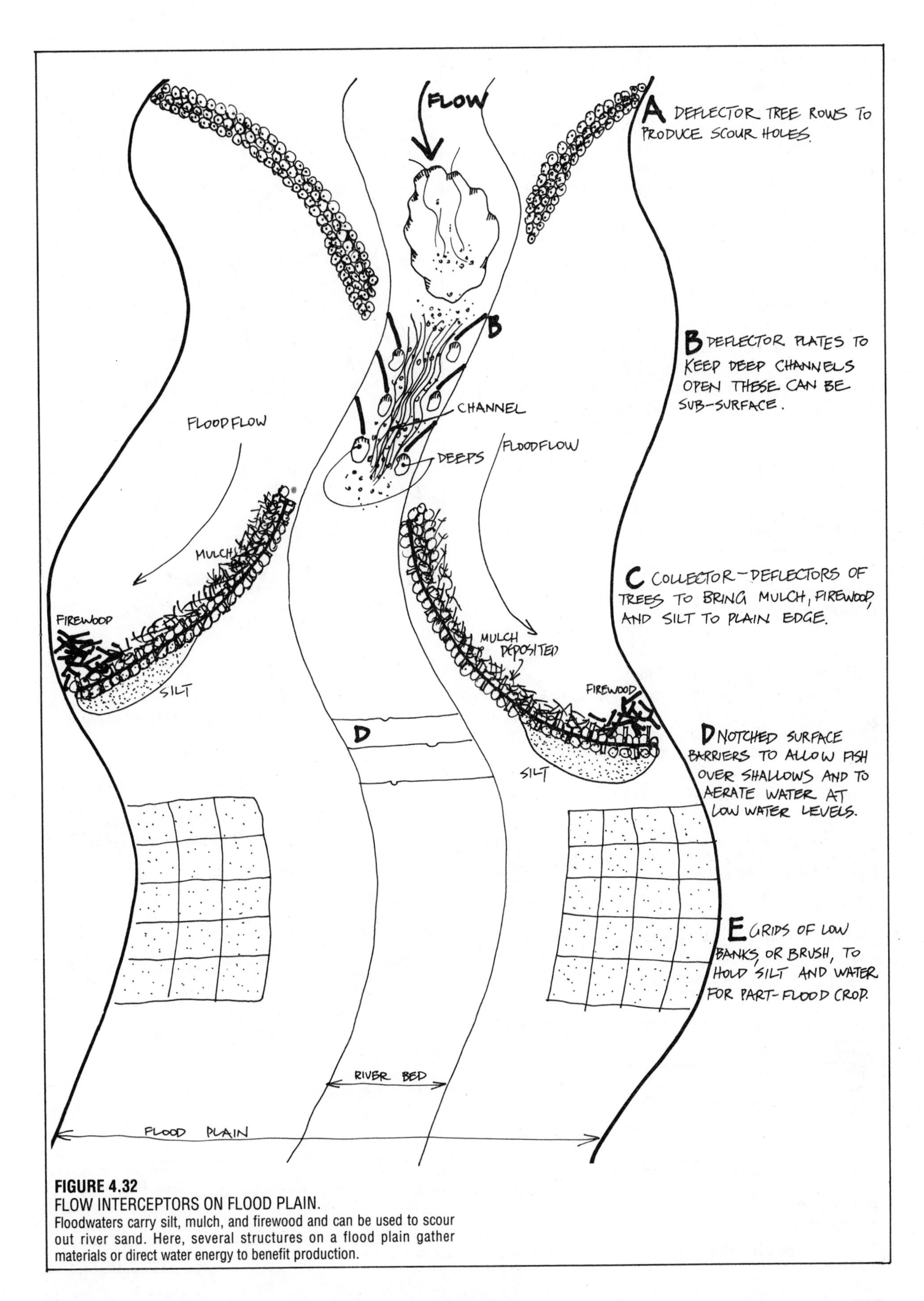

FIGURE 4.32
FLOW INTERCEPTORS ON FLOOD PLAIN.
Floodwaters carry silt, mulch, and firewood and can be used to scour out river sand. Here, several structures on a flood plain gather materials or direct water energy to benefit production.

Birch, and Cobb,*The Liberation of Life*, Cambridge University Press, 1981.

Bohm,*Wholeness and Implicate Order,* Routledge and Kegan Paul, 1980.

Capra, Fritz,*The Tao of Physics*, Fontana Press, 1976.

Cook, Sir Theodore Andrea, *The Curves of Life,* Constable, London, 1914 (reprint)

Escher, M. C. and J. L. Locher, *The World of M.C. Escher,* Harry N. Abrams Inc. New York, 1971.

Goold, J. *et alia, Harmonic Vibrations and Vibration Figures,* Newton & Co. London, 1909.
(see also: *Model Engineer* 3 May '51, 8 Sep '60; *Hobbies* Nov 1966; and*New Scientist* 22/29 Dec '83).

Illert, Christopher, *Sea Shell Mathematics,* self–published, 1984: 76 Seaview Rd., West Beach, South Australia 5024.

Lovelock, J. E.,*Gaia: A New Look at Life on Earth,* Oxford University Press, 1979.

Leapfrogs, *Curves,* Tarquin Pubs. Stradbroke, Diss, Norfolk, U.K., 1985.

Mandelbrot, Benoit, *The Fractal Geometry of Nature,* W.H. Freeman Co. New York, 1982.
(The basic book on fractals, computer graphics)

Murchie, Greg, *The Seven Mysteries of Life,* 1984, self–published: Marlborough, NH, USA.

Pearce, Peter, *Structure in Nature as a Strategy for Design,* M.I.T. Press, 1979.

Schwenke, Theodore, *Sensitive Chaos; the Creation of Flowing Forms in Air or Water,* Schocken Books, N.Y., 1976.

Thompson, D'arcy W., *On Growth and Form,* Cambridge University Press, 1952.
(Multiple examples of forms in nature, spirals)

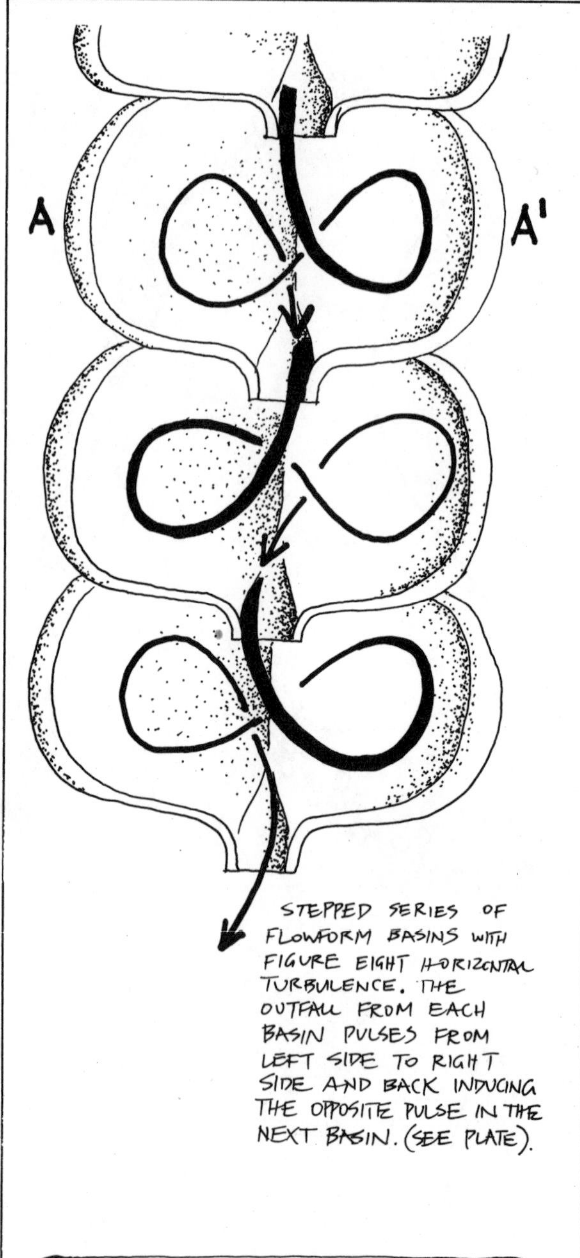

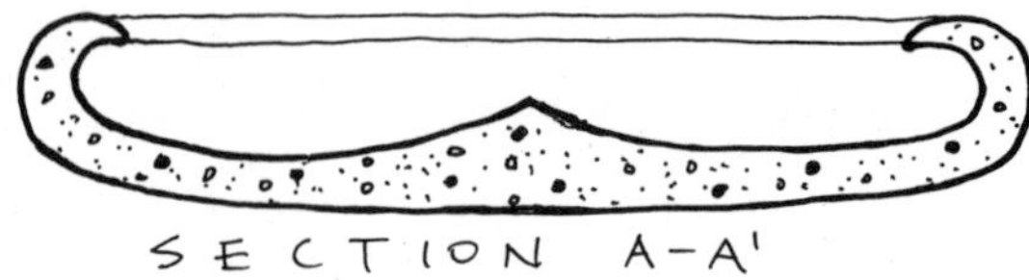

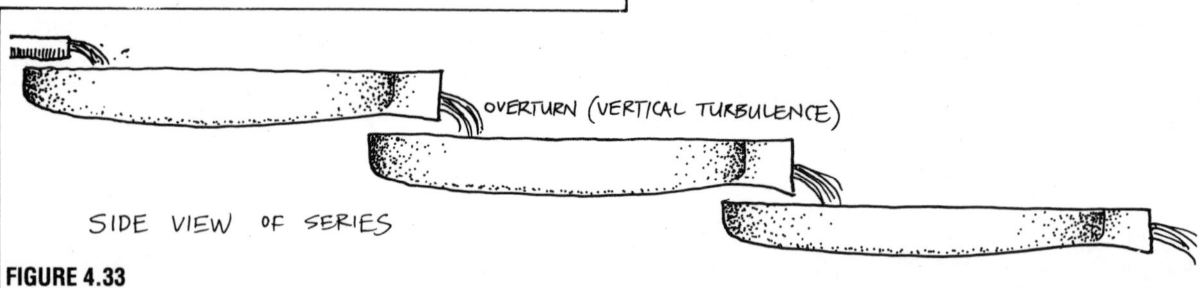

FIGURE 4.33
FLOWFORMS.
Of ancient usage, and natural occurence, can be neatly fabricated in concrete or glass reinforced plastic to aerate water in the case of a constant (or little varying) flow.

Tweedie, A. D., *Water and the World*, Thos. Nelson (Aust) Ltd., 1975.
(On the order of stream flow and stream patterns)

Virbela Institute, Emerson College, Forest Row, East Sussex RH18 5JX (Flowform designs and research, posters.)

Vogel, Steven, *Life in Moving Fluids; the Physical Biology of Flow*, Willard Grant Press, Boston, 1981.
(A sensitive and scholarly study of life forms in flow)

Weyl, Hermann, *Symmetry*, Princeton University Press, NJ, USA, 1983.

New Scientist, New Science Publications, Commonwealth House, 1–19 New Oxford St. London WC1A 1NG:

5 Oct '78 : On pulsers and biological clocks
15 Nov '84 : On the gravity anomalies in the geoid
31 May '84 : The Robinson Congruence
7 Jun '84 : Recombination of DNA in a plus–torus
20 Mar '79 : The swirling of water
14 Jun '79 : The spiral classification of galaxies
5 May '83 : Symmetry, geometry, fractals
26 Apr '84 : Fractals
5 May '83 : Inversion and reflection of forms
17 Nov '83 : On impact craters
21 Apr '77 : "The great tennis ball of earth"
11 Oct '84 : The pineal gland as a timepiece
4 Apr '85 : Fractals, pulsars, black holes

4.28

DESIGNER'S CHECKLIST

Read in pattern analysis, and study the relationships of ORDERS and FORMS in nature. Patterned systems must be of appropriate size, or of the right order (i.e. note that *small* systems operate for things like frost protection and water conservation in crop).

When designing gardens, ponds, or access ways, try to minimise waste space by using spiral, keyhole, and least–path systems, clumped plantings, and sophisticated interplants.

Study and use edge effects, especially in relation to intercrop and in the construction of plant guilds, pond production, and fail–safe species richness in variable climatic regimes.

Use appropriate patterns to direct energies on site, and to lay out the whole site for zone, sector, slope, and orientation benefits. This approach alone creates the most energy savings.

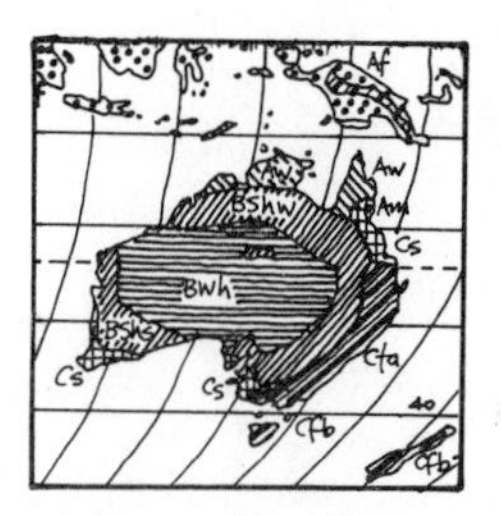

Chapter 5

CLIMATIC FACTORS

If I go out shopping, a glance is sufficient to predict if I am likely to need an umbrella. However, long-term prediction of the weather, over a scale of more than about 10 days is a thankless task. This is because the dynamics of the atmosphere form a system whose behaviour is usually chaotic. The surface of the earth absorbs heat, and so heats the atmosphere from below, and this warm air rises. Heat is lost from the upper atmosphere, and this cooled air falls. A roughly hexagonal cellular array of vortices forms, with the ascending warm air feeding the descending cool air.

(Arun Holden, *New Scientist*, 25 Apr '85.)

The glass is falling hour by hour
The glass will fall forever
But if you break the bloody glass
You won't hold up the weather.

(Louis McNeice.)

5.1 INTRODUCTION

Climatic factors have their most profound effect on the selection of species and technology for site, and are thus the main determinant of the plant, animal, and structural assemblies we can use. There is an intimate interaction between site and local climatic factors, in that slope, valley configuration, proximity to coasts, and altitude all affect the operation of the weather. Such factors as fire and wind effects are site and weather related. It is the local climate that inevitably decides our sector strategies.

Although we will be discussing the individual weather factors that define climate, all these factors interact in a complex and continuously variable fashion. Interactions are made even more unpredictable by:

- longer-term trends triggered by the relative interaction of the orbits of earth, sun, and moon;
- changes included in the gaseous composition of the earth's atmosphere due to vulcanism, industrial pollution, and the activities of agriculture and forestry; and
- extra-terrestrial factors such as meteors, the perturbations in high-level atmospheric jet streams, the oceanic circulation, by fluctuations in the earth's magnetic field, and by solar flares.

There is a general consensus that world climatic variation (the occurrence of extremes) is increasing, so that we can expect to experience successively more floods, droughts, periods of temperature extremes, and longer or very intense periods of wind.

We have separated climatic studies from that of earth surface conditions, and there are climatologists who know little of the effects of forests, industrial pollutants, agriculture, and albedo (albedo is the ratio of light reflected to that received) on the global climate. There is no longer any doubt that our own actions locally greatly affect global and local climate, and that we may be taking unwarranted and lethal risks in further polluting the atmosphere.

Because climatic prediction may forever remain an inexact science, we should always *allow for variability* when designing a site. A basic strategy is to spread the risk of crop failure by a mixture of crop species, varieties, and strategies. This fail-safe system of mixed cropping is basic to regional self-reliance, and departure from such buffering diversity brings the feast-or-famine regime that currently affects world markets.

In house design, the interactions of thermal mass (heat storage) and insulation (buffering for temperature extremes) plus sensible siting permit us to design efficient and safe housing over broad climatic ranges. Strategies such as water storage and windbreak modify extreme effects. Many plant and animal species show very wide climatic tolerances, and local cultivars are developed for almost all important food plants. The variety of food grown in home gardens varies only slightly over a great many situations.

As designers, we are as interested in *extremes* as in means (averages). Such measures as "average rainfall" have very little relevance to specific sites. Of more value are data on seasonal fluctuation, dependability, intensity, and the limits of recorded ranges of any one factor. This will decide the practical limits that need to be included in a design.

People who are called on to design or instruct over wide climatic ranges would do well to read in more general treatments such as Eyre (1971) and James (1941), or in modern biogeographical texts. These treatments deal with world vegetation patterns and climatic factors.

Total site factors related to land configuration will impose specific limits to any design; soil data will also be specific to site. There is, therefore, no substitute in any one design for local observation, anecdotes, detailed maps of local factors, lists of locally successful plant and animal species, and analysis of local soils.

It is obligatory for any designer to study the regional long–term human and agricultural adaptations to climate. Above all, we should avoid introducing temperate (European) techniques and species to tropical and arid lands on any large scale. Aboriginal peoples were never so "simple and primitive" as we have been led to believe by the literature of their invaders. Native agricultural and pastoral management practices are often finely tuned to survival, are sometimes very productive, and above all are independent of outside aid.

5.2
THE CLASSIFICATION OF BROAD CLIMATIC ZONES

Most global climatic classifications are based on precipitation–radiation interactions as formulated by Vladimir Koppen (1918), and subsequently modified and updated by authors such as Trewartha (1954). **Figure 5.1** is from the latter reference. More closely–defined plant lists can be given by reference to the "Life Zone" matrix developed by Holdridge (**Figure 5.2**), which has enabled James Duke and others to annotate plant lists with concise climatic keys. Many plant compendia attach "zones of hardiness" to plant listings, commonly used in the USA. As given in *Hortus Third*, the zones are in **Table 5.1**.

Measures or cut–off points are usually chosen that approximate the limiting boundaries for life forms, and are mainly good approximations of lethal or optimum ranges. The main qualifying factors on the broad classification of climatic factors are:

- special mountain conditions;
- the modifying effects of coasts (and the extremes of continental interiors);
- local energy transfer by winds and oceanic currents; and
- long–term cyclic factors.

Some problems in this area are:

1. Instruments for accurate measurement are expensive, and often specific to a narrow range of the total spectrum of effects.

2. Averages in such areas as precipitation and radiation often refer only to one part of the total spectrum. We have few long–term records of fog precipitation, dew, long–wave radiation, ultraviolet incidence, or gaseous atmospheric composition.

3. We are aware that rain, sun, and wind interact in a dynamic and continuous fashion, so that averages mean little to a plant or animal subject to *the normally changeable effects* that may cover wide ranges of interactive measures.

In this chapter, we are concerned only with the very broad climatic zones (design specifics for each climatic zone are given in later chapters). These have been grouped as follows:

- TROPICAL: no month under 18°C (64°F) mean temperature, and SUBTROPICAL: coolest months above 0°C (32°F) but below 18°C (64°F) mean. In effect, frost–free areas.
- TEMPERATE: coldest months below 0°C (32°F), warmest above 10°C (50°F) mean temperature, to POLAR: warmest month below 10°C (50°F) or in perpetual frost (8°C or less) mean.
- ARID: mean rainfall 50 cm (19.5 inches) or less to DESERT: mean rainfall 25 cm (10 inches) or less. Includes sub–humid, or any area where evaporation exceeds precipitation.

5.3
PATTERNING IN GLOBAL WEATHER SYSTEMS; THE ENGINES OF THE ATMOSPHERE

Dense cold air flows continually off the polar ice caps. This high–pressure or down–draught air spirals out of the polar regions as persistent easterlies which affect high latitudes (60–80°) near the ice–cap themselves. Long spokes of this air curve outward to Latitude 30°.

As the spiral itself is caught up in (and generated by) earth spin, these cold cells of air drive a series of contra–rotating low–pressure cells (turning clockwise in the southern hemisphere and anti–clockwise in the northern). These in turn mesh with rotating spirals of high–pressure air which have risen at the equator, and are falling at Latitudes 15–40°.

The high–pressure mid–latitude cells turn anti–clockwise in the southern hemisphere and clockwise in

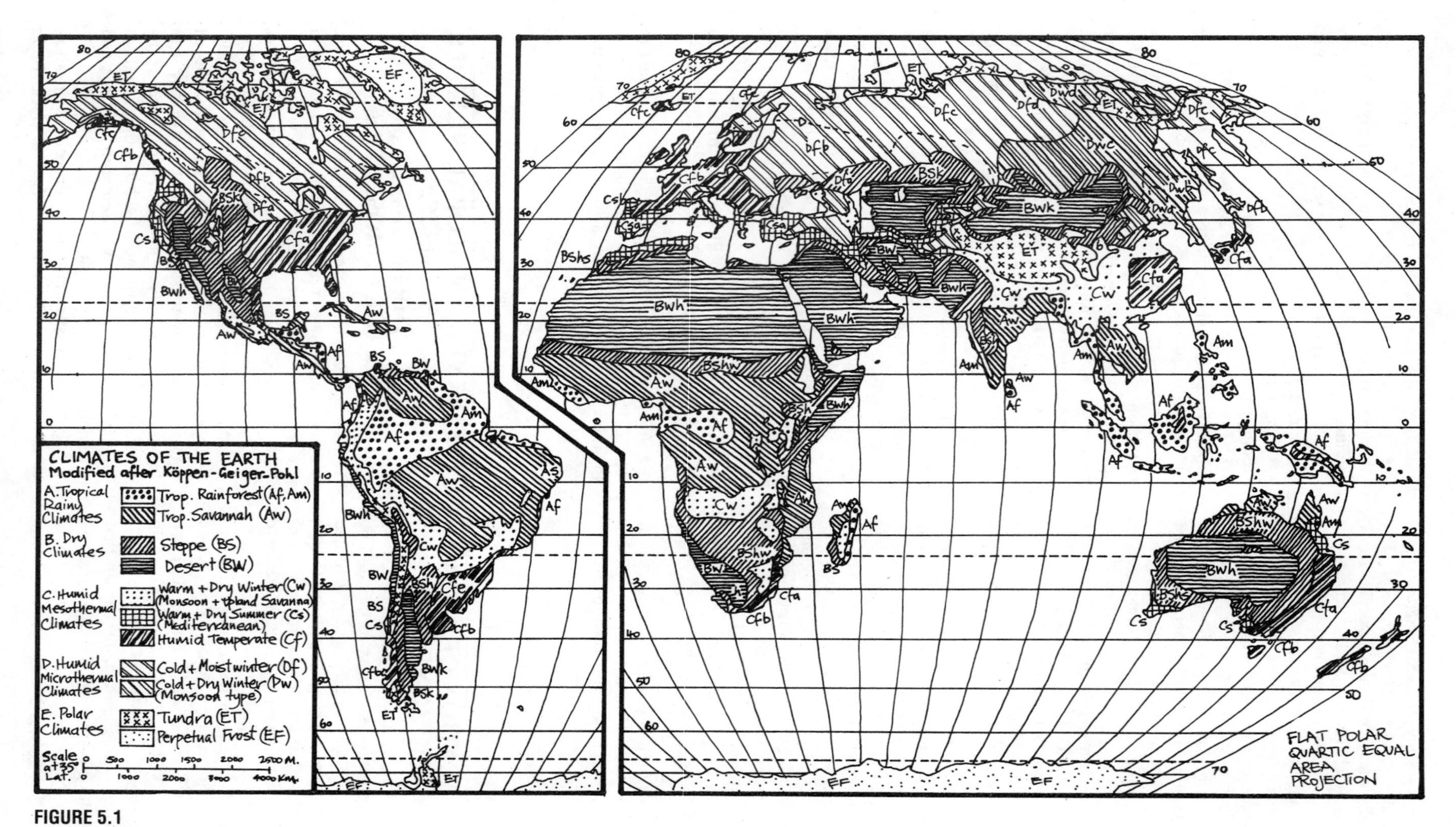

FIGURE 5.1
KOPPEN CLIMATE CLASSIFICATION.
A basic world classification; mionor subdivisions are specified in detailed maps or basic references.

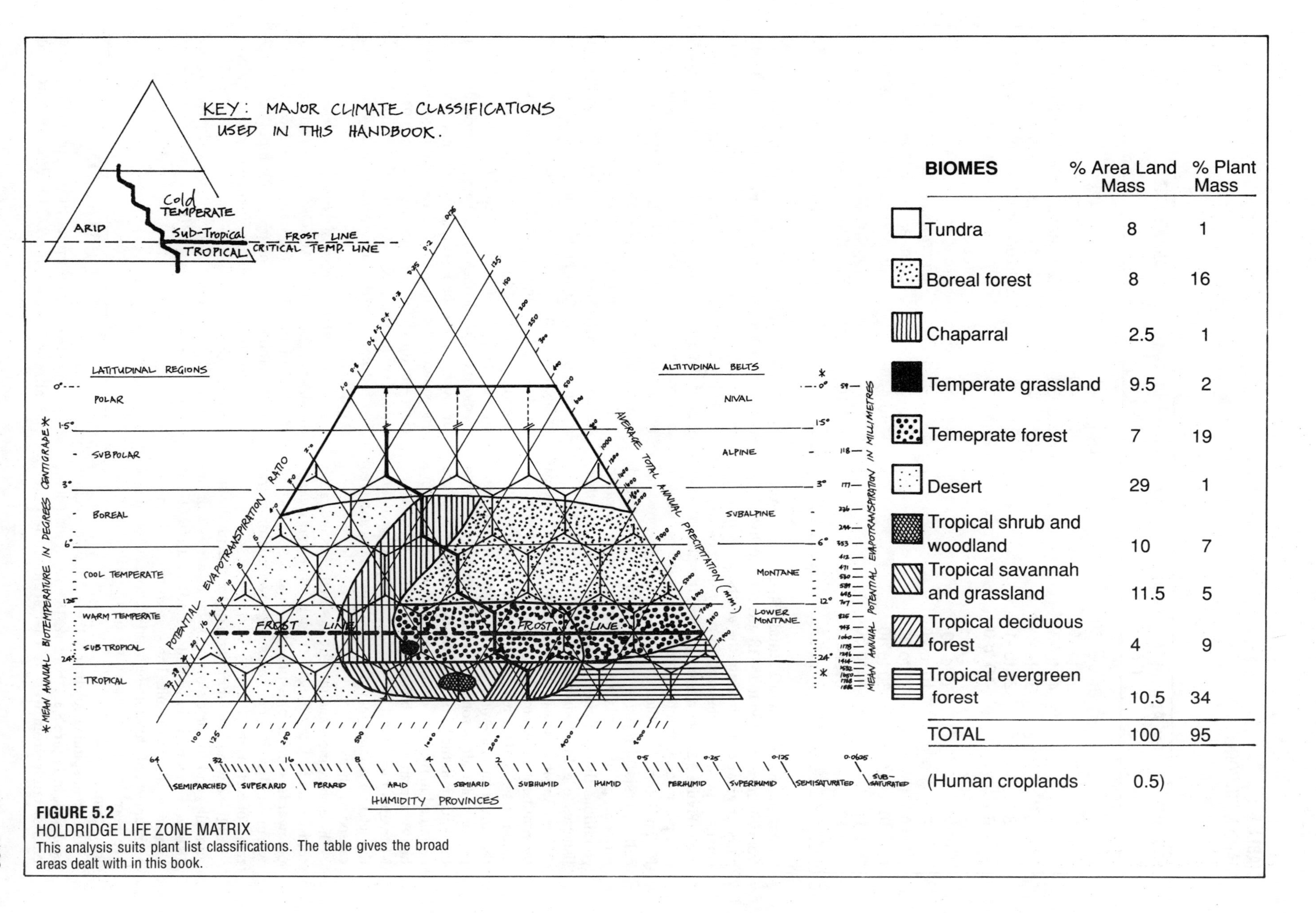

BIOMES	% Area Land Mass	% Plant Mass
Tundra	8	1
Boreal forest	8	16
Chaparral	2.5	1
Temperate grassland	9.5	2
Temeprate forest	7	19
Desert	29	1
Tropical shrub and woodland	10	7
Tropical savannah and grassland	11.5	5
Tropical deciduous forest	4	9
Tropical evergreen forest	10.5	34
TOTAL	100	95
(Human croplands	0.5)	

FIGURE 5.2
HOLDRIDGE LIFE ZONE MATRIX
This analysis suits plant list classifications. The table gives the broad areas dealt with in this book.

TABLE 5.1
HARDINESS ZONES

ZONE	Av. Annual Min. Temp. (°F)	Av. Annual Min. Temp. (°C)	COMMENTS
1	Below -50	Below -45	Arctic tundra
2	-50 to -40	-45 to -40	Cold prairie and conifers
3	-40 to -30	-40 to -34	Conifers and mixed forests
4	-30 to -20	-34 to -29	Cold interiors of continents
5	-20 to -10	-29 to -23	Mixed forests, cool prairies
6	-10 to 0	-23 to -18	Broadleaf and deciduous forests
7	0 to 10	-18 to -12	Broadleaf forests
8	10 to 20	-12 to -7	Arid grasslands, savannah
9	20 to 30	-7 to -1	Semi-arid coasts and basins
10	30 to 40	-1 to 4	Sub-tropical, palms, coasts
11	40 to 50	4 to 10	Tropical forests, deserts
12	over 50	over 10	Equatorial rainforests, monsoon

the northern. Thus from Latitude 50–20°, and in the "roaring forties", about 15–18 alternating high–low pairs of great cells circulate the earth, all of them as smaller spiral systems around the great polar spiral itself (**Figure 5.3**). On westerly coasts, the alternation of cold polar and warm high pressure air arrives at about 10–day intervals, although some great high–pressure cells persist in place, thus blocking westward movement of winds and creating static oceanic conditions that can affect oceanic over-turn, and thus fisheries (e. g. the *el Niño* effect).

These great processions are disturbed and deflected by continents, stubborn high–pressure cells over cool land masses, and the relative intensity of the air cells, so that irregular cold–warm fronts arrive at any one site. Just as polar air is sometimes drawn strongly towards the equator in the lows, so warm tropical air masses are entrained in the outer circulation of the highs and bring heavy warm rains towards the poles. High level jetstreams may speed up or block this procession and the jetstream itself may also break up under stresses caused by shear.

The disturbances and impedences in the system cause cold fronts to pile up against each other and deflect polewards at high–pressure cells, and a sequence of warm– and cold–front rains (the cyclonic or spiral rains) of earth results.

All these wind belts shift north or south with the sun annually, and to some slower extent as a result of the 18.6–year moon cycle, so that periods of drought and excessive rain can result. The system appears chaotic, and subject only to short–term prediction, but of late we are learning to assess some of the effects of the long–term cycles.

The great spiral circulation of the south polar regions is shown in **Figure 5.3**. About 12–18 cold fronts (cloud bands) circle from west to east around the poles, arriving as "cyclonic fronts" every 10 or so days on coasts in that region. They affect areas up to 30° south, with four or so large fronts continuous with (and probably driving) cloud up to 10° south or north latitude, mostly along the western margins of South America, Africa, and the south Atlantic. It is now clear that it is the *oceanic* circulation that drives the air masses, rather than the opposite.

The fronts are dragged in a curve to the west as the earth spins to the east. Each cloud front is a result of the meeting of cold polar and warm sub–polar air masses or high–pressure cells. The low–pressure areas rotate clockwise, the highs anti–clockwise in a series of cog–like spirals or tori that travel every 3–4 months around the poles. Rotation is in the opposite sense in the northern hemisphere. It is the cold, dense, dry polar air sweeping off the ice–caps, and the hot rising air of the equatorial calms which drives these great wheels; clear–air (descending) intrusions are of hot–dry and cold–dry continental air (Australia, Africa) or air descending from the equatorial (rising) congruence (**Figure 5.4**).

In the next sections of this chapter, I will be discussing CLIMATIC FACTORS under parts, as below:

- Precipitation (rain, fog, dew, evaporation–**5.4**);
- Radiation (light, heat, frost, solar input–**5.5**);
- Winds (normal winds, hurricanes and tornadoes - **5.6**);
- Landscape effects (altitude, valleys, slopes–**5.7**); and
- Latitude–altitude factors (**5.8**).

5.4 PRECIPITATION

There are two basic inputs to precipitation: that of rainfall, snow, and hail (WATER FALLING from the clouds), and that of CONDENSATION (water condensed or trapped from sometimes clear air or fogs by cool surfaces). Although the latter may be of critical importance on seaward slopes and at higher altitudes (cloud forests), the only reliable and widespread measures we possess are of "rainfall". World rainfall averages about 86 cm (34 inches). While we may take 50 cm (20 inches) of rainfall or less as semi–arid, and 25 cm (10 inches) or less as arid and desert, we can locally experience seasonal or relative aridity due to long–term cycles and weather effects caused by periodic fluctuations in jetstreams or oceanic currents *in any climate.* Longer periods of increased aridity can also be caused by deforestation on a broad or local scale.

It is because of the potential for changes in precipitation that we give so much space in later chapters to water storage strategies and the conservation of water. Water promises to be the main limiting factor for survival and growth, and the major future expense of food gardens and agriculture. Thus, any strategy we can adopt to generate, conserve, or store water is critical to our design approach. Any gardener knows that climatic averages are at best a very general guide to precipitation effects in the garden or orchard. It is a much safer strategy to see to it that both the species chosen and water strategies developed ensure some yield in "drier than usual" conditions. After all, a fish population out of water for an hour is as dead as if a year–long drought were in effect.

Our annual gardens and crops are also susceptible to short–term changes in available water. People live, and garden, in average annual rainfalls of 10 cm (4 inches) or less, and they manage to both exist and produce crops. Exotic (non–local) water enters dry regions as rivers and underground aquifers, and this enables us to make judicious use of that water and to implement a great variety of local strategies to cope with the lack of actual rainfall.

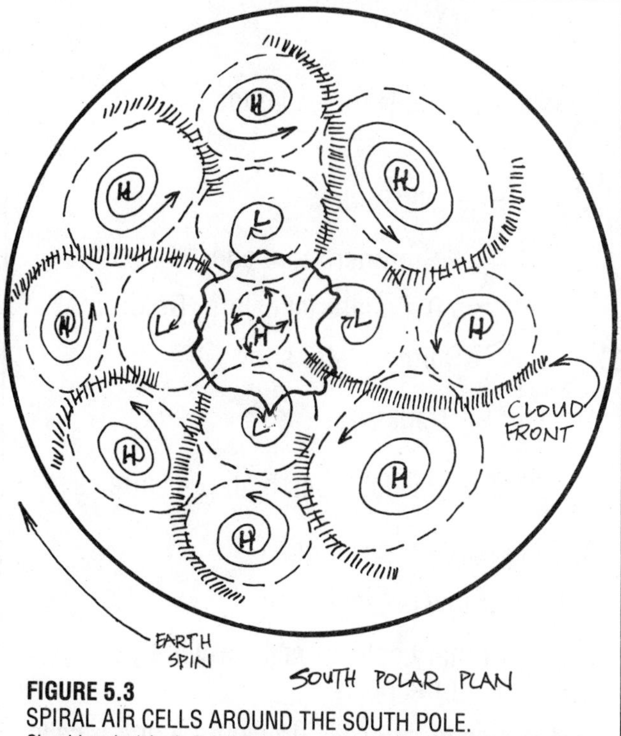

FIGURE 5.3
SPIRAL AIR CELLS AROUND THE SOUTH POLE.
Cloud bands (shaded) bring rains on the E–NE sides of high pressure, and the S side of low pressure cells. Earth spin produces a drag effect. This pattern affects climate to 25° Latitude, when cyclones feed the system.

Rainfall averages are best used as broad indicators rather than as definable limiting factors. Of far more use to us is the expected DISTRIBUTION of rainfall (including extremes such as 100–year flood records) and data on the INTENSITY of rains, as these factors are a limiting influence on the size of road culverts, dam spillways, and the storage capacity needed to see us over dry periods. Flooding histories of sites and districts often indicate the real limits to the placement of plant systems, fences, and buildings, so that attention to flood records avoids future costs and disappointment If flood data is omitted, life itself can be at risk in intense periods of rain.

As precipitation rises, available light decreases. Thus, in extremely cloudy industrial or fog–bound humid climates, *light* becomes the limiting factor for some plants to ripen or even flower. At the dry end of the rainfall spectrum (as we reach 50 cm or 20 inches mean rainfall) sun is plentiful and *evaporation in excess of precipitation* becomes the limiting factor. That factor determines our arid–land storage strategies, just as the depth of seasonally frozen soils and ice cover determines water reticulation strategies in cold climates.

Rainfall is conveniently distinguished by the processes causing rain as:

- OROGRAPHIC: the cooling of air as it rises over mountains or hills.
- CYCLONIC or FRONTAL: the over–riding of cool and warm air masses of the polar circulation.
- CONVECTIONAL: columns of hot air rising from deserts or oceans into cooler air.

Apart from rain, we have dew and fog. DEW is a common result of clear nights, rapid radiation loss, and a moist air mass over coasts and hills. It occurs more frequently in clear–sky deserts than in cloudy areas, and a slight wind speed (1–5 km/h) assists the quantity deposited. Both still air and strong winds reduce dewfall. Intensity of deposition is greatest 3–100 cm above ground level; the highest deposition due to areas of dry ground, the lower due to wet earth, which chills less quickly.

Not to be confused with dew (a radiation heat loss effect from earth with *clear* night skies) is the moisture found on leaves above warm damp ground on *cloudy* nights. This is either GUTTATION (water exuded from the leaves) or DISTILLATION from rising ground vapour; it represents no net gain to total precipitation. The waters of guttation cling to the tips of leaves, dew to the whole leaf area.

Only in deserts is the 4–5 cm (1–2 inches) of dew per year of any significance in precipitation. Dew in deserts can be regarded as an accessory to, rather than a replacement for, trickle irrigation. Dew may be captured by building piles of *loosely–stacked* stones, where low night winds cool rock surfaces and dew can accumulate to dampen the ground below. In the Negev desert and other dry areas, some plants are associated only with these dew condensers. Each mound of stones may suffice to water one tree (**Figure 5.5**). Very large radiation traps, such as those on Lanzarote in the Canary Islands (**Figure 5.6**) may grow one grape vine in each hole.

The most efficient dew–collectors are free–standing shrubs of about 1–2 m (3–6 feet) in height. Groups or solid stands of plants and grasses do less well in trapping dew, and this may help to explain the discrete spacing of desert plants, where perhaps 40% more dew is trapped on scattered shrubs than would be caught in still air, or on closed vegetation canopies.

It is possible to erect metallic mesh fences 1 m (3 feet) or so high, and to use these as initial condensers in deserts, growing shrubs along the fence drip–line, and moving the fence on after these plants are established. In Morocco such fences are proposed for deforested coastal areas.

FOG forms where warm water or the vapour of warm rain evaporates into cool air, or where cold ground chills an airstream and condenses the moisture. Chang (1968) concisely differentiates between:

1. RADIATION GROUND FOG: where, on clear nights, hollows and plateaus cool rapidly and fog forms, often in much the same pattern as the frosts of winter.

2. ADVECTION FOG: where cold offshore currents condense the moisture in warm sea airstreams. These are the coastal and offshore fogs that plague many coasts such as that of Newfoundland and parts of northwest Europe.

3. UPSLOPE OR OROGRAPHIC FOG: where warm, humid airstreams are carried up hill slopes, and condense as the air cools.

Unlike dew, fogs can provide a great quantity of moisture. Chang gives figures of 329 cm (128 inches) for

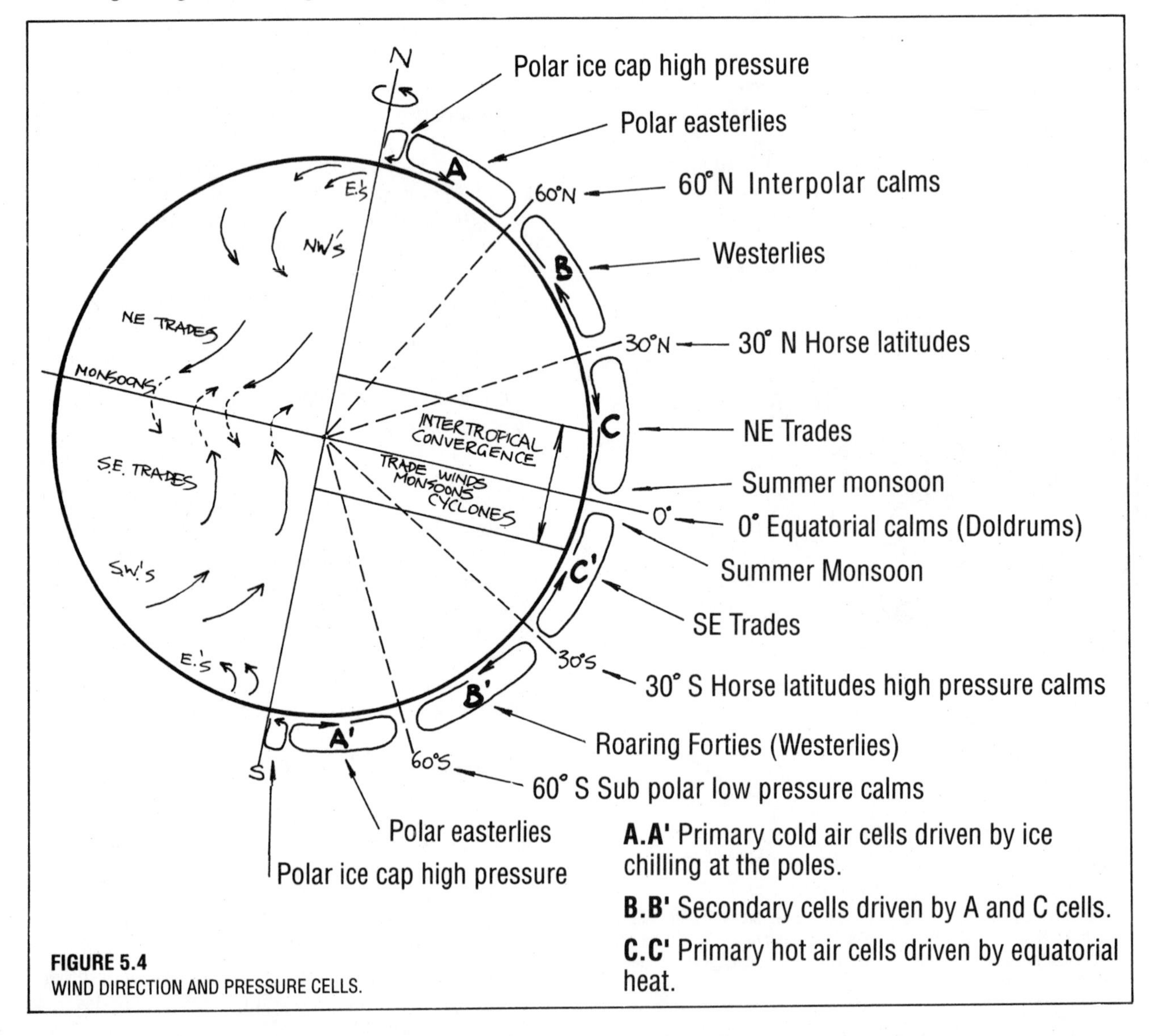

FIGURE 5.4
WIND DIRECTION AND PRESSURE CELLS.

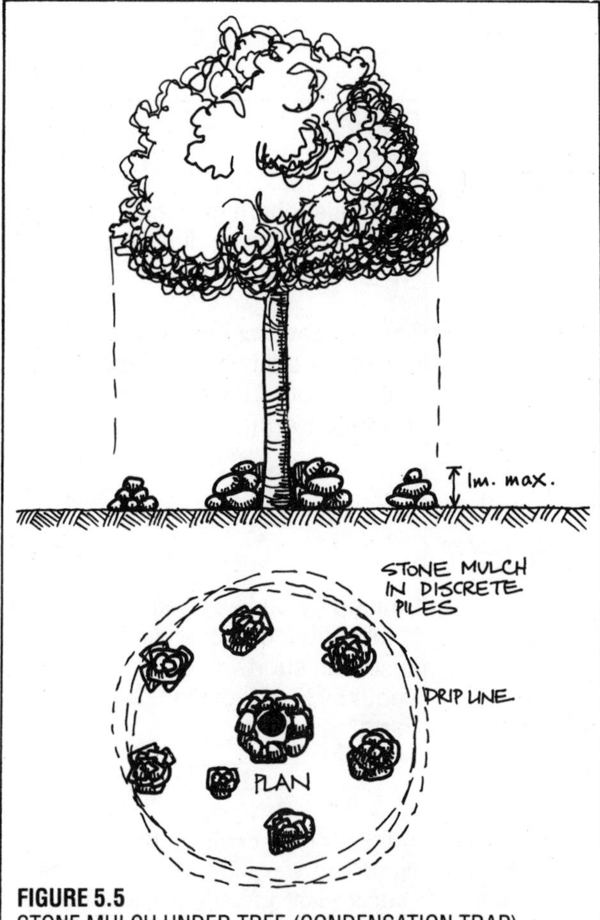

FIGURE 5.5
STONE MULCH UNDER TREE (CONDENSATION TRAP).
Loose piles of stone of less than 1 m high condense moisture from night air movement. Free air flow is permitted between stone piles.

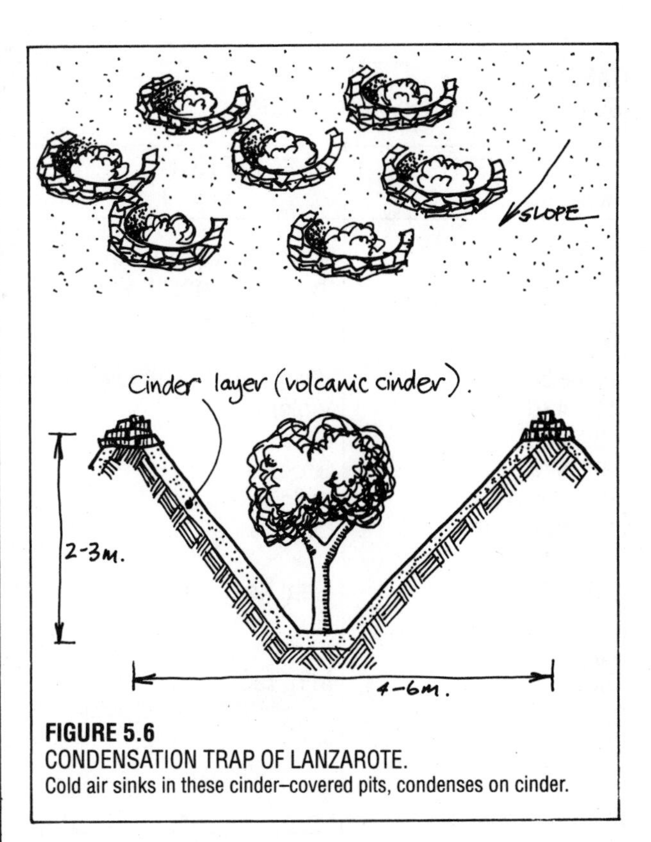

FIGURE 5.6
CONDENSATION TRAP OF LANZAROTE.
Cold air sinks in these cinder–covered pits, condenses on cinder.

Table Mountain, South Africa, and 127 cm (50 inches) for Lanai (Hawaii) from fog drip alone. In such areas, even field crops may thrive without irrigation. Typically, bare rock and new soil surfaces are colonised with lichens and mosses on sea–facing slopes, while rainforest develops on richer soils. Much of New Zealand experiences upslope fog precipitation, and unless burnt or cleared to tussock grasslands, dense forests will develop; the irregular canopy of such forests are excellent fog condensers. Even with no visible fog, trees will condense considerable moisture on sea–facing slopes with night winds moving in off warm seas over the land, and encountering the cool leaf laminae of forests.

In the very humid air of fog forests, giant trees may accommodate so much moisture, and evapotranspiration is so ineffective if fogs and still air persist, that more large limbs fall in still air than in conditions of high winds (which tend to snap *dry* branches rather than living limbs). It is an eerie experience, after a few days of quiet fogs, to hear a sudden "thump!" of trees in the quiet forests. Almost permanent condensation fogs clothe the tops of high oceanic islands, and hanging mosses and epiphytes rapidly develop there, as they do at the base of waterfalls, for the same reasons (free moisture particles in the air).

5.5
RADIATION

SOLAR RADIATION

Incoming global radiation has two components: DIRECT SOLAR RADIATION penetrating the atmosphere from the sun, and DIFFUSE SKY RADIATION. The latter is a significant component at high latitudes (38° or more) where it may be up to 30% of the total incoming energy. Near the poles, such diffuse radiation approaches 100% of energy. We have reliable measures only of direct solar radiation, as few stations measure the diffuse radiation which occurs whenever we have cloud, fog, or overcast skies.

Light and heat are measured in WAVELENGTHS, each set of which have specific properties. We need to understand the basics of such radiation to design homes, space heaters, and plant systems; to choose sites for settlement; and to select plant species for sites. **Table 5.2** helps to explain the effects of differing wavelengths.

A minor component of terrestrial radiation at the earth's surface is emitted as heat from the cooling of the earth itself. The greater part of the energy that affects us in everyday life is that of radiation incoming from the sun.

Of the incoming or short–wave radiation (taken to be 100% at the outer boundary of the atmosphere):

• 50% never reaches the earth directly, but is scattered in the gases, dust, and clouds of the atmosphere itself.

TABLE 5.2
SOME WAVELENGTHS AND EFFECTS

WAVELENGTH (Millimicrons)	**DESCRIPTION**	**COMMENT**
0-400	Actinic or Ultra Violet: only 1.5-2% reaches the Earth, most being absorbed by the ozone layer.	Causes sunburn, skin cancers. May be increasing due to ozone layer destruction.
400-626	Visible light (white light) composed of:	The rainbow colours visible as differentiaited by water vapour or a prism. About 41% of radiation reaching Earth.
400-435	Violet	
436-490	Blue	
491-574	Green	
575-595	Yellow	
595-626	Orange	
627-5,000	Heat (long wave radiation) and radio waves	
627-750	Red	
751-3,000	Far red	Emitted by bodies heated by combustion, or those which have absorbed short wavelengths. About 50% of radiation reaching Earth.
3,001+	Infra red	

Of this 50%:

— half is reflected back into space from the upper layers of cloud and dust.

— half converts (by absorption) into long–wave or heat wavelengths, within the dust and clouds that act as a sort of insulation blanket for earth.

• 50% reaches the earth as direct radiation, mostly falling on the oceans. Of this 50%:

— 6%, a minor amount, is again lost as reflection to space.

— 94% is absorbed by the sea, earth, and lower atmosphere and re–emitted as heat or converted to growth.

Of the outgoing, or terrestrial, radiation (absorbed solar radiation and earth heat; including the added heat released by biological and industrial processes and condensation), the heat that drives atmospheric circulation:

67% is re–radiated to space, and lost as heat. In the atmosphere, therefore, most heat is from this re–radiated heat derived from the surface of the earth.

• 29% is released from condensing water as sensible heat.

Ozone in the upper atmosphere absorbs much of the ultraviolet light, which is damaging to life forms. Carbon dioxide, now 3–4% of the atmosphere, is expected to rise to 6%, and cause a 3°C (5.5°F) heating of the earth by the year 2060. This process appears to be already taking effect on world climate as a warming trend, and will cause sea level changes.

The effect of radiation on plants is different for various wavelengths, as in **Table 5.3**.

Other sources of light for the earth are the moon (by reflection of sunlight) and star light. Although weak, these sources do affect plant growth, and even fairly low levels of artificial light affects animal and plant breeding. The major effects of radiation overall are:

• PHOTOSYNTHESIS in plants, the basis of all life on earth.

• TEMPERATURE effects on living and inorganic substances, much used in house design.

• FLOWERING or GERMINATION effects in plants, of basic importance to the spread of specific plant groups; this includes the day–length effect.

Plants actively adjust to light levels by a variety of strategies to achieve some moderate photosynthetic efficiency. They may keep the balance between heat and light energies by adopting solar ranges to suit the specific environment (silvery or shiny leaves where heat radiation is high; red leaves where more of the green spectrum is absorbed and less heat needed). Leaves may turn edge–on when light and heat levels get too high, or greatly enlarge their surface area under a shady canopy. Trees have larger leaves at the lower layers.

COLOUR

When we look at any object, we see it by receiving the wavelengths of the light it REFLECTS or screens out. Thus, many plants reflect green/blue wavelengths, while flowers reflect a wider spectrum of light, becoming conspicuous in the landscape. About 10% of light penetrates or is transmitted by foliage, although the canopy of rainforests in very humid areas (tropical or temperate) may permit only 0.01% of light to pass through to the forest floor. Absorbed light, as heat, is re–radiated or used in growth.

In addition to leaf colour, plants have bark surfaces ranging from almost white to almost black, the latter good absorbers and heat radiators, the former good reflectors. Leaf surfaces may vary from hard and shiny to soft, rough, and hairy. Typically, waxy leaf surfaces are found in coastal or cold areas, and in some understory plants, while woolly leaf surfaces are found in deserts and at high altitudes. The waxiness often gives a greater reflection of light regardless of colour, while dark or rough surfaces absorb light, so that dark evergreen trees become good radiators of heat.

All of these factors (colour, reflection, heat radiation) are of as much use in conscious design as they are in nature, and can be built in to gardens or fields as aids to microclimatic enhancement.

ALBEDO AND ABSORPTION

The albedo (the reflected light value) of plants and natural surfaces determines how they behave with respect to incoming radiation. The light *reflected* goes back into the atmosphere, or is absorbed by nearby surfaces and by structures such as greenhouses. The light *absorbed* is converted into long–wave radiation, and is re–emitted as heat **(Figure 5.10)**. Soils and similar dense materials normally absorb heat from

TABLE 5.3
THE EFFECTS OF RADIATION ON PLANTS IS DIFFERENT FOR VARIOUS WAVELENGTHS.

WAVELENGTH (Millimicrons)	**DESCRIPTION**	**EFFECT**
0-280 (UV)	UV or actinic	Kills plants and animals. Germicidal.
281-315 (UV)	UV or actinic	Detrimental to plants, growth.
316-400	Violet	Plants shortened, leaves thickened.
410-510	Violet-Green	Strong absorption and growth in plants. Effective photosynthesis. Transmitted by fibreglass, several plastics.
511-610	Green-Orange	Low growth and photosynthetic effect.
611-720	Orange-Red	Strong absorption and photosynthesis, photoperiodic behaviour=day length effect.
71-1,000	Red-Far Red	Plants elongate, important for seed germination, flowering, photoperiodism, fruit colouration.
1,000+	Infra Red	Absorbed and transpired into heat by plants; no strong growth effects.

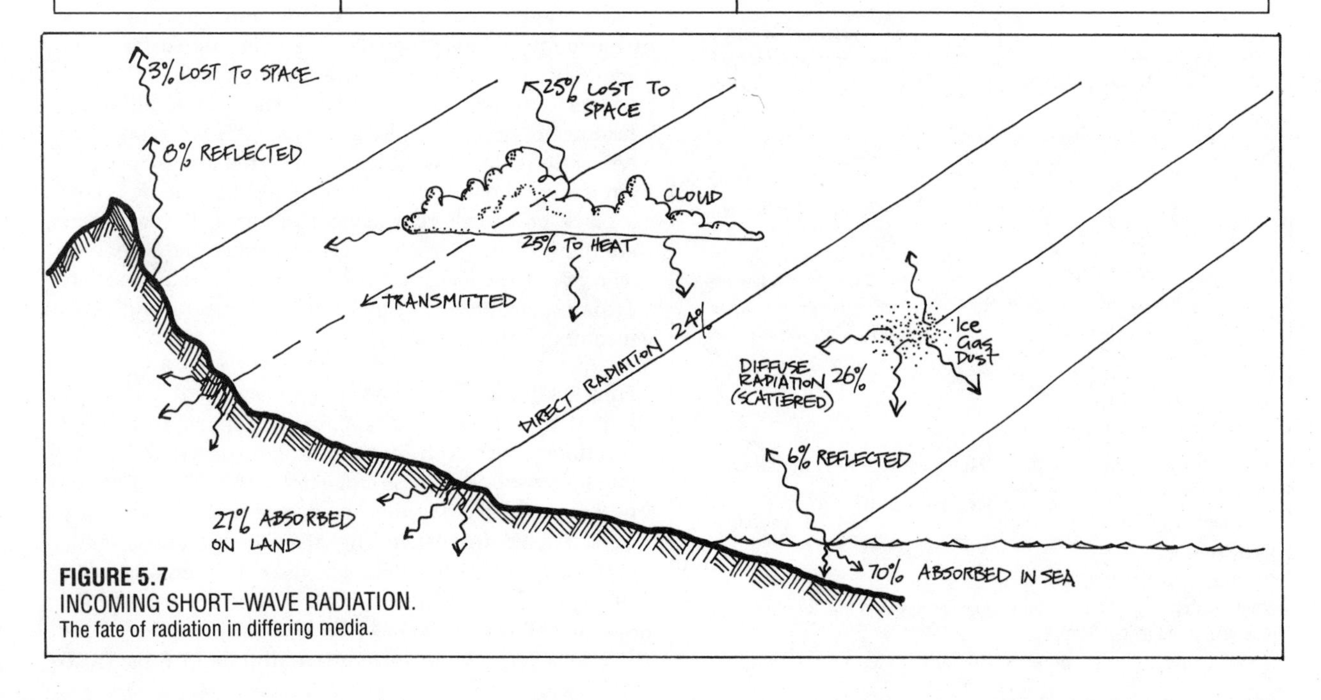

FIGURE 5.7
INCOMING SHORT–WAVE RADIATION.
The fate of radiation in differing media.

daytime radiation to a depth of 51 cm (20 inches) or so. As this takes time, the build–up of soil heat lags a few hours behind the hourly temperatures. Re–radiation also takes time, so that such absorbing surfaces lose heat slowly, lagging behind air temperatures. Thus we have our lowest soil temperatures just after dawn. The radiation loss at night produces frost in conditions of still air [in hollows, on flats, and in large clearings of 9–30 m (30–100 feet) across or more in forests]. Some frost (ADVECTION frost) flows as cold air down hill slopes and valleys to pool in flat areas. Frost forms rapidly on high plateaus. Dense autumn fogs often indicate the extent of winter frosts, and are clearly seen from high vantage points.

As designers, we use water surfaces, reflectors, and specific vegetational assemblies for forest edges. **Table 5.4** gives an indication of the value of diffuse reflectors, as **albedo**. A perfect reflector refuses 100% of light (mirrors); a perfect absorber is a BLACK BODY that absorbs all light and converts it to heat.

The fate of incoming waves encountering an object or substance is either:

- REFLECTED: turned away almost unchanged, as light off a flat mirror or off a white wall.
- REFRACTED: sharply bent or curved, as is light in water, images in curved glass, or sea waves around a headland.
- ABSORBED: soaked in, as when a black object soaks up light. This changes the wavelength (light to *heat* or short to *long* wavelength). All absorbed light is emitted as heat.
- TRANSMITTED: passed through the object.

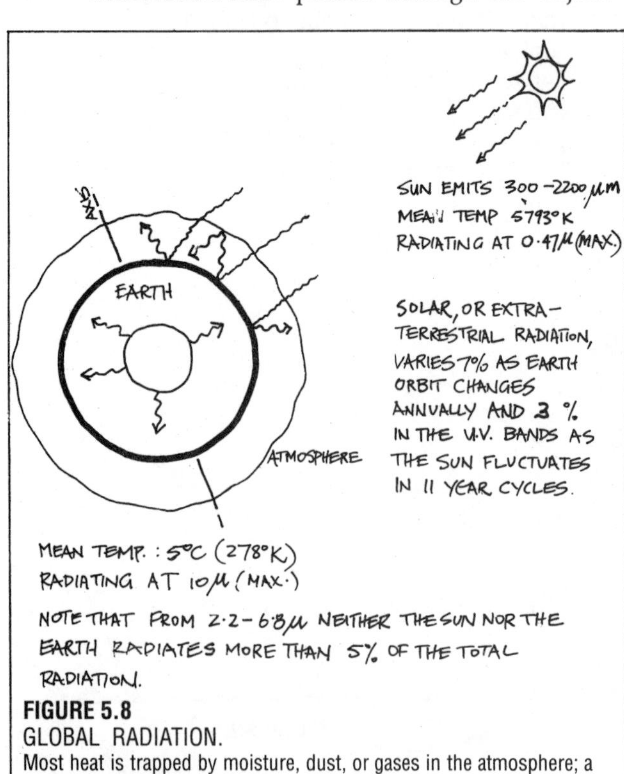

FIGURE 5.8
GLOBAL RADIATION.
Most heat is trapped by moisture, dust, or gases in the atmosphere; a little is emitted by the Earth itself.

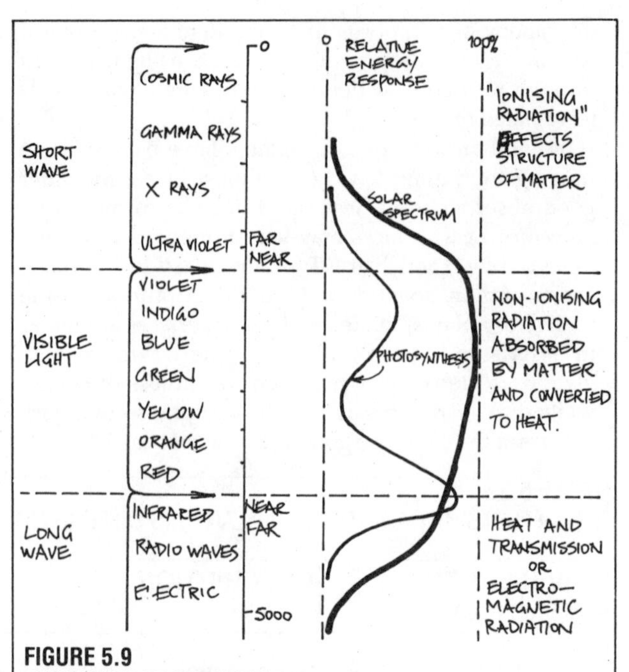

FIGURE 5.9
THE SUNLIGHT SPECTRUM (0–5,000 A)
Plant growth (photosynthesis) relies on two narrow bands of light, is damaged by extreme ultraviolet or heat wavelengths.

Different substances pass on, or are "transparent" to, different wavelengths due to their molecular structure.

Thus it is by our choice of the materials, colours, or shapes of fabricated or natural components that we manipulate the energy on a site. We can redirect, convert, or pass on incoming energy. The subject of radiation ties in with areas of technology as much as with natural systems, and this section will therefore serve for both areas of effect.

The earth itself acts like a "black body", accepting the *short* wavelengths from the sun, and emitting after absorption the *long* wavelengths from the surface and atmosphere. **Table 5.2** deals mainly with the short wavelengths, as they are those coming in as light and heat from the sun. The long wavelengths we experience are those re–radiated to earth from the atmosphere, or emitted by the hot core of earth. Curiously, snow is also a black body in terms of heat *radiation*. Black objects such as crows or charcoal can become effective *reflectors* if their shiny surfaces are adjusted to reflect radiation (a crow is black only at certain angles to incoming light).

HEAT (Longwave radiation)

It is difficult to store heat for long periods in field conditions, although it can be done in insulated water masses or solids such as stone and earth. There is some heat input every day that the sun shines or diffuse sky light reaches the earth. The mean temperature of the earth is 5°C (41°F), of the air at or near ground level 14°C (57°F), and of the outer layers of the atmosphere –50° to –80°C (–90° to –144°F). Normally, we lose about 1°C for every 100 m increase in altitude (3°F per 1000

TABLE 5.4
SOME ALBEDOS

	Reflected (%)	**Absorbed** (%)
"The perfect reflector"	100	0
White, smooth paint	96	4
Clean fresh snow	75-95	5-25
White gravel	50-93	7-50
Dense white clouds	60-90	10-40
Calm water (Sun 15° elevation)	50-80	20-50
Adapted desert shrubs	30-38	62-70
Sand dunes	30-40	60-70
Sandy soils	15-40	60-85
Dry hay	20-40	60-80
Wood edges	5-40	60-95
High sun, rough water	8-15	75-92
Young oaks	18	82
Young pines	14	86
Dark soils	7-10	90-93
Fir forest	10	90
"The perfect black body"	0	100

feet). In most conditions we experience a reduction in temperature with increasing altitude, but in many valleys, or on plains surrounded by mountains, cool air from the hills or cold air generated by rapid radiation loss from soils creates a condition where layers of dense cooler air are trapped below warmer air, and we have a TEMPERATURE INVERSION. It is in such conditions that fog, smog, and pollution can build up over cities lying in valleys or plains, where wind effect is slight. Such sites must be carefully analysed for potential pollutants.

As in the case of precipitation, it is advisable to research temperature extremes for site. Poultry (and many wild birds) do not survive temperatures greatly in excess of 43°C (109°F), nor do plants survive transplant shock from nursery stock when soil temperatures exceed 36°C (97°F), whether in deserts or in compost piles. Many plants are frost-affected at or below 0°C (32°F), and below this, sustained periods of lower temperatures will eliminate hardier plant species (even if well-established). Thus, the very widespread and sometimes economically disastrous black frosts that affect whole regions should be noted by site designers as much as flood periodicity. Livelihoods should not depend on broadscale plantings of frost-susceptible crops in these situations.

CONVECTION LOOPS AND THERMOSIPHONS

For building and garden designs, we should be aware of just how heat is stored and transmitted. First, we need to distinguish between lowgrade heat transmitted by CONVECTION, or the passage of air and water over slightly heated surfaces. It is this effect which operates in valley climates, and which creates valley winds. Cool air is heavier (more dense) than heated air; the same factor holds true for water or liquids, and other gases (and fluid flow generally).

Thus, providing heated air or water is contained in pipes or ducts, a closed loop circulation can be set up by applying heat to the lower part of that loop, providing that a least rise or height difference of 40 cm (about 18 inches) is built in to the loop; any greater height is of course also effective in producing a **thermosiphon** effect (**Figure 5.11**). This is the effect used in refrigerators driven by flames or heat sources.

In the atmosphere, columns of heated air over land ascend as an "Overbeck jet" (**Figure 4.13**), and at the top of this column, condensation and rain may occur as the air is cooled in the upper atmosphere. Such convectional rains are responsible for the mosaic of rainfall that patterns the deserts.

Convection loops will *not* occur in closed rooms, where hot air [at 8–10°C (1518°F) higher temperture] sits in a quiet or stratified layer below ceilings. As air is difficult to heat, and stores little heat, air convection is not an efficient way to heat building interiors, although it is the main "engine" of atmospheric circulation in the global sense.

Thermosiphons are useful in transferring heat from solar ponds or flat plate collectors to home radiators or hot water tanks; we should, wherever possible, site these heat collectors 0.5 m (1.6 feet) below the storage or use points so that they are self-regulated thermo-

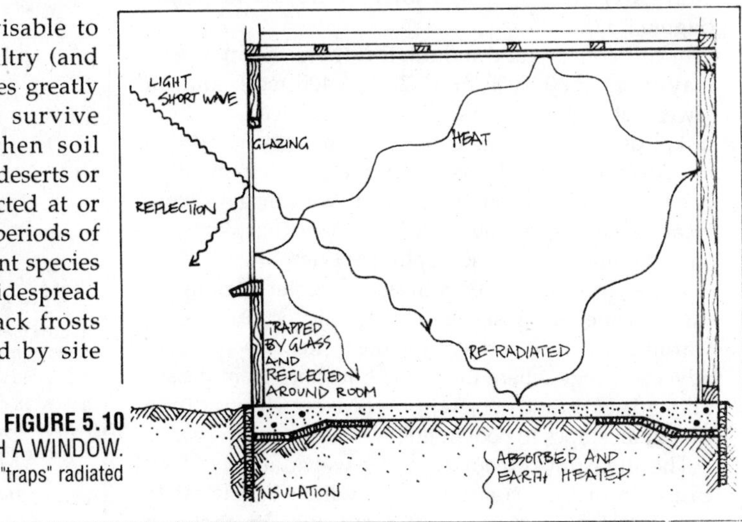

FIGURE 5.10
EFFECT OF LIGHT PASSING THROUGH A WINDOW.
Glass is less permeable to long (heat) radiation, thus it "traps" radiated heat but transmits short-wave light.

siphons.

HEAT TRANSFER
Heat flows from warmer to colder bodies, and just as warm air transfers heat to cool solid bodies by day, so warm bodies can heat large volumes of air at night. Bodies that are heated expand, decrease in density, and (where there is freedom to move) heated air or water rises as CONVECTION CURRENTS.

The common heat unit is that needed to raise one gram of water from 14.5°C to 15°C. In terms of incoming radiation, gram calories per square centimetre ($g/cal/m^2$) are termed LANGLEYS; the sun provides about 2 Langleys/minute to the outer atmosphere.

The quantity of heat received on earth is greatly affected by:

- latitude and season (the depth of atmosphere);
- the angle of slopes (which in turn affects reflection and absorption); and
- the amount of ice, water vapour, dust, or cloud in the air above,

This means that the Langleys received at ground level vary widely due to combinations of these factors. Nevertheless, most homes receive enough sunlight on their sun–facing areas to heat the water and space of the house, if we arrange to capture this heat and store it.

However, even when the sun is directly overhead on a clear day, only 22% of the radiant energy penetrates the atmosphere (1 atmosphere depth). In polar areas, where the slanting sun at 5° elevation passes obliquely through at a distance of 11 atmospheres, as little as 1% of the incoming energy is received! Slope has similar profound effects, so that slopes facing towards the poles receive even less energy from radiation.

It follows that siting houses on sun–facing slopes in the THERMAL BELT is a critical energy–conservation strategy in all but tropical climates, when siting in shade or in cooling coastal windstreams is preferred. Sun–facing slopes not only absorb more heat, but drain off cold air at night; they lie below the chilly hilltops, and above the cold night air of valleys and plains **(Figure 5.12)**.

In hill country and mountains, these thermal belts may lie at 1000–5000 m. (3,280–16,400 feet), and on lower hill slopes at 100–200 m. (330–650 feet), whereas in hot deserts the frost levels may only reach to 10–15 m. (33–49 feet) up the slopes of mesas. Each situation needs specific information, which we can gain from local anecdotes, the observation of existing plants, or trial plantings of frost–susceptible species.

Winds travelling from warmer to cooler regions, or the opposite, bring ADVECTED (exotic, or out of area) warmth and cold to local regions. Thus we speak of advection fogs where these come inshore from coasts, and advection frosts when cold air flows down mountain slopes to pool in hollows.

The invasion of cool areas by warm advected air causes moisture condensation, which is critical to precipitation in forests, but a nuisance in enclosed buildings. Thus, we should attempt to bring only *dry* warm air into wooden houses, or provide ways to direct condensation moisture to the house exterior.

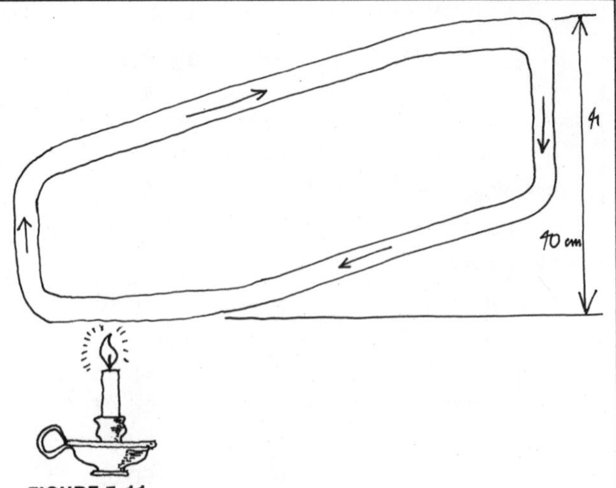

FIGURE 5.11
CONVECTION LOOP OR THERMOSIPHON.
Heat applied at "A" in a closed loop causes fluid flow in the loop if the top of the loop is 40 cm or more above the heated section of the loop.

Intermediate grades of heat can be transmitted by CONDUCTION, as when solids are in contact. It is in this way that we heat an entire floor or wall by heating it in one place, and this is the basis of the efficiency of the slab–floored house, where the floor is previously insulated from surrounding earth. In open (uninsulated) systems, conduction effects are local, as

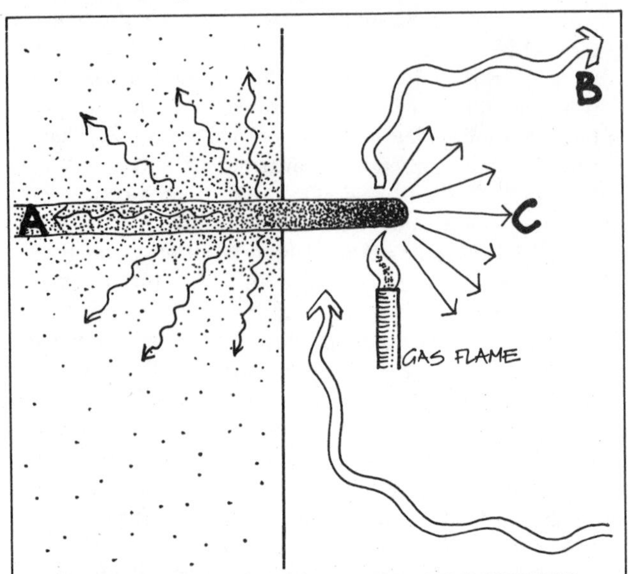

A HEATED BAR EMBEDDED IN CONCRETE DEMONSTRATES HOW HEAT DIFFUSES TO COLDER AREAS.
A "Low grade" heat is conducted from solid to solid or fluid to fluid by contact. Insulation is effective in trapping such heat.
B "Medium grade heat " is convected" by the movements of fluids or gases, as in air, wind, or water. Draught–proofing.conserves this heat. Heated fluids rise.
C "High grade heat" travels by straight–line radiation in all directions and can only be conserved by reflective (dust–free) surfaces or mirrors.This is how the sun heats the Earth.

heat is fairly rapidly radiated from solids or soil surfaces. Pipes buried in hot solid masses have heat conducted to their contents, or hot water pipes conduct heat to slab floors in which they are buried; such heating is most efficient in homes.

Intense heat trapped in solids and liquids is RADIATED, which is the effect transmitted across space by the sun. Radiant heaters affect air temperature very little, but radiation heats other solids and liquids (like our bodies) or dust in the air. Thus, we can keep very warm even in a draughty or cool room by the use of radiant electric, gas–fire, or wood–heated massive stoves; these are very efficient space heaters. As radiation crosses space, and is nondirectional, focused radiation can produce very intense heat locally.

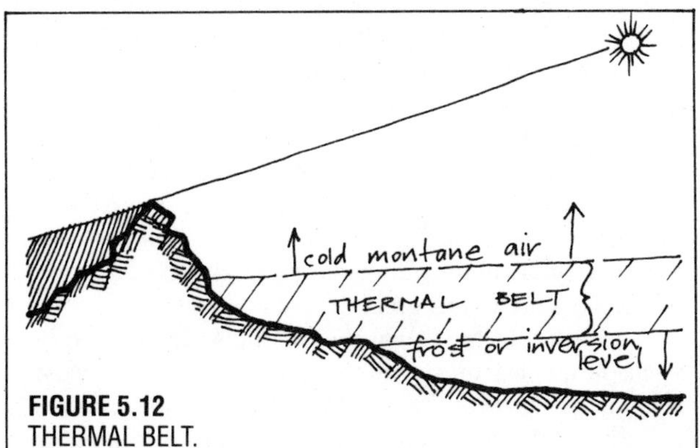

FIGURE 5.12
THERMAL BELT.
A midslope, moderate zone suited to gardens and housing; the shaded side of the hills accentuate cold, low evaporation, and suit very different species.

PLANTS AS HEATERS

Most or all Arum lilies, and species such as *Philodendron selluum* store fats which are "burnt" to create heat, so that the flowers heat up. Philodendrons may register 46°C (115°F) when the air is 4°C (39°F), and crocuses heat up to 15°C (27°F) above the ambient air temperature. The warmth generated is probably used to attract flies and heat–seeking insects to the pollen. Some plants (skunk cabbage, *Symplocarpus foetidus*), however, may use their heat to melt a hole in the spring snow, and so protect the blooms from cold [at 20–25°C (36–45°F) extra heat] as well as to provide a cosy incubator for the rest of the plant's growth" (*New Scientist*, 9 May '85) and to scatter odorous scents that attract pollinating flies. More amazingly, the shape of the first leaf of this species creates a vortex (from wind) that is contained within the hot leaf and carries pollen down to the unpollinated lower flowers, thus achieving fertilisation, in cold winds, without the presence of insects!

As all these "heaters" may have unpleasant smells, we should use them with caution. Understorey clumps of such species may assist frost–tender, fly–pollinated, or heat–starved plants, just as tall interplant systems may assist general heat requirements for some ground crops.

RADIATION AND GERMINATION OF SEED

The effect of soil temperatures alone on germination of a wide range of vegetable seeds can be profound. Between 0°–38°C (32°–100°F) the time to germinate (in days) can be reduced to one–tenth or one–fourth of that in cold soils by increasing soil temperatures. At the extremes of this temperature range, however, we find many plants have limiting factors which result in no germination. While almost all vegetable seed will germinate in soils at 15–20°C (59–68°F), such oddities as celery refuse to germinate *above* 24°C (74°F), and many cucurbits, beans, and subtropicals do not germinate *below* 10°C (50°F). Thus, we are really talking about waiting until 10°C is reached, or warming up the soil in greenhouses or with clear or black plastic mulch in the field before planting. Sometimes just the exposure of bare earth to the sun helps. A simple thermometer inserted 2.5 cm (1 inch) in the soil suffices to measure the soil temperature, or a special soil thermometer can be purchased. For specific crops, we can consult such tabulations as are found in Maynard and Lorenz (*Knotts Handbook for Vegetable Growers*, 1980, Wiley, N.Y.)

A second effect on germination is light itself, e.g. carrots need a definite quantity of light, and are usually surface–planted to effect this. We can surface–scatter such seeds, or first *soak* them overnight and then subject them to a day under a low–wattage light bulb or in the open before planting and covering them lightly (they react to this light *only if wetted first*). Larger seeds usually accept burial and germination in the dark, while some weed seed and desert seed will germinate deep–buried. For a few weed species such as wild tobacco, a mere flash of light (as when we turn over a clod of soil) suffices to start germination.

Next we come to *cold*, and we speak of the STRATIFICATION or VERNALISATION of seeds. Cold–area seed, and specifically tree and berry seeds from boreal or cold areas, should spend the period from autumn to spring in a refrigerator when taken to warmer climates. Apple seeds stored in sand or chestnuts in peat sprout in this way, and can be potted out as they shoot. This in fact reproduces the exposure to cold [at about 0°–5°C (32°–40°F)] that they normally experience at the litter level in cold forests or marshes. Wild rice and other "soft" aquatic seeds are stored in open ponds, or under water in an ordinary refrigerator.

Stratification can often be accomplished by keeping such seeds in sand or peat (or water for aquatics) in cold shaded valleys, or under open cool trellis in warm climates. They can be checked on late in winter and spring for signs of germination. The opposite of this is heat treatment, such as we can give to many tree legume seeds, by heating in an oven at 95°C (200°F) for a 10–20 minute period, or by pouring very hot (near–boiling) water over them, or by burning them in a light straw fire.

Many older gardeners will also feed seeds to themselves (in sandwiches), or their animals (chicken or cattle), collect the manure, make a slurry of it, and sow such seeds as tomatoes, berries, and tree legumes. The voyage through the digestive system is a compounded process of acid/alkali, hot/cold, mechanical cracking in teeth or in bird crops, and packaging in manure to which a lot of seeds are adapted.

FLOWERING

Day length (in fact, night length, but we will take the day side) varies over latitudes, and flowering plants are adapted to bloom and set seed in response to specific day lengths and the change of seasons. While many plants are DAY NEUTRAL and will flower if other factors are satisfactory, some will not flower at all in shorter or longer day–lengths than those to which they are adapted. This can be put to use, as when we transfer a tropical (short–day) corn to a temperate (long–day) hot–summer climate, and get a good green–leaf crop as fodder, or take tobacco from temperate to cool areas and get leaf rather than seed production. The same goes for some decorative foliage plants. But this effect is in fact the reason for choosing varieties from local growers, or selecting for flowering in new introductions so that a local seed source is available for all those crops we want in seed.

In New Guinea highlands (short days), cabbages from long–day climates may never flower, and some *Brassicas* reach 1–3 m in height, the leaves being plucked off at regular intervals for vegetable fodder, and the plant cut down only when too tall to reach!

Latitudes have specific day lengths as follows:

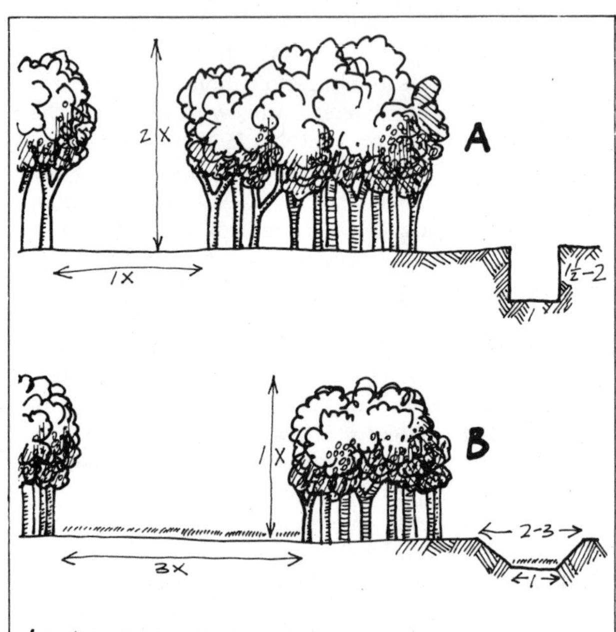

A: THIS CLEARING AND PIT SHOULD REDUCE FROST
B: THIS CLEARING AND PIT SHOULD INCREASE FROST

FIGURE 5.13
CLEARINGS IN TREES FOR FROST PROTECTION. (or pits in soils)
Frost loss is less in small steep–walled clearings that are about half the width of tree height.

• LOW LATITUDES (0–30°): Usually tropical climates, with colder mountain climates; equal or near equal days and nights.

• MID LATITUDES (30–50°): Cool to temperate climates with boreal mountain regions; long summer days and short winter days.

• HIGH LATITUDES (>50°): Very long summer days, and probably good radiation from diffuse light all the growing season. No plants grow in winter.

FROST

Frost is caused by radiation loss (rapid cooling) of the earth on clear nights, in still air. To reduce frost on any site (or in a small pit), it is necessary to have a *steep–sided* clearing or pit so that radiation is restricted to a small area of the sky. In such clearings, we have two effects: radiant heat from the vertical edges *plus* the obscuring of the horizon (hence less radiant heat loss at night). The proportion of heat loss on a cold night is proportional to the area of the night sky that is visible to the object losing heat. For example, a mouse in a cardboard tube in the ground loses very little heat, but a mouse on a mound on a flat sight is exposed to the whole sky and loses a great deal of heat.

The second factor is that the pit or clearing should be *small*; large clearings will create or contain *more* frost. The rule is to make the clearing (or pit) about one–half as wide as high, and to keep the sides trimmed to vertical. In forests, such clearings should not exceed 30m across (**Figure 5.13**).

It is necessary, therefore, to try to build up a complete crown cover to prevent frost on a site, and this is best done in stages. For example, we could plant the whole area to frost–tolerant legume like silver wattle (*Acacia dealbata*), then plant semi–hardy fruits in the shelter of these, eventually cutting back the *Acacia* as the frost–sensitive, protected trees gain height. It is obviously necessary to assist this process by supplying water to the selected trees, and this may also help ameliorate the frost effect on nights of high risk.

The effect of trees on soil moisture and frost may be profound at edges and in small clearings, as the tree crowns obviously create their own water distribution on the ground. Crown drip can direct in excess of 100% of rain to a "gutter" on the ground, and for some tree species with down–sweeping limbs and leaves, this is a profound effect. At the rain–shadow edges of forests, dry areas are to be expected. What makes this effect more pronounced is that the "wet" edges are more often than not also away from the sun (most rain comes from the polar side of sites). **Figure 5.14**.

The sunny edges of the forests help protect seedlings from frost, and these and small clearings are used to rear small trees, or to plant them out in frosty areas.

Some implications for designers are as follows:

• IN COMMUNITY AND PLANT HEALTH: Areas of severe direct or diffuse radiation, and especially where the atmosphere is thin (on mountains), where albedo is high (in snow, granite, or white sand areas, or over hayfields in summer) can produce severe radiation burns, skin cancers (very common in Australia), and temporary or longterm blindness.

Plants, too, must be screened against sunburn by partial shade and by white paint on their stems in conditions of severe radiation (especially young plants). Older plants may suffer bark damage, but will survive.

• FOR AUTOMATIC TRANSFER thermosiphon effects are best achieved by:

- placing heat sources below storage and use points;
- inducing cross–ventilation by building **solar chimneys** to draw in cool air;
- actively fanning heated air to underfloor gravel storages where solar attics or trapped ceiling heat is the heat source; and
- eliminating heat–induced condensation through the use of heat exchangers.

• IN HEAT STOVES: Massive earth, brick, stone, or concrete heat storage masses must be insulated to retain heat that is otherwise lost by *conduction* to the ground, or by radiation to the exterior of houses. Conduction is prevented by solid foam or air–trap insulation (straw). Radiation loss is prevented by reflection from double–glazed windows, or reflective insulation hanging in air spaces.

Reflective insulation doesn't work if it is dusty, dirty, or pressed against a conducting surface, hence it is of most use as free–hanging sheets, or ceiling sheets looped loosely across rafters. It can be kept clean (and effective) only in such situations as solar attics. Plain white paint is an excellent reflector for everyday use on walls or in concentrators.

• PLANT CHOICE: All plants with high biomass (e.g. trees) store heat in their mass (which is mainly water). Thus, fairly small clearings may be frost–free in cold climates. Dark evergreen trees absorb (and radiate) heat effectively; white–barked, shiny, or light–coloured trees reflect heat in cool districts, on forest edges, and where light itself is a limiting factor.

• WATER AND STONE are good heat storages, having a high specific heat. Thus, bodies of water are good heat storages. Air has a low specific heat and is a very poor conductor of heat, hence a good insulator. Many insulation systems work simply by trapping air, or by being poor conductors (cork, sawdust, wood).

These short examples, and some of the tabulated material, give the essential features of radiation that are applicable to everyday design. A preliminary design choice is to choose house sites for the maximisation of solar radiation in subtropical to cool climates and to shelter from radiation where excessive heat is a problem. Excess heat in one area of a house can be used in arid and tropical areas to "fuel" a cross– ventilation system, also essential in the humid tropics for cool dry air intake to the home.

Designers should always be aware of opportunities to convert light to heat, to reflect more heat on to cool areas, to light dark areas by reflection or by skylight placement, and to store heat below insulated slab floors.

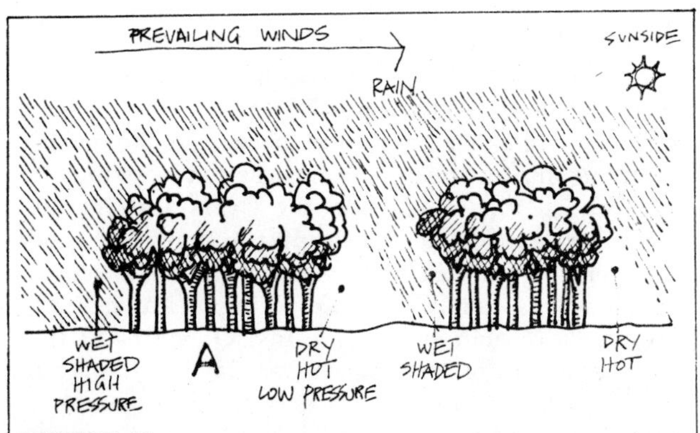

FIGURE 5.14
WET AND DRY SIDES OF A FOREST CLEARING.
Microhabitats develop due to combined sun and rain effects.

5.6
WIND

Both wind and water transport can influence (by impedence or reflection) the quantity of light and heat on any site. Particles or molecules carried in air have a profound effect on available light, and heat can easily be transported about our system by air and water, or by substances mixed in with them.

Of all of these elements, we have least control of wind in terms of storage or generation, but we can control its behaviour on site by excluding, reducing, or increasing its force, using windbreak and wind funnels to do so.

As a resident of a bare, cleared (once–forested) peninsula, when I speak of wind I know of what I write! The very table on which this book is penned rocks to gusts from the "roaring forties" and when the wind blows from the east, spicules of salt form on beards, clothes lines, and plants, burning off leaves and killing plant species; some hardy plants that withstand years of normal gales can die in salty summer winds.

When it comes to crop, winds of 8 km/h are harmless. Those of 24 km/h reduce crop production and cause weight loss in animals, and at about 32–40 km/h, sheer mechanical damage to plants exceeds all other effects; in fact, I have seen my zucchini uproot and bowl along like a tumbleweed. Trees are severely wind–pruned by a combination of mechanical damage and salt burn near coasts, and by the additional factor of sandblast in dunes (desert or coasts) and iceblast in cold climates. Wind transport of sands in deserts and incipient deserts buries fences, buildings, trees, and crops.

Although many sites are little affected by wind as a result of fortunate local conditions, or a general low level of wind effect in a region, very few coastal, island, sub–tropical or exposed hill sites can afford to ignore this factor.

There are broad categories of wind speed and effect,

just as there are for rainfall and temperature. The Beaufort Scale is the normal way to report winds, and equivalents are given in **Table 5.5**.

More severe than mechanical damage are the minute sulphur and nitrogen particles carried by wind. In Colorado, Virginia, Utah, the Urals and in fact anywhere downwind of nuclear waste stores, tests, and accidents we can add plutonium and other radioactives to the wind factor. Dry sulphur, falling on leaf and soil, converts to acid in misty rains. On parts of the northeast coast of America and Canada, these rains can burn gardens and forests, or make holes in garments and tents in a few days. Near acid production factories, even paints and roofing are pitted and holed; this factor has become general in industrial areas and for many miles downwind.

Windbreaks may mean the difference between some crop and a good crop, but in severe wind areas, the difference is more absolute and may mean that susceptible plants will produce no crop at all. Thus, a list of wind–tolerant (frontline– trees) is a critical list for food production and animal husbandry.

DIRECTIONS AND SEASONS

Winds are fairly predictable and often bi–modal in their directions and effects in local areas. For the landscape designer, **wind–flagging** or older trees and wind-pruning tell the story; the site itself has summed total wind effects over time.

From Latitudes 0° to 35° north and south in oceanic areas, winds will be bi–modal and seasonal, southeast or northwest in the southern hemisphere, southwest or northeast in the northern hemisphere. Locally, the directions will be modified by landscape, but the phenomena of windward and leeward coasts are almost universal. From Latitude 35° to the limits of occupied coasts, westerlies will prevail in winter, easterlies more sporadically as highs or lows pass over the site and remain stationary to the east or west. Cold winds will

BEAUFORT NO.	KNOTS	km/h	MPH	DESCRIPTION
1	1	1	0.5	Calm
2	1 - 3	1 - 5	1 - 3	Light airs
3	4 - 6	6 - 11	4 - 7	Light breeze
4	7 - 10	12 - 19	8 - 12	Gentle breeze
5	11 - 16	20 - 28	13 - 17	Moderate breeze
6	17 - 21	29 - 38	18 - 24	Fresh breeze
7	22 - 27	39 - 49	25 - 31	Stormy breeze
8	28 - 33	50 - 61	32 - 38	Near gale
9	34 - 40	62 - 74	39 - 46	Gale
10	41 - 47	75 - 88	47 - 55	Strong gale
11	48 - 55	89 - 102	56 - 64	Storm
12	56 - 63	103 - 117	65 - 73	Violent storm
13	64 +	118 +	74 +	Hurricane

BEAUFORT NO.	EFFECTS	
0 - 2	Slight, no damage to crops or structures	
3 - 4	Damage to very susceptible species	Useful energy produced
5 - 6	Mechanical damage to crops, some damage to structures	
7 - 12	Severe structural and crop damage. Damage to windmills.	

THE REDUCTION OF WIND VELOCITY IN FORESTS:

Penetration (m)	Remaining Velocity (%)
30	60 - 80
60	50
120	15
300 - 1,500	Neglible wind.

TABLE 5.5
THE BEAUFORT SCALE.

blow from continental interiors in winter, and warmer but still chilling winter winds blow in from the seas.

Islands and peninsulas from Latitudes 0° to 28° eperience two main wind modes, those of the winter winds or trade winds (southeast in the southern hemisphere, northeast in the northern hemisphere) and those of the monsoon. In effect, we look for two seasons of winds and two short periods of relative calms or shifting wind systems in these latitudes. These are the main winds of tropical oceanic islands.

In summer, the cross–equatorial monsoon winds deflect to blow from the northwest in the southern hemisphere, and as southwest monsoons in the northern hemisphere. Southeast Asia and the Pacific or Indian ocean islands are most predictably affected by these bi–modal systems. Although many sites are also affected by only two main strong wind directions, these are rarely as strictly seasonal in effect as they are closer to the equator.

WIND LOADS

On warm sea coasts, where onshore winds not only carry salt but also evaporate moisture, salt deposits on vegetation are the limiting factor on species selection, and only selected hardy species with fibrous or waxy surfaces can escape death or deformation by salt burn.

As well as inorganic materials, wind may transport organisms ranging from almost impalpable spores of fungi and ferns to very weighty insects such as plague locusts which are swept aloft by heated air columns, and carried as frozen or chilled swarms to down–draught areas. Here, they thaw out and commence feeding, or perish in oceans far from land. Mosquitoes, fruit flies, wasps, and spiders deliberately spin aerial floatlines and also migrate over mountain and oceanic barriers on windstreams. Flocks of migrating birds also take advantage of windstreams as they circle the globe.

Flow of air (wind) over leaf surfaces promotes rapid transpiration, as does high light intensity (Daubenmire, 1974). When we have both effects together, shrubs and trees may lose too much water, and trees guard against this combined factor by presenting whitish undersides of leaves to the light as the wind blows, thus carrying on a dynamic balance between the light and wind factors. Vines and trees may alter leaf angles to reflect light, trap air, or to reduce the area of leaf exposed to light or wind. Thus, both pigmentation and leaf movement are used to balance the effects of variable incoming energy, and leaf pores close down to prevent moisture loss.

WIND HARMONICS

In a radio programme on sailing (Australian Broadcasting Corporation, 19 Dec '84), Frank Bethwaite, a New Zealand–born Australian boat designer, pilot, and sailor outlined some of the characteristics of ground winds. Such winds do not blow steadily, but vary as gusts and calms in a predictable and locality–specific way; that is, the common winds of any one site have regular pulses.

He states that such regular variations are easily timed; a 49–60 minute frequency of gusting is typical of mid–latitudes, with gusts 40% stronger than lulls. In lulls, the wind *direction* also changes, as light crosswinds at about 15° to the main wind direction.

The variation in wind speed and direction is systematic and regular, and both frequencies, durations, and amplitudes can be obtained by combinations of stopwatches, anemometers, and wind vanes (or all of these recorded on automatic equipment).

Such "waves" of wind are made visible in grass-lands, or on the surface of waters viewed from a cliff. They are also, at times, reflected in clouds as "rank and file" systems. The lulls show as spaces between cloud ranks, and in these spaces, light clouds of different alignment represent the change of direction typical of lulls.

The gusts are ponderous, representing vortices; the lulls are of light crosswinds. Some periods are short (Bahrain, 5.25 minutes; Sydney, 6–12 minutes; Toronto, 10 minutes). Wave "fronts" on grasslands may come at every 14 seconds, with gusts at longer intervals. Sea waves themselves have a characteristic periodicity and speed, usually about 5–12 per minute, the period lengthening in storms.

In the westerly wind belts, we can distinguish between the PREVALENT WINDS of from 8–24 km/h, which blow for five out of seven windy days, and the ENERGY WINDS of from 16–40 km/h which blow on the other two days. The energy winds come from between 15°–20° off the direction of the prevalent winds (Michael Hackleman, *Wind and Windspinenrs*, Peace Press, California, 1974).

WINDBREAKS

It is the chill factor—the removal of heat from surfaces, and evaporation of fluids—that creates cool to cold climates in the tropics at lower altitudes than adiabatic or altitude factors would indicate. This chill factor retards plant growth, and lowers the efficiency of solar devices and insulation. In cyclonic or hurricane areas, catastrophic winds may become the over–riding design modification, around which all other factors must be arrayed.

We are not much concerned with sheltered and low wind–energy sites, except to choose them for our dwellings in exposed landscapes, but close attention must be paid to shelter strategies in exposed sites.

On sites with predictable wind patterns, revealed either by trees, derived from local knowledge, or indicated by wind records over time, we can plan directional, patterned windbreak of earthbank and trees. On sites where severe winds and sandblast may come from any direction (as in some deserts), the strategy is to impose a close rectangular or network pattern on windbreak.

However the windbreaks are arranged, buildings, gardens, and animal shelters can be arranged to face the sun and benefit from solar impact.

Essentials of a Windbreak

The essentials of windbreak are fairly well known and local lists of species for windbreak are often available from forestry and agricultural advisors or departments. Essentials are:

- Good species selection to be used as pioneers (easily grown);
- Initial protection of planting from mechanical or wind damage (bagging, fencing);
- Periodic or trickle irrigation to reduce desiccation;
- Anchoring by stones or mulch; and
- Species with 40–50% penetrability in the front line or as dominants.

Many fire–resistant plants are also wind resistant, and in addition to these, some drought–resistant but fire–prone species (pines) will withstand wind. What they have in common are ways of resisting desiccation and sandblast. Such plants have, as common features:

- Fibrous stems (palms).
- Fleshy leaves (aloes, agaves, *Euphorbias*).
- Hard, needle–like leaves or stems (pines, tamarisks, *Casuarinas*, some *Acacias*).
- "Furry" or hairy (tomentose) leaf covers, or waxy leaves (*Coprosma*, eucalypts, some pines, some *Acacias*).

Initial protection can come from:

- Individual open–ended plastic bags around stakes (a common and effective establishment method).
- Earth mounds, or side–cast earth banks of greater length than the tree line.
- Brush fences, even wire–mesh fences, or staked fences with 40% wind penetrability.
- Tussock or tough unmown grass to windward (leave if already present).

All of these can be used in combination in very hostile areas. It is usual for the windward rows of trees to be heavily wind–flagged, and for taller species to be placed in their lee. On coasts and in deserts, it is not until after the fourth or even fifth tree row evolves that wind–prone fruit or nut–bearing trees will yield, so windbreak is the *first* priority for gardens in these situations.

Substantial trellis is a more immediate alternative, but care should be taken to make this sinuous (if of brick or mud brick) or zig–zagged (if of timber), as it has to withstand persistent and severe forces until shelter grows on either side of it. Earth mounds can be better streamlined, being less sensitive to windthrow. The hollow from which the earth is taken to make the mound can be made to hold water or to give protection to young plants.

Tyre walls are sometimes feasible, and create great warmth inside the tyres, but are scarcely aesthetic unless very regularly arranged and planted. They have the advantage of being cheap, and can be removed once effective. Mesh fences, if stoutly built with a heavy top rail, can be the basis for fedges (fence–hedges) of thick–leaved vines, which on coasts may completely mound them over with tough semi–succulents such as *Rhagodia*, *Tetragonia*, *Carpobrotus*, or *Mesembryanthemum*. Rock walls and tyres may be similarly mounded with scramblers or cacti, some of which provide bee forage

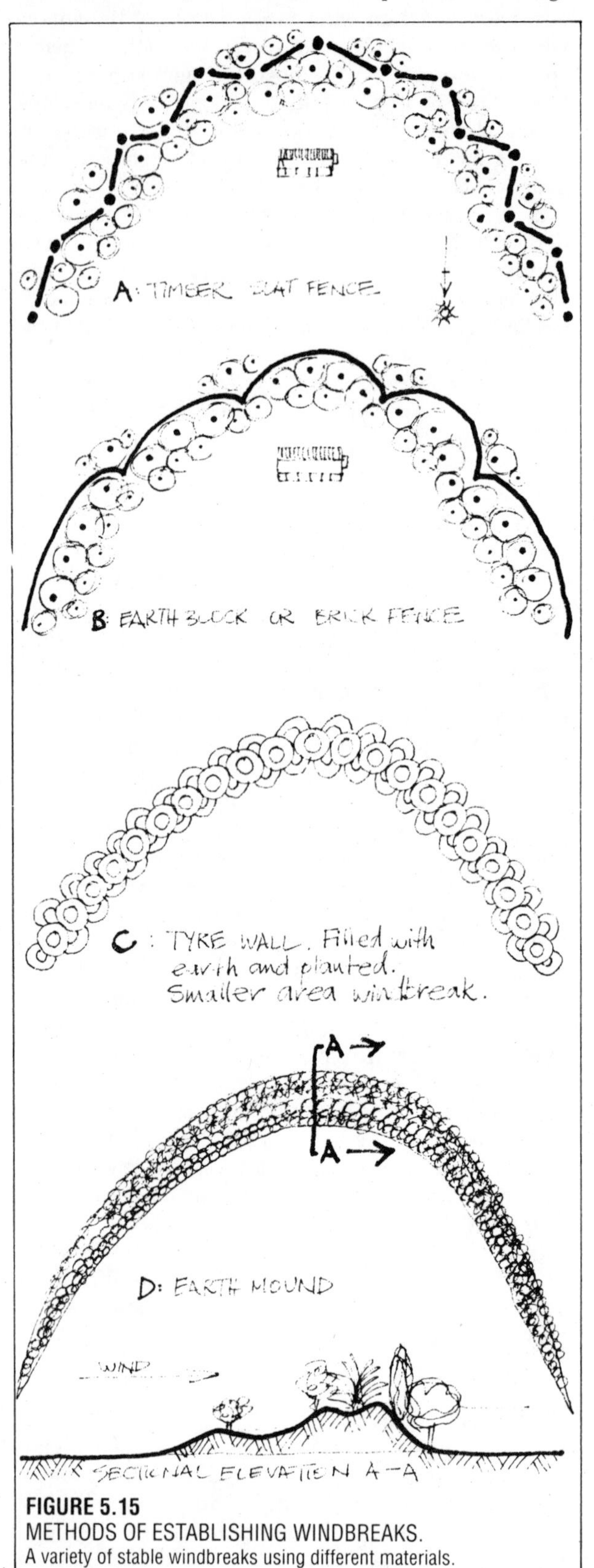

FIGURE 5.15
METHODS OF ESTABLISHING WINDBREAKS.
A variety of stable windbreaks using different materials.

and berries or edible fruits.

It is rare for tree canopies on dry saltwind coasts to gain more than 46 cm height in 1 m width (18 inches in 3 feet), so considerable width must be given to pioneer windbreaks in these situations, unless those hardy pioneers such as the Norfolk Island pines can be nursed to grow to windward. However, as this slow climb to height commences from *ground level*, a fence, building, earth bank, or barrier gives it a great start for far less spread (**Figure 5.15.D**).

Even a 46-62 cm (18-24 inch) high "fence" or mound earth will grow a sweet potato, strawberry, or cabbage in the lee, while hard–pruned canopies need not be barren, as many dwarf fruit, vine, and flower crops will grow below these if mulch and water are provided. In windbreak forests near coasts, small openings of 6–9 m (20–30 feet) provide garden shelter and admit light.

There is, in fact, a special charm about those 3–3.5 m (10–12 feet) high dense coastal shrubberies in which nestle small shacks, through which wander sandy paths, and in which people create small patches of scattered garden using wastewater and mulch. Once shaped, fruit trees in this situation seldom need pruning, and at times one wonders if the wind is not an advantage in that it forces compact and careful work, punishes carelessness, and promotes wastewater use.

Across the whole of the flattish peninsula of Kalaupapa on Moloka'i, the Hawaiians had built tiny stone fences of 25–50 cm (10–20 inches) high and only 4.5–5 m (15–18 feet) apart, behind which they grew a basic sweet–potato crop, and in which grew tough fern for mulch. All are now abandoned, but on the seaward coast, wild date somehow struggles to 4.5 m (15 feet) or so in the teeth of the tradewinds, and would have made a grand windbreak had the Hawaiians retained their land against tourism and graziers. Just to windward, the strong winds bring so much salt spray ashore that it crystalizes out in pinkish ponds, mixed inexorably with the red of the volcanic earth on which it forms. Even today, it is gathered as "Hawaiian salt", and is further mixed with the roasted kukui nut and chilies for a delicious raw–fish condiment.

Some Benefits of Shelterbelt

1. *Shelterbelt effects on house design*. For glazed areas and hot water (flat plate) collectors, wind chill factors remove 60% of heat alone. Shelterbelt (including thick vine trellis) around a house can effect a 20–30% saving in heating fuels in moderate to severe winters. Thus, in cold areas earthbanks plus shelterbelt, and a sun–facing aspect, is a critical design strategy. In deserts, where advected (wind–carried) heat is the most severe effect on human comfort, shelterbelt trees serve to reduce ground temperatures up to 15°C.

2. *Effects of exposure on livestock*. Blizzards will kill livestock and newborn lambs, and even hardy and adapted animals can lose 30% of their bodyweight in 3 days of blizzard. As well as shelterbelts in fields, we need to be very careful to design fences so that they do not form downwind or downslope traps, as herds escaping blizzards will pile up against them and smother in fenced corners. All moorland and high plateau fences should allow easy downwind escape to woodlots, sheltered valleys, or lower elevations.

In less severe conditions, sheep weight in unsheltered fields in New Zealand is 15% less than that of sheltered areas. Australia attributes 20% of all lamb losses to wind chill factors, and issues regular wind chill warnings at shearing time to prevent adult sheep loss.

Cattle fed winter rations on exposed sites will eat 16% less of this food, so that winter hay and concentrates need to be fed out in shelter for animals to obtain full benefit. Both heat and cold have similar effects on weight gain, and shelterbelt is one of the most effective ways of increasing livestock production, and conserving rations. Thus, in designing for livestock, fences, shelter, access to shelter, and feeding and watering points all need sensible placement, so that animals are not exposed to extreme temperatures. In the tropics and subtropics, a ridge planting of pines or *Casuarinas* with a wind gap left below the crowns affords both shade and an induced breeze that discourages flies and mosquitoes. Such ridges are also rich mulch sources for lower slopes.

3. *Civil construction*. Snowdrift across highways is more effectively and permanently blocked by hedgerow of hardy *Caragana* and *Eleagnus*, which are estimated to be 50% cheaper than stout fences, and of course outlast them. Juniper in high country actually grows better in areas of snow drift (below the sharp ridges where snow forms cornices), and swales at such places enable more snow melt, therefore more available root moisture for trees in spring and summer.

Wind shear on exposed highways or at caravan parks can cause casualties and property damage, so that we need to design windfast median strips and highway shelterbelt in areas of known hazard, but especially on mountain passes and near exposed coasts subject to gales.

4. *Shelter in and around croplands and orchards*. For croplands, a matrix of shelterbelt species 10–16 m in height and 33–66 m apart (*Casuarina*, poplar, Matsudana willow, trimmed eucalypt) affords wind protection for such crops as kiwifruit and avocado, and give the greatest increases in yield while reducing wind damage to fruit and leaf. For instance, citrus culled as damaged is 50% of the crop in unsheltered areas, versus 18.5% in shelterbelt systems, cotton yields are 17.4% higher within five times the height of the shelterbelt, and fall off to a 7.9% advantage at ten times the height of the belt.

Effects of shelterbelt are compound, and include more meltwater from snow, much greater fruit or seed set in bee–pollinated crop, the preservation of good shape in the trees, hence less pruning. Species selection of shelterbelt trees is essential, and a set of factors can be the criteria that assists the farm enterprise sheltered. These include:

- Nitrogen fixation or good mulch potential from leaves and trimmings;

• Hosting of predatory insects or birds that control crop pests;

• Least moisture competition with crop (although roots from the shelterbelt can be ripped or trenched at the edge of crops);

• Excellent forage yields or concentrated foods for livestock; and

• Natural barriers to livestock (thorny plants, or woven hedge).

Shelterbelt is planted as a succession from a tall grass to a taller legume to a long–term, tall, windfast hedge of e.g. *Casuarina*, poplar, willow, eucalypt, oak, chestnut. All this complex can be set out at once, and managed as it evolves to maturity. Quickset (by cuttings) hedges of poplar and *Erythrina* are popular because of their fast windbreak effect, but species must be chosen to suit a particular climate.

Where space is ample and winds strong, the profile of a windbreak can be carefully streamlined, and up to six rows of tree and tall grass lines established, giving a mixed yield of forage, timber, fuel, mulch, honey, and shelter. In more constricted areas, a matrix of single–tree lines is usual, and effective if close–spaced. However, there is no such thing as a standard shape or windbreak, and very different configurations are needed for different sites, functions, and as accessory species to the enterprise sheltered, the wind strength, and the wind load (salt, sand, dust).

5. *Effects of windbreak on soil moisture.* Windbreak is very effective in snowy areas, increasing soil moisture 4% to four times the height of the break, and that to 1.2 m depth in soils. Obviously, the benefits to trees in cold deserts are as a reserve of soil moisture that is rare in cold dry climates. The same effect occurs locally in the lee of tussock grasses, and can be used to establish a tree.

In foggy climates or facing sea coasts, we must add the effect of sea air condensation, which can be from 80–300% of rainfall as leaf drip. In hot deserts and hot winds, the advected hot winds are the major factor in soil moisture loss. Such effects are produced over large treeless areas of dry grain crop as well as in deserts.

The effects on grain crop of windburn and seed shattering in hot winds is insignificant for up to 18 times the height of the windbreak.

6. *Less soil loss due to windstorms.* Very serious soil losses of up to 100 T/ha/day in duststorm episodes (usually followed by torrential rain) are prevented by windbreak and soil pitting with tussock grasses. Approximately 50–70% of dusts settle out of the air 100 m into tree clumps, so that treelines are the essential accompaniment to any pastoral or crop system in arid areas.

On coasts, removal of mangroves and coastal dune vegetation results in a sudden acceleration of wind erosion on beaches and coastal soils, and following deforestation, up to 30% more silt per annum flows into and reduces the useful life of water storages.

7. *Windbreak and hedgerow as accessory to crop and livestock.* Quite apart from the above effects, windbreak species can be chosen to provide excellent crop mulch (*Prosopis, Acacia, Erythrina, Melia, Canna*) and fodders (all the foregoing species plus *Leucaena*, Fig, *Pennisetum*), and also to fix or recycle nitrogen and phosphatic fertilisers, or to mine trace elements (*Casuarina, Banksia, Eucalyptus camaldulensis*).

Dry or cold–deciduous species and monsoon deciduous trees give a natural leaf fall in crop, automatically adding growth elements to the crop. In every crop and orchard it is advisable to interplant leguminous trees for mulch, soil building, and in–crop windbreak or frost cover. Trees like avocado and crops like papaya can be grown on sub–tropical frosty sites providing there is a high canopy of hardy palms or light–crowned legumes (e.g. Butia palm, *Jacaranda*, *Tipuana tipu*). Such sites do not frost, as there is no bare–ground radiation at night, and advected frost is impeded.

Finally, forage and firewood from windbreak provides excess fuels to cook crop products, which is an important factor in the third world. In summary, well–chosen and designed windbreak can occupy up to 30% of the total area of any site without reducing crop yields, and if windbreak species are chosen that aid the crop itself, there will be an increase in total yield, soil quality, and moisture available.

Hedgerows and Shelterbelts

Shelterbelt species must be carefully selected to give multiple uses, to either ASSIST the crop yield, or ADD TO the end use yield (e.g. forage trees in pasture). This ensures the area occupied by shelterbelts adds to the

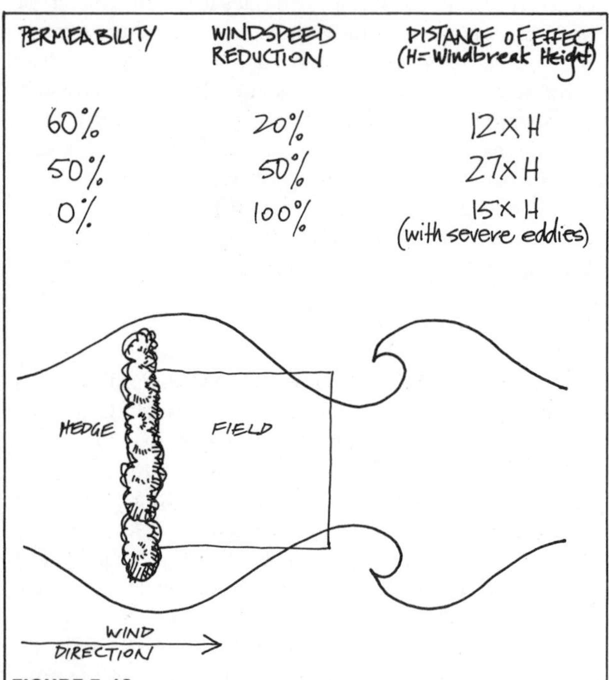

FIGURE 5.16
PERMEABILITY OF WINDBREAKS.
This figure shows how winds flow around a windbreaks which needs to be wider than the field for less permeable hedgerow.

total crop yields, rather than deducts from them. In general, we would gain in crop *or* pasture yield using nitrogen fixing and browse–edible shelterbelts species, and lose crop yield by using high water-demand, non-leguminous, and inedible shelterbelt species.

However, where we experience severe sea or desert winds, which greatly reduce all yields, we must select salt–resistant or sand–blast–resistant windbreak no matter what the intrinsic yields of the shelterbelt. It is rare for sea–front trees to bear effectively (e.g. the outer 4–5 rows of coconuts on exposed islands yield little crop), so that choice of frontline seacoast plants for seed or fruit yields is often irrelevant when considering species for multiple function.

For isolated trees, or trees whose canopy lifts above the general forest level, wind of even low speeds may increase the transpiration rate, sometimes doubling water use. The effect is greatest on water–loving plants, and much less on dry–adapted species which have impermeable leaf cuticle and good control of stomata, or a cover of spines and hairs.

Hot, dry winds, and winds laden with salt have the most damaging effect on plant yields (hence, animal yields), although at high wind speeds mechanical damage can occur, which prevents or reduces yields no matter what the humidity or salt content of the winds. Damaged crop plants such as corn or bananas suffer photosynethic inefficiency ranging from 20–85% when the leaf laminae are torn or frayed, or the midribs are broken (Chang, 1968).

Plants show different resistances to wind damage:

- *Wind tolerant (*and wind–fast). These are the many short or creeping plants at the boundary layer of still air near the ground, or the front–line plants of sea coasts, e.g. *Cerastium, Araucaria heterophylla* Yields are little affected by strong winds.
- *Exposure tolerant*, e.g. barley, some *Brassicas, Casuarina* and *Coprosma repens.* Yields are reduced in strong winds, but dry matter yield is less affected than in wind–sensitive plants.
- *Wind sensitive*. These are the many important crops such as citrus, avocado, kiwifruit–vines, many deciduous fruits, corn, sugar cane, and bananas. Both plant height and yields rapidly decrease with increases in wind speed. For these species, very intensive shelterbelt systems are essential.

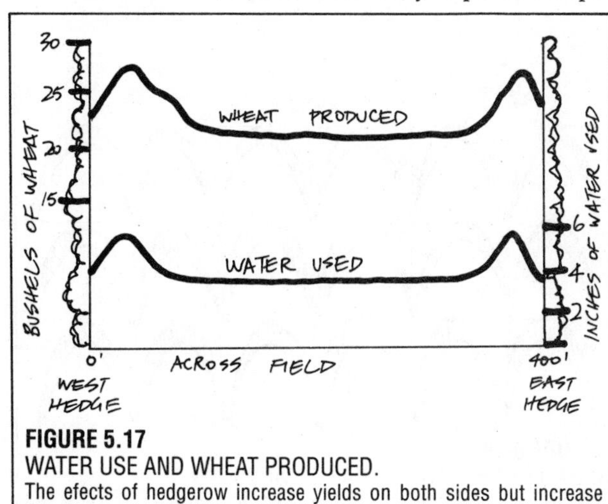

FIGURE 5.17
WATER USE AND WHEAT PRODUCED.
The efects of hedgerow increase yields on both sides but increase water use.

Problems arise when the plants used for shelterbelt (e.g. poplar) are themselves heavy water–use species with invasive roots. An annual root–cutting or rip–line may be necessary along such windbreaks to permit the crop sheltered to obtain sufficient water, but it is best to choose more suitable species in the first place.

Windbreak Height and Density

The height and density, or penetrability, of windbreak trees are the critical shelter–effect factors. Some configurations of windbelt may causes frost–pockets to develop in the still air of sheltered hollows. PERMEABILITY is an important factor if we want to reduce frost risk or to extend the windbreak effect (**Figure 5.16**).

Briefly, we need windbreaks spread at *no more than* 20 times the hedgerow height in any severe wind. In the establishment of wind–sensitive tree crop we may actually need continuous (interplant) windbreak. The length of windbreak needs to be greater than the length of the field protected, as wind funnels around the end of windbreaks in a regular flow pattern.

Windbreak Configurations

In general, species chosen for windbreak should permit 40–70% of the wind through, which prevents the formation of a turbulent wind overturn on the leeward side. Windbreak height is ideally one–fifth of the space between windbreaks, but is still effective for low crop at one–thirtieth of the interspace.

Sensible configurations are shown in **Figure 5.18.** Note that some wind shelter systems are placed throughout or within the crop or fruit area.

A. *Dense windbreak* with bare stem area below. Effects: good summer cool shade for livestock; poor to useless winter shelter. Clumps of such trees on knolls allow animals to escape heat, and flies and mosquitoes are much reduced.
Sample species: *Cupressus, Pinus, Casuarina.*

B. *Alternate* (zig–zag) planting of very permeable trees. Effects: good "front–line" seafront systems to reduce salt burn and provide shelter for more dense trees on islands and coasts. Species: *Araucaria, Pinus, Casuarina.*

C. *Compound windbreak* of high density. Effects: The best protection for eroding beaches, lifting the wind smoothly over the beach berm and trapping sand. Also effective in dust–storm areas as a dust trap. Species: Ground: *Convulvulus, Phyla (Lippia), Mesembryanthemum.* Low shrubs: *Echium fastuosum*, wormwood. Shrubs: *Coprosma repens.* Trees: *Lycium, Cedrus, Cupressus,* some plants.

D. *Permeable* low hedgerow of *Acacia* or legumes. Effects: Good effects on grass and crop growth, allows air movement to reduce frosts. Species: *Acacia, Leucaena, Prosopis, Albizia, Glyricidia,* tagasaste and like tree legumes.

E. *"Incrop "windbreak*

1: Savannah–style configuration of open–spaced light–crowned trees in crop or pasture. Effects: Excellent forage situation in arid areas, especially if trees provide fodder crop; pasture protected from drying winds. Species: Several fodder palms, *Inga, Acacia,* tagasaste, baobab, *Prosopis.*

2: Complete or almost–complete crown cover *in* tree crop. Effects: Excellent frost–free sub–tropic and tropic lowland configuration where fruit trees (F) are interplanted with leguminous trees (L) as shelter and mulch, with *Casuarinas* (C) as borders. Suited to humid climates, or irrigated areas. Species: Fruits (F) from palms, avocado, *Inga,* banana, citrus. Legumes (L) of tagasaste, *Acacia, Albizia, Inga, Glyricidia, Leucaena.* Borders (B) of *Casuarina,* low palms (*Phoenix canariensis*), *Leucaena, Prosopis,* and other wind–fast trees and tall shrubs.

The partial list of windbreak configurations given above covers only some cases, and in every case a designer must select species, study suitable total conformation, and allow for evolution or succession. As with all permaculture designs, general known principles are followed but every actual site will modify the design, as will the purposes for which shelter is intended.

Windbreak is essential for many crop yields, particularly in orchards. As discussed, wind causes mechanical damage, salt–burn, and may transfer (advect) heat and cold into the crop. Unless conditions are very severe, single–line windbreak spaced at 15 times height may have a satisfactory effect on ground conditions, and this is recommended for crops and grasslands. However, severe montane and coastal winds need more careful design, and a complex windbreak of frontline species able to buffer the first onslaught of damaging winds is needed (**Figure 5.19**).

For both tree crops and orchards, we have a very different potential strategy in that the windbreak may be composed of trees compatible with the protected forest or orchard system we wish to shelter, and can then be integral with the crop (**Figure 5.18 E1** and **E2**). Great success with such strategies has been demonstrated both for wind and frost moderation in susceptible crop such as citrus, avocado and macadamia nuts or chestnuts, using a protective interplant of hardy *Acacia, Casuarina, Glyricidia,* tagasaste, or *Prosopis* spaced *within* the crop. As all of the windbreak species mentioned fix nitrogen or phosphates, provide firewood, radiate heat, and shelter crop, it is sensible and beneficial to fully interplant any susceptible tree crops behind barriers of front–line windbreak. Windbreak in this instance is integral with the crop (as it is in natural forests).

The importance of windbreak extends to SOIL CONSERVATION. In dry light soils, windbreaks can reduce dust and blown sand to 1/1000th of unsheltered situations (Chang, 1968) within 10 times the height of windbreak. Thus, in crops in arid or windy areas, it is necessary to plant windbreaks closer together for the sake of soil conservation. The loss of soil at 20 times windbreak height is 18% of open situations, which is still too much when we can lose 8–40 t/ha in windstorms!

Similarly, WATER EVAPORATION can be halved in strong winds (32 km/h or more) for distances up to 10 times height. Over 24 km/h, 30% gain in soil water conservation is achieved. Only in still–air conditions is evaporation loss about the same for sheltered and open field conditions.

SNOW MOISTURE is increased by a windbreak of type A or B (**Figure 5.18**) when the snow is trapped on fields. The snow depth in winter bears a close correlation to dry matter yields in spring and summer,

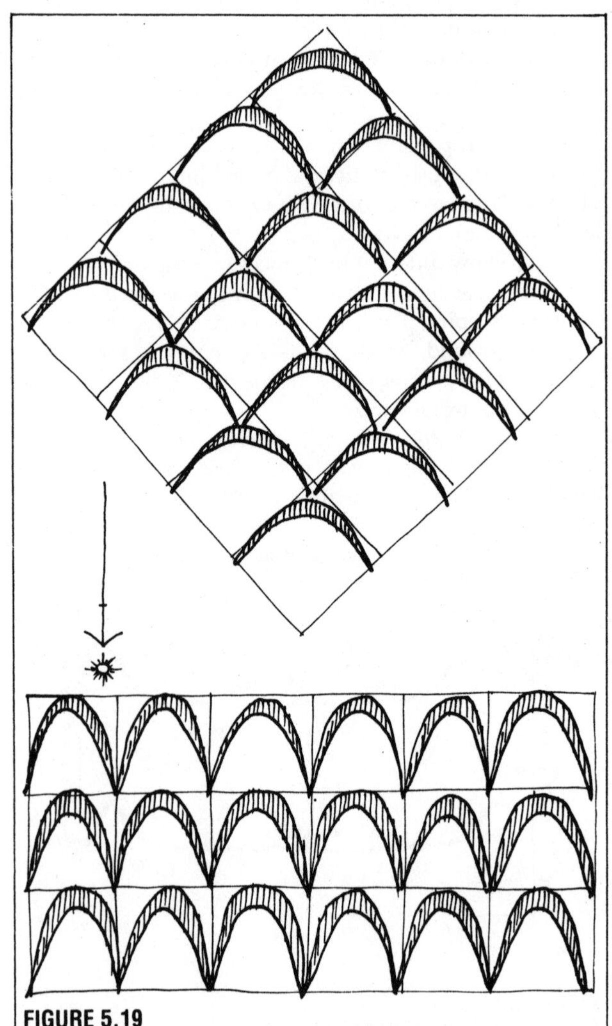

FIGURE 5.19
COMPLEX NETS FOR VARIABLE WINDS.
Two examples are given to suit straight–line fencing. Crescents are bowed towards the most damaging winds.

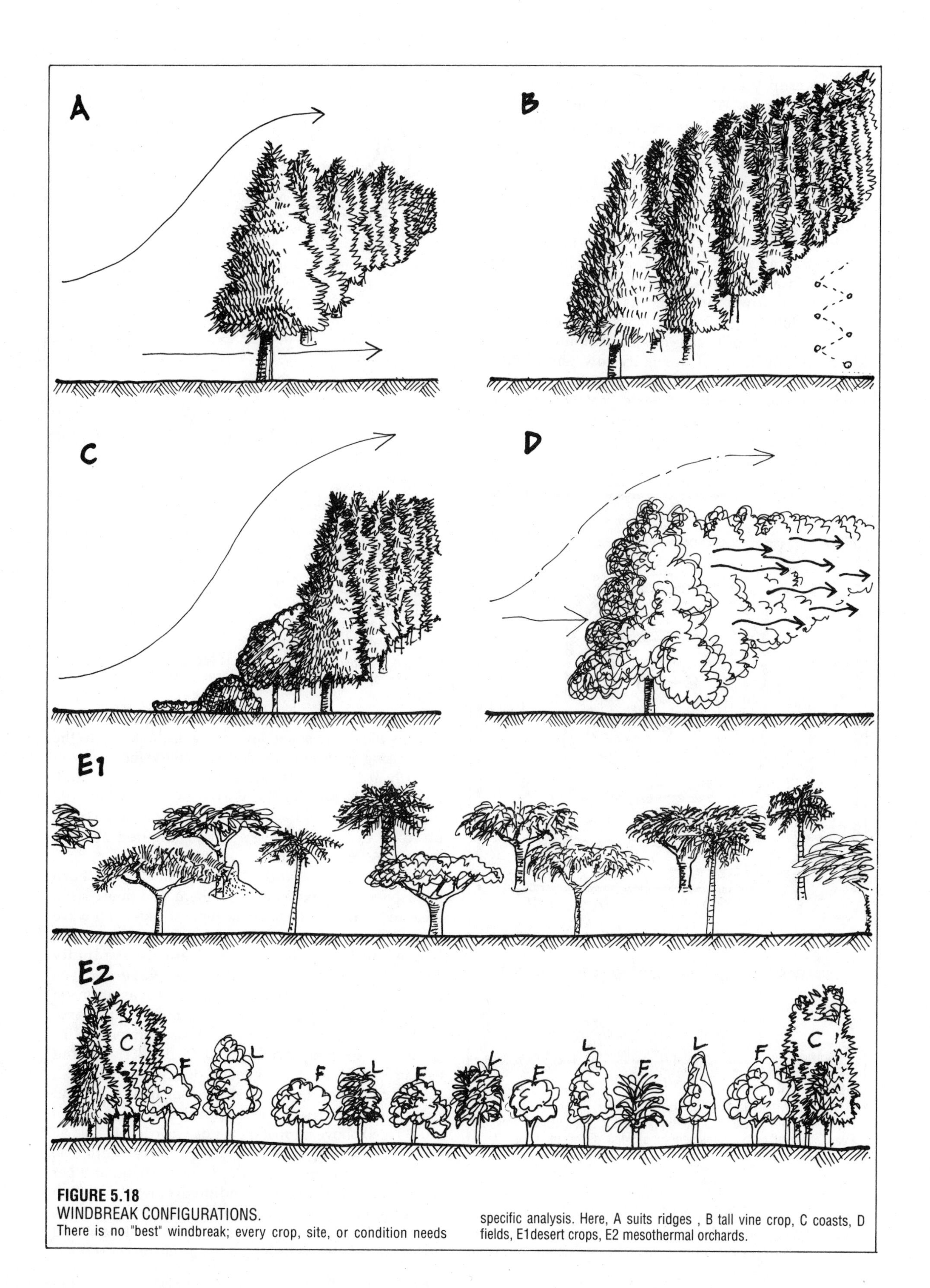

FIGURE 5.18
WINDBREAK CONFIGURATIONS.
There is no "best" windbreak; every crop, site, or condition needs specific analysis. Here, A suits ridges , B tall vine crop, C coasts, D fields, E1desert crops, E2 mesothermal orchards.

so that crop yields are highest downwind from windbreak areas. Wherever snow blows across the landscape, windbreaks of savannah configurations create spring soil moisture traps. It is also possible to do this by using open swales in snow–drift areas. Windbreaks in exposed snowfield areas can be better established in the lee of earth bunds or in natural cornice just polewards of ridges.

CROP YIELDS vary in increase from the 100% increase in such crops as avocado to 45% in corn, 60–70% in alfalfa, 30% in wheat, and lesser gains (7–18%) for low crops such as lettuce. All these increases follow windbreak establishment on exposed sites. Effects are of course less in naturally sheltered situations or areas of normally low wind speed. However, almost all normal garden vegetables (cucurbit, tomato, potato) benefit greatly from wind shelter. For this reason, a ground pattern similar to that in **Figure 5.20** is recommended for such crops and wind–affected pastures.

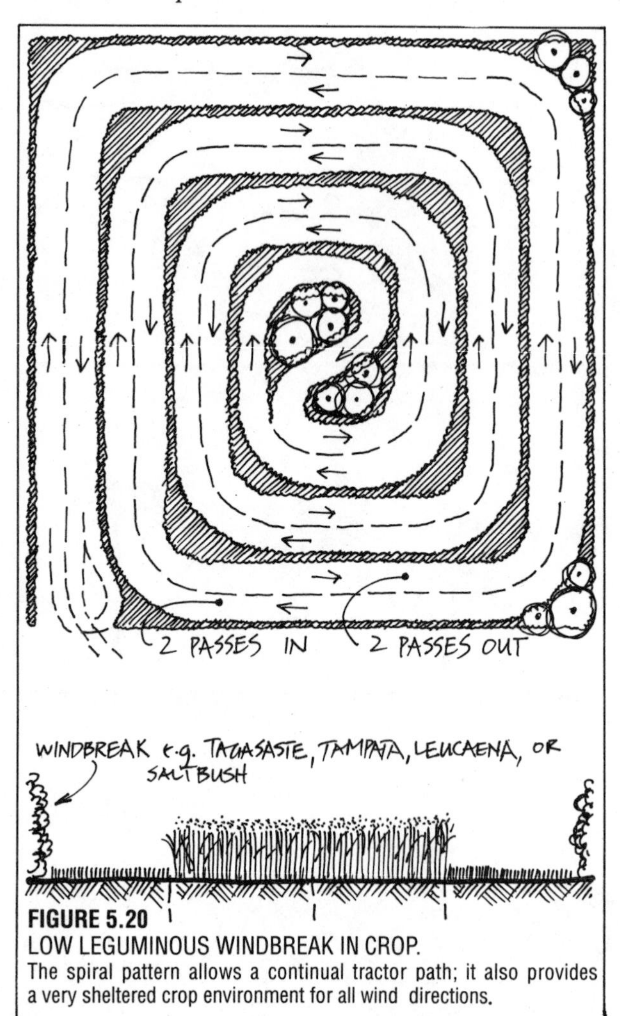

FIGURE 5.20
LOW LEGUMINOUS WINDBREAK IN CROP.
The spiral pattern allows a continual tractor path; it also provides a very sheltered crop environment for all wind directions.

HURRICANES (CYCLONES, TYPHOONS)

Very stable and still–air calms near the equator may produce fierce updraughts of air over warm oceanic areas, which over some days or weeks build up to the great rising spirals of hurricanes. As these move slowly (usually at 24–32 km/h) across the ocean towards land, wind speeds around the vortex can reach 128–192 km/h, while within the vortex itself a "tidal bulge" rises up to 2.7 m (9 feet) above sea level (**Figure 5.21**); this water bulge causes a tidal surge at coastlines.

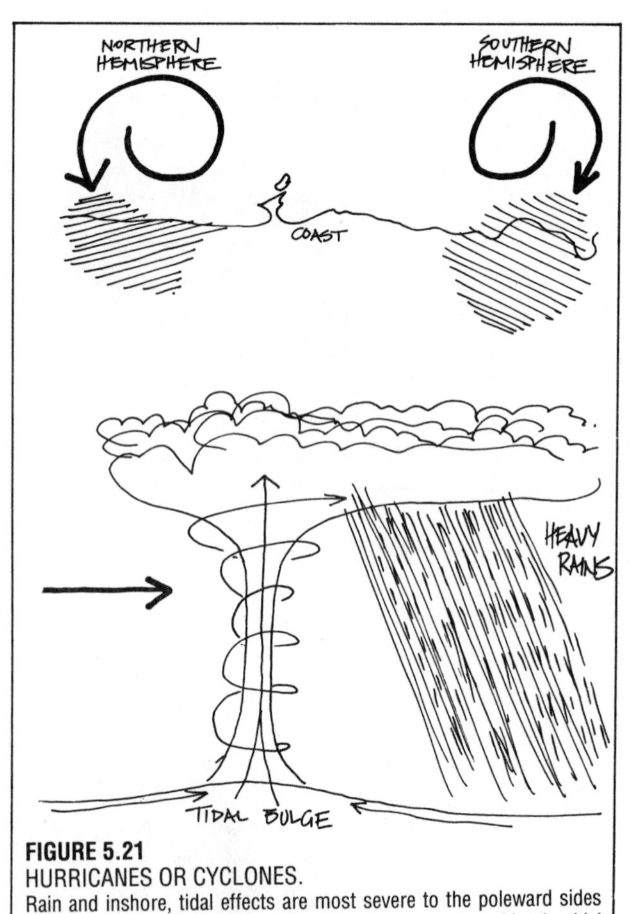

FIGURE 5.21
HURRICANES OR CYCLONES.
Rain and inshore, tidal effects are most severe to the poleward sides of the low–pressure cell, as is inland flooding, accentuated by the tidal bulge.

Vortices revolve anti–clockwise in the northern hemisphere, clockwise in the southern, and thus coastal areas to the north side have the highest water and wave levels in the northern hemisphere, and to the south side in the southern. The combined effects of rapidly fluctuating pressures, tidal bulge, wave and sea pile–up, and wave backwash create devastation on coasts. Although hurricanes cannot persist far inland, as the sea itself generates the vortex, the intense rains generated do reach well inland to flood rivers and estuaries, adding to the general destruction. With all effects combined in a "worst case" of high tides and prior rains, destructive wave attack can reach 6–9 m (20–30 feet) above normal high–tide wave levels.

As wind strength increases at sea, wavelength also increases, so that normal wave fronts arriving at 8 per minute in calm Atlantic conditions slow down to a storm frequency of 5 per minute before great winds. These wider–spaced waves travel fast, are larger, and create severe backwash undermining of shorelines.

Storm waves may therefore arrive long before a cyclonic depression or hurricane, and the change of wave beat gives warning to the shore crabs, birds, fish, and turtles, who either take shelter inland or go to sea to escape the approaching hurricane.

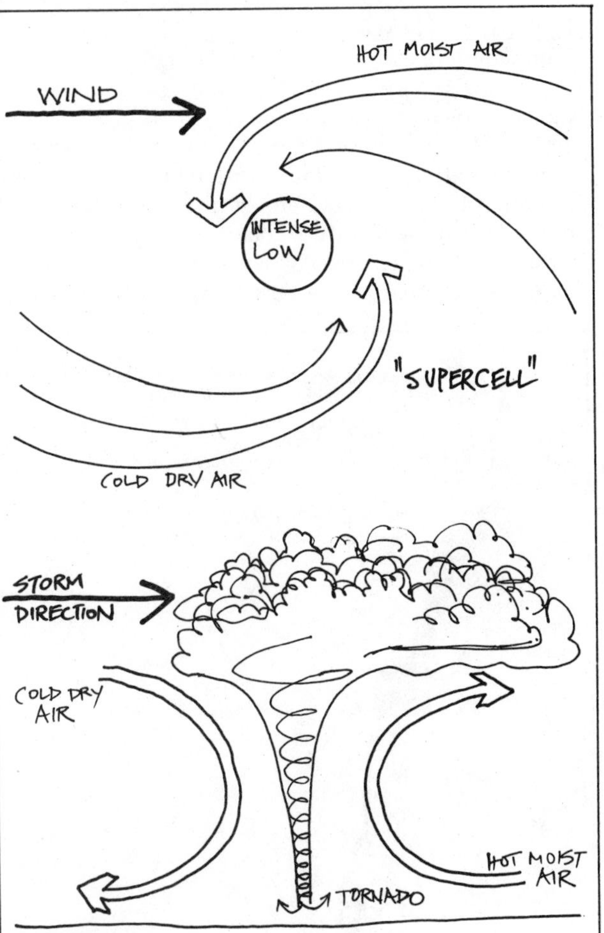

As modern satellite photographs are used to track the hurricane, there is usually a few days' warning for coastal areas, and evacuation is sometimes ordered. Well–built towns (such as Darwin, Australia, after its cyclonic devastation in 1972) can withstand cyclones with minimal damage, but such stoutness is usually only built in after an initial (and sometimes total) destruction. It is possible to strictly regulate and supervise buildings to be safe in hurricanes, and in areas where flimsy constructions are normal, to dig refuge trenches and caves for emergency shelter. All such shelter must be in well–drained hillside sites.

FIGURE 5.22
TORNADOES.
Unlike hurricanes, tornadoes occur over land and sea from shear effects at the junction of hat and cold air masses, creating intense low–pressure vortices.

TORNADOES

Hurricanes are large, slow phenomena covering hundreds of square miles, and mainly confined to coasts facing large stretches of tropical seas, with very large heat cells. Tornadoes, however, may occur in quite cold inland areas, last only seconds or minutes, and affect only a few square kilometres. Thus, they usually escape detection by satellite and ground sensors.

Nevertheless, the stresses placed on buildings, civil constructions, chemical or nuclear facilities, airfields and villages can be disastrous. Wind speeds may reach 120 km/h, at worst 280 km/h; these speeds can exceed hurricane winds. The conditions for tornadoes are:

- Thunderstorms with fast–growing cumulonimbus cloud;
- A persistent source of warm moist air to feed the updraught side of the front;
- An input of cold dry air entering the system from another direction; and
- A vortex formation in the resulting storm; this reaches the ground as a tornado, caused by wind–shear effects at the border of the conflicting system, or as a frontal dust storm in deserts.

Effects: Trees twisted off and broken; people and

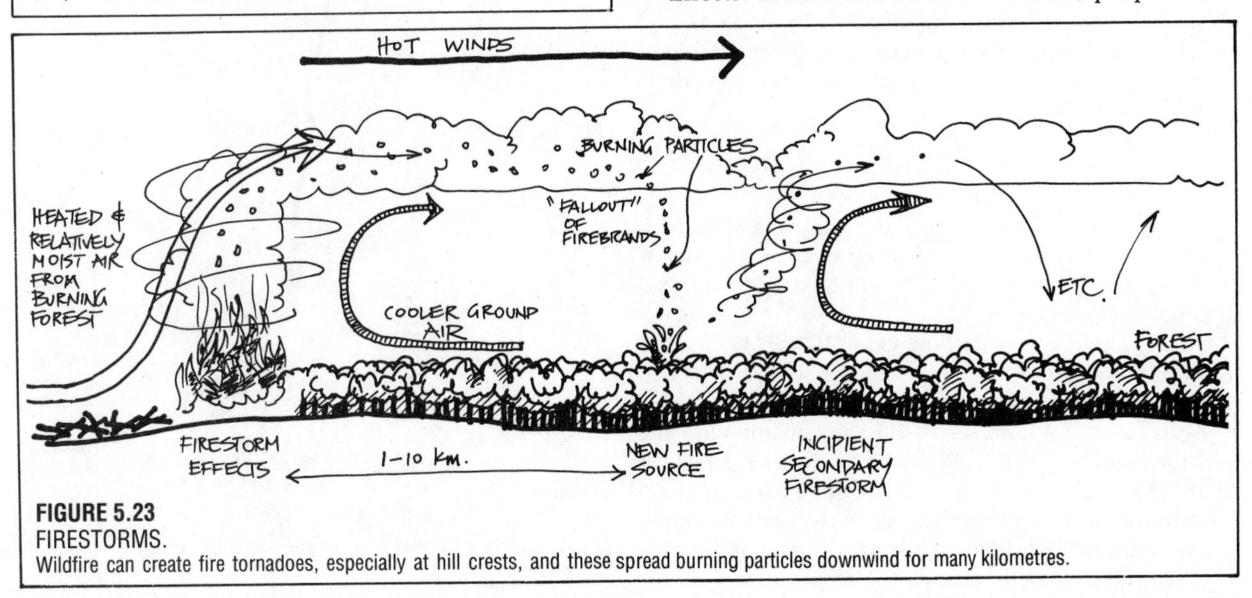

FIGURE 5.23
FIRESTORMS.
Wildfire can create fire tornadoes, especially at hill crests, and these spread burning particles downwind for many kilometres.

objects sucked out of cars and buildings; "rains" of soil, fish, frogs may fall out ahead of the disturbance.

FIRESTORMS

Intense wildfires (urban and rural) fanned by dry winds will create powerful vortices due to conditions very much like that of the tornado. The mass ignition of large areas of forests and buildings feed a powerful updraught. Colder dry air rushes in to replace the air consumed in burning, and fire tornadoes (firestorms) result, carrying large burning particles aloft on "smoke nimbus" clouds. Whole house sections pinwheel across the sky to drop out ahead of the main fire front, where they in turn set up secondary firestorm conditions. The effects on people and property are very much like those of tornadoes, but with the additional danger of intense heat.

5.7

LANDSCAPE EFFECTS

CONTINENTAL EFFECTS

Heat is transported on a world scale by two great circulations: that of the air masses, and those of oceanic currents. Of these, air masses are more wide-spread in their effect, and are least limited by land masses. Oceanic currents, or indeed proximity to any large body of water, have their greatest moderating effect on down–wind shorelines. Such effects may have little inland influence. The concept of continental climates was evolved to describe those extreme and widely fluctuating inland climatic zones that are not buffered by the effects of sea currents, and which demonstrate periods of extreme heat and cold, all the more marked on high mountains.

Thus, the third complication on the simple temperature–rainfall classifications is CONTINENTALITY. After this, only one special factor remains, and it is the effect of hills or ranges of mountains on local climate; these effects are very like the latitudinal effects on a global scale.

LATITUDE AND ALTITUDE

An average measure of temperature fall with altitude is: 9.8°C/km (5.4°F/1000 feet) in rainless or dry air; or 4–9°C/km (2.2–5°F/1000 feet) in humid and saturated conditions.

As a rough approximation, every 100 m (330 feet) of altitude is equivalent to 1° of latitude, so that at 1000 m (3300 feet) on the equator, the temperatures are about equivalent to a climate 10° off the equator with the same humidity. At 10° latitude off the equator, a plateau at 1850 m (6000 feet) has a climate more like that at 30° latitude, with a probability of wind chill to below freezing. For high islands or ranges of mountains, this altitudinal factor is crucial to design strategies for homes and gardens. Altitude effect alone enables us to grow a wide range of plant species on a high island, using the area from ocean to mountain– top.

High Altitude Effects

Mountains are not in fact strictly "latitude equivalents", as the air is more rarefied, air pressure less, and radiation therefore higher. On very high mountains of 4000 m (13,000 feet) and more, people may experience oxygen deficiency (mountain sickness), snow or radiation blindness, and suffer from the extremes of day–night temperature fluctuations. The mountain sickness of oxygen stress is not felt by locals, but can cause extreme fatigue, insomnia, and laboured

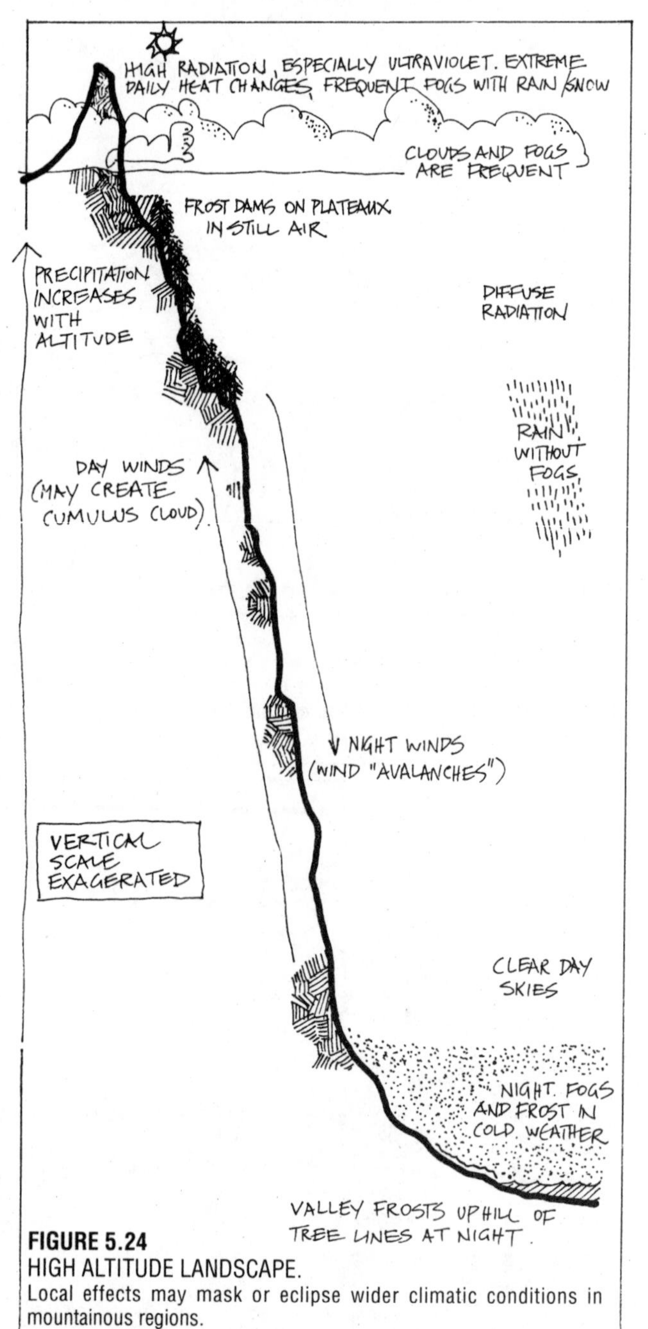

FIGURE 5.24
HIGH ALTITUDE LANDSCAPE.
Local effects may mask or eclipse wider climatic conditions in mountainous regions.

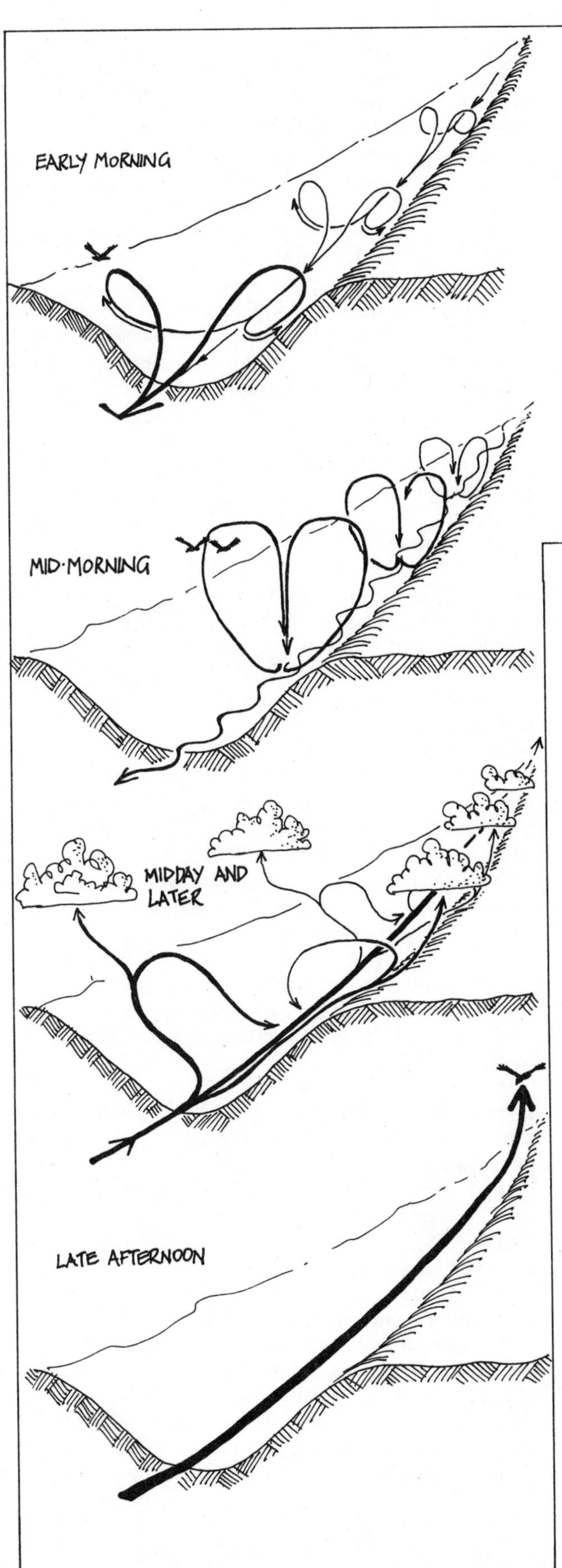

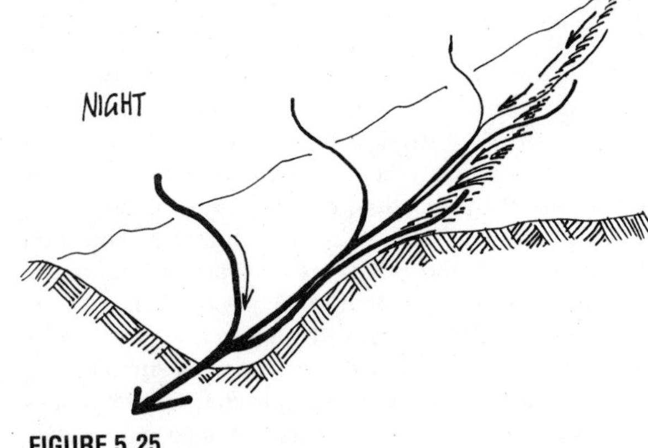

FIGURE 5.25
VALLEY WINDS
These follow a daily cycle. Many bird species use these winds to follow a daily migration (downhill at dawn, to ridge forests at evening). Both wind and temperature effects are local.

breathing in visitors. In the Peruvian Andes, day temperatures remain at 16–19°C (61–66°F) all year, but night temperatures fall rapidly to –10°C (14°F). Shade temperatures are lower than at sea level, due to the less effective heat transfer and insulation effects of the rarefied air. Water boils at lower temperatures due to the low atmospheric pressure, and snow may sublime directly to water vapour rather than melting. High mountains reduce the range of foods available.

Snow cover may serve as an insulating blanket, and prevent early spring thawing, or even autumn freezing if it covers unfrozen ground. Snow cover also causes intense reflection, and raises air temperatures just above the snow by day. At night, radiation from snow causes an extremely cold ground air layer, so that any plants protruding from snow suffer these extremes of diurnal temperature.

As great as the effect of altitude is, the effect of slope is even more pronounced. Daubenmire (1974) records that slopes of 5° towards the poles "reduce soil temperatures as much as 168 km distance" towards the poles, so that even a gentle slope away from the sun creates very much cooler conditions locally. The effect of cold ravines in near–permanent shadow is extreme indeed, and one may stand in hot sunlight in the Himalayas and gaze into icy depths where only the hardiest life forms exist, and where ice may permanently cover rocks and spray zones caused by waterfalls or rapids.

VALLEY CLIMATES

We have referred to the chill of narrow, shaded, high altitude gorges, but an opposite effect occurs in sun–facing wider valleys, sheltered from winds. Here, hot air builds up rapidly, soils are drier, and strong winds may be generated (upslope and up valley by day, downslope and down valley at night). **Figure 5.25** demonstrates this effect in moderate mountain areas of

3,000–4,000 m (9,850–13,000 feet).

In large valleys, and especially in cool moist climates, the upslope wind may result in the generation of a chain of cumulus clouds at the valley head, trailing off as a succession of clouds from mid-morning to evening. In more tropical humid climates, the cloud may be continuously held on the mountain tops; this forms part of the standing cloud of high islands. Such cloud (and rainfall) effects are accentuated by forest on the valley sides and ridges, as trees actively humidify the air streams by transpiration in hot weather.

Valleys in tundra and desert support tree populations absent from the plain or peneplain areas surrounding them, but the reasons may differ in that tundra valleys are likely to be protected by (driven) deep snow cover. This preserves warm or sub-lethal soil temperatures in winter (as well as providing excess summer melt moisture). Valleys in deserts remain moist due to the deep detritus which fills their floors; the shaded soils lose less moisture to evaporation. Lethal soil temperatures are also avoided by partial shading. Both ice-blast and sand-blast are modified or absent in valley floors, so that unprotected seedlings can survive high winds in the shelter of valleys.

Thus, valleys (or wadis) are preferred growing sites in deserts, and provide tree products in otherwise treeless tundras, although the latter sites are rarely occupied by human settlement.

In the field, we often notice a sudden coldness just before dawn in valley areas; this is the time of the greatest depth of cold air, and hence the greatest intensity of cold. Air flowing down from the mountains has pooled all night, and just before the sun rises, we (and many animals) are at our greatest exposure and lowest ebb. It is at this time that winds off glaciers flowing down cold valleys reach their maximum speed.

Without wind or air flow, radiation frost can form, as it does in sheltered hollows and tree clearings. In these areas, opening up the clearings or draining them of cold air may help reduce frost, if that is the aim (**Figure 5.13**).

When, some years ago, I grew such crops as tomatoes and cucurbits inside open tree canopies, I did prevent frost, but lost crop due to low light levels and a lack of wind or insect pollination. In such cases, a shade-side screen of reflector plants facing the sun would help to keep light levels up, and plants to attract bees need to be placed around the clearing. Arboreal or ground browsers within forests are also worrisome in gardens (possum and porcupine for example). Green-leaf vegetables, however, are not usually eaten, and can be successfully grown in small forest clearings or in open forest in frosty areas.

5.8 LATITUDE EFFECTS

Despite the weak light and short growing season at high (sub-polar) latitudes, the very long summer days provide more than a sufficient quantity of light for vigorous plant growth. The daily total for late summer (July) is 440 Langleys at Madras, India (13°N); 680 Langleys at Fresno, California (36°N); 450 at Fairbanks, Alaska (64°N). The average radiation in temperate areas is 1–5 times that of the tropics (Chang, 1968).

This is accentuated, on or near coasts, by the moderate temperatures from the convection of air over warm currents in such areas as Alaska and the northwest coastal regions of Europe. The benefit of these areas is that the generally lower temperatures, which suit photosynthesis, confer a photosynthetic efficiency that makes a considerable production of cereal, berries, tomatoes, potatoes, and vegetable crops tolerant of short season/long day conditions. The often deep periglacial soils provide the basis for the production of gigantic lettuce, cabbage, spinaches, and root crops, so that these areas are very favourable for agriculture in summer.

Such conditions prevail in Alaska, Ireland, Scotland, and parts of Norway. Shelter and added nutrients from seaweeds and manures yield rich meadows and heavy vegetable production during the long summer days. The small stone-walled fields of Ireland produce abundant sweet hay, root crops, and greens for storage during winter.

Conversely, the ample light at low (equatorial) latitudes is inefficient due to the extremely high temperatures there, and the excess light may mean that plants are light saturated. Photosynthesis may actually decline in the intense light, and the energy built into the plants may be less in sunlight than in partial shade. Shade (down to a level of 20% sunlight) is of great benefit in tropical deserts and sunny equatorial climates. Trials of shadecloth with 50–70% light transmission may greatly increase plant bulk and production, e.g. of sugar beet, thus the importance of tree shade and shadecloth in deserts and cleared-area tropics.

Similarly, temperatures above 25°C (77°F) sharply decrease photosynthetic efficiency, so that the normal desert or equatorial condition of high light and temperature is very inefficient for the production of plant material. In the arctic or high latitudes, 15°C (59°F) is *optimum* for adapted species and cultivars, and 20–24°C (68–75°F) for many useful food plants. Tropics are noted for a low production of those crops which can be also grown in temperate areas; light shade may be the essential component for increased yields.

In bright sunlight, leaf temperatures often exceed air temperatures, so that the diffuse light of overcast or cloudy days in high latitudes helps plant growth, especially after midday as temperatures would then also rise above optimum in direct sunlight.

Photosynethetic efficiency is limited by the ability of the leaf to obtain carbon dioxide or by low levels of available carbon dioxide. At high light intensity, we need to supply carbon dioxide (to the saturation level of 0.13% to obtain a 2–3 times increase in photosynthetic rate. Carbon dioxide can be supplied by composting or by housing animals in greenhouses where light is more than sufficient.

It follows that the summer periods of the high latitudes are ideal for biomass production, while equatorial regions evolve biomass mainly as a result of a year–round (inefficient) growth and perennial crops. The ideal of steady low light/low temperature conditions may be at times achieved below the closed forests of tropical mountains, but these sites are very limited in extent, and carbon dioxide concentration is also low. Rice, for example, yields 4–5 times better in temperate areas than in tropical ones, although up to three crops per year in the tropics helps to increase local yields over the year.

It should be feasible to assist tropical crop yields by spacing permeable–crowned trees throughout crops to reduce both light and temperature, e.g. using *Prosopis* trees with millet crops in India, or partially–shading taro in Hawaii. Grass growth in temperate areas also increases with shelterbelt, but this may reflect the *warmer* conditions and lack of mechanical wind damage that such trees as tagasaste provide. Trials of light–transmitting or thin–crowned palms and legume trees would quickly show results, and there are a good many observations to suggest that (if water is sufficient) crops under leguminous trees do much better in the tropics than a crop standing on its own.

Part of the problem in tropics (both for biomass production and nutrition) is that non–adapted temperate crops are persistently grown there. True tropical plants can not only stand much higher levels of light before saturation, but can also maintain photosynthesis at low (0.10%) carbon dioxide.

In summary, we do not have to accept the climatic factors of a site as unchangeable any more than we do its treelessness or state of soil erosion. By sensible placement of our design components, we can create myriad small differences in local climatic effects on any site. In the technical field, we can create useful conversions of energy from incoming energy fluxes such as wind and sun, and produce energy for the site. In the patterning of a site with trees, ponds, earth systems, or hedgerows, we can actively moderate for better climatic conditions, or to eliminate some local limiting factor.

5.9
REFERENCES

Chang, Jen–Hu, *Climate and Agriculture,* Aldine Pub. Co., Chicago, 1968.

Chorley, R.J . (ed.), *Water, Earth, and Man,* Methuen & Co., London, 1969.

Cox, George W., and Michael D. Atkins, *Agricultural Ecology,* W. H. Freeman & Co., San Francisco, 1979.

Daubenmire, Rexford F., *Plants and Environment,* Wiley International, 1974.

Eyre, S. R., *World Vegetation Patterns,* Macmillan, London, 1971.

Gaskell, T. F., *Physics of the Earth,* World of Science Library, Thames & Hudson, London, 1970.

Geiger, Rudolf, *The Climate Near the Ground,* Harvard University Press, 1965.

Holford, Ingrid, *Interpreting the Weather,* David & Charles, Newton Abbot, UK.

James, P. E., *Outline of Geography,* Ginn & Co., Boston, 1941.

Trewartha, S. T., *An Introduction to Climate,* McGrawHill, New York, 1954.

Twidale, C. R., *Structural Landforms,* Australian National University Press, Canberra, Australia.

Staff of the L. H. Bailey Hortorium, Cornell University, *Hortus Third: a concise dictionary of the plants cultivated in the United States and Canada,* Macmillan Publishing Co., 1976.

5.10
DESIGNER'S CHECKLIST

Check data on average rainfall, temperature, and wind speed and direction for the region (often found by contacting the Bureau of Meteorology).

Ascertain the general "hardiness" zone for plants and animals. This is based on temperature, with frost being the limiting factor. Make a survey of the plants that grow in the area, noting special circumstances surrounding plants that are "marginal"; what is the technique or microclimate that allows them to grow?

Find out about flood locations and periodicity, rain intensity, temperature extremes, and the *seasonal* rainfall pattern. Allow for extremes (e.g. no rain in summer) when designing.

Consider *total* precipitation (snow, hail, rain, fog, condensation, and dew) so that your design can include ways to trap and store moisture (dry climates) or ways to dispose of too much moisture (wet climates).

Consider *light availability,* especially on foggy coasts; light becomes the limiting factor for flowering plants.

Continental climates mean more temperature extremes, while maritime climates buffer severe heat or cold.

Altitude effects: approximately every 100 m of altitude is equivalent to 1° of latitude, so that a variety of plants can be grown if the property contains hills and flats. In the sub–tropics, even temperate–area plants can be grown on high islands or hills.

Note where frost is produced (in hollows, on flats, and in large clearings) and where it is absent (the "thermal belt" on hills, under tree canopies).

Note tree flagging on the site; this shows the direction of *persistent* winds (although winds, sometimes severe, may blow from other directions). You can put tall stakes with coloured cloth or plastic streamers at different locations and observe them seasonally. (**Figure 6.2**).

For accurate temperatures, you can have several maximum/minimum thermometers in different locations. These thermometers record the highest and lowest temperatures reached during 24 hours, and are helpful in locating microclimatic areas such as thermal belts (if on a sun–facing slope), cold drainage areas, frost hollows.

Site house and garden on the thermal belt if possible.

In minimal–frost areas, plant light–canopy trees *in* the garden for frost protection (tree canopies help keep rapid cooling of the earth to a minimum). Or plant into a steep–sided clearing or pit.

In houses, design so that you use light and radiation to best effect, particularly in temperate climates. Particular use should be made of the thermosiphon effect of heat, so that heat sources are placed below storage and use points.

Use the principle that white reflects, dark absorbs, heat. Plant shrubs and trees needing heat and light in front of white–painted walls.

When planning windbreaks, consider:

- Trees that give multiple function, e.g. mulch (*Casuarina*), bee nectar (dogwood), sugar pods for animals (carob, honey locust), edible leaves (*Leucaena*, tagasaste), berries for poultry (*Coprosma repens*, Russian olive).
- The windbreak planting itself may need initial protecion and care (nutrients, water, weeding, or mulching).
- If the winds are *very severe*, look around the area to see what stands up to it, and plant it whether it provides multiple function or not. Plant more useful plants in its lee. Protection includes fencing, earth banks, tyre walls, etc.
- Choose a windbreak configuration that is effective for the particular design situation. In tropical and subtropical areas, a thin–crowned windbreak in crop can be used to advantage, providing shade and mulch for vegetable crop.

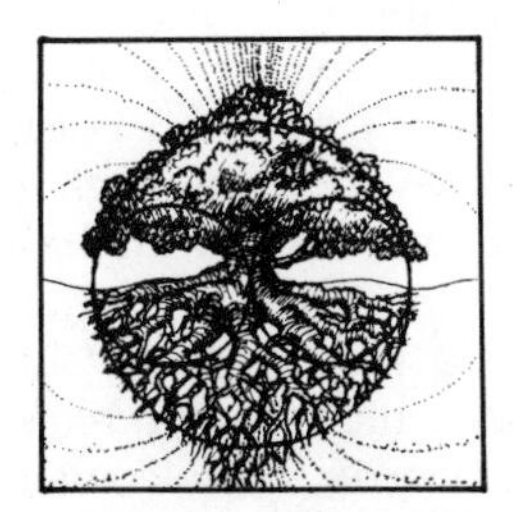

Chapter 6

TREES AND THEIR ENERGY TRANSACTIONS

On the dry island of Hierro in the Canary Islands, there is a legend of the rain tree; a giant 'Til' tree *(Ocotea foetens)*, "... the leaves of which condensed the mountain mists and caused water to drip into two large cisterns which were placed beneath. The tree was destroyed in a storm in 1612 A.D. but the site is known, and the remnants of the cistern preserved ... {This one tree} distilled sufficient water from the sea mists to meet the needs of all the inhabitants."

(David Bramwell)

For me, trees have always been the most penetrating teachers. I revere them when they live in tribes and families, in forests and groves... They struggle with all the forces of their lives for one thing only: to fulfill themselves according to their own laws, to build up their own forms, to represent themselves. Nothing is holier, nothing is more exemplary, than a beautiful strong tree.

(Herman Hesse, "Trees", *Natural Resources Journal*, Spring 1980)

I am astonished to find whole books on the functioning of trees which make no mention of their splendid mechanical and aerodynamic performance.

(Vogel, *Life in Moving Fluids*, 1981)

A point which is often overlooked is the effect of trees in increasing the total precipitation considerably beyond that recorded by rain gauges. A large proportion of the rime which collects on the twigs of trees in frosts afterwards reaches the ground as water, and, in climates such as those of the British Isles, the total amount of water deposited on the twigs from fogs and drifting clouds is considerable, and most of it reaches the streams or underground storage, or at least replaces losses from subsequent rainfall.

Of more importance, however, to hydraulic engineers is the effect of woodlands in modifying the run-off. The rush of water from bare hillsides is exchanged for the slower delivery from the matted carpet of the woodland, losses by evaporation may be much diminished and the melting of snow usefully retarded. In catchments from which flood waters are largely lost, woodlands may increase the available runoff by extending the period of surface flow. The maximum floods of rivers are reduced, and the lowest summer flow increased. Woodlands are usually much more effective than minor vegetation, such as gorse and heather, in preventing the soil from being carried from the land into an open reservoir.

To protect a reservoir from silting, it may be unnecessary to plant large areas, the silt being arrested by suitable planting of narrow belts of woodland, or by the protection of natural growth, along the margins of the streams.

Some engineers consider that in the case of small reservoirs the shelter afforded by a belt of trees along the margins is of value in reducing the amount of scour of the banks caused by wave action. Afforestation over considerable areas in large river basins would, in many cases, reduce the amount of silting in navigable rivers and estuaries.

A matter which does not receive sufficient attention in connection with hydraulic engineering is the effect of judicious planting or woodland conservation over small areas. A narrow belt of woodland along the foot of a slope will arrest the soil brought down by rains from the hillside. The encouragement of dense vegetation along the bottom of a narrow valley may check the rate of flood discharge to a useful extent. The planting of suitable trees along ridges and for a little way down the slope facing the rain bearing and damp winds, will produce the maximum of certain desired effects, in proportion to the area occupied. Suitable tree and bush growths in swampy areas and around their margins will increase their effect in checking flood discharge, and may prevent these areas from contributing large quantities of silt to the streams during very heavy rains. Areas of soft, cultivable soil liable to denudation may similarly be protected. Generally, a country which is, in the ordinary English sense of the words, 'well timbered' is, from the point of view of the hydraulic engineer, a favourable country; and in the development of new lands the future effects of a poposed agricultural policy should be considered from this point of view, and in consultation with hydraulic engineers.

(R.A. Ryves, *Engineering Handbook*, 1936)

6.1 INTRODUCTION

This chapter deals with the complex interactions between trees and the incoming energies of radiation, precipitation, and the winds or gaseous envelope of earth. The energy transactions between trees and their physical environment defy precise measurement as they vary from hour to hour, and according to the composition and age of forests, but we can study the broad effects.

What I hope to show is the immense value of trees to the biosphere. We must deplore the rapacity of those who, for an ephemeral profit in dollars, would cut trees for newsprint, packaging, and other temporary uses. When we cut forests, we must pay for the end cost in drought, water loss, nutrient loss, and salted soils. Such costs are *not* charged by uncaring or corrupted governments, and deforestation has therefore impoverished whole nations. The process continues with acid rain as a more modern problem, not charged against the cost of electricity or motor vehicles, but with the inevitable account building up so that no nation can pay, in the end, for rehabilitation.

The "capitalist", "communist", and "developing" worlds will all be equally brought down by forest loss. Those barren political or religious ideologies which fail to care for forests carry their own destruction as lethal seeds within their fabric.

We should not be deceived by the propaganda that promises "for every tree cut down, a tree planted". The exchange of a 50 g seedling for a forest giant of 50–100 tonnes is like the offer of a mouse for an elephant. No new reafforestation can replace an old forest in energy value, and even this lip service is omitted in the "cut–and–run" forestry practised in Brazil and the tropics of Oceania.

The planting of trees can assuredly increase local precipitation, and can help reverse the effects of dryland soil salting. There is evidence everywhere, in literature and in the field, that the great body of the forest is in very active energy transaction with the whole environment. To even begin to understand, we must deal with themes within themes, and try to follow a single rainstorm or airstream through its interaction with the forest.

A young forest or tree doesn't behave like the same entity in age; it may be more or less frost–hardy, wind–fast, salt–tolerant, drought–resistant or shade tolerant at different ages and seasons. But let us at least try to see just how the forest works, by taking one theme at a time. While this segmented approach leads to further understanding, we must keep in mind that everything is connected, and any one factor affects all other parts of the system. I can never see the forest as an assembly of plant and animal species, but rather as a single body with differing cells, organs, and functions. Can the orchid exist without the tree that supports it, or the wasp that fertilises it? Can the forest extend its borders and occupy grasslands without the pigeon that carries its berries away to germinate elsewhere?

Trees are, for the earth, the ultimate translators and moderators of incoming energy. At the crown of the forest, and within its canopy, the vast energies of sunlight, wind, and precipitation are being modified for life and growth. Trees not only build but conserve the soils, shielding them from the impact of raindrops and the desiccation of wind and sun. If we could only understand what a tree does for us, how beneficial it is to life on earth, we would (as many tribes have done) revere all trees as brothers and sisters.

In this chapter, I hope to show that the little we do know has this ultimate meaning: *without trees, we cannot inhabit the earth*. Without trees we rapidly create deserts and drought, and the evidence for this is before our eyes. Without trees, the atmosphere will alter its composition, and life support systems will fail.

6.2 THE BIOMASS OF THE TREE

A tree is, broadly speaking, many biomass zones. These are the stem and crown (the visible tree), the detritus and humus (the tree at the soil surface boundary) and the roots and root associates (the underground tree).

Like all living things, a tree has shed its weight many times over to earth and air, and has built much of the soil it stands in. Not only the crown, but also the roots, die and shed their wastes to earth. The living tree stands in a zone of decomposition, much of it transferred, reborn, transported, or reincarnated into grasses, bacteria, fungus, insect life, birds, and mammals.

Many of these tree–lives "belong with" the tree, and still function as part of it. When a blue jay, currawong, or squirrel buries an acorn (and usually recovers only

80% as a result of divine forgetfulness), it acts as the agent of the oak. When the squirrel or wallaby digs up the columella of the fungal tree root associates, guided to these by a garlic–like smell, they swallow the spores, activate them enzymatically, and deposit them again to invest the roots of another tree or sapling with its energy translator.

The root fungi intercede with water, soil, and atmosphere to manufacture cell nutrients for the tree, while myriad insects carry out summer pruning, decompose the surplus leaves, and activate essential soil bacteria for the tree to use for nutrient flow. The rain of insect faeces may be crucial to forest and prairie health.

What part of this assembly is the tree? Which is the body or entity of the system, and which the part? An Australian Aborigine might give them all the same "skin name", so that a certain shrub, the fire that germinates the shrub, and the wallaby that feeds off it are all called *waru*, although each part also has *its* name. The Hawaiians name each part of the taro plant differently, from its child or shoot, to its nodes and "umbilicus".

It is a clever person indeed who can separate the total body of the tree into mineral, plant, animal, detritus, and life! This separation is for simple minds; the tree can be understood only as its total entity which, like ours, reaches out into all things. Animals are the messengers of the tree, and trees the gardens of animals. Life depends upon life. All forces, all elements, all life forms are the biomass of the tree.

A large tree has from 10,000 to 100,000 growing points or MERISTEMS, and each is capable of individual mutation. Unlike mammals, trees produce their seed from multitudinous flowers. Evidence is accumulating that any one main branch can therefore be an "individual" genetically. Some deciduous poplars may produce a single evergreen branch. "Seedlessness" in fruit, or a specific ripening time, may belong only to one branch. Grafts and cuttings perpetuate these isolated characteristics, so we must look upon the tree itself as a collection of compatible genetic individuals, each with a set of persistent characteristics which may differ from place to place on the tree, and each of which may respond differently to energy and other stimuli. Like ourselves, trees are a cooperative amalgam of many individuals; some of these are of the tree body, but most are free–living agents. As little as we now know about trees, they stand as a witness to the complex totality of all life forms.

6.3 WIND EFFECTS

Vogel (1981) notes that as wind speed increases, the tree's leaves and branches deform so that the tree steadily reduces its exposed leaf area. At times of very high winds (in excess of 32 m/sec) the interception of light, efficient water use, and convective heat dissipation by the tree becomes secondary to its survival.

Vogel also notes that very heavy and rigid trees spread wide root mats, and may rely more totally on their weight, withstanding considerable wind force with no more attachment than that necessary to prevent slide, while other trees insert gnarled roots deep in rock crevices, and are literally anchored to the ground.

The forest bends and sways, each species with its own amplitude. Special wood cells are created to bear the tension and compression, and the trees on the edge of a copse or forest are thick and sturdy. If we tether a tree halfway up, it stops thickening below the tether, and grows in diameter only above the fixed point. Some leaves twist and reverse, showing a white underside to the wind, thus reflecting light energy and replacing it with kinetic or wind energy. In most cases, these strikingly light–coloured leaves are found only in forest edge species, and are absent or uncommon within the forest.

As streamlines converge over trees or hills, air speed increases. Density and heat may also increase, resulting in fast low–pressure air. To leeward of the obstruction, such streamlines diverge, and an area of slower flow, higher pressure, and cooler air may result. If rain has fallen due to the compression of streamlines, however, the latent heat of evaporation is released in the air, and this drier air can be warmer than the air mass rising over the obstruction. The pressure differentials caused by uplift and descent may affect evaporation as much as wind drying or heat.

Apart from moisture, the wind may carry heavy loads of ice, dust, or sand. Strand trees (palms, pines, and *Casuarinas*) have tough stems or thick bark to withstand wind particle blast. Even tussock grasses slow the wind and cause dust loads to settle out. In the edges of forests and behind beaches, tree lines may accumulate a mound of driven particles just within their canopy. The forest removes very fine dusts and industrial aerosols from the airstream within a few hundred metres.

Forests provide a nutrient net for materials blown by wind, or gathered by birds that forage from its edges. Migrating salmon in rivers die in the headwaters after spawning, and many thousands of tons of fish remains are deposited by birds and other predators in the forests surrounding these rivers. In addition to these nutrient sources, trees actively mine the base rock and soils for minerals.

The effect of the wind on trees is assessed as the Griggs and Putnam index (**Table 6.1**), and the accompanying deformations in both crown shapes and growth (as revealed in stems) is given a value which is matched to wind speeds with an average 17% accuracy.

Such scales and field indicators are of great use in design. When we go to any site, we can look at the condition of older trees, which are the best guide to gauge wind effect. Trees indicate the local wind direction and intensity, and from these indicators we

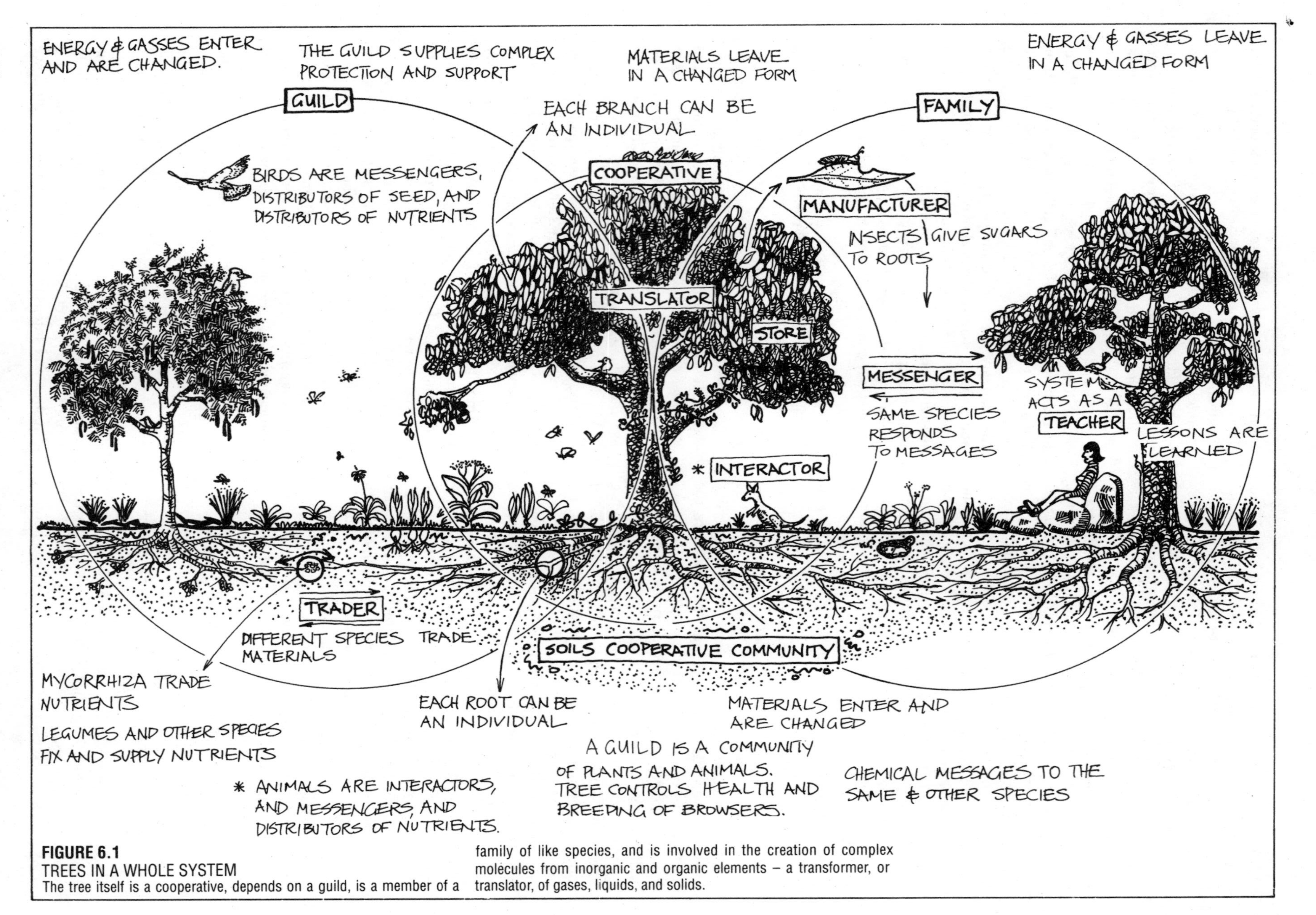

FIGURE 6.1
TREES IN A WHOLE SYSTEM
The tree itself is a cooperative, depends on a guild, is a member of a family of like species, and is involved in the creation of complex molecules from inorganic and organic elements – a transformer, or translator, of gases, liquids, and solids.

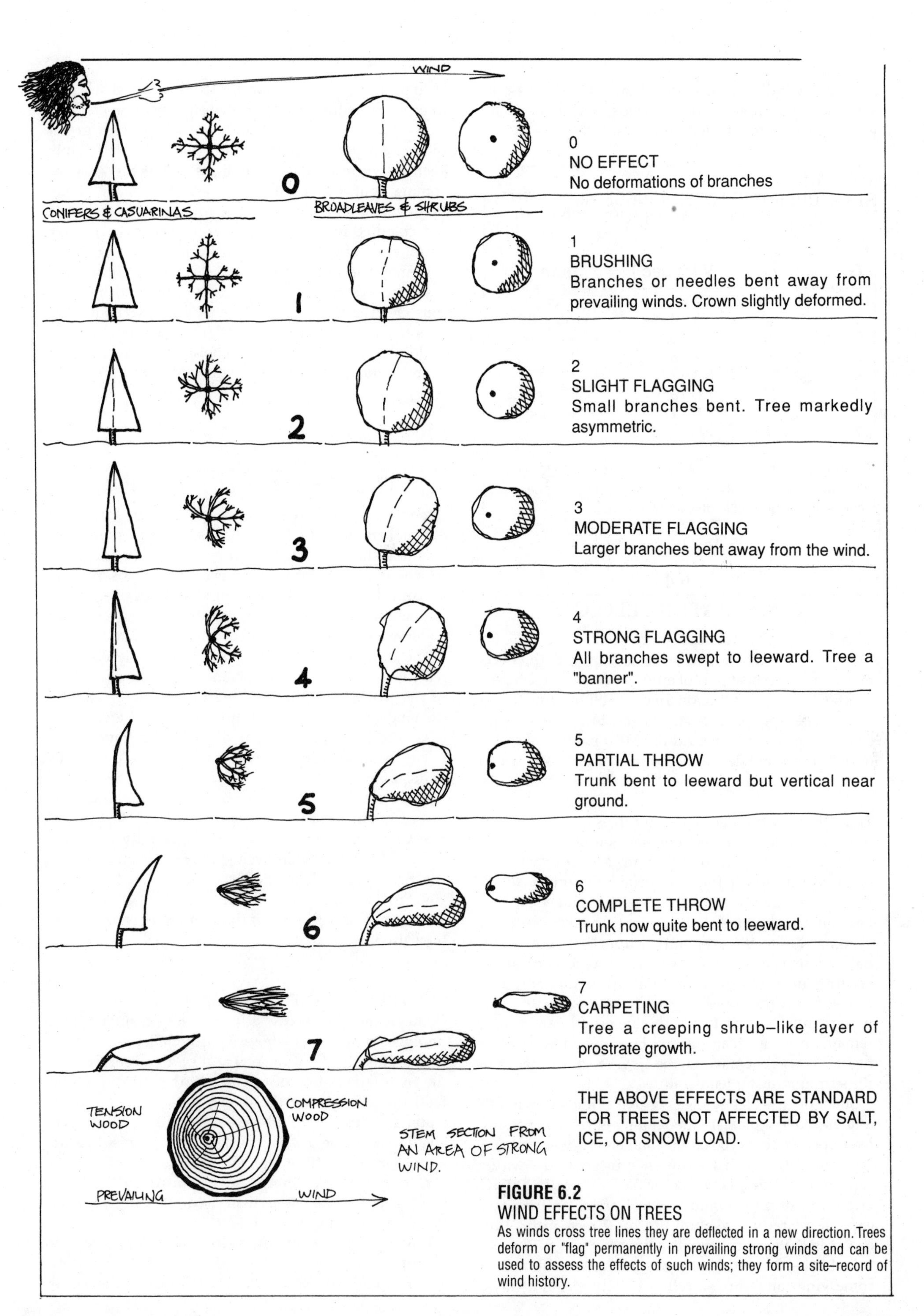

FIGURE 6.2
WIND EFFECTS ON TREES
As winds cross tree lines they are deflected in a new direction. Trees deform or "flag" permanently in prevailing strong winds and can be used to assess the effects of such winds; they form a site–record of wind history.

can place windbreak to reduce heat loss in homes, to avoid damage in catastrophic winds, and to steer the winds to well–placed wind machines.

TABLE 6.1
RELATIONSHIP BETWEEN GRIGGS AND PUTNAM INDEX (G) AND ANNUAL MEAN WIND SPEED (V) IN m/sec.

G	V (m/sec.)	mph
0	<3.3	<7.5
1	3.3 - 4.2	7.5 - 9.5
2	4.3 - 5.1	9.6 - 11.5
3	5.2 - 6.2	11.6 - 14
4	6.3 - 7.5	14 - 17
5	7.6 - 8.5	17 - 19
6	8.6 - 11	19 - 24
7	>11	> 24

[From Wade, John E., and Wendell Hewson, 1979, *Trees as indicators of wind power potential*, Dept. of Atmospheric Sciences, Oregon State University.]

6.4
TEMPERATURE EFFECTS

EVAPORATION causes heat *loss* locally and CONDENSATION causes heat *gain* locally. Both effects may be used to heat or cool air or surfaces. The USDA's Yearbook of Agriculture on Trees (1949) has this to say about the evaporative effects of trees: "An ordinary elm, of medium size, will get rid of 15,000 pounds of water on a clear dry hot day" and "Evapotranspiration (in a 40 inch rainfall) is generally not less than 15 inches per year."

Thus, the evaporation by day off trees cools air in hot weather, while the night condensation of atmospheric water warms the surrounding air. Moisture will not condense unless it finds a surface to condense *on*. Leaves provide this surface, as well as contact cooling. Leaf surfaces are likely to be cooler than other objects at evening due to the evaporation from leaf stomata by day. As air is also rising over trees, some vertical lift cooling occurs, the two combining to condense moisture on the forest. We find that leaves are 86% water, thus having twice the specific heat of soil, remaining cooler than the soil by day and warmer at night. Plants generally may be 15°C or so warmer than the surrounding air temperature.

Small open water storages or tree clumps upwind of a house have a pleasant moderating effect. Air passing over open water is cooled in summer. It is warmed and has moisture added even in winter. Only water captured by trees, however, has a DEHUMIDIFYING effect in hot and humid tropical areas, as trees are capable of reducing humidity by direct absorption except in the most extreme conditions.

Reddish–coloured leaves, such as are developed in some vines and shrubs, reflect chiefly red light rays. Sharp decreases in temperature may result by interposing reddish foliage between a thermometer and the sun, up to 20°C (36°F) lower than with green pigmented plants (Daubenmire, 1974). Whitish plants such as wormwood and birch may reflect 85% of incoming light, whereas the dark leaves of shade plants may reflect as little as 2%. It follows that white or red–coloured roof vines over tiles may effectively lower summer temperatures within buildings or in trellis systems. Additional cooling is effected by fitting fine water sprays and damp mulch systems under trellis, thus creating a cool area of dense air by evaporation. This effect is of great use in moderating summer heat in buildings, and for providing cool air sources to draw from by induced cross–ventifilation.

6.5
TREES AND PRECIPITATION

Trees have helped to create both our soils and atmosphere. The first by mechanical (root pressure) and chemical (humic acid) breakdown of rock, adding life processes as humus and myriad decomposers. The second by gaseous exchange, establishing and maintaining an oxygenated atmosphere and an active water–vapour cycle essential to life.

The composition of the atmosphere is the result of reactive processes, and forests may be doing about 80% of the work, with the rest due to oceanic or aquatic exchange. Many cities, and most deforested areas such as Greece, no longer produce the oxygen they use.

The basic effects of trees on water vapour and windstreams are:

- Compression of streamlines, and induced turbulence in air flows;
- Condensation phenomena, especially at night; Rehumidification by the cycling of water to air;
- Snow and meltwater effects; and
- Provison of nucleii for rain.

We can deal with each of these in turn (realizing that they also interact).

COMPRESSION AND TURBULENCE EFFECTS

Windstreams flow across a forest. The streamlines that impinge on the forest edge are partly deflected over the forest (almost 60% of the air) and partly absorbed into the trees (about 40% of the air). Within 1000 m (3,300 feet) the air entering the forest, with its tonnages of water and dust, is brought to a standstill. The forest has swallowed these great energies, and the result is an almost imperceptible warming of the air within the forest, a generally increased humidity in the trees (averaging 15–18% higher than the ambient air), and air in which no dust is detectable.

Under the forest canopy, negative ions produced by life processes cause dust particles (++) to clump or adhere each to the other, and a fall–out of dispersed

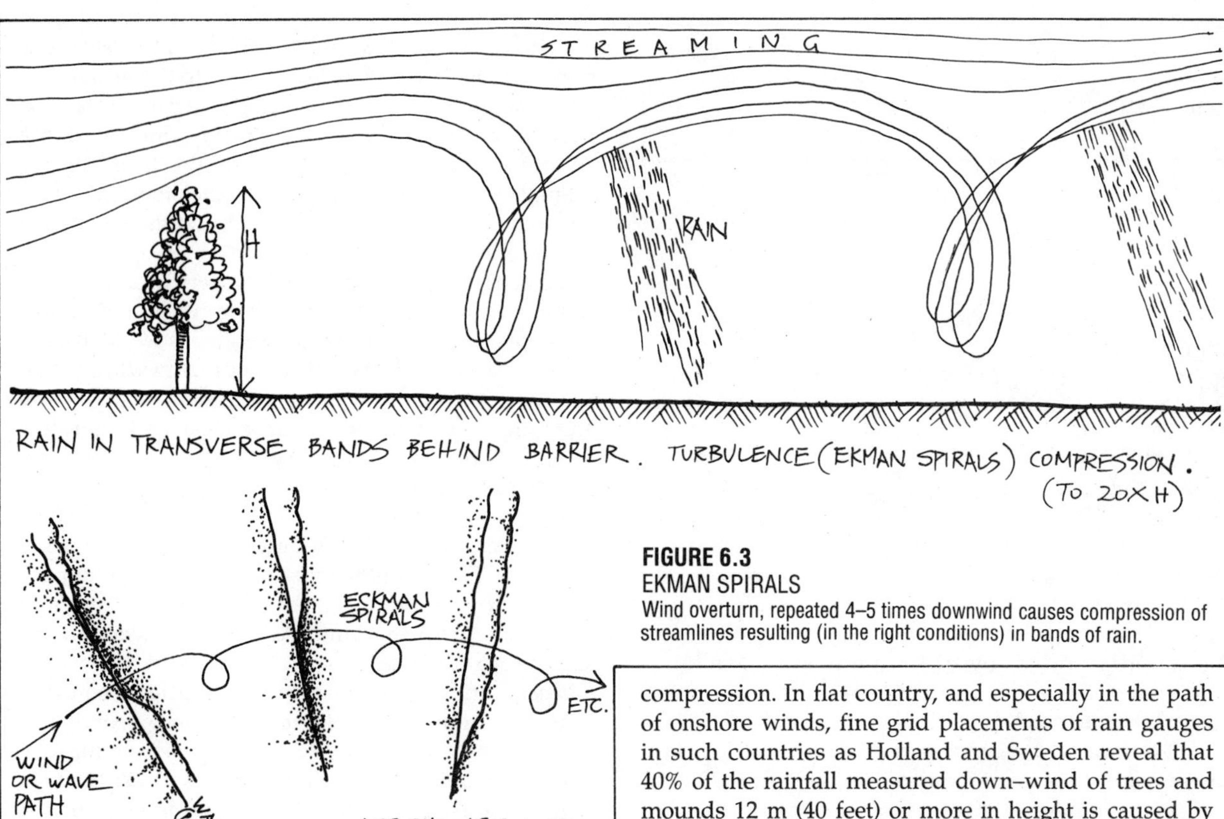

FIGURE 6.3
EKMAN SPIRALS
Wind overturn, repeated 4–5 times downwind causes compression of streamlines resulting (in the right conditions) in bands of rain.

dust results. At the forest edge, thick–stemmed and specially wind–adapted trees buffer the front–line attack of the wind. If we cut a windward forest edge, and remove these defences, windburn by salt, dust abrasion, or just plain windforce may well kill or throw down the inner forest of weaker stems and less resistant species. This is a commonly observed phenomenon, which I have called "edge break". Conversely, we can set up a forest by planting tough, resistant trees as windbreak, and so protect subsequent downwind plantings. Forest edges are therefore to be regarded as essential and *permanent* protection and should never be cut or removed.

If dry hot air enters the forest, it is shaded, cooled, and humidified. If cold humid air enters the forest, it is warmed, dehumidified, and slowly released via the crown of the trees. We may see this warm humid air as misty spirals ascending from the forest. The trees modify extremes of heat and humidity to a life–enhancing and tolerable level.

The winds deflected over the forest cause compression in the streamlines of the wind, an effect extending to twenty times the tree height, so that a 12 m (40foot) high line of trees compresses the air to 244 m (800 feet) above, thus creating more water vapour per unit volume, and also cooling the ascending air stream. Both conditions are conducive to rain.

As these effects occur at the forest EDGE, a single hedgerow of 40% permeability will cause similar compression. In flat country, and especially in the path of onshore winds, fine grid placements of rain gauges in such countries as Holland and Sweden reveal that 40% of the rainfall measured down–wind of trees and mounds 12 m (40 feet) or more in height is caused by this compression phenomena. If wind speeds are higher (32 km/h or more), the streamlines may be preserved, and rain falls perpendicular to the windbreak. However, at lower wind speeds (the normal winds), turbulence and overturn occurs.

Wind streaming over the hedgerow or forest edge describes a spiral section, repeated 58 times downwind, so that a series of compression fronts, this time parallel to the windbreak, are created in the atmosphere. This phenomena was first described by Ekman for the compression fronts created over waves at sea.

The Ekman spirals over trees or bluffs may result in a ranked series of clouds, often very regular in their rows. They are not perfectly in line ahead, but are deflected by drag and the Coriolus force to change the wind direction, so that the wind after the hedgerow may blow 5–15 degrees off the previous course. (One can imagine that ranks of hedgerows placed to take advantage of this effect would eventually bring the wind around in a great ground spiral.)

Winds at sea do in fact form great circuses, and bring cyclonic rains to the westerly oceanic coasts of all continents. These cyclones themselves create warm and cold fronts which ridge up air masses to create rain. In total, hedgerows across wind systems have a profound effect on the airstreams passing over them, and a subsequent effect on local climate and rainfall.

CONDENSATION PHENOMENA

On the sea–facing coasts of islands and continents, the relatively warmer land surface creates quiet inshore airflows towards evening, and in many areas cooler

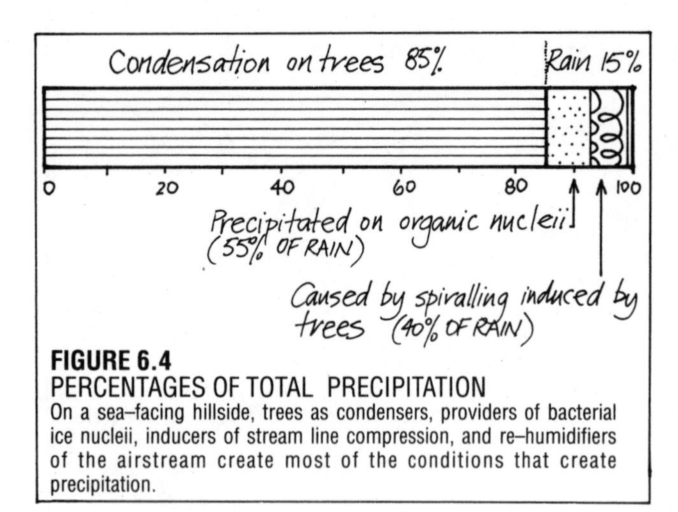

FIGURE 6.4
PERCENTAGES OF TOTAL PRECIPITATION
On a sea–facing hillside, trees as condensers, providers of bacterial ice nucleii, inducers of stream line compression, and re–humidifiers of the airstream create most of the conditions that create precipitation.

water–laden air flows inland. Where this humid air flows over the rapidly cooling surfaces of glass, metal, rocks, or the thin laminae of leaves, condensation occurs, and droplets of water form. On leaves, this may be greatly aided by the colonies of bacteria (*Pseudomonas*) which also serve as nucleii for frost crystals to settle on leaves.

These saturated airstreams produce seaward–facing mosses and lichens on the rocks of fresh basalt flows, but more importantly condense in trees to create a copious soft condensation which, in such conditions, may far exceed the precipitation caused by rainfall. Condensation drip can be as high as 80–86% of total precipitation of the upland slopes of islands or sea coasts, and eventually produces the dense rainforests of Tasmania, Chile, Hawaii, Washington/Oregon, and Scandinavia. It produced the redwood forests of California and the giant laurel forests of pre–conquest Canary Islands (now an arid area due to almost complete deforestation by the Spanish).

A single tree such as a giant Til (*Ocotea foetens*) may present 16 ha of laminate leaf surface to the sea air, and there can be 100 or so such trees per surface hectare, so that trees enormously magnify the available condensation surface. The taller the trees, as for example the giant redwoods and white pines, the larger the volume of moist air intercepted, and the greater the precipitation that follows.

All types of trees act as condensers; examples are Canary Island pines, laurels, holm oaks, redwoods, eucalypts, and Oregon pines. Evergreens work all year, but even deciduous trees catch moisture in winter. Who has not stood under a great tree which "rains" softly and continuously at night, on a clear and cloudless evening? Some gardens, created in these conditions, quietly catch their own water while neighbours suffer drought.

The effects of condensation of trees can be quickly destroyed. Felling of the forests causes rivers to dry up, swamps to evaporate, shallow water to dry out, and drought to grip the land. All this can occur in the lifetime of a person.

Precipitation from clear air is much less than that from fog, from which the precipitation by condensation often exceeds the local rainfall. Advection fogs are most noticeable where cold currents such as the Oya Shio off East Asia and the Labrador current off northeast America cause humid inland airstreams in spring and summer. Southfacing coasts near Newfound-land get 158 days of fog per year. Wherever mountains or their foothills face onshore night winds, fog condensation will probably exceed rainfall. On Table Mountain (South Africa) and on Lanai (Hawaii), fog drip has been measured at 130–330 cm, and in both cases condensation exceeds rainfall. Redwoods in California were once restricted to the fog belt, but will grow well in areas of higher rainfall without fogs

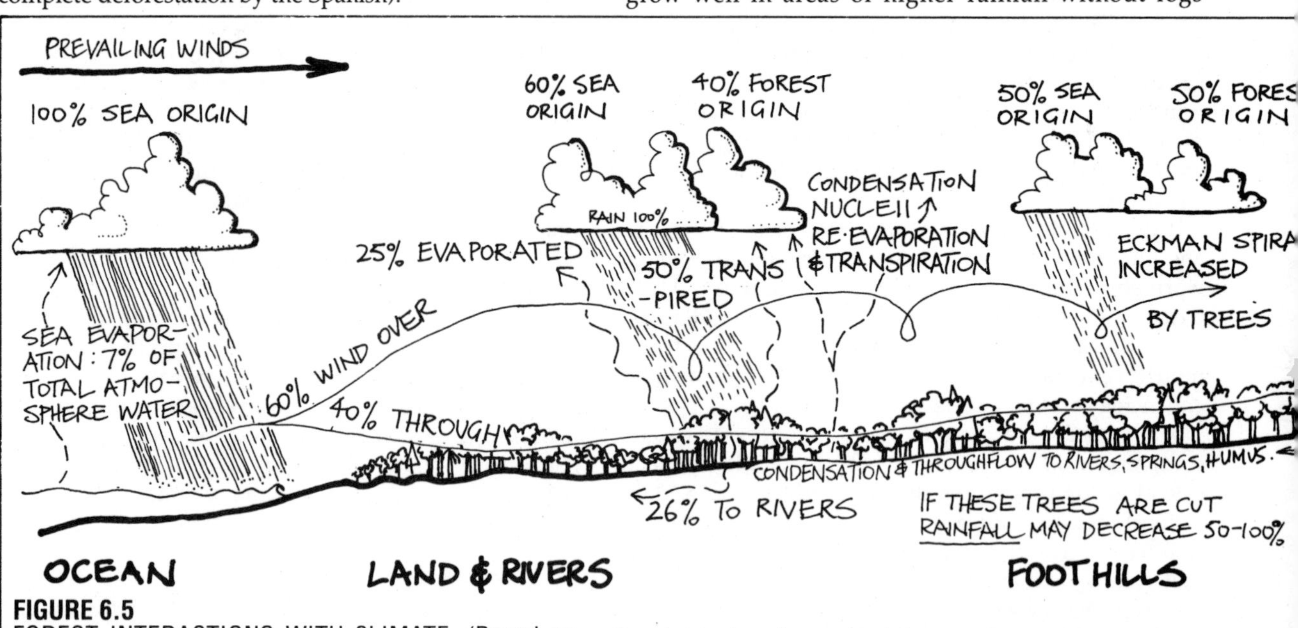

FIGURE 6.5
FOREST INTERACTIONS WITH CLIMATE. (Based on work in Brazil).

Forests inland produce most of the water for subsequent rainfall; recycled water is repeatedly transpired to the airstream.

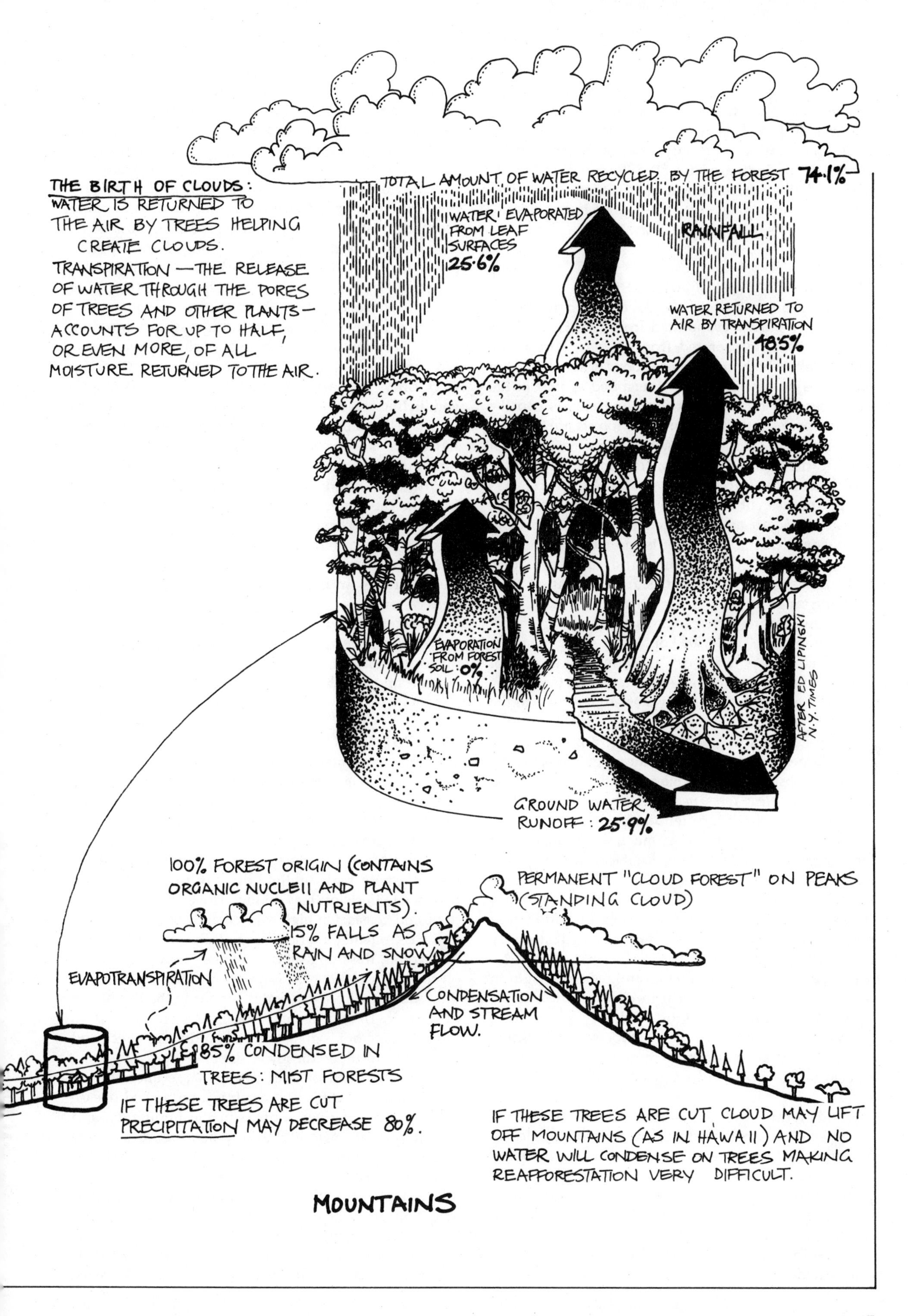
THE BIRTH OF CLOUDS: WATER IS RETURNED TO THE AIR BY TREES HELPING CREATE CLOUDS.
TRANSPIRATION — THE RELEASE OF WATER THROUGH THE PORES OF TREES AND OTHER PLANTS — ACCOUNTS FOR UP TO HALF, OR EVEN MORE, OF ALL MOISTURE RETURNED TO THE AIR.
TOTAL AMOUNT OF WATER RECYCLED BY THE FOREST 74.1%
WATER EVAPORATED FROM LEAF SURFACES 25.6%
RAINFALL
WATER RETURNED TO AIR BY TRANSPIRATION 48.5%
EVAPORATION FROM FOREST SOIL 0%
AFTER ED LIPINSKI N.Y. TIMES
GROUND WATER RUNOFF: 25.9%
100% FOREST ORIGIN (CONTAINS ORGANIC NUCLEII AND PLANT NUTRIENTS).
15% FALLS AS RAIN AND SNOW
PERMANENT "CLOUD FOREST" ON PEAKS (STANDING CLOUD)
EVAPOTRANSPIRATION
CONDENSATION AND STREAM FLOW.
85% CONDENSED IN TREES: MIST FORESTS
IF THESE TREES ARE CUT PRECIPITATION MAY DECREASE 80%.
IF THESE TREES ARE CUT CLOUD MAY LIFT OFF MOUNTAINS (AS IN HAWAII) AND NO WATER WILL CONDENSE ON TREES MAKING REAFFORESTATION VERY DIFFICULT.
MOUNTAINS

(Chang, 1968). In Sweden "... wooded hills rising only 3050 m (9,500 feet) above the surrounding plains may cause precipitation {rain only} during cyclonic spells {fronts} to be increased by 50–80% compared with average falls over the lowland." In most countries, however, the rain gauge net is too coarse to detect such small variations (Chorley and Berry, 1971).

REHUMIDIFICATION OF AIRSTREAMS

If it rains again, and again, the clouds that move inland carry water mostly evaporated from forests, and less and less water evaporated from the sea. Forests are cloud–makers both from water vapour evaporated from the leaves by day, and water transpired as part of life processes. On high islands, standing clouds cap the forested peaks, but disappear if the forests are cut. The great bridging cloud that reached from the forests of Maui to the island of Kahoolawe, remembered by the fathers of the present Hawaiian settlers, has disappeared as cutting and cattle destroyed the upper forests on Maui and so lifted the cloud cap from Kahoolawe, leaving this lower island naked to the sun. With the cloud forests gone, and the rivers dry, Kahoolawe is a true desert island, now used as a bombing range for the U.S. Air Force.

A large evergreen tree such as *Eucalyptus globulus* may pump out 3,600–4,500 l of water a day, which is how Mussolini pumped dry the Pontine marshes of Italy. With sixty or so of these trees to the hectare, many tens of thousands of litres of water are returned to the air to become clouds.

A forest can return (unlike the sea) 75% of its water to air, "in large enough amounts to form new rain clouds." [Bayard Webster, "Forests' Role in Weather Documented in Amazon", *New York Times* (Science Section), 5 July '83]. Forested areas return ten times as much moisture as bare ground, and twice as much as grasslands. In fact, as far as the atmosphere itself is concerned, "the release of water from trees and other plants accounts for half, or even more, of all moisture returned to air." (Webster, *ibid.*) This is a critical finding that adds even more data to the relationship of desertification by deforestation.

It is data that no government can ignore. Drought in one area may relate *directly* to deforestation in an upwind direction. This study "clearly shows that natural vegetation must play an important role in the forming of weather patterns" (quote from Thomas E. Lovejoy, Vice–president of Science, World Wildlife Fund).

Clouds form above forests, and such clouds are now mixtures of oceanic and forest water vapour, clearly distinguishable by careful isotope analysis. The water vapour from forests contain more organic nucleii and plant nutrients than does the "pure" oceanic water. Oxygen isotopes are measured to determine the forests' contribution, which can be done for any cloud system.

Of the 75% of water returned by trees to air, 25% is evaporated from leaf surfaces, and 50% transpired. The remaining 25% of rainfall infiltrates the soil and eventually reaches the streams. The Amazon discharges 44% of all rain falling, thus the remainder is either locked into the forest tissue or returns to air. Moreover, over the forests, *twice as much rain falls than is available from the incoming air,* so that the forest is continually recycling water to air and rain, producing 50% of its own rain (Webster, *ibid.*). These findings forever put an end to the fallacy that trees and weather are unrelated.

Vogel (1981), applying the "principle of continuity" of fluids to a tree, calculates that sap may rise, in a young oak, fifty times as fast as the leaves transpire (needing only 7% of the total trunk area as conductive tissue, with an actual sap speed of 1 cm/sec). It is thus certain that only perhaps one–fiftieth of the xylem is conducting sap upwards at any one time, and that most xylem cells contain either air or sap at standstill. Perhaps too, the tree moves water up in pulsed stages rather than as a universal or continuous streamflow.

With such rapid sap flows, however, we can easily imagine the water recycled to atmosphere by a large tree, or a clump of smaller trees.

It is a wonder to me that we have any water available after we cut the forests, or any soil. There are dozens of case histories in modern and ancient times of such desiccation as we find on the Canary Islands following deforestation, where rivers once ran and springs flowed. Design strategies are obvious and urgent—save all forest that remains, and plant trees for increased condensation on the hills that face the sea.

EFFECTS ON SNOW AND MELTWATER

Although trees intercept some snow, the effect of shrubs and trees is to entrap snow at the edges of clumps, and hold 75–95% of snowfall in shade. Melting is delayed for 210 days compared with bare ground, so that release of snowmelt is a more gradual process. Of the trapped snow within trees, most is melted, while on open ground snow may sublime directly to air. Thus, the beneficial effects of trees on high slopes is not confined to humid coasts. On high cold uplands such as we find in the continental interiors of the U.S.A. or Turkey near Mt. Ararat, the thin skeins of winter snow either blow off the bald uplands, to disappear in warmer air, or else they sublime directly to water vapour in the bright sun of winter. In neither case does the snow melt to groundwater, but is gone without productive effect, and no streams result on the lower slopes.

Even a thin belt of trees entraps large quantities of driven snow in drifts. The result is a protracted release of meltwater to river sources in the highlands, and stream–flow at lower altitudes. When the forests were cleared for mine timber in 1846 at Pyramid Lake, Nevada, the streams ceased to flow, and the lake levels fell. Add to this effect that of river diversion and irrigation, and whole lakes rich with fish and waterfowl have become dustbowls, as has Lake Winnemucca. The Cuiuidika'a Indians (Paiute) who live there lost their fish, waterfowl, and freshwater in less than 100 years.

The cowboys have won the day, but ruined the future to do so.

PROVISION OF NUCLEII FOR RAIN

The upward spirals of humid air coming up from the forest carry insects, pollen, and bacteria aloft. This is best seen as flights of gulls, swifts and ibis spiralling up with the warm air and actively catching insects lifted from the forest; their gastric pellets consist of insect remains. It is these organic aerial particles (pollen, leaf dust, and bacteria mainly) that create the nucleii for rain.

The violent hailstorms that plague Kenya tea plantings may well be caused by tea dust stirred up by the local winds and the feet of pickers, and "once above the ground the particles are easily drawn up into thunderheads to help form the hailstorms that bombard the tea–growing areas in astounding numbers... Kenyan organic tea leaf litter caused water to freeze in a test chamber at only –5°C, in comparison with freezing points of -11°C for eucalyptus grove leaf litter, and –8°C for the litter from the local indigenous forests" (of Colorado). That is, tea litter "is a much better seeding–agent than silver iodide, which requires –8°C to –10°C to seed clouds." (*New Scientist*, 22 Mar '79). Thus, the materials given up by vegetation may be a critical factor in the rainfall inland from forests.

All of these factors are clear enough for any person to understand. To doubt the connection between forests and the water cycle is to doubt that milk flows from the breast of the mother, which is just the analogy given to water by tribal peoples. Trees were "the hair of the earth" which caught the mists and made the rivers flow. Such metaphors are clear allegorical guides to sensible conduct, and caused the Hawaiians (who had themselves brought on earlier environmental catastrophes) to "tabu" forest cutting or even to make tracks on high slopes, and to place mountain trees in a sacred or protected category. Now that we begin to understand the reasons for these beliefs, we could ourselves look on trees as our essential companions, giving us all the needs of life, and deserving of our care and respect.

It is our strategies on–site that make water a scarce or plentiful resource. To start with, we must examine ways to increase local precipitation. Unless there is absolutely no free water in the air and earth about us (and there always is some), we can usually increase it on–site. Here are some basic strategies of water capture from air:

- We can cool the air by shade or by providing cold surfaces for it to flow over, using trees and shrubs, or metals, including glass.
- We can cool air by forcing it to higher altitudes, by providing windbreaks, or providing updraughts from heated or bare surfaces (large concreted areas), or by mechanical means (big industrial fans).
- We can provide condensation nucleii for raindrops to form on, from pollen, bacteria, and organic particles.
- We can compress air to make water more plentiful per unit volume of air, by forcing streamlines to converge over trees and objects, or forcing turbulent flow in airstreams (Ekman spirals).

If by any strategy we can cool air, and provide suitable condensation surfaces or nucleii, we can increase precipitation locally. Trees, especially crosswind belts of tall trees, *meet all of these criteria in one integrated system*. They also store water for local climatic modification. Thus we can clearly see trees as a strategy for creating more water for local use.

In summary, we do not need to accept "rainfall" as having everything to do with total local precipitation, especially if we live within 30–100 km of coasts (as much of the world does), and we do not need to accept that total precipitation cannot be changed (in either direction) by our action and designs on site.

6.6
HOW A TREE INTERACTS WITH RAIN

Rain falls, and many tons of rain may impact on earth in an hour or so. On bare soils and thinly spaced or cultivated crop, the impact of droplets carries away soil, and may typically remove 80 t/ha, or up to 1,000 tonnes in extreme downpours. When we bare the soil, we lose the earth.

Water run–off and pan evaporation, estimated as 80–90% of all rain falling on Australia, carries off nutrients and silt to the sea or to inland basins. As we clear the land, run–off increases and for a while this pleases people, who see their dams fill quickly. But the dams will silt up and the river eventually cease to flow, and the clearing of forests will result in flood and drought, not a long–regulated and steady supply of clean water.

When rain falls on a forest, a complex process begins. Firstly, the tree canopy shelters and nullifies the impact effect of raindrops, reducing the rain to a thin mist below the canopy, even in the most torrential showers. There is slight measurable silt loss from mature forests, exceeded by the creation of soils by forests.

If the rain is light, little of it penetrates beyond the canopy, but a film of water spreads across the leaves and stems, and is trapped there by surface tension. The cells of the tree absorb what is needed, and the remainder evaporates to air. Where no rain penetrates through the canopy, this effect is termed "total interception". INTERCEPTION is the amount of rainfall caught in the crown. It is the most important primary effect of trees or forests on rain. The degree of interception is most influenced by these factors:

- Crown thickness;
- Crown density;
- Season;
- Intensity of rain; and
- Evaporation after rain.

Broadly speaking, interception commonly falls between 10–15% of total rainfall. *Least* interception occurs in thinned and deciduous forests, winter rain, heavy showers, and cloudy weather conditions, when it is as little as 10% of rain. *Most* interception occurs with dense, evergreen trees, light summer rain, and sunny conditions, when it may reach 100% of the total.

However, if more rain falls, or heavy rains impact on the trees, water commences to drift as mists or droplets to earth. This water is called THROUGHFALL. Throughfall depends on the intensity of rain, and there is little interception effect in heavy downpours. As an average figure, the throughfall is 85% of rain in humid climates.

At this point, throughfall is no longer just rainwater, any more than your bathwater is rainwater; throughfall contains many plant cells and nutrients, and is in fact a much richer brew than rainwater. Dissolved salts, organic content, dust, and plant exudates are included in the water of throughfall (**Table 6.2**). "The results show that rain washes large amounts of potassium and smaller amounts of nitrogen, phosphorus, calcium, and magnesium from the canopies to the surface soil. Litter adds organic matter, and is a rich source of calcium and nitrogen and a moderately rich source of magnesium and potassium." (Murray, J. S. and Mitchell, A., *Red Gum and the Nutrient Balance*, Soil Conservation Authority, Victoria, Australia, undated).

Nor can throughfall be measured in rain gauges, for the trees often provide special receptors, conduits, and storages for such water. The random fall of rain is converted into well–directed patterns of flow that serve the needs and growth in the forest. In the stem bases of palms, plantains, and many ephiphytes, or the flanged roots of *Terminalia* trees and figs, water is held as aerial ponds, often rich in algae and mosquitoes. Stem mosses and epiphytes absorb many times their bulk of water, and the tree itself directs water via insloping branches and fissured bark to its tap roots, with spiders catching their share on webs, and fungi soaking up what they need. Some trees trail weeping branches to direct throughfall to their fibrous peripheral roots.

With the aerial resevoirs filled, the throughfall now enters the humus layer of the forest, which can itself (like a great blotter) absorb 1 cm of rain for every 3 cm of depth. In old beech forests, this humus blanket is at least 40 cm deep, and the earth below is a mass of fungal hyphae. In unisturbed rainforest, deep mosses may carpet the forest floor. So, for 40-60 cm depth, the throughfall is absorbed by the decomposers and living systems of the humus layer. Again, the composition of the water changes, picking up humic exudates, and water from deep forests and bogs may then take on a clear golden colour, rather like tea.

TABLE 6.2
Nutrient content of litter, canopy drip, and rain in the open of a naturally regenerating stand of red gum (*Eucalyptus camaldulensis*), Gringegalgona, VIC, Australia. [Source: Murray, J.S., and A. Mitchell, Red Gum and the Nutrient Balance, Soil Conservation Authority, Australia (undated)].

SOURCE	PERIOD*	TOTAL RAIN (in.)	NUTRIENT RETURN (LB/A.)							TOTAL LITTER (lb.)
			N	P	K	Ca	Mg	Na	Cl	
Old trees: (5% of Total)										
LITTER	1*	-	19.0	1.2	6.0	25.0	6.0	4.0	ND	2,800
CANOPY DRIP	1	30.67	6.0	1.1	28.0	13.0	11.0	71.0	143.0	-
TOTAL	1	30.67	25.0	2.3	34.0	38.0	17.0	75.0	-	
Regrowth: (95% of Total)										
LITTER	1		38.0	1.9	10.0	49.0	15.0	5.0	ND	5,400
CANOPY DRIP	1	30.67	3.0	0.7	16.0	6.0	5.0	29.0	51.0	
TOTAL	1	30.67	41.0	2.6	26.0	55.0	20.0	34.0	-	
Rain:										
Nearby	2*	9.25	0.5	0.1	0.7	0.8	0.7	4.2	7.0	
Coleraine§	3*	33.61	0.5	ND	1.5	3.0	3.0	21.0	38.0	
Cavendish §	4*	21.75	ND	ND	1.0	3.0	2.0	14.0	20.0	

*1: 5/5/60 - 4/5/61. *2: 22/11/60 - 4/5/61. *3: 1/9/55 - 1/9/56. *4: 1/9/54 - 1/9/55.
§ From Hutton and Leslie (1958). ND = Not Determined.

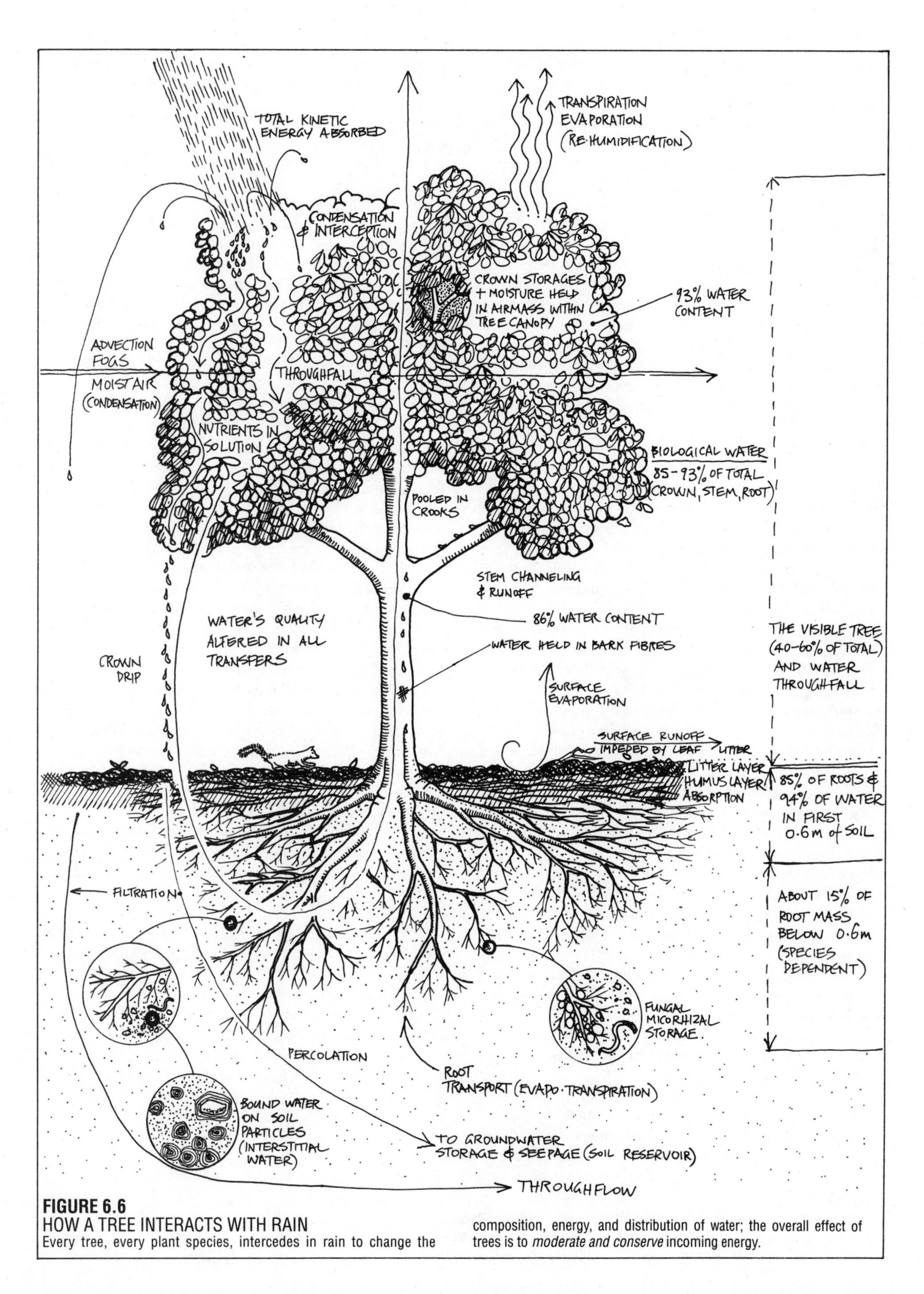

FIGURE 6.6
HOW A TREE INTERACTS WITH RAIN
Every tree, every plant species, intercedes in rain to change the composition, energy, and distribution of water; the overall effect of trees is to *moderate and conserve* incoming energy.

pH can reach as low as 3.5 or 4.0 from natural humic layers, and rivers run like clear coffee to sea. Below the humus lies the tree roots, each clothed in fungal hyphae and the gels secreted by bacterial colonies. 30–40% of the bulk of the tree itself lies in the soil; most of this extends over many acres, with thousands of kilometres of root hairs lying mat–like in the upper 60 cm of soil (only 10–12% of the root mass lies below this depth but the remaining roots penetrate as much as 40 m deep in the rocks below).

The root mat actively absorbs the solution that water has become, transporting it up the tree again to transpire to air. Some dryland plant roots build up a damp soil surround, and may be storing surplus water in the earth for daytime use; this water is held in the root associates as gels. *Centrosema* and *Gleditsia* both are dryland woody legumes which have "wet" root zones, and other plants are also reported to do the same in desert soils (*Prosopis* spp.)

The soil particles around the tree are now wetted with a surface film of water, as were the leaves and root hairs. This bound water forms a film available to roots, which can remove the water down to 15 atmospheres of pressure, when the soil retains the last thin film. Once soil is fully charged (at "field capacity"), free water at last percolates through the **interstitial** spaces of the soil and commences a slow progression to the streams, and thence to sea.

At any time, trees may intercept and draw on these underground reserves for growth, and pump the water again to air. If we imagine the visible (above–ground) forest as water (and all but about 5–10% of this mass *is* water), and then imagine the water contained in soil, humus, and root material, the forests represent great lakes of actively managed and actively recycled water. No other storage system is so beneficial, or results in so much useful growth, although fairly shallow ponds are also valuable productive landscape.

At the crown, forceful raindrops are broken up and scattered, often to mist or coalesced into small bark–fissured streams, and so descend to earth robbed of the kinetic energy that destroys the soil mantle outside forests. Further impedence takes place on the forest floor, where roots, litter, logs, and leaves redirect, slow down, and pool the water.

Thus, in the forest, the soil mantle has every opportunity to act as a major storage. As even poor soils store water, the soil itself is an immense potential water storage. INFILTRATION to this storage along roots and through litter is maximised in forests. The soil has several storages:

- RETENTION STORAGE: as a film of water bound to the soil particles, held by surface tension.
- INTERSTITIAL STORAGE: as water–filled cavities between soil particles.
- HUMUS STORAGE: as swollen mycorrhizal and spongy detritus in the humic content of soils.

A lesser storage is as chemically–bound water in combination with minerals in the soil.

As a generalisation, 2.5–7 cm (1–3 inches) of rain is stored per 30 cm (12 inches) depth of soil mantle in retention storage, although soils of fine texture and high organic content may store 10–30 cm (4–12 inches) of rain per 30 cm depth. In addition, 0–5 cm (0–2 inches) may be stored as interstitial storage.

Thus the soil becomes an impediment to water movement, and the free (interstitial) water can take as long as 1–40 years to percolate through to streams. Greatly alleviating droughts, it also recharges the retention storages on the way. Thus, it almost seems as though the purpose of the forests is to give soil time and means to hold fresh water on land. This is, of course, good for the forests themselves, and enables them to draw on water reserves between periods of rain. (Odum, 1974).

6.7

SUMMARY

Let us now be clear about how trees affect total precipitation. The case taken is where winds blow inland from an ocean or large lake:

1. The water in the air is that evaporated from the surface of the sea or lake. It contains a few salt particles but is "clean". A small proportion may fall as *rain* (15–20%), but most of this water is CONDENSED out of clear night air or fogs by the cool surfaces of leaves (80–85%). Of this condensate, 15% evaporates by day and 50% is transpired. The rest enters the groundwater. Thus, *trees are responsible for more water in streams than the rainfall alone provides*.

2. Of the rain that falls, 25% again re–evaporates from crown leaves, and 50% is transpired. This moisture is added to clouds, which are now at least 50% "*tree water*". These clouds travel on inland to rain again. Thus trees may double or multiply *rainfall* itself by this process, which can be repeated many times over extensive forested plains or foothills.

3. As the air rises inland, the precipitation and condensation increases, and moss forests plus standing clouds may form in mountains, adding considerably to total precipitation and infiltration to the lower slopes and streams.

4. Whenever winds pass over tree lines or forest edges of 12 m (40 feet) or more in height, Ekman spirals develop, adding 40% or so to rainfall in bands which roughly parallel the tree lines.

5. Within the forest, 40% of the incident air mass may enter and either lose water or be rehumidified.

6. And, in every case, rain is more likely to fall as a result or organic particles forming nucleii for condensation, whereas industrial aerosols are too small to cause rain and instead produce dry, cloudy conditions.

Thus, if we clear the forest, what is left but dust?

6.8 REFERENCES

Chang, Jenhu, *Climate and Agriculture*, Aldine Pub. Co., Chicago, 1968.

Daubenmire, Rexford F., *Plants and Environment: a textbook of plant autecology*, Wiley & Sons, N. Y., 1974.

Geiger, Rudolf, *The Climate Near the Ground*, Harvard University Press, 1975.

Odum, Eugene, *Fundamentals of Ecology*, W. B. Saunders, London, 1974.

Plate, E. J., *The Aerodynamics of Shelterbelts*, Agricultural Meteorology 8, 1970.

U. S. Department of Agriculture, *Trees*, USDA Yearbook of Agriculture, 1949.

Vogel, Stephen, *Life in Moving Fluids*, Willard Grant Press, Boston,1981.

Chapter 7
WATER

Water is the driving force of all nature.
(Leonardo da Vinci)

In an animal or a plant, 99 molecules in 100 are water... An organism is a pool in a stream of water along which metabolites and energy move through ecosystems.
(W.V. Macfarlane)

The sustained flow of rivers is truly remarkable, considering that precipitation is an unusual event in most areas of the earth. Localisation of precipitation in space or time is striking {e.g. at Paris, it rains for 577 hours a year, or only 7% of the time. *B.Mollison*}. Few storms last more than a few hours, so that even storm days are mainly rainless. Yet rivers flow throughout the year. The sustaining source of flow is effluent groundwater... The amount of soil water is about fifteen times the amount in channel storage in rivers.
(Nace *in* Chorley, 1969)

7.1
INTRODUCTION

Very little of the world's total water reserves are actually available for present human needs. Many areas of earth, particularly dryland areas, over-developed cities, and towns or cities surrounded by polluting industry and agriculture, face an absolute shortage of useable water. Millions of city-dwellers now purchase water, at prices (from 1984 on) equivalent to or greater than that of refined petroleum. This is why the value of land must, in future, be assessed on its yield of potable water. Those property-owners with a constant source of pure water already have an economically-valuable "product" from their land, and need look no further for a source of income. Water as a commodity is already being transported by sea on a global scale.

The PRIMARY SELECTION FACTOR, when choosing a cropland property to develop, should be an adequate, preferably well-distributed, and above all reliable rainfall. "Adequate" here is about a minimum of 80 cm (31 inches) and upwards. An equally important factor is the ability of the area to hold water as dams in clay or clay-loam storages; any stream flow within the boundaries is a bonus. All other factors (soil type, present uses, number of titles, market potential, access, and forested areas) are secondary to water availability.

Little of the lands now used for crop agriculture have such fortunate characteristics. Few farmers have invested in "drought-proofing" their land by creating gravity-fed irrigation systems of **Keyline systems**[5]. Specific strategies of water conservation and control are given in this book under their appropriate climatic and landform sections.

While there are no economically-feasible strategies or technologies for freshwater creation from the sea or from polluted sources, there are several currently neglected strategies for recycling, purification, conservation, and increased storages of rainwater. In particular, the construction of tanks and dams have been neglected in built-up areas, as have earth storages on farms and in rural areas. Waste usage ranging from over-irrigation, non-recycling in industry, inappropriate domestic appliances, and unnecessary uses (on lawns and car washes) have not as yet been adequately costed by legislators or by householders.

Tables 7.1 and **7.2** show abstract figures of the global and local water cycles. These should not be regarded as fixed or even sufficient representations of water in relation to actively designed or rehabilitated landscapes. There are ways in which we can constructively reverse past trends in water deficits, waste, pollution, and misuse. There is plenty of water for the world if we define the ways in which we store and use it carefully.

7.2
REGIONAL INTERVENTION IN THE WATER CYCLE

CLOUD SEEDING

Silver iodide, and no doubt other ice nucleii such as tea dust can be "seeded" into cumulus, cumulonimbus, or nimbostratus clouds (by plane, ground burners, or rocket) in order to initiate local precipitation.

Until recently these attempts to make rain were assessed as ineffective because no one had, at the time, realised how far and for how long the seeding effects spread and persisted. More recent analysis show that rain in fact increases over a very wide area, and that secondary effects last for months, so that varying wind directions and speeds carry the induced rain effects for hundreds or thousands of square kilometres (*Ecos*, 45, Spring 1985). It also seems probable that ground burners or ground release of ice nucleii could have a similar effect. On the ground, silver iodide is absorbed into coal dust, and this is then burnt when clouds form on hill crests. Strategic downwind hills can generate clouds and rain over large areas of land.

Once initiated, however, such effects cannot quickly be stopped, and even in places like India or Ethiopia may create a little too much rain if ground storage systems are not previously developed to cope with the extra water. In arid or semi-arid areas, flood retardation basins, oversized swales, large sand dams, water spreading systems, pelleted seed of fast-growing plants, and in fact any sensible civil strategy to preserve soil and people from any effect of increased precipitation is a necessary prelude to cloud-seeding.

Initial precipitation (due to increased bacterial or ice nucleii stimulus) increases can be as much as 30%, and subsequently averaged in Australia at 19% (17% in Israel) over weeks, falling to 8% in months. The cloud-seeding system promises to help increase monsoon or frontal rains in areas where suitable clouds occur without sufficient precipitation. This system can be very cheap for large land areas. For more data or references, contact the Cloud Physics Laboratory,

TABLE 7.2
FRESHWATER LOCATION.
Freshwater is only 3% of all water on earth, and very little is in circulation, most being locked up in storages.

STORAGE	% OF FRESHWATER
Ice and glaciers*	75.0
Groundwater more than 800 m deep	13.5
Groudwater less than 800 m deep	11.0
Lakes	0.3
Soils	0.06
Atmosphere (in circulation at any one time)	0.035
Rivers	0.03

*Frozen ground or permafrost is not assessed in this table. It represents a considerable storage (about 40% of the landmasses of Canada and the Soviet Union.

TABLE 7.1
RENEWAL TIMES OF ALL WATER IN BASIC STORAGES (seawater and freshwater)[From: Southwick, C.H., *Ecology and the Quality of our Environment*, Van Nostrand Reinhold, NY, 1976.]

LOCATION IN STORAGES	DISTRIBUTION (% of total water)	RENEWAL TIME (Turnover rates, cycles)
Ocean	93.8	37,000 years
Glaciers and permanent snow	1.986	16,000 years
Groundwater (to 5 km depth)	4.1	4,600 years
(Actively exchanged)	0.274	300 years
Lakes	0.0051	13 years
Atmosphere	0.000959	9 days
Rivers	0.00008	13 days
Biological water	0.000005	3.4 days

Division of Atmospheric Research, CSIRO, Canberra, Australia.

As similar effects (thunderstorms, rain) have been noted for tea leaf dust downwind of Kenya plantations, more homely strategies may also be developed if the underlying nucleation causes can be established. It may even be that the fires and dances of the old "rain makers" on a high hill were, in truth, effective. Certainly, fires of specific vegetation and dances with the "right" dust plume could help seed ice nucleii in clouds; quite local rain falls near some factory smoke emissions.

Windward slope forests, cross–wind tree lines, and even slight earth rises of a modest 4–6 m have been observed to induce air humidity, cloud formation, and even rain, by orographic (uplift) effects on windstreams. Thus, we are not powerless in the matter of increasing local moisture by a series of sensible ground strategies based on providing trees, mounds, and cloud ice nucleii, and perhaps a serious attempt to induce these changes will in the near future bring relief to areas such as the Indian Deccan, the Sahel, and large areas of Australia and the USA subject to rainless cloud masses.

OROGRAPHIC AND FOREST EFFECTS

Strategically–selected cross–wind ridges of even modest height [3–20 m (10–65 feet)] are ideal sites for the planting of known tree "condensers" and cross–wind tree–lines. These ridges are most useful when lying in the path of the summer afternoon sea breezes that flow inland, or located where the air drifts in at night, such as on the Californian and sub–tropical trade–wind coasts. The clearing of trees from such sites may well induce long–term drought and create a drying effect for hundreds of kilometres inland.

It is long past time that we also assessed vegetation for some of the following effects:

- Ability to provide rain nucleii, as bacteria and natural sulphur particles, and also to effectively condense water from air at night.
- The rainfall effects from forested ridges, where forests exceed 6–10 m (19–32 feet) in height, on rainfall induced by streamline compression effects. This effect is credited with up to 40% of rainfall where now assessed in Sweden and Australia (Tasmania and Victoria).
- The total effect of forested catchment area. Historical and recent evidence suggests that rainfall, streamflow, and cloud may all be seriously depleted by upland deforestation. Such effects are never assessed or costed against deforestation or wood–chipping. The soil erosion and salted land effects are, however, well known in deforested areas.

Any conservationist policies of future effective and informed regional governments would first research

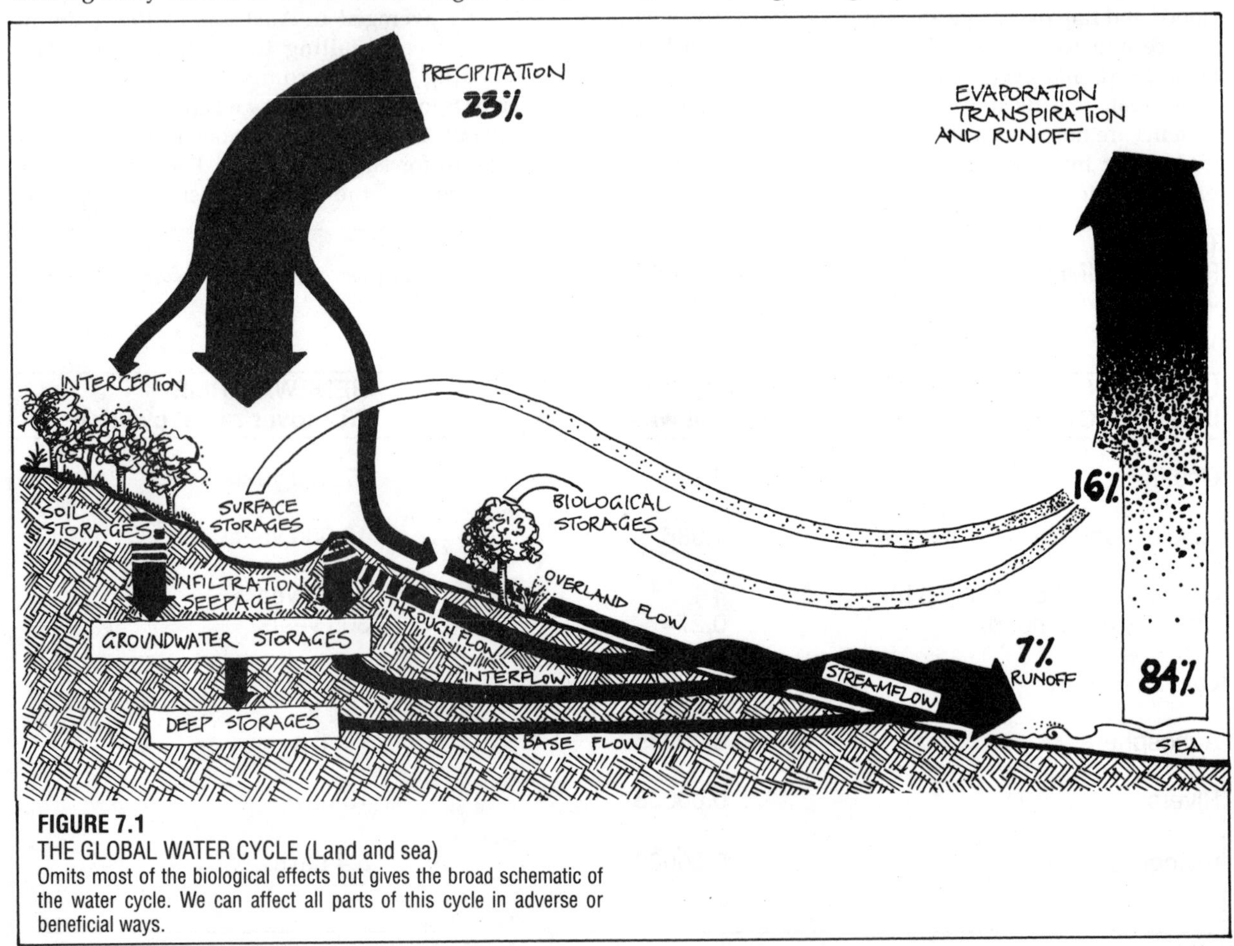

FIGURE 7.1
THE GLOBAL WATER CYCLE (Land and sea)
Omits most of the biological effects but gives the broad schematic of the water cycle. We can affect all parts of this cycle in adverse or beneficial ways.

such effects, then quickly establish national forest and watershed management or restoration policies based on such research. For ourselves, as designers, the proper approach to land planning infers that we recommend permanent forests and the preservation of older forests on cross–wind ridges, and on steep (18° slope or more) sea–facing slopes. The preservation of alpine or upland absorption areas is also essential.

SOIL STORAGES

Soil conditioning or "ripping" (see Chapter 8), providing it is followed by tree plantation, trace element additions, and a non–destructive agriculture of well–managed natural yields, sparse grazing, and conservation farming certainly increases (by factors of up to 70–85%) the ability of soils to hold and infiltrate water. Areas of up to 85% run–off can be converted to zero overland flow by a combination of soil conditioning, swales, and water spreading to forests.

As soils can contain many times the water of open storages or streams, then both the throughflow, baseflow, and water available for plants also increases. It follows that the CYCLING of water via evapotranspiration and rainfall also increases. Soil treatments now need to precede tree planting over almost every area that has been used by contemporary agriculture. In particular the barren areas used for constant cropping in dryland areas need soil treatment to initiate water absorption. Trees are essential to prevent water–logging of soils and soil salting in the long term.

INFILTRATION VIA EARTHWORKS

Cheap broadscale earthwork systems, and many minor forms of earthworks can aid the infiltration of overland water flow. PITTING, SWALES and WATER SPREADING are the main aids to getting fresh water to deeper storages for long–term use, and also to increase base flow. Diversion of surface flow to sand basins, dune fields, swamps, and soakage beds in earth–bermed fields all ensure resident water reserves for crop and trees, and longer–term storages for use in dry seasons.

Diversion drains and their associated valves, slides, cross–walls, intakes, and irrigation systems enable effective water harvesting, dependable storage, and fast emergency use in normal rolling lowlands, hill country, and drylands. They can also recharge sand basins and swales from otherwise wasted overland flow, and damp our wildfires.

POND AND FARM DAM STORAGES

Wherever precipitation exceeds the demands of transpiration and evaporation, small dams, wetlands, and swamps can proliferate. All of these act as long–term water and wildlife reserves in the total landscape, and many Australian farms are now "drought–proof" due to sensible investment in Keyline or similar water conservation systems by the owners. Excellent technical manuals exist (see references at the end of this chapter).

Dams and ponds are potential aquaculture sites, and the production of a diverse plant and fish or waterfowl protein product should also be considered during their construction. In humid areas, therefore, these water storages can occupy up to 20% of the landscape with great benefit in providing fish and a great variety of aquatic product, while at the same time moderating the effects of drought and flood.

BIOLOGICAL STORAGES

The great forests and the biological water storages in the form of fruits and nuts (such as the coconut) are the basis for the proliferation of life forms where no "free" water otherwise exists. In particular, browsers, insects and fungi draw on these biological tree reserves year–round, and perform a host of useful functions in any ecosystem. On atolls and arid islands of free–draining sands, the biological reserves are the main water reserves. This is often overlooked, except by those inhabitants dependent on the waters contained in fruits or nuts. Many plants such as cactus, palms, and agaves have specific tissues or organs to store water.

In the local microclimate, the water in vegetation greatly moderates heat and cold excesses, and both releases to, and absorbs water from passing air streams. Essential crops such as cassava will produce crop as a result of the humidity provided by surrounding vegetation, so that even this side effect of vegetation is of productive use (*New Scientist* 29 May '86).

TANK STORAGES

Water can be captured off roof areas, roads, and other paved areas, and used for both drinking water and shower or garden water, providing it can be stored. Roof water is least polluted, or most easily treated for drinking and cooking in houses, while absorption beds are sufficient for the muddy or polluted run–off from roads and parking areas.

7.3

EARTHWORKS FOR WATER CONSERVATION AND STORAGE

For the serious small dam and earth tank builder, there is no substitute for such comprehensive texts as that recently compiled by Kenneth D. Nelson (1985). This small classic deals with catchment treatments, run–off calculations, soils, construction, outlets, volume and cost estimates, and includes detailed drawings for most adjunct structures.

However, like most engineers, Nelson concentrates mostly on valley dams (barrier or embankment dams), and less on the placement of dams in the total designed landscape. Few dam builders consider the biological uses of dams, and the necessary modifications that create biological productivity in water systems.

A second essential book for water planning in landscape is P. A. Yeomans' *Water for Every Farm/The Keyline Plan* (1981)[5]. This very important book, written in 1954, is without doubt the pioneering modern text on landscape design for water conservation and gravity–fed flow irrigation. As it also involves patterning, tree planting, soil treatment, and fencing alignment, it is the first book on functional landscape design in modern times.

There are two basic strategies of water conservation in run–off areas: the diversion of surface water to impoundments (dams, tanks) for later use, and the storage of water in soils. Both result in a recharge of groundwater. As with all technologies, earthworks have quite specifically appropriate and inappropriate uses. Some of the main productive earthwork features we create are as follows:

- Dams and tanks (storages);
- Swales (absorption beds);
- Diversion systems or channels; and
- Irrigation layouts, and in particular those for flood or sheet irrigation.

SMALL DAMS AND EARTH TANKS.

Small dams and earth tanks have two primary uses. The minor use is to provide watering points for rangelands, wildlife, and domestic stock; such tanks or waterholes can therefore be modest systems, widely dispersed and static. The second and major use is to contain or store surplus run–off water for use over dry

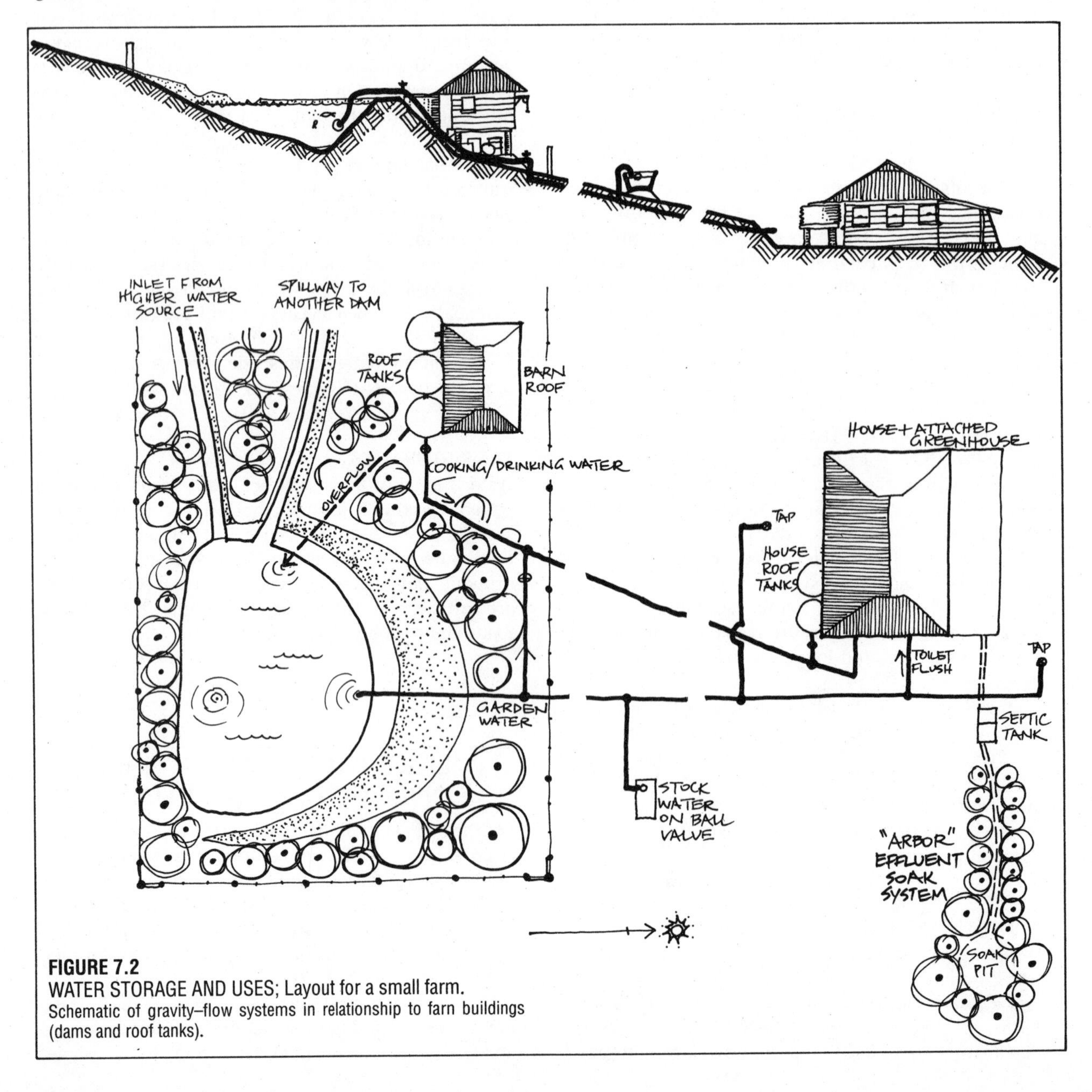

FIGURE 7.2
WATER STORAGE AND USES; Layout for a small farm.
Schematic of gravity–flow systems in relationship to farn buildings (dams and roof tanks).

periods for domestic use or irrigation. The latter storages, therefore, need to be carefully designed with respect to such factors as safety, water harvesting, total landscape layout, outlet systems, draw–down, and placement relative to the usage area (preferably providing gravity flow).

A separate category of water storages, akin to fields for crop or browse production, are those ponds or wet terraces created specifically for *water crop* (vegetation or mixed polycultural systems of aquatic animal species).

Open–water (free water surface area) storages are *most appropriate in humid climates,* where the potential for evaporation is exceeded by average annual rainfall. There is a very real danger that similar storages created in arid to subhumid areas will have adverse effects, as evaporation from open water storages inevitably concentrates dissolved salts. Firstly, such salty water can affect animal health. Secondly, the inevitable seepage from earth dams can and does create areas of salted or collapsed soils downhill from such storages. And in the case of large barrier dams, so little water may be allowed to bypass them in flood time that agricultural soils, productive lakes, and estuaries may lose more productive capacity by deprivation of flush–water and silt deposits than can be made up (at greater cost) by irrigation derived from such lakes.

Dryland storage strategies are discussed in Chapter 11. What I have to say here is *specifically addressed to humid areas and small dams* unless otherwise noted.

Earth dams or weirs where retaining walls are 6 m (19 feet) high or less, and which have a large or over–sized stable spillway, are no threat to life or property if well–made. They need not displace populations, stop flow in streams, create health problems, fill with silt, or block fish migrations. In fact, dams or storages made anywhere *but* as barriers on streams effectively add to stream flow in the long term.

Low barrier dams of 1–4 m (3–13 feet) high can assist stream oxygenation, provide permanent pools, be "stepped" to allow fish ladders or bypasses, and also provide local sites for modest power generation. While almost all modern assessments would condemn or ban large–scale dams (and large–scale power schemes) on the record of past and continuing fiascos, a sober assessment of small water storages shows multiple benefits.

Given the range of excellent texts on small dams (often available from local water authorities, and by mail order from good bookstores), I have therefore avoided specific and well–published construction details, and have here elaborated more on the types, placement, links to and from, and function of small dams in the total landscape. Yeomans (*pers. comm.,* 1978) has stated that he believes that if from 10–15% of a normal, humid, lowland or foothill landscape were fitted with small earth storages, floods and drought or fire threat could be eliminated.

Not all landscapes can cost–effectively store this

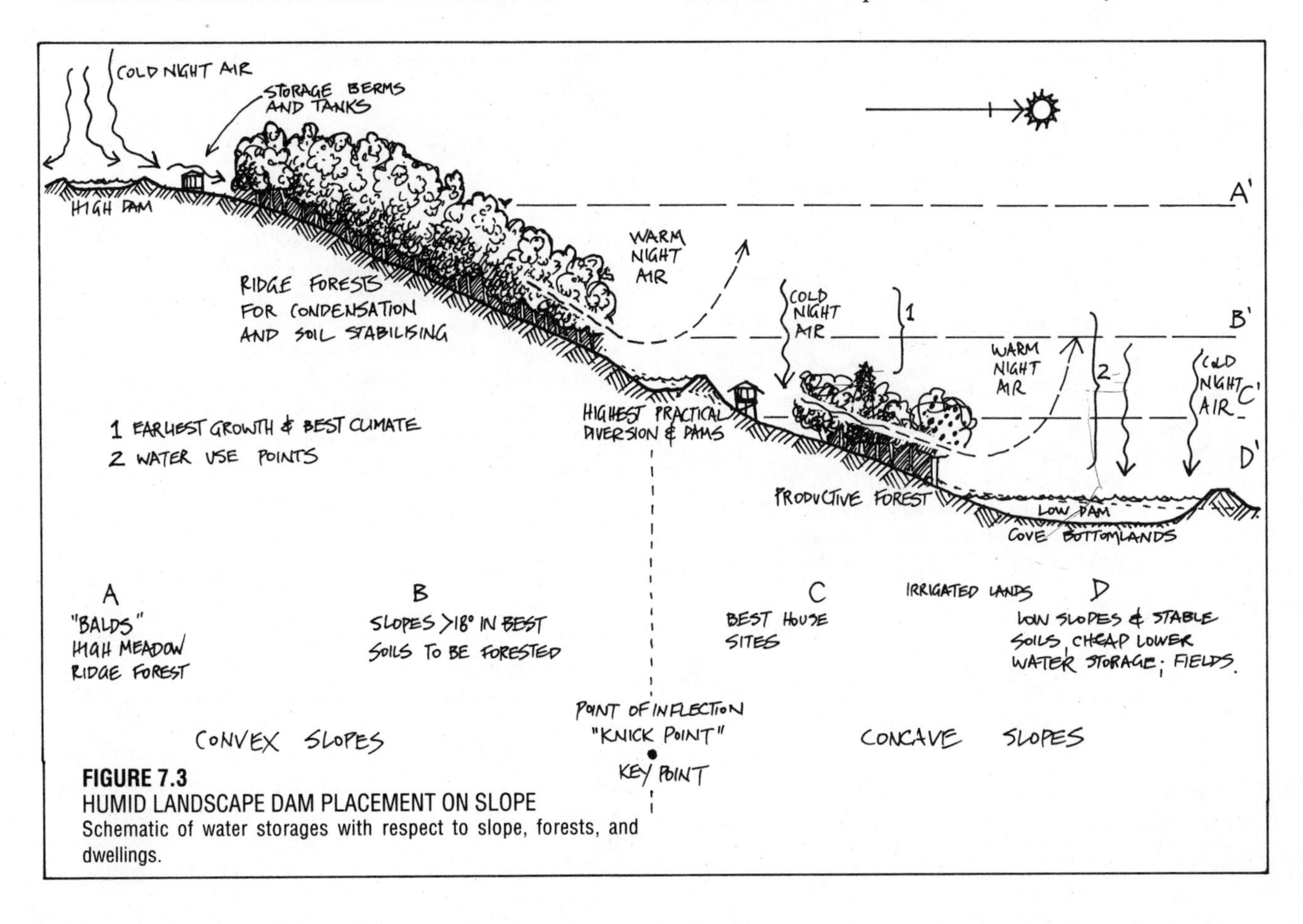

FIGURE 7.3
HUMID LANDSCAPE DAM PLACEMENT ON SLOPE
Schematic of water storages with respect to slope, forests, and dwellings.

proportion of free surface water; some because of free–draining soils or deep or coarse sands. Other areas are too rocky, or of fissured limestone, and yet others are too steep or unstable. But a great many productive areas of clay–fraction subsoils (40% or more clay fraction) will hold water behind earth dams, below grade levels as earth tanks, or perched above grade as "turkey's nest" or ring dams. There are very few landscapes, however, that will not store more soil water if humus, soil treatment, or swales are tried; the soil itself is our largest water storage system in landscape if we allow it to absorb.

Almost every type of dam is cost–effective if it is located to pen water in an area of 5% or less slope. However, many essential dams, if well–made and durable, can be built at higher slopes or grades, made of concrete, rock–walled, or excavated if water for a house or small settlement is the limiting factor. Each and every dam needs careful soil and level surveys and planning for local construction methods.

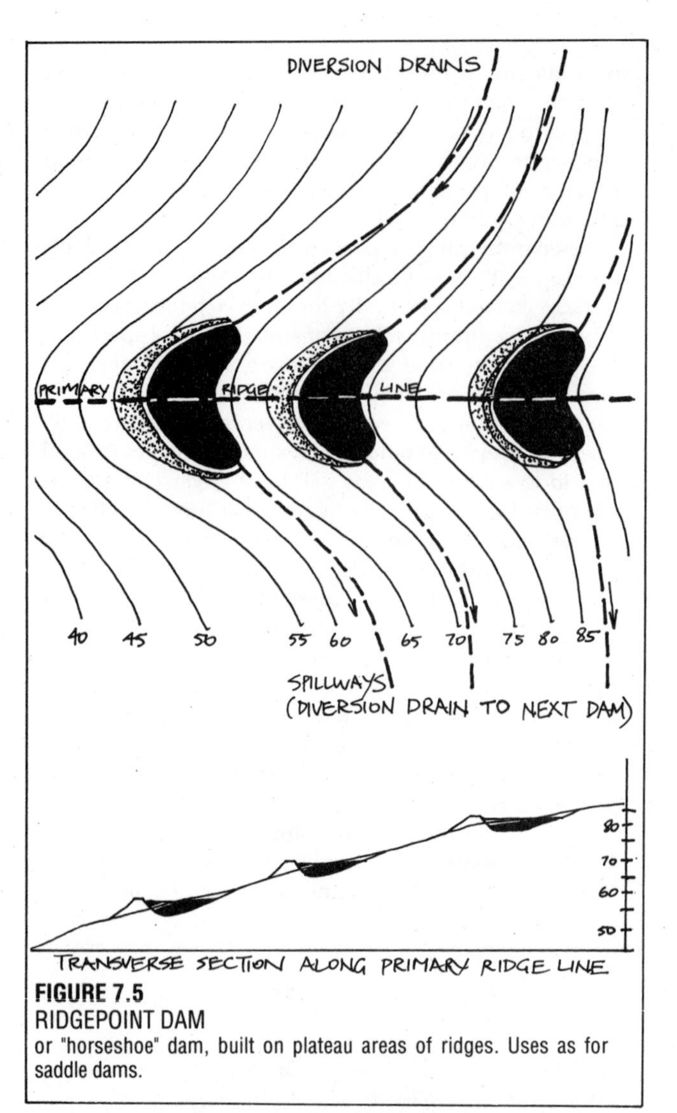

FIGURE 7.5
RIDGEPOINT DAM
or "horseshoe" dam, built on plateau areas of ridges. Uses as for saddle dams.

DAM TYPES AND LOCATIONS

There are at least these common dam sites in every extensive landscape:

SADDLE DAMS are usually *the highest available storages*, on saddles or hollows in the skyline profile of hills. Saddle dams can be fully excavated below ground (grade) or walled on either side of, or both sides of, the saddle. They can be circular, oblong, or "shark egg" shaped with horns or extensions at either end (**Figure 7.4**).

Uses: wildlife, stock, high storage.

RIDGEPOINT DAMS or "horseshoe" dams are built on the sub–plateaus of flattened ridges, usually on a descending ridgeline, and below saddle dams. The shape is typically that of a horse's hoof. It can be made below grade, or walled by earth banks (**Figure 7.5**).

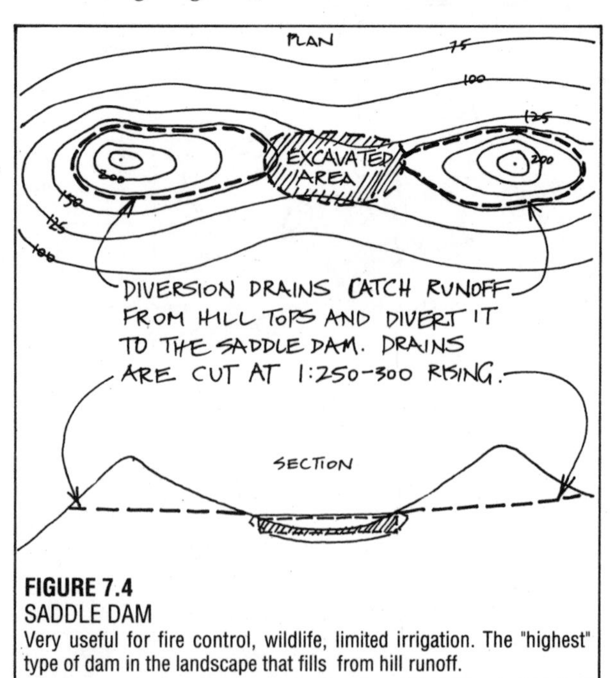

FIGURE 7.4
SADDLE DAM
Very useful for fire control, wildlife, limited irrigation. The "highest" type of dam in the landscape that fills from hill runoff.

Uses: As for saddle dams. Only of limited irrigation use, but very useful for run–off and pumped storages. Note that both saddle and ridge dams can act as storages for *pumped* water used for energy generation.

KEYPOINT DAMS are located in the valleys of secondary streams, humid landscapes, at the highest practical construction point in the hill profile, usually where the *stream* profile changes from convex to concave; this place can be judged by eye, and a descending contour will then pick up all other keypoints on the main valley (**Figure 7.6**).

Uses: Primarily to store irrigation water. Note that a second or third series can be run below this primary series of dams, and that the spillway of the last dam in a series can be returned "upstream" to meet the main valley, effectively spilling surplus to streams.

CONTOUR DAM walls can be built on contour wherever the slope is 8% or less, or sufficiently flat. Contours (and dam walls) can be concave or convex to the fall line across the slope (**Figure 7.7**).

Uses: Irrigation, aquaculture, or flood–flow basins in

semi–arid areas.

BARRIER DAMS are always constructed across a flowing or intermittent stream bed. These dams therefore need ample spillways, careful construction, fish ladders on biologically important streams, and are made most frequently as energy systems, but are also used for irrigation if they are constructed well above the main valley floors where crops are grown (**Figure 7.8**).

TURKEY'S NEST DAMS or above–grade tanks; water has to be pumped in to these, often by windmill or solar pump. They are common in flatlands as stock water tanks or for low–head irrigation (**Figure 7.9**).

CHECK DAMS. There are many forms of barrier dams not intended to create water storages, but to

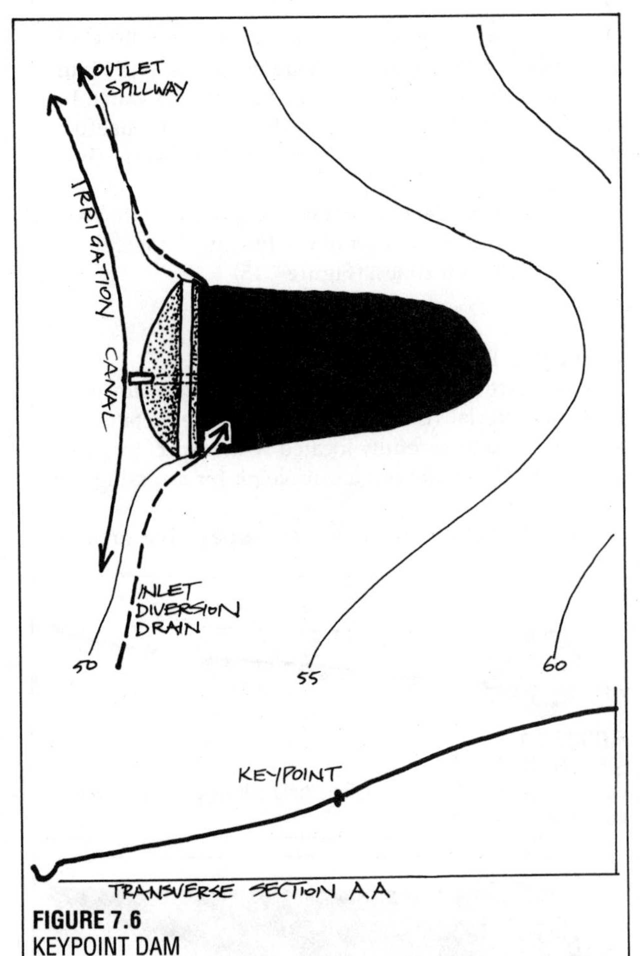

FIGURE 7.6
KEYPOINT DAM
A. If used in series, no spillway is built and the overflow goes to the next dam, and eventually to a stream. Fitted for irrigation on lower slopes.
B. The keyline (heavy dashes) links keypoints in primary valleys.

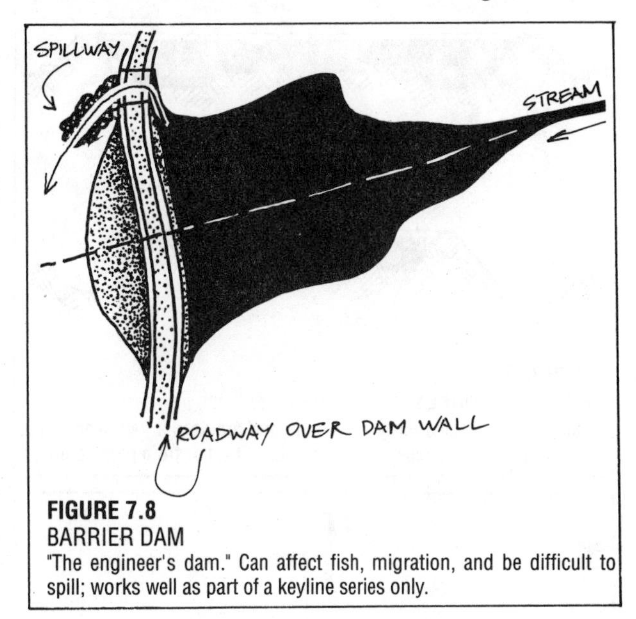

FIGURE 7.8
BARRIER DAM
"The engineer's dam." Can affect fish, migration, and be difficult to spill; works well as part of a keyline series only.

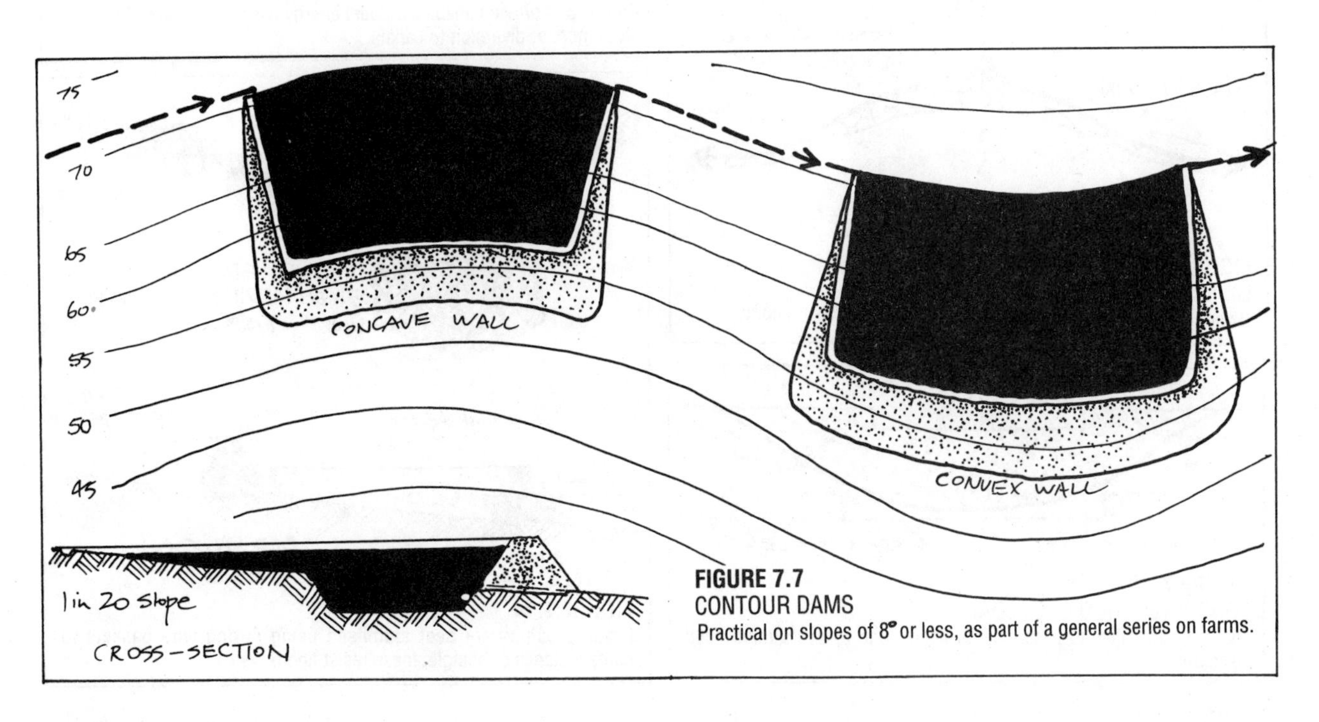

FIGURE 7.7
CONTOUR DAMS
Practical on slopes of 8° or less, as part of a general series on farms.

regulate or direct stream flow. Even a 1–3 m (3–10 foot) wall across a small stream gives enough head to drive an hydraulic ram, to fit a waterwheel, to divert the stream itself to a contoured canal for irrigation, or to buffer sudden floods. Dams intended to regulate flood crests may have a base pipe or fixed opening in the streambed which allows a manageable flow of water downstream while banking up the flood crest behind the dam itself, so spreading the rush of water over time. The base opening allows silt scour and so keeps the dam free of siltation (**Figure 7.10–13**).

GABION DAMS. In drylands, permeable barriers of rock–filled mesh "baskets" (**gabions**) will create silt fields and water–spreading across eroding valleys. The scale of these dams varies, but for farm construction, walls 0.5–2 m (2–6.5 feet) high are usual. As with **Figure 7.12**, the purpose is not to store free surface water, but to create a flat area where silt loads can usefully deposit, and so form absorption beds in flood conditions.

We can see the landscape (as though sliced into layers through contours) as a set of catchment, storage, usage, and revitalisation zones. (**Figure 7.15**)

BUILDING DAMS

Although we can build dams or tanks on *any* site, given enough material resources, commonsense dictates that storage dams be carefully located with respect to:

- Earth type (core out a sample pit for assessing clay fractions);
- Grade behind wall (lower slopes give greatest

FIGURE 7.9
RING AND TURKEY'S NEST DAM.
Hold water pumped in by a windmill and provides a low head in flat landscapes; can be pipe–filled from a large roof or from parking areas.

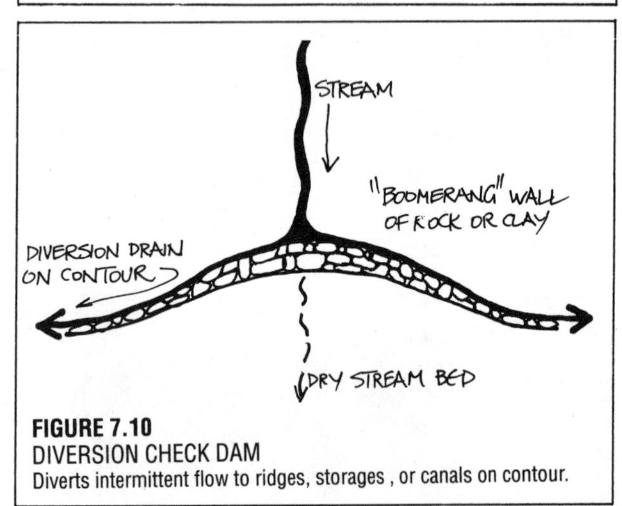

FIGURE 7.10
DIVERSION CHECK DAM
Diverts intermittent flow to ridges, storages , or canals on contour.

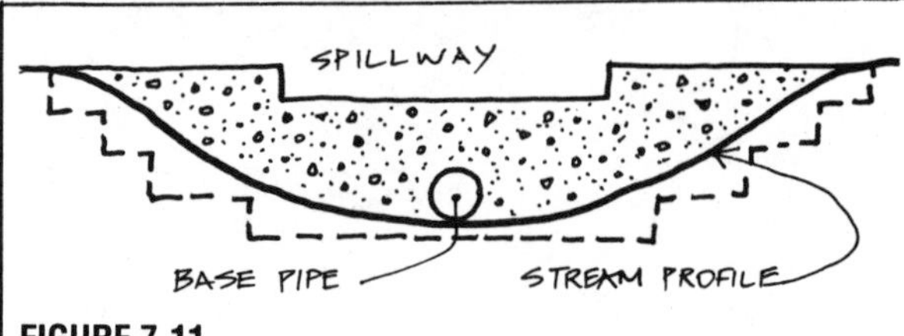

FIGURE 7.11
CONCRETE FLOOD CHECK DAM
Allows normal flow to pass, retards floods, and prevents rapid flood discharge.

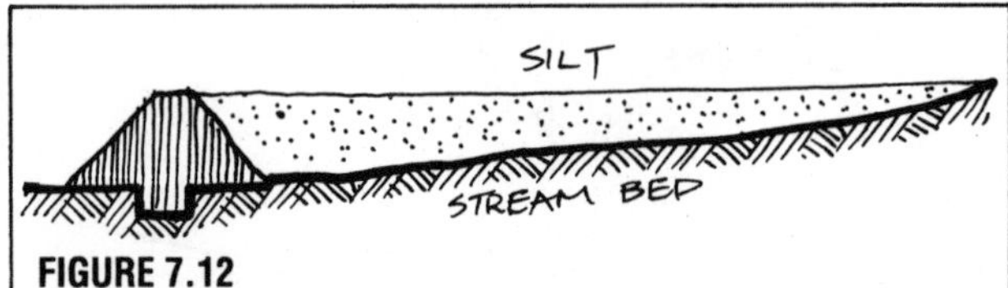

FIGURE 7.12
SILT CHECK DAM
Earth or concrete walls or gabions hold silt fields, spread water, reduce silt load in streams.

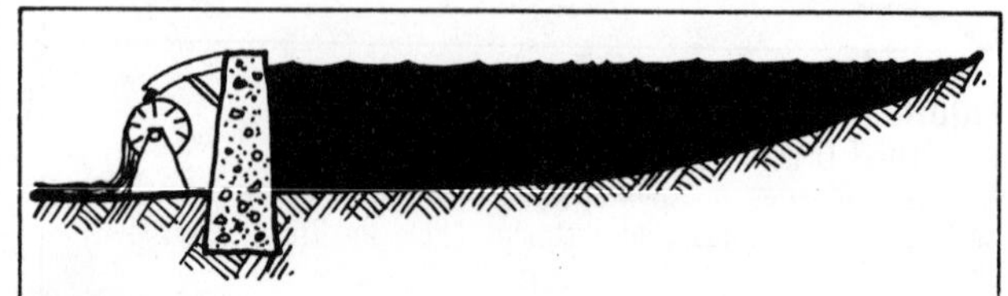

FIGURE 7.13
CHECK DAM FOR RAM PUMP OR WATER WHEEL
Only 1–3 m of head enables modest energy use for mechanical power, lift pumps, or diversion to canals.

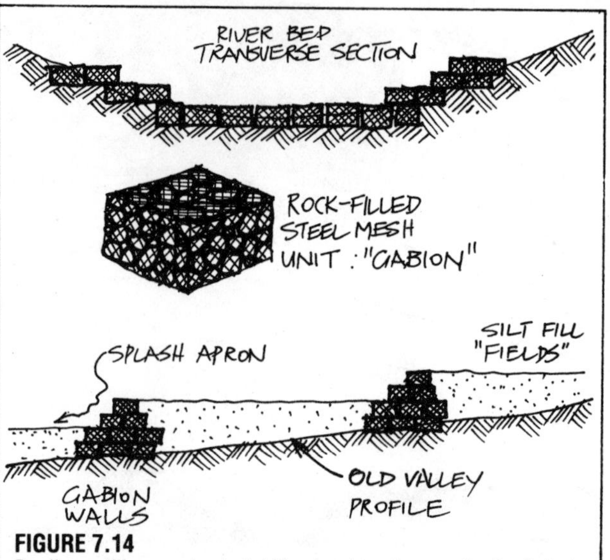

FIGURE 7.14
Eroding gullies are best stabilised using strong wire baskets to contain stones or shingle; these resist floods well.

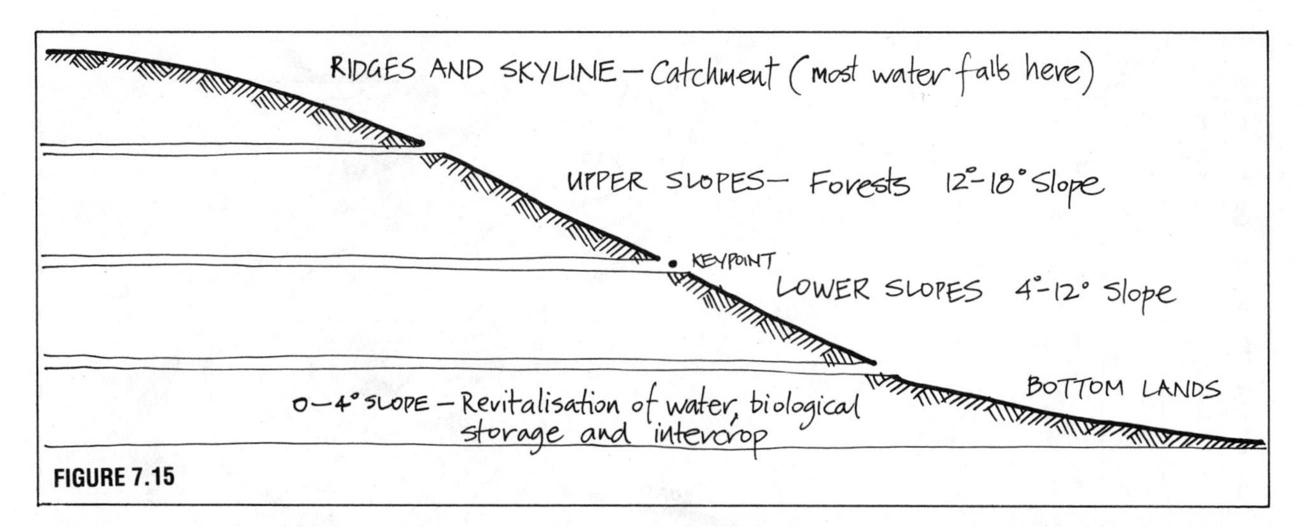

FIGURE 7.15

capacity);

- Downstream safety of structures and houses (a key factor in large dams);
- Height above use points (gravity flow is desirable); and
- Available catchment or diversion.

Tamped earth with some clay fractions of better than 50% is a waterproof barrier up to heights of 3.6 m (12 feet), not counting the holes behind such walls caused by their excavation. Therefore we speak of depths of 4.5–6 m (15–20 feet) for small earth dams. Few of us will want to build farm dams higher, and we must get good advice if we wish to do so.

Slopes to crest should be concave, and every 25 cm (10 inches) a machine such as a roller, or the bulldozer tracks themselves, should ride along and tamp down

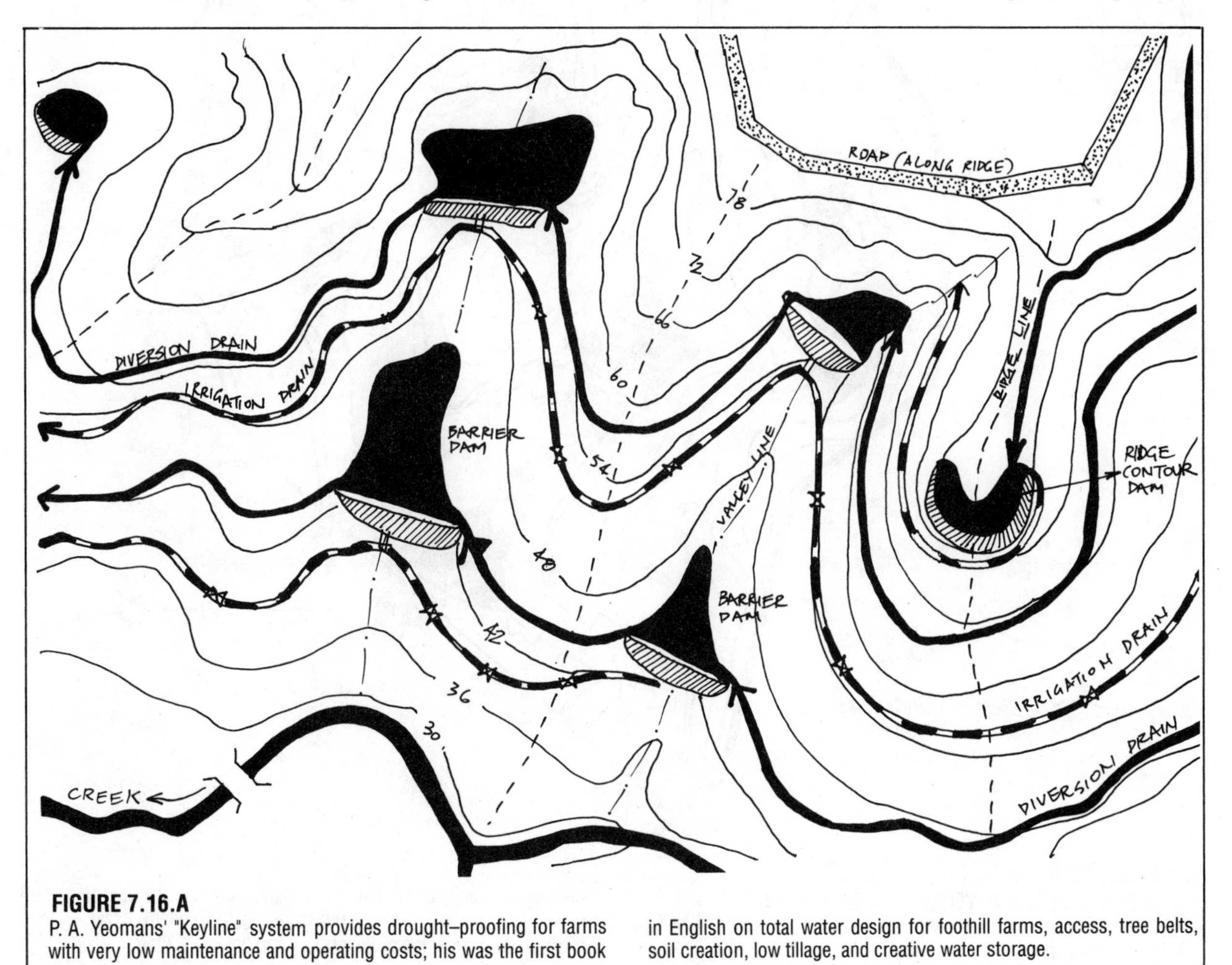

FIGURE 7.16.A
P. A. Yeomans' "Keyline" system provides drought–proofing for farms with very low maintenance and operating costs; his was the first book in English on total water design for foothill farms, access, tree belts, soil creation, low tillage, and creative water storage.

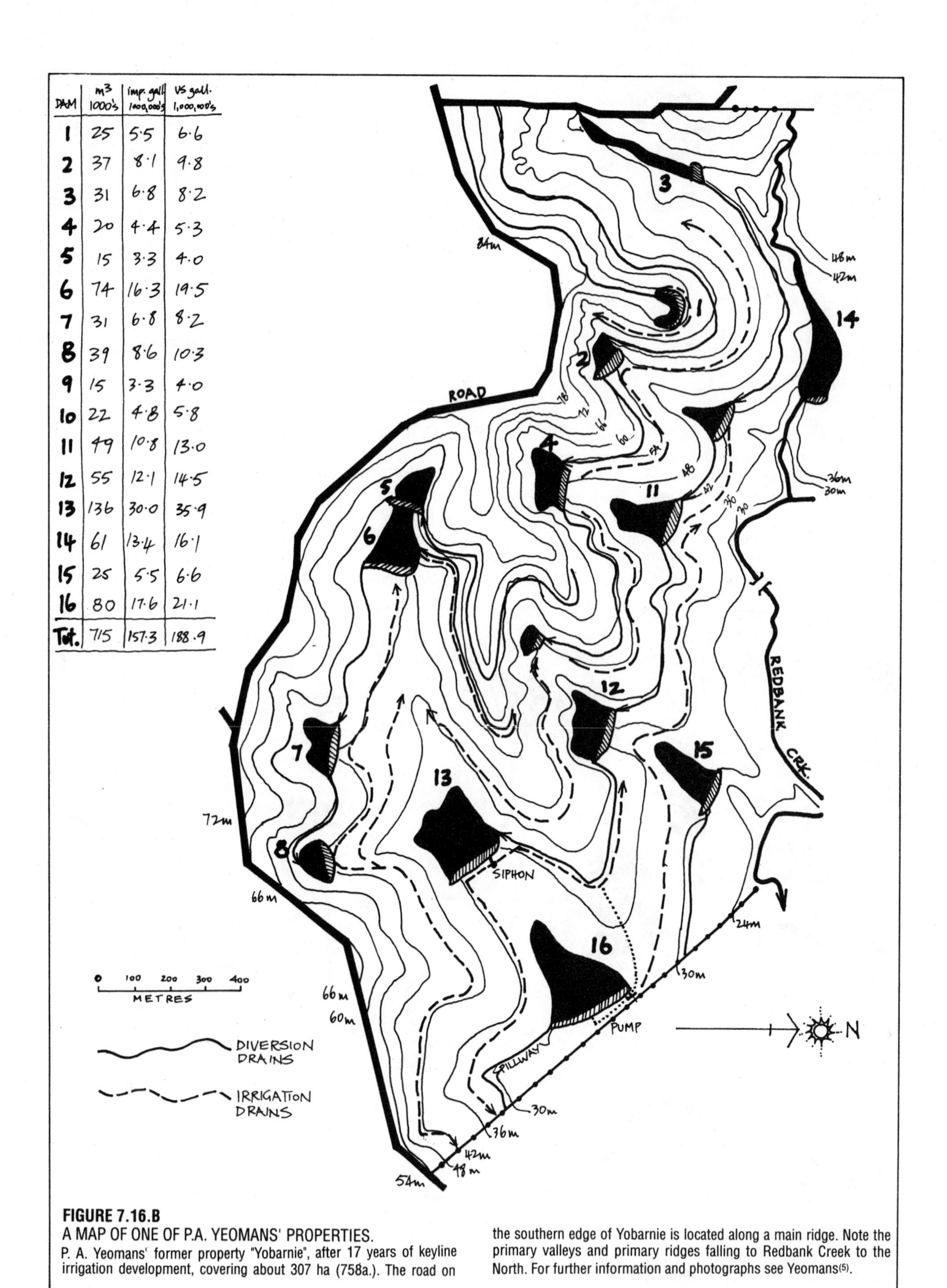

DAM	m³ 1000's	imp. gall 1000,000's	US gall. 1,000,000's
1	25	5·5	6·6
2	37	8·1	9·8
3	31	6·8	8·2
4	20	4·4	5·3
5	15	3·3	4·0
6	74	16·3	19·5
7	31	6·8	8·2
8	39	8·6	10·3
9	15	3·3	4·0
10	22	4·8	5·8
11	49	10·8	13·0
12	55	12·1	14·5
13	136	30·0	35·9
14	61	13·4	16·1
15	25	5·5	6·6
16	80	17·6	21·1
Tot.	715	157·3	188·9

FIGURE 7.16.B

A MAP OF ONE OF P.A. YEOMANS' PROPERTIES.

P. A. Yeomans' former property "Yobarnie", after 17 years of keyline irrigation development, covering about 307 ha (758a.). The road on the southern edge of Yobarnie is located along a main ridge. Note the primary valleys and primary ridges falling to Redbank Creek to the North. For further information and photographs see Yeomans(5).

the earth. This, like the exclusion of boulders and logs, grass clumps and topsoil, is critical to earth stability (shrinkage of well–compacted dams is less than 1%). Earth so rolled should be neither so dry as to crumble nor so wet as to slump or squash out under the roller.

A key should be cut to prevent shear and cut off any base seepage. This is needed on all walls 1.8 m (6 feet) or more high, otherwise the base should be on a shallow clay–filled ditch. Slopes are safe at a ratio of 3:1 (inner) and 2 or 2.5:1 (outer), freeboard at 0.9 m (3 feet), key at 0.6–0.9 m (2–3 feet) deep. In suspect soils, the whole core can be of carted clay (**Figure 7.17**).

The wall can curve (out or in), but if carefully made as diagrammed and provided with a broad spillway, should be stable and safe forever, barring explosions or severe earthquakes.

The SPILLWAY base should be carefully surveyed at 1 m below crest and away from the wall or fill itself (don't try to judge this, measure it), and a SIPHON or BASE OUTLET pipe fitted with baffle plates placed to draw off water (**Figure 7.18**).

The efficiency in capacity of dams depends on the flatness of the area behind the wall. A "V" valley or "U" valley, plateau, or field should be as level as can be chosen for greatest efficiency. The key to efficiency is the length of the dam wall, compared with the "length" of water dammed. If the back–up is greater than wall length, then this is a measure of increasing efficiency of energy used or earth moved for water obtained. A careful survey of grade plus dam length gives this data before starting the wall. Some dam sites are very cost–effective, especially those short dams at constricted sites where the valley behind them is flattish.

Small dams of this nature are a jewel in the landscape. Fenced and planted to 30–60 m of forest and fruit surround, they will provide biologicially clean, if sometimes muddy, water, and if the topsoil is returned,

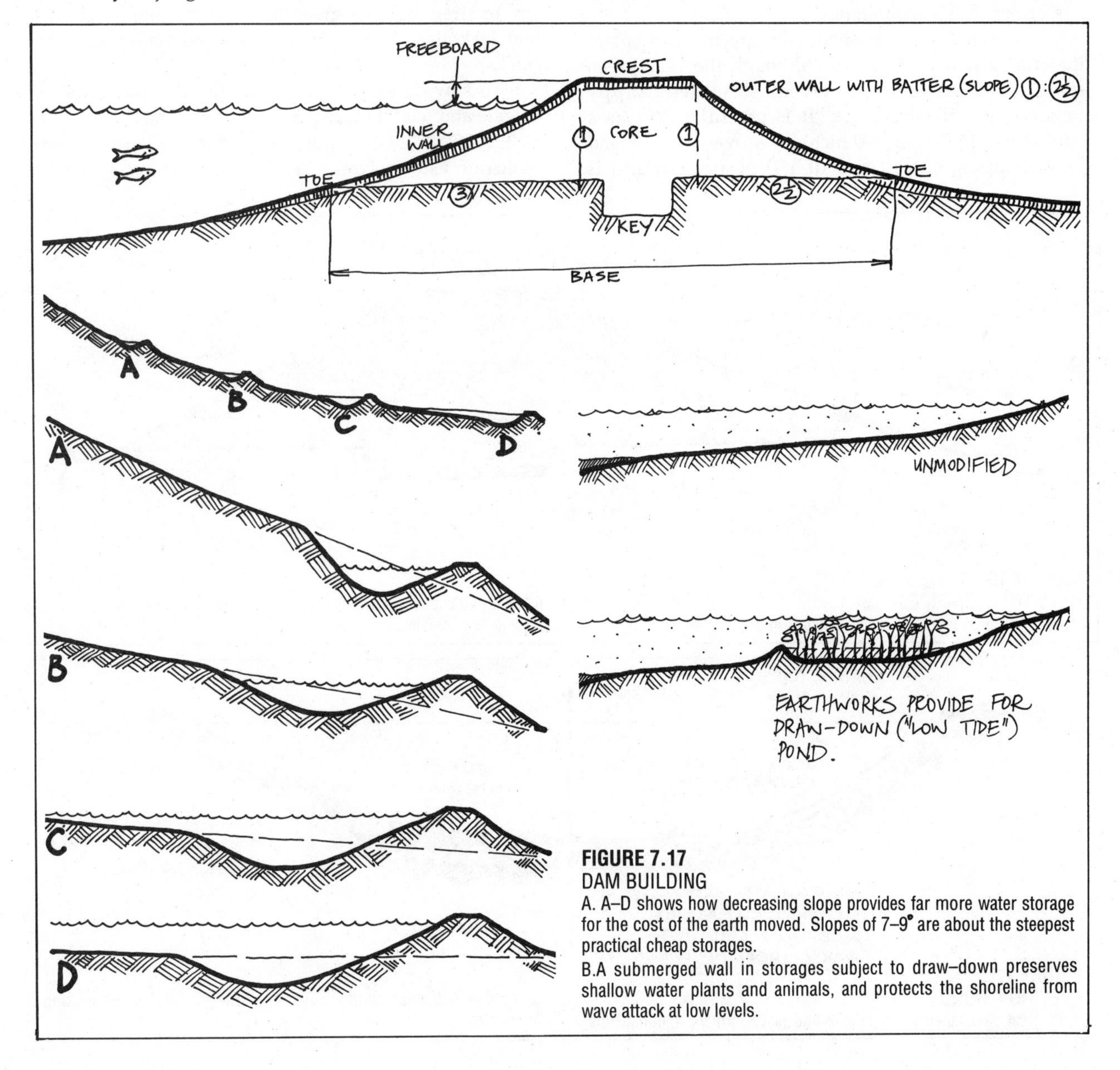

FIGURE 7.17
DAM BUILDING
A. A–D shows how decreasing slope provides far more water storage for the cost of the earth moved. Slopes of 7–9° are about the steepest practical cheap storages.
B. A submerged wall in storages subject to draw–down preserves shallow water plants and animals, and protects the shoreline from wave attack at low levels.

lime used, and edges planted, mud will decrease and eventually clear. For water cleanliness and parasite control, cattle, sheep, and other animals should be watered at spigots or troughs, not directly at the dam. Troughs are easily treated with a few crystals of copper sulphate to kill snails and parasitic hosts; dams stocked with fish will do the same job.

Crests can be gravelled and safely used as roads to cross valleys or bogs, and special deep areas, islands, peninsulas, and shelves or benches made inside the dam for birds, plants, and wild–fire–immune houses.

SEALING LEAKY DAMS

There are several ways to seal leaking dams:

- Gley;
- Bentonite;
- Explosives;
- Clay; and
- Impermeable membranes.

GLEY is a layer of mashed, wet, green, sappy plant material sealed off from air. Although the very green manure of cattle is preferred, shredded, sappy vegetation will also work. It is carefully laid as a continuous 15–23 cm (6–9 inch) layer over the base and *gently* sloping sides (ratio of 1:4) of a pond, and is covered *completely* with earth, cardboard, thick wet paper, plastic sheets, or rolled clay, and allowed to ferment anaerobically. This produces a bacterial slime which permanently seals soil, sand, or small gravels. Once ferment occurs, the pond is pumped or hosed full of water, and the paper or plastic can be later removed. I have used carpets and odd pieces of plastic sheets overlapped with good results. In cold areas, ferment can take a week or two, in tropics a day or so. Lawn or second–cut grasses, papaya and banana leaves, vegetable tops or green manure all serve as the base layer. I believe that in very good soils, especially in the tropics, it may be possible to grow the gley as a mass of *Dolichos* bean and just roll it flat before sealing it (**Figure 7.19**).

Modifications are:

- To pen and feed a herd of cattle in the dry dam until the bottom is a manurial pug; occasional watering assists this process.
- To strew bales of green hay and manure on ponds that leak slightly, producing algae which seal minor cracks.
- To sow down green crop in the dry dam, spray irrigate and feed it off regularly with cattle.

BENTONITE is a slippery clay–powder derived from volcanic ash. It swells when watered and will seal

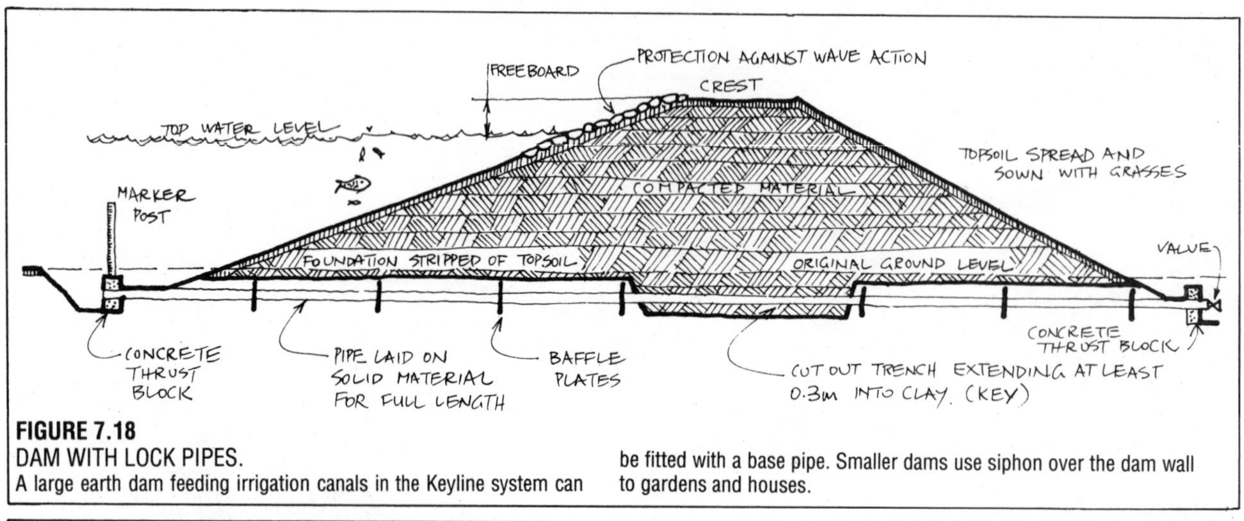

FIGURE 7.18
DAM WITH LOCK PIPES.
A large earth dam feeding irrigation canals in the Keyline system can be fitted with a base pipe. Smaller dams use siphon over the dam wall to gardens and houses.

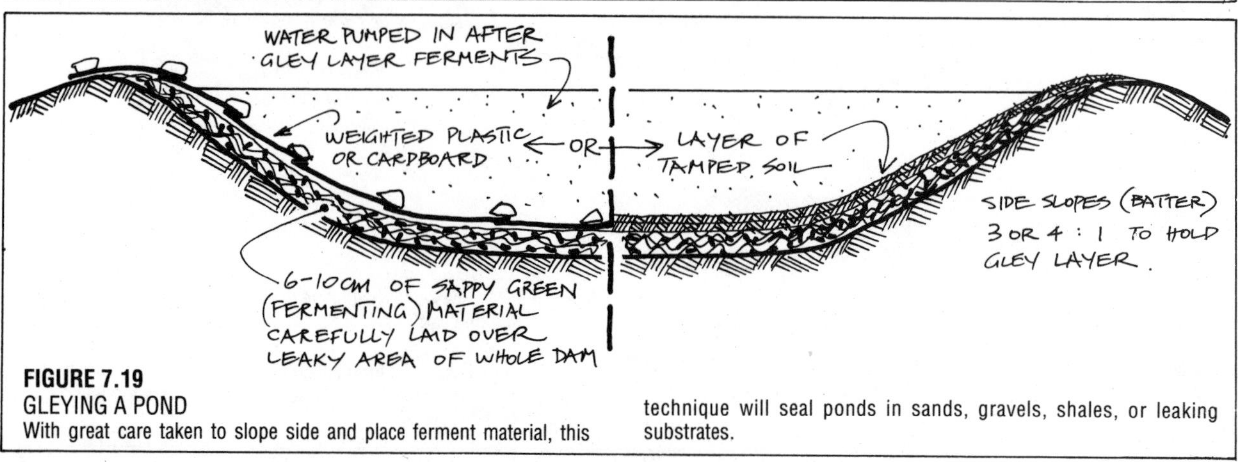

FIGURE 7.19
GLEYING A POND
With great care taken to slope side and place ferment material, this technique will seal ponds in sands, gravels, shales, or leaking substrates.

clay–loams if rototilled in at 5–7 cm (2–3 inches) deep and rolled down. However, it is expensive and doesn't always work. Cement and tamping plus sprinkling might be preferable, or a bituminous spray can be rolled in after tilling. In clay soils, salt or sodium carbonate can have the same effect.

EXPLOSIVES are sometimes used to compact the sides of full dams, and consists of throwing in a 3–5 stick charge of dynamite. This works well at times, but is dangerous if you own a retriever, or if the dam wall is poorly compacted to start with.

CLAY is expensive if it has to be carted in, but it is often used to seal dams near a clay pocket. The clay is spread and rolled 23–30 cm (9–12 inches) thick over suspect areas.

IMPERMEABLE MEMBRANES can be of welded plastic, neoprene, or even poured concrete. Impermeable membranes are too expensive to use on any but critical dams, which may mean a guaranteed water supply to a house or garden in very porous areas. Using membranes enables banks to be steeper than in any other earth–compaction or gley system, so that more water can be fitted into smaller space. It is not "biological" unless a sand or topsoil floor is also added over the sealing layer, when fish or plants can be added.

Earth storage is now the cheapest, easiest, and most locally self–reliant method of water conservation. Unless both cities and farms use such methods, clean water will deservedly become known as the world's rarest mineral, ill–health will be perpetuated, and droughts and floods alike become commonplace. None of these are necessary.

Costs vary greatly; as a rough guide, water stored in soil and humus is the cheapest and of greatest volume, surface dams next cheapest, and tanks dear, but still much less expensive than piped water from mains supply. I can only urge all people of goodwill to promote, fund, and investigate water and water storage, water energy and water cleanliness, as the chlorinated, metallic, asbestos–fibred, poisonous water of modern centralised systems is producing such epidemic disease and illness as cancer, bone marrow failure, and gastrointestinal disorder.

If a 22,500 l tank costs 20 units of money, the same units in a sensible eath storage pays for 2,500,00 l, or about 100 times as much water. Up to 135,000–2,500,000 l tanks get cheaper, as less concrete is used for more water. That is, a large tank is relatively cheaper than a small tank. Above 22,500 l, such tanks are usually poured on site; below this, they are carted from a central manufacturing site.

Dams, in contrast, begin to cost more as the height of the wall rises. About 3 m (10 feet) of retaining wall is the limit of cheap dams. Above this, costs rise rapidly as greater skills, more expensive and massive materials, more complex controls of levels, and much greater environmental risks take their toll.

As noted, "cheap" water in dams depends on the choice of site, so that very low dams on well–selected sites impound 20–100 times more water than the same earth used on steeper sites, where every unit of earth moved equals a unit of water. However, even earth tanks excavated below grade are at one–tenth the price of concrete tanks above grade.

Where are tanks, modest dams, and massive dams appropriate? Tanks are appropriate on isolated dwellings, in flatlands, and everywhere in cities and urbanised areas. Dams of from 22,500 to 4.5 Ml are best built on any good site in country and parkland areas. Massive dams are appropriate hardly anywhere but the the rock–bermed or glaciated uplands of solid and forested hills, subject to low earthquake risk and then only for modest domestic (not dirty industrial) power generation.

TANKS

Cultural and historical precedent may determine how earth is moved and used, or even if it can be moved at all. Thus, an Australian, accustomed to a great variety of surface storage, is astounded that there are no

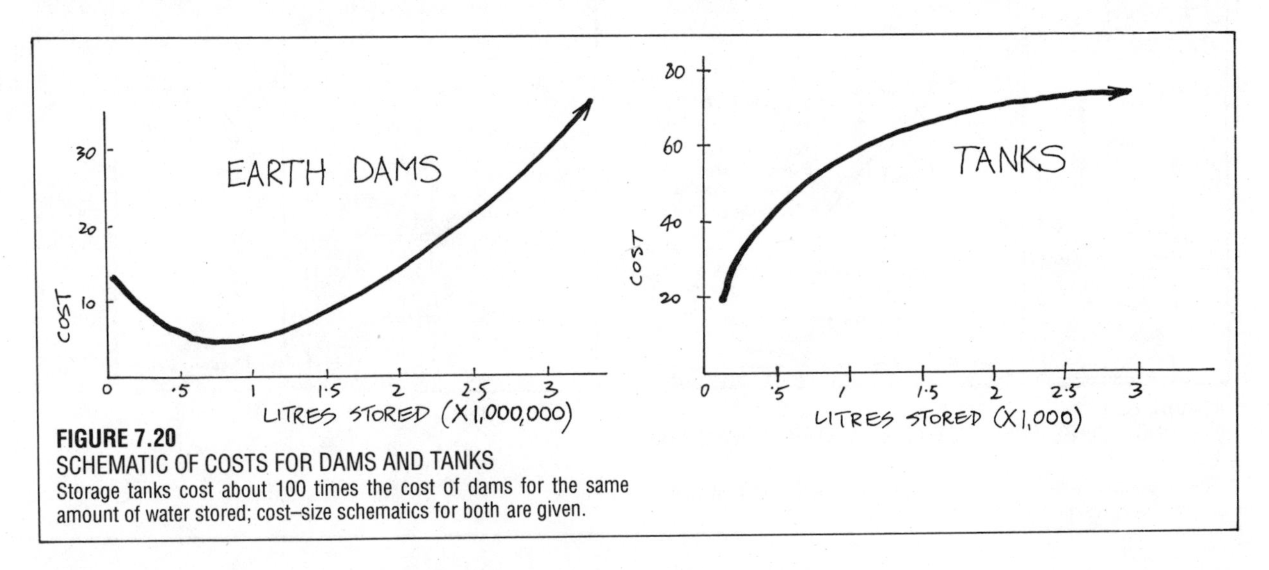

FIGURE 7.20
SCHEMATIC OF COSTS FOR DAMS AND TANKS
Storage tanks cost about 100 times the cost of dams for the same amount of water stored; cost–size schematics for both are given.

significant domestic rainwater tanks in Europe, the USA, or India (where clean drinking water is rare), that British, American, and Brazilian farmers rarely use multiple earth storages of water, and that expensive pipelines and bores are the preferred "alternative", even where local rainfall often exceeds local needs.

The simple forms for making concrete tanks cost a few hundred dollars, and may be used hundreds of times. About 22,500 l provides a family with all needed water (drinking, showers, cooking, modest garden water on trickle) for a year; tank water is renewed by rain at any time of year. Every roof, whether domestic or industrial, would fill many such tanks, and simple calculations (roof area x average rainfall in millimetres or inches) and conversion to litres or gallons gives the expected yield.

Granted that roof areas themselves can be contaminated by birds, dust, or industry, the first precaution is to reject the first flow–off of water, and use it on gardens or in swales. Two methods for doing this are shown in **Figure 7.21**.

As for the entry of insects, birds, or rodents to tanks (and this includes mosquitoes), a "U" pipe entry and exit, a sealed tank roof, and an overflow pipe emptying to a gravel–filled swale all effectively exclude these potential nuisances. If birds persistently perch on roof ridges, a few very fine wires or thread stretched along the ridge as a 10 cm (4 inch) high "fence" will discourage them.

Gutters on roofing can be cleaned out regularly, or "leaf–free" gutters or downpipes fitted (about 3 or 4 types are commercially manufactured; some systems are illustrated in **Figure 7.22**).

Given that most dust and leaves are removed, residual organics are usually harmless. These "fix" as an active biological velvety film on tank walls and bases. Taps or outlet pipes are normally fitted 15–20 cm (6–8 inches) above any tank floor to allow such a film to remain.

Finally, a net or bag of limestone, shell, or marble chips is suspended in the tank. This creates hard (alkaline) water, preventing heavy metal uptake from the water and decreasing the incidence of heart attacks in those using the tank. Washing and shower water can be soft (acid) but the water we drink is best made alkaline for the sake of health.

It makes far more sense to legislate *for* such tanks on every roof than to bring exotic water for miles to towns; it will also ensure that clean air regulations are better observed locally, that every house has a strategic water reserve, and that householders are conservative in their use of water.

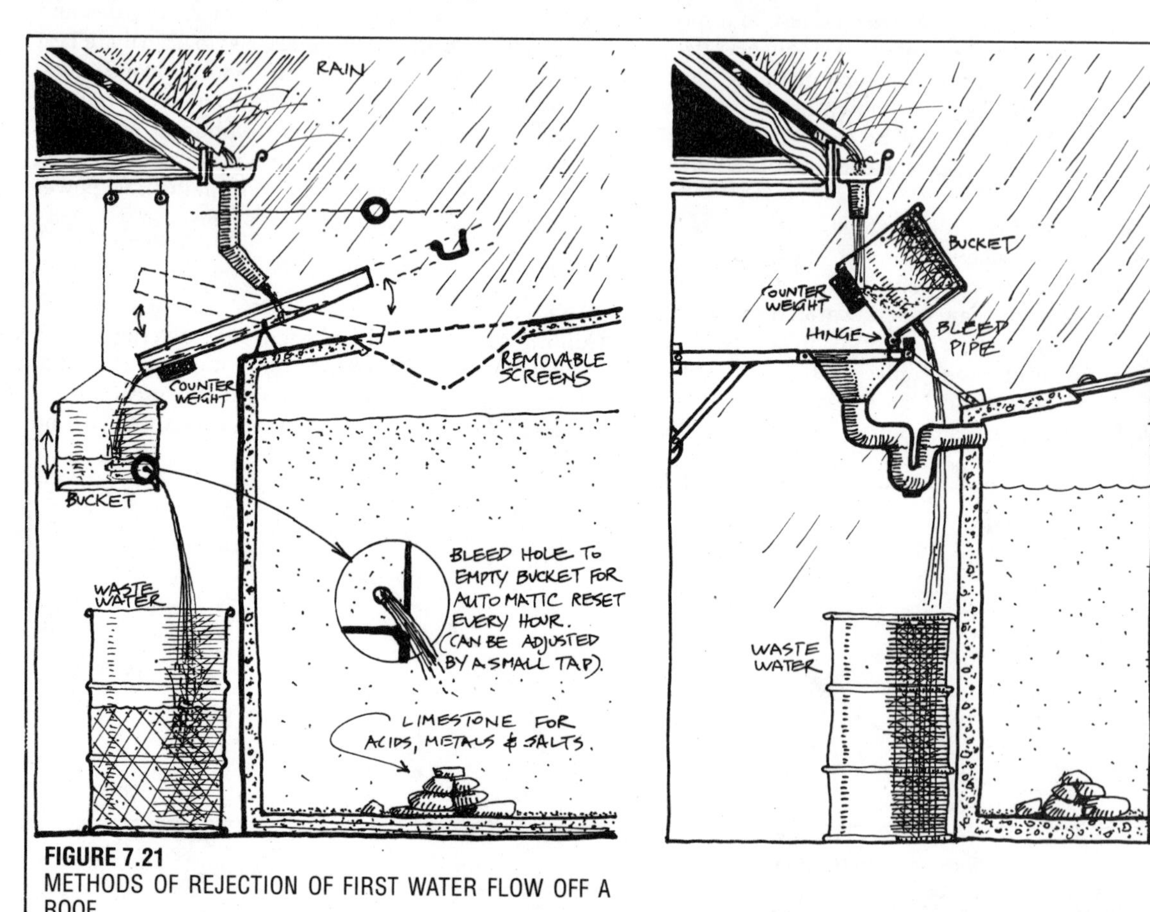

FIGURE 7.21
METHODS OF REJECTION OF FIRST WATER FLOW OFF A ROOF.
The first rains wash the roof, and are rejected; these systems automatically reset when empty.

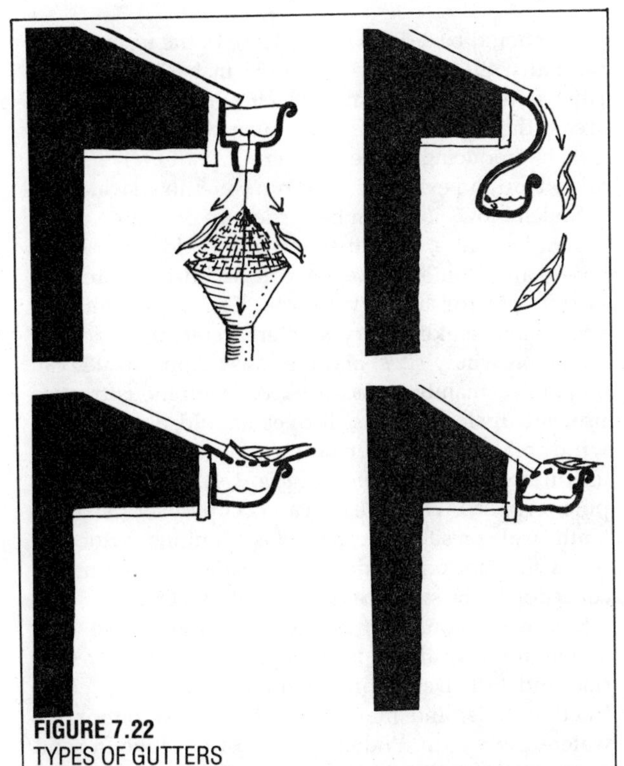

FIGURE 7.22
TYPES OF GUTTERS
Self–cleaning and leaf–free gutters or downpipes are useful where trees overhang a house; cone to pipe is also useful to collect water from rock domes.

SWALES

Swales are long, level excavations, which can vary greatly in width and treatment from small ridges in gardens, rock–piles across slope, or deliberately–excavated hollows in flatlands and low–slope landscapes (**Figure 7.23**).

Like soil conditioning or soil loosening systems, swales are intended to store water in the underlying soils or sediments. They are, simply, cross–flow dry channels or basins intended to totally intercept overland flow, to hold it for a few hours or days, and to let it infiltrate as GROUNDWATER RECHARGE into soils and tree root systems. *Trees* are the essential components of swale planting systems, or we risk soil waterlogging and a subsequent local rainfall deficit caused by lack of evapotranspiration of the stored water. Thus, tree planting *must* accompany swaling in arid areas.

Swales should ideally not exceed in width the total crown spread of the fringing trees planted to use the stored water absorbed into the swale sediments. Trees overshade and cool the soils of swales, further reducing the risk of evaporation and dissolved salt concnetration, or water loss. Although swales can also be grazed, few grasses can effectively remove the absorbed water to re–humidify airstreams.

Swales are therefore widely used in arid to sub–humid, even humid areas, on both fairly steep slopes and flatlands, and in both urban and rural areas. They are appropriate to road and other silty or contaminated run–off harvest (where the dust or tar oils washed off have no adverse effect on tree growth).

The essentials of swale construction are simple: they are all built on contour or dead level survey lines, and are neither intended nor permitted for water flow. Their function is just to *hold* water. Unlike dams, swale banks and bases are never compacted or sealed (although small tanks can be sunk in swale bases for watering livestock or trees). Conversely, the swale soils can be gravelled, ripped, or loosened to assist water infiltration. The swale depth and width can be varied to cope with the speed of infiltration locally, so that wider and shallower swales are made in sands, narrower and deeper swales in clay–fraction soils.

After an initial series of rains that soak in a metre or more of water, trees are seeded or planted on either bank or side slopes of the swales. This can take two wet periods. Thereafter, it takes about 3–5 years for tree belts to overshade the swale base, and to start humus accumulation from leaf tissue. (Humus will accumulate, however, by wash–down and wind movement from bare or uphill areas.)

Early in the life of an unplanted swale, water absorption can be slow, but the efficiency of absorption increases with age due to tree root and humus effects. As this happens, it is possible to admit water to swales from other areas, leading it in via DIVERSION DRAINS DRAINS. This "exotic" water from unused road or rock surfaces or overland flow can enable the planting of high–value trees of higher water demand or a new set of swales to be constructed.

Every sub–humid and arid townscape can, with great energy gain and much reduced cost for roading and water use, fit all roads and paved areas with swales, along which tree lines shade pavement and reduce heat oases while they produce fuel, mulch, and food products. Every roof tank overflow, and some greywater wastes can be led to swales (if boron detergents are not used).

Swales interpenetrating the suburban development of Village Homes in Davis, California (Michael Corbett, designer) accept all road and excess roof run–off, and

FIGURE 7.23
Swales on contour do not flow; they first stop and then infiltrate overland flow. Swales on hillsides are part of access or production systems.

support hundreds of productive trees in settlement. Water penetrated soils to 6 m (19 feet) deep after a few years of operation, and swales were self-shaded after 3–4 years of tree growth. In Hawaii and in central Australia, swales I have designed produced fast growth in trees in volcanic cinder and sandy soils.

Swales in Australian drylands have consistently grown larger and healthier water run-off fed trees than have open plantings. In arid areas, it is imperative to plant trees on swales, or we risk salt concentration and soil collapse downhill. All swales are therefore temporary events, as trees supplant their function; they are precursors to rehabilitation of normal forests in their region. Natural swales in humid forests (Tasmania) not only generate much larger trees and provide level access ways, but support a thick humus and specialised plants on the swale floors. Orchids, fungi, and ginseng do better in swales.

Most swales should be adjusted (by widening, gravelling, or ripping) to absorb or infiltrate all water caught in from 3 hours to 3 days. Fast absorption will not harm most tree species, although trees such as chestnuts and citrus may need to be planted on nearby spoil banks for adequate root drainage. I believe swales to be a valuable and greatly under-used earth form in most climates, including upland and plains areas of snow-drift in winter.

In summary, a swale is a large hollow or broad drain intended to first pool, then absorb all surplus water flow. Thus, the base is ripped, gravelled, sanded, loosened, or dressed with gypsum to allow water INFILTRATION. Trees ideally overshade the swale. The base can be uneven, vary in width, and treated differently depending on the soil type. The spoil is normally mounded downhill or (in flat areas) spread. Water enters from roads, roof areas, tank overflows, greywater systems, or diversion drains.

The distance between swales (the run-off or mulch-planted surface) can be from three to twenty times the average swale width (depending on rainfall). Given a useful swale base of 1–2 m (4–6 feet), the interswale space should be 3–18 m (12–60 feet). In the former case, rainfall would exceed 127 cm (50 inches), and in the latter it would be 25 cm (10 inches) or less. In humid areas, the interswale is fully planted with hardy or mulch-producing species. In very dry areas, it may be fairly bare and exist mainly to run water into swales.

Mulch blows into, can be carried to, or is grown and mown in swales. Fine dust and silts build up in swale bases, and domestic wastes can be buried here as a mulch-pit for hungry plants. The swale and its spoil-bank make a very sheltered starting place for plants on windy sites, and the lower slope swales can be planted mainly to *Casuarina* or leguminous trees to prevent upslope winds. Ridges should always have windbreak and condensation plants of hardy and useful species (*Casuarina, Acacia, Leucaena*, silky oak, pine, cypress). Windbreaks can occupy every sixth to tenth swale on sites where wind is a limiting factor. It is better to plant on the downslope side to allow mulch collection in the swale base for use elsewhere.

Swale sections can be over-deepened, so that although the swale lip is surveyed level, its floor may rise and fall. Deepening is most effective in clay-fraction soils, and may result in shallow ponds for water-needy crop. Widening is most effective in sand or volcanic-fraction soils, and readily admits water to the ground table.

Two other pit systems are useful in swales: one mulch and manure-filled for heavy nutrient feeders (yams, bananas, etc.) and the other to hold oil drums, plastic liners, or tyre ponds as a sealed water reserve for watering young plants. These can be planted with lotus, kangkong, watercress, Chinese water chestnut, or like crop.

Keeping the swale width to the tractor, donkey cart, foot track or wheelbarrow access width that one has planned, sections can be widened at regular intervals to take assemblies of plants, to dig ponds or mulch pits, and to plant trees of higher water need. This leaves access open and enables many assemblies, species, and constructs to be built along the swale as need dictates.

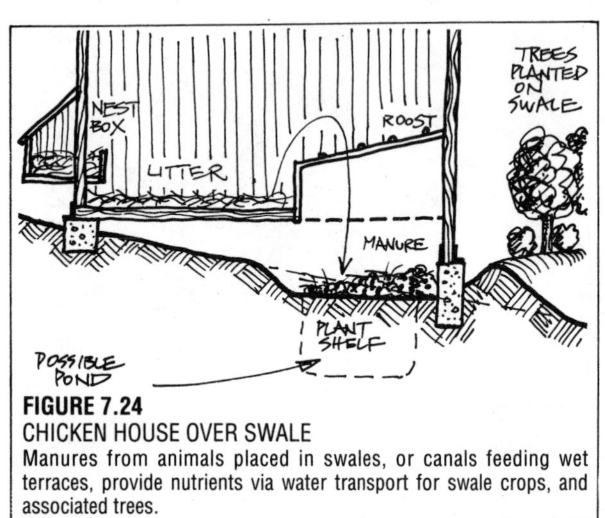

FIGURE 7.24
CHICKEN HOUSE OVER SWALE
Manures from animals placed in swales, or canals feeding wet terraces, provide nutrients via water transport for swale crops, and associated trees.

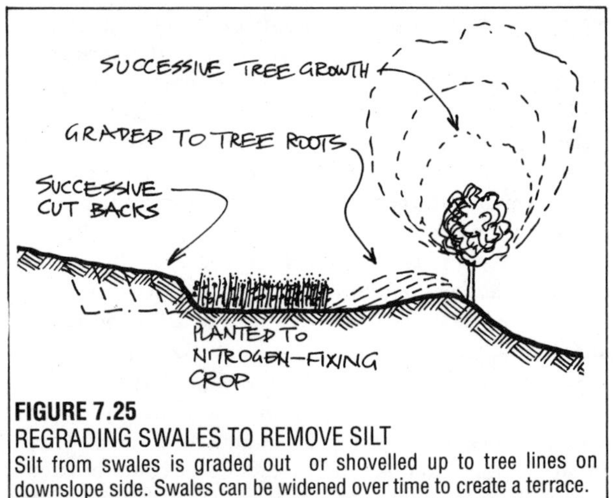

FIGURE 7.25
REGRADING SWALES TO REMOVE SILT
Silt from swales is graded out or shovelled up to tree lines on downslope side. Swales can be widened over time to create a terrace.

DIVERSON BANKS AND DRAINS

Diversion drains are gently sloping drains used to lead water away from valleys and streams and into storages and irrigation systems, or into sand beds or swales for absorption. If low earth–walls are raised across the flow channels of larger diversion drains, these then act as a series of mini–swales for specific tree sites, while surplus water flows on to storages. Diversion drains differ from swales in that they are built to *flow* after rain (from overland flow or from feeder streams). They are the normal and essential connectors of dam series built on the Keyline systems, so that the overflow of one dam enters the feeder channel of the next.

Such diversion systems need careful planning and survey, with drain bases sloping as little as 1:6000 in fine sands. Dam crests and irrigation drains need equally careful placement. Even without a stream intake, diversion drains will gather water from overland flow in as little as 1–1.5 cm (1/2 inch) of rain over 24 hours, so that isolated dams are normally fitted with diversion drains even in quite dry country. Diversion drains can be led to broad level swales in drylands, or made of simple concrete or stone walls across solid rock faces. In the Canary Islands, these gather rock run–off which is led to underground cave storages or large open tanks. Simple sliding gates across or in the downhill banks of such drains allow controlled flow, controlled irrigation, and (in floods) by–passing of dams.

Spill gates can be FIXED (concrete slide holders and aprons) or MOVEABLE (plastic sheets weighted with a chain sewn into the foot, and supported across the drain by a light pipe sleeved into the top). These latter are called "flags" (**Figure 7.27**).

Slide gates can open on to ridge lines, and water then spreads downhill, while plastic flags can be placed and taken up at any point along a drain. This enables one or two people to water 200–240 ha (400–600 acres) in a morning. It is also an effective wildfire control system in forested areas. For sophisticated wildfire or irrigation control, both slide gates and dam–base gate valves can be remotely operated, by radio signal and storage batteries or buried electrical conduits, to power small motors or hydraulic slides. A complete wildfire control can be achieved by dams and sheet irrigation, and using infrared sensers and automatic spill–gates. Such systems remove the risk for fire–fighters, and allow forests to regenerate in semi–arid areas.

Interceptor drains. These drains–or rather, sealed interceptors—act in the opposite tense to diversion drains. These earthworks are specifically designed (by Harry Whittington of West Australia) to prevent overland waterflow and waterlogging, which has the effect of collapsing the dryland valley or downslope soils of a desert soil catena. Thus, they ideally totally intercept overland flow, and direct it to streams or valley run–offs. They can be cross–slope as are swales, but they differ from diversion drains and swales in that their construction always involves the ramming (by bulldozer blade) of subsoil layers hard against the downhill bank. This effectively prevents or impedes water seepage through the downhill wall of the interceptor bank. Moreover, they are always 1.5–2.5 m (5–8') deep, carefully spaced, and effectively stop not only overland flow but also salt water seepages in shallow sand seams. In effect, they isolate large blocks of soil from waterlogging and salt seepage.

After this preparation, trees can be planted in previously desertified soils.

Where deeper sand seams carrying salty water are located, these can be trenched out and stopped with a vertical plastic barrier, backed by compacted clay on the uphill side.

Made to flow at 1:600 to 1:1500 cross–slope interceptors effectively cut up incipient or degraded croplands subject to desertification into blocks of 100 m (330 feet) to a maximum of 300 m (985 feet) wide, isolating each block from flooding and the salty "cascade" flow from uphill. Surplus water is carried off in streams. Interceptor banks also cut off seepage from salt lakes, and can divert early (salted) overland flow around saltpans, letting later fresh floodwater fill the pans or shallow lakes.

Spreader drains or banks are intended to spill a thin sheet of surplus (overflow) water down a broad grassy slope, either for irrigation or (in deserts) to prevent channel scour and gullying. They are normally made to

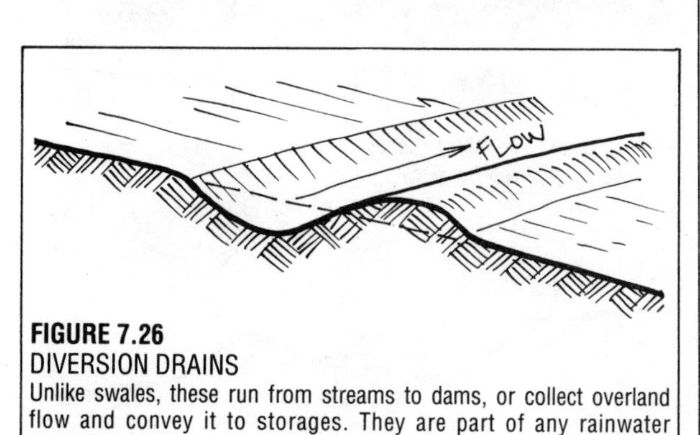

FIGURE 7.26
DIVERSION DRAINS
Unlike swales, these run from streams to dams, or collect overland flow and convey it to storages. They are part of any rainwater harvesting system.

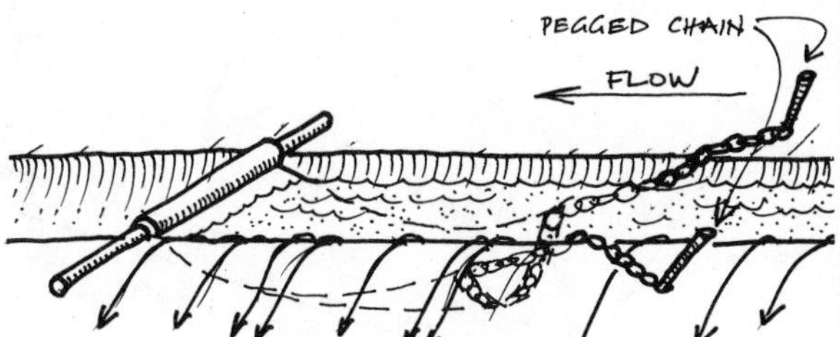

FIGURE 7.27
FLAG
A plastic sheet, one end supported by the channel banks, the other wighted by a chain, forms a temporary dam, causing the channel to flood out and irrigate land downhill.

take any overflow from swales and dams, and may be tens or hundreds of metres long. Spreader banks have the lower side dead level, and compacted or even concreted. Water enters from a dam, swale, or minor stream and leaves as a thin sheet flow downslope. Spoil is piled *uphill*, preferably in mounds, or removed to allow downhill sheet flow to enter the spreader ditch (**Figure 7.28**).

For irrigation areas of flatlands, the bank is often pierced by a series of dead–level pipe outlets, each feeding an irrigation bay (itself planed flat), to which water is confined by low side walls (STEERING BANKS). At the lower end of that bay, a TAIL DRAIN—a surplus water drain—leads off excess water to a stream or secondary storage dam. In such cases, the uphill or primary feeder drain is called the HEAD RACE, and has cross–slides that block flow; this causes the levelled pipes to flood out into the bays. Sometimes the pipes are replaced with level concrete sills and in old–established spreader banks, the whole of the lower lip may be concreted to give a permanent level spill immune to breakage by cattle or vehicles.

Spoil lines uphill from spreader banks are best piled or at least broken by frequent openings to allow downhill overland flow to enter the drain without carrying silt loads.

In *any* landscape, a subtle and well–planned combination of dams, drains, spreader banks, swales, and appropriate pipes, gate–valves, spill gates, flags, or culverts will harvest, store, and use surplus flood or overland flow waters. These are used to flush out salts from soils, spread water evenly over crop, put out wildfires, and modestly irrigate land. It is a matter of first deciding on, then surveying any or all these systems, and above all of considering the long–term effects on the immediate landscape and the soils.

All earthworks can be regarded as REHABILITIVE and remedial if they replace salted and eroded lands with perennial browse or forests. They can be seen as DAMAGING and exploitative if used to irrigate high water–demand crop (like lucerne/alfalfa) in drylands, to cut off flow to dry areas, or to run water for unessential uses to large urban centres devoted to lawns and car washes.

FIGURE 7.28
SPREADER BANK
Water overflow is evenly spread by carefully levelling the downhill bank as a sill.

7.4
REDUCTION OF WATER USED IN SEWAGE SYSTEMS

In cities, water is chlorinated and fed back into the system, sometimes mixed with seawater or "treated" waste water to give it that city taste so typical of, e.g. Las Palmas in the Canary Islands. Surplus sewage,

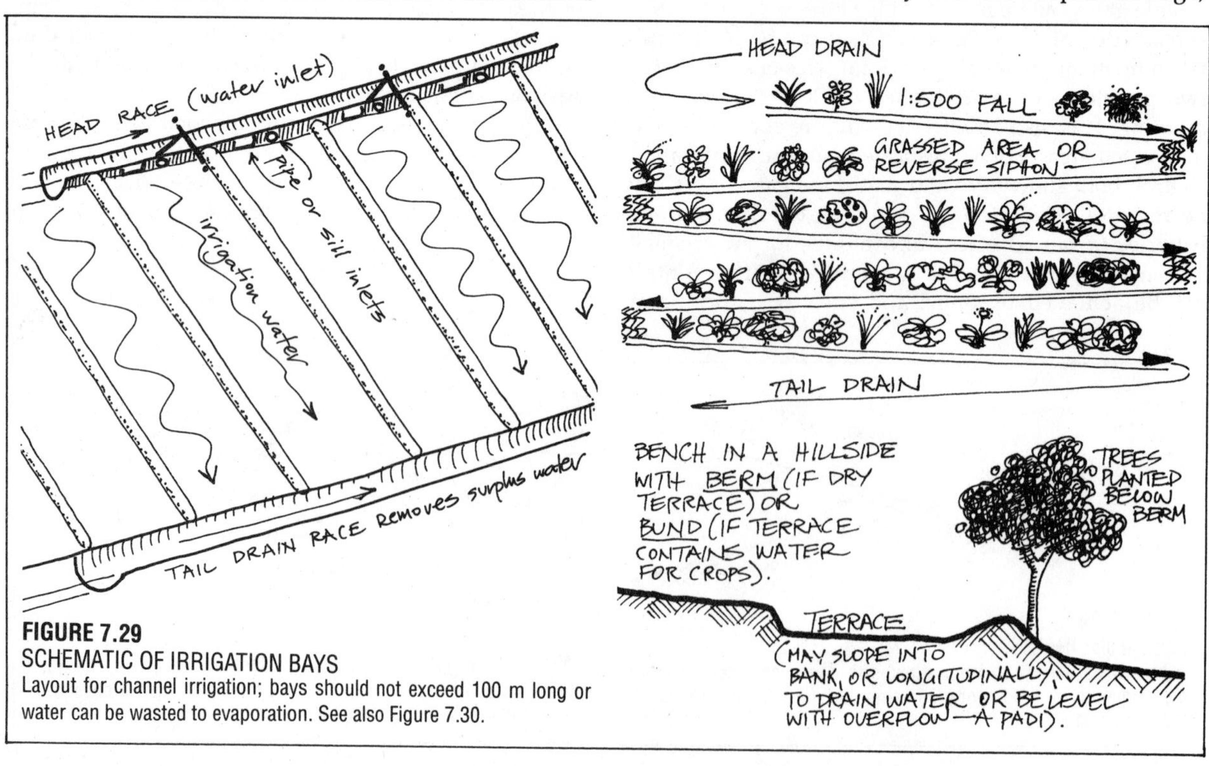

FIGURE 7.29
SCHEMATIC OF IRRIGATION BAYS
Layout for channel irrigation; bays should not exceed 100 m long or water can be wasted to evaporation. See also Figure 7.30.

untreated, is often passed to sea, with bacteria, viruses, and parasites intact for bathers to wallow in.

All of this arises from the frequent, wasteful, and unneccessary flushing of toilets by those of us living in the effluent society. In Sweden, it is compulsory to use dry toilets in remote, unsewered, or unsuitable areas. In the USA, UK, and Australia, one has to fight hard to get permission to use these, as it is the vested interest of industry and town clerks to supply and charge for sewerage systems.

However, *no* clean water need be used to flush toilets if there is a diversion from a hand–basin to the toilet tank. In Australia at least, hand–basins moulded in to toilet flush tanks are available (**Figure 7.31**). It is essential to use low–flush toilet bowls with such systems, as they otherwise flush incompletely, and build up heavy pathogenic bacteria populations.

This is a simple solution to 40% of domestic water misuse, and encourages hand cleanliness rather than the false cleanliness/tidiness of toilet flushing for its own sake.

Dry toilets are not always appropriate, except in cities and other water–critical areas. They are unnecessary on farms or in well–drained soils, or wherever sewage is used to produce methane by anaerobic digestion in tanks. In fact, dry toilets reduce the potential uses of sewage, just as compost is a reduction in the potential use of mulch. Dry toilets are quite specifically useful where:

- No methane system is used;
- Sewage is not used in the production of plants;
- Soils do not suit septic tanks; and
- Cities have critical water supply problems.

In using wastewater from kitchen, bathroom, and laundry, it is wise to establish just what chemicals, and at what concentration, are being released to gardens and soils (or waterways). A typical analysis of a powdered detergent or a soap could include: sodium or potassium salts or polysulphates, silicates, sulphates or bicarbonates, borates, residual biocides (concentrated in animal fats) e.g. DDT, Dieldren, Hexachlor from dairy cattle, additives such as resins (hardeners), scents, dyes, and brighteners, faecal bacteria and viral or worm pathogens from washing (in showers or via clothes). This data is from Kevin Handreck, CSIRO Division of Soils, (*pers. comm.* 1979).

Of these, most can be dealt with by soil organisms, but if the basic water supply is already saline, sodium and potassium salts can add to this and deflocculate soil clays or damage leaves (at >1,000 ppm), while

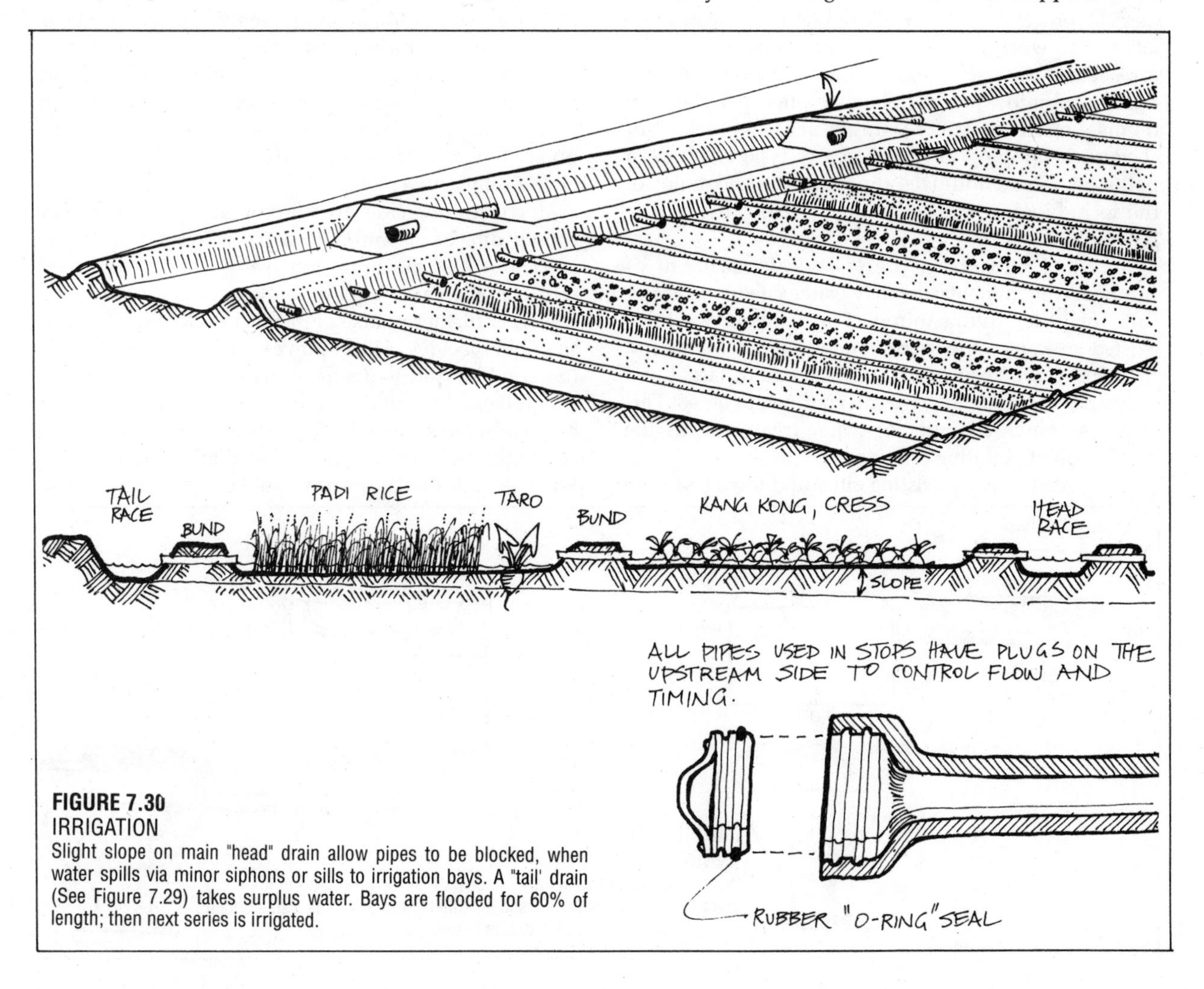

FIGURE 7.30
IRRIGATION
Slight slope on main "head" drain allow pipes to be blocked, when water spills via minor siphons or sills to irrigation bays. A "tail' drain (See Figure 7.29) takes surplus water. Bays are flooded for 60% of length; then next series is irrigated.

borates at >0.5 ppm can create excessive boron concentration in soils, and above 1.0 ppm is harmful to soil life and plants. Thus, we need to use plain soaps on crop if possible, and route more complex pollutants to tree systems (well–monitored), as woody perennials can cope better than garden vegetables, and allow more time for decomposition of long–term pollutants.

In critical areas, and especially in arid or delicate environments, we may need to create both special soaps (unpolluted oils, potash or sodium) and plant special crops which remove excesses (many water plants) before passing on greywater to the soils and streams. There is no blanket policy, only specific cases where we can expect to gain yield and also clean up water *if* we know the composition of soils and soaps.

7.5
THE PURIFICATION OF POLLUTED WATERS

The only long–term insurance of good water supply to a settlement is by rigorous control of a forested catchment, including a total ban on biocides and metallic processing. As there are few such clean areas left in the world, house roof tanks must do for the foreseeable future. The 30–40 additives commonly introduced into water supplies are often pollutants in themselves to that increasingly sensitive sector of society developing allergies to any type of modern pollutant. These additives represent the end point of the technological fix: pollution is "fixed" by further pollution.

Herein, I will stress the *biological* treatment of common contaminants; the only water safe for us is also safe for other living things. For millenia we have existed on water supplies containing healthy plants and fish, and if we keep natural waters free of faecal and industrial contaminants, we can continue to do so. This is not so much a matter of water treatment, as the prevention of polluting activities.

However, for many existing cities and towns, sewage and stormwater supplies must continue to represent a "disposal problem". As the wastewaters of upstream settlements are the drinking waters of downstream areas, our duty is to release from any settlement only water of sufficiently good quality to be safely useable by others.

The problem contaminants most likely to affect drinking water are:

- TURBIDITY: silt and fine particles suspended in the water.
- BACTERIAL or ORGANIC pollution from sewage, and as decay products, e. g. *E. coli*, disease organisms and viral or protozoan pathogens, parasitic worm eggs and so on.
- METALLIC POLLUTANTS such as chromium, cadmium, lead, mercury.
- BIOCIDES, e.g. Aldrin; Dieldrin; 2, 4–D; 2, 4, 5–T; dioxin; PCB, etc (organophosphates, halogenated hydrocarbons).
- EXCESSIVE FERTILISER, especially nitrogenous compounds, phosphates, sodium and potassium salts.
- ACIDS or acid–forming compounds (a pH less than 5.5 increases metallic pollution).

Many of these factors interact. Acid rain dissolves out of rocks and soil poisonous forms of aluminium, mercury, lead, cadmium, and selenium, or other metals such as copper, nickel, and lead from drinking tanks, tea urns, and hot water tanks. Organisms may convert inorganic mercury to organic forms (as happened in Minamata, Japan) which are readily absorbed by the body. Sewage in water aids such conversion to biologically active metals.

Mercuric fungicide dressing on seeds has not only caused direct poisoning of people who have eaten the seed, but also poisons the soil. Excessive artificial fertiliser increases aquatic biological activity, which results in further uptake of metals in acidic waters, and so on. In biocides, Aldrin prevents DDT being excreted; the combination is deadly (one can buy this mix in Australia and the third world, or farmers will achieve it by successive sprays. DDT is a stable residual poison co–distilling with water, so that distillation will not help). An additional threat to public health comes from

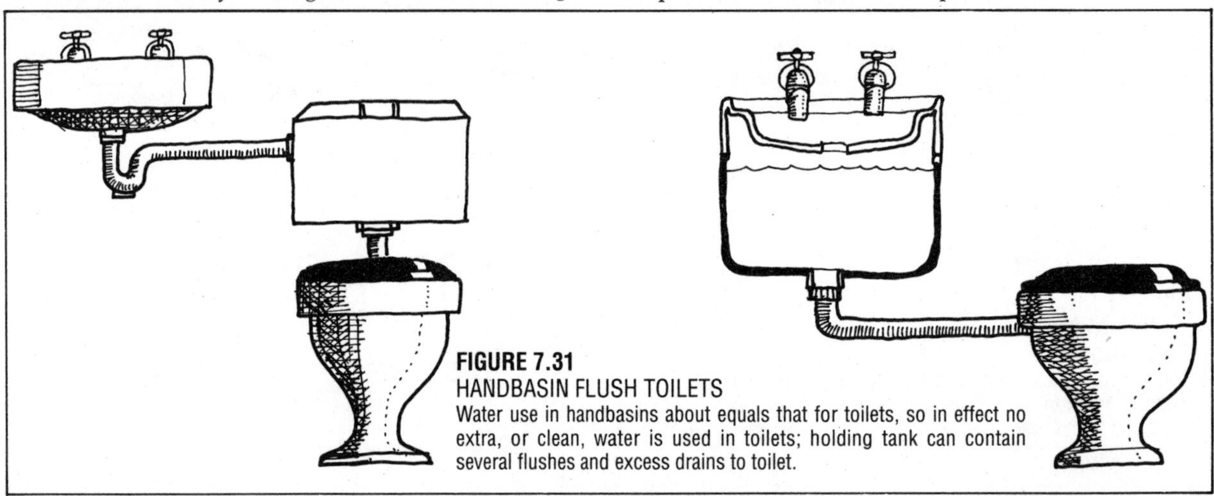

FIGURE 7.31
HANDBASIN FLUSH TOILETS
Water use in handbasins about equals that for toilets, so in effect no extra, or clean, water is used in toilets; holding tank can contain several flushes and excess drains to toilet.

the many miles of asbestos pipe used in public water supply systems; there is a definite threat of both stomach and bladder cancer from asbestos particles in water supplies.

Ferric and aluminium sulphate, salt, and lime are all added to water to cause fine particles to flocculate and settle out as clay. In England, as pH increases due to acid rain, and in fact wherever acid rain occurs, aluminium goes into solution, and with lead and cadmium may bind to protein in vegetables and meat, especially those boiled or steamed. Even if salt is added to cooking water to decrease these effects, levels far exceeding the 30 μg/l allowable for those with kidney problems are experienced. Cooking may increase the water content of metals by a factor of 5 due to this protein binding, and as well make the metals so bound easy to assimilate in the body. Cooking acidic substances in aluminium pots simply worsens the problem. Aluminium from *acidic* rain leaching is now thought to be a major cause of tree and lake death. Ferric sulphate may be safer to use, especially if water is initially or reasonably alkaline. Obviously, these effects need more study and any inorganic salt or metallic salt deserves very cautious use.

WATER TREATMENTS COMMONLY USED

- AERATION (oxygenation) by wind, mechanical aeration, or by increasing turbulence in flow. Aeration is also achieved by trickle columns and vegetation, phytoplankton, or injected air.
- SETTLING: spreading flow in still–water ponds or rush beds to allow particles to fall out, filter out, or flocculate.
- SKIMMING and SIEVING to remove large organic particles.
- FILTRATION via sand beds or charcoal–fibre columns, soils, the roots of aquatic plants.
- COAGULATION or FLOCCULATION by using chemical additives (lime, salt, ferric sulphates) or organic (bacterial) gels.
- BIOLOGICAL REMOVAL by bacteria, phytoplankton, and higher plants.
- pH ADJUSTMENT by adding calcium (as lime) or sulphur compounds as needed.

Filtration

A classical and widely–used filter is sand. Britain and many cities use sand filters followed by chlorination to clean settled and treated raw sewage water sedimentation. Filtration by slow drip through 1.2 m (4 feet) of sand (top half fine, bottom half coarse) is used even in temporary rural camps for water filtration. For cities, fixed sand beds with brick bases are used, the top 1 cm (0.5 inch) or so of sand periodically swept, removed, and dried or roasted to remove organic particles before the sand is returned.

Activated charcoal, often from bones or plants such as willow or coconut husks, is also used as a fine filter in homes and where purity is of the essence. Fine dripstone (fine–pored stone) is used in water cleaners and coolers to supply cool water in homes.

Trickle filters through sand and gravel columns actually feed resident bacteria which remove the surplus nutrient. In less polluted environments, a similar task is carried out by freshwater mussels.

Carbon is essential for the removal of nitrogen or for its conversion by bacteria to the gross composition $C_5H_7NO_2$, and it is generally added as carbohydrate, which can be liquids such as methanol, ethanol, or acetic acids, many or all of them derived from plant residues. This is a bit like "adding a little wine to the water" to encourage the bacteria to work. Surplus nitrogen is released by bacteria to air. Unless bacteria are encouraged and allowed to work, nitrates move easily through sub–soils in which no plants or bacteria can live, and can emerge in wells and streams.

In ponds intended for drinking, light exclusion and surface water stabilisation reduce both turbidity, and thus algae, to a minimal quantity. The stabilisation of banks by grasses and clump plants helps considerably. Pond surface stabilisers are water lilies, *Azolla*, and water hyacinth. Bank stabilisers are *Juncus, Scirpus,* various grasses and clovers, *Phyla nodosa (Lippia)*, and bamboo and pampas grass clumps.

With turbidity much reduced, filtration loads are likewise reduced. Liming will further reduce turbidity if pH is 6.0 or less. This is as simple as placing crushed marble or limestone as a layer in a tank, or casting burnt lime over a pond before filling and (if necessary) after filling. Crushed shells or even whole shells in water tanks and ponds have the same effect. Lime flocculates particles, causing them to settle out of the water.

There are several techniques for filtration, some or all of which can be used in series. First, trickle filters of

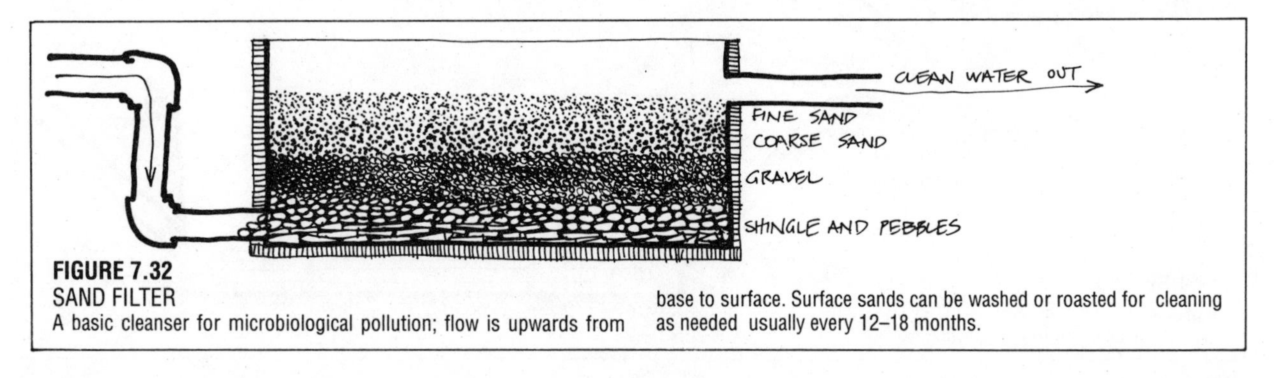

FIGURE 7.32
SAND FILTER
A basic cleanser for microbiological pollution; flow is upwards from base to surface. Surface sands can be washed or roasted for cleaning as needed usually every 12–18 months.

loose pebbles (2.5–10 cm) can be used to form an active bacterial surface layer to absorb nutrients, then a sand filter can be used to absorb bacterial pollution. Water rising through a sand column is fairly clear.

The shells of water mussels can be substituted for pebbles, and the living mussels in the pond or tank not only monitor acidity (dying at pH 5.5 or thereabouts) but filter, individually, up to 100 l/day, digesting bacteria and depositing wastes in the mud base. Mussels and crayfish are not only susceptible to low pH but are also very sensitive to biocides such as Dieldrin, so that their living presence is a constant monitor on life–threatening pollution.

Water, now fairly clean, can be passed through a bed of watercress to remove dyes and nitrates, and the cress cut and fed to animals or dried and burnt to ash. As a final process, the water can be trickled through a column (a concrete pipe on end) of active carbon (10%) and silicon dioxide (90%), otherwise known as burnt rice, oat, or wheat husks.

The results should be clear, sparkling, safe water to drink. No machinery is involved if the system is laid out downslope to permit gravity flow

Lime (freshly burnt) is often used to remove phosphorus and sludges in a primary settling lagoon, and then water is passed to a trickle tower for ammonia removal by bacteria. In towers, of course, the bacteria are not further consumed, but in open lagoons a normal food cycle takes place, with myriad insect larvae and filter–feeders removing bacteria, and frogs, fish, and waterfowl eating the insects. In small towns, the water can be passed from filter towers to sewage lagoons, which in fact may become rich waterfowl and forest sanctuaries. It can then be routed to field crop such as forest, pasture, and to crops to be distilled or burnt, which does not directly re–enter the food chain.

Sewage Treatment Using Natural Processes

Raw sewage is a mixture of nutrients, elements, heavy metals, and carbon compounds; it also contains quite dangerous levels of bacteria, viruses, and intestinal worm eggs. A typical analysis is given in **Table 7.3**. Units are as mg/l; samples are of 30% industrial, 60% domestic wastes at Werribee, Victoria, Australia (Hussainey, Melbourne Metropolitan Water Board Pubs., 1978).

Melbourne is a city of 2,700,000 people and its sewage lagoons cover 1,500 ha (3,700 a.) Thus, there is one hectare of pond (in total) to 1,800 inhabitants (or about 1 a. for 820 people). In the ponds, raw sewage is run into about 724 ha (1,790 a.), where it settles out. Each of these primary settling ponds rarely exceed 7 ha (17 a.) in area, so about 100 ponds receive and settle all raw sewage. Scaled down, this means 1 ha (2.5 a.) of settling pond to 3,800 people.

All these settling ponds are anaerobic, and give off biogas, a mixture of methane (CH_4), carbon dioxide (CO_2) and ammonia gas (NH_3), with traces of nitrous sulphide or marsh gas (NO_2). Biogas is, of course, a useful fuel gas for engines, or a cooking gas for homes. However, it is also a gaseous component of the atmosphere that is creating the "greenhouse effect" and thus should be used, not released to air.

The next set of ponds is faculative (as described below) and the last set aerobic. These, in total, slightly exceed the area of the anaerobic or settling ponds. Most

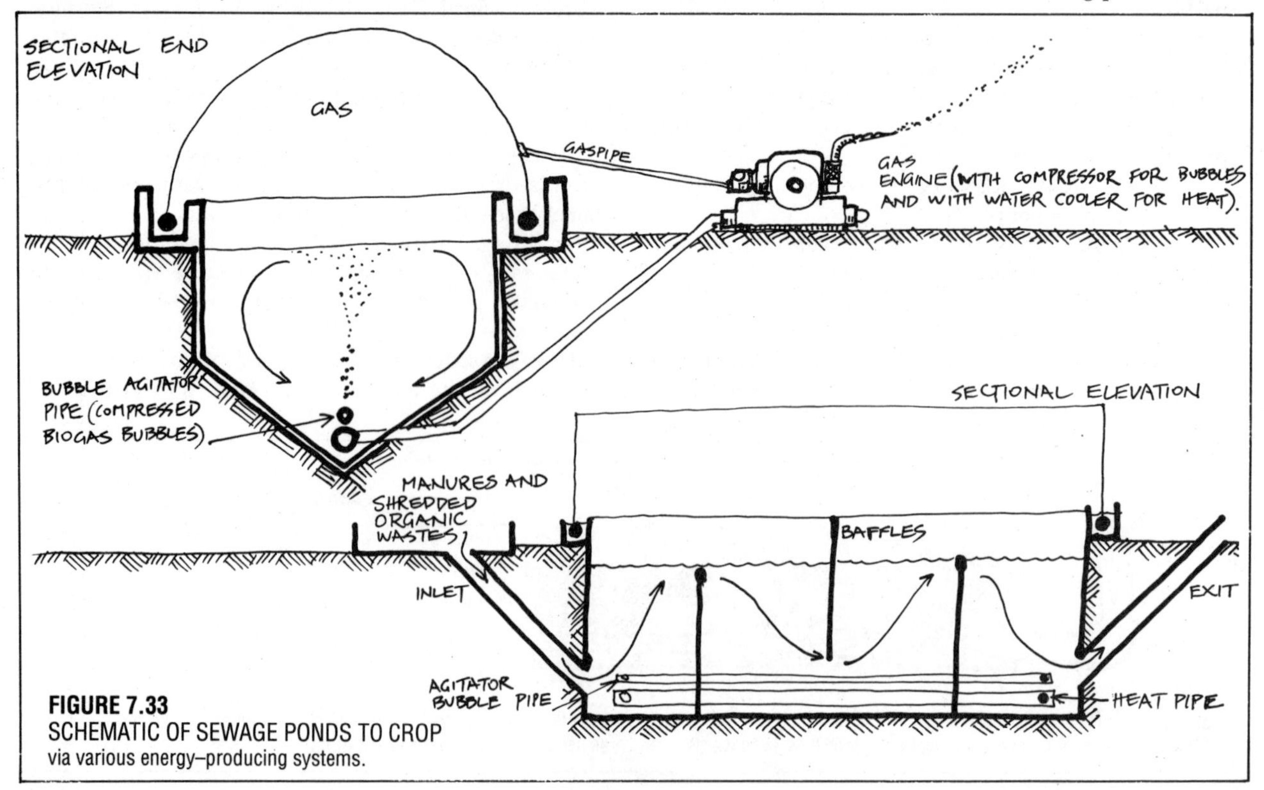

FIGURE 7.33
SCHEMATIC OF SEWAGE PONDS TO CROP
via various energy–producing systems.

are 7–10 ha (17–25 a.) in size.

Ponds can be built (as they are at Werribee) to fall by gravity flow from one to the other. In the first series of (settling) ponds, the sludge creates an ANAEROBIC condition. In the next series of ponds, some sludge passes over and becomes anaerobic at the pond base, while the surface water in the pond (due to wind or algae) is AEROBIC (oxygen–producing). The final series of ponds is totally aerobic. Thus, from intake to outlet, we have the terms:

- ANAEROBIC, or methane–producing (digester ponds).
- FACULATIVE, or part methane, part oxygen–producing.
- AEROBIC, or oxygen–producing ponds.

Ponds at Werribee are only an average of 1 m (3 feet) deep. Deeper, and the sludge breakdown and wind aeration effects are less.

One thousand townspeople and their associated industries therefore need as little as 270 metres square of settling pond 1 m deep. We could, in fact, achieve this as a "long" pond (or series of ponds) 3 m wide x 90 m long, or 3 side–by–side ponds 30 m long and 3 m wide, or any such combination. We can halve the length by doubling the depth to 2 m, and get a pond 3 m x 45 m long; or treble the depth and condense the pond area to a 3 m deep x 3 m wide x 30 m long "digester" pond.

Such a long and narrow pond is easily made *totally* anaerobic by fitting water seals and a weighted cover over the top (which can be of plastic, metal, butyl rubber, or fibreglass). Note that for these deeper digester ponds we would need to *artificially agitate* the sludge (using pumped biogas to stir it), otherwise it settles and becomes inactive (**Figure 7.33**).

Sludge is "active" only in contact with the semi–liquid inputs of the sewer; thus when we stir up the sludge, the better we break down the sewage to biogas. Another (critical) benefit in sealed and agitated digesters is that no scum forms on the pond surface, which can slow the breakdown process further and cause an acid condition.

Of the total dissolved solids (or influent) entering such a digester, over a period of 20 days and with a temperature of 25–30°C (77–86°F), a very high percentage of the mass is transferred into methane; a small proportion is also passed on to other ponds, some as living cells (bacteria or algae). As methane forms, so the oxygen demand of the effluent falls; about a cubic metre of methane generated removes about 2.89 kg of solids, reducing biological oxygen demand (B.O.D.) to that extent.

In the digester, 90–94% of worm eggs are destroyed, as are many harmful bacteria. Useful energy is generated, and can be used at that location to run a motor for electricity, or to compress gas for cooking or machinery (or both, as power demands vary). This motor both supplies the heat for the digester process, and also compresses the gas for digester agitation, and for energy supply.

What happens in the digester? The marsh gas produced, hydrogen sulphide (H_2S), combines with any soluble forms of heavy metals to produce sulphides, which are insoluble in water above pH 7. A little lime can also achieve or assist this result.

TABLE 7.3
ANALYSIS OF RAW SEWAGE

ANALYSIS	MG/L
SOLIDS	
Total dissolved solids	1,200 (TDS)
Bilogical oxygen demand	170 – 570 (BOD)
Suspended solids	160 – 620
Volatile liquids	180 – 510
Total organic carbon	110 – 360
Anionic surfactants	1.0 – 3.6
NUTRIENTS	
Nitrite as **N**	0.05
Nitrate as **N**	0.1 – 0.3
Ammonia as **N**	5 – 32
Organic **N**	7 - 24
Total N	9 – 56.2
Orthophosphate as **P**	1.5 – 6.0
Total phosphorus	1.5 – 9.0
METALS	
Copper	0.09 – 0.35
Chromium	0.25 – 0.4
Cadmium	0.015
Iron	1.6 – 3.3
Lead	0.3 – 0.4
Mercury	0.003
Nickel	0.15
Zinc	0.4 – 0.8
COLOUR	
(as Pt/Cp Units)	100 – 300
pH	6.9 ± 2.0 (near neutral)

Of the total sewage input, from 45 – 60 % of the volume builds up as sludge in settling ponds.

Hussainy found that the following result occurred in anaerobic ponds (see original metal content, **Table 7.3**):

- Copper is removed 97%, of which 78% was removed anaerobically.
- Cadmium is removed 70%, all anaerobically.
- Zinc is removed 97%, 83% removed anaerobically.
- Nickel is removed 65%, 47% aerobically.
- Lead is removed 95%, 90% anaerobically.
- Chromium is removed 87%, 47% anaerobically.
- Iron is removed 85%, 47% anaerobically. (Up to 92% of iron was removed by the faculative pond process, but some iron was partly dissolved in the aerobic pond again, to give the 85% quoted.)

The results are that solids, metals, and disease organisms are very greatly reduced by the first (anaerobic) treatment of sewage. What, in fact, happens to the sludge? It becomes methane. In an anaerobic shallow pond, or a deeper agitated pond, the more sludge, the more active the pond. Thus, *a self–regulated equilibrium condition soon establishes* where input balances gas output. If we remove the sludge, the process slows down or stops. This is a clear case of leaving well alone, of active sludge becoming its own solution; rather than being a problem, it generates a resource (methane).

In the anaerobic pond, there are few algae, but there are some specialised sulphur–loving bacteria of the genera *Thiosporallum, Chromatium,* and *Rhodopseudomonas*. These (in open ponds) may appear pink and give this colour to the ponds. They use hydrogen sulphide as a hydrogen source for carbon assimilation; their by–product is therefore elemental sulphur (S), which binds to the metals present. About 1.8–2.0 mg/l of heavy metals are precipitated as sulphides at 1.0 mg/l of elemental sulphur. The bacteria help in this process.

Passing now to the faculative ponds, we see both the life forms and the biochemical processes change. Here, algae blooms; four almost universal sewage lagoon algae are forms of *Euglena, Chlamydomonas, Chlorella,* and *Scenedesmus*. The total algal and bacterial flora (of many species) are called PHYTOPLANKTON (plant plankton).

Bacteria are also phytoplankton, the bacteria benefiting from the oxygen produced by the algae. Typical bacteria in the open ponds are *Cyclotella, Pinnularia, Hypnodinium,* and *Rhodomonas.* The sulphur–loving bacteria may linger on in the sludge base of faculative ponds, but are absent or rare in aerobic ponds. The algae fix carbon, releasing oxygen to the bacteria.

With such a rich algal food available, ZOOPLANKTON now thrive: most are rotifers (*Brachionus, Trichocerca, Haxarthra, Filinia*); cladocerans (*Daphnia, Moina, Chydorus, Pleuroxus*); copepods (*Mesocyclops*); and ostracods (*Candanocypris, Cypridopsis*). Among these are protozoan flagellates, ciliates, and some nematodes. On this rich fauna, waterfowl and fish can flourish.

Some of the remaining metals are gathered by the zooplankton. In mg/l (dry weight) they contain 1,200 of iron, 152 of zinc, 37 parts of copper, 28 of chromium, 12.2 of nickel, 10.3 of lead, 1.7 of cadmium—almost a mine in themselves. Harvested, both zooplankton and algae can be added to foodstuffs for poultry. Pumped into forests or fields, they provide manures and trace elements for growth. In rich algal growth, blooms of such forms as *Daphnia* can be as dense as 100 mg/l.

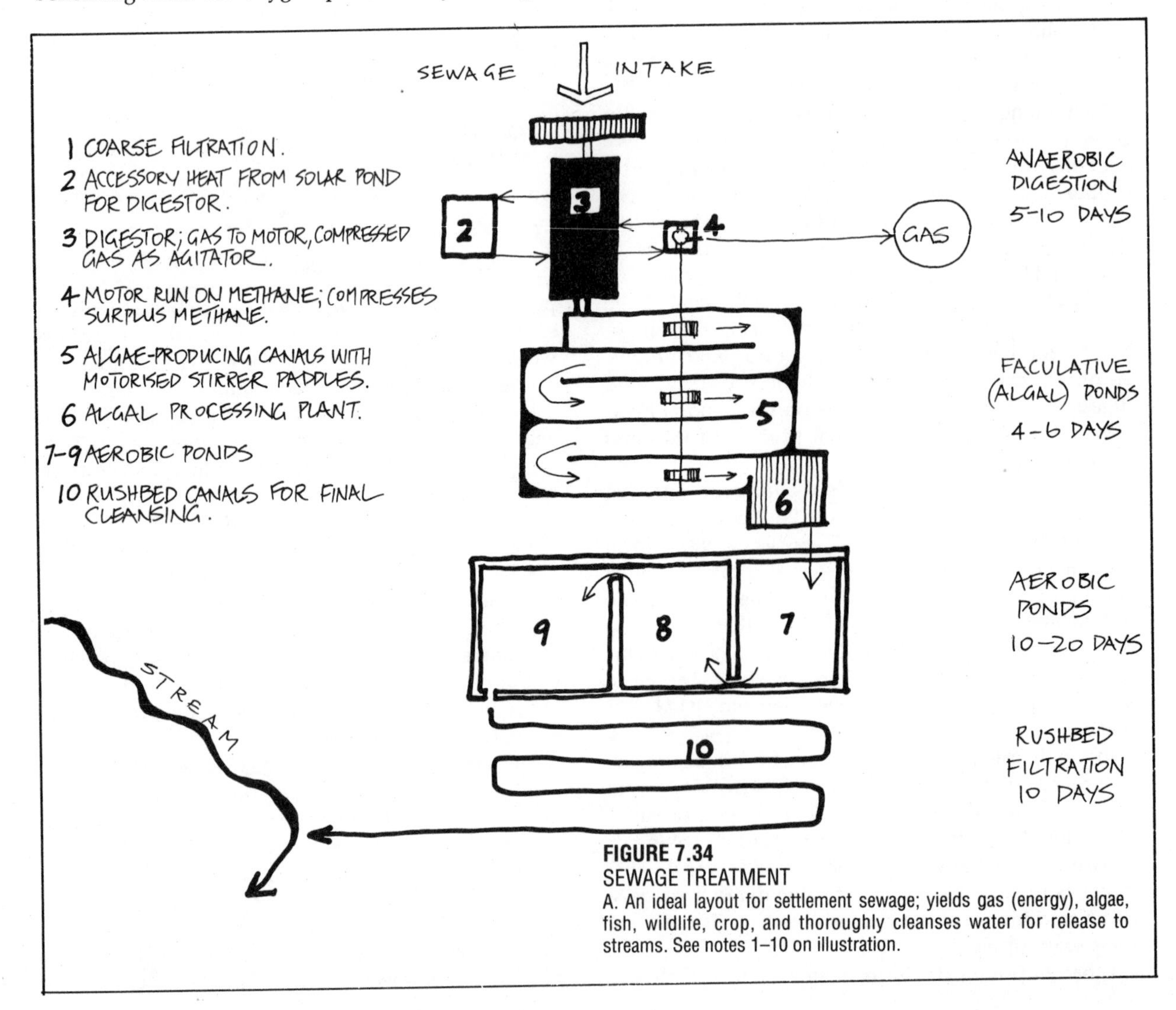

FIGURE 7.34
SEWAGE TREATMENT
A. An ideal layout for settlement sewage; yields gas (energy), algae, fish, wildlife, crop, and thoroughly cleanses water for release to streams. See notes 1–10 on illustration.

These zooplankton masses are self–controlled by eating out their algal foods, and can in their turn be eaten by fish in subsequent pond systems.

Of the pH, which varies both long–term and in 24–hour cycles, it too increases from stage to stage: anaerobic pH 6.2–7.8; faculative pH 7.5–8.2; and aerobic pH 7.5–8.5. In clogged algal waters at night, it may climb higher.

At the aerobic stage, the B.O.D. is only 3–57 mg/l, due mainly to nitrogenous compounds, the suspended solids 32–50 mg/l (now mainly algae and zooplankton). About 80% of these have been removed and incorporated into life forms, and the metal levels are now down to World Health Organisation standards. The water can be used for irrigation, or filtered via rush beds to streams.

Seasonal changes are noticeable. In winter, more hydrogen sulphide is given off by anaerobic ponds (8–15 mg/l compared with summer's 2–5 mg/l), and winds may contribute more to oxygen levels in open ponds than do algae; in winter too, more ammonia (NH_3) is released to the atmosphere.

Summer sees residues oxidised to nitrates. The oxygen being provided more by algae than by wind, less hydrogen sulphide is given off, and there are greater ranges of temperature. In winter (10–15°C), decomposition slows and sludge levels build up, only to be more actively converted in the summer warmth of 18–22°C (64–72°F). B.O.D. is 495 kg/ha/day in winter, 1034 kg/ha/day in summer (at optimum pond conditions), showing that activity almost doubles as temperature increases. Consequently, almost twice as much gas as methane is given off in summer (or in *heated* digesters). In winter, the cooling water of methane–powered engines can provide the essential heat to digesters via a closed loop pipe.

In all, this simple lagoon series produces a very beneficial effluent from heavily–polluted influent. However, there are even more sophisticated biological treatments omitted—those effected by the higher plants. As outlined below, some genera of rushes, sedges, and floating plants can greatly assist with removal of heavy metals and human pathogens, but perhaps more importantly, some plants can also break down halogenated (chlorine, bromine) hydrocarbons synthesised as herbicides and pesticides.

Israel (*New Scientist*, 22 Feb '79) leads sewage waters to long canalised ponds, agitated by slowly–revolving paddle–wheel aerators. Ponds are 0.5 m or less deep. Under bright sunlight (or under glasshouse covers) dense algal mats form, and these are broken up by the addition of aluminium sulphate (a pollutant!), skimmed off, drained, centrifuged, steam–dried, and fed to either carp or chickens (although I imagine that carp could self–feed on aquatic algae). Algal protein replaces 50% of soya bean protein in feed rations to poultry. Total treatment by these methods takes about 4 days. The water is alkaline and somewhat anaerobic, needing more agitation in winter or on cool days. Holland runs sewage to similar canals, and reaps reeds or plants as green crop or for craft supplies.

It has been found (*Ecos* 44, Winter '85) that the artificial aeration of faculative ponds is most efficient if run at intervals of two hours in six (30% of the time). The faculative bacteria follow two digestive modes, and operate best if a rush of air is supplied after a four–hour anaerobic period, excreting carbon dioxide and thus reducing the bulk of sludge. There are corresponding reductions in energy costs for aeration. Nitrogen was reduced from 20 mg/l to less than 5 mg/l, phosphorus from 8.5 mg/l to less than 1 mg/l when ferric chloride was supplied. The process has been dubbed A.A.A. (alternating aerobic and anaerobic) digestion.

Thus, *agitation* of anaerobic systems by bubbling with compressed methane, and A.A.A. of faculative ponds can be used to obtain useful yields of methane and high–protein alage from sewage. As for the aerobic ponds, such higher plants as water hyacinth removed residual metals, surplus nutrients, and the *coli* group of bacilli (*New Scientist*, 4 Oct '79, p. 29). Microwave radiation can also be effective at breaking up algal mats, and sterilising algal products, eliminating toxic aluminium salts.

As for temperatures, solar ponds used in conjunction with compact anaerobic ponds can supply the low grade heat necessary for efficient sludge digestion, and methane will drive any motors needed for both aeration and the gas compression used for the agitation of sludge. The whole processing system can be made very compact, and at the aerobic pond level, throughflow can be led to firewood or fuel forest systems, to irrigated grasslands (as at Werribee), or via trickle irrigation to crops in arid areas.

Final treatment, now in use in Holland and recommended by scientists at the Max Planck Institute in Switzerland, can be released via a sinuous, sealed canal of a variety of rushes and floating water plants.

Waters polluted with metals, biocides, or sewage can be cleaned by travelling through reed beds of *Scirpus, Typha,* and *Juncus*; or by harvesting off floating plants such as water hyacinth. The rushes and sedges can be mown and removed periodically for mulch or cellulose. For *untreated* sewage, a holding time of 10–12 days is necessary, or travel through a series of maze–like gravel–filter canals with floating weeds and sedges. For swimming pools and less polluted systems, a pumped "cycle" of water through ferns, rushes, and watercress suffices to remove urine and leaves. Such pools need a 23–30 cm (9–12 inch) coarse river gravel base, with intake pipes below, and a skimming notch for leaves.

Species recommended are:

- *Phragmites communis* and spp., *Typha* spp.: Flocculate colloids, dry out sludges, eliminate pathogens.
- *Schoenoplectus* spp.: Takes up copper, cobalt, nickel, manganese; exudes mould antibiotics.
- *Scirpus* spp.: Breaks down phenols, including toxic pentachlorophenol.

Low to zero populations of *E. coli*, coliform bacteria, *Salmonella*, and *Enterococci* are found after water is

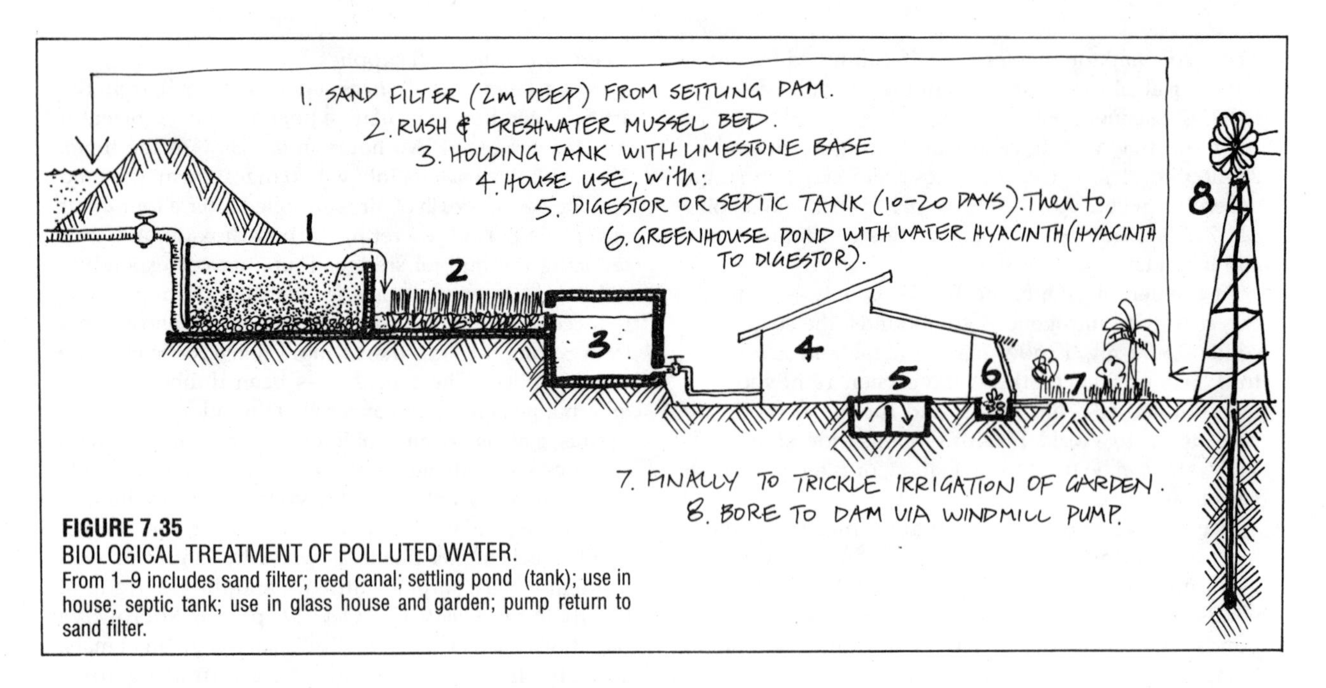

FIGURE 7.35
BIOLOGICAL TREATMENT OF POLLUTED WATER.
From 1–9 includes sand filter; reed canal; settling pond (tank); use in house; septic tank; use in glass house and garden; pump return to sand filter.

treated via the above species. Virus and worm eggs are also eliminated.

Also active in pathogen removal are (although these species must be tested and selected for specific problems): *Alisma plantago–aquatica, Mentha aquatica, Juncus effusus, Schoenoplectus lacustris, Spartina* spp., *Iris pseudocorus*.

For chlorinated hydrocarbons, use rush types with large pith cells (*Aerenchyma*), e.g. *Juncus* spp., especially *Juncus effusus; Schoenoplectus* spp.

Cyanide compounds, thiocyanates, and phenols were treated in fairly short flow times (7+ hours) with *Juncus*.

Systems must be carefully tended and monitored in field conditions. Water can flow through a gravel base planted to purifying species, or for longer rest times, passed through lagoons and ditches.

Domestically, a comfrey bed is one way to absorb the faecal products of animals, where wash–water from yards or pens is available. Comfrey can stand heavy inputs of raw faeces in solution, and the crop may then be used for fodder or trenched for "instant compost" under other species of plants such as potatoes. Flowthrough systems for methane production take little plant nutrient from faecal matter, and comfrey or algae ponds deal with the residues, while producing useful by–products for compost and stock feed.

The water from sewage lagoons has been safely used to rear beef cattle at Werribee for 35 years, and at Hegerstown (Maryland, USA), sewage waters supplied

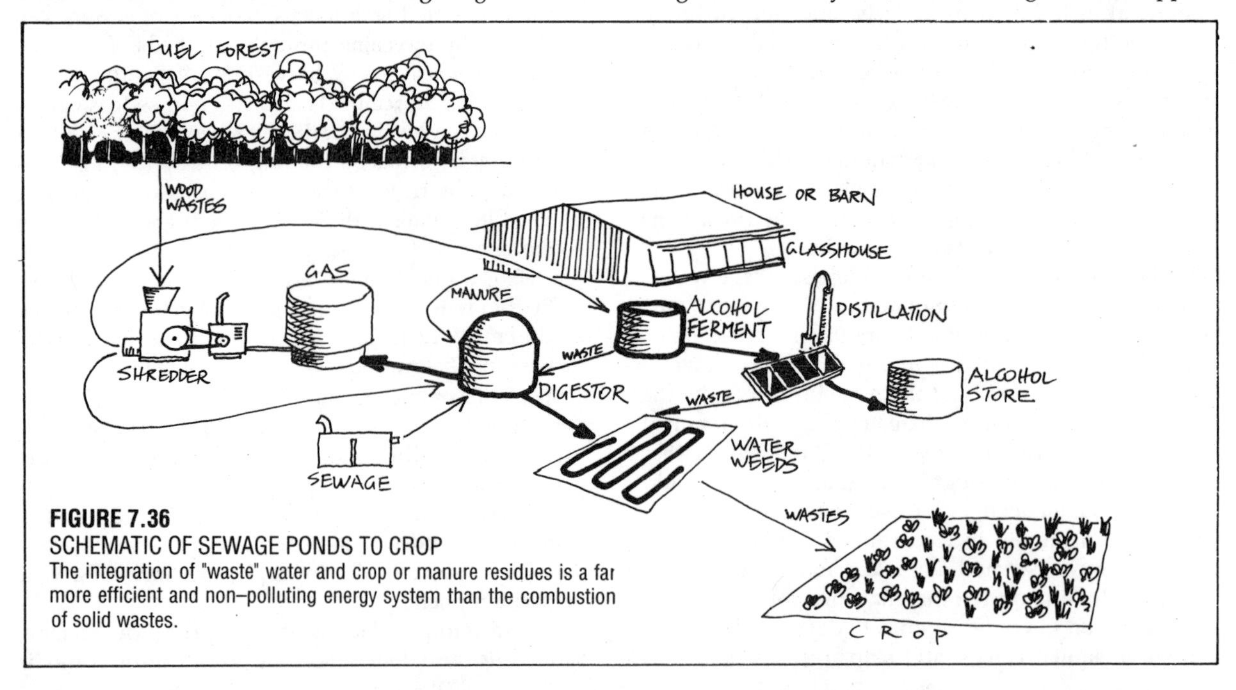

FIGURE 7.36
SCHEMATIC OF SEWAGE PONDS TO CROP
The integration of "waste" water and crop or manure residues is a far more efficient and non–polluting energy system than the combustion of solid wastes.

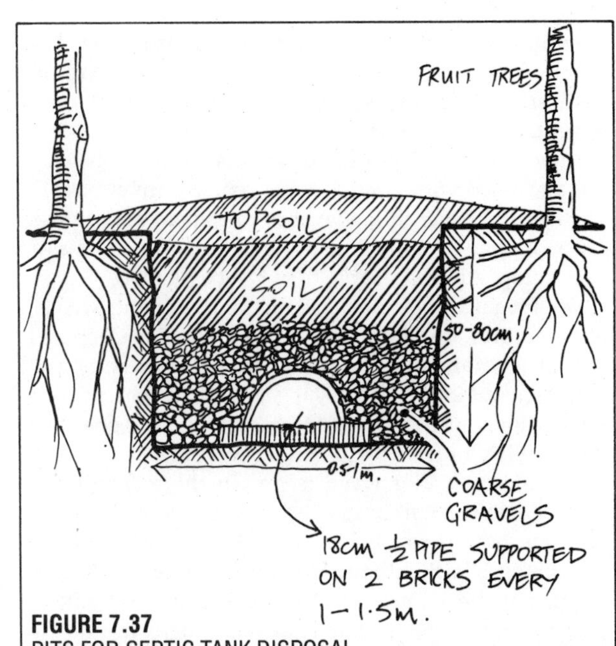

FIGURE 7.37
PITS FOR SEPTIC TANK DISPOSAL
Developed in South Australia as the "Arbor" system. Supported half-pipe never clogs with tree roots, enables trees to remove waste water from trench which has cross-supports every 1.2 m to create "pools".

to selected coppiced poplar plots can produce (as wood chips) some 60% of town energy use. Obviously, water saved from reducing the extent of urban lawn systems can supply the remaining deficit *plus* food crop for any town.

As waters pass through towns, it may gain from 300–400 ppm in salinity—a grave factor in usage in any dryland area (*New Scientist*, 13 Oct. '77). Saline waters can cause problems in irrigated systems, but algae and plant production and removal will reduce this surplus salinity. Discharge of sewage to subsoils does not remove nitrogen compounds from sewage or farm run-off. Again, it is necessary to use productive pond production of algae to reduce nitrates to safe levels for discharge to soils, or we risk pollution of wells and bores, as has occurred in Israel and the USA.

As with garbage, separation of sewage into solids and liquids at the domestic level has productive advantages; 2% urea sprayed on the foliage of rice plants in padi has increased grain protein yields to 40% (11% protein by weight; *New Scientist*, 1 Sept. '77). Such separation can also be used to recover alcohol and chemicals from urine wastes. Urine diluted with water to a 5% solution controls moulds on cucurbits, and aids garden growth or compost activity generally.

In summary, it has long been apparent that modestly–designed sewage treatment systems based on

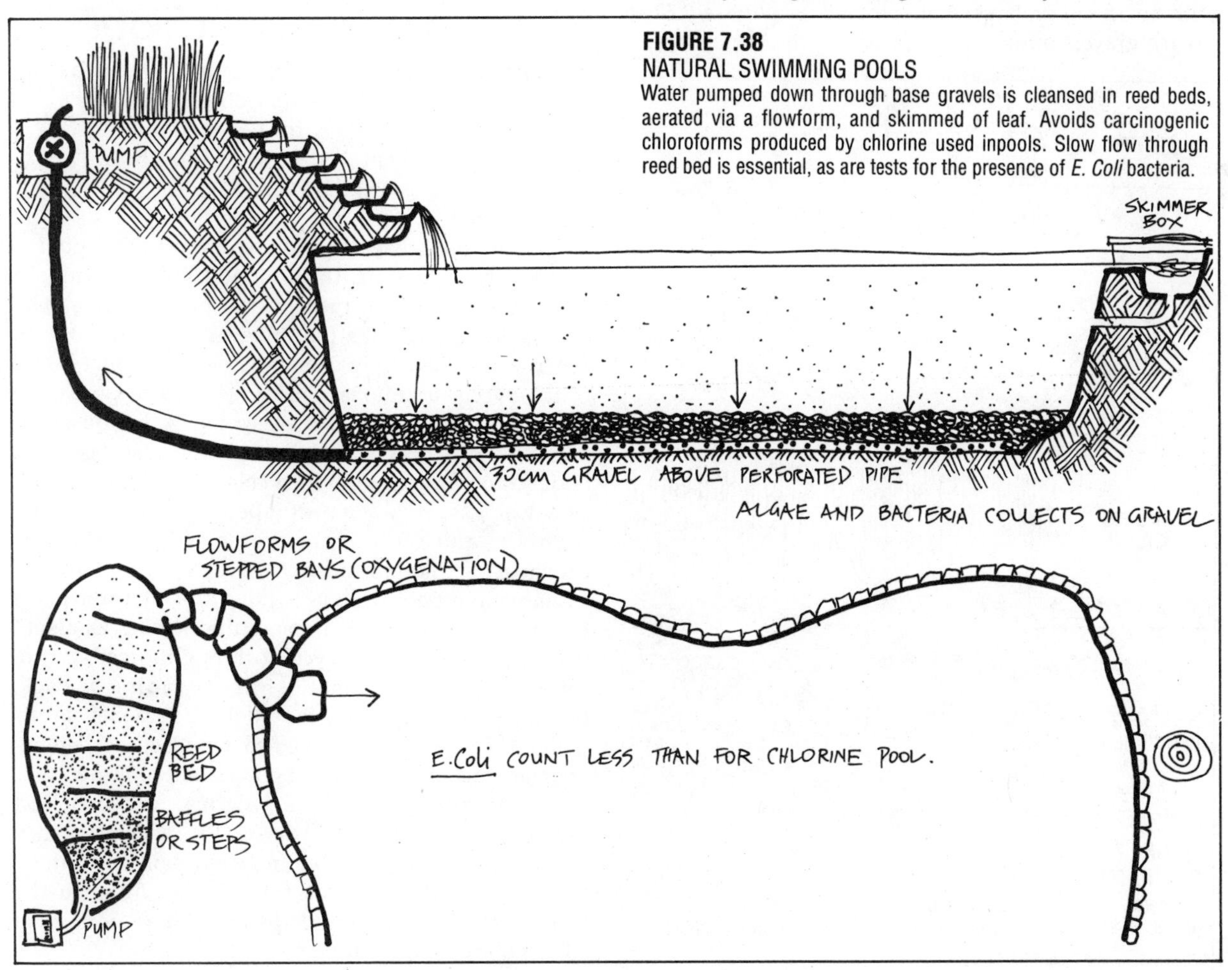

FIGURE 7.38
NATURAL SWIMMING POOLS
Water pumped down through base gravels is cleansed in reed beds, aerated via a flowform, and skimmed of leaf. Avoids carcinogenic chloroforms produced by chlorine used inpools. Slow flow through reed bed is essential, as are tests for the presence of *E. Coli* bacteria.

sealed (not leaky) lagoons and their associated biological systems not only function to recycle water efficiently, but to create a variety of yields from the 'wastes' of society. There are simply no modern excuses for continuing with the dangerous disposal of such wastes to seas and subsoils, where they inevitably turn up as pollutants in wells, streams, and on beaches, or add considerably to the greenhouse effect of atmospheric carbon dioxide. It is possible to design small and large systems of water treatment systems which are both biologically safe and productive.

Creative Disposal of Septic Tank Effluent

There are two basic productive disposal systems for septic tank effluent:

- Underground and surface leach fields around which trees are grown.
- Biogas conversion, followed by a pond growing aquatic crop for biogas feed stock, then a leach field.

A leach field is a trench or open gravelled soakage pit through which sewage wastes from a septic tank flows. In clays and clay–loams, tank water from a family home will stimulate fruit tree growth (without other irrigation) for 20 metres or more. The system follows normal procedures in that a long trench with a 1:12 ratio base slope is dug away from the septic tank outlet pipe. Topsoil is put to one side, and the trench is fitted with an 18 cm or larger half–pipe as per **Figure 7.37**. Coarse gravels or stones are placed in the trench, and over all this a strip of plastic or tarpaper is placed. The trench is then back–filled and trees planted 1–2 m off both sides as 2–6 m spacing. All fruit and nut trees benefit.

Square or round pits about 25 m square can be dug out and filled with graded stone (coarse 6 cm at base to 2 cm at top). Over this, a layer of cardboard and a thick layer of straw is spread, and the latter sown to oats or green crop. Around the pit, trees can be planted.

For biogas applications, septic tank effluent, weeds, and manures are loaded into a tank 2 to 1.5 m deep and 3–4 m in diameter. A loading chute for weeds and wastes about 20 cm and 30 cm slants to the base. Septic tank effluent also enters at the base. Overflow goes to a pond with baffles, and *Pista*, watercress, or any rampant soft water weed is grown there. These are returned to the tank every week. A perforated pipe at the tank base is worked by a small gas compressor to "bubble" gas back into the tank for 1–3 hours daily on a timer. This breaks up the scum on top of the ferment. Gas caught in an inverted tank is fed to the house cooking range, lights, refrigerators. Surplus from the pond is fed to a leach field.

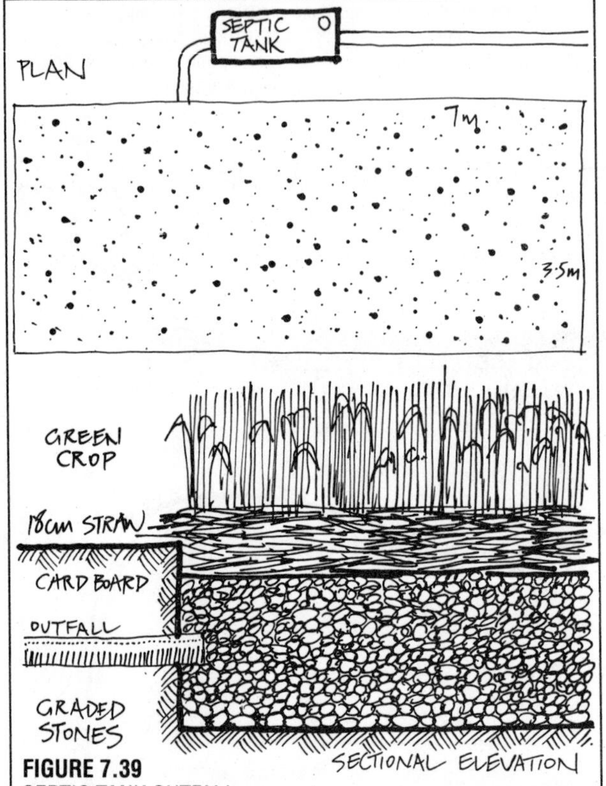

FIGURE 7.39
SEPTIC TANK OUTFALL
For semi–arid areas goes to a 8 x 8 m disposal pit with 0.5 m of gravel, straw mulch, trees around pit. Widely used in rural Australia.

7.6
NATURAL SWIMMING POOLS

When thou wilt swimme in that live bath
Each fish, which every channel hath
Will amorously to thee swimme
Gladder to catch thee than thou him...
(John Donne)

Swimming pools have crept across the affluent suburbs so that, from the air, these ponds now resemble a virulent aquamarine rash on the urban fringe. The colour is artificial, like that blue dye that imitates an ocean wave obediently crashing down the toilet bowls of the overly–fastidious. Chemicals used to purify the water are biocides, and we are biological organisms; if fish can't live in our pools, we should also keep our bodies out of the water. When chlorine isn't being used as a war gas, it is being dumped into our drinking, bathing, and swimming water, where it forms carcinogenic chloroform.

Innovative pool designers now filter natural pools below a base pebble bed, using the pebbles as algal/ bacterial cleaners, then cycle it through a reed–bed to remove excess nutrients before cascading it back, freshly oxygenated, into the pool. Such pools can be delightful systems with tame fish, crayfish, rock ledges, over–arching ferns, and great good health (**Figure 7.38**).

They are also reserves for fire–fighting, potential heat sources for heat pumps, barriers to fire, and emergency water supplies rechargeable from the roof, and can be recycled by photovoltaic pumps. Goodbye to the endless servicing, and perhaps hello to an occasional lobster or overgrown trout!

7.7

DESIGNER'S CHECKLIST

On any property, identify sources of water, analyse for quality and quantity, and reserve sites for tanks, swales, or dams. Wherever possible, use slope benefits (or raise tanks) to give gravity flow to use points, and detail plant lists that will grow (as mature plants or trees) unirrigated.

In the general landscape, soil samples (for 40% or more clay content) will reveal sites suited to earth–dam construction; such sites need to be reserved for future storages. A sequence of primary valleys may enable a Keyline system to be established for downhill fire control and irrigation.

Where evaporation exceeds precipitation (arid areas), make sure all water run–off is infiltrated to soil storages via soil conditioning (rip–lines), swales, pits, or sandfield soakages. In humid areas, open–surface dams dams can be used.

Define water "pathways" in use, so that water use is economical in houses, and that greywater is used in gardens (via filtration beds), forests, or (for villages) design for clean–up on site through a common effluent scheme based on maximum use (methane, plant production, irrigation).

Get good advice on (and supervise construction of) all dams. Wherever possible do not impede normal stream flow or fish migration, and site houses out of the way in case of dam failure. In particular, allow adequate stable spillway flow for "worst case" rain intensity.

Make sure that all earth storages, and in particular swales, are planted with trees, to remove infiltrated water and (in arid areas) to prevent salting problems.

Before recommending cloud–seeding, make sure that the area to be affected is warned, and that dams and swales are designed to cope with any increase (up to 30%) in rains.

Design for forested ridges, and maximise forest on strategic uplands; do not lend your skills to high–country deforestation (or any deforestation).

Windbreaks and in–crop trees are essential to reduce water loss in croplands.

7.8

REFERENCES

Chorley, R. J. (ed.), *Water, Earth, and Man,* Methuen and Co., London, 1969.

Nelson, Kenneth D., *Design and Construction of Small Earth Dams,* Inkata Press, Melb., Australia, 1985.

Seidl, Kathe, *et al, Contributions to the Revitalization of Waters,* Max Planck Institute, Krefeld–Hulserberg, West Germany, 1976.

Yeomans, P. A., *Water for Every Farm,* Second Back Row Press, Leura, NSW, Australia, 1981.

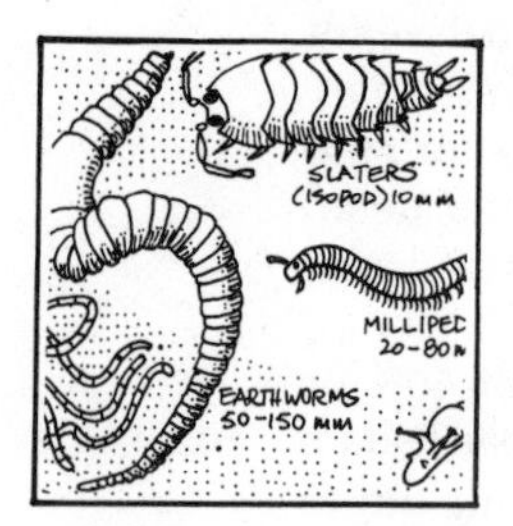

Chapter 8

SOILS

The best fertiliser is the footsteps of the gardener.
(Chinese Proverb)

8.1 INTRODUCTION

The properties and treatments of common soil types in specific climatic areas are summarised in later chapters, so that in this chapter we deal more with soil as a material, including some of the ways to stabilise and assist soils in retaining their productivity (soil conservation and soil conditioning).

Soils defy precise treatment, as their structure (and permeability), organic content, gaseous components (some derived from the atmosphere, some from processes within the soil, and some exhaled from the sediments below), minerals, pH, and water (or rather solute status) changes from hour to hour with soil depth and treatment, and in response to micro–elevations. Added to this is the fact that many soils are originally complex mixtures derived from a variety of rock types and that they may have had a very long and varied history.

A final factor is that despite all our knowledge, in spite of soil services and soil analysis, and despite the best attempts of people to care for land, we are losing topsoil at an ever–increasing rate. Australia, where I live and write, has perhaps 30% of its original soils in fair condition. The rest are washed or blown away, or sadly depleted in structure and yield; this is true of most countries of the world where extractive agriculture and forestry occurs.

The closer soils are defined, it seems, the less likely we are to know them. Notwithstanding this specific uncertainty, there are some sustainable approaches to soil maintenance and to soil rehabilitation. It is in these areas that we will outline strategies. Keep in mind, however, that we are always dealing with a matrix or mosaic that is in constant variation in place and time. Nobody can be dogmatic about any natural system.

Soil science concentrates very much on *what is there* (classifications), but not on *how to evolve soil*. Often it is left to amateurs—gardeners and farmers—to create good soil by water control, modest aeration, and plant and animal management. Farmers and gardeners seem to be so often the practical, innovative, experimental, successful group (while often ignored by academics) that I despair of esoteric knowledge ever preceding effective action. Very few farmers can persuade a group of scientists to assess their apparently successful soil trials. It is past time that we assessed whether more "science" is not being done by outdoor people than by scientists who (like myself) more often collect the results of others than generate them by example. Science is good at explaining why things work, and thus making skills teachable. It is not so good at initiating field work, or in training people already in the field to work effectively.

It is hard to say if scientists lack the means to get into the real work, or if they choose science to escape from field problems. It *is* hard out in the field of erosion, landlords, foreclosures, poverty, greed, malnutrition, and exploitation, but that's where all the action is. A field approach means choosing values, getting involved with people, and inspiring broad scale change. Farmers' field days advance knowledge far more effectively than scientific papers, and local educational sessions more than either of these. However, both scientists and farmers have much to "give and take". This sort of coalition is slowly starting to happen as a result of joint concern on a private level.

Soils can be RESIDUAL (resting in place over their present rock) or TRANSPORTED by water, ice, gravity, and wind. In their formation the key factors are rock type, climate, and topography (or landform). Water has a key role in rock breakdown, combining with such common minerals as felspars in rock to swell and fracture the rock, then to hydrate the felspars to kaolin, clay, and potassium carbonate. The carbonates released make of soil water a stronger carbonic acid than it is as rain.

Atmospheric oxygen dissolves in rain to oxidise iron minerals (pyrites) and forms both haematite and a quantity of sulphuric acid, which again dissolves metals, so that water makes an effective rock decomposer, even without invoking the expansion of water as ice, or the power of ice (as glaciers) to grind rocks to flour. Plants too are wedging open rocks and mineral particles, recreating acids, and transporting minerals in their sap to other locations. All of this work creates a mantle of topsoil, which is estimated to build at about 2–4 t/ha per year as uncompacted topsoil, but which we remove at a rate of from 40–500 t/year in cropping and soil tillage. Even the most ideal tillage just keeps pace with the most ideal conditions of soil formation, and in the worst cases we can remove 2000 years of soil in a single erosion season, or one sequence of flood or strong wind over cultivated soils.

The *only* places where soils are conserved or increased are:

- In uncut forests;
- Under the quiet water of lakes and ponds;
- In prairies and meadows of permanent plants; and
- Where we grow plants with mulched or non-tillage systems.

These then are the core subjects of sustainable societies of any conceivable future. They are not, you might notice, the subjects most taught in the agricultural colleges or forestry courses of the recent past, nor do they occupy the minds of politicians, investment bankers, or TV stars.

Before starting on the complex subject of soils (and I am *not* a soil scientist), it is wise to draw back a little and consider the question of soils from some very different viewpoints, or sets of values. They are broadly these:

- Health (both human and plant). We must be careful and conservative in approach, *especially* in the area of biocides and high levels of artificial fertilisers. After all, our ancestors lived to a ripe old age on home-grown produce without the benefit of herbicides, pesticides, or artificial fertilisers. We must therefore improve on, not lessen, that factor of long-tested vitality that was, and is, integral to good gardeners.

Not that I believe that their health was purely due to diet. Several other factors associated with gardening may be one day better assessed, including:

— *mild but regular exercise:* gardening is a sort of steady and non-stressing *tai chi*.

— *meditation*: we can sit and look as much as work, and banish all cares. I found my grandfather as often just looking as working (and he fed about 20 families from his market garden).

— *meaning* in gardening, there is a very conscious sense of doing a job that is worthwhile, and of *direct* value to others. Gardening, especially food gardening, is "right livelihood".

— *life interest,* perhaps derived from the above. Every day, every season there is change, something new to observe, and constant learning. Permaculture greatly adds to this interest, and has the dimensions of a life-oriented chess game, involving the elements, energy, and the dimensions of both life-forms and building structures (also with political, social, financial, and global implications).

- Yield. Here, we come to a grave impasse. There is no doubt that the once-off yield of a ploughed and fertilised monoculture, supported by chemicals and large energy inputs, can out-yield that of almost every other production system. *But*: at what public cost? for how long maintained? with what improvement in nutrition? with what guarantee of sustainability? with what effect on world hunger? on soils? and on our health? There is now abundant proof that such forced yields are temporary, and that plough cultures destroy soils and societies.

These are some very awkward questions to ask of the agricultural establishment, for very few, if any, modern agricultural systems do not carry the seeds of our own destruction. These systems are those that receive public financial support, yet they destroy the countryside in a multitude of ways, from clearing the land of forest, hedgerow, and animal species to long-term soil degradation and poisoning. We are thus obliged, by entrenched bureaucracies, to pay for the destruction of our world, regardless of the long-term costs to be borne by our children and our societies.

- Life in soil. Soil organisms are a major soil factor, and have myriad perceptible and profound effects on pH, mineral content and availability, soil structure, and erosion.

- Ethics. It is *not* the purpose of people on earth to reduce all soils to perfectly balanced, well-drained, irrigated, and mulched market gardens, although this is achievable and necessary on the 4% of the earth we need for our food production. Thus, what I have to say of soils *refers to that 4%*, with wider implications only for those soils (60% of all agricultural soils) that we have ruined by the plough or polluted by emissions from cars, sprays, radioactives and industry.

Our largest job is the restoration of soils and forests for the sake of a healthy earth itself. It is most definitely not to clear, deforest, or ruin any more land, but first to put in order what we have destroyed, at the same time attending to the modest area that we need for our survival and full nutrition.

Without poorly drained, naturally deficient, leached, acidic or alkaline sites, many of the plant species on earth would disappear. They have evolved in response to just such difficult conditions, and have specialised to occupy less than perfect soil sites.

Colin Tudge (*New Scientist* '86) muses on the proportion of the British Isles that could be given back to nature. He comes down with a very conservative estimate of perhaps 60%. And at that, without letting go of the misconception that it is agriculture (not individual and market gardening) that will actually provide the future food we eat (a common fallacy). John Jeavons estimates (on the basis of *gardens*) that we could return perhaps 94% of land to its own purposes. Not that I think that we will get there this next decade, but we can start, and our children can continue the process, and so develop new forests and wilderness to explore. A reduction of the ecological deserts that we have called agriculture is well overdue, as is a concomitant reduction in the twin disasters of newspapers and packaging derived from ancient forests.

8.2

SOIL AND HEALTH

As long as we live, we will be discovering new things about the soil–plant–animal relationship. Soils harbour and transfer both diseases and antibiotics; plants will take up from the soil many modern antibiotics (penicillin, sulfa drugs), and we might then ingest them at concentrated levels. Animals will retain residual antibiotics (therefore new and resistant strains of disease organisms), and will contain residual hormones and biocides. Plants and animals may concentrate, or nullify, environmental pollutants. Most of these pollutants are in fact concentrated by both plants and animals, but the degree to which this happens varies between species.

Both dangers and benefits arise from our food. Natural levels of soil antibiotics may sustain us, and natural resistance to disease is in great part transmitted to us via food. It is certainly the case with vitamins and trace elements that they maintain the function of many metabolic processes, in minute but necessitous amounts.

When people lived as inhabitants of regions (as many still do) they adapted to local soils, plants, and nutritional levels or they died out. Today, we bring in global food to global markets, and so risk the global spread of "agricologenic" (farm-caused) diseases. Like the home water-tank, the house garden represents a limited and localised risk *under our control*, and of little risk to society generally. Public water supplies and commercial foods are a different matter, distributed as they are throughout many modern societies.

In soil rehabilitation, we are forced to start with what is now there. Only rarely have we a soil containing all the nutrients a plant may need to grow. Most gardeners and farmers who have developed sustainable soil systems allow 3–4 years for building a garden, and 5–15 years to restore a devastated soil landscape. This applies only to the physical restoration of soils and to the development of appropriate plant systems. There are far more lengthy processes to be undertaken where past chemical pollution has occurred.

THE POLLUTION OF SOILS AND WATERS

Orchards, sugarcane areas, pineapple, cotton, tobacco crop, and banana plantations (to name just a few well–known cases) have had such an orgy of mineral additives, arsenicals, Aldrin, DDT, copper salts, and dioxins applied that even after 18–20 years of no chemical use, a set of apple orchards in West Australia produces unsafe levels of Aldrin and Dieldrin in the eggs of free–range chickens. Attempts to grow prawns in ponds on such lands have failed on the basis of residual Dieldrin levels in soils.

When we come to assess the total environmental damage caused by persistent misuse of chemicals, we will find many farms (as well as bores and rivers) that will need to be put into for non–food production (as fuel forests or structural timber) for decades to come. The same may already hold true for soils within 100 m of roadways where leaded petrol is used (and where 800–1000 cars pass daily).

We face lock–up periods of tens or at times thousands of years for the radioactives blowing off, or leaking from, waste dumps and strategic stockpiles of yellowcake Uranium (Iowa, Kentucky, Russia and the UK or France), or from "accidents" such as those occurring at atomic power plants, or from their wastes. Cadmium and uranium–polluted soils of chemically–based and heavily–fertilised market gardens, waste dumps of industry, and the long–term effects of nitrate–polluted soil waters can be added to those lists of already–dangerous areas. Even now, *applied* health levels would close down many farms and factories, and (as awareness rises) this will be done in the near future by public demand.

The costs of rehabilitation (as for acid rain) already far outstrip the profits of degradation, and may in fact be prohibitive for areas that were developed for farming from 1950 to the present (the age of agricultural pollution).

Large quantities of lead, arsenic, copper, and persistent biocides are applied on most apple orchards. Data is available for some metals (*ECOS* 40, Winter '84). Copper and lead stay at or near the surface of soils in high concentrations. Arsenic may also stay at this level in clay soils, or wash down to subsoil (50–60 cm deep) in acid sands. Leaching from clays or organic profiles (10–20 cm) is unlikely, although phosphate application *may* dislodge arsenic to deeper levels. In pasture plants under such orchards, copper can reach 50 ppm (poisonous to sheep). Excess copper in the diet causes toxaemic jaundice (liver poisoning) and blood in the urine. (People in Australian deserts often show high copper blood levels and blood in urine.)

Molybdenum, zinc and sulphur may buffer copper uptake in sheep at least; uptake by plants increases with temperature and acidity (for lead and copper), as for vegetables. Arsenic uptake is not related to acidity; silver beet (Swiss chard) fed with nitrogen lose high arsenic concentrations, but may then be unsuitable for children due to high nitrate levels.

Several substances have now polluted soils. There

are no easy remedies for polluted soils, but the following strategies may help:

• LEAD (from car exhausts and lead paint, pipes, battery burning). Worst cases are in urban areas of older buildings. Lead at 1,100 ppm can be present, and is both taken up by and dusted on the surface of vegetables. However, it is possible to garden by:

— Cracked bricks or gravel as a base.

— Building up beds to 30 cm deep, and making up a rich composted soil of over 40% organic content.

— Growing vegetables and having leaf analyses done; washing in dilute vinegar if lead is still used locally in petrol.

• PERSISTENT BIOCIDES, especially DDT, Aldrin, Dieldrin, BHC, etc. If you buy or inherit an old orchard, canefield, or plantation (banana, pineapple, cotton, tobacco) it is unlikely that any animal product (milk, eggs, meat) will be free of high levels of biocides. There is no choice but to go into forestry, and to produce non-food crop until other methods are developed. Also, test your own vegetables for residual toxic materials.

• GROUNDWATER can contain 80–90 biocidal substances below farms, including those derived from fertilisers, sprays, and fuels. Near industrial waste dumps, dioxins, radioactives, and heavy metal wastes (cadmium, chromium, mercury) can be added. Do not use untested wells or bore waters for *any* purpose. Drink tankwater and try to harvest surface run-off for gardens. It is estimated that several decades may be needed to clear most aquifers of pollutants. Almost every state of the USA has serious problems. Rainwater harvest and strict water conservation is indicated for the long-term future. Several substances are added to town water supplies—and these may include chlorine, fluorine, alum (aluminium sulphate) and other metabolic poisons.

We must not add to this mess. Avoid all biocides, high levels of nitrates, and watch on-farm disposal of oils and fuels.

HUMAN HEALTH AND NUTRITION

The subject of human nutrition is complex, and under fairly constant assessment by scientists from many disciplines. Four very broad statements can be made:

1. A normally-mixed diet (omnivorous) has been exhibited by most human groups. Excessive dietary simplicity, reliance on too few foods, or a restricted dietary range has its dangers, while a mixed diet of local foods, plus an active life, has usually proved healthy, providing good hygiene is also observed (public, personal, and domestic).

2. In the developed countries, refined and processed foods, too much inclusion of animal fats, and a plethora of food additives has certainly resulted in malnutrition and in degenerative diseases (obesity, high blood pressure, heart disease). There is now a general move away from smoking, heavy drinking, and many fatty or processed foods, and a rising demand for lean meats, fish, and clean vegetables, fruits, and nuts.

3. In areas subject to famine, semi-starvation, or where very low levels of critical vitamins (commonly C, A, B-complex) or minerals (iron, zinc) exist, it is necessary to take great care with human and soil health, and to have a very sound knowledge of the possible results of any new dietary change (whole grains and pulses may strip out the little zinc the body retains where there are starving people in alkaline desert areas; zinc tablets may then been needed if such foods are commonly used). Traditional diets, long maintained, are a guide to local food tolerances.

4. There is a complex and constant interaction between food, soil, trace elements, pH, biocides, and fertilisers. Too often, product yield, weight, or processing suitability is the only reason given for using biocides and fertilisers; nutrition is rarely mentioned in plant breeding programmes. Heavy use of fertilisers—the macronutrients—can cause a deficiency of micronutrients [*ApTech* 6(1), 1984].

We are *individually different* in our ability to metabolise and tolerate foods. For some of us, specific food allergies are very real, and at least in part (e. g. lactic acid intolerance) arise from little exposure to certain food groups in our racial history. Individuals can test for ill effects by becoming conscious of headaches or other symptoms, and experimentally eliminating some foods or beverages from their diet, or even a class of food (e.g. dairy products, grains), still leaving a very wide range of foods from gardens, farms, ponds, and nature.

Another basic individual variable (apart from metabolic efficiency) is bodyweight itself. Dosages of any substance vary with body build, fats, and dosage per kilogram weight, so that alcohol and anaesthetics (for instance) may have a very different metabolic effect on two people of equal weight, one of whom is fat and one lean. So it is with specific foods.

The sane procedure in health is to maintain basic hygiene, grow and eat healthy plants and animals, avoid biocides and pollutants, take easy exercise, drink clean water or beverages, and stay as cheerful as this world permits (or adopt a positivistic lifestyle). The rest is up to chance, traffic accidents, or megalomaniacs and wars, and these demand commonsense changes to the social systems, plus a little good luck.

8.3

TRIBAL AND TRADITONAL SOIL CLASSIFICATIONS

From *The Ecologist 14* (4) 1984:

Tribal and traditional people classify their soils on a great variety of characteristics based on:

• Colour (indicates humus content).
• Taste (agrees with our pH measures).
• Moisture capacity and water retention.

- Sand content.
- General texture.
- Firmness.
- Structure (dry soils).
- Wet–season structure.
- Vegetative indicators ("health" of a specific crop).
- Drainage.
- Slope.
- Elevation.
- Animal indicators (e.g. termites: the shape and size of their mounds).
- Plant indicators (for acidity, drainage, fertility).
- Catena (types of soils based on slope relationships).
- "Hot" and "cold" soils — relative fertility (*not* temperature); can also indicate water retention.
- Usage, e.g. pigments, pottery, salt extraction from reeds.
- Work needed for crop (an *energy input* classification).
- Suitability for specific crop (e. g. yam soil, taro soil). Soils can be ranked for up to twelve crop types, giving a complex classification.
- Organic content (apart from colour).

Thus, sophisticated assessments of soils are available from most agricultural societies. However, modern classifications are more complex in terms of nomenclature and physical categories, and use standard colour charts (Munsell and others), standard comparisons, standard sieves and so on.

8.4
THE STRUCTURE OF SOILS

Soil is a complex material, and if it has enough plasticity (usually clay), or glue or fibre from organic sources, it can be pressed or compacted into mudbricks, hard pise, or baked to clay or stoneware. In any of these forms, it is of little use to plants.

Uncompacted soils are open, crumbly, or soft unless concreted by chemical solutes or compacted by ploughs, hooves, or traffic. Crumbly soils have nevertheless a definite structure. The soil particles are in nodules or clumps held together by roots, clay minerals, and chemical bonds. If we speed a plough or drag harrows through these fragile assemblies, they may powder up as they do in a potter's ballmill, or on outback roads. ("Bulldust" is the term used in the Australian outback. A soil scientist might speak of "snuff".) Dryland soils with a high salt content are particularly susceptible to loss of crumb structure, only partly relieved by application of gypsum.

The mantle of soil and subsoil that covers the earth is as thin as the shine on the skin of an orange, and this mantle extends as living mud below the waters of earth as well as on land. It is composed of these elements:

- MINERALS, mainly silica, oxides of iron and aluminium, and complex minerals.
- SOIL WATERS, fresh, saline, with differing pH, and dissolved minerals and gases.
- GASES, some from the atmosphere, others emitted by the breakdown of rocks and the earth's interior.
- LIFE FORMS, from fungal spores and bacteria to wombats and ground squirrels, from massive roots to minute motile algae.
- ONCE–LIVING REMAINS; the humus of the earth; decayed, compressed, and fossil organic material.

Soils rarely extend much below 1–2 m, and are more often a living system 6–12 cm deep. Subsoils, lacking the life components, and buried soils or deep washed silts are rare and confined to valley floors and deltas, or glacial mounds.

To estimate the proportion of clay, silt, sand, and coarse particles in soil, a sufficient first test (for judging the suitability of soil for dam building, mud brick construction, and crop types), mix a sample of soil from a few typical sites, and pour a cup of soil in a tall jar, filled almost with water. Shake vigorously and let the soil fractions settle out over a day or a week (clay can remain in suspension for up to a week). 40% or so clay is needed for dam walls, and less than that for good mud bricks (without lime or cement added).

Of these fractions, the coarse particles are inert, although useful in fine soils as a wind–erosion deterrent. Sands are 0.05–2 mm, silt 0.05–0.02 mm, clay particles less than 0.002 mm (1 g of clay has a surface area of up to 1000 times that of 1 g of sand).

Soil crumb structure, aided by lime (calcium) aids the bonding together of these fractions and creates 20–60% pore space. The organic materials and gels hold the structure open in rain, and where plant nutrients can become soluble for absorption by roots.

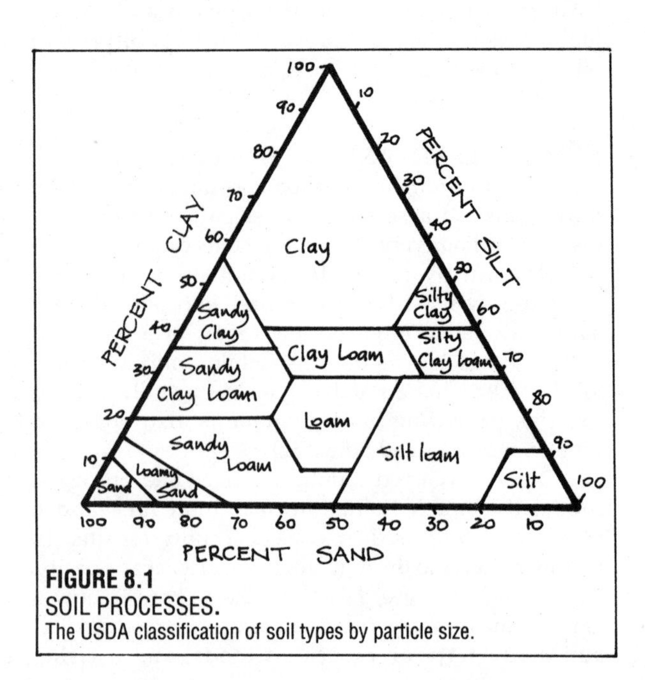

FIGURE 8.1
SOIL PROCESSES.
The USDA classification of soil types by particle size.

8.5

SOIL AND WATER ELEMENTS

Of the 103 known elements, only some are commonly dealt with in the literature on soils and water. Soil itself is predominantly composed of aluminium, silica, iron, and (in certain shell, sand, or limestone areas) calcium. Only mineral ore deposits have large quantities of other elements.

The modified form of the periodic table is given in the centrefold of the colour section and a more extensive annotation on specific elements is given in **Table 8.1**. It is essential that designers in any field have a basic knowledge of nutrients, poisons, and tolerable or essential levels of trace elements in food, water, and the built environment.

The health of plants, animals, and the environment of soils and buildings are all dependent on the balance of elements, radioactive substances, radiation generally, and water quality. Of particular concern in recent times is the level of radioactives in clays, bricks, paints, and stone, and both the emissions from domestic appliances (TV, microwave ovens) and microwave radiation from TV transmission and power lines.

Thus, the annotations in **Table 8.1** cover a range of topics, including short comments on human health. In this case we are using the periodic table not as an aid to chemistry but as a guide to understanding the role of the elements in soil, water, plants and animal tissue, and nutrition.

The colour–coding and the dots on the periodic table (**see centrefold**), plus the following annotations of elements give some of their known effects on soil, plant, and animal health, or their special uses.

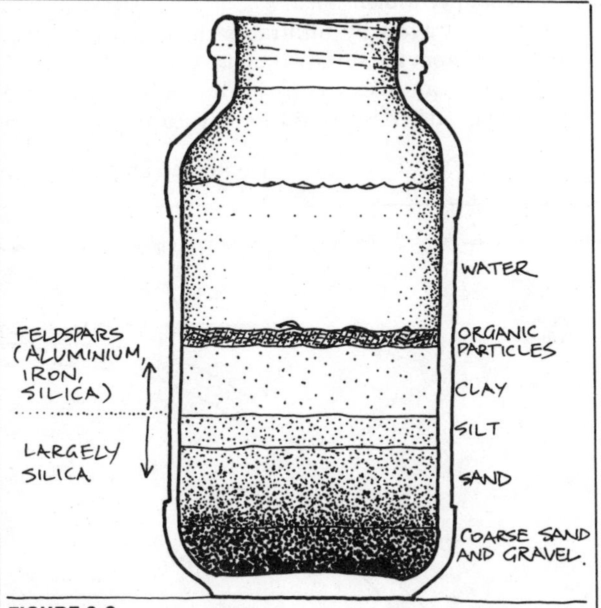

FIGURE 8.2
JAR METHOD
of assessing crude soil composition; useful for classification, uses for mud brick or pisé work. Soil sample is shaken in water and allowed to stand until layers form (1–20 days). The volume of each fraction determines uses and a texture classification (see **FIGURE 8.1**).

8.6

PRIMARY NUTRIENTS FOR PLANTS

Soils are often analysed as deficient in both PHOSPHATES and POTASH in heavily leached areas. Phosphates are supplied either by guano (bird manures) from dry islands, or from older deposits found in sedimentary rock. Potash occurs in the mineral kainite, formed in areas of evaporated waters. Desert salts usually contains 20–25% potash. Phosphatic rock is restricted in distribution, and contains 8–15% phosphorus in various combinations with oxygen or water (hydrated). There are large reserves of potash in common minerals like orthoclase (a major constituent of granite).

NITROGEN can be supplied by water or land plants inoculated with rhizobia, or fixed by algae and water plants such as *Scirpus* or *Azolla*. We can create the conditions for fixing nitrogen by growing these nitrogen–fixing plants, inoculated with the appropriate rhizobia. Much higher levels of nitrogen than were previously thought to be available are fixed by land plants, in a series of zones extending from the roots. Even after nitrogenous plants such as *Acacia, Albizzia*, and *Eleagnus* are cut, the root zone will continue to release nitrogen for up to 6 years, so that pioneer legumes or nitrogenous trees serve as cover crop for trees, and release nitrogen during their lifetimes and for some years after. Legumes may not be needed in older forests, and typically die out under canopy. Only a few larger leguminous trees (*Samanea, Acacia melanoxylon*) persist as forest trees in a mixed forest.

Both phosphate (concentrated by seed–eating birds) and potash (from burnt and rotted plants or compost) can be locally produced if birds are plentiful and their manures are used. The phosphates mined from marine guano, however, may contain concentrated levels of cadmium and uranium, either or both of which (and other heavy metals) can be taken up by the oceanic fish and shellfish used by marine bird colonies. Continual heavy use of such resources is likely to become polluting to soils. Our only ethical strategy is to use just enough of these resources, and to conserve them locally.

SOURCES OF MINERALS IN SOILS

The Sea

It makes sense to assume that as soils are leached, and so made mineral–poor, these minerals later become more concentrated in the sea, in marine organisms, or in inland saltpans. Seaweeds, seagrasses, and fish residues have always been part of agricultural fertilisers, and have maintained their place even in modern times. As seawater evaporates, first calcite and dolomite, then gypsum and anhydrite separate out; all are used for soil conditioning, pH adjustment, or to restore soil crumb structure.

Next, rock salts crystallize out, but only wet tropical uplands may actually lack this common nutrient, although even there specific plants (often aquatic) concentrate salt which can be gathered or leached from their ashes.

Lastly, potash, magnesium salts, and a host of minor elements remain; the evaporites (those already deposited) being the most soluble and therefore earliest deposited. The liquid that remains after the common salt content deposition is a rich source of minor minerals and trace elements. It is, in fact, sold as "bitterns" (bitter oily fluids) for dilution and incorporation in crop soils, or in low concentrations (diluted 100–500:1) used directly as foliar sprays in strengths varying from 1–20 l/ha. Very corrosive, bitterns (which include bromine and many of the early elements of the periodic table, plus some rare minerals), are safely held and distributed only via non-corrosive vessels and pipes (today, polyethylene pipes and drums).

Bitterns are cheap, and easily transported to leached areas, but their effects must be established by local trials on specific crop. As these evaporites are so easily dissolved, they are also those most likely to be carried to sea in rains.

Rocks and rock dusts

Granites contain felspars yielding potash or sodium salts. Limestone and dolomite yield calcium and magnesium, and mineral deposits or their ores give traces of the basic minor elements. Of these, calcium (in all but highly calcareous areas) is most needed, dolomite (except where magnesium is already in high ratio) is next; phosphates and felspars follow, along with trace elements in small quantities (as low as 5–7 kg/100 ha for zinc, copper, cobalt, and molybdenum).

Field trials have established that cheap ores, finely ground, are as effective as more refined sulphates or oxides (Leeper 1982). Sometimes such minerals are given to animals as salt licks, in molasses, in water, or as injections or "bullets" of slow-release elements (cobalt) in a pellet which lodges in the rumen. Some mineral elements also reach plants via urine, but foliar sprays are more rapid-acting and effective.

Fine rock dusts of a specific rock suited to local needs are often cheaply available from quarries or gravel pits. Basalt dusts are helpful, for example, on leached tropical soils. Rock phosphate contains 8–15% phosphorus, but is very slow to release nutrient, and may in fact be absorbed completely on to leached clays and clay-loams. Super-acid phosphate added to compost, or to plants used in compost, may be necessary under such conditions. Rock dust as an unselective category can do as much harm as good on soils, adding excessive or poisonous nutrients in some cases, or excessive micronutrients.

8.7 THE DISTRIBUTION OF ELEMENTS IN THE SOIL PROFILE

There seem to be two important determinants of the concentration of elements or nutrients through the soil profile (that vertical column of soil from the surface to 2 m or so deep). The first is the penetration of water. Water is a universal solvent, enabling compounds to dissociate into ions, and transporting them, firstly, to various deposits at microsites in the soil. This effect works in three dimensions:

- Water travels by infiltration to varying depths, soaking in from the surface down.
- Water can also rise from the soil water table upwards, either by flooding, by capillary action

{continued on page 195…}

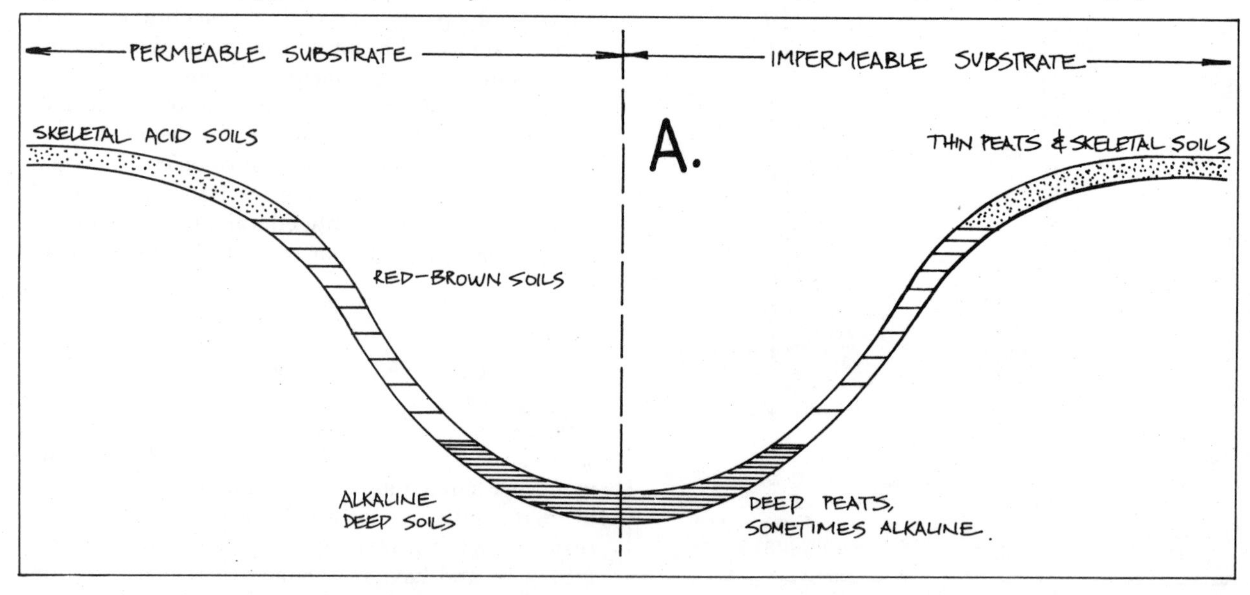

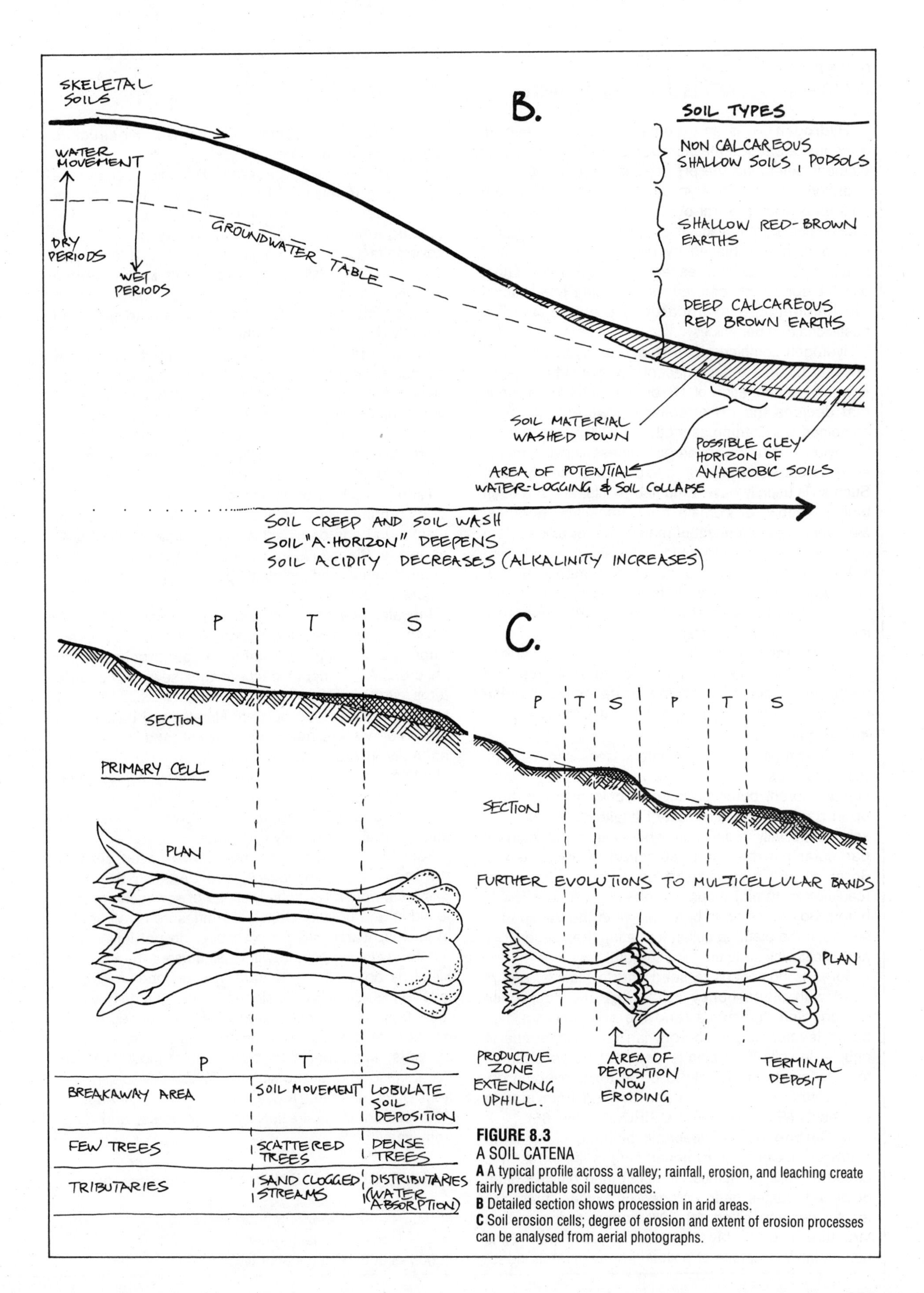

P	T	S
BREAKAWAY AREA	SOIL MOVEMENT	LOBULATE SOIL DEPOSITION
FEW TREES	SCATTERED TREES	DENSE TREES
TRIBUTARIES	SAND CLOGGED STREAMS	DISTRIBUTARIES (WATER ABSORPTION)

FIGURE 8.3
A SOIL CATENA
A A typical profile across a valley; rainfall, erosion, and leaching create fairly predictable soil sequences.
B Detailed section shows procession in arid areas.
C Soil erosion cells; degree of erosion and extent of erosion processes can be analysed from aerial photographs.

TABLE 8.1
KEY TO PLATES 12 AND 13: THE PERIODIC TABLE

1. **Hydrogen** (H) is an extremely mobile and reactive gaseous element; the number of free ions of this element determine the pH of soils, together with the hydroxyl (OH^{2-}) radical in alkaline areas. Hydrogen combines with several other elements and organic substances to form acids. It is a potentially inflammable light gas, now replaced by helium in balloons and airships. Some plants, especially algae and rushes, can transpire hydrogen, and in so doing break down the halogenated hydrocarbons that are used in pesticides and herbicides.

Hydrogen combines with oxygen as hydroxyl (OH^{2-}) or water (H_20), the basic gaseous and liquid elements. It is the concentration of H^+ or OH^{2-} ions in solutions that decides the pH of soils and water. Hydrogen combined with carbon as methane (CH_4) is emitted by decaying humus in anaerobic (airless) environments such as under water or in compacted or boggy soils. Such soils usually have a mottled profile, and are often bluish, yellow, or contain iron stains and nodules. Methane is a component of marsh gas, or biogas from digesters, usually associated with carbon dioxide and sulphur dioxide (the gases of decomposition). A sulphurous smell in subsoils is a guide to wet–season water–logging, and should be noted for plants intolerant of stagnant waters.

3. **Lithium** (Li), the lightest metal, is prescribed medically as tablets in cases of hyperactive and disturbed people; it may moderate nerve impulse transmission across synapses. Lithium is found in plant and animal tissue.

5. **Boron** (B) is a trace element, necessary to (e.g.) the brassicas and beets. As many detergents also contain boron, boron pollution or poisoning can build up in gardens and affect plant health adversely. Use soap, especially in drylands, where citrus and grains in particular can be boron–poisoned in dry seasons. Borax at 160 g/ha is used on beet and *Brassica* (cauliflower, turnip) crops where soil levels are low in boron. Boron seems to be essential to the transport of sugars in the plant, to pollen formation (hence, fertility), and to cell wall structure (like calcium).

Borax is poisonous to seed and to insects, and is used with dilute honey as an ant and cockroach poison. Sea sediments (and the sea) contain high boron levels. Surplus boron can cause anaemia in people. In soils, 0.75 ppm is ideal, but 1.0 ppm can be toxic to plants. It is likely to be at toxic levels in dry years, where soils have been derived from marine sediments (*Rural Research*, CSIRO, Autumn '86).

6. **Carbon** (C) is the basic building block (with hydrogen, oxygen, as hydrocarbons) of life forms. It is added to soils as humus, compost, and mulch. Pitted or buried wastes need nitrogen or oxygen, therefore *air*, for decomposition. 10–20% humus ensures good structure in both clays and sands. More than 50% humus inhibits uptake of polluting heavy metals (lead) cities; less than 7% may not improve soil structure *unless* calcium is added.

Carbon combines with oxygen to form carbon dioxide gas, which is released from agriculture, forest felling, and industry to create the greenhouse effect and subsequent earth heating; this may be a critical adverse factor for human survival on earth. Plant trees, do not use bare fallow, and add humus to soil. Carbon in soils is about 58% of organic matter (combustible), existing mostly as colloids (assess nitrogen and multiply times 20 to obtain the carbon content of soils).

7. **Nitrogen** (N) is a major plant nutrient, 80% of the air itself. Legumes and many non–legumes (alder, *Casuarina*) that have root associates (fungi, bacteria) will fix nitrogen from air if molybdenum is present as a catalyst to convert nitrogen to ammonia. Nitrogen as a foliar spray of urea can, however, increase the protein content of grains such as rice by 40%.

Nitrogen is part of all amino acids (hence proteins), of chlorophyll, and of enzymes. It is very mobile in plants, and has dominant effect on other nutrient use, or uptake. Plants absorb nitrogen in the form of nitrates or ammonium, and the plant coverts this to ammonia to create proteins.

In water or with bacteria, nitrogen forms NITRATES which at low levels are benefical, but at high levels (more than 80 ppl) are lethal to young animals and children. Also, mouth bacteria and sewage bacteria turns nitrates into NITRITES, hence NITROSAMINES which are cancer–associated. Nitrates are building up in all soil waters below agricultural land, and occur naturally in many desert bore waters. Thus, fertilise plants in the garden at *minimum* nitrogen levels; do *not* use heavy doses of nitrogen fertilizer, but rather a modest legume interplant. Check water levels for human health, especially in deserts.

Too much animal manure also releases surplus nitrogen. Thus, *reduce* cattle and run them on pastures, *not* on feedlots. Dispose of sewage in forests to aid in tree growth. Constantly measure nitrate levels in leaves, water, soil. In particular, do not inflict high nitrate levels on people with malnutrition; they can be killed by an excess of nitrate in their food. Accepted levels in Europe (E.E.C. standard) are 50 mg/l in water (50–80 ppm). These levels are being exceeded where nitrate fertilisers and even natural manures are spread on soils; ammonia from cow dung is now 115,000 t/year in Holland, accounting for 30% of the acid in rain from Dutch sources (*N.S.* 6 Feb '86). In humid areas, 70% of human intake is from vegetables, and 21% from water.

SOME NITRITE (OR NITRATE) LEVELS (ppm):

0–50	Safe for human consumption
500	Spinach grown from compost
2,000–3,000	Chemical fertiliser–grown spinach

8. **Oxygen** is a necessary for respiration (energy release in blood). Soils also need oxygen (air) and thus good open structure and soil pores as 12–30% of the soil bulk. Aeration in soils is achieved by using calcium, trees, worms, ripping, humus, or sometimes by blowing air through sub-surface irrigation pipes. WATER–LOGGING reduces pore space and thus root oxygen. Crops such as walnuts, oranges, chestnuts, and potatoes require deep loose soil and plenty of oxygen. Pear trees and marsh grasses (Yorkshire fog, Imperata grass) require much less oxygen and can grow in more compacted or wet soils. Soil loosening and draining (conditioning) improves oxygen supplies to soils. Soils need to have good open pore spaces to 1–2 m (3–6 feet) deep. Large soil pore spaces are only achieved by soil life and perennial crop.

9. **Fluorine**. Although beneficial in small amounts, as in seaweed fertiliser, it is a serious poison at high levels. It is commonly found in bore water, or in downwind pollution plumes from metal processing works (zinc, aluminium, copper), causing bone deformation in animals, stunted growth, and serious metabolic disturbance. Check water, especially bore water, for high levels and use tanks for drinking water. Fluorine is useful at very low trace levels **only**. Surplus can cause anaemia; high levels are found in (e.g.) marine phosphate deposits, hence fluorine can be high in fertilisers derived from these.

The fluorine level sufficient for normal growth and bone develoment is about 1 mg/day. Tea drinkers ingest this from tea leaves; as the chief source is in water, 1 mg/l (1 ppm) is sufficient. Seafoods (fish, oysters) can contain 5–10 ppm fluorides.

Problems of excess flourine occur where dusts from mineral processing fall out on vegetation, where "hard" waters (common in deserts) contain more than 1 ppm fluorine (actual levels can reach 14 ppm) and where diet is mainly seafood. The bones, spine and ligaments can become calcified in areas where excessive flourine is present in the diet. Teeth may be "case–hardened" by flourine, and show little outer sign of decay while the tooth itself rots.

11. **Sodium** (Na). With chlorine, it makes up common salt. At low levels, it is necessary to plants and animals, cell health, and function of the nervous system. At high levels, sodium causes collapse of soils and loss of pore space, displaces calcium and disperses colloids. Beets and some other crops need adequate levels, likely to be absent or low in wet tropical uplands only. Elsewhere, enough sodium comes in with rain as salt crystals in raindrop nucleii. Real problems occur when soil salts are mobilised by irrigation, potash fertiliser, or soil flooding in dryland areas. Excessive salt causes circulatory and kidney problems in people, especially those in sedentary occupations. Salt in vegetable cooking water blocks uptake of some heavy metals on plant proteins, so if lead pollution is suspected, add salt to cooking waters.

12. **Magnesium** (Mg) With calcium it is found in DOLOMITE. It is needed by plants, and common in most subsoils. It forms the central atom of the chlorophyll molecule, activates enzymes, and concentrates in seeds. Like nitrogen, it is very mobile in the plant. Magnesium is likely to be deficient only in sands and sandy soils; it is present in the clay fraction of soils, and is released by soil acids and humus. It is supplied by dolomite application on farmlands, as tablets of dolomite for animals, and mixed in animal foodstuff. Magnesium carbonate from dolomite buffers plant and soil acids, as does lime, but it can cause plant poisoning if not balanced by high lime (calcium) levels.

Magnesium is essential for chlorophyll and in the mitochondria of human cells, hence energy metabolism. It is present in all green plants. Deficiencies do not normally occur where people are healthy and are not subject to diarrhoea; when required by people it is given as an intravenous fluid at 10 minimoles per kgm bodyweight.

13. **Aluminium** (Al). A "Jekyll and Hyde" element, it forms a large part of soils and red clays (with iron), or as poorer bauxitic dryland deposits. It is poisonous as aluminium sulphate, formed in acid rain or sulphur–polluted soils, e.g. in industrial areas, and is also released when cooking onions in aluminium cookware! Heavy aluminium silica uptake in humans may assist the development of senile dementia, thus *calcium* in the soil and as tablets is a partial defence. Some areas are naturally high in acid–aluminium soils. An adequate level of calcium is needed in all gardens, and even in rainwater tanks (as marble or limestone chips, whole sea shells, gravels of lime).

Avoid aluminium utensils (use iron, enamel, stone, steel) if cooking salty, acid or sulphurous substances (vinegar, onions). Soluble aluminium in concentrations of 0.1 to 1.0 ppm is tolerable, but above this level susceptible plants start to die. At 15 ppm or higher, many plants die.

14. **Silicon** (Si) is an important part of cell walls in many grasses, bamboo, and is essential in soil cation exchange capacity of deep, red, heavily–leached tropic soils (an application of cement dust can assist this exchange). Pine trees and conifers generally are poor nutrient recyclers, and can produce nutrient–deficient silica soils under their litter, thus losing calcium and other elements to leaching. Therefore, use grasses and broadleaf, leguminous, or soil–building trees with conifers. Many trees deposit salt, phosphorus, manganese, zinc, potash, etc. at high topsoil levels due to good nutrient recycling *via* leaf fall. Silica is normally at 20–40 ppm in soil waters, and in highly alkaline, wet, warm areas can be leached away altogether, hence the use of cement (calcium silicate) to restore some silica to plants; in such areas, **tree crops** are the only sustainable solution.

Silicates make up much of the bulk of normal soils, but very high silica in rock may produce acid soils in high rainfall areas (silicic acid). Bamboos are good sources of calcium and silica as a garden mulch in the tropics. In ponds, diatoms need silica to proliferate; these are an excellent fish food.

15. **Phosphorus** (P) (as phosphate) is an essential, common plant element, recycled by many trees and fixed by the root associates of several trees (*Casuarina, Pultenea, Banksia*), by algae, in the mud of ponds, in bones, and in freshwater mussels. High phosphate levels in bird manures are derived from fish bones and seeds. Phosphorus is essential to the energy metabolism of plants, hence photosynthesis and respiration. Cell divison, root development, and protein formation are regulated by phosphorus. It is highly mobile in plant tissue.

In some oceanic guano deposits, phosphates can be contaminated by cadmium, mercury, uranium (40 ppm) and fluorides. All heavy cropping demands phosphate for production, and there is a general world soil phosphate deficiency, especially in poorer countries. Most western soils are now over–supplied, with a very large unused soil bank of phosphorus, hence with the pollutants of phosphatic rock. Cadmium levels in inorganic market garden crop may commonly exceed health limits.

Bird islands, areas of recent volcanic ash, and some soils over phosphatic rocks are not phosphate–deficient. Sandy, bare–cropped, wet, and water–logged soils, old soils, and alkaline soils may show deficiencies. On the latter, try sulphur to adjust availability. Super–acid phosphates are used to make phosphate available to plants, albeit expensively. Most soils (superphosphated since the early 1950's) may have 750 ppm in the top 4 cm, and most of the rest of the applied phosphate is bound up and unavailable in the top 20 cm. Only on deep, coarse, leached treeless sands with heavy rains or in bare–soil fallows is a lot of phosphate lost. Drainage water contains 0.2 ppm in clay or loam soils, a minute loss. In natural systems, phosphates are supplied by fish and bird manures.

Calcium, iron and aluminium immobilise phosphate; a pH of 6–7.5 releases it. Basic superphosphate (phosphate and lime) finely ground is available (soluble) to plants. The home gardener can use bone dust, phosphate and lime, and mulches, which are all effective. About 45 ppm phosphate in soil is needed for grains (optimum pH 6.0–6.5). Pelleting seeds with basic superphosphate provides phosphorus. In soils, phosphorus combines with iron, aluminium and calcium as insoluble compounds, and with living materials or humic compounds as "available phosphorus". Despite folk stories to the contrary, superphosphate has not been found to acidify soils (unlike ammonium sulphate).

Phosphate levels ("native phosphate") fall with depth, and are low in most subsoils. Generally, to increase phosphate availability, try humus in warm wet areas, sulphur in drylands, and adjust pH elsewhere with calcium (lime or dolomite). For trees, apply light phosphate dressings regularly in sands. Phosphorus deficiency reduces growth in animals by depressing their appetite for herbage.

Of all the elements of critical importance to plants, phosphorus is the least commonly found, and sources are rarely available locally. Of all the phosphatic fertilisers used, Europe and North America consume 75% (and get least return from this input because of overuse, over–irrigation, and poor soil economy). If we really wanted to reduce world famine, the redirection of these surplus phosphates to the poor soils of Africa and India (or any other food–deficient area), would do it. Forget about miracle plants; we need global ethics for all such essential soil resources. As long as we clear–cultivate, most of this essential and rare resource will end up in the sea. Seabirds and salmon do try to recycle it back to us, but we tend to reduce their numbers by denying them breeding grounds.

Unpolluted phosphate deposits are found only in limited areas, in sedimentary rock. Trees do mine the rare phosphorus released by igneous rocks, and they are responsible for bringing up phosphorus to the topsoil wherever it is rare in more shallow–rooted plants.

About 15–20% of the inessential use of phosphorus is in detergents. The older potash soaps are the answer to that sort of misuse, or the recovery of phosphorus from greywater rather than a one–way trip via a sewer to the sea. Uncut forests may lose 0.1 kg/ha/year, while clear–cropping can lose 100 kg/ha/year or more, a 1000 times increase in lost minerals, never accounted for in logging and clearing, or added to the cost of woodchipping and newspapers.

Phosphorus is found concentrated in seeds and in the bones of vertebrates, especially fish, in mussels (freshwater) and the mud they live in, and in the manures of animals eating fish, seeds, and shellfish. Bone meal from land animals is a traditional source, and most farms (up to 1940) kept a flock of pigeons as their phosphate factory, while in aquatic cultures, phosphorus was recovered from the mud of ponds stocked with mussels and from fish and waterfowl wastes. Bat guano is also a favoured source of phosphate in Holland. Even a modest perch in a bare field will attract a few perching birds to leave their phosphates at a tree or along a crop line.

Conservation farming (Cox and Atkins, 1979, p. 323) loses about one–half to one–third the phosphorus of contemporary agriculture, even without non–tillage. Non–tillage farming would lose even less, but is rarely assessed except in dollar yield terms. Bioregional farming, and home gardening with wastes returned to soils would lose even less; and finally, regional food supply, waste recycling, and a serious consideration of devoting 30% of land surface to trees might just be a sustainable system. Next to clean water, phosphorus will be one of the inexorable limits to human occupancy on this planet. We must not defer solving these problems or conserving our resources any longer, or we betray our own children.

16. **Sulphur** (S). Many anaerobic bacteria (thiobacilli) fix sulphur, which is why anaerobic ferment of plant and animal materials is rich in the sulphur–based amino acids, and of high nutritional value. The sulphur oxidised by anaerobic bacilli also removes, as

insoluble sulphates, most heavy metals from such flow–through systems as sewage digesters. Some thiobacilli occur in all warm wet soils, and many can operate down to pH 1.0! With ammonium sulphate, soils tend to become acidic and so reduce plant yields. Sulphur is used in drylands to reduce pH and so make iron, zinc, and trace elements available. Clover, in particular, may show sulphur deficiency in the sub–tropics (wet summers).

Sulphur is part of all proteins, and is present in the body as the two amino acids methionine and cystein. The vitamins thiamin and biotin contain sulphur. Food intake by people is from the amino acids, and deficiencies do not occur if meat protein is sufficient, or if anaerobic ferment of leaf materials is part of food preparation. Many vegetarians get their animo acids from yeasts or bacteria rather than from fresh plant material.

17. **Chlorine** (Cl) was used as a war gas, and today to "sterilise" waters; it is a dangerous gas to inhale. Chlorine is used by plants, and is normally available as salt. It is a trace element, and used only in minute amounts. It concentrates in crop, e.g. 350 ppm in soil gives 1000 ppm in crop. In water, and in contact with organic materials, it releases chloroform, a carcinogenic gas. Avoid chlorinated water if possible!

19. **Potassium** (K) is used in large quantities by plants. It is usually plentiful in arid areas. Potassium is deficient on sandy, free–draining coast soils. Not much is removed by livestock, but potatoes, beans, flax, and the export of hay may remove soil reserves to below plant needs. It is readily absorbed on colloids, and is usually plentiful in clays, especially illites, **not** in kaolin. Gardeners add ashes, bone, natural urines and manures or green crops for supply to heavily cropped ground.

Earthworm castings commonly concentrate potash (at 11 times soil levels). Excess potash fertiliser can greatly increase soil sodium, and so block calcium uptake; beware of this in alkaline or dryland soils.

20. **Calcium** (Ca) is needed in all soils, and is removed by sodium in drylands. Even where calcium exists in an alkaline area, sodium may suppress its uptake by plants. Sometimes, gypsum is applied (30 tonnes/ha) and the excess sodium then removed by flushing out as sodium sulphate. Plants need large amounts of calcium. It is an essential part of cell walls, enzymes, and in chromosome structure.

The proportion of the four major ions in ideal agricultural soils should be about: Calcium: Magnesium: Potassium: Sodium (50:35:6:5)

Note that adding potassium can increase sodium, that sodium is antagonistic to (displaces) calcium, and that magnesium ions should always be less than calcium ions. Do not add too much dolomite if soils (as clays) already contain adequate reserves of magnesium.

Many peoples have lactose intolerance and cannot get calcium on a milk diet, so lime or dolomite on gardens or as tablets may be needed. Calcium is lost (excreted) in stress, and needs replacement after periods of prolonged stress. Low calcium areas produce predominantly male farm animals and humans (as a primary sex ratio), and lack of calcium produces skeletal and metabolic malfunction.

Calcium phosphate is the chief mineral constituent of bones. Bony tissue is always losing and gaining calcium, but older women, in particular, suffer bone fractures from loss of calcium in the ageing process; immobilised limbs also lose calcium. Gross calcium and vitamin D deficiency results in rickets.

22. **Titanium** (Ti) is not a nutrient, but in sands and the presence of sunlight it acts as a catalyst to produce ammonia for plants, often combined with iron (TiFe) (*N.S.* 8 Feb 79, 8 Sep '83). Ammonia is provided at 50–100 kg/ ha/year. Rain and moist soils make this available to plants—a useful desert strategy.

24. **Chromium** (Cr) is a poison to plants and animals, occurring in serpentine rock. It is used in the preservation of timber, and is to be guarded against from electrolytic and leather works as it poisons active biological agents in sewage works. It is easily removed or recycled at the source.

As some chromium occurs in all organic matter, it is in fact a trace element, related to glucose tolerance in humans, and necessitating insulin if absent or in very low quantities. Very little firm knowledge of the metabolic function of chromium is available.

25. **Manganese** (Mn) is a readily available trace element on acid soils, except in sands. It may cause manganese poisoning below pH 6.5. On alkaline soils, or when pH exceeds 7.5, it may be deficient in grain crop or vegetables, but seed soakage, seed pelleting, or foliar sprays supply this nutrient. Even flooding at periods will mobilise manganese. A typical deficiency situation occurs on poor sands heavily dressed with lime.

Bacteria fix insoluble manganese, and can create problems in pipes and in concrete water raceways even at 2–3 ppm manganese. Aluminium sulphate (from acid rain on soil) may mobilise manganese, mercury, cadmium to lethal levels. Manganese leaches out of acid soils, deposits in alkaline horizons as manganese–iron concretions, and on sea floors as larger nodules.

26. **Iron** (Fe) is normally plentiful in soils, and is a critical element for plants and animals. Plants can show iron deficiency symptoms in alkaline drylands or on heavily limed garden soils. Sulphur in gardens, or iron chelates, remedies this situation; iron can also be added as small amounts of ferric sulphate. Deficiency causes interveinal leaf yellowing in plants, and low blood iron in people (anaemia). Ferrous iron spicules in soil assist the formation of ethylene. Ferric iron (oxides) are "unavailable" (F_2O_3). Iron concretions and iron staining in sands are often associated with microbial action along plant root traces. Iron–deficiency anaemia in people is a common problem in deserts, or after blood losses in menses or dysentery.

In humans, blood iron is as important in energy

transfers as is copper in invertebrates. Deficiencies are normally rare but must be watched in women of reproductive age, when lack of iron is widespread and causes ill–health and lassiitude; anaemia also occurs in infants and children. Meats such as veal, liver, and fish eaten with soya beans, lettuce, parsley, maize, etc. can provide dietary iron. Meat and ascorbic acid (vitamin C) increase iron absorption from food. Iron in bone marrow is needed for new blood cells. Clinically, ferrous salts ingested can be used in cases of excessive blood loss.

People brewing beer and wine or cooking cereals in iron pots often ingest excess iron. Alcholism and diabetes exacerbate iron excess, although excessive iron is an unusual condition.

27. **Cobalt** (Co) Copper and cobalt can be deficient on poor coastal sands. Cobalt is necessary to many legumes, and especially needed by livestock, who are "unthrifty" if it is deficient, as it can be in highly alkaline areas. Only an ounce or two per acre is needed, and this can be supplied as a cobalt "bullet" in the rumen of sheep.

Cobalt is necessary for synthesis of vitamin B_{12} in ruminants; if deficient, further metabolic problems then occur. Often diagnosed via B_{12} tests on blood levels. Deficiency can cause anaemia. Toxicity can occur if excess is ingested. Radioactive cobalt is a serious poison.

28. **Nickel** (Ni) is not noted as a plant nutrient or poison, but with copper can intensify poisoning in people, e.g. in office rooms using urns, or in mining areas. Many people demonstrate skin reaction (rash) if nickel is used as bracelets.

29. **Copper** (Cu) is a necessary micronutrient which can become a poison at higher concentrations. Copper in plants is an enzyme activator, and is concentrated in the chloroplasts of leaves. Copper is deficient on shell–grit dunes, acid peaty soils, coastal heaths over poor quartzite sands, and in deeply weathered basalts. Black sheep are good indicators, and if deficient develop whitish or brown wool, or white–coloured sheep will grow "steely" wool. Copper sulphate at 1 kg/ha every 5 years is a cure. Old stockmen put a few crystals into water–holes. Deficiency (with iodine) can lead to goitre (thyroid inbalance) and anaemia in animals. Levels above 12 micromoles/l are excessive in drinking water. High blood copper levels in deserts are often as a result of zinc deficiency.

While copper deficiency is rare or absent in adults, copper toxicity is of more concern especially where copper sulphate is used to clear algae from drinking water, and also from foods or water standing in or boiled in copper utensils and pipes. Copper poisoning is serious, with liver deterioration (cirrhosis) and brain effects (tremors, personality changes). Once hopeless, such cases can now be somewhat alleviated by chelating agents.

30. **Zinc** (Zn) is deficient on leached sands, dunes, alkaline sands, and supplied by zinc sulphate spray on plants or zinc salts in seed pellets. It should be tested for in all desert gardens. Zinc is critical for both tree establishment in many deserts or dunes, and to human health, and is cheap to supply. In plants, zinc is essential to the growth control hormones.

Zinc deficiency causes gross metabolic imbalances and stunting of growth. People with malnutrition often have low zinc blood levels and poor wound healing. Diabetes is often linked with zinc deficiency. Excessive zinc acts as an emetic, causes vomiting and weakness.

Zinc is necessary to several enzymes in the body; prostate (hence seminal) fluids are high in zinc. Diets of coarse grains, unleavened bread, and low meat intake contribute to zinc deficiency in poor people, especially in high–calcium soils. Oral zinc sulphate can be given clinically. Alcoholism and diabetes, feverish sweating and stress all lower zinc levels. Severe deficiency causes hair loss, moist eczema of the mouth area, impotence, apathy, diarrhoea.

33. **Arsenic** (As) sometimes seems to be needed by horses, and is included also in chicken pellets, but is a poison to animals even at slight concentrations. Polluting sources come often from gold processing areas, or from chickens fed on pelleted foods containing high arsenic levels.

34. **Selenium** (Se) deficiency in lambs causes "white muscle" or "still lamb" disease which is cured by selenium and vitamin E injection. It also causes "illthrift" in animals in many countries where a severe dietary deficiency occurs. Five mg/year is sufficient for sheep; more can cause toxicity. Seaweeds may supply sufficient selenium to gardens.

38. **Strontium** (Sr) is becoming, as a long–term and very poisonous radioactive element (Sr90), a widespread danger from atomic plants. It now pollutes many areas, milk, and is being used as a "blackmail", e.g. in New York water supply in 1985. One of the great pollutants of the future, it causes cancer at absorption site, leucaemia in children, and is a "hidden cost" (very hidden) of the atomic age. It is excreted in urine of breast–fed babies, and concentrates 4–8 times in cows' milk, thus levels in cows' milk must be monitored after atomic fallout. There are no safe levels. Ordinary strontium is a trace element.

42. **Molybdenum** (Mo) is a trace element for clover establishment, and needed by all plants, rhizobia. A few ounces per acre is used on many acid soils, and rarely again needed. It is locked up not by alkali or lime but by sulphuric acid, and may therefore be deficient in plants subject to acid rain. Legumes need molydenum for nodulation, but non–legumes deficient in molybdenum can concentrate dangerous levels of nitrates, causing "leaf burn".

48. **Cadmium** (Cd) is a poison that is concentrated by green leafy plants and shellfish. It is derived from traffic (tyres) and superphosphate. It may already be in very high levels in acid soils of market gardens using artificials (as it is in Canberra). Cadmium causes painful human disease *itai–itai* (Japan) and permanent deformations.

50. **Tin** (Sn). Ores and wastes create plant

establishment problems. It is not noted as toxic at low levels in diets, but can become toxic at high levels in canned food.

51. **Antimony** (Sb) As for arsenic.

53. **Iodine** (I). Deficiencies occur in weathered basalts, causing growth problems and goitre; these are remedied by fish and shellfish, seaweed diet. I131 now a common radioactive fall–out from atomic plants and tests; it poisons milk over wide areas, and affects thyroid function. Iodine can cause cancer, death of children from hyperthyroidosis; a real risk from atomic establishment and tests.

80. **Mercury** (Hg) is a common poison released by mines, metal processes, acid rain, and is more active in organically polluted areas. It creates very serious coordination and sanity problems, central nervous system malfunction, bone deformity, insanity.

81. **Thallium** (Tl) is a toxin, causing birth deformities, thus very dangerous.

82. **Lead** (Pb) is a common poison from petrol, old paints, battery burning. It is a serious urban soil pollutant, needing heavy organic soils to block uptake, or removal of lead–concentrating vegetation for disposal. Earthworms may concentrate lead to lethal levels in polluted soils.

86. **Radon** (Rn) is a gas from the decay of uranium. Seeps up in most soils, especially over volcanic areas, igneous rock, and can pollute super–insulated buildings if the air is not exchanged. Several million homes in the UK and USA have high levels of radon. Avoid building poorly ventilated homes in areas of radioactive ores, granites, and basalts, or the use of such rocks or crushed metals in buildings. If houses are to be super–insulated or draught–proofed, then adequate heat exchangers must be fitted to bring in clean air.

{...from page 188}

(soaking), or by being pressured upwards by an aquifer flow in permeable sediments.

- Water travels by throughflow, traveling by the effects of gravity downslope in the profile.

This latter effect produces soil types called *catenas* specific to slope, drainage, rock type, and landscape (**Figure 8.3**).

INFILTRATION effects are most marked in soils where, for very long periods, a succession of light intensity rains have alternately wetted and then dried out in the soil. At the wet–dry boundary we find concentrations of easily soluble salts and deposits of minerals and nutrients (sometimes in over–supply).

The second set of effects on element transport are biological. Elements are actively sought out or selected, and either concentrated or dispersed by living organisms. These effects are myriad in total, but some of the ways this is known to happen are:

- CONCENTRATION BY SELECTIVE SPECIES of fungi, bacteria, and invertebrates which can seek out, assemble, and change specific compounds to stable new forms as concretions or nodules. Iron, iron–manganese, calcium, phosphate, zinc, nickel, copper, selenium, cadmium, phosphate and nitrogen are all so selected and concentrated by one or other form of mycorrhiza (root fungi), bacteria, molluscs, or algae.
- CONCENTRATION BY ACCUMULATION OF DETRITUS. Diatoms, swamp peats, whole forests, sponges (and their spicules of silica), molluscs, and vertebrates are at times buried by vulcanism, sedimentation, or deposition in oceanic deeps to form specifically concentrated sediments, and eventually ore bodies or rock types (coal, rare earths, or manganese deposits).
- DISPERSAL BY TRANSPIRATION. Many water plants seem to be able to dissociate and transpire a great variety of substances to atmosphere. Reeds do this with mercury, hydrogen, and other elements from phosphates to chlorine. Not all of these are vapourised to drift off in the winds; many are deposited in special leaf repositories, or evaporated to a wax, dust, or efflorescence on the leaves and stems of trees, from where they are washed down again to earth by rain throughfall. In this way, both major plant nutrients and minor elements are concentrated in the top 4 cm (1–2 inches) of soil below trees. Metals, oxides, halogens, acids, alkalis and salts are also concentrated.
- CONCENTRATION BY METABOLIC PROCESSES. We build our bodies up by ingesting a large range of complex foods, as do all living things. From these ingested bodies, materials are selected to build our bones, flesh, blood, and brain tissue or organs, nails and hair, fats and milk. Thus, our own bodies and those of other animals are complex storages of elements; even our faeces package very different concentrations of potash, nitrates, and pollutants from the different concentrations of foods that we eat. And, by our behaviours, these are variously disposed of in a personalised, culturally–determined, or species–specific way in the environment. Plants and animals ceaselessly ingest and defecate, refect and exude over their whole lifetimes, thus altering the concentration of nutrients in their immediate environment.

8.8
pH AND SOILS

Of the parent rocks of soils, we speak of ACID rocks as containing 64% or more of silica (SiO_2), INTERMEDIATE rocks at 50–64%, BASIC at 40–50%, and ULTRABASIC at less than 40% silica. Of soils themselves, we speak of acidity and alkalinity in terms of a logarithmic scale, in which each point is 10 times the concentration of hydrogen ions less than the scale point below it, so that pH 8 (alkaline) is 10 x 10 x 10 x 10 or 10,000 times *less* acidic than pH 4, and pH 3 ten times *more* acidic than pH 4.

Table 8.3 serves to portray the availability to plants of some important elements with respect to pH value.

ACID AND ALKALINE WATERS AND SOILS

We commonly speak of "hard" (alkaline) or "soft" (acid) waters. The latter used with soap lathers easily, and is desirable for washing; the former (including seawater or other alkaline waters) is difficult to use for washing, as soaps and detergents are themselves alkaline and so do not easily dissolve in other alkalis. Soaps, based on sodium or potassium (ash) and fats will lather in soft acidic waters, and detergents based on phosphates or sulphur, lather in hard waters. Hard waters contain calcium (Ca^{+}) or magnesium (Mg^{+}) ions. Soft water contains hydrogen (H^{+}) ions.

The properties of alkali (soaps or carbonates) and acids (vinegar, citrus juice) have long been recognised and used medicinally or in village chemistry. The word for deserts in Arabic is *al khali* ("the salt").

Acids and alkalis arise from the solution of oxides, hydroxides, sulphates, or carbonates of metals and non–metals. In water and soil water, the common rock

TABLE 8.2
NUTRIENTS IN SOIL AS MINERAL CONTENT.

PRIMARY		SOURCES (Compost generally, rock dusts)	COMMENTS by crop	Removed in Soil (Kg/ha)	Available in soil (kg/ha)	Insoluble (Kg/ha)
N	Nitrogen	Legumes, water plants, urine,	Basic to crop growth.	100	20-200	1,000-10,000
P	Potassium	Ash, leafy materials, kainite, bone meal.	Basic to crop growth.	100	40-200	5,000-50,000
K	Phosphorous	Bones, bird manures. Found in super-phosphate (acid-treated bone or rock phosphate.	Basic to crop growth.	20	20-100	1,000-10,000
S	Sulphur	Elemental Sulphur, Volcanic mineral deposits, swamps.	Adjusts pH towards acid.	30	50-100	100-10,000
Ca	Calcium .	Limestone, some crops (e.g. buckwheat), dolomite.	Adjusts pH toward alkaline.	40	100-5,000	10,000-100,000
Mg	Magnesium	Dolomite		20	100-1,000	2,000-100,000
MINOR (TRACE) ELEMENTS		(Seaweed concentrate generally, sea algae)				
Fe	Iron	(All of these are needed in minute amounts, and either	Likely to be deficient (with Mg) at high pH.	0.5	10-200	2,000-100,000
Zn	Zinc	made available by adjusting pH or by adding oxides, sul-	Likely to be deficient in dunes.	0.2	2-200	100-10,000
Cu	Copper	phates, or sodium salts of the element itself. Most are	Likely to be deficient in coastal plain pastures.	0.1	1-20	2-200
Se	Selenium	present im seaweeds, or	Frequently deficient.	0.01	0.002-1.0	0.5-5.0
Bo	Boron	seaweed concentrate.)	Frequently deficient.	0.2	1-5	4-100
Mb	Molybdenum	Deficient in deep, volcanics.		0.01	0.002-1.0	0.5-10
Co	Cobalt			0.001		

MULCH OR COMPOST OF MIXED MATERIALS SUPPLY ALL NUTRIENTS, plus if available, 200g dry pulverized poultry manure per square metre. Gross deficiencies are likely to occur in the primary nutrients; calcium and phosphates are usually low in old or leached or overworked soils. Rock dusts add potash, limestone adds calcium, dolomite for magnesium, and tree or clover legumes for nitrogen. Manures add most of these except calcium.
N:P:K is ideally added in a 5:8:4 ratio, for sandy loams with 5 parts of lime to balance acidity. In a 5:6:6 ratio for clays.
Fertilizer can be added as manure teas sprayed on leaves, with seaweed concentrate, and in humid weather this is absorbed in from 24-56 hours.

and soil constituents are:

- Metals: Sodium, potassium, magnesium, calcium, and minerals (iron, zinc, aluminium, copper).
- Non–metals: Silicon, sulphur, traces of phosphorus, boron, fluorine, chlorides. Carbon is also found in organic soils.

In solution, metals release positive (H) ions. Nonmetals release negative (OH) ions. Chalk, limestone, calcite, dolomite, magnesite, and gypsum are rocks and minerals giving rise to hard water (from air and water which gives carbonic acid). All of these are carbonates, sulphates, or oxides of calcium or magnesium, or both, e.g. dolomite.

All these below are used to raise pH values in soils:

- CHALK and LIMESTONE are calcium carbonates.
- GYPSUM – calcium sulphate.
- MAGNESITE – magnesium carbonate
- DOLOMITE – calcium magnesium carbonate

In the table of elements (Centrefold), groups I and II

TABLE 8.4
SOIL AMELIORATION: A BASIS FOR TRIAL PLOTS

SOIL	LOCATION CLIMATE	pH 11<--------------------------------------- 7 ALKALINE	---------------------------------------> 4 ACID
SANDS	General	Add magnesium sulphate as spray at 10 g/L, or at 45 g/m^2.	Add Dolomite (for Ca, Mg) at 100-300 g/m^2 (Neutralise with lime $20g/m^2$)
	Humid	Add ammonium sulphate or fine dolomite (for Ca). Add zinc every 7 years at 1 g/m^2. Add water-retaining polmers.	Add water-retaining polymers, add gypsum, sulphur. Use copper-based sprays sparingly. Add neutralised superphosphate (for P).
	Non-wetting	Add 10 g/m2 bentonite or monmorillinite clays, polymers,	Add bentonite, polymers.
	Deserts	Add ferrous sulphate (for Fe) at 23 g/L as a foliar spray. Add polymers.	Add dolomite, gypsum, polymers.
	Coarse	Trial Cobalt sulphate at 0.11 g/m2 or foliar sprays of cobalt.	Add potash . Trial neutralised cobalt.
	Wet areas	Use rock phosphate, dolomite.	
	Coastal	Add copper sulphate (for Cu) every 7-14 years, if cropped, at 1 g/m2. Add zinc sulphate (for Zn) every 7-14 years at 1 g/m2. every 7-14 years.	Add neutralised copper sulphate (for Cu) every 7-14 years, if cropped, at 1 g/m2. Do not add sulphur unless absent. Add Zinc sulphate 2:1 with lime at 1 g/m2
LOAMS	Tropics	Add Silicate or spray silica solutions. (Calcium silicate) at 100-400 g/m^2.	Coarse textures in tropics. Add cement.
	Deserts	Add dolomite, gypsum.	
	Loams	Add potash if cropped. Add zinc sulphate	Addpotash if cropped.
	Coasts	Add copper sulphate 1 g/m2 every 7-14 years.	Add neutralised copper sulphate every 7-14 years at 1 g/m^2.
CLAYS		Acidify with sulphur , add manganese sulphate, at 2-3-g/L as spray. Molybdenum needs trials at 350 g/ha.	Add gypsum to improve drainage, reduce salinity in dry areas.
GRAVELS	Iron oxides present	Add copper sulphate at 1 g/m2. Immobilize phosphates.	Molybdenum at 350 g/ha every 10-15 years.
ALL SOILS		Blood and bone, manures, compost, acidic phosphate, urine for potash. Foliar sprays of seaweed concentrate.	Blood and bone, manures, compost, dolomite, seaweed, pebble dust from cement works.

contain the non–metals of which lithium, sodium, potassium, magnesium, calcium, strontium, and barium compounds give up alkaline (OH) ions to soils.

The usual soluble (solid) bases for alkalis are magnesium, calcium, potassium, barium, sodium, and selenium. The usual soluble (liquid or gaseous) compounds for acids are silica and sulphur based, or derived from humus.

The measurement of acidity–alkalinity (pH or hydrogen ion concentration) is basic to soil and water science, as it affects the availability (solubility) of other key or trace nutrients, and (at its extremes) the ability of life forms to obtain nutrition, or even to live.

The pH scale ranges from 0 (acid) to 14 (alkaline), although in nature we rarely find readings below 1.9 (lime juice) or above 11.0 (alkali flats). In the presence of air, and in ploughed and aerated soils, both metals and non–metals form oxides, and these dissolve in water or soil water.

The METALLIC OXIDES (bases) form alkaline solutions. The litmus reaction is *blue*. There is an excess of hydroxide ions (OH^-) present in solution.

The NON–METALLIC OXIDES form acidic solutions. The litmus reaction is *red*. Excess hydrogen ions (H^+) are present in solution.

Common acidic substances are citrus juices and battery acid. Common alkaline substances are sodium bicarbonate (baking soda) and washing soda.

As rain falls, carbon dioxide in the air combines with water to form weak carbonic acid (as in soft drinks); this helps to dissolve metallic oxides in soils, and to bring the minerals of rocks and soils into solution. Sulphur from industrial processing, or from pyrites in rocks or soil can form sulphuric acids, and these also aid the solubility of metallic oxides. Phosphorus and nitrogen can form phosphoric and nitric acids with water. Silica in soils dissolves to silicic acids, and chlorine to hydrochloric acid. Weak nitric acid is a plant growth stimulant (*New Scientist* 22 May 86). All of these acids, in *moderation*, are helpful in nutrient supply to plants.

pH is not a constant for soil or water. Not only does it exhibit diurnal or seasonal changes due to rain, growth, and temperature changes, but it is essentially a *mosaic* in soil crumb structure, on the surface of colloids, and at microsites. Further, pH exhibits vertical soil gradients, being more acid in surface mulches and more basic or alkaline where evaporation, wormcasts, and capillary action draw up bases to the surface of the soils (dry or wet–dry areas). Mosaics on a larger scale are imposed by slope, and both rock and vegetative types. As long as people are aware of this, and realise that root hairs can both create and seek out ideal pH environments if there are no gross imbalances, then gardens are likely to contain every sort of pH level *somewhere* in the soil.

Only by grinding mixed samples of soils to damp pastes, or measuring conductivity, can we "average" pH and obtain some idea of net balance, but (as usual) our measuring methods alter the thing we measure. If

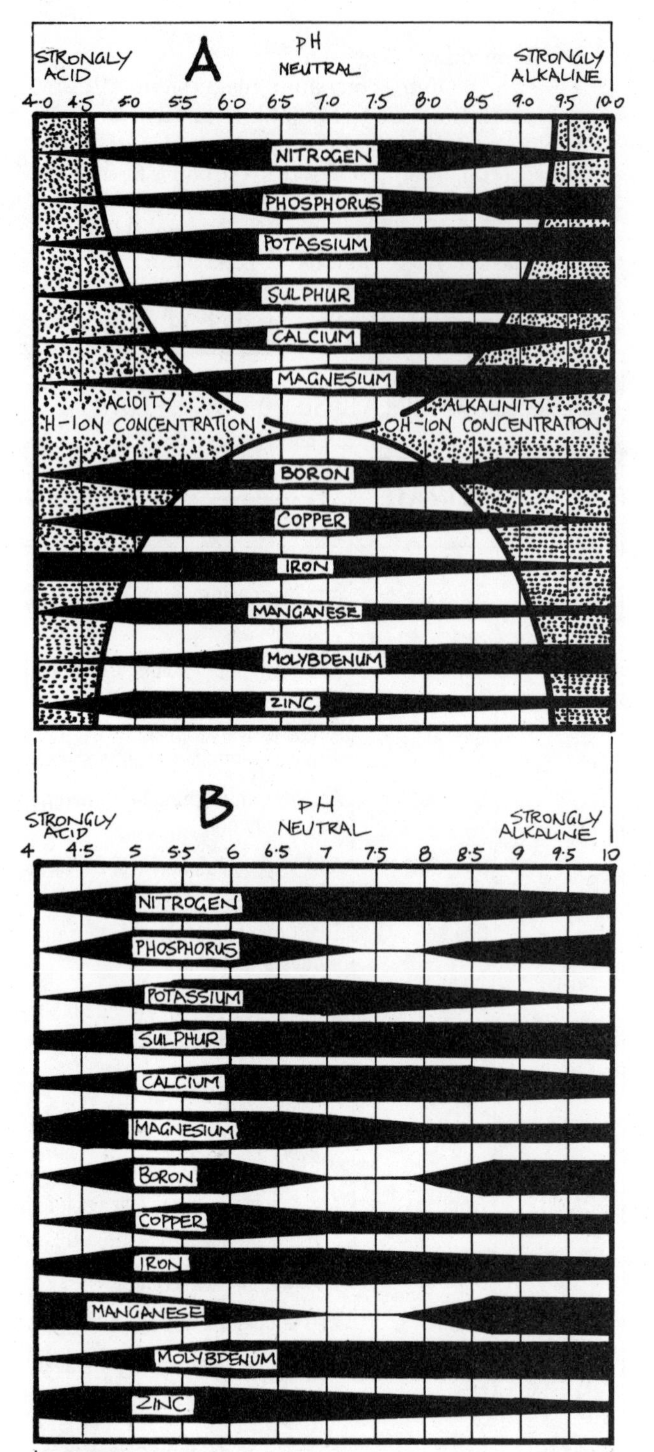

FIGURE 8.4

INFLUENCE OF pH ON AVAILABILITY OF PLANT NUTRIENTS

A In inorganic (mineral) soils. The widest parts of the black areas indicate maximum availability of each element. The curves represent pH values.
[After Nelson, L. B., (Ed.), *Changing patterns of fertiliser use,* Soil Science America, Madison, WI (1968)].

B In organic soils. The widest parts of the black areas indicate maximum availability of each element.
[After Lucas, R. E., amd J. F. Davis, "Relationships between pH values of organic sols and availability of 12 plant nutrients", *Soil Science,* 92:17-182 (1961)]

we keep the average pH value to between 4.5 and 10, we can grow a wide range of plants and rear aquatic organisms *if* calcium is present. If we narrow the range to pH 6.0–7.5, most vegetables grow well in gardens. Outside these ranges (less than pH 4.0 or more than pH10.0), only specialised bacteria or higher organisms can cope. Moreover, soil humus is itself a buffer as is calcium (lime); humus will grow plants at satisfactory levels even if the pH changes, so that limed and mulched gardens rarely show plant deficiency symptoms.

8.9
SOIL COMPOSITION

We think of soils as mineral compounds, but the pie diagram in **Table 8.4** will alter that impression; it does not, of course, represent any one soil, but is an average sort of figure for a good loam with adequate humus. Outside the pie, I have noted some of the possible ranges of variation from peats to sands in old dunes.

THE HUMUS CONTENT OF SOILS

Soils in nature can vary from a humus content of 2% to close to 100% (as peats). In gardens, 40% or more humus helps block heavy metal uptake by plants, holding heavy metals bound in colloids. Many compost–fed or mulched garden soils contain 10–30% humus (some much more). The effect of adequate soil humus is both physical, in effecting good water retention and in preventing erosion, and chemical via colloid formation. The breakdown products of humus, including the mineral content of the donor plants, form the readily–available and biologically active components of soil that are of value to newly–established plants. If these plants are perennial or (in part) mulch–producing, and if we return food wastes to the garden (including urine and well–composted faeces) then very little humus loss occurs.

On the broad scale, humus can only be provided by the root and the above–ground mass of grasses, trees, and plants. Prairie grasses and broadleaf trees are particularly effective at this job. When we aerate (plough) soils we turn up humus and oxidise it to carbon dioxide, thence to atmosphere. It is lost. Burning the vegetation is worse, with a host of additional pollutants (terpenes, creosotes, nitrogen, and dust particles), and a rapid loss of soil humus. Where soils are not tilled or burned, soil humus lasts a long time (hundreds or even thousands of years) and provides for a complex soil life.

ORGANIC SOIL ADDITIVES

Mulching is here defined as covering the soil surface with 15 cm or more of organic material, as a loose (uncompacted) mulch; 8 cm of tight–rolled sawdust does not qualify! Mulching is more generally applied to loose dust "mulches", plastic sheet mulches, and so on, and these may have specific local value in soil amendment, heating, sterilisation, weed suppression, or pest reduction, but as here considered, the *object* of mulching is to add plant nutrients, buffer soil temperatures, prevent erosion, promote soil life, and restore soil structure.

Plastic mulches, soil gels (polyacrylamides), herbicide–treated soils, and organic or natural mulches may all achieve the result of preventing erosion and helping soil crumb structure develop. Only long periods of natural mulches stabilise nutrient supply, and complex the soil life. None can be judged over one or two seasons, as it can take 3–5 years to create a balanced soil under mulch from a compacted or mined–out soil. Even longer periods are necessary to develop humus in permanent crops assessed for yield on the broad scale, where added mulch is not carried to the site, but derived from tree wastes and specially–sown crop (green manures) produced on the site itself.

Used in areas such as wet tropics and arid lands, or on dry coarse sands, mulches may prove to be ephemeral (even if their effects continue), as ants, termites, and leaching reduce the mulch to humic acids or underground storages in fungi and bacteria. In particular, water absorption is improved under mulch, both as field crop mulch and imported garden mulch, thus water needs are reduced. Jeanette Conacher, in Western Australia's *Organic Gardening*, 1979, reporting on extension trials in Nigeria, records 11% better water infiltration on low– to no–tillage and mulched plots. Under mulch, excessive soil temperature ranges are buffered, being cooler by day and warmer at night or in winter. Seed germination is enhanced, and *over the long term*, major nutrients (N, P, K) remain at satisfactory levels. Only under mulch does the population of important soil organisms, such as earthworms, increase.

Mulches need some selection for minimal weed seed, minimal residual biocides, and for best effect on specific crops (tested as row–by–row comparisons).

Plastic mulches (black for heat and weed control, silver for aphid repellancy) have a more limited role, important in the short–term, but often expensive or impractical in poor countries, or rejected by growers who suspect that many plastics release persistent chemical polymers of unknown effect on the soil life.

Mulch is an excellent way to add nutrients to soils; the "cool" decay loses little nitrogen, while stimulating soil life generally. There are some problems with compost, where "hot" (aerobic) heaps heat up and nitrogen (ammonia) losses are severe. Cold heaps (pitted or silage) do not lose nitrogen, but neither do they kill weed seeds. One percent of superphosphate added to a hot compost heap prevents ammonia escape. Chinese scientists get the best of both worlds by first building an aerated heap with bamboo poles as holes to create air tunnels. This is then covered with mud and the heap heats up to 55–60°C (130–140°F) for a few days. Then all holes are sealed, and the rest of the decay is anaerobic. With sealed boxes, either hot or cold processes can take place.

Compost or mulch is critical to preserving soil crumb structure, buffering pH, and (in taste tests) improving sugar content and the flavour of vegetable product. The gums and gels produced by soil organisms create crumb structure, aerate the soil, and darken it so that it heats up faster in spring. The humic acids assist root development dramatically even at levels of 60 ppm carbon. Inorganic (chemical) and mechanical farming can as easily destroy soil structure.

In the U. K. (*New Scientist* 3 Nov. '79), liquid manures sprinkled on straw in silos or tanks, together with a forced air draught, produce compost in about a week (efficient open piles encased in straw need 10 days). The liquid effluent system plus straw is suited to treatment of a manurial sludge (still full of seeds) such as that we get from biogas digesters. It is best to use

TABLE 8.6
PIECHART OF TOTAL VOLUME COMPONENTS OF AN "AVERAGE" SOIL.

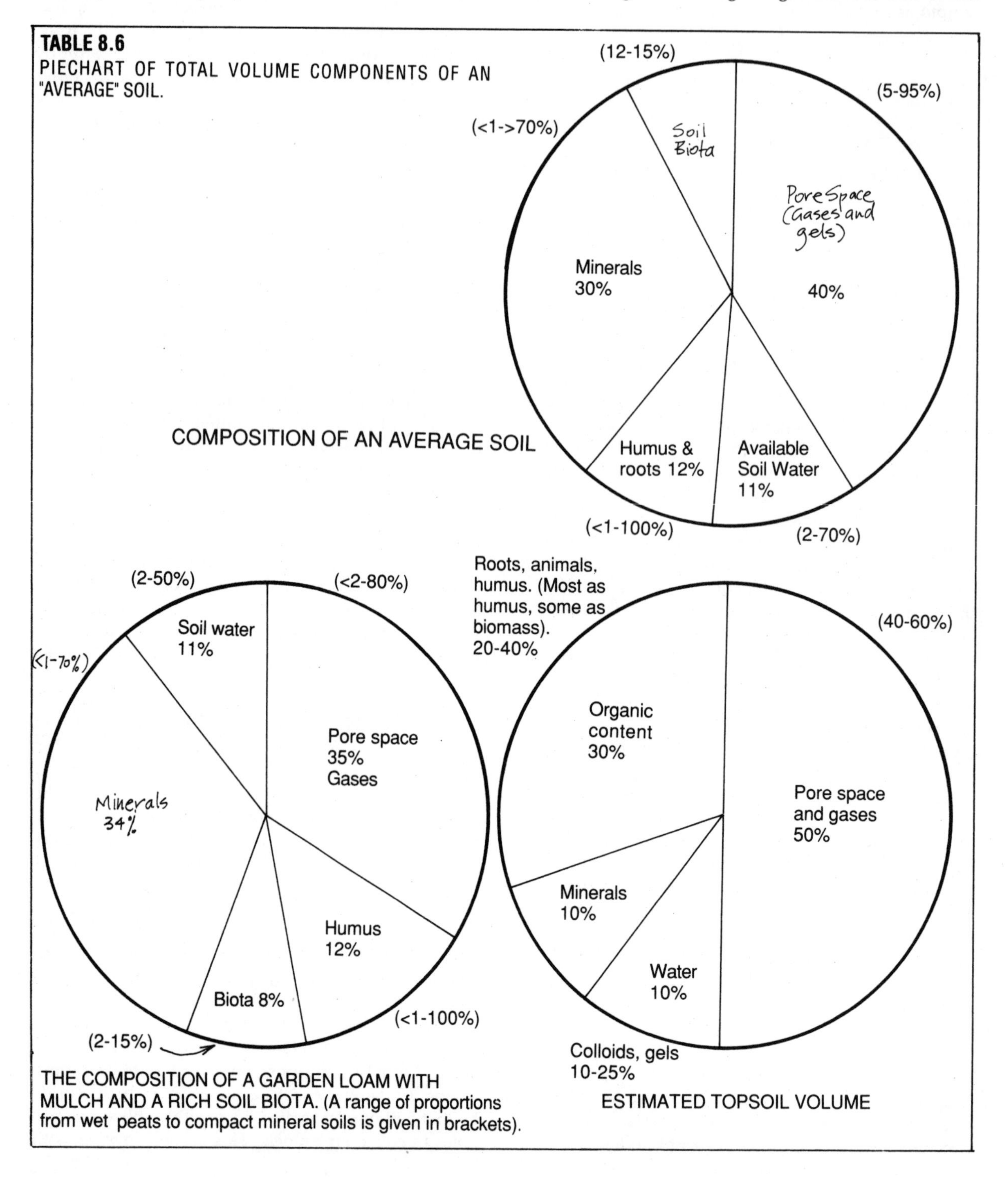

COMPOSITION OF AN AVERAGE SOIL

THE COMPOSITION OF A GARDEN LOAM WITH MULCH AND A RICH SOIL BIOTA. (A range of proportions from wet peats to compact mineral soils is given in brackets).

ESTIMATED TOPSOIL VOLUME

this with dry twiggy or straw material as hot compost to both kill weed seeds and to produce useful heat, after the compost pile has been made. However, in severe winter areas, such "efficiency" is counter–productive as a slower heat release from large compost piles of 10–50 cubic metres (12–59 cubic yards) can provide heat over a long period in winter and greatly reduce glasshouse and house heating costs, while the compost itself is best applied to soil in spring.

CARE IN SELECTION OF COMPOST MATERIALS

Even using organic residues, or "natural" wastes, soil problems can arise from a concentration of nitrates in manures, or toxic mineral residues. Leaf material cut and mulched (or composted) green will contain more nutrients than fallen leaves, although the latter are still useful for humus production. Kevin Handreck (*Organic Growing*, Autumn '87, Australia) has identified potential mineral contamination from these sources:

- Galvanized, copper, or brass containers containing wet residues of manures, or manurial teas; clay, stainless steel, glass, or iron are preferable. Zinc and copper are produced in excess from galvanized or copper/brass containers.
- Manures from pigs, poultry, cattle, or sewage sludges can add excessive copper, zinc, nickel, boron, lead, or cadmium to soils (and especially acid sandy soils).

Where such elements are naturally deficient, such manures may initially help, but heavy or constant application will build up a toxic soil condition. Excess zinc can be built up by using earthworm casts from contaminated pig, sheep, or domestic wastes. Animals penned in galvanized areas can produce excess zinc in pen wastes.

It is in poultry mixes that such contaminants show up, with excess zinc inhibiting plant growth and health, so that the urban gardener should test and use safe mixes, or proceed via plant trials. Handreck recommends a limit of 10% worm casts in a potting mix. In such cases, nitrogen and potash can be supplied by dilute urine.

Note that copper, arsenic, and other minerals are commonly added to stock feeds, and therefore try to buy clean natural feeds from organic sources. After initial soil treatment, and a continuing watch for foliage deficiency symptoms (see Deficiency Key, this chapter), the safest course is to grow our own green manures as the foliage of leguminous hedgerow, windbreak, or intercrop, and to use these as mulch and compost materials.

8.10

SOIL PORES AND CRUMB STRUCTURE

The structure of soil (whether compact or open) depends on the soil composition itself, the way we use it, and the presence or absence of key flocculating or ionic substances (synthetic or natural). Crumb structure in well–structured soils permit good gaseous exchange and free root water penetration without the creation of excessive anaerobic conditions by waterlogging. In free sands, and in the kraznozems developed over deeply–weathered basalts, crumb structure is either not a factor, or else is so well developed that it permits leaching (to immobile clay sites) of almost all applied fertilisers.

However, in most other soils, we would like to see a good crumb structure develop as in our gardens or crop soils. Where crumb structure is poor, we can use a great variety of coulter and rip–tine machines, soil additives, and deep–rooting plants such as trees or lucerne to re–open and keep open the soil structure, which in turn allows adequate water penetration and drainage, and eventually develops the oxygen–ethylene processes that make bound nutrients available.

Soil crumbs of 0.2–2 mm diameter can form as little as 10% of the total soil volume, and still produce crop; below this, crop is greatly reduced and impoverished. The same soils can, when not ploughed lifeless, contain 92–95% of such crumbs (Leeper, 1982). We destroy crumb structure by destroying permanent vegetation, flooding soils for long periods, using high–speed or heavy vehicle cultivation, stocking with sheep or cattle in especially wet periods, using fertilisers that deflocculate the soils (e.g. too much potassium where soil sodium levels are already high), or by burning in hot periods.

The whole set of disasters outlined above can collapse soils to the cemented, dusty, hydrophobic, salted and desertified areas typical of wheatlands on desert borders. It is a question of improper use, disastrous planning (or no planning), and a total lack of applied goodwill to earth.

Given a good structure, pores develop for the diffusion of gases and exchange of ions, provided that we make the transition to perennial, low cultivation systems of forest and crop.

COLLOIDS AND GELS

Colloids are stable aqueous gels or suspensions of clay, organic, or long–chain polymer particles in a finely–diffused aqueous state in soils. These are particles so fine that they stay in suspension unaffected by gravity, and become active sites for ionic bonding and interchange. Colloids also form gels which hold soil water reserves, and are in part formed by or derived from natural (and more recently, artificial) substances some of which form hydroscopic gels by water

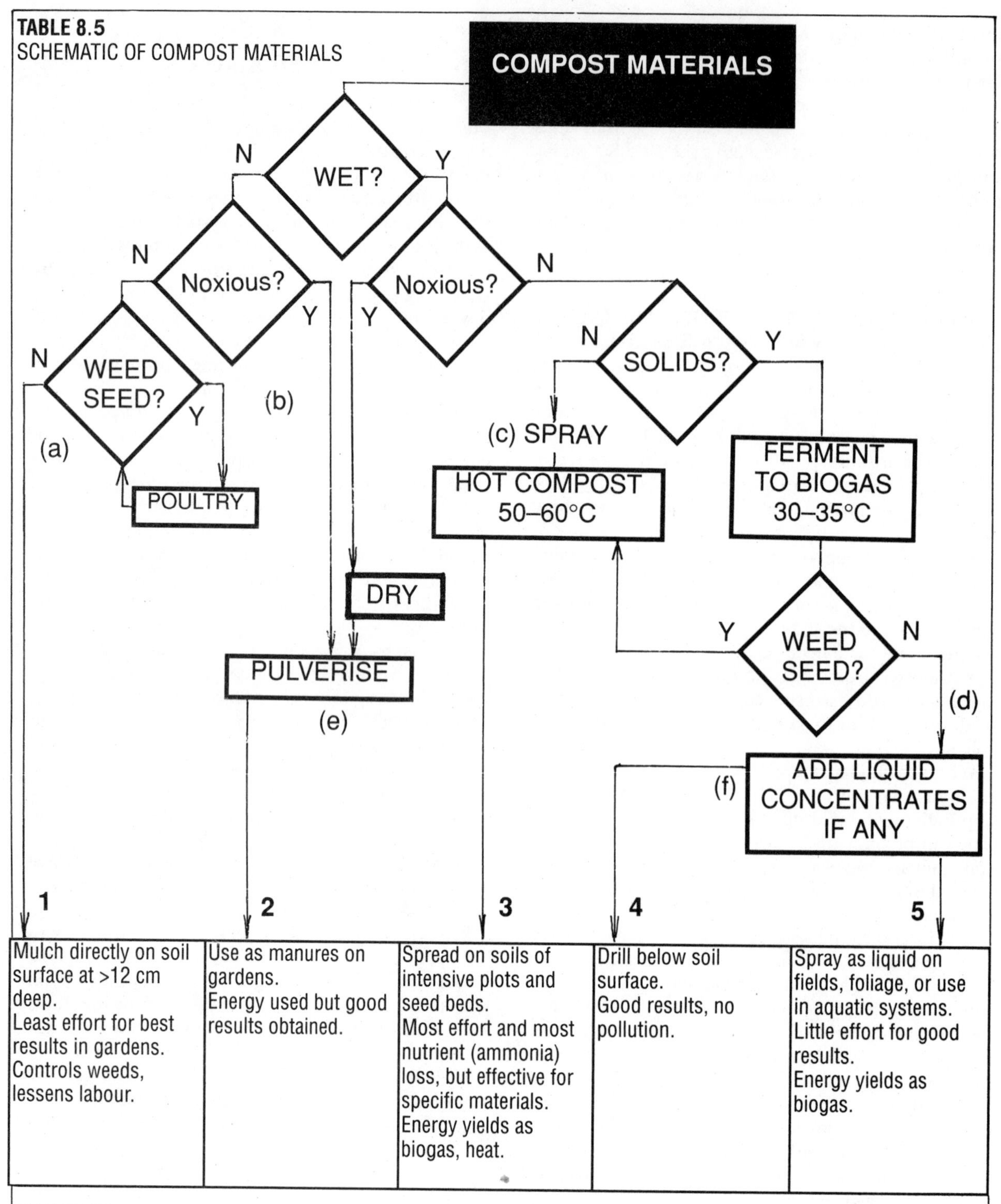

TABLE 8.5
SCHEMATIC OF COMPOST MATERIALS

EXAMPLES:
(a) Nut husks and shells; coffee, teas, and cocoa residues; shredded paper and branches; bark, woodchips, and sawdust; and old carpets, underfelt (*not* pesticide treated ones), bags, canvas(*all made of natural materials*).
(b) Hay with seed heads, weeds in flower, bulbils or roots of weeds.
(c) Sewage and sullage, liquid manure and urine, meat and animal paunches and trimmings, general household wastes. Add lime and superphosphate (1%) to hot compost.; "teas" of seaweed and manure.
(d) Sludge from digesters and weed–free manures.
(e) Chicken and bird manures, litter from animal sheds, blood, bone, feathers, hide scraps, seaweeds.
(f) Dissolved minerals, urine, seaweed and manure "teas".

TABLE 8.6
RECOMMENDED SCALE OF SOIL PARTICLE SIZES (mm).
[After McDonald *et al*]

Fine clay	< 0.002
Clay particles	0.002
Silt	0.002 - 0.02
SAND	**0.02 - 2.0**
fine	0.02 - 0.2
coarse	0.2 - 2
GRAVEL	**2 - 60**
Fine	2 - 6
Medium	6 - 20
Coarse	20 - 60
COBBLES	60 - 200
STONES	200 - 600
BOULDERS	>600

absorption.

Natural
Humus or humic decay and bacterial products provide colloids, as do finely ground graphite and silica, fine clays (particles of 2–200 millimicrons) such as illites, bentonite (usually deposited on clay pans from evaporation of clays in suspension), grain flours and animal (gelatinous) flours, e.g. from hooves, horns, sinew wastes, albumen and so on. Bacteria also secrete polysaccharides which form soil gels.

Synthetic
Soil conditioners which form stable colloids are sold under a variety of trade names (ask a local agricultural supplier), e.g. Agrosoke®, Ikedagel®, Terrasorb®, which are *hydrophilic* (attract water) as are many natural gels from grain products and seaweed. Water conservation of up to 50% is claimed for some of these additives. Most artificial gels are acrylic–based (acrylamid polymers) and supplied as granules able to absorb hundreds of times their weight in water. They are used to great effect in nursery plants, row crops, new tree plantings in deserts, and in transplanting. The acrylics are applied to soils at 6 kg/ha or more, and conserve irrigation water by preventing evaporation (*Small Farmer*, New Zealand, Aug. '84). These substances are of value in seed pelleting.

Colloids also form *hydrophobic* (repel water) substances such as those of clays and some soil fungi products. The surfaces of colloid particles are usually negatively charged, as are root hairs, and thus attract positive ions to their surfaces, e.g. the positive ions of sulphates, nitrates, and of metals (sodium, calcium, magnesium, iron). About 99% of such ions are so held in soils (Leeper, 1982). Colloids from humus are hundreds of times more effective than clay colloids in ion exchange in soils.

Ammonia and sodium can dislodge these ions, and they can become available to plant roots, which attract them by producing negative (H) charges. With too much flushing by sodium, the calcium and metallic ions can be lost to leaching processes and carried to streams, hence to seas or lakes.

Colloids used in water softening capture calcium ions and release sodium ions, allowing soaps to lather. Ferric oxide colloids have *positive* charges and give water a brown stain, typical of waters issuing from acid peats and pine forests. The colloid particles can be flocculated (aggregated) and settle out if aluminium salts (sulphates) are added; and this may happen as a result of acid rain, or can be induced with the acid forms of any negative particles. Salt also flocculates colloids up to the point where excess sodium *deflocculates* clays, with calcium and other ions, which is the effect of high salt concentrations in dryland soils (over 1–5 ppm salt).

Burning destroys the colloidal properties of surface clays, and is another reason why desert soils leach out after fires. Burnt clay particles *no longer form colloids*, and become poor in mineral nutrients, which are then found only in deep soil profiles. Organic humus forms black slimy pools on top of the collapsed soils.

It is the clay particles and colloids (organic and inorganic colloids) that bind water and nutrient in soils. In the tropics at least, most of these colloids are in the biomass of the soil as cellular gels, or are produced as sheathing material by soil bacteria. Without colloids, soil minerals rapidly leach out and become poor in nutrient. Fire, clearing, ploughing, and cultivation destroy such colloids and soil structure, as does excess sodium ions. Thus, to hold and exchange nutrients, we as gardeners need to develop natural or artificial colloid content in soils, from where plant roots and soil fauna or flora derive their water and essential nutrients. Very little clay is needed in sandy gardens to create a colloidal soil environment, and life forms are encouraged and developed by humus, mulch, and perennial plant crops. Good soil structure (to hold the colloids) is developed by careful earth husbandry, together with flocculating additives such as gypsum and humus.

SOIL WATER

The water content of soils is a soup of free–living organisms, dissolved gases and salts, minerals, gels, and the wash–off from throughfall in trees (waxes, frass, tree "body wastes"). Organic and inorganic particles are held in soil water. Soils have a widely variable water–holding capacity dependent on their composition and structure, so that sands absorb and retain water more quickly than clays, but clays hold more water per unit volume. Available water thus varies from 2% (surface sands) to 40% or more of soil volume. We can assist the quantity held in dry sites by swaling, contouring or terracing, loosening soils, adding flocculants, introducing artificial or natural gels (seaweeds or plastic absorbers), or by placing a clay or plastic sheet layer 30 cm below the surface of gardens.

In soils that become waterlogged, we can resort to raised beds or deep drains to reduce infiltrated water build–up, or plant trees to keep active transpiration going and so reduce soil water tables and re–humidify the air.

Two factors at least affect soil water availability:

- The strength of molecular bonding of water to particles in the soil (ionic bonds). Plant root hairs cannot remove this bound–water at pressures above 15 atmospheres.
- The salt content of the water. Too many salts, and the soil water exerts reverse osmotic pressures on plant roots, and will take water from the roots.

8.11
GASEOUS CONTENT AND PROCESSES IN SOIL

Soils are permeated, where not waterlogged, by the gaseous components of the atmosphere (80% nitrogren, 18% oxygen). This enters the soil via pore spaces, cracks, and animal burrows, and diffuses via pore spaces to plant roots. The exchange of gases, atmosphere to soil and soil to atmosphere—the breathing of earth—is achieved by a set of physical and biological processes, some of which are:

- EARTH TIDES. The moon tides, much subdued on continental masses, nevertheless affect groundwaters in cobbles or boulders, the earth itself (about 25 cm rise and fall across the continental USA), and of course the soils of estuaries and mud flats (where air can be seen bubbling up for hours as the tide rises).

Everywhere, low or high pressure cells and turbulent wind flow creates air pressure differentials that draw out or inject air via crevices and fissures, or burrows. These same effects assist or retard the diffusion of water vapour across soil surfaces, and just as the wind dries out surface soils so it also, by fast flow, draws out other gases from soil. The gases of soil respiration (carbon dioxide), oxidation, aerobic and anaerobic metabolic processes, radon from radioactive decay, and simple or complex hydrocarbons from earth deposits or humus decay pass to air.

Much of the ammonia and carbon dioxide in the atmosphere, and at least 16% of the methane, is supplied by soil processes.

- BIOLOGICAL EXCHANGE. A single large broadleaf tree, actively transpiring, may increase the area of transpiration of one acre of soil by a factor of forty; a forest may do so by hundreds of times. So oxygen, carbon dioxide, water vapour, metallic vapours, ammonia, and hydrogen or chlorine gases are transpired by algae, rushes, crops, trees, and herbs or grasses. Plant groups vary tremendously in the volume and composition of gases transpired. Although we owe much of our atmospheric oxygen to trees, a great many non–woody plant species consume more oxygen than they produce.

Thus, the plant is a gaseous translator, trading both ways with air and soil. Some specifics of this trade are given in the section below on oxygen–ethylene processes.

Animals, too, are very active in opening up soils with small or large burrows; these act as pump pistons (like a train in a tunnel) to draw in and exhaust both their waste gases and atmospheric gaseous elements, which are then diffused to roots via soil pores. Many burrowers (ants, crabs, termites, prairie dogs, worms, land crayfish) raise up mounds or chimneys which then act as Pitot tubes for air flow or to create pressure differentials which draw air actively through their burrows. Or they erect large surface structures of permeable sediments across which waste gases diffuse (e.g. termite mounds). By a great many such devices, animals contrive to live in aerated or air–conditioned undergrounds, and increase gaseous exchange in the soils.

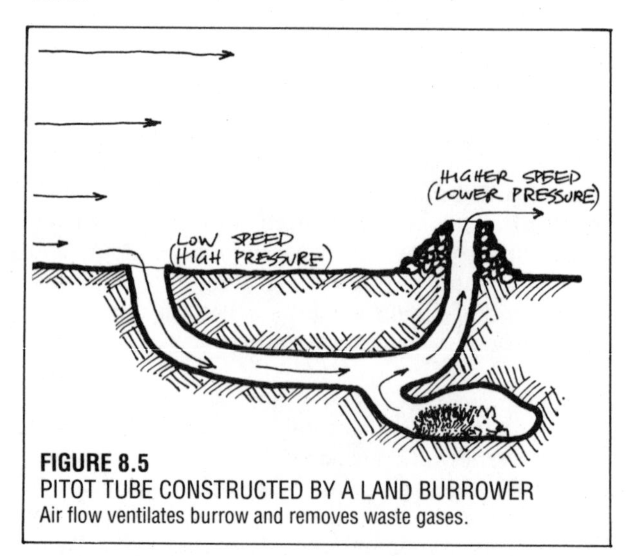

FIGURE 8.5
PITOT TUBE CONSTRUCTED BY A LAND BURROWER
Air flow ventilates burrow and removes waste gases.

GASES MANUFACTURED IN SOILS

By way of titanium or rutile (TiFe), ammonia is manufactured in sands in the presence of sunlight. By way of ferrous iron, ethylene (C_4H_4) is manufactured in anaerobic soil microsites. In anaerobic soils and waters, carbon dioxide, methane, sulphuretted hydrogen, ethylene, and sulphur dioxide are formed, and escape to air as biogas or marsh gas. The same products are present in the mottled soils of hydrophobic clays in winter, soils where crumb structure has been destroyed by misuse, where salt has deflocculated clays and caused soil collapse, or where water periodically floods the soil. Many of these gases are found as a result of humus decay and thiobacillus (sulphur bacteria) action.

Ammonia is released from actively nodulating legumes (trees and herbs), and is used in unploughed soils as a plant nutrient. Thus, gaseous compounds are continually made in the soil itself by process of metabolic growth and decay in the presence of metallic catalysts and micro–organisms. Moybdenum, vanadium, and zinc all assist root bacteria in the creation of available soil nitrogen (as catalysts).

GASEOUS MOSAICS OVER TIME;
SOIL MICRO–SITES

Like pH, cation exchange, and structure, natural soil is always a mosaic of aerobic and anaerobic patches called micro–sites, where either oxygen (aerobic) or ethylene (anaerobic) sites develop. Ethylene inhibits (in the sense of suspending), microbial activity, and like carbon dioxide, is present as 1–2 ppm in soils (Smith 1981). As the ethylene at an oxygen–exhausted site diffuses out, oxygen floods back and re–activates the site. Under natural forests and grasslands, this cycle, or dance, of oxygen–ethylene is continuous, and most nitrogen there occurs as ammonia, useful to plants and plant roots.

When we cultivate and aerate, nitrogen becomes a nitrate or a nitrite (which then inhibit ethylene production), and *ferric* rather than *ferrous* iron forms, thus making ethylene formation difficult (the process from decaying leaf to ethylene production requires a *ferrous iron* catalyst). Also, plant nutrients are tightly bound to ferric iron and become unavailable for root uptake. It follows that the production of ethylene is essential to plant health and the availability of nutrients.

It is important to realise that the aerobic condition of soils such as we get from ploughing or digging not only creates a condition of "unavoidable" nutrients and ferric iron, but also oxidises humus, which goes to air as carbon dioxide. Also, most plant root pathogens require the aerobic condition. As well, the nitrate form of nitrogen, which is highly mobile, leaches out when bare soils (not plants) occupy the site. In all, ploughing and earth turning create a net loss of nutrients in several ways, thus atmospheric pollution, stream pollution, and low soil nutrient states.

Smith (*ibid.*) therefore recommends *least* soil disturbance, the use of surface mulch (*not* incorporated) as an ethylene precursor (old leaves are best for this), and very small but frequent ammonia fertiliser, until soil balances are recovered. The ideal conditions would be:

- Permanent pasture;
- Forests;
- Orchards with permanent green crop as mulch;
- No–dig or mulched gardens;
- No– or low–cultivation of field crop, or field crop between strips of forest to provide leaves and nutrients; and
- The use of legumes in a similar proportion to that occurring in natural plant associations in the area, at all stages of the succession.

Under these conditions, soil mineral availability is made possible and soils do not lock up nutrients in oxides or produce pollutants.

When we achieve this balance, soil loss and mineral deficiency become yesterday's problems. And when *plant leaf* (not soil) deficiency can be adjusted with aqueous foliar sprays, nature then starts to function again to obtain nutrients from soils via microbes and root mycorrhiza at the microsite level. (Smith, A., 1981, "The Living Soil", *Permaculture Journal #7*, July '81.)

8.12

THE SOIL BIOTA

On semi–arid and poor pasture, it is difficult to keep sheep at a stocking rate of 3–6/ha. The very same pasture may support 2–5 t of pasture grubs, or up to 6.5 t of earthworms/ha , so that (like grasses) most of the animal biomass or yield is underground, out of sight. Even where wheat cropping is carried on *continuously* for 140 cycles (Rothhamsted, UK) the plough layer supports 0.5 t of living microbial biomass (*New Scientist*, 2 Dec '82). About 1.2 t/ha of organic carbon is returned annually to the soil as root and stalk material from grain crop.

So large is the soil biomass that its growth must be very slow, sporadic, and based on a turnover of humus/food *within* the soil rather than a food input from the wheat crop wastes. Humus in this soil has a mean age of 1,400 years, and probably derives from forests that long preceded the wheat; it yields up its nutrients very slowly, and is resistant to bacterial attack. However, it is equally clear that there are periods of sudden food supply from root masses at harvest, and from root exudates during the growth of wheat (30% of plant energy may be lost as sugars or compounds released to the soil via roots). However the soil biota achieve it, they exist on a very meagre food supply for such a biomass, rather like an elephant eating a cabbage once a day! Of the total biomass at Rothhamsted, 50% is fungi, 20% is bacteria, 20% yeasts, algae, and protozoans, and only 10% the larger fauna such as earthworms, nematodes, arthropods and mollusc fauna (the micro– and macro–fauna), and their larvae. Such classes of organisms are found in soils everywhere, in different proportions. Anderson (*New Scientist*, 6 Oct. '83) gives some idea of the complexity below ground, where every square metre of forest topsoil can contain a thousand species of animals, and 1–2 km of fungal hyphae!

Very small animals are able to live a basically aquatic life in soil, in the water film attached to soil crumbs, while larger species are confined to pore spaces and the burrows of macrofauna.

A wheat field is not the place most likely to produce high levels of soil biota, and plough cropping has (in Canada) reduced humus levels to 1% of the original levels over much of the wheat country.

Climatically, the balance and proportion of soil biota varies greatly, with the acid soils of coniferous and oak forest yielding few earthworms, and the humic peats even less, so that the soil recycled by worms also varies from 2–150 t/ha (0.5–25 cm depth of soil/year).

Ecological disturbance or imbalances by predation, aeration, disturbed soils, low oxygen levels, and compaction favour the bacteria over the fungi. Fungi are certainly more effective in wood or large plant material breakdown, and can transport materials not only from place to place (e.g. move nitrogen into decaying wood) but also move nutrient into higher plants via their intimate contact with the root cells of the host plant. Usually, such translocations are modest (a few metres), but occasionally a fungal species can send out many metres of hyphae to invade a tree, as a pathogen and decomposer. Thus, it pays higher plants to give energy to their fungal root associates as sugars, and to gain minerals or nutrients in return.

Over time, the death of these soil organisms returns nutrient to new cycles. Even termite nests die out in 20–30 years, and new colonies start up. Larger animals can have a profound effect on primary litter breakdown (millipedes, woodlice) and are typically plentiful in mulch, but rare in compost. As one can imagine, any accurate account of the relationship between such dynamic mass of species awaits decades of work, but some broad facts are emerging; for example, turnover of nitrogen by earthworms exceeds that of the litter fall of plants. Few species fall into clear-cut classes of food relationships, and the chain of events of predation, faecal production, and burrowing are further complications.

As it is probably impossible to research at a species–specific level, and as the gross compartmentalisation of ecosystem analysis is inappropriate, Anderson (*ibid.*) suggests a more possible study based

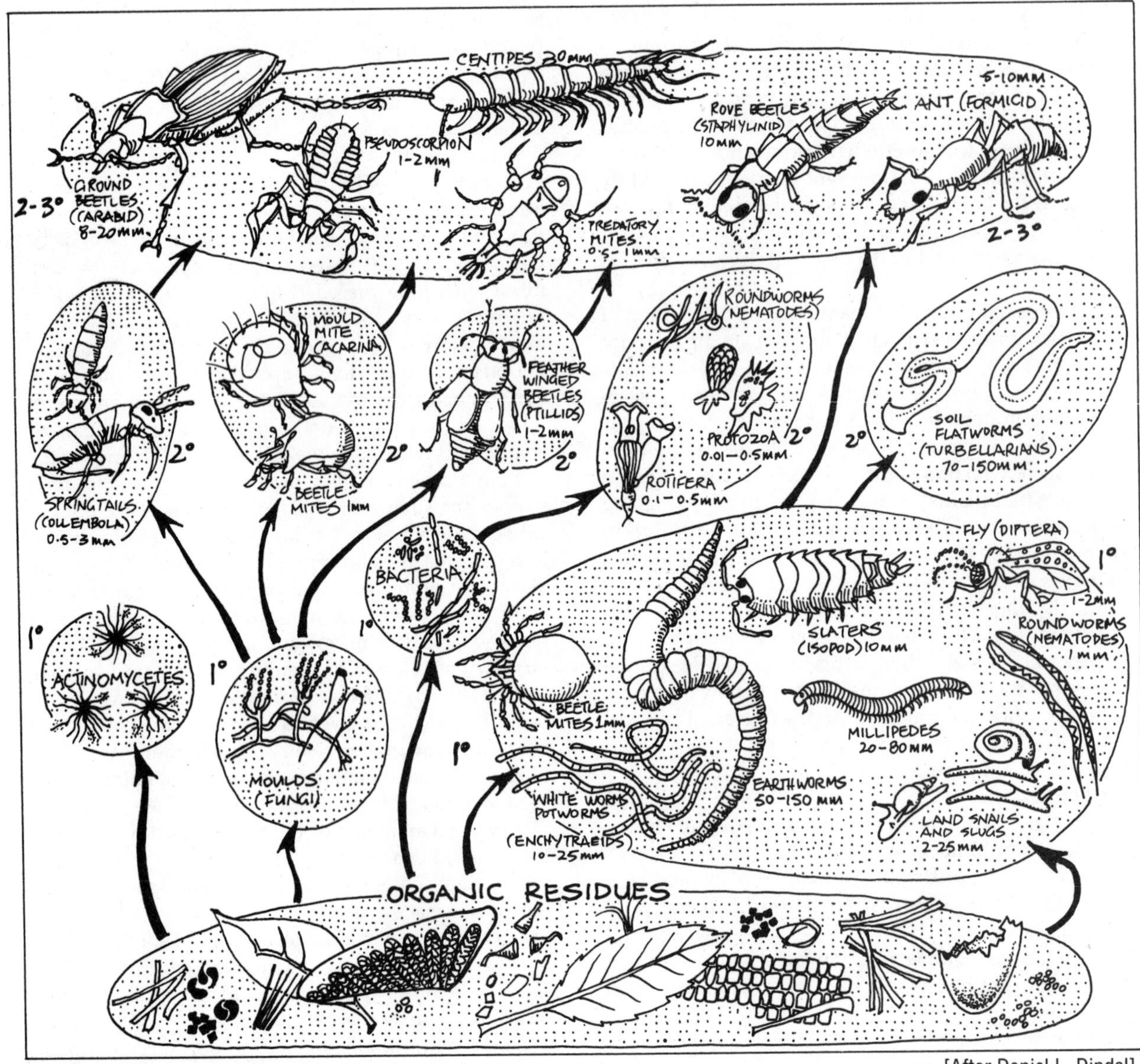

[After Daniel L. Dindal]

FIGURE 8.6
FOOD WEB OF A COMPOST PILE
Energy flows in the direction of the arrows; lengths in millimetres. 1 = first level consumers, 2 = second level, 3 = third level.

on the interactions between *the broad functional groups* of organisms, or a size–food community. Here, he laments, "we know so little about so much."

Large animals (earthworms to wombats) can create major changes in soils locally, by burrowing, soil turnover, faecal production from vegetation, waste products, and even alterations to forest successions after fire. In general, gross disturbances by colonies of larger fauna as in deserts (where rodent biomass can reach 1,000–10,000 kg/ha) shift the balance from soil fungi to bacteria.

We can think of the soil biota as a reserve of otherwise easily leached nutrients (nitrogen, sulphur), both of which elements they gather, store, or concentrate. Their cycle of life and death, which in turn depends on soil temperature and season, releases small or large amounts of these essential elements at multiple microsites. Termites, in addition, may store calcium from subsoils in their mounds, and bacteria store a number of soil minerals. These are held in the mobile living reserves of the soil biota, and are released by their death for slow uptake by plant root associates. Many plant forms directly eat bacteria (algae in water, fungi) or insects and nematodes, so that plants are either direct predators of the soil fauna, or scavengers of the bodies of the soil organisms.

A useful classification of soil biota based on size is as follows (*New Scientist*, 6 Oct '83):

- MICROFLORA and MICROFAUNA: Size range 1–100 millimicrons, e. g. bacteria, fungi, nematodes, protozoa, rotifers.
- MESOFAUNA: Size range 100 millimicrons–2 mm, e.g. mites, springtails, small myriapods, enchytraeid worms, false scorpions, termites.
- MACROFAUNA: Size range 2–20 mm, e. g. wood–lice, harvestman, amphipods, centipedes, millipedes, earthworms, beetles, spiders, slugs, snails, ants, large myriapods.
- MEGAFAUNA: Size range 20 mm upwards, e. g. crickets, moles, rodents, wombats, rabbits, etc.

In terms of sheer numbers per square metre, nematodes (120 million), mites (100,000), springtails (45,000), enchytraeid worms (20,000), and molluscs (10,000) greatly out–number any other species in temperate grasslands. Fungi, however, may be 50% of the total living biomass.

While the tradition of soil science has been to treat and analyse soils as mineral matter (the living component being carbonised or burnt off in analyses as "C" or humus content), the preoccupation of sound farmers, biodynamic groups, mulch gardeners, and "no–dig" croppers has been the quantity and quality of soil life, both as indicators of soil health and as aerators and conditioners of soil. Another factor which deserves more treatment in science is the mass, distribution, migration, and function of roots and root associates, and the role of burrowers (not only earthworms, but larger mammals, reptiles, and a host of insects).

The soil (if not sterilised, overworked, or sprayed into lifelessness) is a complex of mineral and active biological materials in process. No soil scientist myself, I rely on soil life and the health of plants to indicate problems. Diseases and pest irruptions can be the way we are alerted to such problems as over-grazing, erosion, and mineral deficiency. Removing the pest may not cure the underlying problem of susceptibility. Certainly, strong plants resist most normal levels of insect attack.

Soil analysis, helpful though it is, can help us very little with soil processes. Until very recent years, we have underestimated the contribution of nitrogen by legumes *or* soil microfauna. In addition, the measures of soil carbon has rarely been related to the soil biota, whose lives and functions are not fully known. It seems curious that we know so much about sheep, so little about those animals which outweigh them per hectare by factors of ten or a hundred times, and that we do not investigate these matters far more seriously. Our most sustainable yields may be grubs or caterpillars rather than sheep; we can convert these invertebrates to use by feeding them to poultry or fish. We can't go wrong in encouraging a complex of life in soils, from roots and mycorrhiza to moles and earthworms, and in thinking of ways in which soil life assists us to produce crop, it itself becomes a crop.

EARTHWORMS

> Worms have played a more important part in the history of the world than most persons would at first suppose. In almost all humid countries they are extraordinarily numerous, and for their size possess great muscular power. In many parts of England a weight of more than ten tons (10,516 kg) of dry earth annually passes through their bodies and is brought to the surface on each acre of land; so that the whole superficial bed of vegetable mould passes through their bodies in the course of every few years... Thus the particles of earth, forming the superficial mould, are subjected to conditions eminently favourable for their decomposition and disintegration...
>
> The plough is one of the most ancient and most valuable of man's inventions; but long before he existed the land was in fact regularly ploughed, and still continues to be thus ploughed by earthworms. It may be doubted whether there are many other animals which have played so important a part in the history of the world, as have these lowly organized creatures.
>
> (Charles Darwin, *The Formation of Vegetable Mould Through the Action of Worms*, 1881)

From the time of Darwin (and probably long before), copious worm life in soils has been taken as a healthy sign, and indeed more modern reviews have not reversed this belief (Satchell, 1984). Worms rapidly and efficiently recycle manure and leaves to the soil, keep soil structure open, and (sliding in their tunnels) act as an innumerable army of pistons pumping air in and out of the soils on a 24–hour cycle (more rapidly at night).

Of themselves, they are a form of waste recycling product, with a dry–weight protein content of from 55–71% built up from inedible plant wastes. Only a few peoples eat worms directly, but a host of vertebrates from moles to birds, foxes to fish depend largely on the worm population as a staple or stand–by food. Cultivated worms are most commonly used as an additive to the diets of livestock (fish, poultry, pigs).

However, as processors of large quantities of plant wastes and soil particles, worms can also accumulate pollutants to extraordinarily high levels; DDT, lead, cadmium, and dioxins may be at levels in worms of from 14 or 20 times higher than the soil levels. Eaten in quantity by blackbirds or moles, the worms may become lethal. That is, if the "pests" that are moles, blackbirds, and small hawks abound on farms, there is at least some indication of soil health. Where these are absent, it is an ominous and obvious warning to us to check the soil itself for residual biocides.

As non–scientists, most gardeners deprived of atomic ray spectrometers, a battery of reagents, and a few million research dollars must look to signs of health such as the birds, reptiles, worms, and plants of their garden–farm. For myself, in a truly natural garden I have come to expect to see, hear, and find evidence of abundant vertebrate life. This, and this alone, reassures me that invertebrates still thrive there. I know of many farms where neither birds nor worms exist; and I suspect that their products are dangerous to all life forms.

All modern evidence agrees on the value of worms in fields, as decomposers and manure recyclers. They may be even more valuable as garbage disposal systems, and as fish or poultry food, providing a mass of high–protein food from vegetable wastes.

8.13
DIFFICULT SOILS

CONCRETIONS AND PANS

Several types of concretion or cemented particles occur in soils. These are commonly the following:

- CALCRETE (*caliche, platin, kunkar*) is a hard, mainly level subsurface concretion about 0.5–1.0 m below a granular or sandy topsoil, typical of coral islands (calcium triphosphate), and the downwind areas of desert borders. Calcrete must be broken open to plant trees, or the roots will spread out laterally, allowing wind–throw to occur. On atolls, fresh–water deposits develop below the caliche.

Broken caliche can be used as a building material and also forms a safe roof for tunnels or dugouts. Calcium/magnesium concretion, worsened by the addition of superphosphate, is whitish to creamy. In acid (vinegar), calcrete releases bubbles of carbon dioxide.

- SILCRETE (*cangagua*) is a grey to red shiny hard layer developed below some tropical forest soils, which gives a glassy surface if forests are cleared. The soil is concreted by silica deposits. If such deposits lie below forests, it is unwise to clear the forest itself. Durian: Silica–cemented non–wetting horizon, earthy, brittle, found only in volcanic areas. Red–brown hardpan: occurs in many soil types, not volcanic, semi–arid, and is 10 cm–30 m (4 inches to 98 feet) thick.
- FERRICRETE: Iron–cemented pans and soil layers of varying thickness, sometimes as thin sandy layers of 5–10 mm; also alumina–iron laterites (often capping desert hills with veins of silcrete) or iron–manganese nodular horizons in soils. Ferricrete may lie over pale bauxites, and is also called ironstone, plinthite. Ortstein occurs in podzols as iron–organic hard B horizons. Coffeerock is a thick sandy coffee–coloured horizon, low in iron and easily broken; it is a common horizon in humic podzols. Duricrusts form hard silica–iron caps on hills in deserts.
- PLOUGH PANS are usually clay–based compacted layers developed below croplands in wet periods; these can be caused by mouldboard ploughs.

All of the above need ripping, explosive shattering, or deep mulch pits to establish trees. Sodium in soils may develop a "collapsed" cemented, greyish, gravelly pan (SOLCRETE) impermeable to water. Only deep–rooted trees, reduction of salt, and humus relieve these cemented conditions. Deep drainage of 1–2 m is essential for salted soils.

Concreted soil layers (calcium or silica–cemented) are the calcretes (*caliche, platin*) of dry islands and coasts, or the ferricretes (iron–cemented) of deserts. Any or all may form duricrusts (hard layers) in eroded areas. Under some tropical rainforest (e. g. in Ecuador), an iron–silica pan which follows hill contours, locally termed *cangagua*, lies 3 m below the forests; it is a daunting sight to see this glassy and impermeable surface after the removal of forest and a consequent loss of topsoil. Where *cangagua* is known to exist, perpetual forests used for products other than their wood (honey, fruits, medicines) are the only sustainable use of land.

NON–WETTING SANDS AND CLAYS

Some classes of very fine blackish sands, and sands invaded by hydrophobic soil fungi, are difficult to wet; the water sits on top as droplets. There are several remedies for this in gardens:

- Ridge soil to make basins.

Every square metre, core out sand and drop in a loam or clay–loam plug (4–10 cm by 30 cm deep).

- Compost thoroughly and build up organic material to 8% of surface soil.
- Add a handful of bentonite per square metre, or powdered clay from clay pans.
- Mulch thoroughly, and plant. Keep surface mulch

supplied.

On the broad scale, deep ploughing in autumn (to 45 cm) is used, followed by rotary plough or chopper, mixing of the top non–wetting profile of 10–20 cm (4–8 inches) with subsoils. A cover crop is immediately sown to prevent erosion, and this used as a cover crop or green manure for deep–rooting crop or tree species. The successful establishment of trees permanently curbs the problem.

CLAYS which seal on the surface in light rains are often sodium–rich and "melt" in rain. Remedies are:

- Make low banks across run–off.
- Add gypsum at 2–3 handfuls per square metre, and if possible flush out with fresh water (removing sodium as sulphate).
- If practical, place sand over the surface to 4 cm deep.

For deep cracking clays and lumpy soils add a sand layer, scatter gypsum at a handful per square metre, and mulch.

For acid or deep silica sands it is best to add clay and mulch, and to lay plastic at 0.5 m deep in garden beds (**Figure 8.7**).

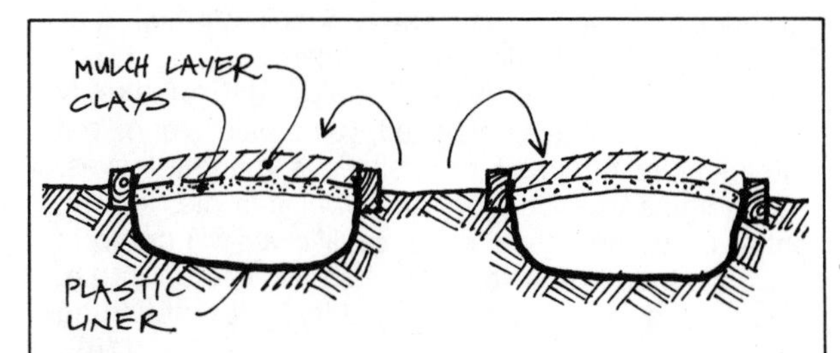

FIGURE 8.7
LAYING PLASTIC IN DEEP SANDS
Plastic sheet prevents deep leaching , clays hold water at root level; mulch prevents surface evaporation.

8.14

PLANT ANALYSIS FOR MINERAL DEFICIENCIES; SOME REMEDIES

Figure 8.8 (after Clark and Smith, "Leaf Analysis of Persimmons", *Growing Today*, New Zealand Feb. '86, pp. 15–17) illustrates the *seasonal* levels of important plant nutrients (as 90% dry matter) and micro–nutrients (as micrograms/gm) in leaf. The figures illustrate several things:

1. SEASONAL CHANGE IN LEVELS. Thus, the potential for early intervention in adjusting levels (before it is too late to save the crop).

2. VERY DIFFERENT BEHAVIOUR OF NUTRIENTS:

Type 1: Zinc, Iron, Copper: high in new spring growth, falling over summer, and finally (as leaves are lost to the tree) becoming more concentrated in the last leaves, as uptake by roots concentrates these elements in the last leaves.

Type 2: Boron, Manganese, Calcium: increase throughout the whole season of growth. Not "mobile" once in the plant.

Type 3: Nitrogen, Phosphorus, Sulphur: rapid early uptake, then a gradual decline over the season.

Type 4: Potassium: remains steady over the year, then declines as wood storage and root reserves build up.

As persimmons are not atypical plants, these findings have implications for *pre–emptive adjustment* (e.g. by foliar sprays), and *selective mulch* (e.g. the season at which leaves are taken for compost).

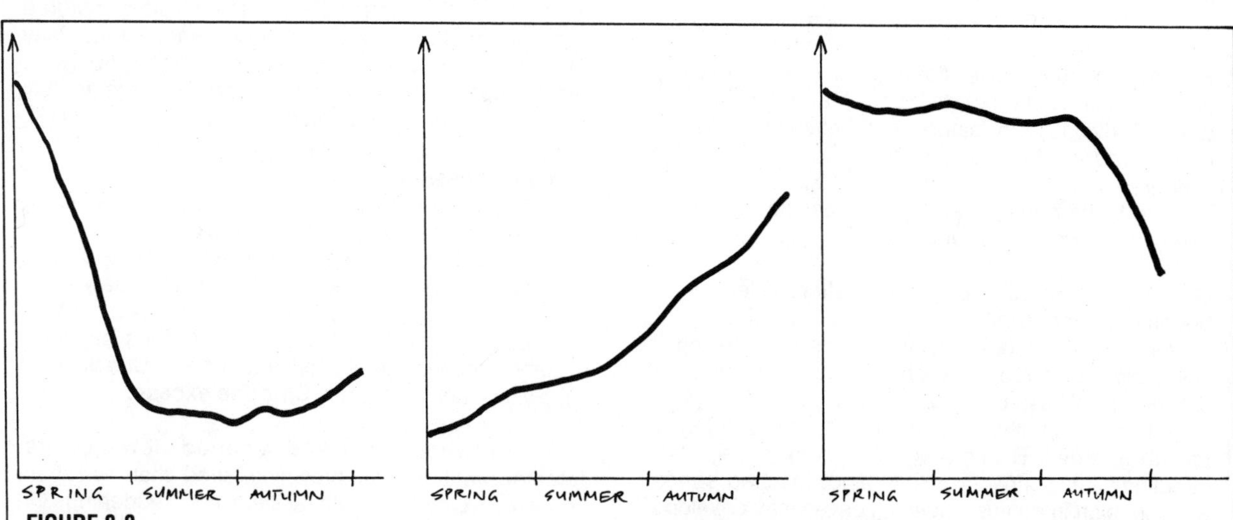

FIGURE 8.8
A 'TYPOLOGY' OF THE ELEMENTS IN LEAF TISSUE (in winter, trees are bare).
Behaviour of basic nutrients in the leaf analysis of a deciduous tree (here a persimmon). Levels vary across the growing season.
[After *GrowingToday*, NZ Tree crops Association, Feb. 1986 (See pp16–7 for actual graphs)].
A Cu, Fe, Zn, P, N, and S. Similar curves, slight increases in late season.
B B, Mn, Mg, and Ca. The non-mobile elements increase steadily in concentration; Mg is not normally regarded as non-mobile.
C K. Remains fairly steady, and is "withdrawn" from leaves in autumn.

TABLE 8.7
KEY TO MINERAL DEFICIENCIES

How to use: Read the first set of choices (A), make one, and follow on to the letter group in the righthand column. Then make a second choice (or find the answer). Some remedies follow, and are given under the numbers in brackets { }.

First Choices: **Go to**

A. • Leaves, stems, or leaf stalks are affected ....**B**
• Flowers or fruits are affected ....**M**
• Underground storage organs (roots, bulbs, tubers, etc.) are affected ....**N**
• Whole field or row shows patchy or variable yields ....**O**

B. • **Youngest** leaves show most effect, or early effects ....**C**
• **Oldest** leaves or later whole plant affected...**I**

C. • Pale yellowish or white patches on the leaves ....**D**
• Pale patches not the worst symptom, but death of tips or growing points, or storage organs affected ....**H**

D. • Leaves uniformly colour–affected (yellowish or pale), even the veins, poor and spindly plants (especially in heavily cropped or poor leached sandy areas, acid **or** alkaline ....**{1}**
• Leaves not uniformly affected, veins or centres still green .... **E**

E. • Leaves wilted, then light–coloured, then start to die. If onions, crop is undersized. If peas, seed in pods barely formed, matchhead size. Coastal sands. Black sheep in flocks may show a brown tinge to wool and are often used as "testers"...................................**{2}**
• Wilted and dying leaves not the problem......**F**

F. • At first, colour loss is interveinal (between veins), and only later may include veins. Mature leaves little affected, dying not a feature, and common on calcareous or coral atolls, desert soils. Distinct yellowing (See also J) ....**{3}**
• Veins remain green, pale areas not so yellow, often whitish or lack colour. ....**G**

G. • Areas near veins still green, affected leaf areas become transparent, brown, or start to die. Young leaves first affected. Peas and beans germinating in soil show brown roots and central brown area on leaf cotyledons. pH usually >7.0.................................**{4}**
• Leaves smaller than normal, stems shortened, growth retarded. Beans and sweet corn, several tree seedlings most affected. Soils acid, leached sands, alkaline, high in humus, coastal. Leaves may develop a rosette appearance, bunchy tops ....**{5}**

H. • Plants brittle, leaves die or are distorted, growing points die, stems cracked, rough, short between leaves, split lengthwise (cabbages), cracked (celery). Probable on acidic sands, or on heavily–limed high–humus soils..**{6}**
• Plants not brittle, but stunted, tips dying, feeder roots die, and leaf tips and terminal bud margins dying. Cabbage or cauliflower have young cupped or dead margins; old leaves all right. Young infolded leaves brown–edged, rotting (jelly–like decay). Check on over–watering, excess Na, K, Mg in water, or in or dolomite. Tomatoes show blossom end rot. ......**{7}**

I. • Plant with marked yellow (chlorosis)............ **J**
• Yellowing not the main problem; leaves brown–edged or purple ....**L**

J. • Yellowing between veins or on margins of leaf...**K**
• Yellowing affecting whole plant, ranging from light green to yellow; plant gets spindly, older leaves drop off. Prevalent in cold peaty soils, leached sands, soils subject to waterlogging. Turnips show purpling on leaves. Plants flower or mature early....................**{8}**

K. • **Margins** yellow, or blotched areas which later join up. Leaves can be yellowed or reddish, purple, progressing to death of leaf area. Later, younger leaves affected. Affected areas curl or become brittle, brownish. Common on acid sandy or soil with high K or Ca readings. Growth slow, plant stunted. ....**{9}**
• **Interveinal** yellowing, looks at first like N deficiency . Old leaves blotched, veins pale green, leaf margins rolled or curled, progresses to younger leaves. Leaf margins of cabbage, cauliflower can die, leaving central tissue only ("whiptail"); cauliflower will not form curds. Common on acid or leached alkaline soils, e.g. shellsand dunes, corals. Difficulty in establishing clover, legumes..**{10}**

L. • Leaf margins brown, scorched, can cup downwards, dying target **spots** appear in leaves; spots have dark centres, yellow edges; general mottled appearance. Growth reduced, first on young matured and then on older leaves, finally to young leaves. May appear late in plant's growth if a root crop (K is translocated to roots). Leached acidic or organic clay soils. Tomato leaf margin pale. ....**{11}**
• Leaves wilt, droop, die at tips and edges: **Sodium excess.**
• Leaves dull, dark green or red–purple, especially below (under–surface) and at the mid ribs. Veins and stems may also purple, growth is much reduced. Common in very acid, alkaline, dry, cold, or peaty soils ....**{12}**
• Leaves at tips wilt early as soil dries out, then become bronze, then die. Not often seen. Check water supply, salt content of soil: **Chlorine excess.**

M • Fruit rough, cracked, spotted, few flowers. Tomatoes with internal browning, seed chamber open, uneven or blotchy ripening, stem end reddening. On acid soils, leached sands, humus–rich and limed soils. Terminal buds may die and laterals then develop. Top leaves thicken, can roll from tip to base. ....**{6}**
• Fruits rot on blossom end (opposite stalk), or show sunburnt dark areas there. Affects tomato, peppers, watermelons ....**{7}**

N. • Internal dying or water–soaked areas, uneven in shape (in beet, turnip, rutabaga if soil acid, leached, or with free lime. ....{6}

• Cavities in root core, then outside collapses as pits; common in carrots, parsnips on acid leached soils. Roots may split open. ....{7}

O • Areas of affected crop test acid: soils may be sandy; pH < 5.5: Acid: **Try lime**

• Areas of crop test alkaline: pH > 7.5 : Alkaline: **Try Sulphur.**

• Soil at depth mottled, smells of sulphur: waterlogged: **Arrange drainage.**

• Leaves tattered and dying at crown. Salt winds: **Try shelter.**

• Check for viral disease in grasses: **Try a plant pathologist.**

REMEDIES FROM THE KEY (FOR GARDENS)

First, keep a fertiliser **diary** for your garden, and leave it for the next person. Tell them what you have done for the soil.

• If you are on leached (washed out) sands, dig up your garden beds, place a plastic sheet liner below, then add a bucket or so of clay and a handful of dolomite per square metre. Also, try a soil gel. Then add compost, and a complete fertiliser like blood and bone. Mulch thickly and replant, then return to the Key if symptoms recur.

• If you are on peats, or have piled on the compost, add some urea or blood and bone, raise your beds, and lime the area.

• If you have lots of lime in the soil, or are on coral sands or dry desert coasts with calcrete, spray weak zinc and copper sulphates **on** plants, iron sulphates in **very** dilute solutions (12 g/10 square metres with lots of water), or add it to a liquid manure. Make pits of compost and grow on the edges of these, use sulphur at about a handful per square metre, or add trace elements and sulphur to compost pits. Or, lay a sheet of plastic on the ground, build up logs around this to 25 cm high, and fill the area with humus (compost plus 50% sand), then mulch heavily. Add blood and bone. On atolls, dig down to near water table and then mulch thickly (make a big growpit 3–4 m wide by 10–12 m long by 3 m deep). Use **any** mulch, especially *Casuarina*, palm, house wastes.

{1} **Sulphur**. Add plain sulphur (not of medical quality) at one handful per square metre. If you are near a city, you could have enough from fallout!

{2} **Copper**. Add as fine–crushed ore, or in water as copper sulphate at 7 kg/hectare or spread (1 g/square metre) every 5–7 years.

{3} **Iron**. Try sulphur first, then if necessary add iron sulphate or spray foliage with very dilute iron solution. Bury old iron in humus pits near trees (e.g. pieces of galvanised iron, old wire or car parts).

{4} **Manganese**. Try sulphur first, then use very dilute foliar spray of manganese sulphate.

{5} **Zinc**. Add zinc oxide in acid areas, sulphate in alkaline, also sulphur in alkaline areas. Zinc at 7 kg/hectare or equivalent every 7–10 years.

{6} **Boron**. Be careful not to add too much; it is poisonous in large quantity. First, lime acid areas and peats, and add sulphur to alkaline areas. If this doesn't work, add borax (sodium borate) at 1 gram/square metre and try cabbages to test reaction. Try not to buy detergents with "borates"; they can poison your soil. **Boron excess** (poisoning) can occur on sea sediments, and are common in reclaimed marine areas (Holland). Raise garden beds, lime, and flush out with fresh water.

{7} **Calcium**. Use lime as limestone in areas where manganese is plentiful, dolomite if not, or as cement powder in deep red hot tropical soils, then continue to add mulch and use lime only if deficiencies occur. Use gypsum in alkaline salty soils, then flush with fresh tank water and continue to use lime. Bone, bamboo mulch, buckwheat straw are all calcium sources.

{8} **Nitrogen**. Make sure the soil is well drained to 0.5 m for vegetables, 1–2 m for trees. Check for cobalt levels, deficiency. If legumes are used, make sure they are innoculated, and that manganese levels are not too high. If all this is satisfactory, add dilute urine (20 parts water:1 part urine), ammonium sulphate in alkaline areas, or use legume mulches or interplant (about 48 small *acacia* or tagasaste trees per one fourth acre will do). Use compost, then surface mulch. Build up **worms** and soil life, use dilute bird manure. Don't overdo it, or nitrates will build up in green plants and kill your kids or piglets with bluebaby syndrome. Just relieve the symptoms, then get good soil life going.

Use cobalt for severe nitrogen deficiency, poor clover growth or establishment in peaty or coastal soils. If manganese is high, just add lime to balance this soil (one handful per square metre), or in **alkaline soils** spray on at very low dilutions at 1 g/10 square metres every 10 years or so.

{9} **Magnesium**. Check if potash is not too high, or add clay to sandy acid soils (plenty of magnesium in most clays). Use **dolomite** for first dressing, then limestone. Epsom salts were used around citrus by old–timers. Or dilute it in water for foliar spray in very severe deficiency situations.

{10} **Molybdenum**. Get some sodium molybdate, about 10 g, and mix well with 5 kg of sand. Take 1/100 of this (weigh the sand), and put it on per square metre every 10 years.

{11} **Potassium**. Use ashes on green crop, diluted urine in early growth, then build up mulches, including dried or fresh seaweeds, flue dusts from cement works (fly ash), also "teas" of bird manures, comfrey. Potassium is found in the mineral kainite (20–25%) potassium) in evaporite deposits of deserts.

{12} **Phosphorus**. Bring pH to 6–6.5 or thereabouts, using lime in acid soils and humus in alkaline. Use bone meal, bury bones, or use tested rock phosphate free of cadmium or uranium. **Stop deep digging** and start mulching with least soil disturbance (build up narrow beds). Encourage soil life, add mulch on top, water with comfrey "tea", dilute bird manure on the **leaves** of plants. Keep this up each time symptoms appear; they will eventually disappear if you have clay in the beds (add some if not). If you have high–iron clays, you will need a lot of bone meal to start with, but it will slowly release later on. If desperate, use a few handfuls of superphosphate per square metre, then continue with other sources. Feed a patch of comfrey with bird manure and make a comfrey tea in a drum of cold water. Water the plants with this. All animal manures (including yours) contain some phosphorus. Calcined (roasted) rock phosphate is effective on acid soils in high rainfall.

{13} **Chlorine poisoning**. Sue your Council, or let tap water stand with a handful of lime in it for a day, then use on the garden. Don't take a shower!

{Developed and modified after a format developed by English, Jean E. and Don N. Maynard, *Hortscience* **13**(1), Feb. '78, and with data from the author and Handreck, Kevin A., 1978, *Food for Plants*, CSIRO Division of Soils.}

MINERAL FERTILISERS OR SOIL AMENDMENTS

The present testing method used on specific soil types is to sow down mixed legume (clover), *Brassica*, and grass crop (or any important crop that may be grown). This sowing is then divided into TRIAL PLOTS which are treated at varying levels, and with soil or foliar spray amendments, to test plant health and response, based on a soil test for pH and mineral availability, or on a leaf analysis such as given in **Table 8.7**.

For the home gardener, or keen observer, a deliberate wander through the system, and a good key to mineral deficiencies may be all that is needed to spot specific problems. Problems are in any case rare in well–drained garden beds using composts and organic moulds, and where one–species cropping is not constantly practised.

On a broader scale, as in prairie or forest re-establishment and erosion control, land reclamation, or plantation, every practical farmer and forester uses TEST STRIPS of light to heavy soil treatments (from soil loosening to fertiliser, micronutrient, and grazing, cutting, or culling trials). When such field trials (as side–by–side strips) are run, it is wise to include typical areas of soil and drainage, and to avoid areas under trees, on the sites of old stockyards or hay-stacks, intense fire scars, watering points, and gateways and roads (all of which have minor but special features and need a separate assessment from the open field situation). I have often noted, for instance, the colonisation of chicory, thistles, and tough and deep–rooted weeds on the inhospitable areas of old roads and trafficked areas; this sort of data is of use for some cases, but does not need to suggest that we compact a whole field in order to grow chicory, rather that chicory is a useful pioneer of compacted soils.

Plant response on the test strips, which can be as little as 1% of the total acreage, may quickly indicate how modest and innovative soil treatment, minute amounts of micronutrients, or the timing of grazing or browsing can be managed to give good effects at least cost. There is no assurance as certain as the actual, assessed plant response. To see two small plots of pines, coconuts, or cabbages side by side, the one healthy, vigorous, and productive, and the other (lacking a key nutrient or on compacted soils) stunted, sickly, and unproductive, is a definite guide to future treatments. The same sort of trials are applied to plant mixtures or polycultures, pest controls, and the benefits or otherwise of mulch for a specific soil or crop.

Assessment can be casual (in clear–cut cases), or analytic and careful where only slight differences appear. Such test strips are best securely marked by stout pegs for long–term visits, as effects of some treatments persist, or become evident, over several seasons.

Not until trials are assessed is it wise to widen the area treated, although in commonsense it may always be wise to add humus or manures to non–peaty soils, or dolomite to acid sands. In alkaline and heavy clay soils, trace elements may become insoluble, and these are best added as foliar sprays to mulch or green crop, or to trees.

8.15

BIOLOGICAL INDICATORS OF SOIL AND SITE CONDITIONS

In any local area, the composition, shape or size, and distribution of the plants give many clues to soil type, depth, and extrinsic factors. Some specific factors indicated are:

SOILS:
1 Depth
2 Water reserves
3 pH
4 Mineral status (see preceeding section)

SITE:
5 Fire frequency
6 Frost
7 Drainage
8 Mineral deposits and rock type
9 Overgrazing and compaction of soil
10 Animal (macrofauna) effects

1 SOIL DEPTH: Shallow soils dry out quickly and hold few nutrients. A very good indication of soil depth is to look at one species of tree (e.g. *Acacia*, *Prosopis*, honey locust) over a range of sites; a "height and spread" estimate will reveal areas of deeper soils where the largest specimens grow. The same species will be dwarfish on shallow soils of the same derivation or rock type.

2. WATER RESERVES. Deep–rooted trees which need water—the large nut trees and candlenuts (*Aleurites*) are good examples which occur naturally only in well–drained but water–conserving sites—often show water–lines not associated with valleys, and stand over springs or aquifer discharge areas.

In sands, a great variety of deep–rooted shrubs and trees indicate where a clay base lies at 1–2 m down. This situation is common on desert borders and hills in drylands. In brief, large tree stems reveal well–

watered sites, small stems drier sites. Armed with these observations, we can create sites by water diversion and select sites for large trees or shrubs.

3. pH: Sorrel and oxalis in pastures may indicate compact or acid conditions, whereas several fen and limestone species establish in alkaline areas; large snails and dense snail populations occur only over alkaline soils or in alkaline water. No snails or minute species occur in acid water (pH < 5.0). In the garden, our cultivated plants demand acid or alkaline soils, e.g.

Alkaline intolerant (pH 4.5–6.0):
- Blueberry
- Chicory
- Chestnut
- Endive
- Potato
- Fennel
- Tea
- Coffee
- Rhubarb
- Shallot
- Watermelon

Alkali tolerant:
- Oats
- Kale
- Rye

Acid intolerant (pH 7.0–8.5):
- Cauliflower
- Cabbage
- Asparagus
- Green peas, bush beans
- Celery
- Leek
- Beet
- Onion
- Chard
- Parsnip
- Spinach
- Lucerne
- Broccoli

Acid tolerant:
- Lupin
- Oats
- White clover

This will have a profound effect on our home garden planning, but providing garden soils are mulched, and a little lime is added to compost, *all* plants thrive in high humus soils supplied with some lime at modest levels. It is the perennial species that may need more care in site selection, or with mulch and compost in alkaline areas. Almost all our pollutants, and many of our fertilisers, tend to make soils acid, as does continued cropping or over–grazing.

5. FIRE FREQUENCY. East–west ridges often reveal abrupt species changes at the ridge wherever fire occurs. Fire produces dry, scrabbly, summer–deciduous, thick–seeded species; lack of fire develops broadleaf, winter–deciduous, small–seeded plants with thin seed capsules and a deep litter fall.

Cross–sectional cuts of trees will reveal fire scars as gum pockets or charred sections, and these can then be counted to get the "fire frequency" of the site (**Figure 8.9**). If tree stem sections are marked for directions before sampling, the direction of fires can also be judged.

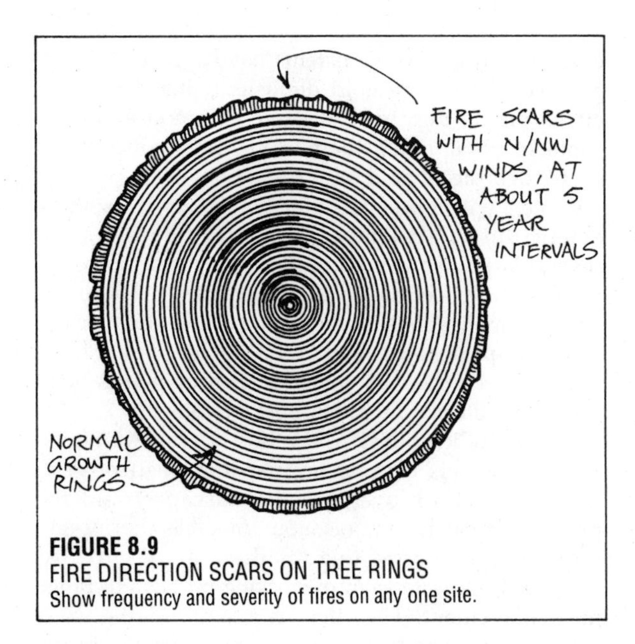

FIGURE 8.9
FIRE DIRECTION SCARS ON TREE RINGS
Show frequency and severity of fires on any one site.

6. FROST. Many species of trees and plants will indicate frost–lines; it is a matter of observing local flora, or planting frost–susceptible species down a hill profile to measure frost intensity. Tomatoes, bananas, and potatoes are all frost–sensitive and will reveal frost–lines on hills in subtropics and deserts.

7. DRAINAGE. Mosses, sundews, and fine–leaved heaths indicate poorly–drained soils, as large trees such as chestnuts (which require 2 metres of well–drained soil) indicate good drainage; these indicators assist survey before pits are dug or drainage measured.

8. MINERAL DEPOSITS. Davidov (*Sputnik*, 12 Dec. '79) gives data on plant systems over mineral deposits (for Russia). The analysis of plant residues often indicates concentration of ores in the underlying soil or rock. Lead and copper–molybdate are so indicated. Leaves or humus from birch, cherry, honeysuckle, St. John's Wort, wormwood, juniper, and heather reveal the above lodes plus tungsten and tin concentrates. "All purpose" plants so discovered are:

Rue or violets.......... zinc
Catchfly................ cobalt
Asters..................... selenium
Milk vetch.............. selenium and uranium
Russian thistle........ boron
Alyssum.................. nickel
Honeysuckle............ silver and gold
Horsetail................ gold and silica

General plant ash analysis may reveal more *specific* plant–ore associations. This has further implications for the rehabilitation of mine waste areas, and also to select plant sources for the supply of trace minerals in compost. As plants have the ability to both concentrate and tolerate unusually high levels of specific minerals, there seems to be a field here for the biological concentration (and subsequent removal) of metallic soil pollutants like lead or uranium, and the use of concentrator plants to mine or collect locally rare trace elements. In fact, some patents have apparently been granted for mining gold deposits using banana or citrus plants deprived of some common elements (potash, phosphate); their leaves then concentrate sparse deposits of gold.

Oysters will concentrate zinc (to 11% dry weight, an emetic dose), abalone concentrate cadmium, and several large fish concentrate mercury and biological poisons from corals (to inedible levels). There are obvious implications for the removal, collection, or use of such species—element relationships, and lead, cadmium, or mercury levels in fish or plants need careful monitoring for public health reasons.

9. OVERGRAZING AND SOIL COMPACTION. Both the levels of grasshopper and pasture grub activity (high on overgrazed landscapes) and the presence of patches of poisonous, inedible, thorny, and unpalatable plants (e.g. Sodom apple, oxalis, capeweed) indicate an over-stocking problem or range mismanagement. The effect is a synthesis between changing soil conditions, plant stress, and the heavy selection by livestock of palatable species, so favouring the survival and spread of spiny or inedible species. Too often, the pastoralist blames the weeds and seeks a chemical rather than a management solution; too seldom do we find an approach combining the sensible utilisation of grasshoppers and grubs as a valuable dried-protein supplement for fish or food pellets, and a combination of soil conditioning, slashing, and de-stocking or re-seeding to restore species balance.

10. MACROFAUNAL EFFECTS. The site of a sea-bird rookery, a rabbit warren, the ground nest of a goose or eider clutch, the pellet-pile of an owl, or the decay of a large carcass will cause a sudden and often long-term change in the immediate vegetation, as will termite mounds and harvester-ant colonies. Once such sites are recorded, and the plant assembly identified, similar sites can be located and recognised. The data can be used as an aid to conservation, an indication of soil drainage (rabbits choose good drainage), as a result of specific nutrient supply (guano on seabird rookeries), or as a way to establish tree clumps following natural indicators.

A large proportion of wind-blown, nitrogen-loving, and inedible plants, or plants carried by birds as seed, depend on the specific habits of birds or mammals, on their dung, or on soil disturbances. The role of animals in the distribution of plant seed, and plant root associates, is well-recognised; their role in soil change, less commonly noted.

8.16
SEED PELLETING

In pioneering the rehabilitation or stabilisation of soils, many of the local deficiencies in soils can be overcome by seed pelleting, which is a process of embedding seed in a capsule of substances that give it a good chance of establishment despite soil deficiencies in local sites, or microsites.

• SEED PRETREATMENT. If seeds have thick coats, or need heat or cold treatment or scarification to break dormancy, they must be treated before pelleting.

• INOCULATION. Purchase and inoculate legume seed with their appropriate microbial or fungal spores. Soak the seed in inoculant solution, then dry the seed. Mix dried seed with a primary coat as below.

• PELLETING. Use a lime, clay layer, and a trace of fine rock flour, calcium, or phosphate mixed into a damp but plastic slurry around the seed. This is then extruded (e.g. via a meat mincer with the cutting blades removed) to a shaker table or tray covered with dust, and on a slight incline. Dust is added as needed to dry and shape the pellet, or to set a desirable size of pellet (**Figure 12.18**).

The dust, or outer pellet coat, should incorporate a soil conditioning gel or polymer, a colloid-forming substance (fine graphite), a bird repellent (green dye helps repel birds), an insect repellent such as powdered neem tree leaf (*Azadirachta indica* or *Melia azedarach*) or diatomaceous earth, and perhaps some swelling clay such as bentonite.

Pellets are now dried and scattered, drilled, or sown on sites to await rain. The protected seed germinates when the pellet absorbs water, and the emerging root finds its nutritional needs satisfied, while the root associates also become active in nutrient transfer to the plant.

The same vibrating table that we use to pellet seed serves, when fitted with screens, to clean and sort seed from the soil below trees or from seed and husk mixtures, and the mincer can be returned to the kitchen none the worse for wear. Fukuoka achieves the same result by pressing seed-clay mixes of grains through a coarse sieve, onto a dust-filled pan which is shaken to round off the pellets.

8.17
SOIL EROSION

As all else depends on a stable and productive soil, soil creation is one of the central themes of permaculture. Soil erosion or degradation is, in fact, the loss of production and hence of dependent plants and animals. Soils degrade in these ways:

• Via wind: by dust storm and the blow-out of dunes and foreshores.

• Via water flow by sheet erosion (a generalised surface flow off bare areas and croplands), gully erosion (caused by concentrated flow over deep but unstable sediment), and tunnel erosion (sub-surface scouring of soils below).

• Via soil collapse or deflocculation following increased salt concentrations in clay-fraction soils.

Thus, the placement of windbreaks, tree crops, and fast-spreading grasses stabilises erosion caused by wind, while permanent crop, terracing forestry, and (in the case of gullies) diversion and spreader drains plus gabions help reduce or heal scoured areas. Tunnel erosion may call for de–stocking, contour drainage, and the establishment of deep–rooted plants, while the problems of desalting need the combined factors of reafforestation (to lower groundwater tables) following deep interceptor drains to cut off salt seepages in surface soils (see Chapter 11).

Erosion follows deforestation, soil compaction, disturbed soil–water balance (increased overland flow and rising water tables or salt seepages), overgrazing, plough agriculture on the broad scale, episodes of high winds or rains in drought periods, or severe disturbance caused by animal tracks, roading, and ill–advised earthworks.

Insofar as landscape design is concerned, soil erosion repair is the priority wherever such erosion occurs. Apart from the physical factors, no designer, or nation, can ignore the economic or political pressures that inevitably create erosion by requiring or permitting inappropriate land use and forcing production or over–production on to the fragile structure of soils. Third world debt and western world over–production are both primary factors in soil collapse. In a conservative society, the very basis of land use planning would encompass the concept of permitted or restricted use of soils, carefully plotted in regions following analyses of slope, soil stability, minimal forest clearing (or reafforestation), and permitted maximum levels of crop production, or livestock density, following the procedures of good soil husbandry.

In assessing erosion in the U.K., Charles Arden–Clarke and David Hodges (*New Scientist* 12 Feb '87) point out that "many of the recent outbreaks of severe erosion are clearly linked to falling levels of organic matter in the soil... the more organic matter there is in the soil, the more stable it is." This stability is because of good soil structure and infiltration of water, whereas an inorganic soil may break down under rain. With the following increase in overland flow, most soils will then erode as rills or gullies, or the destroyed surface can powder and blow away without organic matter to bond it.

On many delicate soils (over chalks) the only answer is to replace crops with pasture or forests. Intensive arable use and winter cropping both create more erosion. The very radical conclusion is that mulching, green manure, grass leys on rotation, hedgerows, and minimal cultivation are not only urgent but imperative. Thus, "the time to examine the organic (farming) approach has passed, the time to adopt it has arrived." (*ibid.*) At long last, some scientists are saying that enough evidence is enough; we need to turn to known effective land management based on permanence and organic methods. This will take the combined good will of farmers, scientists, financiers, and consumers.

8.18
SOIL REHABILITATION

Careful gardeners take care not to break up, overturn, or compact their valuable soils, using instead raised beds and recessed paths to avoid a destruction of crumb structure. Responsible farmers try to govern the speed and effect of their implements in order to preserve the soil structure, and can get quite enthusiastic about a dark, humus–rich, crumbly soil. We seldom give farmers time or money to create or preserve soil, but expect them to live on low incomes to serve a commodity market, whose controllers care little for soil, nutrition, or national well–being.

No matter on what substrate we start, we can create rich and well–structured soils in gardens, often with some input of labour, and always as a result of adding organic material or green manures (cut crop). No matter how rich a soil is, it can be ruined by bad cultivation practices and by exposure to the elements: wind, sun, and torrential rain.

Worms, termites, grubs, and burrowers create soil crumbs as little bolus or manure piles, and they will eventually recreate loose soils if we leave them to it in pasture. But we also have other tools to help relieve compaction; they can be explosives, special implements, or roots.

We use the expansive and explosive method rarely, perhaps to plant a few valuable trees in iron–hard ground by shattering. People like Masanobu Fukuoka[3,4] are more patient and effective, casting out strong–rooted radish seed (daikon varieties), tree legume seed, and deep–rooted plants such as comfrey, lucerne, *Acacias*, and eventually forest trees. Much the same subsurface shattering occurs, but slowly and noiselessly. The soil regains structure, aeration, and permits water infiltration.

A measure of the change wrought by green manures, mulch, and permanent windrow is recorded by Erik van der Werf (*Permaculture Nambour Newsletter*, Queensland, Dec. 1985 and Mar/Apr 1986). Working in Ghana at the Agomeda Agricultural Project, he reports on the improvement of crumb structure by measuring the bulk density {weight per volume ratio (g/cc) of soil samples} is given in **Table 8.8**.

TABLE 8.8
Improvement in Crumb Structure

Soil Treatment g/cc	Bulk Density
Annually burnt bush	1.35
Bush left 2 years without fire	1.27
Farmland, cultivated 2 years	1.29
Farmland, permanently mulched and cropped for 3 years	0.92*

*Even with cropping, the mulched soils show how humus alone restores good aeration; soil temperatures were lower by 10°C, and both crop grain yields and a three times increase in organic matter production were noted.

We can use rehabilitative technology on a large scale, followed by the organic or root method, by pulling a shank and steel shoe through the soil at depths of from 18 cm (usual and often sufficient) to 30 or even 80 cm (heroic but seldom necessary unless caliche or compacted earth is all we have left as "soil").

In field or whole site planning, a soil map delineating soil types can either be purchased or made based on local knowledge and field observation. In designing, it helps future management if uses, fencing, and recommendations for soil treatment and crop can be adjusted to such natural formation as soil types. An aid to SOIL TYPING can be found in basic books on soils. These publications give practical guides to landform, floristics (structural) typing, and soil typing and taxonomy (categories or classes of soils).

We can recommend low–tillage systems, pay close attention to water control during establishment, and get soil or leaf analyses done. We can also make careful trials of foliar sprays, the additions of cheap colloids to sands, the frequency and timing of critical fertiliser applications (often and little on sands, rarely or as foliar sprays on clays). Crops suited to natural pH (it is often expensive to greatly modify this factor) and rainfall should be selected for trials. Close attention needs to be paid to the soil stability and thus the appropriate use for soils on slope.

In particular, priorities should be set for erosion control in any specific soil or on specific sites or slopes, and earthworks or planting sequences designed to establish soil stability, for if we allow soil losses to continue or worsen, all else is at risk. The next stage in the design is to assess the capacity of soils for dams, swales, foundations, or specific crops (this may need further analysis, test holes by auger, or soil pit inspection).

Thus, if we have adopted a pre–determined set of values based on soil and water conservation and appropriate uses of sites versus erosion and high energy use, any site with its water lines and soil types noted starts to define itself in usages.

How we need to proceed in soil rehabilitation is roughly as follows:

1. WATER CONTROL. Drainage and sophisticated irrigation are needed to rehabilitate salted areas, and soil mounding or shaping to enable gardening in salted lands (as explained in Chapter 11 on arid lands). We need to rely much more on natural rainfall and water harvest than on groundwaters. Drought is only a problem where poor (or no) water storage has been developed, where tree crops have been sacrificed for fodder or fuel, and where grain crops are dependent on annual rains.

Although many sands and deeply weathered soils are free–draining, waterlogging can occur wherever soil water lies over an impermeable soil layer or where water backs up behind a clay or rock barrier; anaerobic soil results. Remedies lie in any of three techniques:

1. Raised garden beds: paths are dug down for drains, and beds raised; in very wet areas give paths a 1:500 slope to prevent erosion. **Figure 8.10.A**

2. Deep open drains every 10–80 m (clay–sands) upslope and downslope or on either side of garden beds. **Figure 8.10.B**

3. Underground pipes (tile drains; fluted plastic pipes are best) laid in 1.5 m deep trenches and backfilled at 1.5 m (4.5 feet) deep and from 10–80 m (32–262 feet) apart, starting on a drain or stream and with a gentle fall (1:1000–1:600) to the ridge. **Figure 8.10.C**

Water retention in soil is now greatly aided by long–term soil additives. These are gels which absorb and release water over many cycles of rain. This is a practical system only for gardens or high–value tree crop (where the cost amortises).

2. SOIL CONDITIONING. Compacted, collapsed, and eroded soils need rehabilitative aeration, and a change in land use.

3. FERTILISATION. We can reduce and replace past wasteful or polluting fertilisation by sensible light trace element adjustment via foliage sprays *if* undisturbed soil systems and permanent crop have been developed. Foliar spray of very small amounts of key elements greatly assists plant establishment, as does seed pelleting using key elements deficient in plants locally. We may then be able to utilise much of the phosphate that is locked up in clays, and using legumes, create sufficient nitrogen for food crops from sophisticated interplant and green manures.

4. CROP AND PLANT SPECIES SELECTION. Many older varieties of both annual and perennial crops will yield with less fertiliser and water applications than will more recently–developed varieties. There is a growing trend amongst farmers and gardeners to preserve and cultivate these varieties not only for the reasons above, but also for flavour. Many older apple varieties, such as some of the Pippin and Russet types, are more flavourful than, say, the market–variety Red Delicious. There is still a large diversity of food crops left in the world; the key is to grow them and to develop a regional demand. Many older apple or wheat species are not only pest–resistant, but have higher nutritive value, and can produce well in less than optimum conditions.

There are different species of plants that can live in almost any type of soil, starting the process back to rehabilitation. It is often the case that so–called noxious weeds will colonise eroded landscapes, beginning a slow march towards stabilisation; these can be used as mulches.

Soils can be created or rehabilitated by these basic methods:

- Building a soil (at garden scale);
- Mechanical conditioning; and
- Life form management (plants and macro– or micro–fauna).

BUILDING A SOIL

Gardeners normally build soil by a combination of

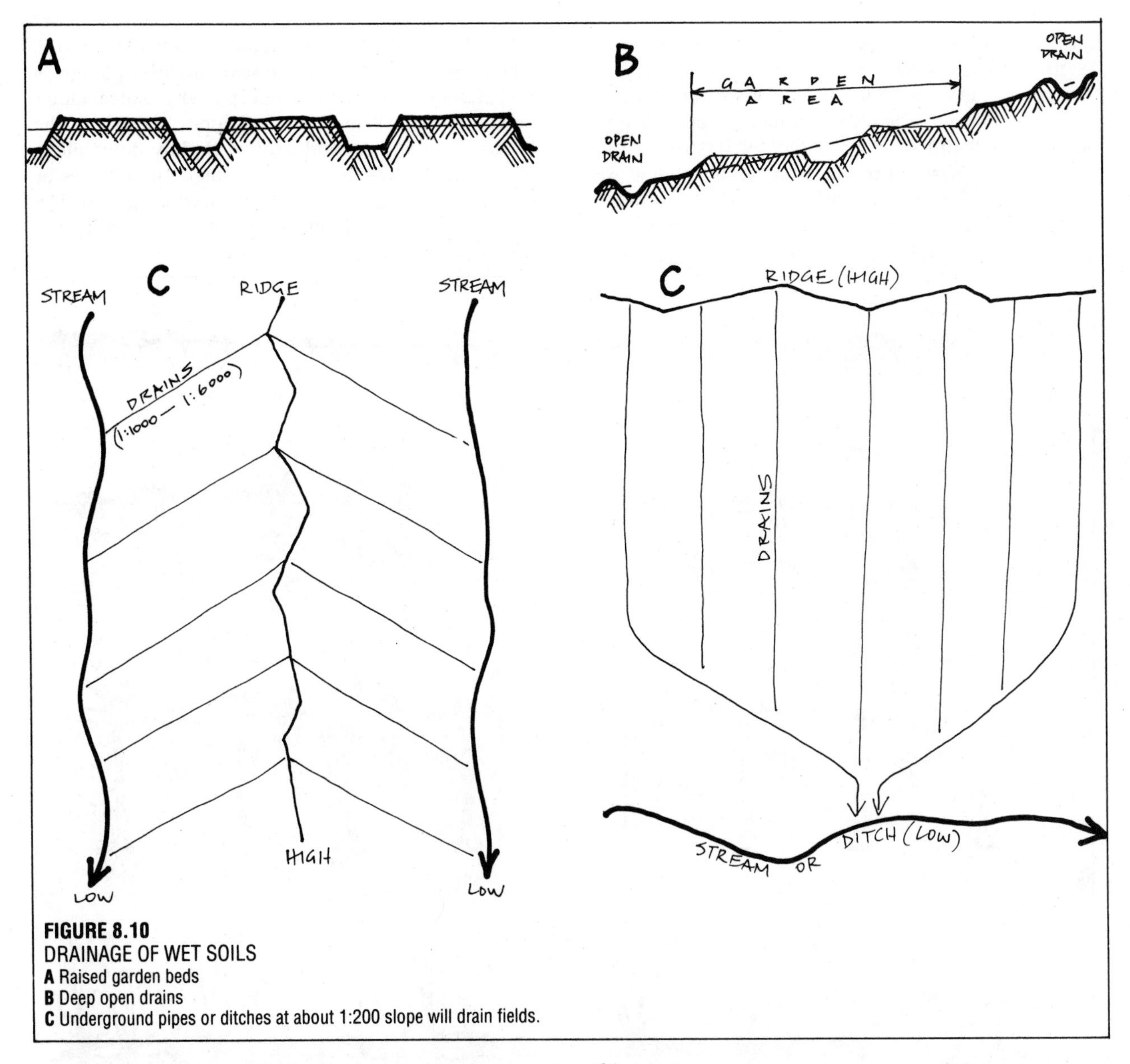

FIGURE 8.10
DRAINAGE OF WET SOILS
A Raised garden beds
B Deep open drains
C Underground pipes or ditches at about 1:200 slope will drain fields.

three processes:

1 Raise or lower beds (shape the earth) to facilitate watering or drainage, and sometimes carefully level the bed surface;

2 Mix compost or humus materials in the soil, and also supply clay, sand, or nutrients to bring it to balance; and

3 Mulch to reduce water loss and sun effect, or erosion.

Gardeners can, by these methods, create soils anywhere. Accessory systems involve growing such compost materials as hedgerow, herbs, or soft–leaf plots, or as plantation within or around the garden, and by using a combination of trellis, shadecloth (or palm fronds), glass–house, and trickle irrigation to assist specific crops, and to regulate wind, light, or heat effect.

By observing plant health, gardeners can then adjust the system for healthy food production. Many gardeners keep small livestock, or buy manures, for this reason.

Large–scale systems (small farms) cannot be treated in the above way unless they are producing high–value product. Normally, farmers create soils by broadscale drainage or by soil "conditioning". As most degraded soils are compacted, eroded, or waterlogged, they need primary aeration (by one of the many available modern machines, or by biological agents), then careful plant and livestock management to keep the soil open and provided with humus.

Daikon radish, tree or shrub legumes, earthworms, root associates for plants (rhizobia) all aerate, supply soil nutrient, or build soil by leaf fall and root action. The management of livestock for least compaction and over–grazing is part of the skill of soil building and preservation. Many organic farmers introduce worm species to pastures as part of their operation, and sow deep–rooted chicory, radish, or comfrey for green manures.

SOIL TREATMENT ON COMPACTED SITES—SOIL CONDITIONING

On the common degraded soils of marginal areas, we can observe compacted, eroded, lifeless soils; they are overgrazed and often invaded by flatweeds and non–forage species of plants. They are boggy and wet in winter, and they are dry, cracked and bony in summer, having little depth. The reconstitution proceeds as follows:

At the end of winter, or in autumn after some rain, when the soil will carry a tractor, a chisel plough is pulled 5–10 cm (2–4 inches) deep over the area, either on contour parallels or on low slopes, starting in the high valley bottoms and driving slightly downhill to the ridges. Unless there are absolutely no legumes or grasses already growing, no extra seed is applied. The response is increased penetration of roots, germination of seed, and a top–growth of pasture.

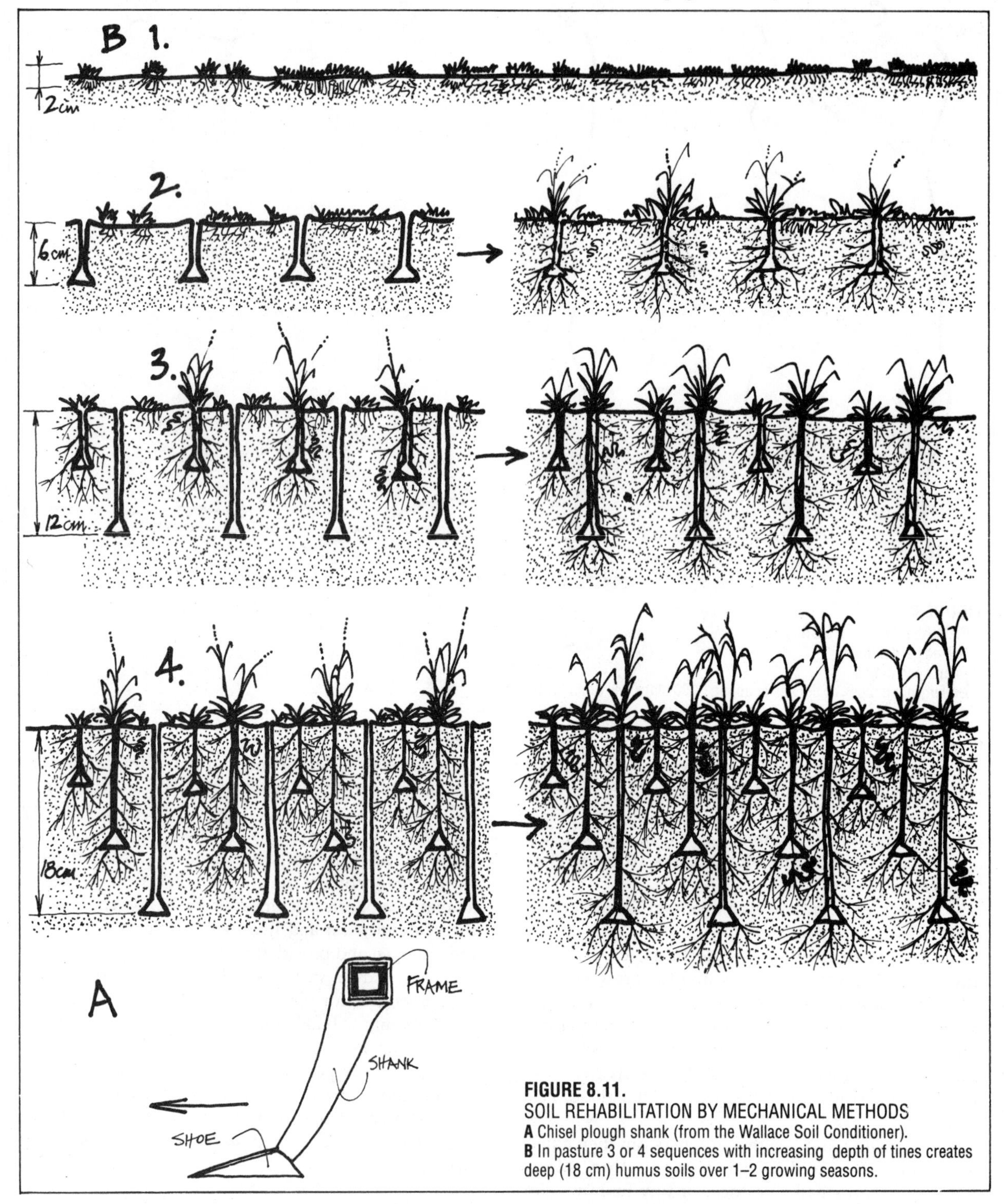

FIGURE 8.11.
SOIL REHABILITATION BY MECHANICAL METHODS
A Chisel plough shank (from the Wallace Soil Conditioner).
B In pasture 3 or 4 sequences with increasing depth of tines creates deep (18 cm) humus soils over 1–2 growing seasons.

A chisel plough or soil conditioner is a rectangular steel frame (tool bar) towed by tractor or draught animals, to which a number of shanks are attached. These are narrow–edge (axe–edged) forward–curved vertical flat bars to the point of which a slip–on steel shoe is attached. The shanks clamp to the tool–bar frame, and the points to the shank. Even one implement of 5 shanks (25–50 b.h.p. tractor) covers a lot of country. There are now at least six or seven makers of soil–loosening machines, in the USA, Europe, and Australia.

Geoff Wallace has produced a soil conditioner of great effectiveness. A circular coulter slits the ground, which must be neither too dry nor too wet, and the slit is followed by a steel shoe which opens the ground up to form an air pocket without turning the soil over. Seed can be dropped in thin furrows, and beans or corn seeded in this way grow through the existing grass. No fertiliser or top–dressing is needed, only the beneficial effect of entrapped air beneath the earth, and the follow–up work of soil life and plant roots on the re–opened soil.

This new growth is then hard–grazed, or cut and left to lie. The plants, shocked, lose most of their root mass and seal their wounds. The dead roots add compost to the soil, as does the cut foliage or animal droppings, giving food to the soil bacteria and earthworms, and softening the surface. As soon as the grazing or cutting is finished, chisel again at 23–30 cm (9–12 inches), on the same pattern as before. Graze or cut again, chisel again at 23–30 cm. Graze or cut.

During this process, often a matter of a one–year cycle, the pasture thickens, weeds are swamped with grasses and legumes, myriad roots have died and added humus, and thousands of subsurface tunnels lead from valley to ridge, so that all water flows down into the soil and out to the ridges. Earthworms breed in the green manure, bacteria multiply, and both add manures and tunnels to the soil. A 23 cm (9 inch) blanket of aerated and living soil covers the earth.

Dust, deep roots, rain, and the bodies of soil organisms all add essential nutrients. The composted soil is, in essence, an enormous sponge which retains air and water, and it only needs a watchful eye and an occasional chiselling in pasture (or a forest to be planted) to maintain this condition.

If tree seed, soybeans, millet or other crop is to be planted, the sequence is as follows: after a few hard grazings or mowings, a seed box is mounted on the chisel plough frame, and the seed placed in the chisel furrow; the grazing or mowing follows germination of the seed. These new plants (sunflowers, millet, melons) grow faster than the shocked pasture, and can be let go, headed, or combine–harvested before the grasses recover. There is never any bare cultivation, and grain growers can move to a minimum tillage method of cropping, with fallows of pasture between crops.

Soil temperature is greatly modified, as is soil water retention. Geoff Wallace (*pers.comm.*) recorded as much as 13°C (25°F) increase on treated versus untreated soils in autumn. This increased temperature is generated both by the biological activity of the soil and the air pockets left by the chisel–points at various depths, and enables earlier and more frost–sensitive crops to be grown.

Nodulation (of nitrogen–fixing bacteria) is greatly increased, as is the breakdown of subsoil and rock particles by carbonic acid and the humic acids of root decay. Methane generated from decay aids seed germination, and water (even in downpours) freely passes into, not off, the soil. After a year or so, vehicles can be taken on the previously boggy country without sinking in. Drought effects are greatly reduced by soil water storage.

Water, filtered through soil and living roots, runs clear into dams and rivers, and trees make greatly increased growth due to the combined factors of increased warmth, water, root run, and deep nutrients.

Fukuoka[3] (in Japan) uses radish and *Acacia*; Africans use *Acacia albida* or *Glyricidia*; New Guineans use *Casuarina*; and Mediterranean famers use *Tamarix* for biological "chisel ploughs" where land is too steep and stony for implements. Otherwise the "graze or cut and let lie" method is still followed. On such difficult terrain as boulder fields, dunes, steep slopes, and laterites, forests of mixed legume/non–legume crops (citrus, olive, pine, oak) are the best permanent solution to soil conservation.

No matter how we aerate soil (or condition it), whether with humble implements like a garden fork levered slightly, by planting a daikon radish, or by sheer mechanical power, we can soon lose the advantage of looseness and penetrability by over-stocking, cropping, heavy traffic, or heavy–hooved animals stocked in wet weather. All of these pug or compress the soil into a solid state again. Final solutions lie only in following on with permanent and deep–rooted plants (forests or prairies), and by maintaining good management (minimum tillage) cropping.

Any reduction in cultivation saves energy and soils, and wherever no–tillage systems can be devised, and heavy hoofed animals kept to a minimum, soil structure can be repaired.

Intense fire, intense stocking, intense cropping, and intensive production all threaten soils. Thus, mechanical soil rehabilitation can be a one–time and beneficial process, or another way to waste energy every year. It is the usages that follow on re-habilitation that are beneficial or destructive to soils in the long term.

Mechanical loosening of soils is appropriate (on the broad scale) to almost all agricultural soils that have been compacted. Soils with coarse particles, of cinder, or dunes do not benefit from or need loosening, and very stony or boulder–soil mixtures are appropriately rehabilitated not by mechanical but by organic (root penetration) methods, as are soils on steep slopes. Some soils (like volcanic soils with permanent pastures) may never lose structure, and will maintain

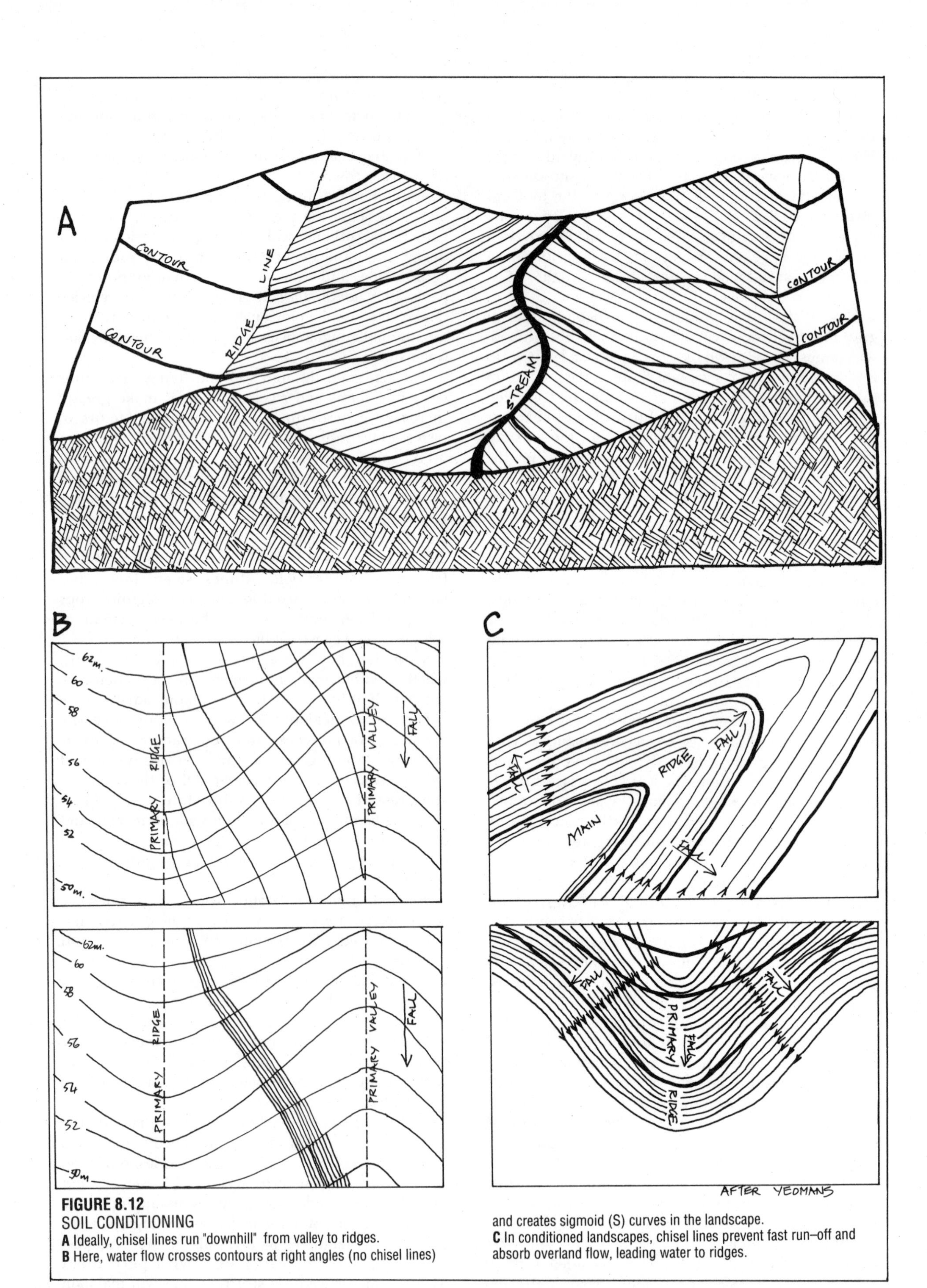

FIGURE 8.12
SOIL CONDITIONING
A Ideally, chisel lines run "downhill" from valley to ridges.
B Here, water flow crosses contours at right angles (no chisel lines) and creates sigmoid (S) curves in the landscape.
C In conditioned landscapes, chisel lines prevent fast run–off and absorb overland flow, leading water to ridges.

free internal drainage after years or centuries of grazing. Thus, we use rehabilitative energy only where it is appropriate.

Soil conditioning can be sequential, allowing a year between treatments, or all–at–once at 20 cm or so, in order to prepare for tree crop planted immediately. The time to use implements is also critical, and early spring or at the end of a gentle rainy period is ideal, as the soil is not then brought up to the surface as dry clods, nor collapses back as being too wet.

There is only one rule in the pattern of this sort of ploughing and that is to drive the tractor or team slightly downhill, making herring–bones of the land: the spines are the valleys and the ribs slope out and down–slope (**Figure 8.12**). The soil channels, many hundreds of them, thus become the easiest way for water to move, and it moves *out* from the valleys and below the surface of the soil. Because the surface is little disturbed, roots hold against erosion even after fresh chisel ploughing, water soaks in and life processes are speeded up. A profile of soil conditioned by this process is illustrated by **Figure 8.11**.

There is no point in going more than 10 cm in first treatment, and to 15–23 cm in subsequent treatments. The roots of plants, nourished by warmth and air, will then penetrate to 30 cm or 50 cm in pasture, more in forests. For disposal of massive sewage waste–water, Yeomans[5] recommends ripping to 90 cm or 1.5 m, using deep–rooted trees or legumes to take up wastes.

I have scarcely seen a property that would not benefit by soil conditioning as a first step before any further input. Pasture and crop do not go out of production as they do under bare earth ploughing with conventional tools, and the life processes suffer very little interruption.

In small gardens, the aeration effect is obtained in two ways:

• By driving in a fork and levering gently, then removing it.

• By thick surface sheet–mulch; worms do the work.

To summarise briefly, the results of soil rehabilitation are as follows:

• Living soil: earthworms add alkaline manure and act as living plungers, sucking down air and hence nitrogen;

• Friable and open soil through which water penetrates easily as weak carbonic and humic acid, freeing soil elements for plants, and buffering pH changes;

• Aerated soil, which stays warmer in winter and cooler in summer;

• The absorbent soil itself is a great water–retaining blanket, preventing run–off and rapid evaporation to the air. Plant material soaks up night moisture for later use;

• Dead roots as plant and animal food, making more air spaces and tunnels in the soil, and fixing nitrogen as part of the decomposition cycle;

• Easy root penetration of new plantings, whether these are annual or perennial crops; and

• A permanent change in the soil, if it is not again trodden, rolled, pounded, ploughed or chemicalised into lifelessness.

Trees, of course, act as long–term or inbuilt nutrient pumps, laying down their minerals as leaves and bark on the soil, where fungi and soil crustacea make the leaves into humus.

8.19

SOILS IN HOUSE FOUNDATIONS

Soils cause perhaps 60–80% of all house cracks and insurance claims for faulty construction and "tree damage". About 20% of the soils we build on will subside or heave depending on water content. Specifically, black cracking clay, surface clays, and red–brown clay loams are subject to swelling and shrinking. Solid stone and brick houses are most subject to structural failure, with wood–frame and veneer less so.

Over–irrigation of gardens, causing the water table to rise, is a primary cause of soil swelling. The removal of trees assists this process, as do paved areas, and burst or leaking sewage and water pipes. Some notorious white or yellow clays collapse as dam walls when wetted. It is as well to consult your local soil expert for large constructions as trials can be expensive.

While the effects are most noticed to 2 m deep, probes to 10 m deep need to be monitored for groundwater levels. Soils subside and shrink with excessive drying (too many trees too near the house) and swell and heave with excessive watering and no trees. Adelaide (Australia) is an area where most damaged houses are on blacksoil clays, but several other areas also suffer these effects, and in some, large buildings need to be built on foundations capping deep piles (to

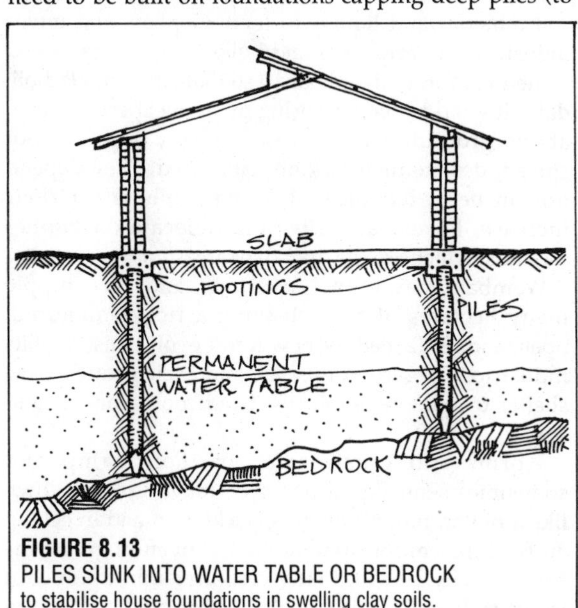

FIGURE 8.13
PILES SUNK INTO WATER TABLE OR BEDROCK
to stabilise house foundations in swelling clay soils.

20 m) sunk to the bedrock or deep into the permanent water table (**Figure 8.13**).

Most Australian native trees have an efficient water removal via roots, so that eucalypts remove 2–3 times the water of pines or pasture (to 10 m radius). Generally, householders should keep large trees at least one half the mature height from the house when building in high clay–fraction soils; sands and sand–loams are usually stable, as are rotten or fractured rock and sandstones.

8.20
LIFE IN EARTH

Before we ever learned to cut open the soil, it was thoroughly dug, aerated, and overturned by multitudes of industrious burrowers. The unploughed meadows of Europe and America are as soft as a great mattress, and are well aerated due to the moles, gophers, worms, prairie dogs, rodents, and larvae eternally at work below ground, even under the snow. Termites, ants, and crustaceans all do their part. The results are obvious from the good soils and great productivity of unploughed ground which has not been compacted by hooves or machines.

Termites and ants are the earthworms of the deserts and drylands, carrying tons of organic material to underground compost piles, in some of which they may grow fungi to feed their colony. The upthrown earth, whether from ants or moles, forms a specific niche for annuals to seed on, and wind–blown pioneer trees to occupy. If birds are the seedscatterers of the forest, burrowers are the gardeners.

Underground and beneficial fungal spores eaten by squirrels or wallaby and activated by their digestive enzymes break hibernation to occupy new ground and help the new roots of acorns and eucalypts to convert soil minerals and liquids to food. Gophers and moles industriously carry roots and bulbs to secret stores and sometimes forget their hoards, so that sunroot, gladioli, daffodils and hyacinths spring up in unexpected places above ground. This is how comfrey and sunroot spread, despite their lack of viable seed. They depend not on bees, but on moles and gophers for their increase. Foxes eat fruits, and defecate on gopher mounds, which are the dug–over areas for new trees.

Wombats may tunnel, overturn, and even topple many hectares of trees, leaving a richly–manured, open, and fertile bed for new forest evolutions. Rabbits industriously garden thistles, and their tunnels give shelter to possum, squirrels, bandicoots, snakes, and frogs.

Worms and crustaceans, in their damp and sometimes semi–liquid burrows, move up and down like a billion pump plungers, sucking in and expelling air (and thus nitrogen) to roots, and in effect giving the soil its daily breath. Many creatures mix up special mudbrick soils with body secretions, and from

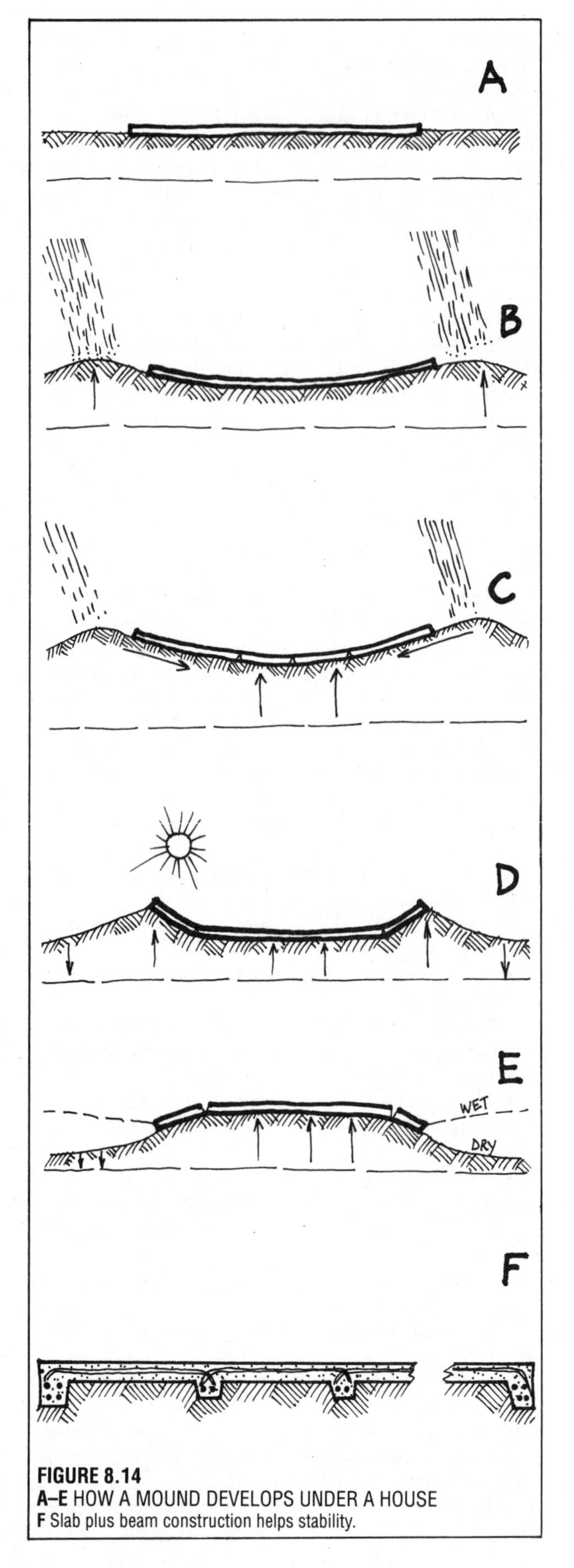

FIGURE 8.14
A–E HOW A MOUND DEVELOPS UNDER A HOUSE
F Slab plus beam construction helps stability.

swallows to termites create homes from stabilised soils. Seeds and spores are buried, excavated, hidden, activated, and forgotten by burrowers, and recycled to life or humus as chance and nature dictate.

Roots die seasonally, invade and retreat, and leave minute or massive tunnels for animals, fungi, and new roots to follow. I once tried to dig a parsnip out of newly drained swamp ground, following it along an old root trace, but gave up after 2 m. I did persist in following a 4 cm long engaeid "land crab" or earth lobster to 2 m down and 2 m along, and have often dug out rabbit and mouse burrows to find their nests, mating circuses, air vents, nursing chambers, disposal chutes, and escape hatches. Many old gopher burrows are filled with plant remains and faecal wastes.

A few dedicated souls in the history of science, from Sir Albert Howard to modern ecologists, try to excavate and discover something of roots, but as a simple bluegum (*Eucalyptus globulus*) can easily embrace an underground 1.5 ha, a forest is so complex and even intergrafted below ground that the canopy seems simple. Many desert plants lead a long and sturdy underground life while thin, straggly, and ephemeral in air. Some insects, like swift moths (*Hepialidae*) spend 7–8 years underground as large bardi grubs (a succulent treat for Australian diggers), with only a few days of nocturnal foodless life in air, mating and laying eggs before disappearing again to the root sheaths and soil, as their near cousins the ghost moths and witchetty grubs do in aerial stems. The bardi grubs, too, open thousands of shafts to the air, and cycle tons of nutrient underground, as do their relatives in air and sunlight. Their predators follow these hoarders and burrowers below the soil, and hunt them in darkness and secrecy.

The implications for designers are that many of these effects may be put to use, or their uses appreciated; it is as valid to plant an *Acacia* for the considerable by-product of swift moth or ghost moth larvae as it is plant a mulberry for silkworms, and to use moles instead of mole ploughs, or gophers as daffodil gardeners (unpaid). Even on shores and the bottom of lakes and seas, the burrowers work to carry nutrients below to roots and to bring up fresh minerals for decomposition, while assisting the flux of liquids and gases across the surfaces of mud media.

Roots have their own PENETRATIONS (depth), PATTERNS or spread, SCHEDULES, seasonal MIGRATIONS to or from the surface, and equivalents of deciduous drop or bark decortication, dying off and sloughing off root branches and bark. It follows that there is a topography of plants underground that parallels that of plants in air. There are also basic differences, in that special storages or fire–resistant organs found underground as ligno–tubers, tubers, bulbs, and rhizomes are very common.

Some species secrete phenols or creosoles to inhibit other plants (bracken, tamarisk, *Juglandaceae, Brassicas*); others encapsulate or surround hapless competitors (*Eucalyptus*, willows, tamarisks). Some trap nematodes and other would–be predators, or poison them out (marigolds, fungi, *Crotalaria*). While agricultural crops exploit from 0.6–4 m (2–12 feet) below the earth, some trees may penetrate to 50 m (164 feet) in deep desert sands. Around the roots of dune trees, calcium and other minerals are deposited as stone–like secretions by root–associated fungi and bacteria. Root space sharing is also scheduled, so that spring bulbs have fed, flowered, and died before the tree roots begin their upward thrust for nutrients and water. Tap–rooted late starters such as thistles and comfrey reach deep for late summer moisture, while a very few plants and fungi take advantage of the autumn rains for flowering and dispersal.

Where there is no season of cold death, as in the low latitudes, aerial roots may develop, or strangler figs send down roots from high in the crotches of other trees to the earth, there to build great buttresses as they strangle their host tree in a root well.

For designers, the diversity of roots in soil can be used as effectively as the diversity of crowns and canopies. Unstable slopes are pegged with the great root "tree nails" of chestnut and pine, oak and walnut. Even after a hundred years, the steep slopes of this island (Tasmania) are only just starting to collapse as the roots of the cleared forests rot, and could still be saved by pines, *Acacias*, or chestnuts. Bamboo not only holds landslides, but for light structures provides an earthquake–proof mattress of roots. The root mats of swamp vegetation save bulldozers from watery graves, and the fibrous web of the prairie defeats the wind.

GEOLOGY AND LIFE FORMS

Many rocks and strata on earth arise from the actions of living organisms. Whether it is the nodules of manganese in oceanic depths, deposits of diatomaceous earth, coal, or limestone, or opals and amber, all were once the products of living organisms. Much of the strata we see, except much–changed granitic and volcanic deposits, were formed from or modified by life. All soils are life–created, as are the corals and coral sands of many oceanic islands.

Life is also busy transporting and overturning the soils of earth, the stones, and the minerals. The miles–long drifts of sea kelp that float along our coasts may carry hundreds of tons of volcanic boulders held in their roots. I have followed these streams of life over 300 km, and seen them strand on granite beaches, throwing their boulders up on a 9,000 year old pile of basalt, all the hundreds of tons of which were carried there by kelp. Round stones are dredged from great depths in the mid–Atlantic; this does not mean that they were formed there, but more likely that drift kelp carried them there in their roots. Before they fly to Japan and Alaska, some millions of petrels (*Puffinus tenuirostris*) annually fill their crop with Tasmanian pebbles, seeds, and charcoal, which will be voided somewhere in the Pacific.

Life moderates every erosion process, every river basin, every cliff and rock fall. It shapes and reshapes earth in a thousand ways.

The hydraulic weight of great forests, such as were once in the Americas, would have exceeded any water catchment weight we can now afford to build, and dispersed it over a greater area. This greatly moderates climate, and with it geological processes. It is possible, in Iran, Greece, North Africa, USA, Mexico, Pakistan, and Australia to see how, in our short history of life destruction, we have brought the hard bones of the earth to the surface by stripping the life skin from it for ephemeral uses. We can, if we persist, create a moon–landscape of the earth. So poor goatherds wander where the lake–forests stood and the forest deities were worshipped. The religions of resignation and fanaticism follow those of the nature gods, and man–built temples replace trees and tree spirits.

8.21
THE RESPIRATION OF EARTH

All of the skin and organs of the earth breathe; it is a regular respiration. The "diaphragm" or energy for this may be provided by the moon tides in water, earth, or air. Locally, the filling up of soil by rainwater forces an exhalation of air; the drying–out an inhalation. Fast winds disturb boundary layers, create low pressure and soil exhalation; slow winds and high pressures force inhalation. Millions of earth animals open breathing tubes, and arrange them (for their own sake) to force an exchange between the atmosphere and the waters, the soils, or sea–sands in which they live. Water is as much breathed as air.

Deeper respirations come from deeper flows and fissures, and radon gas or methane seeps out from the earth. When the earth itself expands, great flows inward and outward must occur through the multitudinous fissures that open up in rigid sediments. This earth respiration transports and transforms fluids and their associated loads, solutes, states, and ionic potential from earth to atmosphere to ocean, setting up the potentials that create thunderstorms or hurricanes. We are of this same respiration. The burrows of spiders, gophers, and worms are to the soil what the alveoli of our lungs are to our body. We can assist this essential respiration by assisting life and natural processes in soils.

> If you kill off the prairie dogs, there will be no one to cry for rain.
> (Navajo warning)

Amused scientists, knowing that there was no conceivable relationship between prairie dogs and rain, recommended the extermination of all burrowing animals in some desert areas planted to rangelands in the 1950's "... in order to protect the roots of the sparse desert grasses. Today the area (not far from Chilchinbito, Arizona) has become a virtual wasteland." Fierce run–off, soil compaction, and lack of fresh seedbed have carried the grasses away (Barre Toelken, in *Indian Science Quest '78*, Sept/Oct).

Using prairie dog burrows as water sinks, and causing water run–off to flood down them, thus germinating stored underground seed, had the opposite effect on the Page ranch, now a dryland

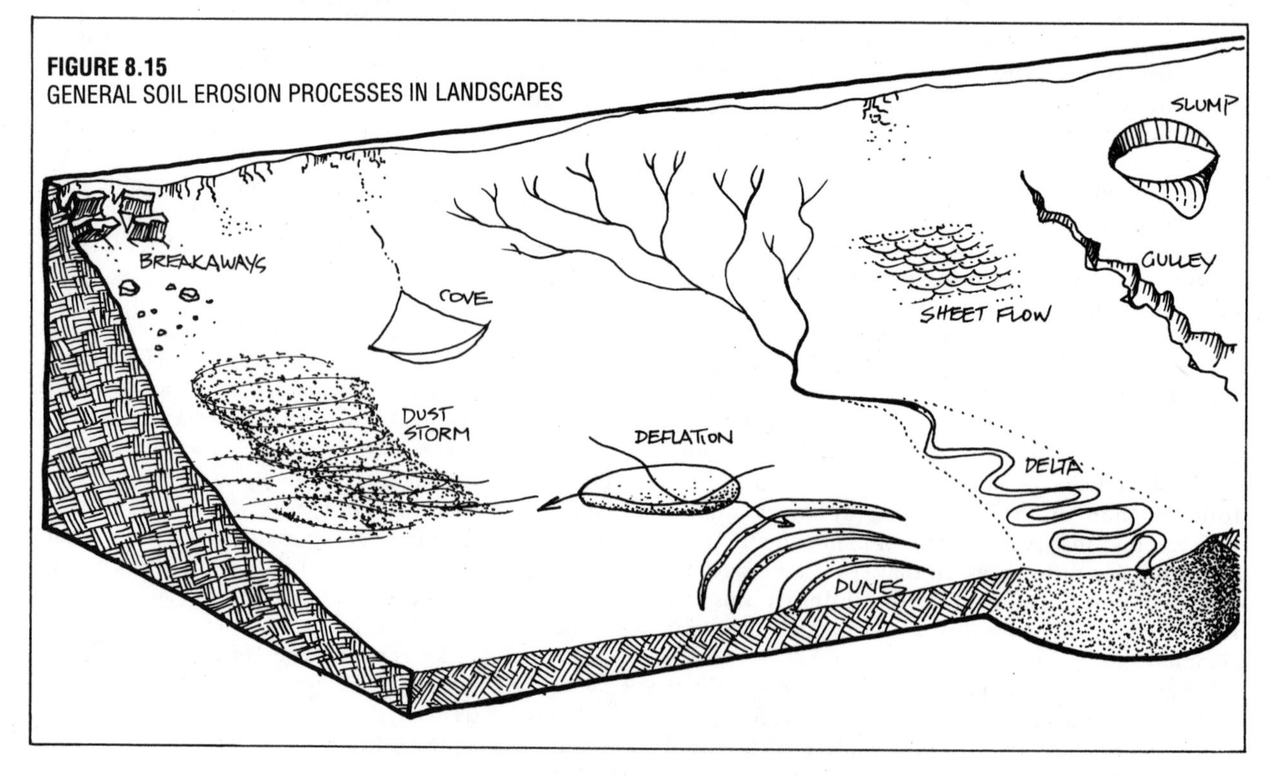

FIGURE 8.15
GENERAL SOIL EROSION PROCESSES IN LANDSCAPES

rehabilitation exhibit of the University of Arizona. Here, prairie dogs and a new and thriving patch of permanent bunch grasses thrive in an area where overgrazing, ploughing, or soil compaction has ruined other grasslands.

> Water under the ground has much to do with rain clouds.
> If you take the water from under the ground, land will dry up.
> (Hopi elder in *Tellus*, Fall '81)

At Black Mesa, near the Four Corners area of the Hopi Indians (USA), a scientist studying thunderstorm occurrence (using computer analysis) noticed an unusual number of storms occurred in that area. She was told by the Hopi of the area where the earth breathed, emitting air as the moon affected the groundwater tides. This air proved to be heavily charged with negative ions, which may have initiated the thunderstorms and consequent rain.

Of the breathing of the earth, there has been little study, although it was regarded as a known phenomena to tribespeople. The earth must breathe, by at least these processes:

- The movement of burrowers in their tunnels;
- The movement of groundwater by tides or replenishment of aquifers (often seen in wells, especially near coasts and lakes); and
- The evaporation of moisture from soil surfaces by the sun.

8.22

DESIGNERS' CHECKLIST

1. It is a *primary* design strategy to prevent topsoil losses and to repair and rehabilitate areas of damaged and compacted soils.
2. Permanent crop, soil bunds, terraces, and low–tillage systems all reduce soil and mineral nutrient loss.
3. Soil rehabilitation and pioneer green crop should precede other plant system establishment.
4. Adequate soil tests, plus test strips of crop examined for deficiency or excess symptoms, leaf analysis, and livestock health should be assessed to guide soil treatments.
5. If soil types can be specified, fencing, cropping, and treatment should coincide with these specific soil assemblies, and specific crops for such type researched.
6. Soil life processes need to be encouraged by provision for green crop, humus, mulch, and the root associates (mycorrhiza) of plants. A useful earthworm may need to be introduced.
7. Drainage, hence pH and soil water capacity, need specific treatment or assessment, and will largely determine crop and tree types.
8. Minimal use of large livestock and heavy machinery is to be recommended on easily–compacted soils, as is burning and clearing.
9. Use pigeon and animal manure where major elements are scarce, as in third world areas (also use of greywater and sewage, or wastes).

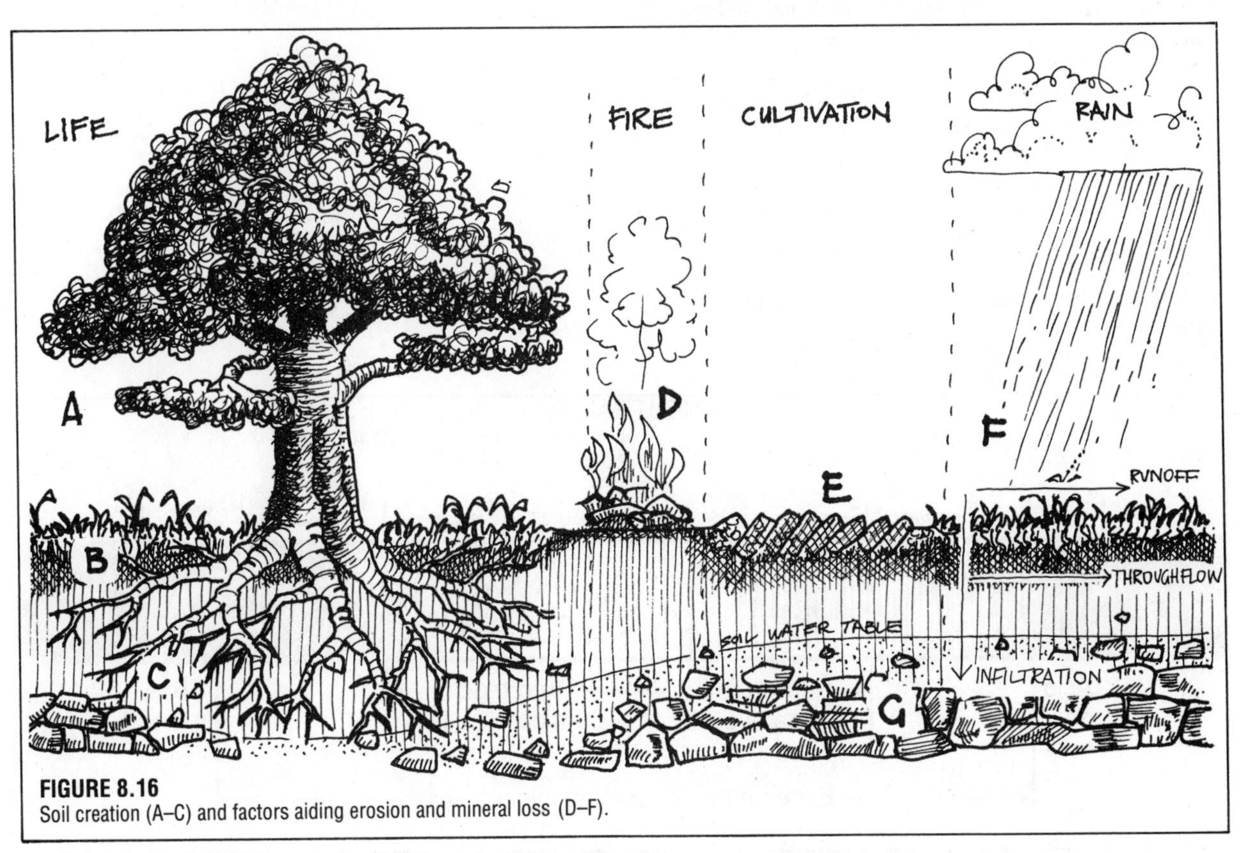

FIGURE 8.16
Soil creation (A–C) and factors aiding erosion and mineral loss (D–F).

10. Before draining waterlogged soils, recommend crops to suit this condition. Never drain wildlife habitats, fens, or bogs which are species–rich.

11. Choose the right soil–shaping or earthworks to suit crop, drainage, and salt threat.

12. Using an auger, check soils for house foundations. Using a (wetted) soak pit, time the absorption of greywater for sewage disposal at house sites.

13. Preserve natural (poor) sites for their special species assemblies; pay most attention to human nutrition in home gardens, and select species to cope with poor soil conditions on the broadscale.

14. Fertilise plants using foliar sprays containing small amounts of the key elements, or pellet seeds with the key elements which are deficient locally. Pelleted seed and foliar sprays are economical ways to add nutrients to plants.

8.23
REFERENCES

Cox, George W. and Michael D. Atkins, *Agricultural Ecology: an analysis of world food production systems*, W. H. Freeman & Co., San Francisco, 1979.

Davidson, Sir Stanley, *et. alia.*, *Human Nutrition and Dietetics*, Churchill Livingston, N. Y., 1979.

Dineley, D. *et.alia.*, *Earth Resources; a dictionary of terms and concepts*, Arrow Books, London, 1976.

Fairbridge, Rhodes W. and Joanne Bouglois, *The Encyclopaedia of Sedimentology VI*, Dowden, Hutchinson & Ross, Stroudesberg, PA, 1978.

Handreck, Kevin A.,
1978, *Food for Plants*, CSIRO Soil Division, Australia, #6.
1978, *What's Wrong With My Soil?*, CSIRO Soil Division, #4.
1979, *When Should I Water?*, CSIRO Soil Divison, #8.

Leeper, G.W., *Introduction to Soil Science*, Melbourne University Press (Aust.), 1982.

McDonald, R. C. *et. al.*, *Australian Soil and Land Survey Field Handbook*, Inkata Press, Melbourne, Australia, 1984.

McLeod, Edwin, *Feed the Soil*, Organic Agriculture Research Institute, Graton, California 95444,1982.

Radomsky, S. W., Heidi Kass, and Norman J. Pickard, *Elements of Chemistry*, D. Van Nostrand Co. (Canada) Ltd, 1966.

Satchell, J. E. (Ed.), *Earthworm Ecology*, Chapman and Hall, London, 1984.

van der Werf, Erik, "Sustainable Agriculture in South–East Ghana and The Agomeda Project" *Permaculture Nambour* Newsletter, Dec. 1985 and March/April 1986. P. O. Box 650, Nambour, QLD 4560 Australia. $3.00 per newsletter.

Weir, R. G., *The Plants' Nutrient Requirements*, Seed and Nursery Trader (Australia) July 1979, pp. 55–57, 1979.

Yeomans, P. A., *Water for Every Farm*, Second Back Row Press, Leura, Australia, 1981.

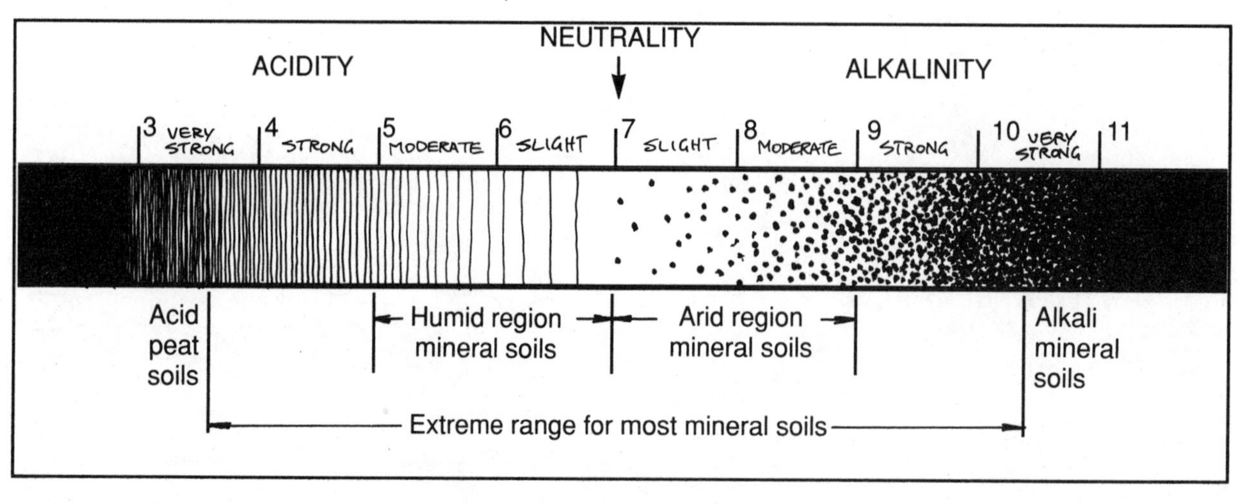

Chapter 9

EARTHWORKING AND EARTH RESOURCES

Moving of the earth brings harms and fears
Men reckon what it did and meant
But trepidation of the spheres
Though greater far, is innocent
(John Donne)

And did the earth move for you?

9.1 INTRODUCTION

Few people today muck around in earth, and when on international flights, I often find I have the only decently dirty fingernails.

For the soil scientist, soil has endless classifications; to engineers, it is a material; to potters, their basic resource; and to housekeepers, footprints on the floor, or part of the eternal dust. However we look upon earth, we will all return to it, and help create the soils our ancestors made, ruined, or form part of. Climate, vegetation, animals, and soil are intimately connected, and each will have influences on the other. It is a great subject, like that of water, and this chapter will concentrate more on its use in structural design than on uses as a growing medium.

People, and other animal species, have mined soils and earth deposits for specific earth resources since their inception; examples are animal migrations to salt licks, and specific dusting sites in which birds and mammals roll or bathe to rid themselves of parasites. People have used silica minerals (obsidian, chert, chalcedony) as tools, and have ground iron oxides or graphites as pigments for many thousands of years. Soils rich in iron oxides have long been mixed with acorn meals to fix the bitter tannins as insoluble ferric tannates ("black" breads), or to supply missing elements in the diet.

There are several clays (illite, smeltite, kaolin, ferrous oxide "red" clays) traditionally used by tribal and modern peoples to absorb poisons (from *Solanum spp.*, yams), to reduce diarrhoea and digestive upsets, or to relieve feelings of nausea (kaolin clay). Although little work has been carried out on this factor, we should record all uses when specific clays are so used, and how they are combined with specific food to reduce or eliminate poisonous or unpalatable toxins from foods. Some earths may be eaten to supply trace elements, as clay and clay–salt licks are commonly visited by deer, kangaroo, cattle, and antelope. Earth–eating (geophagy) is a widespread and seemingly natural habit of children and tribal peoples. As well as clays, the white, mineral–rich ash of trees is widely used as a food dip at campfires by Australian Aborigines. Cooking food in clay is prevalent throughout the world.

Earthworking for agriculture and monuments has existed for at least 17,000 years. Mineral smelting, pottery, and peat or coal mining for fuels increased the scope of earthworks, and this was quickly followed by the development of large machinery intended to remove vast amounts of ores and fuels for modern industry, which built quickly from 1800 on, so that the last 200 years of our existence has seen the greatest development of earthworking and mining.

The common use of self–transporting machines really dates from the post–war years (1947 and on), following the wartime development of tracked tanks, earth scoops used for airfield construction, and large pneumatic tyres for this type of equipment. While civil engineers have to some extent kept pace with these developments, neither the public at large nor those in architectural or agricultural fields have fully realised the potential of earthworking machines in the modern sense.

Our power to move the earth with modern machines is now almost unlimited; we can, if we wish, raise new hills or plane off existing ones, and create or obliterate minor landscape features. Small earthworks are so immediately effective, cheap, and permanent that it continually amazes me that people will suffer local drought, seawinds, noise, erosion, or even flooding without spending the few hundred dollars on a well–built and planted earthbank that would solve the problem. They will build expensive tankstands or towers rather than a cheap hill, and suffer death by storm and fire rather than make a very safe earth shelter for their families (it serves as an outdoor cellar at other times).

Earthworks are necessary and ethical where they:

- reduce our need for energy (underground housing in deserts);
- diversify our landscape for food production (fish culture ponds);
- permanently rehabilitate damage (contour banks, interceptor banks);
- save materials (house site design); or
- enable better land use, or help revegetate the earth.

As with all techniques, it profits us to make as many uses of earth shaping as we can; it is shameful to see quarries, mines and roads serving a single purpose, and usually left as a sterile system, when they could be shaped or planted to assist landscape diversity.

A whole set of skilled and well–tried waste or soil reclamation strategies has developed as methods of stabilising devastated landscapes, both for natural instability and the carelessness of engineers. An excellent handbook for those involved in such painstaking work, and one covering many climates and areas, is that of Dr. Hugo Schiechtl (*Bioengineering for Land Reclamation and Conservation*, University of Alberta Press, 1980). I cannot too highly recommend this book for would–be earthmovers.

Earth can be moved for productive reasons, many of them classifiable as landscape restitution:

- To create shelter; to assist with foundations and to make areas level for floors;
- To terrace hill slopes for stable padi crop, wet terraces, or gardens;
- To raise banks or to dig ditches as defences against flood, fire, attack, or wandering vegetation–eaters;
- To drain or fill areas (to direct water flow or run–off);
- To create access roads to those places we commonly visit;
- To get at earth materials (ochres, clays, minerals, fuels);
- To make holes for any number of reasons and of greatly varying sizes from fence–posts to dams, wells to deeply drilled bores.
- To create special storages and enlarge living space (cellars and caves);
- To stop erosive forces carrying off soils (soil conditioning and erosion control);
- To prevent noise pollution (embankments); and
- To permit recharge of groundwaters (swales and ripping).

We also move the earth to play and to plant. For all these reasons, we have devised hand–held and mechanical diggers, ditchers, augers, drills, blades, buckets, shoes, rakes, ploughs, rippers, delvers, scoops, earthplanes, loaders, rock–cutters, draglines, excavators, and dredgers. We also move earth with explosives, hydraulic jets, and as an unintentional result of erosive processes generally.

Erosion has itself been used to build soil terraces on lower slopes in more than one culture, but it is debatable if the terrace idea was not as a result of attempts to stop erosion before the idea of using erosion to create terraces later developed.

Until the Second World War, earth was moved by sheer numbers of people, by hand or horse and car, or by a few people working with wheelbarrows or baskets over a long period. All this has changed. Why put up thousands of mud bricks when a machine can compact a 6–8 m thick wall immune to flood, fire, and earthquake in a few hours? Or labour long hours over a hole when we can blast a fence–post in a hard shale base for a few cents?

In this section, I will not attempt to deal with large, complex, or precise civil works, those necessary for sewage layouts, the landscape excavations for large buildings, or large dams, ports, or aerodromes. I will instead limit myself to those on–farm, private, useful, relatively small, and rehabilitative or sustainable earth–moving systems that an individual might employ to shelter a house, or to control water in a productive landscape.

9.2
PLANNING EARTHWORKS

It is best to plan all aspects of the earth–moving process *before* the machines or labourers arrive on site.

1. Make an initial decision where you would like to place the (e. g.) road, dam, house site, drains, etc., using a contour map and plan if necessary.

2. Test the soil by auger holes, soil samples, and soil pits to determine if the soil is good enough to suit your plans (a good clay soil for dams is essential). Seek professional advice or do more research before deciding conclusively on placement.

3. Peg out the site, using a level (which can be simple or complex), a measuring tape, and a good many stakes with red or white cloth attached (so that the earth–moving machine can follow them).

4. Plan a place to store all the TOPSOIL removed during the excavations. Never allow topsoil to be mixed with subsoil, but *carefully* remove it, to be later returned to the site as a growing medium and to stabilise subsoil erosion.

To stabilise the site immediately, have on hand as many seeds and plant materials as needed. These can be purchased from a nursery or grown in pots on the site several months before the planned earthworks (see

9.3 Planting After Earthworks).

When soil is moved, it becomes loose, so that soil air space and hence total volume may increase to 145% of the original. Even when compacted by machine, the fill may occupy a space 10% larger than the cut it came from. Although conscientiously compacted clay in dam walls may settle as little as 1% over time, loose fill will eventually settle to 75% or less of its uncompacted volume; this has great relevance to house foundations and wall stability.

When topsoil is replaced over fill, the area should not be over–compacted, or we risk waterlogging, but when it is replaced over deep or solid subsoil in a cut, we will probably need to first rip or loosen the base subsoil to allow root penetration, just as we need to rip old roads, quarries, parking areas, or heavily–trafficked fill before planting trees and meadows.

To prepare a house site (with drainage), we can proceed as follows:

- Careful survey. Place pegs *outside* the site. Call in the bulldozer.
- Strip off topsoil carefully and mound above and at either end of the site.
- Cut house/garden level; use subsoil fill for access roads only.
- Call in ditcher or backhoe to cut foundation and drainage trenches. Pour foundations and slab, paths; place drains and pipes.
- Call in small blade and bucket machine (a Bobcat or wheeled tractor) to replace topsoil and neaten the site where needed. Some soil can be mounded to the windward side for hedges.
- Plant or seed all topsoil to prevent erosion.
- Fine–tune with barrow, rakes; check drainage.

9.3 PLANTING AFTER EARTHWORKS

Every time we move soil, we should be ready to follow straight on with planting or seeding. That is, we need to have planned the planting and stabilisation of the area, and to have the plant materials on standby to implement our plan *as soon after disturbance as possible.* There are two reasons for this:

1. To prevent erosion, which can be severe on bare slopes at only 2% slope, especially in rains; and

2. To prevent invasion by unwanted volunteer plants, which may become difficult to displace later.

If a full set of ground covers, pioneers, and long–term plants can be set out in new earth, a great deal of time and work is saved. A broadscale scatter of mixed seed, raked in, will prepare the way for permanent placements. We are most fortunate if we can immediately mulch bare soil sites with hay, hessian (burlap), or woodchips, to break the force of rain and to suppress unwanted weeds.

If you have prepared for bulldozing, you have seed, divisions, cuttings and potted plants ready to go before the machine pulls out. Seed can be garden–collected or purchased; just scatter and if possible rake in. Some mixes that work in most soils:

- Sunflower, or mixed parrot seed with sunflower, millets, pulses, chard, parsley, lupin, and clovers.
- Parsnip (fresh seed), salsify, daikon radish, radish, turnip; all of these "spike" the soil.
- Bulbs of lilies, grape hyacinth.
- Roots of sunroot, comfrey, chicory, horseradish, ginger, sweet potato, tumeric.
- Divisions of bamboo, banna grass, pampas grass, aloes, agave, New Zealand hemp.
- Seedlings or sets of elephant garlic, asparagus, globe artichoke.
- Cuttings of small fruits; elderberry, willow, poplar, mulberry, Pride of Madeira (*Echium fastuosum*).
- Tubed seedlings of *Acacia*, *Prosopis*, tagasaste, New Zealand mirror plant (*Coprosma*), pines, eucalypts, shrubs.

All compete very well with self–sown weeds, and with very little help or none at all establish a varied and useful early and perennial crop system; a few annuals self–sow and seed down another year. Many species can be further divided or cuttings set out, and stakes can be set at the perennials so they are easily located. Excess grass is cut back in the following years.

Clover can be late–sown to allow vegetable seed to get away. Grasses are not sown, but will invade if spaces appear, and can be grazed by geese in a few years. Failed plants can be replaced with successful

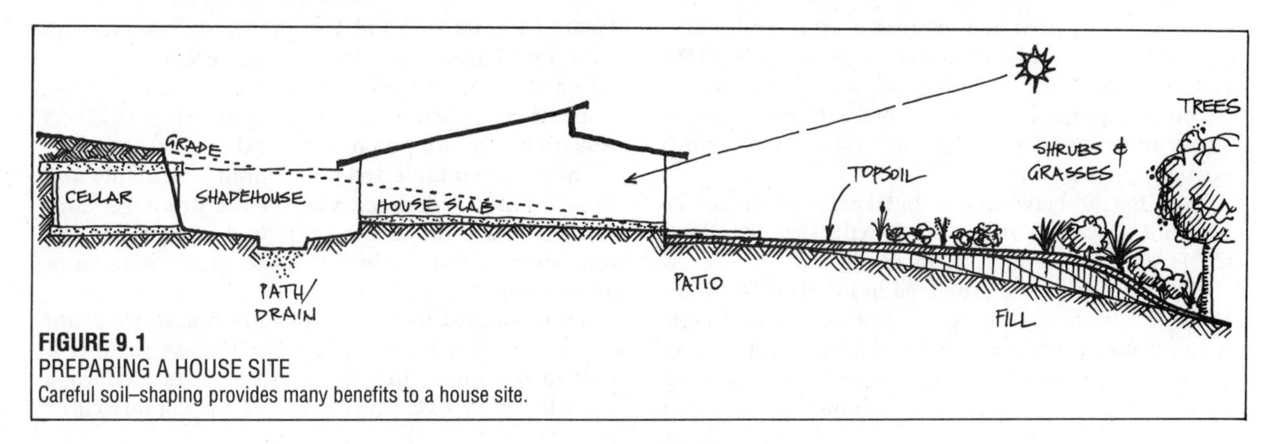

FIGURE 9.1
PREPARING A HOUSE SITE
Careful soil–shaping provides many benefits to a house site.

plants late in the cycle.

It is always an advantage to smooth–finish banks and surrounds so that a mower or scythe can be used until the selected plants take over. Trees are a danger on dam walls; if they fall (and they often do in those conditions) they take part of the wall with them, but bamboos, ginger, sweet potato, pepino, and clump grasses assist bank stabilisation. Trees at the *base* of walls are advantageous in shading, removing water, and reducing weeds.

The "net and pan" planting pattern of **Figure 11.84** is an effective control in overgrazed, eroded, mined or bulldozed sites. If tyres are available, the "pans" can be made from these, filled with mulch, and the diversion drains led in above the tread level. For people with access to logs, these can be staked cross–slope, on a slight downhill grade so that water is made to zig–zag across the erosion face, and hence absorb into the ground.

Even small logs and branches, pegged across erosion channels, build up a layer–cake of silt and leaves, beside which can be planted willow, *Acacia,* or any other fibrous–rooted and hardy species, which then act as a permanent silt trap. Mulch behind logs and barriers quickly stabilises the seed bed for planting. Fallen leaves and scattered dung also accumulate in these mini–deltas to provide plant nutrients. Small wire netting fences, with stone–weighted hay uphill, will trap silt and spread water, as will cross–swales of lemongrass or Vetiver grass.

On very steep slopes, there is often no recourse other than to plant pampas, bamboo, lemongrass, and root–mat pioneers and to make upslope plantings of chestnut, *Acacia,* carob, olive or other large species which will cascade seed downslope over time. Where implements such as chisel ploughs can be used, these are effective in erosion control. Planned chiselling and planting makes a permanent and stable change on hillside. Details on gully control in drylands and in tropical humid areas are given in subsequent chapters.

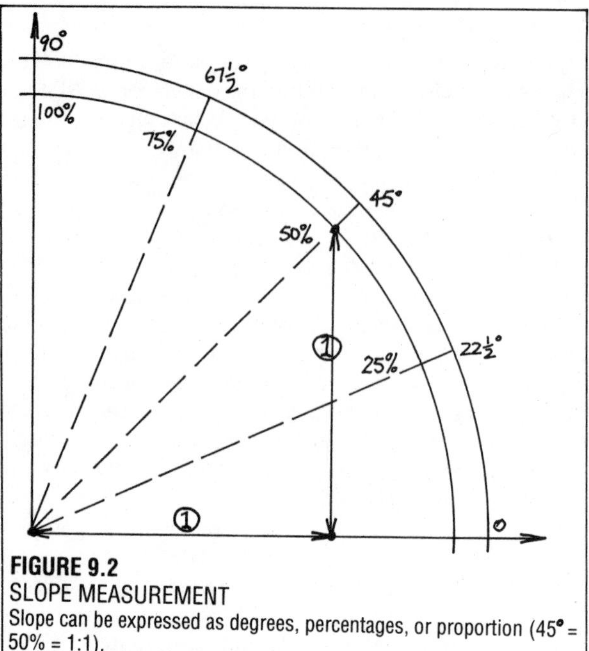

FIGURE 9.2
SLOPE MEASUREMENT
Slope can be expressed as degrees, percentages, or proportion (45° = 50% = 1:1).

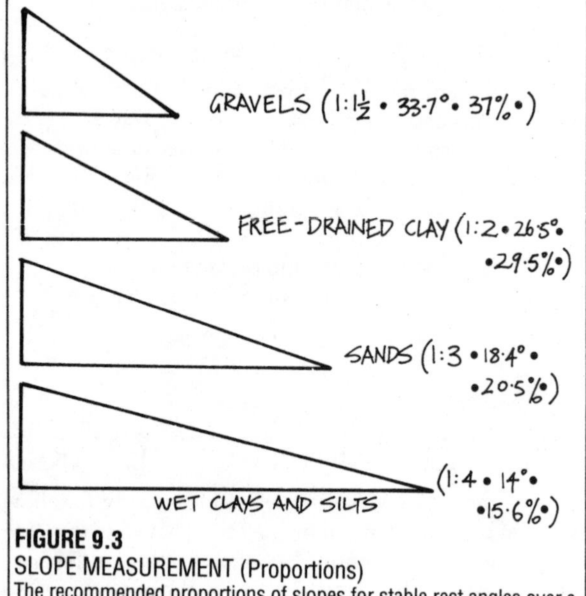

FIGURE 9.3
SLOPE MEASUREMENT (Proportions)
The recommended proportions of slopes for stable rest angles over a range of soil materials.

9.4
SLOPE MEASURES

Slopes are measured and expressed in several ways. First, the slope can be measured as DEGREES FROM THE HORIZONTAL, or the base angles of **Figure 9.2**.

This is expressed as "a slope of 15 degrees", and is sometimes used by geologists to describe the dip (angle) of rock strata.

Next, the 90° between horizontal and vertical can be divided into 100 parts, and expressed as PERCENTAGES, a measure used by engineers.

Finally, slope can be expressed in PROPORTIONS or ratios of base to height, e.g. 1:4 is a slope 1 unit high, with a base 4 units long. This is the rise of slope over distance, often used for drains and pipes, in units used by hydrologists. Many irrigation drains use slopes of 1:500 to 1:2,000, or a fall 1 m in 500 to 2,000 m, and these are referred to as "grades" (**Figure 9.3**).

Coarse slopes are often expressed in degrees, low–slope drainage across slopes as a proportion (1:2,000). Any units serve for proportional or degree measures.

There are suitable irrigation grades for soils and drains. Fine sands need very gentle drainage slope (1:,2000) and earth cuts in sand need an easy angle of rest, about 1:4.5. **Table 9.1** makes clear some slope implications.

Stable natural foothill slopes are not straight but down–curved (concave). When building a road, it is well to remember that they will achieve a concave curve by slump over time, so it is best to cut this curve

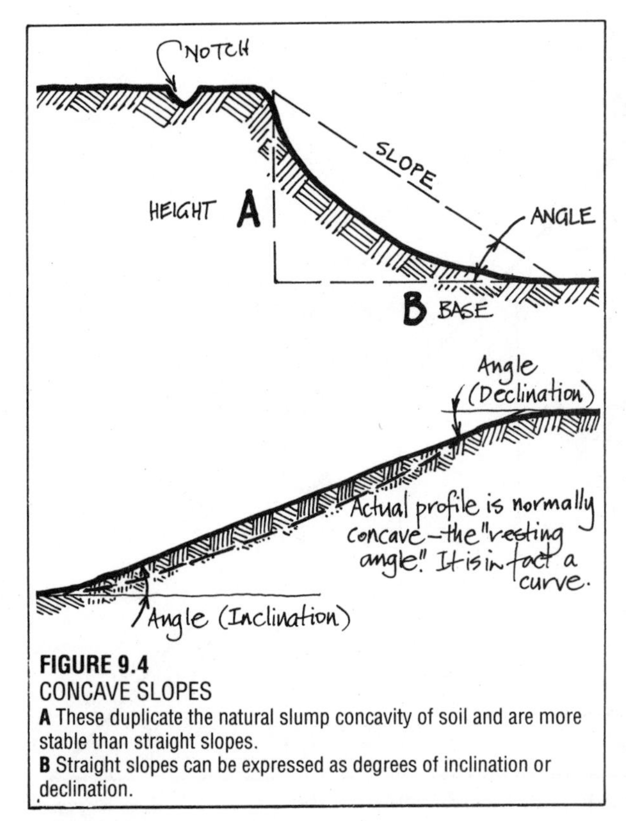

FIGURE 9.4
CONCAVE SLOPES
A These duplicate the natural slump concavity of soil and are more stable than straight slopes.
B Straight slopes can be expressed as degrees of inclination or declination.

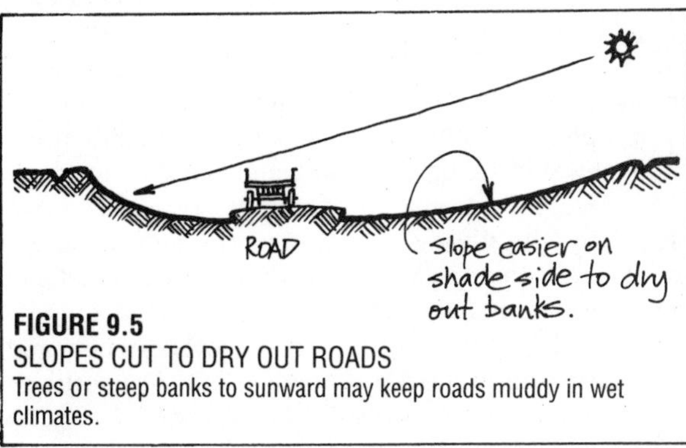

FIGURE 9.5
SLOPES CUT TO DRY OUT ROADS
Trees or steep banks to sunward may keep roads muddy in wet climates.

into the slope to start with. Natural lower slope faces in humid areas are also down–curved, and should be made to curve down or to be concave. Earth banks will assume this profile in time (indicated by a dotted line in **Figure 9.4**). Soils have type–specific resting angles with concave profiles.

Slope angle (the dotted line) is therefore a straight–line approximation of real soil slopes. The notch at the top of a cutting runs water off the face to safer slopes, and is a standard feature in embankment stabilisation. Average safe slopes used by engineers are:

- Gravels 1:1.5
- Clay (well–drained) 1:2
- Clay (wet) 1:4

To dry a road, the sun side should be cut further back and so allow sun in; conversely, to shade areas, banks can be as steep as the soil will stand (**Figure 9.5**).

When we cut or "bench" into a slope, the natural erosion processes try to re–establish the original slope,

Table 9.1
SOME FEATURES OF COARSE SLOPES (given as percentages)

100	Vertical bank; safe only in stone or very dry stable sediments, soft rock.
90	Vertical bank; the walls of dry desert dugouts in stone can be quite stable at this slope.
80	Precipitous. Safe only in stable rock or loess.
70	Precipitous. Safe only in stable rock or loess.
60	Very steep; a bulldozer can be used up or down, not on the traverse (cross–slope).
50	Very steep; 1:1 proportion or 45°. Strictly not for tree clearing, needs permanent forest.
40	Steep; the maximum a track machine (bull dozer) can safely traverse and work.
30	Steep; as above; can be cautiously harvested for firewood or coppice, also benched for access.
20	Moderate; usually accepted as the maximum slope for safe cultivation and erosion control; wheeled tractors are used at about 18%. The maximum slope of good road surfaces is accepted as13%; anything over this should be concreted or sealed.
10	Moderate; may need contour banking and careful use. 4–6% is considered ideal for railroads, without special cogs or cables, to assist the grip of the wheels.
0–10	Low; can erode if soils are cultivated, but these slopes are usually stable for crop agriculture, terraces, and swales.

and if left to their own devices will do so. Benching at the base of sloping and unstable tilted strata will therefore bring down the whole hill, and incur endless costs in road–clearing and slump removal. Slumps in unstable hillsides will carry with them, or bury below them, houses and people.

A great many film stars perched on unstable ravine edges in the canyon systems of Los Angeles will, like the cemeteries there, eventually slide down to join their unfortunate fellows in the canyon floors, with mud, cars, and embalmed or living film stars in one glorious muddy mass. We should not lend our talents to creating such spectacular catastrophes, but to the avoidance of obvious problems and by re–location before disasters occur. Local government often delineates areas of instability where building is not permitted due to mudslides, fire, and earthquake.

In the dry hill scrublands of California and France, or on desert borders, intense fire upslope can generate sufficient heat to bake soils. When such fires occur in steep, unconsolidated shales and mudstones, the baked earth (heated to 6 cm deep) forms water–sealed slab surfaces (such hydrophobic crusts will no longer form crumb structure).

Valleys in such areas can then expect catastrophic mud flow unless early misty rains fall to restart vegetative growth, and to that end the ashes can be sown down with oat, lupin, and shrub seeds, but if

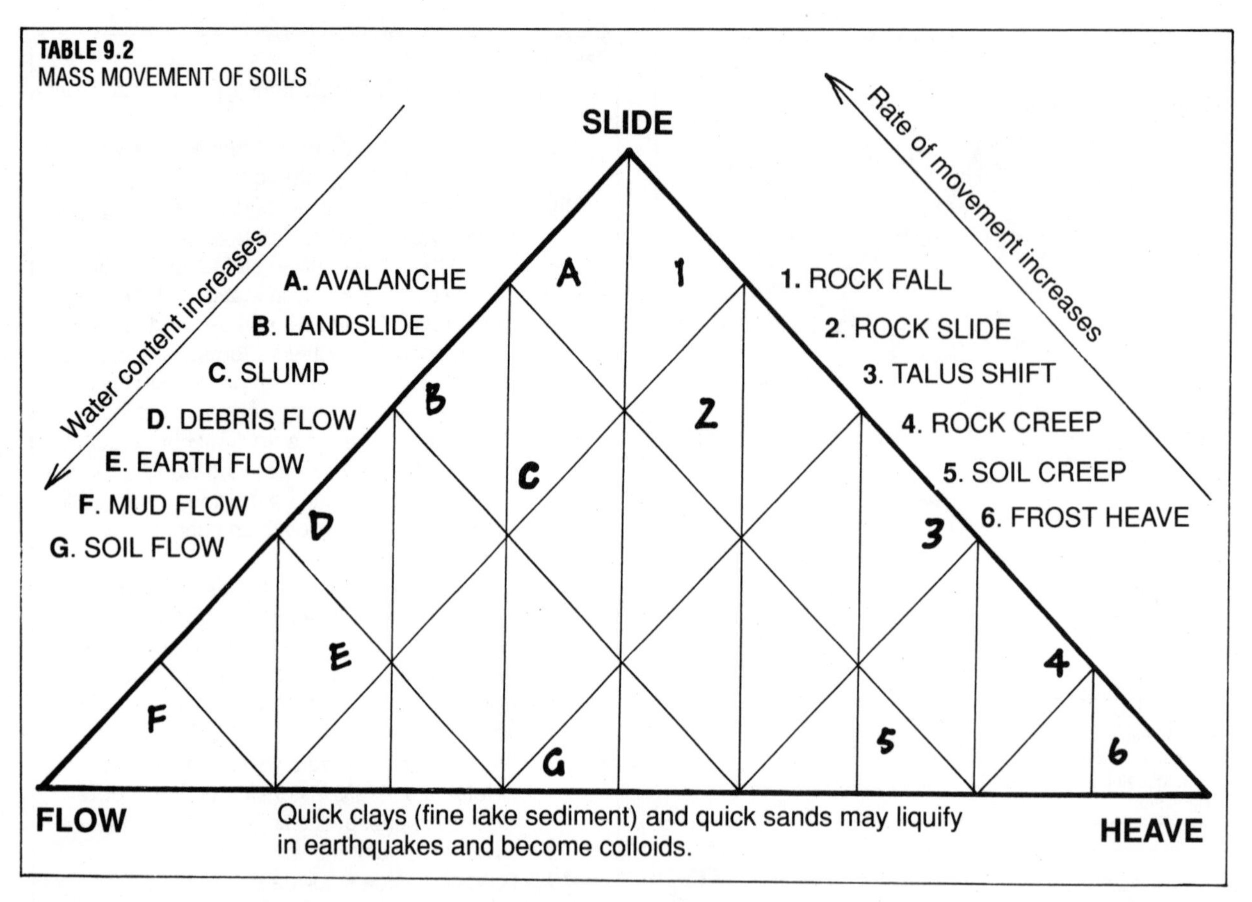

heavy rains fall before plants take hold, water can penetrate below the glazed surface, and mass movement (bulking) of mud and glaze will then occur. The wall-of-mud effect, and the inevitable cascade downhill may well block roads in any such mud-flood emergency, threatening resources in the valleys. Mud flows are a feature of unstable sediments in wet-dry climates.

Thus, the water run-off after fire (as in Adelaide, Australia in 1984) can be catastrophic, and hot fires can have a long-term effect on soils and soil loss. The combination of fire in late summer, followed by a sudden autumn onset of rains is typical of Mediterranean lands, and of devastated soils. For a designer (and town planner) areas of known soil instability should be avoided as settlement sites, and kept in permanent forest; most are designated on soil maps, or can be researched.

9.5
LEVELS AND LEVELLING

Levels are taken to ensure that spillways work, drains run, and houses sit level enough so that one can play at marbles without them all falling towards one corner.

There are tools of levelling, of which the most ancient and even the most reliable is that level assumed by the surface of water. One can always check a horizontal against the sea horizon, or make sure a drain or gutter works by using a little water to flow down it. Even sophisticated instruments like the theodolite use a bubble level to set up the base plate, and then to check the telescope against this with its bubble before preceding with tilt or swivel measures. There are a few simple levels that can be handmade and are of great use to farmers and homeowners:

HOSE ("BUNYIP") LEVEL. A hose level can be made from easily-purchased materials, and is useful to survey some metres or even kilometres of diversion drain, level a house site from corner to corner, or set a wall or spillway at a pre-determined level. The materials needed are (as put together as in **Figure 9.6**):

- 20 m (65 feet) or so of clear plastic hose, diameter about 10 mm (1/2 inch).
- 2 corks or stoppers that fit into the ends of the clear plastic tube, each with a breather hole.
- 2 small cork balls that freely fit inside the clear plastic tube below the corks (optional)
- 2 stakes (2 m or 6.5 feet), marked with the measurement of your choice (centimetres or inches); dressmakers' tapes glued or screwed on to the stakes serve this purpose.
- 4 clamps or tape to connect the clear hose to the stakes.

To prepare for use, uncork one end and fill the entire hose with water. Be sure *all* the bubbles are out of the hose; this is only certain with a completely clear hose.

Place both stakes together, tops level, on a piece of *level* concrete or wood. Keeping the stakes *vertical*, mark the water level in the clear tubes by unclamping and moving the measuring strips until the water level (or both cork floats) are at precisely the same measurement marked on the stakes. Recork the tubes.

In the field, one person walks past the other alternately, driving a stake every 6 or 15 m (20 or 50 feet) to guide diggers or machine operators, and allowing for the rise (or fall) of drains by allowing one cork ball to be an inch or so lower than the other at the next peg, moving the pole up or downhill to achieve this. The stakes must be kept vertical when measuring.

1 cm in 5 m is 1:100 x 5 equals 1:500 fall, ample for most diversion drains to carry run–off to dams. Dead level lines can also be run across landscape, and dam walls checked for level or (if used for roads) a gentle rise. I recently saw a 12 km drain flowing gently all the way in sand at 1:2000, laid–in not by a skilled surveyor but by two women hired for a week and trained in 30 minutes on a hose level.

THE PLANE TABLE is a sturdy tripod which can hold a U–shaped piece of plastic tube or glass, corked and de–bubbled as for the hose level, except this can now be swivelled about to sight in all directions (**Figure 9.7.B**). Lacking tubing, a bowl of water on which floats a piece of painted wood with two nails of equal length as "sights" will do (**Figure 9.7.A**).

Sealed ends with small hole
Clamps
Pith ball or plastic bubbles
Clear tubes
2 Equal metre sticks
garden hose say 20m.

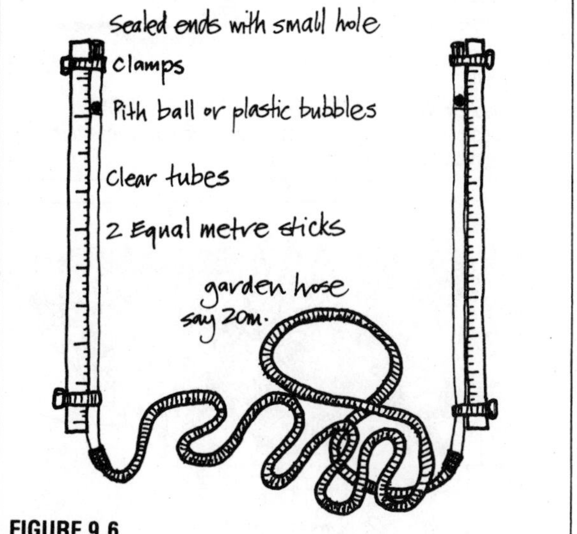

FIGURE 9.6
"BUNYIP" HOSE LEVEL
Used by two people (about 15 m apart) to peg out a contour line around a hill.

Although Model B is self–levelling, it may be necessary to tap in one nail a bit on Model A by way of adjustment.

To check a plane table, set up two marked stakes (with easily–visible white masking tape) a little up hill from the point you will stand at, driving the tops to level using the table.

Viewing the stakes from position B with the plane table, check that the level shows new level marks (b) to be exactly the same distance (d) below the stake tops (measure this). One can then proceed to level, from position B, anything that can be staked, painted, or nicked by another person (**Figure 9.8**).

This is a very handy across–valley level to set in siphons, or to find out where the outlet of a dam will send water by pipe and gravity flow, or to see if a hill on which you want to place a tank at a distance is higher than your shower head at home.

Levelling need only be done once if done right. If in real doubt, buy a dumpy level or theodolite, and read the instructions, or if wealthy and incompetent, hire a surveyor (whose real job is mapping). Thousands of miles of drains have been surveyed, and levels set, by very simple tools. Plane tables of various designs are used by surveyors, but these usually include a telescope for long distance work.

Ralph Long of Australia has devised a tractor–mounted level which he has successfuly used to cut

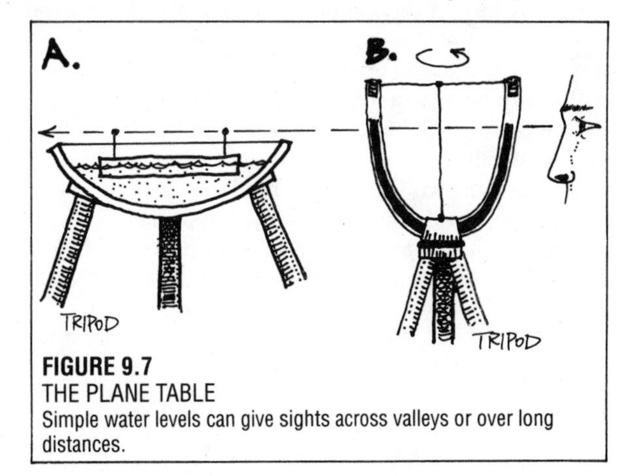

FIGURE 9.7
THE PLANE TABLE
Simple water levels can give sights across valleys or over long distances.

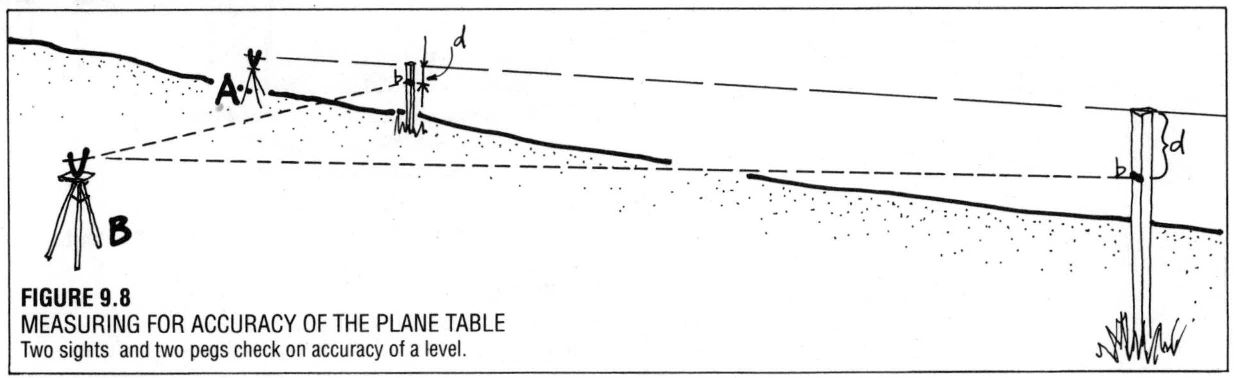

FIGURE 9.8
MEASURING FOR ACCURACY OF THE PLANE TABLE
Two sights and two pegs check on accuracy of a level.

drains, using the tractor itself as a check on levels. This is a bubble level which can also be fitted to any vehicle on farms, and is called the Long Inclinometer® (*Permaculture Journal* #22, 1985).

THE "A" FRAME. This is made from two wooden or metal legs, nailed or welded firmly together at the apex, and fixed about two–thirds down their length by a welded or nailed and glued cross–piece. A plumb bob or weight hangs from a cord at the centre of the apex, and a mark is made on the cross–piece at the cord line. The "A" frame is then reversed, and another mark made where the cord touches the cross–piece. The mid–point between the two marks is that point where the cord hangs when the feet of the frame are level (**Figure 9.9**).

In use, a person swivels the frame about one leg, and sets in the second leg to the place where the cord hangs vertical. A peg is put in, temporarily, to judge where to place the next leg. Many metres of drain have been set in using A frames, often made on the spot from lumber, cord, and a weight for the plumb bob.

9.6
TYPES OF EARTHWORKS

BANKS

It is cheaper to construct a bank than to step or retain the earth, *if* the bank can be made stable. Wherever a four–way bucket (Bobcat or drott) or power shovel can be used to dish the bank, they will remain stable at greater slopes than a straight cut, as we have frustrated soil slump by giving it the profile it would achieve had it slumped. Banks or cuts of more than 4 m high need careful stabilisation.

Unlike dam walls, cuts for roads are not normally compacted, nor do they necessarily have much pressure on them. The greatest cause of slump is water flow through or over banks and the lie of strata in shaly

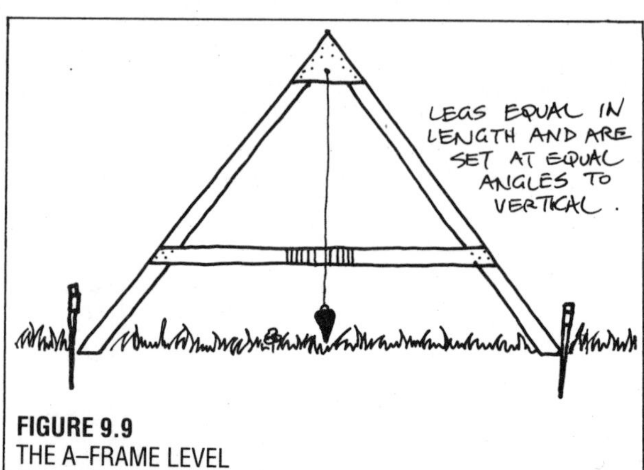

FIGURE 9.9
THE A–FRAME LEVEL
About 2 m wide; legs are reversed so that plumb line cuts "level" mark on crossbeam. Used for centuries to lay out contours and drains at desired slope or flow rate.

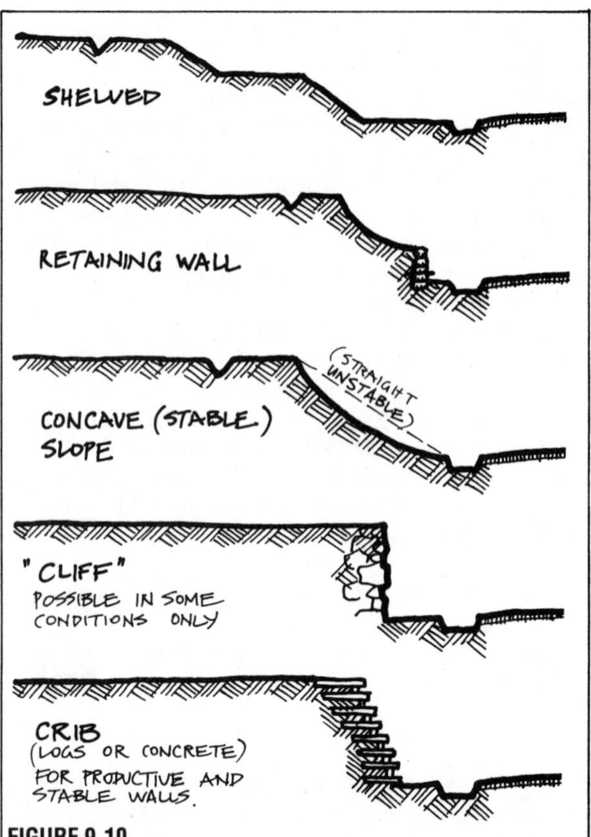

FIGURE 9.10
ESSENTIALS OF WATER AND SOIL SLUMP CONTROL.
Many methods can be used to prevent slump from cuts above terraces or roads; vegetation helps all these systems.

FIGURE 9.13
KICK–DOWN SYSTEMS AND STEEP TERRACES
Using steep slopes to assist the movement of nutrients and water.

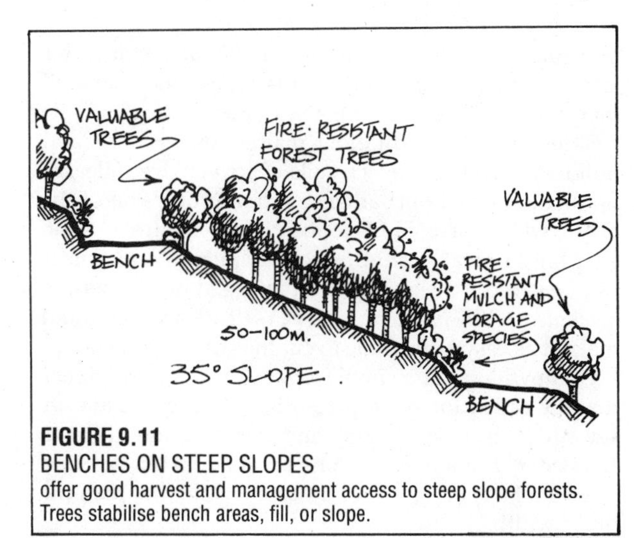

FIGURE 9.11
BENCHES ON STEEP SLOPES
offer good harvest and management access to steep slope forests. Trees stabilise bench areas, fill, or slope.

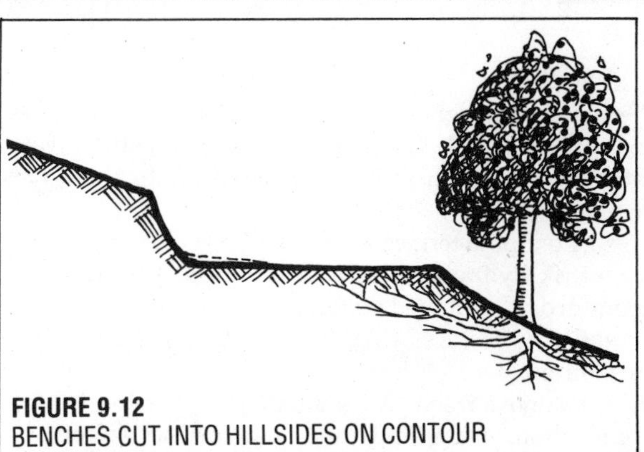

FIGURE 9.12
BENCHES CUT INTO HILLSIDES ON CONTOUR

ground. All normal soils with some clay and stony banks are fairly stable, given that the angle of rest of these materials is preserved, the banks cut somewhat concave, and drainage fitted. In severe slump areas, re–routing of roads or expensive drainage and concrete retaining walls may be last resorts, but **Figure 9.10** shows the essentials of water and slump control; only some may be needed but all can be used.

BENCHING

A bench is a flat, near–contoured cut made in a slope. Very severe slopes of 30–40° can be bench–cut if a bulldozer can start on safe ground (typically, from a ridge of 20° or so). Benches are used to make roads and house sites, and are very useful in long–term forestry if a steep hill is benched every 100 m or so (**Figure 9.11**). Benches greatly aid access, planting, and eventually harvest. The first fast–growing trees to plant are those on the lower (loose soil) side of the benches; these can be nut or fruit trees, for harvest, or for seed crop.

Small–holders on very steep hillsides must work with the slopes, or expend much energy on carrying water and fertiliser. Orchard and mulch *above* poultry *above* garden is the easiest system. All the better if the ridge or hill above that system is planted to mulch–producing trees such as *Casuarina* and pine, oak and beech. Thus, mulch and bedding is thrown down-hill as greenfeed and seed for chickens, and they kick–down to the lower fence where the gardener accepts the manured and shredded mulch for the essential (terraced) garden

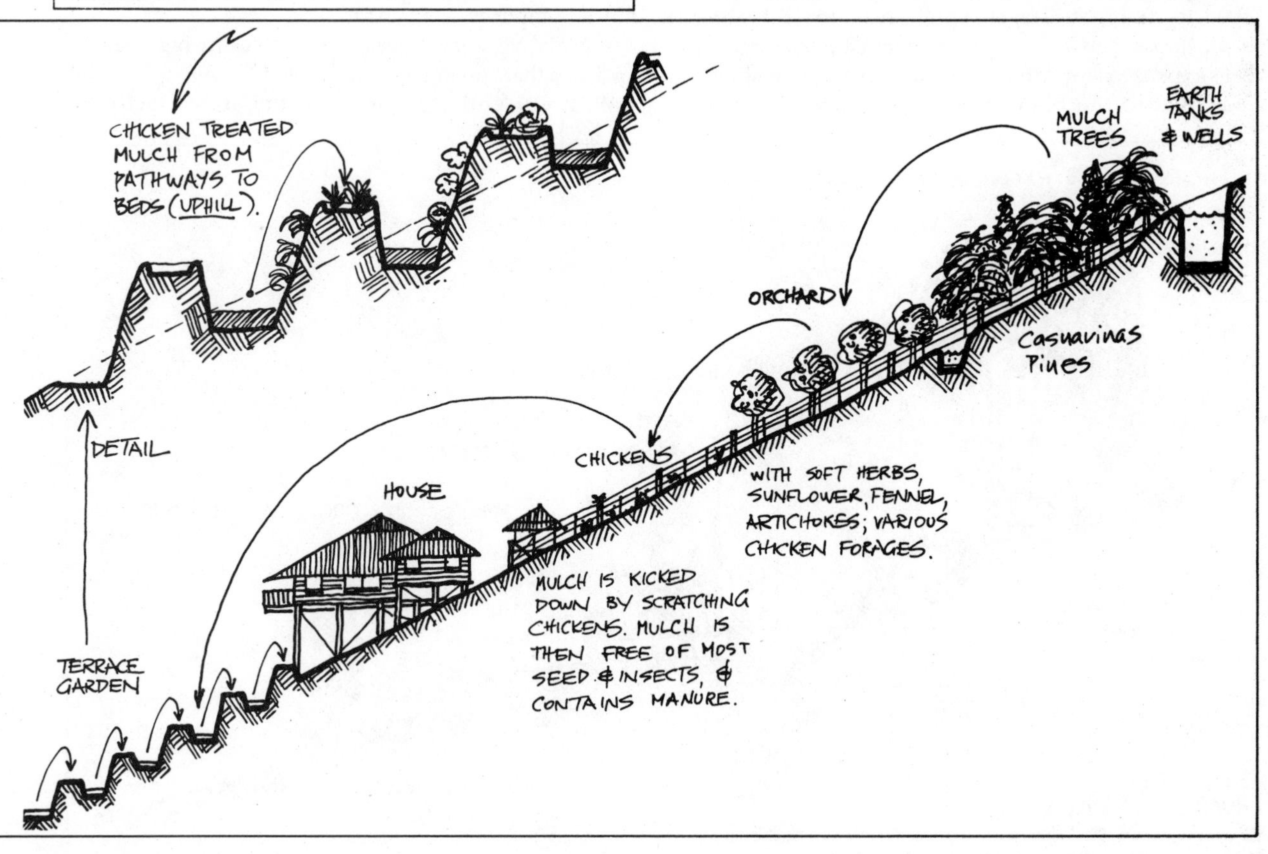

beds (**Figure 9.13**).

Benches can be fertilised, ripped, and sown to legumes or soft–leaved mulch plants, and periodically cut or grazed off. If cut, mulch can be raked to trees on either border. As benches on very steep slopes will rarely be travelled upon, they can slope slightly outwards to drain. Benching should not necessarily continue across water runnels unless pipes, gabions, or culverts are fitted in the watercourses; small watercourses can be bridged.

Benches are quickly made to a survey line pegged every 6 m (20 feet) or so followed by a bulldozer with blade set to side–cast (angled). A run with a raised blade will shave back or slightly step the uphill bank. In *stable* soils, benches can slope *into* the hill and so form swales to infiltrate water for trees below the bench. Cross–walls can be made every 20–30 m to prevent gutter flow. Such benches are not so much roads as tree shelves, and are a blessing in steep country. A quiet traverse with a donkey or cart serves to distribute plants and fertiliser (or mulch) and to gather in the crop. Narrow (1.5–3.0 m) benches are quickly over–hung by tree crop, shaded, stable, and accessible (**Figure 9.14**).

Wherever benches cross drainage channels, pipes or rock–lined swales must be made to carry water across, and these (and their maintenance) are the greatest long–term expense, so they need to be well made and durable.

Although the true finesse of earthworks is best entrusted to the experienced machinery driver, designers need to designate stages, spoil areas, topsoil stores, bank slopes, and so on. Also, they must remember that it is expensive for a bulldozer to travel far pushing a load before the blade to create banks (about 6–9 machine lengths is maximum), and that much faster and better work can be done cross–slope and downhill than upslope by any machine but a bucket digger.

Side–casting is a different matter, and graders and bulldozers both can bench soils effectively for miles, as no dirt is carried but rather cast out to one side of the blade, which is ANGLED to the direction of travel to do so. Thus, a great deal of time, fuel, money, and timber can be saved if a site is carefully pegged out *to suit the machines called in*. Imagine yourself as a driver, and make it as easy as possible to cut the shapes you want.

On anything but a simple job, two or more machines may be called for, as in preparing a house site and its sewage lines, or in digging and carting a clay deposit. Bulldozers can loosen and pile up material but not load it, bench but not trench, or at least not smaller than the blade width.

TERRACING

Terraced lands, given a reserve of local green manures or composts and adequate water, are potentially very stable production systems. Exceptions to this arise when we:

- Attempt to terrace in unstable soils or sediments;
- Risk hydraulic pressures on hill slopes from impounded or infiltrated water;
- Create terraces that are unstable at the bund or wall face;
- Extend terracing as annual crop over too large a proportion of the landscape, and so lose leaf or tree nutrient input to crop; and
- Make very large series of terraces in high rainfall areas, so that run–off is concentrated.

With a useful assessment of the above factors in

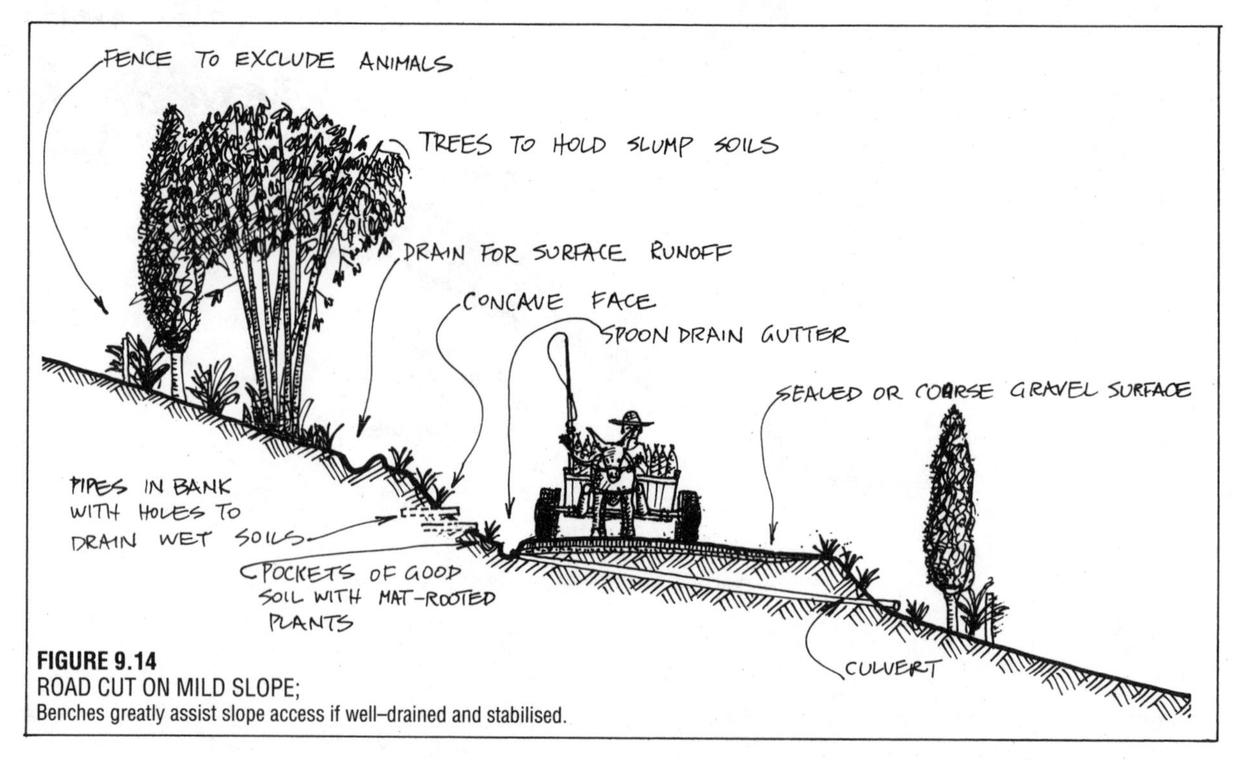

FIGURE 9.14
ROAD CUT ON MILD SLOPE;
Benches greatly assist slope access if well–drained and stabilised.

mind, we can gain long–term production from short series of polycultural terraces (wet and dry crops), with stable bunds (either rock–walled or at a 1:3 slope). Trees, on bunds and between, above, and below terrace series, should form 40–60% of the total landscape plan, and both soils and installed water inlets and outlets should permit safe and controllable irrigation.

The great benefits of terraces are these:

- Very easy crop access on slopes;
- Easily controlled and effective irrigation procedures;
- Minimal soil loss due to overland water flow, or to slope cultivation; and
- A potential gain in silts or nutrients in irrigation or run–off waters and from leaf fall.

As with dams, terracing is most effective where slopes are least, as earth moved versus area of cropland developed becomes impractical or inefficient as slopes steepen. At about 30° slope, but preferably at 10–18°, terracing becomes worthwhile.

Terrace construction always begins on the lowest terrace level, with the removal and stockpiling of topsoil over the whole area of the lower terraces, and proceeds uphill as each terrace is made, so that the topsoil of the next highest level is cleared on to the preceding lower terrace. Stockpiled soil at the lowest terrace is finally carted or lifted to the last of the series uphill (**Figure 9.15**).

Every terrace system is ideally designed to allow perennial bund and terrace wall plants, specifically for *wall stability* and *green manure crop*.

As for the extent and series size in a terrace system, it is wise to limit both on the basis of:

- Heavy rainfall (hence expected run–off) in tropics; and
- Expected rainfall harvest in arid areas, where total terrace areas should not exceed one–twentieth of the catchment harvested, and where perennial or adapted crop (never water–demanding crop) should form the selected species.

Thus, all terrace systems should aim to occupy no more than 30% of tropical or 5% of dryland areas, and in the tropics tree crop should be developed to maintain fertility of the terrace areas. Water catchment areas should be developed to do the same for arid areas, so that run–off brings leaf mulch to terraced slopes.

In humid cool or tropical areas, wet terraces (10–20% of all terrace areas) can be devoted to a fish–plant polyculture, giving yields of fish, shellfish, and water plant products. Yields of protein from water cultures can exceed all land–based systems if managed at the same levels of husbandry and care.

Wherever people occupy *very steep* sites (slopes of 20° or more), especially in areas of high rainfall, it is preferable to abandon broad terracing for a series of 4–6 narrow production terraces, each series carefully drained to spill excess rain down permanently vegetated slopes (**Figure 9.12**).

Thus, by ridging the terrace tips, mulching paths, and staggering path spills in short series, we can get the advantages of terrace on quite steep slopes without risking erosion and soil loss. Needless to say, machines are inappropriate for such construction, and steep slope terracing of this nature is always hand–cut.

9.7 EARTH CONSTRUCTS

Wherever earth is dug, banks are raised. There is a great concentration on the holes, and far less on the mounds, so that "spoil" in mining becomes a pollutant. This need not be the case, providing topsoil is first removed

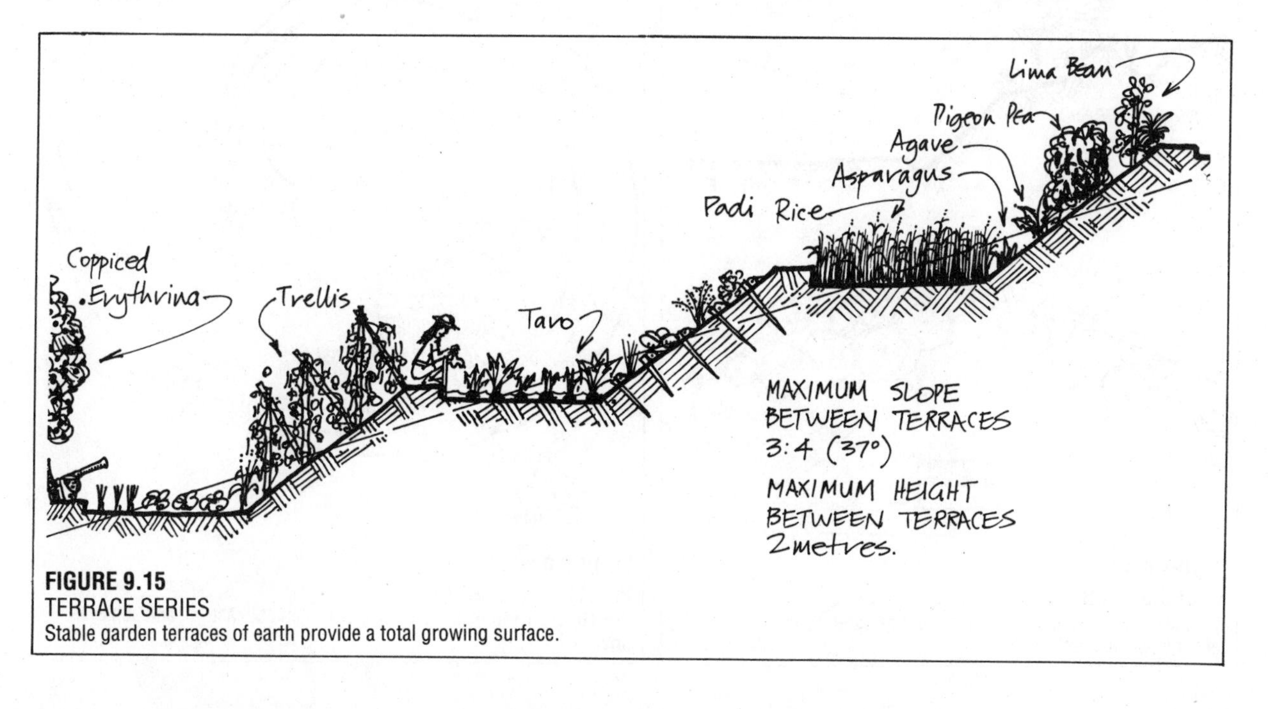

FIGURE 9.15
TERRACE SERIES
Stable garden terraces of earth provide a total growing surface.

and then returned to cover the spoil, in stages or as a whole job.

Just as land can be hollowed out, so roads and banks can be raised above marsh or flood levels, and to deflect winds, noise, or water. Banks raised by machines can serve the following purposes:

1. Shelter for houses and fields;
2. Plant sites for windbreaks (on the crown of the bank and in its lee);
3. Walls of houses or storage barns;
4. Containments for large inflammable fluids, and fireproofing;
5. Noise deflectors and absorbers, e. g. at airports and along highways;
6. Tracks and plant sites in marshes and on clay soils subject to flooding;
7. Patterns to deflect and direct wind and water to storages or energy systems;
8. Flood or tide control systems (polders and levees);
9. Railways and road grade adjustment;
10. Earth ramps and stands; and
11. Earth walls or "ha ha" fences.

These are further elaborated as below:

1. SHELTER FOR HOUSES AND FIELDS. Either in flatlands or on high exposed sites, an earth crescent bowed into the cold winds is the fastest way to create shelter and warmth on its lee side. Even a few blade sidecasts run along a field boundary provides cover for hedgerow and windbreak species and may act as a swale for root water collection. High banks of 2.5–3.5 m (9–12 feet) close behind house sites, whether excavated or raised, create instant shelter over the occupied site, shelter which may be further reinforced with trees (**Figure 9.16**).

This is a long term and considerable heat energy saving in cold climates, and a shading or cooling wind director in warm climates. **Figure 9.17** shows an actual ground design for a flatland site with cold southwest, hot northwest, and cooling northeast winds, all controlled by one earthbank and the pond formed by the excavation.

2. PLANT SITES FOR WINDBREAKS. Low side-casts heap up topsoil, create shelter, and catch or delay run–off, as well as reduce root competition for a year or two, all good reasons to sidecast for long windbreak sequences, and in wetter ground to establish willows, poplars, and tamarisk above water–logged ground (**Figure 9.18**).

3. HOUSE OR BARN WALLS. Unrealised by most architects and home builders, machines exist which can raise and compact a complete house or barn wall in a morning's work. All we need to add is floor and roof (another two days work) to be in a long–term, fireproof, silent, energy–conserving and sheltered house. This technique is suited to open–space situations, cheap barns and large outbuildings. As the walls are raised, a smaller tractor and roller can compact them. Almost any earth will do, providing the compacted rest angle is watched. This technique is not suited to sands unless wall corners are bagged (stabilised with soaked bags

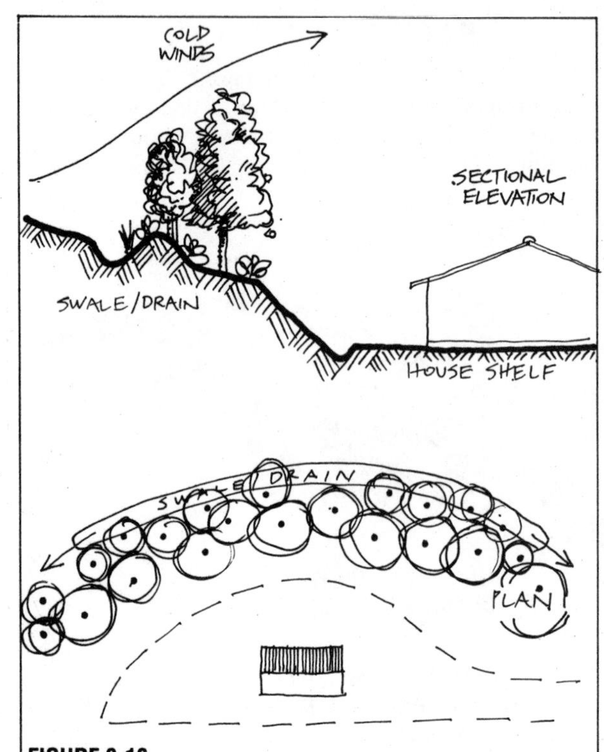

FIGURE 9.16
RAISED BANK BESIDE HOUSE
In areas of heavy rains, house sites must be carefully designed to deflect downhill overland flow.

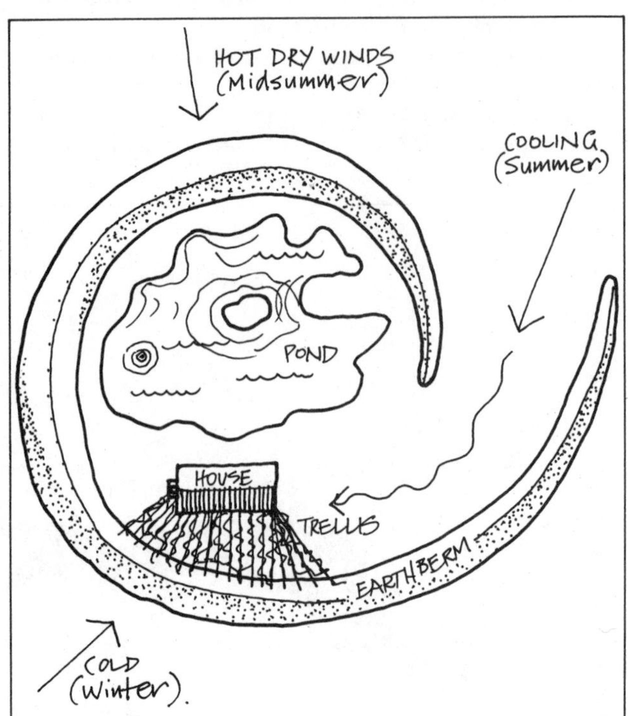

FIGURE 9.17
RAISED BANK AROUND HOUSE;
this structure modifies several external climatic factors, and achieves privacy.

FIGURE 9.18
BANKS FOR WIND BREAK ASSISTANCE,
shelter, flood–proofing, fire–proofing, noise reduction, and drainage adjustment.

filled with cement and sand or sandy soil). **Figure 9.19**.

4. CONTAINMENTS. Whether for inflammable fluids or as fire–proofing, and especially effective against radiant heat, earth walls can surround tanks. Above–ground wildfire radiation refuges in bushland are earth–covered and may be built off forested roads and near isolated bush houses.

5. NOISE DEFLECTORS OR ABSORBERS. Traffic noises are effectively blocked from housing by earth walls, and plants on these banks decrease noxious fume effects. Deflection of noise is effective for long wavelength noise only.

6. TRACKS AND PLANT SITES IN MARSHES. Earthbanks, islands, and mounds in marshes give multiple opportunities to access and place trees and structures. Mounds in ponds isolate useful but rampagious species such as thorny blackberries and runner bamboo, while banks allow foot or vehicle access across and into marshes to place and service duck nest boxes, to harvest fruit and vine, and to attend to fish ponds (**Figure 9.20**).

The raising of a bank in marsh is an interesting operation; the best tool is a very light "swamp" tractor with wide tyres or tracks, and a swivel bucket (an excavator). This unit, perhaps using "mats" in very soft or silty ground, can dig and raise an earth bank where the ground is too soft for the operator to walk. Equipment can be light for the first crossing, the aim

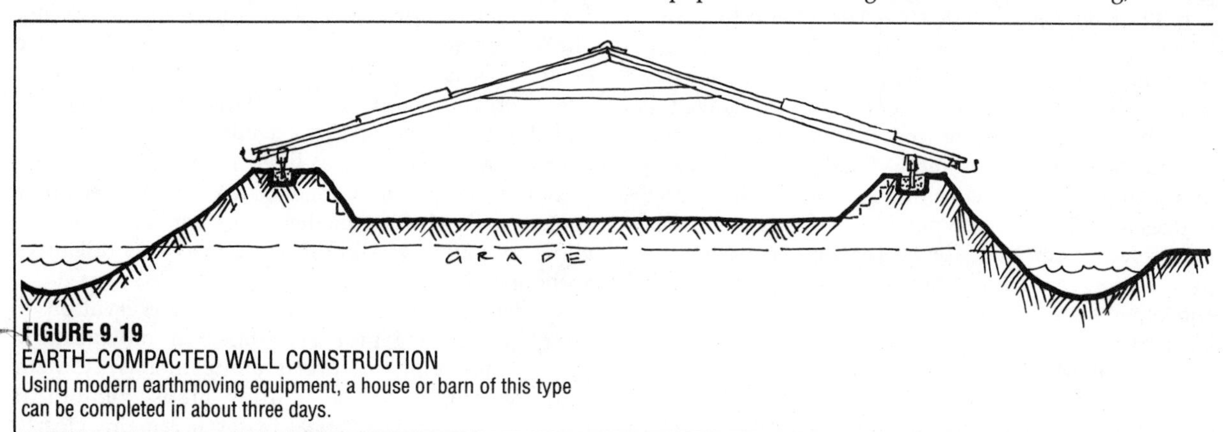

FIGURE 9.19
EARTH–COMPACTED WALL CONSTRUCTION
Using modern earthmoving equipment, a house or barn of this type can be completed in about three days.

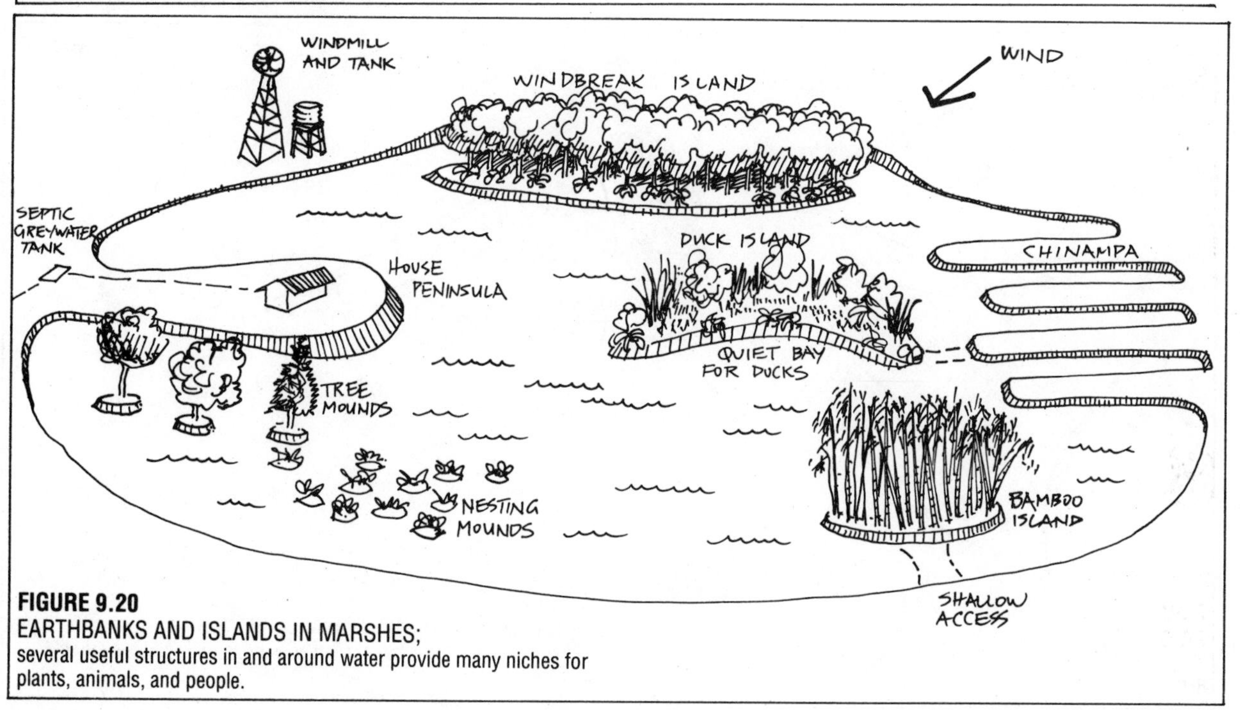

FIGURE 9.20
EARTHBANKS AND ISLANDS IN MARSHES;
several useful structures in and around water provide many niches for plants, animals, and people.

being mostly to throw up a broad earth mound to dry out, so that a safe transit of heavier equipment can be made at a later date and a substantial bank raised for a road–bed or planting. It is easier, of course, and less harmful, to first raise banks, and then create a marsh by building a low berm. Natural marshes need protection for their unique values and for waterfowl.

"Mats" are three bulky wooden gratings, made of 10 cm by 15 cm (4 inch x 6 inch) timbers, which the bucket machine stands on. At each move the machine itself recovers one from behind and places it ahead. Once the first mound is built, larger excavator machines can reach out to dig deep channels and build high banks of great solidity, or the smaller mound can be levelled by light tractor for foot traffic. Draglines are not really an alternative in mound–building, and in timbered swamps are impractical.

7. WIND AND WATER DEFLECTION. **Figure 9.21** illustrates a design based on older Afghan mills of a saddle or ridge wind tunnel for mill power. Systems for water diversion are discussed in Chapter 11.

8. FLOOD AND TIDE CONTROL (polders and levees). Well–maintained earth banks are the only protection of houses and villages in flood plains and below tide level. They are, in effect, reverse dams, the inhabitants living inside and the waters contained outside. Mounds can be built as flood refuge islands in many deltaic areas, containing storage barns and refuges in fields subject to periodic flooding, dangerous to people or animals. Causeways of earth are often used to access low mounds or islands near shore. Bangladesh and other deltaic areas need wooded refuge mounds just for the survival of people.

Such mounds can also be made in *fire–prone* sites, where villages and houses are safe in lakes or on moated islands. Peninsulas should be included in dam construction for this purpose by designers (**Figure 9.22**).

9. RAIL, CANAL, AND ROAD GRADES. These common banks and cuttings are all well–known to all of us.

10. EARTH RAMPS AND STANDS. Earth ramps and stands are of great use for at least these reasons:

- To unload trucks from the side or back (heavy objects and wheeled vehicles);
- To load cattle to trucks, at various levels;
- To unload hay or bales into the upper floors of a barn for use at lower levels; and
- To load and unload boats into water.

Once built, they are permanent in use, and sometimes a whole district can use the facilities of one loading–unloading ramp. The loading face itself must be stabilised with stone, concrete, or beams (**Figure 9.23**).

11. EARTH WALLS OR "HA HA" FENCES. The "ha ha" is a below–grade ditch, which acts as a fence. It is used in classical vista gardens where views are to be uninterrupted, and in zoos for direct viewing of large and potentially dangerous animals. It is also a defence for villagers against stock where the resources do not allow wire fences, but labour or a machine can be obtained. It is essentially a deep pit, dry or wet, with one steep wall faced by stone (**Figure 9.24**). It can be scaled to size for the species excluded.

The same machines that build roads will also build wetlands, swales, and small dams suited to wildfire control and wildlife. Underpasses and guide fences allow migrating wildlife to cross road and rail ways without accidents.

Roadsides can be an area for the preservation of bunchgrasses, sagebrushes, rangeland and meadow plants, and remnant forests along their way, with pull–over areas and vegetation maps, geological features, and archaeological or fossil remains clearly

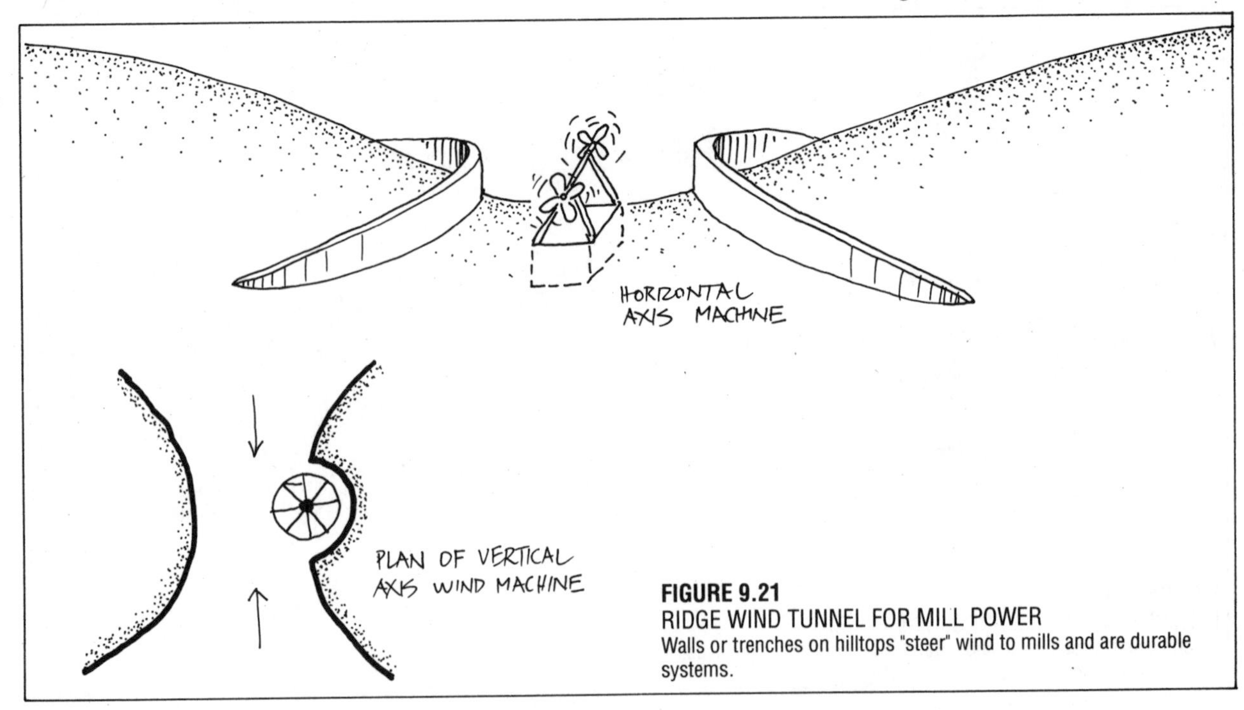

FIGURE 9.21
RIDGE WIND TUNNEL FOR MILL POWER
Walls or trenches on hilltops "steer" wind to mills and are durable systems.

indicated. Such planning must precede actual construction.

We pour kilometres of black bitumen surface, but fail to lay under it pipes for heat pumps that would heat the towns through which the road passes, and we fail to harvest and store road run-off for local irrigation and wetlands. All this will change only when well-intentioned people govern the spending of public monies, or when road engineers are trained in permaculture, and look upon roads as a community resource!

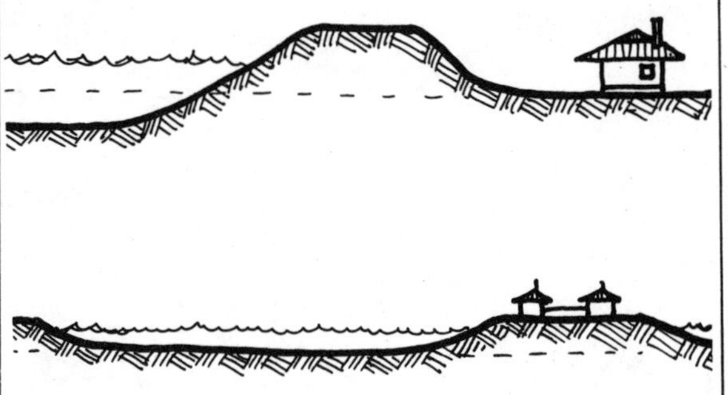

FIGURE 9.22
POLDERS AND LEVEES

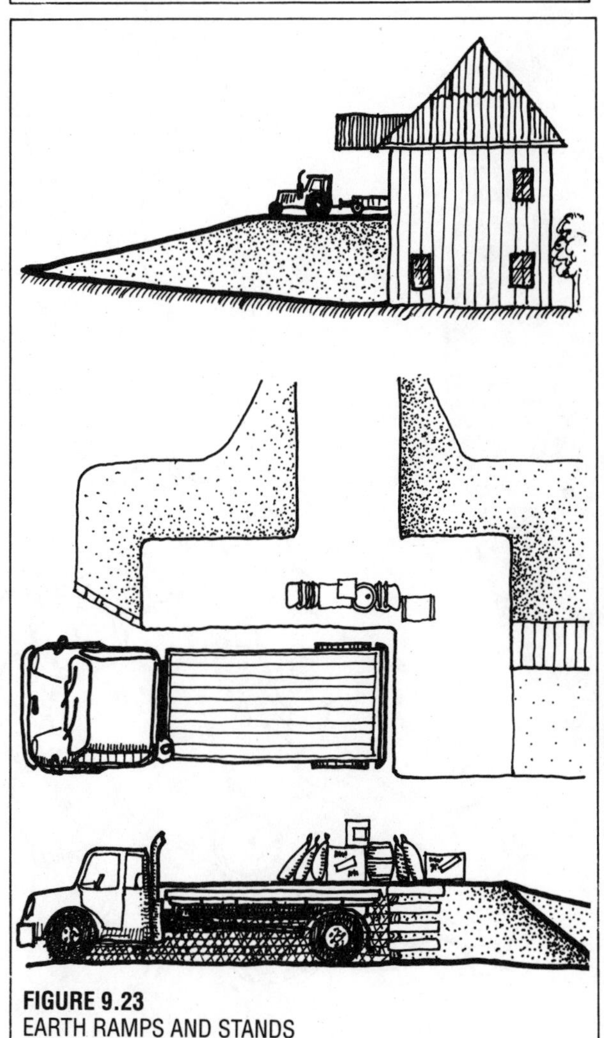

FIGURE 9.23
EARTH RAMPS AND STANDS
Cheap and permanent ramps ensure easier load and unload operations.

9.8
MOVING OF THE EARTH

People have always moved earth: to reach water in dry river beds, to mine pigments for their decoration, to excavate food as bulbs, grubs, or fungi, and to bury their faeces or their dead. The archetypal tool is the digging stick, which gets food of greater variety and more nutrition than the spear. The basic hoe, rake, and shovel exist today in most cultures.

Hand tools have moved most of the earth we see today shaped into mountain rice terraces. There are some very useful and cooperative ways to dig, effective in making miles of low irrigation banks or unloading gravels. The two-person shovel is useful here, one digging in, the other pulling over in a see-saw motion (**Figure 9.26**). Try it, and be surprised at how rhythm and cooperation will move mountains. Sing a little tune to pass the time and get the rhythm:

> **Down** among the **dead** men
> **Let** me **lie**.... etc.

However, the exigencies of education have meant that designers and architects have seldom been personally involved in earth moving. A brief description and nomenclature of machinery and modern tools is not only needed but essential to earth design, given that when we move earth we should do so for permanent and beneficial ends. We can revolutionise eroded and arid landscapes by commencing the process with tools and consolidating it with life forms, especially trees.

EARTHMOVING MACHINES

Just as there are hand tools suited to particular ways of

FIGURE 9.24
"HA-HA" FENCE
Much used in areas where post and wire fencing is unavailable or expensive.

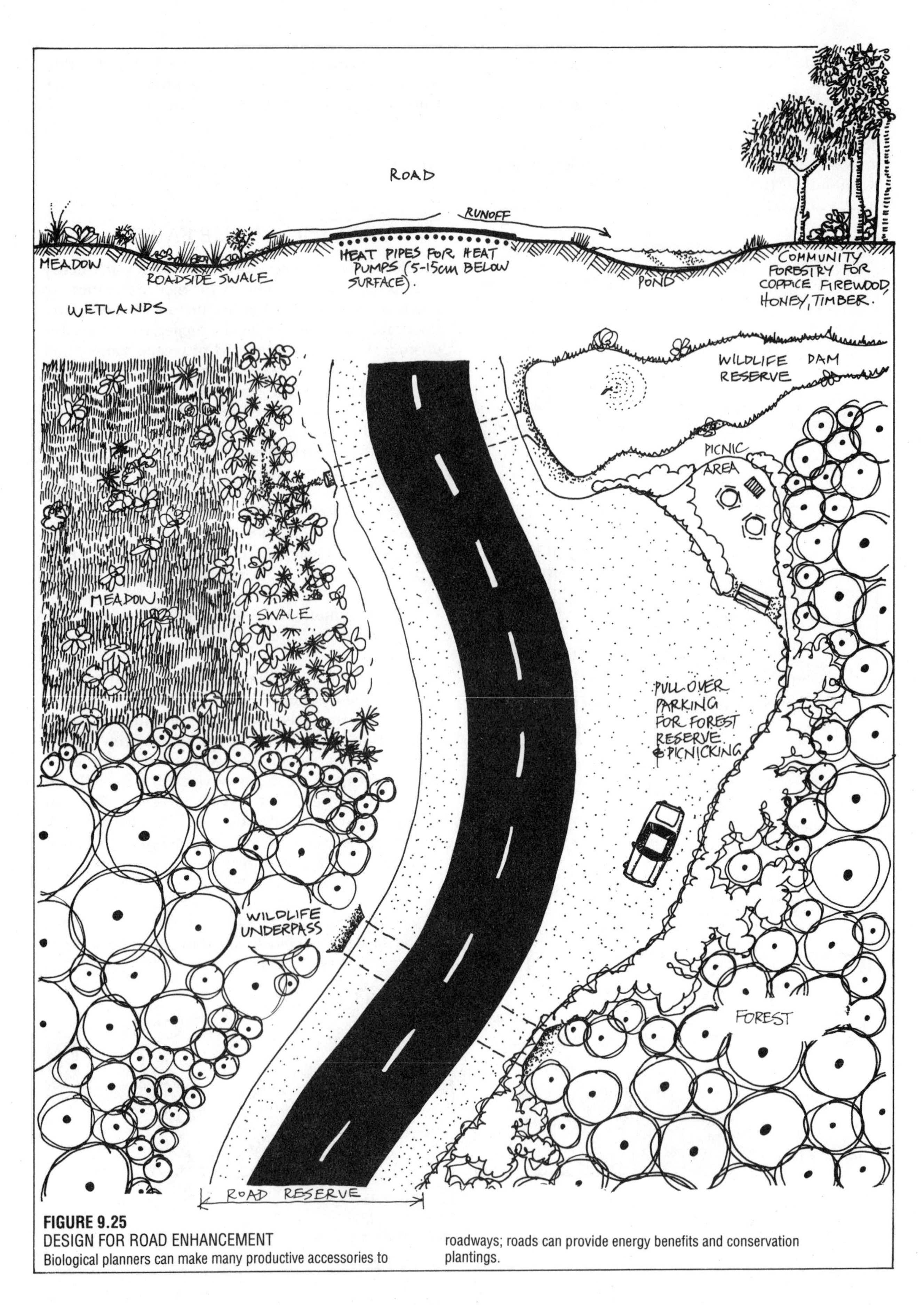

FIGURE 9.25
DESIGN FOR ROAD ENHANCEMENT
Biological planners can make many productive accessories to roadways; roads can provide energy benefits and conservation plantings.

digging, so there are large machines suited to special landscape tasks. Any of these can be supplanted by human labour where it is plentiful. The basic earth–moving attachments are these:

- BLADE (can be mounted on almost any vehicle, or towed); includes "V" blades or delvers.
- BUCKET (for lifting and loading loose material); narrow and toothed for hard ground, or of special shapes for drains.
- BUCKET CHAIN (for foundations, pipelines, narrow deep ditches, underwater dredging).
- SCOOP (can be horse or bullock–drawn, or articulated on an hydraulic or telescopic arm).
- RIPPER (for breaking up compacted soils); usually towed or rear–mounted on tractors.
- DELVER (for one–pass drains; often mounted on a grader, or towed on a frame behind a bulldozer).
- SPINNER (rear–mounted on a special tractor); a fast–revolving disc of about 2–3 m diameter with peripheral buckets.
- BORERS AND DRILLS (holes, fences, explosives, pipes, wells, bores, and the like)
- JET PUMPS (to pump out silt and sand in wet places).

As well as these, there are explosives, which are of special use in marshes and swamps, and some specialised tools for rock and swamp work, for mines, and for massive tunnelling.

FIGURE 9.26
THE TWO–PERSON SHOVEL
A simple, non-tiring digging system for drains, gardens, and terraces.

Our concern here, however, is for the common tools of landscaping and water storage or water channeling. Each has appropriate uses, although most can do something of what the others do. To take them one at a time:

The Blade Machines

The blade earthmovers are the machines for levelling and benching or terracing, and ideally they should be able to lift (and drop), angle, and tilt. They can be mounted forward (bulldozer), in the centre (grader) or at the rear (wheel tractors, for levelling). Tilted, the blade can make shallow V ditches, or put crowns on roads. Angled, it spills earth to one side as steering walls, or makes long shallow drains.

Blades are ideal for levelling house sites, for terraces, for benches on hillside, for side–cutting roads, and for pushing up earth walls. Even a small tractor (17–25 h.p.), patiently worked, can make very large dams and terraces at less fuel cost (but greater time cost) than a bulldozer, while large bulldozers are the most economical of time. They can work difficult, steep, or stony sites and move very large objects such as boulders.

The bulldozer is a blade machine, mainly for pushing and planing earth. It is of greatest use in roading and for dams. It is an excellent machine to put up, roll solid, and spread earth, to dig large shallow holes, move small hills, and bench. The blade TILTS for road crown slopes and ANGLES for casting aside windrows of earth; it also LIFTS to spread and release loads, and DROPS to delve out drains and ditches.

Angled blades are used in long runs to cast earth out continuously to one side (called side–casting). Tilted blades are used to cut channels or to bench slopes, and lift is used mainly for piling up or levelling loads. Blades are normally forward–mounted for sight and control reasons, but on that special road machine, the GRADER, the blade is mid–mounted to give even spreading, and on small farm or wheeled tractors are often rear–mounted. Some small blades rotate about their axis (180° swivel). The grader can be used to make long drains, of shallow angle; these are miscalled spoon drains, but are effectively more angled than spooned in section.

Thus, for all normal bench work, a bulldozer is useful, and for long flat road or levelling runs a grader is better. A SCRAPER is a large, self–filling bucket or land dredge which can both fill and empty itself to plane off or dig out large areas. All large machines can now be laser–guided to accurately level and grade fields at a pre–set slope. Lasers can also automatically work the hydraulics to lift and drop earth, but land forming is mainly restricted to large flattish irrigation areas or civil works, and is normally contracted out to specialists with large machines.

The Four–Way Bucket

This machine (sometimes called a drott) combines all four motions of lift, dig, push, and pull, and is a

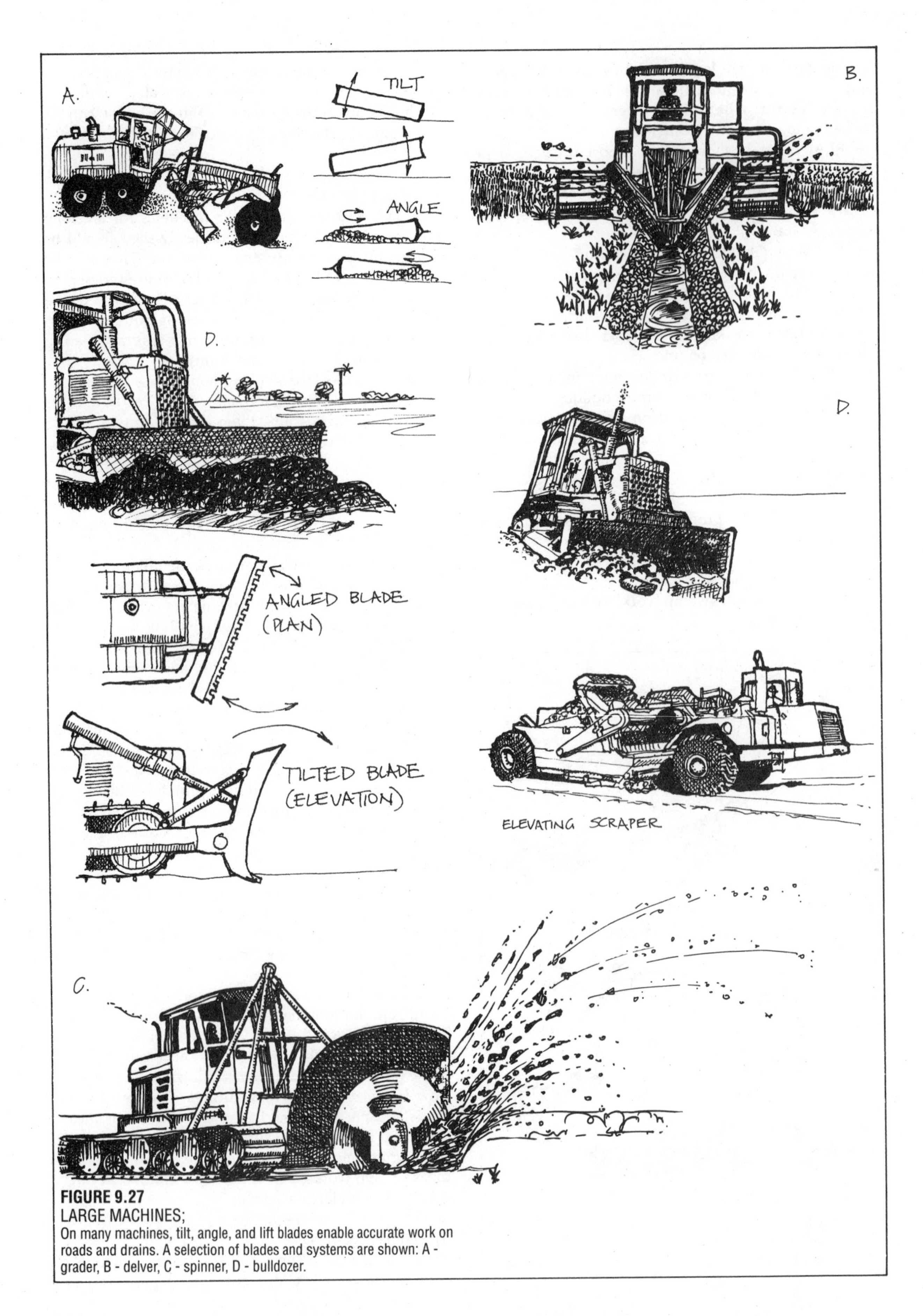

FIGURE 9.27
LARGE MACHINES;
On many machines, tilt, angle, and lift blades enable accurate work on roads and drains. A selection of blades and systems are shown: A - grader, B - delver, C - spinner, D - bulldozer.

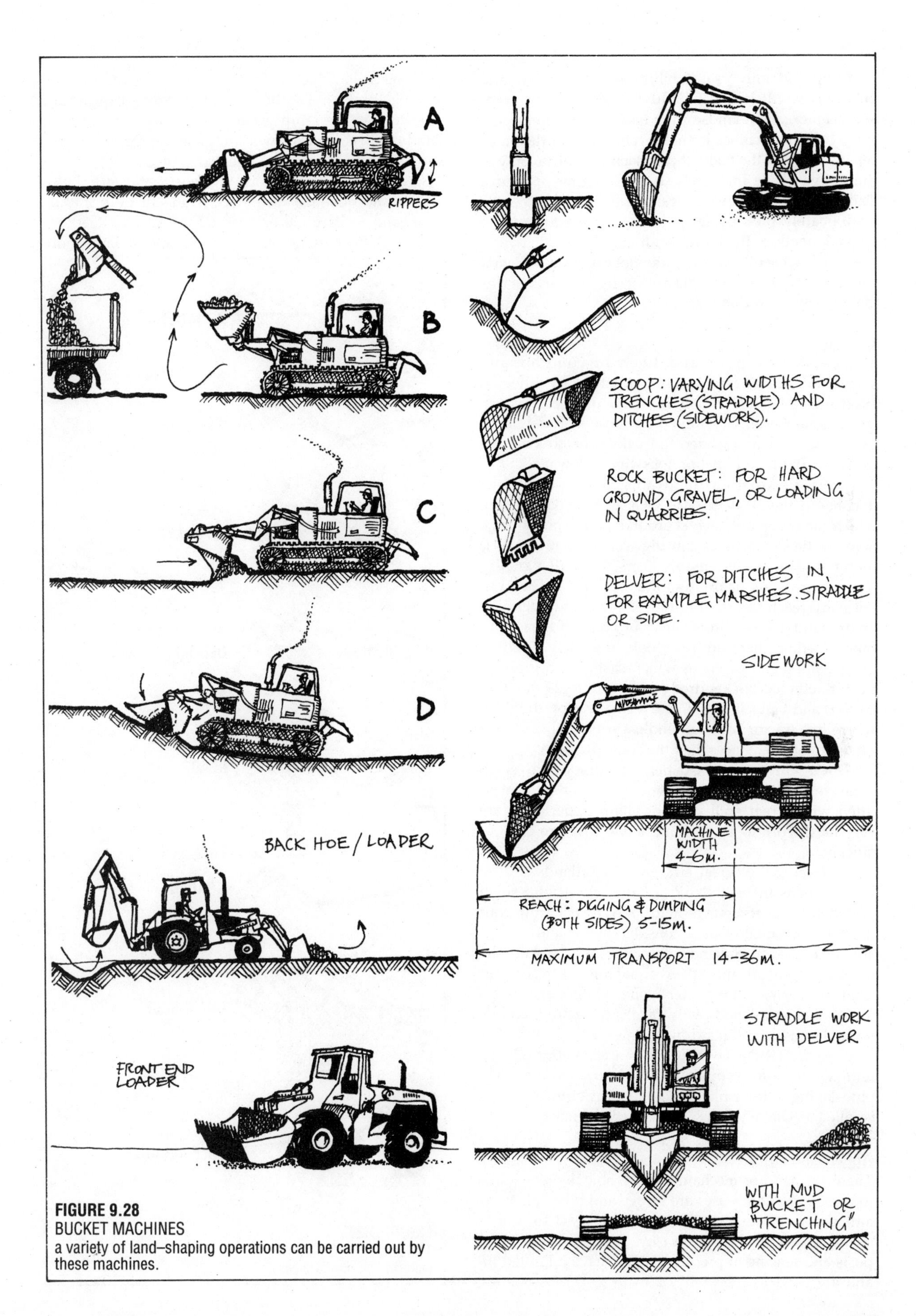

FIGURE 9.28
BUCKET MACHINES
a variety of land–shaping operations can be carried out by these machines.

bridging and universal machine between blade and bucket types. It is usually fitted to a bulldozer body, and is an excellent landscaping machine. It differs from some bucket machines in that it cannot *swivel* the blade separately from the body of the machine, nor can any of the above blade machines. It is sometimes called a clamshell bucket because it can close on loads of earth, or delicately pick up large stones, shave a curve in embankments, or fill a truck with soil.

A small wheeled machine, the Bobcat, is an excellent finishing tool or used alone for light work or for making swales in Zones 2 and 3.

Buckets

These can be simple and fixed or complex and swivelled; the former are often fitted to wheeled tractors as LOADING BUCKETS of great use in quarries, nurseries, and anywhere loose material needs to be picked up and loaded to trucks at heights up to 4 m. They can lift and tip, but not swivel sideways.

A very useful tool for drains, and as a tool in marshes, is the SWIVEL or SCOOP BUCKET machines, which can scoop out, swivel around, and deposit loads. Some of these machines can dig wells to 9 m (30 feet) using hydraulic extension arms, and all can dig sharp-edged pits or drains for special uses. They can normally reach out 6–10 m (20–32 feet), and (standing on dry land) take silt from canals or ponds. To reach out much further, there are two tools, the DRAGLINE, a giant fishing rod or crane which casts out a loose or tractor-tethered bucket dredge up to 18 to 24 m (60 to 80 feet) and hauls it in full, and a two-tractor dragline, where one tractor hauls an endless rope with a bucket across greater distances than the crane dragline.

SCOOPS are often used behind horses or oxen to excavate small ponds or clean out ponds of silt. Unless fitted with a rear spill-door or chain tipper, they are very hard work to tip by hand alone, and tire one out quickly.

A special tool of great efficiency in flatlands and on low slopes is the SPINNER, a large wheel with small peripheral buckets spun at high speed, which throw out the earth as small clods a considerable distance. They are much used in Holland to drain polders, and are the most economical and speedy machines for flatland (shallow) drains. The spinner is one tool that distributes the spoil as small clods as it travels; there are therefore no banks beside spinner drains.

A special TRENCH DIGGER is a chain-bucket system which continuously digs narrow trenches for pipe-laying or to insulate foundations. Commercially, it is called a "Ditch-Witch®" or "Trench-Wench®".

Augers and Drills

These are hand or mechanical post-hole diggers suited to most soils. Other uses are in tree and tuber planting, although care must be taken not to compact soils or to create water-filled holes in clay. The prime use is for posts and fencing at present. Augers have a limited lift and are used to (normally) 2 m (6 feet) or so depth, whereas drills which have extension tubes can operate to great depths. The borings are removed by liquid flushing or are simply compacted and tube-lined. Purely hydraulic drills work to great depths, and bores and hydraulic nozzles are often combined.

Some of the largest vertical and horizontal earth holes (mines and quanats) were once hand-cut by primitive tools, and wells to 30 or even 60 m (100–200 feet) are still hand dug throughout the world. Special large-bore

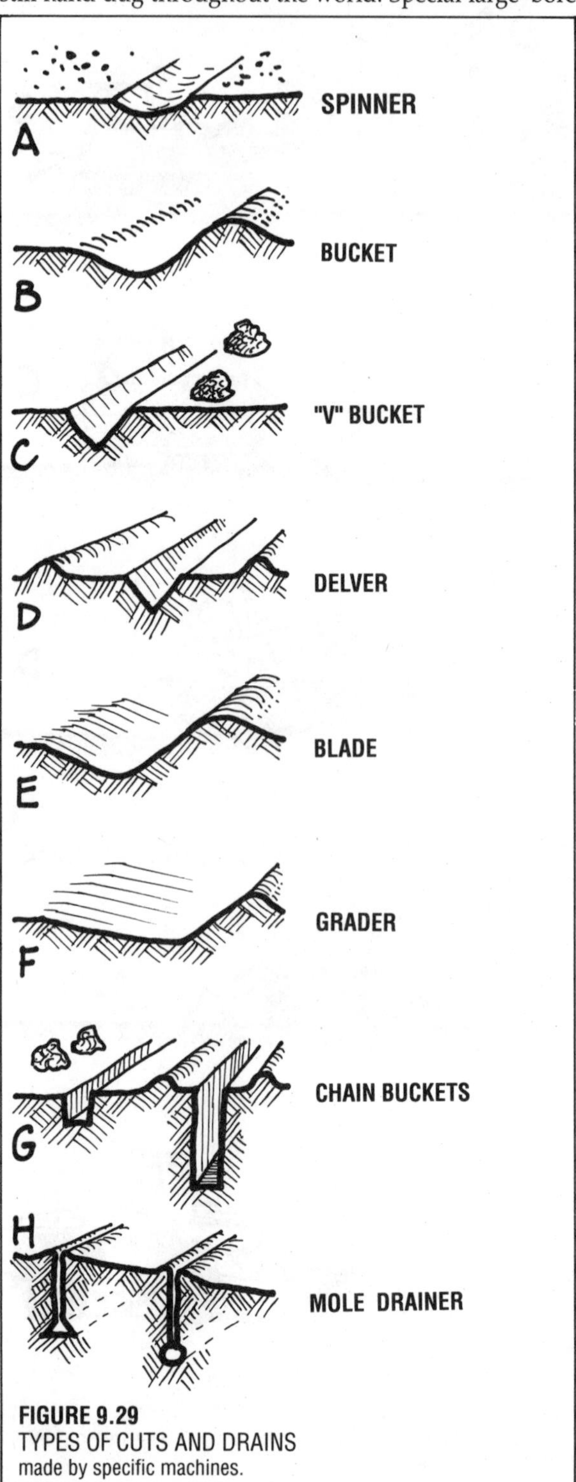

FIGURE 9.29
TYPES OF CUTS AND DRAINS
made by specific machines.

augers to 1 m (3 feet) are used at mines, and most carry extensions to make a 2 m wide shaft. Very large pneumatic blowers are crane–mounted to remove loose materials, and are often locally made as at the Coober Pedy opal mines in Australia. Large–bore drills there are called Caldwells after the inventor and are used for underground housing, excavation, shafts, and tunnels up to 1.5 m (5 feet) diameter.

Explosives

Explosives are of most use to assist auger holes in hard ground, to make holes in otherwise unstable ground such as marshes, or to loosen rock in quarries. With the advent of the swamp tractor, marsh blasting is less common, but the basic ease and effectiveness of explosives should not be overlooked where they can solve some otherwise intractable problems.

Nitroglycerine (cellulose treated with acid, in glycerine, absorbed into wood dust or clay earths) as gelignite, dynamite, or plastique is an inexpensive way to solve some earth–moving problems. Even cheaper is the mix of ammonium nitrate fertiliser and dieseline known in the trade as "chickenshit" (nitrates have from ancient times been gathered from manures, around the soil of toilet pits, or extracted by washing and evaporation from guano).

Old "recipe" books give dozens of reliable recipes for cheap explosives, and even custard powder, flour, or face–powder will blow a room apart, as will the fumes from ether or domestic gases. I once worked as a scientific glassblower, and managed to remove all doors and windows from my room by allowing ether to evaporate from a bottle. Earlier, as a baker in my father's business, I created some spectacular flashes using plain flour near open flames.

Today, we must take a course to obtain a "powder monkey" certificate in order to set our own explosives, or we should hire skilled people. Yesterday (pre–terrorist), we simply made the stuff up and let it go, with unpredictable results and often too much effect.

Hydraulic Jets

There are several ways in which the quiet but powerful action of water jets assist our endeavours. Firstly, a simple jet nozzle fixed inside a pipe and connected to a garden tap may well serve to drill a water bore (a water spear) in a few minutes or hours in sands, gravels, or deep soils. Secondly, an old hose–pipe fastened to a sharpened pile or post and connected to a hose will serve to sink jetty pilings in silt or sand in very little time.

A jet pump (commercially available) will remove silt, sediment, mulch, and even large gravel from a dam, and deposit these behind a retaining wall as rich terrace soil, while the water flows back to the dam. Lastly, jets fired at loose sediment washes down gravels, sands, and soils for mining, as terraces, or to remove land slips from roads and fields. Quite precise control of such action is achievable.

Hydraulic nozzles, "drills", and jet pumps are all made by industry to serve these purposes, and also to steer deep drills in boreholes to specific locations underground when tapping oil, thermal energy, or for venting mines and caves. Again (as with explosives) it is enough to know the range of uses, keep them in mind, and to get good advice if the need arises. Inadvertent jet drilling has a powerful pulling action, and jet pumps with their motors, rafts, or platforms can be dragged below water or across the ground if one is left unattended. Occasionally, a garden hose will bury itself in sandy soils in this way and so be lost.

Powerful air jets now take gravels from mines in much the same way, and a whole series of turbines and powerful pistons operate from pressure applied by water, air, or oil (the science is applied hydraulics, or pneumatics). A jet plus a cutting drill will cool the area and remove waste, in one operation.

Finishing

Rakes and rollers are the tools for finishing excavated earth; the former to fine level, the latter to compact. Compaction is only necessary where traffic or water retention is intended, but where it is necessary may need to be accompanied by water sprinkling in dry conditions. Care must be taken to compact only 15–30 cm (6–12 inches) of soil thoroughly at a time. Even very heavy machines can seldom compact depths greater than 38 cm (15 inches), and impact or hand compaction is effective for less than half that depth. It is necessary to strictly adhere to this careful compaction for dams and house foundations. Modern compacting machines often vibrate, using a pneumatic or mechanical eccentric weight to impact and shake soil as it is compacted.

9.9

EARTH RESOURCES

Whenever earth is moved, it seldom reveals a uniform composition; in fact, most excavation reveals materials already sorted by nature, and of specific use. To mix them up is to set back the clock a few thousand years, and the supervising designer or property owner will do well to follow every excavator (animal or technological) and put aside the following:

TOPSOIL. Topsoil is the often dark, root–filled living surface of the earth. It is to be carefully stripped off, piled aside, and later returned over other material onto banks, over padi or fields, and even below water in ponds. Topsoil can be very thin and rare, and is usually no more than 6–18 cm deep. Where deeper, it can be spread on poorer soils to help them produce, or pocketed for trees in poor soil areas.

PEATS. Excavation in marshes, bogs, or lowlands may reveal 1–9 m (3–30 feet) of semi–compacted fibrous plant material. Most can be stockpiled, otherwise 0.6 m (2 feet) or so returned to the surface as for topsoil. The peat stockpile is most valuable for mixing with sands and loams for making even more topsoil. It can also be used as a fine growing medium in nurseries, as an insulator in buildings, and only in desperation as a fuel, for those who burn peat are near the base of life on earth; the next step is into barrens and rocks. Peatlands throughout the world now urgently need preservation as threatened habitats of unique vegetation.

Peat preserves timber, animals, and such unexpected treasure–troves as hoards of acorns and firkins of beech butter from the forests which preceded the bogs. A whole archaeology may very well lie in peat, and the pollen record may reveal past history. At the base of Irish bogs the *Fir Bolg* (the little people), their axes, bridges, butter, and forest life are well preserved. They and their forests were banished, as if by magic, by the *Tuatha de Danan* (the Children of Diana) who now dig the peat. Diana was displaced in turn by Mary, mother of God. But all are mixed in the peat and the tongue of Ireland.

CLAY. Good clay is very useful stuff; a depth of it can extend 0.3–6 m (1–20 feet). This resource can be stockpiled, and preferably covered with plastic sheets. Both dried and baked brick can be made of it, dams and ponds sealed, and pots shaped and fired. Some types of clay make good cricket pitches, and other types make fine porcelain or special filters and insulators. Fireclay is 58–75% silica, 25–36% alumina, 0.25–2% iron oxide; silica bricks 95% silica, no alumina, and <2% of lime.

SAND. Clean, it may make grinding powder or silica chips, or if yellowish have enough calcium to counter acidity. If fine and black, it may be of use in casting metals or as a material for a black–sand solar collector. If white, it reflects light like snow and can reflect heat to houses and walls. Fat sand (containing some clay <20%) makes ovens and mortar; sharp sand (or mostly silica) makes glass and cement, or can be mixed with peat to grow trees or other plants in nurseries (the best use). Sieved sand gives special sizes for special use as a grit or polish in mortars and grinders.

GRAVELS. Heaped up, it makes good roads, drains, and soft sun–bathing patches much better than lawns (no green stains). Angular, gravel makes good concrete, and smooth, good water filters or enzyme columns. It is a mulch for trees, as is cinder and crushed pumice. 19 mm–size angular gravels are the best for heat stores, where air from solar attics or glasshouses is blown into a heat store wall or tank filled with such gravels. Heat stoves can be placed under floors to heat gravel beds.

SHINGLE. Shingle is good under gravels on roads, for drains, and coarse filters. It is an excellent mulch for condensing water in deserts or dry places, a refuge for snails and decomposers, a filter bed for swimming pools *au naturel*, and so on.

SLATE. Floors, tablets, and billiard tables are of slate, as are roofs. Any rock that can be split can be utilised for walls and houses, roof areas, garden paths, and floor tiles.

BOULDER. Boulders can be used as a coarse mulch, wildlife refuge, and walling and windbreak material which gains and radiates heat. Some boulders are very good as pounders, others excellent mortar and pestles, weights, anchors, and ballast in boats.

To me at least, a bulldozer or excavator is a source of information, and the making of a pond is a rare chance to explore, explain, and store useful information and material relating to the site. On a complex site (old volcano, shoreline, swamp, or desert pediment), one can confirm or deny theories on formation, geology, and geological history. Best of all, experience on many sites gives a predictive capacity on similar sites, and more confidence in finding the right sites for ponds, excavated houses, silos or silage pits, and in finding more earth resources. All this, and just for a careful look at actual operation. It would not be uncommon for excavation to pay for itself on any one of the factors listed, and for yourself or your client to benefit long after the water storage is built, from materials put carefully aside during excavation. If you are wise, *you never leave the site* when earth is being moved; it is very expensive to bring machines back for fine touches, to adjust the work, or to sort out mixed materials.

TRACKS AND SIGNS

Like the snow and windblown sand, newly moved earth and puddled clays reveal to the observer a section of earth life, and makes an imprint of those secretive and nocturnal birds and mammals on site, from beetles to elephants, mice to plovers. After rain, it is a good time to take a notebook and observe the footprints, scrapes, and tracks of any site where earth was freshly moved. A list of species obtained this way adds to basic information resources and may warn of troubles in store from ground foragers. Many cryptic animals avoid traps, but can be detected by their tracks.

The earth hides a great variety of life forms, from secretive larvae and worm–like legless lizards to redolent fungi revealed only on the surface by their garlic–like scent or fruiting bodies. Mice, gophers, and larger mammals leave tunnels, nests, and food stores underground, and all these and other secrets, such as the extent of root penetration and spread, can be revealed for our analysis.

SHARDS AND FOSSILS

Under the earth, almost everywhere, lie the mute stories of prior life. These have more lessons for us than do futurologists, who after all can only look forward by looking back first (or they would not know which way is forward). The camps of our ancestors are revealed in soils by shells, tools, shards, ashes, bones (and can be so dated). Buried trees give growth rings to tell the story of older climate, air quality, and to accurately date their own burial. Pottery shards may encode astronomical or technological data, and bones reveal past diseases and wars. As the Ganges River today bears its children's bodies, their clothes, adornments, and possessions to sea, to race blindly over the continental shelf to the ooze below, and as we carelessly drop broken pottery on the ground, so in the far future other people (or other minds) may want to know about us, how we lived and perhaps when and why we died. Today's disposal pits may well become tomorrow's mines. They are ourselves in past cycles, an expression of life preserved for our education and guidance.

9.10
REFERENCES

Bradshaw, A.D. and M.J. Chadwick, *The Restoration of Land*, University of California Press, 1980.

Schiechtl, Hugo, *Bioengineering for Land Reclamation and Conservation*, University of Alberta Press, 1980.

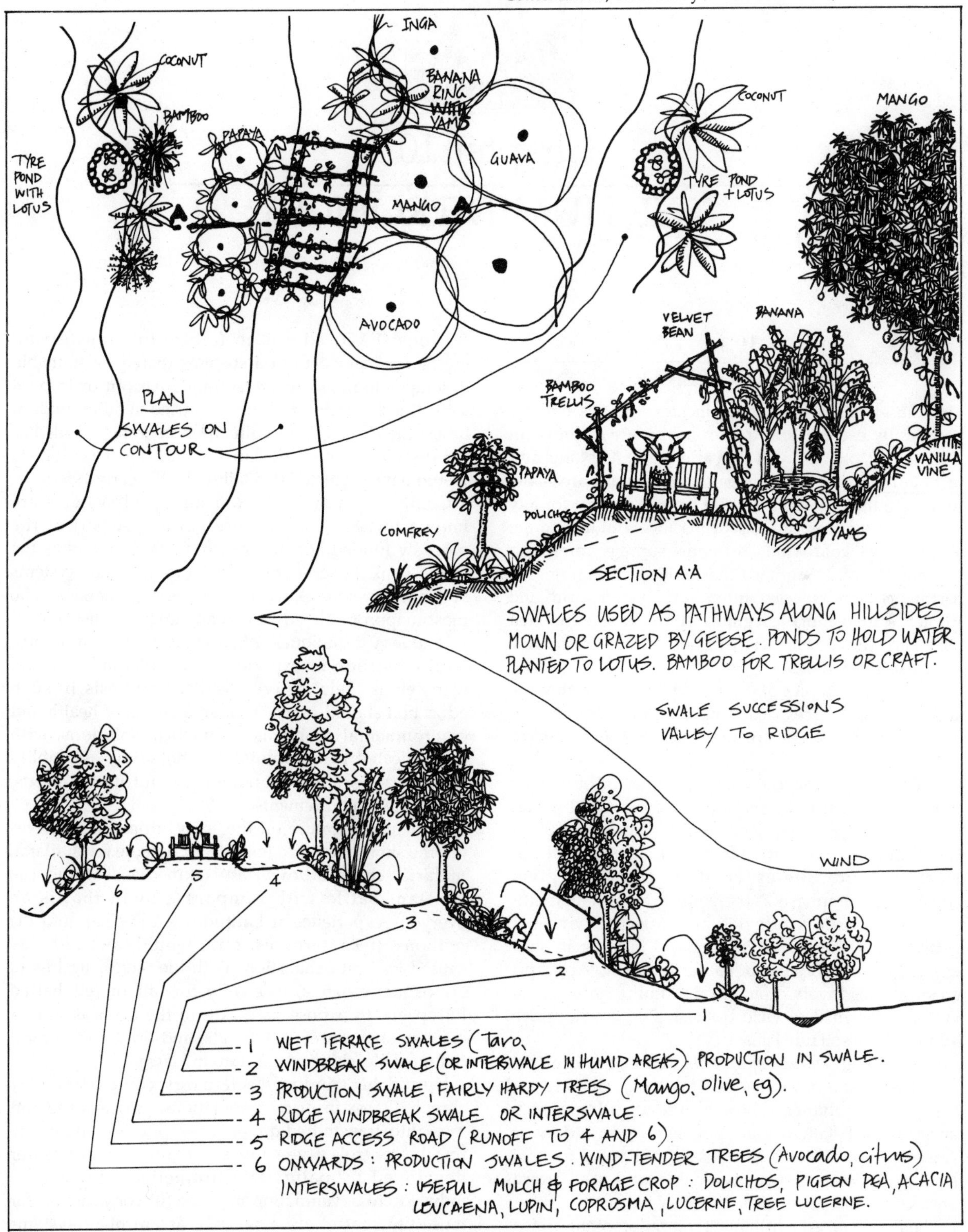

Chapter 10

THE HUMID TROPICS

10.1 INTRODUCTION

The subsets of climatic zones included in this chapter specifically exclude arid tropics, which are included in Chapter 11, together with cold arid areas. As plants and techniques do not split off neatly into climatic areas, the following three chapters should be read in total for any one site. A subtropical site, for instance, can have quite severe frosts, cold winds, torrential summer rains, and 7–9 months of drought, so that it needs the strategies, earthworks, and species suited to temperate, arid, and tropical humid climatic regimes. However, it is true that soils and climatic characteristics do dictate the specific broad design responses.

Some special topics of the humid tropics are those of soils, mulch sources, planning for polycultures, and appropriate house construction, each of which is given a section.

In the wet tropics, heat and high rainfall would leach most mobile nutrients from soils, except for the biomass of the great variety of plants, which contain 80–90% of the available nutrients. Humus is an essential soil fraction, and humus creation must be given considerable emphasis as prerequisite to sustainability.

Inappropriate strategies are those of bare–soil cultivation, or intensive clearing and burning in short cycles (less than 8 years or so) for cropping. Appropriate strategies involve complex and multi–storied plant systems designed to yield basic staples, create mulch, and preserve soil nutrients.

James Fox in *Harvest of the Palm* (1977), has been one of the few who have analysed the social changes and loss of self–reliance following the abandonment of ancient and balanced palm polycultures in Indonesia. Ancient tropical civilisations have been noted for their stability, indicating that sustainable land use patterns are an essential prerequisite for social harmony.

The Food and Agriculture Organization of the United Nations (FAO) admits to failure in transferring mechanised monocultural sustems (barely sustainable in temperate moist areas) to fragile African or tropical soils. It is a wonder that this was tried at all in modern times; the clearing and cultivation of tropical soils has for decades proved disastrous, and most ecologists would have predicted this failure by the early 1950's.

Complex perennial fodder and food systems are known to be stable, but are not as yet part of the officially funded agricultures from such sources as the World Bank. Deserts present even more fragile systems and need greater skills to stabilise and manage. The most inappropriate advisor is an agriculturalist trained in "modern" techniques. What is needed is continuous, local education of experienced people and a lateral transfer of their evolved skills. Emphasis in such education should range from an analysis of health and environmental problems to practical solutions, with sophisticated plant/animal/technological assemblies adapted to local food preferences, nutritional needs, and cultural requirements.

Romantic literature on the "easy" tropical life leaves out the skin cancers, rodent ulcers, dengue fever, filaria, malaria, chronic bowel and skin disease, and the constant battle with rampant growth that is an everyday experience at Latitudes 0–25°. That, and the pythons, ticks, termites, rain, mould, and lethargy caused by heat exhaustion. With the increasing loss of atmospheric ozone, it is folly for fair or red–haired Europeans to expose bare skin to the tropical sun—a cause of skin cancer in Australia and "*haole* rot" (a form of fungal bleaching of the skin) in Hawaii.

Humid heat induces a lethargy compounded by chronic illness in many populations. Water–borne and mosquito–transmitted diseases are almost impossible to totally control, given the aerial resevoirs of water developed by palms and bromeliads. In houses, induced cross–ventilation and careful construction for mosquito control are essentials, as are plant systems

based on a tree–species polyculture; the two combine very well to reduce climatic extremes.

We can largely emulate the tropical forests themselves in our garden systems, establishing a dominant series of legumes, palms and useful trees with a complex understory and ground layer of useful herbaceous and leguminous food and fodder plants; vines and epiphytes can complex this situation as it evolves. In the wet–dry tropics, more open palm polycultures are appropriate; the excesses of heat, light and rain are best modified by an open canopy of palm fronds and the fern–like leaves of tree legumes.

10.2 CLIMATIC TYPES

(Based on Trewartha, 1954)

WET TROPICS

These are the river basins and wet coasts from Latitudes 0–25°. Major localities are the Amazon and Congo basins, Central America, Sri Lanka, Malaya, Borneo coasts, and New Guinea. This climate covers about 10% of the earth's surface (5% of the human population). It is heavily occupied only where terraced alkaline volcanic soils enable sustained cultivation (Java). It contains remnant tribes of hunter–gatherers (often pygmoid) in remaining forests, and is rapidly being ruined by over–exploitation of forests, mining and extensive cattle rearing, mostly developed by large corporations. Hence, there is a recent tendency for catastrophic wildfires to develop in logged areas such as Borneo, and for soils to be leached to low fertility, or eroded to ferricrete or silcrete subsoils.

The sun is mostly overhead, with temperature fluctuating little at about 21–32°C (70–90°F). Humidity is

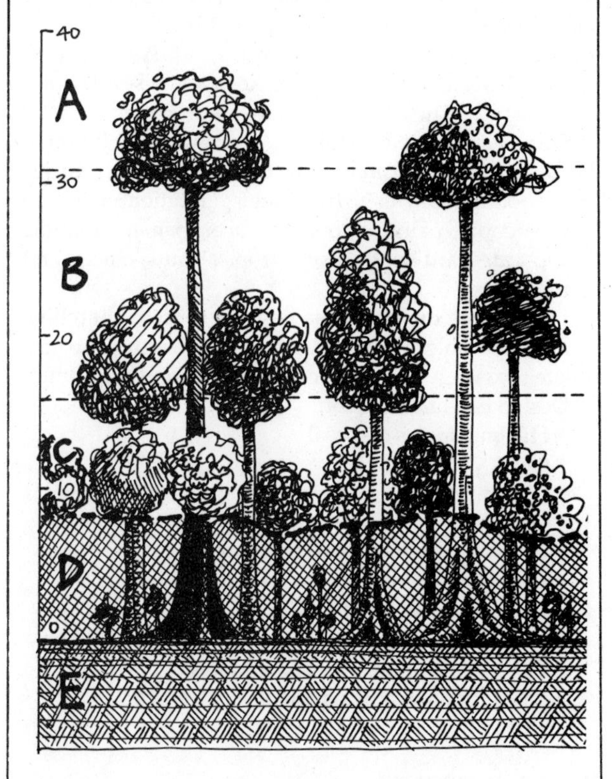

FIGURE 10.1
STRUCTURE OF A WET TROPICAL FOREST
Dense planting is possible, and beneficial near villages; species assemblies are simplified on the broadscale. Levels in natural forests are also indicated (after Moore, *New Scientist*, 21/8/86).

constantly high, frost unknown. Rainfall is from 152–328 cm (60–129 inches), with rain most days and frequent thunderstorms (75 to 150/year), usually towards evening or late afternoon.

The landscape features perennial streams, deeply weathered and rotten rock over bedrock (regolith), and rounded hills. There is rapid water run–off and evaporation, with swamps confined to coasts and lowlands. Rivers usually have flood–plains and extensive alluvial plains and deltaic deposits. Access inland is often by rivers and tributaries.

The vegetation is luxurious, best developed as broadleaf rainforest with lianas and epiphytes, very much mixed as to species—up to 800 tree species per square kilometre. The shaded floor of the forest contains little growth and has subdued light. Mangroves are extensive on appropriate coastal and estuarine sites. Growth is rapid, uninterrupted and continuous. Insects and birds are plentiful and varied. Most fauna is nocturnal and arboreal, and there are abundant fish and aquatic species. Ground grazers are rare and large herd species do not occur.

About 85% of nutrients are held in plants or animals, so the soils themselves are infertile especially if clear–cultivated, tending to erode and leach to insoluble oxides of iron and aluminium. Only terraces, flood–plains and new volcanos keep some soil fertility replenished or held if land is cultivated.

Staple cultivated foods are plantain and banana, cassava, yams, coconut, corn, taro, paddy rice, ducks, pigs, poultry, and fish. Trade and plantation crops are spices, copra, palm oil, cacao, rubber, banana, manilla hemp, rare hardwoods, balsa, tropical nuts, chicle, and drug plants.

Housing is usually raised, steep roofed, thatched, with permeable walls, and screened. Health problems relate to sewage disposal, insect vectors, and skin fungi.

Design essentials are for:

- Hygenic faeces disposal.
- Clean water sources.
- Integrated and benign insect control techniques.
- Gradual replacement of ground crops by trees.
- Preservation of natural stands of trees.
- Development of river versus road traffic.
- Evolution of natural products .
- No–dig (mulch) techniques on root crops.
- Domestic foragers for snail and insect pests.
- Appropriate medicinal plants.

WET–DRY TROPICS

These adjoin the wet tropics but are poleward of them. They take up about 15% of earth's surface from Latitudes 0°–25°, unbalanced in favour of the southern hemisphere. The Campos, Llanos, Gran Chaco areas of South America, parts of Central America, encircling the Congo basin, and many central Pacific islands (Hawaii) are all wet–dry tropical areas, most now developed to grazing.

Winter, the low–sun period, is the dry time, when clear skies and intense sunlight take day temperatures to 38°C (100°F) or more. Humidity is low, and strong dessicating winds may blow. Summer, the high–sun period, is like the wet tropics, but episodic flooding is more common and natural erosion therefore greater. There are no frosts, and temperatures range 21–27°C (70–80°F) in the wet season, 32–38°C (90–100°F) in dry season. Rainfall is 25–152 cm (10–60 inches), decreasing towards desert margins. Windward inland slopes and coastal mountains may receive excessive rain to 1016 cm (400 inches), but rain is erratic and least predictable towards the desert margins. Rain shadows evolve on leeslopes or in the lee of mountains.

The landscape is of intermittent streams, some wadis and flood plains, karst (limestone) areas with sinkholes, cenotes, and absence of surface water (Yucatan, Mexico). Hills are rounded, but gully erosion can develop rapidly on slopes. Extensive inland swamps may develop in flooded areas, and lakes in rifts are common (Africa). Rivers often have dangerous barways of silt and sand due to active erosion sequences.

These regions contain the vast savannah grasslands of the tropics, with thorn–bush and flat–topped *Acacia* trees (Africa), evolving to steppe grassland on plateaus, with baobabs and dry–deciduous trees. Grasses reach 1–6 m (4–20 feet) in the wet season, and are often burnt off. African areas contain enormous numbers of herd species: zebra, gnu, antelope, and therefore large carnivores. Arboreal species occur only within tree islands and the gallery forests of valleys. There are termites, ostrich, rhea, locust, and large numbers of reptile species and insects.

Soils are generally more fertile and alkaline than the wet tropics, especially where they are less leached by rain towards the deserts. Cultivated land is still at risk from erratic rain and erosion, leaching, and wind effects. Due to overgrazing and fire, erosion may extend the desert into these areas, or into dry–summer subtropics. Serious soil erosion results from short–term shifting cultivation (less than 15 years fallow period).

Staples are corn, millet, wheat, beans, potatoes, cucurbits, peanuts, cattle and goats, sheep, and game products. Herding of low–yield large herds is a major erosion hazard. Plantation crops are sugar, cotton, peanuts, pineapple, sisal. Exports are cattle and sheep products, and hardwoods from gallery forests.

Houses and granaries are generally mud or pise with thatch.

Design essentials are for:

- Small domestic water storage and reticulation.
- Hedgerow against winds.
- In–crop tree legumes such as *Acacia albida.*
- Improved stock varieties and stock management.
- Natural herding system of local herd species.
- Mulch use of grasses.
- Increased tree crop of high forage value.
- Decreased fire frequency.
- Tree stands for fuel and structural timber.
- No–tillage (cut and mulch) grain techniques.
- Low bunds for water retention.

• Chisel plow and sod–seeding techniques.
• Greater reliance on in–village tree crop near wells and ponds.
• Reclamation of eroded lands using pioneer species.
• Keyline techniques of flood control.
• Soakage pits and impoundment of run–off by low bunds or swales across slopes.
• Tree forage and tall grass hand–fed to domestic stock .
• The use of manures in gardens.
• The development of domestic fuelwood systems near villages.

MONSOON TROPICS

These are really a sub–type of wet–dry tropics, but influenced by nearby continental land masses and oceanic winds onshore. They are confined to the Indo–Thailand region, northern Australia, East Java, Timor, southern New Guinea, and extend Latitude 0° to 35° north in India. Despite only about 8–10% of the world's land surface, monsoon areas contain large human populations.

Late summer heating of the continents causes onshore sea winds and (with luck) heavy rains. The dry (winter) season reverses winds from cool interiors to coasts, giving a cool period not experienced in the wet–dry tropics (temperatures: 13–21°C—55°–70°F). Temperatures rise, and dry hot winds develop (to over 38°C—100°F) in spring, with heat increasing until the onset of the monsoon . About 60% of the rain falls in summer, but rain is erratic and varies from 102–1016 cm (40–400 inches), depending on topography and distance from the coast. Floods and droughts are equally unpredictable, but common. Most activities are determined by the monsoon rain (transport, fishing, farming).

Tropical forests once clothed the hill slopes and river plains, and grasslands extended towards deserts as savannah. Population pressure, deforestation, and marginal agriculture has devastated this ecology in India. Dry–deciduous broadleaves are common, teak and bamboo once extensive. Tree canopy is less dense than in wet tropics, so that dense understory is also developed. Mangroves occupy river mouths and low coasts. Large native animals are now rare in the Indian sub–continent, but reptile life is abundant, as are feral or native deer, buffalo, and primates. Monsoon Ausralia is better vegetated, with scattered eucalypt and *Acacia* trees, riverine forests, and very low human populations to date. Large marsupials, feral buffalo, and marsh waterfowl are abundant.

Soils are lateritic, often very hard in the dry season, and of low nutrient status. Some are cracking clays. Housing is often mud–pole structures, thatched, steep–roofed, with wide eaves and good drainage for wet period.

Design essentials are similar to those of the wet–dry tropics.

10.3
TROPICAL SOILS

Special problems arise with tropical soils, in that except in areas of recent vulcanism such as Indonesia, soils are old (not renewed by glaciation) and deeply leached. Most of the silica and calcium is in low supply. In clays, aluminium ions substitute for some silica ions, giving soil particles a net negative charge. Especially in the oxidic kaolinitic soils (common in weathered volcanics) only kaolin clays and oxides of iron–aluminium remain. In these soils, the charge or cation exchange capacity (CEC) of the soils is *affected by pH*. Once cleared, the humic particles leach out to about 30% of prior levels, and infertility appears in such crops as banana and sugar cane. There are a few ways to restore the soil's ability to hold nutrients (Wayne Ralph, "Managing Some Tropical Soils" in *Rural Research* No. 117, pp. 15–16):

• Restore humus with green crop, and especially perennials such as *Leucaena* and tree legumes generally Any cultivation loses humus as carbon dioxide, so try to grow plants with intercrop.
• Now, add *small* quantities of superphosphate at frequent intervals so that plants can take it up before leaching. If possible, add fine crushed basalt, a scatter of cement powder, and use shredded bamboo or cane mulches for silica and calcium. Increase pH with lime after trees and green crop are growing well.
• Whatever is added or available as fertiliser, give as a light spread *all year at 6–week intervals*, until plants are well grown. If at all possible, substitute perennial for annual crop, and *never* practice frequent cultivation.

Basalt, cement powder, coral, and bamboo mulch supply essential nutrients and increase soil pH, hence increase the negative charge on soil particles and their ability to hold calcium, sodium, phosphates against leaching.

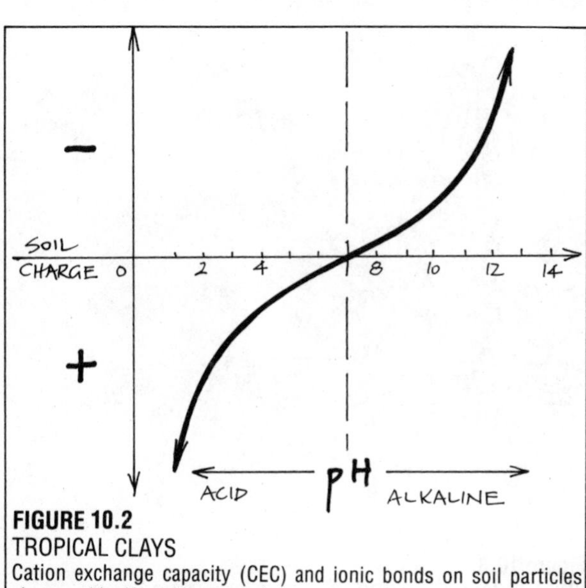

FIGURE 10.2
TROPICAL CLAYS
Cation exchange capacity (CEC) and ionic bonds on soil particles change with pH as diagrammed.

On coral cays, the calcium–rich sands bind to phosphate to form insoluble calcium tri–phosphate, so that a sort of cement (*platin* or calcrete) forms. This may be naturally evolved from the guano of seabirds, but superphosphate rapidly forms the *platin* by its greater solubility. In calcium rich tropical soils, fine rock phosphate yields more slowly and is therefore more likely to provide long–term benefits. A return of crop wastes as mulch is also essential, which can *reduce* pH in coral sands (pH 8–9) to a level nearer to pH 6.5 or 7, which is suitable for gardens.

In fresh volcanic areas, or areas with volcanic dust deposits, soils are sufficiently rich to sustain intensive agriculture without such aids, but constant cropping will exhaust even these soils.

There are many excellent tropical soils such as the alkaline volcanic soils of Indonesia, which support rich terrace and palm polyculture systems, and many tropical high–island soils where dolomite tops, or forms a mosaic with, recent volcanics. Apart from testing for minor elements, the addition a of mulch–manure mix creates excellent gardens on such soils, and plants show only minor nitrogen deficiences. These deficiencies can be eliminated by legume intercrop and manures.

In the long term we must rely on tree and ground legumes to keep up soil health in the tropics. Destructive approaches (now very well demonstrated) combine forest clearing, bare–soil cropping, and careless water run–off management to make desolate baked clays, brick–like and hostile, out of once–rich tropical forests. We have (as yet) no categories of "crimes against nature", but these will prove, in the future, to be some of the worst.

DEEP GRANITIC SANDS

In soils over rotted granites, such as are found on the high islands of the Indian Ocean, the Deccan in India, and where granites are left as inselbergs (domed hills), a peculiar problem arises in that open, coarse, granitic sands, often very deep, will not retain mulch beyond one growing season. There are two approaches to these free–draining and low–nutrient soils:

Broadscale

Palms, *Albizzia* spp., *Inga* spp., *Acacia* spp., a general planting of adapted leguminous trees, and other native vegetation will establish a light canopy of leaves if small amounts of nutrient are added at regular intervals. Palms planted in mulch–filled pits will then establish a high crown cover, and the detritus from legumes can be used to establish more valuable fruit trees, always retaining a *complete root web* of legumes. Palm trunks are ideal trellis for vanilla and passionfruit crop. Niches and clefts in the granite mass itself will hold pockets of soil and mulch for valuable trees, and vines will establish there to cover the granite slabs, which maintain heat and ripen crop effectively.

It is the mycelial web of the pioneer legume roots which enables us to maintain the benefit of applied nutrients, to reduce water use, and to establish fruiting trees (mango, cashew, pomegranate, persimmon, citrus, tamarind, lychee, custard apple, avocado) in such sparse and drought–prone soils. Drainage is, of course, excellent, and deep–rooting trees thrive. Palms suited to these sites are date, coconut, doum, and *Borassus* palms.

Gardens

It is effective to excavate long trenches in the loose sand (1–2 m wide, 1–1.5 m deep), to lay in a sheet–plastic base (upturned at one end only) or to line the trench thickly with cardboard, paper, carpet, and leaf and then to backfill with sandy loam. The deep sealed layer holds water and leached mulch, and household waste water can be led into these trenches to provide root water and nutrients.

Otherwise, we can build log–boxes above ground level, carpet the box base with plastic or thick paper,

Figure 10.3
BROADSCALE TREE CROP IN LOOSE GRANITIC SANDS

and fill with humus–sand mixes for green crop and vegetables, top–mulching as needed. Large domestic water tanks, gleyed ponds, solid granite dams, and underground plastic–lined cisterns back–filled with sand will hold water.

SOIL LIFE IN THE TROPICS AND SUB–TROPICS

Termites, ants, and some worms are the obvious soil mesofauna of many arid and humid subtropical areas. Both ants and termites are very active in the transport of rotted rock and subsoil to the surface, in opening up galleries for the infiltration of water, and in the breakdown of woody and leafy plant material. Some species create large mounds, others build underground compost heaps for fungal culture, and all are active burrowers and builders.

Termites may have a decisive role in the dynamic and delicate balance between the erosion of surface soil and the replacement of the soil by subsoil and rotted rock particles. They certainly have an important role in plant succession and distribution in savannah areas, or where termite and ant mounds are the only well–drained or elevated sites in a landscape subject to floods, or where impermeable clays underlie thin peats (usually with acid anaerobic soils). In these situations the spoil heaps present an ideal site for pioneer vegetation or adapted crop planting.

Harris (1971) records that both the leaf–cutter ant in South America and termite mounds in Uganda assist forest spread or establish islands of taller vegetation in grasslands on their mounds or colonies. I have also observed this in the granite country in Hyderabad, India, and on acid peatlands in Tasmania. Such mounds protect soils from fire, waterlogging, and poor aeration. In humid areas, some such sequence as tall grasses (*Pennisetum, Eragrostis*) are followed by shrubs such as castor oil bean, *Prosopis*, and thorny legumes. Finally, an understory and forest may develop from larger trees such as tamarind, *Vitex, Sapium*, or palms dominant.

Thus, we can start this process or a modest version of it by seeding into termite or ant mounds using similar species; even low ant–heaps may present a site for ground cover pioneers in grasslands. Harris records crops such as sisal, cotton, and tobacco deliberately cultivated on large mounds in grasslands. Palms and coffee can have much of their outer bark removed by termites without suffering loss of production. Termites may greatly assist the primary breakdown of logs, coarse stems, and hard leaf material used as mulch in plantations of coffee, tea, or bananas.

It is a matter of specifying (by observation and local report) which useful crops or trees are left alone on mounds, which are attacked but remain productive, and which actually benefit by association with a local termite or ant species. Planting in ant or termite mounds is a particular example of niche gardening widely applicable to the tropics. I have successfully germinated daikon radish in ant heaps in grasslands as part of a changeover to crop production.

Ant and termite mounds present a rich deposit of calcium and potash, better aeration of soil, and a faster infiltration of water to release minerals from such rocks as granites, which noticeably rot or erode faster when buried in a free–infiltration soil environment.

SOURCES OF HUMUS FOR TROPICAL SOILS

As humus in the soil provides a good CEC, we must look to the provision of such humus as a priority. Some sources are:

- Logs and branches of trees. These are sometimes rotted in wet terraces or piled up as rough mulch around new tree plantings. Palm wastes are often plentiful.
- Detritus from stands of bamboo, pines, *Casuarinas*.
- Aquatic weed mats and emergent water weeds.
- Crop wastes and manures, household wastes.
- Hedgerow, forage, and specially–grown mulch plants.
- Green mulch and ground cover.

Logs and Branches (Rough Mulch)

The rapid breakdown of wood under the combined influence of rain, heat, termites and fungi means that we can lever whole logs together, or in line cross–slope to act as planting sites. This technique is most useful on bare clay soils, eroded areas, and isolated atolls (using the trunks of old palm trees).

In Hawaii, a traditional strategy is to rot the logs of

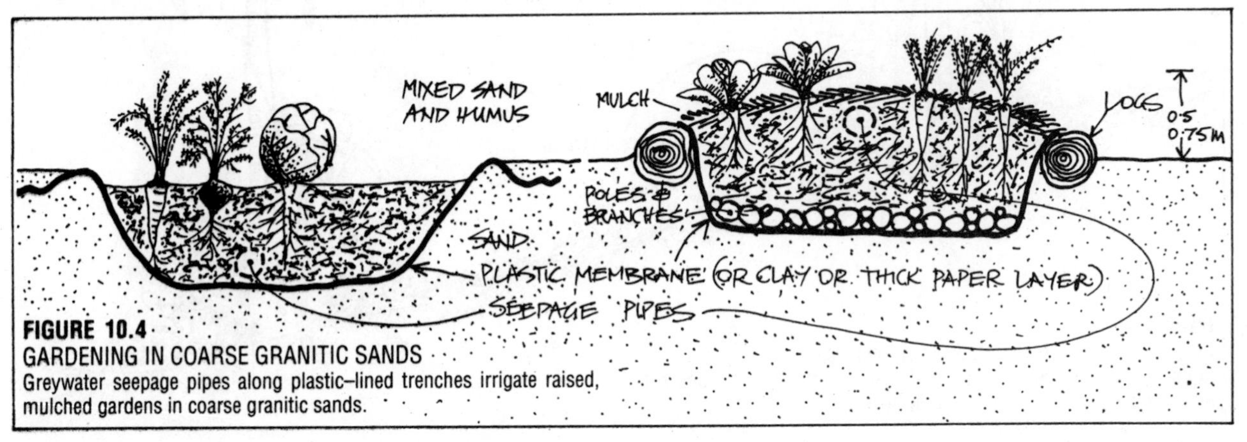

FIGURE 10.4
GARDENING IN COARSE GRANITIC SANDS
Greywater seepage pipes along plastic–lined trenches irrigate raised, mulched gardens in coarse granitic sands.

the kukui tree (*Aleurites moluccensis*) in the shallow water of taro terraces. As logs rot, an edible fungi appears which is taken off as crop. The remaining log is then crushed and spread in the taro terrace. Leaves and branches of kukui and other forest trees were gathered for the same purpose (terrace mulch).

Marjorie Spears, in Queensland, Australia, has successfully built temporary roughwood terraces across a deforested slope using rejected logs, and created a complex and rich garden based on this strategy and green legume mulch. Logs are available from palms, although fast–growing *acacia* species can be previously close–planted for this purpose (**Figure 10.5**).

Detritus

Several plant species (most palms, bamboo thickets, *Casuarina*, and many *Acacia* species) provide silica–rich mulch, as do grain and nut husks and residues from copra operations. This can be applied as shredded or chopped mulch to crop or to the base of newly–planted trees. The silica is released for growth, and has the secondary effect, in alkaline island soils, of reducing pH (from 8.5 to 6.5 in my trials on coral islands).

Of particular value are the fronds and spathes of palms, shredded or whole, and the stems and spathes of bamboo. Both have essential structural uses, and larger stands can be used to produce mulch.

Aquatic Weeds

Floating aquatics, including the water–fern *Azolla* (several species), the water–lettuce (*Pistia*), water hyacinth (*Eichornia*), and algal mats or fern fronds, plus reeds and rushes gathered from ditches, are excellent crop mulch. *Azolla* has largely replaced kukui (*Aleurites*) as a taro mulch in Hawaii. *Pistia* has been successfully used in Africa, and water hyacinth in many areas.

Azolla and algae such as *Anabaena* (one species of which "nests" in the glutinous sacs in *Azolla*) provide nitrogen. In the dry–wet tropics, shallow flood–water bunds collect or produce these plants, which can be gathered as rolls of dry material when the water dries out in the winter period, thus providing garden mulch in abundance from temporarily impounded water. John Selman of Cooktown (Australia) has used algal rolls in this way for his garden plants.

Crop Wastes and Manures

The husks of corn, kitchen wastes (including bones), and human and animal manures are invaluable tropical garden mulches and nutrients, and most need only burial near tree roots or in growing mounds for safe disposal. Cardboard and newspaper, where available, are valuable grass–suppressing weed mulches, and a cover of nut husks completes the job. In the Seychelles, cinnamon leaves and branches from pollarded stumps are considered an excellent mulch for vegetable crop; and the bark is a valuable spice.

Hedgerow and Mulch Plants

All hedgerow (*Hibiscus*, *Casuarina*, banna grass, palms, leguminous trees such as *Gliricidia*, *Acacia* and *Prosopis*) are almost continual mulch sources. The legumes provide in–crop shelter (see later section on avenue cropping). Lower garden windbreak, especially lemongrass (*Cymbopogon citratus*) and comfrey (*Symphytum officinale*) are as useful in preventing kikuyu grass intrusion as they are for repetitive cutting for mulch in the vegetable garden. Many people now use both these species as a combined kikuyu barrier and mulch crop (**Figure 10.6**).

FIGURE 10.5
LOG BARRIERS ON SLOPES (Marjorie Spears)
Logs cross–slope hold mulch until tree–lines establish; logs rot quickly but mulch is renewed annually by banana clumps.

Green Manures and Ground Covers

In and around gardens and trees, soft herbaceous plants such as nasturtium, comfrey, marigolds, tobacco plants, and the tops of mature taro plants and other *Araceae* not only suppress grass, but provide a constant source of "slash" mulch. Even more valuable are such soft legumes as *Sesbania*, vetch, Haifa clover, cowpea, lablab bean, soya bean, *Desmodium*, *Suratro*, and *Centrosema*. These can be slashed or (in wet–dry tropics) interplanted with grains to give a nitrogenous ground cover, aiding in the suppression of grasses. Lablab dies down just before grains ripen in the winter dry season.

Thus, a combination of growing and gathering mulch enables us to create a rich humus for gardens over clays or sands, in loose volanic cinder, on *a 'a* lava, and in loose coral atoll sands. Each of these situations can successfully produce mulch.

Some difficult mulch such as *hibiscus, Lantana,* and weeds which tend to resprout from cuttings or seed if mulched (several grasses and hedge species) can be routed to gardens via poultry or cattle pens (where seeds are removed and foliage eaten). They can also be shredded for anaerobic digestion in biogas plants, bagged in large plastic bales exposed to the sun (where they "cook" to a weed–free silage), or simply bundled and immersed to rot in covered water pits. In fact, some such re–routing is ideal for the primary processing of plant wastes that promise to infest gardens if untreated. Pigs eliminate or eat the nut–grasses, rhizomes, bulbs, and sedges that resprout from compost.

All else failing, even a plastic sheet mulch has excellent effects on row crop, preventing rain splash and nutrient leaching, and at the same time condensing groundwater at night. It does not, however, add to the humus content of soils, nor to the cation capacity of soil structure, and may even release unwanted chemicals to the soil.

The value of surface mulch in weed suppression is a major factor in lowering garden work. For this reason, any mulch should be thickly applied 20–25 cm (8–20 inches) deep when first establishing home gardens. Later mulch can be derived from green herbage and borders or windbreaks.

TABLE 10.1
CROP YIELDS UNDER SELECTED MULCHES

MULCHES	Maize	Cowpea	Soyabean	Cassava
Rice husks	3.7	1.1	0.8	**28.3***
Pennesetum Straw	3.3	1.2	**1.4**	14.2
Elephant grass	3.3	0.9	1.3	16.6
Millet straw *(Panicum)*	3.6	**2.1**	**1.5**	15.5
Legume wastes	**4.3**	1.0	1.1	15.5
Sawdust	3.7	0.9	**1.9**	20.5
Bare ground	3.0	0.6	0.6	16.4

*Heavy yields are emphasized.
[After B. N. Okigboand R. Lal, *Residue mulches and agriculture.*]

B. N. Okigbo and R. Lal in *Residue Mulches and Agrisilviculture* (International Conference on Ecological Agriculture, Montreal, 1978), in mapping strategies to cope with increasing land pressures, found that no–tillage systems maintained or gained yields for maize in Nigeria, and increased yields from mulched crop for cowpea, soya bean and cassava.

I have selected out some natural mulches from the more extensive original table given. Mulch trials are compared with bare ground (on the last line of **Table 10.1**)

Maize had a marked positive response to legume straws or waterplant (*Pistia*) mulch, while the legumes themselves responded well to grass and sawdust mulch, and cassava to both legume and grain husk wastes.

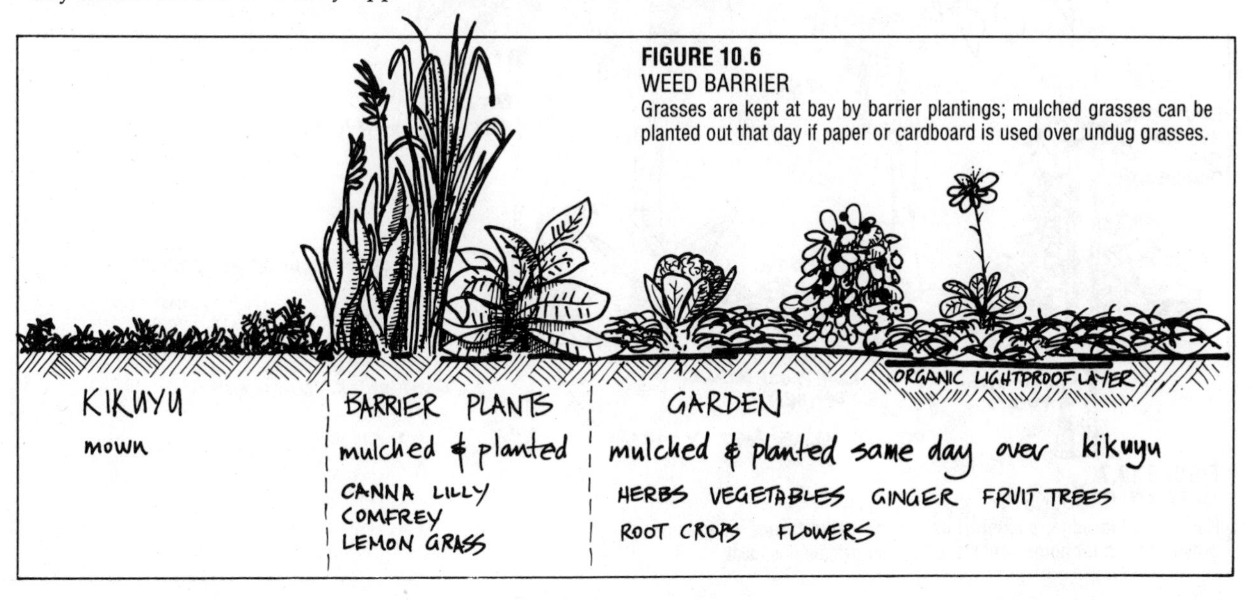

FIGURE 10.6
WEED BARRIER
Grasses are kept at bay by barrier plantings; mulched grasses can be planted out that day if paper or cardboard is used over undug grasses.

Although plastic mulch has a good effect on all crops, it does not add humus to soils, and is therefore not as appropriate to a remote village situation where soils must be built up from wastes and from mulch.

However, every type of organic mulch increases yields, and we should therefore use all available materials for soil restitution. Mulch provision is the cornerstone of tropical home gardens, and green mulch and tree legumes the essential accompaniment of main crops and tree crops.

Special mulches may be used in tropical areas, grown to provide N, P, K (legumes, comfrey, *Pultanea*), and to increase or decrease pH. Pine and legume mulch may benefit the growth of bromeliads (pH 4–5), buckwheat and nut husks serve to raise the pH of garden soils, as do many bark mulches.

For fire control, too, it pays to rake under bamboo and clump canes, and re–route the leaf mulch through animal bedding or poultry strawyards. Branches of legumes and forage trees may also be used in the same way, on their path to the garden.

In the rampant grasslands that replace fallen forests, there is little else we can do than to strip–mow and mulch while trees re–establish. The timing of slashing is important, as seed–free mulch (called second–cut grass) is best for placing around valuable crop. Seed–head mulch should not be placed in gardens or areas where it might be a nuisance. All species of weeds and grasses give weed–free mulch when not in seed. Many useful tree species provide leaf mulch, and so are excellent also for interplanting with crop, for example tamarisk in dry areas, *Casuarinas* in sand, and legumes in all areas.

There is absolutely no excuse for burning any organic wastes in the tropics, as even large logs quickly rot under the onslaught of fungi, termites and beetle larvae. At the same time, logs provide cross–slope barriers against monsoon erosion, until new trees take hold.

Coconut husks have a variety of uses, not the least of which is as mulch for a valued crop such as vanilla orchids. Their one drawback is that they hold small sections of water which will breed mosquitoes, but on many islands they will also be available (with palm fronds) to shred to a first–class mulch of high potash value, to burn and steam to activated (filter) charcoal, or to be used as a solid fuel. Shredded bark and broken shells are ideal mulches for ginger, tumeric, and vines.

I have not found any crop or tree suited to the specific locality that does not grow, produce, and thrive in mulch, nor any widespread pest that grossly affects a total polyculture yield. Ginger, taro, beans, bananas, palms, fruit trees, flowers, yams, sweet potato, melons, etc etc. have been trialed in thick mulches of straw, fronds, nut husks, cardboard, and sawdust. A thick mulch almost totally eradicates kikuyu grass and other persistent grasses. In the field situation, extensive mulching is often impractical, particularly if it is carted in from off the site. However, a pioneer crop of quick–growing tree *Acacias*, bananas, legumes such as lablab, deep–rooting comfrey, and a grove of bamboo and palms will provide continuous mulch for gardens and main crops, fruit trees and valued plants.

Growing in exhausted or poor tropical soils is possible, but the early work of rehabilitation takes hard work, seed, essential fertiliser resources, and a strategy

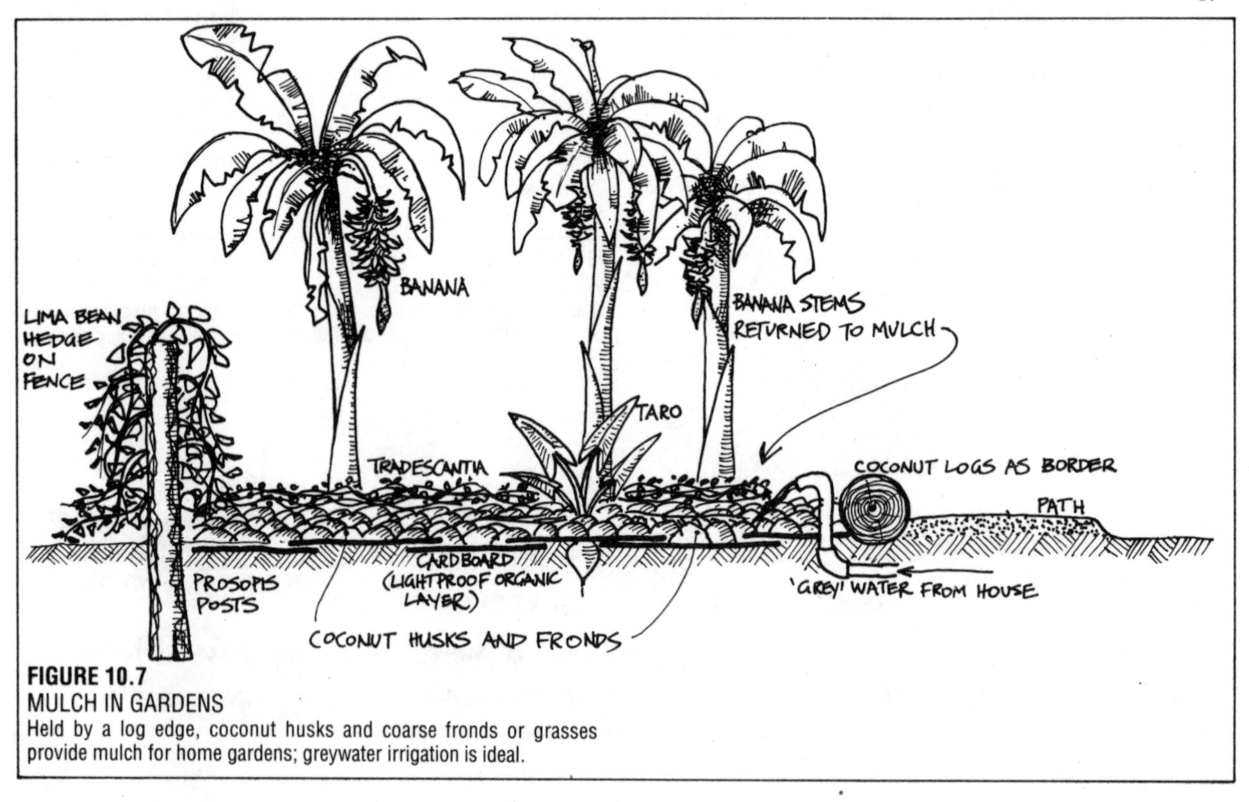

FIGURE 10.7
MULCH IN GARDENS
Held by a log edge, coconut husks and coarse fronds or grasses provide mulch for home gardens; greywater irrigation is ideal.

of starting small and expanding the system at the periphery. Dense planting of nucleus areas plus mulch is the key strategy.

10.4
EARTH–SHAPING IN THE TROPICS

On level ground or gentle slopes (2–8°) in the wet–dry tropics, a series of large contour banks or swales have an excellent soil preservation effect. Coupled with the gradual development of a terrace, the retention of wet–season water, and mulch–providing hedgerow, this ensures a stable situation. Between the main hedges, mulch hedgerow and borders can be developed in crop, or the terraces can be flooded seasonally for irrigated crops (**Figure 10.8**).

On very flat sites (less than 4°), a series of raised mounds or ridges can operate to drain crops in very wet areas, or to impound water for absorption in drier areas. Pits can also be used only where rainfall is less than 76 cm (30 inches), or where soil drainage is good. Thus, cassava, yam, and cucurbits are mounded in areas where drainage is a problem and rainfall intense, and pitted in dry areas or savannah–dry seasons. Pits retain mulch and moisture, as they do in desert areas.

Almost every slope benefits from earth–shaping for soil conservation. Hand–made slope terraces need to be narrower (to 3.5–6.5 m—12–15 feet) than machine–made systems.

Garden terraces on *very* steep humid slopes must be kept narrow, and in sets of 6–8 downslope, otherwise instability may result. Borders can be kept vegetated with trees (**Figure 10.11**).

Classical wet rice and taro terrace has water continuously led into the top terrace of the series, and each has a drain and sump to regulate water level. Fish

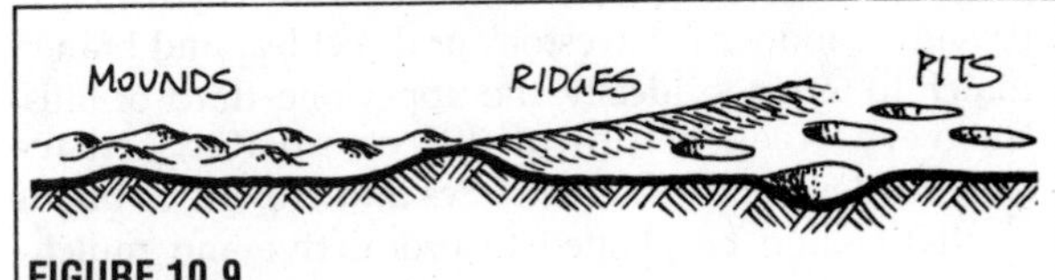

FIGURE 10.9
MOUNDS, RIDGES, PITS
Mounds increase yields of yams; ridges of cassava and sweet potato; pits for taro, arrowroot, and mulch grasses. Terraces need such detailed earthworks.

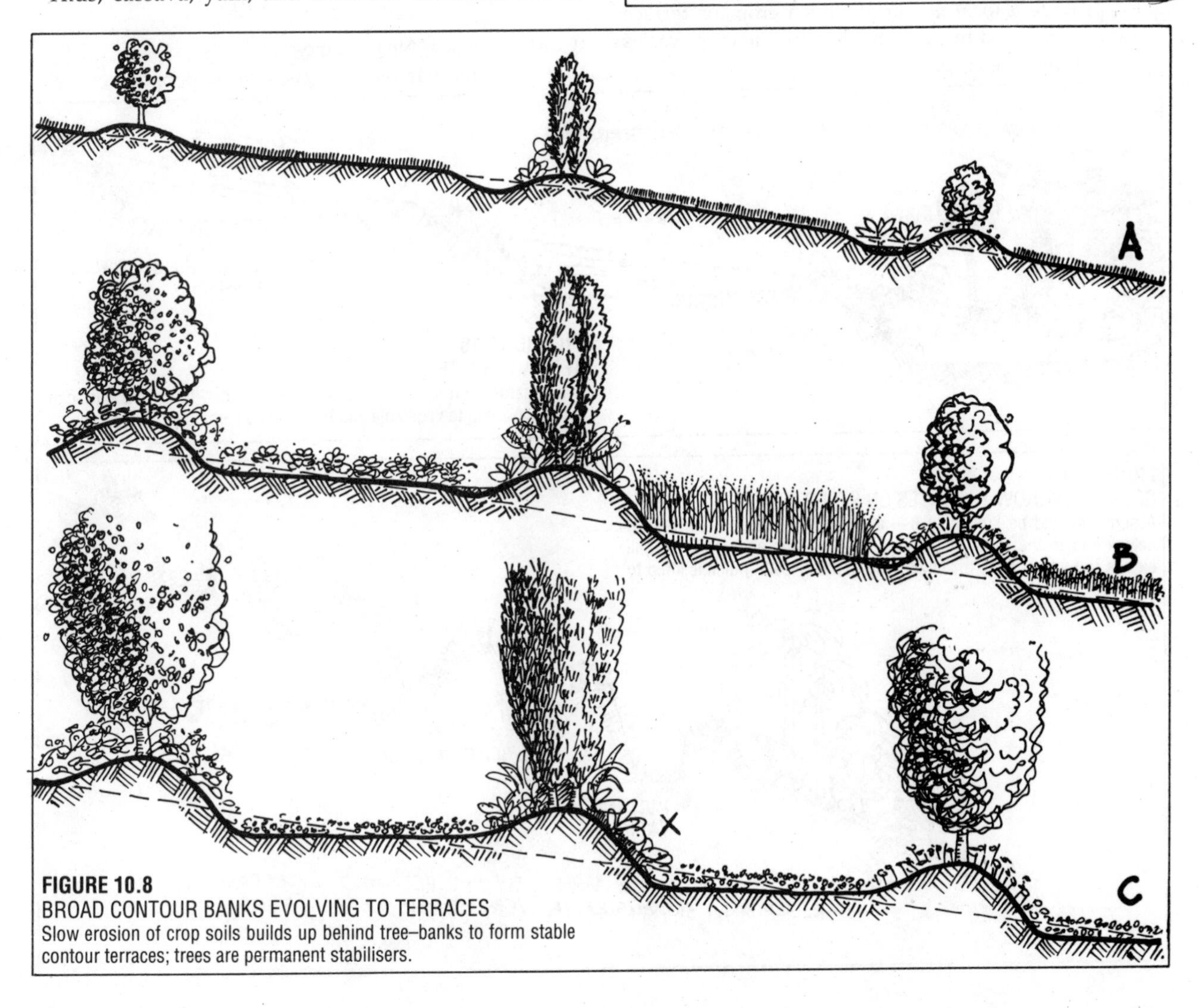

FIGURE 10.8
BROAD CONTOUR BANKS EVOLVING TO TERRACES
Slow erosion of crop soils builds up behind tree–banks to form stable contour terraces; trees are permanent stabilisers.

may be grown in the deeps of such terraces (**Figure 10.12**).

In stable clay or clay–loam soils, terraces not only hold and infiltrate water, but permit mulch application with minimal leaching losses. Where no streams exist to feed the terrace system, DRY TERRACE holds the soil against erosion in cropped areas. Lacking streams, a deep mulch keeps terrace soils moist. Where a stream, or part of a stream, can be led to upper terraces, wet crop such as rice, taro, watercress, kangkong, and water chestnuts (Indian or Chinese) can be cultivated in water–level controlled padi. This is the rich WET TERRACE culture of Asia and Oceania.

Essentials are: about one–half to one–third of the total terraced area should be devoted to *mulch tree crop* providing fodder for livestock or direct leaf and branch mulch to terraces. Ideally, the upper one–third of hills, the very steep slopes of 30° or greater, terrace side–borders, and the outer faces and crowns of bunds (walls) should be planted to productive and mulch–producing tree and ground crop. This not only adds to the terrace stability—many of which have existed in production for up to 5000 years, e.g. the wet terrace cultures of the Ifugao people of the Philippines—but will also provide a local manurial–mulch crop for terrace cultures. Included in such mulches are the crop wastes of the preceeding crop.

Specific Growing Situations on Terraced Lands

- Banks and bunds: the rim of the terraces, and stepped bunds made for tree crop.
- Slope faces and walls.
- On trellis out from bunds.
- In and around ditches and drains.
- On steep (unterraced) slopes.
- The flat area of the terrace itself.

Dry Terrace Crop Species

- Millet: summer or dry periods.
- Dryland rice: spring–summer.
- Barley, wheat, rye: winter and cool periods, spring wheat varieties, *Brassicas*, fava beans.
- Amaranth: summer grains, spring greens.
- Quinoa: summer grains.
- Rape/mustard: winter oils and oil seed.
- Lentils, peas: intercrop and nitrogen fixing grain legumes.
- Grams and pulses: intercrop and nitrogen fixing.

Fodders

- Tagasaste, banna grass, comfrey, *Leucaena*, crop wastes and straw.

Garden Terraces Near Homes

- Banana, papaya, melons, chilies, peppers, cucur-

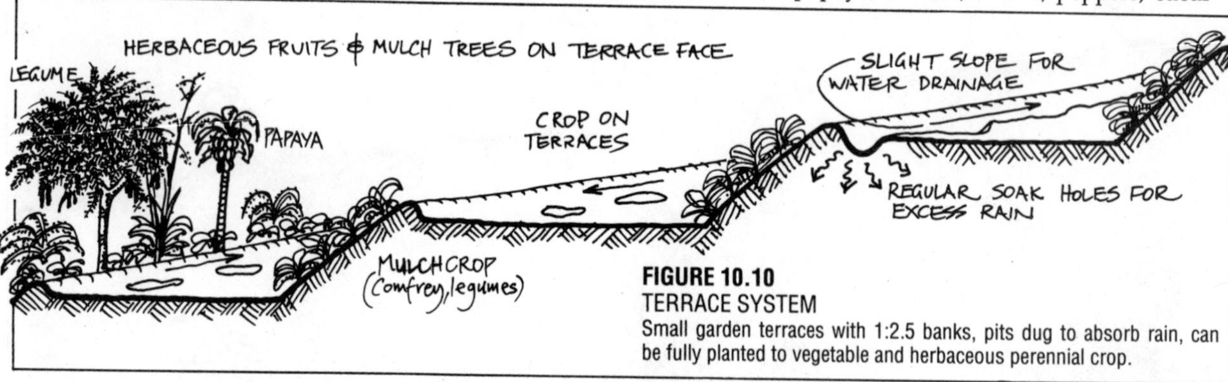

FIGURE 10.10
TERRACE SYSTEM
Small garden terraces with 1:2.5 banks, pits dug to absorb rain, can be fully planted to vegetable and herbaceous perennial crop.

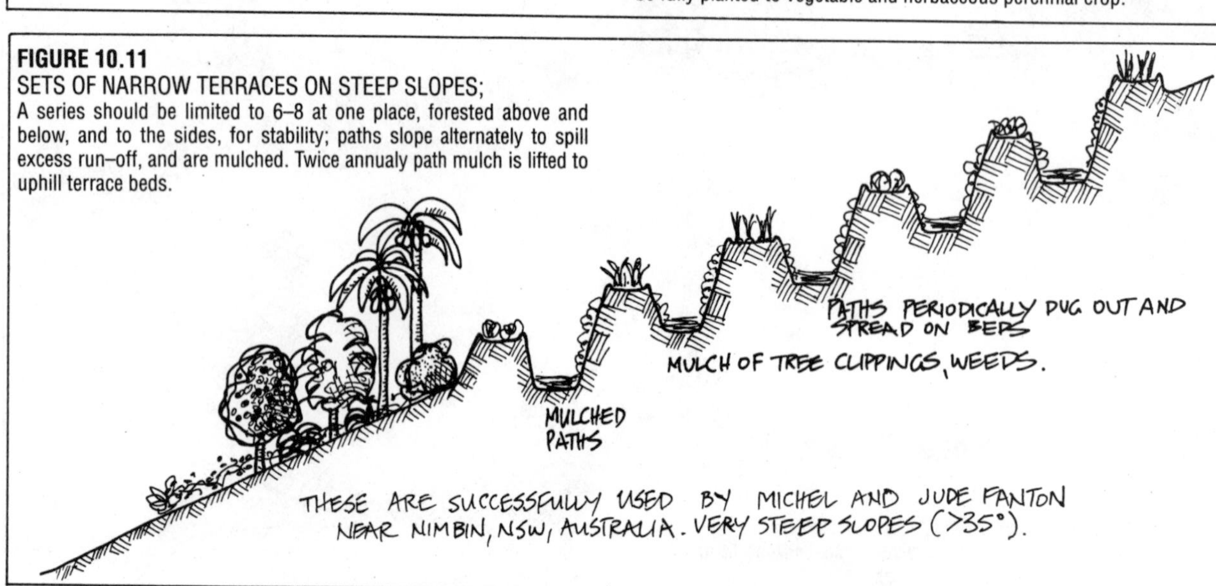

FIGURE 10.11
SETS OF NARROW TERRACES ON STEEP SLOPES;
A series should be limited to 6–8 at one place, forested above and below, and to the sides, for stability; paths slope alternately to spill excess run–off, and are mulched. Twice annualy path mulch is lifted to uphill terrace beds.

rbits, maize, beans, sugar cane, cultivated green–crop *Brassica*, edible *Chrysanthemum*, edible *Hibiscus*, rosella, horseradish tree (*Moringa*), coconut, mango.

Vine Crop Off Bund Faces

- Chayote, cucurbits, beans, passionfruit, kiwifruit.
- Bamboo on borders provide trellis material, as do rot–resistant timbers.

Mulch Crop and Fodders Above to the Borders

- Tree legumes, banna grass (*Pennisetum*), lemongrass, Vetiver grass, comfrey, bamboo and palm fronds, *Aleurites* spp., *Cinnamonum* spp.

Slope Stability

- Contour strips of Vetiver grass, lemongrass, banna grass with tree legumes not only replace contour ridges but trap soil particles, and is a cheap way to "terrace", even on steep slopes. These strips provide mulch for trees and intercrop (**Figure 10.13**).

Essentials and Variations on Terrace Systems

- Borders; and uphill, steep–slope, forest crop planted and selected for mulch value and fodder, or:
- Animal sheds (ducks, pigeon, poultry, pigs, bees) over top terraces; manure on a "washdown" system.
- In–crop mulches such as beans, *Azolla*, clovers.
- Staggered, short sets of terraces for steep slopes and high rainfall, compared with more continuous and longer series for winter–dry irrigated terrace.
- Deep areas in terraces for fish/crayfish/ shellfish refuges.
- Vines over all or part of the terrace to aid such crops as taro.
- Bunds planted in clover, beans, comfrey, lemongrass, fruit crop.
- Splash stones or splash plates for falling water; methods of draining terrace.
- Border drains in terrace to keep soil dry for mid–season crop.

10.5
HOUSE DESIGN

Optimum comfort levels for people are at dry–bulb temperatures of 20°C (68°F) in still air (winter), and 25°C (77°F) in summer, subject also to individual preferences. Above relative humidity levels of 40%, we effectively add 1°C to dry–bulb temperature for every 4% increase in humidity. As average summer humidity in wet–dry tropics commonly exceeds 50%, and long periods of humidity of 70% or so are experienced, there are times when sensible temperature exceeds 30°C (86°F), and heat stress results.

In homes, a useful indicator is a wet–bulb thermometer, where the mercury bulb is kept damped by a cotton wick drawing from a beaker of water. Below 18°C (65°F) wet–bulb temperature, we can remain fairly comfortable.

Factors that accentuate heat are nearby radiant surfaces and lack of air movement. For a nearby radiant heat source (wall or pavement) that exceeds 38°C (100°F), we can add 1°C (per degree radiated) to the air temperature, and conversely, we can subtract 1°C for any air flow above 1 m/second.

Evaporative cooling in dry air greatly reduces heat, but as the high humidity periods of tropics do not enable us to cool by sweating, we must therefore use every strategy available to de–humidify air (mainly by cooling), to cast shade, to develop cool surfaces, and above all to induce cool air currents in houses. Overshading trees, attached shadehouse, white exterior and interior surfaces, and clear–path breezeways are essential design strategies both in equatorial and sub–tropical climates.

In many continental subtropical locations, we are faced with dual problems of quite intense winter cold, with some frosts (and rare snows), and very humid and hot summers. Thus, the sort of house we need to build has several unusual characteristics, and needs perhaps more careful planning than either equatorial houses (where reducing heat is the only problem) or temperate and boreal housing (where providing heat is the only problem).

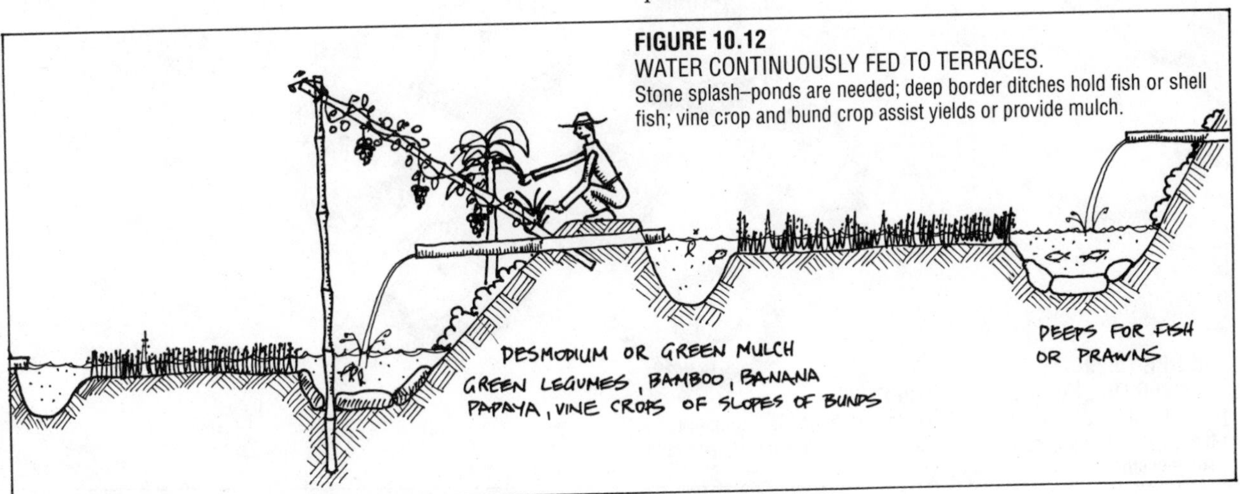

FIGURE 10.12
WATER CONTINUOUSLY FED TO TERRACES.
Stone splash–ponds are needed; deep border ditches hold fish or shell fish; vine crop and bund crop assist yields or provide mulch.

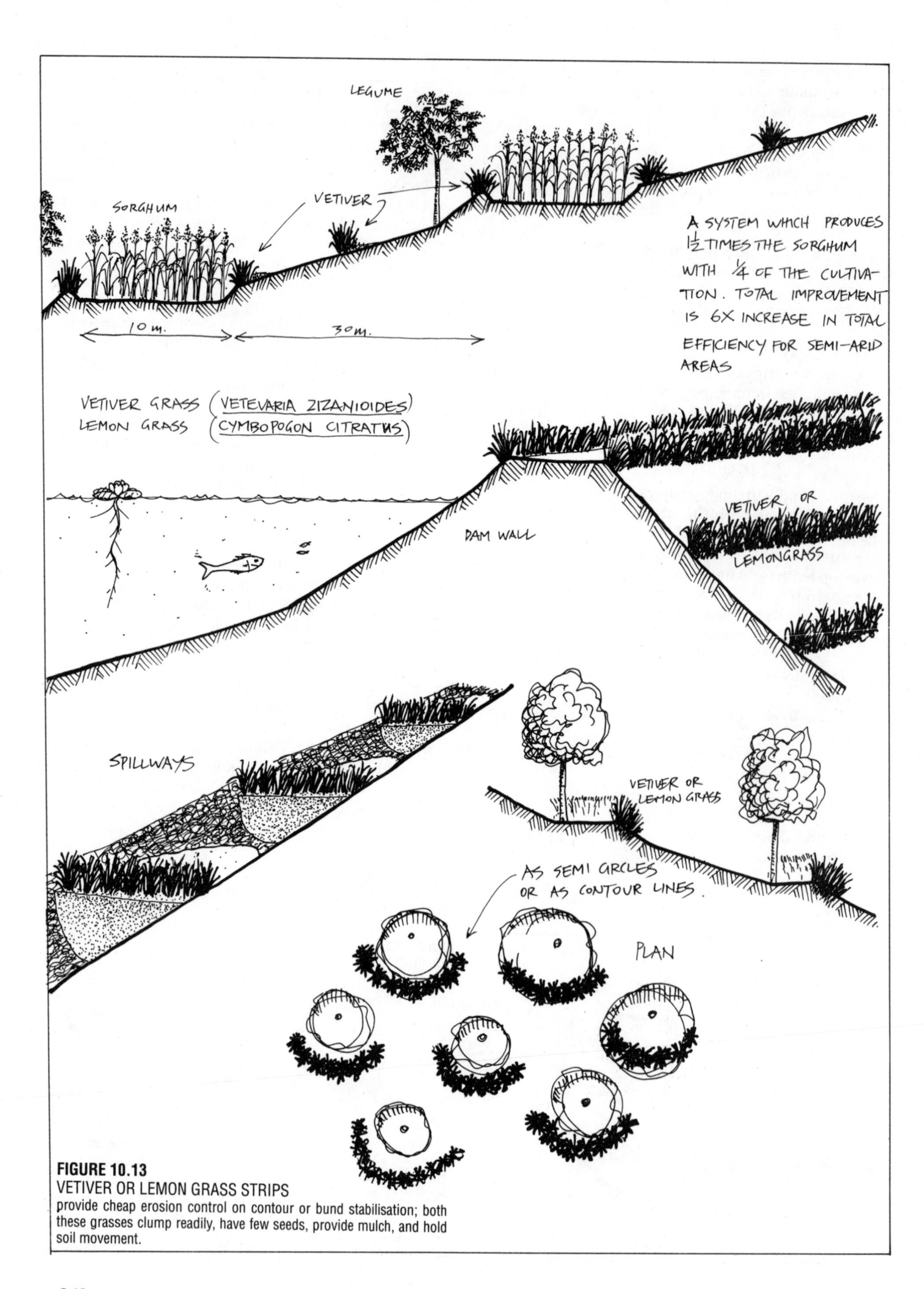

FIGURE 10.13
VETIVER OR LEMON GRASS STRIPS
provide cheap erosion control on contour or bund stabilisation; both these grasses clump readily, have few seeds, provide mulch, and hold soil movement.

The subtropical house needs to both heat and cool. For heating, it needs to have an insulated slab floor or trombe wall, and for the cooling system it needs induced or forced cross–ventilation from a cool or shaded area to an updraught area.

The secondary effects of high humidity range from the merely annoying (salt will not pour) to far more serious effects (clothes, food, film, and books mildew). Thus, we face two sorts of problems in house construction:

- Human comfort.
- Safe storages; for these we need both cool and warm storages, but both need to be *dry*.

Human comfort is greatly aided by these factors:

- SHADE: Light and heat are both excluded as incoming radiation in shade. Shade is particularly critical on massive walls or over water tanks close to homes.
- TRANSPIRATION: Plants can assist cooling by transpiring. Partial shade helps this factor in understory species except in extremely wet conditions.
- COOL BODIES: Large heat resevoirs used as water tanks and relatively cool blocks of (shaded) stone, concrete, and mud brick absorb heat from the air and from warm bodies. Conversely, hot radiant bodies adversely affect us.
- AIR FLOWS: Even low air flows from shaded areas greatly aid both transpiration and evaporative cooling. To create such air flows we need to develop both relatively hotter and colder air sources and to provide a

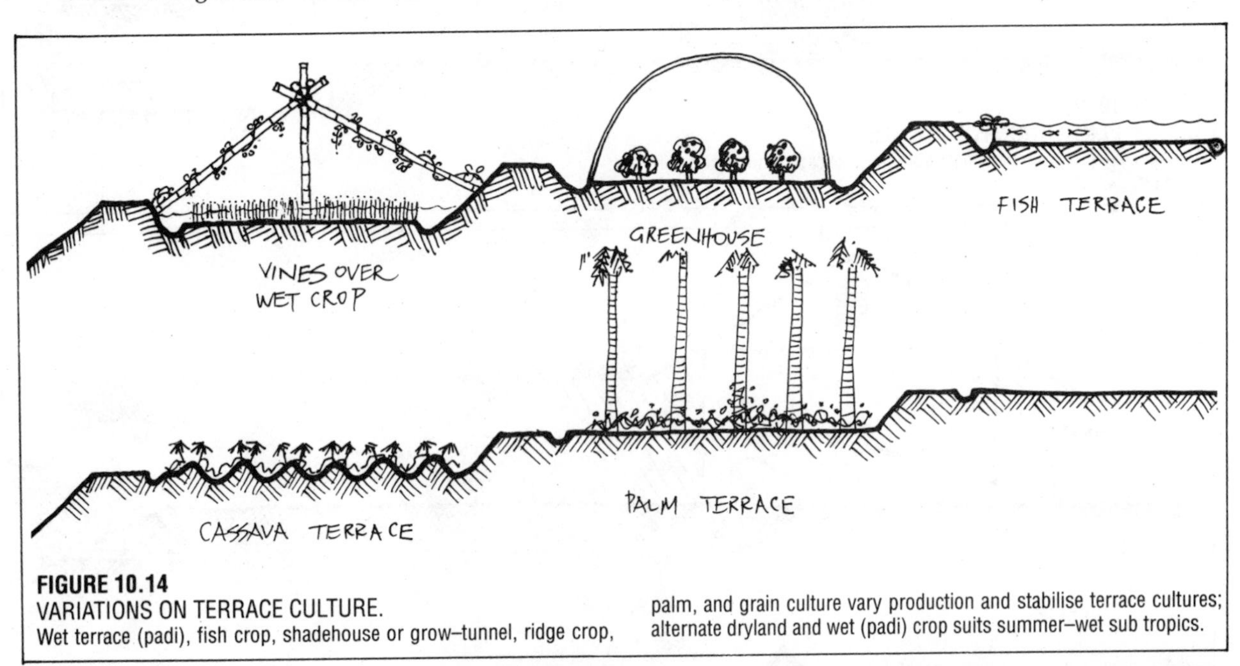

FIGURE 10.14
VARIATIONS ON TERRACE CULTURE.
Wet terrace (padi), fish crop, shadehouse or grow–tunnel, ridge crop, palm, and grain culture vary production and stabilise terrace cultures; alternate dryland and wet (padi) crop suits summer–wet sub tropics.

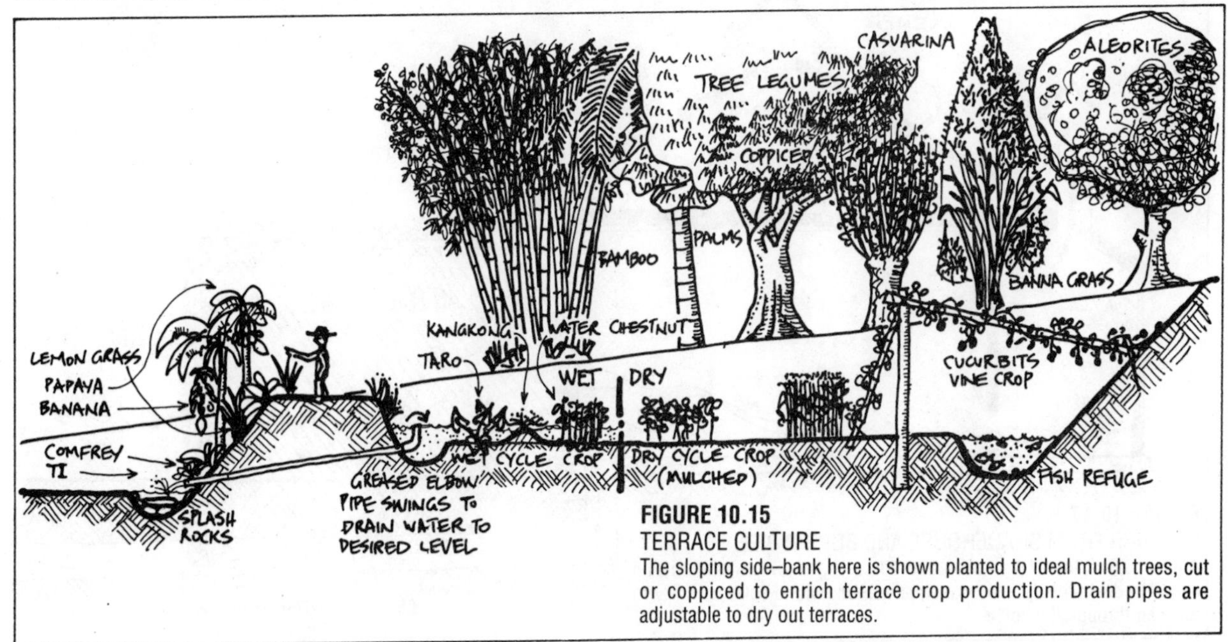

FIGURE 10.15
TERRACE CULTURE
The sloping side–bank here is shown planted to ideal mulch trees, cut or coppiced to enrich terrace crop production. Drain pipes are adjustable to dry out terraces.

cross–flow airway. Even a fan, simply stirring the air, aids in human comfort.

• REMOVING HEAT SOURCES: All massive cook–stoves and hot–water systems are best placed in a semi–detached kitchen in the tropics. Commonly, these are reached via a vine–covered shade area, are themselves shaded by palms or trees, and have wide eaves and ceiling vents for hot air escape.

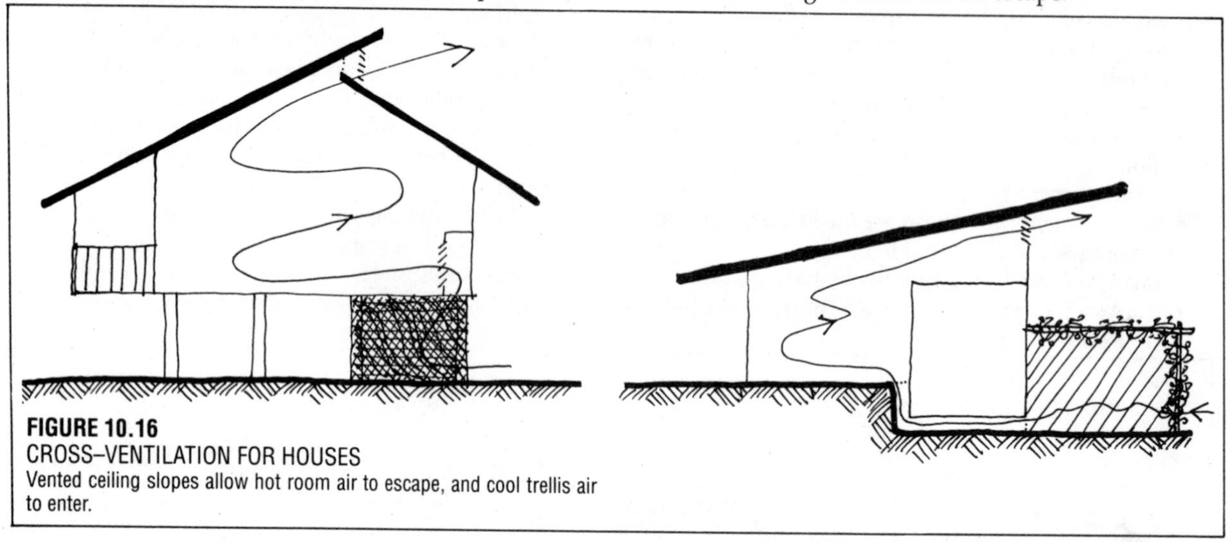

FIGURE 10.16
CROSS–VENTILATION FOR HOUSES
Vented ceiling slopes allow hot room air to escape, and cool trellis air to enter.

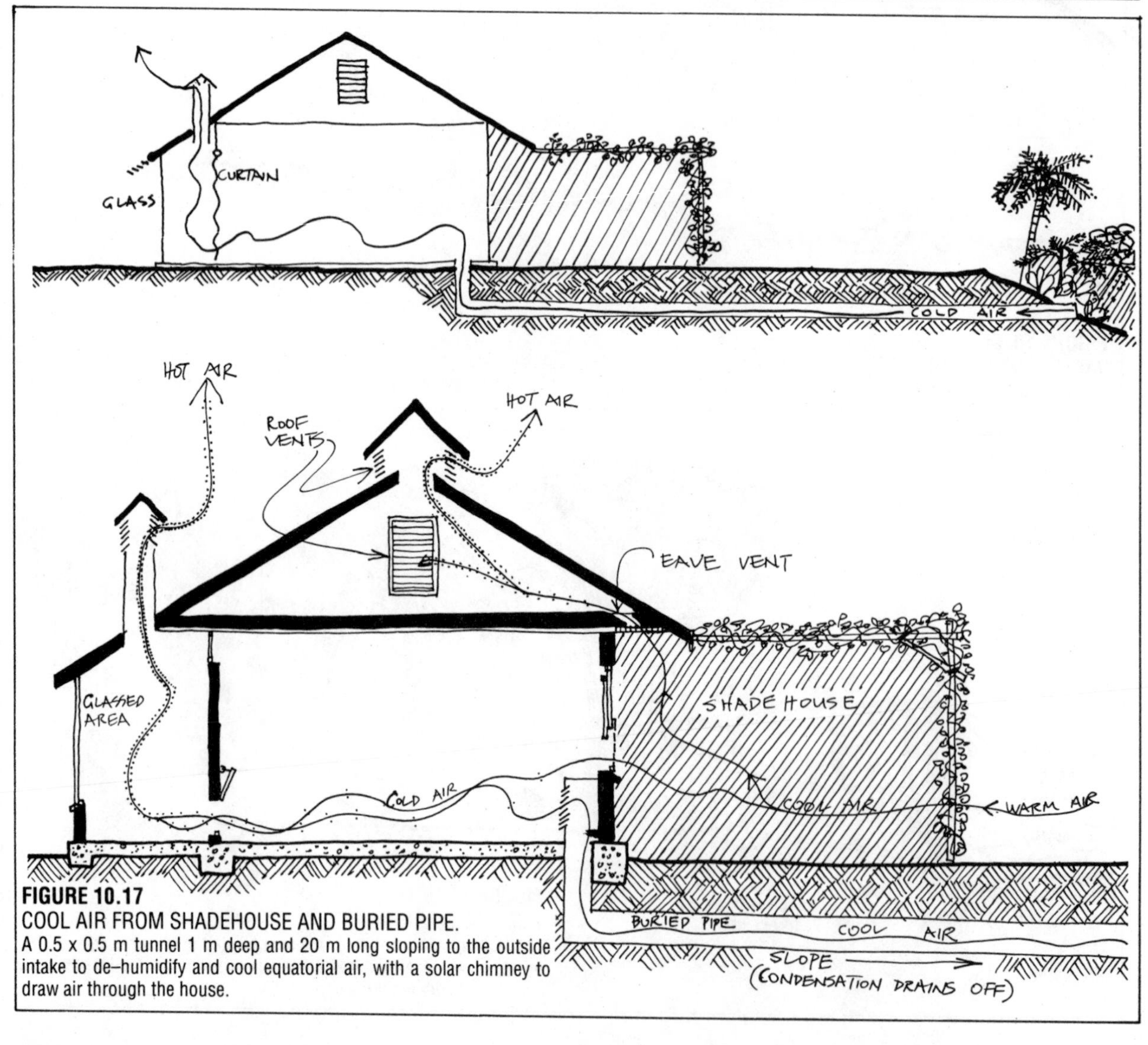

FIGURE 10.17
COOL AIR FROM SHADEHOUSE AND BURIED PIPE.
A 0.5 x 0.5 m tunnel 1 m deep and 20 m long sloping to the outside intake to de-humidify and cool equatorial air, with a solar chimney to draw air through the house.

Heat can be used via metal roof areas, hot–water storages, and attached glass–houses or solar chimneys to vent hot air and create updraught, which in turn provides a heat engine to draw in cool air. The essentials of good cross–ventilation are that the flowing air has a simple pathway to follow (no unnecessary corners to turn), and that large vents are used to allow a good volume of air through workrooms and storage areas (**Figure 10.16**).

Probably the best cooling systems in tropical houses are those which use a hot roof or metal chimney to draw in cold air from earth–cooled underground tunnels or pipes. As this cool air is dense, it will naturally flow downhill or sink to lower levels; this cool air can be drawn into houses via a positive exhaust system, or actively fanned into rooms.

To cool a pipe and lead off the heat continously, we need to construct a trench 1 m deep and 15–20 m long cut in the earth, drain off the condensed water (ideally making it self–draining to a lower slope), and provide this trench or pipe with a sloping floor. The intake end can be box–screened to keep out mice, and shaded by plants. Outlets can be floor grills or a louvred "cupboard" opening to the pipe or pipes in the trench (**Figure 10.17**).

Natural cross–ventilation can occur if a well–sealed room has a roof vent or chimney to create an updraught. Some forms of air scoop may help this process

The cold tunnel solution is very effective, and can be used together with evaporative cooling in desert housing, but it is also expensive, and difficult to fit to an existing house. For this reason, many homes can be sufficiently cooled by the use of vertical shutters acting as air scoops—a satisfactory solution on subtropical tradewind coasts. Or a shadehouse can be added to the poleward side of a house and cross–ventilated to a well–vented GREENHOUSE on the sun side of the building.

BASIC ESSENTIALS OF THE EQUATORIAL HOUSE

Some essentials of truly tropical housing (no cold season) are:

Site Choice

- Orientation is to *prevailing winds*, not to the sun. Cooling is by cross–ventilation.
- Shaded valley sites greatly aid cooling and shelter.
- Induced ventilation is essential, achieved by siting in palm groves or overshaded by trees, which should be permeable to wind at ground (house) level. Palms and trees can be pruned up the trunk.
- Site sheltered from hurricanes, *tsunami*, and vulcanism, sited on stable soils that resist mudflow in heavy rains.

House Design

- Walls white or reflective, overshaded by wide eaves and palms or trees.
- Heat sources such as stoves and hot water systems detached from the main structure (e.g. outdoor kitchens).
- Wall–material light, even permeable to wind (woven matting and mosquito screens).
- Mass, if any, internal to rooms, smooth and white–painted. The whole house can be of light construction on the outer walls.
- *Vertical* louvres and window shutters aid in

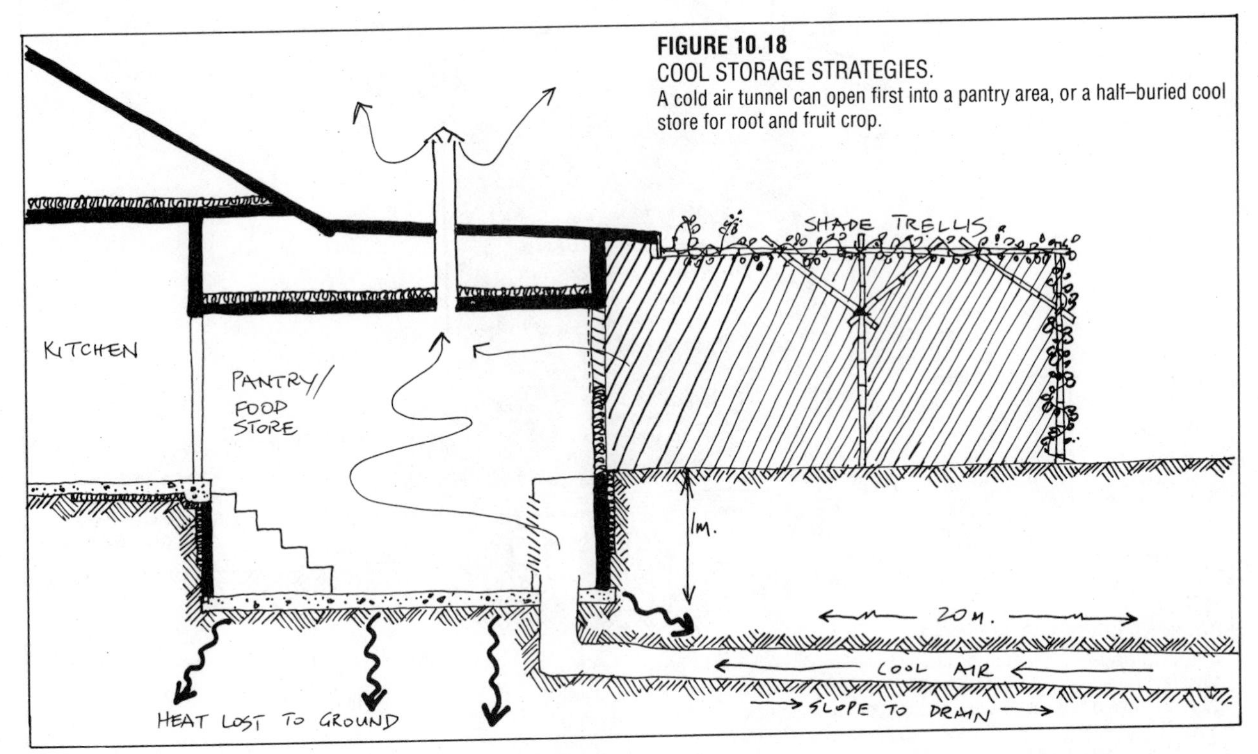

FIGURE 10.18
COOL STORAGE STRATEGIES.
A cold air tunnel can open first into a pantry area, or a half–buried cool store for root and fruit crop.

cross–ventilation.

• In hurricane areas, a strong central core or refuge may be needed, or an earth–bank shelter raised to protect the house; cellars should be entered above ground due to flooding danger in hurricanes, or well sealed against flooding.

• Very strong cross–bracing, deep ground anchors, and strapped timbers may be necessary if powerful winds are known to the area. Large bamboo groves placed to the windward will bend to the wind without breaking, protecting the house.

• Where thatch or tile is impractical, use a vented sheet metal roof. In this case a thin (12 mm board) ceiling is necessary, and soffit lining can be of permable netting or screened to allow an air flow to the roof space and thence to the exterior via high roof vents.

10.6
THE TROPICAL HOME GARDEN

SEASONS: The wet season is the "hungry gap", where plants are growing, but too young to harvest. Early in this season the soil is soft enough to plant and establish trees, but plants must be well–timed as the dry season is long–lasting. Planting too late is to risk drought before ripening of the crop. Vegetable crop is started at either end of the wet season. Water storages (Chapter 7) are essential, no matter how modest, for garden, tree crop, and diversity in yield. Moulds, mildews, and root fungi are encouraged by humidity, and it is best to use resistant plant varieties or root-stocks.

SPECIES: Mango, papaya, sapote, banana, limes, coconut, cashew, macadamia nut, breadfruit, mound–planted avocado and pineapple, durian, and so on are the garden and orchard framework, as are any productive palm crops. Large legumes such as *Inga, Gliricidia, Leucaena, Cajanus*, and so on are essential interplants.

In the vegetable garden, yam and sweet potato yield better than, or in place of, potato. Adapted small–fruits and tomatoes of wilt–resistant strains grow well. Amaranth is a good green and grain crop. Lima, velvet, and *Dolichos* beans trellis on tree legumes. Forage and ground legumes provide green mulch and help suppress grasses, as do comfrey and lemongrass. Chilies, peppers, and the range of tropical vegetables are preferred to temperate species.

Bamboos, balsa, teak, palms, and mahogany provide structural and craft materials, rattans can be encouraged along waterways and in mangrove edges. Oil palm, jelly palm, *Bactras, Maurantia*, salak palm and doum palm provide trusses of useful fruits.

PESTS: Large insect pests (locust, cicadas, sucking bugs) are plentiful; guinea–fowl or chickens on range are some defense. Native rodents and pigs can be damaging, and pythons rather than foxes take poultry. Termites and ants largely replace worms in soil–building, and buildings must be constructed to resist them. Geckoes in houses eat many insect pests, as do wolf spiders.

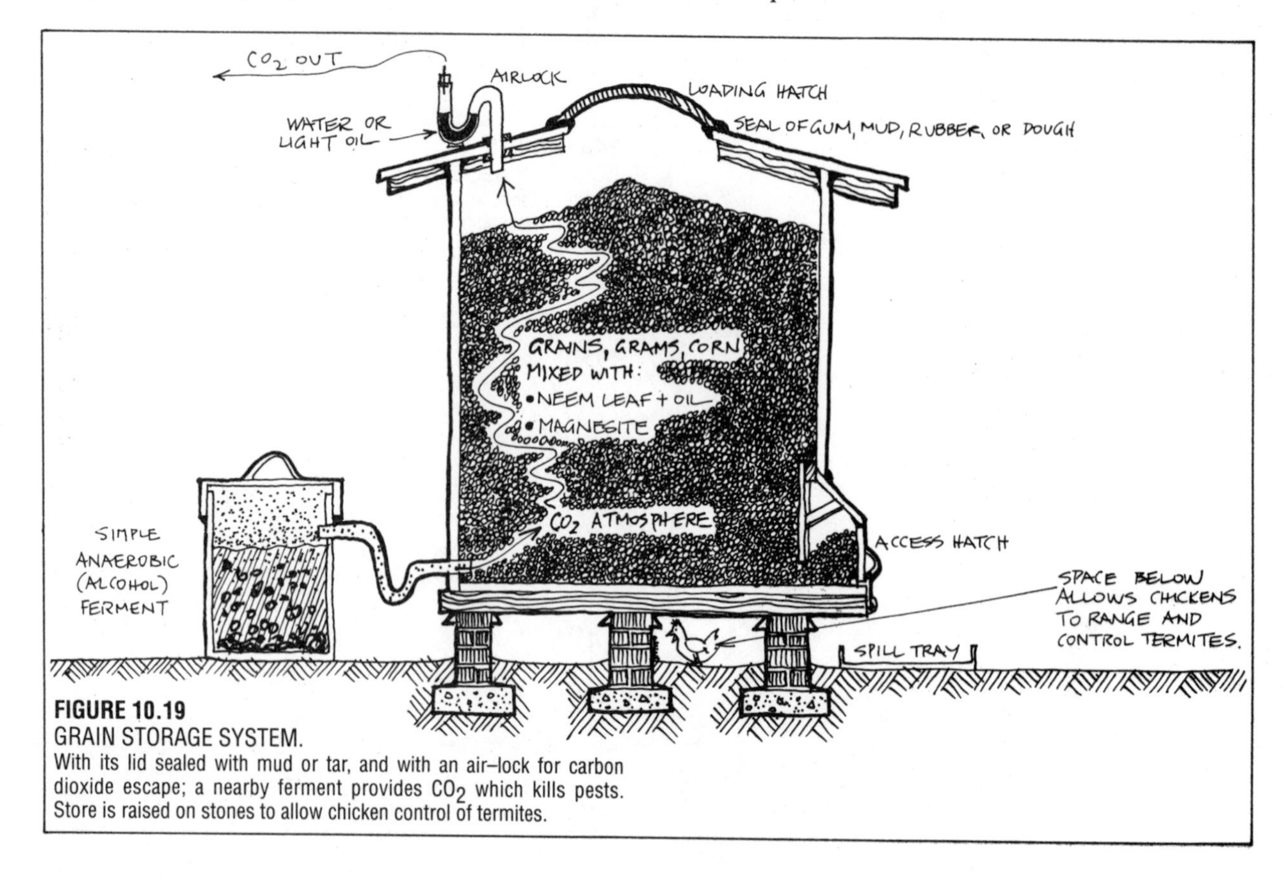

FIGURE 10.19
GRAIN STORAGE SYSTEM.
With its lid sealed with mud or tar, and with an air–lock for carbon dioxide escape; a nearby ferment provides CO_2 which kills pests. Store is raised on stones to allow chicken control of termites.

DOMESTIC ANIMALS: Pigeons and bees are most easily protected from predators by elevation on pole structures, or over shallow ponds. Guinea fowl, francolin, pheasant, and bantams provide essential foraging and insect control services. The guinea–pig aids small tree establishment as they "chip" the base of young grasses, and small pigs of Taiwanese strains provide orchard–fruit garden scavenging duties. Waterfowl and aquatic species add yields to water storages and assist in grass control.

HURRICANE DAMAGE: can be limited by raising large earth banks, selecting valley garden sites, screening plantings with bamboo groves, establishing a general tree canopy through garden and plantation, or a combination of these strategies. Oversize swales aid wet–season water run–off control and diversion to storage.

SPECIALIST CROP: There is a wide range of specialist crop potential, from rubber (*Hevea*), betel-nut, chalmougra oil, and chicle to essential oils and medicinals. Many are suited to primary processing in remote locations, or conversion to commercial–quality end–products. The high value of processed product enables smallholders or cooperatives to pool research and processing facilities, and to select high–yielding varieties.

INTERPLANT: As well as the essential legumes, a scatter of *Banksia, Casuarina, Gigasperma calaspora*, and *Pultenea* with their mycelial associates will fix phosphate and return it via leaf mould. Several plants "pump" sugars or carbohydrates into soils, while leaf–sucking insects and scale insects exude sugars

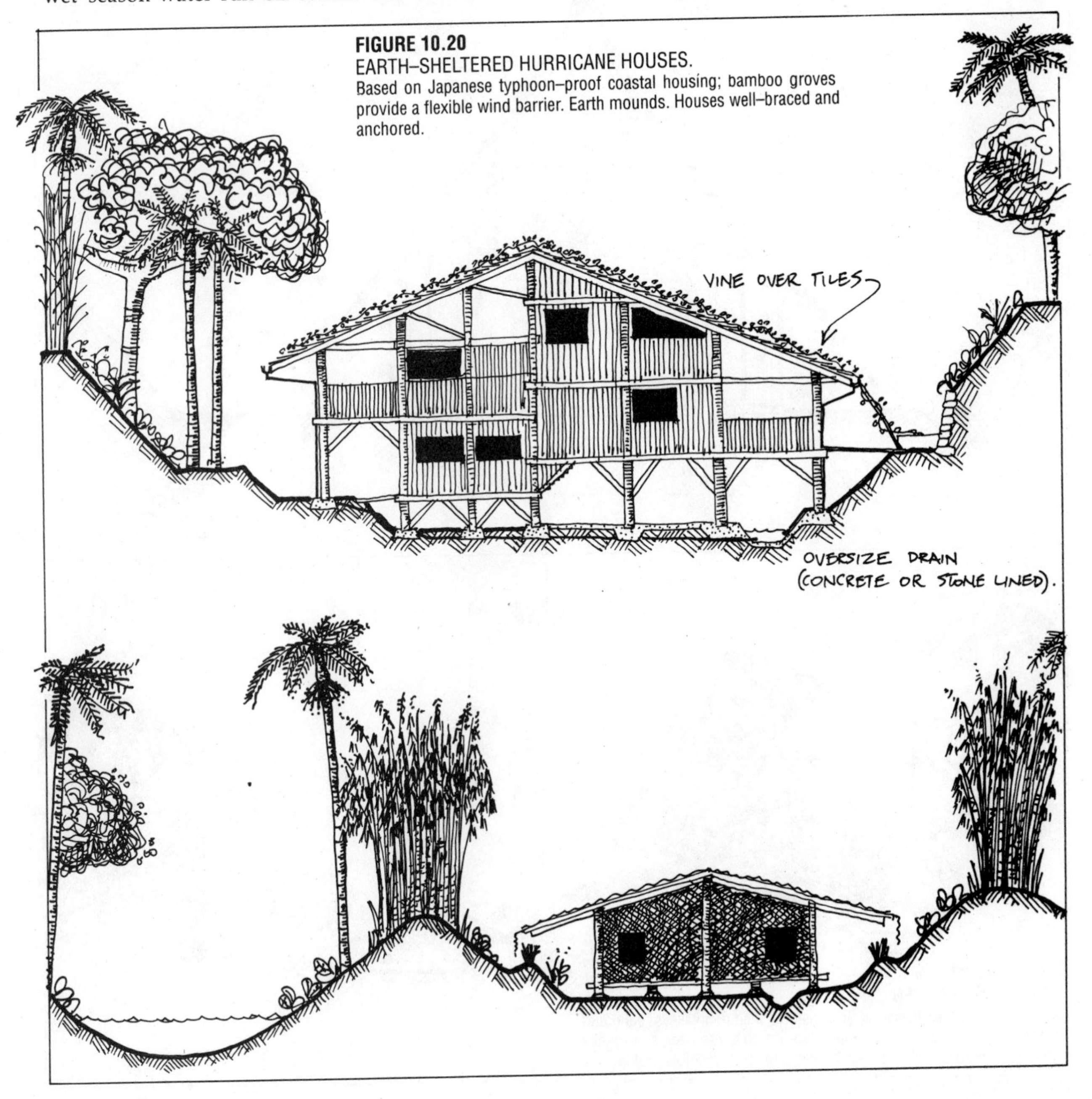

FIGURE 10.20
EARTH–SHELTERED HURRICANE HOUSES.
Based on Japanese typhoon–proof coastal housing; bamboo groves provide a flexible wind barrier. Earth mounds. Houses well–braced and anchored.

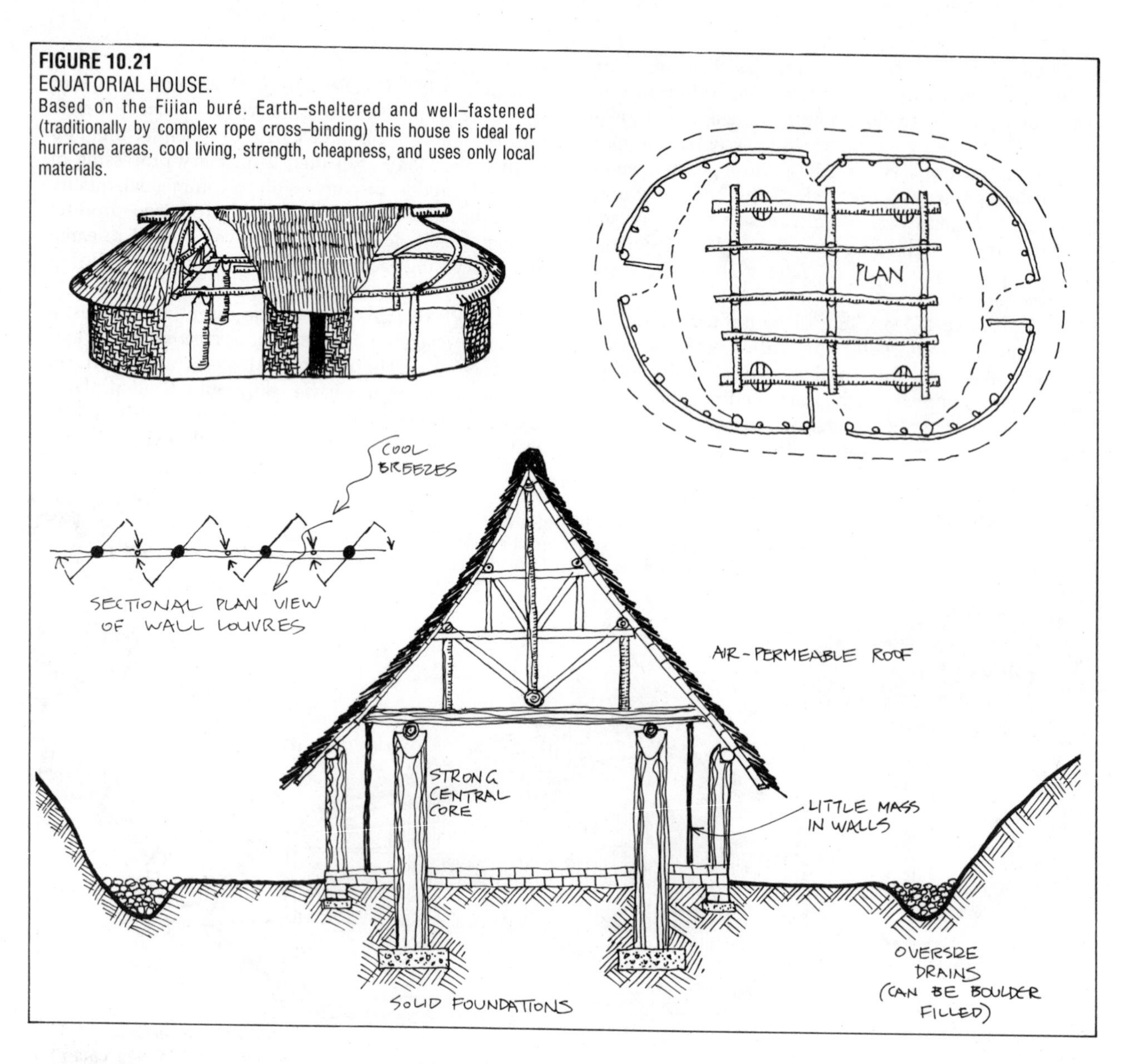

FIGURE 10.21
EQUATORIAL HOUSE.
Based on the Fijian buré. Earth–sheltered and well–fastened (traditionally by complex rope cross–binding) this house is ideal for hurricane areas, cool living, strength, cheapness, and uses only local materials.

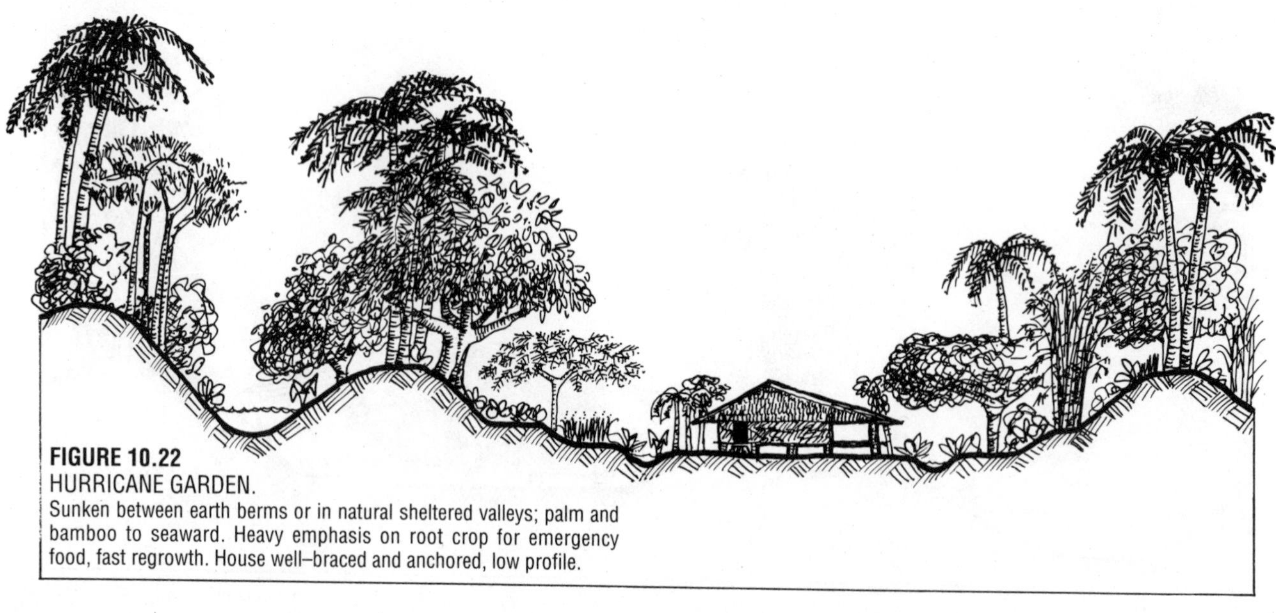

FIGURE 10.22
HURRICANE GARDEN.
Sunken between earth berms or in natural sheltered valleys; palm and bamboo to seaward. Heavy emphasis on root crop for emergency food, fast regrowth. House well–braced and anchored, low profile.

from stems. Dilute molasses or cane and sorghum sugar juices and stems also activate soil fauna. Marigolds, neem tree leaves or berries, and pyrethrum daisy control soil pests and provide insecticides or water insect control. The neem tree is often planted to overhang ponds, so that the berries that drop control water–flies and mosquitos.

Established tough grasses of the savannahs resist gardens, and need to be mulched or overshaded with tree canopies. Essentials are:

- To mound or raise garden beds for good drainage in the wet season.
- To use mulch and mulch–tree species to create topsoil for gardens.

Where logs exist, they make ideal garden bed edges to hold mulch and soil.

Gardens have been devised for many tropical areas, usually containing the following:

- Designed for full nutrition for an average family.
- Water conservation and safe water disposal (hygiene) a necessity.
- Species chosen to suit local cultural preferences.
- Sufficiently varied to survive reasonable climatic change, or seasonal irregularity.
- Protein sources, livestock; their forages, or grain/legume replacements for meats.
- Water routes and use.
- Basic foods or staples
- Fresh vegetable and fruit for vitamins, minerals, varied uses.
- Some fuels, medicinals, flowers.

The elements in **Figure 10.23** are those that make up the house structure itself, and those that make up the garden, hedgerow, livestock and path access structures. The best way to use this section is to read it through very carefully, study the plans and diagrams, and then improve it, or better it to fit to a specific site.

THE HOUSE

Room size and number is adjusted to family size, but is basically a simple, easily–heated and cooled structure, preferably on slab or raised pise floor, and preferably edge–insulated. The induced cross–ventilation acts to cool and heat as per **Figure 10.16**. In addition, vertical sashes or shutters to each room help to scoop air in.

In hot periods, the main living area is outside rear, or under a similar porch trellis to the front if the people prefer to be seen from the road (as is the case in most close–knit societies). The side trellises are seen in **Figure 10.17**. Materials can be local, as can any insulation. Glass is needed, as are some pipes or drains of stone, and a tank.

THE GARDEN

How the Garden Works (See **Figure 10.23)**

First, it accepts all water and wastes of use. Only plastic and glass or metal are not used, although some cans *may* be buried for slow zinc and iron release). Second, it provides most mulch and a lot of fodder or forage, which when bulked out by house–scraps should feed rabbits, guinea pigs, some poultry, and even fatten a small pig. Next, it is very accessible and well–designed on a need–to–visit–and–tend basis. It is also very natural in appearance and function. If no septic tank is present, a dry toilet will do, and the manure can be put under trees (in pits). Even "toilet paper" can be built into the hedges (E) (*Nicotiana* is great, as is *Leucaena*).

Accessories

Hot water for showers can be achieved by a hot–water collector or at least a coil of black pipe on a roof or bare area. With a couple of oil lamps or a photovoltaic array, a solar oven on castors, an efficient cooker, and a small solar food drier, life should be fairly cheap and healthy. Adventure can be sought in teaching neighbours how to do it, writing novels, or joining an adventure–camp group—or even a permaculture group (they behave *very* diversely!)

GANGAMMA'S MANDALA

In Taiwan and the Philippines, small intensively–planted home gardens are planned to feed a family of five all year. I have added to these designs my own permaculture "least–path" layouts to give a very concise and effective model of sustenance garden design for tropical and subtropical regions. These can also be adapted to temperate regions, using suitable species. The overall pattern can be altered to fit almost every site form, but is presented here as a flat site pattern. Although the building of such a garden is fast and simple, its design is sophisticated.

The whole design owes much to the work of the East–West Institute in Hawaii, and the Samaka gardens of the Philippines, but the layout is purely permaculture. I have named it "Gangamma's Mandala" after one of our Karnataka (India) permaculture design graduates.

Steps in the process are:

1. At the centre of a 100 square metre (1075 square foot) or larger area, describe a circle 2 m (6 feet) across and excavate the topsoil (or subsoil) to a dish shape, ridged on the perimeter, and about 0.6–1 m (2–3 feet) deep from hollow to rim. This is the banana/sweet potato/papaya circle garden, as per **Figure 10.26.1**.

The whole circle is then covered with wet paper or wet cardboard, banana leaves, or any mulch material, and the hollow is filled (or over–filled as a dome) with rough mulch of short logs, coarse twigs, hay, rice husks, and sawdust or indeed any humus–creating materials. A little scatter of manure, ash, lime, dolomite, or fertiliser can be added. If stones are available, bank them to the outside of the rim.

The rim is then planted to 4–5 papaya (a tall variety), 4 or so bananas (dwarf types), and 8–10 sweet potato. If available, yams or taro can be placed *inside* the rim. Later, beans can be planted to climb the papaya and banana stalks. In the banana circle, we can place a grid

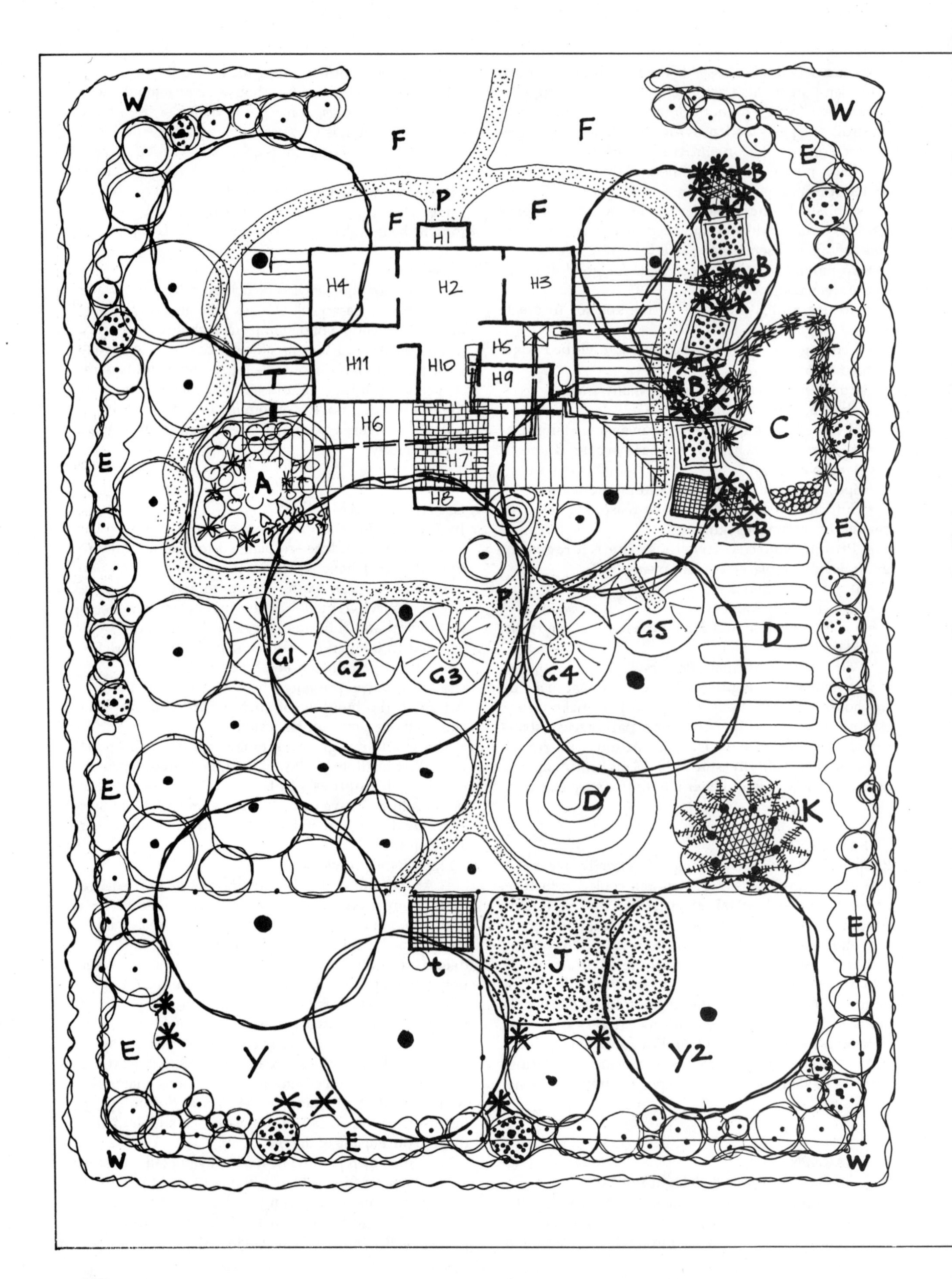

W
F
F
W
P
E
F
F
B
H1
H4
H2
H3
B
H5
H11
H10
H9
T
B
H6
C
E
H7
A
H8
B
E
P
D
G5
G1
G2
G3
G4
E
D'
K
E
t
J
Y2
E
Y
E
W
W

FIGURE 10.23

ELEMENTS OF A TOTAL DESIGN FOR A GARDEN.

Based on a Brazilian design by Margrit Kennedy and the writer; climatically appropriate house of local brick and tiles amortizes in 4–7 years if a garden is developed. All water is caught or directed to gardens.

H1: An entry; in the subtropics this is a glasshouse, fitted with a wide ceiling vent and drawing cold air in from **H6**—the shadehouse. Note the straight–through airway via living area (**H2**) and winter kitchen (**H10**).

H3, 4, 11: Bedrooms and study/library if needed or possible. All open off or into the central kitchen (**H10**)/living axis (**H2**).

H5: The bathroom/toilet. From here, shower water goes to the "wet" area of garden (**A**) or to mulch/ forage crop (**E**); toilet and sink water goes always to (**C**). Handbasin water goes to (**B**), banana circles.

H6: Is the shadehouse, which can extend around three sides of the whole, and has entries, trees (or their trunks) in it, minor trellis resting against it, and a crop below.

H7: is the part–paved area of the summer kitchen, and **H8** the outdoor cooking stove and bulk–cooking stove, the ashes of which go to **A**, **C**, and the garden.

H9: Mudroom/pantry off the kitchen, with outside basin/sink for primary vegetable processing.

Table scraps go to livestock *or* below mulch in boxes (**B** series) of the garden; sweepings to mulch.

T is the tank, catching roof water. There is a small tank (**t**) on any animal shed. The tank can also be filled from a reticulated water system (on a float–valve). All tank overflow goes first to (**A**), the wet garden bed, which is in part under the house trellis, in part in the open. There is a path (**P**) around the whole house and to the outdoor kitchen, animal house, etc.

The garden has the following main elements:

A is a wet food patch with earth bank edge (detail **Figure 10.24**). This receives clean water and grows wet crop, e.g. taro, rice, kangkong, watercress, Chinese water chestnut, etc. The bank is planted to dry staple foods, e.g. yam, sweet potato, cassava, etc., which benefit somewhat from the wastewater. Some of this crop is under partial shade, which taro appreciates, as do some other sub–aquatics.

B are banana circles (see **Figure 10.26**). They can also be papaya circles or mixed papaya/banana/pigeon pea/tomato/yam circles, or any such combination. They are watered from the hand basin and kitchen sink and mulched inside from the hedge (**E**).

C is the "dirty" water patch (**Figure 10.25**). It contains *no direct food plants* and *no root crop*, only vigorous, damp–tolerant, manure–tolerant green forage crop, e.g. comfrey, banna grass, lemongrass, legumes. These are regularly cut and removed either to the animal pens as green feed *or* to the little potato boxes (**B**) *or* to the garden beds (**G** series). The area is sunken and surrounded by a bank on which can grow pigeon pea, papaya, banana, or all of these and more.

D & D' are avenue crops of maize, corn, or some main crop of value. Two layouts (spiral and zig–zag) are shown (detail **Figure 10.27**). They consist of self–mulched and stable intercrop systems such as maize/ beans, yam/beans, cassava/beans, or melon/beans, with plants of *Leucaena, Gliricidia, Sesbania*, or some such productive legume as a fixed hedge. Sticks from these go to the fire; ash returns to the garden. Hedge clippings from **E** or exotic mulch (paper, carpet, wastes) can be put on these beds.

E is a low, raggedly–trimmed hedge of forage–mulch species, and is used for privacy, as a windbreak, and a sometime food source. Just *outside* it is a very dense weed barrier (**W**). In two places this hedge is *inside* two chicken pens (**Y** and **Y2**). Where there are "spots" it means a few well–chosen clumps of sympodal, edible–shoot, solid bamboos form part of the hedge. These are partly for food, partly for mulch, partly for the trellis repairs.

F is the front yard of traditional or showy flowers, if the culture demands it. It has a few basic medicinals, pleasant scents, a fibre plant or two (to help the trellis), a few ornamental food plants (cacti, agave), and is rarely watered so all plants there are hardy.

G are "keyhole" garden beds (**Figure 4.10**); these can be extended to either side of the house for large families, and more to the rear if necessary. All the preferred vegetables, especially greens, are here. The nearest bed to the kitchen has the usual bulk culinary herbs, but some of these are also on the small spiral herb bed (**Figure 4.34**) beside **H8** (the outdoor stove), and a few wet–area herbs are in pots under the trellis in the A pond (mint). Some rare dry–area herbs may be at or near the front door (**F**).

The two trees near **H8** are a lemon or lime and an allspice or caper tree (shrubs, really). They placed near the house as they are the most–used trees in the system.

The rest of the **G** beds are carefully–planned seasonal vegetable crop. They are mulched from the hedge and the (**C**) pit. Some food scraps, ashes, mulch and shredded paper or plant waste are dug in to them in little pits, as are their own crop wastes and the cuttings from the windbreak plants (**W**). They are always in action, full of plants. Water is bucketed or reticulated from a mains system. Shower or bath–water can also be used, if run along a length of low–pressure leaky pipe.

J is a Fukuoka–style grain plot, mostly for people and partly for poultry. It has a low surround of dry–tolerant mulch such as *Centrosema*, but is also mulched from the hedge **E** and the palm clump **K**.

K is a circle or clump of 6–12 palms, of 1–4 species depending on site, tastes, availability, and needs. They ideally would be 2–3 coconuts, 2 oil palms, 2 dates, and a sugar palm, but may be simpler. The inside circle is mulched with their chopped fronds, their husks and wastes, ashes, and green hedge material or mulch brought in from other sources. In this deep mulch, a yam or vine will grow in the wet season.

Every dense black spot is a tree trunk; these are carefully–selected trees, and their crown spread is indicated. They can overshade the house roof, and may even be placed inside the house trellis, with holes in the roof to allow the trunk out. These few trellis trees have light crown and/or fruits that fall or that can be brought down by a child off the roof. They are intended to shade the roof area and can be tough, spreading legumes (*Prosopis* is good) yielding a food crop for livestock, or they can be palms, or a thin foliaged tree.

Other trees are producing staple fruits, a few avocado (on mounds 1 m high and 5 m across in wet areas), citrus, a few nut trees, and an oil palm. They have soft ground–cover below (nasturtium, soft legumes, comfrey) and this is sometimes slashed as mulch for the tree, sometimes fed to livestock.

The little hatched areas are the roofs of rabbit hutches and chicken roosts or pigeon lofts, quail houses, or guinea–pig rest areas (depending on the culture and preferences). A lot of the hedge (**E**), the wet mulch (**C**), the tree herb layer, and the spare corners are devoted to plants chosen just for the animals. Each little roof catches some water and leads it to a small tank or the garden or both.

Y–Y2) These are two fenced areas. They are changed over to suit the animals, and while one is rested, it can be cropped if space allows. Both are well–shaded animal runs. The fence can be of woven bamboo, palm rib, banna grass stalks, or of wire netting. Both areas are limed if sour.

The house trellis areas are very carefully fitted up with vine crop, spaced to let *some* light through (30% is fine). Basic vines such as chayote, kiwifruit, beans, cucurbits, and grapes are carefully chosen for the house trellis, which can extend 1–1.5 m above the whole roof area if the roof is sound and solid. One or two non–bearing vines can, in fact, be let rampage over the roof with great benefit to the cooling of the house. Note, however, the type of gutters needed in this case to catch the roof run off (**Figures 7.21** and **7.22**).

On the sides of this trellis, more ephemeral bean and cucurbit crop can be grown on leaning trellis of bamboo, or separate bamboo trellis can divide the G beds. There is no vertical limit to some vine crop, and if water permits, every palm in the K clump can eventually carry a high vine (granadilla, grape, or kiwifruit), as can any large leguminous tree (grape; kiwifruit is too dense). *Under* the trellis a few shade–tolerant plants will grow (coffee, taro).

or platform of wood over the mulch, and this then becomes an outside shower or wash–up area.

2. A circular sunken path 0.6–1 m (2–3 feet) wide is covered with sawdust or gravel around this central circle garden, and off it, 5–6 "keyholes" or indentations are made. The system now looks like **Figure 10.26.2**.

3. Around each keyhole a bed 1.5–2 m (5–6 feet) wide is first edge–banked with soil 100–200 cm (4–8 inches) high to prevent water run–off, and the beds are then papered and mulched (as for the banana circle). The whole garden now looks like **Figure 10.26.3** in plan. The thick lines represent low earth ridges.

Thus, we have six major keyhole beds, each of which is separated from the next by a thin strip of lemongrass *(Cymbopogon citratus)* or Vetiver grass *(Vetiveria zizanioides)*. Just outside the periphery ridge, strips of lemongrass, comfrey, and arrowroot *(Canna edulis)* form a kikuyu grass barrier, and behind that, a taller border of cassava/banana/papaya/pigeon pea/*Leucaena*/*Crotolaria* forms a hedge or windbreak. All these borders give mulch, forage, barrier effects, or food. The whole mandala is fenced or has a spiny woven hedge boundary for cattle exclusion, if necessary.

The mandala has now been earth–shaped and mulched to prevent water run–off and to conserve moisture. We now proceed to plant, using buckets of good soil to place the following zones of plants:

A. On the track edge border of the central path and keyholes, within stoop–reach of the path, plant those frequently–plucked or everyday greens of high value. Here, the placement and selection criterion is that *all the plant, or most of it, is picked for much of the year*. These are the PATHSIDE GREENS; they include all the chive and shallot species, plenty of parsley, coriander, thyme and sage, celery, broccoli, edible chrysanthemum, chard, and any such long–bearing or perennial greens (e.g. various perennial spinaches). This is therefore a narrow border to the inner side of the keyhole beds, planted in the ridge soils there.

B. Behind or outside the pathside plants, we plant a 1 m (3 foot) wide strip of species which are *frequently picked over a short to long season*, e.g. tomatoes, eggplants, bell peppers and chilies, bush or staked beans and peas, kale, corn, okra, and so on. These are the NARROW BED plants, all within reach of a path or keyhole. As yet, we do not need to step on any beds to harvest.

C. Just out of reach, on the outer borders of the keyhole beds, we place most long–term root crop (potatoes, sweet potatoes, carrots) or any crop we CUT AND REMOVE (cauliflower, head lettuce, and cabbage). Thus, for this crop we step (once) on the bed to harvest and replant, following root crop with fava beans or dahl (dried beans or lentils).

All beds are replanted as they are harvested, and a top mulch of straw, sawdust, bark, dry manure, or chips is added annually. Rabbits, guinea–pig, chickens, or small livestock are fed from weeds, waste vegetables, household scraps, and forage greens from the border hedge (comfrey, cassava, *Leucaena* or lemongrass). Vetiver grass, lemongrass etc. are cut 3–5 times annually for mulch. The roots of Vetiver grass prevent rodent burrowing from outside the system, as do the root masses of *Euphorbia* species.

All trees, shrubs, and tubers are planted before papering and mulching, then about 30 cm (1 foot) of trodden–down and wetted mulch is added. Tray seedlings 8–10 cm (3–4 inches) high, and large seeds such as peas or beans, are each planted in a hole burrowed in this mulch, with a good double handful of soil to each hole. Small seeds (lettuce, carrot) are scattered thickly on lenses of soil 50 cm (1.5 feet) across and 5–8 cm (2–3 inches) thick, placed and firmed on top of the trodden mulch, followed by dusting of a 1 cm (1/2 inch) thick layer of fine soil. All seed can be pre–soaked. The whole bed needs a good soak with a sprinkler at each stage.

If no weed seeds are included in the mulch, the beds are weed–free. It takes 9–15 months to build up worms and a good soil. Any surplus compostables can be pushed under the top mulch layer.

A larger system, designed for a community kitchen at a rural centre in Karnataka state (India), uses a core assembly of 4–5 banana circles, and has 8–12 keyhole beds. In this case, a keyhole accesses the banana circles (**Figure 10.26.4**).

Any one of these banana circles can contain a small

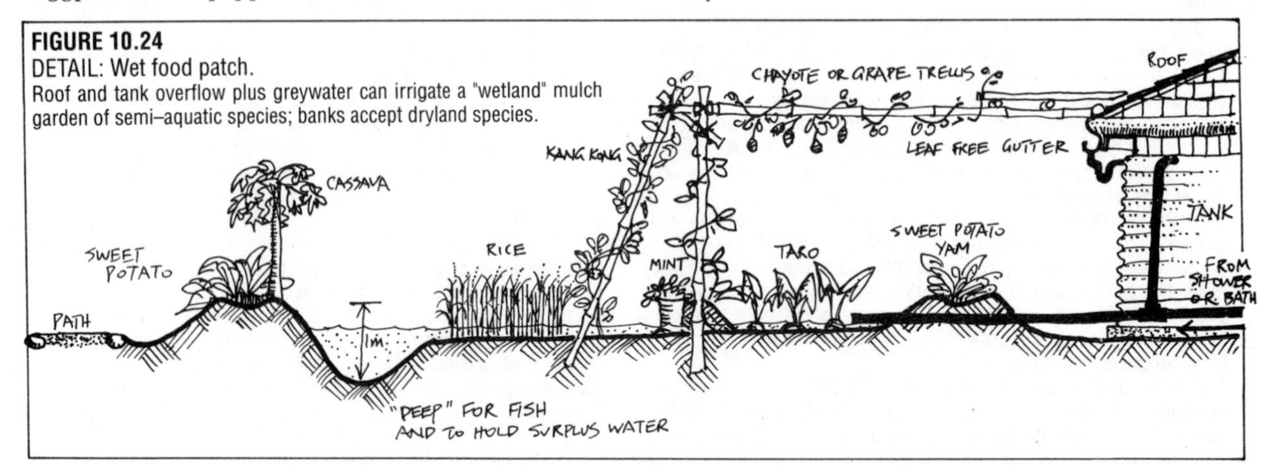

FIGURE 10.24
DETAIL: Wet food patch.
Roof and tank overflow plus greywater can irrigate a "wetland" mulch garden of semi–aquatic species; banks accept dryland species.

pond for frogs and water convolvulus, or taro. We can plant a *Leucaena* or palms for high shade and mulch, at the junctions of the keyhole beds. The hedge surround eventually provides the annual mulch (**Figure 10.26.5**).

This garden is intensively–planted, has very little path per bed area, is easy to build and maintain, provides everyday greens, minerals, vitamins, allows no water run–off, and can be built on *any* substrate (rock, concrete, roof areas).

We can here combined basic nutrition, soil building, rainwater harvest, eventual self–mulching, various weed and animal barriers, small livestock fodder, overhead shade, non–dig gardening, "least–path" access, direct waste water disposal, and a pleasing design.

AVENUE CROPPING

The system described here is adapted from Ray Wijewardene, 1981, *Conservation Farming*, IITA–Sri Lanka Program, Dharmapala Mawatha, Columbo 7, Sri Lanka. It is a deliberate fuelwood/mulch/soil improvement crop integration of great use in the tropics but adaptable to any climate, with good species selection (**Figure 10.27**).

The essentials are simple enough:

- Correct spacing to shade the ground fully but to allow cropping between the tree legumes.
- Lopping to leave stems 0.5–1.0 m above ground (a suitable tree legume is tagasaste or *Leucaena*).
- Sowing of crop just before rain or irrigation, to beat the growth of weeds.
- Possible two–crop sequence, with mulch also returned from crop wastes.

A very similar system is used in Africa with *Gliricidia*, and more recently in New Zealand with tagasaste as the green crop. A parallel system (regular or irregular in ground plan) is successful in establishing small trees, which otherwise perish from frost or in open grass competition.

At least two problems arise in sustained coppicing of legume trees. Firstly, such coppice should be confined to warm wet periods, allowing mature leaf to carry over into dry, cool, or frosty periods. Secondly, constant coppice weakens trees over 5–8 years, and replanting is necessary. Very few legume trees will sustain constant coppicing, and other strategies are called for. Perennial thin–crowned leguminous trees can be spaced throughout the orchard and garden, or fast–growing and short–lived legumes can be allowed to grow and die, or can be ring–barked or felled on a 2–5 year cycle.

However, with permanent high legume cover, mulch can be obtained all year from a variety of non–legume hedge and understory. Such species as *Nicotiana*, *Echium fastuosum*, *Lantana*, cinnamon, even clumps of daisies, wild ginger, lemongrass, Vetiver grass, *Pennisetum*, and crop wastes from maize, *Sesbania*, and soft ground legumes or comfrey provide constant mulch under high legume cover, so that the coppicing of susceptible legumes themselves is reduced or eliminated.

In any evolved system, avenue cropping or mulch provision can be sustained by a carefully–planned system of mixed non–coppiced tall legumes giving a seasonal leaf drop (*Erythrina*, *Tipuana tipu*), and a row series of non–legumes for ground mulch.

BARRIER PLANTS

There are several reasons to erect barriers of plant species:

- As a block around the annual garden to resist invasive grasses (such as kikuyu and buffalo grass, knotweeds, etc).
- As a corral for animals, or to keep grazing animals out of a compound, or to guide animals to a gate or

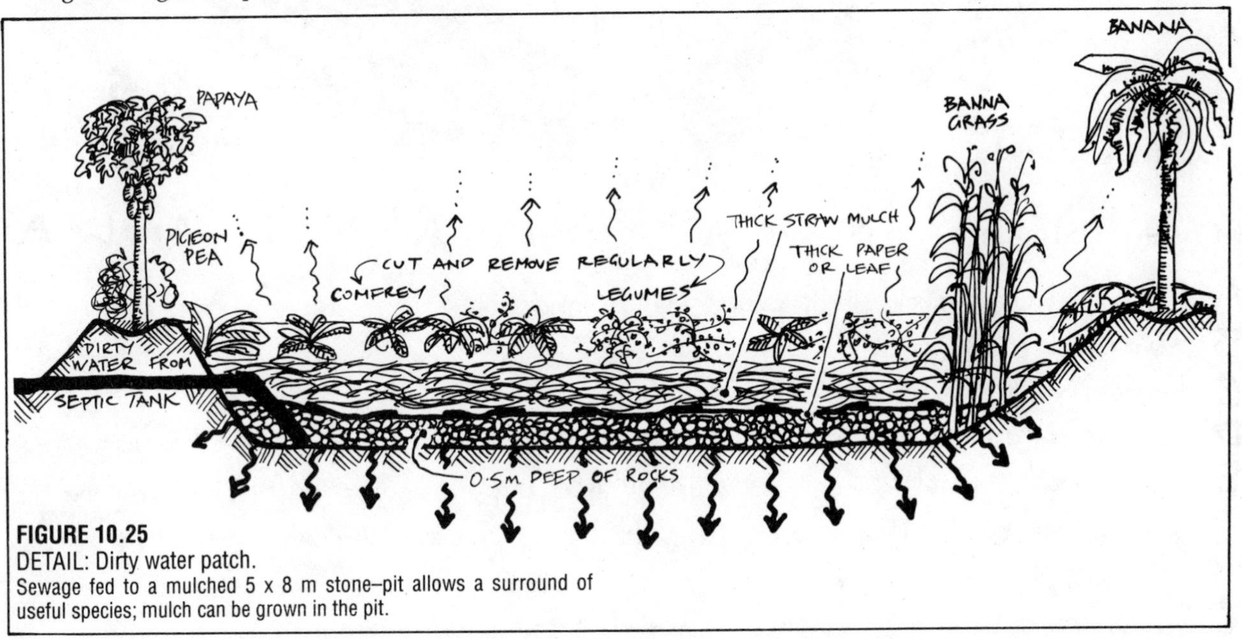

FIGURE 10.25
DETAIL: Dirty water patch.
Sewage fed to a mulched 5 x 8 m stone–pit allows a surround of useful species; mulch can be grown in the pit.

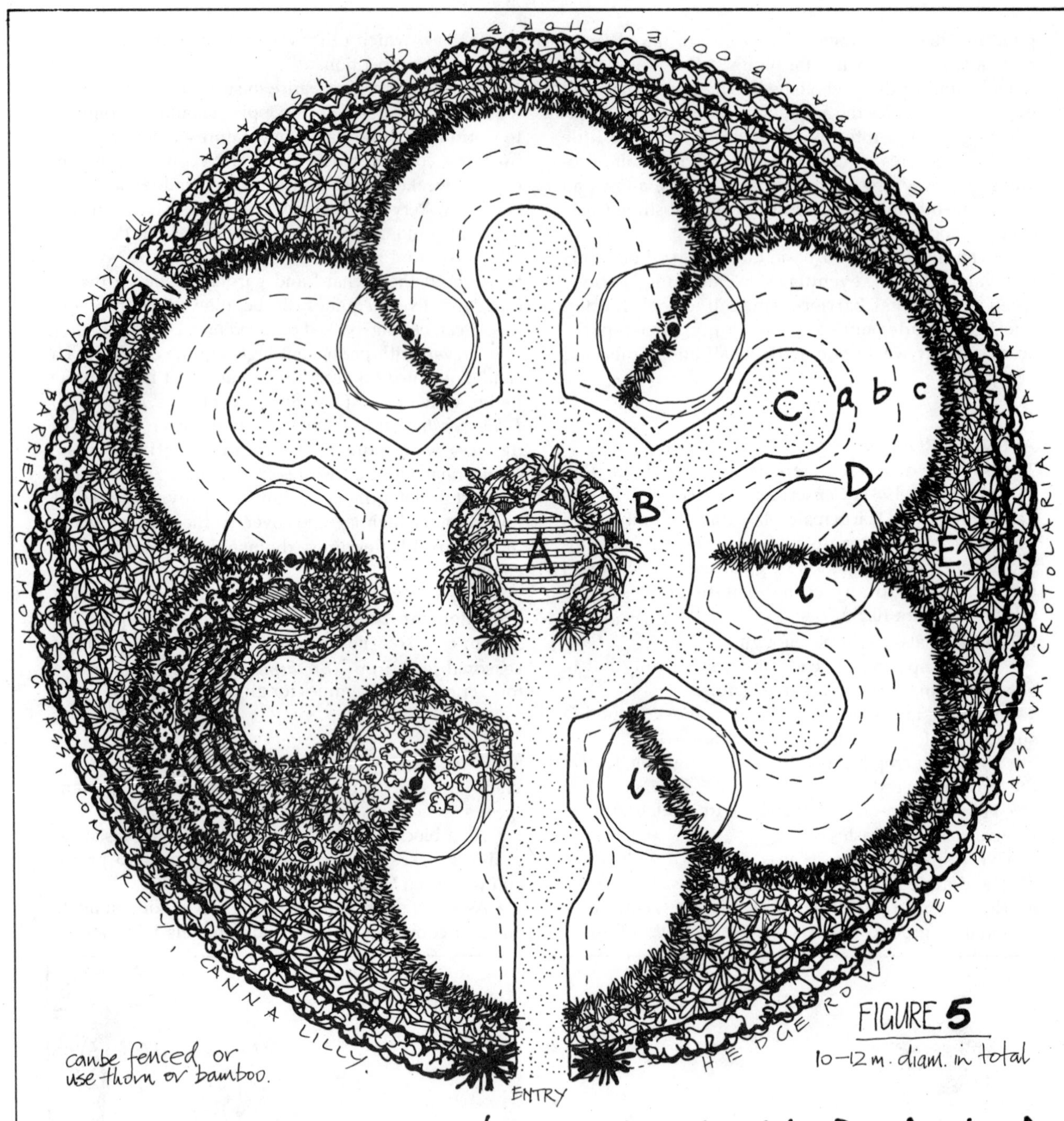

·G A N G A M M A'S · M A N D A L A·

A Banana– Papaya circle with shower-wash grid fitted.
B Sawdusted, rice husks, or gravel paths
C Keyhole paths as B above
D Keyhole beds **a** pathside plants **b** narrow bed plants **c** one visit plants
E Weed, wind, animal barrier hedge sequence, e.g.: (inner → outer) Vetiver or lemon grass, comfrey, arrowroot; taller hedge of cassava, papaya, crotalaria, leucaena, pigeon pea, and banana.
l In garden trees are leucaena or palms for shade in hot regions.

FIGURE 10.26
GANGAMMA'S MANDALA.
An ideal "least path", zero–runoff, acessible layout for tropical home gardens. FIGURE 10.26.4 shows how this mandala can be scaled up for group gardens at schools, villages, etc. Gangamma is an Indian permaculture student.

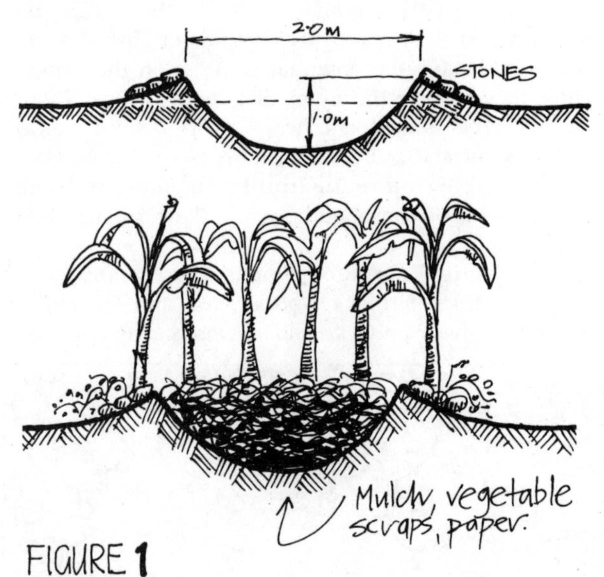

FIGURE 1
Sectional elevation of banana/papaya circle.

FIGURE 3
D Keyhole beds about 1·5m across.
E Hedgerow & barrier plants.
ZONES IN KEYHOLE BEDS:
a Pathside greens.
b Narrow bed plants.
c One visit crops.

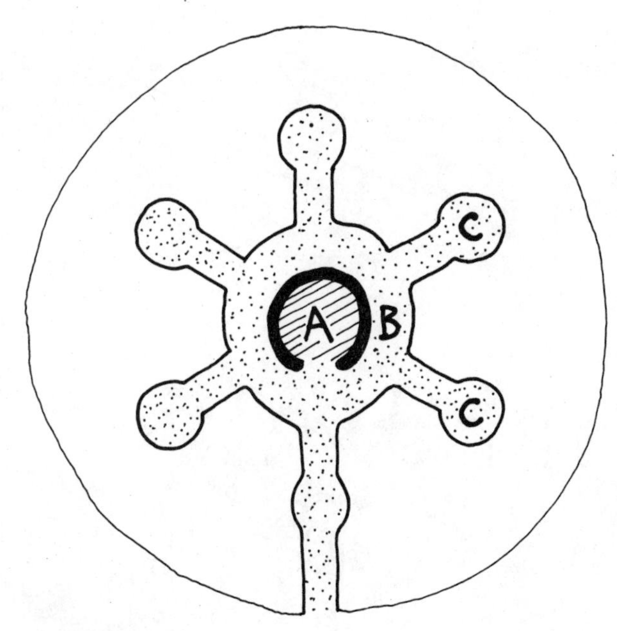

FIGURE 2
A Mulch or circle garden = "banana circle."
B The annular path.
C The keyhole paths.

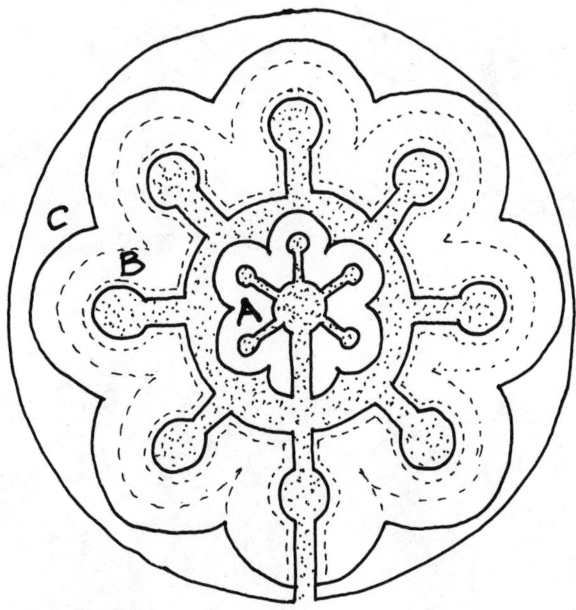

FIGURE 4
A 5 "Banana circles"
B 8 Keyhole beds
C Hedge and barrier.

corral.

• Against hot, dry, or salt–bearing winds, on seacoasts, cliffs, or exposed sites.

• As contoured strips to disperse overland water flow, to catch silt, and to prevent erosion of soils.

Garden Barriers

Around annual, mulched gardens laboriously freed of weeds, a band of grass–barrier plants prevents weed re–invasion. There are 4–5 forms of plants which are effective, and all of them can be used if space permits:

• A deep–rooted broadleaf (e.g. comfrey).

• A clump grass which does not seed down or is not browsed (e.g. lemongrass, Vetiver grass).

• A carpeting plant such as sweet potato, nasturtium, or *Impatiens*.

• A dense low shrub (*Oncoba, Corposma, Echium*).

• A bulb such as *Canna, Agapanthus*.

In total, the same plants can form a fire barrier, provide ample mulch for the garden, and if initially cared for, establish in one season. When first placed, they need to be mulched, manured, and watered.

Animal Barriers

Corrals or cattle–goat–sheep barriers can be strengthened with one or two strands of barbed wire, but should have the potential to resist on their own. Thus, they involve thorny or distasteful shrubs. These can either be planted as a hedge, or as a coppice crop which is cut and built into a thorn fence (boma). The latter enables more flexibility in changing the arrangement of compounds, while the former is less trouble.

Species ideal for such bomas are *Lycium ferocissimum, Acacia tortilis, Oncoba spinosa,* and *Euphorbia tirucalli*. Of these, only *Lycium* and *Oncoba* may not need

FIGURE 10.27
AVENUE CROPPING.
Coppicing legumes (*Leucaena, Cassia, Acacia, Glyricidia*, etc) cut annually provide nitrogen and mulch for intercrop such as maize, sweet potato. ginger, pineapple; crop wastes are returned to field; firewood is a by–product.

1. PRINCIPLES AND DESIGN

2. PT. AUGUSTA, SA, AUSTRALIA.
At first, acid rains from cars and power stations supply sulphur and nitrogen to forests and fields. This "fertiliser", in the long–term, becomes a fatal pollutant to life systems. The first signs of collapse are plagues of pests like this tent moth caterpillar which invests stressed trees. (Principle of Disorder No. 1).

1. MAUI, HAWAII.
Lower slope clearing for cattle lifted the cloud base here and on Moloka'i. This loss of cloud cover reduced precipitation and lead to the desertification of the verdant offshore island of Kahoolawe, now unoccupied, desolate, and used for a bombing range by the US Air Force. Our ability to change the face of the earth increases at a faster rate than our ability to forsee the consequences of that change. (Birch's Principle No.5).

3. CANTERBURY PLAINS, NEW ZEALAND.
This apparently ordered landscape is maintained only by excessive work and at great energy cost. Such systems are disordered and represent chaos in life systems and produce a variety of pollutants in air and water. (Principle of Disorder No. 2).

4. OVER MONTANA, USA.
Great energies spent on centre–pivot irrigation can in 3–4 years produce desert by evaporation effects. Water runoff harvesting and perennial plant systems could lead to harmony in this desert landscape. (Principle of Stress and Harmony).

5. WILTON, NH, USA.
Careful site observation of natural processes. Here the author assesses woodlands on SAMUEL KAYMAN'S farm in early spring for forages, microsites, snow melt effects, wildlife refuges, and leaf litter distribution leading to appropriate design strategies. (Methods of Design No. 2 Observation).

6. WOMIKUTTA, CENTRAL AUSTRALIA.
Following observation of runoff in deserts a rock can be used to direct water to one useful tree. Such strategies arise from observation of rain effects and tree distribution around inselbergs. (Methods of Design No. 3–Deductions from Nature).

2. PATTERN

7. BUFFALO, WY, USA.
This section of a cottonwood limb cut by a chainsaw resembles the "general model".

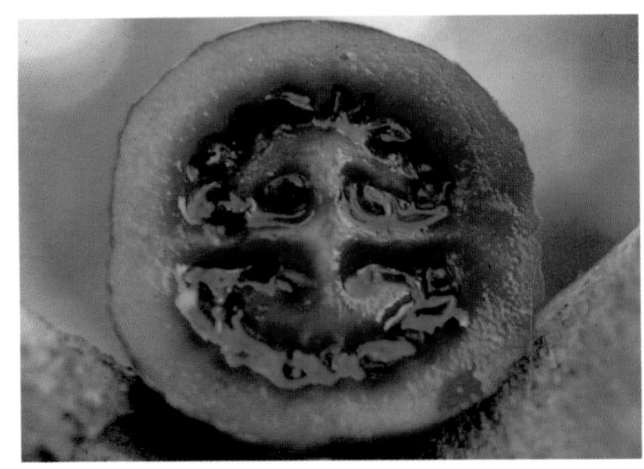

8. STANLEY, TAS, AUSTRALIA.
This section of a tamarillo also resembles the "general model".

9. STANLEY, TAS, AUSTRALIA.
The double spiral of seed in a sunflower head; counts of seed reveal the "summation series" or Fibonacci numbers along each spiral band.

10. TUCSON, AZ, USA
An aberrant suguaro erupts into an "Overbeck Jet", and the stem reveals increasingly turbulent growth.

11. STANLEY, TAS, AUSTRALIA.
A spiral of spirals of spirals. This range of orders of similar forms is basic to fractals; it can be seen in many phenomena here, a Romanesque broccoli.

12. ERNABELLA, SA, AUSTRALIA.
Pitjatjantjara women transfer traditional patterns to new materials, here using a batik (wax–dye) process to print fabrics. Older patterns were made in sand.

13.
A sea–snail egg seen from its underside. Conical form, crenallations, and spiral slots all ensure excellent stability in flow and self–induced flushing due to differing flow velocity near the sand surface.

14. VIC, AUSTRALIA.
An estuary reveals the nature of flow here, a Von Karman trail.

15.
Another Von Karman trail in a lava flow.

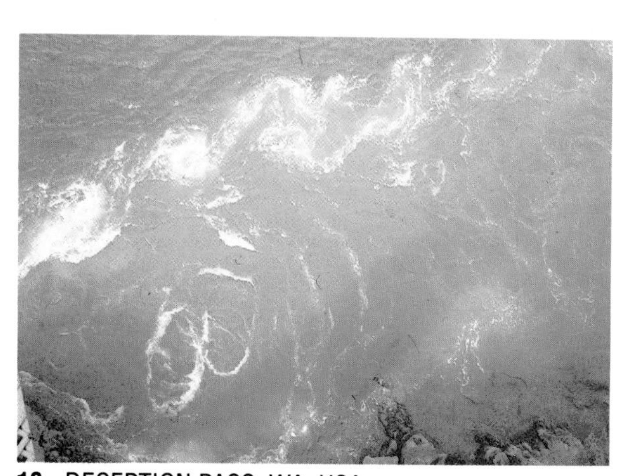

16. DECEPTION PASS, WA, USA.
The tidal flow in the narrows between Whidbey and Orcas Islands shows another Von Karman trail.

17.
The unfolding of agave leaves reveals the imprint of other leaves. Von Karman trails show very similar increasing (or decreasing) magnitudes of flow.

18 & 19. BORE FARM, KENT, UK.
A flowform system. The paired photos show water "switching" naturally into alternate basins; excellent for water aeration in a situation where volume is fairly constant.

4. TROPICS
Developing a Food System

20. HAWAII ("the Big Island"), HAWAII.
Fresh *a 'a* lava flow has been crushed and levelled by tractor, spot-mulched with compost and *Macadamia* nut husks, and irrigated. See **21** for growth in 1 year. Steve Skipper inspecting. USDA trial.

21. HAWAII ("The Big Island"), Hawaii
Growth of trees in one year on *a 'a* lava site; more extensive mulch enables interplant. Numbers **20** and **21** reveal that mulch plus water can establish valuable trees on apparently hostile sites, here mango, citrus, macadamia, guava.

22. MOLOKA'I, HAWAII (Lat. 28°); Sub-tropical chicken tractor system.
Developed by Dano Gorsich. Foreground is 6 weeks of vegetable growth following the clearing of the chicken tractor (5 pens in 0.5 ha). Dano and his wife, and members of the Busby family stand at the front edge of a papaya system which is 2.5 years old. Chickens (Australorps) are moved from pen to pen over the year to clear weeds and grasses; they get most of their food from fallen papaya, seed, insects, and greens. Income from their eggs and papaya at the local cooperative *plus* home garden largely support the Gorsich family.

23. MOLOKA'I, HAWAII.
Traditional Hawaiian garden combines banana, papaya, chili, and taro. Pigs fed from surplus, chickens range below system. Developed by IMU and RACHEL (RAHELA) NAKI. Together with fish and home garden, this is a self-reliant and commercial system.

24. MAUI, HAWAII.
Kikuyu grass. Here, people have planted *Tagetes* (marigold) around their avocado tree to kill, and form a barrier to, the grasses.

25. KAIWAKA, Nth Is., NEW ZEALAND; Instant garden or jungle
About 18 months previously, this was pasture. KAY BAXTER first set up a small nursery, then set out (at one time) a windbreak of eucalypt and blackwood *(Acacia melanoxylon),* and an interplant of nasturtium, comfrey, sweet potato (ground covers); banana, tagasaste, taro (understorey); and a scatter of *Acacia*, guava, mulberry, citrus, apple, peach, etc. as fruit crop. Grass is suppressed and the system co-evolves. Mulch from comfrey, tagasaste is thrown around fruit trees.

5. GETTING RID OF WEEDS, GRASSES, RAMPANT SPECIES

See also Plate 22 -Small Livestock

26. KULA, MAUI, HAWAII.
A sea of kikuyu lawn is held at bay by a low barrier of canna, lemon grass, and comfrey. Part of the lawn (Number **27**) has been converted to a no–dig garden by JACK and SANDY LEWIN.

27. KULA, MAUI, HAWAII.
The reverse side of Number **26**'s barrier planting is an "instant garden" of mulch over newspaper and cardboard which, laid directly on the kikuyu lawn, obliterates that grass; the garden was built as a class excercise at a Permaculture Design Course based on field observations of kikuyu behaviour.

28. KUALAPUU, MOLOKA'I, HAWAII.
Electric fencing restrains pigs which root out and eat kikuyu and grasses, leaving *Prosopis* trees unaffected; this can now be planted and the pigs moved on. 100 pigs in 2 ha pens will "tractor" 40 ha in 2 months. (Hawaiian Homelands area).

29. MULLUMBIMBY, NSW, AUSTRALIA.
David McIlraith has "grenaded" an area of lantana with choko *(Sechium edule)*, which has killed out (by shade) the lanatana (and covered his truck!). This patch yields $120/week in fruits. Choko is easily removed by chickens or pigs, and the area can be planted to orchard or crop.

30. MOLOKA'I, HAWAII.
Chickens clear all grasses and weeds for gardens, papaya, and are ranged on the greens outside fence; papaya provides most of food.

6. TROPICS
Beaches, Islands, and Windbreaks

31. OAHU, HAWAII.
Profile of a stable sand beach with beach convolvulus, and casuarina. This shore can withstand severe storms.

32. MAUI, HAWAII.
Once–stable coast eroding due to beach vegetation destruction; "recreation" can destroy coasts, destabilize snowfields, and erode dunes (sand buggies).

33. NAITAUBA, FIJI.
A hurricane has almost cut this island in half, following deforestation of the shore–line. A fringe of mangroves *(Rhizopora)* is re–stabilising the sea edges; on low coral islands and shores, the removal of coastal vegetation can destroy the island quickly in cyclones.

33. MOLOKA'I, HAWAII.
At the Moloka'i Plant Research Centre tropical windbreak trials indicate the importance of windbreaks in reducing stresses on plants and animals especially on islands with salt laden winds, and consistent monsoons and trades.

35. MOLOKA'I, HAWAII.
A "liveset" (cuttings) hedge of *Erythrina* ("willi–willi") and banna grass *(Pennisetum)* shelters crop against coastal winds; the grass is cut for mulch, and can later be removed as the *Erythrina* bushes out;. Casuarina, Leucaena, or Grevillea is often used as permanent hedgerow.

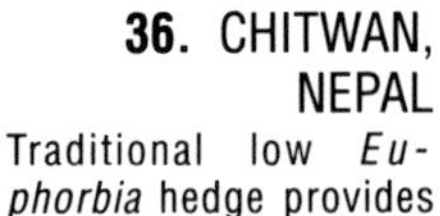

36. CHITWAN, NEPAL
Traditional low *Euphorbia* hedge provides windbreak, a habitat for spiders, a section of bamboo, and demarcates a cattle lane-way from mustard crop; *Euphorbia antiquorum* is often so used, and *E. tirucalli* in Botswana; all need goggles when cuttings are taken to prevent eye injury from the milky sap.

7. TROPICS
Patterned Systems

37. KURANDA, QUEENSLAND, AUSTRALIA.
Some years of work by MARJORIE SPEAR have evolved a very productive garden over 2.8 ha; sited at the edge of a coastal–facing cliff. Night condensation of mist is heavy. Marjorie has stacked the garden with Acacia, coconut, citrus, ferns, flowers, and mixed fruits to produce maximum yield in least space, with minimum grass or weed maintenance. Systems of this maturity demand management more than development work. Only access tracks are in grass.

38. NAITAUBA, FIJI.
Clumped coconuts (here, 8 together) yield as well as spaced coconut (12 x 12m), free up more ground, enable spot mulch serving all trees, assist harvesting, and concentrate watering. Several plant species (e.g. banana, Macadamia) can be clumped in this way, maximising space for intercrop.

39. KUALAPUU, MOLOKA'I, HAWAII.
CHUCK BUSBY has developed mulched circle gardens—some with trellis fitted—to form nucleii in exhausted pineapple farm soils. Mulching and watering is compact. Shelter is of banna grass. Note that several yams and beans can use banna grass for support or trellis.

40. DETAIL OF NUMBER **39**.

41. NIMBIN, NSW, AUSTRALIA.
MICHEL and JUDE FANTON use lemon grass planted on contour to hold steep slope soils, provide mulch, and create upslope planting sites for *Alectrytes*. Lemon grass is also used in beverages and cooking. Vetiver grass is used in this way in Africa whenever stability is needed on slopes, *Euphorbia tirucalli* in deserts.

42. CHITWAN, NEPAL.
UMA and M. C. PEREIRA have developed a more structured stacked system than in Number **37**, with pineapple and ginger main crop edged with leucaena and lychee trees. Leucaena provide fodder, firewood, nitrogen in soils, and mulch. This system too needs minimal maintenance, and is no–tillage. Unirrigated, but sustains yield, provides surplus firewood, mulch, and supports 3–5 adults on 1 ha.

8. TROPICS
Irrigation and Weed Control

43. MALANDA, QUEENSLAND, AUSTRALIA.
BOB KINGWELL has devised a simple manifold to lead water in 3 directions from a head drain; pipes are plugged or opened to allow flow in channels, using simple concrete plugs with rubber "O" rings.

44.
As for Number **43**, here Bob has fitted a head drain with cross–stops to allow water to back up and flood the field to the right; again simple pipes are used, carefully levelled.

45. GUINEA PIG SHELTER.
3–4 guinea pigs manure, and keep kikuyu from choking, young litchee trees; they forage below the canopy, and are fed food pellets. Owls and pythons eat excess populations.

46.
In lieu of guinea pigs, *Desmodium* restrains kikuyu grass. This uses more water than the guinea pig solution but does not need food pellets.

47.
Flood irrigation of papaya on slight ridges also waters green mulch. *Desmodium* carpets the papaya mounds, and comfrey provides a mulch interplant.

9. DRYLANDS
Grazing and Range

48. COOMA, NSW, AUSTRALIA.
Desertification due to overgrazing. Collapse of soils and forests is evident.

49. WILMINGTON, SA, AUSTRALIA.
Goats reduce ground cover and make regeneration of trees impossible. The area needs swales, long–rotational grazing, intensive cut forage for milk goats (in swales); a scene that could be of any semi–arid area.

50. LAS PALMAS, CANARY ISALNDS.
A tethered milk goat (fed on cut hay and cactus) creates a well–manured niche in bare granite rock. After this, an almond, fig, or chestnut tree is planted; oats regenerate for hay. In this way, extensive orchards are developed in eroded granite country. (See Number **51**).

51.
As for Number **50**. A general view of an almond orchard developed in granite using goat and donkey manures. Small hay fields lie below this slope. Cattle or goat pens can be used in this way in any arid area.

52. WILUNA, SA, AUSTRALIA.
Caves part–excavated by wallaroo (a rock kangaroo) offer shelter below a scarp edge; such shade sites determine the number and type of wildlife (e.g. lizards, wallaroo, pigeon) able to occupy and breed in drylands. Author takes refuge from the sun.

53. PYRAMID LAKE, NV, USA.
GERARD and GUS JAMES of the Pyramid Lake Páiute tribe examine Indian rice grass re–seeded in sands. The seed is acid treated, buried 0.3 m, and the glumes flamed to free seed for cooking. *Oryzopsis* is a staple of the Shoshone in this region, and a good perennial range forage for cold deserts. It spaces itself to about 0.6 m between bunches.

10. DRYLANDS

54. PORT LINCOLN, SA, AUSTRALIA.
Sand dune on hard pavement absorbs water and grows larger trees; the range is in fair condition. Sessile trees prevent sand erosion or trap sand on dune crests and sides.

55. POVERTY FLAT, CEDUNA, SA, AUSTRALIA.
Raised beds, sunken paths, perforated pipe irrigation, mulch, sulphur, and bentonite, combine to produce a rich organic garden on an eroded sandy soil site (water 1,100 ppm salt). Aboriginal students taught by PETER BENNETT produce excellent crops in desperate soil conditions.

56. NEAR FLINDERS, SA, AUSTRALIA.
BARRY PILTZ gathers sheets of guano from below tree roosts beside a dryland dam—pigeons and parrots visit for water and enrich area with guano, below roost trees.

57. NEAR OODNADATTA, SA, AUSTRALIA.
Hard bank (on skyline) constricts flood water to create a scour hole in river sand beds. Such effects can be created.

58. ALICE SPRINGS, CENTRAL AUSTRALIA.
Gabions—rock–filled wire basket structures—stabilise a floodway near Alice springs reducing flood flow, and silt burden in the stream.

59. CHARIKOT, NEPAL
Gabions stabilise this steep hillside gully on the road to Charikot. A Swiss design.

11. DRYLANDS
Housing

60. LAS PALMAS, CANARY ISLANDS.
A rocky arroyo (wadi, canyon, gully) floods periodically bringing down large boulders. (See Number **61**.)

61. LAS PALMAS, CANARY ISLANDS.
An arroyo is part–blocked by concrete/stone barrier on right, and bled off via a coarse trash screen to tunnels *(galeria)* and cave storages linked to irrigation systems at lower levels.

62. KOONIBBA, SA, AUSTRALIA
Aboriginal gardeners dig an absorption trench from a kitchen sink outlet, in brick–hard ground. Trench is plastic–lined, manured, refilled, and planted. (See Number **63**).

63. KOONIBBA, SA, AUSTRALIA.
A few months growth in greywater produces abundant vegetable food, establishes fruit trees, cools area, and utilizes nutrients previously lost to a waste lagoon.

64. LAKE WINNEMUCCA, NV, USA.
Recessional boulder benches on the shoreline provide a natural swale system and infiltration, determining vegetation (range) distribution. Gibber deserts (Reg) can be swaled in this way for range species, saltbush, and useful crop.

65. SEROWE, BOTSWANA.
Setting out swales in thorn acacia to infiltrate overland flow for vines and valuable trees. Sorghum and melon crop can be interplanted in thorn forest. (See Numbers **66** *et seq* for swale growth in similar rainfall conditions).

PERIODIC TABLE KEY

1	Brown triangle	THE MAJOR ELEMENTS of which soil is composed.
2	Dot	Elements occuring in minute quantities in plants and marine and freshwater algae.
3	Green triangle	MAJOR NUTRIENTS in plant tissues and soils.
4	Dot	TRACE ELEMENTS necessary to metabolic functions in plants.
5	Dot	CATALYTIC ELEMENTS involved in metabolic function and hydrocarbon processing.
6	Blue triangle	MAJOR ELEMENTS necessary to metabolic function in animals.
7	Dot	TRACE ELEMENTS necessary to metabolic function in animals.
8	Dot	Elements critical or limiting to phytoplankton, aquatic organisms.
9	Orange triangle	POISONS; radioactive isotopes taken up by plants and animals e.g., Rn222, I131, Cs187, Th232, U235, Pu239.
10	Dot	POISONS; metabolic poisons, elements or their compounds dangerous to life forms.
11	Dot	Elements specifically concentrated by bacterial action, root associates of plants, anaerobic bacteria, or thiobacilli.
12	Diamond	STATES– Light blue: gas at room temperatures. Dark blue: liquid at room temperatures.
13	Name	Element's name.
14	Number	Element's number.

I	II								
H 1 HYDROGEN									
Li 3 LITHIUM	Be 4 BERYLIUM								
Na 11 SODIUM	Mg 12 MAGNESIUM								
K 19 POTASSIUM	Ca 20 CALCIUM	Sc 21 SCANDIUM	Ti 22 TITANIUM	V 23 VANADIUM	Cr 24 CHROMIUM	Mn 25 MANGANESE	Fe 26 IRON	Co 27 COBALT	
Rb 37 RUBIDIUM	Sr 38 STRONTIUM	Y 39 YTTRIUM	Zr 40 ZIRCONIUM	Nb 41 NIOBIUM	Mo 42 MOLYBDENUM	Tc 43 TECHNETIUM	Ru 44 RUTHENIUM	Rh 45 RHODIUM	
Cs 55 CAESIUM	Ba 56 BARIUM	La 57 LANTHANUM	Hf 72 HAFNIUM	Ta 73 TANTALUM	W 74 TUNGSTEN	Re 75 RHENIUM	Os 76 OSMIUM	Ir 77 IRIDIUM	
Fr 87 FRANCIUM	Ra 88 RADIUM	Ac 89 ACTINIUM							

PERIODIC TABLE

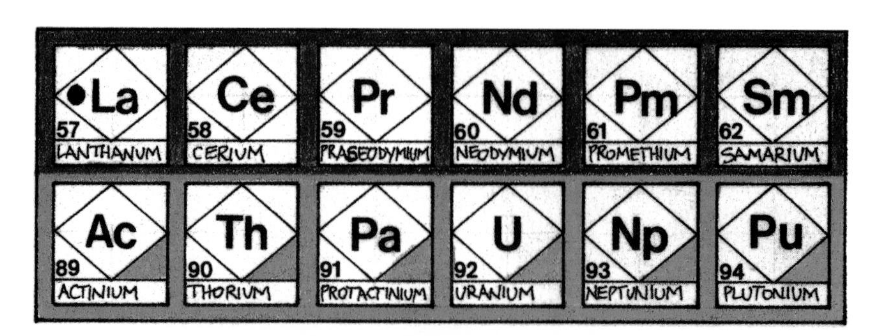

	BORDERS
Green	Hydrocarbons, amino acids (plant and animal tissue) C, H, O, N, S, P.
Yellow	Non-metals.
Dark Blue	Alkali metals.
Light Blue	Inert gases.
Crimson	Major catalysts.
Brown	Rare earths.
Orange	Radioactives.

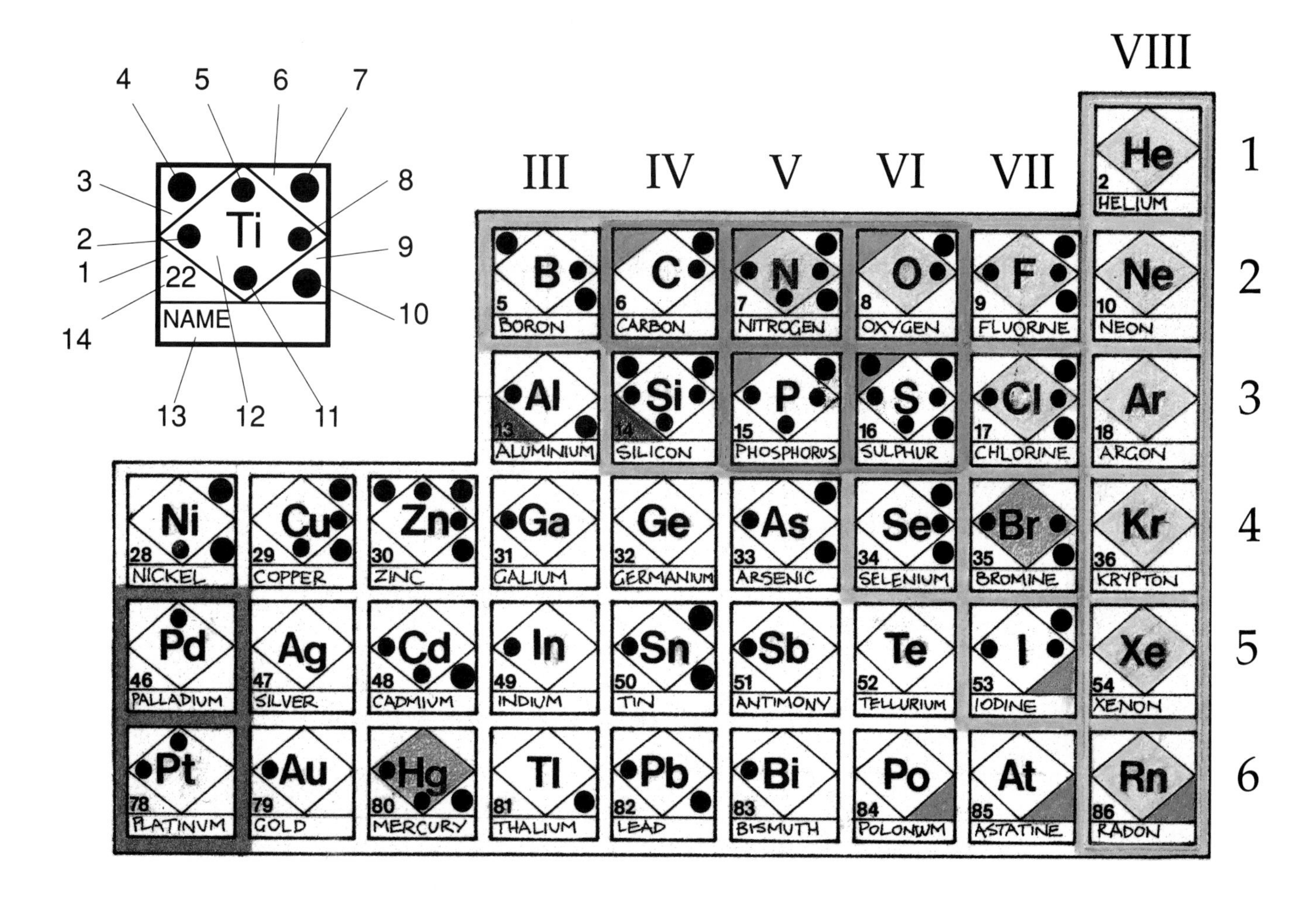

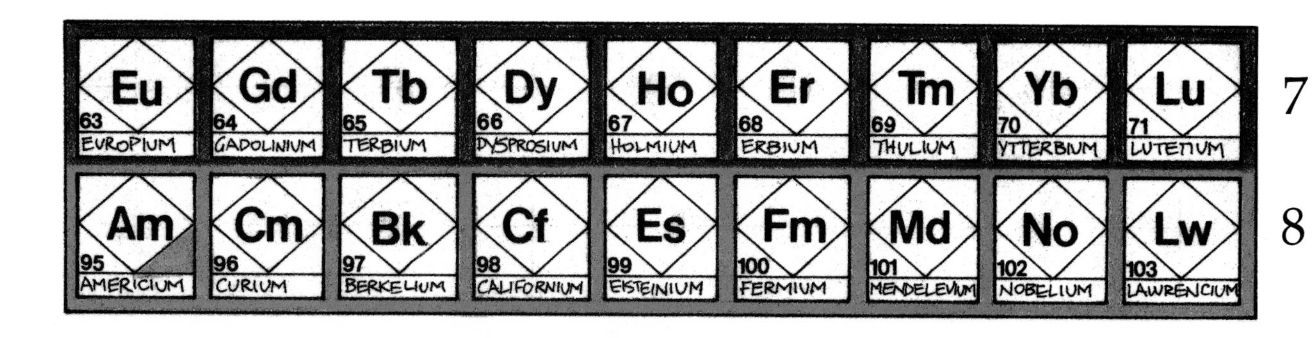

14. SWALES

66. KOONIBBA, SA, AUSTRALIA.
A swale under construction using a side–casting grader in the drylands of central southern Australia.

67. PUMPENBIL, NSW, AUSTRALIA.
Swale under construction.

69. VILLAGE HOMES, DAVIS, CA, USA.
Michael Corbett, designer/developer. New swale area being planted.

68.
Swale in Number **67** six months later with watermelon, sorghum, corn, cucurbits on the bank, and taro in the base.

70. VILLAGE HOMES, DAVIS, CA, USA.
Another swale, heavily shaded by trees which are nourished by the swale soakage. ANDREW JEEVES surveys the lush vegetation growing in what would otherwise be a desert.

71. VILLAGE HOMES, DAVIS, CA, USA.
A gravel–based swale collects water from a suburban road. Because of extensive swaling Village Homes sits on its own deepening, soil–stored, water resevoir in a surrounding desert.

15. DRYLANDS
Building

72. CANYON DE CHELLY, USA.
Mud houses cluster like swallow's nests in a notch at the scarp base of the canyon. Navajo fields and cottonwoods occupy the canyon floor.

73. TAFIRA ALTA, CANARY ISLANDS.
The facade of cave houses is all that reveals the dwelling; excavated in soft tufa (volcanic ash). The rooms need good ventilation to reduce radon levels.

74. KOONIBBA, SA, AUSTRALIA.
Dust storms in settlement carry infectious diseases, bacteria, encephalitis, and cause about 40% of sinus or asthma problems, and sore throats. These broad N–S streets need paving or re-alignment, windbreaks, or zig–zagging. Houses are dust–filled by winds.

75. ALICE SPRINGS, CENTRAL AUSTRALIA.
Extensive disc pitting over 600 square km has reduced dust storms around the town.

76. DETAIL OF NUMBER **75**.
Detail of grasses in pits.

77. NEAR ASPEN, CO, USA.
Traditional earthed–up barn in high cold desert affords winter hay storage and shelter for livestock; such barns are insulated by snow.

16. DRYLANDS
Building And Settlements

78. FRESNO, CA, USA.
Underground orchards; hand–dug caves open out to shaded pits containing fig, grape, citrus, and pomegranate. This house covers 2.8 ha and was hand–dug by M. FORESTIERI.

79. AS FOR NUMBER **78**
Looking up from underground, the sky is covered by oranges; roots in pits are cool, easily mulched. The ground surface can be funnel–shaped to catch water and direct it to plant containers.

80. GREENOCK, SA, AUSTRALIA.
Traditional buried dairy shed keeps milk and food cool. Vine, earth, or tree shade assists cooling.

81. OLD SNOWMASS, CO, USA.
Early experimental version of The Windstar Foundation's "Biodome" uses solar electricity, heating ponds, domed greenhouse to grow tender crop year–round in high (7,500' ASL), cold, dryland where the growing season averages 60 days. Frogs and finches help with pest control in greenhouse, *Tilapia* and hydroponic vegetables are grown in the ponds.

82. VILLAGE HOMES, DAVIS, CA, USA.
Earth–sheltered house with solar hot water, turf roof, and slab floor, provides good comfort control in the house.

83. Roof of house in Number **82**; ice plants and herbs provide active cooling by transpiration and shade.

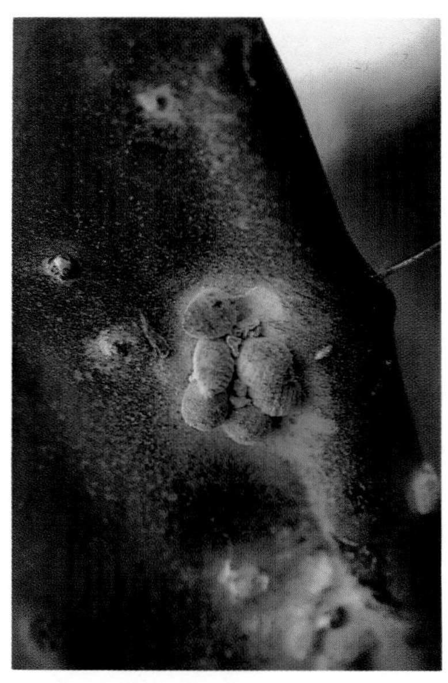

84. LAS PALMAS, CANARY ISLANDS.
Opuntia provides fruit (Tuna), vegetable food, cattle forage, hedge, and is here hosting the Conchineal scale, used locally for food dyes.

85. ST. THERESA MISSION, CENTRAL AUSTRALIA.
House verandah is screened with Bryonia (a cucurbit), Lab–lab bean, and tomato vines. Such screens cool west and sunward walls, and are sustained by waste wash–water and grey water.

86. LAS PALMAS, CANARY ISLANDS.
Apple and pear crop is sustained by a deep (0.3 m) mulch of *Enarendo* (deep cinder) technique using volcanic cinder to produce fruit, potato, and vine crop in an area receiving mainly night fog and condensation moisture.

87. ST. THERESA MISSION, CENTRAL AUSTRALIA.
Gidgee *(Acacia cambagei)* creates a desert mound or *Nebka* about 1.5 m high and 12 m across. Many sessile or weeping trees in deserts will cause dust accretion in this way. *Acacia spp* in the Kalahari do the same, providing deep absorbant soils and mulch for vines, other trees, and burrowing animals.

88. LAS PALMAS, CANARY ISLANDS.
Many desert *Acacia spp* provide abundant poultry seed, as do *Lycium spp* and *Solanum spp.* Here, *S. lidii* provides abundant fruit and seed suitable for chicken forage; many woolly *Solanum spp* provide dryland fruits or pot herbs.

89. TUCSON, AZ, USA.
A spiny *Ocotillo* hedge guards an O'o'dam (Papago) garden. This species is cultivated for fencing, can be liveset to grow if irrigated, and is widely used as a reinforcement or base for mud wall construction. It is a useful crop in itself for desert structures.

18. DRYLANDS
Salting and Interceptor Banks

90. BODLINGTON AREA, WA, AUSTRALIA.
Deforestation, grazing, and wheat cultivation have produced widespread salting, here visible as collapsed soil at seepage lines. The brownish barley greas *(Hordeum maritimum)* is also indicative of soil collapse.

91. NEAR QUAIRADING, WA, AUSTRALIA.
A closer look at a once–forested marsh in wheatlands, now a saltpan.

92. QUAIRADING, WA, AUSTRALIA.
A farmer develops interceptor banks to run surface water and throughflow off fields to a natural valley, thus isolating soil blocks for rain leaching and preventing a "cascade" effect downhill of salty overland flow or throughflow.

93. QUAIRADING, WA, AUSTRALIA.
A classic interceptor bank, ending on the streamline as marked by the distant trees. A bulldozer has rammed subsoil on the downhill wall of the bank. Depth = 1.5 m, width 4 m, spacing not more than 100 m apart, or (on slopes) 1 m vertical separation. Soil pit below this bank shows no throughflow, no rising salted groundwater.

94. BEERMULLAH, WA, AUSTRALIA.
An interceptor has prevented soil collapse by flooding (note rushes on upslope side), thus protecting crop below the bank. Spike rush *(Juncus)* is another indicator of imminent soil collapse, surface flooding, and anaerobic soils. Subsoils are dry, cemented, or subject to groundwater rising.

95. BEERMULLAH, WA, AUSTRALIA.
A series of interceptor banks in wheatfields isolate soil blocks from overland flow and salting effects.

19. COOL TEMPERATE
Establishing Systems

96 and **97**. COLLINGWOOD RIVER, NEW ZEALAND.

Rampant gorse covers 280 ha; lines are rolled down or slashed and eucalypt, pines, and acacia are planted by DICK NICHOLS. "Zone 1" is fully cut and planted to pioneer trees; Tamarillo cuttings can be planted directly into the slash.

98. AS FOR NUMBERS **96** AND **97**.

After 2-3 years, the remaining gorse is slashed, and mature rainforest species self-seed into the forest floor. After 5-6 years the gorse has rotted, and rainforest establishes under acacia (which in turn is felled); native birds carry in rainforest seed.

99. NELSON, NEW ZEALAND.

DICK ROBERTS uses gorse as "rough mulch" to protect fruit and forage trees in steep sheep pasture; reduction of a "weed" species protects the succeeding crop. Tree crops will shade out gorse if thickly planted as in Numbers **96-98**.

100. KATJIKATJIDARA, CENTRAL AUSTRALIA.

Fallen trees, here near a desert inselberg, provide protection from feral animals for young trees and for Andrew to establish a food tree in the branches.

101. MAHALAPYE, BOTSWANA.

Fierce thorny mulch from *Acacia tortilis* protects new trees from intense browsing pressure in compounds, here planted by DOROTHY NDABA.

102. TAHEKE, NEW ZEALAND.
Avocado in" hard" frost area thriving in a clump of tagasaste. Kikuyu is slashed to provide additional mulch at JIM and MIRIAM TYLER'S farm. The tagasaste yield upto 7m of mulch trimmings per annum.

103. CANTERBURY PLAIN, NEW ZEALAND.
Tagasaste drilled with turnip for future sheep forage; seedling trees establish through turnip (brassica) crop at MATTHEW CARPENTER'S property.

104. BANKS PENINSULA, NEW ZEALAND.
1 m in row, 2 m between rows, tagasaste provide summer forage from coppice and short–period browsing by sheep. Grasses thrive between rows (DSIR trials by DOUG DAVIES).

105. BANKS PENINSULA, NEW ZEALAND
Pampas grass is both fast shelter for lambing ewes, and a preferred forage for most livestock; propagated by divisions. Combines well with tagasaste strips.

106. STANLEY, TAS, AUSTRALIA.
Tomatoes in a keyhole bed sheltered by a windbreak of sun root (Jerusalem artichoke) in old tyres on a cold and wind swept site.

107. STANLEY, TAS, AUSTRALIA.
A herb spiral in the form of a ziggurat provides ample culinary herbs at the kitchen door; drainage and aspect suits most cool area species of herbs (spiral 2 m across, 1 m high).

108. BLACK FOREST, WEST GERMANY.
Traditional integration of home and winter barn; mature mixed orchard. Even the bees are housed in this region in winter. Note earth ramp to upper floor of barn to assist hay storage.

109. NEAR CHRIST-CHURCH, NEW ZEALAND.
Willow coppice crop, here intended for basketry, can also be used for medicinals, forage, "stickwood" for radiant (mass) heaters, active charcoal filters, artist's charcoal, and so on. "Willow water" (fresh chips soaked in water) provides gibberelic acid for promoting root growth in the striking of cuttings.

110. NEAR CHRIST-CHURCH, NEW ZEALAND.
The willow coppice field as for **109**.

111 PERTH, TAS, AUSTRALIA.
Here, willows are pollarded for propagation cuttings at a nursery.

112. NELSON, NEW ZEALAND.
Matsudana willow provides strict windbreak for a kiwifruit vine crop on the plains near Nelson. The strict form allows minimal space loss on the ground and needs little maintenance to control shape.

22. COOL TEMPERATE
Small livestock

113. FOREST, TAS, AUSTRALIA.
A rabbit cage "tractor" cleans clover and grasses from mounded strawberry crop at MICHAEL MANGEVELAKIS' farm. Rabbits are part of the yield of the weed system, and provide manures for the strawberrries.

114. STANLEY, TAS, AUSTRALIA.
Lycium ferrocissimum hedge carries vines of hops, banana passionfruit, and shelters main crop pototatoes. Its berries are loved by poultry and being a sharp thorny shrub (used to keep out lions in Africa) provides ideal shelter.

115. LAKE CHELAN, OKANOGAN, USA.
Chickens "tractor" grasses for small fruit (background crop), "weed" small fruits over the autumn period, and provide manure. At MICHAEL PILARSKI'S (Friends of the Trees Coordinator) farm.

116. ASPEN, CO, USA.
Chicken–heated greenhouse and solar–heated seedling cloche enable early planting in short season high mountain areas.

117. WILTON, NH, USA
Turkey forage below autumn olive *(Elaeagnus umbellata)* with abundant berries. Siberian pea shrub *(Caragana arborescens)* and autumn olive are important tree forage legumes of cool and cold climates.

23. COOL TEMPERATE

118. CALDER, TAS, AUSTRALIA.

JIM BASSETT confines his rampant thorny blackberry (35 kg/year yield) in a straight-jacket of old oil drums which also makes for easier harvesting.

119. KIEWA, VIC, AUSTRALIA.

GEOFF WALLACE observed a free-seeded apple in blackberries on his farm. His cattle demolished the blackberries to get to the apples. He now plants apples in the brambles (part of a natural guild in England), they prosper in their protected hole, fruit, and soon the brambles are trodden down by cattle eager for sweet apples.

120. CLE ELUM, WA , USA.

Clearings in fir forest maintain high production of berry fruits, fungi, and nettles all an important crop to the SALISH indians. Here, filberts, thimbleberry, black raspberry, huckleberry, wild strawberry thrive.

121. WHIDBEY ISLAND, WA, USA.

Stumps of firs grow many berry species, and slabs from fir stumps and logs generate berry clumps on lawns or in clearings, as do fallen fir logs. Birds bring seed and manure to stumps and they establish protected from deer browsing.

122. TUMBARUMBA, NSW, AUSTRALIA.

NEIL DRUCE and JASON ALEXANDRA reduce fuels in a "Jean Pain" chipper. Chips are mulched, composted, and fermented to methane. Mechanical removal of forest fuel reduces the fire risk and enriches soils unlike "cool burns" which impoverish soils and often lead to wild fires.

123. EVERGREEN COLLEGE, WA, USA.

Cloche raises early tomatoes in leaf mulch; sides are hinged to allow late spring growth of crop. RENY MIA SLAY.

24. URBAN

124. THE BRONX, NY, USA.
The Green Guerillas transform vacant land in urban wastelands to productive, beautiful, and recreational usages as city farms.

125. WEST GERMANY.
The productive Schreber gardens adjoin almost every settlement and provide food and much–needed recreation in cities. People are permitted to spend weekends on these mini–farms. (One example of greening the cities).

126. GLIE FARM, SOUTH BRONX, NY, USA.
This urban farm provides 8% of New York's herbs, and provides employment for many people in an area of urban decay.

127. THE BRONX, NY, USA.
The only example of productive use of New York's food waste is this small composting business operated by "The Bronx Frontier" group. Small mountains of compost are sold off site to create meadows over rubble–filled wastelands.

128. BLACK FOREST AREA, SOUTH GERMANY.
West or sun–facing walls in cities and towns provide radiant heat for marginal crop here an espalliered pear grows well and "softens" the wall of an urban barn. We can therefore regard most urban buildings as *additional* agricultural space as are trellis systems over alleys, shopping complexes, up building walls, etc.

129. LOUISVILLE, KT, USA.
Project WARM volunteers retrofit and insulate old folks' homes by the thousand. This sharply reduces energy costs and deaths from the cold. The cost amortizes in two winters. Only minimal funds are directed to such sound expenditures in western society. It is always cheaper to conserve rather than gernerate energy. Such projects make many dangerous or dirty power generating systems obsolete (e.g. nuclear power).

trimming. All of these protect diversion banks and compounds from grazing animals. *Euphorbia* takes root from cuttings in even arid conditions, and *Acacia tortilis* is ideal for cut– and–build fences. When cutting *Euphorbia*, be sure to protect the eyes with goggles, and keep the skin covered if allergic to the milky sap.

Windbreaks

Primary tall windbreak of *Araucaria, Cupressus, Casuarina, Pinus,* hardy *Phoenix* palms, and even mangroves may be needed in front–line locations, followed in the lee with such hardy quickset species as *Euphorbia tirucalli, Coprosma repens, Echium fastuosum* and so on. It is always best to find local plant species that do well in the district.

Erosion Control

Contours at 10 m on medium slopes (2–7°) and at 5 m on steep slopes can be planted out with root sets of *Canna,* Vetiver grass, lemongrass, or pampas grass. These are set out at 0.3–0.6 m spacing and form an unbroken cross–slope hedge, or a crown on earth walls or dam banks at spillways (**Figure 10.13**). They both disperse water and create silt traps; behind such self–perpetuating walls, soil is deeper and trees can be planted or crops grown. The system is cheap, effective, and provides mulch. Some of the yuccas, agaves, and aloes may provide the same structural effect in desert areas.

10.7 INTEGRATED LAND MANAGEMENT

The Maori *marae* or Hawaiian *ohana* were geomorphic and sociological units in which land and people were integrated for sustenance. They may have evolved out of early errors of over–clearing, excessive burning, and the extinction of useful animals before reaching equilibrium. We can only hope that the modern world also has time to take stock and come to its senses, but that will rely on determined change by many thousands of us within society.

Presuming a hill–to–shore profile (often a volcanic cone profile, **Figure 10.28**), the stable tropical landscape may require some or all of these features:

• PROTECTED SKYLINE AND HILL FORESTS. These not only protect soils and waters, but both mine and release plant nutrients from the upper (sometimes steep) slopes. They can be used as limited forage resource and mulch provision, but should have iron–clad protection. Their clearing brings compound catastrophes ranging from landslide to loss of nutrient in water and crop, desertification, and consequent severe social disruption. At the base of these forests, as the slope eases to 15° or less, water can be diverted or harvested to replenish groundwater and irrigate terraces and crop.

• MIDSLOPE OR KEYPOINT. A diversion of stream water here will lead water out to ridges for terrace crop and village use. Thus, cropping commences below this critical point. Human occupation and complex cultivated forests and gardens can now be established downslope. The stable plateau, the hill rising above the valley, and bench sites above the reach of flood and sheltered from hurricane and tsunami are prime cluster settlement (village) sites, with some scattered housing higher on ridges and the forest edge.

• LOW SLOPES (2°–15° slope) are well suited to earth–shaping as terrace and padi, with limited grazing and innovative forestry. These are the sustainable agricultural areas, where attention to sub–contoured agriculture, windbreak, and access will help direct run–off and water to crop.

• COASTAL FLATS AND VALLEY FLOORS. Rich and often deep humus soils can accumulate on valley floors. Greywater and processed manures from settlement and livestock add to fertility, where extensive aquacultures or rich forest/orchard crop can be developed.

• SHORELINES are preserved for 100 m (330 feet) or so inland for essential windbreak, shore stability, and cloud generation over land. Selected quiet harbours and refuges accept fishing settlements based on sea resources. Shores with 15–20 year *tsunami* history need sacrificial palm/*Casuarina* barriers to reduce storm damage ashore. There may be lightly–built and temporary shore housing, which is used only when working the coast, with more substantial homes and sheltered gardens safely inland or behind earth embankments. If shallows or a coral–based lagoon/reef system stretches offshore, the following categories of use also apply.

• LAGOON AND ESTUARY SEA IMPOUNDMENTS. Ancient and yet innovative sea–walls (usually of semi–permeable coral blocks with entries, flood–tide gates, and a series of inner pens) are an excellent use of the shallows. Restricted mangrove plantations help stabilise the banks and provide sea mulch while catching any silt that washes down from above. These warm–water lagoons are rich in algae and are good sites for cultivated seaweed. They also impound and fatten mullet and mangrove fish, oysters and eels, shellfish and prawns. In the sea lagoons, mobile phosphates are fixed in plants and mud in a few days, and are then available for growth. Such rich estuaries are invaluable as managed maricultures.

• MARINE CONSTRUCTS. Artificial reef systems of tyres (on sand), coral blocks, and boxes of palm trunks greatly enlarge the habitat for sedentary fish and lobsters; within a year, these constructs are consolidated or cemented by corals. Weak electroysis on metal mesh creates an artificial but permanent coralline deposit more rapidly. In extensive sandy shallows, log barriers full of sand and stabilised with mangroves provide shore barriers, lagoon walls, and "edge" in the sea.

It is as well to have an integrated concept of appropriate land use in mind for both broad policy

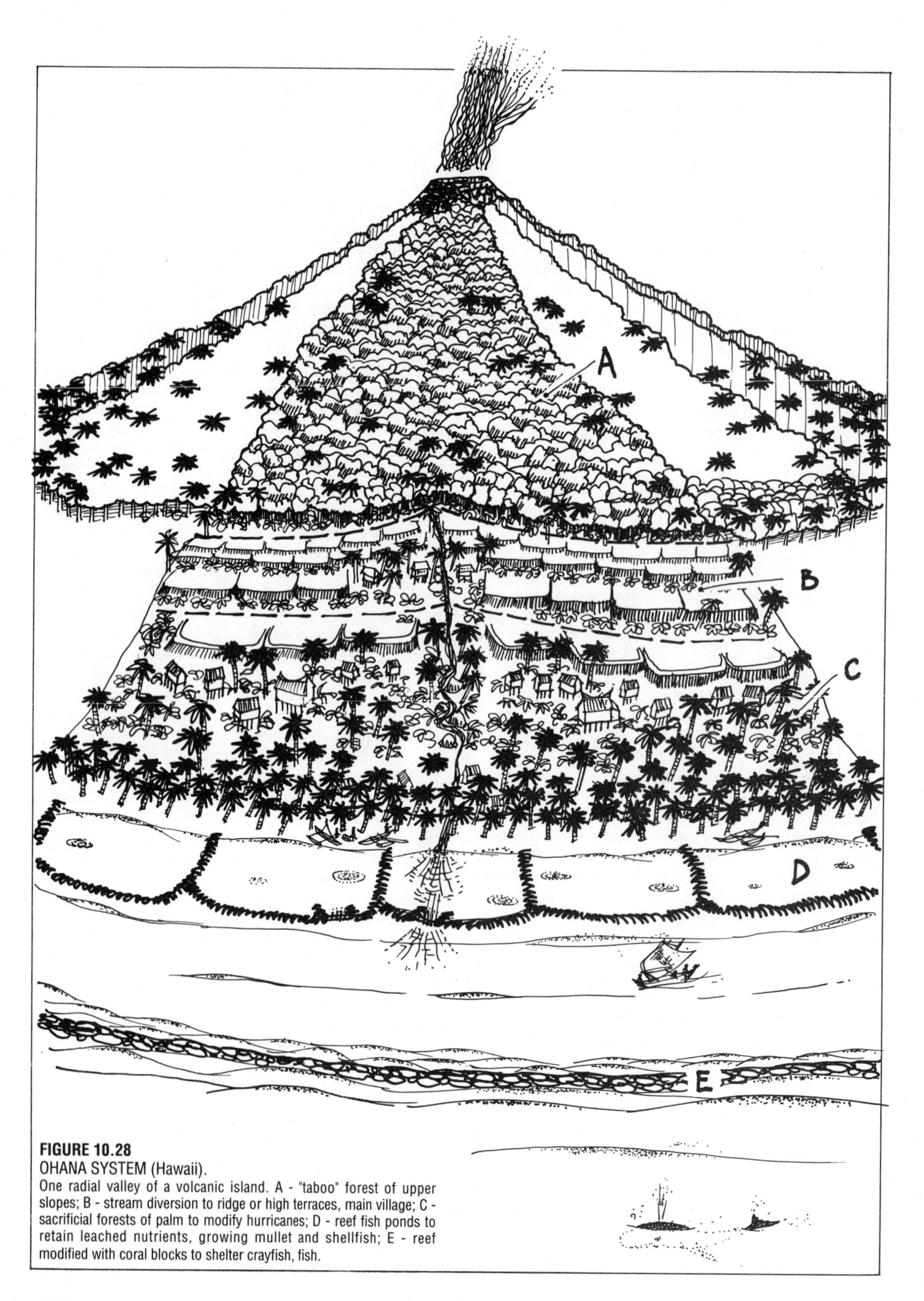

FIGURE 10.28
OHANA SYSTEM (Hawaii).
One radial valley of a volcanic island. A - "taboo" forest of upper slopes; B - stream diversion to ridge or high terraces, main village; C - sacrificial forests of palm to modify hurricanes; D - reef fish ponds to retain leached nutrients, growing mullet and shellfish; E - reef modified with coral blocks to shelter crayfish, fish.

planning and future land redistribution. There are slow moves in present government circles to institute some sanity into life and landscape, as (like agriculture) government must change towards an environmentally sound policy.

10.8
ELEMENTS OF A VILLAGE COMPLEX IN THE HUMID TROPICS

Based on traditional and modern villages, a complex can be built from the following checklist:

• A well-arranged array of housing, often grouped around a compound.

• The compound has processing and storage areas, threshing floors, dancing areas, meeting house, cooperative or retail store, bulk fuel depots, firewood and mulch depots.

• A well, piped water, tanks, or dam must be sited and integrated for clean water (depends on local skills, resources, water available, and site characteristics).

• A plant nursery to serve gardens and forest.

• At each house, 0.25–1.0 ha of home garden and orchard, based on self-reliance.

• At borders of gardens, ranges for domestic species, e.g. chickens, guinea pigs, ducks, rabbits, small pigs, pigeons; these are a manure resource for the home garden.

• In areas of plentiful water, fish ponds over which some animals are housed.

• Sector or zone of fuelwood plantation, potentially integrated with windbreak.

• A strong yard and sheds for cattle and pigs at a commercial level.

• A careful zonation of tree polyculture as per **Figure 10.29**.

• Special facilities such as a log-trimming or boat-building area, fish net drying racks, canoe or small boat landing, wharf, freezer, solar pond installation, power house, large community drying shed, craft work areas, and vehicle or draught-animal park area.

10.9
EVOLVING A POLYCULTURE

If, as is often the case, we start to evolve a permaculture on grasslands or compacted soils, then the very first step is to thoroughly plan the site, and rip, swale, pit, or dam every area to be planted, thus ensuring maximum wet-season soil water storage to carry over to dry periods. This process should commence at the highest point of the property, and around the house or village site.

On these loosened soils and in mulched swales, a mix of tree legumes, fruits, bananas, papayas, arrowroot (*Canna*), cassava, sweet potato, and comfrey can be co-planted. There should be one such plant every 1–1.5 m, with *Acacia* at 3 x 3 m spacing, banana at 2 x 2 m, fruits 5 x 8 m, palms 10 x 10 m, and the smaller species as gap fillers.

As well, all larger planting holes should be seeded with nasturtium, *Dolichos*, Haifa clover, broad bean (fava), buckwheat, *Umbelliferae* (dill, fennel), lupin, vetch (hairy or woolly), dun peas, chilies, pigeon pea, or any useful non-grass mix available and suitable to climate and landscape. The end aim is to completely carpet and overshade the ground in the first 18–20 months of growth.

Ideally, dense plantings of this type should be grass or hay mulched, using monsoon grasses, swamp grasses, and later the tops of arrowroot, comfrey, banana, *Acacia*, and green crop. Later still, shade-loving species such as coffee and dry taro can be placed in any open spots.

Paths for access, openings for annual crop, bee plants on the edges, flowers, and fire-resistant "wet" ground covers such as comfrey, *Tradescantia*, *Impatiens*, and succulents can be placed as time goes on, while the fruit trees are kept free of grass and mulched by cutting out crowded *Acacia* and banana as mulch.

It is far better to occupy a quarter hectare thoroughly than to scatter trees over 2 ha as production is higher and maintenance less, moisture is conserved, and frost excluded.

As for species richness, or species per hectare, this can be very complex and dense near the village or home, and simplify to well-tested species of high yields as distance from the home increases. Any trellis crop should be first placed to shade the home, livestock, or to make fences, and only later placed on *Acacia* or other legumes as they age.

The number of productive, managed and effectively-harvested species in a polyculture is decided by a complex of these factors:

1. The number of people responsible for managing one or two of these crops or animal species (labour).

2. The proximity of the complex to village or settlement (zoning).

3. The relative cost-benefit balance on increasing inputs to optimum levels (fiscal economics).

4. The need for effective plant guilds (harmonious ecological assemblies).

5. The method of marketing and processing (whether these can cope with complex product).

6. The total area which is controlled. Larger areas demand increasing simplicity, at a cost to factor 4 above.

7. The maturity or stage of evolution of the plant system. Older systems provide more niches, younger systems more regular product.

In practice, the gardens, walls, roof areas, trellis systems, and compounds of villages are the most complex and rich areas of cultivated species. We can manage, and find uses for some 200–400 species in such situations, of which the following usage classes are dominant (some very useful species fall into 3 or more of these classes):

Potential Species	
• Basic food species	70–90
• Mulch and fodder provision	10–30
• Medicinals and biocides	20–50
• Structural and craft	5–20
• Culinary herbs	20–30
• Beverages and export specialities	10–15
• Fuelwood and coppice	12–30
• Special uses, sacred uses	8–15
Total cultivated species	155–280
Plus wild-gathered species	40–80
Total species utilised in a complex village situation.	195–360

While a complex polyculture of many hundreds of species delights both the naturalist and (in food plants) the householder or villager, and the benefits to settlements are numerous, it becomes difficult to control an *extensive* rich polyculture and collect its Our very complex polycultures work best at small scale and with close attention from people. The depopulated, dehumanised, and now almost deserted wasteland of modern agriculture is unable to cope with any but the most basic and simple intercrop systems, thus sacrificing yield, quality, stability, and inevitably people.

Thus, if we analyse the dollar economics of such systems, there will be an optimum number of species

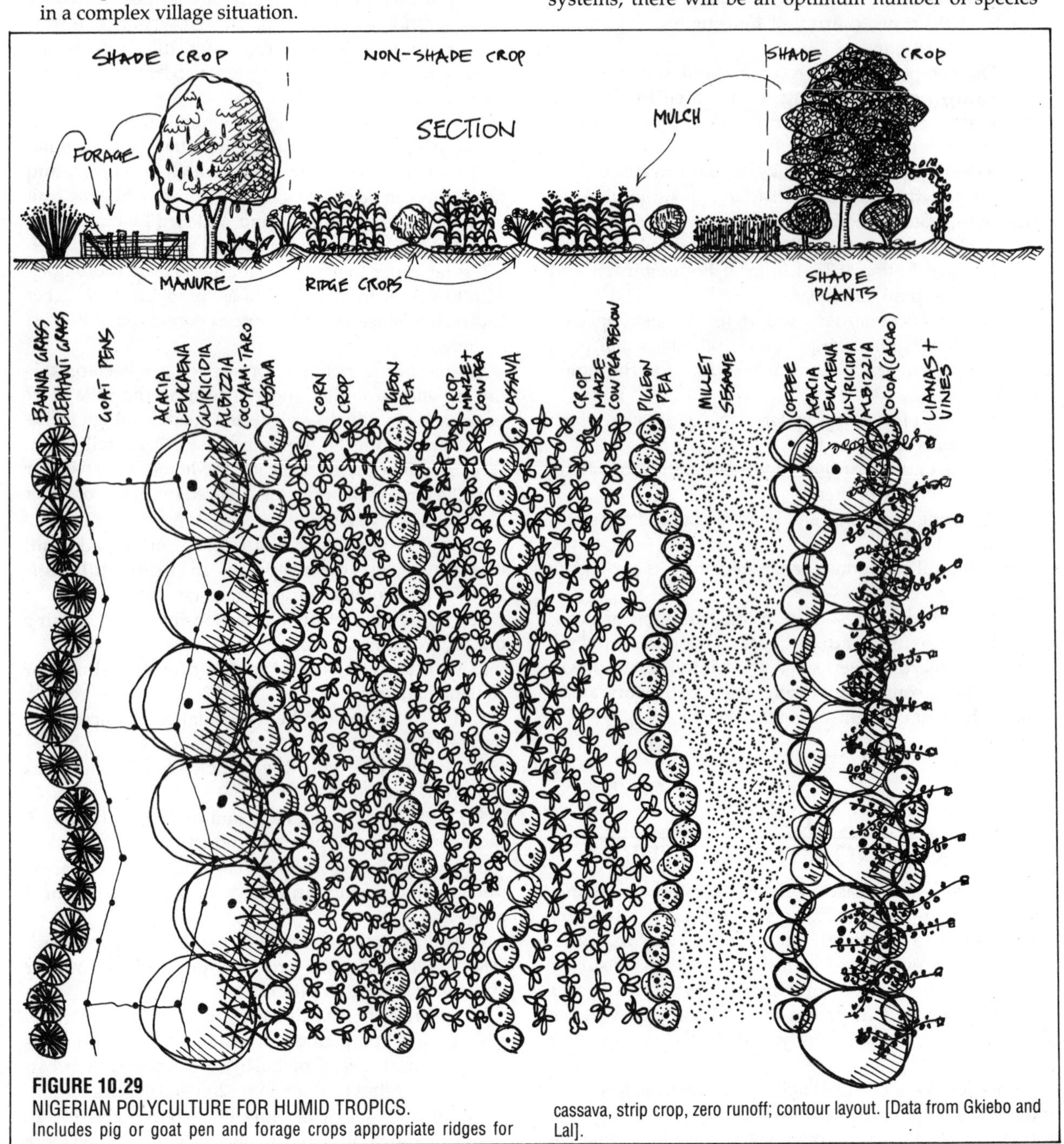

FIGURE 10.29
NIGERIAN POLYCULTURE FOR HUMID TROPICS.
Includes pig or goat pen and forage crops appropriate ridges for cassava, strip crop, zero runoff; contour layout. [Data from Gkiebo and Lal].

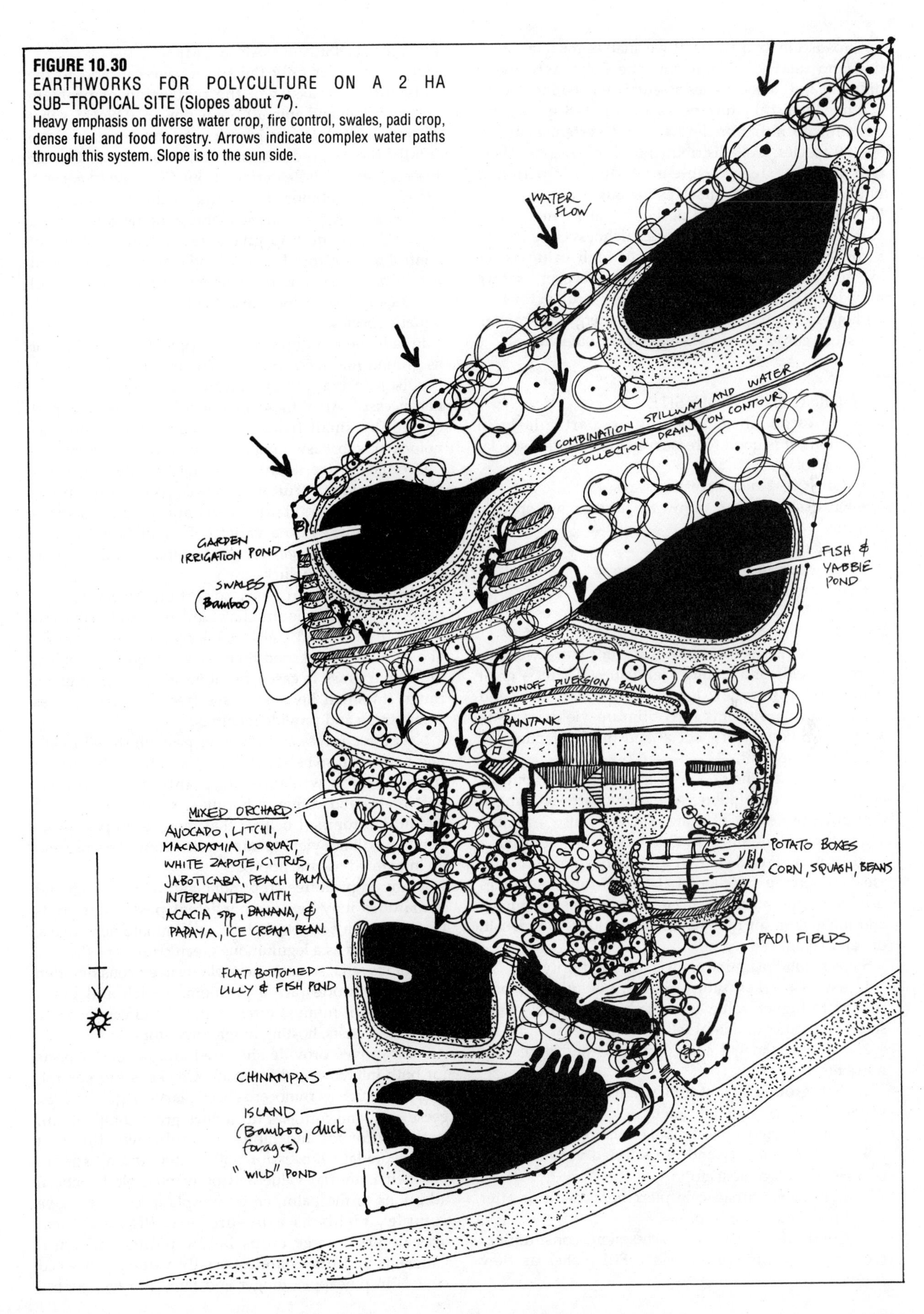

FIGURE 10.30
EARTHWORKS FOR POLYCULTURE ON A 2 HA SUB-TROPICAL SITE (Slopes about 7°).
Heavy emphasis on diverse water crop, fire control, swales, padi crop, dense fuel and food forestry. Arrows indicate complex water paths through this system. Slope is to the sun side.

for broadscale cash yields. If we analyse for total nutrition and total yield (ignoring the dollar returns), a different and richer species assembly will be indicated.

While the fiscal return peaks at about 6–8 species in a system, the nutritional–total yield system peaks at 50–100 species, well–distributed over all seasons. These two factors (extensive–intensive, fiscal-nutritional) must be defined by ourselves for our needs, and will have a profound effect on design. What we may arrive at is a sensible zonation of species richness close in, and a concentration on less species of high value as we extend the system. It is in the garden, however, that we may learn the value of such successful extension without sacrificing large amounts of energy and capital. Thus, our gardens are trial areas for the outer zones.

PLANNING THE WHOLE SITE

Even in established polycultures, particularly in plantations, it is good to re–survey the site with special attention to:

- Main access and harvesting ways.
- Earth–shaping for rainwater harvest and specific crop.
- Sufficiency of mulch.
- Best water and irrigation strategies.
- Better village planning.
- Improved or more sophisticated site processing for market.

These are the main factors that can reduce work or increase yields and commercial values. There is great benefit in testing new legumes, tree varieties, and earth–shaping systems for optimum yields, and in assessing labour, work, social and market factors for future development.

On a new site, the same considerations hold, but the establishment of windbreaks and any earth–shaping is a priority, preceding planting. There are also essential soil tests for plant nutrients and trace elements, as it is a modest amount of these that give early vigor and early yields. Intercrop selection is also a priority, sometimes used to shelter a more delicate crop, but also as mulch and nurse crop for nitrogen fixation and for wind, salt, or sun damage reduction.

Steps in total planning are roughly in priority:

1. Assess market; future; prices; potential for processing to higher value; labour; shares, legal systems; social necessities and local self–reliance needs.
2. Analyse and get advice on soils and necessary nutrients.
3. Plan ground layout and windbreak, access, and water. Detailing can follow later.
4. Plan and carry out essential earthworks.
5. Establish nursery and use selected varietal forms for new or replacement crop.
6. Commence broadscale placements with or after windbreak and nurse crop.
7. Continue by constant assessment, consultation, feedback and innovative trials. Fill niches as they evolve.

PEST AND DISEASE MANAGEMENT

Here, I consider only the species we can add to the polyculture to assist in the regulation of problem species (plant and animal). Some powerful biocides that are found in plants are harmless or short–term and are totally bio–degradable, natural substances. Classic insecticides are those derived from *Chrysanthemum* spp, *Derris* spp (rotenone) and the neem tree (*Melia* or *Azadirachta*). A few of these plants in home gardens and small clumps in crop give a ready source of insect control, or control of invertebrates, nuisance fish, and amphibia in water. Both neem and derris control aquatic organisms; most insecticidal plants are lethal to aquatic species.

Broadscale mosquito control, applied from the air or as ground mists, can combine fats or oils (e.g. lecithin), a poison (neem oil), and an infective agent (*Bacillus thuringensis*). All of these are potentially assisted in pest control by small fish and such insect predators as notonectids (backswimmers) in open water systems.

Ground foragers (chickens, pigs, cattle, large tortoises) eat fallen fruit and larval insect infestations, while leaf foragers (birds, frogs) attend to infestations in the canopy, as do a variety of small skink lizards. Some lizards (*Tiliqua*) forage for snails and slugs at ground level, as do ducks.

Pasture grubs are eagerly sought out by a variety of birds and small mammalian and marsupial insectivores. Tropical land crabs seek larval insects in mulch, and provide useful food themselves. Even the problem of kikuyu grass is eased by domestic guinea–pigs on range (in small houses); these free trees from grass competition and provide manures.

Neem tree leaves and oil deter pests in stored foods, and have been so used for centuries in India. In short, a little research will indicate plants, invertebrates, vertebrates and common harmless substances of great use in the tropics. I believe that there is no pest problem that will not yield to our applied commonsense and an integrated natural approach.

In any tropical tree crop monoculture, soil fungi and nematodes may become persistent pests. Marigolds (*Tagetes*) often serve to reduce or eliminate nematodes, and *Crotolaria* as a leguminous green crop traps them in its root mycelia, so both these plants need consideration where nematodes are a problem. Mulch and green manures (soil humus) often buffer the effects of these and fungal pests, hosting fungal predators.

Palm groves provide sheltered and shaded aspects for both intercrop and livestock. Chickens (for controlling pests such as rhinocerus beetle larvae), guinea pigs, geese and land tortoises (to reduce grass competition), pythons (for rat and mouse control), owls (the best rodent predator), bees for pollination, and all species for their manurial value and other possible beneficial additions to the palm/crop/interplant complex, give complex yields as a by–product. Pigs are ideal scavengers in tree crops below palms and fruit. Chickens and ducks are especially valuable in weed control in pineapple, ginger and taro, and will control comfrey and *Tradescantia* if needed (the latter plants also control grasses if planted as part of a weed barrier).

Perhaps the main function of animal species in the tropics is to "recycle" plant wastes and to help control the rampant growth of ground cover. The usual domestic species, often penned, are pigs, chickens, geese, guinea pigs, rabbits, pigeons, and milking goats, cattle, or buffalo. Horses and bullocks or oxen perform draught animal functions.

Moreover, useful endemic animal species (and especially island species of restricted natural range) need to be more widely examined for their particular uses, and selected for their functions in a wider set of trials out of their normal range.

In the established tropical system, geese control pond–edge grasses, and wallaby or small grazers keep forest clearings and paths open; both are encouraged by feeding–out bran or pollard in the areas to be clipped.

Where no foxes or pythons threaten poultry, flocks of guinea fowl and hens, ducks and bantams perform invaluable pest–control and manurial/scavenging roles. Where predators are a problem, special housing, or pigeons in safe elevated roosts, may be the only way to keep fowl. The effects of electric fences in tropical areas, if not within forest edges, are often nullified by the rank growth of coarse grasses, and become inoperative for repelling python, pig, bandicoots and foxes. For this reason, fence lines must be planted with a dense perennial ground carpet of low herbaceous plants to exclude grasses, or overshaded by tree canopies.

10.10
THEMES ON A COCONUT OR PALM–DOMINANT POLYCULTURE

There are many considerations to bring to the planning of a coconut or palm–dominant polyculture; they can dealt with in the following themes:

- Structure and zoning.
- Species selection.
- Patterning.
- Economics.
- Re–working old plantations.
- The effects of plantation monoculture.

STRUCTURE

Any humid tropical polyculture that duplicates or imitates the normal structure of a tropical forest is likely to succeed. The structure of the system refers to the final cross–sectional appearance of any polyculture. Near large markets, or mass transport systems, it is quite feasible to introduce large tree species into the palm system, and to use the fruits and nuts as a supplementary market crop. Around villages, a far more complex and species–rich approach is needed, with fuel, fodder, structural materials and basic foods, oils, and medicinals in a complex intercrop. Remote from settlement or market, some livestock ranging (pigs, cattle) can be contemplated, with the intercrop selected to assist animals on range over dry periods, as ground forage, or as a drop from fodder trees such as figs, breadfruit, papaya, or *Inga edulis*.

Proximity to village or settlement decides species complexity and (by implication), structure, in that labour–intensive systems are best placed close to the village. Zoning out, we might place:

- Productive trees in palms (total species: 6–12) **Figure 10.31.A**.
- Palms within crop and avenue cropping between palms (total species: 30–35) **Figure 10.31.B**.
- Animal forage and free range in palms (total species: 8–20) **Figure 10.31.C**.
- Village garden and trellis, roof crop, greens (total species: 100–150) **Figure 10.31.D**.
- Fuel–wood in dwarf palm (total species: 3–4) **Figure 10.31.E**.
- Forest and tree reserves.

It is in and around the village that small livestock, fungi culture, padi crop and terrace is appropriate.

CRITERIA FOR SPECIES SELECTION

For any one site a selection of species that go with a palm polyculture has to be made. Some criteria are:

In village: A mosaic of the full range of food, craft, and medicinals is needed. Special intensive crop such as fungi, padi, and even algae can be planned. Experimental plots can be located here.

In palms, species are selected according to a general zoning/use–plan. It is helpful if:

- Species are suited to soils. A mosaic approach is indicated based on soil drainage and nutrient status. If the site is complex, then so should be the main crop.
- Species are locally acceptable, or are very similar to local types. New introductions need trials and instruction as to processing.
- Species chosen have a wide potential for processing. Coconuts have hundreds of known products or uses; this gives market flexibility.
- Species do not become rampant (unless they are controlled by livestock or cultivation).
- There is good information, some varietal types, and assured yield or low management input for the species used.
- Species are compatible, such as those listed herein.
- Species serve a present and future essential use (as thatch, fuels, oils, dry–stored food).

Considerations for Varietal Selection

For coconuts, and many other species (guava, yam, taro, banana, papaya) there are dozens of varieties developed for quite specific sites, soils, microclimates, or uses. There are coconut palms ideally suited to oil production, while others produce very fine quality cup copra for temple use, and others are ideal for shredded coconut, coconut milk, fresh nut markets, and so on.

There are dwarf, medium–height and tall varietes.

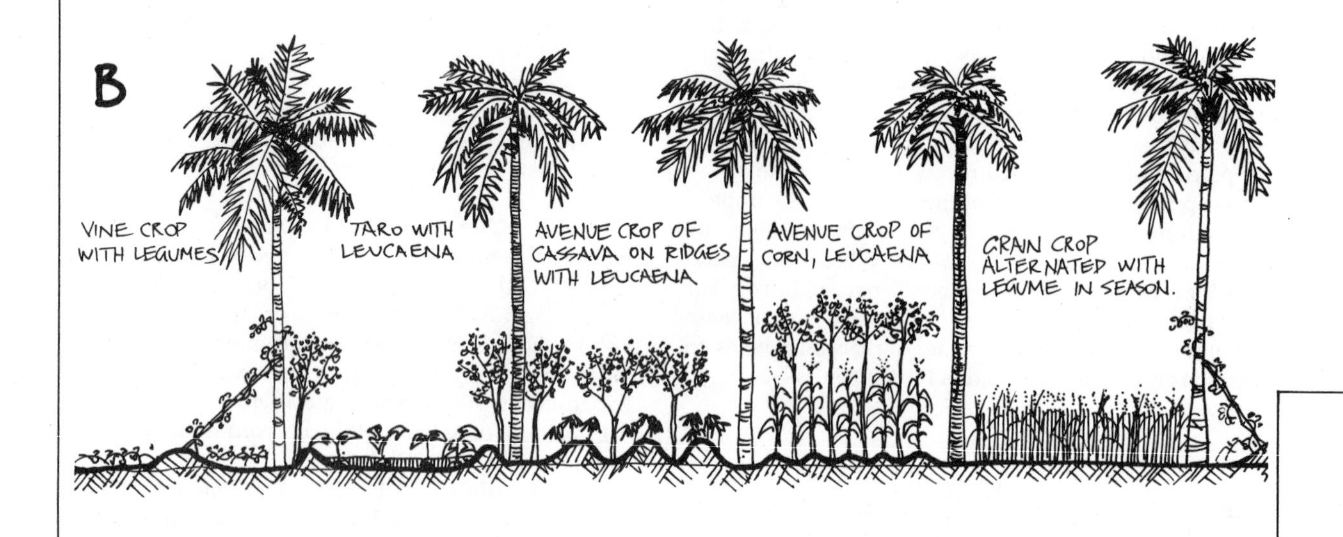

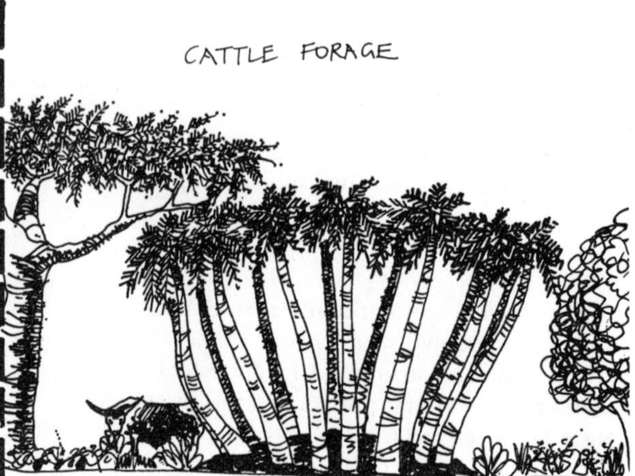

FIGURE 10.31
A–E STRUCTURAL VARIATION IN PALM POLYCULTURE.
A - Palms–fruit layout; B - Palm row crop; C - Palm–pig forage or cattle forage; D - grassland evolution; E - Fuelwood plus palms. All can be zoned around a village.

The former are ideal for village surrounds, especially in windy areas (where there are dozens of deaths each year from falling coconuts). Dwarf varieties (Philippine, Samoan) are easily accessible, have large nuts, good eating characeristics, and will not damage people or buildings if the nuts fall. Pest resistance and soil type must also influence cultivar selection.

In older plantation areas, selected and well–tested local species will be available. For areas with no plantation history, it is perhaps wise to build up a small arboretum of many varieties, and select a range of cultivars suited to the end–product aims. In every country, the cooperation of local agricultural authorities, and their assistance with varietal selection will be needed. Once a nursery is established (either as large containers or open bed planting, later as field plantings where rainfall permits), the site planning can go forward, but every plantation needs a mulched, shaded nursery, no matter how modest. Shade is most cheaply provided by light–foliaged legumes at wide spacing (e.g. *Acacia, Albizzia*).

Natural Variation

As almost all coconuts must be seed–grown, we can expect a variation in all crop characteristics, subject to later selection, culling, and new selections for site. Even if we grow from root tips in tissue culture, meristem and single–cell mutations are very high. In seed–grown crop, we might expect about one in twenty trees to show some very different characteristics, and of these perhaps one–third will be favourable for site, giving a limited set of new characteristics for selection.

Thus, it is unlikely that seed–grown *or* culture–grown palm plantation will demonstrate a very uniform genetic resource, and this will later lead us to "cull and select" options in management. This indicates a need for initial over–planting to allow for a 2–4% cull within the first 7 years (when we can make a fair estimate of vigour, nut production, bearing, and pest resistance) and another 2–4% cull in years 7–14, when the tree is mature. Final culling (14–60 years) should be in the nature of a replant and renewal process. Culling and replanting in palm crop can be a continuous process, so that plantation vigour (and overall design) is updated.

Species Suited to Co–processing

In special planation intended for (e.g.) ethanol or biogas fuel production, the same ferment and distillation equipment will serve a complex of crops that can form a "special use" polyculture.

In alcohol–oriented (fuel) palm crop, interplant of cane sugar, century plant (*Agave*), beet or sorghum sugar may add to the total sugar crop and suit the processing or distillation unit, while oil palms may be interplanted with mustards, sunflower, rapeseed, etc. to take advantage of oil press equipment and to increase honey production for bees, which themselves increase oilseed crop.

Similarly, wetlands suit many swamp palms (*Nypa*,

Maurantia), taro, rice, and *Azolla* fern or blue–green algae complexes, where the fern acts as nitrogenous mulch and the palms as deep nutrient pumps for the padi crops.

Thus, special site conditions, investment in processing equipment, or special end–use may dictate special plant assemblies in the site mosaic.

PATTERNING OF PALM POLYCULTURES

Pattern and Water Run–off

Many sites on clays and clay–loam soils benefit from earth–sculpturing for run–off absorption. On extensive sites, and on clays over limestone or dolomite, absorption swales may be the only practical broadscale irrigation method. In fact, any levelling or subterracing of land helps water infiltration. Palms and

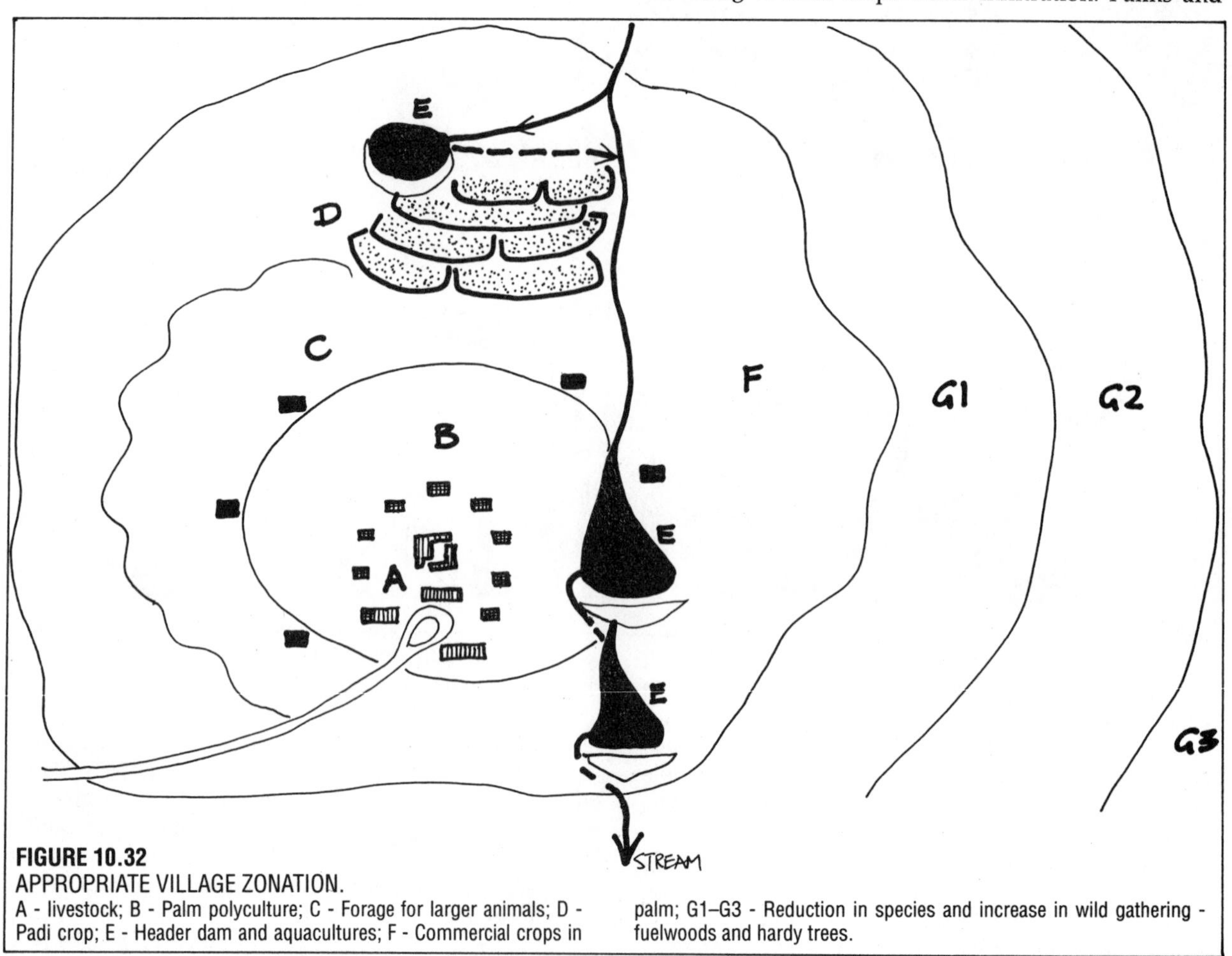

FIGURE 10.32
APPROPRIATE VILLAGE ZONATION.
A - livestock; B - Palm polyculture; C - Forage for larger animals; D - Padi crop; E - Header dam and aquacultures; F - Commercial crops in palm; G1–G3 - Reduction in species and increase in wild gathering - fuelwoods and hardy trees.

FIGURE 10.33
LAND PATTERNING FOR MAXIMUM WATER ABSORPTION.
Cross–slope hollows to hold mulch and runoff should precede palm and tree planting; broad swales also provide access at harvest.

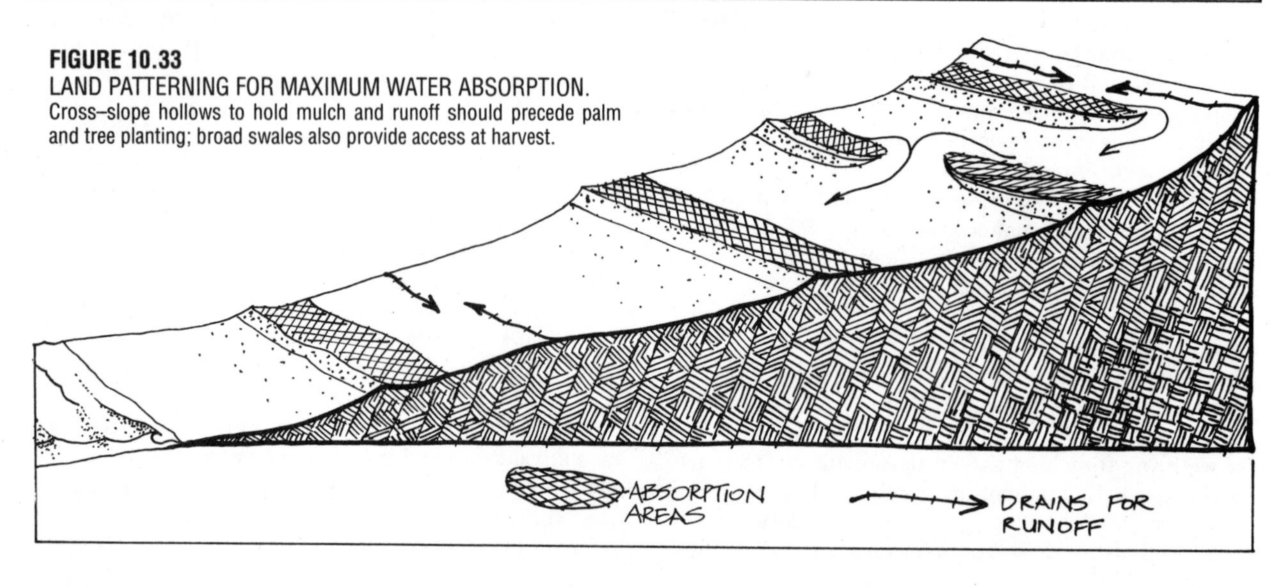

trees appreciate ground water reserves.

A hillside patterned as per **Figure 10.33** suits clump plantings of palms. Swales are illustrated and exemplified in Chapter 9.

Planting Patterns of Palms (Clumps vs Grids)

Without altering too much the appearance, spacing, and amount of coconuts, **Figures 10.34.A–C** illustrate some of the possible plantation layouts. While **A** and **B** are "normal", **C** arises from several independent observations I have made on densely–planted coconut

Ten to twelve coconuts planted in a circle, and each only a few feet apart, do in fact quickly adopt a divergent growth habit something like that in **Figure 10.35.**

Not only do nut counts compare favourably with trees planted on a square grid pattern, and nuts drop cleanly to the ground, but a third (probably more important) factor emerges, to do with mulching. Coconuts in plantation mulched with their own fronds and husks show better growth and bearing, but in normal plantation, husks are left at one tree in 10–30, because

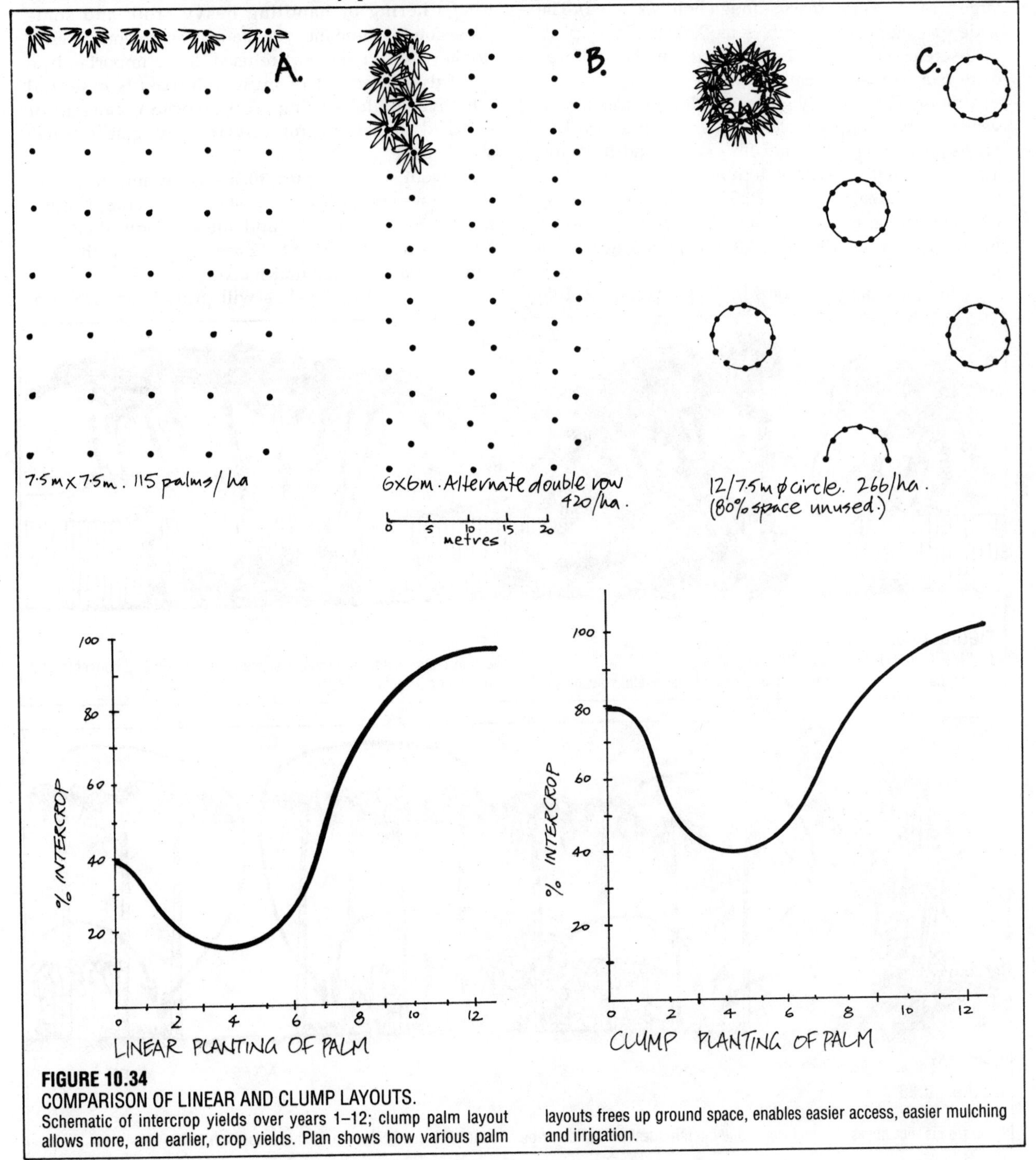

FIGURE 10.34
COMPARISON OF LINEAR AND CLUMP LAYOUTS.
Schematic of intercrop yields over years 1–12; clump palm layout allows more, and earlier, crop yields. Plan shows how various palm layouts frees up ground space, enables easier access, easier mulching and irrigation.

the labour of first gathering and then distributing the mulch is too great. However, with the circle clumps, it is easy to both gather and husk the coconuts in one place, and thus mulch the base of *all* trees, conserving water and returning nutrients to every tree. A little care in turning husks face–down prevents mosquito breeding in these mulch heaps.

Any other nutrient (manure, blood and bone) is equally easily applied to clumps. Clumps also form more suitable trellis for vanilla, black pepper, and other vine crop, are very economical for watering, and leave a large area of ground free (although lightly shaded). The wide spacing of circles enables replanting to take place in discrete sets of 10–12 palms without gross linear disturbance to the system as a whole.

Although I originally saw such clumping as a convenient way to apply mulch, it later became clear that broad areas of clear ground for grazing and intercrop are also available. Such patterning frees up to 60% of the ground area, as against 30% for linear planting. Clump planting is ideal for run–off harvesting of water in circular swales (**Figure 10.33**) or in coconut–circle pits.

As for the spacing of palms in lines, a very good rule is to space at twice the frond length for that site plus two feet. This allows full crown development without abrasion damage to the fronds from wind–sway. Intercrop spacing is as usual: cacao 2 x 3 m, banana 3 x 3 m, cassava 1 x 1 m, velvet beans 0.5 x 0.5 m, maize 1 x 0.5 m in rows, citrus 9 x 9 m and so on (local agricultural people can advise). Coconuts on new sites are normally 6 x 6 m.

Access and Mulch Provision

Sensible roading or grassed access ways are necessary for gathering or handling heavy crop, and some provision for these must be made even where horses or donkeys with panniers are used. More importantly, a careful assessment of mulch sources is essential wherever mulch–loving crop (avocado, banana) or mulched short–term crop (dryland taro, ginger, yams) are planned.

A layout such as **Figure 10.36** ensures mulch sources for the system itself and for short–term crop. Natural fall from palm fronds, and husk or nut shell will line–mulch about one in 8–12 rows of palms with about 2 m wide x 0.5 m high mulch beds.

In clumps, 10–12 palms will provide about 0.5 m

FIGURE 10.35
PALMS PLANTED IN CLUMPS.
Palms, bananas, macadamia nuts and many other useful species yield well in clump plantings, freeing space for intercrop and concentrating mulch and irrigation needs.

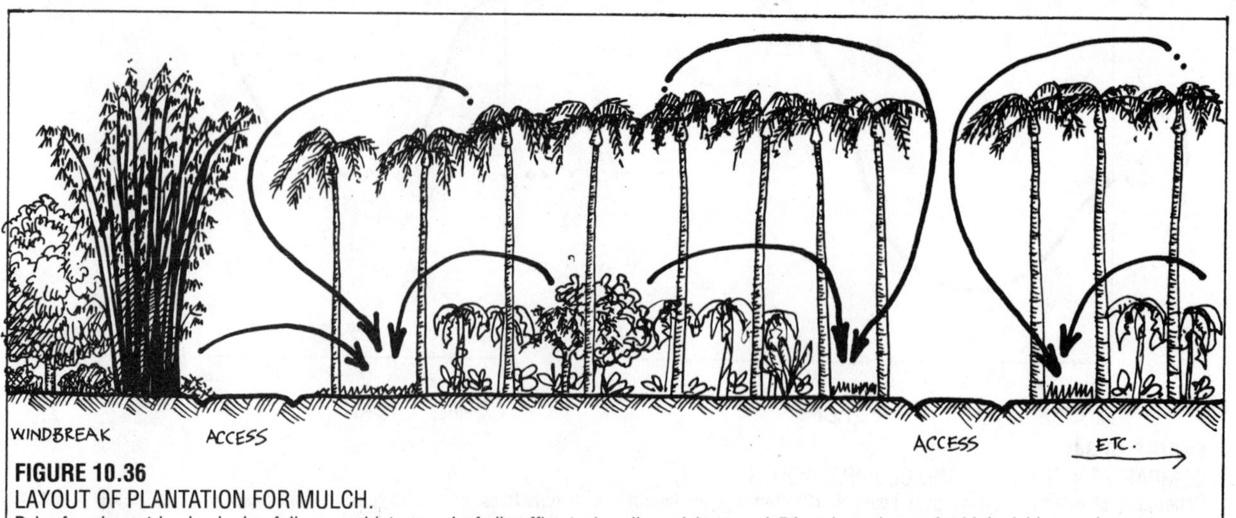

FIGURE 10.36
LAYOUT OF PLANTATION FOR MULCH.
Palm fronds, nut husks, hedge foliage, and intercrop leaf all suffice to heavily mulch every 4-5th strip each year for high yield annual crop.

deep of mulch for the inner circle of mulch. This is easier to gather and keep in place in windy areas. The addition of bananas, especially with avocado, has become standard in many planations, as the banana plants at harvest (with root mass) provide about 25 t/ha of organic matter, a key resource for a healthy fruit and palm crop (Penn, J., *New Scientist* 20 May '85). Small tree legumes (*Cassia, Calliandra, Leucaena*) also help. Bananas in legume crop may be regarded as "pioneer" mulch in grassland reclamation.

The layout in **Figure 10.36**:

- Reduces the labour of harvest by providing regular access.
- Provides sources of mulch for short–term crop.
- Enables mulch accumulation by long–term crop as interplant.

Earth Shaping for Intercrop

Earth shaping is worthwhile for several reasons, not only to assist water infiltration and run–off, but to give a free root run, to retain mulch in wind, to effect better drainage in over–wet areas, and to provide microclimate benefits with respect to wind shelter and ground warmth.

Briefly, earth MOUNDING for root crop and cucurbits is beneficial in humid tropics, and earth TRENCHING is best in dry tropics. Earthworking is discussed in Chapter 9, but some relevant data is given here.

RIDGES. Ridges of 0.5 x 1 m increase yields in cassava, sweet potato, potato, and yam crop. Mulch and green crop can be grown between the ridges. Pineapple and ginger also prefer ridges in wet areas. In **Figure 10.37**, *Leucaena* intercrop for mulch is on mounds, while maize and green mulch (beans) occupy hollows. Ridges permit deep mulching for low crop such as pineapple, the mulch being applied between ridges.

MOUNDS and volcano–shaped mounds with hollow centres are good cucurbit sites if enriched with manures. A stone or two helps heat the earth to germination temperature for cucurbit and melon crop.

FURROWS assist mulch retention for ginger and pineapple in dry areas. They are best covered with mulch, and will carry subsurface water seepage lines.

BASINS, even shallow basins, aid dryland taro and banana, or patches of Chinese water chestnut. Soil is more easily saturated, and deep mulch assists this process.

BOXES of palm trunks are ideal mulch–holders for yams, banana, and vanilla orchid, vines generally, and borders of beds in home gardens. Such log boxes can be 1–3 logs high, and greatly assist weeding if mulch–filled.

Yields Over Time

Plantation can be cropped with short–term grains for a season or two, but by years 2–4, the palm fronds (of linear plantings, not so much of clumps) cause mech-

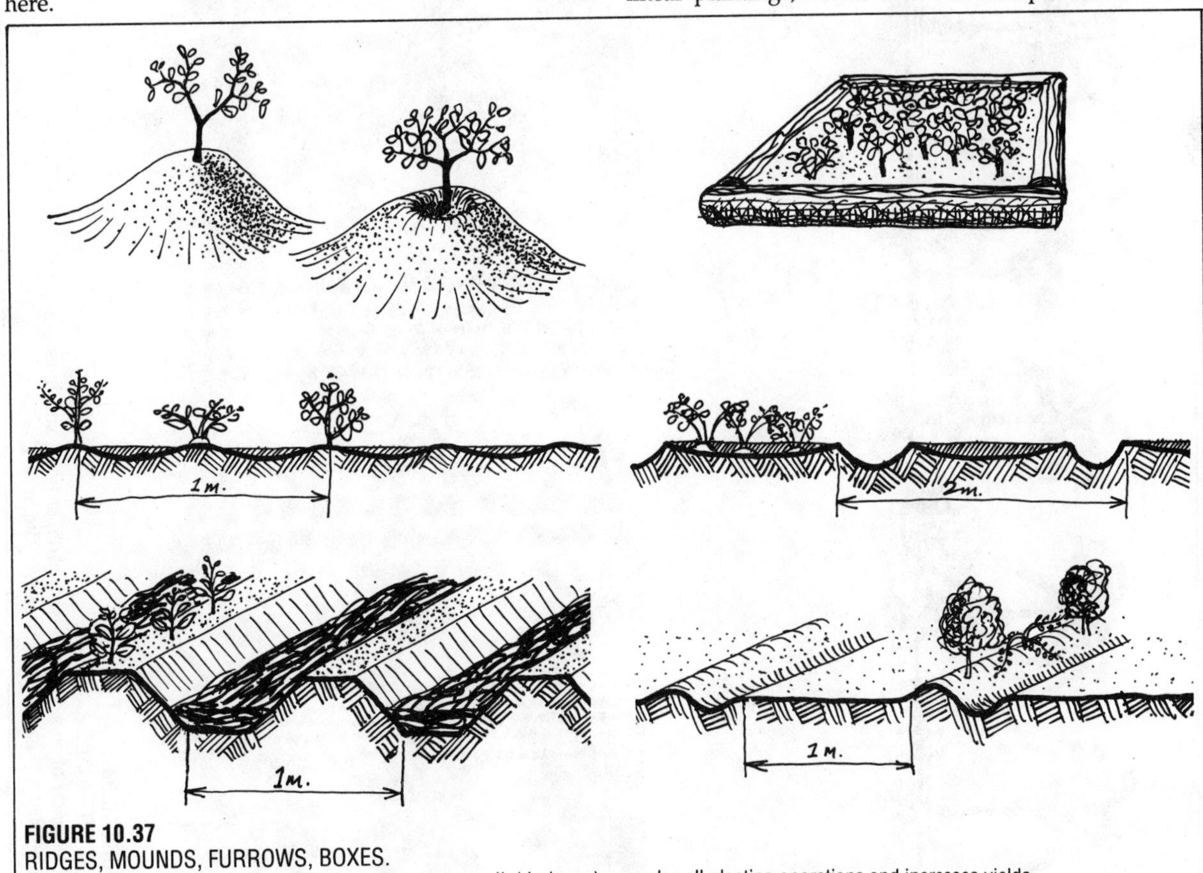

FIGURE 10.37
RIDGES, MOUNDS, FURROWS, BOXES.
Appropriate earth–shaping specific to crop type or soil (drainage) precedes all planting operations and increases yields.

FIGURE 10.38
PALM INTERCROP.
A wide variety of crop, and accessory earth, or trellis, structures can be integrated with palms after years 4–7. [Jörg Schultz].

anical damage and obscure the ground. After years 4–6, a stem forms, and from 6–14 years, complex perennial intercrop (not short–term grains) can be placed in linear systems. In clump systems, the early ground effect is less marked.

ECONOMICS

Nair (1975) gives convincing economic analyses for coconut, showing a 50% increase in yield for irrigation alone and a trebling of the yield for complex intercrop of two or more species, effectively doubling the cash return to the grower on the same area. Costs of irrigation and intercrop (plant or animal) never exceed returns if care is taken to select beneficial plant and animal species for available soil, water supply, and climate. Often, the cheapest irrigation system is to pattern the ground to hold wet–season run–off for tree crop use in dry seasons.

On Nair's anlaysis, where 1 unit = 4 rupees, the net income from coconut was as **Table 10.2**. Adding three species and increasing net yield by 3–9 times increases costs by 3.1 times. This is a clear implication for smallholders that much less area, polycultured, would give as much return (3 to 8 times) for *far less expense* (as expense is also a function of expanded area under crop). Irrigation of any sort is obviously a key factor. There would be a point, however, where more species added, even if very carefully selected, would push labour, harvest, and control costs past sensible limits, as per the schematic in **Figure 10.40**. So it is also clear that a complex polyculture must be managed by many more people if expanded to a wider scale.

RE–WORKING OLD PLANTATIONS

People who inherit or buy old stands of coconut or other palm crop need to undertake clearing and

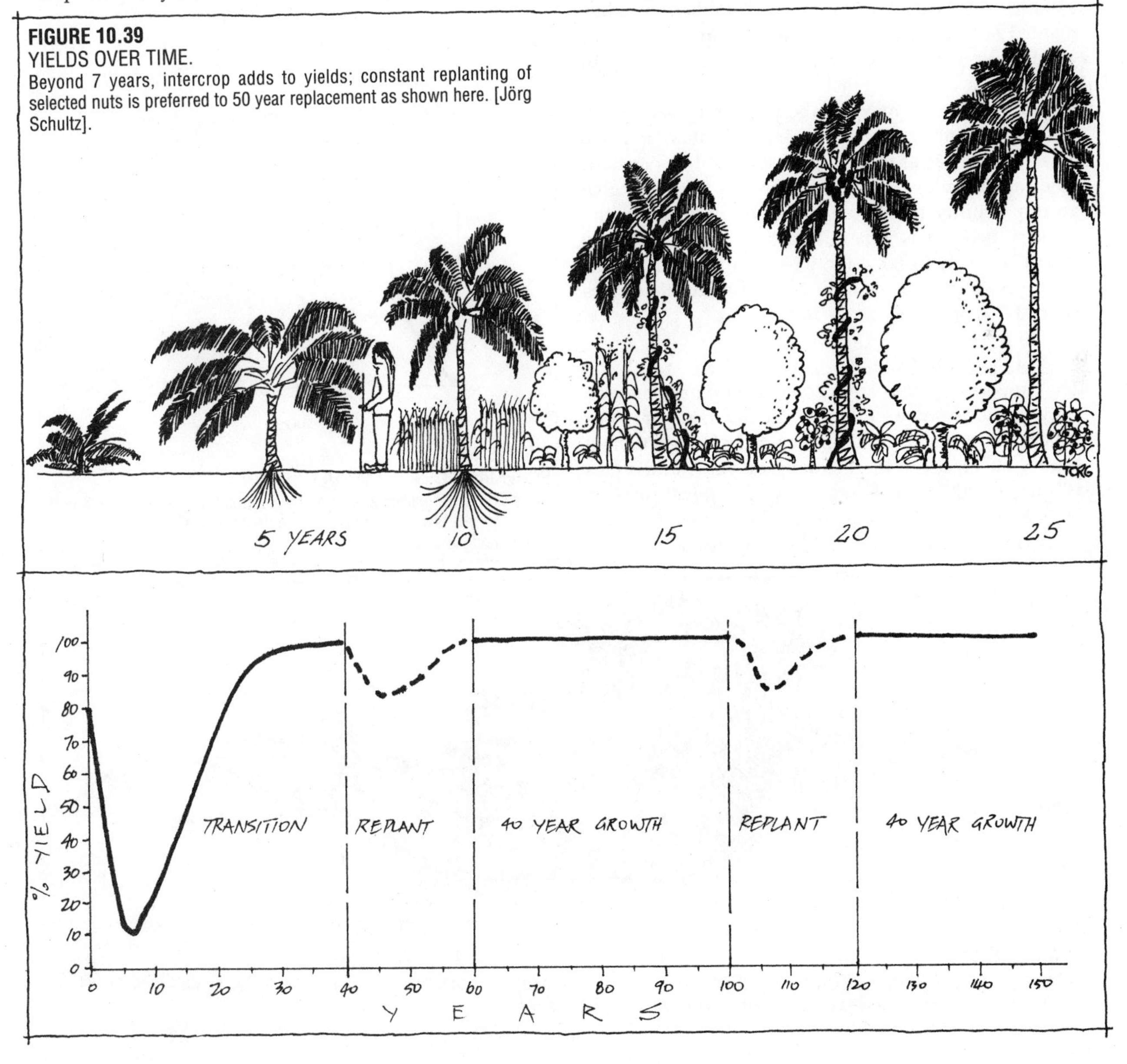

FIGURE 10.39
YIELDS OVER TIME.
Beyond 7 years, intercrop adds to yields; constant replanting of selected nuts is preferred to 50 year replacement as shown here. [Jörg Schultz].

replanting programmes for renewal if the stands are 60+ years of age. This is an ideal time to re–assess the potential for intercrop, to assess local processing potential, and to use the trunks, fronds, sugars, and palm heart products of the over–mature palms for mulch, food, and structural material. If hurricanes have stripped the old crop, it is also timely to assess the placement of windbreak for future plantation, and to place *Casuarina, Acacia, Albizzia*, bamboo, or tough *Prosopis* species to afford greater shelter and to fix nutrient in the crop, or to provide forage for grazers such as pigs, cattle, tortoise, or game birds (turkey, geese).

The clearing of old trees should be carefully planned to give a maximum return and to correct placements in older plantations. As it may take 8–15 years to rework a neglected, old plantation, the process can be staged and tuned as trial systems. In many areas, the old palms are tapped for sugar before removal, and the palm hearts eaten or sold as "millionaire's salad" (although palms planted specifically as young heart crop have a very good yield, can be close–planted, and are a crop in themselves).

Good managers may be about the business of re-placing, replanting, or re–grouping plantation at a rate of 4% or so per annum, giving a slow but constant renewal and culling as needed. The new trees also give an opportunity for field–testing selections from the nursery beds. It is generally agreed that coconut is over–mature in 40–80 years, when nut yield falls from optimum 45 per tree or so to 15 or less. Excellent trees bear 60–100 nuts per year.

Uses of Palm Trunks

The trunks of coconut are a good resource; not only do they provide an excellent building material, but (stacked in open box fashion) they make baskets to hold mulch on land for the growing of yams and vines (**Figure 10.42**). In the shallows of tropical seas or lagoons, they form a frame for coral to cement together, sheltering crayfish, crabs, and fish. The trunks hold silt and sand in reclaiming new lagoon areas, or in creating stable planting ledges in "hurricane garden" hollows cut into coral sands, on the sides of gleyed or plastic–lined surface ponds on islands, or as an aid in retaining bank stability and plant establishment on slopes.

Palm trunks can also be used to create planting benches on pits on coral islands. Pit base: taro, mint, parsley, kangkong. Sides: cassava, papaya, yam, banana. Spoil: sweet potato (mulched). **Figure 10.41**.

THE EFFECTS OF PLANTATION MONOCULTURE

Plantation crop in the tropics may bring with it all the evils of monoculture, and especially those of poisonous sprays, which not only affect the workers themselves but infect all streams and eventually town water. These sprays drift over adjacent properties, making livestock unsaleable, and poison the landscape generally.

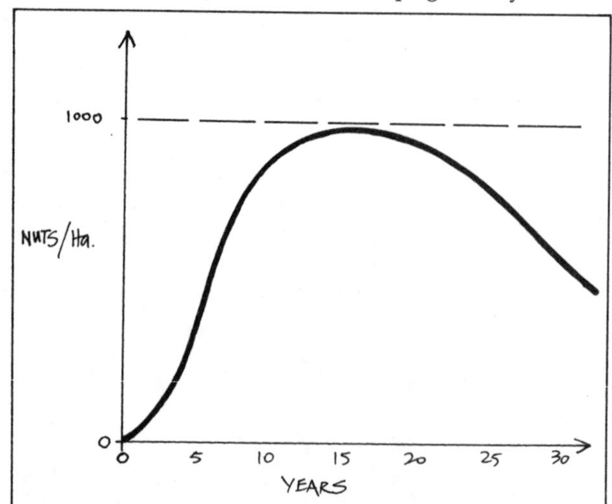

FIGURE 10.40
SCHEMATIC OF A PALM POLYCULTURE.
Nut yields; slow decline after year 15 requires that selected planting of 1/20th of area is a continuous management process which yields logs, mulch, and variable age across stand.

FIGURE 10.41
PLANTING BENCHES IN PITS ON CORAL ISLANDS.
Grow–pits 1–2 m deep and 8–12 m wide are ideal in coral sands; lower levels are close to permanent water table. Mulch reduces pH to 7–7.5.

Plantations almost invariably erode the landscape, pollute rivers, estuaries, and corals with silt and sprays, and exhaust soils. They centralise power and corrupt local politics, often funding repressive politicians, Their products are of low nutrition, and contain high levels of residual chemicals. Perhaps worst of all, plantations almost always displace local self–reliant crop, and replace it with "company store" dependency.

Cures are available. Firstly, plantations can be locally managed by worker cooperatives, as in some Sri Lankan tea plots. Here, at least, the workers have a say in and profit from their labour. This does not necessarily alter many of the ecological factors, however.

Secondly, plantations need not exist, as the same area of crop can be produced by smallholders, and processed centrally. This is a further improvement, as quality can be rewarded and good ecology instituted. Such an approach needs shared research, market, and processing systems. Thirdly, the plantation itself can adopt two reformatory practices:

- Good ecological management through the use of polyculture systems, soil building, organic fertilisers and biological control of pests.
- The "commonwork" approach, where workers lease secondary or tertiary crop in, around, and under the main crop, or lease rights to the processing of the main crop residues.

Modern analyses (computer modelling of intercrop) show that intercrop and polyculture raise employment and income, and plantation itself should be designed specifically to allow intercropping throughout the life of the palm.

It is quite feasible for people to extend specific successful crop mixes from their home gardens to more extensive situations, thus giving a surplus for trade. Money is then gained as a result of extending a stable and tested polyculture rather than by imposing a monoculture on an unsympathetic and fragile landscape. No landscape or soil can maintain long–continued monoculture production of crop, as even tree crop is susceptible to disease in this situation.

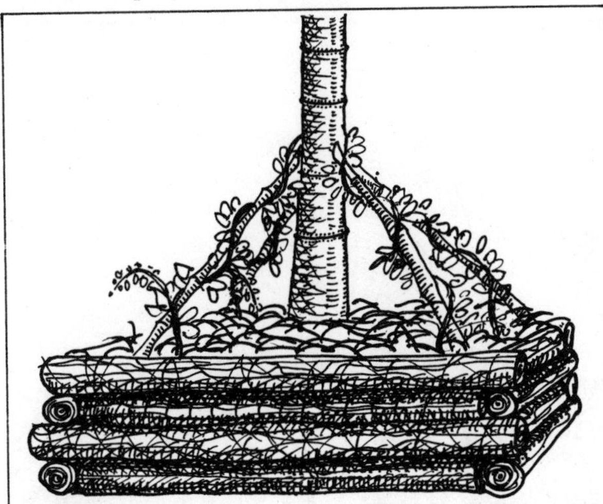

FIGURE 10.42
MULCH BOXES MADE OF PALM TRUNKS.
Palm– or log–surrounded areas filled with mulch are ideal for vanilla, yams, beans, cucurbits in sandy or alkaline soils; also on a larger scale for potato, tomato.

TABLE 10.2
ECONOMICS OF PALM AND INTERCROP (AFTER NAIR).

PLANTINGS	NET (Units/ha)	EXPENSES (Units/ha)
Coconut only (under rainfall)	1,000	600
Coconut only (irrigated)	1,512	817
Coconut and cacao	3,122	1,300
Coconut, cacao, black pepper,and pineapple	3,882	1,880

The approach of extending small and successful trials is basic to success. Broadscale trials have unstable effects (social and ecological) from the beginning, and success is rarely achieved as a result of such an approach. However, it must also be recognised that complex small systems may work well simply because they are close by, and many such systems cannot be scaled up to large acreages as a totality. Size itself creates new factors of cost, control, market, and labour requirements.

Plantation and monocrop have the undeniable advantages of ease of harvest and predictability, neither of which are necessarily the best criteria for human–centred benefits. Malnutrition and low socio–economic status are common factors in the human populations of the wet tropics, and criteria such as full nutrition and enhanced self–reliance are where we should be concentrating for the tropics.

10.11
PIONEERING

If we are going to pioneer in the tropics, the only ethical conditions in which we would contemplate such a process is to rehabilitate:

1. Grasslands developed by burning/grazing sequences and monsoon grasslands.
2. Semi–forested clearings and old monoculture plantations of, e.g. sugarcane, banana, pine, eucalypt, pineapple. (We will suppose some "weed" invasion by *Lantana*, tobacco bush, vines, or shrubs.)
3. Logged and burnt forest with reject logs, branches, stumps, and weedy regrowth.

These are some typical conditions. The end results we would envisage would range from:

- Terrace culture and water absorption systems.
- Extensive aquaculture or substantial dams.
- Polycultural forests.

• Managed forestry or rehabilitative forestry for perpetual yields.

Or, more probably, we would plan for all of these in appropriate combinations for site.

TROPICAL GRASSLANDS

The management of deforested grassland areas is the main problem of the wet–dry tropics: soil erosion, rank grasses in the wet, and inflammable or low–nutrition feed in drought result from burning and over–grazing. Once deforested, the pastures are open to summer winds, and the nutrient cycle of trees/grass/ browsing is broken. Fire, often out of control, only accelerates the process. Although there are very few trees which can survive in tropical grasslands, it is essential to re-establish tree legumes.

Some vigorous grassland cover crop legumes (*Desmodium, Suratro*) will help reduce the grasses and eventually lay down a mulch. Under trees, a short-stemmed *Desmodium* will defeat the grasses, but it is then essential to be able to supply dry–season water, as the legume also competes with the young trees for moisture. Some fast–growing leguminous trees (*Albizzia, Acacia, Inga, Leucaena*) will quickly establish, and can be grown in the shelter of banna grass or elephant grass (*Pennisetum*). If these grow vigorously, they also provide green mulch.

Heavy cattle browsing is a major cause of pasture deterioration and soil loss. Their extensive grazing is probably the most common destructive use of tropical lands. The first step is therefore to relieve the land of the weight of too many cattle. No nation, nor the globe, can support destructive grazing agriculture on the agribusiness/cowboy/pyromaniac model so general in tropical countries, in America, and wherever "cheap" beef is produced. The long term cost makes such systems uneconomic in any terms.

A positive approach is to re–establish either a multi–species system ecology (trees and a variety of browsers), or to intensify cattle rearing. Cliff Adam, Chief Research Officer at Grand Anse, Mahe, in the Seychelles has grown *Pennisetum atropurpureum* (7 parts) plus *leucaena leucocephala*—the low–mimosin type available in Australia—(1 part), and may add the Bocking strain of comfrey. This "pasture", cut and fed to cows, supports seven milk cows to the acre. All manure and washings from stable/dairy are returned to the irrigated field. Imported artificial manures have been reduced to one–tenth, and he hopes to further reduce this import by building soil. Meanwhile, in the same climate in Australia, one cow per square mile is enough to lay waste to the land.

A friend who bought a degraded cattle property north of the Daintree River (Queensland, Australia) gathers a load of coconut from the beaches, and (travelling the ridges of old fields just before the wet season) throws dozens of coconuts at intervals into the stream–lines and gullies. About 4% take root and grow into sheltered and pioneering palms. Not far south of there, another innovator rolls down the monsoon grasses as they begin to die off in the dry season, and broadcasts tall–stalk rye and field legumes (Fava, *Dolichos, Vigna*) into the thick resulting mulch. Enough moisture persists over the dry winter season to grow these crops; after harvest, the monsoon grasses regrow for next year's rye crop.

This clever use of seasons and growth is possible for the establishment of many species, some of which become permanent and grass–defeating pioneers for later evolutions. Consolidation of the area for regenerative forestry, however, proceeds more surely as a scattered set of pioneer tree and herbaceous nucleii; that is, the steady establishment of CLUMPED pioneer trees in open grassland. This is a "natural" process which duplicates the seeding of grasslands by fruit pigeons

FIGURE 10.43
PLANTING IN GRASSLANDS.
A dense (2 x 2 m) planting on "nucleii" of legumes, palms, shrubs, ground covers, and bulbs plus stone or stick mulch quickly shades out grasses and produces a closed canopy. [Jörg Schultz].

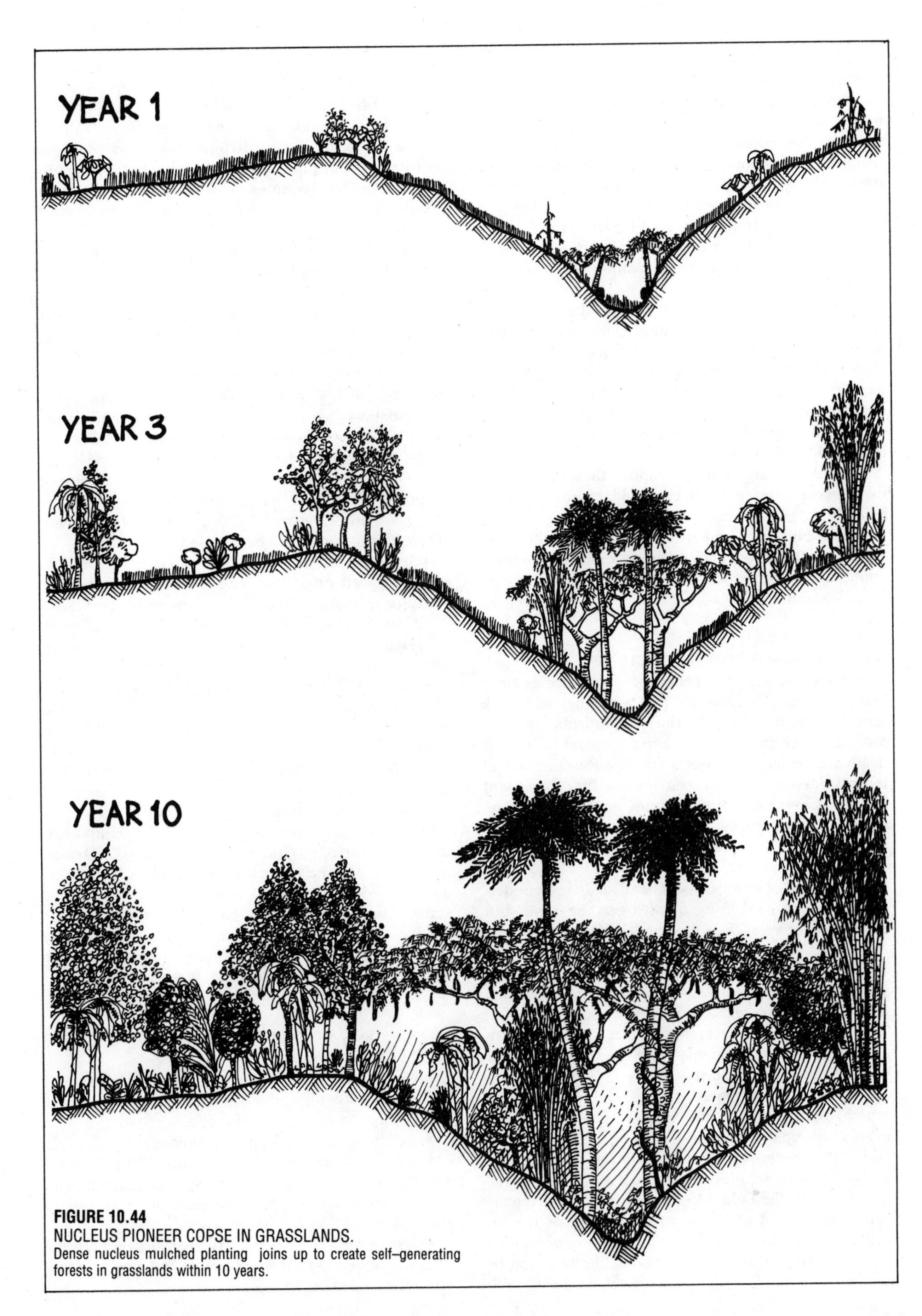

FIGURE 10.44
NUCLEUS PIONEER COPSE IN GRASSLANDS.
Dense nucleus mulched planting joins up to create self–generating forests in grasslands within 10 years.

and fructivorous birds.

In steep moist valleys, there is good reason to plant patches of bamboo and rattan palm for later "wild" harvest, large nut trees of heavy water demand (macadamia, coconut, pecan), wind–sensitive large fruits (avocado), and a selection of high–value timber trees (rosewood, teak, cedar, tropical conifers, balsa, mahogany). On more gentle and accessible slopes, a mixture of productive tree and palm crop can be planted in the shelter of pioneers, and on ridges a protective windbreak of hardy palms, *Casuarina*, and wind–fast legumes.

Excellent nucleii clumps in grassland are built up from a close–planted (1–2 m spacing) mixture of *Acacia mearnsii, A. melanoxylon, Inga, Gliricidia, Nicotiana, Casuarina, Vigna, Tagetes,* comfrey, a box or two of nasturtium (box, soil, and all), a few handfuls of fertiliser, and a visit to slash grass and tall weeds occasionally. Any stones, logs, cardboard, or old carpets help if laid in the clump.

Natural aids are stumps (plant *in* these) and large rocks—keep them central to clump. Plant around boxes, logs, or log piles (pocket soil and plant in the logs) or even old buildings and rock walls. It is within and around these pioneer nucleii that we can commence re-afforestation or productive tree crop, using our nucleii as mulch sources.

A GENERAL NOTE ON THE LEGUMES

Most legumes, and other genera of plants such as alder and *Casuarina* have mycelial root associates which fix atmospheric nitrogen. As these organisms, and the roots to which they attach, are in a constant process of death and replacement over a growing season, much of the nitrogen is also released for use by other plant species. Clover and tree legumes perform the same benefit for pastures. Such trees as the rain tree (*Samanea saman*) can preserve green grass below even in dry seasons.

The amount of nitrogen fixed has usually been under-estimated, at 75–100 kg of nitrogen/ha/year, but efficient legumes such as lucerne (alfalfa) may provide 250–500 kg of nitrogen/ha/year, and tree legumes such as *Albizzia* as much in quite poor sandy soils [Iseky, D, 1982, *Economic Botany 36(1)]*. Every part of such legumes as *Leucaena, Acacia, Albizia, Gliricidia* and *Tephrosia* may contain high nitrogen levels; one can actually smell the ammonia from the trees in rain or when the roots are crushed.

Thus, cut green material from such trees (green mulch) lightly turned into crop, water–mulched, or even as interplant, supplies much of the nitrogen for crops. It is necessary to make sure the trees are *inoculated* as seed with the correct root associates, in the nursery or in the field. Most agricultural departments can supply a list of strains of inoculants, or the inoculum itself. Many firms supply inoculum for legume and other species, or soil from nodulating trees can be washed in around newly planted trees, or mixed with potting soils.

The nitrogen is distributed around root zones as per **Figure 10.46**. Some shrubs and trees lay down about a 9–year supply, and if cut or ringbarked, the slow decay of the roots gives up nitrogen for 6 years or so. Nitrogen, if supplied artificially, quickly leaches in warm rains, so legumes, with their slow nitrogen release, *are of critical importance* in any tropical crop situation.

The management of leguminous tree crop must be carefully assessed for local conditions. Some considertions are:

- SPACING: With shrubs and small trees, 0.5 m apart is the best for foliage production, and a trimming height of from 0.5–1.5 m is recommended.
- SEASON: Only in frost–free tropics can we trim all year (4–5 cuts). Wherever cold is a seasonal problem, two months of growth must be allowed to harden the plant before winter, or it will weaken as the shoots are cold–killed. Trees in drought can be part–trimmed.
- FORAGE: There is some danger that young (coppice) shoots will have higher levels of metabolic poisons than 2–3 year old shoots, and if stock do not thrive, this factor should be assessed.
- SHELTER: Trees can be more widely spaced for root nitrogen, seed production (e.g. for poultry and bees), and for in–crop shelter, as flowering and seed production is better at 2–20 m spacing (depending on tree size). A full canopy may be needed to reduce or eliminate frost.
- REPLACEMENT: Although many trees will coppice for 4–30 years, any sign of loss of vigour should indicate the need to replant. Replant for small shrubs may be necessary every 2–3 years, while some shrubs and ground covers are annuals or become annuals in cold–season areas.

To assess total nitrogen yield, we must assess *soil nitrogen* from mycelia (say 200 kg/ha/year) and leaf and slash nitrogen/ha. In such crops as *Tephrosia*, yielding 135 t/ha in 4 cuts, leaf nitrogen should be about 20–30 kg/t, or 1,000–1,500 kg/ha/year, which is some factors higher than the root nitrogen yield. This is the whole rationale for avenue cropping and legume mulch. Phosphate and potash levels in green mulch are also satisfactory for crop production.

PIONEERING IN SECONDARY FOREST GROWTH AND LANTANA

Lantana is analagous to the rampancy of gorse and blackberry in cooler areas, and the essential process remains the same:

- Roll down, crush, or cut out contour strips.
- Plant advanced, vigorous mixes of *Acacia, Eucalyptus*, vines such as chayote, ground legumes, and local pioneer species. Manure each plant and mark small plants so they can be easily seen.
- Slash every few weeks in the wet season until the trees are above the *Lantana* canopy, and free any new natural tree seedling that comes up.

FIGURE 10.45
COMPONENTS OF THE TROPICAL FOREST TREE POLYCULTURE
Palms, lianas (vines), crown–bearers to the outside and stem–bearers inside clumps; fungi and shade species below.
A. Elevation.
[Jörg Schultz].

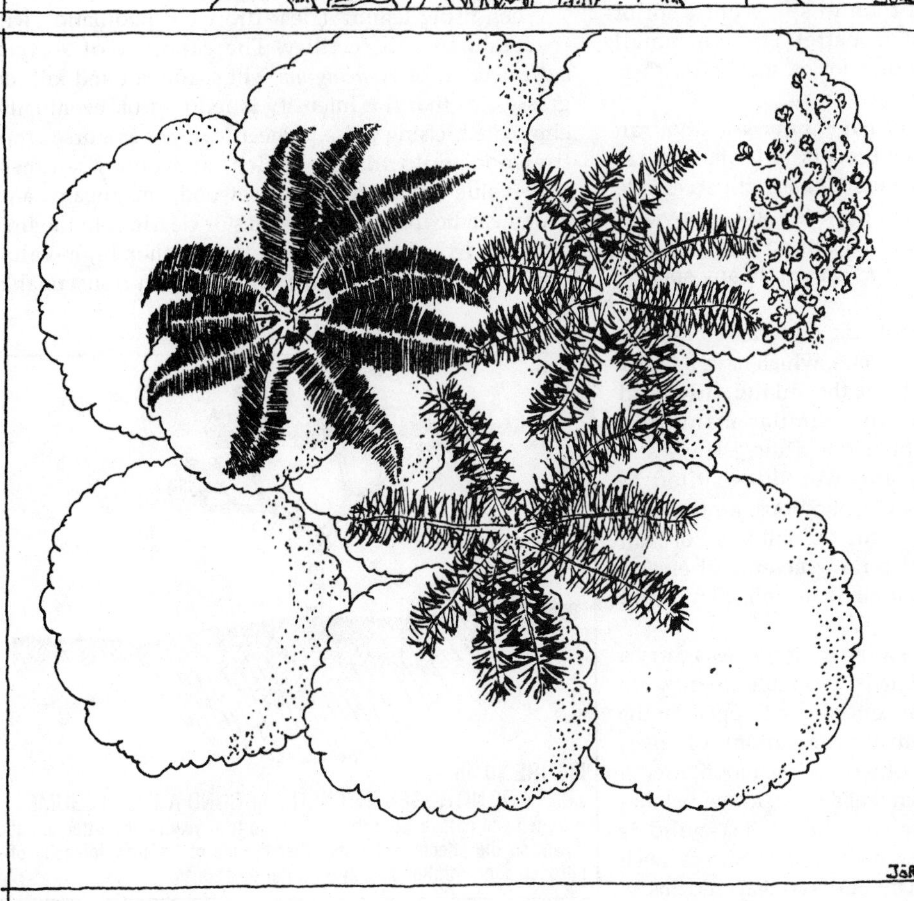

COMPONENTS OF THE TROPICAL FOREST TREE POLYCULTURE
B. Plan.
[Jörg Schultz].

• Do final slash at or about 18 months, and then cut or roll adjacent strips to extend the system.

• Use early pioneers (e.g. *Acacia*) as mulch for selected high–value species as previously planned.

The shading–out of *Lantana* takes from 2–6 years, and only remnant and weak shoots remain under productive forest.

In all extensive hill areas of gorse or *Lantana*, benches or roads cut on contour every 250–100 m is a great aid to regenerative forest processes and subsequent harvesting or slashing of new plantation. These can be kept mowed and cleared, and eventually stabilise as trees grow. Road borders can be of dense, evergreen, wide–crowned trees for track shading and stability. Such deep shade also keeps fences clear of grasses and weed crop.

PIONEERING ON DIFFICULT TERRAIN

On man–made and natural landslide or volcanic areas of the tropics, it is first necessary to pocket the area with soil–mulch mixtures (nut husks from coconut and macadamia are excellent to establish any pioneer species). Thereafter, species such as *Inga edulis, Leucaena leucocephela*, various *Acacias* (*A. mearnsii*), *Scalesia pedunculata, Prosopis pallida*, and like legumes (*Dolichos, Desmodium*) will prepare the area for palms, cacti, figs, and the more useful fruit trees, by providing shelter and mulch for subsequent plantings. It is better to plant small assemblies than to space out a lot of species on their own. It is better to plant small assemblies than to space out a lot of species on their own.

A heavy spiked roller crushes new *a'a* lava (an Hawaiian term for lava which is softish, not far removed from pumice); crushed lava is both accessible and easily rotted to soils. Soil pockets can be provided with trace elements (boron, manganese, zinc, copper, molybdenum) if not analysed as present in any specific location.

Near the sea and on islands, the night air condenses on the sea–facing side of the stones, which have a richer moss–algae–lichen flora than the inland side, and pockets of vegetation act in a similar manner to condense sea vapours for their use. Thus, islands and sea coasts will have dry and wet sides suited to different plant species. In such conditions, a ragged or spiky forest canopy, where palms and tall pines or fruit trees lift above the general canopy layer, will ensure more condensation from sea air than will a level and relatively closed canopy.

The saturated winds that sweep off tropic seas carry a heavy moisture load which is available as dew on grasslands, but is much more effectively trapped on the myriad leaf surfaces of an uneven canopy of trees, hence, the forested slopes of sea–facing mountains in tropic tradewind areas. Even small garden tree patches "rain" softly on clear nights when a sea–wind is blowing, and let down drips in a steady stream to swell dried–out leaves and to channel down leaf midribs to the fibrous trunks of palms and tree–ferns. Once plants are cleared, the effective precipitation falls, rivers cease to flow, and the land becomes truly dry.

SAVANNAH FORESTRY

Wherever overgrazing plus fire or cut–and–burn forestry has ruined native forests, in particular towards the wet–dry tropics, closed grassland species of fire–prone and tough grasslands develop, closing out the tree seedbed and preventing good management practices. Further burning or cultivation may result in a depauperate grassland of low stock carrying capacity over the dry period, and patches of bare and eroded soils, low in nutrient states and at times acidic (pH 4–4.5) may develop. Blady grass (*Imperata cylindrica*) and other tropical forage grasses are stubborn, tough, and almost inpenetrable barriers to gardeners and stock, although they provide good mulch.

Given rains of 60–150 cm, a set of rough pioneer legumes are available for the rehabilitation of exhausted sites (including mine spoils and road embankments). Providing enough seed can be obtained, direct seeding in scratch holes or chiselled strips will result in the fast establishment of some, or all, of the species listed below, to which can be added *Leucaena* and *Albizia* species. Tropical grasses, scythed or mown 5–8 times annually, make good mulch for trees and gardens.

When using legumes, leaf–drop and nodulation will re–establish soil fertility. The canopies of *Acacia auriculiformis* or *A. mangium* will shade out and kill the grasses, so that fire intensity is reduced or eventually eliminated. Using these pioneer legumes as nurse crop, firewood, pulp timber, mulch, and honey sources, high–value timber such as rosewood, mahogany, and ebony can be introduced in lines or clearings in the first crop, and the gradation made to either high–value forestry or to sensible strip cultivation on a sustainable basis.

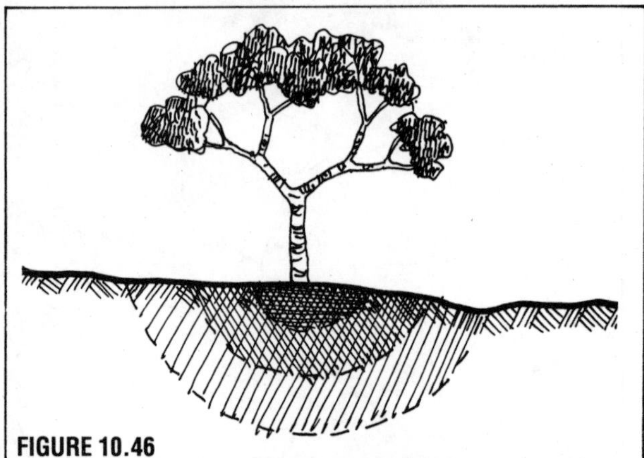

FIGURE 10.46
ZONES OF NITROGEN INTENSITY AROUND A TREE LEGUME.
Nitrogen will diffuse from the soil for up to 6 years after a tree is cut down, so the effects lasts long after the life of the tree. Intensity of nitrogen concentration falls in the outer root zones.

If *Leucaena, Samanea, Prosopis* and *Inga* are planted, a long-term forage system will evolve, providing replanting or rest periods are given for seedlings to re-establish. The only thing preventing or delaying savannah forestry is a lack of tree nurseries and seed sources of appropriate species, and this too presents an opportunity for a pioneering enterprise in the humid tropics. A very good selection of potential species can be found in the National Academy of Sciences publication referenced at the end of this chapter.

Species such as *Pterocarpus indicus* or *P. erinaceus* can be first seed-planted in a nursery stand, then coppiced for 2 m quickset planting in bore-holes in the field. Some species can be set out at 10 cm diameter, and make good timber trees.

PIONEER AND GRASS-EXCLUDING SPECIES, FIREWOODS

Acacia auriculiformis is an important pioneer for exhausted savannah and tropical soils, where over a very wide range of soils and sites it can defeat blady grass (*Imperata cylindrica*), restore fertility, provide firewood, and act as a tree nurse crop. It reduces fire, and provides good paper pulp. It coppices and self-seeds and is widely used in tropics as a shade and street tree. *A. mangium* has similar characteristics but is straight-stemmed and therefore better suited to forestry operations.

Sesbania grandiflora is a fast tropical pioneer, can be coppiced, and is a good forage tree, an excellent green manure in rice, and re-invigorates worn-out land. Exceptional nodulation. Grows to 10 m and provides good firewood. Wide soil tolerance, extensively used for eroded hill sites. Young leaves, pods, and flowers used for human food (36% crude protein). Seeds are 40% protein. Used as light shade crop, vine support. Good in crop. Frost and wind tender, life about 20 years. All food from this tree should be cooked. Exceptionally fast growing.

Calliandra collothyrsus. A stick wood coppicing species which defeats grasses and provides abundant firewood. Repairs exhausted soils and restores fertility.

Dalbergia sissoo is salt and frost tolerant, fast growing, and defeats grasses. It tolerates a wide range of soil types and can be quickset from large cuttings (India).

Enterolobium cyclocarpum is a durable timber tree with large pods, defeats grasses (Central America).

Mimosa scabrella of Brazil is a subtropical pioneer, provides good humus and a living fence.

Samanea saman (rain tree) is a very fast-growing large tree of the tropics and subtropics, with sugary pods. Grass grows well below. Wood is valuable, durable.

10.12
ANIMAL TRACTOR SYSTEMS

Following are two examples of animal tractor systems, either of which can be used to prepare soils and remove grasses or persistent weeds for evolution to garden and tree crop.

CHICKEN TRACTOR

Confined chicken flocks will remove all green ground cover and surface bulbils, depending on how many are confined on how big an area, thus killing out or consuming such plants as *Oxalis*, nut-grass, kikuyu, onion weed, and pasture species of *Convolvulus*.

Dano Gorsich, on a 0.5 ha farm on Moloka'i, Hawaii, has planned and executed a successful chicken tractor/garden system on a stony hillside site. The process is to fence 5–6 plots, and rotate a 40-chicken flock on these plots over a period of 18 months. As each fenced area is scratched bare, it is limed, raked, and sowed immediately to vegetable crop (typically *Brassica*, beans, peas, amaranth, cucurbits, radish, root crop).

The chickens are moved to Plot 2, and in about 6–8 weeks vegetables are in full production on Plot 1. As Dano also needs a cash crop, he has interplanted young papaya in Plot 1 amongst the vegetables. These grow strongly and succeed the vegetable layer, giving high shade, and (from the waste fruit), chicken forage in later rotations.

Thus the tractor system proceeds, with chickens

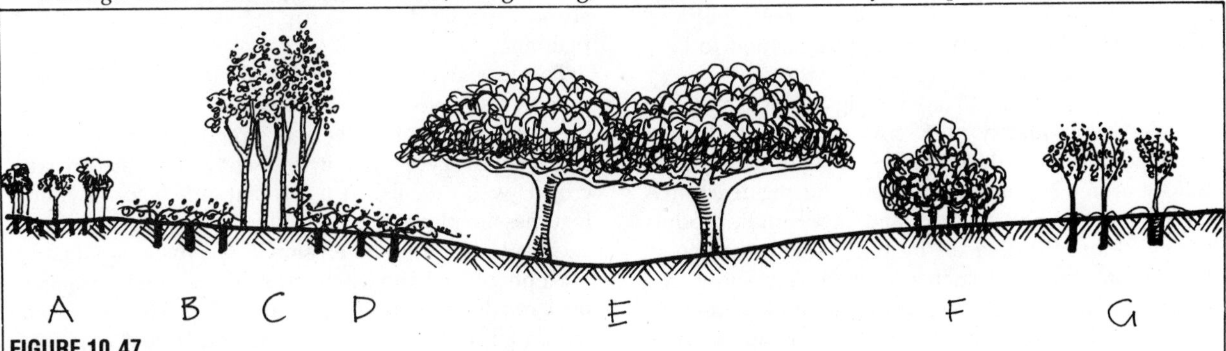

FIGURE 10.47
SAVANNAH FORESTRY.
A. High-value advanced seedlings planted in riplines. **B–D**. Chiselled area sown to Desmodium or lab-lab ground cover. **C**. Acacia strips sown in chiselled ground. **E**. Selected large figs, *Albizia, Inga*, set out in good soils around hollows. **F**. *Leucaena* and beancrop drilled in chiselled area. **G**. Quickset coppice forests of *Bauhinia, Pterocarpus*, some in deep holes.

pioneering the weeds, and vegetables and papaya succeeding them. In about 18 months to 2 years, a more perennial system succeeds the weed layer. Eggs, chickens, vegetables, and papaya are at modest commercial level, and both milk goats and chickens are let out onto the paths to eat greens when the pens are bare.

I have now seen numerous chicken tractors, all different, some with passionfruit fence/trellis crop, some just as vegetable gardens, and some for small fruit or herbaceous orchard. All are remarkable for lack of weeds and high production. In the more mature cycles, buckwheat, comfrey, millet, sunflower, and sorghum can be sown in the pens a few weeks before the chickens are returned, providing greens and grains.

Even rocky or rough country is so prepared for crop by chickens, the main cost being secure pen fencing. Strong fences also support vine crop, and a few larger legume trees provide high shade (*Tipuana tipu*, larger *Albizia*) for pens and crops.

Where chickens are to be the main crop, chicken forage plants replace vegetables and some fruit crop, and the system then provides all food. Near the house, a few small top–netted and secure rearing pens allow broody hens to replace the chickens culled. Normal weeds such as *Oxalis*, cleavers, dandelion, onion weed, nettle, and nut grass are excellent chicken fodders, as are any of the *Solanum* family (huckleberry, black nightshade, pepino, kangaroo apple, tomato, huskberry, Sodom apple, etc.)

PIG TRACTOR

The pig tractor follows the same technique but is more suited to 1–40 ha properties. Larger shrub–weeds (*Lantana*, gorse, blackberry) or deep–rooted weeds (*Convolvulus*, rhizomatous grasses, comfrey) call for a pig tractor. The density of pigs per pen should be at the proportion of 50/ha for full clearance of weeds. In practice, 0.5–2 ha plots are fenced, most economically using permanent electric fencing, which is much cheaper than chicken mesh fencing. Once each pen is bare (6–10 weeks) and rough–plowed by rooting pigs, it is easy to plant lucerne, comfrey sets, sunroot (Jerusalem artichoke), sweet potato, *Inga* trees, papaya, banana, and similar crops for pig forages, and to keep up this rotation until the pigs return to the pen.

On a large scale (20–40 ha), the pig tractor system can pioneer high–quality milk–cow pasture of chicory, dandelion, comfrey, dock, grasses, and clover, and cows follow along 2–3 months behind the pig tractor. A continous rotation is set up, and excess milk product (whey, skim milk) fed to the pigs as accessory food. Piglets ranging over such pasture rarely show iron anaemia deficiency, parasite cycles are broken, and the soil constantly improves in humus. Such large animals as pigs and cows need fenced tree strips, tree guards, and border hedgerow to supply tree forage crop.

Obviously, these intensive animal tractor systems can be a phase followed by tree crop, an accessory to tree crop, or a permanent feature of the mixed farm, or used seasonally to remove crop wastes and fallen fruits. Chickens, I feel, should be a permanent forage system in all mixed orchards.

10.13

GRASSLANDS AND RANGE MANAGEMENT

In view of the prevalence of livestock enterprises in the tropics, some guides to management are required for milk and beef or sheep production. The following management strategies can be implemented:

- THE ADDITON OF FORAGE SPECIES to grasslands. These can be grass legumes or trees; the latter providing foliage, food sugars, seed carbohydrates, or fruits.
- ENVIRONMENTAL CHANGES, particularly in terms of water storage and soil structure, irrigation, windbreak and shelter. Key fertiliser or trace elements can be added, and plant species can be maintained by slashing or light cultivation of pasture.
- CONSERVATION OF FORAGE by rotational or periodic rests from grazing, by using hays and silages, by supplying protein or urea supplements and molasses in drought, and by keeping stocking rates below the worst case conditions.
- STOCK MANAGEMENT, especially by well–planned buying and selling to keep numbers in tune with seasons and longer–term fluctuations, for example timing calving or buying in animals in spring or early summer, and selling them or dry cows in autumn to lessen winter feed demand. At the extreme, stock can be penned and fed harvested fodders. A sequence of species or a species polyculture can be planted to take best advantages of forages.

A mixture of legumes with a selected grass species plus some storable forage is ideal for the tropics. Most grazing systems can extend under palms, between large tree crops, or as a complex with belts of forage tree legumes yielding fodder, fruit, pods, or large seed for food concentrates. Each soil type, location, rainfall area, slope type, and main crop needs assessment and planning.

The leaf swards valued by graziers may also suit green–crop cover for orchardists where regular slashing for easy fruit harvest is practiced (or sporadic grazing). In these cases it is essential that the orchard crop is well established using manures before twining legumes are planted. Soils under slashed pastures are of excellent structure, and erosion is effectively stopped.

Elephant or banna grass (*Pennisetum purpureum*) is best on deep alluvial or coarse flats above 110 cm (alluvium) or 90 cm (coastal) rainfall. It reaches 2.5–4.5 m high, but can be grazed to 1 m or cut to 15 cm for forage and mulch. It needs a vigorous legume, e.g. *Leucaena* interplant, or forage legumes such as *Calopogon*, *Centrosema*, *Glycine* (in high rainfall tropics).

If a cool season is expected, autumn cutting should be later so that cool season regrowth is obtained. Banna grass can be set out as windbreak by burying hard slim pieces of 4–5 nodes horizontally at 8–10 cm depth in summer. Furrows should be manured and kept free of weeds until the stems shoot strongly. Permanent plots can be established, well–suited to feeding selected stall–fed dairy cattle. Accessory plots of bean trees, coconut, banana, etc. for diet variation, and palms for bedding, are ideal. Banna grass can carry 7 milk cows/ha if cut and hand–fed with *Leucaena* and sugar pods.

TROPICAL PASTURE GRASS SPECIES

Guinea grass: (*Panicum maximum*) This is a bunch grass for warm areas of more than 90–300 cm rain. It is drought resistant but yields best in humid areas. Frost–sensitive. Shade tolerant, and suits thin–crowned tree crop (often yields well under trees). Valuable in that growth is maintained in cloudy summer–autumn regimes. Grazed down to 15–20 cm. Combines well with twining legumes which climb on stiff stems. Suits rotational grazing (12–18 fields), interspersed sugar–pod trees and tree fruit forages. Drilled to 6 mm to establish. A first choice for sub–tropic and tropic pastures.

Kikuyu grass (*Pennisetum clandestinum*). Cold tolerant and grown from cool to tropical areas. Valuable for cooler uplands, thinner soils, and for good autumn growth if nitrogen–fixing trees (*Acacia, Leucaena, Prosopis, Albizia*) are established. Prefers light soils, red loams, and can be sown as runners or seed. Excellent for water spillways and erosion control. Few legumes tolerate the tight sward, so that trees for nitrogen are essential. *Desmodium, Glycine,* and white clover sod–drilled in autumn–cut areas can be tried.

Makarikari grass (*Panicum coloratum*). Bunch and spreading types for 40–90 cm rainfall. Tolerates alluvial fans, flood plains, black clay soils, red earths, and even semi–caked salted soils. Needs a year after seeding to establish, so suits rotational systems. Lucerne interplant can succeed in irrigated areas. Drilled at 1.3 cm, 2–3 kg/ha or planted from rooted cuttings. Valuable for winter–green feed, drought resistance. Suits long rotation grazing in open savannah of *Acacia*, sugar–pod trees. On black clay soils, purple pigeon grass (*Setaria porphyrantha*) may germinate better than makarikari.

Para grass (*Brachiaria mutica*). For warm areas of low frost intensity, valuable for swampy soils and at soaks, dams, waterholes. Provides good soil structure due to fibrous shallow roots. Can be grown with the forage legume phasey bean, greenleaf *Desmodium, Centrosoma,* puero. Ideal as a fenced–out reserve food for drought, to finish off animals for sale in poor seasons. Planted from cuttings at 2 x 2 m or seeded if seed is available. Do not plant where clogging of channels can be a problem or where other crops are to be grown. Can reach 4 m in one summer!

Sorghums (*Sorghum almum*), silk sorghum, and Sudan grass (*S. halapense*) are annual, biennial, or persistent from seed, and are of most use as broadcast–sown pioneers in slashed mulch at 50–90 cm rainfall. They can be used as pioneers with the perennials, as mulch in orchard strips, as emergency dry–season fodder, and as a garden mulch source. Easily grazed out, the sorghums provide birdseed, forage, and help control weeds. They are of particular use in early establishment and can be surface–sown.

Establishing perennial grass swards on weedy or eroded areas is a one to three year process. The best way to proceed is:

- Choose a land–forming system such as swaling, interception banks, or pitting. Try to establish some dams for irrigation above good soil types.
- Sow a pioneer grass such as molasses grass, *Sorghum almum*, or silk sorghum mixed with sawdust into slashed weeds, or drill selected grass and legumes after slashing.
- Burn molasses grass, or drill selected perennials and broadcast *Sorghum almum*.
- Concurrently with earth–forming, plant a mixture of leguminous trees along swales, through the area to be grassed at 30–100 metre strip spacing. Allow 2–3 years to grow with light grazing to year 3.
- Commence managed rotational grazing, and drill or broadcast forage legumes into established grasses. About 15–18 fields are necessary for rotation. On irrigated areas, some strip grazing is possible (use electric fences).

On rocky knolls, leguminous tree pioneers followed by kikuyu sward may succeed. Early furrows of banna grass provide erosion and wind control (at 30 m spacing) until tree legumes establish. At every stage, soil analysis and minimal mineral fertiliser amendments may be necessary, and with intensive grazing, sulphur and potash dressings are desirable.

TROPICAL FORAGES AND GREEN CROP

Desmanthus virgatus is a shrub to 3 m resembling *Leucaena* and tolerant of heavy cutting and browsing in the savannah tropics. It is vigorous and seeds are prolific, thus should be on range, not in field crop (7–70 t/ha/year).

Desmodium discolor is a browse shrub to 3 m, yields some 30 t/ha/year green fodder and is sown prior to rain as strips in rangeland. Also compatible with maize.

D. distortum. Perennial to 2 m. Good on acid soils (2–7 t/ha/year).

D. gyroides. Shrub to 4 m. Tolerates wet sites in tropics. Can be cut for forage (stems brittle).

D. nicaraguense. Excellent forage, wide soil range in tropics. Pioneer plant in grassland, for cut forage.

Tagasaste (*Chaemocytisus palmensis*). Tolerant and hardy to tropics, cool areas, widely used in New Zealand in dry areas for cut forage, pioneer, mulch, and nurse crop.

Honey Locust (*Gleditsia triacanthos*). Selected trees bear heavy loads of pods in dry subtropics; frost–hardy.

Thornless forms exist. Deep soil moisture is required in the dry season, but the tree is soil–tolerant and wind–hardy. Best trees are thornless, high sugar types.

Kiawe (*Prosopis pallida*). Staple pod forage on dry savannah sites in subtropics, dense wood, excellent firewood and termite resistant posts. 20% thornless trees on Hawaii. Non–invasive.

THE PASTURE LEGUMES OR FORAGE LEGUMES

Calapo (*Calopogonium mucunoides*). A short–lived twining perennial used mainly as a pioneer of burnt or slashed weed areas to smother weeds before permanent systems are established. It is suited only to low–frost coastal areas of high rainfall (above 125 cm) and is moderately shade tolerant. It reseeds, but shades out or can be grazed or cut out. High seedling vigour.

Centro (*Centrosema pubescens*). A twining perennial used in both pastures and grain crops. Prefers more than 125 cm rain, warm climate between the tropics. Excellent cut forage and soil–builder, tolerant of wide soil range, acid soils, short flooding, some frost. Ideal for guinea–grass permanent pastures, banna, pangola, and para grasses. Climbs to 14 m so is *not* suited to short perennial crops, bushes, small trees. Can be broadcast in burns or slash areas, or drilled. Seed may need hot water treatment, inoculation. Persists well under grazing.

Kenya white clover (*Trifolium semipilosum*). Persists well in shortgrass pastures, dairy strip grazing (more than 100 cm rainfall or irrigated). Flowers autumn and spring. Needs good seedbed, scarification, inoculation.

Haifa white clover (*Trifolium* spp.). Strain adapted to summer heat, subtropics, persists well, reseeds after drought. Good interplant together with woolly vetch.

Greenleaf desmodium (*Desmodium intortum*). Vigorous trailing perennial used as understory in tall orchards (after establishment). Affected by frost, needs more than 100 cm rain, but valuable for soil–building in sandy soils, for early spring and autumn growth. Tolerant of poor soils, and stands some waterlogging. Needs rotational grazing. Seeds need inoculant. Companion legume is *Glycine* for wind control.

Silverleaf desmodium (*D. uncinatum*). Trailing vigorous perennial for mulch in established orchards, rocky sites, pastures, wet (not boggy) areas and acid soils. Pods sticky, and some people get skin rashes if it is used in gardens.

Macro (*Macrotyloma axillare*—was *Dolichos axillaris*). Twining perennial, forming a dense sward. Needs more than 100 cm rain in light frost areas. Valuable in shallow ridge soils, tolerates some dry periods. Establishes readily.

Lab–lab (*Lablab purpureus*—was *Dolichos lablab*) Vigorous annual or short–lived perennial useful for soil–building and weed control. Grown as a forage and mulch legume wherever cowpeas succeed. Will stand sporadic grazing; kept in rotation or strip grazing. Good silage, compost, mulch, pioneer crop. Tolerates acid soil, rough seed–bed. Broadcast at 20 kg/ha, drill at 6–10 kg. Inoculation assists establishment. A good screen plant on trellis for watered dryland gardens. Pods and beans edible.

Glycine (*Neonotonia wightii*—was *Gylcine wightii, G. javanica*). Slender, twining perennial with deep roots. Cycles phosphates from deep soil layers. Resists drought well, but affected by frosts. Useful in cool subtropics and tropics. Good winter growth in pastures; main growth in summer. Often fenced out in late summer or early autumn as a winter reserve. Rainfall ideal at 80–180 cm. Does best on well–drained deep red soils, but also yellow clays, black cracking soils, areas not subject to waterlogging. Needs rotational grazing, rested in late spring. Good silage (with molasses), mulch, fertility restoration of soils. Good seedbed and inoculation desirable.

Lucerne (*Medicago sativa*). Grown from cool temperate to tropics, usually as a pure sward cut to baled hay, but also in well–managed pasture under rotation (allowing a year or so of light grazing). Grows from 55 cm and up rainfall, as it is deep rooted. Combines well with makarikari, sorghum. Regular resting is essential to persistence, and in pasture needs re–seeding every 4–8 years. Cut for hay just before flowering. Reseeded in cut sward by chisel seeding. Inoculation essential, and lime pelleting also essential in acid soils. 6–14 kg/ha sown, lighter on rain–fed areas, heavier if irrigated. Silage with molasses now popular, hay expensive and in high demand. Garden plots used for mulch, rabbit feed, seeds for sprouting.

Phasey bean (*Macroptilium lathyroides*—was *Phaseolus lathyroides*). Self–regenerating annual, long erect twining stems. Needs more than 75 cm rain, heavy soils. Can be sown with para grass in swampy areas, also with glycine.

Siratro (*Macroptilium atropurpureum*) Perennial legume, creeps, good root system. Warm areas of 75 cm or more rain, ideally 90–110 cm. Poor soil tolerant. Excellent contribution of nitrogen to grasses, e.g. Rhodes grass. Ideal for rotational grazing, readily established, resistant to nematodes. The basis of many excellent pastures.

Puero (*Pueraria phaseoloides*). Pioneer green and cover crop, perennial climber. Very vigorous as a smothering summer mat. Used in wet tropics. Palatable, good seedling vigor (can be broadcast). Can be kept in pastures if rotational grazing practiced, but also suits green manuring, orchards, garden mulch crop.

Stylo (*Stylosanthes guianensis*). Perennial pasture legume of warm areas, 90–400 cm rain. Good pioneer of poor acid soils, poor drainage, sands, rocky soils, hillsides. Combined with low grasses (signal grass, pangola). Sensitive to copper and phosphate deficiencies. Excellent mulch in tree systems in such soils, can be cut to silage. Surface–planted, wide range of inoculants. Many varieties. Suited to specific sites and climates. Some shrubby types are an excellent cassava interplant, or also suit banana/papaya once plants of fruits are established, as a slash mulch; often kept as a feed for dry season, suits fenced–off reserves (*seca* variety) Shades out in dense plantation but may be ideal establishment mulch.

Cowpea *(Vigna sinensis)*. A preferred annual cover crop and soil improver. Also with sorghum, maize, millet as a hay or mulch in established orchards.

Lupin (West Australian varieties of seed lupins). Excellent cover crop and seed in acid sandy or good soils. Inoculated, can be broadcast or sod seeded. Good winter green crop (annual) in vines, bush fruit crop.

Mung beans *(Vigna radiata—*was *Phaseolus aurlus)* Vigorous garden green crop, forage annual, hay or grain crop. Suits gardens and low crop systems. Annual.

Note: Serious attempts to establish green crop and productive perennial pasture should be prefaced by research into species. An excellent place to start is: Humphreys, L.R. *A Guide to Better Pastures for the Tropics and Subtropics*, Wright Stephenson & Co. Pub. Australia P/L, P.O. Box 113, Ermington, NSW 2115.

FIGURE 10.48
ROWS OF TROPICAL HEDGEROW AND WINDBREAK.
Evolution from *Pennisetum* hedge to a permanent palm–casuarina–legume windbreak over 2–5 years. Quickset *Erythrina* assists in early establishment as truncheons set in soil.

10.14
HUMID TROPICAL COAST STABILISATION AND SHELTERBELT

If we presume a fairly delicate sandy coastline, then we need to build a complex stable assembly from the wave break to 10–20 m inland. The natural profile of undisturbed beach vegetation is that of a convex profile into the wind, and these uncut shores are very stable.

TROPICAL HEDGEROW AND WINDBREAK

Deliberately–mixed hedgerow is a preoccupation, skill, and literature of the temperate zones (as these were the first to suffer enclosures of common lands), but the rape of the tropics has now proceeded so far and fast that pioneer hedgerow is a priority theme for tropical coasts and hill country. Due to ideal growing conditions in the climate, if not the soil, hedgerows quickly establish. A classical hedgerow for the tropics is given below.

Well–tried procedures are as follows: cultivate, manure, and place dripline along a hedgerow site, and set out (concurrently) a row of:

- Tall grasses or clump bamboo; *Pennisetum* is usual.
- Quickset cuttings of *Erythrina fusca* or *Jatrophe*.
- Seedlings of *Leucaena* or *Acacia*.
- Occasional palms as seedlings, preferably those with spiny trunks or mid–ribs.

The results can be as in **Figure 10.48**.

This is for field conditions. When first setting small orchard crop such as citrus or avocado (both wind tender), first cast up an earth ridge system and plant *Pennisetum* hedges every 30 m (100 feet) crosswind and (if possible) cross–slope. This may result in a series of parallel lines (if wind and slope coincide), or a diamond pattern (wind at an angle to slope), or a series of squares (wind and slope at right angles). Pay particular attention to the top of ridges in wind–prone areas.

As the young orchard grows, the *Pennisetum* at 30 m shelters it. Every second row of *Pennisetum* can be combined with *Leucaena*, and every third and every ridge row with *Acacia* and palms. The evolutions follow. Later, the inner rows can be removed as mulch.

Complicating the Hedgerow

Tomato trellis can be placed on *Leucaena*, and passionfruit on most trees. Mango itself is a good windbreak, *Eugenia* can replace some *Leucaena*, and we are on the way to a mixed hedgerow for wildlife, domestic forage, and food in the tropics. I would never neglect a clump bamboo as a source of structural field material and effective windbreak.

The cross–slope ridges early established become long–term soil and water traps, and accumulate mulch for later evolutions. These are a feature of the Tropical Crops Materials Centre on Moloka'i, and there one can see their uses and long–term evolution (under cultivation) into terraces of undoubted stabilty, as in **Figure 10.8**. Roads should be provided with concrete or stone fill at "X", the *downhill* side of the mounds. Permanent roads can be made after the terraces are formed.

10.15
LOW ISLAND AND CORAL CAY STRATEGIES

All coral sand cays and many low atolls lie within 28° of the equator, as do coral reef areas and sandy alkaline coasts. As many peoples live on atolls, very careful design approaches are needed to avoid the known risks of:

- Hurricane erosion and damage to plantations and coasts;
- Water table (water lens) pollution; and
- Poor nutrition due to a limited diet (usually high in carbohydrate and oxalic acids).

Additional design input is needed:

- To extend vegetable and fruit crop;
- To extend water storage and to conserve water;
- In conserving natural vegetation and unique birds, reptiles, and lagoon or reef fauna;
- To use shallow marine waters for aquaculture and pond fish; and
- In developing energy resources locally.

On atolls, we can expect soil pH values of from 8.0–9.5, sand abrasion, and basic mineral deficiency (especially iron, zinc, molybdenum, boron) in soils and plants. Elemental sulphur, iron sulphates, and humus added to garden soils and planting holes lower the pH. Humus sources are palm fronds, coconut husks, tree trunks, and leaf litter from such pioneer species as *Casuarina equisetifolia*, sea grape (*Coccolobus unifera*), coastal shrubs (*Scaevola, Tornefortia, Pemphis*), mangroves, and *Barringtonia* trees. All yield abundant litter for garden and tree holes.

We also can expect a thin coralline sand over a hardpan of caliche or calcrete. Calcrete is worsened by the application of superphosphate. Usually the guano from seabirds provides sufficient phosphates if colonies of terns, gulls, boobies, frigate birds, and shearwaters are protected or encouraged. Failing this, domestic pigeon, quail, pheasant, geese, ducks, and chickens can be kept as such predators as feral cats and foxes are usually absent.

PREVENTING WATER LENS POLLUTION

The sole natural sources of water on low islands are those biological storages such as we find in baobabs and coconuts (about 12 coconut trees provide for a person's drinking water for the year), and that water trapped in the sand below the caliche, floating on the seawater—the water lens of **Figure 10.49**.

As rain falls on the islands, it quickly absorbs in surface soils and leaks through the caliche to the water lens, which itself leaks slowly to sea between or below tide levels. As the beach berm is 2–4 m above sea level, and the inner atoll about 2–3 m, the delicacy of the situation in the event of water pollution is obvious. For this reason, settlement, toilet areas, processing, livestock, and even gardens should be kept to the periphery of the island, and the interior devoted to dense natural stands of food and native trees, kept free of pesticides or industrial pollutants and fuels. It is obvious that a polluting source at the centre of the island can affect the whole water lens, all plants, and all people.

By what methods can we increase the supply of potable water? First, by using run–off water to house tanks, the latter above or below ground and made of a reinforced cement–coral–sand mix, which is widely used in construction on all but remote atolls.

Secondly, and of most importance to plants, by using deep mulches, and by growing ever–abundant coconut, banana, arrowroot, papaya, legumes, and like plants to provide leaf and trunk materials for gardens and other tree crop.

Lastly, and often successfully, we can try leaf–ferment seals (gleys) of gently–sloped coral pits, where a 20 cm deep mash or shredded mass of banana, papaya, and other soft green leaf is applied to the base and sides of the area, covered with plastic until fermenting (sometimes only 4–5 days), and then pumped full of water from the freshwater lens (**Figure**

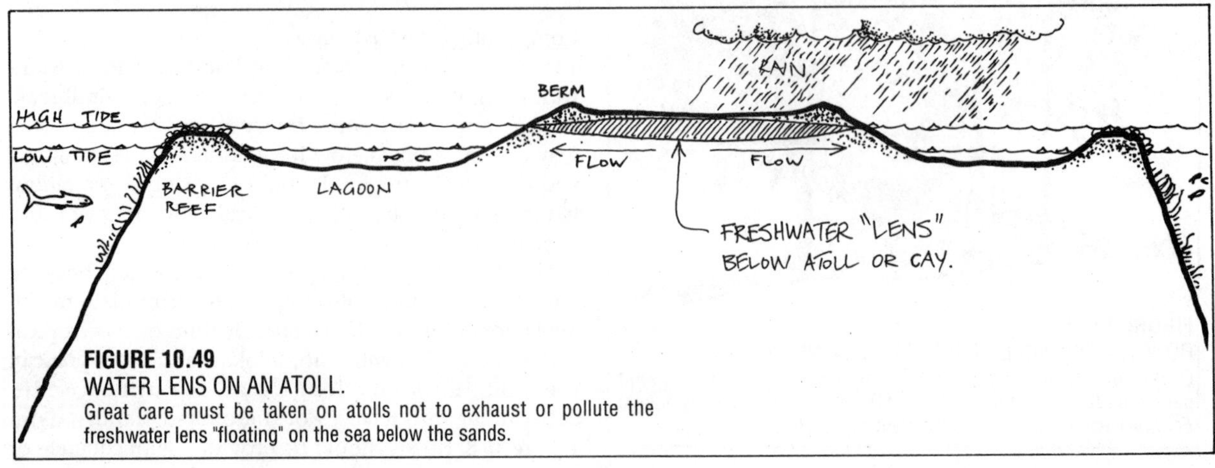

FIGURE 10.49
WATER LENS ON AN ATOLL.
Great care must be taken on atolls not to exhaust or pollute the freshwater lens "floating" on the sea below the sands.

7.19).

Made carefully, such surface ponds will also take roof run–off. Success follows careful trials at small scale to get the system working, then scaling up to significant ponds for ducks and garden water, water leaf crop, and animal drinking water. As such ponds are often a source of cross–infection in children, it is sensible to swim in the lagoon, and to drink tank water. Seawater serves also for many toilet uses.

PLANTING TREES ON CORAL CAYS

To plant trees, we need to clear off a patch of topsoil and break open the caliche, making a pit about 40 x 40 cms and 60–80 cm deep. In this pit we make a humus pile of domestic and plant wastes, plus some sulphur and the mineral trace elements, and plant a coconut or other seedling tree.

The tree roots spread out above the caliche layer, and feeder roots go below the caliche to the water lens, which is usually 0.75–1.0 m down, and 1.0–2.0 m deep as a saturated sand layer. Trees are thus well anchored and easily obtain water.

To plant gardens, we have two or three possibilities. The first (and best) is to open a pit garden of about 8 m x 15 m, and sloped for stability to about 2 m deep, thus often damp or wet on the base (**Figure 10.50**). The sloped sides are stabilised, as steps, with coconut logs or caliche, and the spoil banked around the rim. Mulch is thickly applied to the base and behind the stepped terraces, and when this is rotted, a range of plants are grown. From base to crown, some such sequence as follows is appropriate:

1. BASE (damp, mulch) watercress, parsley, chives, *Brassica*, taro, kangkong, salad greens generally.

2. FIRST TERRACE (18–25 cm above base) tomatoes, peppers, taro, sweet corn, beans, peas, taller crop.

3. SECOND TERRACE(25–60 cm above base). Banana and papaya, sweet potato, cassava (all provide mulch).

4. THIRD, OR HIGHEST, TERRACE. Cassava, sweet potato, banana, dry–tolerant vegetable crop, and mulch trees such as *Leucaena, Glyricidia, Tipuana tipu, Moringa oleifera*, and local tree and shrub legumes to provide leaf mulch and partial shade; palms for frond mulch and high shade, vines to climb on these (passionfruit, four–winged bean).

Secondly, under some light palm–legume canopies, boxes of palm logs will hold thick mulch and household waste for surface gardens. These can be planted as beds of potato, yam, sweet potato, and normal vegetable crop. The thicker the mulch, the less watering needed.

Thus, log boxes, pits, thick mulch, and high shade canopy are the essentials for good atoll gardens. Staples are coconut, root crop, fruits, and normal salad vegetables plus fish and shellfish from the lagoons or coasts.

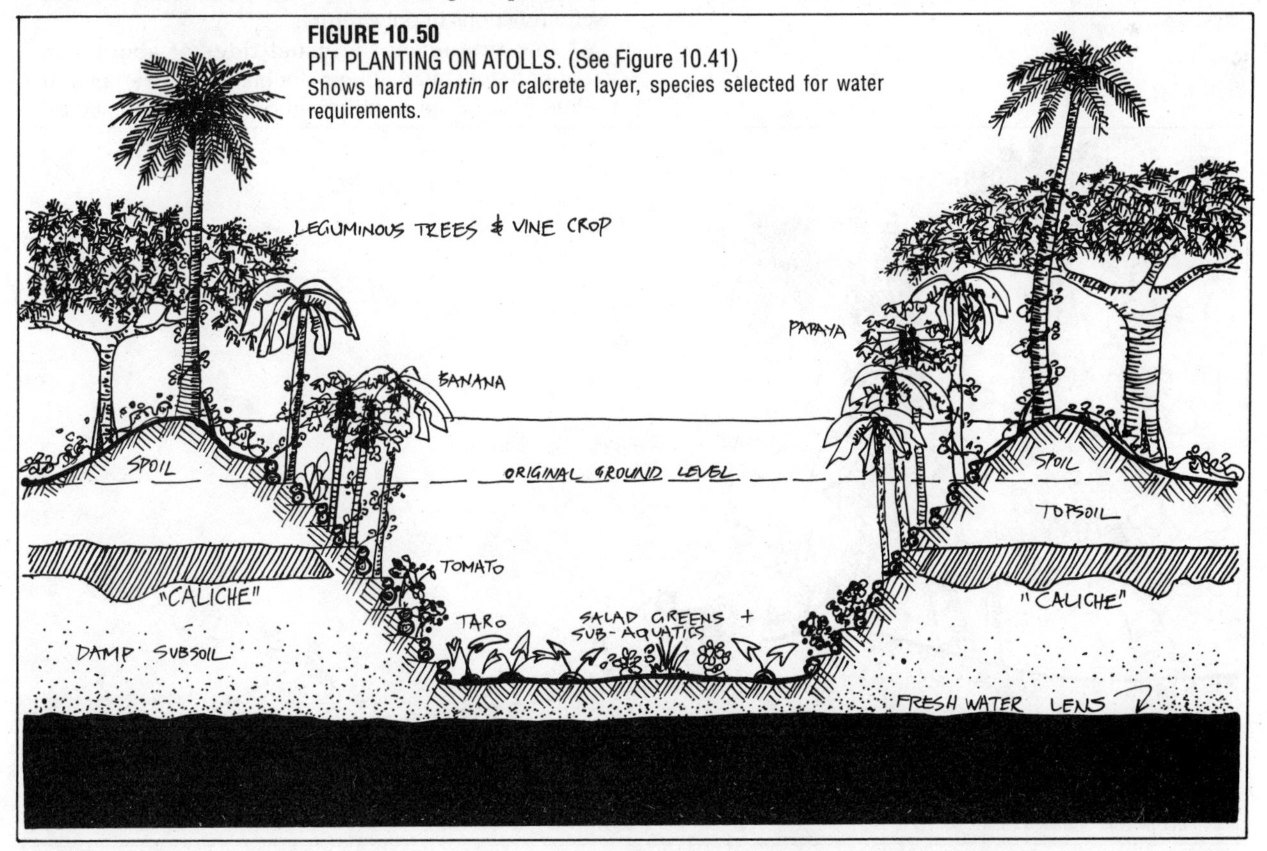

FIGURE 10.50
PIT PLANTING ON ATOLLS. (See Figure 10.41)
Shows hard *plantin* or calcrete layer, species selected for water requirements.

AVOIDING AND REDUCING HURRICANE DAMAGE

Access to atolls is traditionally by boats in reef gaps and today by light planes. When blowing a gap for a reef entry, or clearing a landing strip for a plane, great care must be taken not to open a wave or wind gap to gales, or any atoll can literally wash away. Thus, reef entries are cut on the slant through the reef at the east or west quarters (winds blow southeast to northwest south of the equator, northeast to southwest north or the equator). In fact, reef gaps should be in the most sheltered sector of the reef in any winds, and also just wide enough (6–10 m) to admit a vessel or barge.

Airstrips are also aligned about 20° off prevailing winds, and both ends and sides should be of tall palms and trees, especially those borders on the coast, so that light planes drop in, using their rudders to straighten up below tree crown level. Airstrips carelessly made have destroyed whole islands when hurricane winds have cut them in two following the line of the air strip.

For the same reasons, the sandy coasts of all atolls and cays need a sequence of perennial shelterbelt to hold the shoreline against hurricanes. This starts on the beach as convolvulus (*Ipomoea pescaprae*) and beach pea, rise on the beach berm to a dense shrubbery of vines, *Tournefortia, Scaevola,* and in sheltered bays mangroves, and is backed by a 5–6 tree deep layer of coconut palm, *Casuarina, Coccifera, Barringtonia* and other hardy beach trees (**Figure 10.51**). It is behind this dense frontline windbreak that we site houses, gardens, and productive trees, which will produce in shelter but not as exposed systems.

EXTENDING DIET

People who live on small islands, and indeed small traditional villages generally, may exist eating a very few starchy root foods plus banana, with fish for protein. It is quite probable that mineral deficiencies and low vitamin/high carbohydrate diets can impair health. Thus, a well–mulched pit garden, and a well–selected introduction of tree fruits (guava, citrus generally, vine fruits, and a polyculture of minor fruits and nuts) greatly extends and buffers the diet. The addition of (in particular) zinc and iron to mulched soils, and periodic tests of leaf content of such minerals serves to eliminate problems due to restricted diets and highly alkaline soils. Even on high islands, soils can be devoid of, or have, very limited mineral rocks, and soils may need trace elements.

Almost every island group has unique plant and animal species, some of great value directly, others of value in that they exist and demonstrate new forms and behaviours. Such groups as land crabs (derived from ghost crabs, shore crabs, and hermit crabs) do special work as mulch shredders, scavengers, larval insect eaters, and may form a valued food resource. Giant tortoises are also excellent scavengers of fallen fruits, and keep grasses below palms neatly–trimmed, while putting on a considerable annual growth. Marine iguanas, giant lizards (the Komodo dragon), flightless or specialised birds, and rare plant and animal survivors of older land masses are not only common but usual on islands. All need careful preservation and assessment for their special values, and many provide useful functions in polycultures.

At low tide (even the usual tides of about 1 m variation) atolls may almost double their "dry" area. It is thus possible to modify lagoon and reef for better

FIGURE 10.51
SEA COAST PLANTING
A. Beach convolvulus. **B**. *Scaevola*. **C**. *Tournefortia*. **D**. *Barringtonia*. **E**. Sea grape (*Coccoloba*). **F**. Date or coconut palm, Casuarina behind beach berm. This profile is stable.

conservation and feeding of economically useful fish, shellfish, and marine plants such as mangroves, or to consolidate and protect shorelines with coral–block breakwaters. There are many such marine impoundments throughout the Polynesian world, and new marine breeding techniques are bringing into cultivation such species as trochus, turtles, many inshore mullet, milkfish, and edible seaweeds.

Because of the frequent internal (atoll) or external (annular reef) lagoons in or around low islands, designers and residents have extensive quiet waters in which to trial a wide variety of productive maricultures, shelters for fish, breeding places, and undersea constructs generally. The daily flux of tides through reef outlets brings a regular fish movement well–known to indigenous peoples.

ISLAND ENERGY RESOURCES

Islands in oceanic energy flows behave very much like "bluff bodies" in streams. Tide, waves generated by winds, winds, and the water crashing over low reef areas and flowing out of constricted reef inlets present good opportunities for energy generation locally. Reliable biogas technologies now widely used in Asia, and the less reliable wind–electric systems, can also be used for energy generation.

Solid fuels from coconut husks, fast–grown coppicing legumes, fronds of palms, and *Casuarina* stands are always available on well–planned islands. Climates are usually mild, and the main fuel needs (for cooking) can be much reduced by a vegetable–fruit garden development.

One area all atolls and islands can develop is that of tide–flow turbines (these can be propellers, or "egg–beater" catenary–curve vertical axis turbines). Coral and cement provide strong anchors for such turbines at reef outlets. Both tidal and ocean current flow provides dense energy power at 1kW/square metre/second of flow, so that a few such turbines at selected high–flow sites can provide either electrical or pneumatic power for island workshops and lights.

The above outline should assist island design; but one factor that we cannot design for is that of rising sea levels. Many of today's atolls will simply be overtopped or washed away by a very modest rise in sea levels, which is expected to occur over the next decades. For these sites, early evacuation is the best action!

10.16
DESIGNERS' CHECKLIST

Maximise tree crop, herbaceous perennials (banana, papaya, arrowroot, taro) and plan a multi–tier system integrating windbreak, forest, orchard, understory, and ground cover.

Complete earth–shaping before setting out plant systems.

Choose adapted high–value foods for intensive (mulched) home gardens; allow 30–90 species in Zone 1, but concentrate on 7–20 high–value crops in Zone 2.

Avoid bare soil systems in all areas.

Design a careful plant/animal assembly related to culture, market, processing, available labour, and value to village.

Design houses and villages for low–energy climate control.

10.17
REFERENCES

Davies, J. L. and M. A. J. Williams, *Landform Evolution in Australasia*, Australian National Univeristy Press, Canberra, 1978.

Etherington, Dan M., and K. Karanauayabe, "An economic analysis of some options for intercropping under coconut in Sri Lanka", *Sri Lanka Journal of Agrarian Studies 2(2)*. (Uses MULBUD, a "MULtiperiod BUDgeting" programme for the economic assessment of the intercrops in perennial crop.)

Fox, James, *Harvest of the Palm*, Harvard University Press, 1977.

Harris, W. Victor, *Termites: Their Recognition and Control*, Longman, London, 1971.

Martin, Frank, and Ruth Ruberte, *Edible Leaves of the Tropics*, Mayaguez Institute of Tropical Agriculture, Puerto Rico, 1975.

Nair, P. K. R. *et. alia*, "Beneficial effects of crop combination of coconut and cocoa", *Indian Journal of Agricultural Science 45(4)*, 1975.

National Academy of Sciences, *Tropical Legumes: Resources for the Future*, Washington, D.C., 1979.

Nelliat, E. V. *et. alia*, "Multi–storey: a new dimension in multiple cropping for coconut plantations", *World Crops 26*, Nov/Dec., 1974.

Ralph, Wayne, "Managing Some Tropical Soils", *Rural Research No.117*

Trewartha, S. T., *An Introduction to Climate*, McGraw–Hill, Inc., N.Y., 1954. (See for modified Koppen climatic classifications.)

ACKNOWLEDGEMENTS

Many friends have helped with information and trials in tropical areas. I would like to acknowledge data from Bud Driver, Marjorie Spears, John Selman, Michel Fanton, Chuck Busby, Sandy Lewin, Cliff Adam, Imu and Ahela Naki, Richard Waller, Dano Gorsich, Bob Kingswell, and Walter Jehne.

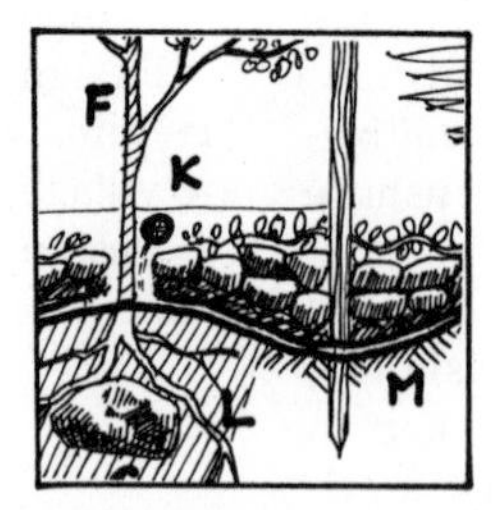

Chapter 11

DRYLAND STRATEGIES

The dunes, like harmonic chords, multiply and repeat the notes of the rustling, shifting sands... you now hear what sounds like the sonorous music of distant drums– roll after roll...

(Quilici, 1969, on the Tobol, or music of the dunes)

11.1 INTRODUCTION

The development of conservative strategies for the preservation of dryland species, and for the responsible human use and management of arid lands is probably the world's most pressing problem in landscape management. Only sporadic world attention is given to famine or drought in arid areas, with such "sudden" emergencies consuming great quantities of resources and finance, much of which is wasted in mismanagement.

Problems of aridity, salted soils, and long–term drought can *always* be expected where we venture into desert borders with pastoralism and cropping, for added to the natural background fluctuations of rainfall due to earth and moon orbits, and solar radiation variance, we as expoloiters add deforestation, soil erosion, and the consequent salting of waters and soils. All desert areas are extending; many dryland areas are being created, and antecedent plant and animal species are thereby brought to extinction.

A great many arid–area species are not so much dry–adapted as drought–evading. Plants dry off, cease growth, or exist only as seeds and tubers in drought. Many trees are dry–deciduous, evading the worst effects of drought in a leafless state, but truly drought–adapted plant species also exist, and use wax, insulation, reduced transpiration, and large water storage organs to withstand droughts. Animals aestivate, migrate to humid areas, or take refuge near oases and wetter areas. Many animals get sufficient water by eating succulent vegetation, by browsing at night to take advantage of condensation, or by predation on insects and other species of animals. Almost all small animals burrow by day to escape soil temperature fluctuations, often to 1 or 2 m deep. Some termite colonies develop deep galleries (to 40 m) to mine water, and also arrange air conditioning by adapted permeable surface nest structures.

Increasing desertification of agricultural lands has focused world attention on salt and drought–tolerant plant species of potential use. Gary Nabham (in *Annals of Earth* IV(1), 1986) has totalled 450 edible plant species for the Sonoran and Great Basin deserts of the North American interior (20% of the total species). The semi–arid or transitional areas are particularly species–rich, thus desertification is taking a rapid toll on potential useful plant diversity. Of the 936 woody plants in the West Australian wheat–belt, 45% are at risk. Large reserves are urgently needed; the present reserves in arid lands are "entirely inadequate to conserve even a usual proportion of the genetic variation..." (Nabham, *ibid.*)

Cold high mountain deserts like the Gobi have some 60 flowering plants, and 360 species of birds (*Geo* 7[4] 1986), while large areas of the arid Sahara have 6–8 plant species in total, and recent deserts such as the Thar some 15–18 species. Most of the world arid species lie in the Kalahari, Sonoran, and Australian deserts.

Arid lands are areas where direct evaporation exceeds rainfall, and where annual precipitation averages fall below 80 cm, and as low as 1 cm (sometimes only as dew). Although the greater part of these areas are hot desert, there are substantial montane hot–cold deserts, and very cold desert areas near the north and south polar regions. Typically, an extensive hot desert area consists of a savannah edge towards more humid monsoon areas, a Mediterranean climate

edge, extensive scattered shrub–forb associations (**forbs** are any non–grass or herbaceous plants), dunes on harder pavement areas, and almost bare pavements of harder rocks with faceted pebbles (gibber plain or reg).

Where over–grazing has not been the dominant feature of land use, a desert may present an eye–level appearance of a low–crowned forest with shrub understorey, and some areas may be as rich or richer in plant species than a more humid forest environment. In detail, however, there will usually be larger or smaller areas of bare soil between all major plant clumps or species. Closer inspection may reveal a fungal–algal–lichen *(cryptogam)* crust on the bare areas, more or less intact depending on the frequency of hoofed animals. This cryptogamic crust is a critical and delicate feature preventing wind erosion; its preservation is essential for soil stability.

Some features of the deserts are:

- Plants produce copious seed with long viability; seed is often wind–dispersed.
- Termites and ants are more effective than worms as soil aerators and decomposers.
- Rain may fall in mosaic patterns, so that vegetation is also a varied mosaic of fire and rain, and ephemeral plants at different response stages from growth to decay are evident. New generations of shrubs may experience favourable seedling conditions as rarely as every 7–20 years. This then becomes the period of **recruitment** of new forests or shrublands.
- Much of the water run–off system may end not in rivers but in inland salt–pans or basins (*endorheic* drainage), from which all water eventually evaporates.
- Normal erosion is by wind, but rare cloudbursts shape the main erosion features and move vast quantities of loose material from the hills in turbulent stream flow. Wind transports materials locally in dust storms.
- Animals burrow, seek shade, or are nocturnal in order to conserve water; many are highly adapted for water conservation.
- Plant associations may be very varied in response to changes in long–term aspects such as slope, soil depth, salinity, browsing intensity, pH, and rock type.

SEMI–ARID areas have steppe, scrub, and low forest vegetation, ARID is steppe and scattered low shrubs, and true DESERT has little but oases and ephemeral vegetation appearing after the rare rains. Deserts have a rainfall classification of:

- Hyperarid: 0–2 cm annual average (e.g. Atacama, Namib desert, central Sahara).
- Extremely dry: 2–5 cm annual average.
- Arid: 5–15 cm annual average (e.g. Mohave, Sonora, Sahara margins).
- Semi–arid: 15–20 cm, maximum 40 cm (much of the Australian deserts, Asian deserts, Kalahari).

Above 40 cm, and to 75 or 100 cm rainfall we have (potentially) dry savannah forests; and it is up to this latter level that we will be dealing with in this chapter. Rainfall is not dependable in arid areas, with a normal 30% variation, and a potential 90% variation in any one year. Potential evaporation can range from values of 700 cm/year in hyperarid areas to 100 cm in steppe.

We can also regard the polar ice–caps as hyperarid (for precipitation) but evaporation levels are very low; wind removal of snow replaces evaporation. Both the very cold and hyperarid hot deserts have one thing in common—they mummify and preserve a great range of organic and fabricated substances, which presents a management problem when dealing with pollution, as breakdown of most organic substances can be very slow in the absence of water.

A paradox in almost all large desert or arid areas is the existence of two types of more humid environments: EXOTIC RIVERS which flow in from better–watered or forested regions, and OASES. There is a third, invisible water resource, that of AQUIFERS (underground waters). All must be used with great caution, as water in all of these resources can be locally exhausted, and aquifers can be depleted over vast areas by immodest use (as they have been in the USA). This misuse can cause widespread subsidence, collapse of the aquifer, and a permanent disappearance of those oases which were in depressions fed by the aquifer.

Instead of concentrating on exotic water in deserts, we should attempt to *increase the input* of water into the aquifers, soils, and streams; and to re–humidify the desert air by planting trees and protecting existing vegetation. It is the presence of trees and shrubs, transpiring rather than evaporating water, that keep the desert salts from evaporating at soil level. Once evaporation alone operates, capillary action quickly brings subsurface salts to the surface as magnesium, sodium, calcium, and potassium compounds (chlorides, sulphates, carbonates) and we can no longer establish vegetative cover.

When we talk of arid lands, we should also remember equatorial low islands, unstabilised sand dunes in midsummer even in cool climates, and whole periods of relative drought even in more humid areas (including the dry winters in many subtropical areas). That is why it is important to specify pioneering and drought–hardy plants for many situations, and why any strategy to get water into soil, and keep it there, is worthwhile. There are large areas of Mediterranean climate on desert borders which are "arid" if they have deep sands and poor water retention.

A profound question to ask about deserts concerns our basic usage. Livestock herding has been traditional—and devastating. Australians, Peruvians and Africans, Arabs, and Tibetans have all used arid areas for herding. Aboriginal Australians managed better, by harvesting the natural abundance of deserts so that the artifical stress of herds was not superimposed on the natural stresses of dry seasons.

We should re–think our strategies of desert use, and the way we occupy arid lands. It is possible to establish a carefully–developed core settlement, to set out hardy plants for many kilometres along favourable areas or CORRIDORS and to take advantage of the rare rains to establish a wide biological resource for dry years. It is

not possible to count on permanent cropping or herding in regions that experience one good year in every 4–9 years. As for desert revegetation, by far the most effective and cheapest strategy is to exclude browsing animals from headwater areas, when after a few years thousands of young plants may establish.

Desert borders are now used for a seasonally and uncertain production of wheat, barley, millets, and sorghums. About one year in four produces a reasonable crop, but severe wind erosion and dust storms make this a precarious use of such delicate soils. Herding and extensive livestock systems also occupy great areas of the deserts, with seasonal or nomadic herding in north Africa and southwest Asia.

If we look at actual income *versus* land use, the Central Desert of Australia allots some 73% of the area to pastoralism, and only 27% to Aboriginal reserves and conservation areas. However, income is 15% from pastoralism, 76% from tourism, and 7–8% from mining in small areas. Even the mining is a "tourist industry" where it concentrates on precious stones, and Aboriginal art is a large part of the tourist interest. Pastoralism and mining are publicly subsidised, and has a grossly unwarranted system of direct and indirect supports, producing a largely surplus product for the world market. No wonder pastoralists must maintain a powerful lobby in the halls of government!

People rarely want to live in the desert—it is an expensive place to maintain a high standard of comfort in modern terms. But they love to visit, and to see wildlife, genuine tribal art, and the landscape itself. To Aboriginal people, the land is life itself; they are a part of the whole, and their art reflects this.

Deserts are inspirational for designers. Not only do the hills reveal patterns, rock types, and processes of erosion, but as the light changes from early dawn to late evening, new insights arise from the shadowed and light areas. What is often unclear on the ground becomes part of a whole pattern if seen from hills or the air. When we understand these many patterns, the distribution of materials and organisms makes sense, and we are able to creatively inhabit the landscape. It is little wonder that desert peoples, in the great silences, beauty, and vastness of nature, arrive at profound mystical pattern concepts. Only the confined oasis– or town–dweller concentrates on the ephemera of finance, commerce, and conduct. In the open, survival demands sensitive reaction to environmental imperatives.

11.2 PRECIPITATION

Rain occurs in deserts as a lesser part of the normal cyclonic, convectional, or orographic rains occurring elsewhere. Both warm, unstable tropical air masses and cold westerly rain may be entrained into air cell circulation over deserts, but there is a significant proportion of convectional rain due to local heating over sands, rocks, and bare soils. Only in some deserts does rain fall fairly reliably in a seasonal distribution, in areas affected by monsoon borders or westerly coastal belts. Elsewhere, rain is episodic and averages are meaningless in that many years may pass at any one place without rain. There are the natural deserts (the Namibian coastal areas and the Atacama) where offshore winds and cold sea currents ensure that rain is rare or virtually absent. Such areas are treeless, and depend on fogs and dew alone for plant growth.

As dew may be critical to plant survival, dew traps of stone, scattered shrubs, or even vertical metal screens to 1 m high are strategies to catch moisture. Moroccan foresters are contemplating such metal screens to condense dew and to thereby establish shrubs, which in turn will become moisture condensers.

OPPORTUNISTIC RESPONSES TO PRECIPITATION

A rainstorm 8–12 mm produces run–off sufficient to start headwater stream flow, and only the occasional downpour of torrential convectional rain, or the heavy rains of monsoons and tradewinds over headwater streams remote from the desert produce river flow down the braided or sandy water networks that thread through the desert.

For tribespeople and the mobile fauna of the desert, distant rain is the trigger that sets off a whole sequence of migration and perhaps a subsequent intensive breeding programme. Just as we can write a dissertation on the phrase *Om mani padme hum*, so we can write essays on such simple concept–words as "walkabout" for Central Australian tribes. Rather than being an arbitrary, willful, or unpredictable movement that the calendar–regulated Europeans see it to be, walkabout translates as something like:

> Our scouts saw thunderstorms to the far north; if we go now we can arrive in time to harvest some of the birds, animals, and plants that will respond to rain, and to follow the water to the waterholes that will fill with freshwater and give us fish, turtle, and frogs for a few months. We must go now, in time to celebrate the cycle of plenty that comes from the rain.

Thus we can see such movements as being sensible, planned, and appropriate to an environment which presents rare opportunities to harvest the varied resources provided by rain, to visit newly–regenerated country, and to lighten the burden of resource pressure on favoured home areas, so that they can regenerate as reserves for hard times.

The response of the desert to rain is truly remarkable. Ephemeral plants carpet the ground, and flowers and seed are produced in abundance. Buried tubers throw out great patches of melon, bean, and yam vines. Shrubs or trees may produce numerous seedlings in areas where few young trees existed. In the streams and

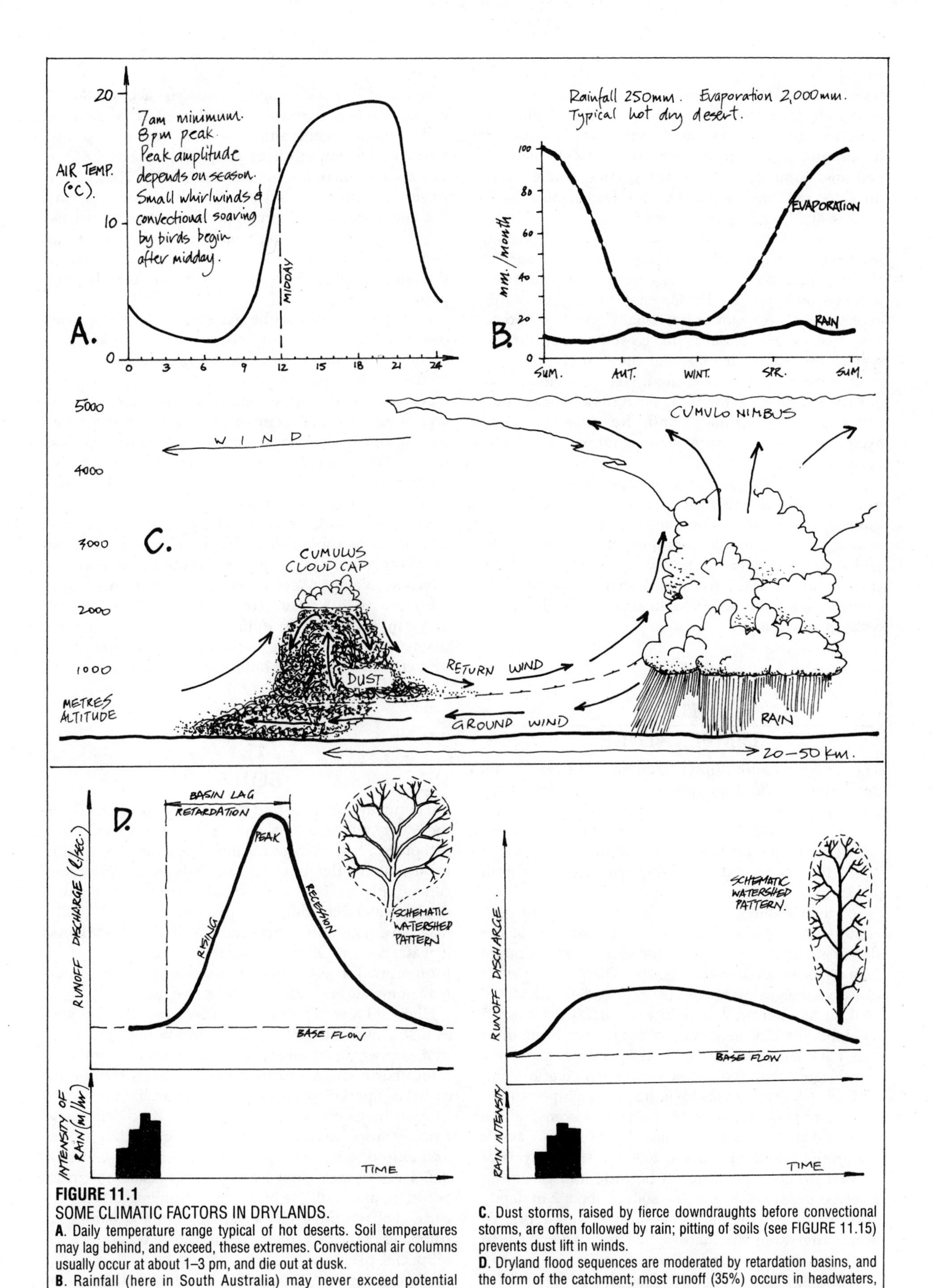

FIGURE 11.1

SOME CLIMATIC FACTORS IN DRYLANDS.

A. Daily temperature range typical of hot deserts. Soil temperatures may lag behind, and exceed, these extremes. Convectional air columns usually occur at about 1–3 pm, and die out at dusk.

B. Rainfall (here in South Australia) may never exceed potential evaporation, hence the importance of infiltrating runoff into soils.

C. Dust storms, raised by fierce downdraughts before convectional storms, are often followed by rain; pitting of soils (see FIGURE 11.15) prevents dust lift in winds.

D. Dryland flood sequences are moderated by retardation basins, and the form of the catchment; most runoff (35%) occurs in headwaters, much less in main streams (0–5%) unless in widespread rains.

pools, frogs and fish appear from mud–packed retreats, or invade from the rare permanent pools. Fish may breed and grow in a few months, and sea–birds migrate inland and nest to take advantage of this. Flocks of pelicans, flamingos, herons, terns, gulls, waders, and ducks occupy the waters, and for a few months life is riotous, and breeding unconfined.

Antelope, kangaroo, and the desert quail, pigeon, finches, and parrots flock to waterholes and spread their ranges over new forage areas, and animal numbers build rapidly. The longer–term collapse of this explosion of life is slower but usually inevitable. Some new tree generations may have established, but the waters turn salty, dry up, and life scatters, dies, or returns to rest in mud and sand. Seeds are blown and buried, and the desert waits again. Birds disperse, as do the migratory animals, and the desert fringes experience a sudden irruption of migrating species. The party is over.

It is part of our strategy to capture some of the estimated 88% of water that either evaporates or rushes unused across the land at these times, and to safely store it below ground for a prolongation of the growth period, and a rehumidification of the desert air itself via transpiration from trees and shrubs. In natural conditions, as little as 0.8% of total rainfall infiltrates to recharge desert aquifers.

11.3 TEMPERATURE

The air temperature (up to 60 m height) over deserts approximates the daily pattern given in **Figure 11.1**; I have not given the 6 a.m. temperatures, which can range from 8–30°C, but show a generalised graph which holds true for most cases, and indicates a rise of about 10°C to 25°C over the day, with a peak at from 12–3 p.m.

Soil temperatures follow this general curve, but peak about an hour earlier. The soil is also colder just before dawn, at about 5–6 a.m., and the effect of temperature falls off as we go deeper in the soil. There is little effect of *daily* changes at 30 cm (12 inches) depth, and the peak surface temperature reduces about 2°C for every 5 cm depth (or 15°C in 30 cm), so that soils at 30 cm deep may have fairly constant daily temperatures some 5°C or so higher than the lowest surface temperature, and 15°C or so lower than the highest surface temperature.

Another way to look at this is that soil evens out the heating that affects air for a considerable distance above the ground. Both air and soil follow a slower cycle of annual range depending on latitude and radiation received in season. This affects soils to about 2 m depth, with about 5°C annual fluctuation at this depth, or about the same fluctuation that we experience in underground houses and cellars.

Soils also can gain and hold much higher levels of heat than air, and at 5 cm depth, in favourable sun slopes and in poorly conducting soils, about 60–70°C can be reached at peak! While this effect is good for solar hot water collectors, it is lethal for young plants. Surface air temperatures frequently reach 30°C in deserts, and there are isolated records of 52°C (125°F). Layers of super–heated air just above the soil may become unstable, and form rapid convectional currents, especially on sun–facing slopes. Cumuliform clouds can arise from this effect as thermals or strong updraughts, or as clear–air turbulence in very hot dry periods.

In order to espcape the extreme surface soil heat, desert animals commonly burrow to below 30 cm in depth, or in some cases seek out high refuges in low shrubs or on stumps or posts in hot weather. The other way that animals escape soil and air heat extremes is to seek shade, become nocturnal, and to develop efficient cooling strategies and water economy. In larger species, shade may be the limiting factor in survival (e.g. the amount of cave shelters in the desert may limit the numbers of large desert marsupials).

Consequently, deserts have a far greater proportion of their fauna as subterranean and nocturnal species. Soil humidity also rises steeply, especially in dune sands. Given an average 4% of water in the top one metre, we can reach 10–20% water content at 2–6 metres, and at 20–60 m, there may be quite saturated soils. For both plants and animals, it is preferable to draw from these deeper, cooler, and more humid layers of soil.

11.4 SOILS

We normally expect to find dominantly alkaline soils in the waterways of deserts, with areas of surface salts and carbonates. It is in these alkaline areas that we usually locate our settlements, to take advantage of water run–off. pH levels of 8.5 and 9.0 are not atypical, and drying waterholes can reach pH values of 10–11. Soils may have high nutrient potential if pH is adjusted and if water for irrigation is available. Acidic sandy soils form around areas of deeply weathered granites, and may dominate large areas of these landscapes.

High and low pH areas have (as a consequence) low available mineral trace elements, and in high pH, zinc, iron, copper, and manganese deficiencies are common, indicated in crops and fruit trees by such factors (in citrus) as interveinal colour loss and leaf thinning, with tip curl in severe cases (manganese). Zinc deficiency causes more severe leaf yellowing, and is often associated with manganese deficiency. Copper deficiency causes giant leaves, gum pockets in citrus branches, and multiple budding or trunking in trees or citrus. Foliar sprays, elemental sulphur, and oxides or sulphates of the deficient elements can be used.

Special problems may be caused by non–wetting sands (a fungus is responsible), and high salt levels in water or soils. Bentonite or humus will ameliorate the "non–wetting" problem, as do swales and raised beds with high edges to prevent water run–off. Salt problems can sometimes be solved by flushing beds with fresh water, but if the water source is itself salty, salinity needs the combined solutions of humus production, perhaps a ponding period of water with algae production and water crop, ionic or distillation treatments, and a choice of salt–tolerant crop. Free–draining sands can be irrigated with salt levels in water much higher than will be tolerated as spray irrigation by plants (to 1500 ppm).

Despite all these problems, we can usually establish home gardens and adapted tree crop systems, and many selected areas (especially near scarps, rivers, hills, or ranges) will grow excellent fruit, vegetable, and tree crop with appropriate water run–off harvesting.

Perhaps the most important thing to remember about any activity in desert is that the natural systems are fragile. Good management and constant appraisal is essential. *Small* systems may be called for, especially where run–off is harvested. Natural yields should be carefully assessed, and broadscale or grand trials avoided until the capacity of the total system—especially of water resources—is assessed.

Excessive mineral content in soil or water can also be toxic, both to plants and animals, and in particular bore waters must be tested for fluorine, sodium, and radioactives. Nitrates are to be rigorously tested where children are consuming water and garden leaf products, and nitrate fertiliser used at minimal levels, or not at all in the absence of mulch or high organic soil content (20% or more humus).

Aluminium, boron, sodium, and manganese can be in over–supply at very low or high pH values, and only humus can buffer the uptake of these excess minerals. Aluminium in acid soil solutions damages roots, and high levels of manganese causes stunting and yellowing in plants. 0.5 ppm copper, or 10 ppm lead or zinc can stop root growth completely (Bradshaw and Chadwick, 1980).

A complex soil nutrient, essential to all plants for growth and enzymes, is phosphorus. It readily becomes insoluble in acid soils, combining there with iron or aluminium. In calcareous soils it forms insoluble calcium compounds. In arid areas, phosphates can be deficient except in humus, in forests, and in the silt of ponds. Manure from seed–eating birds has high phosphate levels. Correct inoculation of introduced plant species with phosphate–mobilising mycorrhiza may be essential to their growth. *Acacias*, eucalypts, legumes generally, pines, *Casuarinas*, and even garden crops can benefit from the root associates that enable their roots to accumulate phosphorus from the surrounding soil. Phosphates must always be applied in small quantities, close to the crop.

Trace elements (zinc, copper) can be applied to soil (as sulphates), and 7 kg/ha of both zinc and copper sulphates added to alkaline South Australian soils enables pasture growth (and the carrying capacity of sheep) to increase by a factor of 40 times. 50 g/ha of molybdenum in more acid soils enables clover production. However, it is also possible to add micronutrients as foliar sprays, to infuse them into irrigation water at root level, to include them in slow–release pellets at root level, or to mix them in bulk soil amendments such as dolomite. Over–supply of these same nutrients will result in toxicity and health problems. No amount of guesswork can supplant careful and skilled soil analyses. Plant tissue from healthy garden plants should analyse about as in **Table 11.1.**

MINERAL SOLUBILITY AND pH; SOIL AMELIORATION

A general solubility diagram is given below for soil minerals; this is, in fact, a measure of the availability of specific minerals to plants. Important soil constituents (iron, cadmium, silica) show a range of solubilities with pH as follows:

- IRON: very soluble at pH 3–3.5 as limonite (yellow) Fe_2O_3, and at pH 7–8 as iron oxide (FeO), which reddens alkaline deserts.
- ALUMINIUM: very soluble at pH 4–4.5 and pH 9.5–10 as Al_2O_3; relatively insoluble over normal garden ranges.
- SILICA: slightly but increasingly soluble over pH 0–8, but then rapidly becoming more soluble up to pH 10 as SiO_2.

Iron and manganese (ferromagnesium minerals) are closely associated in soils, and show the same spectrum of deficiency with pH. We can often therefore ameliorate soils at the higher pH levels with the addition of simple sulphur, and restore iron, magnesium, and phosphate levels to plants if these are "locked up" (and test as being present). We seldom garden at pH 3–4, but if we do, we will need a fairly massive calcium input as crushed or burnt lime, or crushed shells, to increase the availability of nutrients.

In the deserts, the soluble aluminium and silica may be evaporated to the clay minerals allied to illite or montmorillorite soils. With all minerals, higher temperatures in water greatly increase solubility. Aluminium ores are usually formed at the lower pH values, and hydrated to gibbsite or bauxitic ores.

The carbonic acid naturally present in rain (and very much increased in rainwater as it infiltrates soils) removes potash from orthoclase in granites. In the plagioclase of white granites, calcium and sodium bicarbonates are similarly formed, or potassium and magnesium carbonates from biotite rocks (biotite, dacite).

FERTILISERS IN ARID AREAS

Fertilisers, apart from humus and limited animal manures, should be used sparingly. Excessive green growth can subject trees, in particular, to drought stress. Magg suggests a novel method of burying two plastic bags in the planting hole for valuable trees, provided

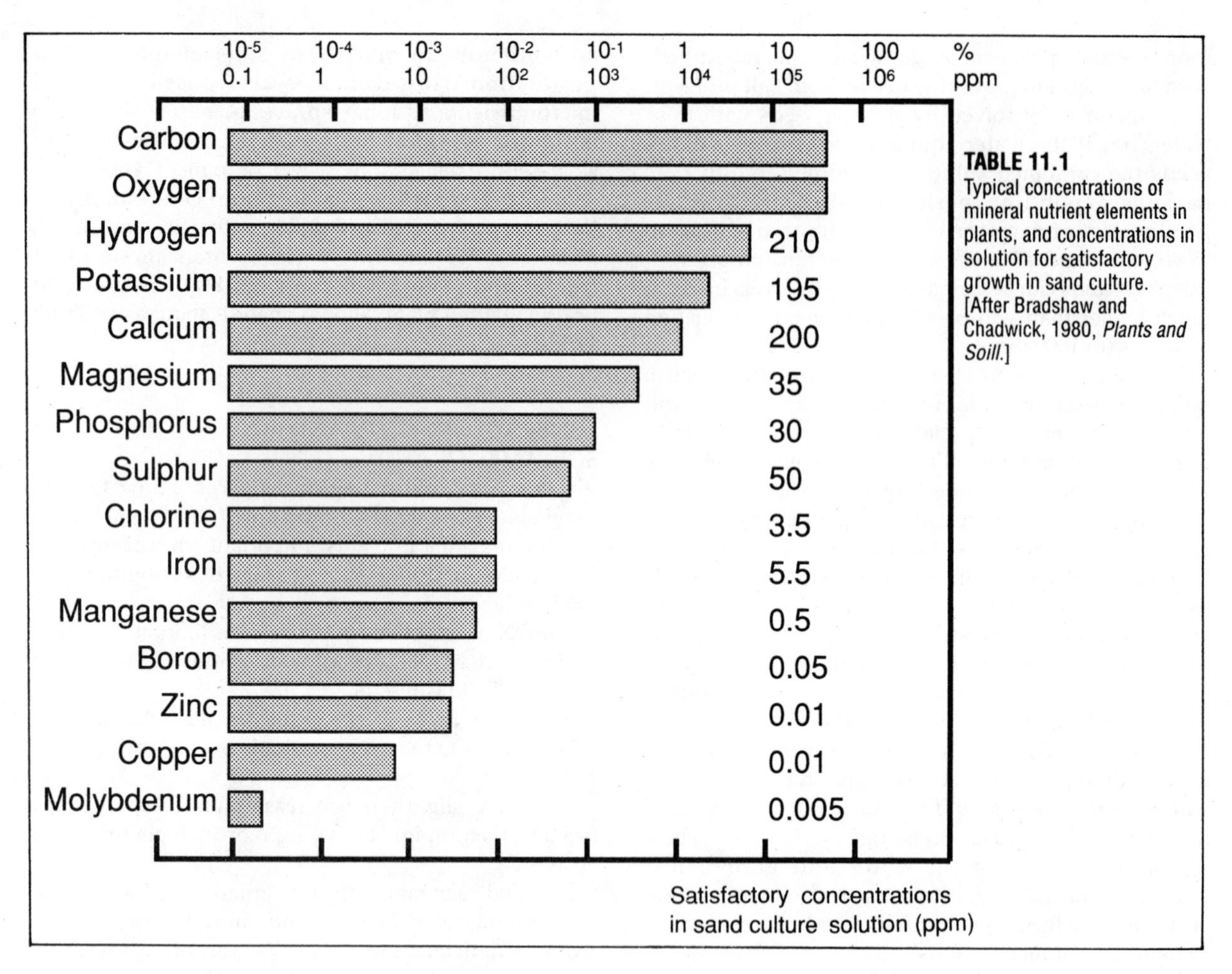

TABLE 11.1
Typical concentrations of mineral nutrient elements in plants, and concentrations in solution for satisfactory growth in sand culture. [After Bradshaw and Chadwick, 1980, *Plants and Soil*.]

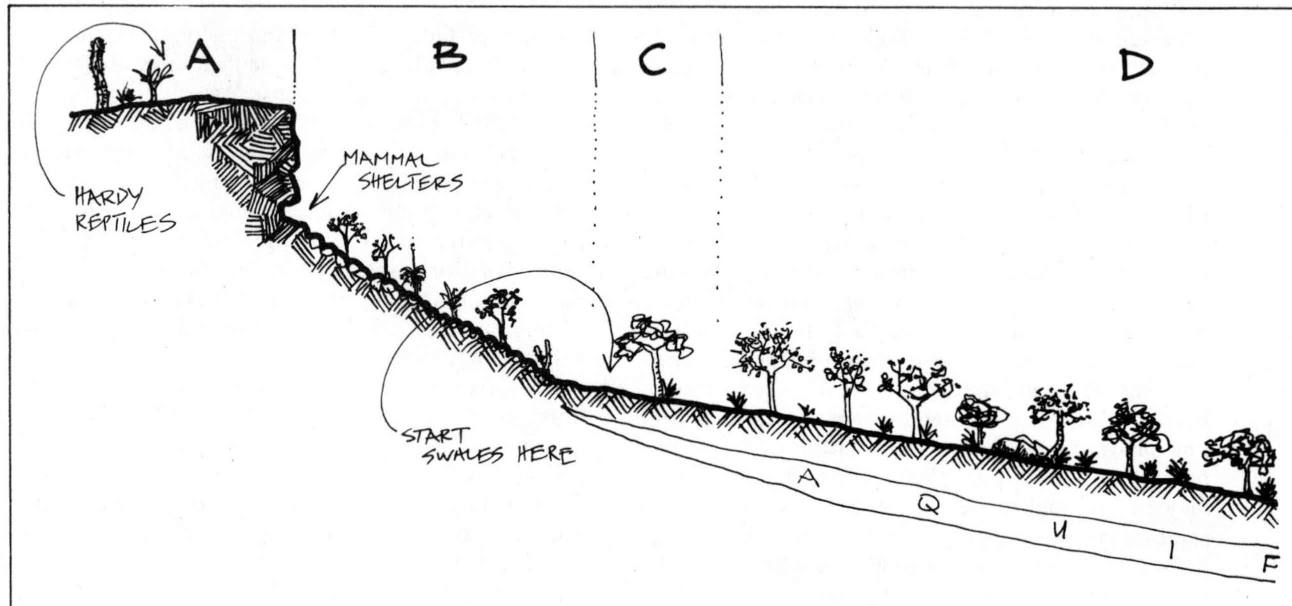

FIGURE 11.2
DESERT VALLEY PROFILE IN FOLD MOUNTAINS.
A. Hard lateritic or ironstone cap in plateaus; skeletal soils, long–lived hardy, woody plants.
B. Loose scree slopes; bunch grasses, hardy plants, some large trees if infiltration is good; coarse sediments; mammal and reptile shelters and burrows.
C. Outwash plains, sandy clays, fast runoff; will grow trees if swales are introduced.
D. Sandier infiltration and"flood out" areas. Trees can be as a scattered

with pin holes. These contain 0.5 kg of phosphate and slow–release pelleted fertiliser. These last the life of the tree (Magg, D. H., *The Potential for Horticulture in Central Australia,* CSIRO Horticultural Science).

Gardeners generally have had good results from shredded bark, manure, and leaf nutrients as mulch, with compost below this, and some sulphur added if pH is high. Others have developed pit composting with continuous mulch layers above. An excellent system where salt in water is not a major problem is to combine a permanent compost pit with drip irrigation from a small sprinkler, or a smaller humus pit with a jar for watering (**Figure 11.69**).

In dunes, phosphate, ammonia or guano, and zinc give good results (for trees). Ammonium sulphate is also used. Sewage waters have everywhere given good results with fuelwood trees (salt content must be checked for, as water gains in salts 300 ppm on its way through towns). Sewage lagoon walls are excellent tree sites, as are raised islands in the lagoons. Shallows grow lotus and water–lily. One feature of such lagoons is the large munber of trees that are bird–carried, as it is of waterholes generally (many seeds are defecated or regurgitated near water).

Most people do *not* add potash, as these salts, and sodium, chlorine, carbonates, calcium, boron, and magnesium may be plentiful in many arid soils, but the best advice is to get thorough analyses of both soils and leaf before and after growing.

THE EFFECTS OF POISONS IN DESERTS

All persistent biocides should be sensibly banned from drylands, as it is the (missing) aquatic plants that break these down most effectively. Herbicides used on lawns will seep slowly to sandy watercourses and down them for miles, travel up and down on groundwater, and kill thousands of plants over time. Water movement in soils is slow, and deep sands do little to offset poisons. Thus, natural remedies are a high priority for deserts and all fragile ecosystems. We should out–think, not poison out, any species that promise to trouble us, as most weeds provide biomass or mulch.

California is in very deep trouble with its groundwater supplies after a 40–year chemical orgy, and many Californians die of these poisons in their water. The few nearby swamps and lagoons are being wiped out, as are the trees near Alice Springs (Australia) where lawns are developed along with their poisons.

The position of settlement, and the disposal of sewage must be carefully assessed in arid lands (including arid islands). In Israel (*New Scientist*, 13 Oct., 1977) sewage and agricultural pollution rises through the sands in summer, and nitrates can be included in bore water pumped for domestic use. Winter rains again carry pollutants down, and they must then be kept clear of the water pumped to resevoirs. As many desert waters already have high levels of dissolved salts, any additional stress from nitrates may cause kidney malfunction. Some 47% of Aboriginal outstation people using bore water do in fact suffer kidney damage, consequent high blood pressure, and excessive intercellular water (a form of dropsy). It is easy enough

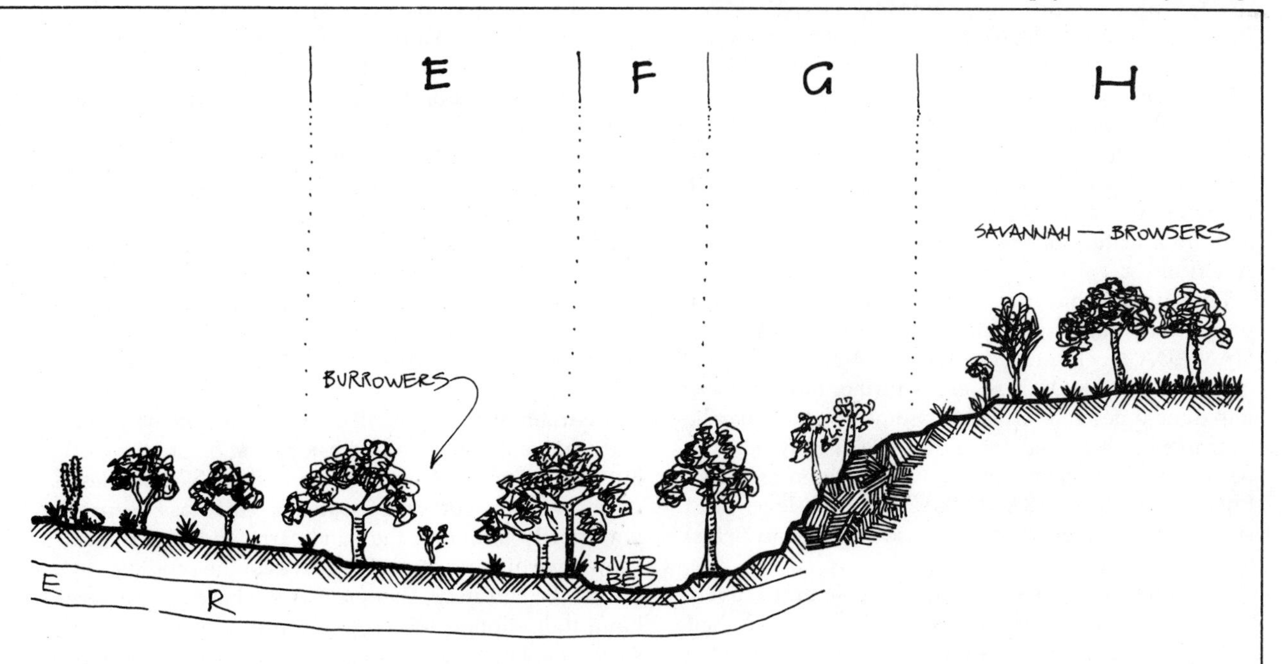

savannah, thick grasses after rains; stick–nest rats and mound–building birds.
E. River deposition of silts; many burrowing reptiles and small mammals; large trees with deep roots, figs, casuarina, river trees (cottonwoods in North America, redgums in Australia).
F. Sandy river bed or wadi; a fringe of dense large trees and vines if ungrazed.
G. Resistant rocky cliffs; can be springs; adapted figs, hardy trees.
H. Dry rocky hills, scattered shrubs and grasses after rain.

to go wrong in desert, and to put whole populations at risk; water in particular must be subject to frequent analysis for pollution and salts, and deep bore or well waters rigorously tested for excessive mineral and radioactive pollutants.

11.5
LANDSCAPE FEATURES IN DESERTS

Occupied and fire-managed arid lands present a total mosaic of vegetation, with very different changes obvious to the traveller. To "read" such a landscape mosaic, a designer needs to note:

- PROCESS: whether wind, water, or infiltration is active locally.
- ROCK AND SOIL TYPE: these decide local response to process and produce characteristic landforms.
- ASPECT: even slight shading by hills changes opportunities and promotes growth.
- FIRE FREQUENCY, and the time since the last fire.
- THE DATE OF THE LAST HEAVY RAIN: (more than 12 mm) which may have been the trigger for a specific age-group of plants (recruitment of species).

In drylands, erosion landforms are both more significant, conspicuous, and more numerous in type than in humid areas. Extensive deserts may show ranges of MOUNTAINS with complex long valleys and shear-sided gorges draining these. Isolated or grouped granitic domes (INSELBERGS) rise steeply out of the desert plain, and the complex SCARP AND PEDIMENT landscapes of fault-lines, mesas, and wadis (box canyons) are also features of the desert. BADLANDS of complex, eroded, softer sediments may develop on unconsolidated areas, and sharp-sided and much branched GULCHES (gullies) develop on steep slopes. Series of folds give a BASIN AND RANGE topography over most desert borders near mountains, and are a feature of the Great Basin deserts of North America.

True desert has the broad primary landscape pavement classification of ERG (sandy desert), HAMADA (rock and boulder pavement), and REG (gravel surfaces). Of these, erg is further broken down into dune types and formations, some of which may lie over the reg or hamada base. Reg is generally taken to be an area from which sand and silt has been removed. Erg is composed of SANDPLAIN as near-level sand sheets over various substrates (also called sand fields) and dunes of several types.

In total, desert landscapes are angular, and actively eroding. Humid areas, and especially those clothed with grasslands and forests, have softer and more rounded outlines, with rare cliff faces at recent fault-lines or shorelines (**Figure 11.2**).

The lower slopes, basins, and playas (pans) of overgrazed and eroded drylands need attention, as do the dry river beds and aquifers that lie below the surface sediments. For this reason, and because a very scattered literature exists on desert strategies with respect to specific landform types, I have tried to assemble here a set of strategies suited to specific sites and landforms (**Figure 11.3**).

SCARPS AND WADIS

Wherever periods of uplift, or scarps of faults fracture desert peneplains, long cliff-lines (some running many tens of kilometres) stand above the lower erosion surface. While these uplift scarps are also visible in humid lands, in deserts they remain less softened, and develop characteristic profiles. These profiles, angular and essentially simple, will quickly emerge from the rounded hills of faulted humid areas once the vegetation is cleared and desertification sets in, as it has emerged in country cleared of trees and eroded in the 1920's in South Australia.

The process develops as per **Figure 11.4**. In true deserts, the profile is typically as per **Figure 11.5** (scarp section above and wadi plan below). Usually, the scarp face is fairly straight or only gently curved, and the wadis are at near right-angles to the scarp face. Within the wadi side, valleys leave again at near right-angles to the main valley; the pattern is that of parallel fault weakness and compensating joints in the rock.

Complex box canyons (wadis ending in cliffs), mesas (isolated pieces of the main scarp), and buttes (cut-off scarp sections) make various scarp-lines. Scarps are capped (the upper erosion surface) with durable ironstones, hard sandstones bands, ferricrete, or silcrete crustal material. Cliffs and scarp faces are generally of softer, sometimes bauxitic (alumium oxide) material, and the lower erosion surface is covered with silts and sands washed out of the wadis and later distributed by winds.

When it rains, the hard rock surface of the upper scarp surface sheds most of the water, which follows gentle valleys behind the scarp and falls as sudden waterfalls over the cliff ends of the wadis. Just before it does so, however, turbulence may scour out deep holes in the upper surface (**Figure 11.6**). Water may rush in torrents through the system for 3-4 days after rain, and the volume discharged largely depends on the catchment area.

A curtain of water, or rills, may fall along the whole scarp (there is usually a downslope just before the cliff) but this is a much more ephemeral flow. Nevertheless, over time, this curtain of water or cliff-base seepage may partly undercut the soft scarp, and **Figure 11.5** shows one of these scarp-base caves in the profile.

Like the cliff itself, the notch at the base can be very large (tall enough to take a 6-storey house) or very small (just large enough to give shade to a kangaroo or a sheep). Notches in the cliff face may be occupied by pigeons, owls, swallows and swifts, rodents, reptiles, and insects or their larval forms. Shade, like water, is a critical resource in the desert.

The wadi profile, or fair approximation of it, is also

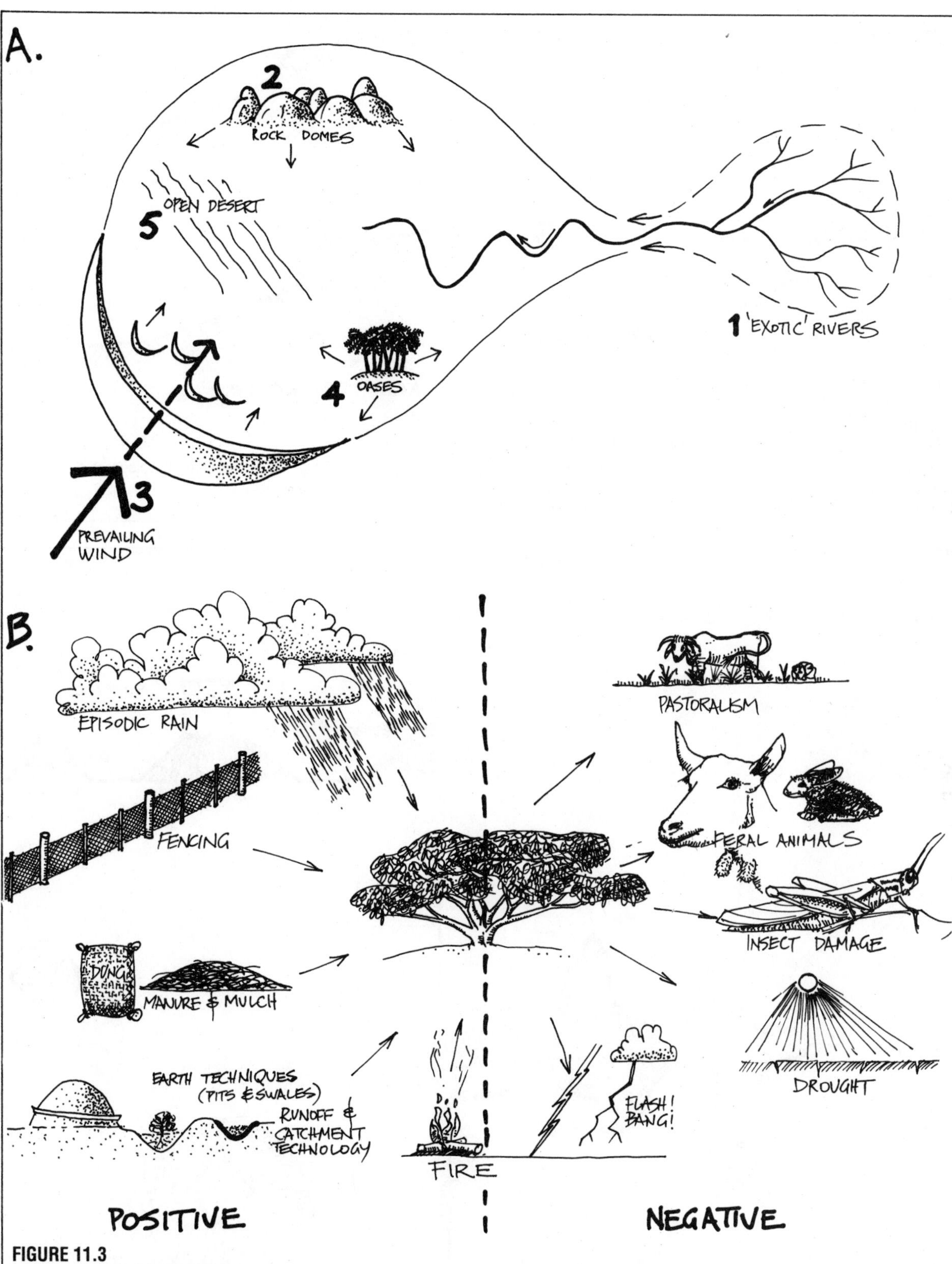

FIGURE 11.3
TOTAL DESERT STRATEGIES.

A. 1. River headwaters; most runoff and best area to create swale forests and generate vegetation downstream.
2. Inselbergs, scarps, and folds give many sites for runoff collection, some dams, foothills flood retardation and forests.
3. Upwind; pelleted seed, soil pitting, dune stabilisation protects downwind areas.
4. Oases, deflation hollows, provide nucleii for settlement and dune stabilisation.
5. "Flood–outs" and pans may yield crops after heavy rains; a general scatter of pelleted seed on the desert awaits rains.
B. Factors affecting desert stabilisation; investment and political policies have the greatest influence on these.

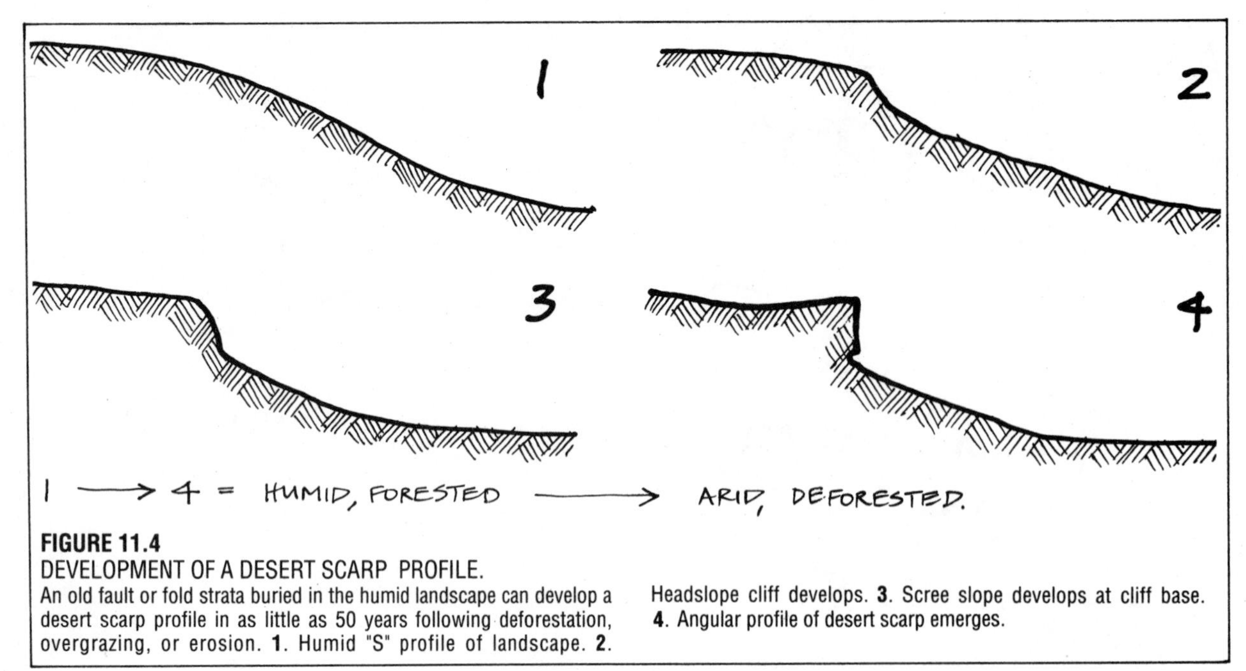

FIGURE 11.4
DEVELOPMENT OF A DESERT SCARP PROFILE.
An old fault or fold strata buried in the humid landscape can develop a desert scarp profile in as little as 50 years following deforestation, overgrazing, or erosion. **1**. Humid "S" profile of landscape. **2**. Headslope cliff develops. **3**. Scree slope develops at cliff base. **4**. Angular profile of desert scarp emerges.

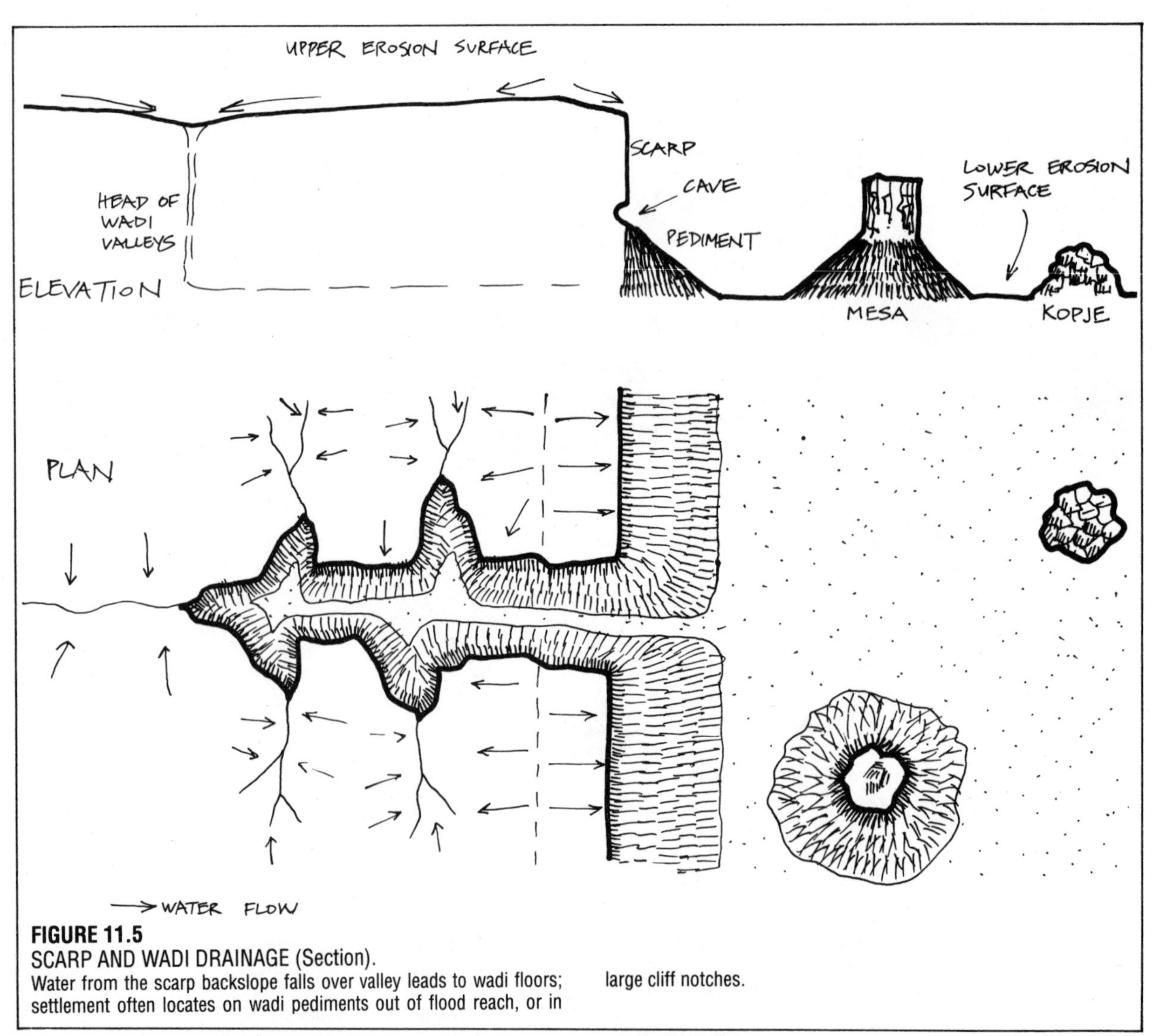

FIGURE 11.5
SCARP AND WADI DRAINAGE (Section).
Water from the scarp backslope falls over valley leads to wadi floors; settlement often locates on wadi pediments out of flood reach, or in large cliff notches.

found in incised sandstone valleys in semi–arid older landscapes, where rivers have cut down as the land rises. In these cases there is a valley floor (often with a river remaining), and the cliff is not a simple scarp but more often a set of complex smaller scarps and caves in bedded sandstones or mudstones, as per **Figure 11.7**. Caves still occur (large or small) at the base of a cliff, or in the cliff face itself. Side valleys may resemble box canyons, but are less regular than those of fault–scarps, and waterfalls may be a series of cascades.

After heavy rain in deserts, water pours off the cliffs and rushes as mixed mud–silt–stone torrents down the wadis. At times, the flood can be devastating, and has caused many fatalities in unwary campers, or has carried away badly sited villages and houses. The iron–clad rules are to keep houses and roads well above the valley floors, to cut them into the pediment of the scarp itself, or to site houses in the base caves of the cliffs.

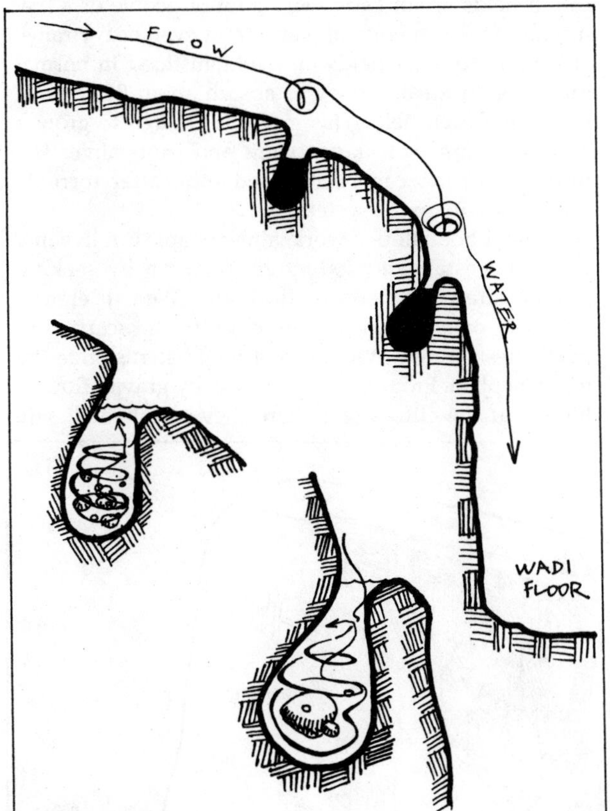

FIGURE 11.6
SCOUR HOLES.
Just before plunging over wadi walls, streams often scour out waterholes; some of these can be enlarged into cisterns; all are import to wildlife, especially birds. Old sandfilled holes can be rock–rimmed for tree planting sites.

Water Storage and Use in the Scarp Landscape

We can start on the top erosion surface to harvest and divert water. Up top, we can erect stone and cement dams across the generally gentle valleys of the upper plateau. These will hold clean water, and can be fitted with plugs or base pipes to let the flood water through more gently. Next, we can cut gutters across the scarp top above useful caves, and prevent too much water splashing over to our cave–house sites.

We can also clear silt from, deepen, and seal the scour–holes in the upper surface, and use these as tank storages. Some of these can lie *inside* dam walls as deep spots. We can also cut gutters away from the main valley head and spill all (or a larger proportion of) the water to the side valleys, again lessening the torrent in any one place. Once our high dam fills, however, we may still receive torrents over the headwater cliffs, and the only way we can further moderate the erosive effect of these is to make concrete or stone–cut chutes leading into excavated pools at the cliff base. All wells, pools, and dams will gather silt, however, and will need periodic cleaning. Dams can be cleaned by leaving base pipes open in alternative flows. Cliff pools self–clean by turbulence.

There is little hope of growing any but the very hardiest trees on the upper surface, except in crevices or deliberately constructed rock–walled containers. It is on the wadi floor that we find deep sands, gravels, and

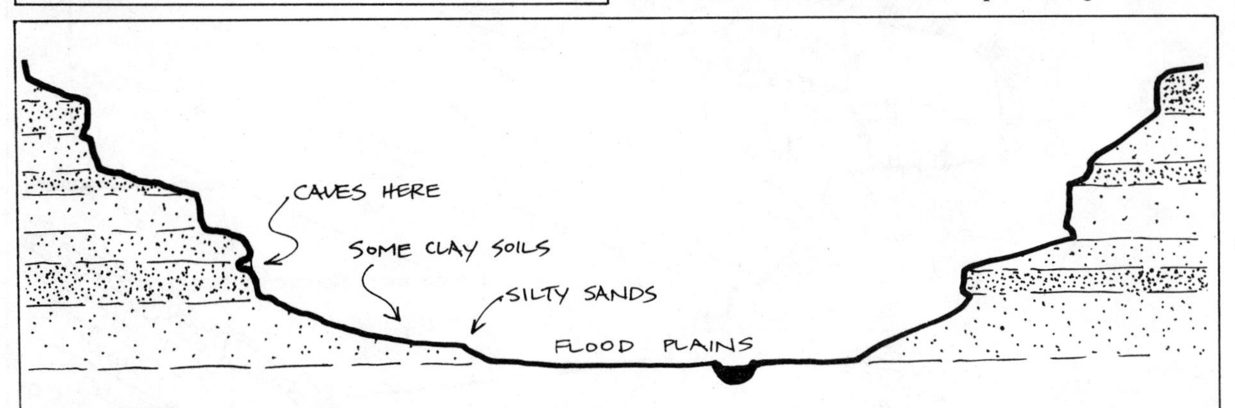

FIGURE 11.7
SANDSTONE VALLEY SCARPS AND CAVES.
Sandstones in semi–arid areas give a succession of scarps, caves, ledges, and a deep sand flood plain. These are more complex sites than true desert scarps, and are open to innovative design; wet seepages are common on lower ledges. Such profiles can arise in more humid landscapes of massive sandstones.

silts, and from there they spill out for a few kilometres onto the lower erosion surface.

In the wadi therefore, we need to erect strong rock walls across the floor, and let silt fields build up behind them. For minor wadis, the whole width may be dammed, and spillways cut in the pediment rock. In wide wadis, a central channel can be left open, and fields built to the sides (fed by side valleys). Once the flood spreads out on the lower plain, however, we can form very broad walls of earth and stone and soak it in to the fields so contained; these banks are best stabilised with unpalatable shrubs.

The most extensive wadi development studied to date is the ancient Nabatean systems of the Negev desert (Michael Evenari and D. Koller, 1956, "Ancient masters of the desert", *Scientific American* 194(4): 39–45) These have been in part restored, but there are also thousands of hectares of fields still abandoned.

The Nabatean systems hold 80–90% of run–off behind dykes in the wadis, and these are spread to lower dykes in the plains. The water is spread by highly coordinated flood and soak regimes to orchards. For every irrigated hectare, 20 ha of run–off surface are reserved as catchment. From the moment it begins to rain until the fields are soaked, the systems around Petra need close attention. In this area of the Negev, some rains fall reliably in winter every year.

The ideal is to let a minimum of water escape as run–off, and to absorb as much fresh rain water as we can in fields or silt beds. With a lot of people or a few machines we can hold all water from even heavy rains, although the outer fields may seldom flood in normal rains. As an ideal, we try to absorb about 0.5–1 m of water into each field. This gives us enough to grow a crop of grains, or to keep palms and fruits alive. The upper resevoirs can be released soon after torrents cease, and top up the system.

Even without all this work, a few people can live in a wadi and establish productive tree crop by seeking shaded sites and deep natural silts. We can erect a windmill over a well or bore close to the scarp, and pump fresh water up to tanks or rock cisterns on to the top peneplain. From there, it will run by gravity flow to houses at the cliff base (where they are shaded, safe

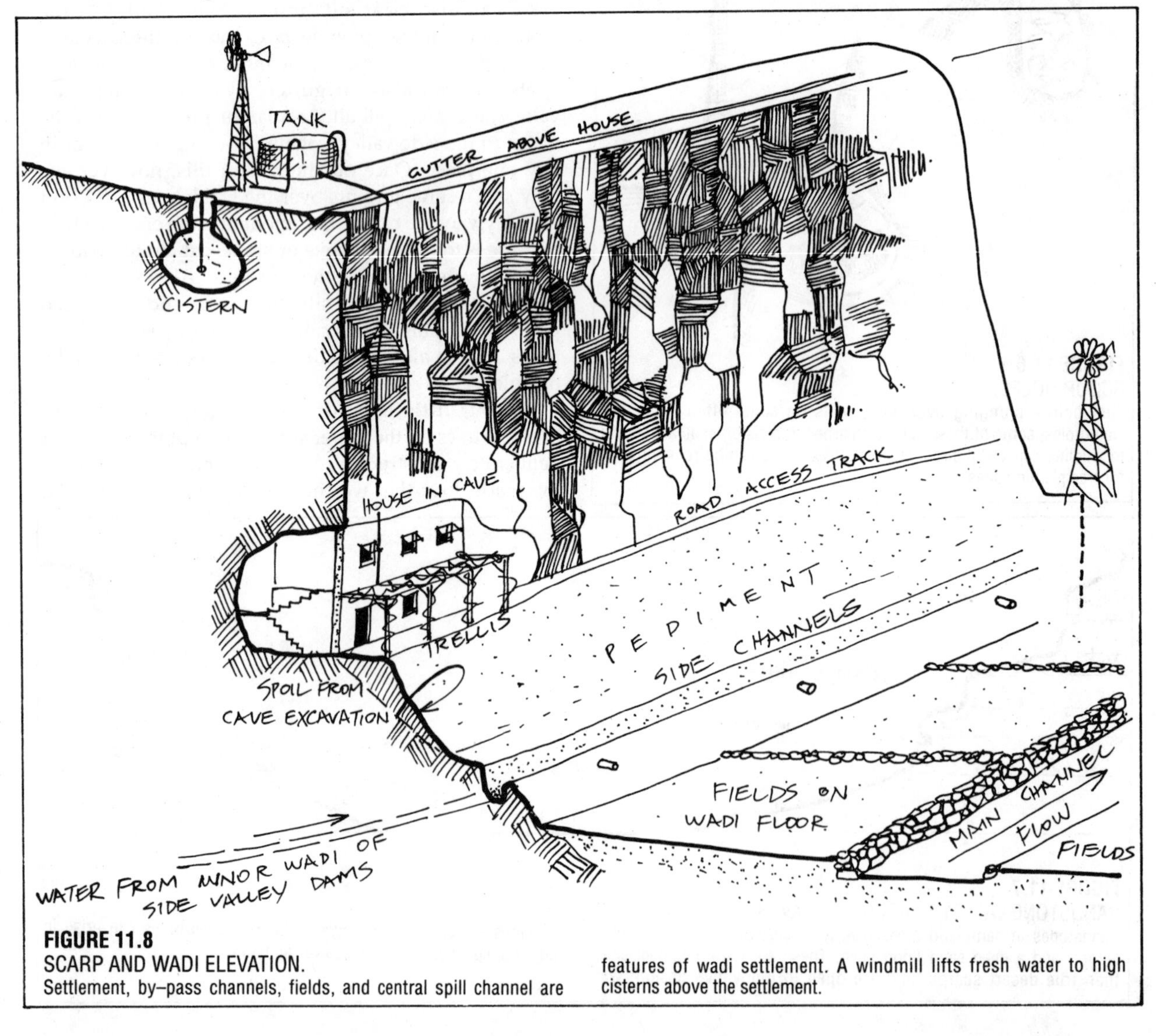

FIGURE 11.8
SCARP AND WADI ELEVATION.
Settlement, by–pass channels, fields, and central spill channel are features of wadi settlement. A windmill lifts fresh water to high cisterns above the settlement.

from flood, and can be in part excavated into the cliff base). **Figures 11.8** and **11.9**.

Fortunately, most peneplain crusts are formed over softer rock, so that the cutting of channels and gutters is fairly simple. There is usually a plentiful supply of rock for field walls on the pediment slope, and it is simple enough to dig a house notch if none exists. It is even easier to dig small animal shelters and pigeon nest sites, and these also form a food source. Fencing, too, is minimal in that box canyons can contain many hectares, most are bounded by steep cliffs, and have only a single entry, so that both human settlements and animal compounds are easily defended or protected from feral animals or predators.

I believe that with modern machines, fencing, windmills, and solar panels or photovoltaics we could creatively occupy many wadi systems (on a modest scale) that are now just grazed or neglected, and plant a desert forestry system.

Large trees already grow in most wadis, shaded and protected by the cliffs and immune from sand blast. By restricting hoofed animals many more trees would grow; overgrazing is the most obvious plague of arid lands. There is a wealth of detailed work that can be

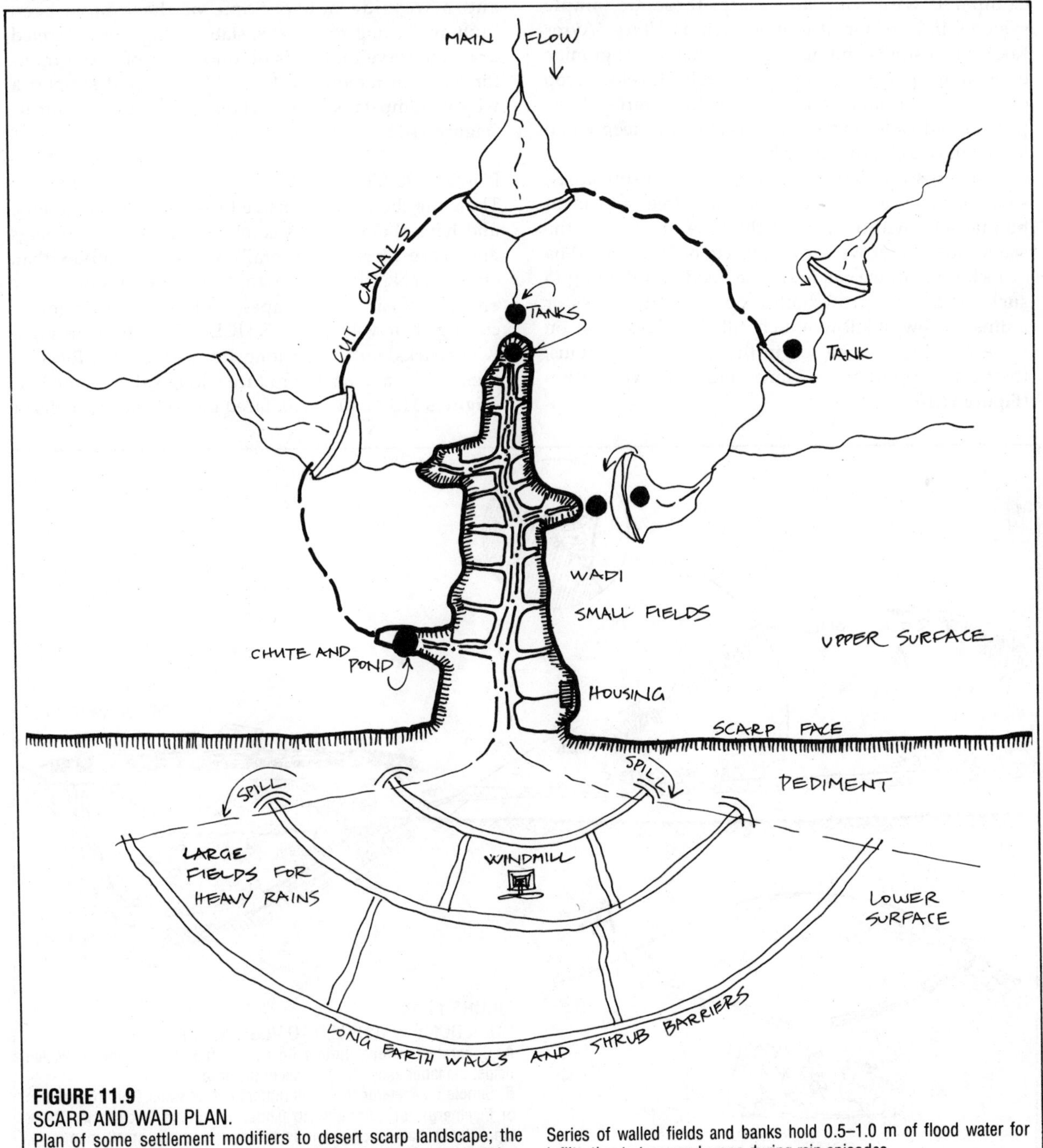

FIGURE 11.9
SCARP AND WADI PLAN.
Plan of some settlement modifiers to desert scarp landscape; the dams on the upper surface act as retardation basins for heavy rains. Series of walled fields and banks hold 0.5–1.0 m of flood water for infiltration to trees and crops during rain episodes.

done in the complex wadi systems, with many special niches utilised, and a rich flora and fauna (in most deserts) to be preserved and encouraged. Just by increasing shade, crevices, and ramp–access water cisterns we can stabilise and support a great many more desert quail and pigeon.

Where grazing animals (camel, goats), have been removed or controlled, as in parts of South Australia, a dense *Callitris* pine forest has regenerated, and palm forests can be established.

RESIDUALS, DOMES AND INSELBERGS

Compared with wadi and scarp, these are simple systems indeed. Great domes such as *Uluru* (Ayres Rock) in Australia and many similar massifs of granitic or metamorphosed sandstones rarely develop deep caves (but do have some rock shelter), rarely have pronounced valley entries, and usually dip steeply into loose sandy soils around the base.

Inselbergs are, however, very solid run–off areas, some of many tens of hectares. Large trees and good humus soils will develop at the base of cliffs on the shade side, and if we allow one acre of "field" to 20 ha of rock run–off, we can lead base floods via fluming to such fields, and grow both crops and fruit trees or palms. A few shallow mulch–filled valleys exist on these massifs, and the soil in them is mainly humus; they can support trees, yams, and other vegetation (**Figure 11.10**).

Occasionally steep and shaded clefts present sites for rock dams, but for the main, partly sealed tanks and soil infiltration must suffice. The larger systems (100 ha or more) have often been occupied in historical times by careful and conservative peoples, usually as base camps for long range foraging after rains. Each one has a special charm, and very sensitive and restricted occupancy is called for, developing and protecting natural resources rather than attempting extensive systems.

Very small rock slabs (1–10 ha) present a useful area for a gutter and tank system, a hectare or two of trees, and a wayside house. Some of these have been "guttered" using rock walls, slab on edge, or concreted drains to serve the needs of a house or of steam trains for water in remote locations. Many would support a wildlife ramp–tank for desert birds and small mammals (**Figure 11.11**).

FOLD MOUNTAINS

These are the most extensive features in many deserts, and have characteristics combining those of inselbergs and scarps, but are generally far more complex than either SYNCLINES (down–flexed sediments) may erode to great canoe shapes, with a few river gorges cutting their way out. ANTICLINES may form great whalebacks, but in the long run a river will form on their spines and again produce a long valley in the hills (**Figures 11.12 and 11.13**), faced on each side by cliffs or

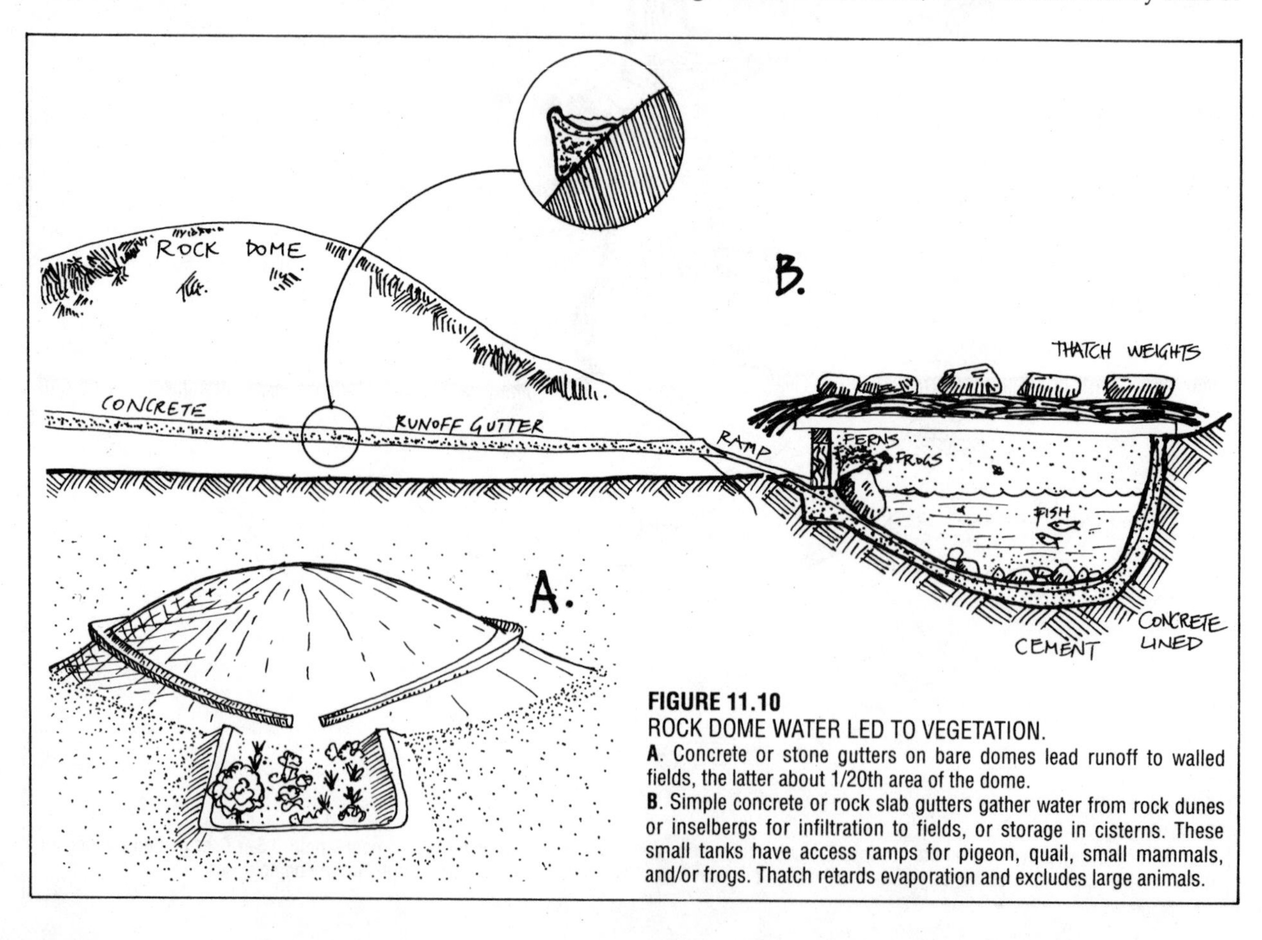

FIGURE 11.10
ROCK DOME WATER LED TO VEGETATION.
A. Concrete or stone gutters on bare domes lead runoff to walled fields, the latter about 1/20th area of the dome.
B. Simple concrete or rock slab gutters gather water from rock dunes or inselbergs for infiltration to fields, or storage in cisterns. These small tanks have access ramps for pigeon, quail, small mammals, and/or frogs. Thatch retards evaporation and excludes large animals.

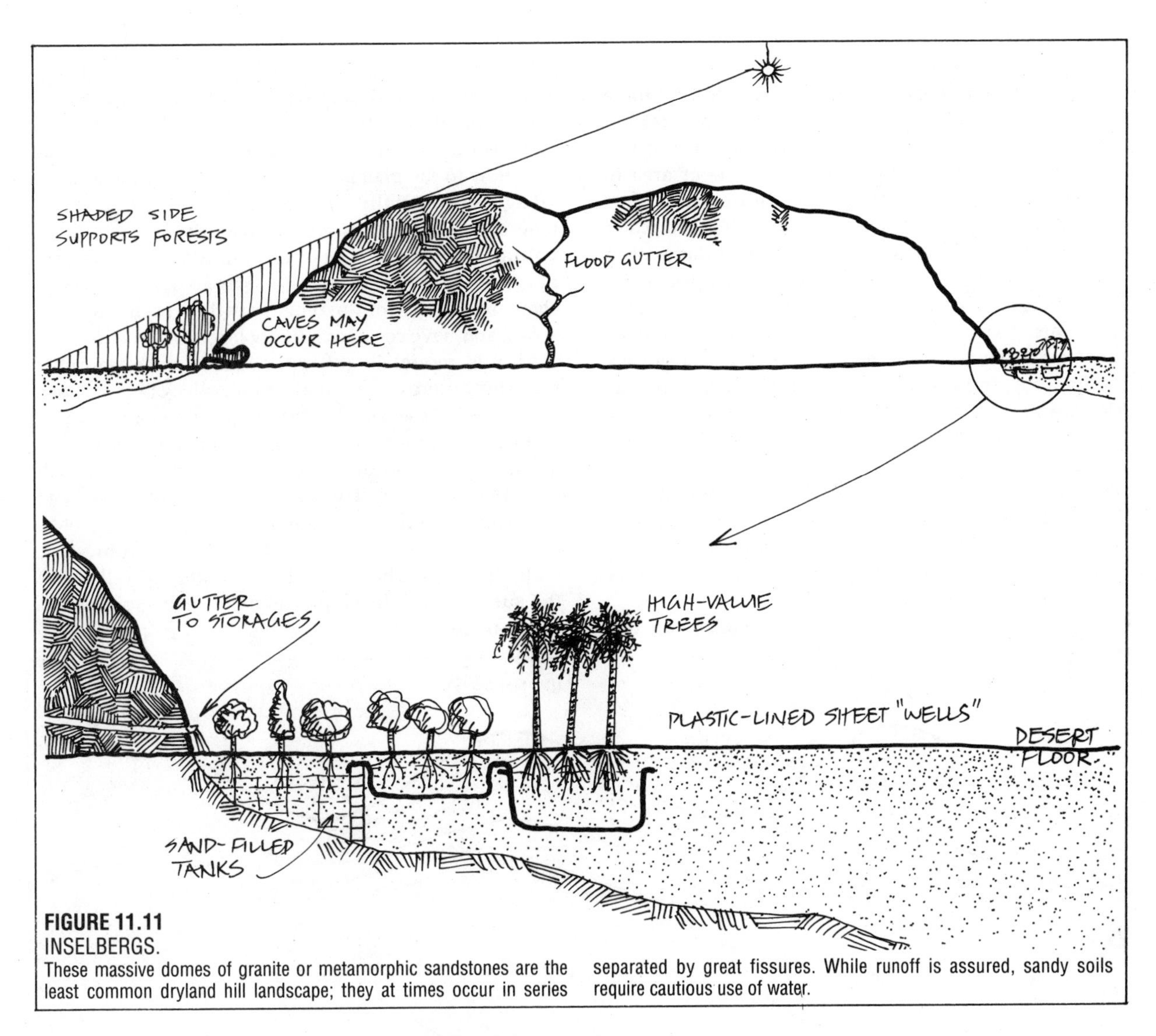

FIGURE 11.11
INSELBERGS.
These massive domes of granite or metamorphic sandstones are the least common dryland hill landscape; they at times occur in series separated by great fissures. While runoff is assured, sandy soils require cautious use of water.

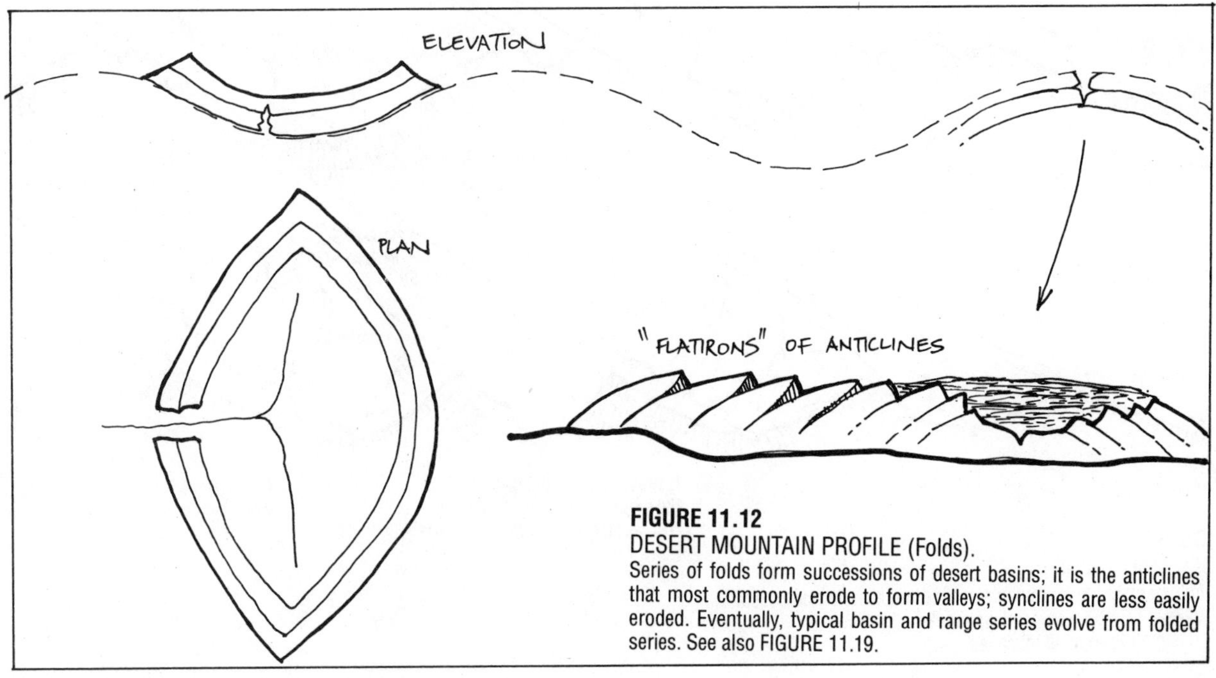

FIGURE 11.12
DESERT MOUNTAIN PROFILE (Folds).
Series of folds form successions of desert basins; it is the anticlines that most commonly erode to form valleys; synclines are less easily eroded. Eventually, typical basin and range series evolve from folded series. See also FIGURE 11.19.

steep tilted sediments.

In many places, a dam at the exit of a river from these ranges has resulted in a large ovoid freshwater lake in the interior of the ranges, but the permanance of this water depends on the proportion of run–off area to storage area (this must be 20–30:1 in deserts). Thus modest and deep storages are more likely to hold water all year than large shallow storages. Such dams need careful hydrological assessment and engineering. However, the opportunity to create successful freshwater dams is greater in fold mountains than in other desert landforms. Wherever massive ranges back the fold mountains (as in Colorado, Nevada, and near Mt. Ararat in Turkey), foothill dams will hold water from a combination of snow melt and rain run–off. Settlements and roads need to be well off any flood path should they burst. As few settlements exist in deserts, this is an open option, whereas large dams in well–settled areas tend to be built above the existing valley settlements, with frequently disastrous consequences in war or earthquake.

A great many more modest dams are available in fold sytems. They range from narrow rock–walled exit valleys to foothill dams where exit waters are led to diversion dams. Some complex freshwater dam systems can be led off low–slope wadi–like streams as they flow to the plains. A vigorous day hunting around village sites usually reveals many possible storages wherever the ranges are sufficiently large to shed volumes of water. These are almost always a few permanent rockholes and streams, as it is a characteristic of fold sytems that they occur in large series, and have correspondingly large catchments.

All fold mountains of sedimentary or metamorphic rock show more or less marked palisades of harder rock series, and these usually run along the slopes like so many sub–contours. Wherever vegetation exists, it is far taller, greener, and denser on the upslope side of such palisades, because it is there that run–off is slowed down and forced to infiltrate. These palisades are not as effective as swales, but they help. Some are wide–spaced, fairly well on contour, project well above the surface, and develop striking tree–lines. Most are more subtle.

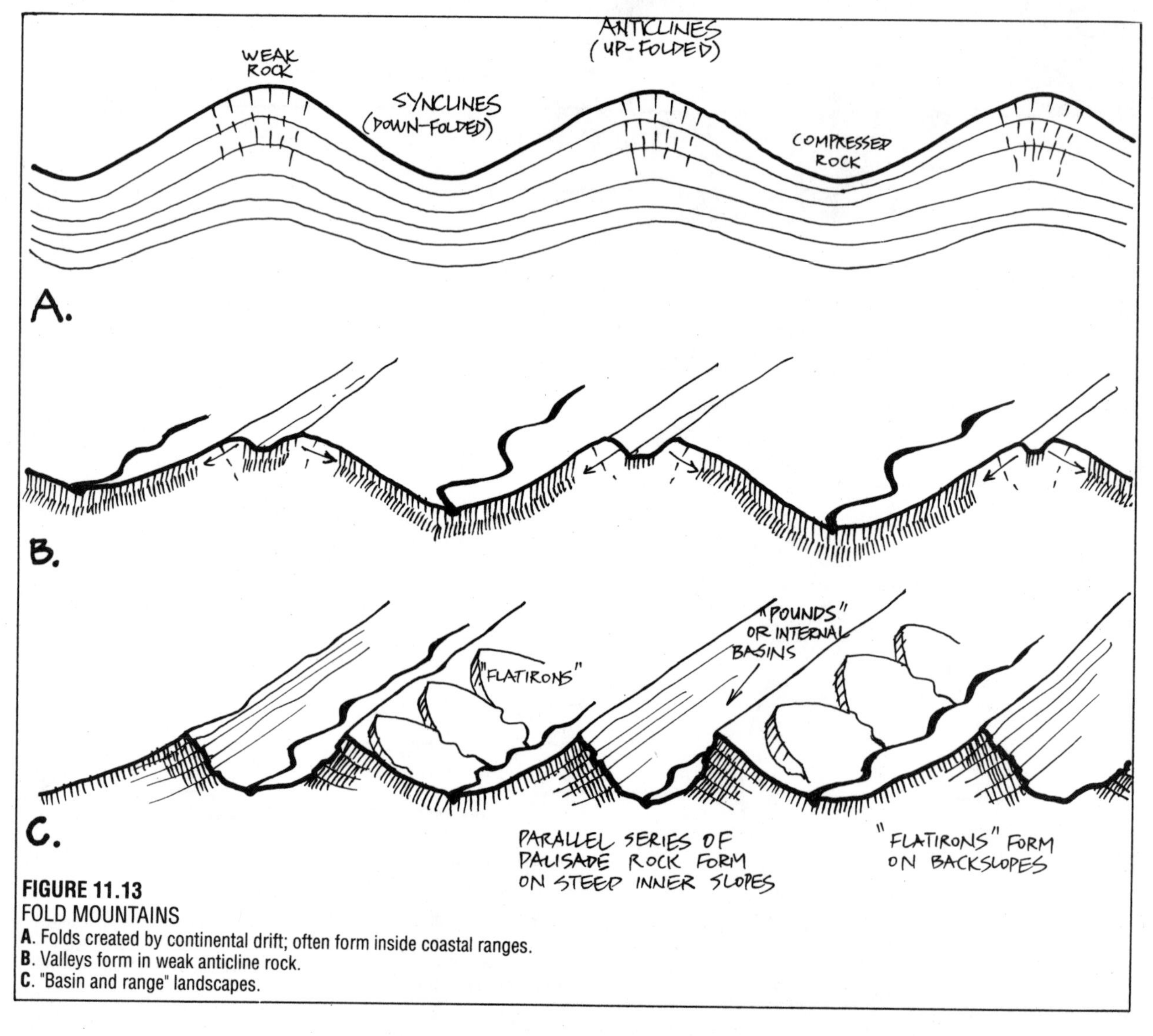

FIGURE 11.13
FOLD MOUNTAINS
A. Folds created by continental drift; often form inside coastal ranges.
B. Valleys form in weak anticline rock.
C. "Basin and range" landscapes.

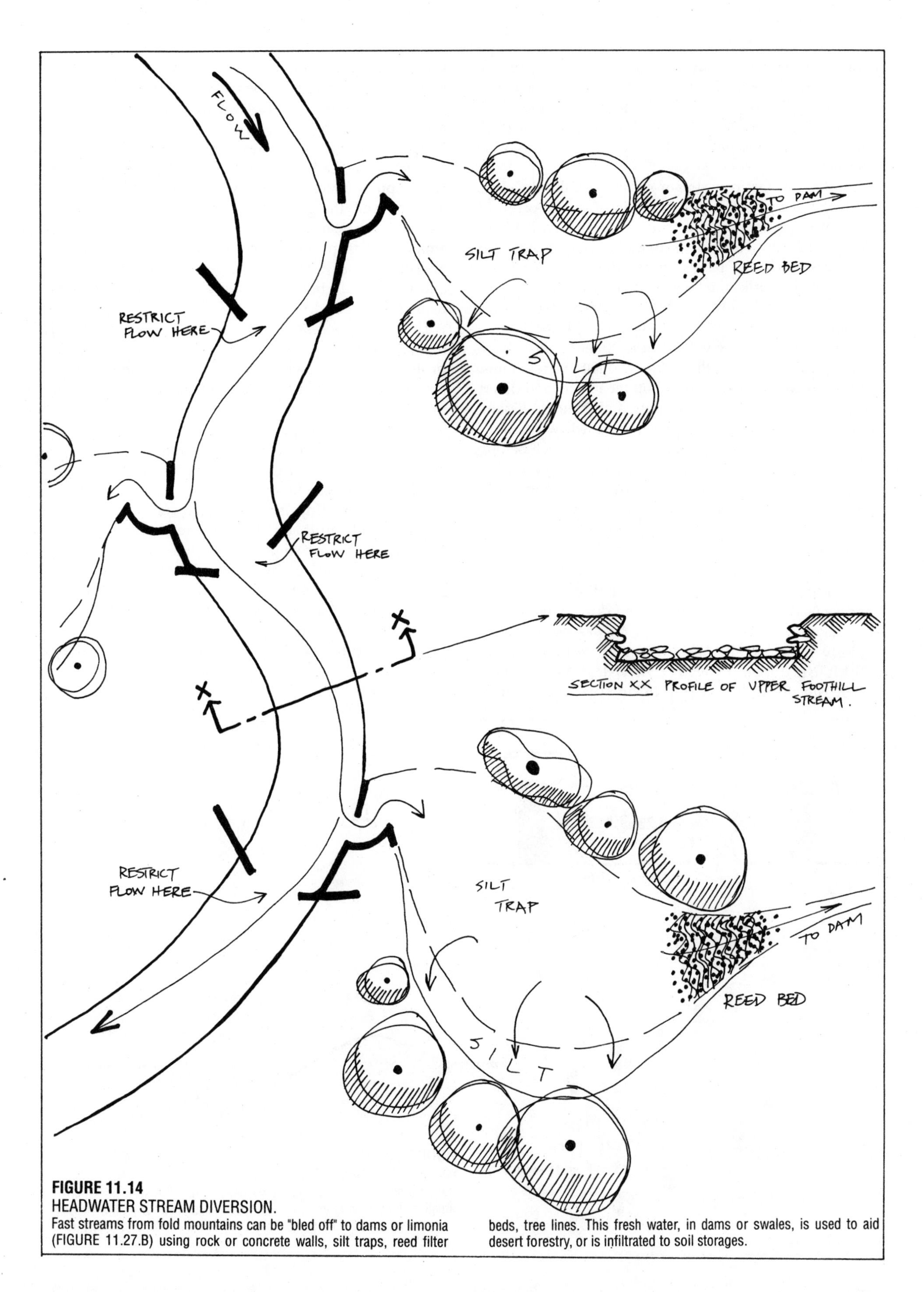

FIGURE 11.14
HEADWATER STREAM DIVERSION.
Fast streams from fold mountains can be "bled off" to dams or limonia (FIGURE 11.27.B) using rock or concrete walls, silt traps, reed filter beds, tree lines. This fresh water, in dams or swales, is used to aid desert forestry, or is infiltrated to soil storages.

Wherever series of these palisades are a feature, a horizontal bore will often enough tap sweet water trapped in the rock strata. Unlike quanats (**Figure 11.37**) such water can be sealed off and used only as needed, and can be be regulated to that amount taken into the aquifer.

A shallow, fast dry–stream system usually develops out of the upper foothill regions of folds, often of stream sequences only 4–15 m wide and with banks 0.5–2 m high. The bed is usually rock–filled, and detritus along the banks shows the extent of flood levels. As in wadis, the flood periods (when they occur) are violent and sudden. The boulders attest to this, along with the smashed lower branches of large trees.

We cannot effectively dam these violent flows, or we would fill that dam with boulder and silt in a few rains. But what we can do is to bleed off the torrent and divert it to a more peaceful contour flow, and thence to a dam. If we lead this flow along an oversized, wide, and shallow contour trench, silt falls out in the trench before ever reaching the dam, and the trench itself presents a planting site *and* a silt trap that is easily cleaned out, whereas a dam is difficult to de–silt (**Figure 11.14**).

DUNE COUNTRY

Sands occur on pavements, or as dunefields (sand seas). Depending on the wind intensity, duration, streamlines and velocity, and the supply of sand downwind, dunes can take up any of a number of characteristic forms, chiefly those of TRANSVERSE and regular ridges, OBLIQUE dunes, LONGITUDINAL dunes aligned with the wind, BARCHANS or crescents (horns downwind), and sand seas of curiously wave–like (rough sea) forms, with lobed advance edges.

Dunes on pavement (isolated dunes) can be planted after rains with a fast–growing grain or oat crop (sorghum, millet, some desert legumes such as moth bean) or yam beans, and a set of hardy *Acacia* seedlings can be put in place, preferably with a mixed fertiliser. If all goes well, the grains flower and the straw lasts for two or more years, the moth beans leave seed, and the small *Acacias* grow to effect permanent stabilisation. Species such as *Acacia victoriae* give copious leaf mulch, and desert yam bean pits can be placed in later years using this leaf mulch.

The whole process depends on a good rain, pest and browser control before sowing, and some minimal protection after sowing. A few larger quickset trees (tamarisk) may succeed if they are deeply set in or near the dune base.

Extensive dune systems need a very different approach, although even here, pelleted seed can be broadcast from the air to await rain. Many such pellets will be buried by sand, and if a heavy rain occurs, some will grow. Pelleting is a relatively simple matter of mixing seed, mud, fertiliser, some insect repellent (neutralised copper sulphate with lime, neem tree leaf powder, magnesite) in a stiffish mass, passing it through a mincer with the blades removed, and then on to a vibrating tray with a slight slope on which dry powder or dusts are shaken. The extruded "rolls" of pelleted seed become round pellets on the shaker tray, and can be spread out to dry. Pelleted seed is not eaten by birds or insects.

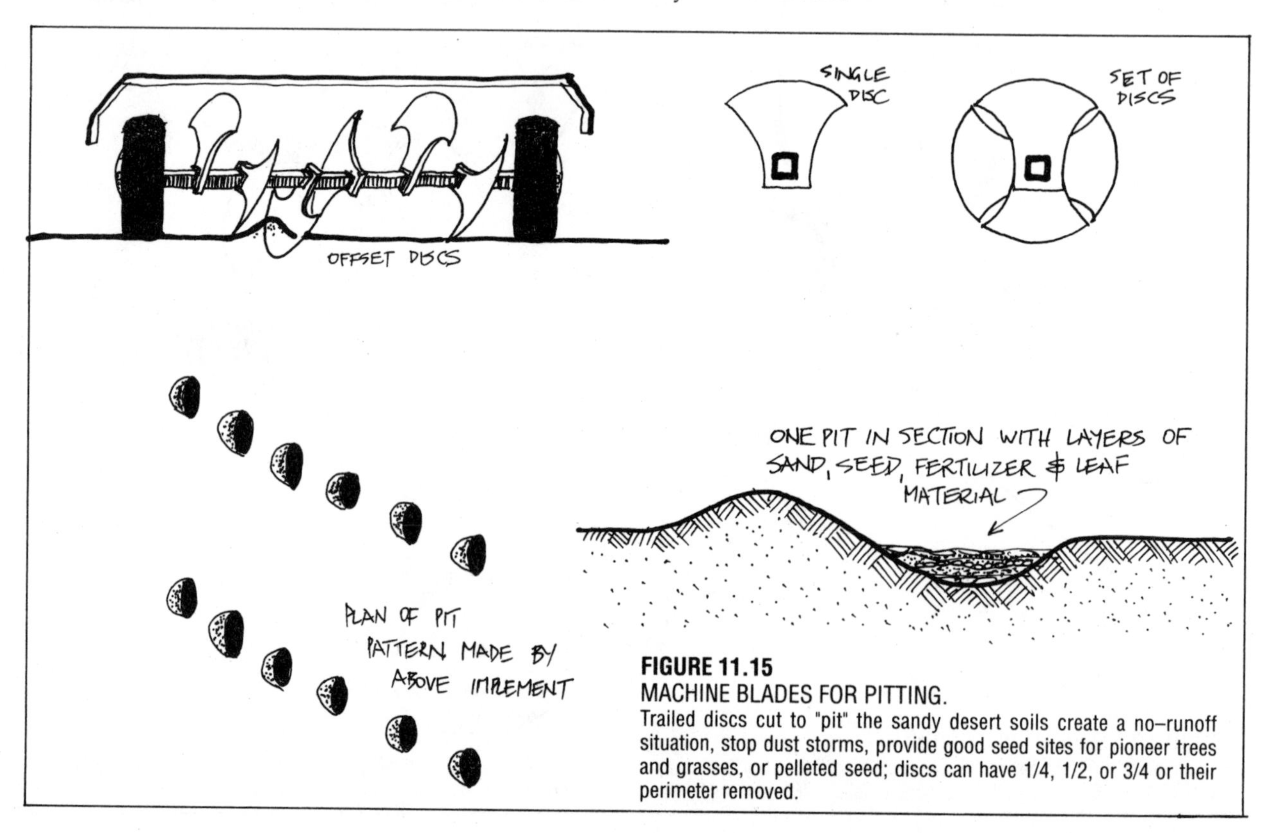

FIGURE 11.15
MACHINE BLADES FOR PITTING.
Trailed discs cut to "pit" the sandy desert soils create a no–runoff situation, stop dust storms, provide good seed sites for pioneer trees and grasses, or pelleted seed; discs can have 1/4, 1/2, or 3/4 or their perimeter removed.

Wherever rain flows or falls occasionally over bare sandy ground, PITTING is an excellent device, Here, a set of large discs with one–quarter section cut off or a small disc fixed to a wheel (**Figure 11.15**) is drawn across–country, and seed and fertiliser spread soon after. Seed, fertiliser, and sand blow into the pits, and these respond very well to subsequent rain. Many hectares of these pits have grown well near Alice Springs in Central Australia.

Severe dune "blowouts" on coasts need more intensive treatment, with hand planting of quickset grasses such as *Ammophila* spp., pelleted moth bean seed, and brush fences in 7 m squares to effect early stabilisation. Such fences are built from stakes 2 m long, driven 1 m deep, and with numerous bushy weaves between. Pits of seed and *Acacia victoriae* may succeed after rain.

We can look on dunes and dune fields as large water tanks. In deep sands, the surface to a few metres may have only 4% water, but as we dig deeper, at about 6 m we find damp sands, and at 40 m saturated sands. Some dune fields (with humus and dust particles) will support quite dense vegetation (50% cover) while others are almost sterile and lack basic plant nutrients.

Any traveller in vegetated deserts will notice that the largest trees (sometimes the only trees) stand in the dune ridges. This is apparent with dunes on harder pavement, or dunes on clay and a strong base pavement, but it is also true of coastal dune hollows and deflation hollows in sands. Dunes represent a resevoir of freshwater, much as the coral sands of an atoll hold rain which "floats" on the salt water, so dunes present a very large surface area of sand grains for water to adhere to, and rapid infiltration of rain is possible (**Figures 11.16 and 11.17**).

At the edge of large dune complexes, it is usual to see water rushes and sedges at the base seepage, and even shallow lagoons which are as much dune–fed as dune–dammed. It follows that dune on pavement or salted ground is the best site for permanent plantings; the problems are how to establish those trees, and to stabilise the dune.

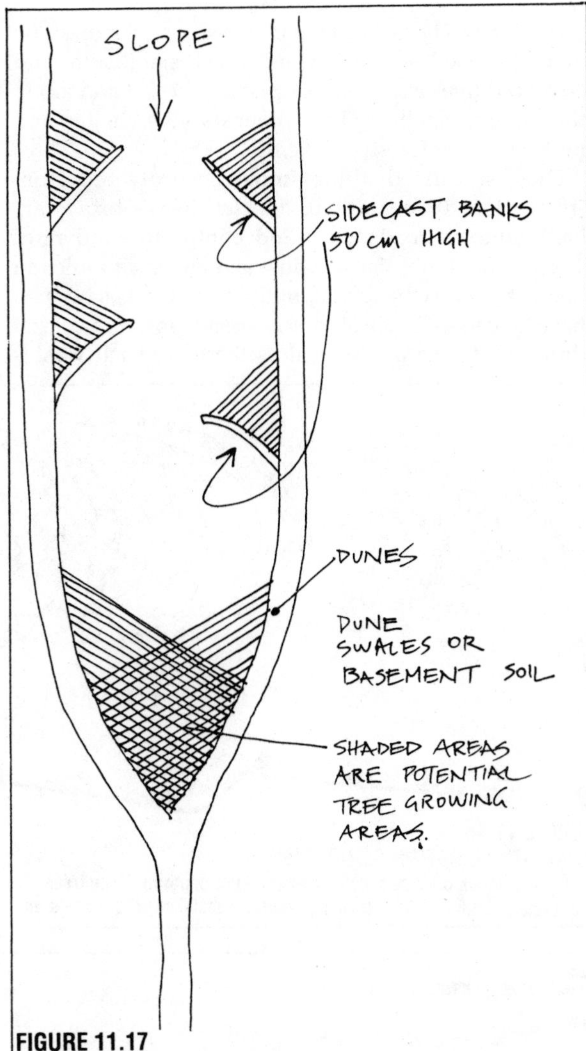

FIGURE 11.17
CHEVRON BANK IN DUNE SWALES.
Side–cast banks halt flood flow down dune swales; hatched areas will grow trees (parallel dune series); swales so chosen must have a slope "downhill".

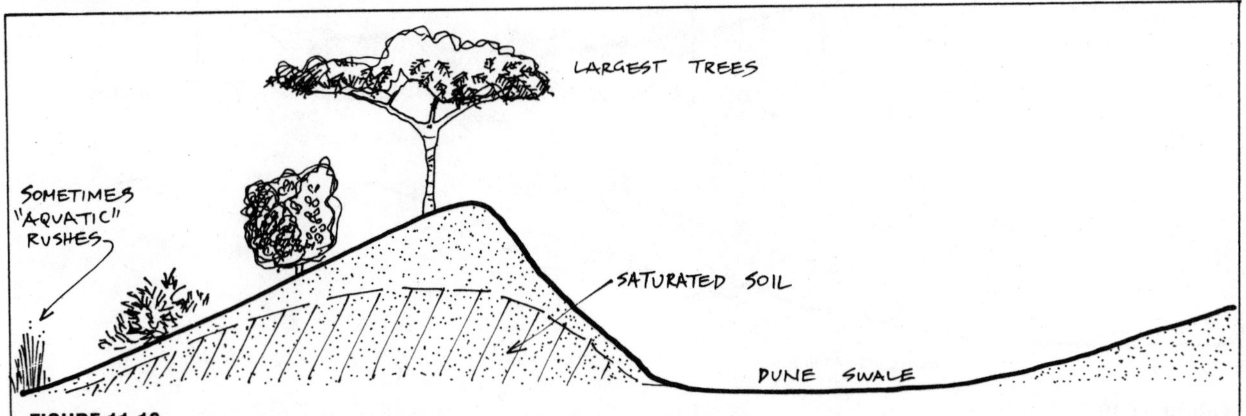

FIGURE 11.16
DUNES, LARGE TREES, FRESHWATER LENS.
Stabilisation by brush fences bring BARCHAN dunes to a halt; sand drift then creates an oval dune which infiltrates water and will carry adapted vegetation. Some sessile trees greatly aid this process (Acacia, Mesquite).

In China (where they take trees seriously), rice–straw mats are used as fencing or surface stabilisers, and advanced trees in baskets are planted (basket and all) in the matting shelter. These pioneers give shelter and mulch for later forestry and crop.

Once stabilised, dunes must be protected from "recreation" (devastation) vehicles, heavy browsing, badly aligned roads, and sand mining to windward. There is no doubt that sawdust and any such mulch in hollows is very beneficial; mill wastes (bark, sawdust) have helped establish an *Acacia sophorae* forest in coastal dunes in Tasmania, but unless there are such wastes locally available, other strategies must be undertaken. All plants in dunes benefit from nitrogenous fertiliser, phosphate, and trace elements, as old dunes are usually deficient in plant nutrients.

DUNE STABILISATION

Apart from the stabilisation of sand and silt surfaces by pebbles and vegetation, any "cementing" system will help. Water creates bonds between particles that call for much higher wind speeds to move the sand, as does salt (especially where dew is present, as salt is

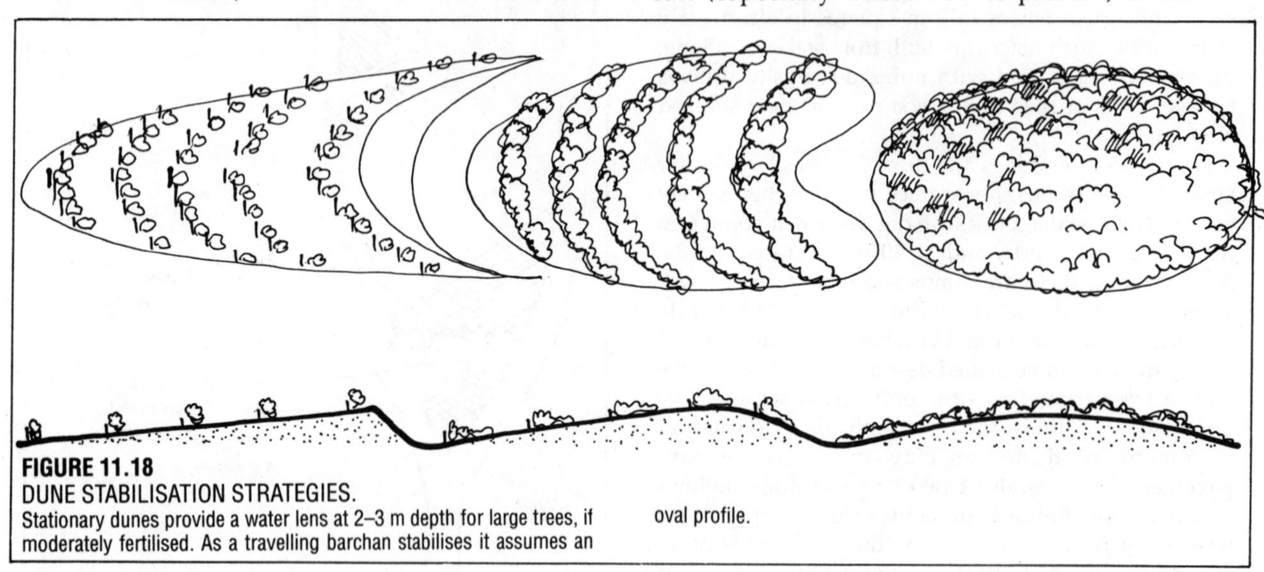

FIGURE 11.18
DUNE STABILISATION STRATEGIES.
Stationary dunes provide a water lens at 2–3 m depth for large trees, if moderately fertilised. As a travelling barchan stabilises it assumes an oval profile.

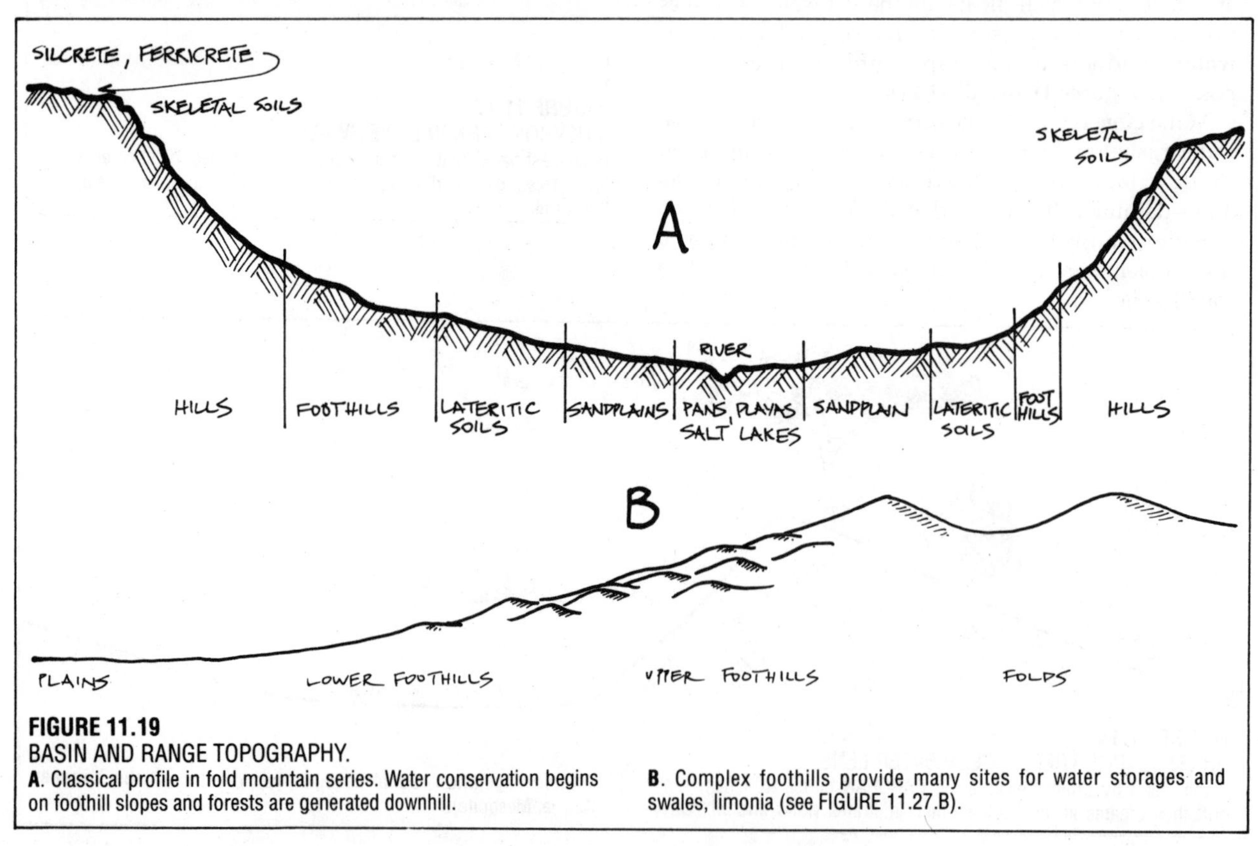

FIGURE 11.19
BASIN AND RANGE TOPOGRAPHY.
A. Classical profile in fold mountain series. Water conservation begins on foothill slopes and forests are generated downhill.
B. Complex foothills provide many sites for water storages and swales, limonia (see FIGURE 11.27.B).

hygroscopic or water–absorbing), or we can use tars, oils, or glues such as latex. Lichens, bacteria, fungi, and algal mats form naturally, as do salt crusts. Thus, no dunes form in some deserts where salt crusts or minute plants cement the surface of the ground. These often delicate desert crusts are critical to stability, and when hoofed animals, fast vehicles, or (worst of all) agriculture is brought into deserts, wind erosion may quickly follow.

Stabilisation in urgent cases can be effected by spraying tar oils, laying down pebble beds, or building brush fences 0.9–1 m high in parallel rows 7 m apart across the main wind direction. These fences must have wind gaps of less than 50% to drop sand out, firm posts driven every 3–4 m, and be combined with planting sequences for permanence. Temperature reduction of 12°C or so have been measured in the shade of fences. Moisture loss there is less, at 1.8–3.0% less than the open areas. Such fences are precursors to the planting of hardy perennials adapted to dune conditions.

As lethal soil temperatures are rarely formed below 15 cm in dunes, it follows that a careful placement of trees in holes this deep or deeper, stabilised from collapse by a woven basket or a cardboard collar, will prevent the new seedling being "cooked" until it can cast its own shade.

Any fertiliser should likewise be thinly placed at this time, as the rapid infiltration of water in rains can take it down below root reach. After the tree has spread a root web, however, more fertiliser can be added, and will then be taken up by the plant. Any plants pioneering the dune must be able to stand stem burial until the area is stabilised.

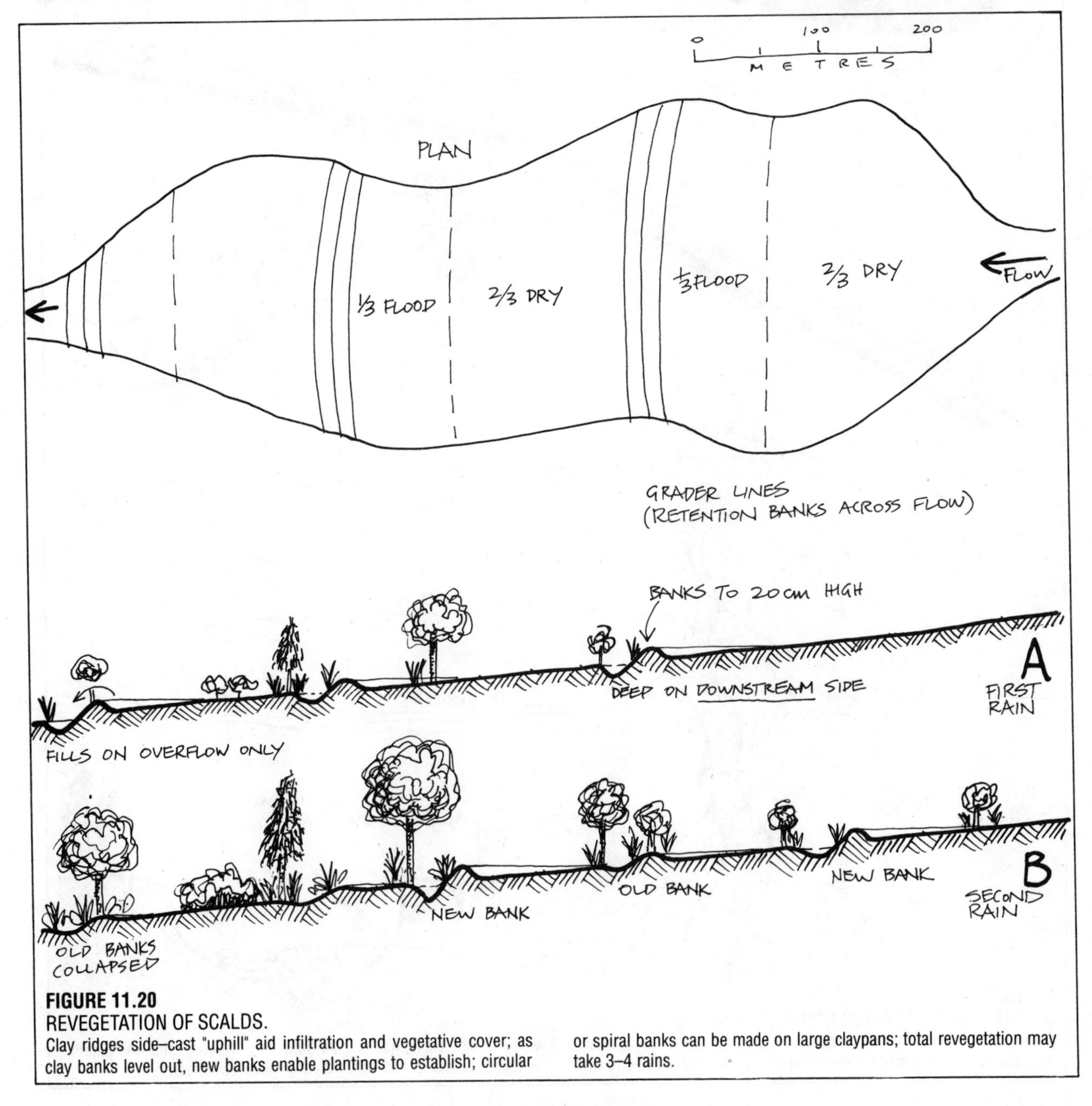

FIGURE 11.20
REVEGETATION OF SCALDS.
Clay ridges side–cast "uphill" aid infiltration and vegetative cover; as clay banks level out, new banks enable plantings to establish; circular or spiral banks can be made on large claypans; total revegetation may take 3–4 rains.

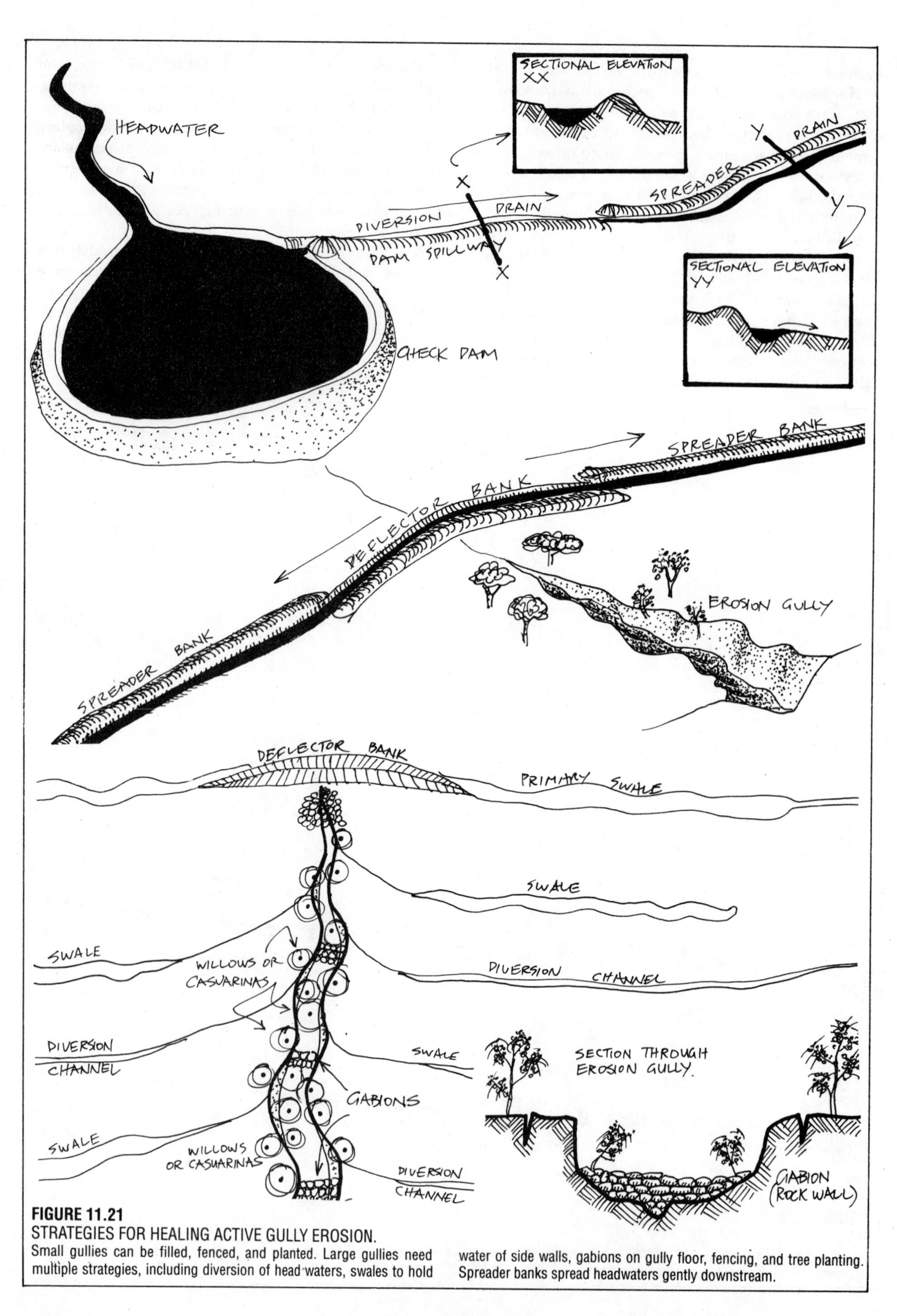

FIGURE 11.21

STRATEGIES FOR HEALING ACTIVE GULLY EROSION.

Small gullies can be filled, fenced, and planted. Large gullies need multiple strategies, including diversion of head waters, swales to hold water of side walls, gabions on gully floor, fencing, and tree planting. Spreader banks spread headwaters gently downstream.

Such strategies as vegetation stabilisation do not apply to hyperarid zones, where only fences and sprayed surfaces can operate to hold sands from roads or settlements (which would be rare in such landscapes).

DEPRESSIONS AND BASINS

Low flat areas or near–circular depressions are called (on the large scale) TECTONIC BASINS. Next in order are fairly large DEFLECTION (WIND) HOLLOWS, and smaller depressions with flat floors are called PANS. Some of these are clay–based evaporation areas and are called CLAYPANS, others are salt–based and are called SALT PANS. Clays can be treated with gypsum to increase their capacity to infiltrate water, or with sodium carbonate or bentonite to seal them where they are to be used to store water. Natural montmorillinite or illite clays swell and form impermeable surfaces after a short period of rain, and this may greatly impede infiltration.

Some small depressions (GILGAIS) form as a result of the swelling and shrinking of plastic clays. These are circular in outline and are useful water–capture systems where erosion has left them as hollows. Sand–filled they cease to function.

It is in the great tectonic depressions that salt lakes and dune seas (DRAAS) form; smaller depressions gather run–off and develop typical base materials of suspended silts or salts carried by the water and left on the pans when the water evaporates. A typical desert profile is given in **Figure 11.19**.

Soils in older deserts usually show some yellow or reddish colour from a complex of iron oxides, but areas may vary in composition from free sands to heavy, compacted, or cracking clays, which becomes notoriously sticky if wet (preventing all vehicle movement). Clays are most typical of flats and evaporation pans, and the softer clay–loams typical of dune swales and foothills.

SCALDS

A scald is the name given to a bare clay–pan or where a duplex soil (originally light sandy loams over a sharply–defined transition area with deep clay below) has lost its upper soil layer. Clays remaining are often *solonetic* (with a high content of sodium ions). As a result, the clays tend to "melt" when first wetted in rain, effectively sealing the surace of the clay base and preventing infiltration. The wet crust so formed on the surface is almost impermeable; the same effect is used to seal clays deliberately for use in leading run–off water to tanks. Scalds are not hollows, but have a flow into and out of the area (as opposed to claypans).

However, if a bank is graded up to 10–15 cm high, with the cut on the *lower* side, water will back up a considerable distance. Let this distance be one–third of the total area to be treated, allowing two–thirds of bare soil as run–off. Seed pre–sown on this surface will germinate, the low bank will eventually "melt" flat, and the next one–third can be treated as a successional strategy (**Figure 11.20**). Over 2–3 years a complete vegetative cover can be re–established. If banks are made too high or too solid, the plants will drown. Spiral earth–casting and ridges sown at their apex are also successfully used to seed claypans.

Revegetation of scalds is important, because in Australia at least, plague locusts lay their eggs and hatch most successfully in the narrow shaded edges of the scalds. Most scalds are produced by over–grazing, so that the desert grazier inflicts plague on whole regions.

CLAYPANS

Unlike scalds, claypans rarely overflow, but receive silty waters from clay soils, and after they pool for awhile, clay settles out and water evaporates. In the wet phase, tall canegrass (*Eragrostis australasica*) can grow, duck and waterfowl nest, and as the swamps dry out, marsupials and rodents take refuge. These swamp areas were frequently burnt by Aborigines in autumn, or as the canegrass dried out, but regenerated in wet periods.

Non–saline eroded claypans have also been revegetated by pitting, by building flood retention banks, and by chequerboard ridging. A low bank of 0.5–1 m is sufficient to support adapted trees in saltpan areas. The trees then assist desalination. Ploughed or ripped clay allows faster water infiltration, and vegetation keeps the initial crevices open if dry–leaf mulch species are avoided, and herbaceous or needle–leaf groups substituted.

SALTPANS AND SALT LAKES

These can hold many metres of water in floods. When this water evaporates, great depths of salt may accumulate, sometimes with a thin clay top. Vegetation of salt–tolerant plants occurs only on the margins.

Salt enters the freshwater zone in sandy deserts and atolls from these sources:

- Salt is washed *down* by infiltration of water through saline surface sands.
- Salt diffuses *up* from deep, heavily–salted waters (usually saltier than the sea).
- Salt *advects* in rains from the sea or saltpans after rain. This effect is most pronounced if freshwater has been overdrawn from the ground.
- Salt is drawn *up and in* by forceful pumping locally, and then pollutes surface soils.

GILGAIS

Gilgais are "puff and hollow" areas formed by patches of clay 3–5 m across, which expand into 6–20 cm deep hollows on a "swell–shrink" regime. The hollows occur in solonetic clay soils which shed their water into the hollows. The hollows can link up in chains or patterns,

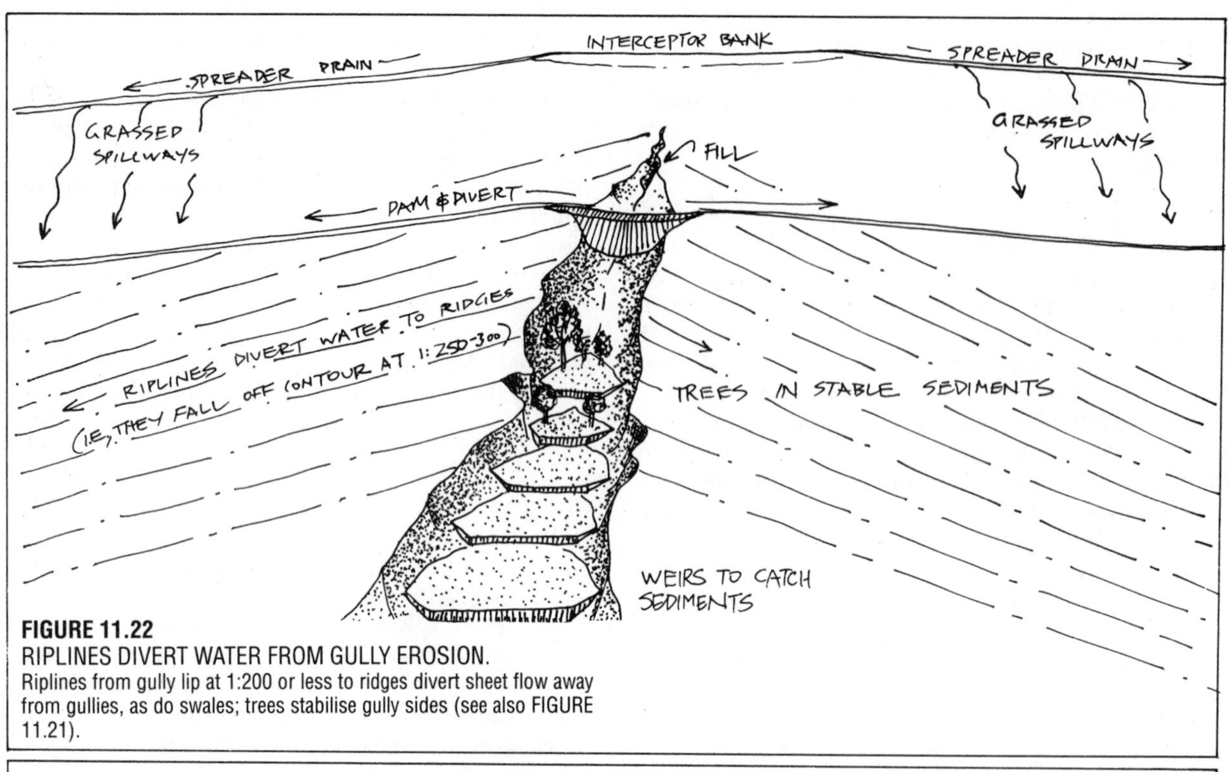

FIGURE 11.22
RIPLINES DIVERT WATER FROM GULLY EROSION.
Riplines from gully lip at 1:200 or less to ridges divert sheet flow away from gullies, as do swales; trees stabilise gully sides (see also FIGURE 11.21).

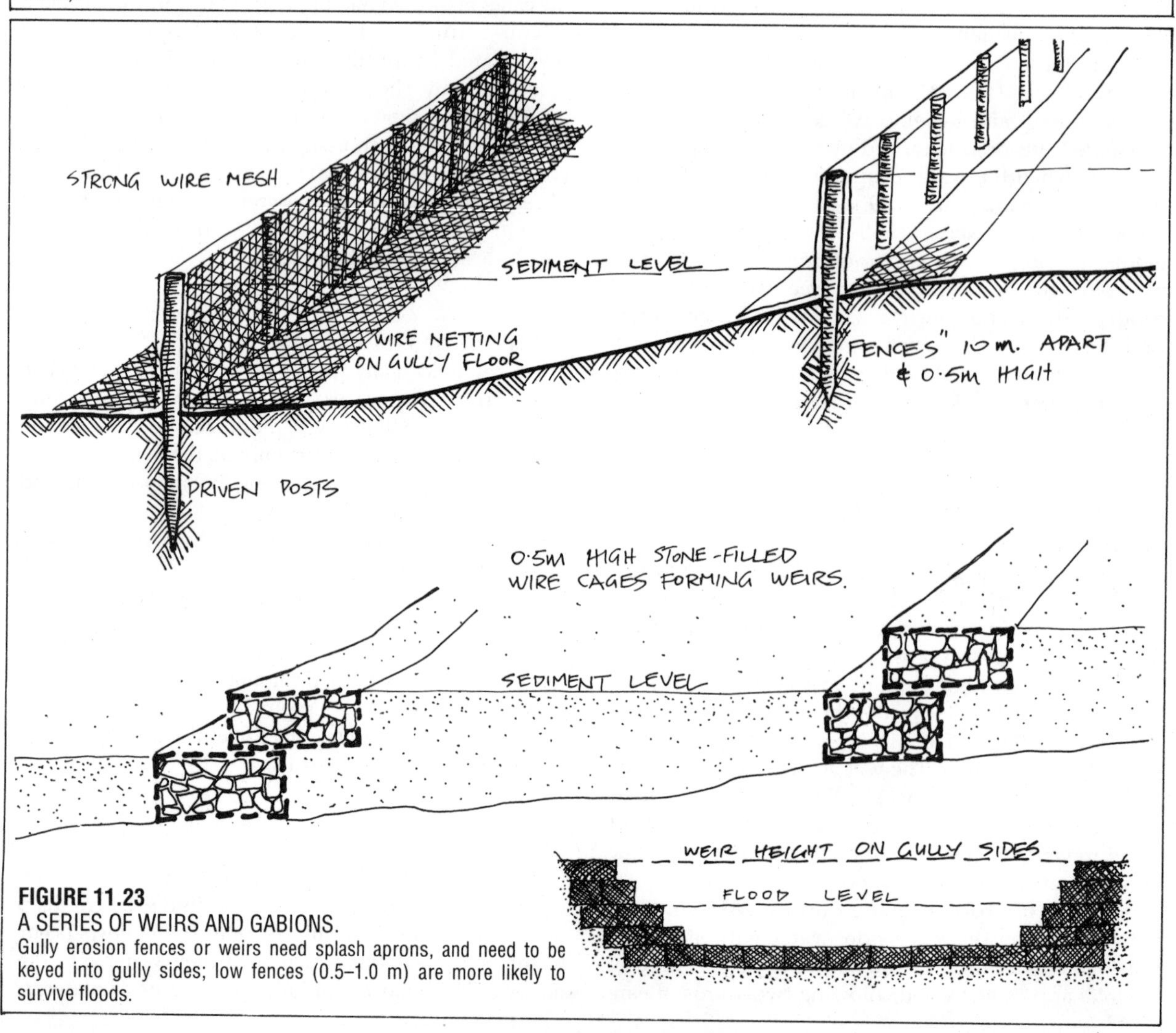

FIGURE 11.23
A SERIES OF WEIRS AND GABIONS.
Gully erosion fences or weirs need splash aprons, and need to be keyed into gully sides; low fences (0.5–1.0 m) are more likely to survive floods.

and the whole area is rich in plant species.

Where the area is overgrazed, the gilgais become sand–filled, the solonetic clays wash away, and a series of sandy mounds evolve. The area loses most of its plants and becomes unstable.

FLOOD–OUTS, GULLIES, BADLANDS

Also called flat–outs, or run–ons, these are the ever–widening flattish floors of valleys as they leave the hills. These "streams" are only 2–8 cm deep and 20 m or more wide, braided, and stable. Water absorbs over a wide area of such plains. Overgrazed, the streams commonly first erode some deeper channels, then become steep gullies. At this stage, diversion of water from gully heads, water spreading from large canals, and gully planting must be applied. Small gullies can be filled, and larger gullies planted behind stone walls or gabion barricades once the feed water has been diverted or spread (**Figure 11.21**).

Controlling gullies is essentially a matter of:

- Relieving the causes, usually those of overuse and lack of water control; and
- Preventing further cut–back of the gullies.

Fragile soils, loose shaley ground, and recent sediments exposed to sudden overland flows may gully in a few downpours. In deserts, this most frequently occurs where foothill flood plains are overgrazed, where rabbits tunnel into loose subsoils, where vehicles or cattle create downhill tracks, or where badly built roads direct sheet flow to culverts that concentrate flow. Such areas need fencing out, light or no stocking of animals, pest control, and above all, spreader banks for the water.

In the valleys above the gullies, stream flow can be diverted to flood spreader channels, and a series of rip–lines drawn downhill from the gully bank towards the ridges (sloping at 1:1000 or so). **Figure 11.22**. Small gullies can be bulldozed full, or interceptor drains made across them to cut off flow. Large gullies can also be filled or dammed if soils are suitable, but extensive gullied badlands are too expensive to treat in this way, and we may need to fall back to creating small silt dams in the valley floors themselves.

As with many dryland techniques, gully retention banks need to be small, frequent, and well–made to effectively spread flow and create absorbent flats where water trapped in silts create beneficial growing media. In many areas, once gullying has developed, there is no alternative to allowing water to flow in the gully systems and to create fields there.

A series of weirs 0.5 m high traps sediment so that tree establishment can stabilise the run–off. Weirs can be made from a series of well–braced wire–mesh fences or low stone bunds across gully floors, or stone–filled wire baskets (GABIONS) placed to spread the flow of floodwater. **Figure 11.23**.

If the weirs are higher than 0.5 m it is necessary to provide a splash apron at the foot of the weir where plunge pools form. However, if interceptor banks can be made above the gully–head, and water diverted, then the lower volume of flow permits weirs of 1.0–1.5 m high without scour damage.

All weirs are started at the gully outlet, and built in sequence upstream. Where no local supply of boulder is available, the lower height of weirs are built of wire mesh only. Where there are plentiful stones, stone seepage weirs are very satisfactory and form roads across the gullies.

Weirs must always extend up the gully sides, well above any flood level. Stone weirs are more easily maintained and restored than wire fences, and more permanent. All weirs need Vetiver grass or some such tough plant barrier to hold erosion.

At projects such as that at Assomada (Cape Verde) in the Sahel (*International Agricultural Development* Jan/Feb 1985), terrace–building and check dams built of stone–filled wire mesh gabions in the stream beds have effectively checked and stabilised soil erosion. Both help to absorb run–off water. A change from sugar cane cropping to tree and vegetable culture also assists soil conservation and aids local nutrition in this area of high unemployment. Some 2000 ha have been reclaimed in two target valleys.

Pigeon pea *(Cajanus cajan)* has supplied food and ground cover, and cooperatives have formed to grow and market food, with credit provided for crop production. Wells and galleries dug to tap groundwater

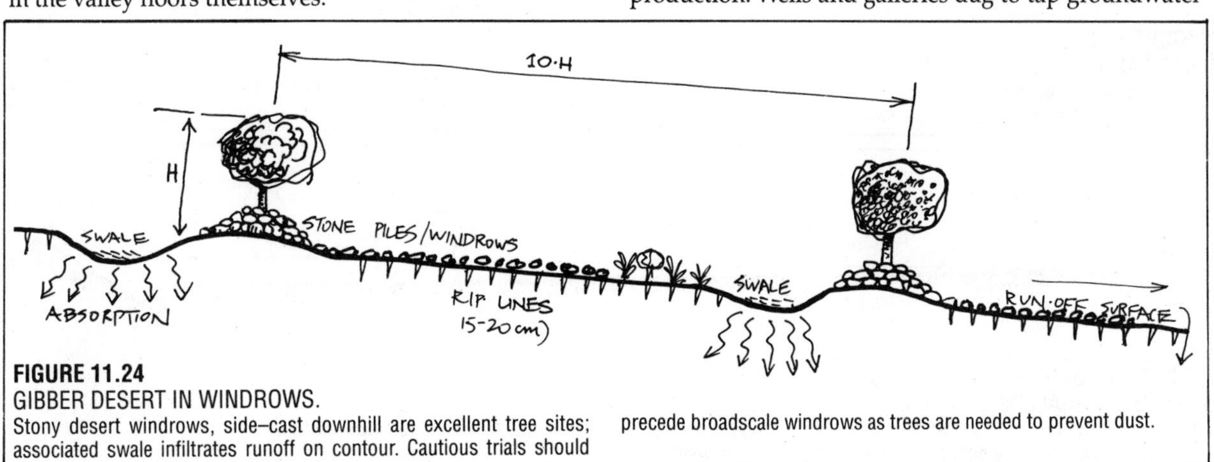

FIGURE 11.24
GIBBER DESERT IN WINDROWS.
Stony desert windrows, side–cast downhill are excellent tree sites; associated swale infiltrates runoff on contour. Cautious trials should precede broadscale windrows as trees are needed to prevent dust.

have served to establish tree and vegetable crops. The silt brought down by erosion has built terraces 6–8 m deep and about 200 m wide across eroded valleys, and copious spillways let floods flow over. A change from goats and sheep to rabbits and pigs have helped replace animals that overgraze the hills to those suported by domestic wastes. In all, the project has integrated several important biosocial strategies to make modest improvements to a degraded area:

- Government aid is available as revolving loans, to plant new crops and to buy in alternative livestock (replacing erosive species).

Community involvement is assured by the employment and education of unemployed people towards projects that relieve their poverty (meaningful local work).

- Education is integrated with development.
- Export crop is being replaced with local nutritional food crop and tree crop.
- Pioneer legumes have helped improve soil and halt erosion, while providing forage and food locally.

STONY (GIBBER) DESERT

Vast areas of desert may develop a stone layer over fairly well–structured soils, the stones being once part of a soil matrix, but now revealed after centuries of wind erosion. Old stony deserts have wind–faceted stones, whose facets reveal the direction and intensity of local sand–storms, or the saltation of sand grains in wind.

Wherever a road grader passes across this country, a windrow of stones results; large sturdy side–rakes are now developed which will also windrow stones in fields or on gibber. It is these windrows that present a great opportunity to soak in run–off and to start tree lines across slope every 10–20 m. The bare area between windrows not only gives greater run–off, but provides a seed bed for smaller shrubs, legumes, and forbs. Insects, reptiles, and birds leave manures and their remains in the windrows of stones. Water is impeded and largely absorbed below the stone piles, and dew forms in aerated heaps. If practical, ripping the pavement below or between windrows helps root penetration, and a touch of fertiliser greatly aids growth. Swales can be made just upslope of windrows (**Figure 11.24**).

Some natural stone windrows evolve below cliff shorelines on receding lakes. Below the stones the soil is cool, moist, and obviously well–structured due to insect occupancy. Many tree species appreciate these conditions.

REG (gibber desert) is stabilised only by the layer of stones. (Active dunes can also be stabilised by laying pebble beds over them.) Thus, if there is any disturbance of the stone layer, erosion can recommence, so that windrowing stones for annual cropping is certain to *increase* erosion. However, windrowing in stages to plant trees at 10–30 m intervals (in stages, because we need to assess the possibility that we would be able to grow trees there) will reduce erosion and also trap dust from outside the area, or bring sand to rest. Like a great many desert techniques, a limited trial is needed before broadscale windrow systems evolve. It is thought that the Nabateans made windrows on steep slopes in order to increase run–off, and to lead water by hill–base channels to fields, but good absorption does occur in

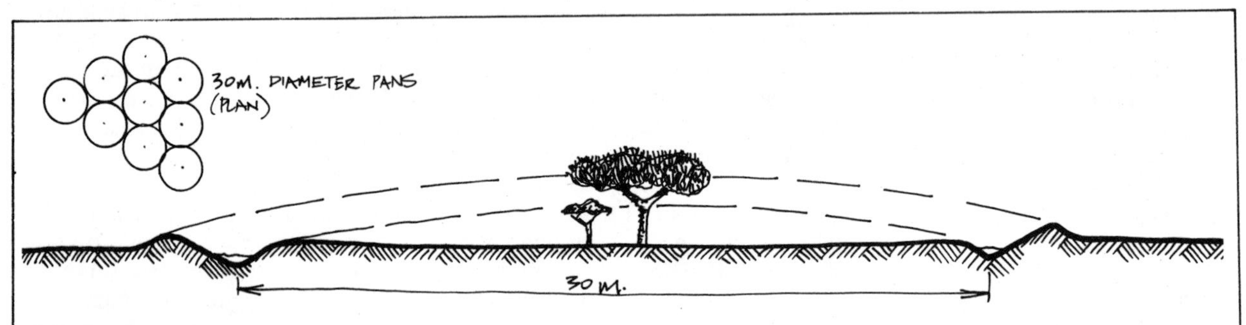

FIGURE 11.25
DITCH AND BANK LANDSCAPE IN PLAINS.
Circular side–cast pans of 30 m or less diameter prevent all runoff and will grow forage species in perimeter swales; ideally, smaller (20 m or less) pans may become partly over–shaded by trees. Pelleted seed in ditches helps revegetation.

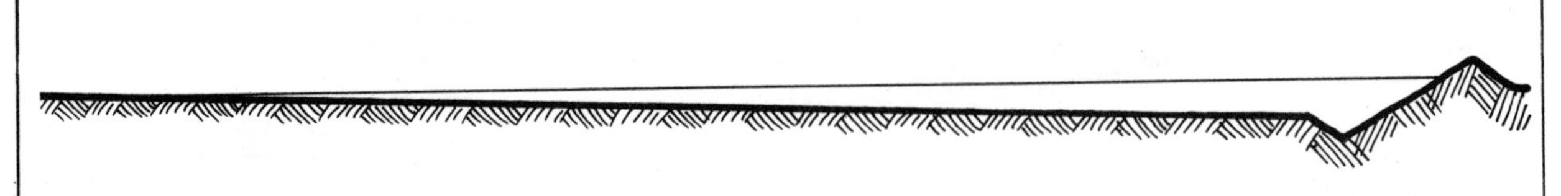

FIGURE 11.26
YEOMANS' SHALLOW SWALES.
Very low (1–1.5 m) banks on flat deserts hold back runoff water for up to 6 km; bled off to fields, and infiltrated, this water is useful for opportunistic crop (sunflower, millet).

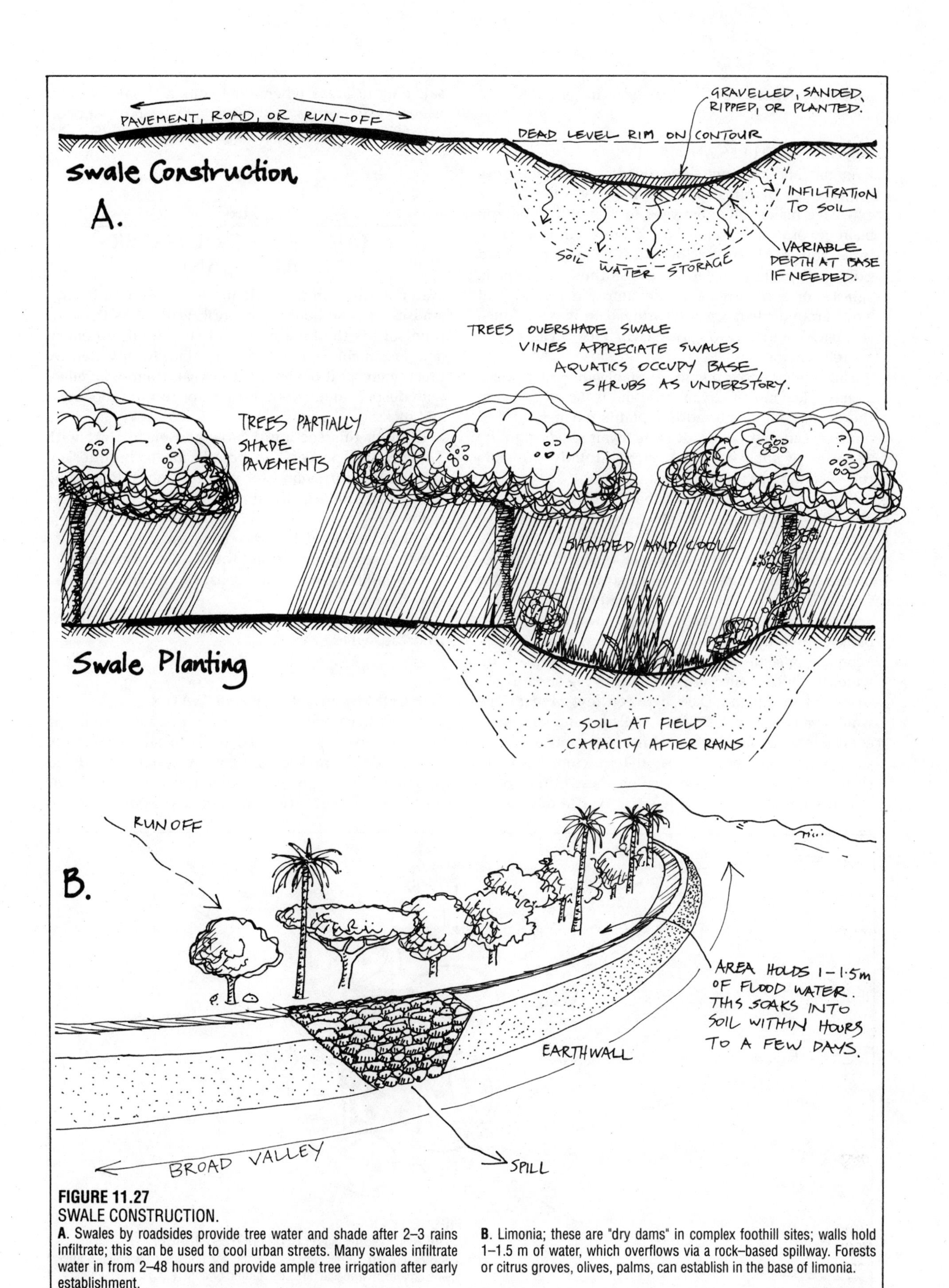

FIGURE 11.27
SWALE CONSTRUCTION.

A. Swales by roadsides provide tree water and shade after 2–3 rains infiltrate; this can be used to cool urban streets. Many swales infiltrate water in from 2–48 hours and provide ample tree irrigation after early establishment.

B. Limonia; these are "dry dams" in complex foothill sites; walls hold 1–1.5 m of water, which overflows via a rock-based spillway. Forests or citrus groves, olives, palms, can establish in the base of limonia.

windrows over clay or clay–loam soils in gentle slopes.

LOWER FOOTHILLS AND PLAINS

Even flat land has some slope (not detectable by the eye but revealed by rains and sheet flows of water). These extensive plains, often browsed to near extinction, can be regenerated or seeded by simple circle–swales, preferably made by tractor and tilt blade, or a road grader. Especially where a few saltbush, bluebush, *Acacias*, or shrubs remain, one simply drives a great circle around them, casting earth *out* to make a "ditch and bank" landscape (**Figure 11.25**). Such circles can be 30–100 m across.

This too has succeeded in re–establishing poor range, as seeds, leaf mulch, dry manure and water all gather in the ditch. Where no mother plants exist, seed and fertiliser can first be broadcast to await wind and rain. A series of such conjoined circles totally prevents run–off and traps seed and dust.

Equally well, graded contours will back up and absorb water for a relatively great distance. I have worked on plains with falls of 10 m/km, and here a 30 cm contour wall will hold water up for a few kilometres in heavy rain. Seed can be sown as soon as the ground is firm, or pelleted seed spread beforehand.

A scale larger than these shallow swales is a 1–2 m bank thrown up by repeated side–casts, used by P. A. Yeomans in semi–desert (**Figure 11.26**). These back up water for miles, especially if the banks are sited to run from hill to hill in a pass between low ranges. The water can be left to soak in, or spilled under slide–gate control into yet smaller walled fields to grow a crop of sunflower, millet, beans, or short–term crop. Trees and shrubs will often establish well on these bunds. Such systems suit fairly permeable soils, but can develop as salt flats in areas where clay soils and salty surface water is used. Crops can be sown as such lakes or pools dry up, in sequences following the receding waters.

11.6
HARVESTING OF WATER IN ARID LANDS

Water is the dominant theme for designers in arid landscapes. The quantity of fresh water (less than 700 ppm salt) is the final arbiter of successful settlement and sustainable agriculture. To obtain fresh water, we must intercept it before it washes salt from soils, mixes with deeper saline groundwaters, or runs off roof areas and rocks.

To open this section, I have chosen to start with TANKS and proceed to broadscale techniques on specific landforms and for specific slopes. Regarding water catchment, our aims can be summarised as follows:

- To store *fresh* water for cooking and drinking;
- To divert sheet flow and waste water to gardens;
- To infiltrate water into soils, and to produce plant growth on that site; and
- To give rainwater run–off time to soak into the landscape where we live.

THE CONSERVATION OF RAINWATER

Here, we have two basic strategies to apply: the fitting of tanks, cisterns, or sealed wells to take house roof water, and the provision of very large public sealed and roofed resevoirs to hold water run–off from public buildings, paved areas, and roads. Many dryland

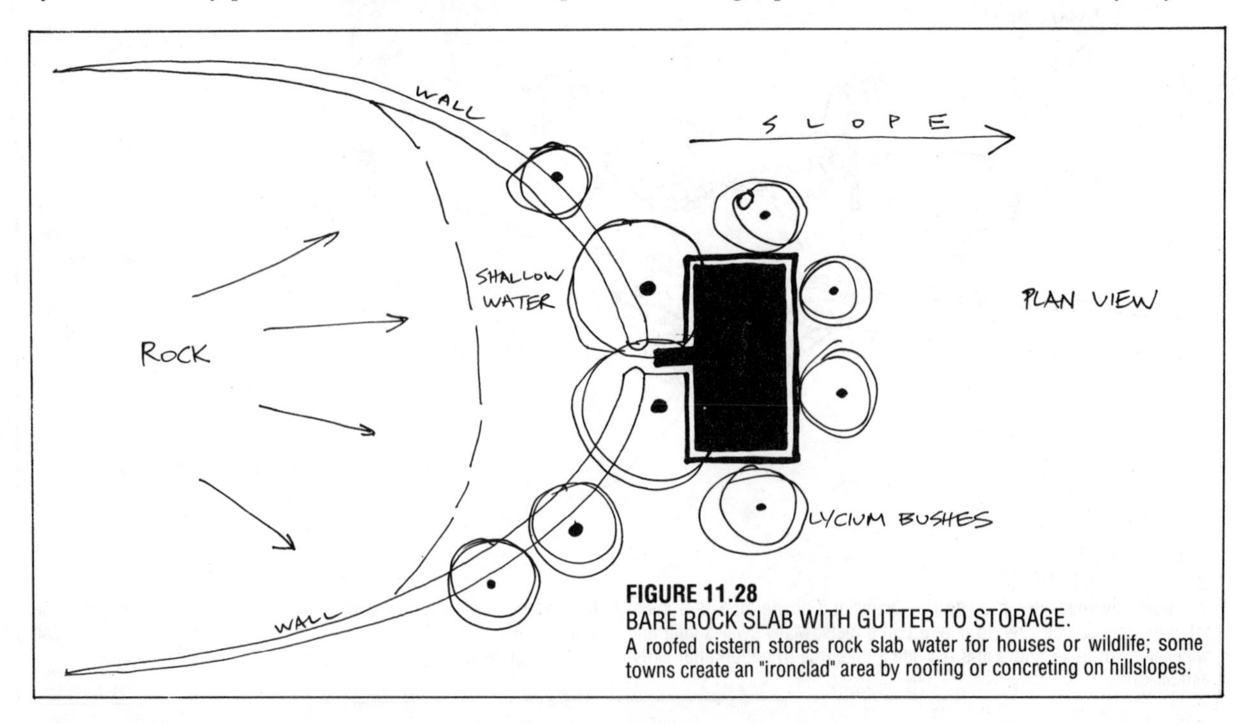

FIGURE 11.28
BARE ROCK SLAB WITH GUTTER TO STORAGE.
A roofed cistern stores rock slab water for houses or wildlife; some towns create an "ironclad" area by roofing or concreting on hillslopes.

settlements in Australia have public cisterns. These are used for domestic or agricultural water, depending on the quality of run–off.

The cheapest water storages for trees are in permeable earth swales (of any shape) alongside the roads, which are essentially large compacted collection areas. A swale is a dead–level hollow built on contour (although the base need not be level) which create temporary pools. Swales are ripped, gravelled, sanded, or planted at the base to assist rapid water infiltration. They must be large enough to take all tank or storage overflow, pavement run–off, and additional harvested and diverted overland water flow from the environs outside the settlement itself. In short, they may need a capacity of 200% of village run–off which then infiltrates locally as soil water storage to 3–5 m down in the earth.

Village Homes (Davis, CA, USA) is fully swaled, and grows its trees on water so harvested, planted in or beside the swales themselves. After a few rains, ample soil water is available for tree growth. In Mexico swales are used along main road sections.

Apart from the greatly reduced cost of eliminating stormwater drains, gutters, and kerbing, swales will grow the very trees needed to shade the pavements, courtyards, and parking or market areas that they serve. Swales also provide dirt spoil to raise road–beds, sand for construction, and topsoil, loam, and clay for gardens.

There are two approaches to the problem of water conservation in homes and buildings; the first (a regular feature of life in arid–area towns) is to allow a strict household ration of water per person. This is achieved by using the simple strategy of fitting float valves to all home roof tanks (of fixed dimension), and pumping them full only once a week.

The second is to strictly meter use—a more generally acceptable approach—and put a sliding increase scale on excess use. Waste then costs much more than conservation. Meanwhile, inside and outside the house, the most water can be saved by:

- Using the least possible in gardens, especially by planting shrubberies instead of lawns, and using sub–surface trickle irrigation, mulch, and shade. This can save about 50% total water use.
- Saving water by less use, especially on toilet–flushing. If hand–basin and raised–floor shower water is fed to a low–head, low–flush cistern (with overflow by–pass), then 40% of the remaining 50% of water use is saved. Further, bathwater diverted to a pre–soak trough for clothes, and laundry water to the garden makes good use of this waste water. Water–saving shower heads and timer taps stop us dreaming in the shower and save 14–20% of water so used.

Whenever kitchen sink water is used, it too can be diverted to the garden and used productively. Every family roof should catch its own drinking water in tanks of from 20,000–50,000 l (5,000–12,000 gallons), depending on family size and regularity of general supply.

In a few sensible towns in Australia, *all* roof areas are run by public pipeways to large storage tanks, and all road run–off to resevoirs to be used on public gardens. Other settlements run all road water to swales where trees and fruit crop are growing .

Much simpler but still quite hygienic systems use "rope and pulley" showers, basins for hands and laundry (often under a lean–to) and earth–closet toilets. I have spent most of my life using such systems in bush camps and in pre–war houses. They work well and save a great deal of water, usually a fraction of that caught on roof areas and stored in house tanks.

Only 1 cm of rain produces 1,000 l for a roof of 100 square metres, or 100,000 l/ha of sealed run–off. Domestic tanks are built at about 20,000 l (to 100,000 l for small settlements). Any multiple of these tanks can

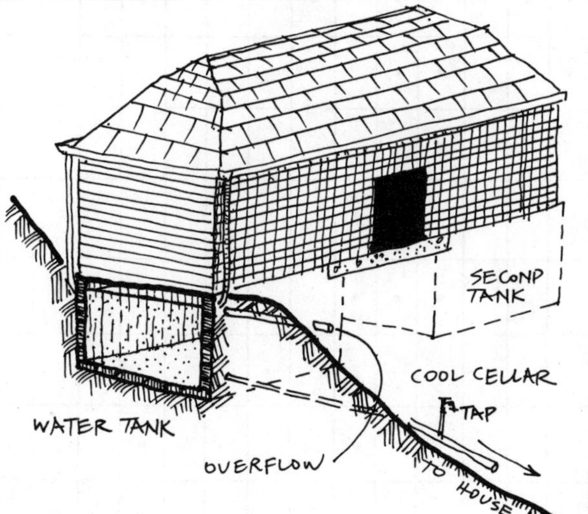

FIGURE 11.29
LARGE TANKS FORMING FOUNDATIONS.
Uphill barns will fill large cellar storage tanks from roof catchment for gravity flow to house and gardens.

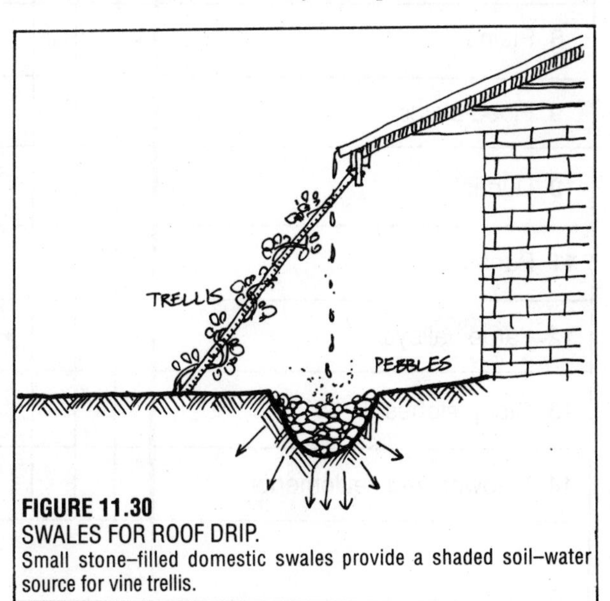

FIGURE 11.30
SWALES FOR ROOF DRIP.
Small stone–filled domestic swales provide a shaded soil–water source for vine trellis.

TABLE 11.2
A GENERAL SUMMARY OF WATER STRATEGIES FOR DESERTS; APPROPRIATE USAGES AND SITES.

SITE	1 Fields (soakages)	2 Open dams, scour holes	3 Check dams	4 Sand–filled dams	5 Swales	6 Spreader drains	7 Catchment basins	8 Chequerboard ridges &bunds	9 Low contour ridges	10 Pits and burrows	11 Net and pan	12 Leach fields	13 Moisture barriers	14 Run–off (paved areas)	15 Boomerang bunds	16 Log and rock check	17 By–pass channels	18 Tanks and cisterns
1 High hard plateau surface			•													•		•
2 Rills in stony country, slab rock			•	•												•		•
3 Shaded rocky valleys		•																
4 Upper valley systems	•				•	•									•			
5 Wadis and wider valleys, dongas	•			•				•									•	
6 Flood–outs	•					•											•	
7 Low foothill slpes (4–8°)	•	•			•	•	•		•		•							
8 Plains	•				•			•	•									
9 Flood plains	•						•	•										
10 Dunes										•								
11 Pans								•	•	•								
12 Large valleys	•	•		•														
13 Steep slopes (8–15°)									•	•					•			
14 In towns and settlements					•							•	•	•			•	•

be built of concrete, using a central vertical pivot and sliding form in which to pour the concrete. It is easy enough to scale up or down from these figures when estimating tank or swale capacity needed for dwellings.

Even in very low rainfalls of 10 cm or so, large rock or roof run–off surface will supply all the drinking water that we have tanks to store it in. Because of the costs of building cisterns, we can seldom afford to store all the rain it is possible to catch.

Roads are commonly sealed and otherwise compacted by traffic, and in desert downpours collect large expanses of water. In other areas, we can seal hill slopes above villages with concrete, wax, bitumen, or plastic, and use the area as a "roof". But the cheapest roof is to locate near a bare rock slab in granite, or a natural sealed cap on hills, around which we run a gutter (rock–cut or cemented) to storage cisterns or tanks.

Tanks (versus dams) can stand on or above the soil surface, and are fully sealed and usually roofed. Large tanks have a roof which is itself a catchment to the tank. Sometimes a tank is excavated and sealed with a tough plastic or rubber membrane. Sites are usually chosen above houses or settlements, and an accessory windmill in foothill settlements (or a solar–powered pump) is used to lift clean groundwater from wells to tanks in any period of higher demand. On very flat plains, the tanks are raised on a stand, and the windmill lifts to the tank for house and stock water—a very reliable and widely–used system even in towns. Some figures (as an aid to storage planning) for arid areas is shown in **Table 11.2**.

Large garages, factories, recreation halls, and shops provide ample roof area for catchment in most communities, and are so used in arid areas. The tanks themselves can be earth–sheltered, partly or fully buried, covered with vines, or they can stand in shaded courtyards. Sited uphill from houses, large ground tanks can form the foundations of barns or garages, and the space between such tanks a cool store for root and fruit crop (**Figure 11.29**).

All the more wonder, therefore, that (apart from in Australia) tanks are a rare event in settlement, and clean rainwater is let run to waste.

There is seldom enough storage for *all* our roof water in tanks, so we can then run the overflow to swales in the garden. Such swales can be simple channels, or (in deep sands) long plastic–lined and sand–filled swales, from which we can pump water, and on or beside which we can grow plants. This surplus garden water can be augmented by shower water from the house. Neat swales for roof drip are made from rock–shingle–filled hollows or channels with vertical concrete edges, but unsealed below to allow infiltration (**Figure 11.30**).

There are some myths surrounding the storage of water in tanks:

1. It is stagnant water (untrue; it tastes fresh for years and remains clear and clean).

2. It breeds mosquitos (not true of *covered* tanks; mosquitos can be totally excluded. In open tanks, mosquito–eating fish can be stocked).

3. It contains dust, etc. from the roof (true, but this settles as a biological sediment and keeps the water clear).

Generations of healthy Australians have been reared on tankwater (myself included), and have escaped cholera, bladder cancer (from asbestos pipes), heart attacks from acid water, and the 46+ additives now used in reticulated water. Get a tank and survive! Tanks can be brick–lined, domed, and rendered in sandy soils, built of "stack sack" (plastic tubes of 1:9 cement:sand, pricked and soaked in water) in plastic–lined holes, built of cement plaster on a wire–netting base, excavated in rock, or poured of concrete in mobile steel forms on site or at the factory. They are also made of (or lined with) galvanised steel, fibreglass, plastic, or plastic resins.

It is beneficial to place limestone, dolomite, or marble gravel (or mollusc shells) in a bag in every tank to offset acidity, to fix metallic salts, and to "harden" the water, as it is healthier to drink alkaline or neutral water (less heart attacks, less metals in the diet). The fitting of tanks in arid areas should be *integral to every building*. At the least, every home can drink its own tank water; and at most a large house roof will grow a home garden and orchard on wastewater and overflow.

WATER HARVESTING ON OPEN SITES FOR TANKS OR CISTERNS

Wherever good clays are found, we can make earth tanks in the ground. The essentials are that:

- Tanks should be deep and narrow;
- They should intercept a run–off area; and
- They should ideally be roofed to prevent evaporation.

By instituting a system of cisterns, tanks, and swales, any desert settlement (wherever average rainfall exceeds 30 cm) can create for itself a reserve of soil and roof water that is, in effect, an oasis. The proportion of water supplied to swales by local run–off and from more widespread interceptor banks harvesting the overland flow of surrounding areas therefore changes as rainfall decreases; more extensive harvests of landscape run–off must be made once rainfall declines below 30 cm, and must be directed to village swales to support large trees.

Both for domestic and livestock use, a variety of deliberately sealed or compacted run–off surfaces have been constructed in Australia and elsewhere. Some of these are at ground level, and range from bitumen–sealed and fenced areas to roaded catchment of 300 m long by 10 m wide rounded and graded surfaces; side slopes are at a ratio of 1:20, drains at 1:100–1:200. The earth can be stabilised by gums, a light cement mix, or simply well–compacted clay soils. **Figure 11.31**.

Other approaches are the sheet metal roofing areas of up to 600 square metres, at ground level, feeding to 100,000 l tanks. Slopes covered with roof metal can be steeper than those of earth. It can, of course, be

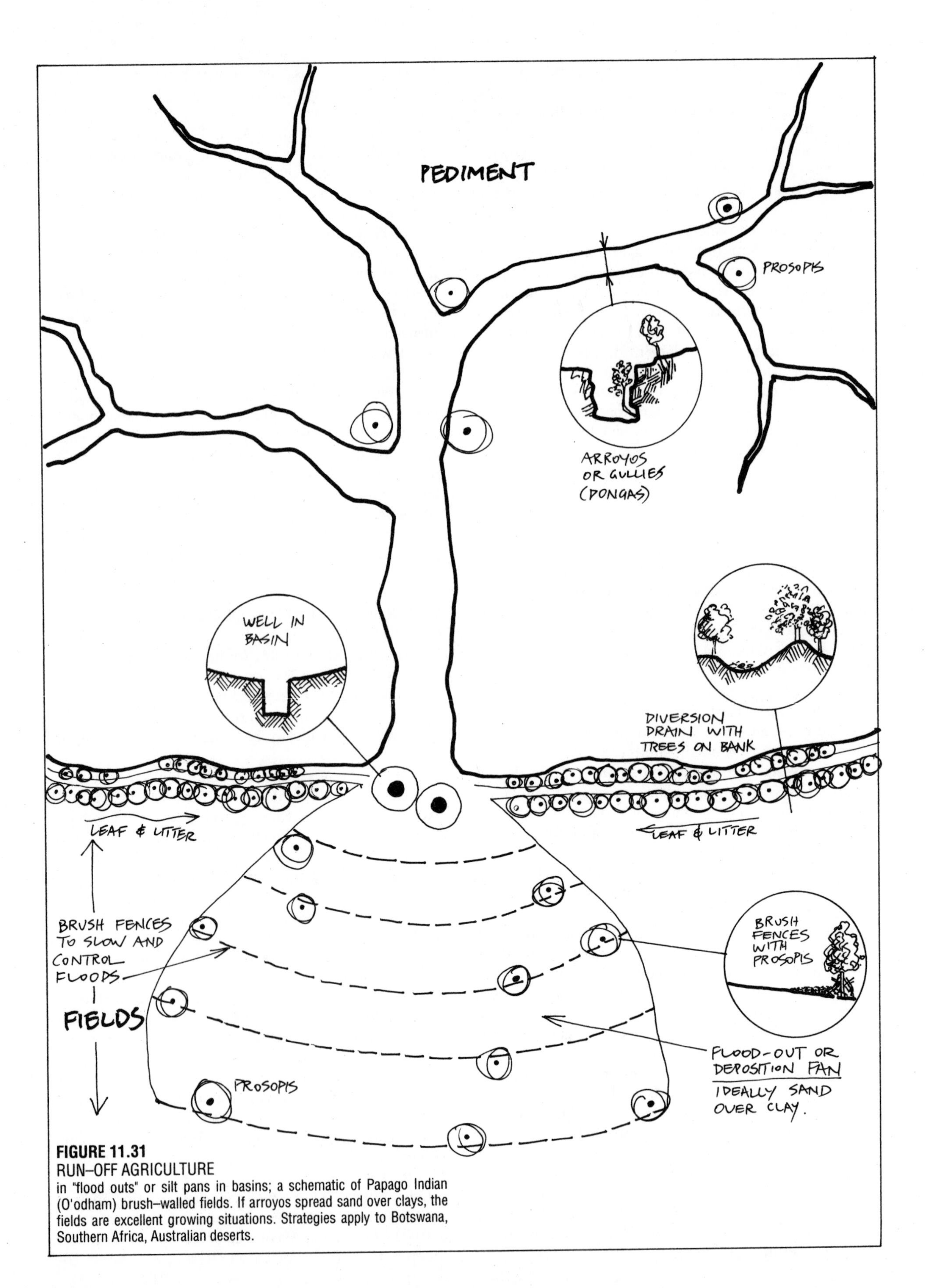

FIGURE 11.31
RUN–OFF AGRICULTURE
in "flood outs" or silt pans in basins; a schematic of Papago Indian (O'odham) brush–walled fields. If arroyos spread sand over clays, the fields are excellent growing situations. Strategies apply to Botswana, Southern Africa, Australian deserts.

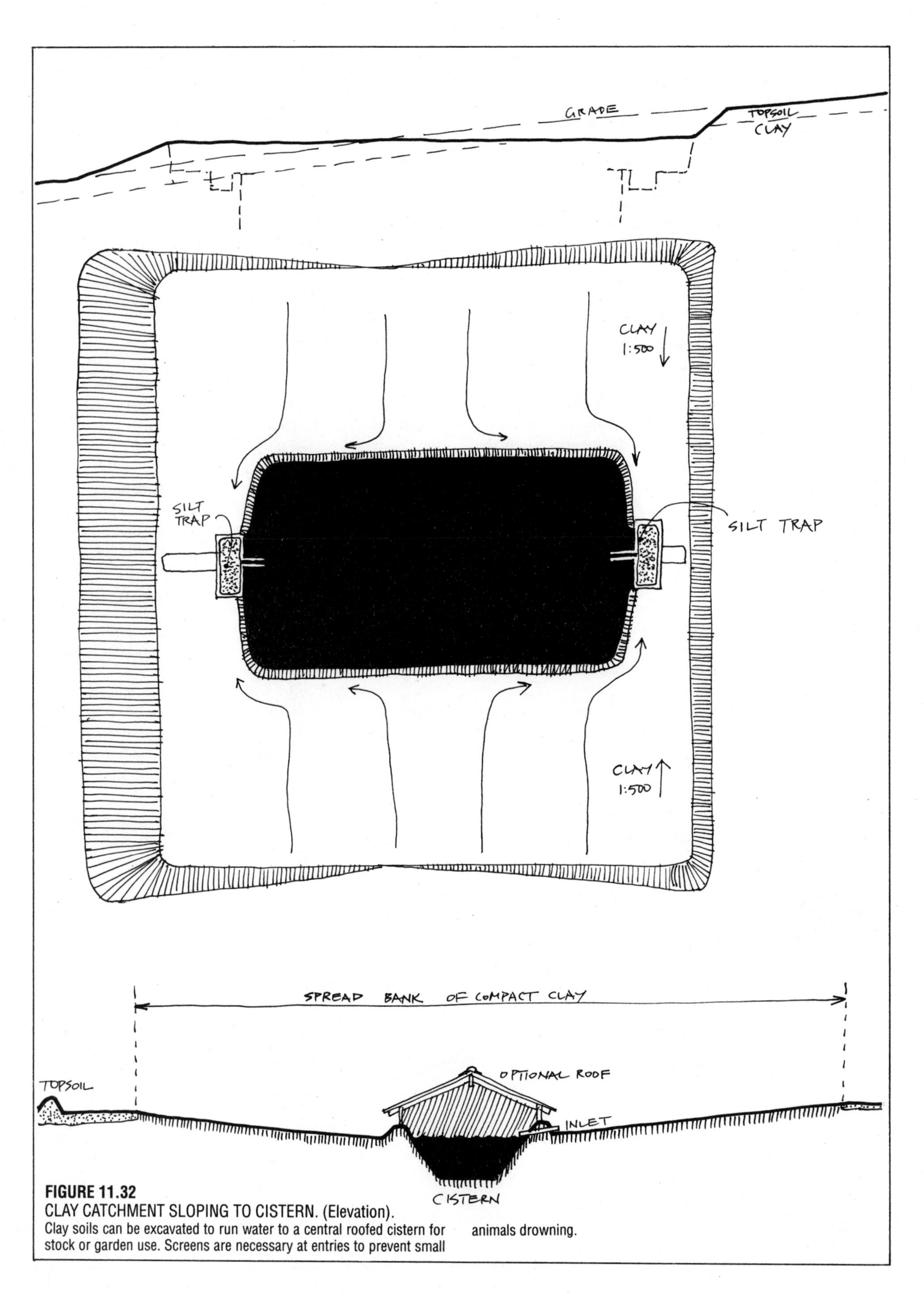

FIGURE 11.32
CLAY CATCHMENT SLOPING TO CISTERN. (Elevation).
Clay soils can be excavated to run water to a central roofed cistern for stock or garden use. Screens are necessary at entries to prevent small animals drowning.

advantageous to actually use galvanized sheet iron to cover extensive roof areas for hay and implement storage, or even for lease as "mini–warehouses" near towns. Many such roofs on factories and warehouses lie unharvested in settlements.

In clay soils, square or circular spread–banks sloping to a central cistern are also made, the topsoil first removed and the excavated central material rolled down to create a water run–off area. Ramp access, base drains, or pumps can be used to access water (**Figures 11.32 and 11.33**).

WATER SPREADING

The fate of rainwater in a thinly–vegetated or overgrazed desert area is usually that of most of the water running off over the land. So little is absorbed (less than 12% in bare hill country) that even modest rains will create sudden and turbulent flood flow in previously dry streams, large boulders will be swept down the steep–sided wadis, and enormous quantities of silt, sand, dry leaf and stick material, dry manures, seed and seed capsules, desert snails and other debris are washed off the hill surfaces and carried out onto the plains. Very little soil water is retained.

As the waters spread out over the plains, they increase in salinity, becoming rapidly more saline the further they travel. In a few days or weeks, the headwater rains reach the salt–pans, and there they may pool for months or years, until they evaporate to salt–encrusted flats. In total, little but ephemeral plant growth results.

If we are to creatively use this rare water resource, we must halt and divert the flows, and make places for water to soak in to the soils while it is still fresh enough to be of use to plants. Once fresh water is infiltrated, it forms a soil reserve or aquifer that travels very slowly underground to the river beds and salt pans, effectively de–salting the shallow water reserves, even though below this fresh aquifer flows very salty deep leads **Figure 11.34**).

Trees and tree lines, swales, deep–ripped contours, silt dams, dams, and lines of stones or palisades of natural rock all help to halt run–off and let it infiltrate into the soil. Trees in particular help recycle the water through growth and transpiration, keeping the deeper salty water below root level. When we cut, burn, and graze off the desert trees, we effectively reduce infiltration and the deep leads rise to the surface due to flooding and bare–soil evaporation.

The forests of Ethiopia have been devastated by grazing and the development of export crop. The rainfall has likewise been reduced, and salted lands increased. If we put as much effort into forest rehabilitation as we put into lawns, or as much capital as we put into war, then we could rehabilitate the earth. China and Taiwan have proved this to be feasible on the broad scale.

As we cut forests everywhere, aridity and fires, salt and sterility will increase. South Australia (with an impoverished 3% of her original tree cover left) has banned all new forest clearing (although many farmers cleared as much as they could just before the legislation was enacted). This is not enough. Any sensible government would allow agriculture as perhaps 20–30% of a total land area; we rely on forests and lakes to support even this area of crop and grazing. It is time for farmers to also become foresters, and so prevent future disasters.

A general profile of water depth and quality from intake to salt pan decides the strategies of settlement, water systems, water quality, water disposal, and placements (hence, design). **Figure 11.34** gives the general pattern. On hill country, the second determinant is frost level and aspect.

Rain in the desert, falling as 2.5 cm or so per day for

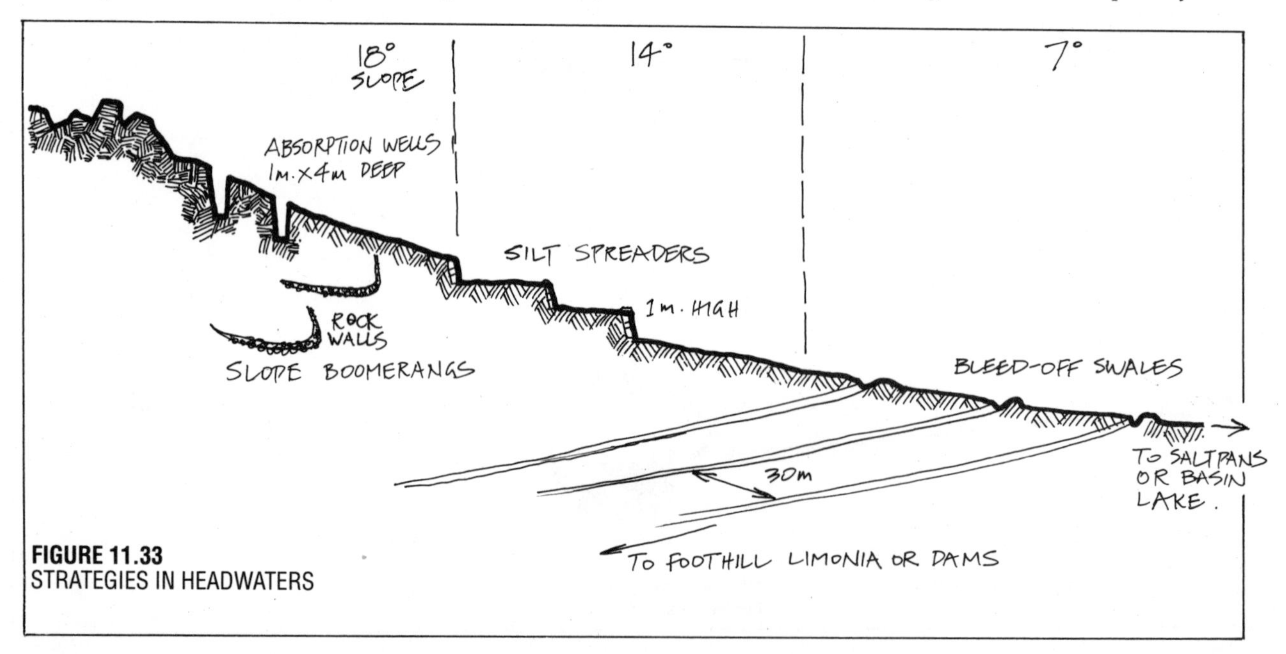

FIGURE 11.33
STRATEGIES IN HEADWATERS

2–3 days (as it does), is likely to produce severe local flooding, and this too must be kept well in mind when settlements are located or built. This flash–flood factor becomes critical when settlements are built in wadis, on the upstream side of restricted gaps in hills, or in hollows near dry watercourses. If we are to store or use the intense, sudden and infrequent falls in deserts, we need oversize channels, drains, infiltra-tion, and diversion systems to cope with such high– intensity storms.

In Australia, 88% of rainfall runs off or is evapor-ated from land and water surfaces. Other hot deserts would have similar losses, with only 12% of rainfall available to plants. Normal estimates are of 12% water run–off for forested areas, 20% for non–forested areas with sandy or friable soils, and up to 80% for the rare concreted, compacted, or clay–sealed bare sites. These few figures enable us to make educated guesses as to the volume of run–off water, and hence the capacity needed in channels or flood by–passes in any arid area, or to estimate flood levels in valleys.

Some precipitation intensity figures (mm/24 hours) are listed in **Table 11.3.**

Given such daily rainfall intensities, what are the typical proportions (percentages) of run–off we can expect from various substrates? It is the percentage of run–off that will decide the proportion of storages needed in arid areas (as soil or swale storage, pits, dams, or cisterns). We presume a rainy period at 10 mm/day. See **Table 11.4** (the data refers to arid lands).

As well as surface conditions, other factors affect water run–off. Some important determinants are:

- CATCHMENT SIZE small catchments discharge a greater peak flood flow than large catchments. Foothill areas will peak to floods more rapidly than plains.
- STREAM GRADIENT. Steep streams discharge more rapidly than downstream areas of gentle gradi-ent. This is related to the factor above, as the smaller headwater streams are, in fact, usually steeper than the larger streams in the plains, which are often sand–clogged and have wide flood plains.
- SIZE OF STORM. Large storms may totally wet a small catchment, causing local floods, whereas the same storm will have little downstream effect. It is common, in deserts or arid areas, to see a flood disappear a few kilometres from the headwaters ex-cept in the case of very widespread rains.
- STORAGES AND RETARDATION BASINS. Dams, retention or interception banks, swales, natural swamps, and created swamps all reduce or delay flood peak. Local soil absorption is much more efficient in small streams of high density, but large catchments have greater dampening capacity for floods, providing such retardation areas are preserved for such purposes as flood control (many of these areas are drained, filled, or built upon)
- CATCHMENT FORM. The normal dendritic (tree–like) drainage patterns of rivers, where branches join at small angles, are subject to faster flooding (greater streamlining) than fold–mountain or fault–area grid catchments with their abrupt turns, constricted valleys, and relatively long runs of middle–order streams.
- DRAINAGE DENSITY (the actual length of stream per square kilometre). Very dense streamlines remove water very efficiently, and create higher and faster flood peaks. Sheet flow across the flat inter-fluves or off slopes in low–density areas retards flood–waters, as does the braided and broad streams of flat desert areas.
- VEGETATION. Run–off increases as vegetation is cleared. The less crown interception, humus storage, and ground interference encountered by rain, the more water runs off. If tussock grass replaces shrubs on dry

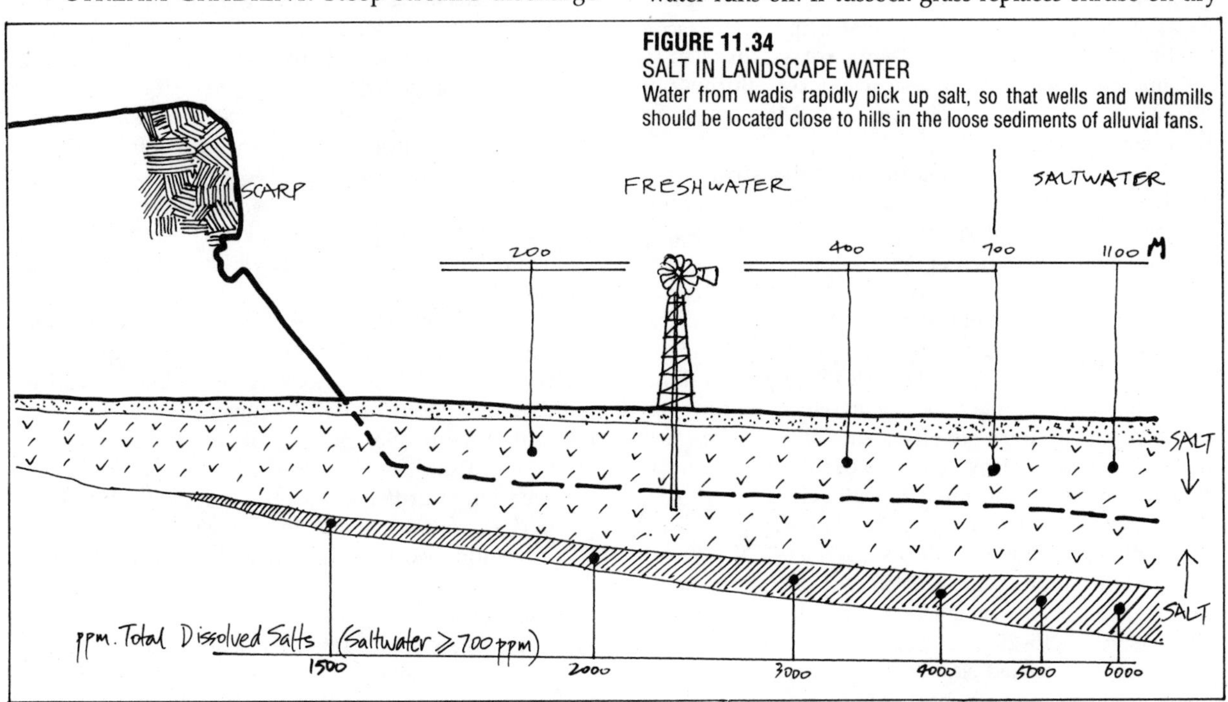

FIGURE 11.34
SALT IN LANDSCAPE WATER
Water from wadis rapidly pick up salt, so that wells and windmills should be located close to hills in the loose sediments of alluvial fans.

TABLE 11.3
PRECIPITATION INTENSITY

mm/day	EXAMPLES AND PHENOMENA
0-10	Light dews (0.8), light rains in semi–arid areas (4), prairie rains (6). The smallest rills commence flow at 6–10 mm.
11–20	All day drizzle (12), sheet erosion can occur at 13, gully erosion can develop at 15. Diversion drains in first and second order streams will flow. Typical of rains of humid coasts, warm frontal air masses.
21–50	Typical of monsoon sloud streets on coasts, or light rain all day in western maritime climates, and periods of convectional rain in deserts.
51–100	The condensation drip from onshore fogs (60 mm); tropical rains exceed 100 mm once a year.
100–200+	Heavy rain all day, and exceeded once a year on monsoon coasts, tradewind coasts. Exceededonce per century in temperate maritime areas.

TABLE 11.4
RUN–OFF FROM VARIOUS SUBSTRATES.

% RUN–OFF	SUBSTRATES
0	Coarse sands, active dunes, fixed dunes in dune seas.
5	Forested.
10	Wooded and grassed areas, tussock grassland.
15	Cultivated sands and sandy loams.
20	Approximate average figure for semi–arid and arid areas.
25	Ungrazed steppe of mixed shrubs, forbs, grasses.
30	Drained and cultivated clays and clay loams.
40	Steppe and shrubland grazed by hoofed animals.
50	Undrained clays, claypans,urban roaded areas.
60	Skeletal soils and scattered vegetation.
70–98	Compacted soils, solonetic soils, non–wetting sands, extensive paved areas, full water storages.

rangelands, run–off decreases. However, it rises rapidly as perennial prairie or tussock grass is in turn replaced by burnt ground or grazed–out areas where only ephemerals or annual grasses can flourish. Burning of clump grasses, therefore, has a sudden effect in increasing water run–off.

• SEASON. Run–off has its most disastrous and erosive effects after long dry summers, broadscale fires, extensive autumn or spring cultivation, and in fact subsequent to any period of stress on the landscape that reduces vegetative cover. It is at these periods that thunderstorms create gullies, flood thin soils, and cause the soil collapse that can create salted soils.

• ARTIFICIALLY SEALED AREAS. Roads, parking areas, house roofs, bare rock, concreted open drains, and compacted areas of settlements all produce flash floods in nearby streams. The area of these total run–off systems in catchment must be calculated if effective flood retardation systems are to be planned.

• SOIL AND ROCK TYPE IN CATCHMENT. Rapid rainfall infiltration occurs only in coarse or open sands, sandy loams, kraznozems (deep, well–structured soils), and soils with 15–20% or more humus content. Non–wetting sands, sealed clay soils, compacted soils, and sheet rock or massive rock domes all shed water rapidly. Areas of each type need to be mapped or noted in the catchment under study.

Gravels and boulder–banks, shattered rock ridges, unconsolidated pediment wastes, limestone outcrops, and eroded fold crests all act as *intake areas* for aquifers. These will remain open and reduce run–off unless (as can happen in deserts), fine blown dusts and sands bury the crevices and boulders, or wash down into rock crevices to seal the aquifer intakes. Once this occurs, springs will cease to flow and overland water flow will increase.

• TIME. Whether as a result of well–dispersed and low–intensity rains (1–2 cm per day), or of impeding surface factors such as tussock, litter, pebble, or artifical pitting and contour banking, water will soak into soils without run–off over periods of from 30 minutes in sands, and from 3–4 days in clay–fraction soils. Thus, we need larger artificial storage capacity in clay areas, or we risk drowning plants in hollows over the period of infiltration. Most plants can withstand 2–3 days of flooding.

Ideally, best plant growth results from 0.5–1.2 m of saturated soil, and infiltration rates of this depth should be timed in a pit (first wetted, then filled to test infiltration time) so that this depth of soil is wetted in something less than 48 hours. Given this data, the width and depth of total interception by swales or banks can be calculated, and retardation basins built to reduce water loss, increase vegetation, and control floods. These artifical intakes are now necessary over very large acres of overgrazed and eroded country.

• WATER VISCOSITY. Water density is little affected by temperature, while temperature has significant effects on water *viscosity*, hence on streamline shear or "friction". Day–time rains on warm surfaces run–off faster than night rains, or rain in cool night periods. As many deserts encounter rain in summer (monsoon) regimes, floods peak earlier in these regions than they do in winter–rain areas. Both the velocity of flow (the main erosive factor) and turbulence are accentuated by warmth and lower viscosity. Viscous water may cause local water level rise in low–order streams at night, and support heavier floating loads of detritus.

In many areas, local long–term stream guaging will not exist. The best guide to past flooding levels is therefore careful observation, based on damage to trees, old detritus levels, flood detritus caught up in trees, local knowledge, and a sensible assessment of the catchment area, its type, and rainfall data from similar regions.

Any designer can, by reference to rain days, catchment area, and a set of the foregoing factors, make an estimate of the potential for run–off (as overland flow) in a specific locality. Thus an estimate of the potential storages needed can also be made, so that loss of water off–site is eliminated or reduced.

Run–off in deserts, depending on the permeability of the surface, ranges from 5–80% of rainfall. Averages may be low, but in flash floods (not uncommon in the total precipitation), higher percentages are obtained. I have seen diversion channels in stable non–wetting dunes flowing as small streams in 1 cm of rain, and these filling large storages in a day. We can accept about 20% as an average of run–off, but in practice it is safer to allow a 20 ha run–off area to 1 hectare of crop, or cultivated vegetation, to ensure sufficient soakage for long–term growth. Only in areas of reliable rains can we reduce this ratio to 15:1 (catchment area to field area or productive crop).

HALTING AND ABSORBING WATER RUN–OFF

The halting and absorption of run–off has two main aims: to convert water from a destructive erosion force into a quieter life–creative energy (to develop forests), and to recharge groundwaters so that freshwater sheets develop above the deeper saltwater leads of the desert, and trees can benefit.

However, we should always do our sums and keep good records, for if we withdraw freshwater from the groundwater storage faster than we can recharge it, we are only temporary inhabitants of the desert. Thus our uses must be frugal, needful, and devoted primarily to establishing deep–rooted and drought–resistant trees, not to supply wasteful lawns, flush toilets, and leaky pipes!

Wherever we establish a surface intake from run–off steams, we may risk silt clogging. If we design a broad intake, with trees on the outer banks, we can use this silt (and any cover crop we sow on it) as a rich side–dressing to trees. For a check on silt depth, it may be wise to place permanent steel pegs with clear graduations in swales, and to sow and grade off at any time that more than 20 cm of silt is deposited. If only hand labour is available, such silt removal requires swales not exceeding 4–5 m wide, but with a tractor and blade (or road grader), swales can be 5–18 m wide. Infiltration can be aided by ripping the swale base, and using gypsum in clay soils.

Should low slope soils themselves be subject to steady erosion, swale banks can be exaggerated in this way to form eventual terraces, with trees now on the outer terrace slopes. Such evolutions occur where the slopes are sandy, or where cultivation takes place on the slopes between swales. As terraces (with treed edges), the slopes stabilise, and silt removal is a less pressing necessity. Such terraces are not a continuous series, but spaced downhill at intervals of 20 times their width.

By a succession of such strategies, large and small, desert groundwaters can be compelled to sustain a recharge. It is quite possible to create springs and to maintain low streamflow if a succession of 5–7 swales are made to behave effectively on run–off slopes; such effects occur naturally where once–full lakes such as Pyramid Lake, Nevada, receded in a stepped succession, leaving 1 m deep and 3–6 m wide bands of "beach boulders" across the slopes, and so halting run–off. Below these bands but above the present lake level, freshwater springs occur.

Bleed–off systems for turbulant and silt–laden streams can tax our ingenuity. In every case, a *restricted* water exit should open into a much *wider* drain (the silt trap), which later leads into a broad reed bed, from which fairly clear water can be led to a dam or swale sited well away from the stream line itself, and not above a valley settlement area (**Figure 11.14**).

While dams store water for village use, swales from these systems can be a few kilometres long (they can reach to the next fast stream) and can therefore grow extensive forest lines with no more trouble than an occasional silt clearing and a few hours with a road grader. Seasonal silt clearing in the silt trap may be necessary until trees are well established.

If heavy machinery (and very large stones) are available, a more substantial dam can be built, allowed to silt up, and a water channel cut from the level area just before the dam.

The aim, however, is clear: it is to lead run–off on contour to absorption or soil storages before it flows out to the salt flats and is made useless. For minor flows, a simple swale suffices to direct and absorb sheet waters. Even dam spillways in deserts can be led from the rear of the dam into long swales before being spilt into minor streams. We can hold thousands of times more

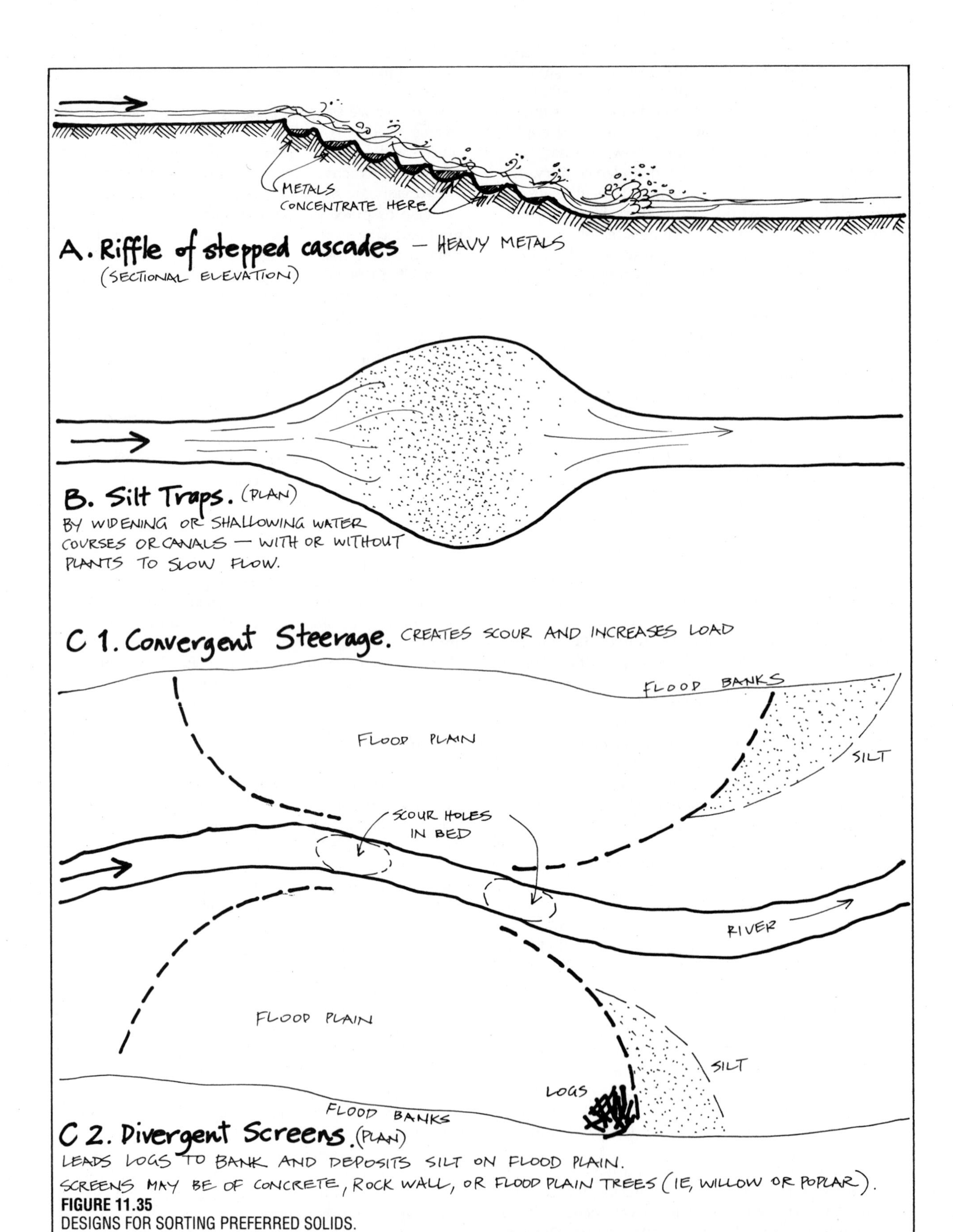

FIGURE 11.35
DESIGNS FOR SORTING PREFERRED SOLIDS.
Rivers, canals, flood plains, and diversion drains can be sculpted to deposit a variety of loads where these are useful, near gardens of houses. Silt traps may be needed in most deserts. See FIGURE 11.40 for details of flood plain fences.

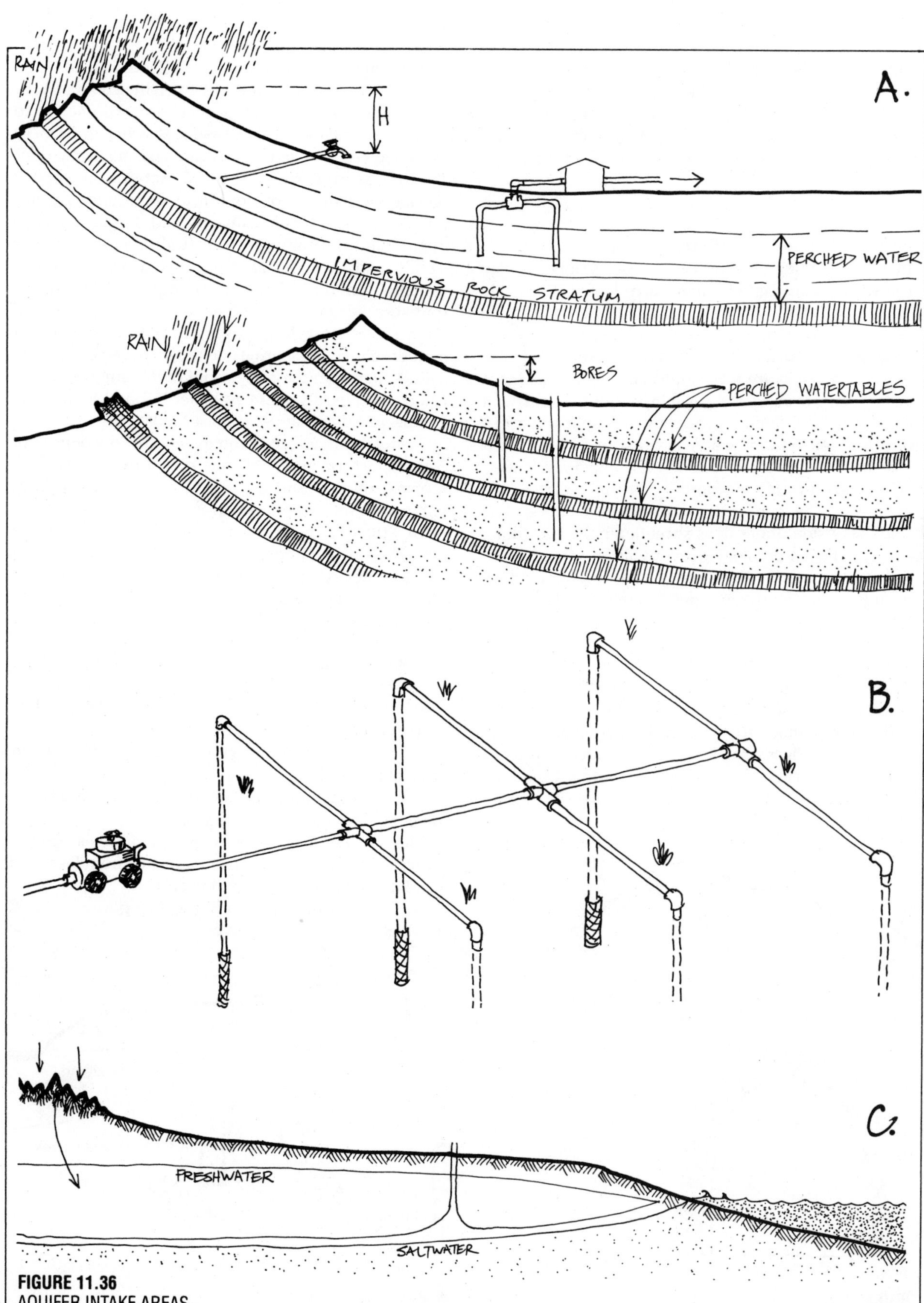

FIGURE 11.36
AQUIFER INTAKE AREAS.
A. Fold sediments may create a series of aquifers above impermeable layers; slant–bores can tap some of these at head by gravity flow.
B. A manifold of 50 mm pipe and filters is ideal for shallow sand pumping.
C. Deep bore wells may draw saltwater table up to pollute surface soils.

water inside the earth than we can in small dams, and with a greater beneficial effect on trees. Two dams of modest size near a village gives a clean water source for houses and a welcome swimming area (these functions need to be clearly separated, even if the latter is also for garden use).

ISRAELI RUN–OFF TRAPS *(LIMANIM)*

In gently–sloping uplands with well–developed valley run–off in rainy periods, and where loess soils enable deep infiltration, Israel has developed what are essentially dry dams of from 0.1 to 0.5 ha, the bottom graded flat, and the dry walls little consolidated. The banks themselves are 1–2 m high and 2–5 m wide. It can be beneficial to support the inside base of the bank with stone. Sills at the ends of the walls will spill water at a depth of 40–60 cm, when it can flow to the next *limanim*. Trees are established in these impoundments.

The run–off area is from 30–500 times that of the catchment. The more gently water can be led in (via low slope diversion drains or as sheet flow), the less silting and sill erosion occurs, but trees will grow well if the soil is kept open by tillage, or compaction by livestock is prevented. Catch–crop is possible in wet years. (From Wilson, G. in *Landscape Australia*, Aug. 1980).

AQUIFER INTAKE AREAS

Only in hill country, and then for the most part on ridges, plateaus, and detritus slopes do we find open subsurface channels for the deep infiltration of water, as loose gravel ridges, shattered or soluble rocks, cappings of loose dune sands, or heavily forested ridges and high slopes of porous materials such as limestone, free–draining soils, or fissured rock (**Figure 11.36**).

From the intakes, water will enter the soil and rock strata, and travel for many metres or kilometers to springs, rivers, or artesian basins. If water from these latter sources is to be utilised, care and maintenance of the primary intake is essential. Satellite telemetry reveals the patterns of flow in underground streams. Most follow the dip of sediments, and the general slope of the land, but aquifers can (unlike streams) flow uphill, and they can also produce a head of water (usually from 0.5 to 7 m) in bore pipes or piezometers (pressure meters).

On flatter land, large areas of sand act as perched aquifers above less permeable desert pavement or clay soils. Water from these seeps out of the dune base, and can be used locally for trees. Water is found from about 2 m deep in dune sands as moist sands.

The intake areas can become ineffective by deposit of volcanic ash or mud, by fine dust from desert dust storms, or by sand and dust blown uphill from cultivated and overgrazed areas. Except for vulcanism, such effects can be minimised by stabilising the landscape, by windbreaks, and by protected reserves. In all cases, deforestation of slopes and ridges reduces the effectiveness of intakes.

We can assist aquifer intake by:

- Planting trees on ridges and slopes;
- Building deep interceptor drains on contour near ridge tops and on plateaus;
- Ripping rock pavement or capstone areas to allow water penetration;
- Leading surface drains to wells cut in shattered rock; and
- Creating banks of loose surface boulder, or opening pits and swales in gravels.

Trees, in the long term, keep such systems open to infiltration and reduce or eliminate dust from loess and sand deposits from surrounding areas.

By a combination of all the foregoing factors, intelligent preparations can be made to utilise this fresh

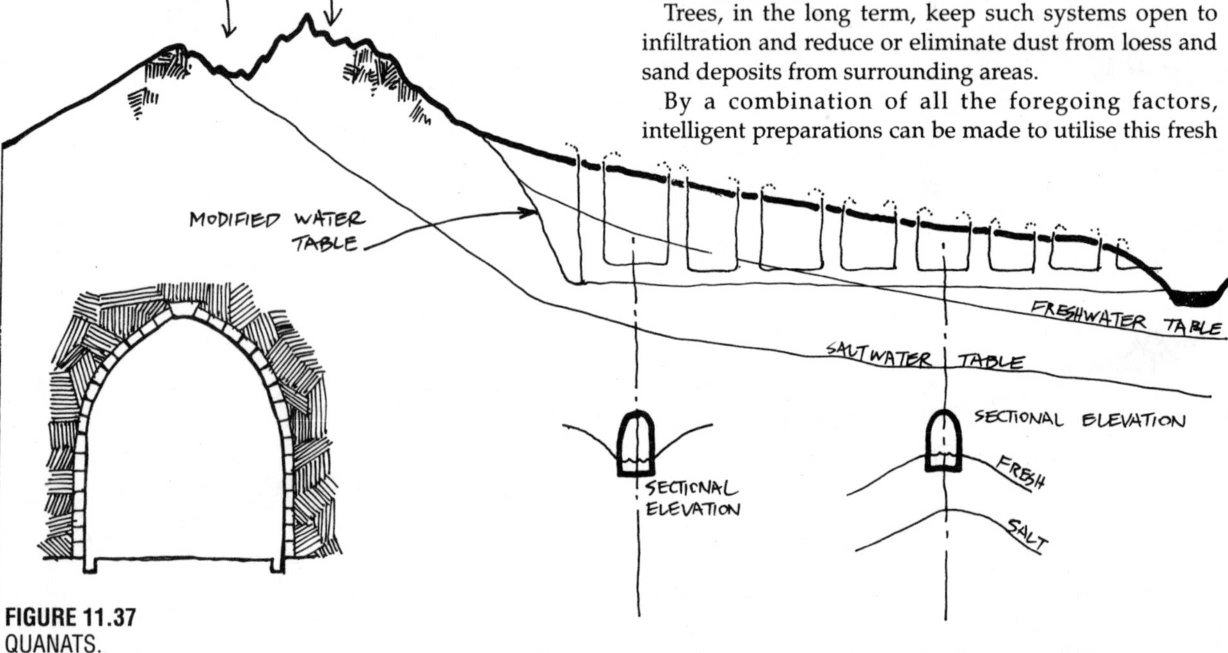

FIGURE 11.37
QUANATS.
Ancient galleries or quanats run for thousands of kilometres in many deserts; they have the disadvantage of being non–stoppable, but are in wide use in S. W. Asia. Air from these tunnels is vented to cool houses; detail gives gallery construction in sediments.

water where it does occur. QUANATS have been used in the Middle East for centuries to tap aquifers (**Figure 11.37**).

When areas were forested, ungrazed, uncultivated, and lacked the hard pavements of roads and settlements, little run–off occurred and throughflow in soils kept streams flowing. Desertified areas lack these characteristics, and we need to intercede in the hydrological cycle to store, infiltrate, and convert rainwater to vegetative growth—in short, to create soil storages to assist forest regeneration.

INFILTRATION

Infiltration of water and the time taken to infiltrate 25 mm of rain (in hours):

- 11–7 mm/hr minimum in sands and dust (loess) 2–3 hours;
- 4–1 mm/hr minimum in clay–loams 6–20 hours; and
- 1 mm/hr in alkaline swelling clays 24 hours.

This has quite a few implications for us, for if we can spread neutral or acidic sands about 0.5 m deep or more over alkaline clay pans, then we have an excellent surface to first infiltrate and then hold rain, and consequently a good garden or tree site, on which we can store up to 1 m of water for long periods. This effect occurs naturally at the discharge end of some gullies or canyons.

Another way to achieve this is to open deep swales in alkaline soils in front of sands and gravels on the move, and let them fill. In many areas, people have (with great labour) buried layers of clay, plastic, thick colloidal green matter (gley) and tar or latex under the deep sands of their gardens; on the broad-scale, drainage slits are left every 100–200 m in these systems. It is obvious that we need to hold water for a day or more for it to soak in to soils.

SLOPE STABILISATION FOR INFILTRATION

Spinifex or porcupine–type grasses help to stabilise slopes and increase infiltration, forming cross–slope sand dams as they age. The normally circular tussock form breaks open to create an open crescent, its horns uphill. Detritus forms deeper sandy rubble banks behind these lunulate clumps, effectively creating a swale. Burning destroys this effect and allows sudden slope erosion to occur.

However, we can create such infiltration aids as swales, wadi dams, levelled gardens, and soakage pits. Of these, the swale is most easily created on a large scale, but wadi dams may also work well for local food forests; neither are a trouble if some basic machinery is available. In cold deserts, tussock grasses and mosses perform the same function for unstable rock slopes and soil flows, and cross–slope Vetiver grass lines on contour greatly assist water infiltration and slope stability.

FLOODWATER HARVESTING

Aboriginal Australians (Stevenson, 1978) made earthwall floodwater dams to increase *Panicum* grass and their seed, as did the Papago and neighbouring groups in the Sonora regions of the USA (Nabham, 1979).

The latter groups worked on river flood–plains and deltaic deposits at the mouth of wadis. Nabham notes the interesting fact that although summer storms bring 40–60% of water, they add only 15% of the run–off to mainstream resevoirs. Most of the run–off is localised in intermittent streams, and is only available for local headwater agriculture. About 8 cm of rain is needed to give useful run–off, over a catchment area 15–27 times the area of fields. Hardy crops such as sorghum are produced with run–off only 4 times that of the field.

Along the Colorado River, floodplain fields were also constructed for major floods. Techniques cover temporary basins, irrigation canals, and field boundaries of stone and earth reinforced with posts, brush, and vegetation. Planted and wild or weed species are grown together as a crop complex over 4 ha or less per field. The system is part of that illustrated in **Figure 11.40**, and needs minimal management.

As headwater streams coalesce and enter broad valley systems flowing out of the ranges, broad braided rivers (dry most of the year) break out across country, dropping silt as they do so, and leaving lines of organic mulch at their highwater level. This mixture of seed, manure, twigs, molluscs, and debris is excellent garden material and can be trapped using a sequence of stout scoop fences (**Figure 11.40**) for use in gardens.

Where the land flooded is flattish, a "grid–iron" of low banks keeps the floodwater and silts settled and absorbed. Such grids need banks only 0.5–1 m high, and those are best stabilised by stone on the upstream side, tussock grasses or tough low shrubs, live–set sticks and brush, or can be made of heavy rock walls from the main river bed. Such systems in Papago Indian areas maintain high levels of the major plant nutrients, and were the basis of the rich Nilotic agriculture.

When and as the river floods, fields of millet, sorghum and legumes can be broadcast sown into the wet silt, and later headed as stored crop. Further away from the main stream, broader deep dykes will hold waters in narrower and deeper channels for tree crops, and the trees should overshade the deep absorption swales so developed.

BRAIDED STREAMS

A close study of braiding patterns in streams, and in the associated "green" areas of vegetation, showing up as darker areas on a black and white photo, reveals that there is an inverse relationship between the land areas wetted or saturated and the channel width. Small multiple streams produce much larger wetted areas than simple main channels (**Figure 11.38.A**). This suggests an as yet untried strategy where (in dry periods) we can build concreted, deeply bedded, deflections in main channels, and set these out to split the flood flow as often as we have space to create small channels (and there is ample space in most arid flatland flood plains).

In effect, we are trying to *widen* and *slow* the floodwaters, and wet as much alluvium as possible. As such floods more frequently arise from headwater run–off, not local rain, it is in our interest to spread and harvest such water before it is lost to saltpans in the desert (**Figure 11.38.B**).

Methods of achieving multiple stream braiding would be best modelled in experimental sand channel flows, but the ground effects on vegetative growth are evident in many arid areas, when the wetted areas support strong tree growth.

The process of braiding depends on a shallow water flow over extensive deep sand sheets; braiding does not occur "underwater". Each braid–form is initially a diamond (rather, a rough diamond) with plunge–pools at each corner, caused by the confluence of side streams. Ideally, the unit forms are as in **Figure 11.39**.

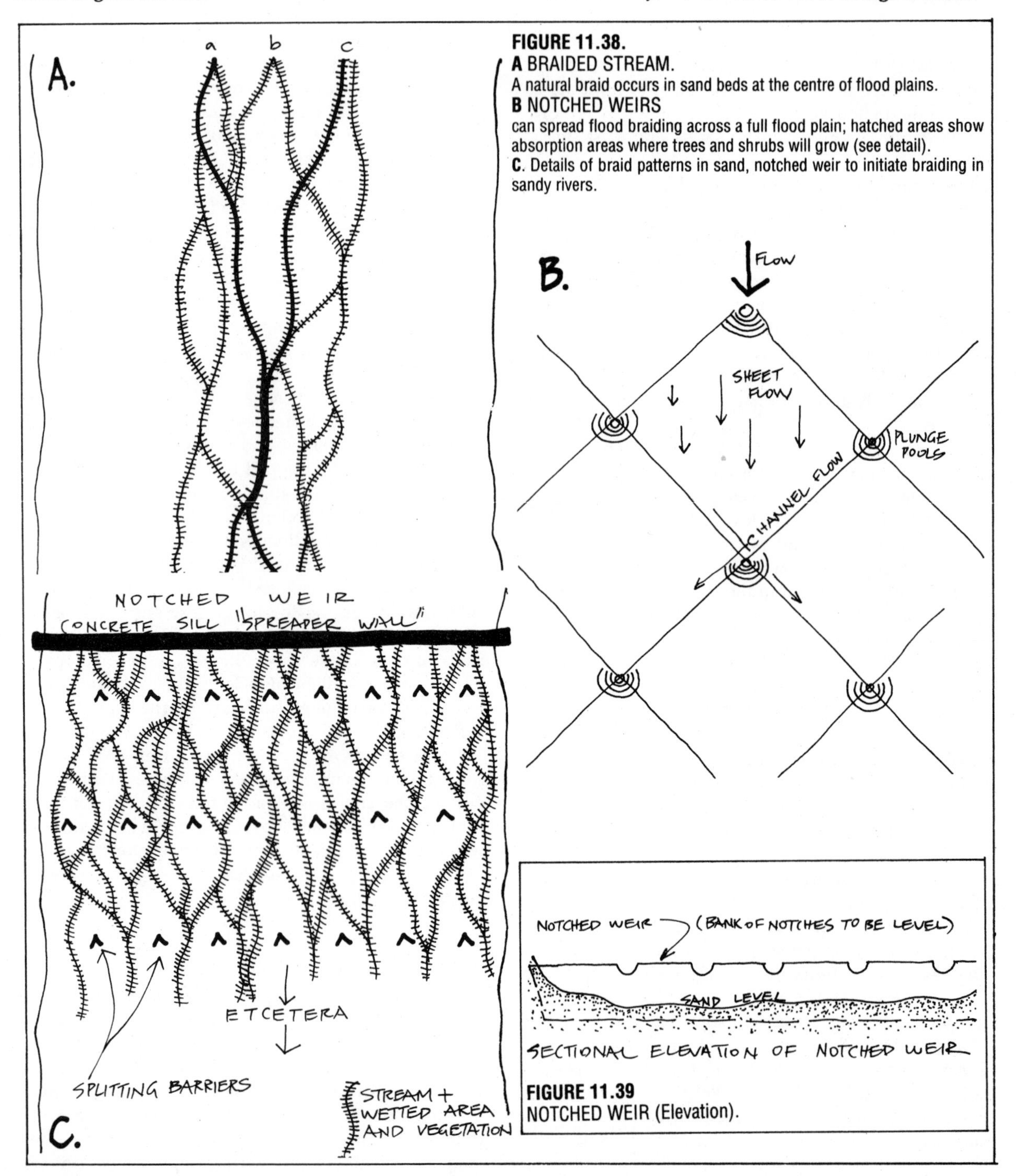

FIGURE 11.38.
A BRAIDED STREAM.
A natural braid occurs in sand beds at the centre of flood plains.
B NOTCHED WEIRS
can spread flood braiding across a full flood plain; hatched areas show absorption areas where trees and shrubs will grow (see detail).
C. Details of braid patterns in sand, notched weir to initiate braiding in sandy rivers.

FIGURE 11.39
NOTCHED WEIR (Elevation).

In low flow conditions, only the channel flow persists, and sheet flow across the interfluve is absent. Just what initiates the pattern is not clear, although similar flow patterns arise from scattered pebbles or shells in sand. However, I believe that we could spread floodwaters more fully across plains, initiate a finer mesh of the pattern, and spread and absorb water effectively if we erected a level, notched wall across the braided area. The notches are in effect the imitation ofplunge pools.

SCOUR HOLES

Within 10–50 km of ranges there may be a few permanent desert lagoons. They are often of great beauty, full of crustaceans, fish, molluscs and water plants such as lotus, kangkong, water chestnut, and submerged weed. Birds and mammals visit by day and night. How do these lagoons form?

In every case, they arise from a strong natural ridge (often unnoticed by visitors) that approaches the river from one or both banks, thus constricting the flood flow and forcing an increase in velocity and a powerful hydraulic digging action. These days, it is simple enough to replace or construct such ridges using a combination of rock–cement walls and a bulldozed bank, stronger and deeper near the river bed itself. The ponds remain full in drought only because the deep sands of the river contain millions of litres of slow seepage (which we can also increase by installing sand dams higher up in the system). Nevertheless, we

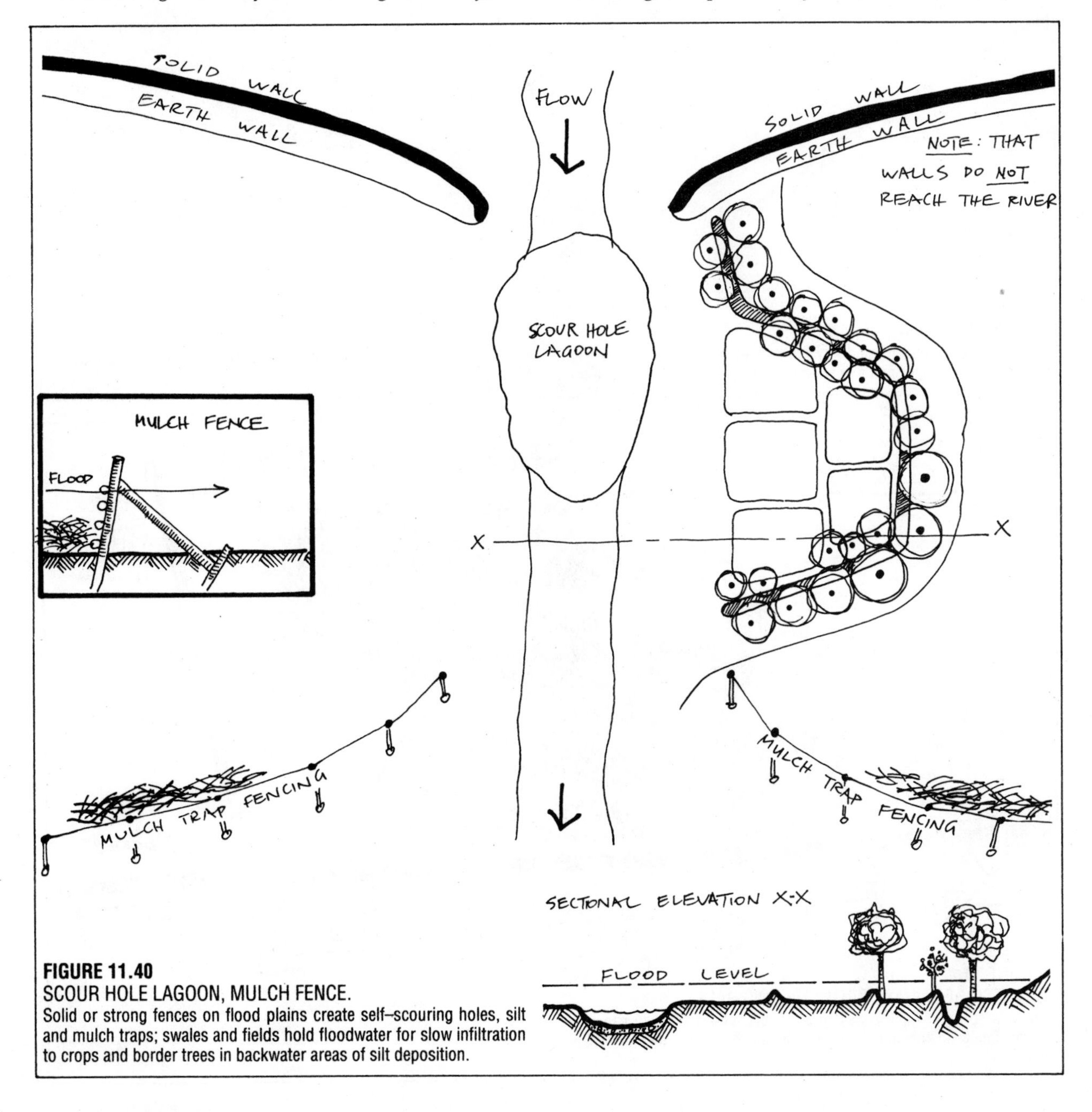

FIGURE 11.40
SCOUR HOLE LAGOON, MULCH FENCE.
Solid or strong fences on flood plains create self-scouring holes, silt and mulch traps; swales and fields hold floodwater for slow infiltration to crops and border trees in backwater areas of silt deposition.

should not create these open–water systems too frequently *unless* we have made silt dams of up to 9 or 10 times the capacity of each pool, as open water will evaporate over time. The profile of such natural scour beds is as in **Figure 11.40**.

SANDY RIVER BEDS

Wherever deep sands fill rivers, palms and large trees will stand in small embayments off the river, where flow damage to their trunks is less likely. A good guide is to observe how native trees survive, and where the largest trees grow (that is, if cattle have not killed out the range).

It is very noticeable that desert river banks in hills will often have quite separate species of trees on either bank; this may be true of all rivers eroding country areas. One species needs the slip–off or silty slope, the other thrives on the harder country rock of the erosion bank. Almost all the burrowing animals use the silted or slip–off slope of the river.

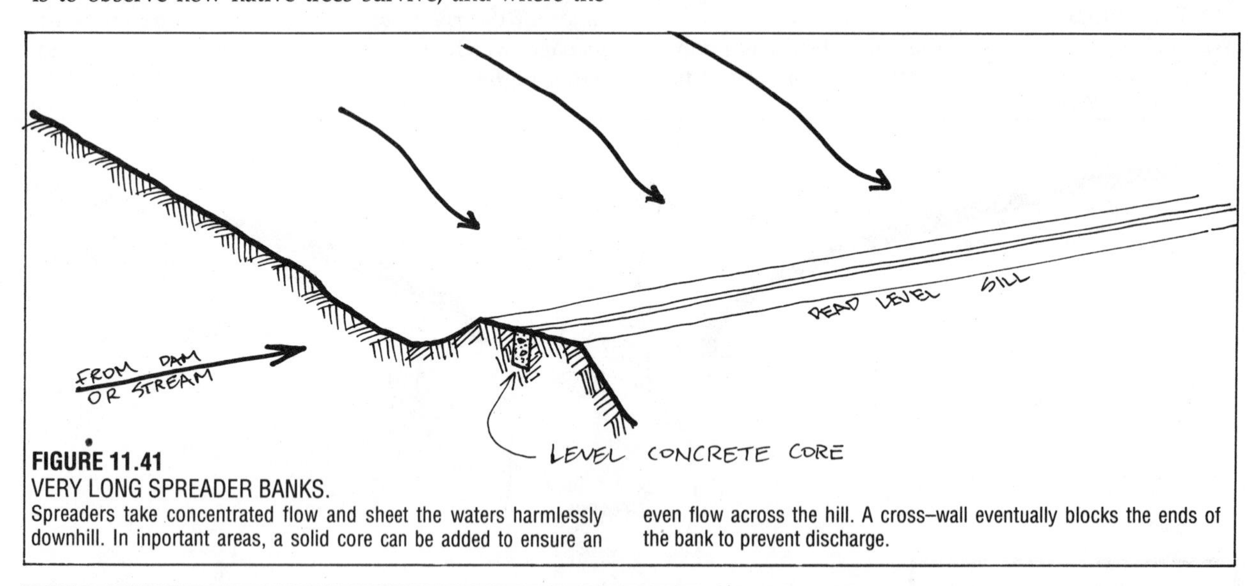

FIGURE 11.41
VERY LONG SPREADER BANKS.
Spreaders take concentrated flow and sheet the waters harmlessly downhill. In inportant areas, a solid core can be added to ensure an even flow across the hill. A cross–wall eventually blocks the ends of the bank to prevent discharge.

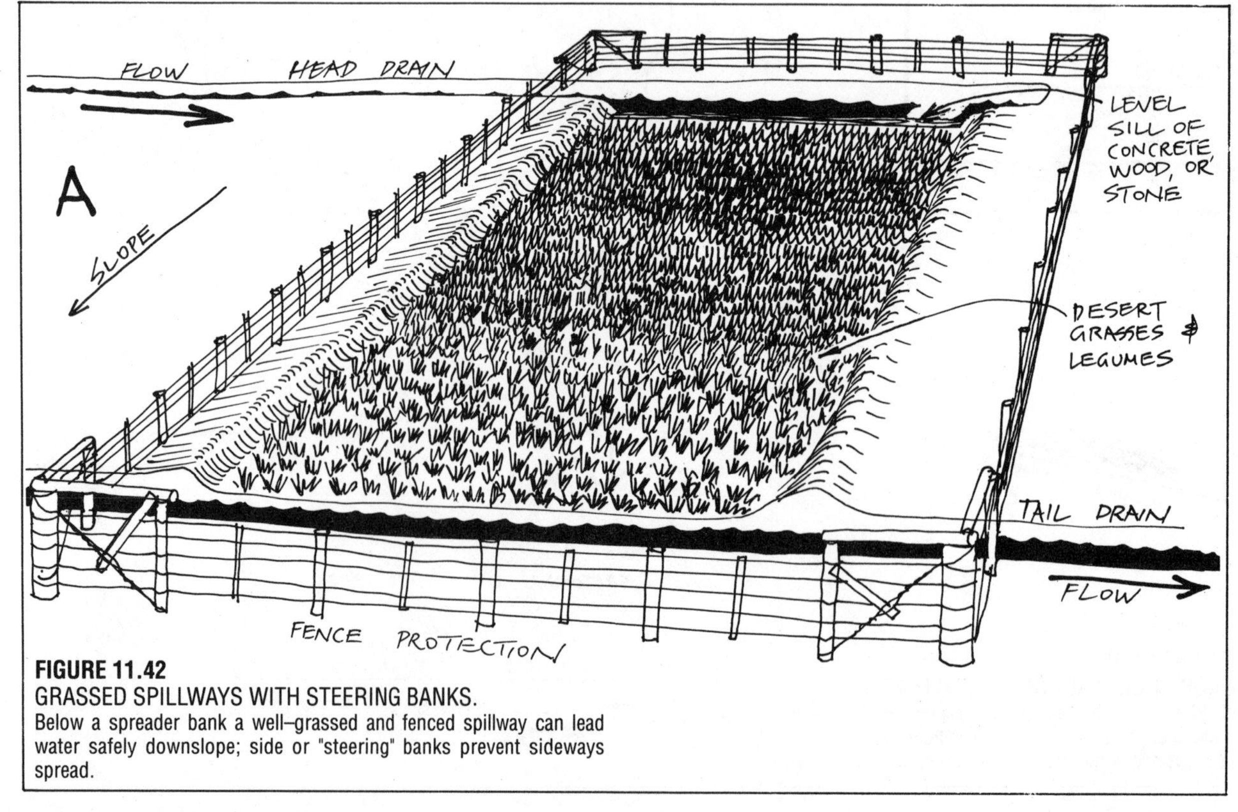

FIGURE 11.42
GRASSED SPILLWAYS WITH STEERING BANKS.
Below a spreader bank a well–grassed and fenced spillway can lead water safely downslope; side or "steering" banks prevent sideways spread.

PITTING IN SANDS AND LIGHT SOILS

By cutting a quarter–section slice off a set of disc blades, "long" pits are made, with a quarter of their length as solid breaks. As discs are drawn along, we get a series of mini–swales, and these too are staggered, like our contour swales.

A great advantage is that such short pits or furrows need not be strictly on contour, so that they can be drawn by eye. No one pit is so large as to spill too much water and cause stream flow. Seed can be added to pits, broadcast, or blown in from residual vegetation. The method is fast, effective, and can be completed on the broadscale; grasses, shrubs, trees, and forbs can be seeded.

These pits effectively become small swales for an area of bare soil. Over many acres, they have enormous storage capacity. They would usually be spaced at 0.5–0.75 m apart, and be from 1.0–1.5 m long and 0.3 m deep at the most. Here, soil is ideally sidecast downhill, or downwind, but this is not a critical factor on near–flat sites.

This form of discing acts as surface roughening, and eliminates or greatly decreases dust storms as soon as it is made. There are many forms of pitting machines, but they can also be hand–dug. Pelleted desert shrub seed plus a light fertiliser dressing creates the condition for permanent soil stability around settlements.

Cordon (1975), working in semi–arid hills, records average annual run–off of 20%, with up to 80% in eroded and compacted soils for individual rivers. He recommends (for slopes in excess of 2%) a series of contour furrows or mini–swales at 2 m intervals, with only every 20–30 contours actually surveyed. The swales are "broken" to allow flooding through to the next series. Vegetation in the furrows remains green in drought. The soil chosen for these systems must have the capacity to absorb the moisture (which mean soils primarily of sand). The scale and area treated can be as large as is required at the 2–4 m spacing.

SPILLING WATER DOWNSLOPE IN FRAGILE SOILS

Steep–sided gullies often result from road culverts, animal tracks out of wadis, or water discharged from diversion drains after fields are flooded.

There are only a few safe ways to spill downslope in deserts; these are:

- Very long spreader–banks (**Figure 11.41**), maintained and inspected in rain.
- Grassed spillways with steering banks (**Figure 11.42**). These take a few years to evolve.
- Downslope sealed spillways or pipes with splash–pools, leading to a swale, one of the above, or a dam (**Figure 11.43**).
- "U"–shaped channels filled with boulder, smaller rock, then gravels, so that water is forced to follow multiple stable paths.These, in fact, imitate in their structure the rock sorting of dryland breccia slopes below cliffs.

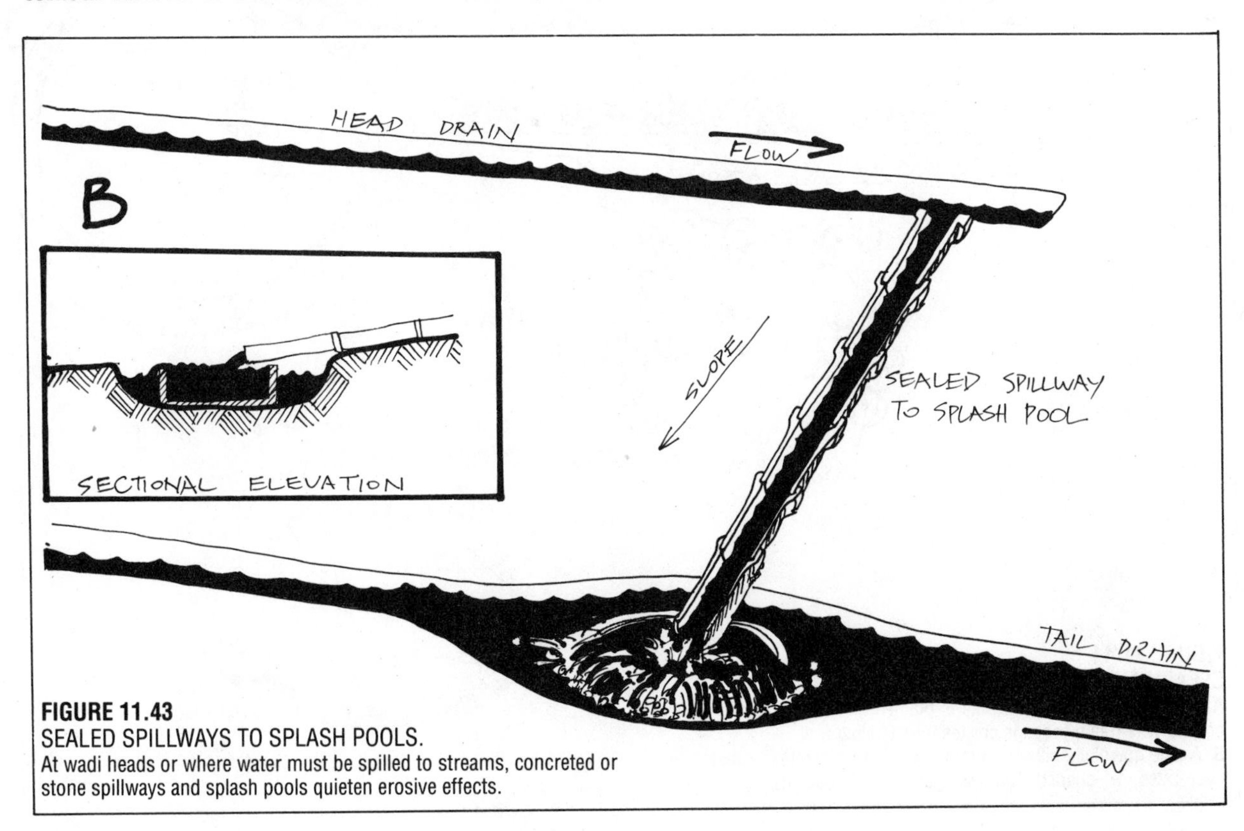

FIGURE 11.43
SEALED SPILLWAYS TO SPLASH POOLS.
At wadi heads or where water must be spilled to streams, concreted or stone spillways and splash pools quieten erosive effects.

SAND DAMS AND CLEARWATER RESEVOIRS

Washes in gently–rolling foothills carry broad flood-water surges. It is very advantageous to dam these so that water backs up 10–20 times the dam width. The wall must be keyed in for 4–6 m to the side banks, or water will scour around the ends. Before this is built, it is necessary to cut a series of 2 m deep bays into the upstream banks, and to have several loads of coarse stone to place against the upstream side of the wall itself. This will form a drain to a clearwater dam, which itself should be less than a ninth of the total area backed up by the wall (**Figure 11.44**).

How it functions is fairly simple. The first floods rush down and over the wall. Silt and coarse sands fall out and in a few rains fill up the storage. The shingle or rocks on the inside of the wall act as a filtered seepage to the clearwater dam. The dam itself also leaks water through the river bed—it is intended to— and some

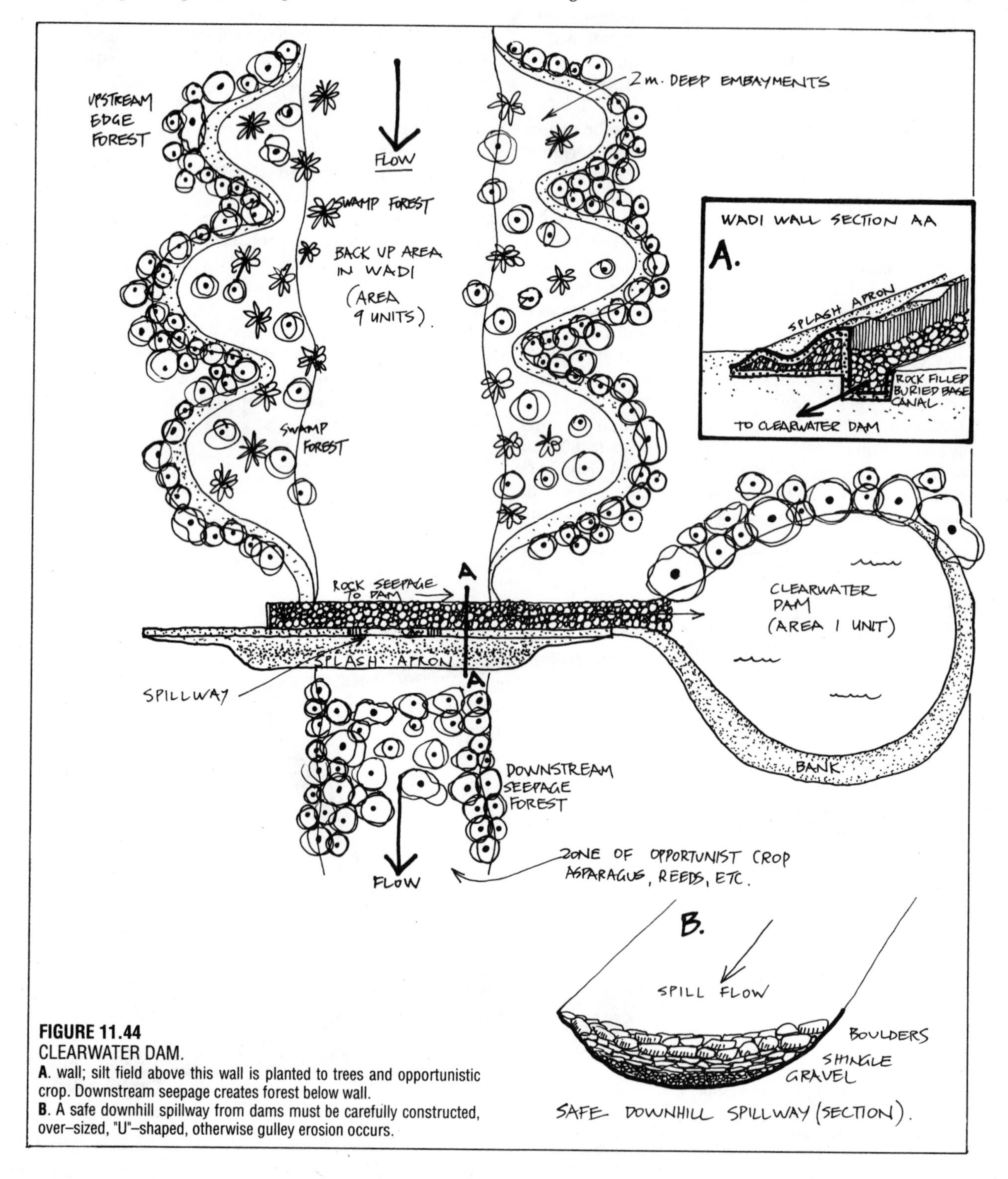

FIGURE 11.44
CLEARWATER DAM.
A. wall; silt field above this wall is planted to trees and opportunistic crop. Downstream seepage creates forest below wall.
B. A safe downhill spillway from dams must be carefully constructed, over–sized, "U"–shaped, otherwise gulley erosion occurs.

water seeps through to the downstream forest. The bays we dug upstream also fill with sand and make a flat water–filled silt bed on the edges of which we plant valuable large trees (pecan, palm, avocado, mango) which make up the upstream forest. Such trees can also surround the clearwater dam (the dam forest).

When the barrier dam is sand–filled, several deep rooted sub–aquatics will grow there, e.g. asparagus, *Mauritia* palms, Chinese water chestnut, swamp cherimoya, etc. Full, the rubble and sand act as a buried water resevoir which keeps the dam full (I know of one such dam that dropped only 8 cm in level over a 7–year drought), and also as a deep reserve of moist sand for trees. The downsteam forest may extend 1–2 km in the deep sands of the river (for a wall only 50 m wide).

There are hundreds or thousands of such sites a few kilometres from foothills in deserts, and this simple but sophisticated system creates oases where there were none before. The figure shows a dam built by a railway engineer at Oodnadatta, South Australia, in the 1920's; it is still functioning well in the 1980's.

Below roost trees around such dams, the guano from desert birds may be gathered as sheets of phosphate fertiliser, and these are of great value in establishing dune or rock windrow trees (as a dilute solution or mixed with 30–40 parts of damp sand). On the silt dam, melons grow very well as the first swampy condition recedes, and a good crop of legumes and melons can be planted at that time. A few such dams support a strong village population, and will keep scour holes downstream recharged with water for swimming holes. All clearwater dams should be 5–9 m deep and of reduced surface area, rather than shallow and wide.

DAMS

Barrier dams across streams in wooded and stable hills can be expected to have a long life of 100–300 years or more, before silt and humus fill the catchment, and even before then a well–designed dam can be silt–flushed from the base. But barrier dams in arid areas must be planned from the outset to fill to the brim with silt and rubble in the first few years. The water they hold will be stored *in* sands and rubble.

Thus, *clearwater* dams in deserts should be built well away from such valley sites, and the turbulent flow from wadis or foothill sand–bed and boulder streams must be bled off into almost–level channels fitted with sand traps, and only after that to a dam. This dam should not itself be sited so that fast water can run in to it over loose surface material. It should be fenced, and the margins planted to rushes and shrubs. **Figure 11.45**.

Some helpful factors for an effective sand dam can be checked before construction begins. Ideally, the rubble and silt in the dry creek bed should be very open and permeable, of stones, pebbles, and coarse sands (this will almost certainly mean that the dam is near the foothills or in a valley close to the headwater rocks). Secondly, the floor of the dam is ideally clay or massive rock, and hence leak–proof. However, even if neither of these factors are ideal, sand dams will still be good growing sites, as we will be slowing down flood water on a long and level site.

An extensive silt dam of 9–12 m in height is always built in 1.5 to 2 m stages, and the wall raised as the earlier stages silt up. Behind each wall, gabions allow water to filter down to a base pipe, which can be led to settlement. The whole area of such catchment needs to be protected from sprays, fertilisers, and browsing animals to protect health in settlements.

ROCKHOLES AND GNAMMAS

Rock basins, often filled with water, arise from a series of origins; some are:

- Rock scouring in stream flow (corrosion) on stream beds.
- Freeze–thaw and wet–dry swelling and shrinking plus rock mineral hydration.
- Salt–eroded ponds in rock, as near the sea.

There are many ways to assist water conservation in rockholes, e.g.

- Fill hole with pebbles to lower evaporation; some water is lost.
- Thatch or cover hole.
- Build gutters to hole.
- Erect a metal mesh fence to direct dew condensation to hole.
- Deepen hole, and plaster it with cement if it leaks.

EVAPORATION AND EVAPOTRANSPIRATION

As we have defined the climatic scope of this chapter, open–pan evaporation (pans of water exposed to air) will always exceed or equal precipation over the year. In fact, almost all of the sub–humid to arid areas tested have a potential evaporation well above that of actual rainfall. This means that for most of the year, there is no longer any free water to evaporate except for the water in dams or other open–water storages. A typical grain crop such as millet needs 50 cm of water to mature. It is unlikely to obtain enough water *unless* water storage from run–off is available, or unless run–off has been infiltrated into the soil.

Direct evaporation in still air (the "oasis effect") will exceed potentially 100 cm per year in most arid areas, but a far more severe effect is that of hot wind evaporation (the "clothesline effect"), which in unsheltered fields most effects the field edge and increases transpiration 10%–30% over the background effects. Chang (1968) records that winds blowing from arid to humid areas may affect crops or trees for 400 m before acquiring a non–evaporative water moisture load. In effect, a 400 m wide evaporation border of trees may be needed between the desert and productive croplands. There is no doubt that a windbreak of trees uses less water than that removed by the hot winds over crop, so that throughout croplands, scattered trees and cross–wind tree breaks are essential.

Needless to say, such trees should be selected to be both dry–adapted and of benefit to the crop. *Acacias*, such as *A. albida*, and *Prosopis spp.*, are some of those commonly used to protect and benefit crops in arid areas. Evaporation from open pans of water outside the crop or tree shelter can be 16 mm/day, while 300 m into the crop, only 5 mm/day is lost. In this case, the crop itself acts as windbreak, with a consequent gain in yield due to less evaporation. Whatever we can do to prevent the drying effects of hot desert winds will be worthwhile, so that although we will always be left with an oasis effect we need not suffer extra loss of water. For restricted areas of crop, narrow canals, or home gardens, we can screen out winds and shade the area either with shade cloth (to 50% or more shade), vine crop, or the canopy of palms and "umbrella" *Acacias*. These strategies plus ground mulch drastically reduce the water lost to evaporation from open storages and crop species.

CONSERVATION OF WATER IN TRANSIT

The transmission of water (as with any fluid or gas) needs great care. More water can be lost in canals, especially in loose soils or shattered rock, than is lost in use. Even open concrete channels can lose great quantities of water to air. There is really no substitute for long runs of pipe, and in small systems extruded PVC (flexible or rigid) pipe is most effective. Runs of 100 m or more can be made in one length; there are now long rolls of collapsible pipe for surface use, which can be unrolled as needed.

There is one other way to store water from opportunistic run–off, and that is in tough plastic or neoprene bags. These can, like swales, operate from fixed–diameter pipe inlets, much as the feeder drains operate from a head drain. The swale in this case is provided with a rolled container, which fills and unrolls as water fills it, and forms a "sausage" along the swale. Precisely the same technique can be used to capture fresh water flowing into salt lakes, so that the freshwater sausage floats, and is available as unmixed water on demand. This can be a much bulkier article, with greater depth/width ratio, as there is no

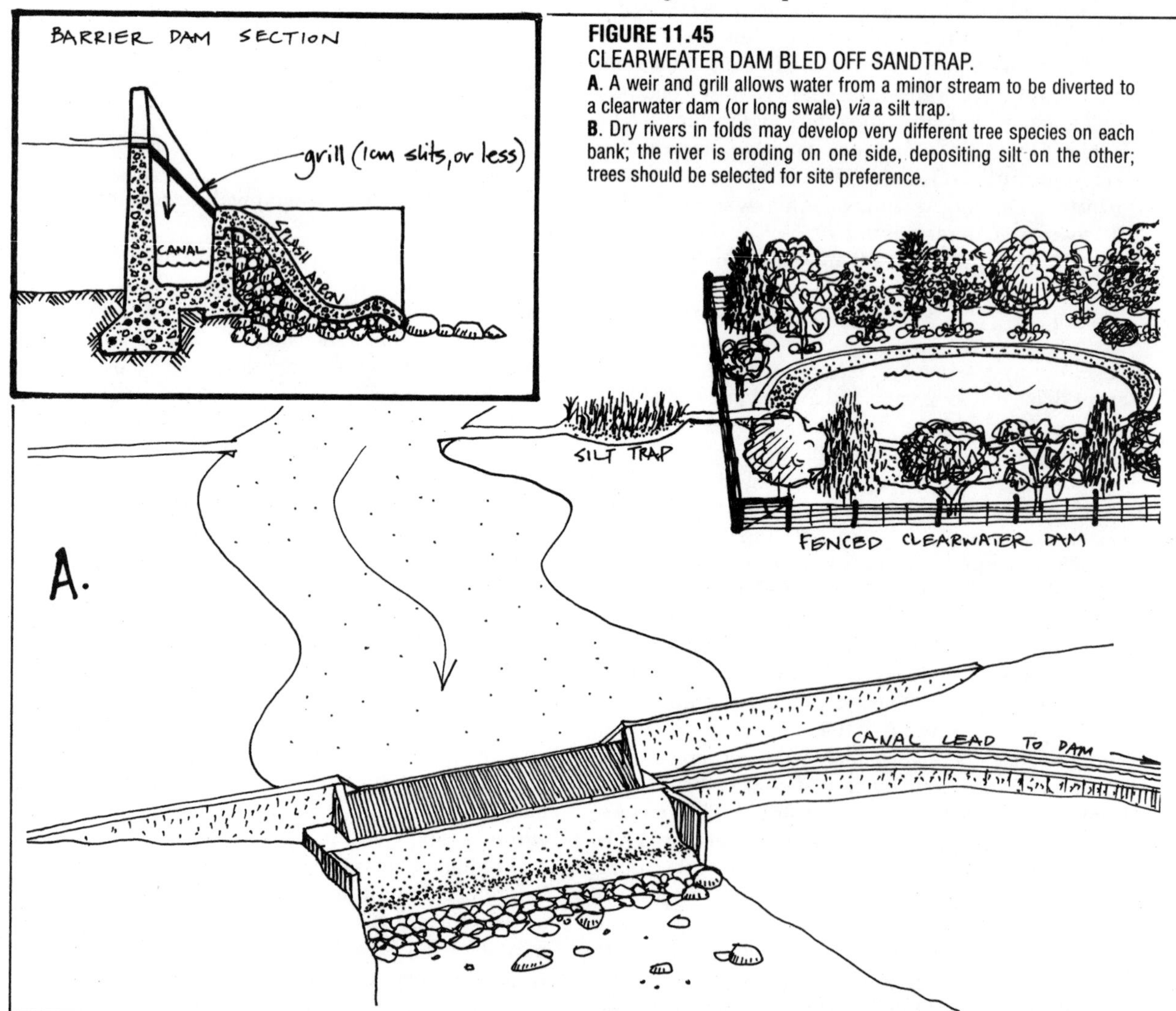

FIGURE 11.45
CLEARWEATER DAM BLED OFF SANDTRAP.
A. A weir and grill allows water from a minor stream to be diverted to a clearwater dam (or long swale) *via* a silt trap.
B. Dry rivers in folds may develop very different tree species on each bank; the river is eroding on one side, depositing silt on the other; trees should be selected for site preference.

substantial head pressure on an immersed bag. In swales, or as a series in swales, such bags would need fencing protection from large animals, but would otherwise be evaporation–free storage, preferably silvered above to reflect light, and trellis or vine–shaded to further reduce light and heat. Bags in salt water can be towed to other locations, or pipelines led across lakes to fill storage bags near villages.

There are these basic approaches to reducing evaporation in water storages in deserts:

- Underground and in–earth storage;
- Surface treatment of dams; and
- Storage configurations.

Any reduction in total evaporation of water is of interest in areas where this factor may reach 180 cm in a rainfall of 30 cm. Water is an expensive resource in any situation, and more so in deserts.

Several cultures in deserts (Peru, Iran, Afganistan, Canary Islands) have avoided gross evaporation loss by conveying water to desert farms via many kilometres of underground galleries (the *quanats* of Iran) and by storing the water in clay or cement–sealed caves. This is certainly an effective strategy, only superseded in modern times by the development of extruded and leak–proof pipes and concrete tanks, by bores in headwater aquifers, and by the well and windmill systems widely used in Australia.

Water absorbed as rapidly as possible into soil is also safely out of the sun's rays, but to be efficiently stored, it is necessary to ensure that just enough water to soak the soil is admitted to swales or fields (that excess does not in fact escape to the groundwater at depth). A loose cover or mulch should be kept on the soil to prevent excess evaporation, and a crop established to use the water effectively. A 10–15 cm surface layer of loose gravel or volcanic cinder is widely used in the Canary Islands for crop and trees; coarse mulch has the same shading and insulating effect. If impermeable clays or rock lies at 2 m or so, trees can retrieve all water so stored, and desert trees penetrate to 30 m in free soils.

In view of the evaporative effect of winds, any exclusion of wind from dams greatly reduces surface loss. The exclusion can be by tree windbreak, artificial mesh windbreak, or even by building trellis or shade structure over small dams, as Australian Aborigines would do with gnammas (open rock pools). Gnammas

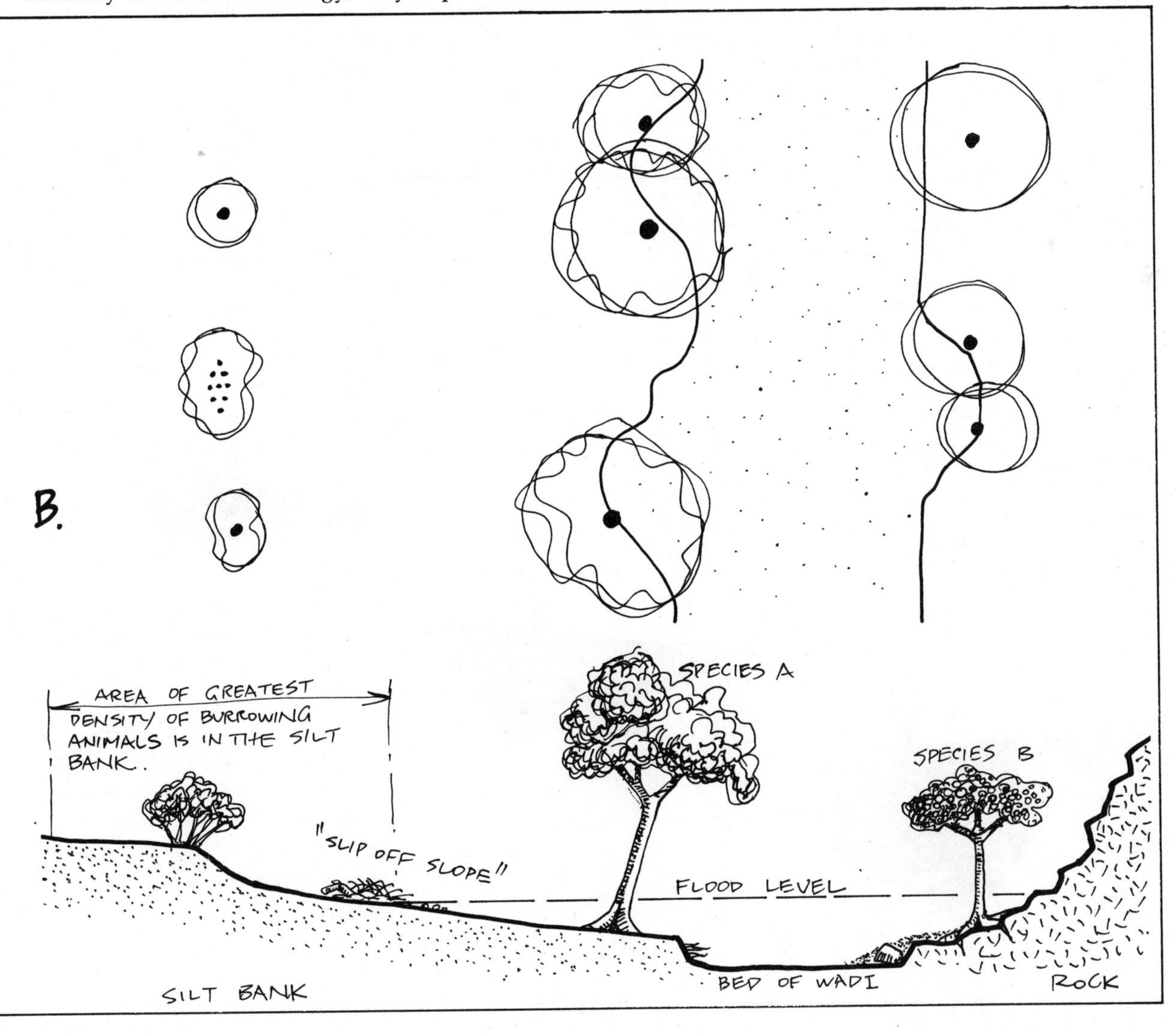

up to a few metres across would be completely covered with a wooden frame covered with thick spinifex, thus providing both shade and wind protection. Covered tanks are likewise protected from wind losses, although vapour pressure can cause some loss in unshaded tanks. Roofed earth tanks are still widely used in Australia.

For more extensive water systems, strategies have ranged from floating rafts of wax, to the more permanent and effective strategy used in South Africa, where hexagonal floats of white–surfaced "light" concrete (a slurry of cement, sand, fine gravel and polystyrene beads) are floated out until the entire catchment surface is covered. This is very effective where water is rare or expensive. Larger and lighter rafts could be made of sealed pipe floats and thin panels, but the concrete is durable and simple; blocks are tar–sealed below, and their specific gravity is 0.8, so that they float deep enough to resist wave effect. Corners are rounded to allow air to the water.

STRUCTURE OF DAMS

There are effective approaches to the structural effectiveness of dams. The first is to make the dam itself as a deep *conical* section, so that as the level falls, surface area is also rapidly reduced (as the square of the radius). Much of the same effect is achieved in "V"–shaped valley storages.

Secondly, deep and shaded valley sites are far better protected from sun and wind than shallow valleys, and if choice is possible a shaded east–west valley is best.

Thirdly, a simple strategy of creating a series of 3–5 storages, each able to be drained by gravity flow to the next lower dam is very effective (far more so than a single storage of the same surface area). The system needs active management so that the use sequence is from the top down; the top dam is drained into the others just when they can take any residual water. As each is emptied, evaporative surface reduces by one–third, one–fourth, one–fifth and so on. Obviously, *the shorter the series the better the result*. It is better to create two series of 3 dams rather than 1 series of 6 dams. Like much else in the desert, the scale of operations decides efficiency, and in this case smaller *is* more efficient (**Figure 11.46**).

Despite all the potential problems with salted soils and water, there are many sites in desert or dryland foothills where freshwater springs and groundwater (less than 200 ppm salt) are available, where shaded valleys can be safely dammed, and where soils are open and free–draining. Endless problems and expense are avoided if such sites are carefully identified and selected wherever local long–term food provision is of an over–riding concern.

Many ideal sites lie in the inner valleys of spaced fold mountains, where deep sands over clay hold many millions of litres of freshwater run–off from the hills, and all that is needed is a reliable windmill or solar pump to raise this water to a mesa or nearby hill. As frost rarely exceeds the 12–20 m level in warm deserts, a raised tank will irrigate "sub–tropical" crop on the upper hill slopes, and deciduous plants needing a chill factor are planted on the lower slopes. This situation is common in hill country with restricted cold uplands.

If all surplus run–off is used in wet years to set out forest trees held in nurseries, such situations quickly become not only sustainable but self–reliant and far less subject to serious drought effect. It is all a matter of careful site choice, responsible behaviour, and modest scale.

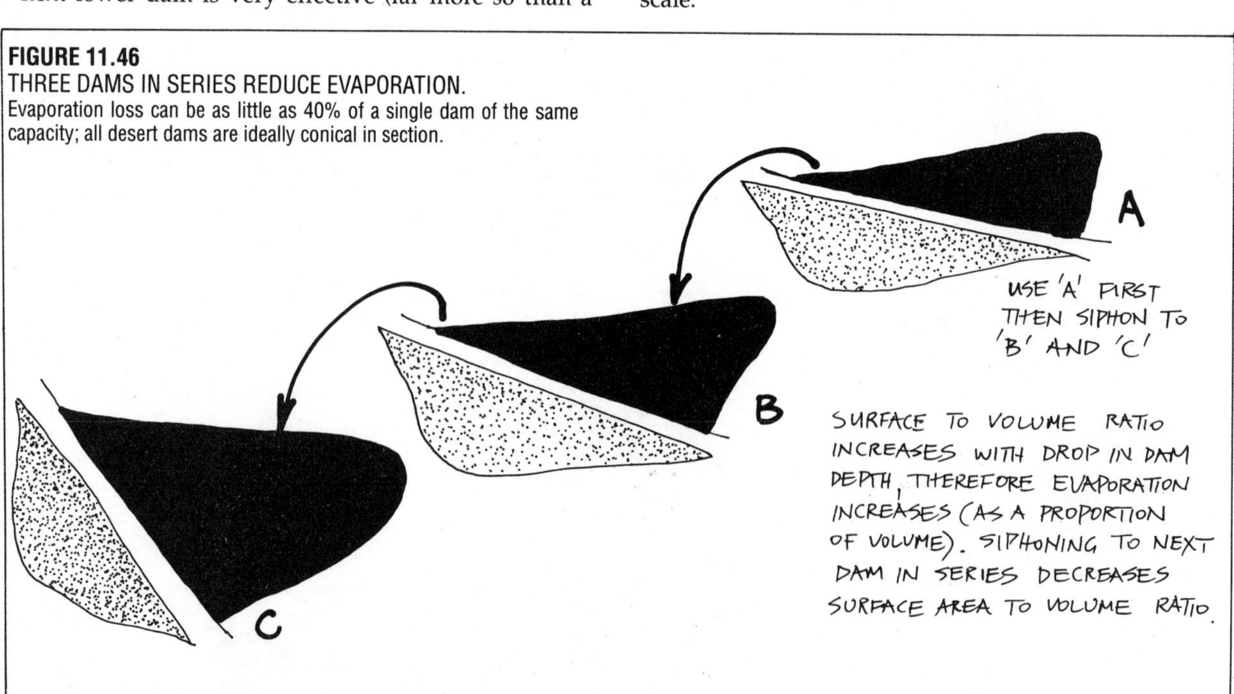

FIGURE 11.46
THREE DAMS IN SERIES REDUCE EVAPORATION.
Evaporation loss can be as little as 40% of a single dam of the same capacity; all desert dams are ideally conical in section.

11.7 THE DESERT HOUSE

Like the sub–tropical house, desert housing needs the twin qualities of summer cooling and winter (or night) warmth. Many features of desert houses resemble those of both sub–tropical and cool temperate housing. Traditional systems are often very sophisticated; the older houses of Iran, Afghanistan, and Rajasthan were all well designed for climate. Some features are:

- Cool courtyards in building interiors; narrow and tall to preserve shade.
- Evaporation strategies from water in tunnels, unglazed pots, tanks, fountains, bark mulch, and coke or hessian "wicks" for water evaporation.
- Narrow east–west steets maximised; broad north–south streets minimised.
- Use of white–painted massive walls (often of mud) as cool surfaces.
- Small windows or stone grilles; most light is indirect from the inner courtyards.
- Towers, vanes, and airscoops for ventilation systems.
- Cooking outdoors under shade trellis.
- Earth sheltered or underground housing of various types.
- Vines on walls, over roof areas, gardens, storehouses.
- Roof area creatively used for drying crops, washing clothes, pigeon lofts.

Houses can be constructed so that they assist other houses. The "colony of swallows nests" appearance of many settlements in arid areas is not a coincidence, as the strategy has independently arisen in Asia, India, the Middle East, in the Mediterranean, and in the Americas. Close–clustered dwellings, with the long axis of their streets east–west, and common or close–spaced walls ensures that neither wind or heat can easily penetrate the fabric of the settlement. If all these dwellings are sun–facing and of more than one story in height, cool air in shaded narrow "wells" is always available in courtyards and streets, and vents at roof level will draw cool air into the rooms of the dwellings. The bottom floors and flat roof areas are used for living, and the upper floors or roof areas for bedrooms.

However, where cultures are accustomed to more space and privacy, a line of bungalows or cabins on an east–west axis, and with dense strips of vegetation to the shade side can achieve much the same result, providing the site itself is correctly landscaped. Village Homes, in Davis, CA, USA, is an ideal model for such design. Thus, for this first factor we can set an ideal of:

- Dense housing, closely placed side by side along an east–west alignment. Few, if any, north–south cross streets are aligned to hot winds. Narrow streets, preferably overshaded by trees or two– to three–storey buildings.

- Multiple storeys are appropriate in restricted sites such as wadi banks, niches at the base of cliffs, and

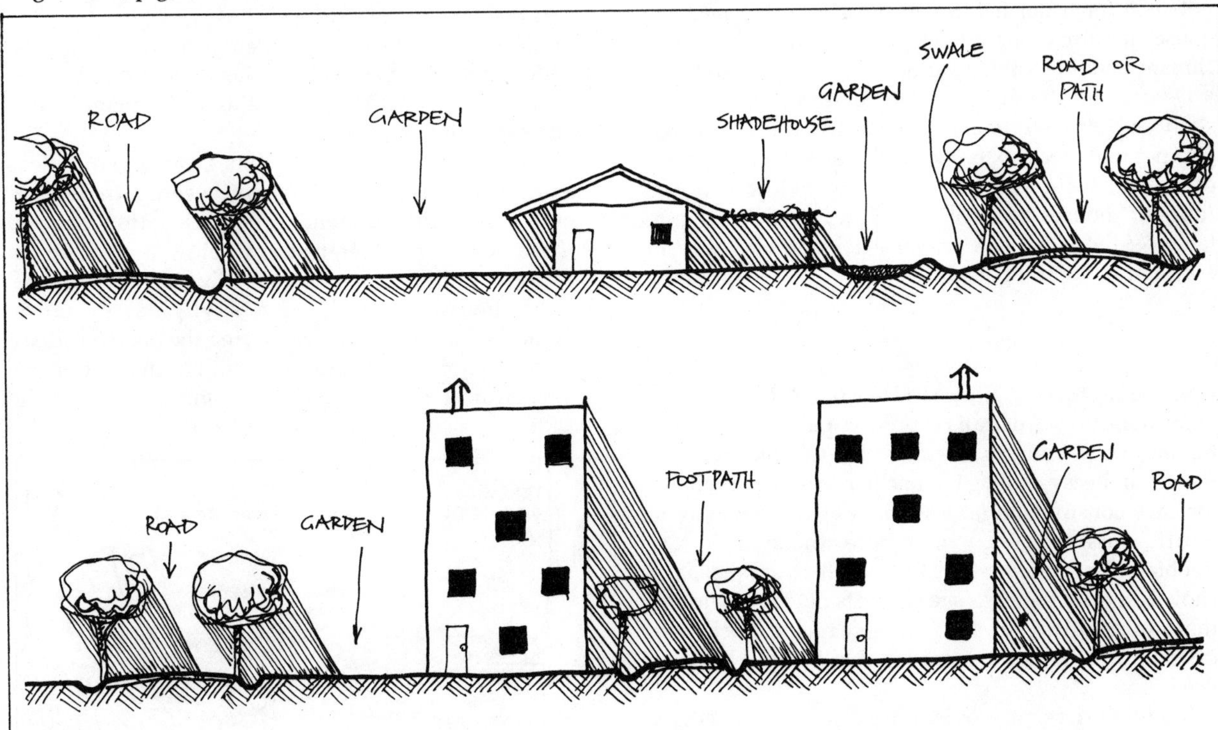

FIGURE 11.47
ISOLATED AND MULTIPLE STOREY HOUSING.
Narrow streets and courtyards can cool multi–storey buildings. Shadehouses are necessary on isolated dwellings. Tree shade reduces ground heat in all situations.

small valley sites. Single storey housing is appropriate on more open sites. but if then needs dense vegetation or trellis on the shade side (**Figure 11. 47**).

Any settlement planning broad–paved (bituminous) boulevards and parking areas will incur areas of uncontrollable local heat. Roads and car parks, or hard paving, *must be shaded* and therefore narrow. In Davis (California), at least 60% of such areas must be shaded, by law. Shading of lanes or streets by trellis and lattice is a very pleasant and practical way to create cool refuges in desert towns. The painting of roof areas or the top surfaces of lattice white reflects heat back to the air above the settlement. The white canvas–covered markets of Istanbul are a model of good design for commercial areas.

SITE CONDITIONS

No house or village site needs more careful selection than that of desert lands. Of the range and basin topography, only 15% is hill country, and of that only 5–10% is foothill or wadi site with adequate run–off. In both hot and cold desert, flat sites can be very cold at night; a thermal belt does exist, usually some 10–20 m above the peneplain, with the frost line sharply defined.

Wadis are excellent sites in very hot deserts, and narrow east–west wadis are particularly well–shaded, but in fold mountains or low foothills a run–off area or spring is needed. The main consideration is the potential for water harvest and storage on or near the house site (excluding the deep groundwaters), equable climate, and if possible a variety of aspects and soils.

Deserts are the ideal sites for underground, earth–sheltered, or cave housing. Arid areas also suit flat parapet roof areas, extensive trellis, and the integration of house, water, and a vine crop. Below ground, "dugout" houses are cool, clean, and need low maintenance; they also suit casual extension and present excellent sculptural potential.

UNDERGROUND AND EARTH–SHELTERED

Due mainly to Australia's preoccupation with opal mining, large–bore (to 80 cm) drills, vacumn earth removal, and easy drilling by machinery is now available for cave construction (at least, it is at Coober Pedy in South Australia). Even where these aids are unobtainable, there remains the wheelbarrow, pick, and shovel. In the latter case, site choice is much enhanced if the proposed dugout site is in soft rock capped with calcrete or a hard and impermeable layer at the roof level.

Mud for bricks, pise, or bulldozed earthwall shelter is usually available, and quite ordinary axes will plane soft rock walls in caves.

Vents, cisterns, toilets, cellars, fireplaces, cupboards niches, beds, and tables can be excavated in carefully planned caves. Chimneys or vents are drilled, cisterns are rendered. (Cisterns can be opened from the upper surface and deliver water at head to the lower cave.) Toilets (in Cooby Pedy) are a vertical drill–hole 20–30 m deep, sited a few feet from the door. Low–flush toilet systems can feed into them, and they are then vented to air. Or they can be dry pits with compost, down draught, and solar chimneys. In either case, they last for years, but are best occasionally pumped out for establishment of trees around the site.

Where water is sufficient, low–flush toilets are led to soakage beds where large trees rapidly transpire the water and convert nutrients. Large shallow leach–fields with thick top mulch are also officially permissible in arid Australia (**Figure 11.48**).

I think there is nothing so peaceful as a cool desert cave with a domed ceiling, the rock trimmed nicely flat, and a sense of safety and silence. For those with deep–seated ambitions, Fresnel lenses and surface mirrors, with reflector mirrors and light guides will illuminate tunnels and rooms for kilometres deep, and "camera obscurae" or lenses will throw scenes of the outer world on walls, where the outside world can be admired but the heat avoided.

Of all desert houses, those who live there prefer caves. One of the more extraordinary underground homes was cut and planted at Fresno, California, under a calcrete ceiling by a miner (M. Forestieri). His underground house covered some 2.8 ha, and included bathrooms, fishponds, and a large citrus garden, of which the only sign on the surface was the top of the citrus trees (**Figure 11.49**).

In the Canary Islands, houses are seen where only the facade stands out; the rest is hidden underground. Even the facade is shaded by grape trellis, as are many of the roof areas of houses in the region.

In both ancient and modern times, caves and underground houses were the preferred dwellings in deserts. Their practicality depends on the location having softish rock, or a softer strata below a calcrete or ferricrete "ceiling". Cave houses (**Figure 11.50**) can be totally below ground, with skylights, or more commonly built with one wall facing the open (shaded) side of a hill. Sunrooms can be built out in front of the underground rooms, or front rooms built on as a facade.

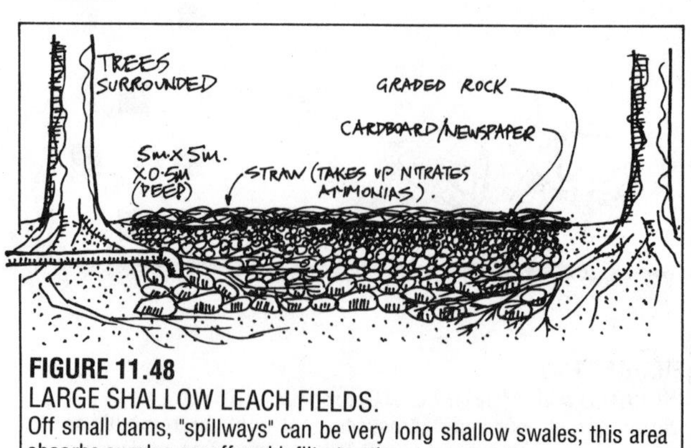

FIGURE 11.48
LARGE SHALLOW LEACH FIELDS.
Off small dams, "spillways" can be very long shallow swales; this area absorbs surplus runoff and infiltrates the water to soil.

At Coober Pedy and in the Canary Islands, decorative facades may be built at a bench–cut entry. Where occasional rains are expected, sections of the hill slope above the cave can be sealed with concrete as a roof or water run–off area for water cisterns; this also prevents water seepage entry into the cave and strengthens the strata above the rooms.

In Coober Pedy, machinery is used to prepare the site. First, a cliff face some 3–6 m high is cut into the selected hillside of soft stone, using a bulldozer. Large–bore drills (1–1.5 m in diameter) such as the "Caldweld" are used to cut corridors, rooms, and storage caves to approximate size, and to sink deep wells for toilets or water cisterns. These or smaller diameter drills cut vents, light shafts, or skylights at strategic places to light the inner rooms. Rooms are then hand–trimmed smooth with an axe, floors sealed with level concrete, and below these the essential electrical conduits and plumbing are laid before pouring the floors.

Large blowers, like vacumn pumps, are used to remove spoil as the drills cut the shafts. The newly–trimmed rooms are left to dry out for a few months before Bondcrete® or a polyvinylacetate sealer is sprayed on the stone to prevent dust. Caves built in this way are dust–free, silent, and easy to maintain.

It is *essential* to ensure good venting, particularly in volcanic sediments or rhyolite rock where radon gas may accumulate in unvented rooms. Capped surface vents or solar chimneys of metal, painted black, ensure positive venting of household heat, steam, or cooking gases. Storage areas built as furniture in surface homes are cut into walls in underground dwellings. In 1985, such houses even when sealed and tiled, cost about 50% of surface dwellings. Comfort levels are easily maintained for both cold nights and hot days.

In many Mediterrarean countries, trellised entries and bright facades satisfy the aesthetic needs of occupants, and the walls themselves are often of veined, colourful, or patterned stone. Such houses were built long before bulldozers or drills by hewing the rock and removing spoil with wheelbarrows, and miners traditionally built them as dwellings in deserts. Temperatures in such houses fluctuate about 5°C, and are fairly constant at about 25°C in the central deserts of Australia.

In North Africa, large pits (20 x 20 m) are cut into the ground, and caves built off them. These courtyard pits, with rooms off them into the soft rock, are cool and protected environments. Good gutters are needed to lead water around the pits, and the walls can be trellised, as can the top opening of the pit. Water cisterns can be cut into the floor of the pit or into the upper surface. The courtyards so formed are drained by a cave exit to a cliff face, and so cliff sites are often selected.

The ultimate refuge in hot deserts are underground, mud–walled, earth–roofed and shaded houses, well–secured against the multitudinous flies and mosquitoes of the desert, and fronting on to a shaded vine arbour, with their rear walls cut into the pediment as caves.

Caves (single entrance shelters) are widely used in deserts as livestock pens (winter, in east Turkey), grain stores, and general storage rooms. They can be built to be "hot" or "cold", as follows:

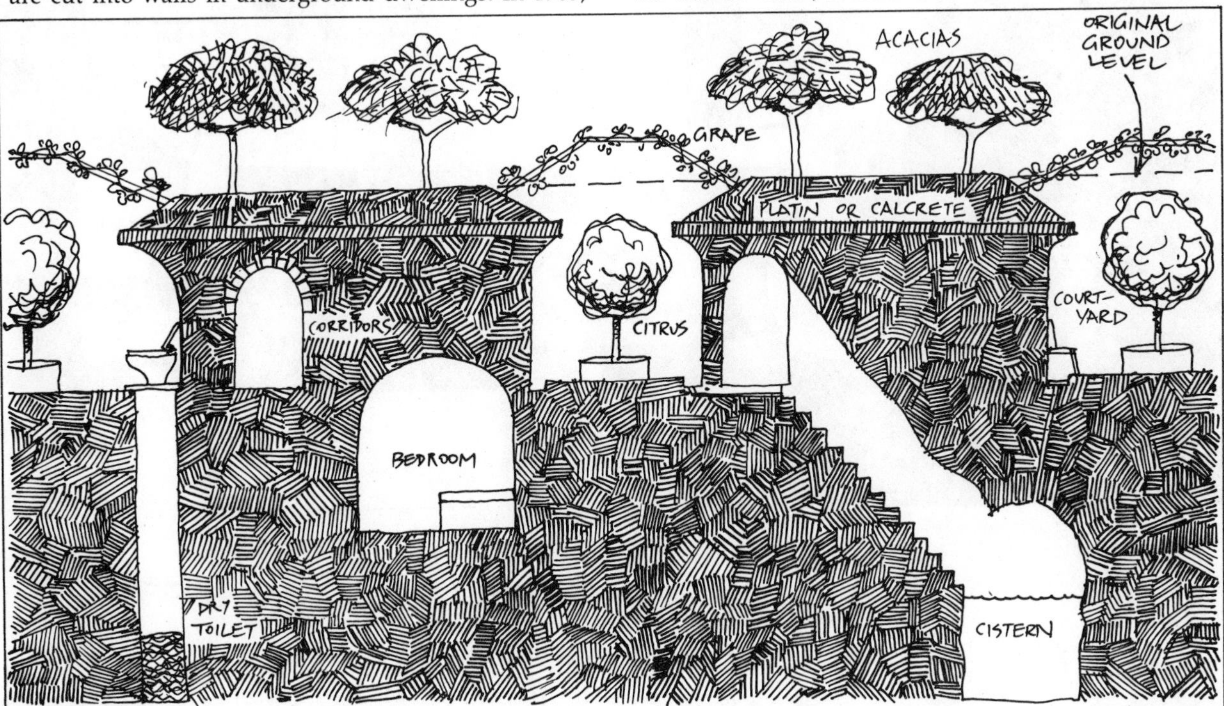

FIGURE 11.49
FORESTIERI HOUSE (Fresno, California).
Complex underground housing; only crowns of citrus trees show on surface; *caliche* forms "ceiling" in upper rooms. House can be endlessly extended, and is very comfortable in desert climate. Good venting for the removal of naturally–occuring radon gas is essential.

• HOT CAVES: A bubble of hot still air gathers in the cave during the day. Dry heat storage suits dried fruits, and mummifies wood and paper. The essential here is that the entry is overhanging to trap hot up-slope air; the cave walls are an effective heat store. Such caves may be unhealthy for living (due to lack of adequate ventilation), but make excellent grain and livestock storages in winter. The cave itself has to be built upslope from the entry in order to hold the hot air. **Figure 11.51.A**.

• COLD CAVES: These are excellent root storage areas, and also general storage areas for books, films, machine parts, and foodstuffs. The essentials are that the entry lies in a dip on a hillside, permitting cold night air to pool in the downsloping cave. A sill to the entry is essential where rains can affect drainage (**Figure 11.51.B**).

Most dwelling caves have level floors and entries, but within these complexes, both hot and cold storages can be built. Many countries have instances of ideal rock, tuff, loess, or bauxitic rock containing complexes of dwellings and storage caves. China, Iran, Turkey, Australia, North Africa, and the Canary Islands all have many examples of very successful and long–occupied dugouts of various degrees of luxury and complexity. Deep cave complexes are now illuminated using surface fresnel lenses, mirrors, and light guides of solid plastic.

EARTH–SHELTERED HOUSING

Where soils or stone will not support caves, surface housing of concrete walls and roofing, with earth–banks built up to the eaves (and if required, over the roof) effectively duplicates cave conditions, without the risk of soil collapse or water seepage. Where stable clay–fraction soils are available, a bulldozer can quickly consolidate a "turkeys nest" dam above ground, and when this is formed and roofed, it too is a cool house for deserts, allowing a variety of treatments on the inner walls, and safe from even severe surface flooding (**Figure 11.52**) .

Intermediates between excavated and surface housing are found in the compact, notch–sheltered pueblo housing of the Great Basin desert (USA), and analagous sites in China and North Africa. Large

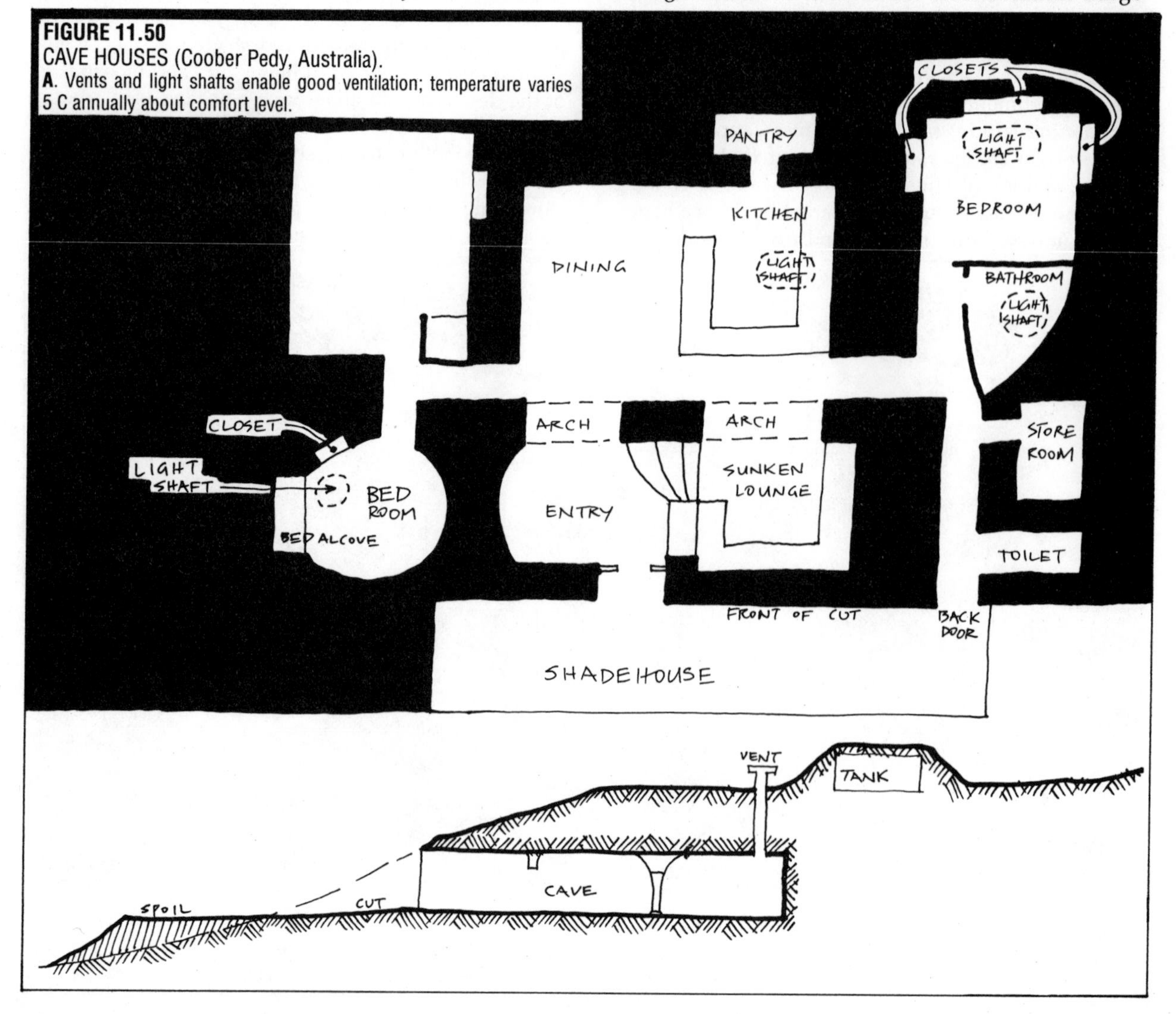

FIGURE 11.50
CAVE HOUSES (Coober Pedy, Australia).
A. Vents and light shafts enable good ventilation; temperature varies 5 C annually about comfort level.

rockshelters at the base of scarps or cliffs are "built in" with mud brick to provide cool and secure storehouses and dwellings, safe from flooding, rain, and sun. Even when abandoned, these pueblos last for hundreds of years in good condition.

SURFACE HOUSING

The desert is also suited to surface–excavated or hill–stepped houses compacted by bulldozer (**Figure 11. 53**). This beats mud brick, and may last for millenia. Thick mud walls can be built out from half–caves, or in very hard country, surface housing can be built. Surface housing is more practical where flooding may occur in thunderstorms, and where sediments are unstable or

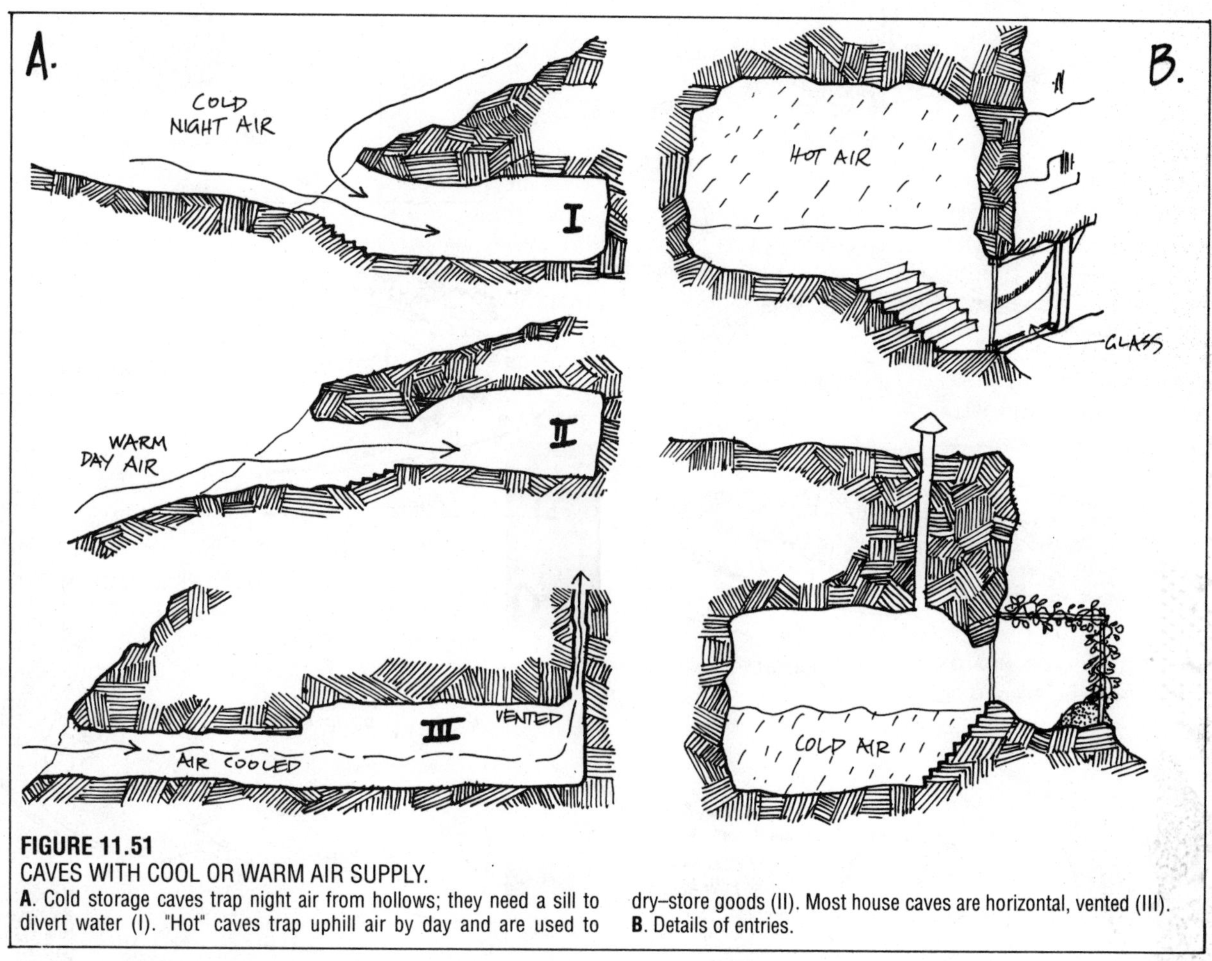

FIGURE 11.51
CAVES WITH COOL OR WARM AIR SUPPLY.
A. Cold storage caves trap night air from hollows; they need a sill to divert water (I). "Hot" caves trap uphill air by day and are used to dry–store goods (II). Most house caves are horizontal, vented (III). **B**. Details of entries.

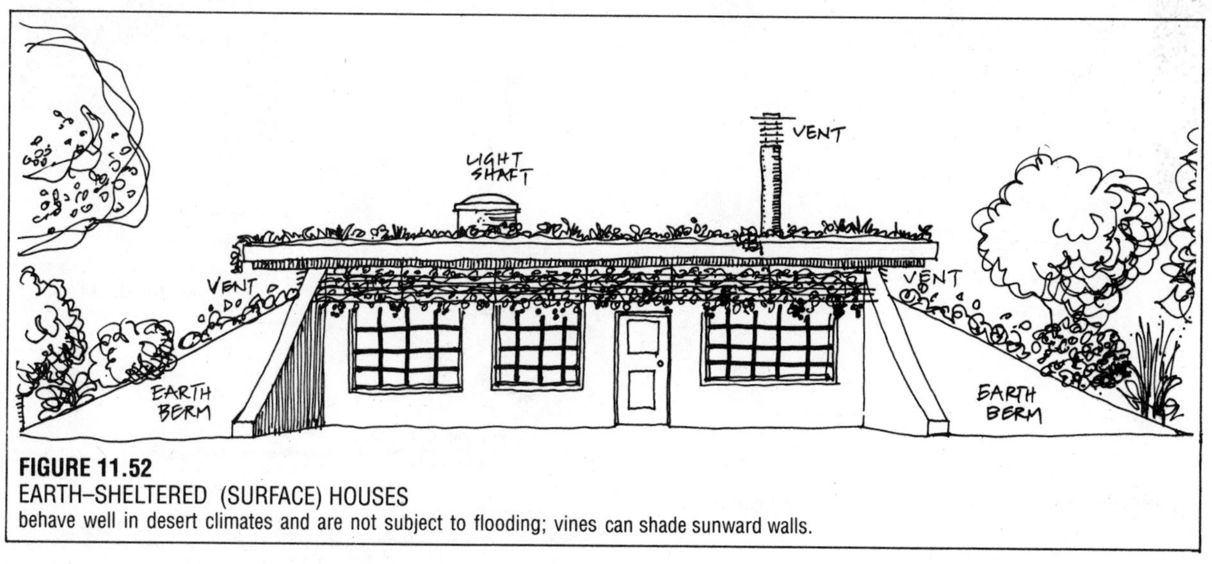

FIGURE 11.52
EARTH–SHELTERED (SURFACE) HOUSES
behave well in desert climates and are not subject to flooding; vines can shade sunward walls.

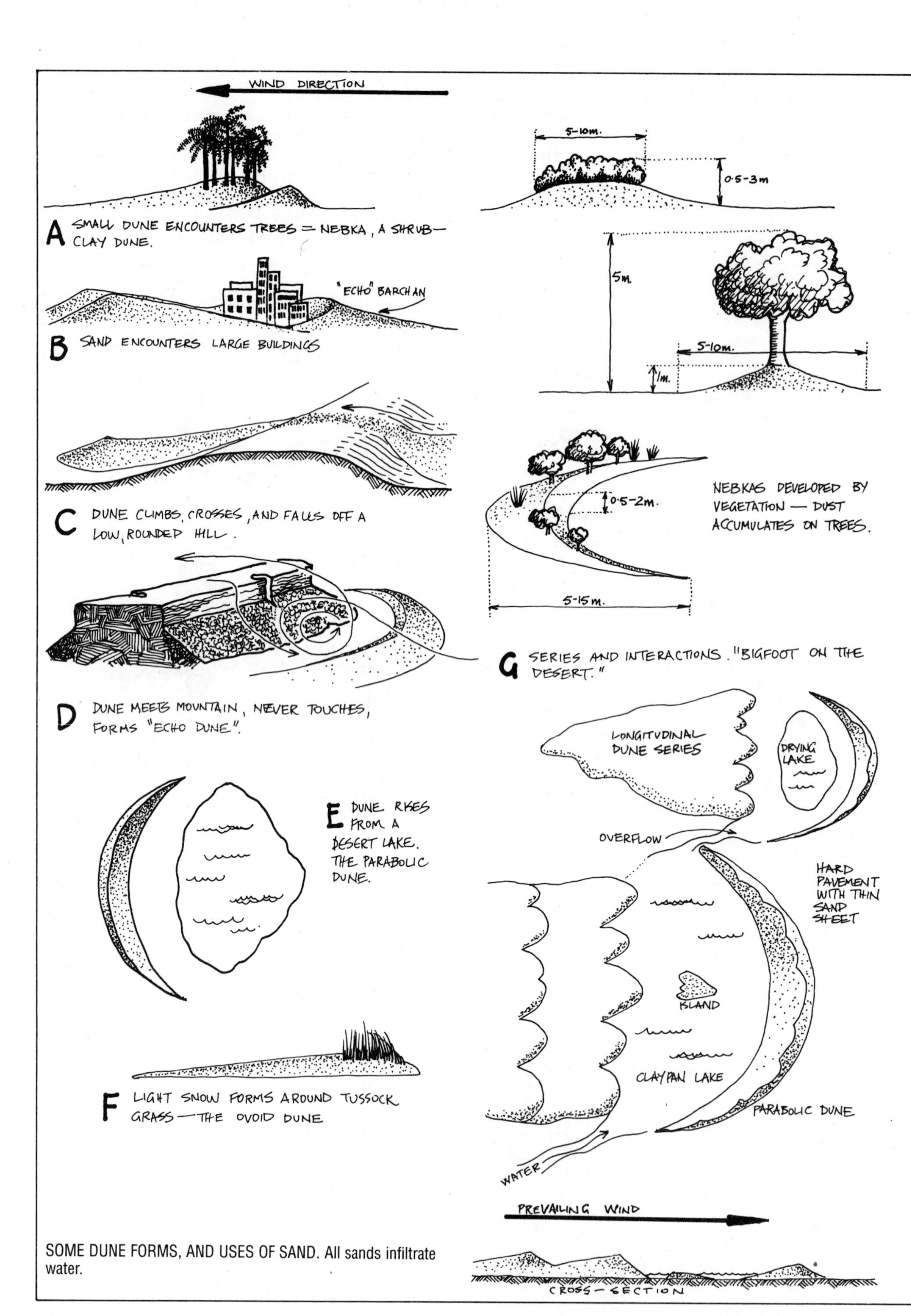

SOME DUNE FORMS, AND USES OF SAND. All sands infiltrate water.

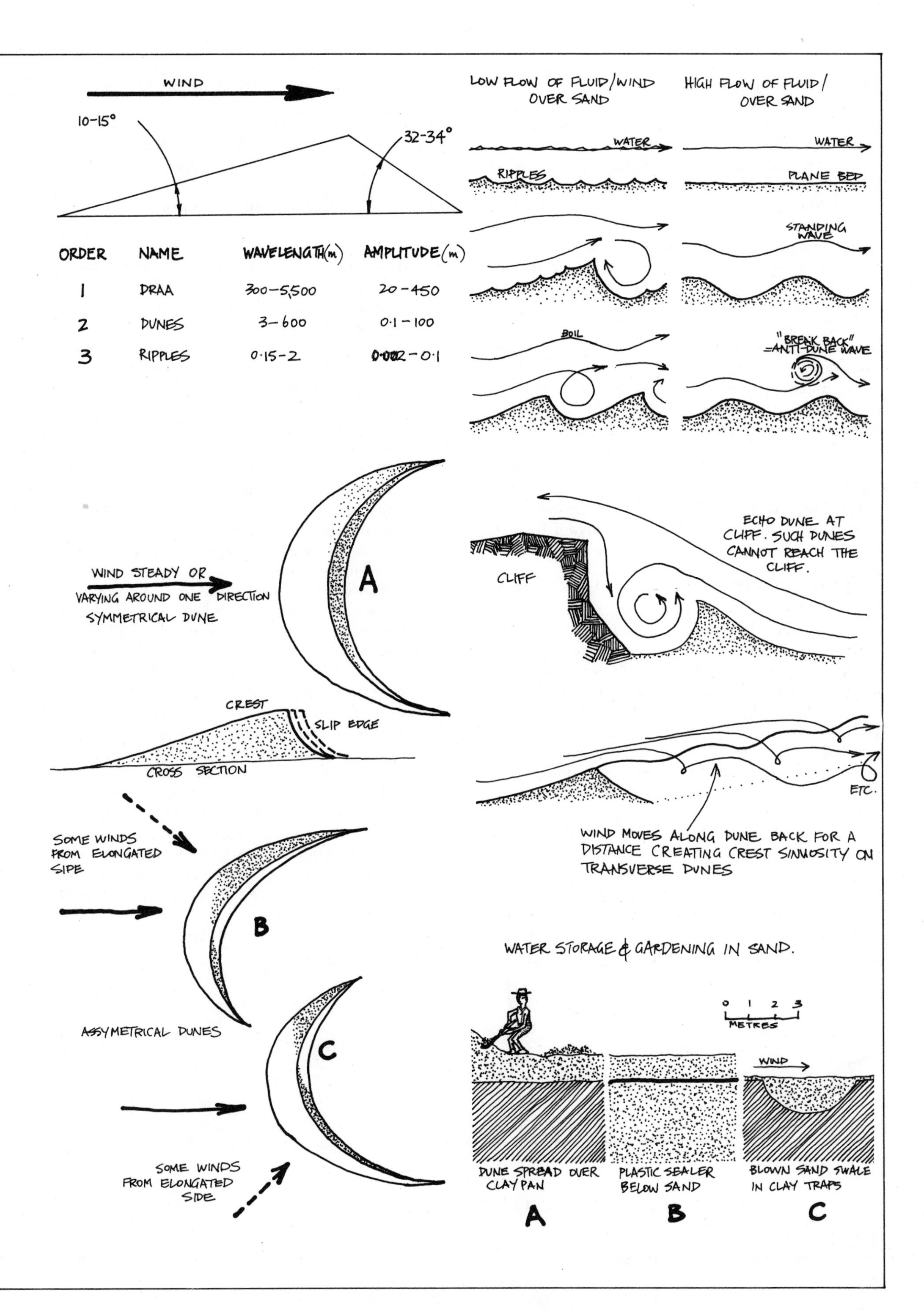

ORDER	NAME	WAVELENGTH (m)	AMPLITUDE (m)
1	DRAA	300–5,500	20–450
2	DUNES	3–600	0·1–100
3	RIPPLES	0·15–2	0·002–0·1

radon–rich (granites and volcanics).

The ultimate cooling device for a surface house, or its pantry and living area, is an earth tunnel; this must be a minimum of 1 m deep and 20 m long, and ideally slopes downhill (the air intake up–slope). In the tunnel, large unglazed pots, pans of wet coke, or curtains of coarse fibreglass weave can be drip–fed to provide evaporative cooling. Even beds of coarse bark kept damp does the job. Cool humid air continually falls through these tunnels to the house rooms (**Figure 11.54**). In swampy areas, such tunnels can be mounded above the surface soil.

Most housing in deserts has been built as surface dwellings. Only a relatively restricted set of sites suits underground housing, unless these are made at considerable expense for water–sealing, excavation and drainage. Thus, the essentials of having efficient housing in desert climates is a critical design strategy. We understandably expect heat to be a major problem. There are at least these ways to provide a *cool* air source for day use:

- INTERNAL COURTYARDS: preferably latticed or shaded overhead, and even more effective if they are two or more storeys high and naturally shaded by the building.
- EXTENSIVE FULLY ENCLOSED VINE ARBORS with mulched floors and trickle–irrigated. These suit single–storey dwellings, although very small courtyards with shade can also be fitted in single–storey houses. Arbors need to be about 30% of the total floor area to provide cool air, and hanging ferns or house plants aid in cooling, as does a water tank.
- TUNNELS, opened as ditches 20 m (60 feet) long and 1 m deep, and with large pipes, half–round cuverts, or evaporative cooling materials provided in their air flow.
- DOWN–DRAUGHTS. Sails, slats, or wind–scoops on roof areas, either fixed or self–steering to force a down–flow of the constant or prevailing winds. At their outlets in rooms, these down–draught inlets can be fitted with damp hessian (burlap), coke avaporation beds, or unglazed pots full of water. These add considerable cooling capacity to the air, and humidify the air indoors.
- INDUCED CROSS–VENTILATION. This is most easily achieved by fitting a black–painted sheet–metal solar chimney to open from ceilings or roof ridges. As these heat up, they effectively draw air into the rooms from any of the above cool–air sources, and create a cool air flow in living areas.

No desert house should be planned or built without its integral trellis and garden systems, as these may not only save most or all climatic energy use (e.g. air conditioning), but provide food and shelter. The attached shadehouse, in particular, must be planned as integral to house design, and in fact as the summer living area. For this reason, the winter or indoor kitchen must open onto the shadehouse (the summer kitchen).

As well as these cooling devices, *heat* sources are needed for the frequently cold winters and cold cloudless nights. Depending on latitude, windows allowing sun to strike an edge–insulated concrete slab will provide all the heat needed. With the advent of rigid foam insulation, trenches cut at or below the foundations (to 0.5 or 1 m deep) effectively insulates the whole of the floor/earth mass under the house. It is this simple strategy that can buffer both heat and cold extremes, without the need of firewood for heating. We have also used plastic bags of straw or sawdust for this ground insulation in Nepal.

Where such slabs have not been built in the original dwelling, a thick vertical wall of mud brick standing 0.5 m inside a room, and faced with a glass window, will act as a heater long after dark (the Trombe wall effect). Finally, in very cool deserts, an attached glasshouse will act both to heat the interior in cold periods, and to vent hot air for cross–ventilation on hot days. At high latitudes (30°–60° north or south) extensive glazing needed, and a courtyard with a glazed roof is an excellent refuge from cold. In lower latitudes (0°–30°), about 20–25% of the sun–facing wall will need glazing to provide sufficient heat on to cement slabs or trombe walls.

A sun–side glasshouse has three potential uses in

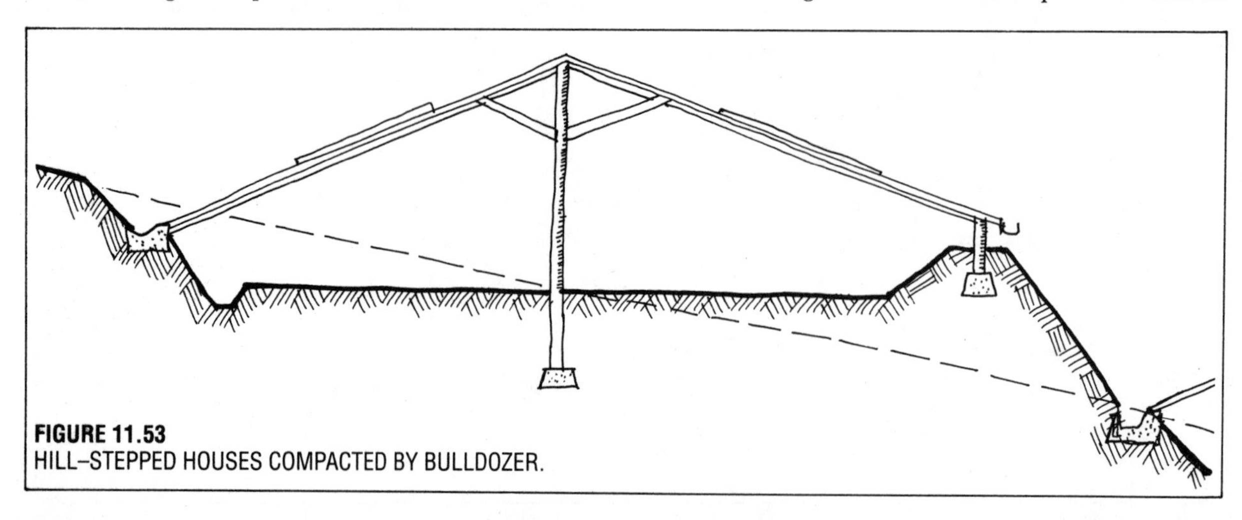

FIGURE 11.53
HILL–STEPPED HOUSES COMPACTED BY BULLDOZER.

deserts:

- To create a winter heat source.
- To draw cool air into the house during the summer.
- To start spring plants early in the year, to ripen late (autumn plants), and to grow greens in winter.
- To dry surplus fruits and vegetables in an enclosed area (unavailable to insect pests).

As these are also the basic functions of the greenhouse in cooler climates, then it is clear that such structures operate effectively over broad landscape and climatic ranges.

For both heat and cold, massive walls, edge–insulated floors, effective draught–proofing, insulated ceilings or roof areas (if necessary trellised or carpeted with thick vines), and efficient cross–ventilation are all essential strategies to moderate the extremes of daily and seasonal temperature that is typical of desert areas. White–painted exterior walls help to reflect excessive heat, and strategic shade trees, palms, vine trellis, and courtyard ponds or fountains assist in buffering heat extremes.

Houses themselves can be very compact, especially where shaded outdoor trellis areas are extensive. It is an excellent design feature to place the winter kitchen indoors, but to also have it open to a screened–in summer kitchen part–roofed under a thickly trellised area, where occupants can spend most of the day out of doors.

Quite small (in cross–section) solar chimneys or attached greenhouses on the sunward side of dwellings can create a cross–draught sufficiently strong to blow out a candle. Thus, if a fully–enclosed and totally vined shadehouse is constructed on the shade side of the house, a continuous cool and humid air cross flow results, providing the cool air can enter the living areas by a fairly direct route, and that some water is available to supply the vines with evaporative cooling (**Figure 10.17**). Commonly, air can be cooled to 10–15°C below ambient temperature by this combination of shade, vines, and induced air–flow. The same cross–flow from earth tunnels also supplies unlimited cool air.

It is a simple matter to close the house at night, or on cold days, and to retain glasshouse heat inside. Such designs save up to 80% of fuel energy, a particularly important factor for low–income groups, where energy can be 30–40% of total household expenditure. As glass and trellis are durable, and as costs amortise in 1–3 years, it makes sense to make these beneficial retrofits to uncomfortable houses.

That we see modern houses built without such aids is a witness to stupidity, waste, and poor design. The results are to be seen in deforestation and reliance on fossil fuels (with continuing pollution). More serious results are the financial strains on poor or low–income families, the ill effects of smoky fires, and illnesses due to cold and tiredness after cold nights.

Roof and Parapet Furniture

Wherever rainfall is slight, flat parapet concreted roof areas are preferred; these roof flats have a wide variety of uses for grain drying, wash–houses, pigeon lofts, and trellised–over recreation areas. In the severe winters of the Caspian area, peasants make their haystacks on the roof, in part to protect the hay from goats, but effectively providing a winter's warmth of thick insulation plus compost heat which radiates from the ceiling to the rooms below.

Features commonly built into parapet roof and trellis roof areas are:

- Pigeon lofts for eggs, squabs, manure.
- Laundry troughs and drying lines.
- Header tanks for 1–2 weeks water supply, fitted with float valves.
- Wind towers, with scoops, slats, lattice, sails, or aerofoils to catch local winds.
- Evening rest areas for outdoor passive recreation;

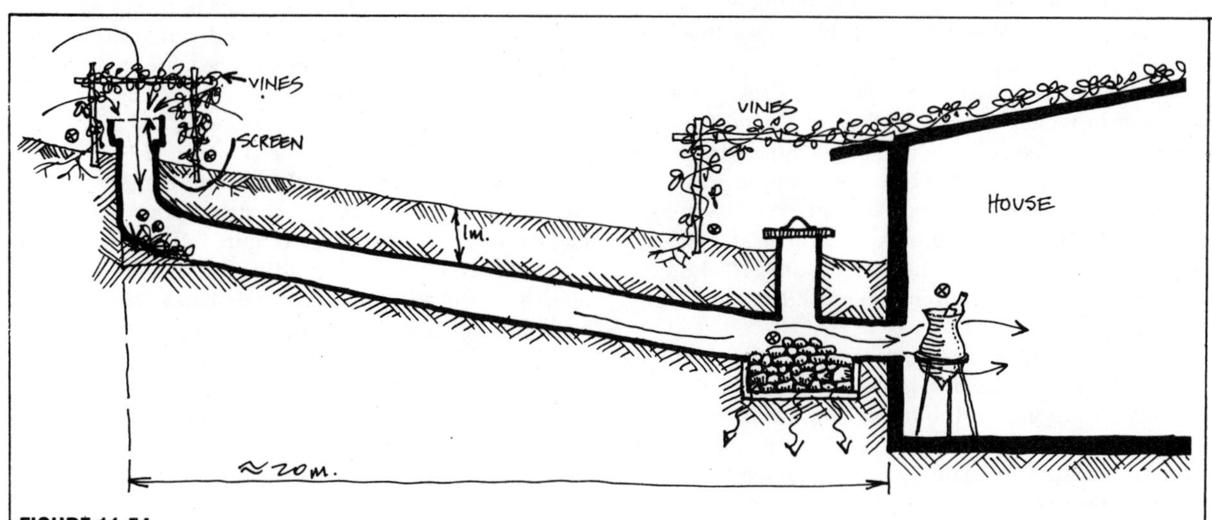

FIGURE 11.54
COOL AIR TUNNEL.
Slopes *down* to house, has shaded intake, moist coke or cinder bed, unglazed pot of water at outlet; length 20 m, depth 1 m. This tunnel provides cool humid air to dry desert houses, food storages, hospitals, etc.

hammocks.

- Grain drying flats, drying racks, bulk food stores or bins.
- Winter hay storage (in cold climates); heat or partial composting is radiated and conducted to the ceiling of rooms below.

THE ESSENTIALS OF THE DESERT HOUSE

- No west windows. West wall painted white and fully shaded, vine–screened, evergreen trees to the west.
- Cold air tunnel or cold air source as a fully enclosed, bark–mulch shadehouse.
- Positive hot air exhaust as a solar chimney or small attached glasshouse to the sun side.
- Unimpeded through–ventilation for cool air.
- Either very thick, white, fully–shaded walls (cool deserts and cool nights) or very light, screened, or matted walls (hot days and nights).
- Evaporative cooling surfaces in a through–draught; as unglazed pottery, coke mounds, ferns in bark, thick vines, or wet hessian (burlap) screens.
- Cool cellars for storages, or deep cool internal courtyards with vines and ferns.
- Ceilings insulated, thick vine mass over roof area.

PLACEMENT OF VEGETATION AROUND DRYLAND HOUSES

It is the low westerly sun that adds most heat to buildings; for that reason, *no* windows are placed in west walls where they incur direct heat gain. Permanent screens of evergreen trees, thick–leaved vines, or turf banks are placed to the west of the house so as to shade and shelter the west wall. The shade side of the house is where enclosed trellis is built. The east side can be part–shaded by deciduous trees or vines, and can have small windows. Glazing is most useful on the sun side where 100% (in cold areas) to 25% (in warm areas) of the wall can be glass.

The house itself needs to be elongated east–west, so that sun in winter can reach all room floors. It is the eave width on the sun side that permits the winter sun to enter the rooms, and the summer sun to miss the walls and windows. *External* (not internal) bamboo, wooden, or aluminium blinds prevent heat entry to windows on hot winter days in deserts. Deciduous vines on the sun side prevent too much heating during the summer months.

The correct placement of vegetation, blinds, windows, and insulation (**Figure 11.55**) means that no extra costs are incurred, but that the dwelling works efficiently. In settlements, such rules can be made part of local building legislation, and this alone will save thousands of hectares of forests in the third world.

Straw, thatch, wool, feathers, cork, woven materials, and even paper and cardboard are adequate insulators, but few of these equal thick vines over wall and roof as active cooling systems; these are affordable by any person. A desert house well screened and shaded by vegetation is an oasis indeed. I believe the emphasis in desert housing should be on developing the arbor for living, and on creating compact and reactive housing mainly for night occupancy, and in the rare case of rains.

HOME ENERGY CONSERVATION

To produce hot water in deserts is easy enough, either from:

- A solar attic with a water tank, a glass cover, and aluminium foil over thick insulation (a bulk water heater);
- A grid of plastic pipe in soot–covered sand, itself under glass; and
- A simple coil of pipe on a metal roof, so that enough warm to hot water is available.

For groups of 5–10 houses a solar pond of 4–5 m sides, and 1–2 m deep (**Figure 11.56**) will supply not only hot water, but house heating. Space heat and hot water together are about 80% of the energy needs of modern housing.

A solitary photovoltaic cell or a bank of such cells provides the little electricity needed for lights and electronics, and it remains to cook using oil, gas, or modest fires. All desert settlements should, as a matter of legislated planning, divert wastewater to firewood plantation. In a modern town such as Alice Springs or new towns, the water from washing, toilets, and sinks could provide *all* the firewood for cooking if it is carefully used in trickle irrigation systems, and stored in non–leaking and covered cisterns. Hardy desert trees on a much broader scale establish firewood reserves.

HOUSE WATER CONSERVATION

Modest water use is easily achievable if efficient shower heads are used for washing, and both shower and handbasin or laundry water is first diverted to the flush tank of toilets (where sewage is provided), or to the garden where dry pit toilets are used. The pit toilet popular in central Australia works well, and is good for years of use without water (**Figure 11.57**). To get shower water to a flush toilet cistern, the shower and handbasin can be raised a few steps above floor level, and a low–level cistern used (**Figure 7.17**).

Wherever possible, spouting from *all* roof areas should be run to covered concrete tanks. These would be a hot sales item in most of the world, but are common only in Australia (like Vegemite, electric jugs, and rotary clotheslines!) It is a rare isolated house in dryland Australia that does not boast of at least 5,000 l of tank capacity. Located on the shade side of the house, under trellis, such tanks provide cool drinking and cooking water.

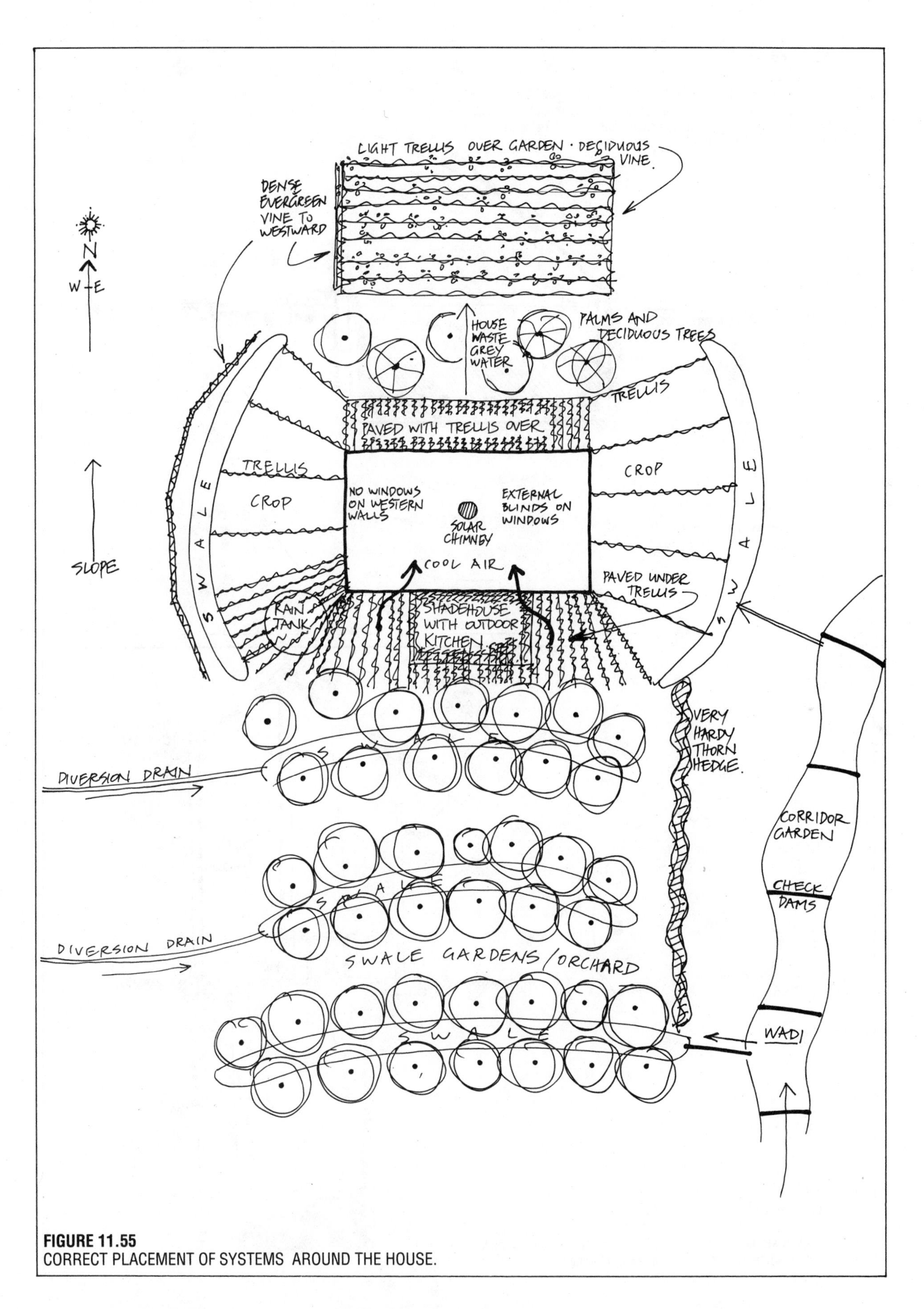

FIGURE 11.55
CORRECT PLACEMENT OF SYSTEMS AROUND THE HOUSE.

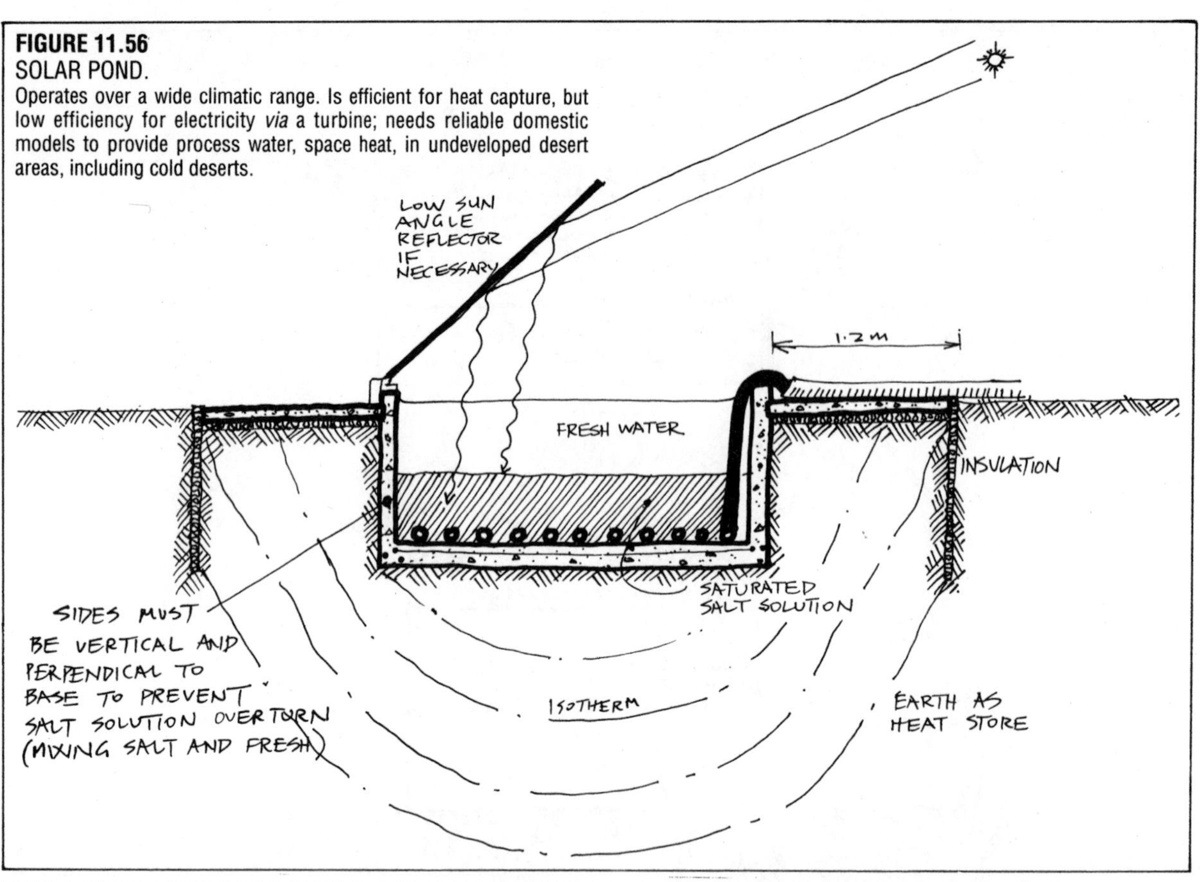

FIGURE 11.56
SOLAR POND.
Operates over a wide climatic range. Is efficient for heat capture, but low efficiency for electricity *via* a turbine; needs reliable domestic models to provide process water, space heat, in undeveloped desert areas, including cold deserts.

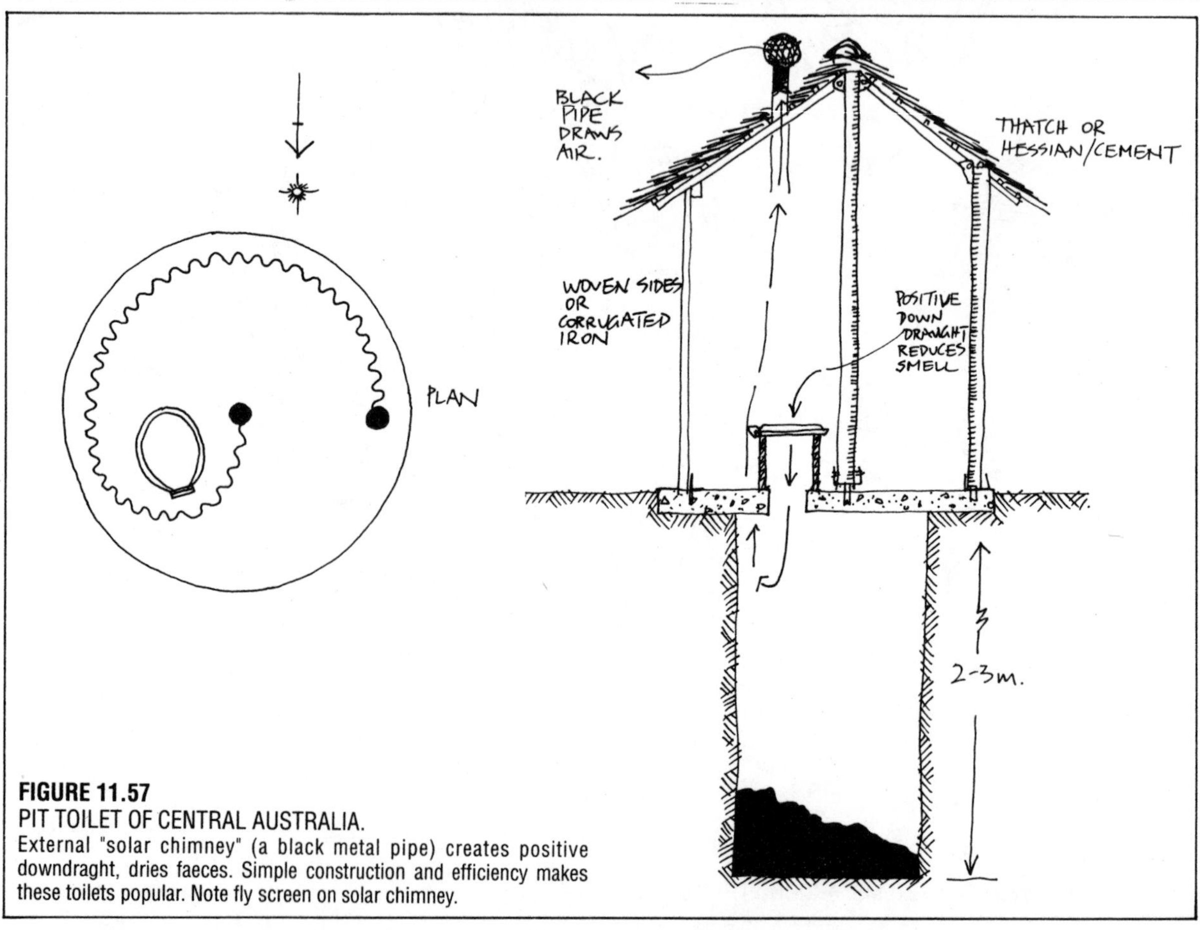

FIGURE 11.57
PIT TOILET OF CENTRAL AUSTRALIA.
External "solar chimney" (a black metal pipe) creates positive downdraght, dries faeces. Simple construction and efficiency makes these toilets popular. Note fly screen on solar chimney.

11.8
THE DESERT GARDEN

It would not be necessary to write a desert garden section were it not for these profound effects that are unique to drylands:

- Water solutes are likely to be factors higher than in humid areas.
- pH and consequent mineral deficiencies can have severe effects on health.
- There may be high nitrate levels in water and food.
- Non–local food is likely to lack vitamins, and to be expensive.
- Water use is necessarily restricted by supply.
- Light saturation of garden plants may create the need for shade to achieve good growth in gardens.
- Nomadic or wild animals are specific problems.

Moreover, because of the general low level of resources in drylands, it is unlikely that ordinary people living there will be able to afford full nutrition if gardens are not plentiful throughout settlements.

To relieve malnutrition, the home garden is our primary strategy. In deserts, gardens (not fieldcrops) are the mainstay against famine. Unlike cereals and grain legumes, garden leaf, fruit, and root products require little cooking, which saves energy in the home. They also contain essential minerals and vitamins, and can make every family food self–reliant.

The desert garden must be planned as a very serious affair. Health in deserts is very much a matter of good nutrition, and imported food often is of poor quality, and expensive. On flat and dusty plains, a few hours with a bulldozer, or many hands, can throw up a ditch and wall to 3 m or run a high barricade that discourages wild cattle and deflects winds. Solid pise and mud brick courtyard walls can be built over time; rock is often plentiful for such protective walls. Inside this protection, beds raised by stones, pise, or logs can be almost totally mulch–filled as mulch sources are developed or gathered. Individual beds can be shaded (to 75%) with shade cloth, palm fronds, or brush supported on frames 1–1.5 m high. Solid pillars of concrete block and mud brick to 2.5 m can be made to support permanent vine trellis over much of the garden.

A careful soil and water analysis is essential. We can reasonably expect phosphorus and zinc deficiency, and probably iron and manganese. Excess boron from detergents can cause plant failure, so that common soaps are best used where waste water is used on garden beds. Water must be checked for excessive nitrates and fluorine, both common dryland pollutants. All rain run–off and roof water must be carefully harvested and used to flush salts if local water exceeds 800 ppm or the garden is not free–drained.

A careful plant sequence for all beds, so that they are always self–shaded by almost constant crop is of great value. Local expertise and agricultural department advice is called for. Deep–rooting and high–yielding perennials (asparagus, globe artichoke) are standby crop, as are drought–tolerant staples such as sweet potato and most cucurbits and the melon family.

The desert garden needs an emphasis on staple trees, adapted to dry periods or able to survive on minimal water. Suggested plants would be 5–6 date palms, 4–5 olives, a doum palm, 2 or 3 citrus, 1–3 avocados, 4–5 apricots, bananas and papayas (climate permitting), and a mass of vine crop. The house should be sited near a run–off area, where water can be soaked into sand dams or swales to keep trees alive.

The garden surrounds the house, using all waste-water either as soakage fields or as underground plastic–lined swales in coarser material. Garden beds for vegetables can be carefully constructed. I have seen them built of compacted mud, filled with better soil, and mulched. Every vegetable we can grow in temperate or tropical lands will grow well in *small* beds (usually 3 x 1 m or so) flooded every 3–10 days, mulched, and part–shaded by slats, vines, or the canopy of a light–crowned leguminous tree.

Wallflowers, marigolds, gladioli, and other companion crops are beneficial to *Brassicas, Solanum spp.*, and onions respectively. Every bed should be planned for succession to fava beans, pea, or tepary beans in season (fava and peas in cool seasons, tepary and moth beans in hot seasons). Even potatoes will grow in deep (0.5–1 m) boxes filled with mulch, pine needles, and household scraps, although yam beans, jicama, sweet potato, and sunroot (*Helianthus tuberosa*) supply more reliable yields than potatoes.

Celery, onion, *Brassica*, carrot, beet, spinach, globe artichoke, tomato, and sweet and chili pepper do particularly well, and surplus vegetables can be dried. The desert is the home of watermelon, melons, and climbing or vine cucurbits gernerally. Every wall should be seen as a vine trellis and the roof covered with dense vine. Even flat roof areas can be shaded with a high trellis of grape.

But the desert garden should also be a corridor plantation down nearby river beds, niche gardens in shaded sites, patch gardens on leach fields, and a spread of very hardy adapted yams, bulbs, and semi–wild fruits, cactus, and palm wherever a site exists or can be made.

Sand–filled waterways always contain some moisture; these are *corridors* of moisture and soil. CORRIDOR FARMING is already a very different idea from area cropping, and we need to choose and place a complex assembly of native and introduced species to give us at least a basic food reserve, fuels and selected structural timbers.

At times, the corridors are rivers, dune series on hostile salted or stony pavement, or the floodplains of exotic rivers. These are the sites for our "semi–wild" plant and animal reserves. We can make some provision for a good rain year by broadcast sowing into areas we have already prepared for infiltration, and taking off a crop of sunflower, millet, sorghum, panicgrass, or melons. These are esentially surplus crop, for long–term storage as such, or for export out of region in trade.

MATCHING UP EARTH BED, IRRIGATION, SOIL TREATMENT, AND PLANT SPECIES

Home gardens return so much in health, cash, and just plain life interest that they deserve intensive, bed-by-bed planning, in which companion plants, seasonal successsion, bed soil treatments, and a permanent watering method are all designed. Rather than do this for every vegetable, I will deal with 3 or 4 classes of each and suggest a method to suit; people can then follow the methods (all of which are working) for other plants of like needs.

Depending on resources, beds can be raised by digging, and edged with pise (rammed damp clay-sands), bricks, logs, or stone walls. Trellis can be of metal mesh, bamboo, hardy and even crooked termite-resistant wood, cane grasses, or living leguminous trees. Deserts have supplies of clays, sands, rubble, and sometimes lime or gypsum for use in gardens.

In gardens, we can experiment with a variety of pits, circles, raised beds, "padi" beds for flooding, ridge and furrow, edge-banked raised beds, "cone and hollow", and mulch baskets or mulch boxes for some vegetables (**Figure 11.58**).

Where plastic irrigation pipes are absent or expensive, adequate substitutes can be made of unglazed pots,

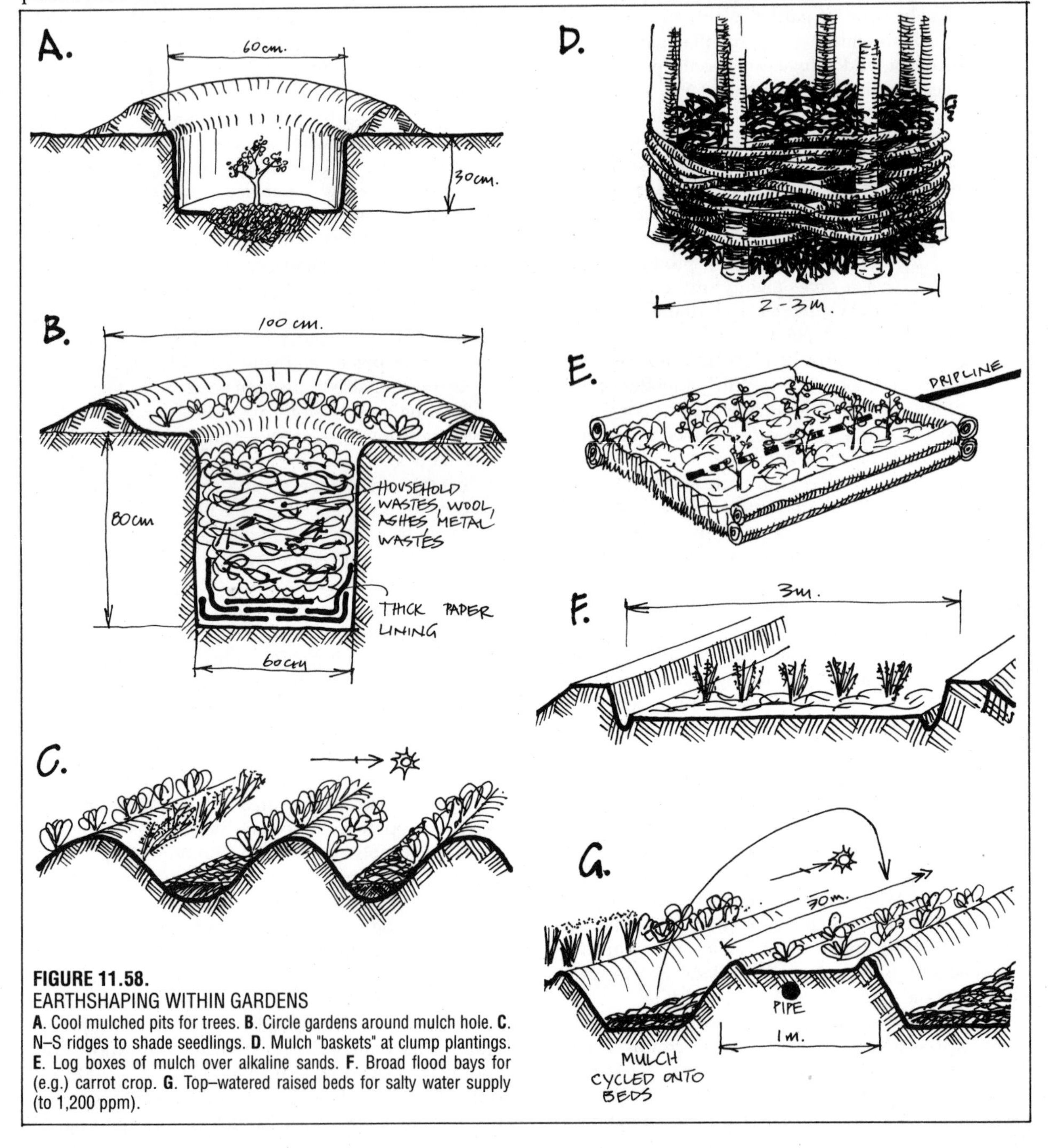

FIGURE 11.58.
EARTHSHAPING WITHIN GARDENS
A. Cool mulched pits for trees. **B**. Circle gardens around mulch hole. **C**. N–S ridges to shade seedlings. **D**. Mulch "baskets" at clump plantings. **E**. Log boxes of mulch over alkaline sands. **F**. Broad flood bays for (e.g.) carrot crop. **G**. Top–watered raised beds for salty water supply (to 1,200 ppm).

broken pipe, gravel beds, or hand–moulded concrete channels on or below the surface, the latter covered with tiles or any flat material. Pipes of bamboo, drilled to drip water, flagons or bottles suspended and the corks perforated to leak, or simple bucket systems can be used, albeit with more labour.

The categories of things we plant are:

• ADVANCED PERENNIALS or trees in pots, e.g. herbs, shrubs, flowers, advanced tomatoes. Keep in pots under shade until rains have fallen, transplant in cool weather. Dig holes to accept full root system; soak this hole and add a small handful of fertiliser or soil improver (lime in acid areas, gypsum in alkaline clays). Turn out the pot and plant the seedling, pushing a shade branch in nearby to shade and shelter the plant.

• SEEDLINGS from trays or pits, e.g. lettuce, *Brassica*, sweet and chili peppers, chives, globe artichoke, cucurbits. Plant in cool weather, towards evening, in well–watered soils. If in compost, dibble a hole for roots. If in thick mulch, pour one–quarter of a bucket (about a litre) of good soil in a hole made in the mulch, and place a flat stone upwind. Plant in the soil hole.

• TUBERS AND BULB, e.g. potato, yam, sweet potato, flower bulbs, sunroot. Greened and if possible sprouted, tubers and tuber shoots can be pushed down into mulch, or buried and then mulched (in damp soils).

• LARGE OR FINE SEEDS. There are two possibilities:

1. Scatter seeds over a fine–tilth bed, ideally of sieved compost, then wet and press down. Sometimes, hessian (burlap) is laid above the bed before watering, and removed on day 3 or 4 after the first signs of germination are seen. This suits larger crop.

2. A lens of fine soil 4–5 cm thick is patted down over a thick (12 cm) base mulch, and some fine seeds sprinkled on this, forming patches of crop in a general mulch; this suits repetitive sowings in home gardens.

Figure 11.63 gives an idea of how each of these classes of plants are established in thick mulch in a home garden. Such instant gardens have commonly been established in a few hours. They give a satisfactory result in the first year, but as with any garden, do not form good garden soils until a few years of continous gardening have passed, when the biological content of the soil is stabilised and crumb structures in the soil have formed.

It remains to design beds and plant guilds to suit a set of vegetables. For carrots and onions, 3–6 beds (4 m long x 1 m wide) can be used, depending on family size. Another 2–3 beds should also be allotted to chives, garlic, and shallots to ensure some *Alliums* throughout the year. Keen gardeners would include beds for leeks, sand leeks, and multiplier onions.

All these beds can be built raised but dead level by levelling with a straight–edge, and either sunken below a built–up rim, or heaped and levelled off. As both onions and carrots prefer free sandy soils with moderate drainage and modest fertility (a pH of 5.5–7.5), these are beds into which old compost and dry fine organic material such as powdered grazing–animal manures can be mixed at ratio of 1:9 or 1:6 (organic material to soil). The beds are then levelled and pressed flat, seed sprinkled over it (previously mixed with sawdust or a coloured sand to show seed coverage) and the bed flooded to the rim with water not exceeding 1100 ppm salt. This then soaks in. If the beds are covered with hessian, and inspected daily for germination, a light daily flood will have good effects on germination.

As crop tops grow (burlap removed as seeds start to sprout), irrigation can be reduced to once every 2–3 days. Carrots have 3 beds allotted, planted mid–summer, and just before and after the cool season. Surplus carrots keep well when picked and layered in cool sand in pits 0.5–1 m deep; the green tops are removed to 1 cm from the crown, and they are pitted without washing or removing any roots. Onions are hung in dry airy plaits.

For carrots, a scatter of small salad radish breaks the soil crust and "thins" the crop as the radish mature in a few weeks and are removed. For onions, gladioli inhibit most pests and provide a small cash crop. For both crops, a nearby border of sunflower, *Crotalaria*, or marigold (*Tagetes*) reduce wind effects, which can devastate onions, and provide a mulch source. *Tagetes* reduces nematodes, as does *Crotalaria*, and the latter provides soil nitrogen and mulch.

An example of a SEEDLING crop are peppers or tomatoes. Although soil tolerant, both appreciate a pH of about 5.5.–6.8. Tomatoes can tolerate about twice the salt in soils of peppers, and thus peppers may need more tank water at levels of salt exceeding 1100 ppm. Both are best arranged for underground or subsurface (root level) irrigation via unglazed pottery or leaky pipes.

Both are *unsuited* to spray irrigation (as are cucurbits). Such sprays aid the spread and growth of leaf mould or mildew, and fungi. For the home garden, an efficient bed is a "keyhole" system holding 20–50 plants in 3 rows deep (1.5 m wide). Both benefit from *Tagetes* interplant and a wind shelter of sunflower, sunroot, or *Crotalaria*. A culinary accompaniament is basil.

These keyhole beds provide easy access with minimum path:bed ratio or wasted space. Both plants do well when set out as strong seedlings in soil pockets in deep mulch. **Figure 11.60.A** and **11.60.B**. For ROOT crop, broad beds, keyholes, or long rows are able. The one essential is thick coarse mulch to shield the tubers and cool the root runs while keeping pH buffered. Potato and sweet potato prefer organic dry soils (pH 5.0–6.8) achieved by compost, perhaps some sulphur, and a thick damp mulch. Subsurface water is ideal, and a few marigold interplants aid root health. A family needs 2–3 successions of each, with areas of 10 square metres or so in each bed, as they are staple crops. Thus, 20–30 square metres in garden needs to be devoted to staple food.

For all these crops, specific beds can be laid out and designed to give ideal conditions. Lorenz (1980) provides an excellent planning guide for growers for many factors to do with spacing, micronutrients, pH, and yields under good conditions.

Planning a garden repays itself quickly, and a good gardener expects to amortise establishment costs (fences, beds, seed, mulch, micronutrients) in 6–8 months, thereafter gaining a profit in health, food, and cash.

The essentials of a desert garden are:

- *Small* raised flooded beds, thickly mulched.
- Permanent hardy trees on leach fields or in swales.
- Semi–wild very hardy bulbs, tubers, and yams in selected sites.
- Every bit of waste water directed to leach fields, also surplus from roof and run–off.
- Vines, their roots in cool mulch or inside shady walls, a major feature of the garden. Every wall, open space, and roof vine–shaded.
- Mulch. Mulch. Mulch. Mulch.

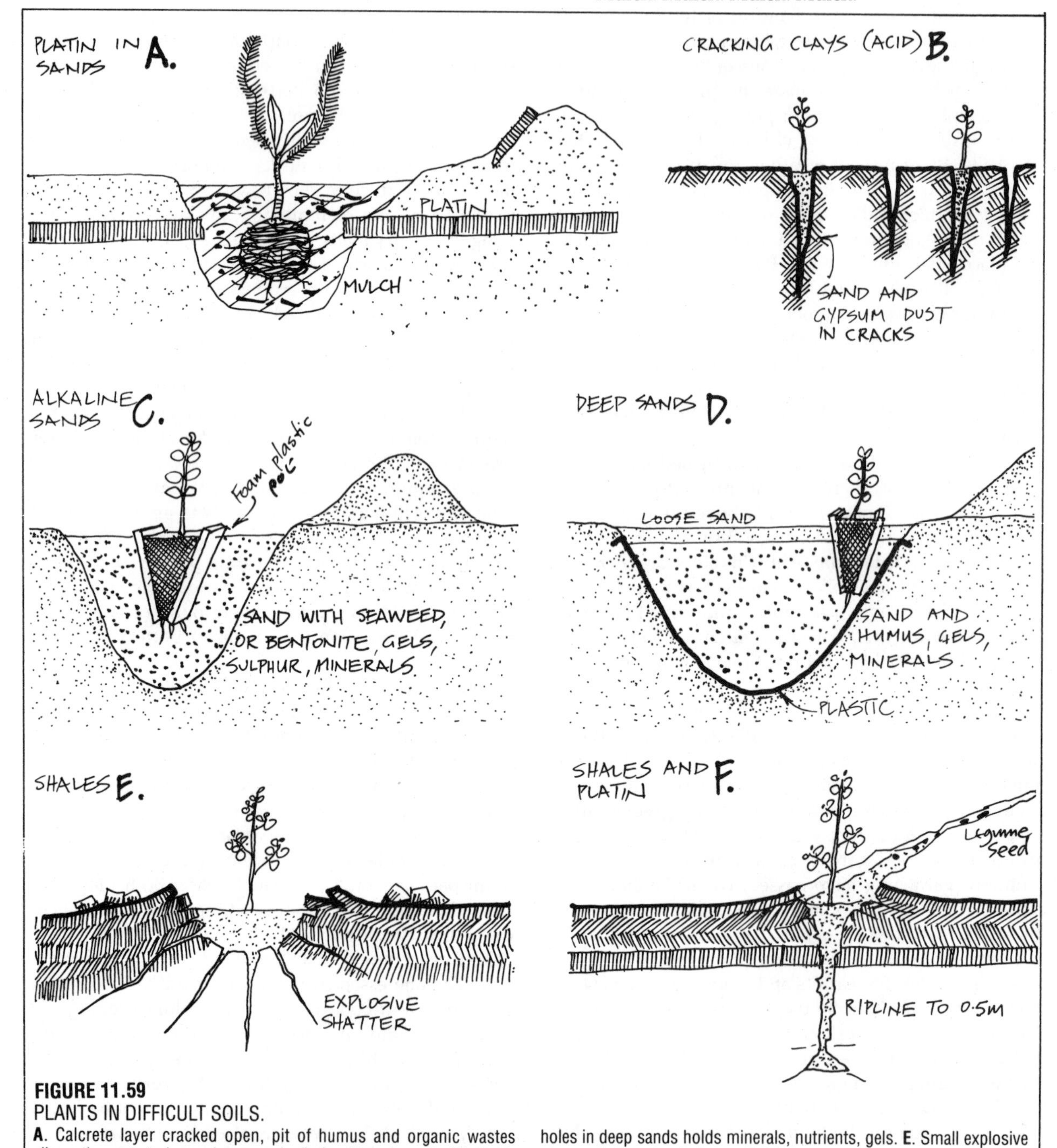

FIGURE 11.59
PLANTS IN DIFFICULT SOILS.
A. Calcrete layer cracked open, pit of humus and organic wastes allows the roots to lower damp sands. **B**. Cracking clay soils filled with sand–gypsum mix before seedlings planted. **C**. Gels mixed with trace elements, sulphur before planting in alkaline sites. **D**. Plastic liner to holes in deep sands holds minerals, nutrients, gels. **E**. Small explosive charge shatters hard shales. **F**. Ripline opens hard pans for tree lines and water infiltration.

STAPLE FOODS

A staple food is defined as one supplying 50% or more of the diet when in season. Today, some 70% of the food eaten in western societies are supplied by 8 staple crops. No European family has 20–30 basic staple foods. Few have a choice of as many as 62–100 foods in all, and a very good home garden and livestock situation would produce about 20 vegetables, 6–9 fruits, and 3–6 meats (well beyond the choice range of an average western family). Of these, 2–4 would be considered staples, depending on the culture.

The very basic question we must therefore ask is how have we improved the Australian desert yield for people? By introducing cattle and sheep, and releasing the rabbit, we have destroyed the bulk of the useful vegetation, and almost every common marsupial or desert–adapted mammal of medium size (although kangaroos and mice survive). We have destroyed the harmless way of living and the profound knowledge of thousands of integrated tribal people for an undeserved ownership by a very few, essentially disintegrated, pastoralists.

This makes no ecological, economic, genetic, informational, or social sense and it is a denial of human rights. We have replaced commonsense with a situation of downright exploitation and rapid loss. Broadscale commercial herding is, to the desert, what broadscale sugar cane cropping is to the tropical rainforest, and broadscale soybeans to floodplain forests—a disaster in every way.

People who managed desert well for 10,000–40,000 years may have known a little about it. We have suppressed this knowledge and its right to continue to exist. Those who were teachers have become fringe–dwellers. Both genetic resources and information on management have been destroyed by "the–right–to–ruin" mentality claimed by private owners and governments.

Even the plants developed by the Papago and Hopi Indians are being lost as store foods replace desert foods. And yet it is clear from desert plant lists that there is a present potential for several staple foods, and an untapped potential for developing better cultivars of many desert plants.

Several hardy seeds were stored to extend the staple over more than one season. In Nevada, Sho-shone Indians gathered, in about 6 days, a year's supply of *Pinus edulis,* and buried unripe cones for later use if needed. Rice–grass seed *(Oryzoides)* was a similar easy staple to gather and store. Thus, staples in deserts (excluding animal foods) seem to be plentiful. It is a question not so much of gardening or farming, but of placing and managing a set of hardy foods in the best situation for later gathering.

As almost every desert garden with modest water can supply 60 or more vegetable species, and as the cultivated fruits of deserts number 30 or more in common use, and as all domestic livestock (including

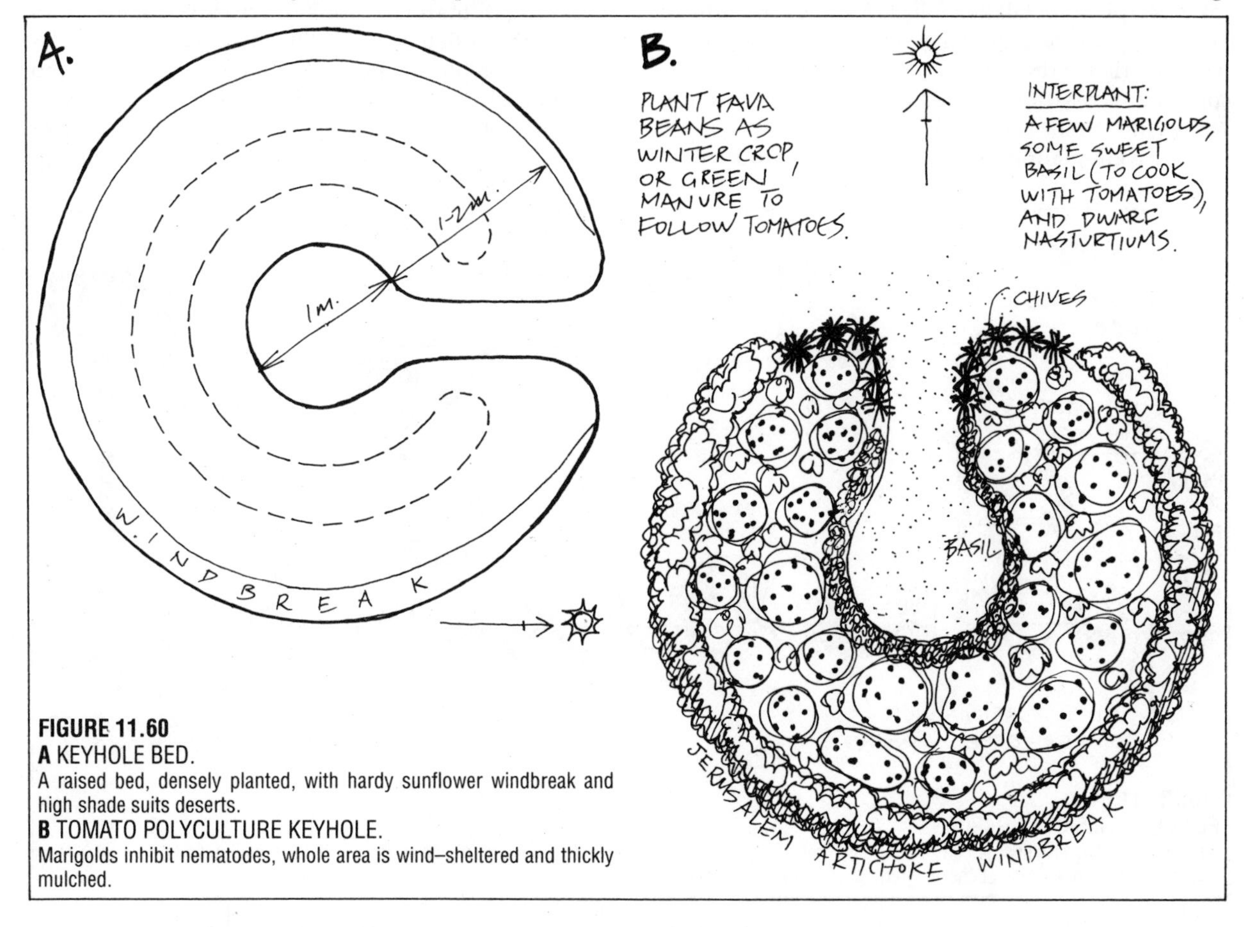

FIGURE 11.60
A KEYHOLE BED.
A raised bed, densely planted, with hardy sunflower windbreak and high shade suits deserts.
B TOMATO POLYCULTURE KEYHOLE.
Marigolds inhibit nematodes, whole area is wind–sheltered and thickly mulched.

ducks and geese) are thriving in arid areas, there is no basic problem for nutrition. Yet adequate food and good nutrition are denied many desert peoples because of the lack of basic seed and plant resources, and the denudation of the total environment by livestock.

For designers, small intensive trial systems within and around settlements are the essential precursors to wider Zone 2 and 3 trials along favourable corridors of better soils and water. The selection of existing plants, and the further addition of new species for the area, greatly assist this "nucleus of small successes" that demonstrates how larger or more extensive systems can also develop, using fewer but well-tested species.

It is always essential to build-in humus and mulch production into crops, and species used for this purpose can range from such edible plants such as *Dolichos*, edible lupin, *Leucaena*, or edible-seeded *Acacias*, to shelter, edge, weed barrier, and hedgerow species such as *Echium fastuosum*, *Acacia*, *Pennisetum*, comfrey, lemongrass, and Vetiver grass. Many trees such as *Casuarina*, some figs, *Acacia victoriae*, *Pongamia* and clump species such as bamboo also provide shelter and leaf mulch, while hardy ground covers suppress grasses and cool the roots of vines and young trees. Good "soft" ground covers outside Zone 1 are vetches, nasturtium, a variety of runner legumes, comfrey, annual lupins, and daikon radish.

The spacing of fruit trees in Zones 1 and 2 can be as usual; it is, however, essential to provide *high shade* from interplants of tall thin foliage legumes or palms, to also interplant fast-growing small and large woody legumes that either die out (pigeon pea) or can be coppiced (tagasaste) or felled or ring-barked to rot to humus (many *Acacia* species).

Such well-planned systems are very productive, drought resistant, cool, and eventually humus-rich. They demonstrate how quickly a barren or hard soil area can be brought into production as a result of intensive biomass production and wastewater use. Thus, the role of teachers and designers is to help plan and place such systems as trials in villages, to locate and provide seeds and propagules of new species, to teach the benefits of soil humus and rainwater harvest, and to stress food preparation and good nutrition.

VINES

Vines have a key role to play in desert gardens. Correctly spaced and pruned they provide both a productive crop and shade cover to mulched and watered gardens. They are also a key element in moderating climate in designed houses, or in retrofitting uncomfortably hot homes. These basic uses can all be applied in the domestic situation. There are special vines for broadscale work, and for selected niches in the desert property. We can deal with each in turn.

Vine Over Garden

Horizontal trellis bars at 1–2 m spacing and furnished with grape vines can throw a shade system over the greater part of the vegetable garden, relieving light saturation. The sides of this trellis can also be more completely closed with herbaceous vines (beans, climbing tomatoes, yams, cucurbits). It is more important to defend from the early (eastern) sun, as dew, distillation, and guttation moisture is important to plants early in the day, and soil temperatures then remain cooler for longer periods. The western side of the trellis needs fleshy vines, as great heat can build up

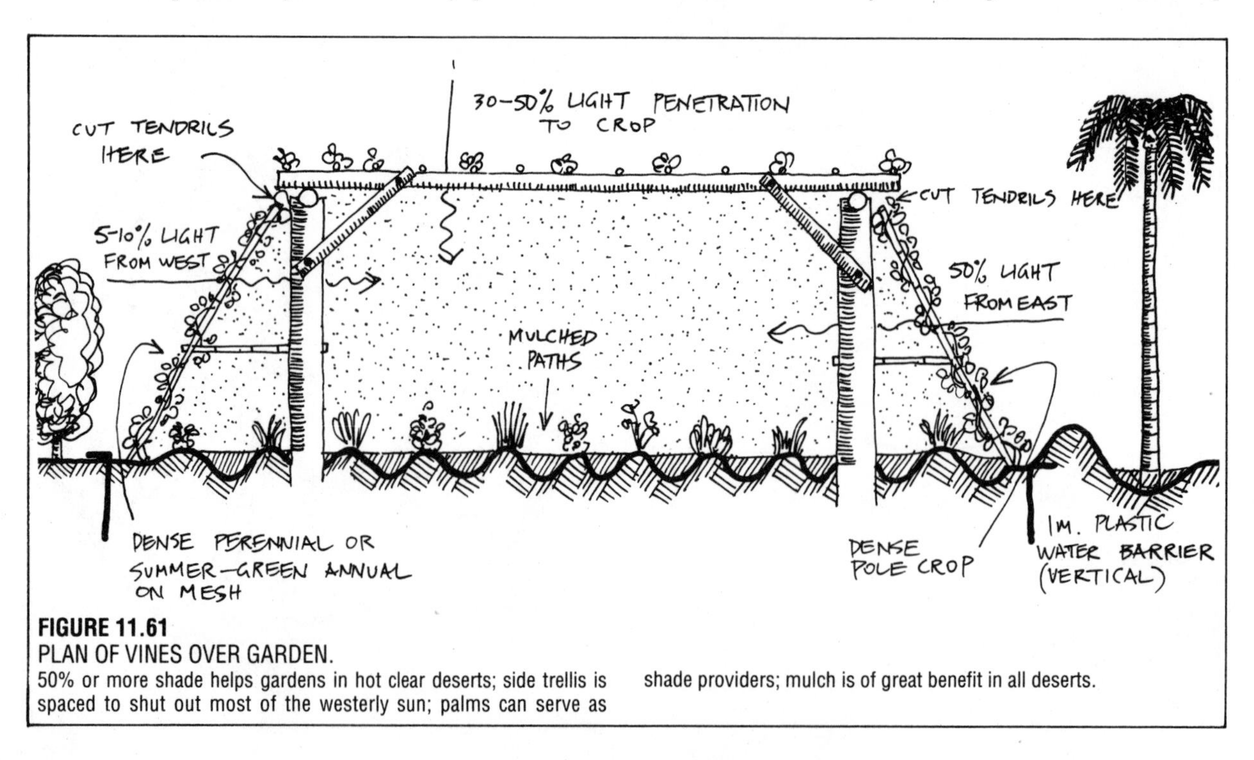

FIGURE 11.61
PLAN OF VINES OVER GARDEN.
50% or more shade helps gardens in hot clear deserts; side trellis is spaced to shut out most of the westerly sun; palms can serve as shade providers; mulch is of great benefit in all deserts.

on the western aspect late in the afternoon. A vine such as *Mikania* (mile–a–minute) is ideal. Only the shady aspect of trellis needs to be left quite open, while the sun side can be of wide–spaced pole crop. A plan is as in **Figure 11.61**.

Vine as a House Retrofit

Considerable comfort can be brought to over–hot homes by attaching trellis. The best results are achieved by:

- Standing dense evergreen vine crop out from west walls (*Mikania, Dolichos, Pelargonium*), and a vine awning above and out from any existing west windows.
- Growing vines (winter deciduous or summer herbaceous) in a screen fixed out from the eastern sun walls. These are well placed at the edge of a verandah if such exists. Washing down the verandah and spraying the vines with water rapidly cools the shaded verandah area.
- Building extensive closed trellis over the rear door, on the shade side to provide a cool air source, preferably with a thick bark mulch below the vine and fine sprays to damp it down. In this case, roof vents via the indoor ceiling or a small sunside glass-house are necessary to draw in the cool air of the trellis.
- Running a perennial non–invasive vine (*Mikania, Pyrostegia*) completely across the roof.
- Placing a water tank under the vine crop to keep the garden air cool, and to cool the water.

Vines as Mulch and Forage Sources

Vines can be vigorous producers of foliage for garden mulch and for feeding small domestic livestock. Well–chosen, shade vines can also provide some stick fuel woods for efficient cook–stoves.

Vines in the Desert Itself

Several vine crops are desert–adapted; if we include the gourds and yam legumes (ground vines), some very basic reserve foods can be set out in shaded, moist, wastewater, or soakage situations. Some vines survive dry, harsh, stony, or dune conditions. All, however, thrive and yield better if a few cubic feet of humus is pitted below the plant when it is set out. Valuable vines grown in large containers can await rains, or the soakage from water harvesting, before being placed out for permanent field growth.

FENCING

Over much of the world's drylands, the great impediment to home garden production of food is the presence of wild, feral, and domestic browsing animals. It is much cheaper to fence out these devastating and ever–hungry animals from settlements and gardens than it is to airlift emergency food aid to starving people, or to withstand the social costs of poverty and famine.

Fencing is thus a primary requisite for intensive food production in deserts. Fenced corridors or "stock routes" can permit milk and draft animals to enter some restricted areas of settlements, but it needs sound fences (preferably electrified) to exclude goats, camels, cattle, donkeys, and sheep from home gardens.

Within fenced areas, sometimes surprising natural regeneration of trees can occur, and hardy food plants can be set out in unirrigated areas if fencing is provided. These form a firewood, mulch, and medicinal resource for hard times. Poultry, domestic rabbits, bees, guinea pigs, and pigeons are all relatively harmless livestock for dryland homes.

Where there is no money available for post–and–wire fencing, more laborious alternatives are used, ranging from ditch and bank systems, either rock–faced or thorn–crowned, to woven or living fences of plant materials (reeds in the Caspian area, cactus in Mexico, *Euphorbia antiquorum* in India, *Euphorbia tirucalli* and *Lycium ferocissimum* in Africa and now parts of Australia). Combinations of stone walls, rock, thorny shrubs, and steel pickets or wires are frequently seen. In affluent societies burnt brick walls, and in Afghanistan and Iran unburnt mud brick walls, are erected around large gardens; even within urban areas of the third world, domestic animals range and destroy vegetation. Where hunting will support dogs, large domestic dogs will defend a house area (at the cost of feeding the dog). It is a matter of adapting to local materials, labour costs, and customs.

SOILS

In drylands, any soil humus can rapidly decompose (in dry–cracked soils) to nitrates with heat and water, giving a sometimes lethal flush of nitrate to new seedlings. Dry cultivated soils exacerbates this effect. Mulches or litter on top of the soil prevents both soil cracking and the lethal effect of rapid temperature gains that cook feeder roots at the surface, so that in subsequent rains there are less roots to absorb water.

Fire is destructive of this protective litter. After fire and cultivation, most of the soil nitrogen, sulphur, and phosphorus is lost, and even a cool fire loses plant nutrients to soil water and leaching. When we know more of the effects of fire in drylands, it is my opinion that we will use any other method (slashing, rolling, even light grazing) to reduce fire litter to soil mulch. It now seems probable that Aboriginal burning has not only gravely depleted soil nutrients, but caused a breakdown in soil structure, and perhaps has been in great part responsible for the saltpans that preceded agriculture. However, agriculture itself is a monstrously effective way to speed up this process and intensify it.

Soil Treatments in Dryland Home Gardens

Where free–draining or non–wetting sand is the problem, bentonite (a volcanic fine clay which swells up and holds water) is a great help in flood–irrigated beds. Conversely, where clay is causing problems with water penetration, the addition of gypsum enables water to penetrate further into the clay particles. Where salted soils or salty waters are a problem, the garden beds must be mounded up or raised, when the salt can wash down into the paths and low places.

Water use is efficient only if all the water soaks in quickly, and watering is stopped once the soil is saturated. Mulch greatly helps this process of fast soakage. Whatever is true of home gardens is also true of swales, and soil treatments can be the same for swales. Ripping swales helps, as do soil additives such as bentonite and gypsum. Where swales are used as tracks, gravel mulch is appropriate; coarse sand (raked to keep swales weed–free) also helps.

Swales are better made about 2–4 m (6–12 feet) wide, so that trees grown along them can overshade the swale base and so help prevent evaporation. As swales develop, salt is carried down to groundwater level. Every desert settlement of 5 cm rainfall or less can direct all road and roof run–off and tank overflow into swales. Domestic water from all sources can also be absorbed into swales, producing fruit, flowers, and vegetables for the house and village, or fuel products for society.

De–salting Soils in Gardens

The Saharawi people of North Africa, as refugees in Algeria, have developed two methods of soil treatment under irrigation that leaches salt. Both methods involve heroic or total soil rehabilitation (*International Agricultural Development* Jan–Feb., 1987).

Method 1: (where shingle or pebbles are available): Salt soil is removed to 1–2 m deep, a layer of shingle laid down, and non–salted soil carted in. **Figure 11. 62.A**

Method 2:

Where no such stones are available, topsoil is removed (if clayey) to 20 cm, salt–free sand is brought in, and deep canals are cut to drain off surplus salt with the irrigation water. **Figure 11.62.B**.

DESERT MULCHES

Organic matter is as invaluable in deserts as it is elsewhere. Sources are:

- The detritus brought down in flood, which can be arrested on strong mulch–trap fences (**Figure 4.33**);
- Tumbleweeds and wisps of plants blown by wind, which will settle in pits, swales, or are trapped on fences;
- Grown mulch in our gardens and orchards;
- Plants such as *Casuarina*, bamboo, tamarisk, comfrey, some species of *Acacia*, forage grasses and legumes can be planted *for* their mulch value or nitrogenous soil fixation in specific situations. Some desert vines also provide good mulch from trimmings, as do hedge species;
- Household and storage wastes; and
- Grazers on range (if regularly penned) for mulch–manure resources.

Categories of mulches are:

- DOMESTIC old skins, blankets, cotton or wool clothes, cardboard boxes, newspapers, hardboard and thin planks. All of these, soaked, can be used to line sandy pits, or tiled over the ground in raised beds before mulch is added.

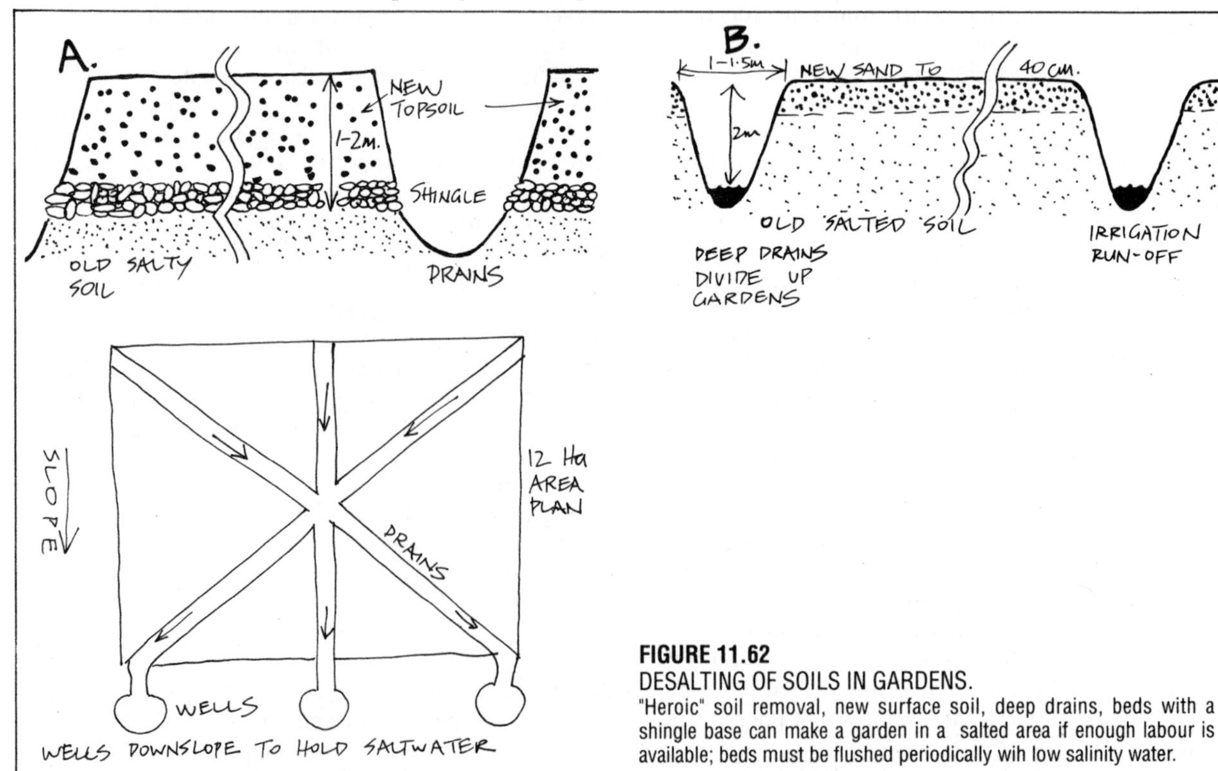

FIGURE 11.62

DESALTING OF SOILS IN GARDENS.

"Heroic" soil removal, new surface soil, deep drains, beds with a shingle base can make a garden in a salted area if enough labour is available; beds must be flushed periodically wih low salinity water.

• COLLECTED ashes, bones, dog, cattle or horse manure, kept dry or pitted in circle pits for in-ground compost. Ashes and dry herbivore manure can be shredded or pounded.

• FINES: chaff, bark, leaves, tea leaves, coffee grounds, flood detritus, rice or grain hulls, rotten wood, sawdust. All these "fines", and shredded paper or wood chips, are ideally combined with shredded manures, sand, and ashes to fill up raised beds; these can be laid as a top or coarse mulch.

• COARSE MULCH: logs, twigs, dry straw, thick bark, old wood. These are used as bed edges, or a 4–6 cm layer over which mixed soils and "fines" are spread to 18–20 cm thick.

• TOPPINGS: Seagrass, woodchips, pine or *Casuarina* needles, cocoa beans. These can be a mulch or cosmetic layer spread over the top of the early layers.

Our bed now looks like **Figure 11.63**.

Good green mulches for grapes and tree crop can be gathered from catch crops of lupins, fava beans, and coppiced *Acacia* or mallee eucalypt; these can be cut after rains, and stored as silage or in earth pits for later use. Lucerne establishes in swales and damp areas, and *Casuarina* leaf or pine leaf can be regularly gathered from trees in the field. We must largely rely on mulch to bring down the pH of alkaline soils and to make minerals available.

In mulch, free-living nitrogen-fixing organisms supply plant nutrients at a steady rate, and leaching is delayed or prevented. In tropic drylands, the greatest losses are in bare soil agriculture, when nitrogen, calcium, and potasium are leached to streams and groundwater. Wherever we use deep humus, mulch, or green manures, we need to provide dolomite, as nitrates prevent copper and manganese mobility if no calcium or manganese is supplied.

When we mulch in deserts, we need to emulate the ant and the termite, who bury their organic materials out of the sun. Wherever possible, we should also place a layer of stone, sand, or soil over our desert mulches, or create hollows where trapped leaves will be later covered with sand.

Drought stress in cultivated plants is greatly ameliorated if these plants are provided with mycorrhiza (root fungi). From the host plants, root associates get shelter (in a safe root environment) and sugars. In return, they scavenge for soil nutrients, and can mobilise phosphorus even at very low water levels in the soil. It is true to say that many plants don't have roots; they have fungal mycelia to explore the minute world of the soil particles.

With almost no exceptions, all the vegetables that

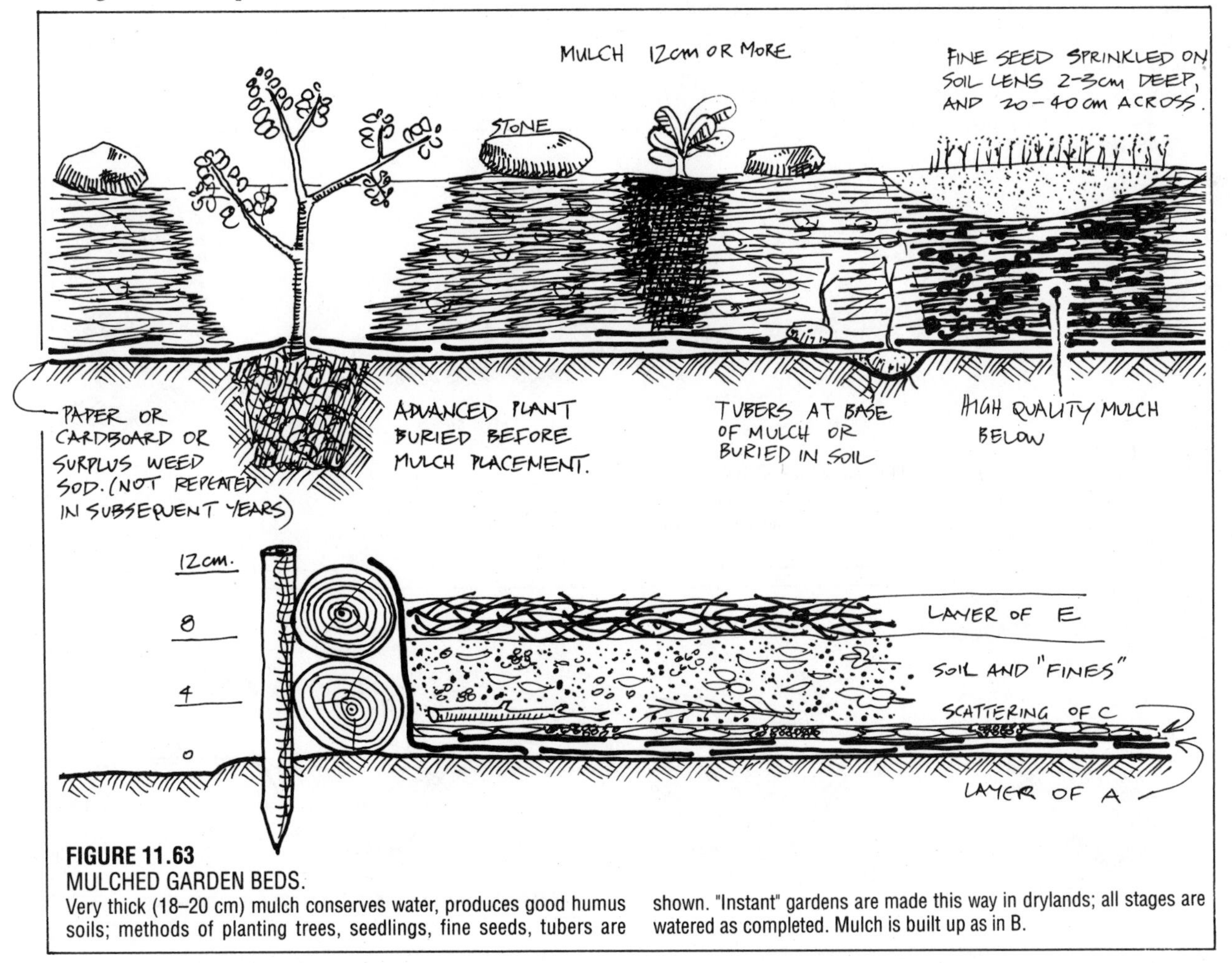

FIGURE 11.63
MULCHED GARDEN BEDS.
Very thick (18–20 cm) mulch conserves water, produces good humus soils; methods of planting trees, seedlings, fine seeds, tubers are shown. "Instant" gardens are made this way in drylands; all stages are watered as completed. Mulch is built up as in B.

grow in humid areas of similar temperatures will grow in deserts or dryland gardens under some form or irrigation or water harvesting. Field crops are more restricted, but grains, oil seeds (sunflower, safflowers) and grain legumes can be grown in suitable soils, after rains. Fields, however, are more expensive to modify and water than gardens, and carry the risks of wind erosion if cultivated. "Fields" in deserts may be possible as cross-wind avenues between perpetual windbreak of *Acacia* species and fieldstone. However, I have not as yet seen this tried, and even in avenues, the width would need to be restricted to 5 or 6 times in tree height, and all crop straw returned (trash or mulch farming) to avoid rain splash and soil loss.

Space is seldom limited in deserts, so that up to 30% of any garden or field can be profitably devoted to permanent or (for gardens) seasonal windbreak, as much for a mulch source as protection from the wind. Unburnt and ungrazed semi-desert can produce very large quantities of grass mulch after rains; it is quite frightening to stand in a chest-high stand of *Danthonia, Stipa, Themeda,* and *Agrostis* dried to a tinder in such conditions, and to contemplate the implications of a fire! I believe that a large area kept unburnt, and forage-harvested in strips after rains, would supply all the mulch that gardens could use. In the (narrower) unharvested strips, I would expect a dense mat mulch to emerge *if* fire could be excluded, even in 15 cm of rain. If this were rolled down between permanent trees, a humus soil would eventually develop. Certainly, all of these strategies should be tested (avenue cropping, forage or mulch harvesting of bunch grasses, roll-down mulching, and trash farming in avenues) on the broad scale.

If such systems are also combined with interceptor banks (**Figure 11.93**) it may be possible to develop a dryland agriculture based on permanence and stability.

The unregulated nature of grazing and wildfire is what prevents this development in many semi-arid areas. Harvested wildlife would, however, appreciate the system for food and shelter. I have tried to diagram such a system in **Figure 11.64**.

On an eroded, alkaline, non-wetting sand dune near Ceduna (South Australia), a group of aboriginal students trained by Peter Bennett of the Soil Association produced in 6 months an excellent, healthy mixed garden of almost every sort of vegetable and many flowers at 1,100 ppm salt in water. The key ingredients were fine-shredded mulch, compost, and coarse mulch on beds, careful watering, and a sound organic approach to soils. Given such a hostile base for a garden, this example should encourage any group to do the same. The graduate gardeners from this teaching garden have extended their studies to home gardens in the area, and to teaching those skills. This is an appropriate way to provide food in deserts.

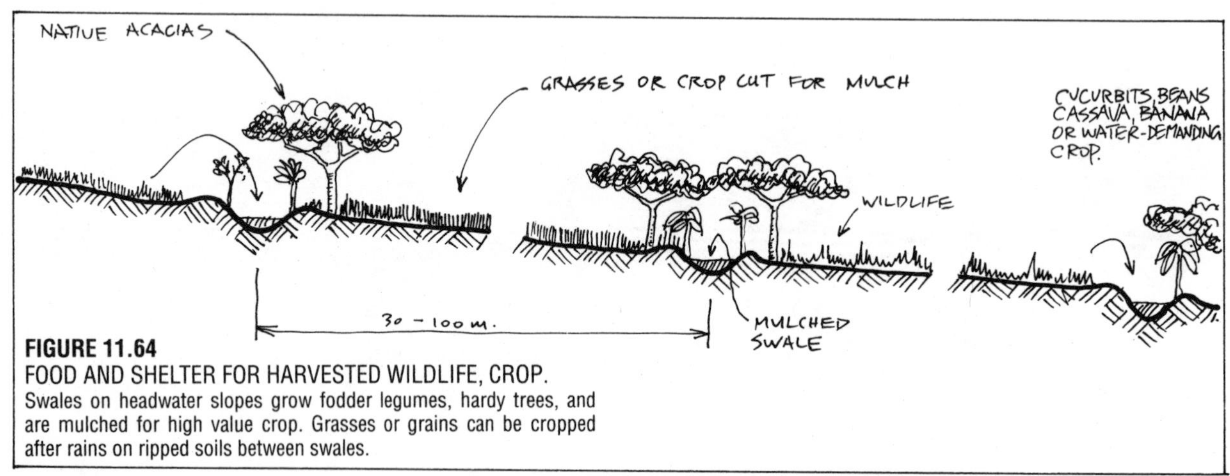

FIGURE 11.64
FOOD AND SHELTER FOR HARVESTED WILDLIFE, CROP.
Swales on headwater slopes grow fodder legumes, hardy trees, and are mulched for high value crop. Grasses or grains can be cropped after rains on ripped soils between swales.

FIGURE 11.65
MICROPORE AND SEEPAGE LINES.

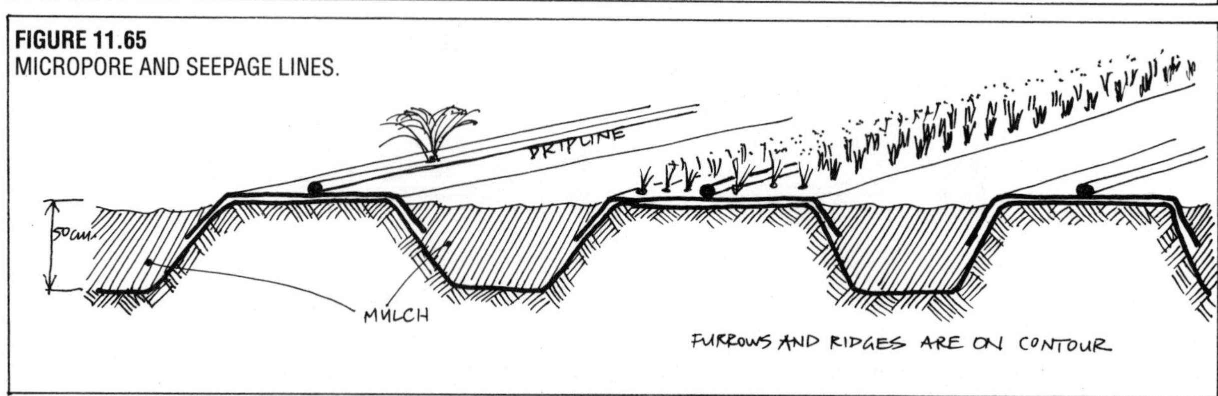

11.9
GARDEN IRRIGATION SYSTEMS

The use of water in unguarded or unsheltered plots is wasteful, as a 10% increase in evaporation is possible on the edges of such plots. There is no doubt that trickle, drip, or seepage irrigation is *the* most effective dryland garden watering method. All we need to argue about is at what level of technology. The most sophisticated systems use automatic timers, 7– or 8–line sequences, and extruded plastic hose with drip emitters of various efficiency and flow rate; there is a vast commercial literature, catalogues, and even training programmes available from manufacturers.

The cost of a good dripline system amortises even in the lowest–paid western households in a few months, given the food produced. However, normal household water use is sufficient to water a good–sized home garden if the wastewater is led to underground seepage pipes and thence to plant and tree roots.

Now popular on a global level, drip–line systems have manufacturers and representatives in almost every country. Water use by drip, especially in orchard or tree establishment, can be from 10–50% of sprinkler use, and if trees are grown, species which are adjusted to climate need water only in the early establishment phase. The particular advantages are that only low head are needed, and that water is placed right at the plants.

There are at least two basic systems, one a solid line with emitters ranging from small sprinklers to drip emitters, and the other a thin, twin–pipe "Bi–wall®" system where a high–pressure pipe is joined and perforated to a low–pressure pipe which is perforated at intervals to suit row crop. Nylon mesh insert filters (or old stockings in a larger pipe) are used as filters, and automatic, soil–sensing, and nutrient addition, or timing systems can be added. This is the technologist's dream system, and installed with small in–line taps, can be laid to completely water the home garden. Even more sophisticated are systems devised to automatically respond to soil probes which sense dryness at root level by measuring soil ionic exchange capacity.

Drip irrigation is ideal for glasshouses and enclosed growing systems, but should be used with caution in waters of more than 800 ppm salt on bare ground, as local salts and carbonates accumulate at the plants. For trees, this effect is overcome by letting drips fall into stand pipes which let water pool at root level, not evaporate to air.

At modest level, people use unglazed earthen jars central to small circle gardens, inverted bottles with leaky tops, and short tubes of pebbles (hand–filled), plus bucket systems. These serve to let special trees survive severe drought conditions in home gardens, or to get windbreak going in deserts. Plastic "pillows" are on sale which can be placed as part–mulch, part–drip at valuable trees. Some pillow systems are designed as circular windshields and small glasshouses.

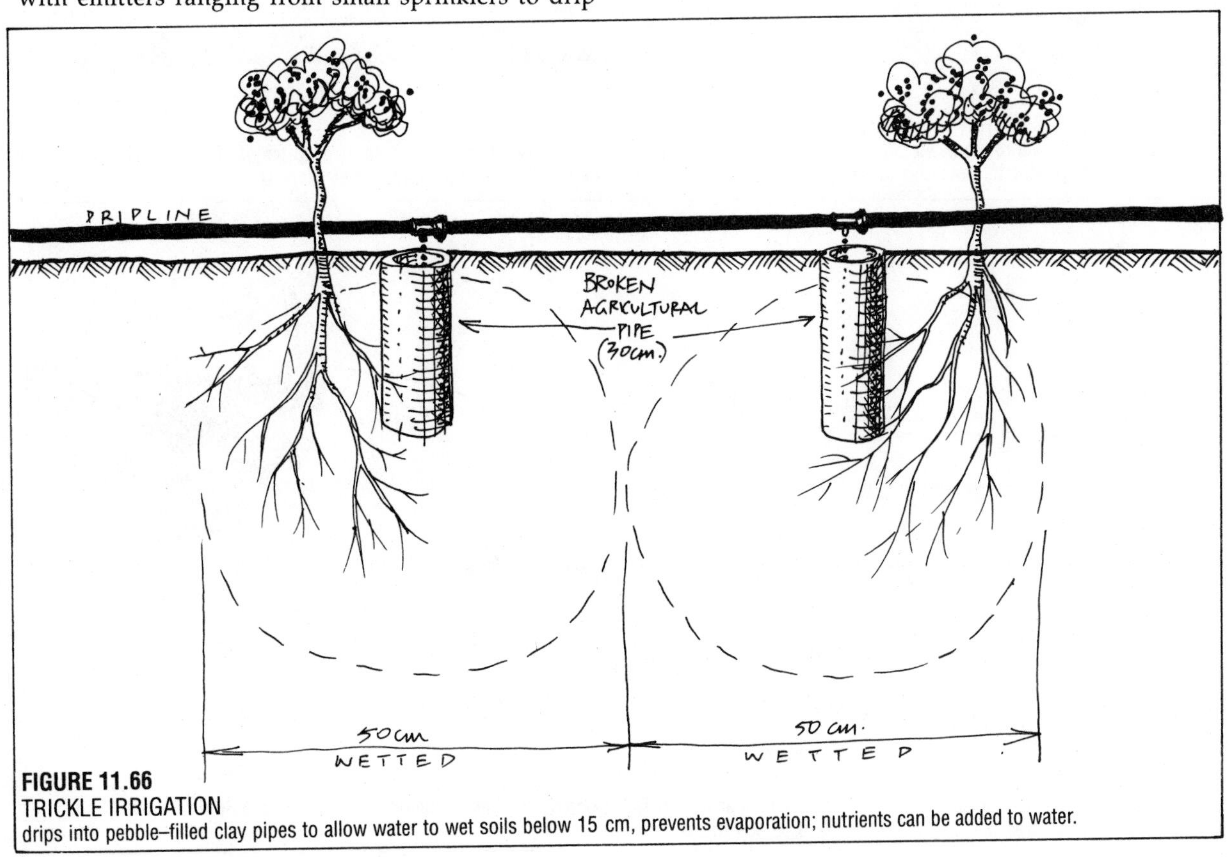

FIGURE 11.66
TRICKLE IRRIGATION
drips into pebble–filled clay pipes to allow water to wet soils below 15 cm, prevents evaporation; nutrients can be added to water.

SUBSURFACE IRRIGATION SYSTEMS

Underground drip and seep systems, if 12 cm or so below the soil surface, are the most conservative method of irrigation. These are best suited to arid sites, where solar evaporation is a factor, and on those sites, the critical use of water is to establish windbreak of desert–hardy plants around orchard or crop. There are good reasons for using domestic wastewater sub–surface, especially sewage water, and these have to do with health and soil filtration. Subsurface may mean below a plastic surface film for some crop like straw–berries, and even deep below the soil surface for desert trees. Swales are the best example of cheap subsurface water development, but there are many special techniques successfully used:

- MICROPORE OR SEEPAGE LINES: these are for valuable row–crop, and have many slits which open under slight pressure. They can be laid below root level or (for shallow roots) at surface, and are worth the effort for such crops as strawberry, medicinals, special seeds (**Figure 11.65**).
- TRICKLE IRRIGATION PLUS POT OR PIPE: a very successful way to water arid–area tree seedlings. Shallow pipe can be replaced with deep pipe to 0.5 m as a tree grows (**Figure 11.66**) and the root zone extends Ideal for primary windbreak or ridge forests in the establishment years, but species that later stand alone should be used.
- DOMESTIC WASTE CHANNELS: used success–fully off desert houses. A sheet of plastic is laid 30 cm below the soil surface from a wastewater outlet, in sands or very permeable soils, preferably 60 cm or less wide. Non–root crop is planted above the plastic, and the channel periodically extended so that no surface

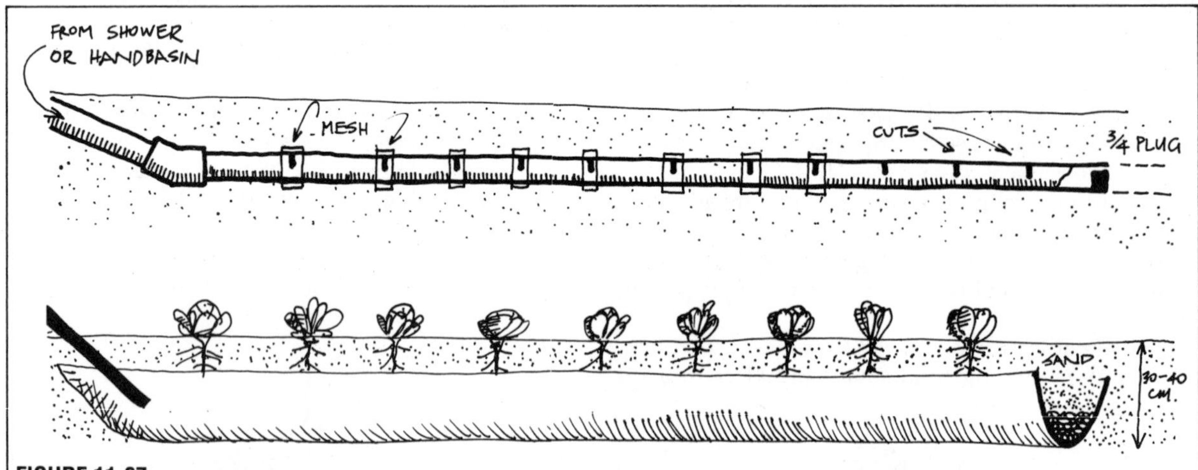

FIGURE 11.67
DOMESTIC WASTE WATER CHANNELS.
A level pipe, half cut through at 40 cm intervals and with screens at cuts allows greywater to seep out at root level of vegetables; a plastic–lined trench can help in deep sands. Rigid polythene pipe is good for this method of greywater disposal.

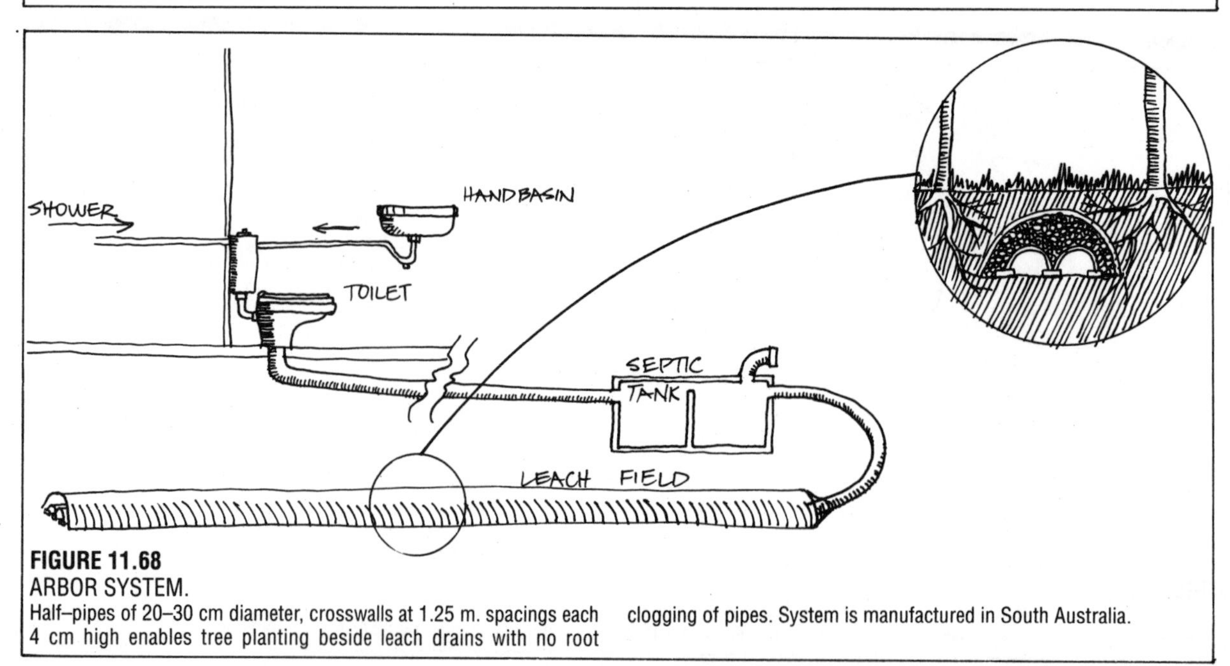

FIGURE 11.68
ARBOR SYSTEM.
Half–pipes of 20–30 cm diameter, crosswalls at 1.25 m. spacings each 4 cm high enables tree planting beside leach drains with no root clogging of pipes. System is manufactured in South Australia.

water appears. Slotted pipe can be used to carry wastewater along the plastic base of the garden beds (**Figure 11.67**).

• ARBOR SYSTEM: commercially available in South Australia, wastewater pipes will not fill with tree roots and can be used without gravel seepage beds in ordinary soils. Ideal for productive trees like citrus, apricot, and palms (**Figure 11.68).**

• POT SYSTEM is excellent for village gardens, with or without dripline. Combines well with circle gardens (**Figure 11.69**).

• HOME SLOT–PIPE SYTEM. The top of the pipe (3–5 cm diameter) has slots cut to one level, nylon mesh or old stockings bound around these, and wastewater led in from handbasins or sinks past a grease trap or crude sieve to catch solid particles. These are excellent too, for small–fruit beds, with the slots at correct spacing for plants. Pipes should be laid to one side of the plants for periodic checks as they do eventually block up. A half–pipe is sometimes used in the same way as the perforated pipe (**Figure 11.67**).

• NUTRIENT FILM AND INJECTION TECHNIQUES. There are at least techniques in use to prevent any water loss by soil absorption; both are essentially hydroponic–intensive. In Israel, direct infusion of water and nutrients into tree or vine stems (a nutrient drip) is in use, and in India a plastic film technique is used to grow Napier grass (*Pennisetum*) for forage, with part of the root mass in earth, part in a nutrient solution in the plastic gullies (**Figure 11.71.A**).

It should also be possible to "wick water" with a slotted pipe, using hessian or nylon wicks to nearby plant roots (**Figure 11.71.B**). In either case, little or no water is lost to soil, and plant roots can explore both the soil itself and the nutrient solution. In view of the obvious maintenance problems of such systems, drip irrigation in open fields and true hydroponics in glasshouses may be preferable if labour is not available for layout, maintenance, and harvest.

Sea or brackish–water can also be used at high tide level in lagoon for garden irrigation (**Figure 11.72**).

• AUTOMATIC IRRIGATION. In areas where electronic sensors are impractical or expensive, some such mechanical apparatus as that sketched in **Figure 11.73** would, in effect, permit automatic watering and self–regulate in rain, and ensure that irrigation is sufficient. Any capable local firm could make such reliable equipment. Rain automatically switches off the tap after 3 cm of downpour. **Figure 11.73** is designed from data from Kevin Handreck, CSIRO, Australia.

CONDENSATION STRATEGIES

Where no piped water is available, and where water is in seriously short supply, trees and gardens need condensation strategies. The aim is to condense water either from night air, from transpired water, or from weeds and trimmings, and return it to root level for re–use. The following methodologies are used:

1. Plant shields of plastic, mesh, or metal.
2. Stone mulches.
3. Sheet plastic sub– or surface mulches.
4. Organic mulches.
5. Pit evaporation systems.
6. Closed recycling systems.

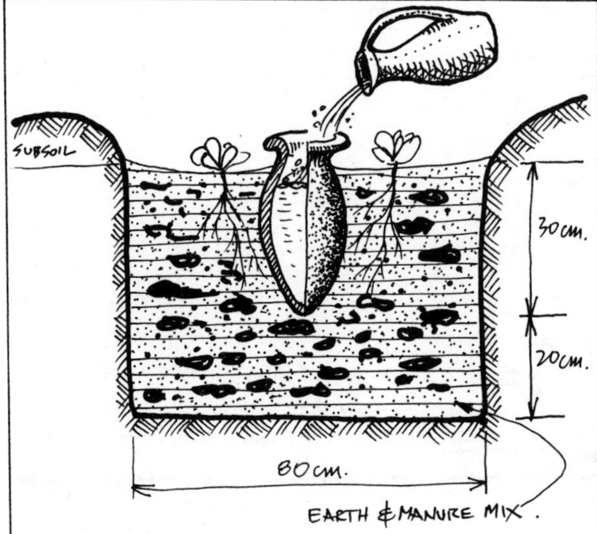

FIGURE 11.69
POT SYSTEM.
An unglazed pot in a humus pit waters a circle of vegetables in circle gardens.

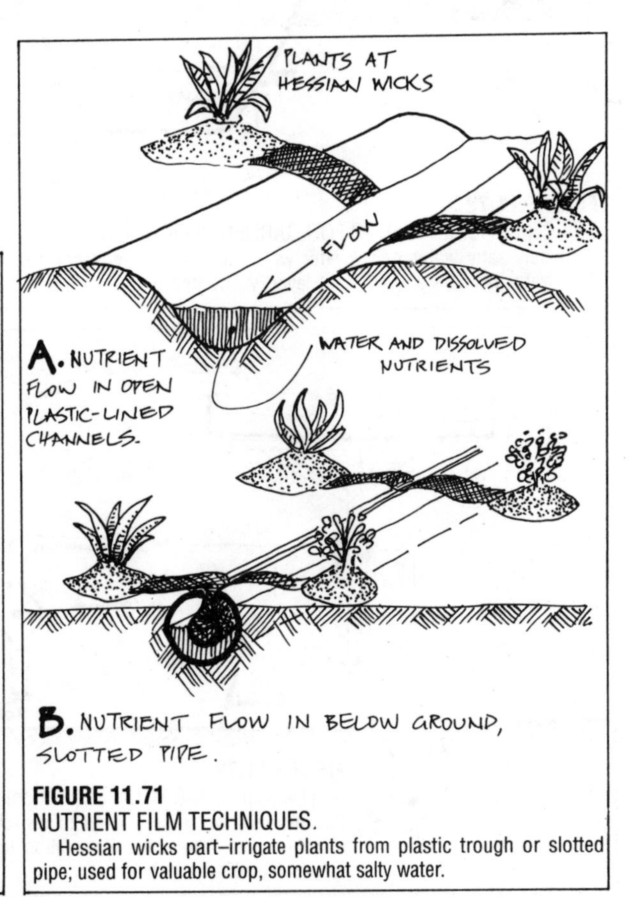

FIGURE 11.71
NUTRIENT FILM TECHNIQUES.
Hessian wicks part–irrigate plants from plastic trough or slotted pipe; used for valuable crop, somewhat salty water.

To take each in turn:

1.Shields of Plastic, Screen material, or Metal

Small trees, planted with a 3 or 4–stake frame over which a bottomless and topless bag is pulled, and around which loose weeds are placed, live in a protected environment in which, at night, soil moisture condenses and runs down the inner bag surface to the roots. Alternatively, coastal ridge areas can condense water from night air using fine metal screens or fences, placed crosswind.

Shields cut down wind, hold mulch, protect from plants from *small* animals, and condense water. My own experience suggests a difference of 80–90% in survival using shields as against unprotected plants. Growth (as light) may double in tall clear plastic bag tubes around a small tree.

2. Stone Mulches

Wherever stone is plentiful, stone mulch acts as a condenser, screen, weed control, windbreak, and root weight against windthrow. Small invertebrate animals take refuge in stone and add nutrient, and the ground below stone piles is always damp. Stones in walled circles on rocky hillsides, with weeds thrown in, are a very successful establishment technique, and can be moved after 2–3 years, or as plants establish. Even a few flat stones at a tree base assists with condensation of soil water.

Linear stone mounds between crops have a similar effect, and natural raised boulder beaches act very much as do swales, so that springs break out below them; they can be used to plant double lines of trees or crop.

3. Sheet Plastic and Sub–Surface Mulches

Sheet plastic, perforated for plants, acts as a ground surface re–condenser as do stones, and a variety of forms are in use (**Figure 11.74**) to trap moisture, reduce weeds, and condense night moisture. They integrate between closed recycling systems and open field mulches, and can be used indoors and out. The common uses are on small fruits and valuable crop, where costs are offset by income. As well, deep *vertical* plastic sheets prevent lateral loss of water around desert gardens, while buried sheet benefits from the weight, sun protection, and additional mulch of a layer

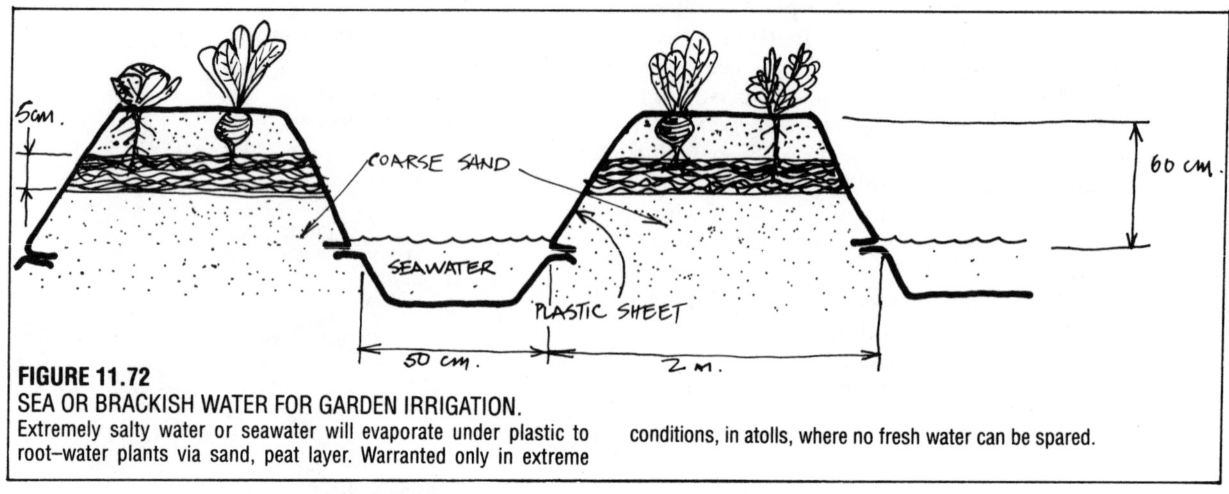

FIGURE 11.72
SEA OR BRACKISH WATER FOR GARDEN IRRIGATION.
Extremely salty water or seawater will evaporate under plastic to root–water plants via sand, peat layer. Warranted only in extreme conditions, in atolls, where no fresh water can be spared.

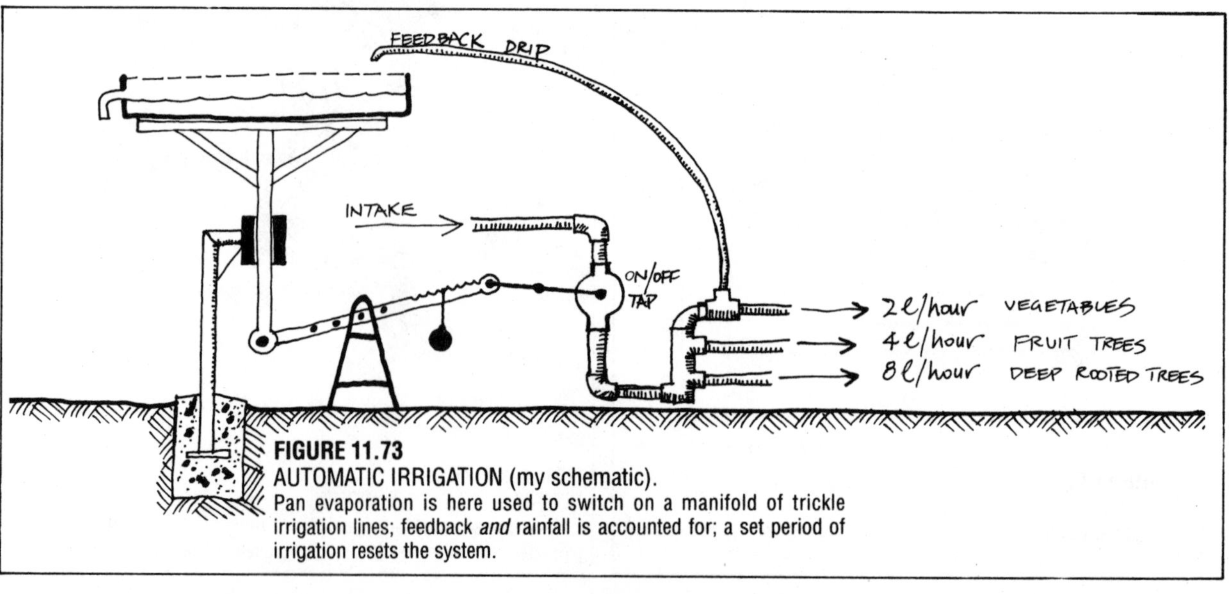

FIGURE 11.73
AUTOMATIC IRRIGATION (my schematic).
Pan evaporation is here used to switch on a manifold of trickle irrigation lines; feedback *and* rainfall is accounted for; a set period of irrigation resets the system.

of dry soil (**Figure 11.75**).

4. Organic Mulches
For very small areas or restricted tree sites, a variety of organic mulches are of great help. In extreme conditions, these need to be 45 cm or deeper to be effective, but in a drought summer, they may mean the difference between survival and death of plants. Special deep mulch boxes are excellent in forests or gardens for root crop such as yams and potatoes. Weed suppression, condensation or release of water, wind protection of soil, and decomposition products are all in effect, so that compound influences aid plant growth.

Combinations with plastic sheets, pits, and drip–line systems make mulches even more effective in home gardens. Small waterholes benefit from being thatched over, and many padi crops (taro, rice) can be grown in mulch with much less water use than open ponds.

5. Pit Evaporation Systems
These are used as survival strategies in deserts, but can be adapted to grow useful plants, using less useful or cropped plants to give up water, or salted water to provide fresh drip to plants (**Figure 11.75**).

6. Closed Recycling Systems
When it is feasible or economic to totally enclose plants, as in glasshouse and plastic tunnel systems, there are some water losses due to ventilation, but little else, as condensation on the inside surfaces can lead back to plants, or salt water flows can be evaporated to air before cooling the vapour to potable water fed to plants.

11.10
DESERT SETTLEMENT–BROAD STRATEGIES

The ultimate safe limit of human occupancy in deserts depends on the capacity of a carefully–balanced wadi or permanent lagoon to support that population. All settlements (ancient to modern) *must be limited by the water supply available to them.*

The majority of dryland settlements have managed, in one way or another, to destroy both themselves and their hinterlands. The common ways they perish are:

- To exceed the capacity of (or to pollute) local water resources.
- To devastate their environment for firewood and fodders for domestic livestock.
- To fail to govern their expansion, or to assess a limit to growth. Consequently (as with many societies) wars and invasions, or refugees and migrations, follow.

All these factors must be taken into account when we chose to live in drylands. I believe pastoralism to be one of the key factors in all arid landscape devastation. A very complete education in rangeland management, and a strict restriction of livestock numbers (constantly assessed) is necessary to enable any settlement to survive. If grazing is controlled, firewood supply and essential windbreak is also ensured, as is a basic food resource from local agriculture. But if we impose the

FIGURE 11.75
PIT EVAPORATION SYSTEM.
Green leafy material in a pit gives off water which condenses at night on the underside of a plastic cone, perforated at the centre to allow rain through; by adding fresh leaves, dirty water, or urine a nearby hardy tree can be sparsely watered, or enough freshwater for a drink obtained.

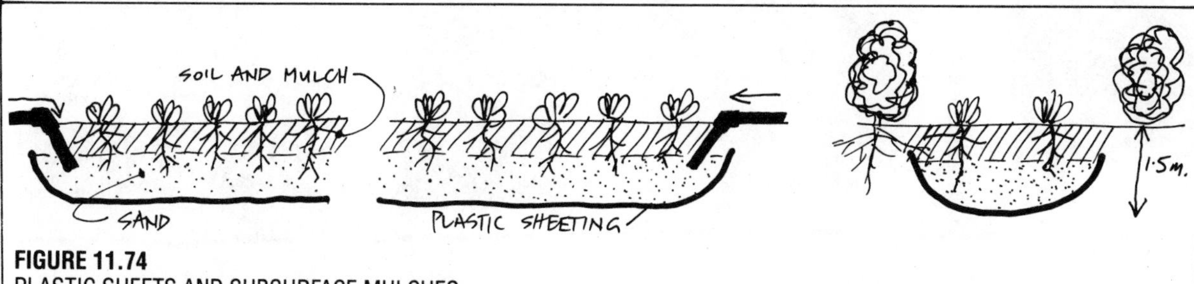

FIGURE 11.74
PLASTIC SHEETS AND SUBSURFACE MULCHES.
In very deep coarse sands, only plastic sheets will hold humus, nutrients, and water at root level of crops; gaps are left every 30 m in broadscale systems.

stress of ungoverned pastoralism on that of expanding settlement, we are inevitably doomed to create real deserts from drylands.

We need a very cautious approach to desert settlements, in that very favoured environments or refuge areas are limited. We must locate near water, or find water resources sufficient to our needs. In practice, this means one of three broad choices:

- The foothills of run–off uplands (about 5% of total drylands).
- Valley or wadi sites, at times under the shelter of cliffs and scarps.
- Around oases, permanent pools, or reliable exotic streams and freshwater wells.

Having selected such a site, we can do a great deal about firewood and water storage in settlement, and eliminate other undesirable factors such as dust and excessive heat.

If a settlement allows itself to expand beyond the minimal resources of dry years, it is simply deferring catastrophe, and at the same time building a larger order of long–term catastrophe. Thus, it is essential, perhaps an iron–clad rule, that any such settlement must be founded to contain only that number of people who can survive the "worst case" scenario.

DUSTSTORMS IN SETTLEMENT

Summer heat over the land can produce dry or rainy thunderstorms, with a dry downdraught just preceding them that can reach 50–65 km/h, and picking up great quantities of dust. Episodes of strong winds are most common under these unstable conditions. The dust devils, or whirlwinds, that precede them are milder examples of instability, occurring with great regularity at about 3 p.m. onwards in bare deserts in summer. Even a small ground movement, such as an antelope running or a car, can set off an upward spiral of superheated air. It needs a great bubble of hot air to rise in order to start a dust storm, which is then sometimes followed by heavy rains. Once a desert storm is initiated, however, the dust itself efficiently heats the air and perpetuates the initial instability, clearing only at nightfall or if clouds obscure the sun.

The effects of such storms near settlement can be greatly reduced by sealing roads, orienting dirt roads across–wind, erecting fences, ploughing lines, or pitting plains for grasses (all strategies to roughen the ground and so reduce effective windspeed).

No settlement should be planned without these attentions, or life can get miserable every time a car drives out of town, and both asthmatic, eye, and sinus problems can become epidemic. Dust carries a host of human pathogens, and dust storms are often followed by episodes of morbidity in settlements. Tree lines at 20–30 m are permanent solutions, and these can be established using town wastewater. Trees are best selected to be useful to the region.

HEDGES AND WINDBREAKS

Within the broad windbreaks of settlement, grown on swales and wastewater, garden hedges serve a multiple set of functions:

- To provide forage for poultry, rabbits, and bees.
- To shelter tall crop such as corn.
- To provide mulch from clippings and leaf fall.
- To exclude rampant grass or weeds from garden beds.
- To help exclude browsing animals or large live-

FIGURE 11.76
WILLIAMS' DESCRIPTION OF NORTH AFRICAN POLYCULTURES.
Palms as high shade over other tree and surface crop.

stock.

Careful selection of a mixed windbreak can achieve these ends. In very dry areas, it is prudent to set out a vertical (to 1 m deep) plastic root barrier between garden and windbreak to minimise water competition, to choose highly drought-adapted windbreak species, and to regularly check on root invasion while trimming hedges for mulch. Soft foliage from *Acacia, Casuarina, Prosopis, Leucaena, Albizia,* and *Coprosma* are ideal compost mulches, and some species provide good leaf fall. An investigation of local flora often yields up many suitable species. Outside such soft forage hedges, cactus, *Euphorbia* and thorny shrubs help exclude livestock.

Hedges within the garden are also of great importance, but can be of smaller hardy perennials (rosemary, lavender), crop species (sunflower, sunroot), or mulch species *(Crotalaria)*. Fleshy vines on fences are ideal in crop windbreak *(Mikania, Dolichos)*.

The total integration of house and garden is important. Even in very poor countries, the space between and close to houses (an area rich in nutrients, close to hand, and easily shaded and watered) is often left unplanted, while people "go to fields to work"!

The grape, passionfruit, a great variety of beans and peas, cucurbits, yams, and vine fruits, forage vines for domestic rabbits or guinea pigs, and vines providing honey and pollen are all desert or dryland adapted. Even in the sub-arctic, the silverberry *(Actinidia arguta)* provides copious fruit crop, and survives hard winters.

Below the thinned or well spaced canopy of vines and palms (date or doum palm) gardens are protected from direct sun and the drying winds.

> The long serpentine trunks of the palm tree rise above every village and about every field. The fibrous palm has entered almost every facet of the peoples' lives. It is the first line of defence against the sun in the open fields, and in its shade grows the olive tree. Under the olive, the fig grows, and under the fig, the pomegranate and vine, then the grain and vegetables. The palm tree's second contribution is dates...
>
> (Williams, C., 1974, *Craftsmen of Necessity*, Vintage Books, N.Y.) See **Figure 11.76**

Defining the desert garden, this strong hedge or fence is essential. If the hedge supplies clippings for mulch pits, compost, and potato boxes so much the better. A fast, but temporary, hedge can be made of banna grass *(Pennisetum)* which can be cut for mulch. It needs irrigation to keep productive, so is furrow-planted on manure. Parallel to this, a slower, hardier permanent hedge of columnar cactus or *Euphorbia* repels animals, and thorny *Acacia* also helps. One of the best in-garden "fedges" (fence-hedges) is a strong rail and wire fence (the post termite-resistant or creosoted) to carry vigorous vines such as *Mikania*, which provide soft mulch and bee fodder. *Dolichos* beans, lima beans, and passionfruit can occupy parts of these fedges, as can vigorous soft vines of other species. At 1.5–2 m high, the fedges protect 9–10 m widths of garden.

Outside this system again, taller slow windbreak of Roman cypress, *Casuarina*, or *Acacia* on drip provide high windbreak, but only if space allows. Olives may also establish; local hardy evergreens can be chosen for these 25–50 m grids.

We need to make a gradation from crops and grazing to tree products and forages, for as trees are the nutrient storages of the humid tropics, they are the groundwater moderators of the drylands. We are not short of tree species with which to do this, and if we have enough trees planted, modest grain and pasture strips can also be developed. Our error has been to develop grain foods at the expense of trees, and to extend grasslands and crops until we create desert, or to destroy trees until the salts in desert soils create surface crusts.

I cannot stress too often that wherever we harvest water, it should be to create forests. Not to plant trees may mean that we create waterlogged and anaerobic desert soils, and thus perpetuate or extend salt problems.

PLANTING AND VEGETATION IN SETTLEMENTS

Rigorous vegetative design of settlement calls for adapted perennials, both for food and shade, hardy and preferably local species in shrubberies, large areas of overhead vine trellis, and only a strictly regulated amount of lawn. Such lawns can be of *Lippia* or other carpeting and drought-resistant plants. Golf courses, extensive grass lawns, car washes, and large open pools or ponds supplied at public cost are dryland disasters which need to be taxed out of existence in arid lands.

There are many very hardy and useful trees needing only a swale nearby to survive years of relative drought. However, even water-dependent fruit trees can be grown in or near swales, in courtyards, and along road fitted to harvest water run-off.

A critical settlement strategy is to develop broad (300–400 m deep) tree parks around or even within the settlement, to eliminate not only the devastating dry desert winds, but also to help provide fuels, mulch, medicines, and to supplement domestic animal forage. This planting is the first line of defence against the desert; within its shelter, gardens and crops can thrive without wind damage or excessive water loss (**Figure 11.77**).

What Jen-hu Chang (1968) calls the "clothesline effect", or the effect of (advected) dry hot air blowing into crop, is of particular concern in desert gardens. A general ring of trees of *50–100 m deep* is needed at the desert borders of cultivated land, or the desert becomes self-propagating. The effect decreases exponentially, so that such a broad guard forest against desert creates a screening which effectively protects crops or cultivated lands from the dessication for a great distance downwind.

In such cases, it is better to establish deep tree belts

around settlements against the prevailing desert winds, and to relax to smaller single tree–width shelters within the area so protected, rather than to place narrow windbreak throughout the crop or garden. Given that we can shelter the crop, and that the advected effect is reduced, we are left only with the "oasis effect" of moisture loss to atmosphere, due to earth heating by trapped resident hot air. In this case, field size is irrelevant as all the field is affected, and we have to accept the water loss, unless fields are so small they can be trellised or overshaded by leguminous trees such as *Prosopis*, or similar high shade trees.

On a national scale, it is imperative to provide tree belts 1–5 km wide on the savannah edges of deserts, or advected hot winds affect crops for hundreds of kilometres into humid areas, and bushfire frequency is then steeply increased.

Finally, very large buffer zones of ungrazed or lightly grazed desert can be developed around villages. Pitted or swaled, these areas completely defend against dust, so ensuring that both public and environmental health is preserved.

The last area of settlement vegetation to plan is that of FUEL FORESTS for cooking and (if necessary) power supply, essential oils, mulch, and other tree products such as honey. If the settlement is sewered, then surplus greywater and sewage can be first ponded in deep (to 3 m), narrow, preferably roofed or trellised, and well–sealed collection ponds, then led by dripline to a carefully selected and designed fuel forest. Such a system is in effect at Yulara, near Uluru (Ayres Rock) in Central Australia; here the river red gum *(Eucalyptus camaldulensis)*, Casuarina *(Casuarina cunninghamii*, and trees adapted to root water in drylands form fuel forests surplus to settlement needs.

Such a fuel plantation needs two elements:

- A grid or matrix of perpetual long–term forest no more than 300 m apart, nor less than 8 trees wide, initially established by trickle irrigation but selected to be very hardy on rainfall alone once established.
- In the sheltered spaces so developed, a grid of closed–spaced trees (2–3 m) on trickle irrigation pipes (preferably on automatic) which are harvested in a 4–6 year rotation as coppice.

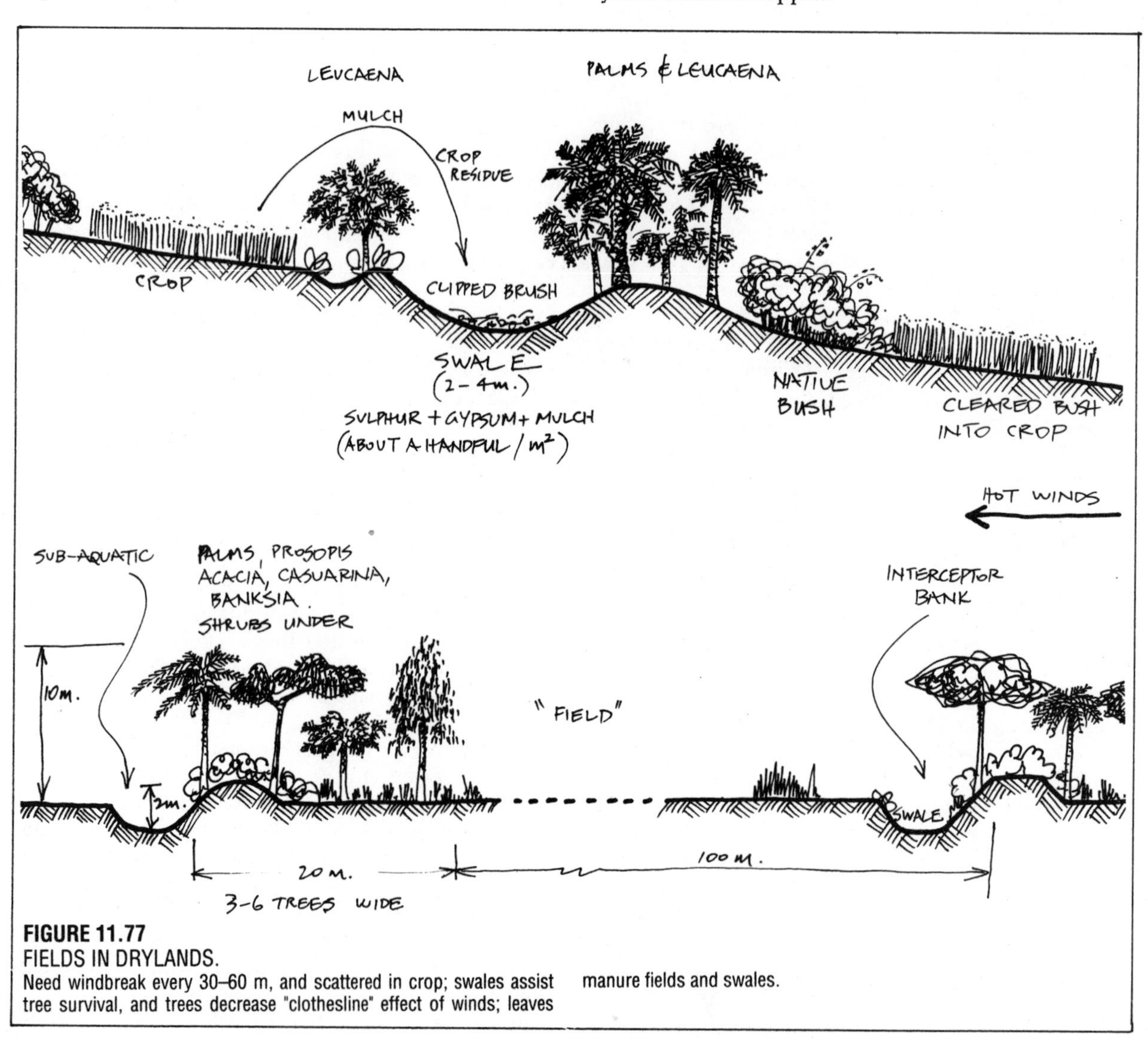

FIGURE 11.77
FIELDS IN DRYLANDS.
Need windbreak every 30–60 m, and scattered in crop; swales assist tree survival, and trees decrease "clothesline" effect of winds; leaves manure fields and swales.

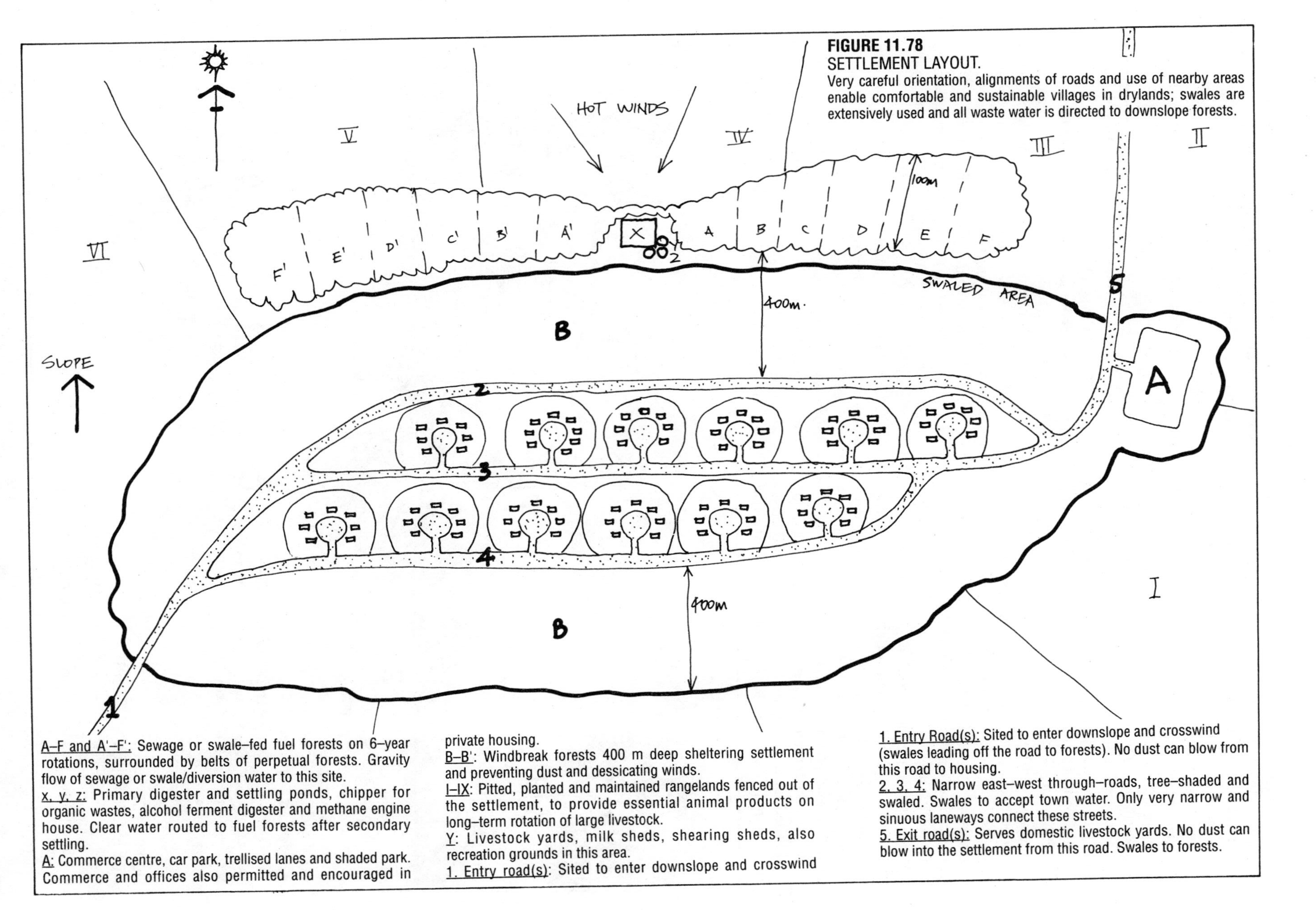

FIGURE 11.78
SETTLEMENT LAYOUT.
Very careful orientation, alignments of roads and use of nearby areas enable comfortable and sustainable villages in drylands; swales are extensively used and all waste water is directed to downslope forests.

A–F and A'–F': Sewage or swale–fed fuel forests on 6–year rotations, surrounded by belts of perpetual forests. Gravity flow of sewage or swale/diversion water to this site.
x, y, z: Primary digester and settling ponds, chipper for organic wastes, alcohol ferment digester and methane engine house. Clear water routed to fuel forests after secondary settling.
A: Commerce centre, car park, trellised lanes and shaded park. Commerce and offices also permitted and encouraged in private housing.
B–B': Windbreak forests 400 m deep sheltering settlement and preventing dust and dessicating winds.
I–IX: Pitted, planted and maintained rangelands fenced out of the settlement, to provide essential animal products on long–term rotation of large livestock.
Y: Livestock yards, milk sheds, shearing sheds, also recreation grounds in this area.
1. Entry road(s): Sited to enter downslope and crosswind

1. Entry Road(s): Sited to enter downslope and crosswind (swales leading off the road to forests). No dust can blow from this road to housing.
2, 3, 4: Narrow east–west through–roads, tree–shaded and swaled. Swales to accept town water. Only very narrow and sinuous laneways connect these streets.
5. Exit road(s): Serves domestic livestock yards. No dust can blow into the settlement from this road. Swales to forests.

All fuel tree species should be at first selected as coppice types, and continuing field selection of seed can then proceed for the varieties, sub–species, or provenance (locality) types that respond very well to this form of cultivation.

Keeping in mind the need to cut one–fourth or one–sixth of the crop annually, and having assessed solid fuel need per household (*and* governed the population level), such coppiced forest areas can provide all essential cooking and public energy needs; more so if the settling pond or tank is anaerobic and harvested as a methane source, and if all organic wastes are shredded and added to this tank. Much of the shredded organic waste will support ferment to alcohol fuels before being routed to biogas production. *No* organic matter can be wasted in deserts, and all can be turned to productive use.

Where sewage is not available, a series of close–spaced swales of from 4–8 m wide, the banks between them only 3–5 m wide and the swales supplied by extensive interceptor banks to harvest broadscale run–off will also support a fuel forest.

Within settlement and nearby, gardens can be established for food, and an area can be reserved (as an outer zone) for a 6 to 9 year rotation of needful domestic rangeland species supplying meat, wool, and other products can be established.

11.11
PLANT THEMES FOR DRYLANDS

TREE ESTABLISHMENT IN DESERTS

As well as pelleted and pit–trapped seed which awaits the rain, valuable fruit, forage, and seed–source trees need to be carefully established in dryland areas, partly as graft, bud, and seed sources for broader reafforestation. There are some essential precursers to success in this endeavour, some of which are:

- Plant in relatively cool periods, and check that soil temperature is not lethally high. Shallow plant roots can be cooked above 30°C.
- Supply mulch in quantity in pits near the plant and around its roots, or stone–mulch the tree root areas.
- Plant in a long swale or in sloping pits to collect, absorb, and retain moisture. For citrus, use larger pits with a central mound so that the stem graft is not mulch–covered.
- Plant a few gourds, legumes, or ground cover crop

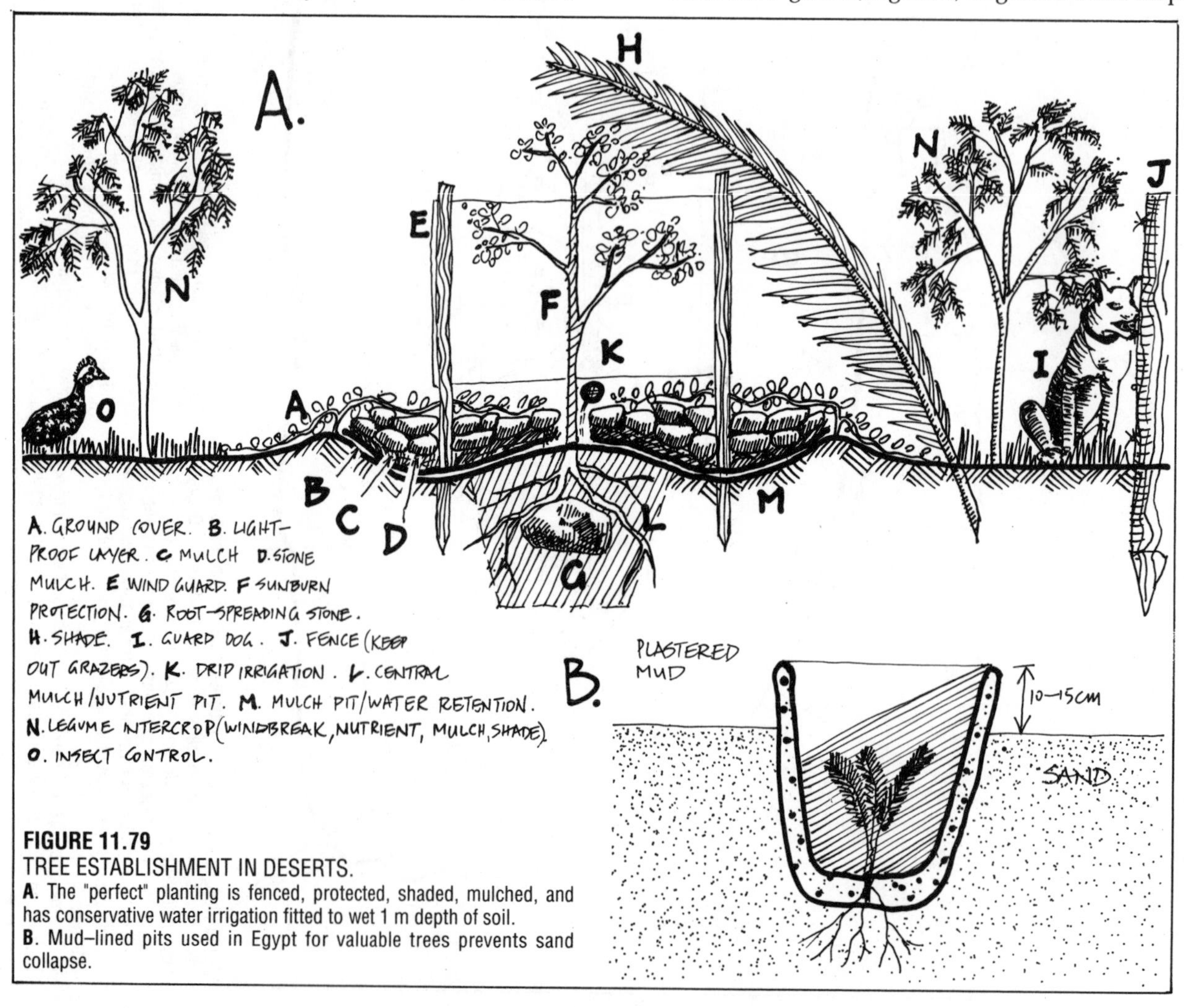

FIGURE 11.79
TREE ESTABLISHMENT IN DESERTS.
A. The "perfect" planting is fenced, protected, shaded, mulched, and has conservative water irrigation fitted to wet 1 m depth of soil.
B. Mud–lined pits used in Egypt for valuable trees prevents sand collapse.

around the tree to cool the root area.

• Supply water as drip for one to two years, or until the tree root area is self–shaded.

• Paint stems white or wrap them in foil to prevent sunburn before the bark has thickened.

• It helps to place a shade such as a palm frond or dead brush on a slant over the small tree.

• Fence the area, or shoot or poison rabbits, hares, and feral goats. They can wipe out a young plantation. Dogs keep all these away.

• Plant hardy tree legume intercrop to aid with wind, sun, fertiliser, and mulch.

Many of these features are summed up in one diagram, as in **Figure 11.79**.

In **Figure 11.80.B**, night condensation stays longer in the morning, enabling plants to use the moisture longer. In **Figure 11.80.A**, the morning sun evaporates night dews and plants suffer longer water deprivation. The heat itself is less a factor than water early in the day.

On Lanzarote in the Canary Islands, pits 8–10 m across and 1–3 m deep are dug, and one vine or tree planted in each. A cover of cinder is then carefully raked over the pit surface. The large pits act as night condensers, trapping cold night air and hence condensation. This method is also used for potato ridges, orchards in sloping country, and home gardens wherever cinder is plentiful. Over clay or loam soils, cinder allows rain in but prevents erosion and soil overheating. As cinder contains many air pockets or gas bubbles, it retains moisture from dews and condensation (**Figure 11.81**).

Apart from dune stability, palms and trees near groundwaters can be pit or slot–planted in a "mud planter" hole with effective avoidance of sandblast, water conservation (in the mud and the sand) and escape from the excessive heat of shallow sands (70°C).

This can be line, drip, or pot–irrigated; the mud (of course) must be locally available, and the whole process is worthwhile only if the trees are themselves of great value.

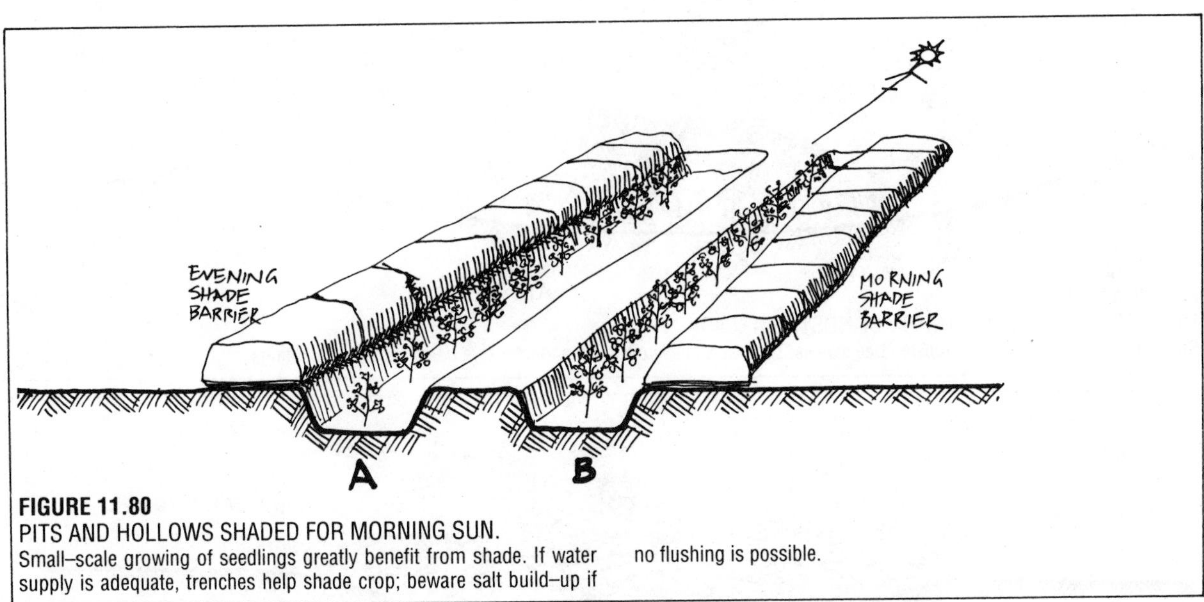

FIGURE 11.80
PITS AND HOLLOWS SHADED FOR MORNING SUN.
Small–scale growing of seedlings greatly benefit from shade. If water supply is adequate, trenches help shade crop; beware salt build–up if no flushing is possible.

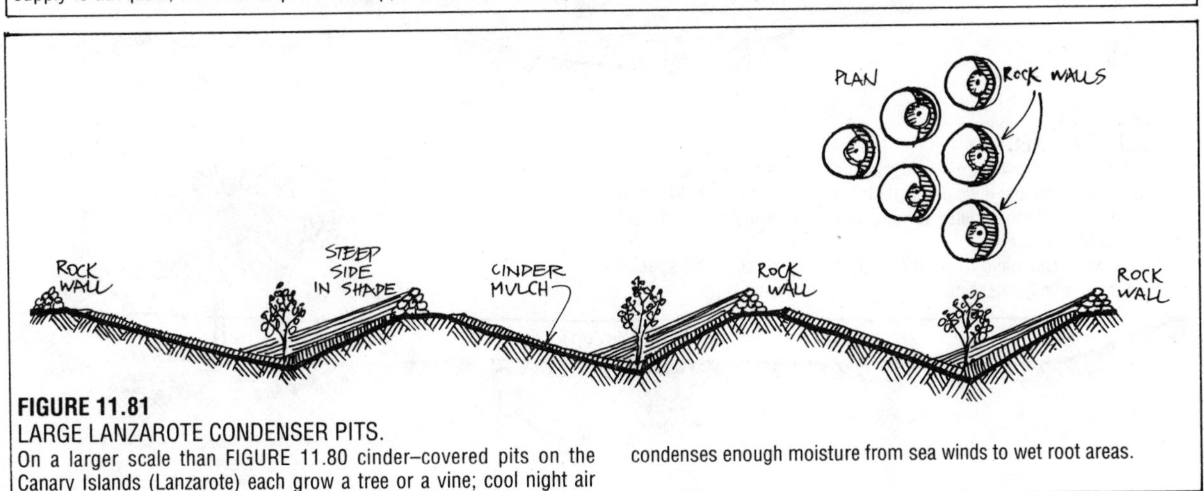

FIGURE 11.81
LARGE LANZAROTE CONDENSER PITS.
On a larger scale than FIGURE 11.80 cinder–covered pits on the Canary Islands (Lanzarote) each grow a tree or a vine; cool night air condenses enough moisture from sea winds to wet root areas.

SPECIAL PREPARATION OF SOILS FOR TREE PLANTING

Deep–cracking clay soils will leave small tree roots in the air, and the tree will die. It is better to collect a modest amount of sand from river beds, wait until rain is expected at the end of a dry season, and having soaked the site, or run rip lines across it, set out the small trees in sand poured into the clay cracks. Gypsum aids root penetration in such clays, which are sometimes acid in the plains areas. For hardy pioneer species, pelleted seed and sand in clay cracks suffices (tagasaste, *Acacias*, pioneer legumes).

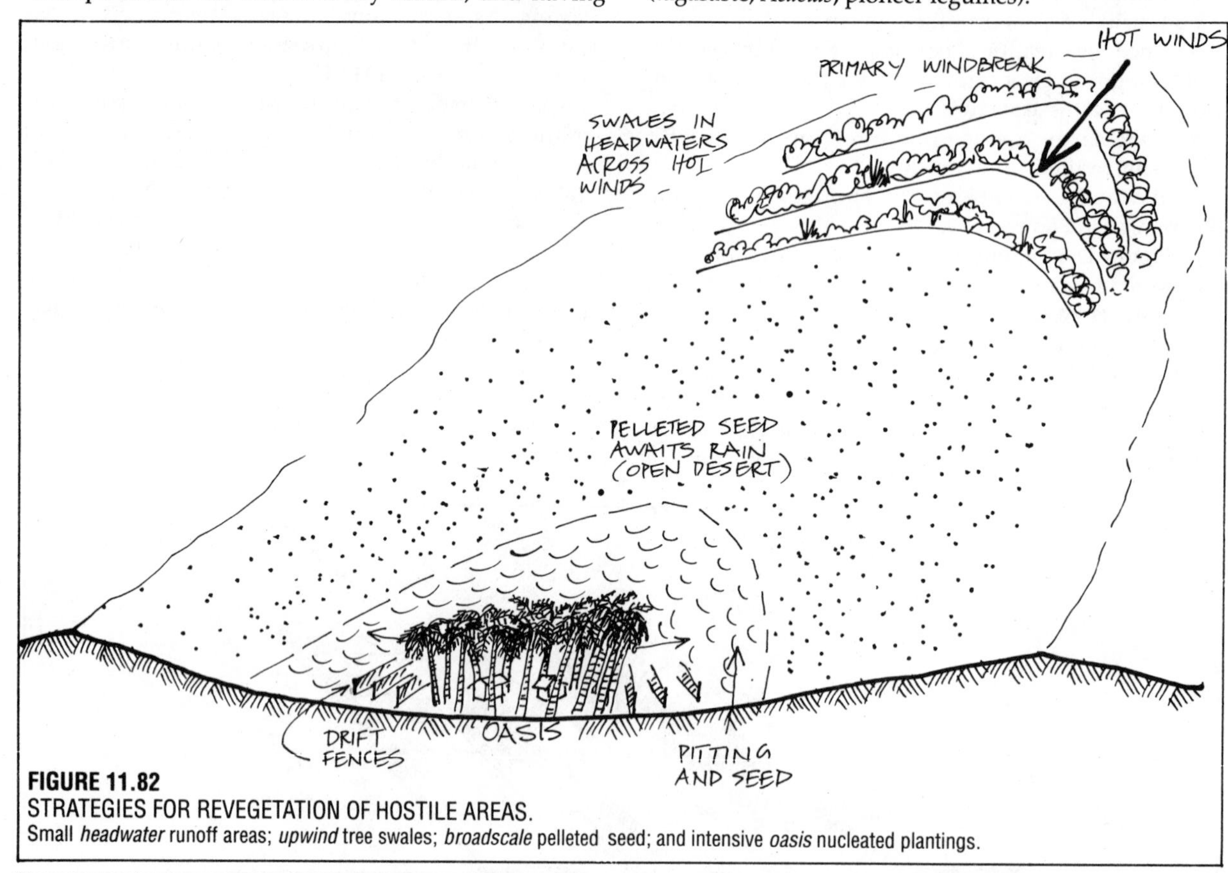

FIGURE 11.82
STRATEGIES FOR REVEGETATION OF HOSTILE AREAS.
Small *headwater* runoff areas; *upwind* tree swales; *broadscale* pelleted seed; and intensive *oasis* nucleated plantings.

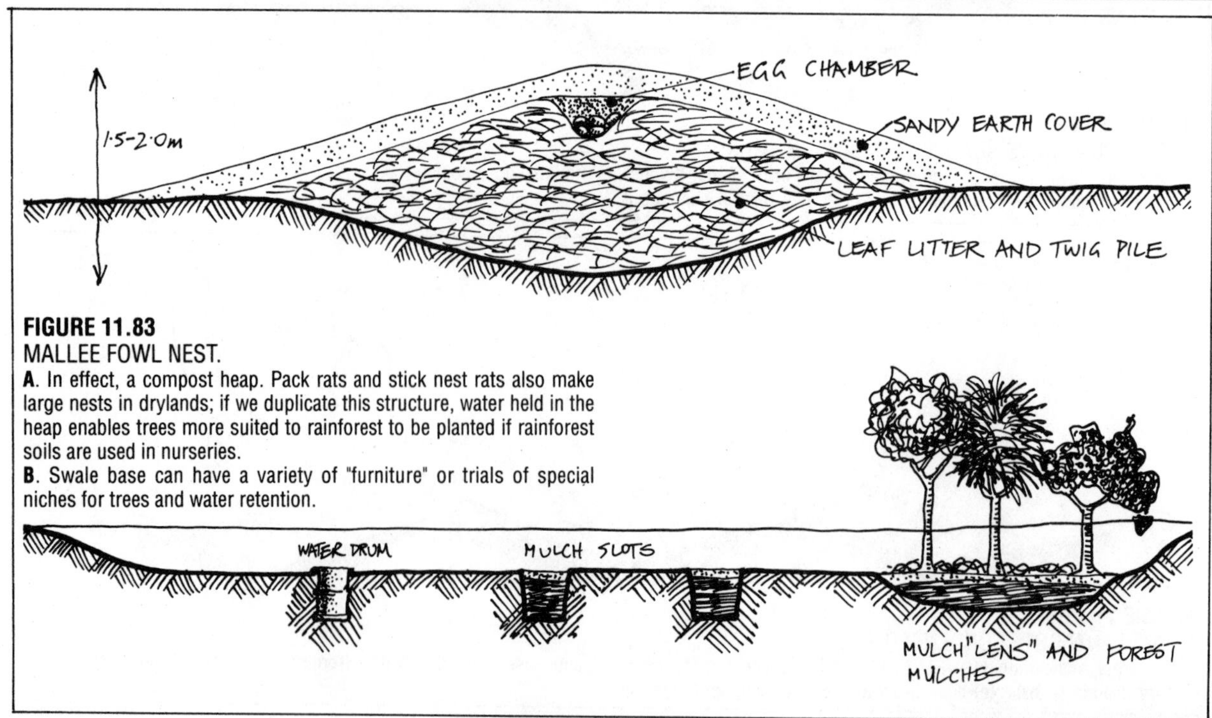

FIGURE 11.83
MALLEE FOWL NEST.
A. In effect, a compost heap. Pack rats and stick nest rats also make large nests in drylands; if we duplicate this structure, water held in the heap enables trees more suited to rainforest to be planted if rainforest soils are used in nurseries.
B. Swale base can have a variety of "furniture" or trials of special niches for trees and water retention.

In the calcrete soils of coasts and islands, the concreted layer must be broken open at every important tree site (as it is with coconut and date crop) and a mulch pit (plus a scatter of elemental sulphur and mineral elements) prepared for the tree. A few fibrous–rooted species will sit on calcrete, but they are then at risk from drought and windthrow. Tree growth keeps the *platin* open and cracked for interplant. A fast way to do this is with an auger and about one–third of a plug of dynamite. The same technique can be used to place fence posts, or to shatter concretions for tree holes in shales and mudstones.

Sometimes a bulldozer is available, and rip–lines can be made for the tree lines. In this case companion crop of small legumes can be seeded between trees along the rip. Desert fenugreek, lucerne, tagasaste, and gourds can be intercropped with palms, *Casuarina*, or jujube.

Just as gypsum helps roots to penetrate clay, bentonite assists sands to hold moisture. Dried seaweed crumble added to planting holes forms a gel in rain, and enables the seedlings to penetrate to deeper levels in the first season. Commercial soil gel additives are also available for adding to the soils of potted plants in the nursery, and some of these function for many dry–wet cycles in field conditions.

THE REVEGETATION OF HOSTILE AREAS

It is certain that we will need to reclaim dry, salted, deflected, and pest–invaded areas in the course of developing a permaculture. Let us return to the practical experience of people who try to re–establish native bush on disturbed areas covered with weeds. The lesson is to start with *small* nucleii and to gradu-

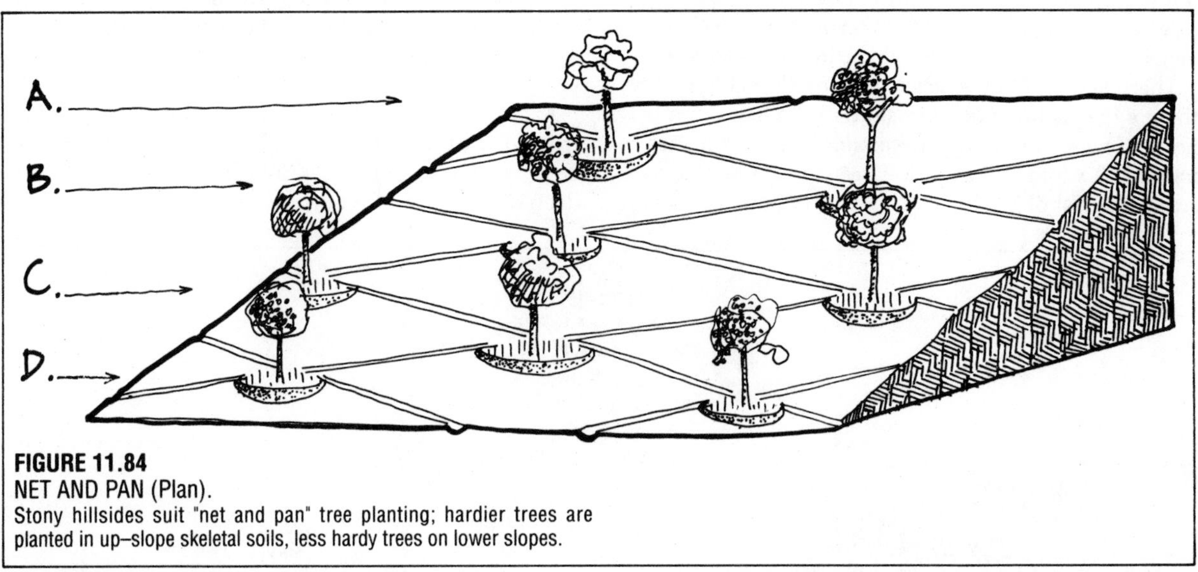

FIGURE 11.84
NET AND PAN (Plan).
Stony hillsides suit "net and pan" tree planting; hardier trees are planted in up–slope skeletal soils, less hardy trees on lower slopes.

A. Crest trees: hardy needle–leaf species and narrow–leaf trees to suit thin soils, e.g. stone pine, olive, *Casuarina*, *Callitris*, *Acacia*, quandong.

B. Hardy trees with known drought resistance, e.g. fig, pomegranate, *Acacia*.

C–D. Midslope and deeper soils suited to citrus, fig, *Acacia*, pistachio.

E–F. Deep base soils with some humus suited to chestnut, mulberry, raintree, citrus.

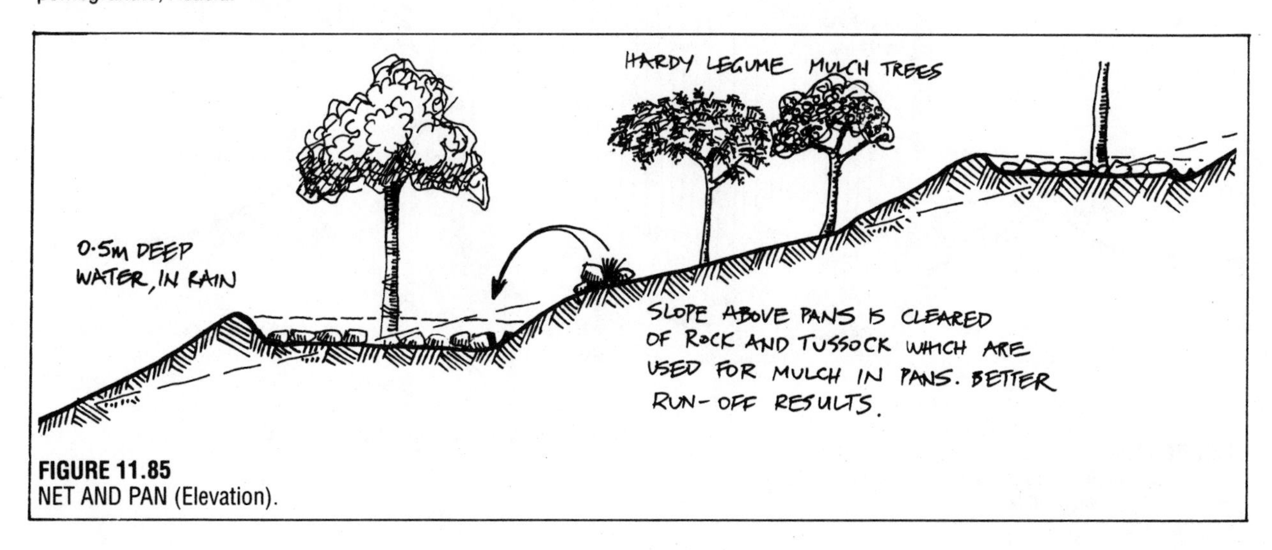

FIGURE 11.85
NET AND PAN (Elevation).

ally expand the perimeter, mulching and returning wastes as we go, and keeping the new system undamaged. Broadscale desert strategies also involve revegetating the upwind and upstream areas, and seeding ahead in readiness for favourable seasons (floods in deserts, salting, fire in grassland, etc.) All strategies are combined in **Figure 11.82**.

CREATING A FOREST IN DRYLANDS

The brush turkey *(Alechura lathami)* in Australia compacts a heap of rain–wet vegetation in an excavated pit, covers it, and allows a female to approach and lay eggs when the mound temperature is 35°C (95°F). The male attends the nest and adds material, or opens cooling vents, to incubate the eggs. A nest section (**Figure 11.83**) is in fact a large inderground compost pile, steady at about 30°C (Von Fritsch, 1975). All megapodes (big–foot) birds behave in some such way.

There are about 20 species of birds that either bury their eggs in hot volcanic sands or make compost heaps. The scrub turkey *(Megapodius freycinet)* makes a mound 12 x 5 m and the arid–area mallee fowl *(Leipoa ocellata)* a nest of 5–6 m x 1–2 m, deeply buried in sand. All of these "compost heaps" present an unusual site for tree species to locate.

In the dry savannahs of Central York Peninsula (Australia) is a rainforest clump about a mile across; at its centre is the compost–mound nest of a megapode (the scrub turkey), which is carbon–dated to about 5000 years old. The rainforest is oldest near the nest and younger at the periphery. We can speculate that the nest provided the soil nutrients for rainforest pioneers dropped by fruit pigeons on migration. In abandoned sheep stockyards near Ernabella (Central Australian desert), a leaking windmill provided some water to a 1 m deep pile of manure. In this stockyard a dense and tall "wet sclerophyll" forest structure is apparent. The desert was once forest; can it be re–forested? I believe so, and this suggests a way to try it out. The process could use the following strategies:

- Grade wide hollows (5–10 m) on contour from outwash (wadi) runnels, sloping these into the hill, and leaving 30–100 m bare (or desert) strips between each swale.
- At intervals of 10–20 m in the swales, tip a truckload of mixed chips, logs, manure, and straw or crop wastes in hollows. Cover these with 1.0–1.5 m deep of sand. Wait for rain.

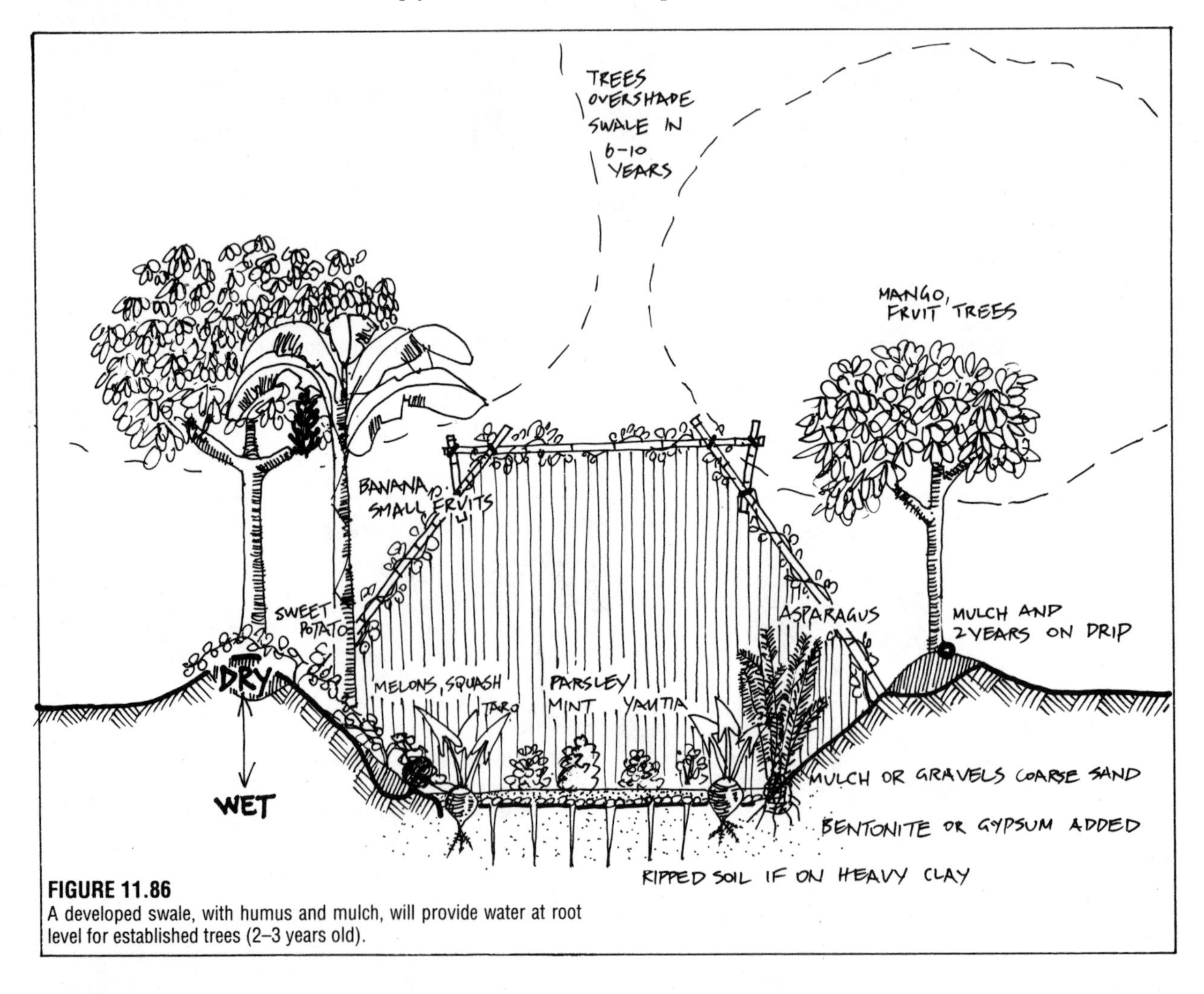

FIGURE 11.86
A developed swale, with humus and mulch, will provide water at root level for established trees (2–3 years old).

• In a nursery nearby, raise a few thousand pioneer legume and nurse crop trees in large pots and containers. All should have rich rainforest soil mixtures or inoculants.

• After a rain, when the swales are just drying out, plant the pioneer crop into the buried manure–chip areas, or around their periphery.

• When (and if) the pioneers grow, plant second–stage forest or very productive tree crop in the pioneers, again after a rain.

• If successful, slash the pioneers to let the rainforest evolve.

The inter–swale desert run–off (supposing 8–20 cm of run–off) should supply each swale with an effective 50–300 cm of water every year. The deep mulch should hold this water at tree roots, and the forest soil provide a mycelial web for roots. I see no reason why forests should not re–establish, and self–perpetuate, under such regimes.

PLANTING TREES ON HARD SOILS, SLOPES, AND MINOR SYSTEMS

High on the steeper slopes of fold mountains, small runnels feed the second and third–order streams. While the slope may be too steep for machines, and of restricted area, there are a few modest systems possible to establish trees.

One system we can characterise as "net and pan" for sheet run–off, and the other as a "boomerang pattern" for absorbing the flow from active runnels. Either can be made by hand or machine.

Figures 11.84 (plan) and **Figure 11.85** (elevation) Net and Pan. Sheet run–off absorption systems. Runnels cut at gentle slope of 1:500 (exaggerated in the diagram).

On a more irregular level, small runnels over hard ground can be blocked using a log, some large stones, and a bundle of straw or spinifex weighted down with these. The grasses trap silt, and leaves and sand soon builds a small delta of detritus into which a hardy *Acacia* or shrub will take hold. A little fertiliser often applied helps to build the system back to natural mulch. The tree roots then become our silt traps.

Figure 11.87 Runnel traps to build silt deltas on bare, hard soils. Used successfully on mined–out land and lateritic soils.

Where larger rills flow, "boomerangs" disperse and absorb flow (**Figure 11.88**).

Figure 11.59 shows various types of tree establishment techniques in difficult soils.

RECRUITMENT

Some of the large trees of the desert live for 200–4000 years. Once we have released animals (feral or domestic) that eat off seedlings—rabbits and hares are devastators, as are sheep, goats, and cattle—we may see no more signs of regeneration, and the trees that are left become less vigorous with age. Likewise, fire at increased frequency in any one area may destroy young or small plants before they seed.

Recruitment (the evolution of a new generation of plants to adult status) depends on an interaction between an adequate rainy season, fire, browsing, and a source of seed. Light fires in spring (winter–wet deserts) may generate new trees, while either cutting, browsing, or fire in early winter can kill out trees, as young shoots suffer from the colder conditions.

Seeds may germinate and plants grow, after rains, but it may need a second period of rain to establish the seedlings (about 6 weeks later). Thus, to reforest desert we need to remove or greatly reduce browsing pressure, to provide seed where none exists, to reduce autumn grazing and fire (to burn lightly in spring if necessary), and to provide some key nutrients (usually phosphates). And as well as all this, to hope for a good rainy season, or a succession of rains. All this may come together every 9–20 years, so that to keep some light stocking on deserts, we would need to have at least

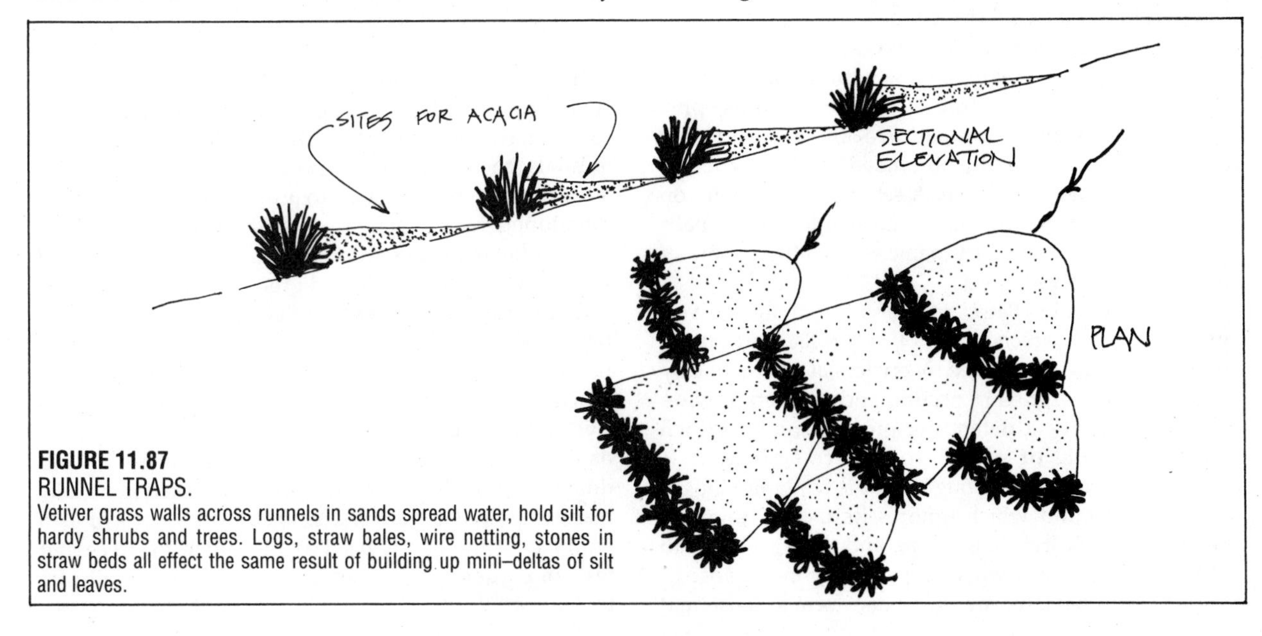

FIGURE 11.87
RUNNEL TRAPS.
Vetiver grass walls across runnels in sands spread water, hold silt for hardy shrubs and trees. Logs, straw bales, wire netting, stones in straw beds all effect the same result of building up mini–deltas of silt and leaves.

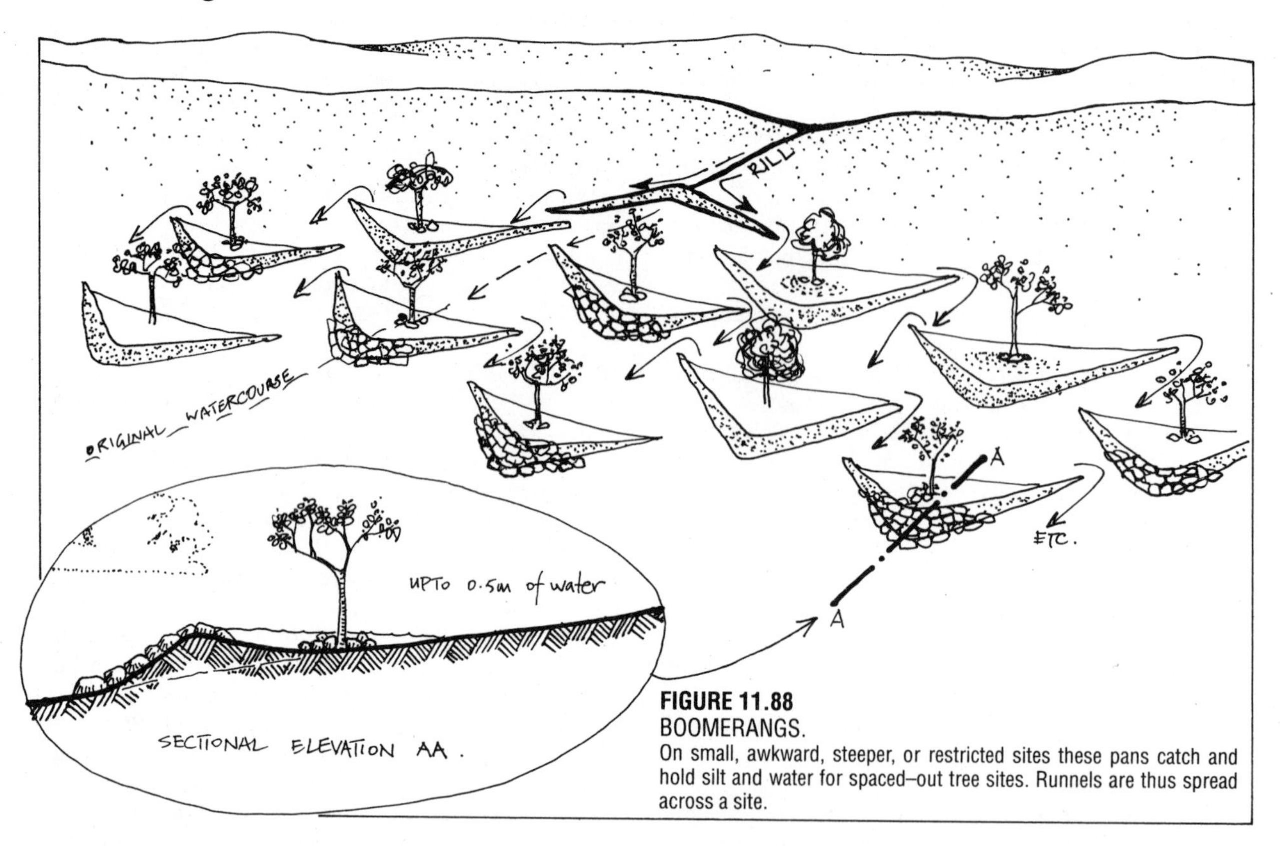

FIGURE 11.88
BOOMERANGS.
On small, awkward, steeper, or restricted sites these pans catch and hold silt and water for spaced–out tree sites. Runnels are thus spread across a site.

one–fourth (some say one–seventh) of the area free of stock at any one time, and to let the area regenerate if rains occur, perhaps assisting with a spring fire, pelleted seed, or broadscale water harvesting.

It is also possible to set up core areas of more carefully tended trees that can expand at their perimeter if seasons permit, which supply seed, and where reforestation trials (and fire trials) can be controlled.

WRAITHS AND GOLEMS

Whole plants, seed–heads, leaves, dried manurial pellets, seed pods, and dust blow with the wind across desert plains, and are trapped in depressions, pits, swales, and against fences and tree–lines. Brush fences can be buried by drift, and need to be vertically extended if sand is unstable. A set of plants depend on the wind for dispersal, seen as substantial rolling balls (golems) or dry, light, airy panicles (wraiths). Some of the species involved are:

Black roly–poly (*Bassia quinquecuspi*), a short–lived perennial shrub which breaks off and rolls, distributing the spiny seeds. A nuisance to graziers, it occurs thickly only where overgrazing occurs, and protects a wide range of soils as a defending pioneer, under which more palatable or useful species can grow. It is "noxious", of course, although what is really noxious is the state of mind which ignores its useful function. Several copperburrs (*Bassis spp*) and salt bushes try to repair scalds (areas where topsoil is gone). Most of these species have thorny seeds, and form a dense mat under rubber sandals, which must be discarded as the seeds get everywhere!

Tumbleweed (*Amaranthus albus*) is a more effete invader, a stiff wraith preferring towns and settled areas, rarely abundant. *A. viridis* (cooked as a vegetable) is also found in arid–area towns as a "weed". A mustard (*Sisymbrium officinale*), of some forage and medicinal value, is usually found only in cultivated (wheat) areas, and also spreads its dry seed stems as a wraith in dust–storms.

ARID–AREA GRASSES AND FORBS

A very careful selection of perennial and annual grasses can greatly increase the number and condition of animals on range. There are good grasses for every situation from sandseas to gilgais (natural swales developed from pockets of expanding clays). They cover a range of uses from human food to poultry and waterfowl forage, thatch, and green forage or hay.

Coarse tussock grasses become unpalatable if left unbrowsed (when dry material accumulates), or at flowering and after, when the food value is low. Thus, mown or managed swale fields are the most productive. Like all green crop, leaf material may contain dangerous levels of nitrates (dangerous to people and domestic animals) if over–manured or grown in heavily manure–polluted waters. This factor does not so much worry us if the hay is used as pit mulch for trees, but tests of garden greens should be made periodically for this factor.

Thus, I would recommend a carefully-selected sowing of some of the best grasses in plots (they can be pot-grown in the first place, and divided), and for quite specific purposes, such as to:

- Reclaim claypans and saltpans.
- Filter out silt and nitrates in diversion drains.
- Provide green hay for domestic stock.
- Give durable thatch and cane fences, mats, and screens or baskets.
- Provide bundles of seed-head for poultry, and to feed wild birds.
- Create a grassed water-way for downhill sheet flow.
- Provide a seed resource for broadscale work.

On the broadscale, there is no substitute for small, intensively managed perennial grass plots, fenced or protected to avoid overgrazing, or managed and encouraged to provide seasonal fodders. Range management is a real skill, and relies on good plant identification and management, a keen eye for animal behaviour, and a modest but sustainable stocking rate.

DESERT AQUATIC AND SWAMP SPECIES

Several aquatic and swamp plants are recorded for deserts due in part to exotic rivers that flow into deserts, and in part to run-off from ranges. There are even more species adapted to the claypans and swales (gilgais) or run-off hollows, where they grow and tolerate very hot and somewhat brackish or salty water.

There are thousands of sites in deserts where ponds and dams discharge into interior flats or flood-outs, some of which may be below sea level (as in the Dead Sea and Lake Eyre, South Australia). Thus, the desert is, in a sense, like a series of small aquatic islands, with no danger of species escaping to infest permanent streams. There are therefore ideal sites for small aquacultural assemblies and experiments.

All growing plants use up salts in growth, and some dense algae will desalinate large volumes of water, while many rushes and weeds remove dissolved salts or unwanted pollutants. It is quite feasible and sensible to set up part-enclosed and part-open natural water filtration systems to remove faecal, nitrogenous, and metallic pollutants from town water supplies, and a model of such a system is given here (**Figure 11.90**). In third-world deserts, some such benign system may greatly reduce transmissible disease, while producing useful forests and biogas materials. In addition, several aquatics provide food and wildlife shelter.

11.12 ANIMAL SYSTEMS IN ARID AREAS

Small livestock, especially chickens, quail, guinea fowl, guinea pigs, even ducks and geese all do well in deserts (however, it is often too hot for rabbits in cages). They need dense shade shelters and access to shallow clean water. They perform a multitude of tasks, especially in the reduction of insects and snails, and provide eggs and small meats not needing cool storage. On range, there are a great many desert seed and grain crops for them to forage, as well as garden and fallen fruit scraps. They will keep termites out of grain storage bins by breaking open the mud tunnels if such bins are supported on rocks.

On the wider range, highly selected and *controlled* meat and milk flocks (sheep, a *few* goats, a *few* cattle, donkeys, and camels) also thrive if they can be herded or penned in 15 or so rotated runs, allowing 2–8 years for each run to recover and re-seed. Some successful Australian sheep graziers allow 7–9 years rest per run, and never suffer the animal stress brought on by drought. It is infinitely preferable to run *small, high-value herds on copious range* than to risk the inevitable collapse of range and flocks by stressing the vegetation.

Meats can be dried in screened and shaded air cupboards, and part-salted or smoked. Strips of meat

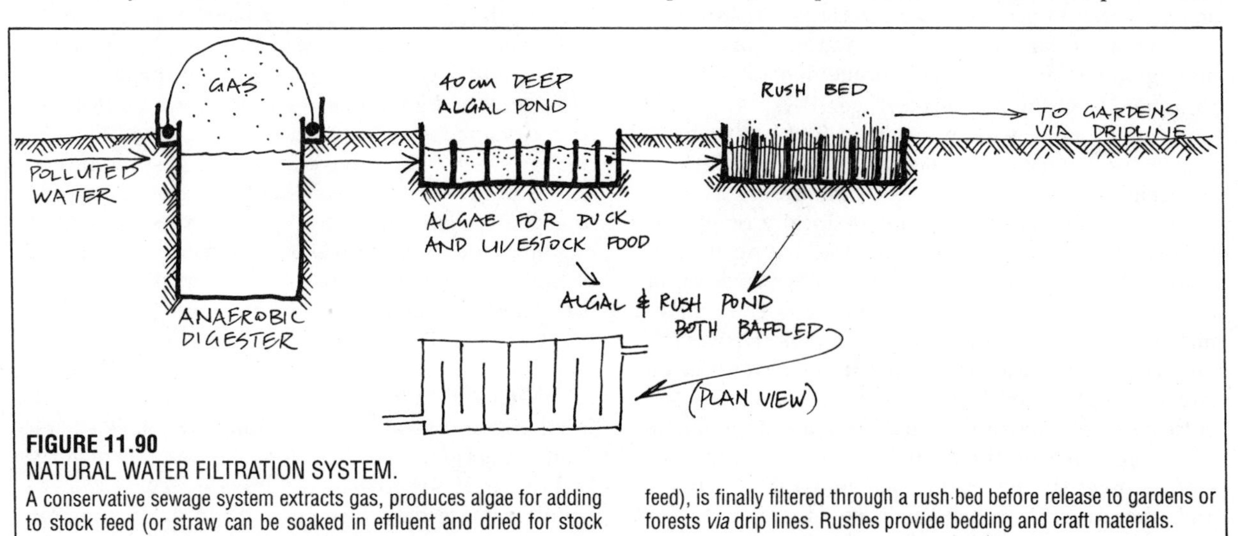

FIGURE 11.90
NATURAL WATER FILTRATION SYSTEM.
A conservative sewage system extracts gas, produces algae for adding to stock feed (or straw can be soaked in effluent and dried for stock feed), is finally filtered through a rush bed before release to gardens or forests *via* drip lines. Rushes provide bedding and craft materials.

beaten with pepper and garlic to 1 cm or less thick and 4 cm wide (as long as possible) dry very quickly and "glaze" in a few hours. This "jerky" keeps indefinitely if stored in insect–proof fine–mesh bags hung from ceilings. Complete drying can take 3–5 days.

One of the sources of animal products in deserts is managed wildlife systems. Here, we can differentiate between resident and migratory species:

- IRRUPTIVE AND NOMADIC SPECIES are large and fairly fast–moving animals such as kangaroo, antelope, and ostrich. They are able to travel long distances to areas of rain—they "smell the wind" and some wet earth and vegetation. There is also a vacuum effect in that as some move to rain, their neighbours find the old areas empty, and move towards "low stress" areas, eventually finding the new growth. Those are the herds from which many *young* animals can be culled—they die in the dry times. Healthy adults need to be preserved.
- SEDENTARY SPECIES are smaller, slower, or more specialised, adapted to drought. They will reside in an area at all times. Their "limiting factor" may be WATER HOLES (for quail and ground birds), SHADE (for large lizards and surface mammals), BURROW SITES (for small mammals and lizards), or FORAGE adapted to dry, salt, or wet areas. Thus, the strategies of small water "ramp dams" (**Figure 11.10**), rock piles and small shaded caves or cliff sites, bulldozed piles of loose earth for burrows, or key plants provided for hungry gaps may result in a local, dependable increase in those stable and staple species.
- PLAGUE SPECIES may lie in either category (rabbits, grasshoppers) but also present a potential for harvest and drying (rabbits) or conversion (grasshoppers) via a domesticated predator (chickens, guinea fowl). We tend to forget that chickens, ducks, fish, and guinea fowl will convert most insects to food and high–value manure, and we need to set up some breeding or attraction systems like the termite breeder to take advantage of this (**Figure 13.32**).
- REPTILES. Large lizards are the "fish" of the desert and are carefully harvested by Aboriginal Australians. They are a constant source of food and need only good management (and some encouragement of insects, fruits, and snails) to do well on range.

In humid areas, we can expect to find most animal species sedentary, or at least of limited and strictly seasonal movements, whereas in the desert we find many animals may become either locally or opportunistically nomadic, with people adjusting to this. However, there are many smaller species which are of necessity sedentary, and those choose "steady state" niches, such as the large grubs of moths (*Hepialidae, Cossidae, Cerambicidae*) which have in the past formed a large part of the reliable foods of people, and are in fact both palatable and nutritious. They are dependable foods because of their humid niche in the arid environment (living in the bark and roots of trees) and—because several species live as grubs for 7–8 years, and have only brief lives as adults—they are always to be found.

It is the same with frogs (*Hylidae, Leptodactylidae*) which also persist in resting states for up to 8 years, and who have fat bodies for their own nutrition, and for their predators or gatherers.

The nomadic tribes of birds and larger mammals provide a "feast and famine" resource depending on rains. At times they retire to refuges or more humid areas to survive long droughts. Such refuges may be "tabu" areas for long–term management of the species, so that they persist as food potential in the long term. These need more skilled and cooperative management, as does any herding system.

Boundaries such as fences make less sense in deserts than elsewhere because of the need of mammals to migrate, and to fall back on reserve areas in hard times. So we find that typical Aboriginal Australian diet mght contain 6–10 insect larvae, 10–12 reptile species, 6–12 birds (some taken in moult), 3–4 large migratory mammals, 9–10 smaller sedentary mammals, 5–8 frog species (mainly the females with large ovaries), the adults of some insects, some aquatic crustaceans and molluscs (3–4 species), and fish (depending on the area, 4–8 species). In all, 48–66 animal foods, and at least that many plants, including aquatics from temporary lagoons, are eaten.

SPECIES MIXES IN VEGETATION

Just as freely nomadic animal species follow mutually beneficial successions in savannahs, so very different species use the desert in a non–competitive and probably complementary way, with some opportunistic overlap. An example is given in **Table 11.5**.

Thus, both species eat dry or green forage not much appreciated by the other at that time. Almost certainly, the eating of coarse dry grasses by kangaroo enable cattle to browse more easily in wet periods (fire also helps). However, kangaroo eat very few grass species compared with cattle, who are less selective in food preference. The cattle prefer annual grasses.

Much the same sort of findings are true for euro (a kangaroo) and sheep. As both euro and kangaroo are valid yields, the total yields are improved with admixtures and good management. Kangaroo meat has been devalued, but is in fact a superior food to either beef or sheep, which have 38–42% digestible protein and saturated fats, whereas kangaroo has 58% digestible protein and low fats, mainly unsaturated. If anything, such situations should be managed in favour of efficient conversion rather than on the basis of a "pre–selected" market, which in any case is failing.

LIVESTOCK IN DROUGHT

In all arid and semi–arid areas, large livestock such as draft animals and milking cattle and buffalo are at risk in drought. If no provison for drought feeding is made, small farmers may lose their basic draft and milk animals, or sell them cheaply for slaughter. From 2–4 of

TABLE 11.5
FROM CSIRO RANGELANDS RESEARCH, BILL LOW, 1979.

CONDITION	TREE STANDS	STEPPE	DEPRESSIONS
	Acacia stands	Open forbes and grasses	Grassy depressions
A **Generally cold** (Wet season)	Kangaroo scattered in small groups of three or less.	Cattle in large groups, group close together.	Cattle in large groups.
B **Generally Dry**	Cattle scattered in small groups.	Kangaroo in large mobs near refuges.	Kangaroo in large mobs.
C **Local storms over grass depressions** (Dry season)	As for B if no local rain.	As for B if no local rain.	Kangaroo and cattle drift in groups for green crop. Some grasses eaten by kangaroo, most by cattle.

these animals are kept by most farmers, and about 17–30 can be fed on a permanent hectare of cut–and–fed forage, whereas free–range animals take from 1–5 ha to browse food in average conditions (up to 40–60 ha in drought and deserts).

Thus, 6–8 farmers need one hectare of emergency forage for drought. Such a survival forage plot needs careful planning. Essentials are:

- A frond–roofed shed for shade, where 15–30 animals can be penned. The floor should be supplied with a mulch of fronds and hard straws from sugar cane, *Pennisetum* grasses, or palms.
- Up to one hectare of perennial forage. This forage must be cut daily and fed as one–third to one–half of the ration. A forage planting layout is shown in **Figure 11.91.A**. Species include large trees such as *Inga* and carob, honey locust, *Prosopis* (for its pods), tagasaste, *Acacia, Glyricidia,* and *Leucaena* for coppice forage, arrowroot (*Canna*), comfrey, and *Pennisetum* for cut forage.
- A careful ground–plan of multiple cross–slope swales to catch and infiltrate run–off water in rains. This is a critical precursor to planting the forages. **Figure 11.91.B**.
- As well, all adjoining fields should be edged and wind–breaked with the same forage species, planted at 20–30 m intervals in rows throughout all other crop, on bunds, and along swales and ditches. The basic survival hectare can be cut and managed in good years, but in bad drought years, all choice or essential livestock need to be penned in or near this forage system for survival feeding. As no crops can be planted in drought, farmers and their families can tend these cattle on rotation.

In drought, cattle can be fed on chopped dry stalk material, small branches, straw, crushed cane, and even cardboard or paper *providing* they have access to a lick of molasses with a little urea added. Such licks are handily made from a petrol drum floating in a half–drum bath of molasses–urea. As the cows lick the floating drum, it revolves and picks up more molasses–urea mixture.

It is the urea–molasses mixture plus high–cellulose cheap bulk food that enables the cattle to break down some of the cellulose in the wood and straw. The rest of the ration is provided from the perennial forages. These are cut in succession and carried to the pen, and all manure and bedding is carried back to the forage fields as mulch, preferably deposited in the swales. This mulch develops cool humus soils with good water capacity over time, and the forage plants thrive on this humus.

DANGERS TO LIVESTOCK ON RANGE FOLLOWING RAINS

Both woody and ephemeral plants in drylands may concentrate any of several toxic substances as they start into new growth after rains. A partial list of these includes nitrates, oxalic acids, cyanides, sodium fluro–acetate, and poisonous alkaloids that cause infertility, liver destruction, spasm, and eventually death in wild and domestic browsers. It is the way dryland plants protect themselves from browsers when they put on new growth. The effect last from 4–6 weeks after rains, and livestock should not be released to range that contains only new growth, at least not for long periods on any one plant stand.

Mature leaf is usually non–toxic, so that the same plants that are toxic in new growth may be good browse in maturity. Cattle on such range must be very carefully herded, as any sudden shock or running will cause death. They must be allowed to move at their own pace, or left still if obviously sick. Dogs, whips, horses, and cowboy tactics generally can be fatal to livestock in these conditions, and any distant droving should be left until forages mature.

Good husbandry demands that livestock are managed in new growth periods to have a wide range of foods, some cut forage available, or some mature leaf from trees fed out.

As browsing is equivalent to severe pruning or coppicing, and wildfire removes nutrients and causes stress in regrowth plants, burnt and browsed plant stands may secrete toxins for some time after such

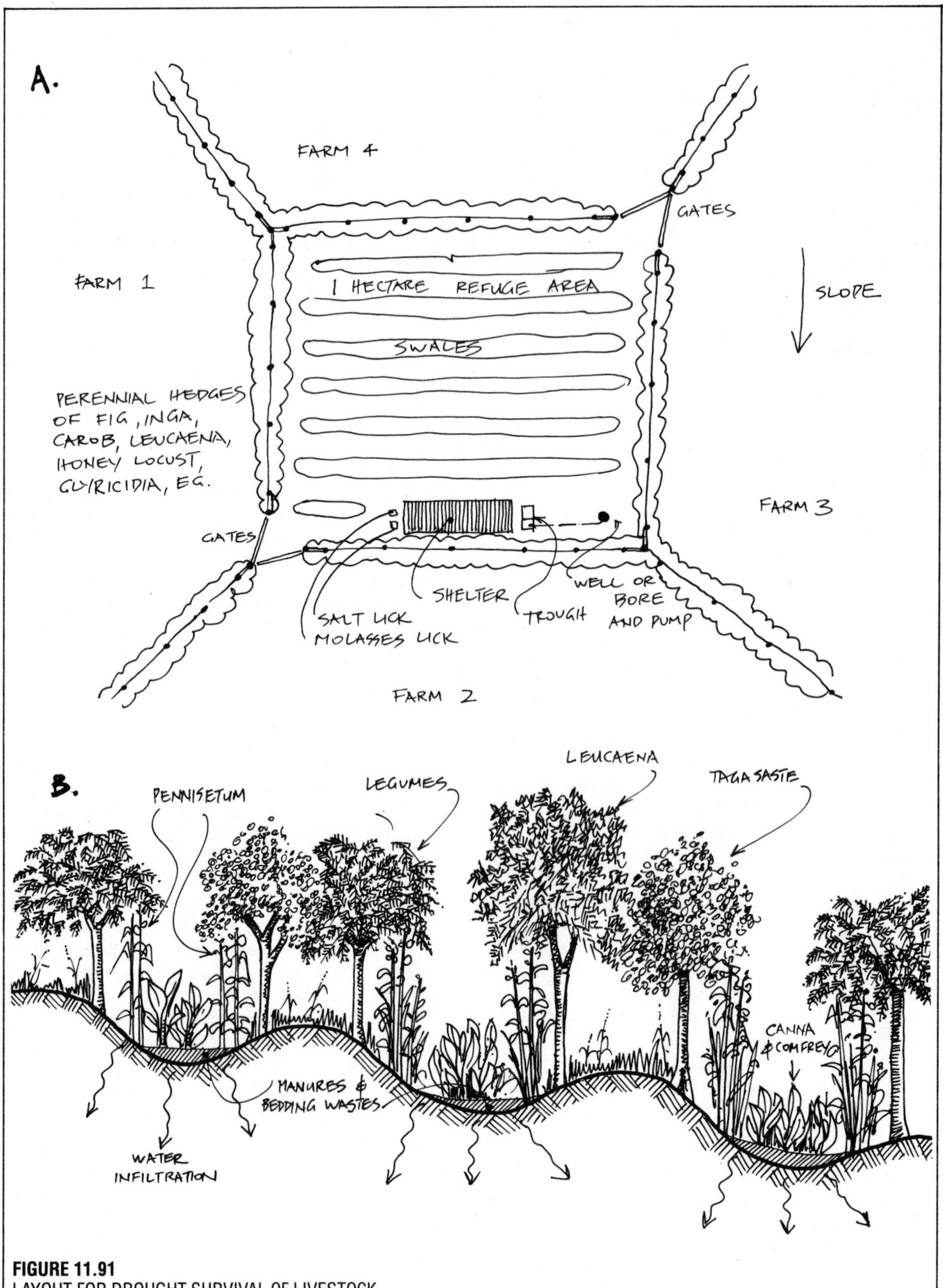

FIGURE 11.91
LAYOUT FOR DROUGHT SURVIVAL OF LIVESTOCK.

A. Thatched shelter, water, and molasses–urea lick, swale forage trees (coppiced), cheap straw or even paper feed, selected stock for survival, reserved area of 1 ha in 20 all enable small farmers to keep and milk animals alive through drought. Animals are kept quiet, bedded, not moved about.

B. Detail of reserved swale forage planting for emergency stock survival.

stresses. This plant response is linked to the other cyclic factors that cause hare numbers to sharply decline on over–browsed range. In hot deserts, fire and browsing, coppicing and firewood gathering can produce such poisonous metabolites in a wide range of both woody and ephemeral plants (Bryant, J. P, 1981, *Natural History* 90[II]).

11.13

DESERTIFICATION AND THE SALTING OF SOILS

DESERTIFICATION

Le Huerou, H.N., 1968 in *La Desertification du Sahara* (Int. Biol. Prog. Sect. C. T. Colloque Hammamet, London) gives the following causes of desertifi-cation, independently of any long–term or cyclic climatic change:

• OVERGRAZING, assisted by well–drilling to maintain higher stock numbers. Herds can not now adjust to climatic factors, and can be held at high numbers until well into drought periods.

• OVERGRAZING; stressing the vegetation beyond recovery, especially near wells.

• FUEL USE; removing timbers cut but not replanted (fencing also takes thousands of trees). Frequent burning.

• SETTLING OF NOMADS local grazing and fuel–cutting stress causes a zone of destruction, usually centred on wells; no regeneration is possible as people are no longer nomadic.

• CROPPING, often as a result of a failing herd support and in response to population increase. This extends down to 15 cm rainfall, wheras 40 cm is accepted as the lower limit for dependable cropping.

• EXTENSION OF CLEARING, following on low crop returns, thus less fallow; this greatly increases soil compaction and topsoil loss.

• EXCESSIVE WATER USE in modern times only; due in particular to deep wells or powered pumps. The upper aquifer dries up, and the local excess water use leads to salination of surface soils.

• SURFACE DRYING; the desert becomes irrevocably abandoned when oases dry up due to groundwater removal.

Damage in hills is devegetation, erosion, denudation. In eroded soils calcrete and ferricrete or silcrete is exposed, preventing productive use. Deep erosion cuts gullies in fields; dust storms do further damage. Once–stable dunes will then start to move. All these are preventable problems if early signs are heeded and if we have goodwill to the earth. Above all, desertification is a land–use, hence political, problem, and reflects the priorities of governments.

While I agree with all of the above, there are other factors affected by and related to them, and necessary to more fully understand the process of desertification. These are:

• WIND EFFECTS: Both the drying effect, wind erosion, and sand blast on plants.

• SOIL EFFECTS: Soil collapse due to deflocculation of clays; hollows and pans develop soil salting due to the above and to the development of hardpans in the B soil horizon.

• WATER EFFECTS: The increase of overland water flow due to deforestation and agricultural compaction of soils, causing gullying and salt transport to lower soils. Added to this, there is a rise in salt water tables due to deforestation.

• WETLAND SALTING: The overuse of mineral and salt–rich waters on clay–fraction soils, and over–irrigation.

These are the processes of salinisation or soil salting.

In 1928 and 1929, W. C. Lowdermilk of the United States Department of Agriculture cast a soil conservationist's eye over what archaeology had revealed of the once–grand civilisations of the Middle East, Israel, and dryland China (U.S.D.A. Bulletin #99 of 1929). He noted the patterns of abandonment of hillside settelments based on agriculture in the semi–arid lands he visited, and the eventual concentration of settlements on the valley floors, in wadis, and in oases. Many excavations in the Saharan and Thar deserts showed that catastrophic sand flows and silting had obliterated first the canals that irrigated these settlements, then the settlements themselves.

Today, thousands of abandoned towns lie buried in desert sands; people planted wheat and reaped salt. After 7,000 years of this history oft–repeated, Australia and India are on the same suicidal path. We lose 30,000–40,000 ha of arable land to salt and desertification each year. In the West Australian wheat belt, 10,000 ha of farmland is lost to salt annually, and some 260,000 ha of salted soils are now noted in croplands.

In the central Australian desert, 74% of the land (bringing in 9% of income, most of which depends on tourism) is devoted to unregulated pastoralism, producing erosion scars that are conspicious on satellite photographs. So we proceed to our own more modern extinction and elimination via an ephemeral export agriculture that will leave Australians as a "third world" people in a few more decades.

SALINITY

Salt is brought in by sea winds, as rain nucleii (CYCLIC SALT), or remains in soils and sediments from marine periods (CONNATE SALT). Cyclic salt provides plant nutrients, but can also form salt ponds or crusts if evaporated, as is evident in pools on dry, windy sea coasts. Salt is also leached from rock minerals by groundwater, which carries large loads of connate salts dissolved from old sea beds. In indisturbed country, the soils of forests have both higher salt and other mineral nutrients, and many desert trees exude or store

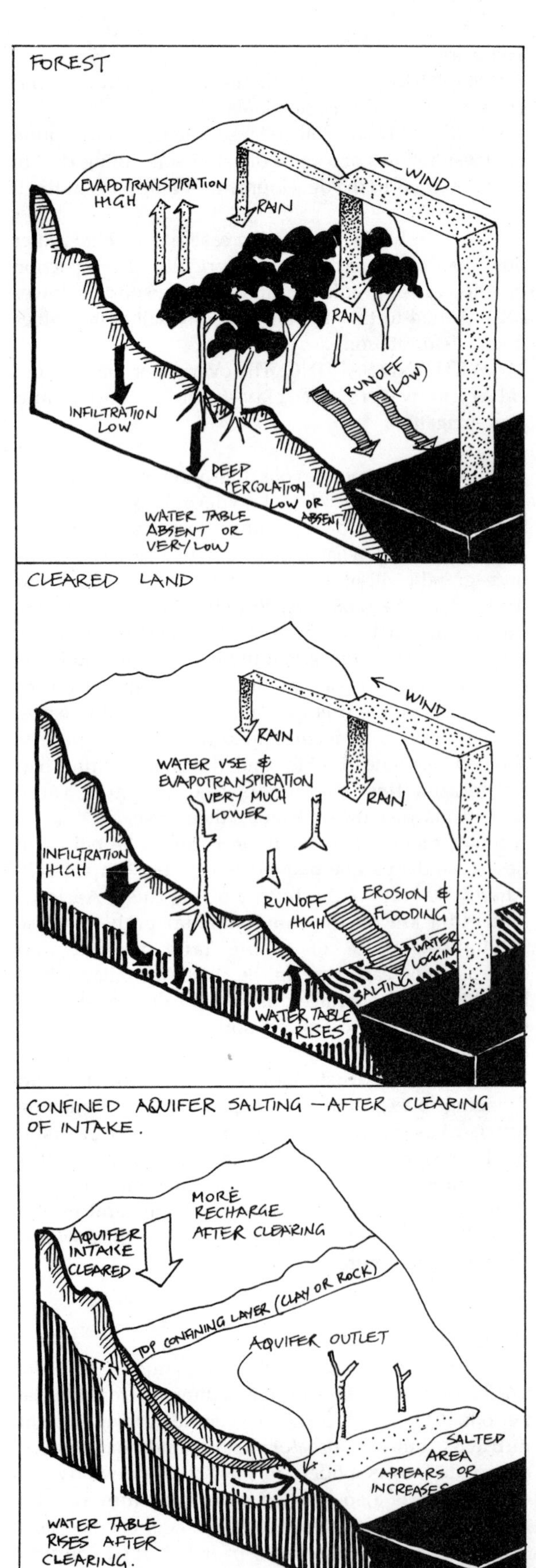

salt in leaves, to release these salts in rain to surface soils.

Salinity is an *induced* problem. That is, it did not exist before clearing and cropping or grazing. Thus, it is theoretically preventable and reducible. Salting is usually confined to lowlands, and is not a problem in sloping upland, forested, or ungrazed regions. The acceptance that the problems of salt are man-made, although later perhaps self-generating, must dictate two early preventive policies:

1. An absolute ban on clearing or tree-cutting in any area subject to salting, which is most arid to semi-arid lands.

2. Practical field work on alternatives to decide effective local strategies.

Another observed widespread and well-attested phenomena is that salted areas actually *increase rapidly* if two or more wet seasons, with flooding, occur. This is an effect, not balanced by any *reduction* in the salted areas in subsequent dry seasons. Although it is counter-intuitive that more fresh or flood water on the land creates salted areas, it is nevertheless a fact. That is, flooded and winter-wet or boggy soil patches are those that later develop salt scalds or salt pans.

Causes of Salting

Broadly speaking, there are two types of salt problem. WETLAND SALTING results from the over-application of irrigation water from bores and canals to clay-fraction soils. The surplus water seals the surface soil, and pools up and evaporates, leaving a salt crust. Below this, slow percolation from rains and excess irrigation may cause the shallow soil-water table of river flood plains to rise to the root level of trees, and so kill off irrigated fruit trees and native plants. This problem can be solved by applications of gypsum, tile drains or deep drainage, and sophisticated sub-surface or tree crown-shaded drip irrigation timed to release just the minimum amount of water required, preferably at root level. However, an intractable problem is the disposal of water from the drains.

The more insidious and widespread problem of desertification and DRYLAND SALTING is not related to arid areas irrigation, but will occur anywhere we crop or graze arid areas. It is self-generating into even sub-humid areas of up to 100 cm average rainfall, and in fact anywhere that we farm country where evaporation equals or exceeds precipitation.

In 1978, Terry White (editor of the *Permaculture Journal*) and I invented the acronym S.A.L.T. (for Salt Action Liaison Team), and set out from Maryborough (Victoria) in an old Volkswagen to convene a meeting of

FIGURE 11.92
"GROUNDWATER RISING."
Winds bring in salt particles; very permeable cap soils infiltrate rain, and if forest is removed, this water causes salted groundwater to rise at the foot of slopes. Often, cap soils are volcanic, slope soils of shale or mudstones (Victoria, Australia). Trees can be replanted on cap soils to stabilise this situation, but lower soils may have collapsed.

farmers (who were losing their land), churchmen (who were losing their congregations), and local government representatives (who were losing their towns and rate incomes). Subsequently, strong action teams formed, on Terry White's initiative, and grants were raised to tackle the Loddon–Campaspe area in Victoria by tree planting.

Our reasoning was based on the premise that salt rises up from deep soil reserves due to tree removal, and that permanent trees pump out these rising groundwaters, keeping the salt water table below damage levels. This is the GROUNDWATER RISING theory, and is widely accepted as the cause of non–irrigation salting in Victoria, where very large areas of country freely absorb rainfall, and where water tables are close to the surface (2–10 m down). **Figure 11.92.**

By planting trees, therefore, we would "pump down" the water and salt, and enable cropping and pastoralism to continue. We had all, I think, accepted this model of dryland salting: cut the trees, and the salt rises; plant trees, and it goes down again. This model presumes that deep waters in the soils are free to rise to the surface by infiltration, followed by evaporation and by capillary action, and it takes no account of the effects of farming and pastoralism *on the soil itself.*

However, late in 1985 I visited West Australia and took the opportunity to speak to many farmers about the W.I.S.A.L.T. scheme (see **Box**), based on a premise not of deforestation alone, but of SOIL COLLAPSE due to two influences. Soil collapse is primarily due to clearing, cultivation, and hoofed animal compaction, and secondarily to swamping of these compacted and damaged soils by surface (overland) flow of rainwater carrying small quantities of salt falling as cloud or raindrop nuclei. In addition, salt is released from the collapsing soils and from the heavier concentration of surface salts which are found under forests and leached when these are cleared for cropland.

Salt (as sodium chloride) releases chloride ions, which are rapidly washed away or escape to air, and sodium ions, which bind on to any clay crumb particles and displace calcium ions, causing a rapid disassociation of the clay crumb structure of the soil if sodium exceeds 15% (known to soil scientists as DEFLOCCULATION). As a result, not only the soil *surface* is sealed, but deeper clay particles and minerals displaced by the sodium migrate, and cement the subsoil into a hydrophobic (water–repellent) block of rock–like consistency, a cement without the air spaces of true soil. This *seals off the surface soil above* and creates an ever–increasing swamping by overland water flow (cascade effect), worsening the initial problems. Rapid deep infiltration is blocked, and the surface soils flood easily in rain (**Figure 11.93**).

At the same time, hot winds begin to take effect on clearings over 500 ha in extent, or more than 5 square kilometres across, drying out crop soils and vegetation as yet unaffected by desertification. Thus, newly cleared land becomes degraded in about 20 years, whereas the older farms (clearings in the bush) can take 100–150 years to develop soil and salt problems.

W.I.S.A.L.T. is an acronym for WHITTINGTON INTERCEPTOR SALT–AFFECTED LAND TREATMENT SOCIETY, founded in March 1978. Harry Whittington of Brookton, West Australia, has evolved (from practical field work) the soil collapse explanation of desertification. This society now has 1,100 farmer–members, and trains consultants and contractors to build interceptor banks correctly. As great care is taken in the size, spacing, grading, and construction of interceptor banks, a period of training with the W.I.S.A.L.T. people is necessary for both farmers and contractors. Even the bulldozers are modified to create a 1 metre blade drop for delving the ditches and ramming the banks to seal them. There is a vast amount of effective (and ineffective) land–forming to be assessed in the wheat belts of West Australia, and a lot of results to hand.

Literature and information is available from The Secretary, Box 154, Quairading, West Australia 6383. Many members can supply data supporting both the explanations and remedial earthworks needed to hold back, or reverse, desertification on farms. They fully convinced me (I peered into miles of canals and ditches) that their theories were correct; certainly their land therapy is working. I believe their efforts are worthy of international recognition. Harry Whittington is yet another unrecognised "great Australian" with a love of land and good husbandry.

My thanks go to Lex Langridge for transport, Gavin Drew of Beermullah, James Gardener of Bungulla, Mac Forsythe of Kellerberrin, Laurie Anderson of Quairading, and of course Harry Whittington of Brookton for explaining their approach to land reclamation. Many thanks also to my friend Terry White in Victoria for his constant inspiration and persistence.

Strong winds, often as a downdraft wind in advance of (dry or wet) thunderstorms, create dust storms that deposit very fine topsoil deposits on high plateaus, on gravel ridges, in the lee of hills, and of course in towns. In Alice Springs, this loess covers the hill graves of the pioneers buried before 1966 to a depth of 1–2 m. Since 1966, soil pitting and grassland growth has stilled the dust in that area, mainly due to the need for a reliably open airport! At the same time, grazing was excluded or rigorously controlled to prevent a recurrence of the dust problem.

The silty dust from wind storms can now trickle or work down into the once–open rock crevices of hills, and thus seal off or reduce the effectiveness of the old water intake crevices that feed underground aquifers, hence springs and rivers. This again exacerbates the effects of overland flow, soil erosion by floods, deflocculation, and soil collapse. Now we have two effects—damaged intake areas and sealed subsoils—that will cause a cascade of run–off rainwater over and through surface soils, while the subsoil remains *dry.* The cemented subsoil also seals off the soil from the parent rock below, created a perched water table in rains.

An aerated soil of 1 m deep in an untouched state can contain 8–20% humus, 10–40% clay, and can be as much as 40–60% air space (Leeper, 1980). Once the soil humus is removed by fire, cropping, and over–grazing, the clay deflocculates to fill the air spaces, and the crumb structure compacts to a cemented surface, or to residual fine sand and dust (which blows away). Little soil humus remains. 1 m depth of soil can by these effects be compressed or eroded into a layer only 10–20 cm or so thick, overlying a cemented "B" horizon of concrete–like consistency.

This is the process of soil collapse. Of course, any salt in that original soil, and previously dispersed through the whole open soil layer, is now concentrated to make a very salty zone around the collapsed area. Desertification has arrived *by a process of soil collapse*, a different but complementary effect from that of groundwater rising as it occurs in Victoria. Tree planting in such conditions will fail to reverse the process as trees cannot long survive.

During the process of soil collapse, crops, pastures, and isolated trees suffer. The collapse produces *anaerobic soils* (without air spaces), and humus decay then becomes subject to anaerobic bacteria producing ethylene, methane, and sulphurous smells, typical of all waterlogged and compacted soils. This is a wide-spread and easily observable phenomenon on our farms and wheat belts. Mottled and waterlogged soils, often with iron stone layers or iron "buckshot" concretions, are found at the base of slopes on many wheat farms, just above a salted area of dying trees and shallow salt pans.

The collapsed soil areas sink 1–2 m, and (in valleys) produce sunken salt pan areas. These, in turn, cause further peripheral soil collapse due to edge swamping and salting, and may develop into extensive winter salt lakes, with a greater and greater collapse effect on their adjacent soils. Salt is also washed downstream and into dams to refer the problem elsewhere. All these processes rapidly become self–generating.

The natural process is believed to take 10,000 or more years of alternate wet and dry seasons, before this argillic (clay) skin, plus iron and silica cements, builds up a layer impermeable to roots and water. The deforestation of soils and the subsequent solubility of clay due to mobile salt can speed up the natural process considerably. While such effects are good for water crops (rice, taro) in humid areas, they become lethal for

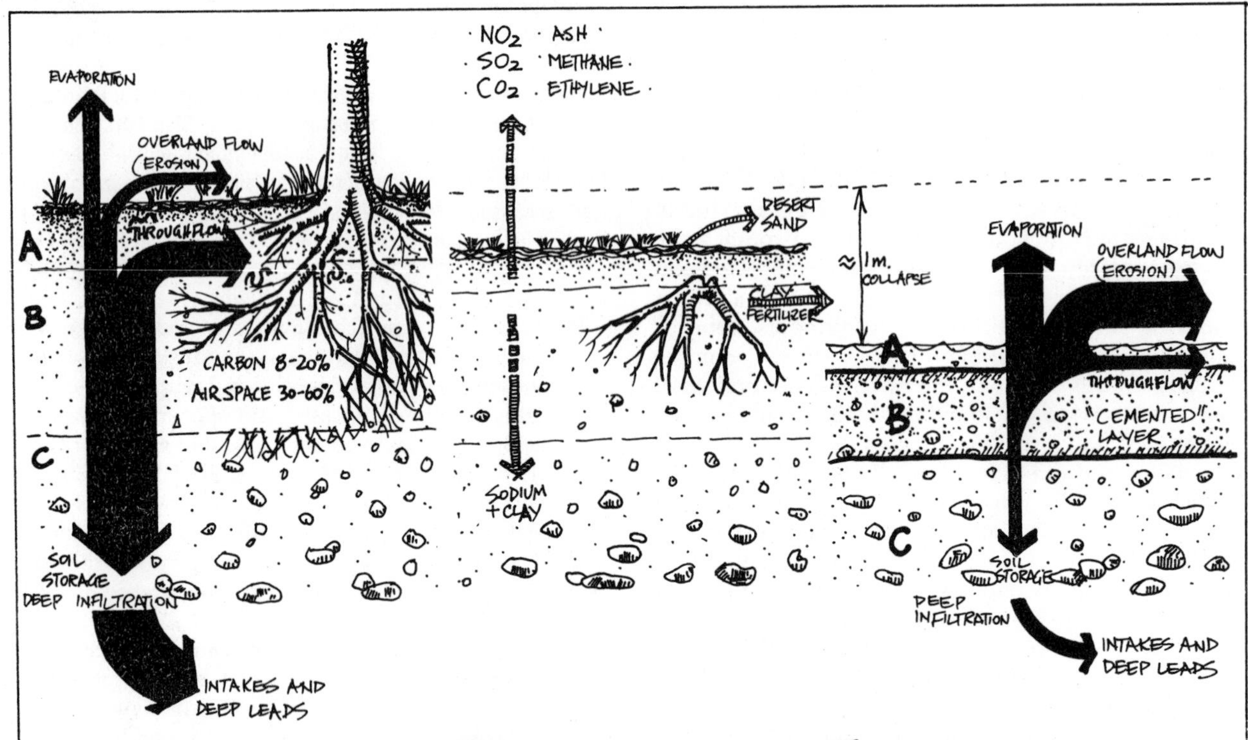

I UNCONSOLIDATED FOREST SOIL PROFILE:
COMPOSED OF HUMUS, WATER, CLAY, SAND, AIR, WITH CRUMB STRUCTURE. 30% OR 50 WATER SPACE; OVERLAND FLOW A MINOR ELEMENT. DEEP WATER–SOIL FIELD CAPACITY;

II PROCESS OF MISUSE:
CLEARING, OVERGRAZING, FIRE, HOOVES, PLOUGHS, MACHINERY (COMPACTION), +RAIN AND WIND LEADS TO NUTRIENT MOVEMENT, COMPACTION. (SODIUM FLOCCULATES CLAY WHICH FILLS AIR SPACES). LEADS TO; →

III COLLAPSED SOIL PROFILE:
"MASSIVE SOIL". B HORIZON NOW HYDROPHOBIC, SALTS CONCENTRATED. 5% OR LESS WATER SPACE. GREATLY INCREASED OVERLAND FLOW (EROSION) AND EVAPORATION. SOIL COLLAPSED ABOUT 1M. ONLY THIN SURFACE SOILS GET WET.

FIGURE 11.93
PROCESS OF SOIL COLLAPSE.
After misuse, sodium ions displace calcium, clays deflocculate and soil pore spaces fill; soil collapses to a cemented hardpan which seals off the subsoil. It is now almost impossible to plant trees without a long rehabilitation process, interceptor banks, and humus development (West Australian soils).

dryland crop.

Foth (1984) notes that deep drainage plus gypsum or (more slowly) sulphur will restore acidic soils, but that the leaching out of salts that follows may effect stream health. Interceptor banks leading to creek or river beds effectively drain and permit leaching of the salted soils. Where the pH is more than 8.5, gypsum is also indicated to assist sodium removal.

Wherever we irrigate with saline waters, we must flush salt out or drain the soils thoroughly, but in the end one wonders if cropping on such soils is inevitably risky, and if diversion to energy production via tree crops, or tree crop and modest animal grazing is not a better use of drylands. In particular, flood irrigation of furrowed fields can result in a 60% evaporative water loss, and a two– to seven–fold increase in salt concentration. Salt in ridge and furrow systems of flood (furrow) irrigation will increase on the ridges where less flushing occurs. Thus, ideally we should flood the ridges by creating beds with raised edges, and flush the salt *down* to the furrows.

Leeper (1982) notes the reduced ability of cropped soils to infiltrate rainwater; in adjoining fields, up to 30 cm can be absorbed by uncropped soils over a period of 5 hours, while in identical cropped soil, only 5–16 cm are absorbed. He notes that these differences are greater than those between different soil types. Wheat cropping is therefore of itself a cause of increased overland flow, and the subsequent flooding of lower soil profiles, as is soil compaction by overgrazing, and the use of heavy or fast farm machinery.

As for pore space, the specific gravity (S.G.) of an open soil is from 1.0–1.6 (equal to or greater than an equal volume of water). The S.G. of soil mineral components is 2.6–3.2. Thus, the average air pore space in the total volume is 50% or more (generally 40–60%). In a flooded soil, all but a few of these pores are water–filled. If that water also contains salts, and deflocculation occurs, the soil crumb structure quickly breaks down, and pores are clogged with dispersed silty particles. In wheat lands, only 12% of the soil may be well–structured (crumbs of 0.25 to 2 mm), while in free–draining uncleared soils, 92% of the soil is composed of such particles.

Thus, it is clear that the structure of soils under dryland cultivation can in truth break down, and that almost all the soil pore space disappears (even though cropping is still possible at 12% crumb structure).

A major factor in soil crumb formation is the presence of polysaccharides (long–chain sugars) and gels produced by humus; this humic material is critical to good soil structure. The widespread practice of burning dryland grasses and wheat stubble further reduces the soil humus. Fire can actually bake surface soils so that they form fine dusts which blow or wash away. Humus as green crop added to neutral or alkaline soils will, however, restore crumb structure if salt water deflocculation can be stopped.

In collapsed soils, tree and plant roots face two inseparable problems; even if not flooded, the reduced pore spaces contain little air, but in rains, soil carbon dioxide levels (waste gases from root respiration) may reach the level of 5% (it is 1% in healthy soils, 0.03% in air). This is a lethal level for plant root growth. As well, sulphur–concentrating (anaerobic) bacteria produce methane and sulphur dioxide (marsh grass) as a result of the decay of humus in these anaerobic conditions.

Finally, the weakened plant roots are subject to rapid drought effect by surface evaporation, and face osmotic stress from the salt water in the soil. A point comes where soil salts prevent water uptake by roots. No wonder we are losing trees and crops throughout the drylands!

It is now obvious that to plant trees in soils already in process of collapse is futile. We need to stop the process of soil collapse itself, and it is here that the W.I.S.A.L.T. group have evolved what I believe to be a unique and successful West Australian hill farm solution to soil collapse and desertification.

<u>Interceptor Banks</u>

In earlier times, shallow (0.5 m deep) CONTOUR BANKS were built to stop soil losses on slopes. Even gentle slopes (3° or less) lose soil in rain as farmers remove tree cover. Because lower slopes get more water, as flow–down, soil erosion is not much less on plains than it is on hill slopes of 10° or so. Contour banks did not, however, have a profound effect on wind erosion, gully formation, tree death, or soil collapse by flooding; indeed, salt eventually causes contour banks to collapse, and run–off floods over them in heavy rains.

Although our salt problems first appear in low flat areas and valleys, the water that causes the problem cascades downslope. The W.I.S.A.L.T. approach is to first tackle the problem at its source—the hill tops. Starting about 1–3 m below the ridgelines and plateaus, a deep (2 m) and wide (3–5 m) ditch is bull-dozed up on contour. Care is taken to first push off any topsoil, to dig 1–2 m down with a bulldozer blade to break up the concreted subsoil (it is often necessary to also rip), and then to compact this clayey and stony layer by ramming it firmly against the *downhill* bank of the ditch. Ditch and bank together are called an INTERCEPTOR BANK (**Figure 11.94**).

If the downslope bank is effectively sealed, all high run–off or overland flow stops at the ditch, as does the slower seepage or throughflow of the topsoil, which perches on top of the compacted "B" horizon as a general seepage line. No flooding of downslope soil results, and further soil collapse ceases. While the higher banks are made on the level as swales, downslope banks slope at grades of 1:3000 to natural stream beds, and thus both salted and fresh water are carried to sea in streams. Within 2–3 years, the strips between the banks commence to regenerate grasses, and dying trees regain health.

More interceptor banks, still on contour, are now built every 3 m (vertical height) downhill. On slopes of 10–15° or more, these form a continuous series of ditches (piped or stone–filled for machinery crossings at intervals) from ridge to foothills, They immediately stop overland flow and seepage, and hold all fresh water after rains (Figure 11.95). Their spacing is carefully calculated to *totally intercept* any water that would otherwise flow or seep downhill. As slopes lessen to 5° or less, the spacing is altered to a maximum of 300 m between ditches, and the ditches are made so as to end on (or rather start at) and spill into natural waterways, carrying off water flow surplus to the capacity of the ditch itself. Wherever such ditches leak or overflow, areas of collapsed soil and grass death are visible, but grasses re–colonise salt pans wherever uphill water is fully intercepted.

Open hill intake areas, even if damaged, still feed deeper streams by slow infiltration into old sand seams, permeable rock strata, or sandy soils trapped below the "B" soil horizon. These shallow aquifers can cause lower slope flooding, and wherever they are located, they too

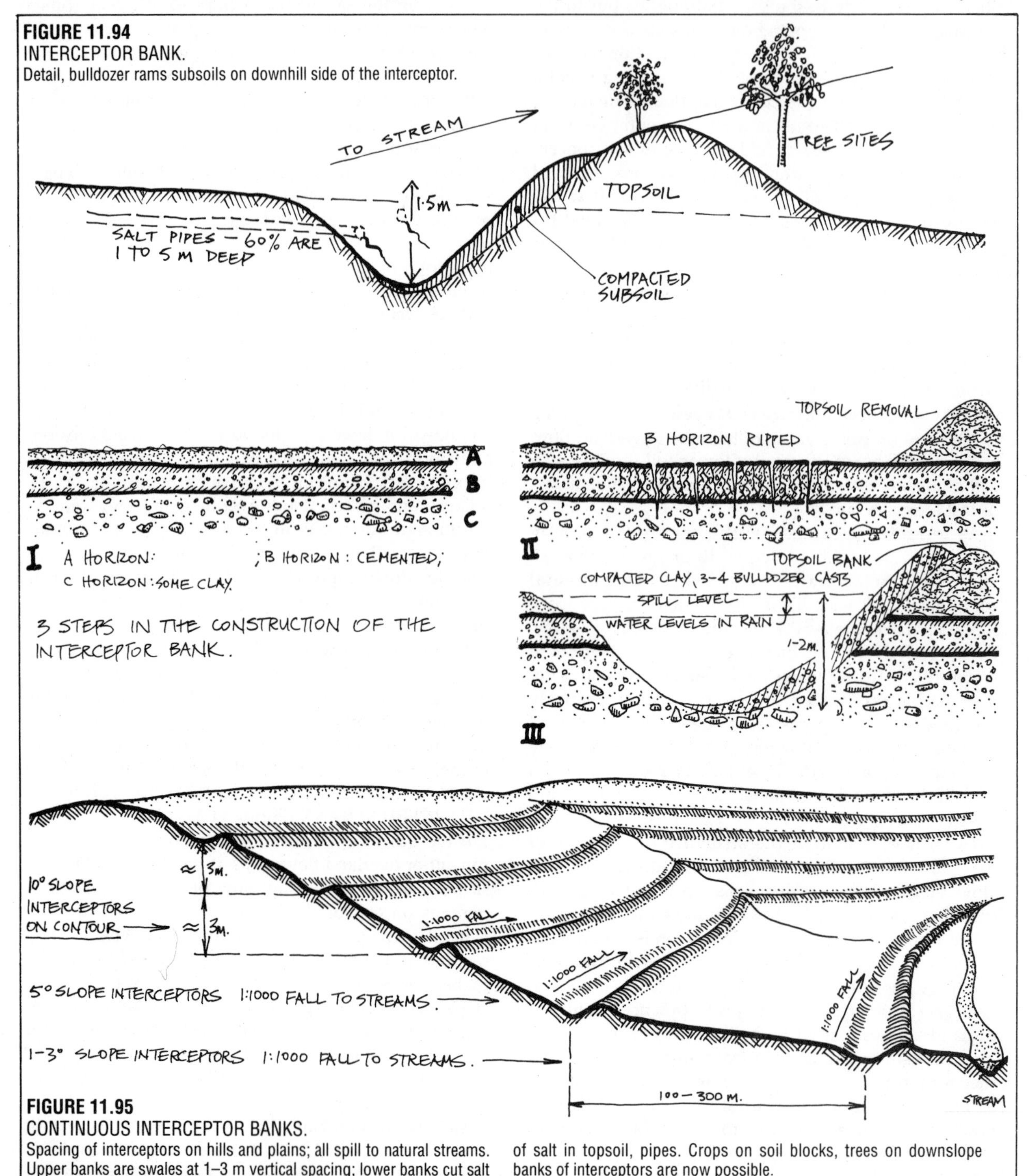

FIGURE 11.94
INTERCEPTOR BANK.
Detail, bulldozer rams subsoils on downhill side of the interceptor.

FIGURE 11.95
CONTINUOUS INTERCEPTOR BANKS.
Spacing of interceptors on hills and plains; all spill to natural streams. Upper banks are swales at 1–3 m vertical spacing; lower banks cut salt "pipes" to 1.5 m down; soils blocks are isolated from "cascade" effects of salt in topsoil, pipes. Crops on soil blocks, trees on downslope banks of interceptors are now possible.

are dug out to base and clay-filled or blocked by vertical plastic barriers so that their water (at head) also flows into the interceptor ditches, and thence to streams, or from uphill ditch areas via pipes to dams or storages. Water is then available to stock and wildlife as waterholes where no such source of freshwater at head previously existed. Most seepage flows occur at less than 2 m depth, but deep sand seams to 6 m depth can be dug out by back-hoe and blocked by vertical plastic sheets buried in a clay backfill. Sand seams are detected by backhoe digging, radar from satellite, or locally by dowsing. When banks are constructed, as many of these deeper seepages as possible are blocked, lessening surface seepage lower in the catchment where the shallow aquifers discharge. In effect, seepage comes to a standstill.

Deep sands or old dunes do not collapse (they are, after all, up to 93% silica sand grains) but instead act as freshwater "tanks", sitting on the landscape and seeping slowly from their base. Interceptor drains downslope from such sandy areas remain water-filled with fresh water as canals all year, and are very attractive to waterfowl, small fish, and water plants. The interceptor banks cut off freshwater seepage water from these, and also from salt lakes or dams that lie above valley farms, directing it harmlessly to streams and thus preventing the winter flooding of surface soils. Salt lakes can have early wet-season floods diverted around them, and only later fresh run-off waters are diverted to fill them. In this way they become suited to fish and waterfowl.

The beneficial effects of interceptor banks are rapidly evident downslope. Wind erosion and flooding in rains ceases on construction. Crops regain health, fertilisers are no longer washed away, dying trees recover, and new trees can establish (they often appear on the banks soon after construction). Long-term soil recovery and gully elimination can, however, take up to 10 years of careful husbandry. Tree-lines 5–8 trees wide *below* each bank stops the drying-out effect of the desert winds, and eliminates wind erosion on the fields. The uphill side of ditches are left free of trees in case maintenance is needed.

I cannot overstress the importance, to every country, of this well-tested present approach to soil and water conservation, and the consequent reversal of salt scalds, gullies, and desertification. In hard cash terms, the farmers I spoke with in West Australia estimate that in 5–7 years, cash returns from increased livestock and crop yields pay for all the earthwork.

We could all help by investing in such beneficial and remedial work that salt-afflicted farmers themselves may not be able to afford. We might also insist that our government co-invests in this work, as much more land has been lost to salt than we can ever replace, If we are ever to get our economic priorities right, this aid to preventing salt must take precedence over thoughtless investments in the stock market and real estate fields that destroy real wealth merely to make money. We must relieve the pressure on the land that caused the collapse of our rural soils in the first place.

<u>Groundwater Rising</u>

Dryland salting is attributed to rising groundwaters derived from deforested aquifers in Victoria; some of the deep sand aquifers sampled contained very salty water (30,000–42,000 ppm), rich in iron, and acidic. Where these eventually surface, 180,000 ppm salt is possible in lagoons or creek seepages. Alkaline limestone aquifers may have less sodium salt ions (960–3000 ppm) and can supply domestic or stock water. In many areas of Victoria and inland Australia, water tables are in any case high, and if these deep leads discharge into them, salt is not far below the surface. Here, the answer is to restore forests to the hills and plains.

While tile drains and pumping can reduce wetland salting, dryland salting via aquifer discharge is too large a problem to solve by mechanical pumping. It is clearly evident that country under forest (in the same areas) is not salted; it is the agricultural clearing and wood-chipping, or net forest loss, that creates salt deserts.

In Victoria, interceptor banks have not been widely trialled, and soil collapse is not invoked as an explanation for the broadscale salt seepages that are obvious above the sealed or cemented "B" horizons of the soils, and that create sunken off-stream salt lakes at the foot of slopes.

In drylands, any rise in water table or compaction of soil can be taken as a warning of salting danger. Surface indicators are brownish patches of sea barley grass*(Hordeum maritimum)* and buckthorn plantain *(Plantago cornopus)*. Crops lose vigor, and the subsoil is saturated and anaerobic.

At a groundwater level of 1.8 m of so, capillary soil action takes over, and surface evaporation of soil water creates salt crusts (efflorescences). Salted areas then spread uphill, rushes evolve *(Juncus)*, and bare soil areas increase. In West Australia, the uphill spread is attributed to the damming of groundwater flow by lower areas of collapsed soils. Wet years exacerbate the problem, which does not necessarily recede in dry years.

It is obvious that evergreen trees, transpiring more water than crops, keep the water table down and permit better rain infiltration and soil pore space. At the same time, forests in drylands excrete salts which accumulate in surface soils, and it is these salts that are further concentrated by soil collapse. In the West Australian soils, over granites and gneissic areas, salt reserves to 20 m deep are 3,400 to 19,000 t/ha, and most of this lies below 2 m, or below the forest root zone. In Victoria, basalt cap soils are good infiltration areas.

Rainfall contains 5–10 ppm salt, derived from the 33,000 ppm of seawater. Salt content falls off inland, and rain deposits 27–150 kg/ha/year in Victoria. It would take 100,000 years at 50 kg/ha to build up the soil salt levels noted for West Australia. Salt loss by stream flow (running to sea) about equals that coming in as rain (hence "cyclic" salt), but where areas are farmed, salt levels in streams can be 20 times higher as connate salt is mobilized by groundwaters and soil collapse.

In West Australia, only 0.7% of original tree cover remains, in South Australia 3%, in Victoria 13%, and in Ethiopia 4%. All suffer increased stream and groundwater salting, raised water tables, and increased artesian pressures from deep salt leads, risking surface salt seepages. It is therefore a critical strategy to forbid further clearing on any land (South Australia has done so), to trial interceptor banks, and to revegetate upland or intake areas with trees.

To sum up the evidence on dryland salting:

- Soil collapse can occur in clay–fraction soils, causing waterlogging, gullying, and crop death from compaction and anaerobic soil conditions.
- In deforested areas with large areas of water intake formed of permeable soils and sheltered rock, salty groundwaters can rise by flood infiltration, as it does in over–irrigated areas.
- Excess soil water from the latter effect is most effectively removed by trees (versus crop or fallow).
- The reduction or elimination of overland flow by interception banks enables collapsed soils to recover, and that until this is done, trees will not survive.

There is every reason, therefore, *to examine the local situation* before recommending either tree planting or interceptor banks; the latter are appropriate in soils with obvious cemented "B" horizons and obvious rainy–reason waterlogging, but where both the "B" horizon and any underlying strata can be *quite dry*. Tree planting is likely to be effective in areas where deep permeable surface soils or strata are found, where no cemented sub–surface layer occurs, and then only where groundwaters have not reached to 2 m of the surface.

Until interception banks have been made in collapsed soils, tree death will occur and new trees live only until they are a year or two into growth; after that, they waterlog and die.

The best way to discover what local problems are occuring is by digging back–hoe pits and observing how they fill—whether water wells up from deep seepages, or trickles down from shallow seepage above the "B" horizon. This decides corrective strategies (trees alone, or interceptor banks plus trees).

Swales to infiltrate rain water are appropriate only if *trees are planted along them* to reduce the net infiltration above deep clay layers. Swales are, however, valuable aids to tree crops and concentrate fresh water at tree roots.

Finally, there is no doubt that any nation which permits over–grazing or cropping which reduces tree cover in areas where precipitation is exceeded by evaporation will, in a few short decades, lose its drylands to deserts and salt. It is long past time to legislate and to divert public monies to both stop and if possible reverse these effects world–wide.

Cautious Approaches

Given the multiple interactions possible between salts, evaporation, soils, slopes, drainage, and land use, and given that problems may not show up for 100 years or more, designers must always be willing to follow a sequence of cautious approaches to ensure the permanent use of drylands. I believe these to be as follows:

- To try a great variety of small systems of earthworks and land use practices on only one slope or soil type.
- To then monitor such factors as infiltration, evaporation, salinity changes, and crop health over a few years.
- To cautiously extend systems that do not show any salt increase in soils over the short term.
- Above all, to store no run–off where tree roots cannot remove the water, or to tree crown shade the storage, unless it is subsurface storage in sand basins or sand dams.
- Where groundwater tables are shallow, concentrate on *lessening* infiltration by floodwater DIVERSION to streams and floodways.
- Where soils are free–draining, concentrate on *increasing* local infiltration by swales to increase tree cover.
- Where gardens are to be made, use ridge trickle irrigation in dry periods, and if possible allow flood–waters to flush out furrows of accumulated salts in rain.
- To favour roof tanks and surface swales rather than wells or bores, which lower total aquifer resources and create local excesses.
- In particular, use back–hoe or deep pit samples to establish the conditions of drainage locally, and then proceed to strategies based on the local conditions (diversion, interception, swale infiltration, and crop type).
- Make every attempt to assess potential yields before doing any of these things. Any yield we can *manage* is preferable to one we *create*; it may be easier for us to manage and increase natural yields (e.g. the saiga antelope in steppe, the kangaroo in Australia, or the 140 species of large animals in Africa), whether animal or plant, than it is to impose an exotic crop or animal species on a fragile environment.
- Observe every case where yields are naturally high; these conditions may indicate a safe way to increase yields at least risk (e.g. patches of forest where dunes converge on a sloping pavement, or large trees fringing a dry river sand–bed). Such conditions are likely to have withstood the tests of time.

Having examined (preferably by digging pits or inserting piezometers in key areas) just what local salting effects are in process, farmers and designers can choose from a set of strategies (or even use all of them if product increase justifies the expense). These are:

1. Swales at high levels to establish trees, which reduce net infiltration.

2. Interceptor drains leading to natural waterways at midslope and on lowlands.

3. Where intake areas can be identified, intensive planting of trees which use plentiful soil water (e.g.

Eucalyptus camaldulensis, E. sideroxylon, E. wandoo [drylands], *E, globulus* [cool wetlands]), and *Casuarina cunninghamii* plus *E. camaldulensis* to use surplus sewage or other waste water in lowlands. The choice of species needs local trials. Natural tea–tree vegetation in swampy valleys (*Leptospermum, Melaluca, Banksia*) are killed by salt, but are re–established if uphill effects are controlled.

4. Subsurface drains ending in natural waterways or interceptor drains are worthwhile where the crop is of high value.

5. Bores and pumps to lower water tables in irrigation areas should be looked on as a last resort, and may be supplanted by trees if headwater infiltration can be controlled.

Whatever the methods used, salt levels will rise in streams as soils are flushed out, and once we have salted lands, fresh water for settlements may need to be supplied by rainwater tanks and hill sand storage dams until streams regain health—and this could take decades.

Interceptor banks and drains leave more land open for cropping, but trees have a very definite value over recharge areas, and as a valuable energy crop in their own right. It is cheaper, however, to reclaim salted lands than it is to clear new country for short–term cropping, so extending the problem. Above all, we should continue to study the chain of effects that cause deserts, and design to eliminate each of these in turn.

Additional Methods of Salinity Control:

- Changes from shallow–rooted annual or pasture crops to forage tree crops and fuel–wood supply is a long–term but currently profitable land–use shift to sustainable culture. Fallow or bare–soil cultivation, stubble burning, and shallow–rooted crop all exacerbate the problem.
- Eliminating leaky channels for impermeable pipes, using sophisticated irrigation at root level, and closely monitoring irrigation water to minimum levels are all essentials to reduce wasteful water use and excess infiltration. In some areas of Australia, only 26% of channel waters actually reach farm, and on–farm use may waste 60% or more of the remainder (e.g. in extensive spray or flood irrigation without close supervision). Piping eliminates land lost to open channels, sand drift problems, and preserves water quality.
- Salt–tolerant plants can be more used to supplant affected crop. Saltbushes *(Atriplex), Acacias, Prosopis, Puccinella* grass, bluebush *(Maireana spp.)*, samphire *(Crithmum, Salicornia)*, tall wheat grass *(Agropyrum)*, Wimmera rye grass *(Lolium rigidum)*, and a variety of seashore and inland plants will thrive in soil of more than 0.4% sodium chloride. However, emphasis on salt–tolerance will simply delay investigation into basic causes and treatment, and so has not been emphasised herein.
- Above all, careful assessment of local factors, a return to more natural yields, and a *true* economic assessment of the costs of farming versus aid to gardeners will almost certainly reveal the benefits of closing down most marginal agriculture, and returning the area to a natural managed system of sustainable yield.

11.14
COLD AND MONTANE DESERTS

In the arid plateau and high valley interiors of continents, are very extensive desert areas with great extremes of temperature, extremely hot clear–sky day radiation, and very cold icy nights. Woolly plants and woolly animals survive these conditions, and silvery hairs protect them from heat and cold. Very solid houses are needed, and fuel is a special problem, as all plants grow relatively slowly. Willows, birch, poplars, junipers, and fast–growing or tuberous summer plants are called for. Damp stream areas grow good tree crops. Snow at high altitudes will melt directly to vapour, to blow away on winds, so it is essential to swale for snow–melt, and to grow trees as snow traps in or near these swales. Even fences help to retain snow for meltwater in streams.

Stone piles and mulch preserve a moderate air layer over plant roots, and hardy grains will grow on small fields, as will root crop for winter storage. Rocky soils are good growing sites, and rocks help preserve ground warmth and moisture. Some good fish species occupy lakes and streams, as do beaver and large aquatic rodents. Chickens and guinea pigs survive, the latter in houses or in a small room near the kitchen to keep them warm. Trombe walls, well–sealed double entries, glazing over massive internal courtyards, and earth sheltering are all ideal house features, together with massive fire–flue walls and efficient stick–wood cookers.

Clothing needs dry, airy, absorbent linings and wind–excluding waterproof exteriors (furs inside out) to be comfortable, and radiation burn must be guarded against by visors or broad hats, dark glasses or snow (slit) glasses.

When we move to cold deserts and arid montane climates, the vegetation changes from dry–dedicious to winter deciduous. Sugars are stored in saps rather than pods, and evergreens are conifers rather than *Acacias*. Berry–eating rather than seed–eating birds account for the distribution of many understory species, and animal migrations are vertical (sometimes daily) or trans–equatorial rather than to rain patches (horizontal). A few plant species (lucerne, grains) are shared, but most differ from the hot desert areas.

Growing seasons are measured in days (40–70 days of growth is common) rather than months, and frosts may occur over most of the year; thus cold hardiness is as important, or more important, than drought hardiness.

There are few succulents at high altitudes; the cold freezes thick–leaved plants and ruptures the cells, although a very few "woolly" cacti can survive the cold as the thick hairs trap insulating air.

11.15

DESIGNERS' CHECKLIST

BROAD STRATEGIES

- Commence operations for freshwater infiltration at the top of catchments, and work downstream as plants establish.
- Give upwind areas the first attention for soil pitting and windbreak trees.
- Establish plant nucleii or oases at favoured sites and follow out on corridor planting.
- Spread pelleted seed over larger areas to await a favourable season.

GARDEN AND FOOD SUPPLY

- Establish shaded and mulched gardens within settlement.
- Establish settlement windbreak and shade trees.
- Extend successful species along corridors of sandy river beds, wadis, water runnels, foothills. Seek out favoured niches in rocks, seepages, shaded areas for high–value trees and vines.
- Establish water–harvesting swales and fields for wet–season storable crop. Swales may support trees and vines after a few rains have infiltrated into the soil.

WATER SUPPLY

- Primary drinking water can be supplied by tanks off roof areas.
- Gardens and firewood plantations are established using a combination of swales and drip irrigation.
- Broadscale crop needs both windbreak and water harvest at a ratio of 20 ha of run–off to 1 ha sown.
- Seek out safe dam sites near settlement, in shaded valleys or off streams diverted to drop silt.
- Store water in sand–filled dams or gabion–stabilised terraces.
- Use water wisely and route greywater to toilet use.
- Never use deep bores or pumped well water beyond the ability of the recharge areas to re–supply usage. Test all such water rigorously for salts, radioactives, nitrates, fluorine, and biological contamination.
- Wherever water is infiltrated into the ground, plant trees to keep salt levels down.

HEALTH

- Reduce dust in settlements via trees and soil pitting to avoid sinus and other problems.
- Check for water–born disease where children swim.
- Supply ample vitamins via home garden vegetables and fruits.
- Check plants, soils, and blood levels for essential minreals, especially zinc and iron.
- Reduce use of imported carbohydrates (sugars, starches) and rely instead on locally–grown crops.

11.16

REFERENCES

Bradshaw A. D., and M. J. Chadwick, *The Restoration of Land*, University of California Press, 1980.

Cribb, J. N. and A. B., *Wild Food in Australia*, Collins, Sydney, 1974. (A basic book on aboriginal food plants in Australia).

Corbett, Micheal, *A Better Place to Live*, Rodale Publications, 1981.

Hall, Norman, *et.al.*, *The Use of Trees and Shrubs in the Dry Country of Australia*, Dept. of Nat. Devel.; Forestry and Timber Bureau, Canberra, Australia, 1972. (The basic book on Australian deserts).

Harrington, H. D., *Western Edible Wild Plants* (of the USA), University of New Mexico Press, 1972.

Hagedorn, H. *et al.*, *Dune Stabilization: a survey of literature on dune formation and dune stabilization*, Germany Agency for Technical Cooperation Ltd., Eschborn, West Germany, 1977. (Excellent summary on this subject).

Newbigin, Marion I., *Plant and Animal Geography*, Methuen & Co., London, 1972.

Indra Kumar Sharma, 1983, *Ecodevelopment in arid and semiarid areas of Asia for food, fodder, fibre, wood, essential bioproducts; shelter against sand storms, parching solar radiation, and droughts*, (Thar desert). Second World Congress on Land Policy (ICLPS), Massachusetts, June 20–24, 1983.

Southwest Bulletin, April 1980 5 (4), P. O. Box 2004, Santa Fe, NM, USA. (Good material on desert plants and agriculture by Gary Nabhan and others).

Weber, Fred, *Reforestation in Arid Lands*, VITA Publications, 1977.

Chapter 12

HUMID COOL TO COLD CLIMATES

12.1 INTRODUCTION

This chapter deals with the cool to cold humid climates, where precipitation exceeds evaporation over the year. These climates differ from drylands in respect to evaporative effect, and from tropics in that frosts can regularly occur in winter. It is in this section also that we will deal with aspects unique to a cold climate (snow, ice, frozen ground, snow avalanche), and the character of settlements, landscapes, and husbandry in these regions.

Cool humid climates include Mediterranean areas, polewards to the boreal forests (the Taiga) that rim the tundra or cold deserts. For the most part, they are winter–wet climates, and suffer fogs and frosts. It is in the European and southwest Asian areas of cool humid climate that what we call contemporary or broadscale crop agriculture developed. This form of agriculture (high capital input, mechanised, energy–intensive, using artificial fertilisers and pesticides) has been inappropriately exported as an ideal to tropics and deserts, where neither soil, water, nor financial resources can support it.

In more sensible times, the cool (mesothermal) climates were a mosaic of mixed forests, extensive hedgerow, small fields, permanent meadows or prairie, and relatively small vegetable plots. Trade and export pressures from a subsidised agriculture have destroyed this stable use of land, and forced large-scale grain crop and feed–lot systems on farmers. Fields have been coalesced, hedgerow destroyed, and few significant forests remain within the farmed areas. Even where forests are preserved, the heavy industrial base of temperate areas, plus overuse of fossil fuels generally, have created widespread problems with acid rain, affecting the foothill forests and the thin soils of older shield areas, with secondary effects on water quality due to a release of excess metals from rocks and soils to streams.

Ozone production and soil losses (people now speak of the "desertification" of such areas as southern England, where wind effects on exposed broadscale soils sown to grains are pronounced) are slowly decreasing production on once–fertile fields. There is widespread pollution from the overuse of nitrogen fertilisers and biocides, compounded by such accidents as Chernobyl and persistent radioactives in acid soils, uplands, and field crops. Groundwater quality is decreased, and potable water supply becoming a problem.

The 40–year thrust for more and more production from temperate farms succeeded in producing surpluses in almost every farm product (grains, meats, dairy, and wines) at enormous cost to the public and the land. Thus, the central problem of the mesothermal climates and their societies is not (as in deserts and tropics) one of subsistence, but the very opposite. It is how to change energy and land use to create a sustainable future for such societies, without the huge public subsidy of modern agriculture.

However, since 1970 or thereabouts, there has been a widespread home garden movement, and an increasing development of urban farms under community control. The production (in dollar terms) of such home gardens now equals or exceeds farm production, and gardening is still increasing in popularity. Public attitudes to industrialised or polluted farm product is also changing due to data almost every day on the health effects from residual sprays, hormones, and excessive nitrogen or mineral content. Radioactive fallout has not helped to inspire consumer confidence in farm product, which promise to create a new category of human morbidity, called agricologenic disease (illnesses caused by farm chemicals).

As a result of better health education, animal fats, carbohydrates, and red meats are falling into disfavour as the effects of diet on diabetes, heart disease, obesity, and immune response are being elucidated. These effects are worsened by the more sedentary lifestyle of a mechanised society. There is a pronounced consumer swing to lighter diets, less fats, starches, and sugars, and a greater consumption of naturally produced vegetables, fruits, lean meats, fish, and poultry.

As many people now view modern agriculture, it has few essential roles to play in a sane future economy. Perhaps the production of lean protein and energy crops (firewood and liquid fuels, biogas and forest product) is the only foreseeable farm future. Conservation forestry for watershed protection and wildlife will certainly be a large part of future land use in the more affluent, leisured societies. As for food production, there is already ample evidence that the existing waste space within urban/suburban aieas can produce most of the essential food of such areas.

The societies occupying mesothernal climates have achieved, for the most part, zero or negative population growth, and thus the pressure on land to produce has lessened while the land itself has been forced into uneconomic overproduction. Perhaps the best indicator of this stability in population is that more people are regarding farms and forests as recreational reserves, and near cities from 60–80% of farm income can be derived from *social* facilities, not farm product.

12.2
CHARACTERISTICS OF HUMID COOL CLIMATES

The cool humid climates are production areas for a great variety of berry crops: gooseberry; red, black, and white currants; blueberry; cranberry; cherry, salal, and service berry; and a variety of bramble berries. The woods contain many species of wild–gathered edible fungi, some of them long since cultivated on logs of oak, poplar, birch, and sassafras (shiitake, *Pleurotus*).

Evergreen and part–deciduous forests, fir forests (northern hemisphere), and species–rich meadows are found in the few areas still uncultivated, as are oaks and maples (for sugar), beeches and alder. In terms of "mast" production (the fallen food of acorns, beechnuts), truffles, and fungi, many natural forests are rich food resources. It is to be expected that the recently glaciated areas of these climatic regions will also be well–provided with lakes.

The waters of these lands can be rich in salmonid fish, suckers, carp, pike, and sturgeon. Pastures and remnant prairie and meadow are uniquely suited to these regions, and because of the often cool or cold and snowy winters, there is much emphasis on solid barns and houses, hay storage, field shelter, and root crop production for winter fodders and storage. Houses must be carefully constructed for winter warmth, and in the colder areas are often built with cellars or pit storages.

In these cold temperate areas of earth, outlined in Holdridge Life Zones (**Figure 5.2**), we are dealing with seasonal factors of frost, snow, ice, and (to the polar extremes), frozen ground. We also lose a great deal of heat energy, hence growth in plants and animals, to cold winds. Even on snow–free coasts, nearby mountains, plateaus, and inland snowfields advect cold air to the fields and forests.

Climatic subsets include the foggy coastal climates of northern California to Vancouver, western Euope, southern Chile and Tasmania, and the winterwet climates of the mid–latitudes. Most of Europe, China, Japan, North America to the northwest and northeast, Canada, and the Andean slopes and southern half of South America, the southeast and a section of southwest Australia, and mountain climates of southwest Asia can all be typified as cool humid areas, as are the outer slopes of the Himalayas and other high foothill areas in more tropical areas. The cool highlands of the tropics lack frosts, and some growth occurs all year.

Original vegetation was, or can be, mixed broadleaf and pine forests (evergreen or deciduous), wet sclerophyll forest, meadows, cool swamps and marshes, fens, and bogs. Periglacial lakes and moraine may occur over vast areas of the northern continents, and cool prairies or steppes develop in treeless areas recently covered by ice.

In general, the areas considered here have their coldest month below 0°C, and warmest above 10°C (mean or average temperatures). There is usually a winter–wet period, although rain can occur in any month, and drought occurs midlate summer. Most plant growth is in spring and summer, with a less marked period of autumn growth before the resting period of winter.

In Mediterranean areas, valley floors are favoured agricultural sites, and on the slopes, deeprooted and drought–resistant trees and vines are often grown Towards the older coniferous forests, soils are generally less fertile, but a band of brown/black clayloams of high fertility lies within the area of broadleaf forests; the centres of contemporary agriculture.

The mesothermal climates lie in the westerly wind belt, with the "roaring forties" (Latitude 40–45°) subarctic air in from polar lowpressure cells. In general, there is a 7–10 day sequence of frontal rains, with occasional easterly gales produced by large stable high–pressure cells to the east. Winds are particularly damaging when they blow off coastal seas in summer, as salt–burn affects trees and gardens.

Thus, except in cooler inland and sheltered sites, windbreak is essential for animal health and crop protection, and *permanent* forest edges must be developed both to retard fire and to prevent blowdown of single–age stands. Oak, willow, blackwood (*Acacia melanoxylon*), *Coprosma repens*, poplar, hawthorn, alder, aspen, and birch are just some of the ideal forest–edge trees for plantation protection.

On coasts, firs, cypress, *Araucaria* spp., *Coprosma*, waxberry (*Myrica*), and *Lycium* in snowfree areas provide frontline protection, with lower hedges of wormwood, rosemary, *Rosa multiflora*, hawthorn, and gorse (*Ulex*) around fields and gardens. Some of the taller grasses (pampas grass) or "fedges" of succulents and hardy vines greatly assist garden protection. Wind is a major determinant of yield over all coastal and upland regions of these climates.

In acidic peaty areas and turfs, fences of living peat or turf species can protect gardens. On cold exposed plateaus or coasts, drystone walls, soil banks, or ditch–and–bank are primary protection (many of these can later be planted to hardy shrubs or bamboo).

Low, flat valley sites, and especially those valleys at the foot of escarpments capped by high plateaus, may be subject to severe winter frosts. In mountain foothills, frost can lie all day in winter on the shaded side of valley slopes. Frosts affect pasture and herb-aceous plants by stopping growth or causing frost death by plant cell rupture. A great many species of plants from bracken fern to tomatoes are frost–killed, and many cool area farmers choose crop plants on the basis of frost–free days in the growing season, or (where frosts can occur in any month) by excluding from field culture any plant which is frost–susceptible over its growing season. Marginal or semi–hardy plants such as bamboo may survive frost if given an autumn dressing of fertiliser salts; high salt values in plant cells prevent cell rupture.

Cold humid air is viscous in flow, and stabilises in valleys as a dense are mass with a near–level upper surface. As most frosts form in calm clear weather, the upper cold air surface is rarely subject to strong winds, and this creates a marked frostline on hill slopes, often revealed by the contours joining frost–susceptible pines, jumipers, or evergreen trees.

Downslope cold air flow may have a distinct pulsing behaviour of about one minute where air flows from highlands, and this pulsing can create a series of surface waves in the cold air masses. Such waves are typical of large areas of the air above frosted ground.

Special frost–free sites, usually high on sun–facing valley slopes, in clearings on ridge forests, or in smaller clearings (less than 30 m across) in tall forest are therefore chosen growing sites in frosty areas.

12.3
SOILS

The striking characteristic of cool temperate soils is their ability to accumulate humus under natural regimes of forest and prairie or meadow. In this respect they differ from both tropical and arid–area soils, pH values in areas of poor drainage can be 3.5–4.5 (humic acids), and only over limestone, dolomite, and chalk deposits do soil pH values normally exceed 7. Traditionally, most soils have been limed to modify this acidity factor.

Another unique feature over vast areas of the northern hemisphere, and to a lesser extent in the southern hemisphere, are sheets of periglacial outwash, compacted till, moraine, and glacio–fluvial (ice and water) outwash, with downwind deposits of rock flour and fine particles (known as loess). Such areas are normally mineral–rich and fertile due to the mix of rock types in glacial debris. Braided wide streams and complex benches of glacial sediment along valleys typify the rivers with glacial headwaters, and low meltwater temperatures (46°C or so) may be maintained for most of the year. Acid waters containing humic acids, tannins, and saponins may issue from bogs and marshes. Even at pH 3.5, such streams (if unpolluted) can sustain good fish populations, although few molluscs occur due to low calcium availability.

It was in these regions that plough agriculture developed, with the old traditions of "high farming", which included the concepts of crop rotation (over 49 years) and a rest period, ley of green crop, or pasture. Because livestock had to be shedded in the colder areas, manure spreading on fields and hay feed stores in barns were also developed, hence ley crop, silage, permanent pasture management, intensive livestock husbandry, and root crop for winter storage (turnip, beet, potato, carrot, parsnip). Traditional grains were barley, oats, wheat, and rye, and traditional foods were meats, vegetables, and breads. Many methods of winter storage of foods were developed, ranging from burial of butters and acorns in bogs to ice storages, pits, cellars, and silage systems for fodder.

Soil, for the most part (excepting some sandy coasts) have good structure and clay fractions, high natural humus, and good CEC capacity. Adjusted for pH, micronutrients, and trace elements they are ideal crop soils, unless derived from acidic rocks (granite, quartzite, gneiss), where peats are developed.

12.4
LANDFORM AND WATER CONSERVATION

Open water storages are peculiarly appropriate to the mesothermal landscape. Soils generally possess sufficient clay fractions (over 40%) to ensure secure dam walls, evaporation does not exceeds precipitation in normal years, and there is a wide range of plant and animal species dependent on water for their production.

It is in the classic "S" profile of the humid landscape that Yeomans[6] developed his Keyline approach to farm water management, which integrates open water storage with soil reconditioning, soil water storage, fencing, and farm forestry in a whole system design. The classic Keyline system of gravity flow irrigation refers only to gentle foothill country in the headwater regions of first and second order streams, but skilled design over a very wide range of soils and of landscapes can achieve the essentials of drought–proofing by gravity–flow irrigation, soil conditioning prior to forest or pasture establishment, fire control by downslope flooding, and integrated forest farm systems.

It is in humid landscapes that the techniques of water storage, diversion, and the various uses of stream or overland water flow can be most developed.

The overriding design input into humid–area landscape is therefore planned on the basis of water management, followed by access planning (roads and tracks), then plant and animal system planning. In water planning, we start from the highest accessible slopes, and work out methods to lead water flow via the longest routes downhill (working on or near contour), creating small dams, swales, rip–lines, and (where appropriate) and energy systems to take advantage of the abundant rainfall.

Water has all of the following tasks or duties in the landscape:

- As soil and pond storage, to irrigate crops and forests.
- In homes, for cooking, drinking, cleaning, and toilet uses.
- As a source of energy via turbines or as hydraulic pressure.
- In growing fish and aquatic plants; as a growing medium in itself.
- As a carrier for nutrients. Nutrients added to upstream or uphill areas are distributed downstream to crops.
- Recreational and aesthetic uses.

12.5

SETTLEMENT AND HOUSE DESIGN

The greatest cost in house maintenance in temperate cold areas is that of space and water heating; together, these may make up 80% of all domestic energy costs. Both settlement design overall and house design in detail is a critical conservation factor in human occupation of wintercold areas, more so than in tropical areas, or in hot deserts.

Not only housing, but barns, outbuildings, and livestock shelters must be solid, well–designed, and carefully assessed for thermal efficiency.

SETTLEMENT DESIGN

Settlement site choice, and the surrounding plantings in landscape, are probably so critical as to provide for 70% of the conservation of heat energy in cool climates; actual structural techniques cope with the rest. Thus, in creating a sustainable settlement, the following factors are important.

- Village or streets aligned eastwest at the mid-slope (thermal belt) of a sunfacing slope, preferably with forests and high water storages above the site.
- Housing closely placed or conjoined at east and west walls, and preferably of two to four stories. These factors reduce insulation costs and create a compact site.
- Careful planning of accessory landscaping to pro-

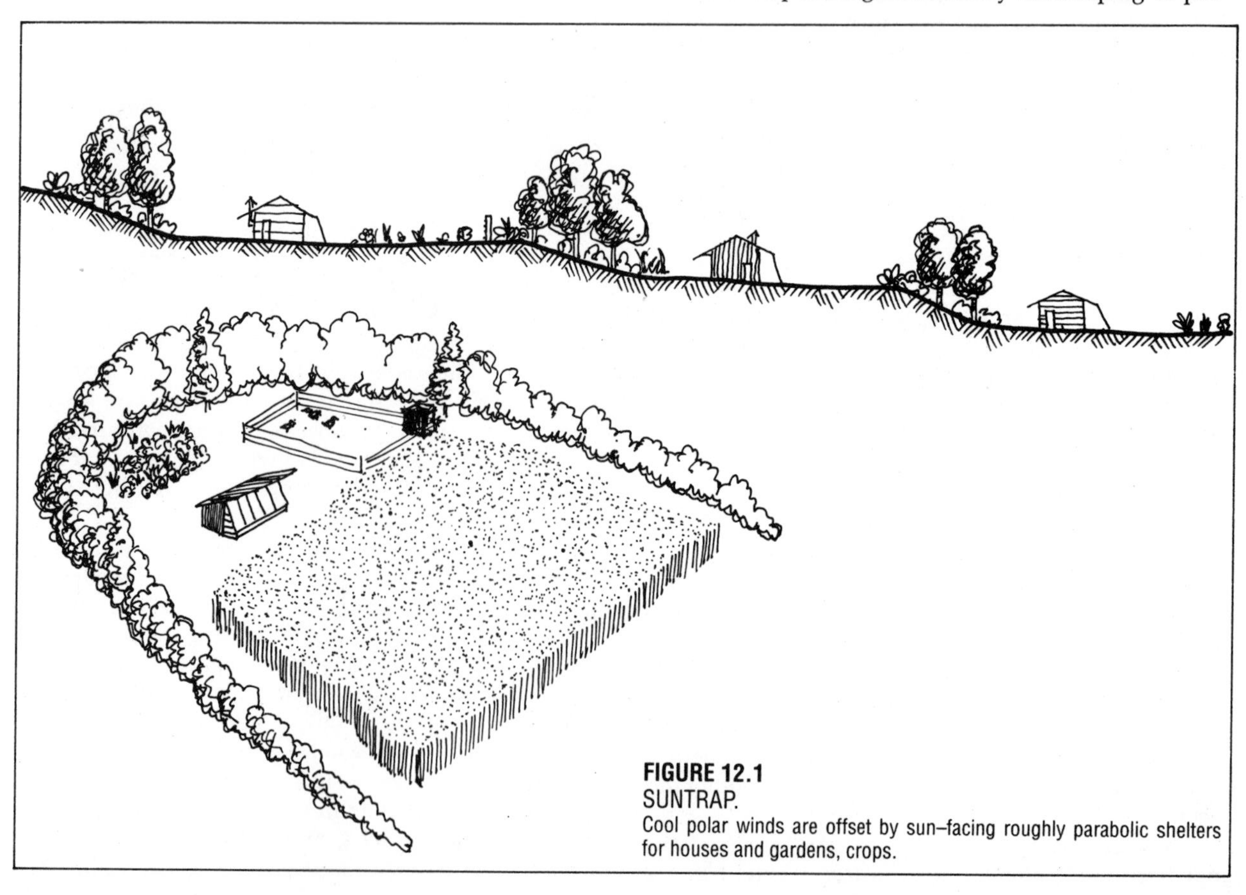

FIGURE 12.1
SUNTRAP.
Cool polar winds are offset by sun–facing roughly parabolic shelters for houses and gardens, crops.

vide for:

a. Dense windbreak polewards or uphill of the settlement, and to the east and west (a "suntrap" structure) **Figure 12.1**.

b. Dense attached vines on all sound masonry walls, or out from wooden walls, to assist insulation.

c. Deciduous trees to sunwards chosen to admit low winter light to all facades to the sunward aspect, especially in mid–winter.

• Siting to avoid unstable soils, avalanche tracks, or flooding areas, and where possible above local frostlines.

• Site to be off radioactive or radio–emitting rocks if possible.

Preferably, settlements in cool areas should present a stepped aspect, so that each dwelling presents a full facade to the winter sun (**Figure12.2**).

HOUSE DESIGN

Glazing

Given that settlement site is carefully selected, and landscaping is as carefully designed, the individual house can self–provide for heat and hot water by good construction technique. Glazing of the facades in any cool–area site should ensure that:

• There is from 30% (Lat 3035°) to 100% (Latitude 60°) glazing on the sunward walls.

• No glazing on the west walls (a window in the east for a kitchen is controllable).

• Minimal and perhaps double–glazing in the poleward walls (above Latitute 40°, in cold continental interiors or alpine sites).

House Proportions

Houses should be no more than two rooms (10 m) deep on the north/south axis, and may be 1.5 times longer on the east/west axis, so that winter sun can penetrate windows to the walls of the poleward rooms. Now, it remains to adjust the sunward roof eave width, and the sill height of windows to admit winter sun to the interior (from early autumn to late spring) or in very cold climates, to attach a solarium or glasshouse for day use in winter. Clerestory or attic windows to the sun side heat rear rooms or roof space.

Floor and Sub–floor Area

As for construction materials, concrete slab or mud, brick, or tiled floors in contact with the ground are ideal to absorb heat into the underfloor earth mass. These can be covered with wooden parquet, carpet, or any preferred surface without long delaying heat transfer efficiency, so that any usual floor covering can be used, including wooden flooring over battens.

The floor area and the earth beneath it are the critical heat stores for winter, radiating solar heat at night, or over cool cloudy periods. As most of this stored earth heat escapes via surrounding soils, and most of that loss occurs within 0.5 m of the earth surface, we must insulate the entire earth mass below the floor by digging a 1 m deep foundation trench, and providing 5–15 cm thickness of insulation around the whole periphery of the floor outline.

Insulation

Given that the subfloor area is insulated, the next most important sources of heat loss are the ceiling and the windows at night or on cool cloudy days. The ceiling should therefore be fitted, concurrently with construction, with 5–15 cm of insulation (as fibre, wool, seagrass, sawdust, feathers, or shredded paper), and the windows provided with oversized floor to ceiling curtains. In areas of severe cold, sliding insulated shutters are appropriate. Insulation adds 5–8% to house costs, but pays for itself in 35 years in energy savings.

Walls

Walls can be massive, of stone or pise, or double–walled (reverse brick veneer is ideal: bricks to the inside, timber or wallboard to the exterior) and the wall cavity also insulated. Light insulation suffices in the east and west walls, mainly as a sound barrier, if houses are conjoined.

In all areas of cold winds and blizzards, double entries are essential, and in alpine and snowy areas, farm housing is usually conjoined to barns, feed storages, and wood or fuel storage so that one need not

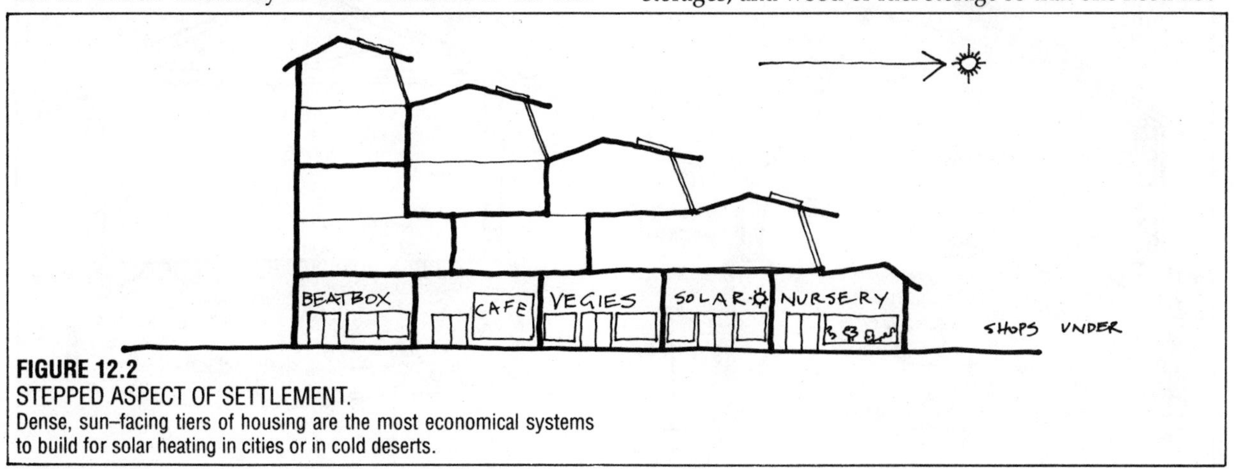

FIGURE 12.2
STEPPED ASPECT OF SETTLEMENT.
Dense, sun–facing tiers of housing are the most economical systems to build for solar heating in cities or in cold deserts.

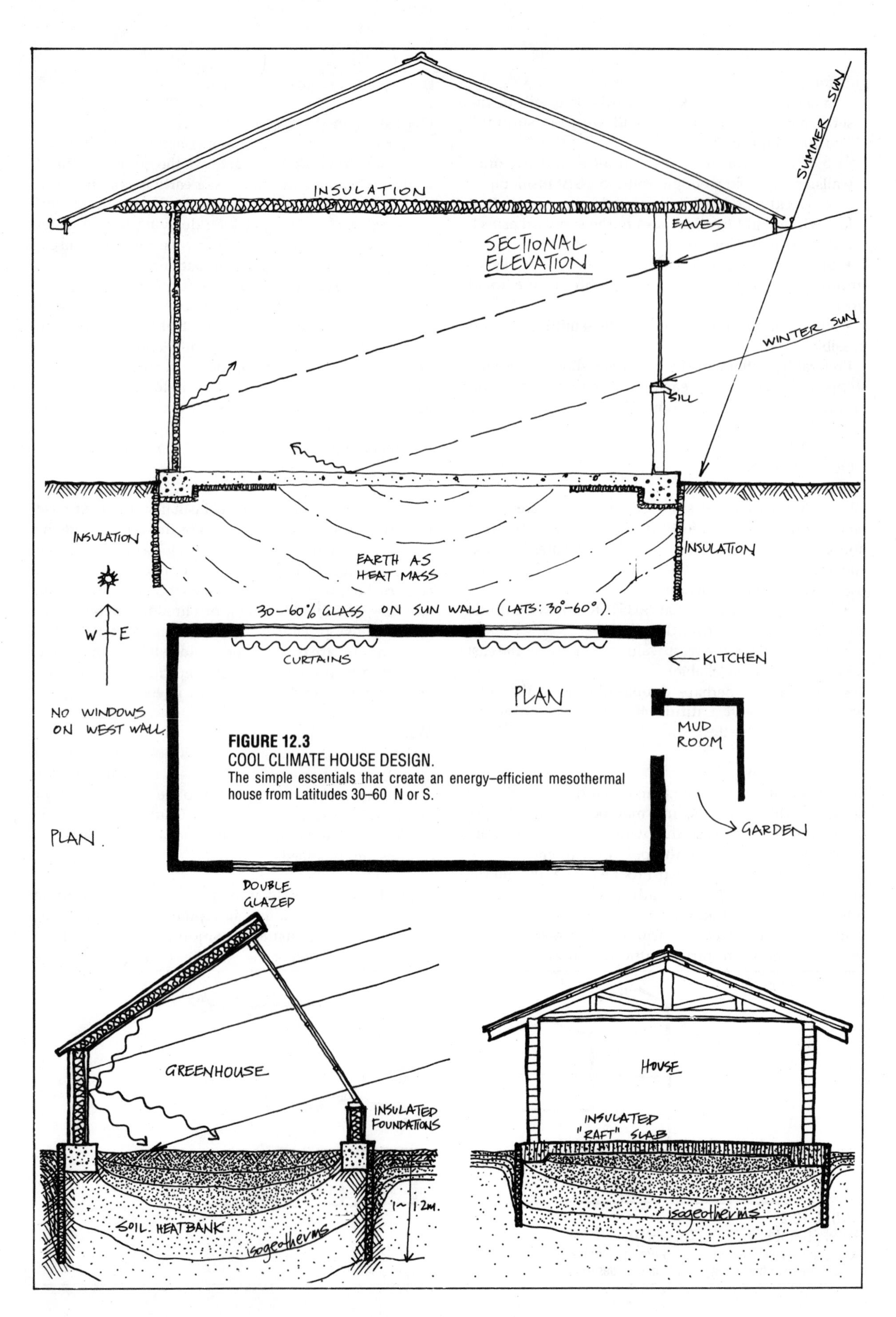

FIGURE 12.3
COOL CLIMATE HOUSE DESIGN.
The simple essentials that create an energy–efficient mesothermal house from Latitudes 30–60 N or S.

go outdoors in blizzards or deep snow to feed livestock or obtain fuels. In such climates cellar storage of house food, and deep placement (1–1.5 m) of all water lines is also essential.

12.6
THE HOME GARDEN

(See also Chapter 10—the Humid Tropics—for mulching, geometry of beds, etc.).

One special feature of cool area home gardens is the need to grow and store crop (in the field or home) over the winter months, as growth effectively ceases from late autumn to early spring. The middle of spring is known as the "hungry gap", when winter crops are finished and new crops are not yet yielding.

The winter food gap is bridged by growing root crop, cucurbits, and subsequently processed crop for preservation. These crops mature in late summer, and are stored or preserved (some crop can also be left in the ground). Staple root crops are the potato, parsnip, carrot, and minor species such as oca, sunroot (Jerusalem artichoke), and bulbs such as onions. Field crops left to over–winter include fava bean, kale, chard (silver beet), parsley, and a variety of cabbage, broccoli, cauliflower, and sprouts under straw or shelters.

There are two main planting periods. The first months of spring are for salad vegetables and all summer crop, and the last month of summer or first month of autumn for all over–winter crop and all root crops of the year. In milder areas, a green crop (green manure) can be sown before winter (oats, tares, fava beans) and slashed or dug–in in the spring.

In cool areas, cellar or pit storage is extensively used to store crop. Potatoes are stored in straw or ash in dark boxes or humid cellars. Carrots and parsnips (tops removed) are in mounds of sand or ash. Apples are separately wrapped or spread out on attic floors (they are not stored with root crop because of their ethylene production, which causes root crop to shoot), and cabbage or kale can be stored, uprooted, by covering with hay in the open, as a ferment (sauerkraut) in large jars, or as sterilised or frozen storage. Today, many crops are stored frozen in deep freezes, and retain good nutritional value.

Planning the garden thus means small beds for the spring planting, with main crops of potato, tomato, and sweet corn, then a reversion to larger mass beds in the autumn planting for winter storage crop (potatoes, carrots, parsnips, turnips, cabbage, fava beans, peas, *Brassicas*). Chard and parsley are important all–year foods and several successions can be sown to reseed at irregular intervals, so that they are always in the garden.

Otherwise, the garden layout is very much the same as that given under the tropics chapter, but omitting any overshaded areas or tree crop interplant, as light levels can be low in cloudy weather and light saturation is rarely a problem. Mulch is, however, appropriate, and the sole exception to this may be for spring carrot planting, when mulch is pulled from the seedbed to let it warm up early. All other crops planted in mulch catch up with bare–soil crop and mature at the same time.

For cool areas, a small glasshouse for seedling trays, and a series of cloches for early transplants are invaluable. Even potatoes can be sprouted on old hessian bags indoors, or in any open shadehouse, and transplanted complete with leaves and roots by cutting up the bags into squares.

There are several ways to perennialise annuals, some of which are:

- Carrots: cut a disc off the top, sprout it on soft wet paper in a shallow tray, and replant; one gardener told me she has been eating the same carrots for 9 years!
- Leeks: let one in four go to seed and dry off, then remove the small bulbils from the base and replant. When harvesting, cut off the leek 2 cm below ground level. The central shoot quickly regrows. There is also a variety of perennial bunching leek (called pearl onions).
- Lettuce: pick single leaves, allow to go to seed, scatter seed. Mignonette variety is a good self–seeder, are some Italian varieties.
- Cabbage: cut the head 8 cm above the ground, then deeply crosscut the stem. Each quarter of the stem regrows a small cabbage.
- Garden fennel: this plant self–seeds, as does chard, parsley, parsnip, etc. if allowed to go to seed. Judicious thinning, weeding, and transplanting is in order.

When saving seed from any umbelliferous plant, collect the strong, mature seed for the terminal panicle only; this germinates best of all. Side panicles contain immature and small seeds, and can be pruned or rejected.

TRELLIS CROP in mesothermal climates is important. Many peas, beans, and cucurbits will trellis, as will varieties of tomatoes. But trellis *over* the garden is inappropriate (unlike in deserts), and either vertical zigzag, inverted "V", or house wall and garden fence trellis is best, aligned north/south where possible for even sun effects (**Figure 12.4.**)

SMALLFRUIT, as currants, gooseberry, raspberry or trellis blackberry, is a feature of cool–area garden. Six to fifteen plants of 78 species yields a great deal of fresh or frozen food of high vitamin C value. Many such rows can act as windbreak in the garden.

SEED AND SEED RESOURCES

There is no need here to restate the evidence that seed, especially the 20–30 basic food crop seeds of grains and grain legumes is today subjected to a conserted effort at total control by a few agro–chemical multinationals, as part of a global power play to control nations (or rather mineral resources) by starving regimes who try to use local mineral resources fairly, or to dislodge the parasitic grip of exploitive multinationals; see, for example, works by Pat Roy Mooney (*Seeds of the Earth;*

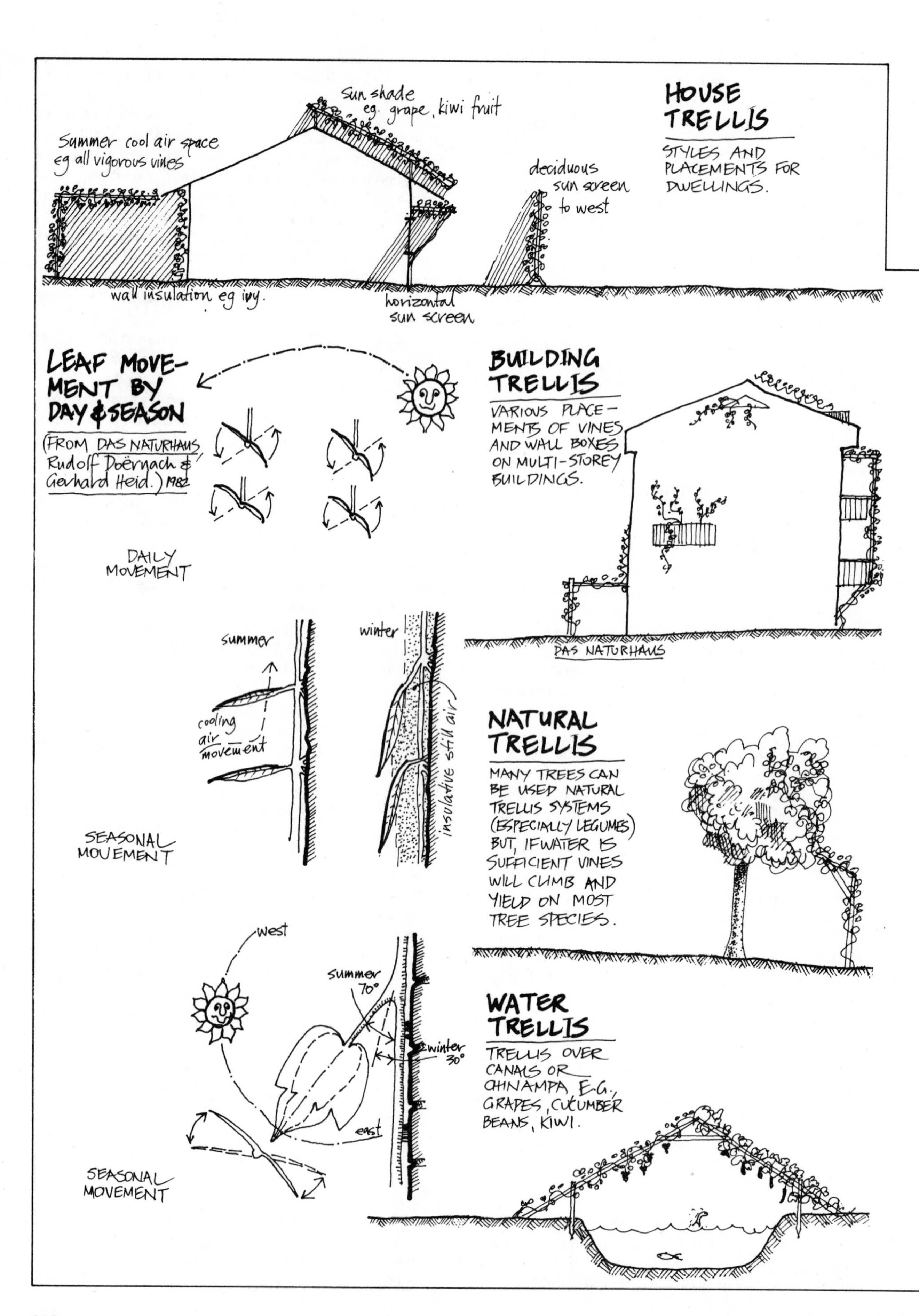
HOUSE TRELLIS
STYLES AND PLACEMENTS FOR DWELLINGS.
Sun shade eg. grape, kiwi fruit
Summer cool air space eg all vigorous vines
deciduous sun screen to west
wall insulation eg ivy.
horizontal sun screen
LEAF MOVEMENT BY DAY & SEASON
(FROM DAS NATURHAUS, Rudolf Doërnach & Gerhard Heid.) 1982
DAILY MOVEMENT
BUILDING TRELLIS
VARIOUS PLACEMENTS OF VINES AND WALL BOXES ON MULTI-STOREY BUILDINGS.
DAS NATURHAUS
summer
winter
cooling air movement
insulative still air
SEASONAL MOVEMENT
NATURAL TRELLIS
MANY TREES CAN BE USED NATURAL TRELLIS SYSTEMS (ESPECIALLY LEGUMES) BUT, IF WATER IS SUFFICIENT VINES WILL CLIMB AND YIELD ON MOST TREE SPECIES.
west
summer 70°
winter 30°
east
SEASONAL MOVEMENT
WATER TRELLIS
TRELLIS OVER CANALS OR CHINAMPA E.G., GRAPES, CUCUMBER BEANS, KIWI.

a public resource, World Council of Churches, Canada).

In short, by contracting seed trade, seed patents, and seed retail outlets, the few powerful state/industrial cooperations are preparing the ground for de facto government by controlling food, via "aid" allocation and market control; a sick and destructive use of power. In response, people and organisations everywhere have set up seed exchanges, seed libraries, open pollinated and non–patented seed companies, and hundreds of thousands of growers have studied home seed saving systems.

Despite the enormous loss of locally adapted seed (estimated at 85% of European varieties) that resulted from the monopoly control of market by seed patenting, the encouraging result of "people power" is that most gardeners or farmers today can assemble more species and varieties of seed than ever before in history, by using the seed exchanges and local

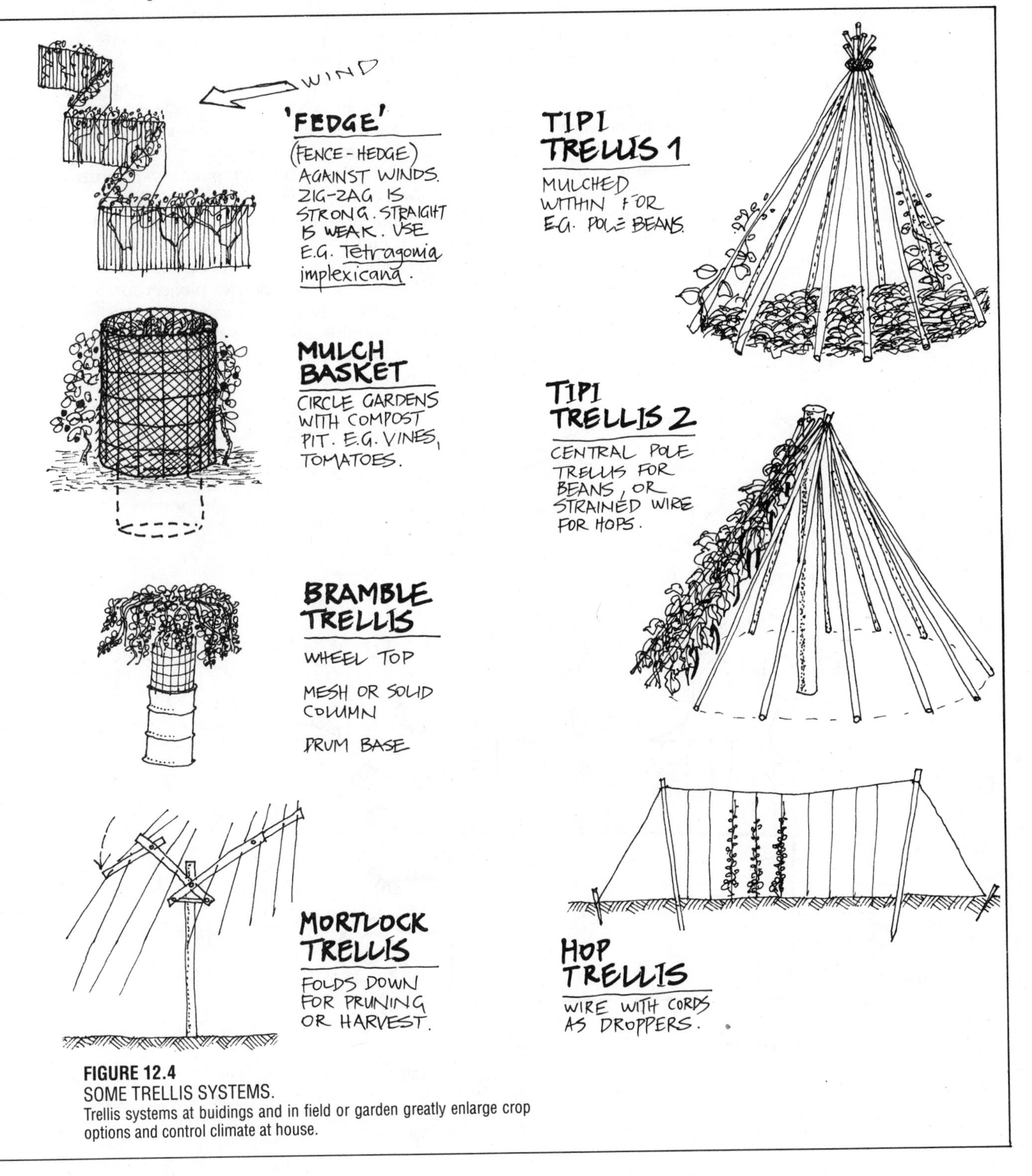

FIGURE 12.4
SOME TRELLIS SYSTEMS.
Trellis systems at buidings and in field or garden greatly enlarge crop options and control climate at house.

collections instead of the patented hybrid seeds offered by the controlled markets.

Every gardener who is opposed to external control (which is *why* we garden) preserves a few valuable varieties: many of these are heirloom or locally reliable traditional food plants specific to site, climate, or regional preference; in total, these species and varieties give any new gardener a vast range of potential product.

Thus, a region in any climate can escape external food control via a local, regional, and national network of seed exchanges and open pollinated (non–hybrid) non-patented seed. It is particularly important for small farmers to grow and share seed, as seed outlets for farmers are bought up by the agro–chemical industry, and seeds altered to suit their purposes (not the purposes of local food production). Seed sources are numerous, and are listed in most organic farming or permaculture publications; note too that perennials and trees providing staple food have largely usurped seed patenting.

12.7
BERRY FRUITS

No region so suits berry production as the cool humid climates. There, one or other berries occupy niches from high montane to seashore sites, and natural stands of berries fill forest clearings and edge roadways and paths. They are food for a variety of birds, for foxes and rodents, and for people.

In the coastal or upland bogs, *Vaccinium* spp. (blueberries, cranberries) thrive in oxygen–poor soil (each has a specific heathfamily root associate to provide nitrogen), and in the clearings of forests bird–carried berry seed germinates on fallen logs and tree stumps, where 46 species may compete for space. So favourable are these stump sites that slabs of fallen trees will proliferate berry mounds on lawns or in gardens, and peat–like bricks of groundcover berries such as salal will convert lawn edges to berry groves.

Berries demand little but a humus–rich and somewhat acid soil, high ammonia nitrogen (provided by birds and rodents in the wild as urea), and a thick mulch to discourage grasses. Berries pioneer for, and protect, seedling trees, so that advancing forest edges often develop bramble and cane thickets. In these thickets oak, chestnut, plum, apple, and birch thrive. Brambles, in particular, protect and nourish young fruit

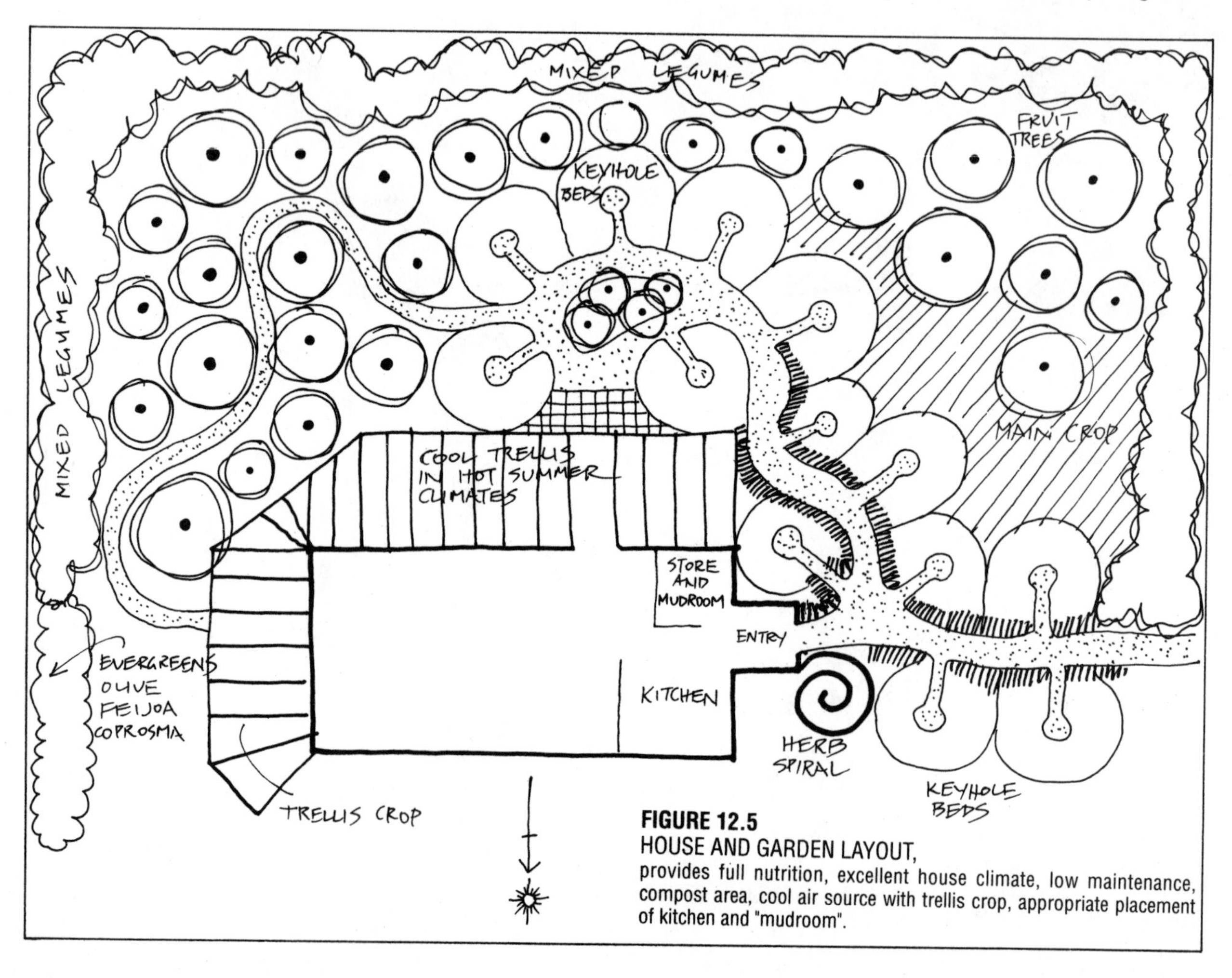

FIGURE 12.5
HOUSE AND GARDEN LAYOUT,
provides full nutrition, excellent house climate, low maintenance, compost area, cool air source with trellis crop, appropriate placement of kitchen and "mudroom".

trees, and on farms bramble clumps (blackberry or one of its related cultivars) can be used to exclude deer and cattle from newly set trees. As the trees (apple, quince, plum, citrus, fig) age, and the brambles are shaded out, hoofed animals come to eat fallen fruit, and the mature trees (7 plus years old) are sufficiently hardy to withstand browsing. Our forest ancestors may well have followed some such sequences for orchard evolution, assisted by indigenous birds and mammals.

BIRD PROBLEMS WITH BERRY FRUIT PRODUCTION.

Production of blueberries, strawberries, and raspberries attracts fruit–eating species such as blackbirds as thrushes, starlings, and parrots. There are several practical aids to reduction of crop losses due to birds:

- Cage Culture. Growers of sweet cherries, blueberry, and table quality raspberry all report economic advantage in completely caging over the area of crop. Such an advantage gives yields of up to 30 times that of open field conditions. In these cages, some about 20 m by, 10 m by 24 m high, using 10 mm mesh, a polyculture of smallfruit plus espalier fruits (nectarine, peach) is safe from birds and even large moth species. One or two such cages of piping plus mesh will support a grower aiming for high quality markets (**Figure 12.6**).

Ground preparation may involve ridging for species such as blueberry (ridges of 1.5 m wide, 0.6 m high, with a drainpipe at base is ideal), the fitting of drip irrigation, a deep mulch for pH adjustment and soil moisture conservation, and a careful schedule of liquid manures selected for crop. It is beneficial, in cage culture, to use species such as tree frogs, small lizards, and insectivorous birds (such as quail) in the cages for pest control, and to select plants of high fruit yield. Given such care, a family livelihood can be obtained. Cage walls may support trellis of bramble fruits (youngberry, boysenberry). Maintenance is mainly that of pruning and adding mulch as needed to maintain

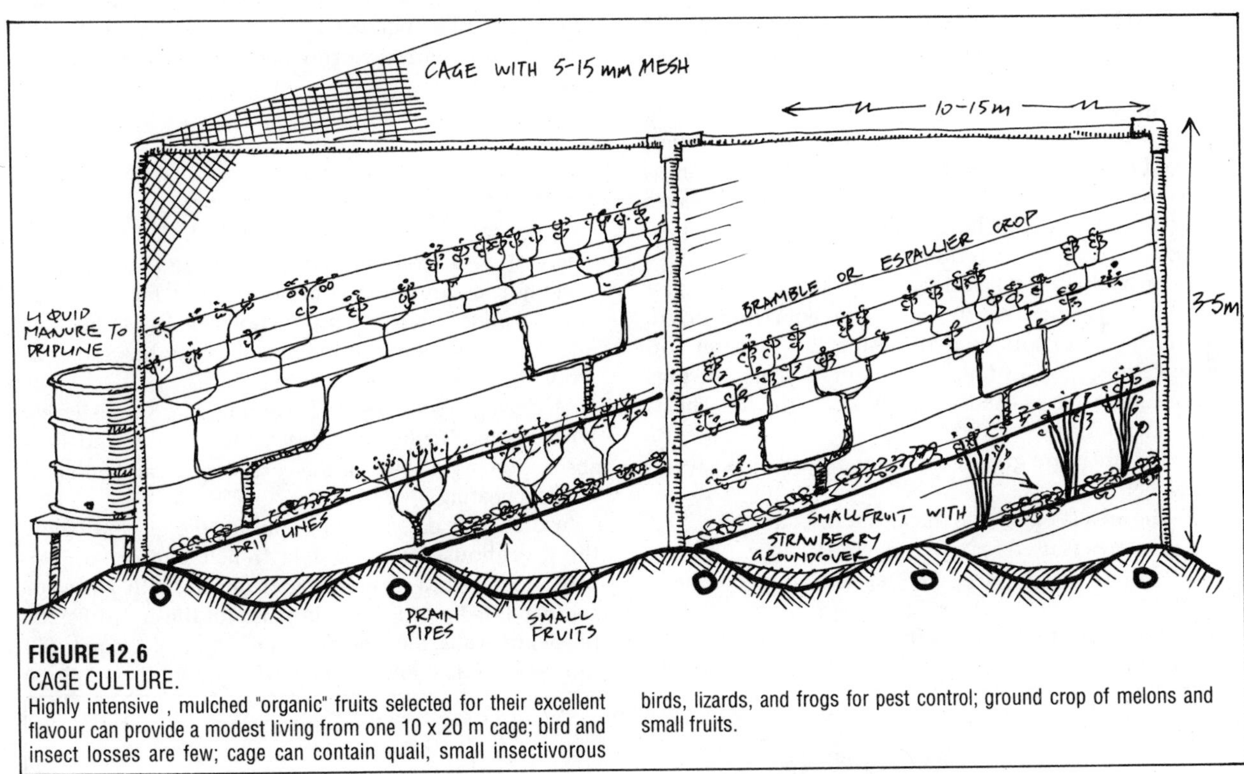

FIGURE 12.6
CAGE CULTURE.
Highly intensive , mulched "organic" fruits selected for their excellent flavour can provide a modest living from one 10 x 20 m cage; bird and insect losses are few; cage can contain quail, small insectivorous birds, lizards, and frogs for pest control; ground crop of melons and small fruits.

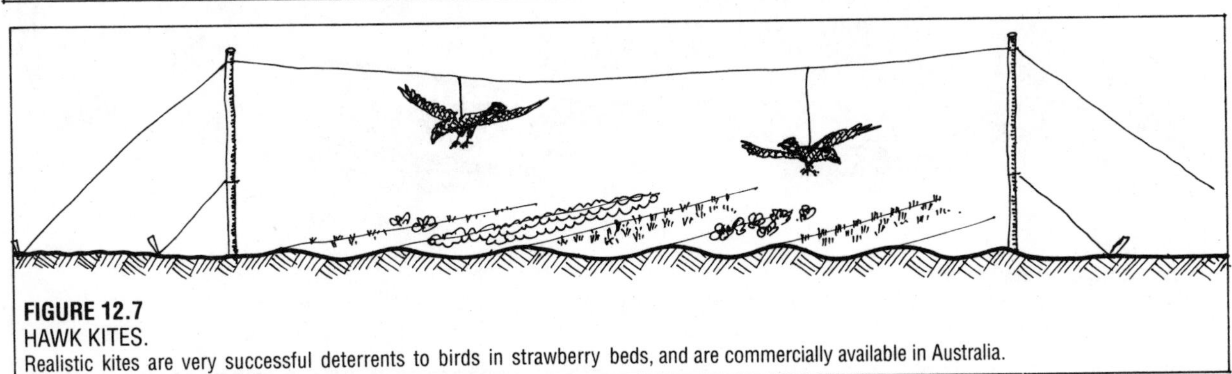

FIGURE 12.7
HAWK KITES.
Realistic kites are very successful deterrents to birds in strawberry beds, and are commercially available in Australia.

soil health.

• Field Culture. In field conditions, losses to birds are unavoidable but may be minimised by using tethered predator hawk kites above the crop at critical ripening periods. Such kites imitate local hawks or eagles, and are very effective bird repellents. They should be removed after the crop is picked so that birds do not get accustomed to them (**Figure 12.7**).

As much of the berry crop cost is in picking, it is preferable to plant for self–pick sales where farms are within 15–20 km of town markets. In this case, the berries (currant, gooseberry, raspberry, strawberry) are best planted double row on contour mounds 1–1.5 m high, with grassed paths and a wider spacing of mounds or ridges than in cage culture, so that pickers have easy access to crop. Selfpick farms need parking and weigh-out centres, buckets for pickers, and produce tubs for crop.

RAMPANT BLACKBERRY CONTROL.

The methods given here do not use any potentially dangerous chemicals, and are suited to the size of the area to be controlled.

1. Individual Plants (up to 0.1–0.25 ha). I have had success in eliminating these using a strong rubber-backed carpet, tearproof fibre, or tough plastic. Clumps are covered and weighted for 710 days, when leaves have rotted and branches blanched. The clump is then uncovered and the roots dug out.

2. Areas of 0.1–2.0 ha can be completely controlled by erecting permanent electric fencing to reduce the area to 0.4 ha (1 acre) lots, and releasing 20 pigs per lot, followed by 12 goats (on rotation). In 24 years, the pigs have eaten the roots and the goats any regrowth. This is permanent if:

• The areas are grazed by sheep or goats later (even geese help);
• The area is planted to thickish forest later; or
• The area is used for hay and regularly cut.

3. Large Areas in Steep Gullies. Two methods have worked:

• Cattle are fed in the winter with *untied* bales of fodder thrown into the blackberry gullies. They trample and destroy the vines, and "sow" grass. Such gullies need reafforestation later. This method has been used successfully by Lance Jones of Tasmania over many acres.

• Apple, fig, quince, pear, and plum trees (ungrafted whips) are planted inside the edge of blackberry clumps (12 metres in from the edge). These grow and fruit in 4–6 years. Cattle come in to eat the fallen fruit and trample out the blackberries. This method has been used over many acres by Geoff Wallace, Kiewa Valley, Victoria. The fruit trees are now up to 20 years old; no blackberry survives.

There are several methods used in confining blackberry brambles:

1. If a field is normally cropped or mown, many farmers leave walls of blackberry along fences and creeks as cattle barriers. These do *not* spread where crop or slashers are used and the fields grazed.

2. Blackberry does not penetrate through (or do well under) 3–5 years of pines or in pine plantations after years 4–7. Blackberry edges have been reported as confined by comfrey, wormwood, *Coprosma repens*, pine, or cypress hedges.

12.8
GLASSHOUSE GROWING

The cool temperate climates are those places where greenhouses are most commonly used for speciality crop or aquatic species production, for house heating, and for the winter production of vegetable and fruit crop.

Modern glasshouses are becoming very sophisticated with respect to heat energy conservation and heat absorption, as 70–90% of the cost of crop can be that of artifical heating.

Insulation is essential not only for the ground below the greenhouse, usually as a trench 1 m or so deep provided with a panel insulation, or with straw or sawdust–filled plastic bags, but also for the endwalls, the poleward walls, and the poleward slope of the roof (all can be solid insulated walls). The sunward roof slope

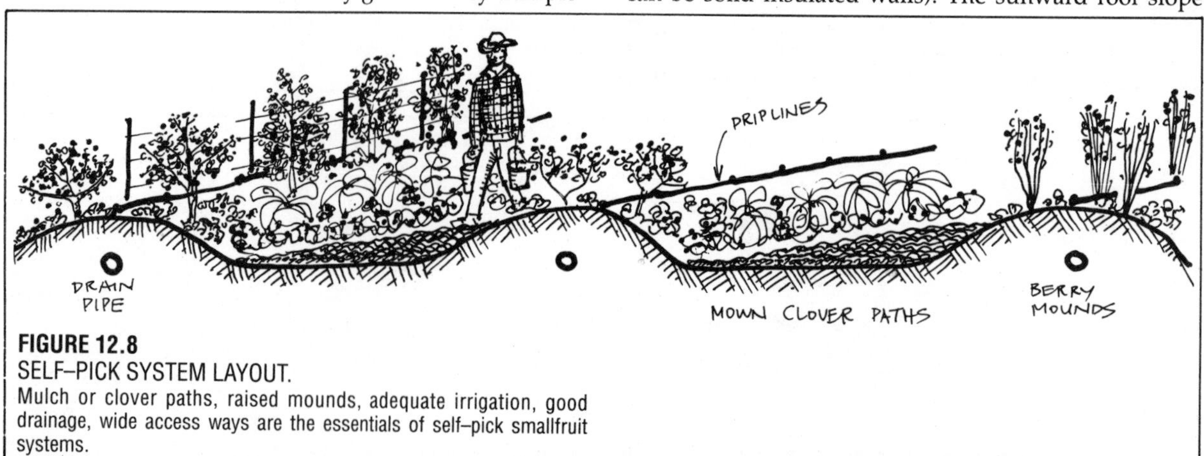

FIGURE 12.8
SELF–PICK SYSTEM LAYOUT.
Mulch or clover paths, raised mounds, adequate irrigation, good drainage, wide access ways are the essentials of self–pick smallfruit systems.

and wall can be air–insulated, either by using trapped bubble insulation as standard plastic sheets, by keeping air at pressure between double plastic walls, or by using insulated shutters at night to cover the glazed areas.

Inside the greenhouse, where the whole earth floor inside the walls is insulated from the cold earth outside, several additional heat masses can be provided, most commonly as plastic or metal tanks or drums of water (some can be fish ponds), but also by providing active heat sources such as domestic animals, compost heaps, or hot water storages filled from panel or trough–concentrator solar heat collectors.

The areas below plant benches, if used to house rabbits, guinea pigs, poultry, or any small domestic animals at night will provide considerable winter heat, as will long compost boxes along the poleward wall, or external to, but below, the level of the greenhouse itself. Dark curtains within the greenhouse also act as heat (long wave) absorbers and radiators.

Lastly, manual or automatic venting of the greenhouse, fans to create air circulation inside, and a close watch on excessive humidity (relieved by a heat exchanger to the exterior, or by venting hot air) completes the design of efficient greenhouse units. Moreover, such devices as glazing over an earth trench in a sunward slope, a narrow street between houses or shops, or the sunward side of a massive stone or brick wall also provides a frost–free and warm plant environment.

All entries to greenhouses and houses need to be porches or covered areas, provided with two successive doors, preventing heat escape as air loss to the exterior.

As greenhouse construction is relatively expensive, then the crop selection in early years should be oriented towards high–value crop such as bell pepper, tomato, snow pea, ginger, kiwifruit, and so on (depending on local preferences and demands). It is manifestly less profitable to grow low–value or bulky crop, but a domestic greenhouse can profitably provide a wide range of family food (animal and vegetable) rather than a commercial market crop.

12.9
ORCHARDS

Cool temperate humid orchard species include pome fruits (apple, pear, quince, medlar), stone fruits (cherry, peach, plum, apricot), nuts (filbert, walnut, chestnut, hickory), and a few hardy evergreens such as olive, loquat, and pineapple guava (*Feijoa*); mulberry and fig may also be grown in milder areas.

In selecting varieties for local orchards, care should be taken with the following factors:

- Adaptation to the Region. This necessitates a search of local nurseries, gardens, town plantings, and older homesteads to list species yielding in region.
- Resistance to Disease. This factor is of specific relevance to such species as apples, where some older hardy varieties show little effect of fungal or insect attack, so that yields of sound fruit may be obtained without using dangerous chemical biocides (fungicides, insecticides).
- Site Selection. The selection of well–drained, sunfacing slopes in cool areas, or shaded slopes where late frosts are expected (to reduce damage by freeze/thaw effects).
- The Selection of Plant Guilds for Orchards. See below for details.

PLANT GUILDS FOR THE POME AND STONE FRUIT ORCHARD.

The enemy of deciduous orchards is grass, thus non–grass crop below tree canopies is ideal. A selection or mix of the following plant groups can be made:

- Spring Bulbs (*Narcissus*, hyacinth): These flower and die back by early summer, as does *Allium triqetrum* (onion weed), and create a grass–free area below trees in fruit, plus a crop of bulbs, flowers, and honey. Iris and tuberous–rooted flowers also assist grass control.
- Spike Roots (comfrey, dandelion, globe artichoke) cover the ground and encourage worms, yield mulch and crop. Soil below their foliage is soft, free–draining, open to roots feeding near the surface, cool.
- Insectary Plants: *Umbelliferae* and small–flowered plants: fennel, dill, Queen Anne's lace, tansy, carrot, and parsnip flowers, and so on. Tachynid and other predatory wasps, robber flies, ladybirds, jewel beetles, and pollinator bees or wasps are attracted to interplants in orchard, e.g.: *Quillaja*, a small tree attracting many wasps, *Photinia*, and some small species of *Tamarix* and *Acacia* species. All of these can be placed in windbreaks around orchards, interplanted in rows, or as clumps in orchard. All bring predatory or pollinating insects into crops.

In the herb layer, catnip, fennel, dill (or any *Umbelliferae*), small varieties of daisy (or any *Compositae*), *Phyla (Lippia) nodosa*, and flowering ground covers generally attract wasps, pollinator bees and insectivorous birds.

- Nitrogen and Nutrient Crop. Clovers, and interplants of tagasaste or *Acacia* provide root–level nitrogen. Some control of root nematodes (*Crotalaria*), as do marigolds (by "fumigation" of soil).

In general, we aim to maximise the floristic or flowering components of orchard interplant, to reduce or even to eliminate grasses, to attract a variety of pollinators and predatory insects, insectivorous birds (using *Kniphofia* spp. *Fuchsia spp. Echium fastuosum*, *Salvia* spp.), and to provide ground cover for frogs and insectivorous lizards (small ponds or troughs throughout orchards will breed frogs for leaf insect control). Soft ground covers such as *Nasturtium* prevent soil drying and give mulch, as does the interplant and windbreak trees, and the herb layer generally.

Thus, pest species in orchard can be reduced to 4–7% of monocrop orchard by a combination of these strategies:

- Selection of disease resistant stock of the main fruit crop;
- Minimal damage by pruning, or no pruning;
- Predation by birds, frogs, lizards, wasp and predatory insects;
- Interplant of leguminous trees and other than main crop species;
- Reducing crop stress with windbreak and mulch; and
- Ground foraging by chickens, pigs, geese to clean up windfalls, or careful collection of windfalls for juice processing or disposal.

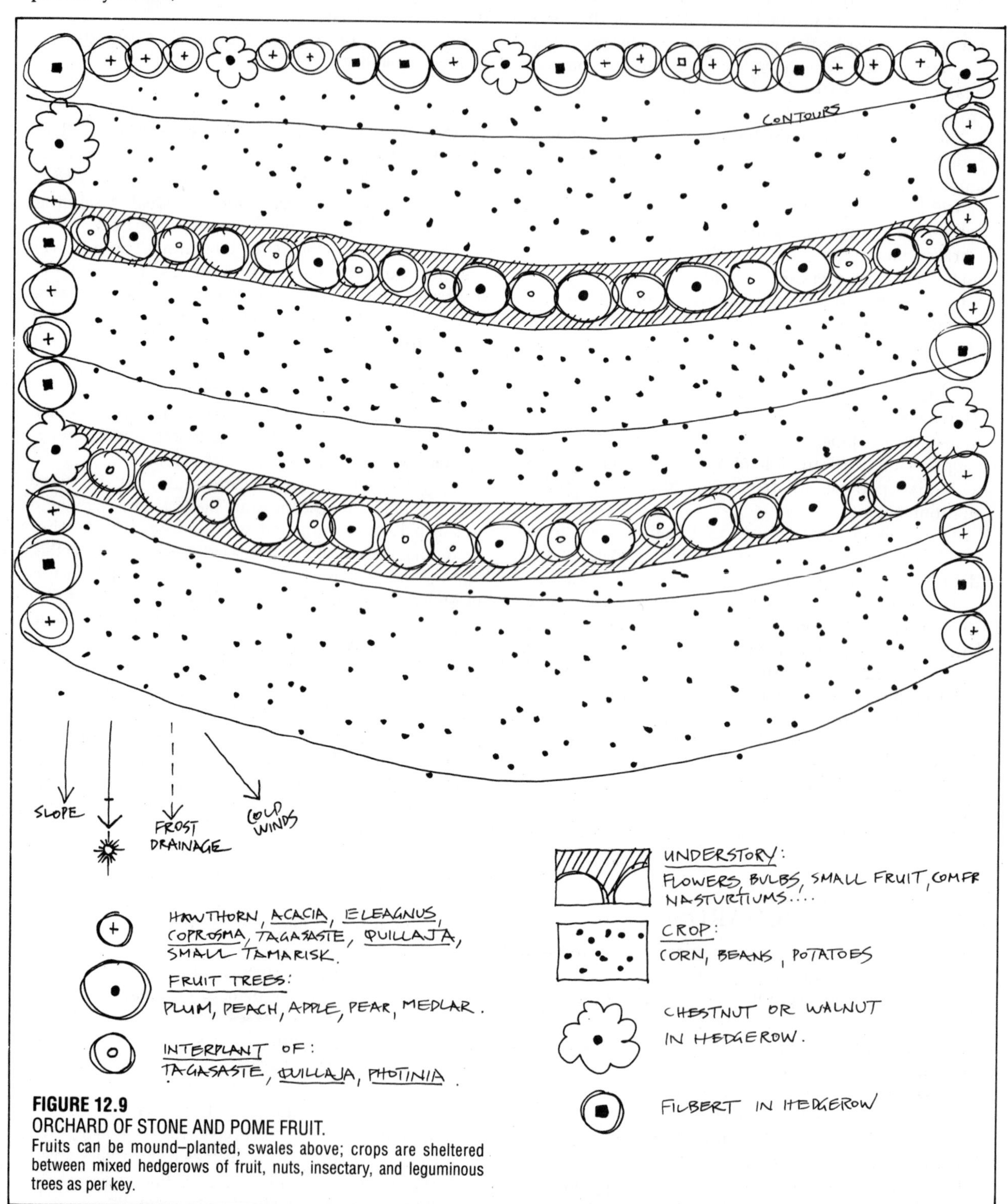

FIGURE 12.9
ORCHARD OF STONE AND POME FRUIT.
Fruits can be mound–planted, swales above; crops are sheltered between mixed hedgerows of fruit, nuts, insectary, and leguminous trees as per key.

ANIMALS IN THE ORCHARD.
Once young orchard trees and their associated plant guild species have established, small livestock can be introduced into the orchard system. Initially, bantams and small poultry breeds can be on range. Poultry have several functions in orchard systems, where they scavenge most soft fruits (and any larvae or pupae of pests), help control ground covers, provide a manurial turnover for the orchard trees, and self-forage seed and greens.

Chickens at 120–240/ha (70–100/a.) do not greatly affect the density of ground covers, and they scavenge most windfalls. They have a well-tested effect on such orchard pests as codlin moth and fruit fly, reducing their incidence to insignificant proportions over a few years of ranging.

When orchard trees are 3–7 years of age, foraging pigs can be introduced as fruit matures to take care of windfall fruits that breed pests. In standard-pruned orchards of from 7–20 years of age, first sheep and later controlled cattle grazing can be permitted. Trees of 10-plus years of age are far less susceptible to grass competition, and thus groundcover guild plants are less needed.

12.10

FARM FORESTRY

In the last decade, or from about 1972 onwards, farmers in the mesothermal climates have commenced to develop tree systems on farm. There are several reasons for this including:

- Loss of farm firewood and building materials due to overclearing of forest;
- Concern for wildlife, especially birds, important to crop pest control;
- Concern about soil erosion on steep slopes (more than 15°), along watercourses, and on steep ridge country;
- The realisation that trees provide forages in hard times for livestock and wildlife.
- Variation of product, which buffers economic changes in prices for crops and livestock;
- A more assured income from tree crops, coupled with a wider market for relatively exotic vine crops, fruits, and nuts;
- Gains in all other yields (crop, orchard, livestock) due to the shelter effects of trees; and
- Transition to a tree-oriented production less subject to drought or cold effect, and away from mass market crop product controlled by large retail or processing firms.

Many farms can select specialty crop, specialty processing, or unique product lines from the many thousands of species of trees, vines, and perennial herbaceous plants recently researched and accessed by tree crop associations, permaculture groups, and agricultural or forestry research centres.

Thus, the last decade or so has seen a very marked transition from cropping to perennial and long-term development of a part or the whole of farms. This is as often as not coupled with ownership changes, so that capital from urban professionals, new village groups, industries, and farm developmental trusts open to public subscription have all contributed to a beneficial set of changes in the farm landscape, often in contrast to, but also in support of, traditional farming practices (those prior to 1940).

A new look at farm design is required (and in demand), not only for forest products but for a rapidly growing increase in aquatic products on farming areas. For such endeavours to succeed, support services are needed, such as combined plant nurseries and new species collections for the provision of seed, bud, and scion materials, vegetative propagules, and grafted, layered, or tissue-cultured plants, as well as processing and market systems.

Pioneer or innovative farmers are now often providers of plant materials to other people requiring sufficient plant materials to commence commercial operations. If a bioregion is integrated and cooperative, such enterprises can cope with very different plant/animal groups, as the short list below illustrates:

- Bamboo nursery, together with canegrasses and large clump grasses. Bamboos are a currently neglected forestry.
- Aquatic and edge plant nursery, including fish forages, insectary species, and marshland perennials for bee fodders, duck forage, and wildlife refuges.
- Berry fruit and vine nursery, a great many species grow in these climates.
- Fish breeding: just a few reliable species is all one establishment can handle for providing farm ponds with stock.
- Poultry species; providing breeds of poultry to suit local soils and conditions on range.
- Bee fodder plants (also, butterfly and insectivorous bird attractors); many such species provide hedgerow and forage crop, fruits, and flowers.
- Hedgerow Tree species; a regionspecific selection for district; also pioneer tall grasses, vines for "fedges" on coasts.
- Conservation and Reforestation Trees for extending natural forests; includes pioneer and selected high-value tree crop species.
- Speciality Nurseries for herbs, salt-affected areas, smog tolerance, medicinals, food dyes, and stabilisers, and so on, are all valid enterprises.

None of the above enterprises are competitive. Troubles occur only where large investors and short-term entrepreneurs create large areas of indentical crops; this is usually the result of out-of-region investment.

Given regional resources of this nature, many farmers can then access unique species assemblies for site, enterprise, and preference. Thus, the second support service that will follow tree and vine establishment must be as processing and marketing systems. Quite a few farmers and cooperatives offer leasehold or hire of planting, harvesting, or primary processing machines to a more general subregion or district. Both nursery supply and primary or final processing are essential to new farm enterprises using new species. However, the range of potential crops and processes is, for all practical purposes, unlimited. Layouts of farm forests can vary enormously, and are for the main part governed by the labour and machinery available, landscape features such as slope, drainage, and soils, and the requirements of sustainability or internal farm self–reliance. Some much–used systems include:

Timber Crop in Pasture

Selected high–value trees are widely spaced in rows, and more closely spaced in line, to permit good pasture development between rows. Ideally, tree rows are on contour, or form a matrix on flat sites. Animals are let in as trees harden (years 36), and pastures are cut for hay over this period. Early grazing is supervised, or the area used by "soft" animals like sheep or geese.

Many trees are hostile to pasture, competing with pasture plants for water or nutrients (forest eucalypts, some pines), while others are less competitive (*Acacia melanoxylon, Eucalyptus camaldulensis, Grevillea robusta,* honey locust, as examples). The system enables a gradual integration of livestock and forestry, but note that many trees so planted may not assist the livestock forage. Trees such as poplar, fig, willow, chestnut, oak, and pine may all provide forage and other products in the medium term. Bamboo may be regarded here as a forest product for many farms, and has multiple end uses.

Woodlot

Woodlots are planted primarily for forest yields, and although they provide shelter and some browse for livestock, are usually less integrated with livestock. Also, woodlots can be themselves very variable in end use. For example:

• *Firewood production* on a 2–7 year rotation (one–half to one–seventh of the plants cut annually). Depending on the culture, firewood can be cut as coppice or stickwood, or grown to 4–10 cm log size. In most cases, firewood species are chosen for persistent coppice (regrowth from stump) and good fuel value (calories/t). Tagasaste, *Acacia* species, *Casuarina*, and eucalypt species are examples of good fuelwoods, but almost any vigorous tree can be used

• *Polewood production.* Polewood is of increasing importance for fencing, house, and furniture construction. In this category, it might be important to define durable timbers such as chestnut, raspberry jam acacia *(Acacia acuminata)*, osage orange, black or honey locust, cedars generally, and eucalypts known to be rot resistant (river red gum, turpentine). Polewood of less durable timbers is widely used for indoor work, furniture, and as scaffolding or formwork support in building. Poplars and *Acacia* species are planted for these less durable uses, as are timber bamboos for scaffolding and house frames, furniture, and household mats or articles.

Special uses of fast polewood production from Chinese elm, poplar, or like species include chipping for wallboard or fuel bricks, crushing for fibre or cellulose production, fermenting for stock feed or alcohol distillation, and chipping for distillation to oils, resins, and chemical products such as creosote and furfural.

• *Longterm fine timbers.* Some farmers reserve off steep land, valley sites, rocky soils, or islands for selected fine timber stands. These can be revalued at every stage of growth, as they add to farm value from their first year. Black walnut, oak, rosewood, fine cedars, redwoods, blackwood, and many other species are planted as potential retirement trusts for the farmers or their families, and also as foundation trusts for schools or limited institutions with long–term aims Council rates in villages can be completely offset by urban forestry operations.

Stands of such fine timbers can have complementary pioneer species interplanted for medium–term yields (leguminous trees and small cedars generally), but are at their greatest value at 40–100 years for fine furniture, inlay, panelling, and plywoods.

• *Hedgerow.* Hedgerow and contour–bank forests, roadside forest, watershed forest, and steep slope forests permit a true polycultural forestry with 6–30 species of trees chosen for fruit and nut yields, forages, honey, special wildlife foods, browse, and both mulch and stickwood production. Unlike some other forest types, hedgerow and conservation forests can contain numerous species, as cropping of product is not on a cut–and–run, but gather on a gather–and–select basis, with constant replant.

There can be 5–6 types of forestry on any farm of 50 ha or more (or in and around villages), including orchard production. Some special forests are possible on specific sites such as swamps or acid uplands. Thus, farm forest design should be oriented to site and purpose to enhance other farm enterprises, to supply local needs, and to give the potential for a wide range of end product. A whole bioregional forestry devoted to one species of eucalypt or pine is the antithesis of this secure approach, and sets up the conditions for several very undesirable end results, some of which have proven to be:

• Glut conditions in market leading to a depressed local economy;

• Land ownership change based on remote ownership and often very unsympathic control;

• Setting up conditions that may lead to catastrophic fire in the area (eucalypts and pines provide fuels for firestorm in settled districts); and

• Displacement of bioregional needs by industrial feedstocks, thus unemployment and social disruption leading to longterm social problems.

All these problems have been developed by industrial, not regional or village, forestry.

For very small farms, trees need to be carefully

selected and placed to maximise short– to medium–term uses (forage, mulch, honey, nitrogen fixation in soils), to assist crop and building efficiency, and to assist the microclimate of the property, so that small farm design (as with small garden design) needs as much detailing as large forest design systems.

A Small Forest Farm for a Cool Humid Area

Large logging operations waste 30–70% of a harvested timber resource, and produce very few forest products apart from wood or woodchips, whereas a small (12–16 hectare) forest site managed by a family or families can yield fungi, seed, smallfruit, furs, meat, wild fruits and nuts, honey, poles, coppice, bark for craft and dyes or tannin, medicinals, and mulches on a sustainable basis. A layout for a tree forest farm is given in **Figure 12.10**, based on observations made in Oregon (USA). The farm is zoned for frequency of harvest.

12.11
FREE–RANGE FORAGE SYSTEMS

To rear animals on range, the following factors must be studied:

- The social behaviour of the animals themselves; preferred sex ratios, ranging behaviour, prefered flock or herd size, and the adaptations of special breeds to soils, pH, site factors, and foraging. This factor decides varieties (breeds) and herd or flock sizes.
- Nutrition over a full year; adequate provision of energy foods, vitamins, and minerals essential to animal health. This factor decides on forages developed or minerals provided.
- The interactions (beneficial or adverse) of ranging animals with crops, forests, other animals, and people. This is the factor deciding the placement and nature of shelters, fences, and water points.

Thus knowing the BEHAVIOUR OF THE ANIMAL is intrinsic to good range management (lessening social stress), and knowing the CHARACTERISTICS OF GOOD RANGE is also critical to factors which include all of those grouped under environmental stress (shelter, nutrition, climate). Livestock management on range is a skilled job, and involves a great deal of observation and monitoring, unlike factory farm systems regulated by recipes and automation. We can choose from three very different animal species—bees, chickens, and pigs (all critical to domestic self–reliance) to illustrate the way to proceed.

BEE RANGE

Bees produce several valuable and unique products: honey, wax, pollen, propolis, and royal jelly. They also

FIGURE 12.10
SMALL FOREST FARM.
For intensive forest management: **1**. house and garden; **2**. smallfruits on forest edge; **3**. coppiced woodland; **4**. high–value managed nut and fine timbers; **5**. mixed age and species stands; **6**. forest clearings; fungi and seed are also crops.

carry out the essential service of pollination for a wide variety of food, oil, fruit, and seed crops (such as mustard, clovers, buckwheat, most smallfruit, apples, and grain legumes). Bees form a hive or clustered hive site range over an area about 2.5 km radius, and commercial registered hive sites are thus sited 5 km apart in order to restrict disease transmission from hive site to hive site.

The hive site itself should be sheltered from extremes of cold, wet, wind, and sun. However, the site should not be atypical of the climate of the range, or bees are tempted out in adverse weather, and perish. As many as 100–150 hives are commonly grouped at one site, depending on the richness of the range resource. Bees prefer to fly 100 m or more to forage, and their flight assists in the evaporation of nectar to honey, so that forage species are planted this distance or more from hives. Bees more efficiently harvest clumped rather than scattered nectar sources, so that hedgerows, fields, or clumps of preferred forage species are better than a scatter of the same species in a mosaic of individual plants.

Cold winds most restrict foraging, so that hedgerow (even low hedgerow of 0.5–1 m) is essential cover, preferably leading from the hive site (bee village) to the forage. Such hedges can be made of rosemary, *Acacia*, or even soil ridges with catmint, capeweed, thyme, or field daisies to assist foraging.

As well as nectar producers, bees need pollen to rear replacement workers, especially early in the season. Thus, willows (especially pussywillow), *Acacias*, pines, and vine crop yielding early pollen are very beneficial on range. Bees also gather propolis, a hard waxy substance, used to plug wind gaps or repaint hives, and some sources of this are *Xanthorrheas*, poplars, and pines; very few sources of propolis are needed in the whole range.

Finally, clustered hives use very large quantities of water, which is used by the bees to cool the hives. The bees ventilate hives as a "super organism", inhaling

FIGURE 12.11
IDEAL BEE FARM.
Cross–wind forage hedgerows shelter bee–dependent seed crop, herbs, vines; adequate water, pollen, shelter, and a well–designed honey house complete this system.

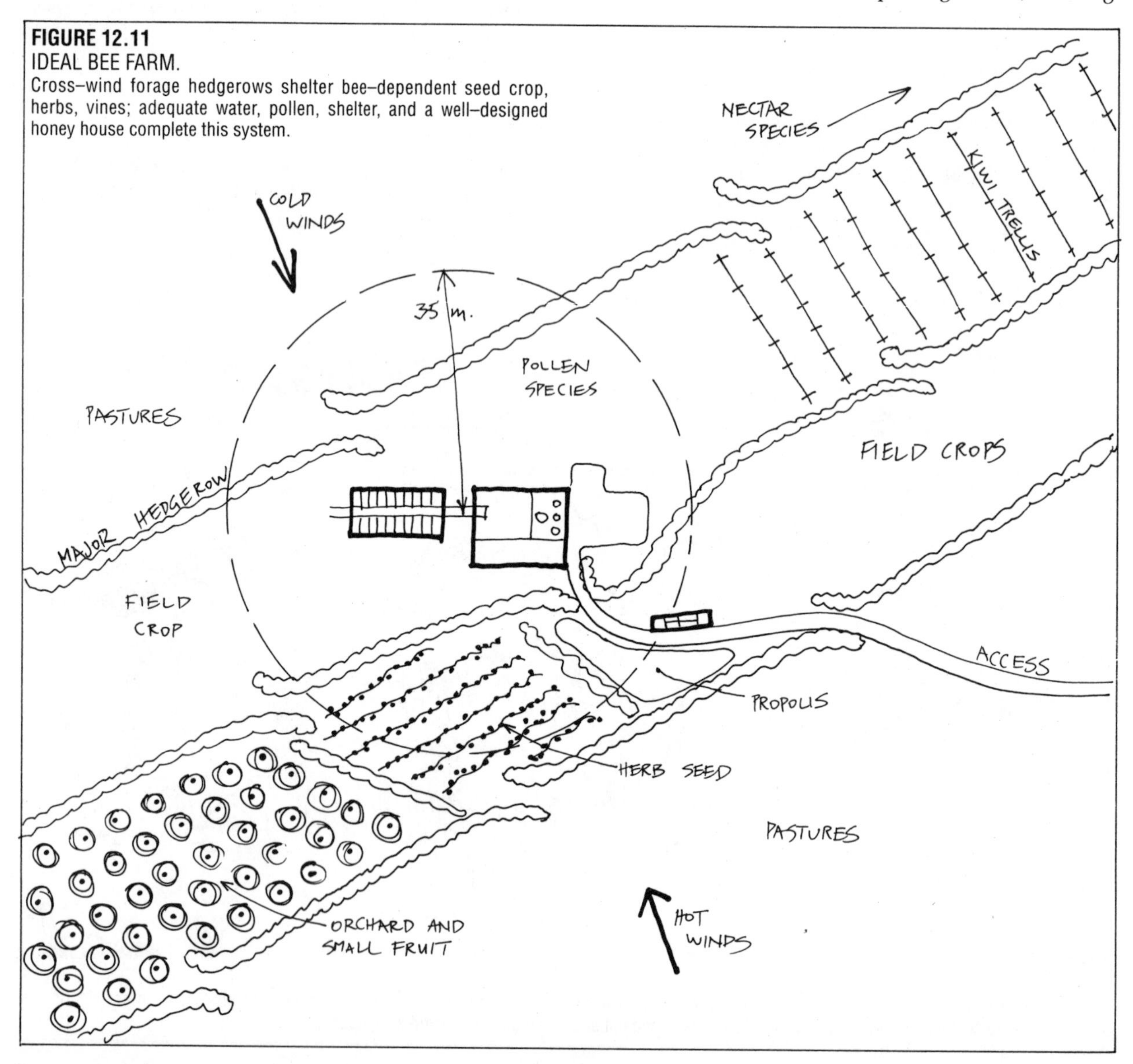

TABLE 12.1
BEE FORAGE SPECIES TABULATED FOR SEASON AND TYPE OF HONEY (COOL TEMPERATE CLIMATE)

	EARLY Season	**MID Season**	**LATE Season**
CLASSIC	Rosemary	Leatherwood (*Eucryphia billardieri*)	
BULK	*Echium fastuosum* Acacia Catmint	White clover Blackberry	*E. fastuosum* Eucalypt
SPECIAL & CROP HONEYS	Gooseberry	Smallfruits	Mustard

and exhaling in about a 1.2 minute rhythm. As hive temperatures are normally high (at about 21°C), so hive insulation in cold weather is important, and in Denmark bees are over-wintered in thatched attic areas, or stored in insulated hive houses in other parts of Europe. Conversely, in areas of hot summers, hives need shade and good ventilation, or honey is lost as energy used in cooling (fanning) by the hive.

Access to water is most safely achieved by providing soaked mats or hessian at pond edges, so that bees cannot drown while drinking, and *small* ponds or troughs free of dragonflies give less losses than large ponds (where dragonflies are efficient predators).

Having designed the physical outline of a bee range (which in truth resemble any well-sheltered farm), then local lists of honey plants can be consulted for pathsides, road verges, hedgerow, and forest. It repays the beekeeper to do careful research into bee forage species. We could handily divide forages into early, mid, and late season for a good spread of yield, and again into classical, bulk, and special honeys and pollens for mixing and specific market. Some classical (highly preferred) honeys are: "heather" honeys (sage, thyme, catmint, rosemary); pine tree or basswood (*Tilia* spp.) honeys in Europe and the USA; marsh tupelo honeys (subtropics), leatherwood (*Eucryphia*) honeys of Chile and Tasmania, and citrus honeys.

Bulk honeys suitable for blending derive from clovers, eucalypts, *Acacias*, field crops such as mustard and buckwheat, fruit crop usually from the family *Roseaceae* (blackberry, raspberry, apples), and "mixed garden honey".

Special pollens are those collected from plants such as some *Acacia* and goldenrod, the pollens of which may cause asthma if inhaled by susceptible people. Eaten, these honeys or pollens (like the young leaves of poison ivy and poison oak) act to prevent allergic reaction.

Thus, the selection of forage species can be entered into a 3 x 3 table, as in **Table 12.1**, for any particular site and climate.

It is best to have 35 main forage species over the whole season, as species such as clover and eucalypts may vary in yield from year to year. Similarly, pollen species are selected on the same basis, and pollen traps fitted to the hives at periods of high yield. Hives should yield about the same weight of pollen as honey, and the latter provides a high-protein additive to any flours or starchy foods.

Given that we have designed a range for bees, then the same areas presents a unique opportunity to grow crops—especially seed crops—dependent on bee pollination. Such crops are: kiwifruit, any bramble berry, smallfruit, mustards and *Brassicas*, clovers, apples or pome fruits, buckwheat, *Acacia*, stone pine, and so on. Crops within a mile of hives will outyield crop in bee-deficient areas by a factor of 3–10 times.

Thus, we can sketch out the essential ground plan of an excellent bee farm, which would also and ideally be a production farm for fruits and seeds (**Figure 12.11**).

POULTRY RANGE

The two basic breeds of modern chicken are the *light* and *heavy* breeds. Light breeds derive from Spanish ancestors, and probably from Indo-Malaysian jungle fowl. They are flighty, poor mothers, lay white eggs in spring and summer, are cold and wind sensitive, and prefer sandy or light soils of high pH. These breeds have long legs, full combs, four toes, are lightly feathered, often possess white ear lobes, and fly well; we know them by their older Mediterranean names of Anconas, Andalusians, Minorcas, Leghorns, and more rarely by northern European names such as Hamburghs.

Heavy breeds derive from Chinese fowl, via ancestors such as the Langshan. They are non-fliers if not chased, excellent mothers, lay brown eggs in summer and autumn (even through winter), are hardy to cold and wind, and tolerate clay and acid soils. Many have heavy feathering, short and sometimes feathered legs, short combs or "rose combs", five toes, and possess names derived from northern Europe such as Scots Dumpy, Orpington, Dorking (of English fame); others have names from breeds developed in European colonies, such as Wyandotte, Rhode Island Red, Plymouth Rock, and Australorp.

Between these breeds, and a few exotic species developed in South America (Araucana), we can select from 60–100 breeds for specific soils, sites, foodstuffs, and either egg or meat production, or both (utility breeds), In laying hens, free-range flock averages of 130–150 eggs per hen per year would be normal for heavy breeds, with rare averages of 180–200, while light breeds such as Hamburghs and Leghorns can usually produce from 160–180 flock average. Unstressed flocks on range will continue to lay from 4–6 years, whereas caged animals wear out under forced regimes of light and manipulation (artificial day length), hormones, and pelleted foods containing several additives. They must be replaced every 18–20 months.

Natural or "wild" flocks of poultry seldom exceed 20–30 individuals, of which 2–3 are cockerels, the rest hens. Surplus cockerels are driven out of the range area or killed, and usually fall prey to hawks or other predators. Larger flocks of 40–60 individuals will break up of their own accord, and form two flocks.

Light breeds are excellent foragers, and breeds such as Hamburghs and Leghorns can obtain most, or even all, of their food on a free range with hedgerow and pasture. Heavy breeds forage well but need supplementary food, especially in winter, and this is generally supplied as grain, grain legumes, and grain mill wastes or root crop mash. All breeds eat domestic and garden wastes, forage fallen fruit and seeds, and pursue and eat insects in the field. Normally, and on correctly stocked range, we would expect poultry to obtain 65% of their food as insects, invertebrates, greens, and grains, and we should supply 30–35% as concentrated grain products (usually weed seed and the cracked or undersized grains from the threshers).

Poultry rearing combines well with a mixed farm economy of orchard, dairy or home milk cow, and some grain crop. In this type of environment, they scavenge many wastes, obtain maggots from manure and cleanse the orchard of the windfalls of soft fruits. In the tropics, abundant fruit like papaya supplies most of their needs. It is true to say that no soft fruit, apple, or citrus orchard thrives without poultry as foragers.

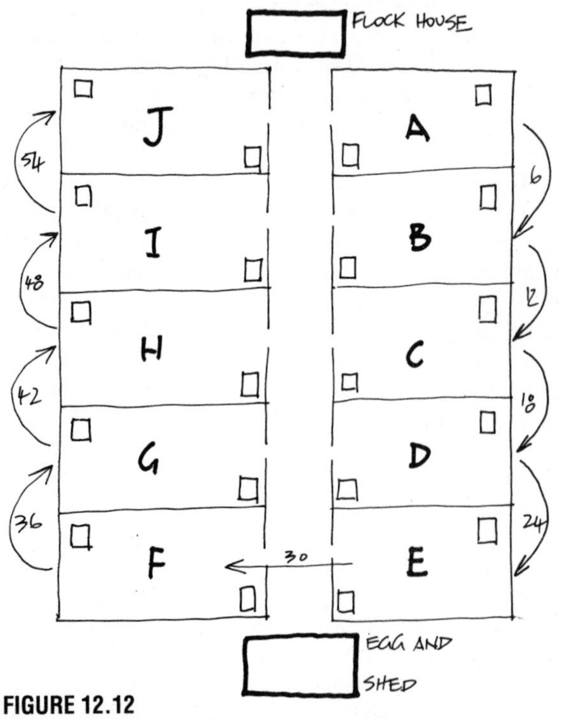

FLOCK HOUSE

J A

I B

H C

G D

F E

54 48 42 36 6 12 18 24 30

EGG AND SHED

FIGURE 12.12
CHICKEN PEN SYSTEM.
Here, flocks are moved every 6 weeks, followed by a vegetable or small fruit crop. Vines on fences add yield; bared areas are limed, raked, sown to crop; figures indicate weeks for a full year's rotation of a "chicken tractor" system.

Population Density

Chickens (chooks in Australia) are unstressed at up to 800–900/ha, providing these are housed as 20 or so small flocks in sheds about 2 m per side (**Figure 12.12**). More ususally, households keep one flock of 20–30 hens as a food supplement, and these may never range on greens where space is limited, but have cut greens supplied to them from garden weeds to roadside grasses.

At 800/ha, chickens forage so well that other livestock cannot find sufficient browse, but at 350–400/ha sheep and cattle can also use the range. Densities of from 120–180/ha are needed just to clean up windfalls and supply fertiliser to orchards. A flock of this size can be rotated around 12–16 pens on one hectare every 18 months, when they remove all weeds, surface bulbs, weed seeds, and grasses and can be used as a "chicken tractor" to clean the ground of nutgrass, kikuyu, onion weed, *Oxalis*, or indeed almost every persistent weed.

As pens are cleared, chickens are moved on, the ground limed, and vegetable or orchard crops placed. Before they return, forage crops of buckwheat, wheat, sunflower, millet, black nightshade, cleavers, chard, and mustard can be sown, and given 2–3 months to mature. The chickens thus plough and clear for their own forage. A separate pen of comfrey, arrowroot, chard, and sunroot can be kept as drought forage, or "throw-over-the-fence" crop.

On range, full nutrition can be obtained by providing seed from such abundant trees and shrubs as *Coprosma repens*, *Caragana*, tagasaste, boxthorn (*Lycium* spp.), pigeon pea, most *Solanum* species with berries, mulberry of 5–6 varieties, and normal fruits. *Acacias* with edible seed (most species), *Leucaena*, clovers, grasses, and "weeds" such as cleavers, chicory, comfrey, *Oxalis*, and dandelion supply the rest of the diet, together with insects and their larvae. Some farmers deliberately pile manure for flies to lay in, and let the chickens in to eat maggots. Others use piles of mulch, cockroaches, termites bred in sunken pits, just loose compost piles to breed insect foods, or mealworms in bran and root mixtures.

Table scraps supplement diets, and large acorns, chestnuts, or starchy seeds can be sprouted or rolled to crack them for chook food, using a garden roller on a hard surface such as concrete or steel. In hot periods, and fed–out grains can be sprouted for 2–3 days, but in winter unsprouted grains and grain legumes are preferred.

Given a free range of such richness, diseases are rare or absent, chooks healthy, and their only needs are dry dust baths with a little diatomaceous earth or dried bracken and neem tree leaves for ectoparasite control, a few wormwoods for intestinal worm control, abundant shell grit or calcium (limestone) gravel for eggshells, hard silica grit to grind food in the crop, adequate predator protection, clean water, and good shade and

TABLE 12.2
BENEFICIAL ANIMAL/ENVIRONMENT INTEGRATIONS.

LIVESTOCK	Market Gardens	Small Fruit	Orchard	Pasture	Wetlands	Range-lands	Flowers & Herbs	Forests
Bees	•	•••	•••	•	••	•	•••	••
Poultry	•	•	•••	••	•	••		•
Waterfowl	••	••	•••	••	•••	•		
Pigs	•		••	••	••	••		••
Sheep			••	••		•••		
Dairy Cattle			•	•••	•	••		•
Beef cattle				••	•	••		•
Game			•	•	••	•••		•••

• Compatible •• Good association ••• Excellent association

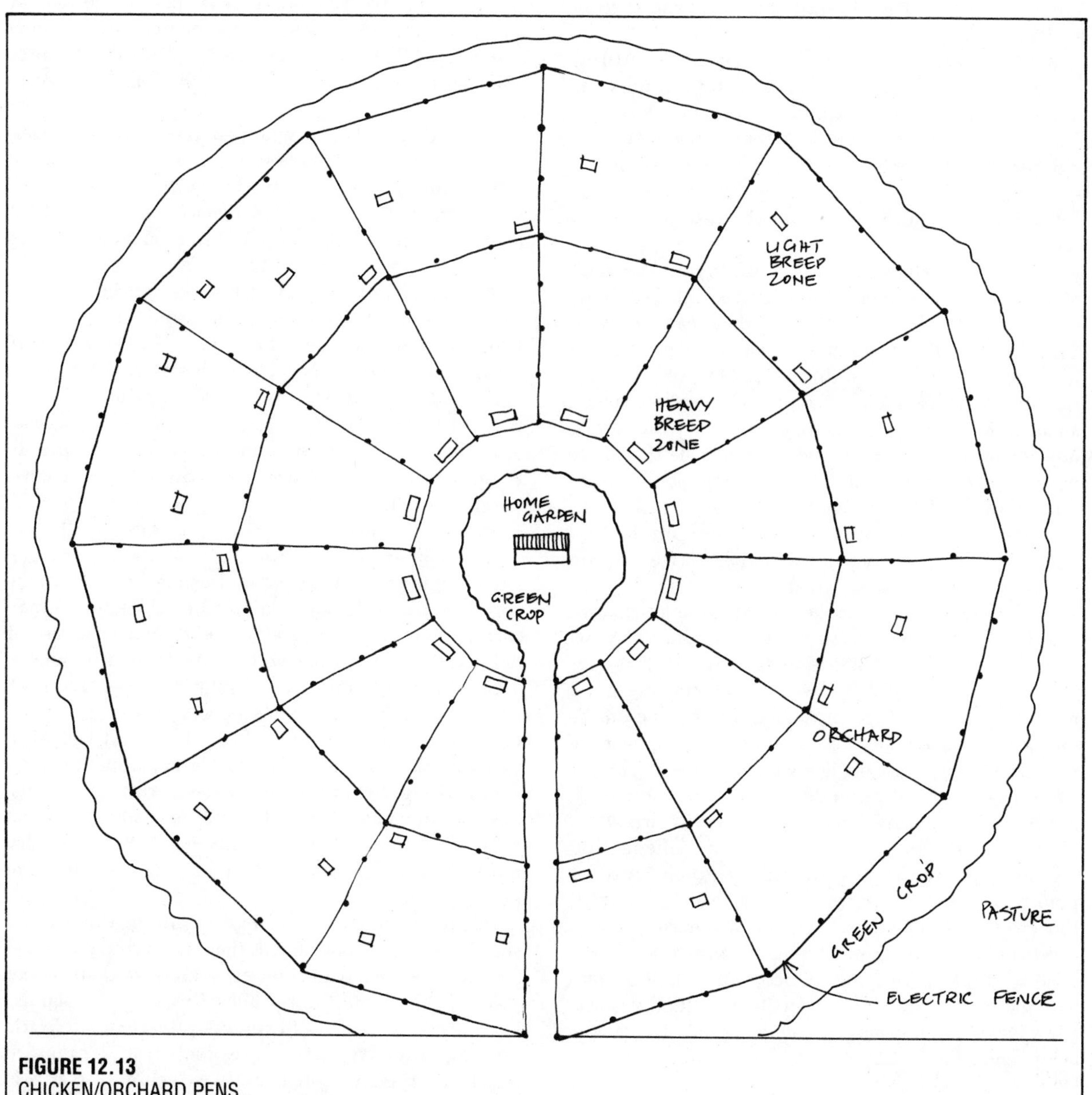

FIGURE 12.13
CHICKEN/ORCHARD PENS.
Idealized free-range layout; heavy breeds close to house, some pens always rested, limed, sown to forage crop. Fruit and forage trees omitted.

night shelter.

People who breed their own chickens in the field keep a small flock of Wyandottes or Silky Bantams as mothers or broody hens, and provide secure pens for hen and chicks. Most hardy bantams breed their own chicks on range, as do older breeds such as Hamburghs, Australorps, or Dorkings.

Specialty breeders are numerous, from those who maintain, select, and breed high–performance layers or growers, to those keeping many pure breeds for show or sale. About 13 ha is plenty for commercial egg production, given a careful range plan. By field selection, tender chickens of up to 3.5–4 kg weight can be produced, and the contrast between these and the soft, fatty, bitter, diseased products of caged flocks fed on commercial pellet foods must be experienced to be believed.

All chicken flocks need observation, culling, replacements, range adjustment, predator control, clean water, and good shelter. Running a 13 ha flock of 3000 birds is a full–time family job, combining well with chicken tractor systems and orchard fruit production. Good husbandry requires close attention, good hygiene, and healthy food, but the results are very worthwhile.

As every livestock keeper knows, it is in the early stages of flock establishment that mortality is high. Once chickens are bred on range, they learn to avoid predators and then settle down to an easy–to–manage and quiet flock, and produce excellent food of high quality. As chickens are a universal domestic meat, small enough not to need refrigeration, they can fit into any mixed farm economy with great benefit. All we have said here about chickens can be said about ducks, guineafowl, turkey, and wildlife such as pheasant and quail on range. The latter species differ only in food preferences, preferring more snails, berries, and a higher proportion of insect foods.

Ducks have always been kept in the same range area as chickens, but need marsh plants, snails, and shallow water for dabbling. They too supply up to 200 eggs each/year, are relatively disease–free, and are good foragers; given predator protection, they thrive on the mixed farm. Geese are the traditional small grazers of mixed orchards, replacing lawn–mowers and tractors to control grasses and groundcovers. In the subtropics, rampant greens such as comfrey, *Tradescantia*, arrowroot (*Canna*), sweet potato greens, cassava foliage, and leguminous tree leaves all supply green crop for poultry and domestic animals. Many also yield surplus root crop for cold periods, fed as a boiled mash.

Whenever we think of keeping small livestock, it is wise to think of running an accessory plant system (aquatic or crop) that both benefits from the livestock, and gives them a benefit. Some beneficial integrations are given in **Table 12.2**.

PIG RANGE

Pigs on range are healthier, cheaper to feed, and have less saturated fats than pigs kept in sheds. The greatest expense is in fencing (now much reduced by the use of electric fencing). The ideal site is preferably wooded, well–drained and dry in wet seasons. Mud and wet soils encourage worms and disease, and can cause trouble with suckling. Winter shedding may be necessary in very cold winter climates, but sheltered ranges and good dry beds help.

Pigs are most economically kept where some dairy, orchard, root crop, or meat wastes are available, and do well on restaurant or household food scraps. As pigs will not eat or do well on coarse grasses, grazing beef or horses may help in reducing these on range, or mowing may be necessary.

A farrowing pen is used by a sow and litter for nearly 3 months. Rested for 6 months, each pen can be used by two sows a year. Various small (or one large) shelter sheds (best thatched or insulated against cold) can be used on rearing ranges, or dense thickets provided in the tropics.

Strong diagonal posts wrapped in hessian make ideal rubbing places, and the hessian can be kept dampened with sump oil, or vegetable oils with some neem oil (made from the berries of *Azedarachta indica* or *Melia azedarach*—white cedar) or pyrethrum oil added to keep the pigs free of lice (**Figure 12.14**).

Restricted wallows and dust baths can be arranged by sheltering or dampening sites, and so add to the comfort of the pigs. Dust baths (a roof over a soft dry area) kill lice and can be laced with some dry neem tree leaves or pyrethrum flowers. Rested mudwallows planted to cattails (*Typha* spp.) provide excellent root and shoot foods at any time of year. Duck potato (*Sagittaria*) can be planted in pond margins for summer–dry forage.

Free–range pigs are not always suitable for bacon, and may need grain–feeding for 2–4 weeks to harden (saturate) the fats. However, for fresh or frozen pork the soft fats are quite suitable and healthier for human consumption. A lot of vegetable oil from avocadoes, oil palms, olives, or oily fish will produce soft fats in pigs.

Good range pasture is of legumes (clover, lucerne), comfrey, chicory, and young grasses (grazed by horses or cattle, or mown). Pigs will eat 11 kg wet weight of this material per day, and have larger appetites than confined pigs. They also need seed, fruit, or kernels. Good quality green feed halves the necessity for protein meal and grains. A sow in milk needs 3.5 kg of dry rations when piglets are new, and 5.5 kg when they are 2–3 weeks old.

To prepare for a free–range planting, the ground should be ripped and limed, then sown down to good grass legume mix, with comfrey, sunroot, and arrowroot pieces pushed into the rip lines. Trees can be planted just outside fences and in corners protected by electric fencing. Any fruit trees are useful, and pigs are beneficial in mature orchards. A selection of plants is given in **Table 12.3** for the tropics and cooler areas.

When considering stocking densities, 20 pigs/4,000 square metres (1 a.) will "plow" (by scratching and

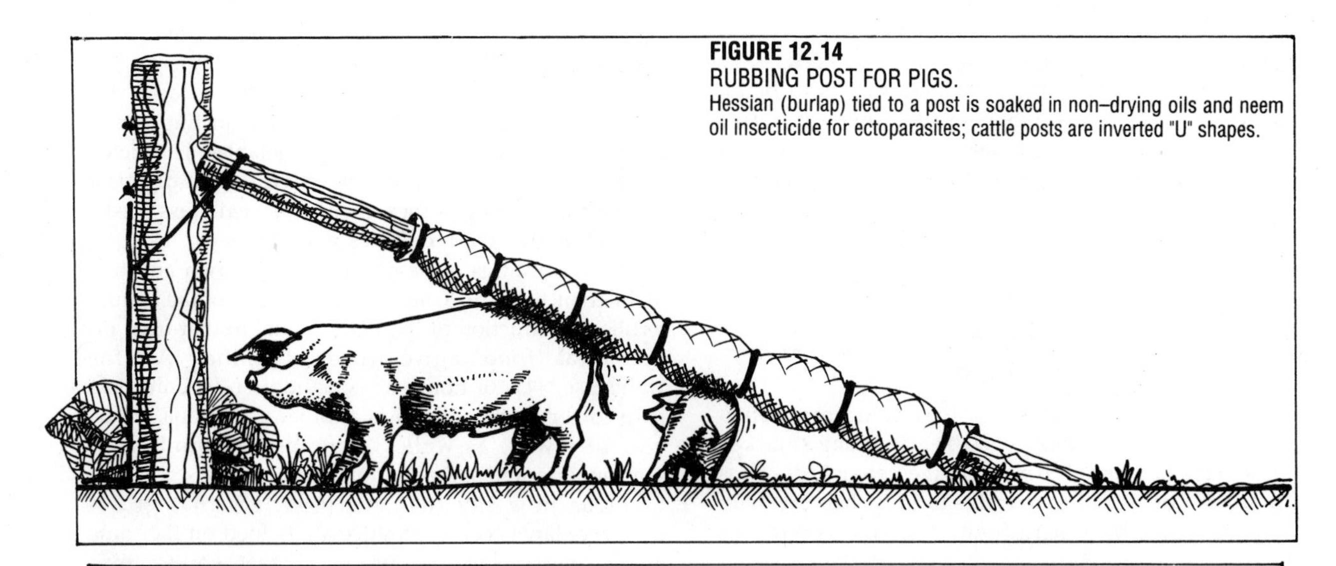

FIGURE 12.14
RUBBING POST FOR PIGS.
Hessian (burlap) tied to a post is soaked in non–drying oils and neem oil insecticide for ectoparasites; cattle posts are inverted "U" shapes.

TABLE 12.3
SELECTED PIG FORAGE SPECIES

TYPE	TROPICS	COOLER AREAS
Trees Planted in protected areas, outside pens	Mulberry, Bunya pine, fig, Inga *(Inga edulis)* Palms: Chilean wine palm *(Jubea spectabilic)* Jagua palms *(Jessenia spp)*, Buriti palms *(Mauritia spp)*, Babassu palms *(Orbignya spp)*, Assai palm *(Euterpe oleracea)*, and *Atallia spp.* Oil palm *(Elais guineensis)*. Date palms (*Phoenix spp*), coconut *(Cocos nucifera)*.	Mulberry, fig, olive. Palms: Chilean wine palm, canary date palm *(Phoenix canariaensis)*.
	Oaks, chestnuts, chinquapins, breadfruit, jak fruit, and *Prosopis.* pods. Honey locust, carob (dry areas). Chinese tallow tree, *(Prosopis)*.	Oaks, especially white oak group, cork oak, turkey oak, chestnut, chinquapin (some need to be collected and shared!) Rain tree *(Samaea (Albizzia) saman. A. dulcis Argania sideroxylon*: fruits.
Roots Planted in rooted–up pens as sets.	Cattails *(Typha)*; arrowroot *(Canna edulis)*; Jerusalem artichoke; comfrey, yams, taro; sweet potato, potato, cassava (boiled), elephant–foot yam, *Marantia.*	Cattails; potatoes; Jerusalem artichoke; comfrey.
Greens As sets and seeds following rooted ground.	Chicory, lucerne, comfrey, white clover, pigeon pea *(Cajanus cajan)*, *Desmodium ovalifolium*, pumpkin, chayote (choko), sunflower (heads), cowpea.	Chicory, lucerne, comfrey, white clover, grasses, broad beans (favas).
Fruits Protected 3 to 5 years if valuable.	Papaya, mango, banana spp., and avocado are staples. Guava; anona; *Inga* in grasslands; sapote; canistel.	Apple, pear, plum; pigs assist orchard health.
Pests	Neem tree oil for mites; the fruit is poisonous to pigs.	Pyrethrum daisy–flowers dried and used in nests or dust baths for mites; also use diatomaceous earth for lice.

rooting) the area for planting comfrey, sunroot, lucerne, chicory, and clover. It then needs to rest. One hundred pigs in 2 ha (5 a.) pens will plow 40 ha (100 a.) in 18 months. They will remove gorse, blackberries, and small shrubs. They can be followed by sowing, then cattle, then pigs again.

Count on at least 1 ha (2.5 a.) per breeding sow, laid out somewhat as in **Figure 12.15**. Areas for tree planting are outside and in the corners to start with; later, pens can be planted out in blocks using electric fencing to keep pigs away from young trees. In the tropics, banana, papaya, and like fruits will be rooted out by pigs, so these need to be outside the pens. As pigs are moved, root crop can be planted in the pens.

The whole site plan can be handled fairly casually for a few pigs on a large range, or a clean pen will hold a family of pigs for using surplus crop, but needs very careful planning and siting if pig–raising is to be an occupation. It takes 3–5 years to develop a full complement of foods on range, and even some of this must be thrown over the fence to the pigs, as is the case for bananas and papayas.

12.12 THE LAWN

"Lawn aesthetics do not permit the chaotic intrusion of vegetable gardens...since vegetable gardening smacked of poverty or peasantry, many of the new middle class (from 1897 onwards) were loathe to sully their turf with mere food." (Bob Schildgen, "Lawns: America's Creeping Carpet", *Pacific Sun*, November 1982).

We can, to some extent, trace the development of the lawn from those shortcropped vistas developed by geese and sheep on rural estates in cool humid climates. "In a photo from the USDA Yearbook of 1897, a flock of sheep grazing in Central Park (New York) are described as 'the lawn mowers or turf makers' of the park." (Schildgen, *ibid*).

In the USA, it is estimated that 16 million acres were devoted to lawn by 1978 (*The Integral Urban House*, Farallones Institute, 1979), and a vast expansion of lawns has taken place there in recent years. At that time, lawns were considered to be the single largest "crop" system in the USA, requiring 573 kilocalories per square metre to maintain—more than the energy used in the production of corn or vegetables. The yields of this agriculture create a massive public disposal problem, consisting as they are of poisoned grass waste, rich in Dieldrin, DDT, biocides, and nitrogen.

Millions of litres of petrol are used in lawn and turf maintenance. By 1978, lawns used 15–20% of the annual fertiliser production of the USA; equal to that used on the total food agriculture of India. As for water use, 44% of domestic consumption in California is used for lawns, which is another enormous public cost of lawns, as well as longterm groundwater, atmospheric, and soil pollution costs.

Let us now say that every society that grows extensive lawns could produce all its food on the same area, using the same resources, and that world famine could be totally relieved if we devoted the resources of lawn culture to food culture in poor areas. These facts are before us. Thus, we can look at lawns, like double garages and large guard dogs, as a badge of willful waste, conspicious consumption, and lack of care for earth or its people.

We can clearly see the lawn as the world's "third agriculture", after food gardens and farms. Few people realise just how large this agriculture has become following the development of automated mowers, slashers, "whipper snippers", edgers, plug–cutters,

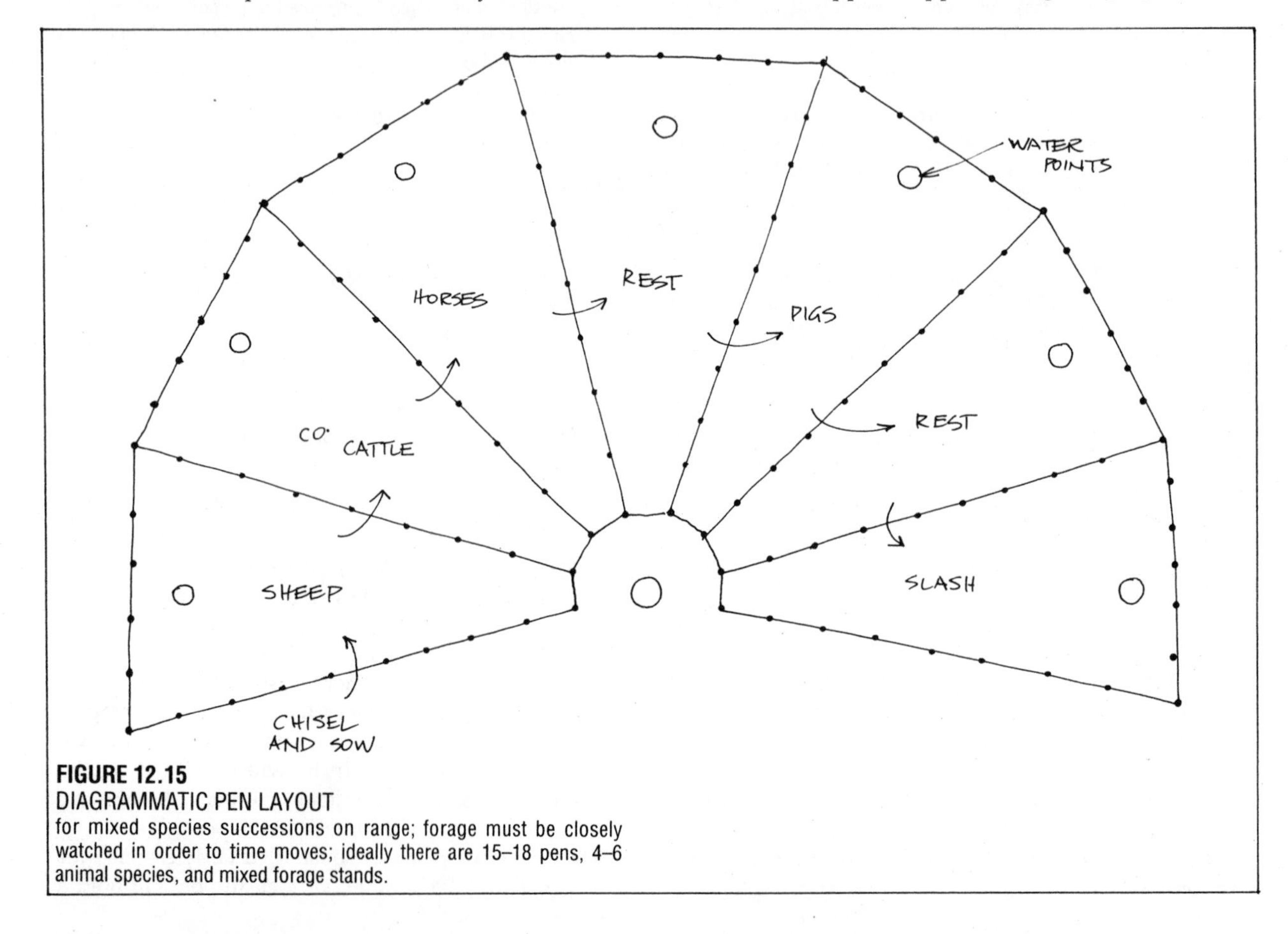

FIGURE 12.15
DIAGRAMMATIC PEN LAYOUT
for mixed species successions on range; forage must be closely watched in order to time moves; ideally there are 15–18 pens, 4–6 animal species, and mixed forage stands.

aerators, sprinkler systems, and the development of teaching institutes, journals, retail outlets, and very large firms to service the turf and lawn industry.

It is now probable that the lawn cultures of affluent nations use more water, fertiliser, fossil fuels, biocides, and person–hours than either gardens or the formal broadscale agriculture of that country, or indeed any agricultural resource of the third world. Of the lawns developed today, perhaps 13% have any use in recreation, sport, or as rest areas. Most lawns are purely cosmetic in function. Thus, affluent societies have, all unnoticed, developed an agriculture which produces a polluted waste product, in the presence of famine and erosion elsewhere, and the threat of water shortages at home. Some actual case histories:

- On the dry island of Moloka'i (Hawaii), golf courses are being developed near resort hotels, and an engineer there estimates an annual cost of some $400,000 as maintenance, interest on capital, and repairs to pumps, water systems, and turf areas. This is more than is spent annually in food gardens on that island.
- In Massachusetts, Connecticut, and Kentucky, many small farms are being converted to lawn systems by city owners; even in towns, the average suburban lawn area is about 650–900 square metres (5000–7000 square feet), and the yearly maintenance cost is about $30 per square metre, with up to six heavy biocidal sprays per year for pests (chinch bug) and weeds (flat weeds).
- In Perth (West Australia) lawns use 254 cm (100 inches) of water per year, in a water scarce area where salted land is extending annually, and the groundwater table is falling from about 90 m to 450 m.
- On the Canary Islands, resort hotels with lawns, swimming pools, flush toilets, and irrigated aesthetic gardens have effectively dried up a productive small farm economy; now atomic power is being contemplated as a method of creating fresh water from seawater.
- In one agricultural college (Orange, NSW, Australia), the energy officer calculates that the fossil fuel and personhours spent on maintaining the 10 ha (24 a) of lawns exceeded costs for the mixed farm of 50 ha (120 a). He recommended shrubberies for all but the 1.5 ha of lawn used by students.

The lawn has become the curse of modern town landscapes as sugar cane is the curse of the lowland coastal tropics, and cattle the curse of the semi–arid and arid rangelands.

It is past time to tax lawns (or any wasteful consumption), and to devote that tax to third world relief. I would suggest a tax of $5 per square metre for both public and private lawns, updated annually, until all but useful lawns are eliminated.

Exemptions would be for non–mown swards grazed by productive animals, short natural turf such as is produced by *Phyla* (*Lippia*) *nodosa*, or chamomile, and handmown areas of lawn that use no water or biocides. The rest is a shameful waste of global resources.

To reform the lawn, new permaculture businesses are evolving, using natural (non–irrigated) ground covers, berry and smallfruit shrubberies (salal, blueberry, cranberry), flowering meadows of native bulbs and perennials, copses of small trees, ponds, marsh or fen areas, and rock gardens or specialty gardens of perennials. Even vegetable plots are slowly becoming respectable as values change from the production of waste to the production of food for the home. A new ethic is arising that will reverse the status of waste producers, and new ethical investment portfolios are eliminating support for many such destructive industries.

Lawn reform, on the basis of health, costs, and ethics alone, is a rich field for innovative design, and needs disadvantageous taxing to bring it under control, or we risk converting all land not in housing to sterile lawns.

12.13
GRASSLANDS

Perennial grasslands (veldt, prairie, meadow, steppe, herbal ley, pasture, and heathlands) are a feature of mesothermal climates. Many, in fact, are treeless areas established during periods of icecap retreat, and have never established as forests, although both soils and climatic factors will permit forest establishment. Other grasslands are maintained by fire (natural, from lightning strikes; or managed, from human interference) against forest invasion.

Many aboriginal peoples so managed their environments as to maintain herd species, berry crop, nettle crop, or some aspect of the fire–succcession ecology of benefit to their food supply or resource base. The management of natural grasslands by fire is a skilled task; too frequent or too hot fires not only destroy soil structure but tip the balance from perennial bunch grasses and herbaceous species to fireloving annuals (wild oat) of low nutrient value and poor yields in the winter or late summer season.

Most natural grasslands lie in the Eurasian land mass, with Africa and North and South America providing veldt and pampas respectively. Areas such as the poa tussock plains of New Zealand and southeast Australia were fire–created and maintained by aboriginal peoples. Except in nature reserves or sparsely settled regions, even the great prairies of North America and the pampas of Argentina and Brazil have been converted to pastures or cropping, as have the Russian steppes and the African veldts. In every case, net protein production has been reduced and soils eroded.

Perennial grasslands (*not* crops of annual grass species) are very stable systems, and while a forest may have a visible "standing crop" of 207 t/ha, and grasses only 7 t, below the ground forests produce 184 t/ha of organic matter to the 345 t/ha of natural prairie in equivalent soil/climate regimes (Foth, 1984, based on other studies). Forests have most of their biomass involved with gaseous exchange in the atmosphere, and also with deep nutrient recycling, while prairies of grasses, forbs, and their associates act more on the near soil surface, and are copious producers of litter and shallow root masses (195 t/ha versus a forest's 106 t/ha in the top 30 cm of soil). Both function to complement the other in different strata of the soil/ atmosphere interface.

The profound difference lies in the fact that the regeneration time after "harvest" or removal of all or part of the standing crop is measured in decades for trees, and seasons or even weeks for grasses, so that yields from the standing crop (where this consists of the vegetative parts of the plants and not the seeds, honey, or fruits) may favour grasses. If the forest takes 80 years to regrow from the ground level, then it "yields" or gains about 2.5 t/ha/year. If the grasses can be cut or grazed twice in a season, then they can yield some 10 t/ha/year. Grasses do this by developing their buds or growth points close to the ground, below cut or browse damage, and keeping a lot of their mass in living roots (97% or so of the total biomass). Trees, when cut, lose most of their meristems (growth centres or buds) and have relatively less root reserves (less than 50% of their total mass).

Yields to people or cattle are not, however, the important factors in the sustainability of systems, as we have often proved to our cost. Trees both shelter, provide nutrient to, and supply offseason or complementary yields when grasses are affected by drought or cold. It is, as ever, a matter of balancing the components of the system so that both trees and grasses can serve the stability of the whole, whether we use them or not. It repays us too, in the long term, to see that a balanced system persists in its yield, rather than our demands run it down to a desertified endpoint of low total yield. Like machines, plant systems are of greatest benefit when they prove durable and reliable, and we should always think of "yield" as complexed by the length of time that yield can be maintained. Grasses, in effect, offer for our use only a little part of what they have, while forests make most of themselves visible and available (sometimes at their peril).

Grasslands (or rather herblands) lie between forests and deserts. Alpine herblands lie above the treeline, and below the barren upper slopes of mountains, as savannahs lie between humid rainforest and deserts, and riverine or shoreline herbfields between the forest and open water. Grasses and herbs can always retreat to the buffered atmosphere of the topsoil and escape cold and heat, while trees must stand and endure the changes of seasons. Every forest has its herbland edges, each characteristic of that place, soil, and history.

Of all of earth's great plant systems, it is grasslands that we most favour by our use of fire and plough. Waiting for the right wind and weather, pre- industrial peoples sent fire across the dry grasses to kill or invade the forest edge. We built many of our plains this way from the edges of water and deserts to the hills and valleys of humid and fire–resistant forests. Maori, Inca, and aboriginal African and Australian peoples explored by fire, and destroyed forests to do so. Behind the fire front, trees (perhaps adapted by developing fire–resistant bark, foliage, buried buds, and hard seed) advance to reoccupy the land, so that many plains have a dark, wet rainforest edge polewards, and a savannah forest of adapted trees (most are fire–dependent) following along from the direction of the hot winds. It is a common pattern to find these "moving plains" in areas still subject to fire management alone. Eventually, we too become dependent on the grasses, the animals that feed on them, and the fire.

We flourish best, however, on the forest–plain edges, where we and our associated animals can profit from both herbland and forest, and we as conscious designers need to complex the agricultural grass deserts with many more herb species, and with clumps, lines, contours, and valleys of trees.

"Grasslands unsuited for farming exist in every continent, and there is surely no better way to use it than to preserve its wildlife." (Eric Duffy, 1979). In fact, most dry savannah veldt areas of Africa and Australia are unsuited for farming, although rich in natural protein yields (e.g. kangaroo). We may never again equal the product yield of the 60 million bison on the American prairies, with their unnumbered associated hordes of pronghorn and mule deer, and a host of minor species. The 80 or so large mammals of Africa, and the 60 associated species of antelope can never be equalled in total biomass, yield, or value to the earth by a propped–up, energy–consuming and essentially retrograde pastoral system of a few species of domestic cattle and goat flocks.

Russia has preserved and encouraged Saiga antelope (from near extinction in the 1940's) to over a million today, and can cull 40% of these annually for food. They are the best–adapted tundra animals. The Lapps likewise manage their reindeer, by travelling with the herds on their migrations, and (until Chernobyll) took a sustainable yield from the sub–arctic meadows and lichens that would be destroyed by cultivation or disturbance.

About 24% of the earth's surface is in some form of grasslands. Originally, these were not the simple grass–legume or mono–crop cereal stands that we have subsequently developed, but contained many hundreds of plant species. Even of the true grasses, which number 7,500 species, we now use only 30–40 species in pastures and less than 20 as grains. A rough classification of grasslands is as follows:

Natural

- *Mountain meadows*. Called alps in Switzerland, they are often flower–rich and least modified by man.
- *Continental interiors*. Variously termed prairie, pampas, veldt, they have been greatly modified and only exist as reserves.
- *Savannahs* or scattered tree systems, in which grasses may be less numerous than forbs (nongrass plants such as bulbs, orchids, various flowers, succulents, annual or biennial dicotyledons, forage shrubs,

and forage trees).

- *Reedbeds* and coastal dunes, coastal or river fringes, swamps.

Artificial

- *Pastures.* Some semi–permanent and relying on 30 or so grasses plus forage legumes.
- *Lawns.* A few species (5–10 common), mainly grass.
- *Cereal crops.* Usually as monoculture or simple mixed crop systems with a legume or legumes. A subset is wet padi crop (rice).

Grasses are of great value in soil protection and erosion control, especially in difficult situations such as on dunes.

A great many of the flowers and vegetables we use today were part of the original flora of meadows, including the dune meadows and seashores. These include almost all onions, poppies, cereals, peas and beans, amaranths, mints, ground orchids; all bulbs, iris and other rhizomes; anemones, sages, herbal *Compositae* such as fennel, carrots, parsnips; the *Brassica* family, comfrey, peonies, asparagus and so on. Alpine meadows still include a great many species, but we have banished flowers and bulbs from the broad "grassland deserts" of our pastures in modern times.

In the Berlin Botanical Garden, there is about an acre of reconstituted European meadow, which includes many nonagricultural grasses, and about 80 species of flowers (some now very rare in the agricultural environment). That this recreated acre is about the only area so rich in native species is a commentary on the sterility of western industrial societies and the chemicalised farm. We have eliminated almost all the ancestors of our herbs, vegetables, flowers, and root crops to make way for beef and a few crops. In Yugoslavia and Ireland, rich meadows and bogs still exist where progress (or misuse) has not overtaken the environment.

Old castles, cemeteries, rail reserves and unsprayed or unmown roadsides shelter the prairie and steppe refuges of the USA and USSR, although belated reserves of their ancient companion plants have been established in both countries. Wherever meadows still exist, a visitor will notice the richness of insect life (especially conspicuous species such as bumble bees and butterlies), and the deep springy softness of the earth itself. It is the deep tunnels of undisturbed burrowers that produces an absorbent mattress of meadow on the earth.

A single Russian ground–squirrel (suslik) will dig some 200 holes per acre, and turns over 95 cubic feet of soil per year. Susliks are eclipsed by woodchucks, moles, wombats, and (as a group) mice and voles in this deep–ploughing effect, and prairie dogs and rabbits may fully tunnel out and manure many acres.

A great many bulbs with poor seed set, and rootset plants such as comfrey are planted by these underground gardeners, who leave forgotten storages of tree and meadow seed, root cuttings, and bulbs in shallow tunnels. A North American gardener can never rely on keeping comfrey in one place if gophers are in the area! Many plants thus came to rely on vegetative rather than seed propagation in rodent–rich meadows, while as many seed eaters became involuntary gardeners. Even the regurgita of owls sprout meadow seed caught in the fur of their victims, while the neglected underground stores of their prey species are left uneaten, to sprout later. As so many of these tunnelers are eliminated by plough culture and many become pests of stored grains or crops, they have largely disappeared from civilised parts, leaving hoofed animals to compact the earth without relief.

In the tree clump savannahs of Africa and Australia, and in the marshlands, only the raised mounds of termites and ants present well–drained and manured tree sites. Everywhere in grassland, the burrow spoil of ants, moles, mice, and rabbits dot the landscape. On these soft bare–soil places, dogs and cats defecate, and blown seed finds a seed bed. Fox and dog dung as often as not contain the seeds of mulberry, plum, loquat, and vine fruits, which rely on a combination of loose soil and manure.

We can recreate meadows and prairies. A little research and gathering will provide the seeds and bulbs, and today many meadowseed mixes are sold. If we do the job properly, however, we must also tolerate the burrowers and their predators. Today, many thoughtful and observant farmers are including more and more herbs in their leys (mints, chicory, dandelion, cleavers, daisies, plantain, and vetches) to the great health benefit of their herds. When we eliminate the bulbs, herbs, mushrooms, copses, hedges, and small animals of the agricultural landscape, we also lose part of the body of the earth and of ourselves. There is nothing so dull, anti–intellectual, and sterilising as the contemporary agricultural scene. We may yet die of the poisonous policy of: "CLEAN, DEAD."

For cool humid areas, perennial pastures usually consist of a sward carefully composed of grasses and legumes. Traditionally, good strains of perennial rye-grass (*Lolium perenne*), cocksfoot (*Dactylis glomerata*), white clover (*Trifolium* varieties) and some lucerne (*Medicago*) would be sown, and carefully managed for grazing on the basis of short–rotation stocking. However, there are several more innovative ways to increase nutrition, drought buffering and mineral balance in such pastures. In high–value landscape (well–structured soils at gentle slope) fenced–off browse lots and hedgerow tree forage strips are obvious ways to extend and buffer the system. Today, permanent electric fencing is cheap and effective in partitioning pastures and protecting new forage hedgerow.

In the contoured hedgerows, an under-sowing or set planting of lucerne, comfrey, chicory, selected dandelion; and a midlevel planting of tagasaste, *Caragana*, *Coprosma repens*, and pampas grass (*Cortaderia*), with a tall overstory planting of willows, poplar (selected high–value forage cultivars), white oak, chestnut, honey locust, and known desirable woody browse (hawthorn and *Rosa* spp e.g.) could be designed to occupy 10% per annum of the area, until year 4, when 40% of the total area would be broad, complex, contoured hedgerow of deep–rooted woody plants and forbs, with evolving tall–tree browse and mast–crop species, some of long–term value as timber. After years 4–5, timed and observed access to such browse could be permitted for sheep, young stock, and cautious harvesting by cattle. From years 6–8, longer browse times can be permitted, and in emergencies species such as willow and poplar cut and fed as drought rations.

In farming, we distinguish between temporary grasslands sown to rest the soil and provide hay or green meadows [leys, or short–rotation pastures (1–4 years)], and the permanent pastures sown for constant grazing (100 or more years). Soil crumb structure, humus, and soil nitrogen all show a slow improvement under permanent grasses, so that 25 years is needed to notice a pronounced effect on these factors. For cropped soils (nitrogen 0.11%), a climb to 0.17% takes 25 years and often 100 years; 0.25% nitrogen is measurable under permanent grassland. In every case, legumes sown with grasses are essential to soil improvement, and arable crops soon after leys show increased yields for 3–4 years. In a traditional cool temperate farm landscape where soil health is valued, some 25% of land will be in 1–4 year leys (red clover as a legume) and 15–25% in permanent pasture (white clovers). The rest (60%) would be in grains, green crop, and root crop. In such a landscape, soil condition is usually good, but modern farming rarely follows such conservative methods.

As livestock must be housed in colder snowy climates for up to 200 days a year, very large quantities of manures and urine are available from barns to keep sward and crop in fertiliser. Manures are more evenly spread as a liquid slurry, or more recently are fed in below the sward using a soil conditioner (**Figure 12.16** shows an illustration of such technology from *New Scientist* Aug '87, p. 30). The latter method prevents loss of nitrogen to the atmosphere, hence deters pollution, and enables maximum use of the slurry by pasture swards. Incidental aeration also results from this method. Slurry is so injected at 140 cubic metres/ha, 13 cm down. Autumn is the best time for least disturbance to the sward (3-4 weeks rest).

Leys are also being used (since the 1950's) in the subtropics, again with grass–legume mixes for 2–4 years, followed by crop. Grass–legume mixes are important soil stabilisers down to 45 cm of rain, especially if sown in wet seasons, fertilised, and planted with scattered or hedgerow–drilled *Acacia* or leguminous trees. Below this rainfall level, opportunistic barley or oats in winter hold soils while leguminous trees and woody shrub seed establishes. Grasses without legumes do not noticeably improve soil reserves. Leys unbalanced towards legumes may produce an acidic soil condition and nutritional problems in livestock. Pastures are kept in balance by strategic browsing or mowing.

MANAGEMENT OF LIVESTOCK ON GRASSLANDS OR RANGE.

Despite the fact that plant breeders have "concentrated on producing grasses...that would respond to high nitrogen inputs...UK farmers now apply nitrogen fertilisers at 20 times their prewar rate, but in 40 years the number of cattle and sheep carried by each hectare of grass has scarcely doubled." This increase, as well, can be attributed more to improved grain feeds and the use of lime on pastures rather than from the application of nitrogen (Harvey, G. "Grassland Production and Nitrogen, *New Scientist* 15 Feb '79).

The great drive to nitrogen fertiliser (aided and abetted by agricultural advisors) was as a result of overproduction in explosives factories, or in fact war surplus dumped on farms. Fertiliser became 30% of energy use on farm but brought no such increase in return. Nitrogen was once supplied free by clover, which has now been discouraged by repeated ploughing (fossil fuel energy) so that grass would grow. Today, scattered leguminous trees on range are known to supply sufficient nitrogen, and free–living soil bacteria and algae also are cultured and inoculated in soils to assist this effort.

No observer of field conditions can fail to see the hunger of cattle on highnitrogen sward for the leaves and bark of trees, so that trees which have stood for years in unfertilised pastures are "suddenly" attacked and killed by cattle where nitrogen is used. Clover can, in fact, supply growth equivalent to 200 kg of nitrogen /ha and is far better utilised by cattle than is the the grass sward recommended by advisors. This energy ratrace is typical of pasture developed without any sound nutritional observations or attention to livestock preferences.

It is evident on any heavily browsed areas that only poisonous, spiny, or very tough plants eventually come to occupy a range, so that noxious weeds are another indicator of overuse of range, although we blindly blame the weeds that protect the soil, not the management or cattle.

Some changes are more subtle; flatweeds (plantain, dandelion) may come to occupy more of grasslands in cool climates or areas of poor drainage. Permanent bunch grasses, succulent in dry periods, may be replaced with ephemeral annuals which drought–off in summer, and low tree browse may be eliminated, as has happened in California and many dry areas.

There are several management strategies which can reverse these trends. For instance, in the above example flatweeds unpalable to sheep are preferred by wallaby. Bunch grasses will remain if heavily stocked, seeded again, and then completely rested (this is a question of fencing strategy). The best strategy is to balance browsing species with plant species, to allow rest periods on range, and to avoid fixed stocking of animals on any one part of the range. Good managers allow a 2–9 year rotation of herds, using 15–18 fields or runs, and pay close attention to observed regeneration of browse. The longer (9 year) period is used on subhumid desert borders, the shorter period on humid lowlands.

There are several ways to crop grasslands and associated woody browse.

- Natural or Managed Wild Systems. A great variety of plants and animals interact; adjustment to available browse is made by migration (herd species), and light stocking (residents), or by very efficient metabolic and reproductory processes.
- Long–Rotation Stocking. Domestic species on extensive range, very light stocking, long rotation (7–9 year). Drought–immune, such range maintains woody and selected browse; there are low returns per hectare but a sustainable yield and no environmental damage. OR, ideally 15–18 fields, 18–20 month rotation, 20–30 days/field, and Keyline irrigation (for beef and sheep production).
- Short–rotation Stocking. On improved pasture, humid areas only. Cows at 15–25/ha for 3–7 days, grass at 15 cm high, rotated for from 3–5 weeks. In good growth periods, some fields are closed–up for hay making or slashed as mulch. *No* grazing before animals are moved (usually 3–5 days). Dry cattle can briefly follow milk cows. In this system 8–12 fields are needed, of which 1–5 are available for mowing in all but very dry or cold periods. Fields not well–grazed can be mown 3–6 weeks before stock are returned. Many good farmers mow and mulch to improve soils.
- Strip Grazing Uses electric fences, permanent or movable, to effect the same result as above, with about an average 2–5 day rotation (grass at 15 cm or better). Intensive, used only on improved pastures, dairy cattle. Akin to *tethering* for a few milk cows or goats on limited improved pastures.
- Cut and Feed. Ideal for milk cows, small herds of 420 cows, tropics, semiarid areas. Shedded cattle are fed from mixed forage (*Leucaena, Pennisetum*, comfrey, browse plants). All manure is returned to the cutover area, preferably as a subsoil sludge (**Figure 12.16**). This is an excellent system for villages in poor arid areas or where large herds are not kept. Hay can be cut from surplus growth in good periods. It is *not* sustainable as broadscale feedlot, due to excessive energy costs, waste of manures, pollution, and high overheads of feedlots; this is essentially a small farmer system.
- Mixed Livestock Rotation. Not much used as yet for more than 2–3 species. This would involve horse/cow/sheep/goose/pig succession (or some such) over a mixed forb/grass/ legume/ forage system. Each animal species prepares the sward or browse for a successor. Pigs are used well in advance to plough the ground for new sowings and root sets of (e.g.) comfrey. This system was used by a farmer (Mr. J. Savage in Victoria, Australia) with success. Comfrey is a key element, also lucerne and grasses. The "pig tractor" system obviates the use of mechanised tractors and improves soils. It is well worth trials in any area, choosing a succession of animal species following each other at optimum densities and intervals. The system attempts to reproduce the high yields of natural wildlife systems.
- Fixed Stocking. This method is dangerous if the land is overstocked, and is the most used and misused system. Livestock are rarely moved, and pasture quality, yield, and soil structure can deteriorate. It is not responsive to seasons unless closely watched, and unless sale–and–repurchase pre–empts environmental damage.This system can work with very light stocking, closed winter range, permanent pastures, 6–8 home fields, and careful observation.

All the above domestic species systems relate to "livestock only" farming, and are far more simple and less sustainable than the usual mixed farm situation of crop, forest, marsh, and a variety of livestock balanced to forage wastes. Inevitably, most intensive broadscale systems call for energy and government subsidy, and are political rather than environmental systems. If the world returns to sanity, most of these broadscale systems would be curtailed, and cattle in forests, semi–arid, and montane cold areas would be removed (4% of all cattle production, and probably 70–80% of cattle damage), as they add far less to national income than ther damge they cause to national assets. If this were to happen a careful harvest of wildlife and water would exceed cattle products by many times their value.

The second factor (after cropping) that has reduced the production of natural grasslands, with their gallery forests in valleys and islands of tree species, is the seemingly intractable and modern concepts of private land ownership, hence fencing and the prevention of migratory movements of herds. We may never live to see such errors totally reversed, but it is a very worthwhile design enterprise to both recreate and to experiment with new forms of perennial and productive grasslands and their associated tree species in livestock managment, and for meadow species preservation.

The net product of a grassland in a cool humid climate has a bimodal growth curve very like that of **Figure 12.18** (derived from pasture yields in southeast Australia).

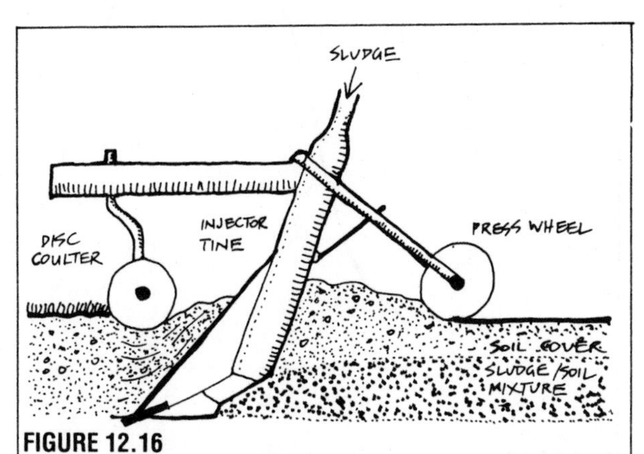

FIGURE 12.16
SLURRY INJECTION IN GRASSLANDS.
Liquid slurry injected 18 cm below sward or crop avoids problems of acid water from waste manures; pastures respond in 4–7 weeks.

Thus, there are two periods of deficit, which is always a feature of open grasslands. The late summer deficit arises from the shedding of grass seed, droughting, and drying–off of grasslands, and the winter defecit ís simply that of slow or no growth, with snow cover or blizzards affecting herd grazing times.

As the figure shows, there is an absence of green feed in summer and winter. For the summer period, green leaf from forage tree plantation largely overcomes this shortage, although evergreen forage trees may not withstand a hard frost if lopped in the last month of autumn. Many species (such as tagasaste) need mature leaf to carry into frost periods, and so should not be winter–lopped. Such species as willow and poplar are also summer leaf forage due to their deciduous habit, and chips of their wood can be fermented to high–value winter concentrates.

Summer foliage from evergreen or deciduous forage can, like grasses, be pelleted, made into silage, or pressed as hay if dried after lopping. Tagasaste (unirrigated) produces the equivalent forage weight of irrigated lucerne. Given a seed source, such forages as tagasaste can be field–drilled with a normal seedbox used for other crop, and a mix of tree seed and daikon radish or turnip is often effective in grass suppression while the trees grow. Seed is most cheaply obtained from the soil under mature trees, which can be thrown on a series of shaker trays to clean seed from the grass sod and soil, using a sequence of three shaker trays in one frame (**Figure 12.19**).

Matthew Carpenter, on the cold Canterbury Plain of southern New Zealand, has successfully field–drilled tagasaste for summer sheep forage. One unforeseen benefit is that grasses between tagasaste strips dry out much later, and produce more growth, than grasses in the open field situation. As tagasaste produces abundant seeds for chickens, and honey for bees, it can be used as a forage crop for these species. For cut forage, spacing at 1.0 to 1.5 m is ideal (the stems kept cut at 1 m high), while for seed and bees spacing at 4–5 m produces better forage (Doug Davies, *pers. comm.*).

In constructing, rehabilitating, or producing a seed resource for mixed meadows, the following broad plant groups are involved:

- Grasses, Rushes, Sedges: mainly perennial, bush, or tussock species, but with some annuals, some of which are of geat value to settlement as food (grains). Examples are the ricegrasses, *Panicum* species, and

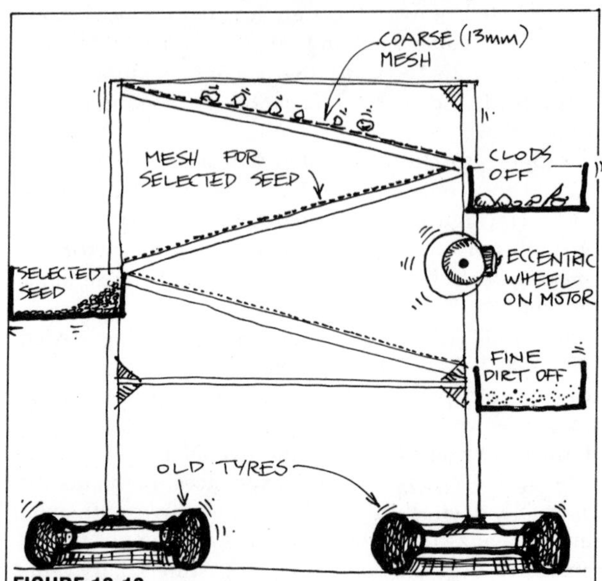

FIGURE 12.19
SEED SHAKER TRAYS.
Seed form below trees is easily gathered by this electrically–operated seed shaker system; soil is simply shovelled onto the top tray.

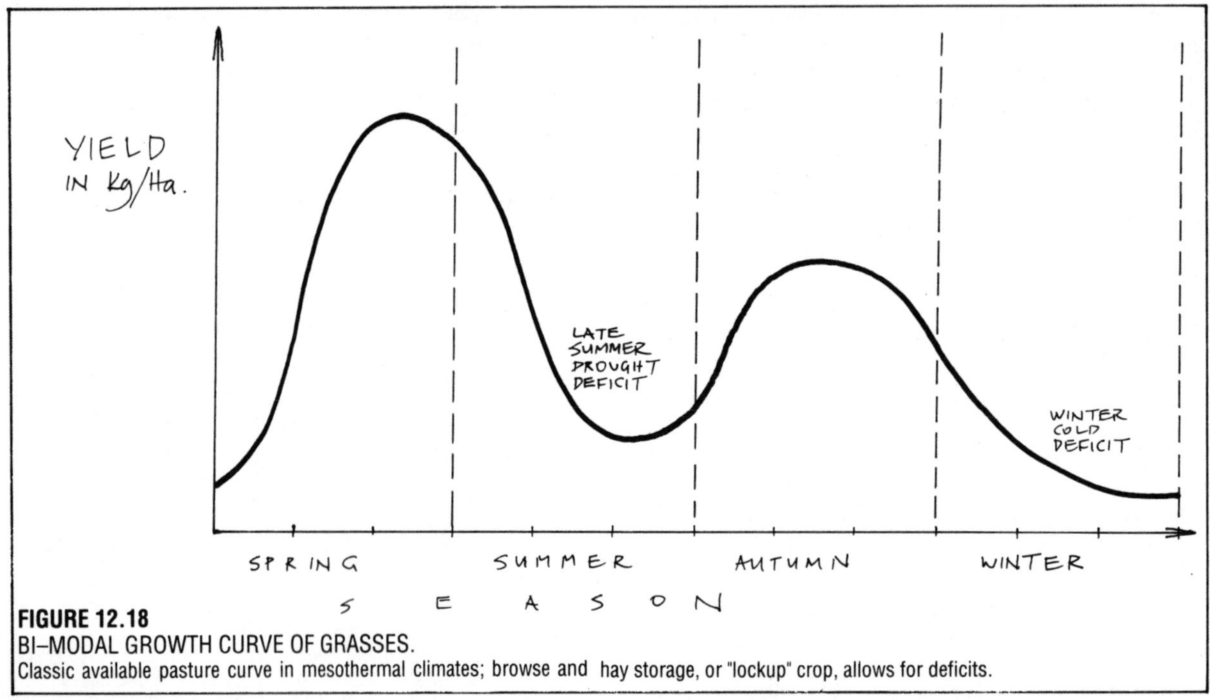

FIGURE 12.18
BI–MODAL GROWTH CURVE OF GRASSES.
Classic available pasture curve in mesothermal climates; browse and hay storage, or "lockup" crop, allows for deficits.

rushes used for thatch and matting.

- Legumes: the grain or pasture legumes, clovers, vetches, lupins, tares, and bulbous or deeprooted legumes, some of use for foods as tubers, some twining species.
- Bulbs, Corms, Tubers: perennial flowering plants as bulbs, rhizomes, or tuberous root masses; many are dug as food in winter. Sunroot (*Helianthus* sp.), breadroot (*Psoralia* sp.), crocus, camass, mariposa or sego lily, and a variety of orchids, are of this group. All the onion group belong to this classification.
- Herbaceous and Perennial Forbs: nettles, daisies, and species of the families *Umbelliferae* (fennel, dill) or *Compositae* generally, as well as many flowering plants of meaows (poppies, forget–me–nots, watercress, buttercups) all belong here. A few are annuals, and many have a wide range of uses.
- Spike–rooted Flatweeds: such as dandelion, thistles generally, plantains, docks, and chicory (all high– value browse and some used as salad plants and vegetables) are features of mixed pastures, meadows, and grasslands.

Whether we are designing or constructing a meadow, or rehabilitating one, it is necessary to study a few of the critical characteristics of each plant group. Such factors are:

- Mode of Occurence: Whether the species occurs as clumps, or are solitary in the system. Some species need to be set out as patches or clumps to persist, others do well alone.
- Method of Propagation. Comfrey, narcissus, and sunroot are all naturally propagated vegetatively by being harvested, stored, and forgotten by burrowing animals. Flatweeds such as thistle and plantain seed on burrow spoils, resist close grazing, and therefore occur near burrow mouths or on cropped areas with soil disturbance. Disturbance also suits annuals and windblown species.
- Preferred Soils and Sites: Drainage (from boggy to freedrained gravels or sands) is one critical factor for meadow species balance, as is soil moisture reserve. Acidity/alkalinity is a second primary factor, so that in initial preparation and placement of meadow sites, some variation in such factors can be built–in so that chosen species have a niche or site to occupy. Although meadow plants prefer free drainage, most sedges prefer poorly drained areas, while chicory will colonise compacted ground.

As a meadow is a perennial or selfseeding system, ground preparation must involve an initial earthshaping and weed removal effort. The collection of meadow species involves searches of seed and plant catalogues, roadsides, and reference books, but the end result of meadow recreation is a very useful and pleasing assembly of plants with a varied product base, producing healthy animals on range.

Up to this point, we have considered only the needs and values of meadow plants, but a meadow is as much maintained by its pollinators, browsers, burrowers, and their predators as it is by plant growth. In fact, many plants only maintain if their animal associates are present, so that owls, field mice, butterflies, bees, and worms are all part of a meadow assembly, and have specific functions in meadow development and maintenance.

REVITALISATION OF COMPACTED SOILS AND WORNOUT PASTURES

Data on soil rehabilitation is given in Chapter 8 of this book, but any soil or compacted pasture can be revitalised by periodic sodseeding, using a broad–flanged chisel point at 6–10 cm depth and 0.5 to 0.6 m spacing to cut the roots of existing grasses and to provide a seed furrow for more vigorous, deep–rooted, more nutritious, or droughtresistant grasses or woody forages. Early spring and from mid autumn to early winter are the usual pasture species sowing periods. With the seed, trace elements, major nutrients, and water–retaining gels can be trickled into the shallow furrows so developed by the tines or chisel points in the old or degraded pasture.

Some typical seed mixes for sod seeding may include a variety of clovers, chosen for site, pH, and drainage; spikeroots such as dandelion, plantain, and chicory; woody browse species and medicinal species such as wormwood, tagasaste, *Elaeagnus*, pines, willows, or poplars, and grasses chosen (as for clovers) for the specific site and soil factors.

If labour or rootset planters are available, root cuttings, bulbs, and plant cuttings of woody browse, comfrey, willow, or poplar can be set out in deep–ripped furrow, either as a pasture browse combination, or as pioneers for farm forestry operations.

A special use of sod–seeding is in sowing down lands where round–rush (*Juncus*) will grow if the land is normally ploughed and cultivated. Here, minimal tillage prevents the spread and growth of rushes and the acidification of soil. This establishes clover and soft pasture or crop without rush competition, or with minimal rush growth (often browsed off by cattle if clover is present). Bert Farquhar of Wyambi (Tasmania) gives this data for coastal pasture establishment. As many coasts and heaths are copper and cobalt deficient, this is also added. Clover at 0.7 kg/ha is sod–seeded after a slash of brush and rushes, and an autumn burn. The sowing is completed in the first month of winter (it can be earlier if the ground was previously fallow). About a ton of dolomite per year is added. Ploughed marshy coasts often result in a pH so low that uneconomical amounts of dolomite need to be used to correct the soil condition—another instance of how high energy use causes cost and work later. The lowland so ploughed also bogs vehicles, pugs badly in winter, and may quickly go out of use. Its best usage, then, is as wildlife marshes, fish ponds, and a source of *Typha* browse or reed thatch for crafts, where acid boggy ground has been created by ploughing and hoof compaction.

12.14 RANGELANDS

Brown (1972) estimates the standing crop of large herbivores on the African savannah to be about 500 kg/ha, or about the same as on improved pastures in humid temperate areas. Instead of one or two species sustained by external oil energy input, the savannah species hosts guilds of browsers (five species), grazers (eleven species) and four species that alternately graze or browse, each having specific food preferences.

This crop estimate specifically excludes the far greater biomass of smaller mammals, birds, insects, and burrowers, reptiles, other lower vertebrates, and plant products, so that it is a minimal yield figure for a natural system. No known cultivated system of livestock can surpass this sort of productivity without external inputs, nor are similar yields achieved without stressing soils, plants, or animals. Our excuses for destroying such systems, therefore, rest not on the basis of health or productive capacity or soil conservation, but on the sequestration of land for private misuse or political gain.

We limit our yields also by not considering the abundant soil life, and by not devising methods to harvest some of this product either directly or indirectly. In a clearance of rabbits, where some 1,100 were caught or dug out per acre (Tasmania, 1954), myself and others estimated yields of close to 700 kg/ha. Such degraded pastures also contain enormous numbers of insect larvae in excess of this figure. We need to assess yields more closely, and to question our direction in trying to raise a few species of animals on lands that will support many more species and much higher yields of protein from hitherto unconsidered species.

In swamps, savannahs, and in semiarid or arid plains, trees may occur in discrete groups. This may be related to natural conditions, or it may be as a result of other (animal) species working on the environment. Such species as alligators in swamps create wallows for their own comfort, as do pigs, buffaloes, elephants, and warthogs. A wallow presents a new clump plant site by altering drainage (as hole and bank), nutrient, and soil structure.

In estimating the effects of animals on browse or grazing, agriculturists have used a "stock unit" system of equivalents based on the weights of the animal. In fact, this system needs more work and modification, as sheer size also affects metabolic efficiency. A second modifier is that of individual metabolic efficiency (as with the factor of water conservation). Even within a species, certain individuals convert food to energy or growth much more efficiently than others. Such differences within species may be greater than differences between species. However, the concept of the stock unit is a useful idea, so that the available food in season or in a particular type of pasture can be assessed over the year in "cow–days" of fodder, and approximate stocking rates for other species, or complex species mixes, can be estimated. In practice, however, ground experience of the effects of stocking rates on pasture need to be adjusted by observation, especially if animals are being selected for efficient growth.

The standard stock unit is a European cow and calf, about 450 kg liveweight or: 2 African cows; 5 European sheep; 20 African sheep; 25 Thompson gazelles; or 125 dikdik antelope; and so on. Ten standard stock units are

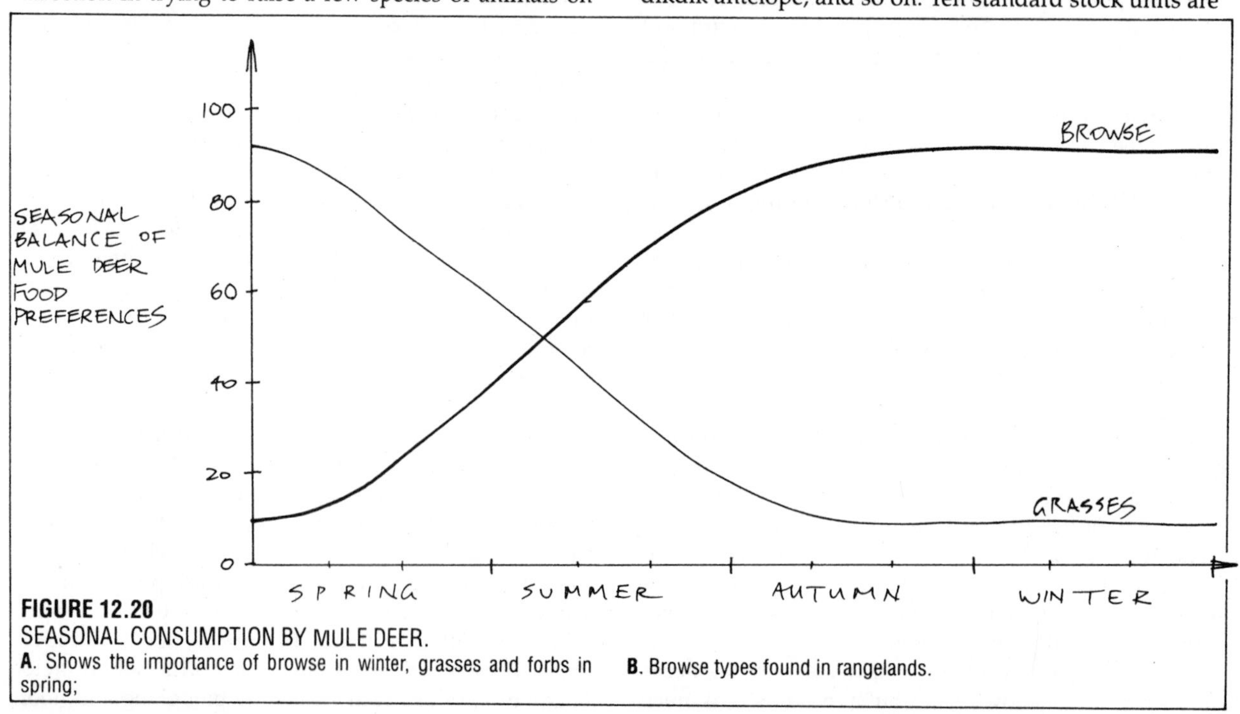

FIGURE 12.20
SEASONAL CONSUMPTION BY MULE DEER.
A. Shows the importance of browse in winter, grasses and forbs in spring;
B. Browse types found in rangelands.

equivalent to 1 elephant; 4 black rhino; 15 zebras; or 27 hartebeest.

Animals on managed range are a valid sustainable system only if the balance of plant and animal species is carefully maintained. In broad groups, range plants are:

- Forbs (annual and perennial herbs and bulbous species)
- Grasses and other monocotyledons, annual and perennial
- Trees and shrubs
- Mosses and lichens
- Fungi and saprophytes

The first two are eaten by grazing species, the third by browsers, and the last two by small herbivores and cold–area species. Animals balance their range diets in very different ways, so that a variety of animal species is needed to utilise the available forage (**Table 12.4**). Further, these are not so much a constant proportion, but a dynamic seasonal change, which is illustrated by **Figure 12.20**).

Some animals are sedentary, some range widely; the latter are able to spread their browsing more evenly across the range. Deer, for example, may commonly utilise 92% of a range area, on which cattle exploit only 52%. Deer may also convert food to flesh at efficiencies greater than cattle by factors of 2 or 3, depending on the species of deer.

This range behaviour is in part species–specific and in part determined by the availability of such critical resources as shelter from extreme heat or cold, the placement of water holes, or the presence of predators such as wolves or cheetahs. Cattle can be expected to range only 3–7 km from water, whereas some deer, kangaroo, goats, and small antelope can exist without free surface water except in extreme drought conditions. For species with clear preferences for fixedsite resources (water), range management consists of providing these critical resources to enable such species to utilise the range.

A judicious selection of animal species enables far greater production from range than the choice of one species alone. Different species will feed at different browse levels (**Figure 12.21.** It is usual for graziers to express range values in terms of "cow browse days" or the number of days a cow can forage on a unit area. This will of course vary seasonally depending on the balance of plant species on range, and it also enables some rough equivalents to other species of animal, e.g.

One European cow browse year = 266 blacktailed jackrabbits; 164 antelope jack rabbits; 18 kangaroo; 6 sheep; or 385 ground squirrels.

As many of these smaller species can be continually harvested and some (kangaroo) have of far higher assimilable protein content (58%) than cattle (35%), then from the point of view of protein production alone, kangaroo or jack rabbits, if well–managed and harvested, are far more productive than cattle. A judicious balance of species will always exceed in yield a monoculture of one species, and also keep the rangeland in better productive condition.

TABLE 12.4
COMPARISON OF RANGE DIETS.

	BROWSE	GRASSES	MOSSES	FORBS
ELK	27	65 (Max: 85)	5	3
ANTELOPE	74	4	—	22

(Percentage average consumption figures)

FIGURE 12.21
DIFFERENT LEVELS OF BROWSE

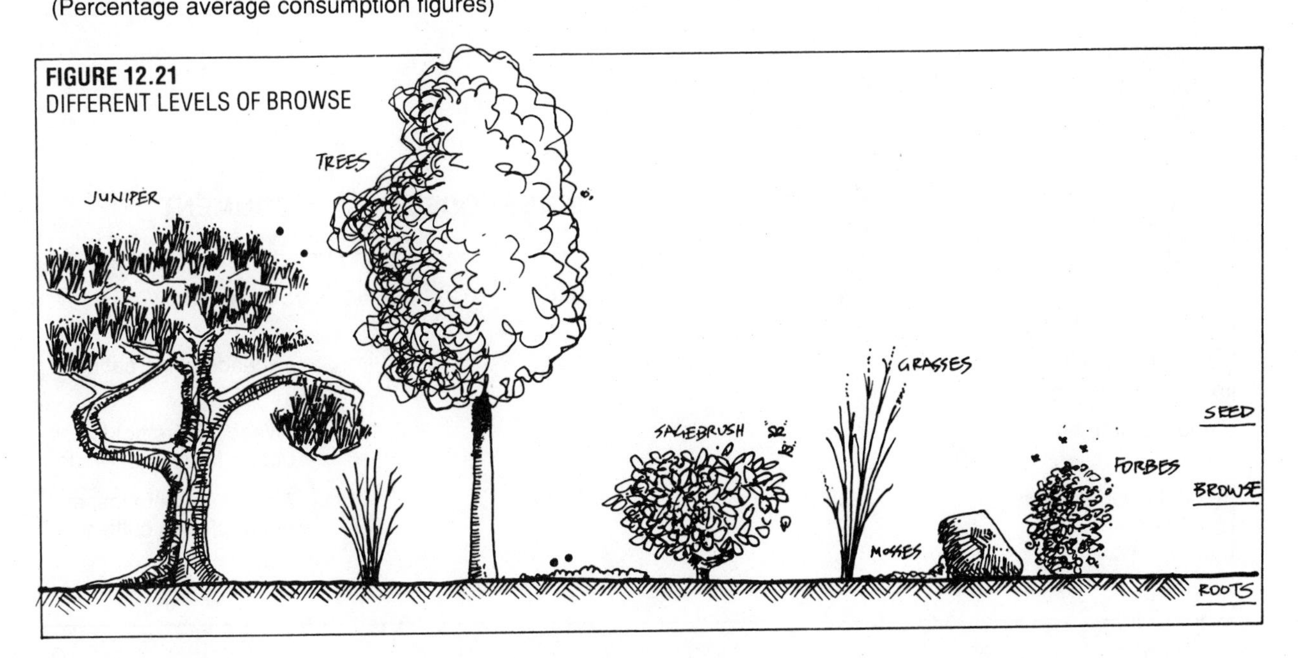

Any species not harvested (or where predators have been removed) can expand numbers so that rangelands are badly damaged. When such damage occurs, insect plagues may result. "Canadian workers found a direct correlation between range depletion and grasshopper outbreaks" (Stoddart, 1955), and rangelands infested with grasshoppers, plague locusts, or mormon crickets can lose 67% of their browse to insects. This is a severe reduction in livestock production. We clearly see that insect plagues are a reflection of poor mangement, as is the disappearance of preferred plant species from range.

Even insects are usable and of excellent protein value, and both fish, guinea fowl, or turkeys on range are efficient converters of plague insects into usable protein. Thus, on suitable range, fish ponds, water-fowl, and insectivorous bird flocks become integral to range management and plague control. Wetlands and ponds need to be created or preserved for this reason alone. If not harvested by range birds or fish, insect populations can themselves be cropped as part of concentrated foods for other acceptable food species. In Botswana, the drought harvest of locusts is accounted (by weight) as four times the food value of stored grains.

TABLE 12.5
SOIL PORE SPACE.

Grazing Condition	Soil Pore Space (%)	Rain Infiltration (%/hr)
Slight (Undergrazed)	68.1	4.14
Overgrazed	51.1	2.16
Depleted by Grazing	46.5	0.82

The plant cover on range has a profound effect on the infiltration (and subsequent river flow) of water. In the first 30 minutes of rain (after which infiltration rates level off):

- 1.2% of rainfall soaks in at 5% vegetative cover.
- 1.5% of rainfall soaks in at 15% vegetative cover.
- 1.7% of rainfall soaks in at 25% vegetative cover.
- 2.4% of rainfall soaks in at 35% vegetative cover.

This is the effect of vegetation alone, but perhaps a more profound effect is the reduction of soil pore space by trampling. In some desert areas of central Australia, I could not find a square meter of soil without a recent deep hoof imprint. After only 9 years of stocking on range, Stoddart (1955) reports that soil pore space was reduced 44% in the top two inches of soil, and 60% in the lower 24 inch deep layer of soils. Infiltration (and river flow) is correspondingly reduced, and runoff correspondingly increased, with resultant soil loss. Actual figures are given in **Table 12.5**.

Thus, up to five times less water is absorbed by eroded and degraded rangelands. This has a profound long–term effect on the general health of the region, and in particular on continuing soil loss, flooding, erosion, and drought.

The proportion of runoff in total precipitation (here given as a percentage) and the allied loss of soil in runoff varies according to range condition, assessed by herbage indicators in **Table 12.6**.

As for total plant cover on range, from 65–70 % is needed to totally prevent erosion. Stoddart (*ibid*) notes that run–off water from rangelands may carry 6–10% silt by volume, and that in a flow measured at 17,500 t water/hour, 500 t (in with the 1,750 t of soil) were of organic matter. This staggering loss is directly due to poor range management, and the effect is so drastic on all areas of the watershed that it may well be a central

TABLE 12.6
VEGETATION COVER, RUN–OFF, AND EROSION.

VEGETATION	RUNOFF % of rain	SOIL ERODED T/ha	COMMENT
Wheat grass (Good range climax)	0.4	0.001	Sustainable
Lupin and Needle grass (early depletion of range)	49.9	0.963	Exceeds soil creation
Annual brome grass	25.5	0.425	"Weed" grass holds soil before final deterioration
Annual weed species	60.8	3.094	Almost equal to losses under plough culture

issue in land/water policy, which itself should be the core of the policy of any nation which hopes to survive. Erosion due to cattle can be minimised or halted by fencing which falls from the valleys to the ridges.

In summary, then, the main function of rangelands may be to preserve the upland soils and water for more intensive use. The careful harvest of low (polyculture) stocking rates of mammals can be permitted, and range predators may assist in controlling browsing species, both insects and mammals. Only heavily vegetated lowland ranges are really safe areas for domestic hooved animal production, and even these may need regular soil rehabilitation to alleviate compaction.

THE MANAGEMENT OF GAME SPECIES ON RANGE.

The first priority in game management on range is to work out a method for harmless live capture of the species. These capture methods range from very large drift (nonreturn) traps for migratory, or moulting, or surfacedriven species (jackrabbits, geese, deer, gazelle, gnu); yard traps at waterholes (goats, horses, kangaroo); large or small baited box traps (most rodents); cage traps (many birds), set net in flyways or at night (many birds, some mammals); to individual capture using fastacting knockout drugs or chemical pellets (large or dangerous species like bear, rhino, elephant). Large species are also culled by shooting where culling is essential, as sexes and ages are easily distinguished in the field.

Many capture methods operate at night, when lights, traps, and nets are used and heat stress is less a factor. Wildlife managers and researchers have worked out safe methods of capture for most game species, and in many cases good culling or management techniques (in the absence of efficient predators such as lions, wild dogs, wolves, tigers, leopards, eagles, or foxes, many of which have been reduced or eliminated locally). Given a good capture and marking method (dyes, collars, ear tags, leg bands, tattoos, ear notches), population estimates can be developed, and tied to other factors such as scat counts or browse effect on range, herd size, or breeding success. Many free– range species are attracted to or benefit from areas of special forage or crop, shade, winter shelter, or water points—all of which can be added to a depleted range to increase wildlife, as can a more general scatter of seed or forage plants of particular value in periods of browse shortage.

These studies are the essential preliminary to the development of true managed wildlife farming, where up to 40 or 80 species of birds, mammals, fish or lower vertebrates are farmed in an integrated product system, which predictably exceeds one–species yields, and may actually improve tree and prairie cover. Even by mid–1960, wildlife researchers had good estimates of yields under managed systems, and could have devised excellent game farms. We may see these developing over a wide variety of species in future, as today the pioneer farms of crocodile, emu, kangaroo, deer, and waterfowl already show economic and social benefits.

Longterm measurements of natural mortalities on range due to drought, predation, plant response, parasitism, or territorial behaviour enables us to predict natural losses with some degree of success, and pre–emptive culling can make use of age/sex classes of animals that are in any case doomed to perish in the field.

Deer (red deer and roe) are now farmed in New Zealand, but are (astonishingly) being raised on improved pasture, and with the usual hay/concentrate system used for sheep, whereas these animals are well suited to managed montane tussock and tree browse systems, and are most efficiently raised in that

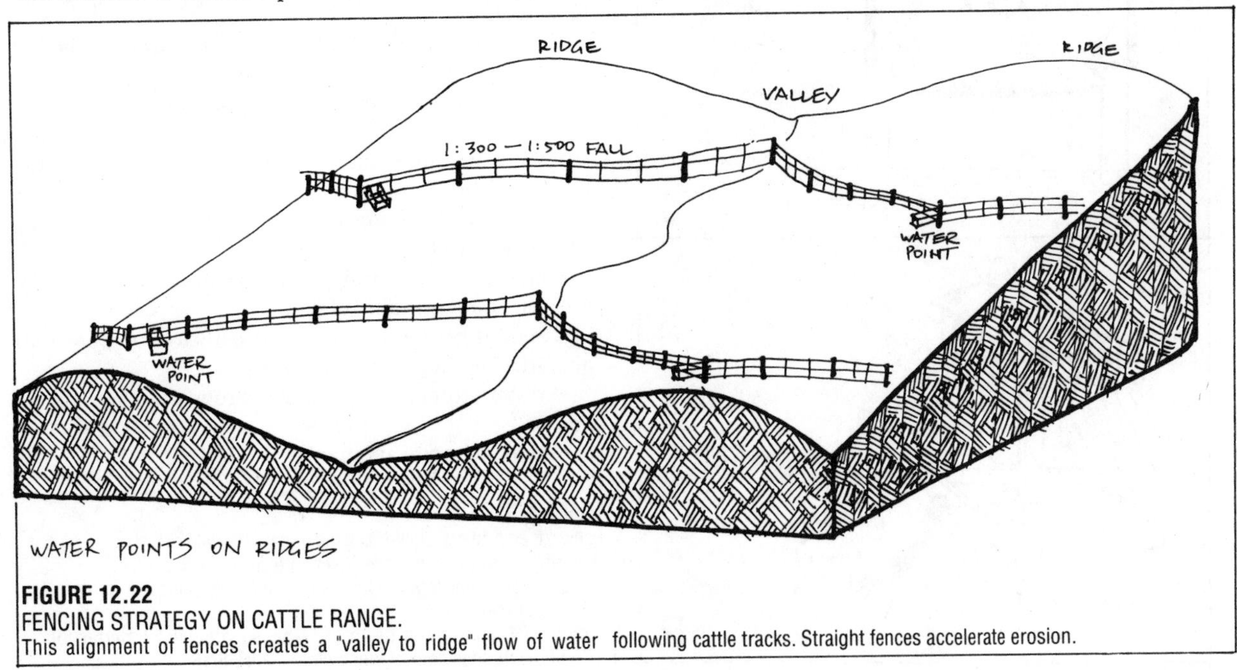

FIGURE 12.22
FENCING STRATEGY ON CATTLE RANGE.
This alignment of fences creates a "valley to ridge" flow of water following cattle tracks. Straight fences accelerate erosion.

environment.

Only in very recent years have we recommenced the intelligent management of game species in their preferred habitat. The benefits are as yet scarcely realised, but in most cases exceed that of "improved pasture" and selected domestic breeds of sheep and cattle, with less capital cost and better feed conversion.

Hopefully, some of the capital now misused in the production of fatty and chemicalised meats will be in future devoted to the intelligent management of healthier rangelands and products. There is no better use of capital than to use it for "increase" in the tribal sense of encouraging nature to show her capacity, instead of dictating directions and species. It has been imported European cultures which have prevented the use of kangaroo and a variety of antelope as farmed species although they are a preferred local food.

12.15
COLD CLIMATES

PHENOMENA OF COLD CLIMATES.

Considerable energy is needed to create snow, hail, and ice, and there are a special set of techniques for really cold areas that can take advantage of the characteristics of cold phenomena, for example:

- Albedo (reflection from) of snow as a heating device;
- Insulation values of snow over buildings;
- Preservation of ice for cold storage, summer cooling of food;
- Use of ice cover on ponds as winter access to deep areas;
- Ice effect in trees along meltwater rivers; debris trapping; and
- Stratification of the seed of cold area crops by refrigeration.

Then, there are special design precautions needed because of cold, and these include:

- Insulation of pipes and burial of pipes below frozen soil levels;
- Wells within insulated earth, preferably within the building fabric;
- Roof construction and steep roof slope in heavy snowload areas;
- General house design for extreme cold, especially for heat storage and insulation;
- Special garden techniques (frostheave on bare soil) and attached glasshouse growing;
- Water storage for summer gardens in earth dams; and
- Avalanche site avoidance.

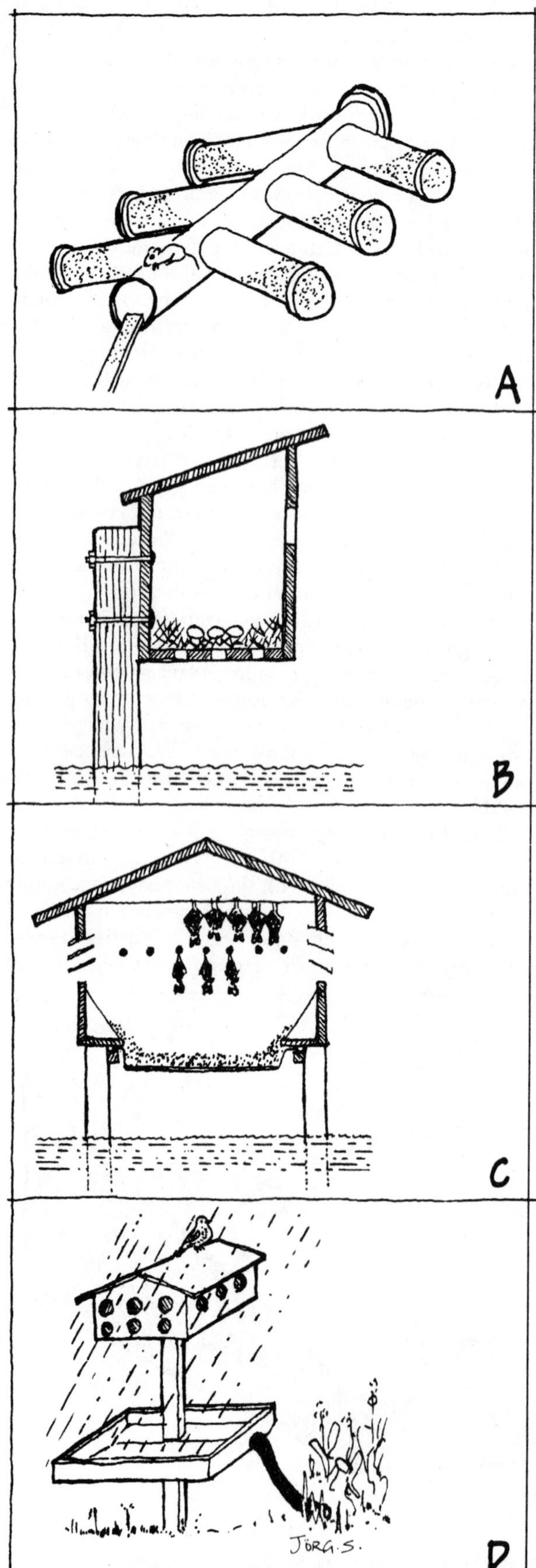

MANAGING WILDLIFE
A. Pack rat; an artificial shelter acts as a store for wild rice;
B. Waterfowl nest safely in boxes fixed to pilings in lakes.
C. Bat colonies in shelters provide rich manure for ponds or gardens, and control insects.
D. Communal blue martin nests yield phosphates and fertilizer for crop, and the martins control mosquitoes. [Jörg Schultz].

There are multitudinous publications on heat, but very few on cold, so that although we can distil alcohol either by heating or chilling water/alcohol mixes, only the former is in common use. Freezing concentrates sugars (maple sugar), alcohol, and salt solutions as efficiently as heating distils water or alcohol from solutions. Open pans of maple sugar can have the surface ice removed regularly (each day) until a sugar concentrate remains. Salts in water, and alcohol in ferment liquors can be concentrated in the same way.

Frost heave on bare, part–eroded or exposed soils leave (in the thaw period) ideal seedbed conditions for clover and other covering seed. Only bare soils show this phenomena, and these are most in need of seed to stabilise soil erosion potential. Frosted and late snow areas present an ideal surface to spread manures in late winter, so that meltdown deposits seed and manure on the surface of the ground for spring growth. Sawdust in the mixture gives a clear picture of seed or manure spread on snow.

In the fall, acorns, filberts, and hickory nuts are gathered by wildlife as winter stores. Field and pack rats bring in smaller seed such as wild rice from the marshes. If storages are provided, these foragers will fill hollow pipes or logs, or smaller pipes, old vehicle engine manifolds, and nest boxes or wall cavities. Seed so collected is sound, clean, and neatly stored. Providing some 15% is left, and given over to winter food for these workers, 85% can be collected for human use. A few people regularly collect their hickory nuts or wild rice in this way, by providing dens for squirrels or pack rats. It is a question of cooperation and provision for others, instead of attempting to kill off the experts and do the job yourself.

Grains and nuts so gathered, and stored in bins in barns or animal houses, feed poultry and form food concentrates for other animals over winter.

As the snow melts from the forest floor and openings, winter acorns begin to sprout. At this time (early spring), pigs and poultry released from winter quarters can forage acorns, as can sheep and cattle. The food value of the sprouted seeds is higher than in unsprouted seeds by factors varying from 10–100 times (for vitamin and sugar content).

In the winter forest, piles of branches and loose compost mounds of leaves form a refuge for small animals and reptiles. These and rockpiles are essential refuges for wildlife, providing deer–immune planting sites, and reducing forest fire hazards.

The pattern of snow–melt indicates insulated, absorbent, reflective,and heat–generating environments or surfaces. Black objects, soot, and leaves sink into the snow. Snow melts rapidly in front of the bare–leaf and white–stemmed stands of birch for about 4 m or so from their stems. Rock walls backed by birch give an early warm site to plant out vegetables. The Chinese use slanted bamboo and straw lean–tos to achieve this early growth of vegetables and to extend their growing season (**Figure 12.23**). The shade side of such shelters accumulate snow for insulation.

Stone walls, embayed, form very warm early sites, as do semicircles of tyres facing into the low sun. Such embayment can be plastic or glass–covered to assit heat retention, or piles of tyres can be topped with glass as miniature grow–holes, especially if the tyres are earth–filled or half–filled with water to retain day heat. Many gardeners use glass cloches, or cover individual plants with bottomless glasss or plastic flagons in early spring.

Ponds begin to melt in time for garden watering. While the ponds are still frozen, bundles of brush can be placed on their surface to provide fish cover,or fixed anchors can be set out to sink later as moorings. Black, glasscovered tubes or tyres keep fishing holes open, as

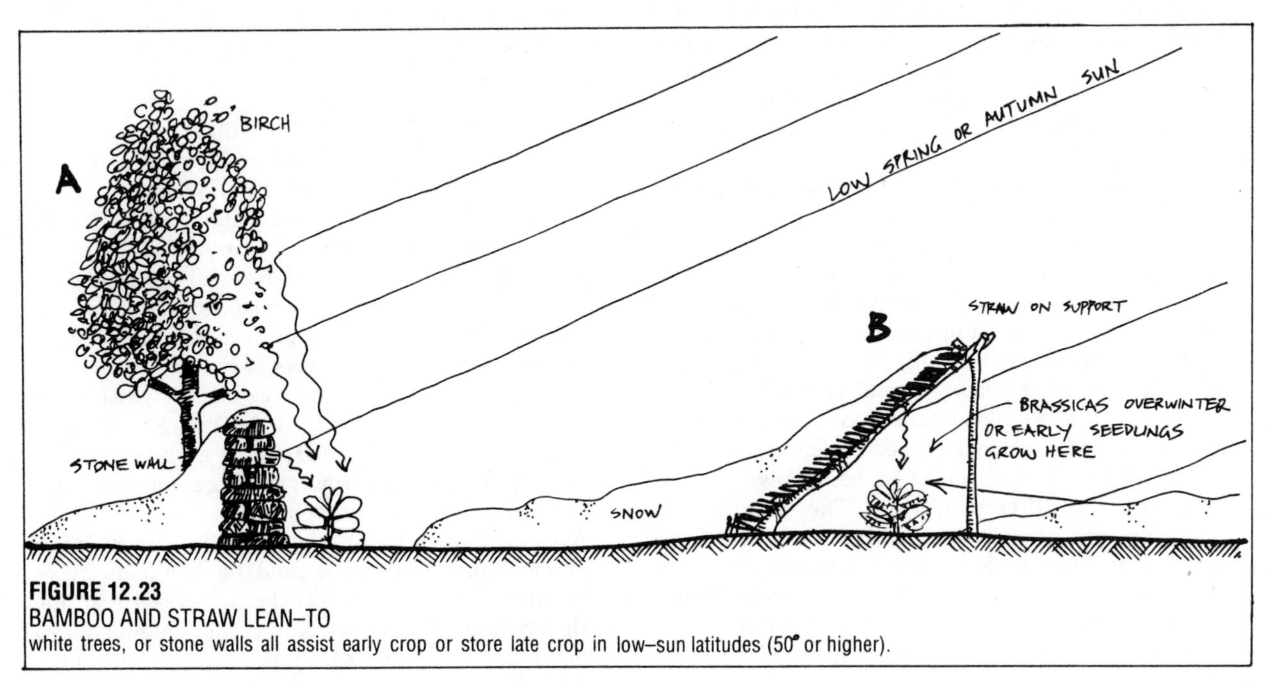

FIGURE 12.23
BAMBOO AND STRAW LEAN–TO
white trees, or stone walls all assist early crop or store late crop in low–sun latitudes (50° or higher).

do rafted glasshouses which act as fishing shacks. These can be towed ashore in spring or used to grow subtropic water plants in the shallows.

Ice, frozen inside insulated boxes in winter, and stored insulated from the ground, will last for a summer in very shaded places, and form an icebox store for summer use. Thick straw or sawdust lids preserve underground ice for summer. Uninsulated ice pits, into which winter snow was packed and then earthed over, were the basis of chilled drinks in this type of climate before 1830, and before the advent of mechanical or evaporative refrigerators.

ICE

Depth is the critical factor in cold waters, especially in stillponds. If ponds freeze to 15 cm–90 cm (6–36 inches), then that depth of water is unavailable in winter for house use, so that where arid areas must calculate depth of storage on evaporation, cold areas must calculate the amount "lost" by winter freezing. Freezing expands water, so that storages need be left unfilled, slope–sided, or open–topped to cope with expanding ice (**Figure 12.24**).

Similarly, pipes must be buried to 1 metre to prevent bursting, or allowed to trickle continuously at 450 l /day. For domestic use, there is no substitute for placing the water storage within the house cellar structure, as a well or cistern below floor level. The same is true of water stored in barns. Glasshouses and houses both can use this water as a heat store with great gains in efficiency. Ice, like ploughs, may produce a plough–sol or hardpan by pressure on soils due to ice weight. Weight is the factor that most differentiates the snow and ice which accumulates from frozen water.

Ice is, like glass and water, a transparent refracting medium suitable for lens construction, so that very cold areas can construct very cheap ice lenses for solar energy concentrators. Ice lenses poured as water in moulds can be turned out cheaply, and focused to direct heat to storages or machines during daylight hours. Components of complex lenses can be glued with water (**Figure 12.25**).

As ice can be moulded in any shape that water can be poured, and reinforcement as bars or fibres added to the mould, a great variety of useful ice objects can be made, or ice so moulded around axles can be towed to summer storages as an ice buggy (**Figure 12.26**).

Snow is a handy form of precipitation, as it is more durable in landscape than mist or rain, thus easier to store. Compacted, it stores as a solid, and endures more than one season. As a semi–solid, it can be stored three dimensionally, caught and heaped on fences and plant barriers. These can be fences or hedgerows, or one replacing the other over time. The varieties of catchment are obvious from studying the way trees store snow. Some snow is stored as sheets, others as clumps or drifts. Thus, snow driving across landscape can be directed to heap in mounds for meltdown into such storages as swales or cisterns in spring.

When it snows, the weather is cold, but snow both insulates and reflects, thus earth sculpturing around either the local landscape, dwellings, or drifts can produce reflected heat of great intensity (sunglasses are required in all reflective landscapes: white beaches, quartzites, still waters, hayfields, and snow fields).

Just as we can build concert bowls, circuses, arenas, and focusing dishes, so we can build earth structures which snow covers to give winter heat concentration. In summer, these surfaces can be of white gravel or sand or clay, but in winter snow is better. So, houses built in earthformed reflectors, or that are themselves reflectors, can use snow for heat and insulation (**Figure 12.27**).

Reflected and concentrated light, even for short periods, gives intense heat which can be stored as hot water, molten metal, steamheated earth, or heated salt solutions.

SNOW

Curiously, snow is almost a "black body" for radiation, losing heat rapidly with radiation values of from 0.986 to 0.962 (dirty snow). Thus, snow cools very rapidly at night, and thin snow cover chills soil to 35°C below air temperature. Snow thicker than 15 cm, however, acts as an insulator for soils, buildings, and permafrost. As for reflection or albedo (20% for muddy slush, close to 100% for fresh dry snow), is it most effective at the low sun angles of dawn, evening, and winter. Radiation (light) penetrates snow 10–15 cm, when most of it is absorbed by the water in the snow mass, but in the hollow ice crystals of hoarfrost, and in dry snow, light penetrates to 30–60 cm. About 30% of outgoing radiation is as heat (long wavelength), but the snow itself is completely non–transparent to incoming long–wave radiation, so that melting more commonly occurs from below due to earth (soil) heat. It is this basal melting that causes poor snow–mass cohesion, and that may trigger snow creep or avalanche.

AVALANCHE AND SNOW INSTABILITY

On slopes of more than 6° and in the valleys of steep foothills, wherever snow builds up to 15 cm or more in depth, there is a risk of avalanche. Snow will avalanche wherever the crust is broken by strong winds (more than 13-18 km/sec. sudden freeze is followed by thaw, warm winds cause snowmelt, or normal spring thaw occurs. The saturation of deeper unstable snow layers by meltwater over layers of buried hoarfrost, or overlying ice or earth layers, weaken the whole snow mass until slides can occur. At this critically unstable point, a gunshot, the collapse of a snow cornice (overhang) or a strong wind can trigger an avalanche, as can a skier, antelope, or rabbit.

Avalanches pick up or trigger off the movement of even more material as snow, mud, boulder and clay, or vegetation. Even so, it is less the mass itself than the high–pressure air wave which precedes, and advances beyond, the avalanche (sometimes carrying on up the

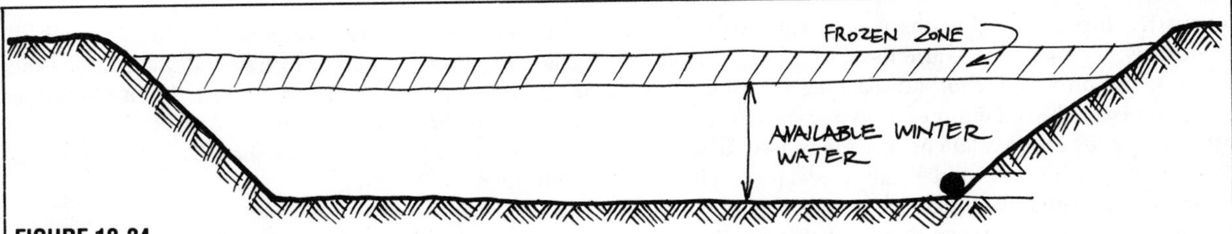

FIGURE 12.24
EXPANDING ICE IN WATER STORAGE.
Sides of storages need to be sloped to allow for freeze expansion; capacity must be judged as unfrozen depth over winter.

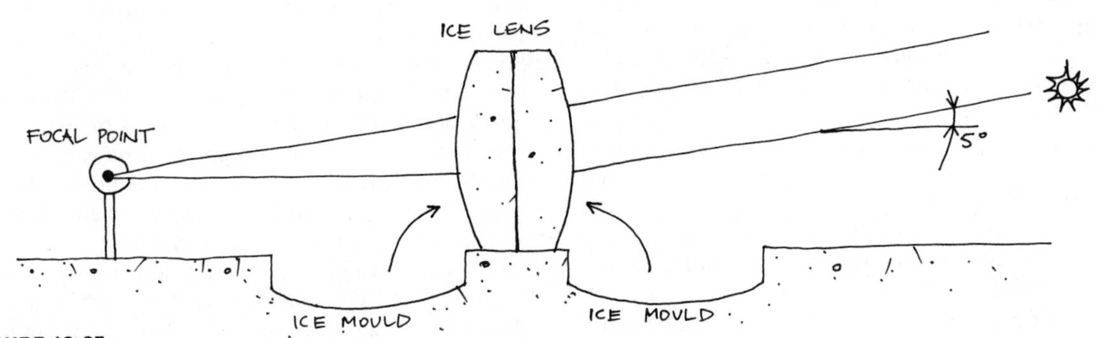

FIGURE 12.25
ICE LENSES.
Ice can be moulded for use in place; here, as a focusing lens.

FIGURE 12.26
ICE TOWED AS A BUGGY.
Ice forms in moulds can be easily moved to cool stores.

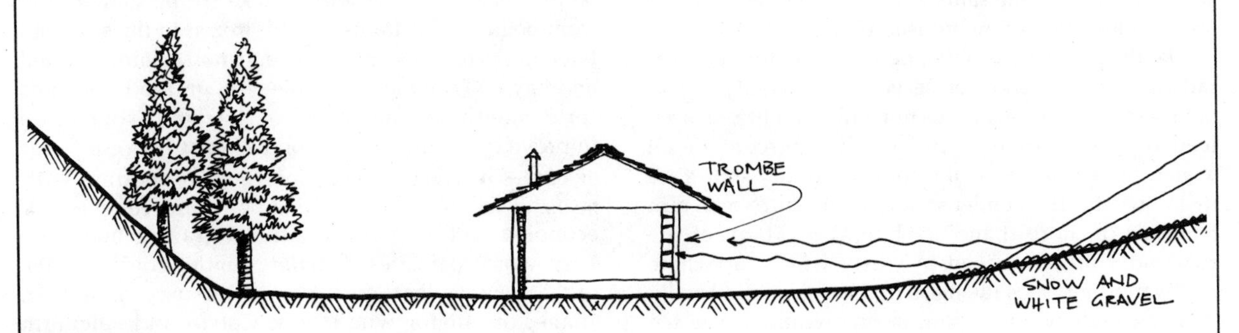

FIGURE 12.27
SNOW AS A REFLECTOR FOR HOUSE HEATING.
Low sun angles "bounced" off snow suraces can add 60–70% more heat to walls of houses.

opposite slope) that can destroy forest, crop, and buildings. Typically, avalanches cut roads, public utilities, and prevent fuel or aid reaching the area. Like other catastrophic phenomena, avalanche areas are often mapped, and avalanche warnings issued when conditions can be predicted locally. Avalanche chutes are often delineated by pioneer species such as aspen.

The best defense is prevention, and this is achieved by a set of strategies ranging from preservation of high slope forests, the placement of essential services in tunnels, the erection of Vshaped barriers to split up the avalanche front, and the prevention of settlement in highrisk areas. A new threat follows on the death of steep slope forests by acid rain damage. In general, crossslope terraces, walls, and barriers have not proved effective, but snow barriers on plateaus *above* the slope can effectively hold snow above the slope and thereby lessen the load .

Much of the data given here is from G.M. Kuaeva (1975) *Physical Properties of Snow Cover of the Greater Caucases*, USDA translation.

Snow, like sand, moves in a variety of modes:

- MICROCREEP, under freeze/thaw and gravity effect advances downslope a few centimetres per day
- SLIP of thick snow masses (SNOW BOARDS) may cover metres per day.
- SNOW MOUND advances from rocky uplands as a streamflow effect.
- SNOW BODIES move en masse down short slopes.
- SNOW SLIDES at velocities of 4–5 m/sec. occur and can be arrested by terraces.
- AVALANCHES of snow can increase in speed down long slopes, reaching speeds of up to 6.5 metres per second. They can be of great magnitude but are usually EPISODIC (every 35 years). SYSTEMATIC avalanches are those that occur every year, but these are unusual. Longer periods of rare heavy snow falls produce PERIODIC or rare avalanches.

Trees are useful avalanche and mudslide movement indicators, and downslope lean, lack of lower limbs and understory, bare "trains" (different age classes or pioneer species) such as aspen in downslope strips, and blown–down forests radiating from avalanche snouts all reveal past instability in valleys. Rock avalanches produce many of the same effects, which persist for at least 20–30 years in coniferous forests.

Like the plant crusts of the desert or tundra, the solar radiation crust on snow preserves slope stability. This is of necessity a seasonal protection only, and breaks up if iced or shattered by wind. Eighty percent of all avalanches occur where old, compact, or dry snows are dislodged by the condensation of moisture and rain from air of more than 70% humidity. Many of the remainder are dislodged by gusty winds, or where meltwater lubricates the mass in thaw periods (usually between midday and 3 p.m. in fine weather). Sudden falls of 50–69 cm of snow followed by long periods of cloud-free weather also cause unstable snow conditions.

Whereas the impact of a snow mass is somewhere from 4–54 tonnes per square metre, depending on the total mass and the load carried, the compressed air blast in front of the mass can travel at 330 m/sec. This produces a severe shockwave effect. Even snowdust avalanches produce waves of 10–100 m/sec and develop pressures of 35 atmospheres, enough to knock down trees and buildings.

PERMAFROST.

Permafrost (permanently frozen ground) occupies 47–50% of such areas as Canada, Russia, and Greenland. Peaty surfaces, which are common over the permafrost, both prevent melting and rapidly admit cold, thus preserving the stability of frozen areas. Either by clearing, fire, or overgrazing plus ice particle scour (ice particles at 40°C are as hard and abrasive as sand), the peaty layers can be eroded or removed. Thaw may rapidly occur, so that bare sunken areas of clayey soils (mollisols) occur, and these thawed areas can reach 2–45 m deep in a few years. The erosion effects (plus 9% volume loss of ice) can be severe, and has destroyed many upland grazing areas in Greenland.

Much permafrost is fossil, at times to 400 m thick, so that normal melting under peat is slow; earth heat from below reduces the thickness only about 1 cm per year. In such harsh areas, soils are formed only as a result of frost shattering of rock, and both soils and rocks are sorted by ice action and thaw (gelifluction). Solar thaw is about 5,000 times more effective than earth heat if the peat is removed.

Loess deposits (windblown soils) are 50–80% silica, 20–25% felspar and low in calcium. Loess fields of deep soils (some deposited over foothill soils) occur throughout Europe, Russia, China, and near continental glaciers elsewhere. They support excellent forests but dry out quickly and have high porosity. Loess deposits are stable as steep walls to 300 m high. They are therefore much used for underground housing in China, and have been occupied for millenia.

COLD AREA GRAZING

When, as in Iceland, frost penetration to 1.2 m in pastures and soil is a common feature, open ditches at 50 m spacing are necessary to take spring thaw runoff from bogs, as tile drains are still frozen in the soil until later in spring. Sheep, cattle, and horses are shedded and hayfed for 6 months of the year, grazed in uplands for 3 months or more, and on lowland (sometimes improved) pastures for 3 months. Frost heave in areas of bare–soil erosion needs levelling, re-sodding with peats, or rest from grazing, and both a mixed–livestock economy and longterm rotational grazing may help keep worm parasites (*Helminth*) under control. Kale, rape, ryegrass, barley or rye can be used as fodder, silage, or cut for winter hay. Cobalt and selenium deficiencies in sheep; and calcium, copper, and phosphorus in pasture must be monitored. Shrub and tree browsing is a factor little developed, although willow, and dwarf birch species are available as browse.

The demand for firewood, here as elsewhere, has reduced forests to 5% of the original cover, and overgrazing has denuded 50% of the total vegetative cover, especially on fragile volcanic upland soils. Historically, better animal growth was achieved in these more wooded and less "improved" highland pastures, but as these break down from overgrazing, the less thrifty lowlands are now being considered as alternatives. Less than 1% of the total area is cropped, and as in most very cold short-season areas, livestock, sea, or forest products are of greatest importance in the human economy.

In sandy, cold deserts, thyme, *Festuca*, lichens, dwarf birch (*Betula nana*), *Rhacomitrium* are typical species of hummock grassland;*Alopecurus*, *Phleum*, and *Poa pratensis* (whitegrass tussock) of hayfields; and sedges (*Carex*) and cottongrasses of bogs. As bogs are drained or areas fertilised, grasses begin to predominate over shrub and tussock species, and worm parasites increase.

Eroded gravel plains develop on overgrazed areas. There has been no serious consideration of high–energy browse as supplied by oaks or the twigs of such sugar storing species as striped maple. It would appear that deforestation and pasture pressures are developing the same spectrum of "limiting factors" as are found in all areas where livestock numbers and clearing expose the livestock to insoluble disease levels, together with a lack of shelter that can reduce weight gains by 16–30%.

Iceland faces the same compound of problems as warmer areas, with deforestation, severe parasite infestation of sheep on lowland or fertilised pasture, and lack of shelter and deep-rooting nutrient–yielding trees. Although no gross deficiencies have been noted for any of the three trace elements tested for (copper, cobalt, selenium), serious intestinal worm infestation on bogs and pastures does reduce the mineral content of bones, and lowland sheep in Norway have about ten times the nematode worm population of highland animals. It is only on the lowlands that high levels of nitrogen fertilisers are supplied, and where consequent lack of energy food in improved pasture species is experienced.

Iceland obviously presents a case for innovative forestry, coupled with a serious attempt to reduce grazing on the eroding soils. The shift to forest products seems to be a much–needed option for farmers.

12.16 WILDFIRE

Wildfire is a feature of many sites and climates, and can be created even in hot humid climates by logging, or by block plantings of eucalypts and pines. It is notoriously violent in summer–dry climates peripheral to large arid areas; "wet" savannah or chaparral scrub will burn fiercely when strong advected heat blows in from deserts.

Periods of high fire danger coincide with periods of strong ground winds from continental desert interiors, and affect many climatic types on desert borders, up to 200 km from the desert edges themselves. These winds are the normal precursors of widespread and catastrophic wildfires, which in the presence of enough local fuel may develop into terrifying firestorms, which themselves generate a type of fire tornado, with fierce ground winds. In the southern hemisphere, fire winds blow anticlockwise, and in the northern hemisphere clockwise.

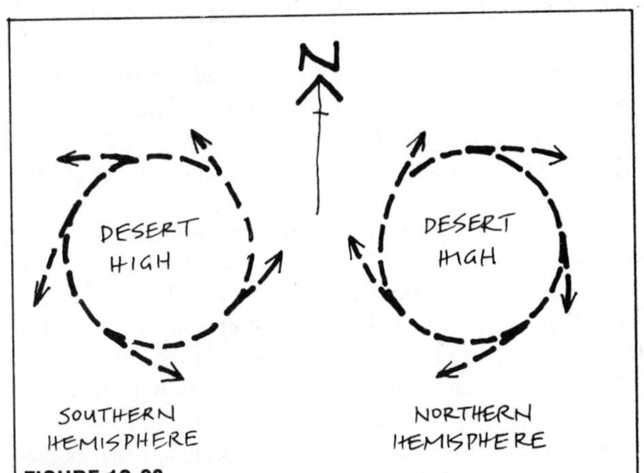

FIGURE 12.28
FIRE WINDS IN SOUTHERN AND NORTHERN HEMISPHERES.
Highs over hot inland areas produce winds on desert borders, as figured. These hot dry winds may trigger wildfires.

The critical factors for firestorm are:

- FUEL SUPPLY; this includes the dryness of available fuels and their distribution and quantity (loose fuels of more than 6 cm diameter).
- OXYGEN SUPPLY as winds to fan the flames, especially hot winds.
- PREHEATING as upslope or radiant heat in front of the flames, or as advected desert winds in unprotected forests.
- UNSTABLE AIR MASSES, so that wind shear, ground whirlwinds (dust–devils), scattered cumulus clouds, and shifting winds all presage fire danger when dry fuels reach to less than 35% moisture. In unstable air, smoke does not level out at low altitude, but ascends to great heights, or is up and down in streamlines, and the air is otherwise clear (no fog or smog before the fire). In some forests, one can smell the volatile oils, terpenes, or resins, and light–blue haze develops over these forests.

A small proportion of fires start from lightning strikes, even in remote forests. This is why many ridge forests show pyrophilous (fire–dependent) species, and on some ironstone ridges, every tree will be scarred by strikes. Such places should be noted as areas where houses need earthing for lightning strike.

However, most fires are deliberately lit, or arise from previous "controlled" burns left smouldering (often lit to *reduce* fire risk!) Freak fires can start from electrical shorts (power lines), backfire flashes from vehicles in grass, heat focussed by bottles or curved glass, and welding, campfire, and cigarette accidents. In all, lightning and accidents are perhaps only 4% or so of total fires; the vast majority are lit by mischievious, psychopathic, or even well–meaning people. A few pyromaniacs light and attend many fires, and even enlist in volunteer firefighting organisations; in aboriginal or tribal peoples, an angry person will sometimes burn out a camp or forest.

FACTORS THAT INCREASE FIRE INTENSITY OR SPREAD.

Once initiated, wildfire can spread with great speed; grass fires spread after 10–11 a.m. (after the dew has dried off); forest fires from midday to 3 p.m. After an initial flareup, an hour or so suffices to develop firestorm conditions, aided by:

• Loose fuels of smaller than 6 cm diameter, grasses and sticks, at less than 20% water content (a chunk of 75 x 50 mm pine, unpainted, can be weighed to judge humidity and wood dryness; where saturated, this wood is judged at 100%). Pine woods erupt at less than 30% humidity or moisture content due to high resin or oil content.

• Winds of from 10–50 km/h accelerate spread, as the square of the velocity. E.g., at 20 km/h, if the spread is 2 km squared per hour, then at 30 km/h, the rate is 4 km squared per hour. At higher wind speeds, tongues of fire break up the firefront. At 80 km/h, ground fires may self–extinguish.

• Winds "backing" (shifting) late in the day may blow out a fire flank into a broad front, or even blow a fire back on itself and make it safer. However, backing winds are unpredictable, and in wildfire the best strategy is to order an early evacuation of a broad area except for teams (in safe refuges) whose job it is to put out minor house fires in the first halfhour after the fire has passed. For this reason, forested suburbs need local refuges (gravelled areas with underground shelters), as do isolated homesteads, and long stretches of roads through inflammable forests.

Wildfire will always occur on arid borders; thus we need to first be able to live with fire, and perhaps only secondarily (over a period of years) design to exclude fire from settled areas by a combination of:

• Altering the vegetation to create more fire–immune systems.

• Designing dams specifically to floodflow over hillsides subject to fire.

• Mechanical or grazing removal of fuels just before fire–danger periods—this includes dead brush, long dry grasses, and the dead lower branches of trees.

In non–inhabited areas, both "cool" fires (damp and cool weather) and "hot" fires (dry periods) are sometimes lit as a management mosaic to preserve fire–dependent flora and fauna; this is unsafe and difficult to control, and often *causes* fires.

Houses, dense surrounds, village surrounds, and intown planting (or the forests at the base of settled slopes) should all be designed to minimise fire damage and mortality. Fire can be expected as wildfire on a more or less regular schedule in specific vegetation types at about 30 years in wet sclerophyll forest, 8–10 years in dry savannah, and even annually in unbrowsed grassland. Thus, fire provides a specific problem for designers and landowners progressing from grassland to forest operations. From 3–5 years, or until forest establshment, the system has high fire risk, and we need to programme planting mosaics to reduce district risk.

FIREBREAK is a way of decreasing fire intensity; roads act as firebreak, as do ponds, marshes, rivers, stony areas, and summer–green or sappy plant crops hedgerows. Horizontal firebreak weakens or reduces the fire front energy. Vertical firebreak, to prevent fire "crowning" in trees, relies on the removal of lower branches, dead tree material, and perhaps on planting sappy ground cover under the forest.

No firebreak (even 10 km of water) is effective in firestorms, as fire tornadoes, with ascent velocities of up to 250 km/h can develop on the lee ridge side of hills, travelling downhill and lifting aloft large logs and branches and pieces of houses, and creating massive aerial gaseous explosions. The ground winds near the base of this column (some tens of metres across) can reach 100 km/hour and will roll people over and over. The noise is deafening.

In these tornadoes, or in more minor whirlwinds, incandescent material is carried aloft and dropped out from 1–30 km downwind to start fresh fires, and fresh firestorm sequences (**Figure 5.23**.)

In towns, fire resistant design (for wildfire) has these features:

• A simple roof and wall outline (no internal roof valleys or re–entrant wall corners to pile up incandescent ash).

• No tarpaper roof lining projecting into gutters. Roof gutters should have either a leaf–free profile, or can be plugged and water–filled in the event of fire (plugs are handily chained to the gutter near downpipes). Roof spaces often catch alight from leaves in the gutter.

• No unscreened windows, underfloor, or wall cavity vent spaces; all need fine–mesh metal screens to reduce spark size. Even beds can catch alight with large embers, and cellars or underfloor spaces may have dry firewood or fuel liquids stored there.

• No inflammable door mats, nor wood piles or shrubs against the house walls. Large cans of petrol, or explosive materials, should be stored in a shed away from the houses, tightly lidded.

SITING OF HOUSES AND BUILDINGS.

In fireprone areas, houses are at most danger from *upslope* fire; few houses survive wildfire on sharp ridgetops, or in hill saddles that have diverging ridgelines creating a wind (fire) funnel effect. The same funnelling or intensification of fire is created by planting inflammable trees (eucalypts) or grasses (pampas grass) along a house driveway; I have seen funnelshaped plantings of this type that would have the effect of a blowtorch on the house, when even concrete will powder, and steel posts behave like spaghetti influenced by Uri Geller (or an Indian snake charmer).

For every 10° increase in the angle of upslope, fire speed and intensity doubles; the effect is that of updraught plus enhanced drying out of the fuel ahead of the front due to upslope flame and wind. That is, if the fire speed is at 16 km/h at 0° slope, it is 32 km/h at 10°, 64 km/h at 20°, and 128 km/h at 30°; thus slope effect alone can wipe out hill ridge settlements. It is critically important therefore that downslope forests are *not* pine or eucalypts, but slowburning deciduous trees, with low leaf oils, and are sappy or thick–leaved forests with a clean floor, or with succulent groundcovers and lily clumps, or succulent vines and crop interspersed.

To reduce the ridge effect, site houses not only off the ridge, and if possible on downslope plateaus, but also excavate the site instead of raising the downslope house wall on stilts or stumps. A house nested into a shelf on the hill is protected from radiation, has no open underfloor area, and can have a rimwall, pond, or earthbank on the edge of the plateau as further protection. Such houses can more easily develop a cave or dugout refuge behind the shelter of the house itself (these are fully earthed over and have a wet blanket door and a dogleg or curved entrance to further escape direct radiation). Each such fire or radiation refuge needs a small (270 l) permanent water tank incorporated, a few old blankets, and a bucket of water or two. This is absolute "fire insurance" for those caught at home (often, women and young children). All these fire aids also apply to barns, livestock shelters, and outbuildings.

Around house and building sites, it is essential to reduce forest and grass fuel to a distance of 30 m (100 feet). This does *not* mean tree removal, but rather the planting of such trees as *Coprosma*, deciduous fruits, figs, willows, poplars (*not* olives, pines, eucalypts), lines and clumps of lilies (*Agapanthus*, spring bulbs, arum, *Canna*) or "summer–green" ground cover (comfrey, iceplants, *Tradescantia*, *Impatiens*, shortgrass sward) to reduce flame and radiation effects.

My own family survived in a dense eucalypt forest only by prior removal of all lower limbs, loose bark, twigs, fallen leaves, dry brush, dry grass, and dead stumps (every year); most of this material was chopped down and stoneweighted into hollows and swales, or burnt as cooking fuels. The tall tree stems not only saved the house from fire wind, but regenerated after the fire. A downslope line of willows and fern–leaved (*not* hardleaved) *Acacia* was firekilled but rejuvenated from the roots when cut down. This sacrificial hedge, or fire barrier, dampened out the ground fire, and even the leaves of the willows did not combust, but shrivelled and gave out steamy ash as the flames reached the trees. The house was blistered, and the zinc roof coating flaked off, but the closefitted boards were unburnt. All walls were whitepainted, and screens fitted; we had ample bucket water stored both indoors and out for the many small spot fires that were left after the front had passed. Apart from sore eyes and some short beards and hair, little damage resulted (Hobart 1967, a firestorm condition). In my street alone, 70 houses were burnt to the ground, with 1,100 houses burnt in the area, and 90 people killed.

The safest house sites are in damp valley mouths, in well–tended buit–up areas, on farms with flood–flow or Keyline irrigation fitted, in irrigated areas, on peninsulas in dams and lakes, and in any plateau site where the design and maintenance criteria are rigorously applied.

It follows that all designers should take fire into account over many climatic regions, and especially where we are developing forest from grazed areas, as long grass or tall stands of straw are the worst fuels for fast fire spread and for ground survival.

Important to human and plant survival are RADIATION SHIELDS; these are solid or reflective (or both) objects that reflect or harmlessly absorb the radiant heat from the fire front. It is radiant heat which quickly kills plants and animals. Most human casualties of fire are not at first burnt, but either smothered by

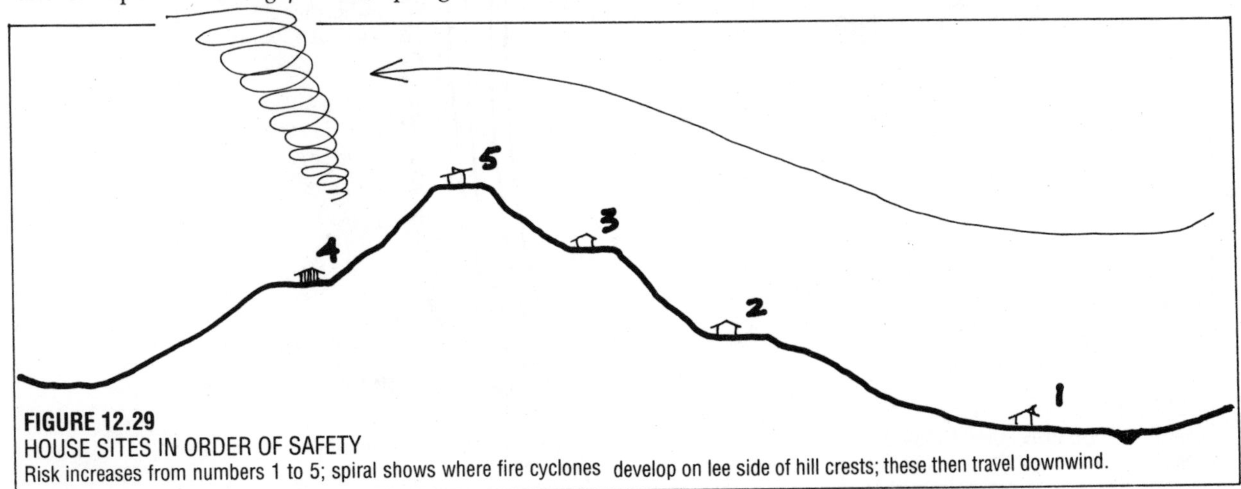

FIGURE 12.29
HOUSE SITES IN ORDER OF SAFETY
Risk increases from numbers 1 to 5; spiral shows where fire cyclones develop on lee side of hill crests; these then travel downwind.

toxic smoke or fumes from furniture or plastics, or killed (unburnt) by radiation.

Thus, radiation shields can be houses, stone walls, thick tree trunks, hollows or caves, hedgerows, and car bodies; a whitepainted brick wall is ideal. White paint on houses reduces radiation absorption, as white roof areas reduce sun heat. Fire–proof or slow–to–burn insulation in houses (mineral wool, seagrass, sawdust, feathers, wool) all keep the interior cool and assist fire control. Wooden panelling transmits little heat, while stone, brick, and mud may convey heat indoors, unless of sufficient thickness to absorb and disseminate it. Thatch and shingle areas must be replaced by tile or metal roof cladding in fireprone areas (by law in some districts).

Note that a fire shadow is tapered to about 4–5 times the height and width of the solid radiation shield, so make shields of trees or walls extend past the house (**Figure 12.32**).

A NOTE ON FUEL REDUCTION

Wildfire will always happen, often every 8–30 years, on many sites. It will not be severe if normal annual fuel reduction is practiced; the most *unsafe* way to do this is to "cool burn". The safe ways are to graze off, slash, compost in swales, use as firewood, or to replace tinder with sappy green plants. Part of bioregional planning must be to keep monocultures of inflammable trees to uninhabited ridgetops, or better to scatter such stands throughout grazed or wetforested areas, or to tend them very well indeed in the matter of fuel reduction.

Ida and Jean Pain (*Another Kind of Garden*, 1982) have clearly laid out a broadscale, beneficial, fire–reduction system of chipping all dry forest fuels, composting or using them for biogas, which in turn fuels the chipping and carting operations, and lastly using compost and sludge to grow gardens, improve soils, and further reduce litter. Every bioregion should, perhaps must, adopt these methods if forests are to be preserved and eventually made fireproof.

Likewise, in the case of scattered suburbs, it should be compulsory for houses to build to fire specifications, have large roof tanks and ponds, and for developers to build fire–damping dams able to operate by radio–control to sheet water over slopes on a Keyline principle. Fire will then be restricted to remote dry–ridge forests, and lightning strikes (as it should be).

FIRE EFFECT IN FORESTS

Fire sharply reduces litter, and leaves a nutrient–rich light ash that can wash or blow away, or wash into lakes and streams as a clear or cloudy nutrient load. Hot fire will remove from the soil in low intensity burns:

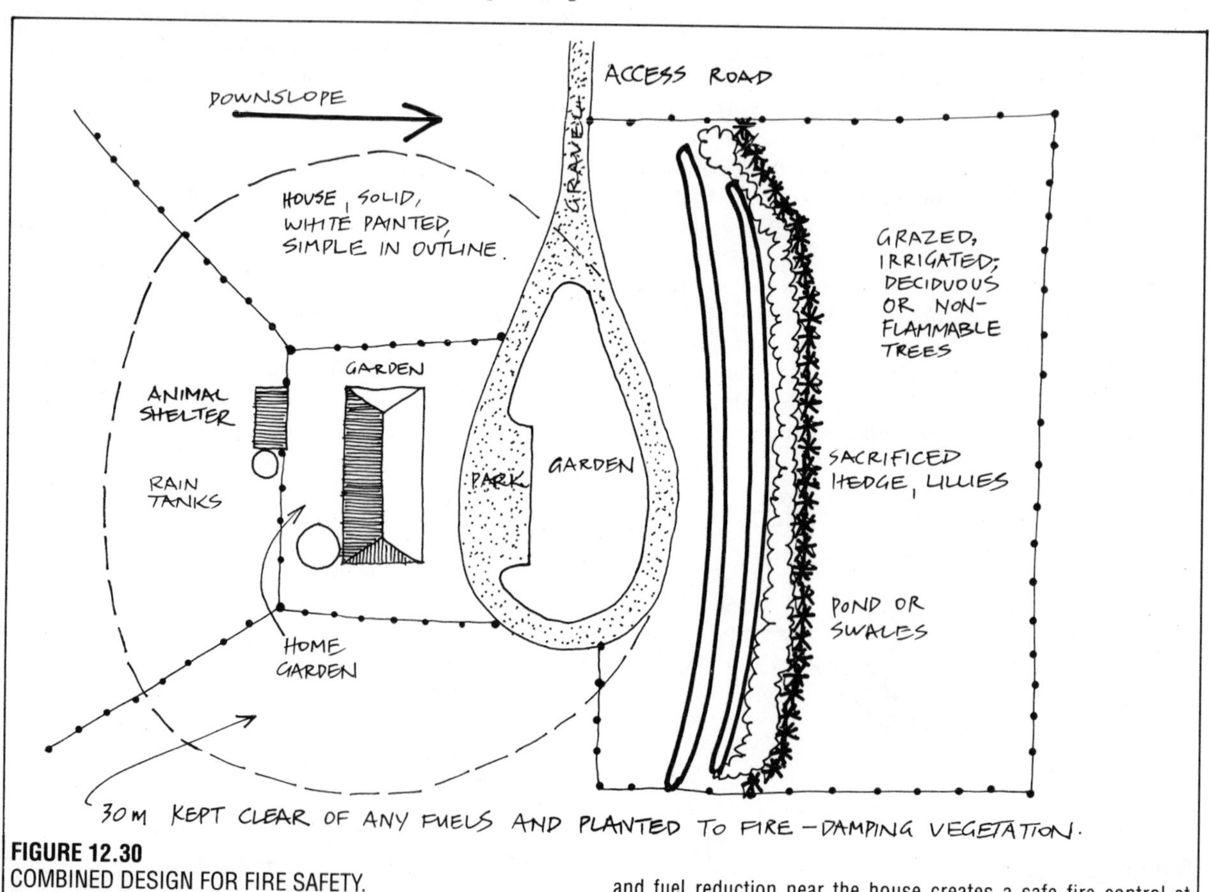

FIGURE 12.30
COMBINED DESIGN FOR FIRE SAFETY.
Radiation shields, multiple downhill firebreak, selected "wet" plants, and fuel reduction near the house creates a safe fire control at settlements.

Nitrogen: 54–75% (109 kg/ha); replaced in 11 years by legumes, rain.
Phosphorus: 37–50% (3.0 kg/ha); replaced in 20+ years by rain.
Potassium : 43–66%
Calcium: 31–34%
Magnesium: 25–43%
Boron: 35–54% (Figures from *Ecos*, 42 Summer '84/85)

Additional losses come from those shrubs and trees where foliage is burnt, and particularly so if the fire occurs early in the growth season. As fires "glow" at 650°C and burn fiercely at 1,100–1,400°C when strong winds blow, sulphur and nitrogen are volatilised, as is carbon. Phosphorus and potassium are volatilised at 774°C and calcium at 1,484°C (cement structures powder at this temperature). However, organic compounds containing these elements may volatilise more easily than soil elements.

Obviously, there is a very slow recovery of soil nutrient after fire, and this depends on trace elements brought in by rain or birds, and minerals recycled to topsoil by deeprooted vegetation. Clearly, fires never improve soil status. Humus loss of 10–12 cm occurs in forest soils, and peats often combust to greater depths. Clays lose structure, and mud flows can result.

STOCK LOSSES IN FIRE

Good stock managers can, by pre–planning, reduce losses in fire. Some ideal situation would be to make sure that some paddocks (small areas) have been closecropped from late winter (or late summer in monsoon areas) to early summer, so that these small fields carry no inflammable fuel and can be used as general refuges in fire. Even more effective is to place a water trough in a deliberately bare soil area, even one where topsoil and shrubs have been bulldozed into a surrounding earthbank (for protection against radiant heat), and to use such areas as always open refuges off 4–6 prepared range paddocks; stock can be confined there early fire danger days. Stock enclosed by a temporary electric fence will clean off rocky knolls of grasses before the fire period; these too can be used as refuges.

Tethered goats or sheep reduce patches of fuel near houses, as do close grazers such as geese, wallaby, and rabbits. Wild wallaby or geese can be "fed in" with

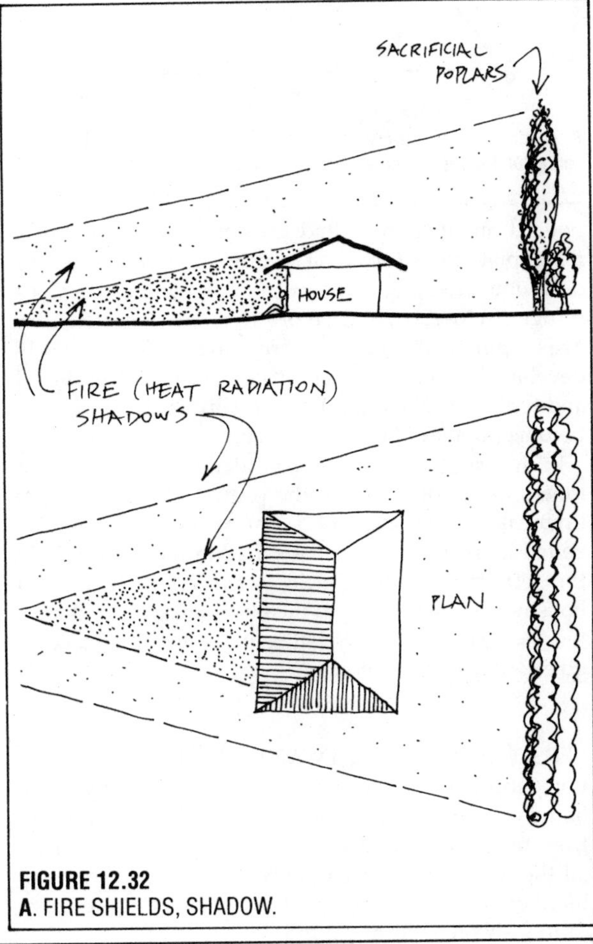

FIGURE 12.32
A. FIRE SHIELDS, SHADOW.

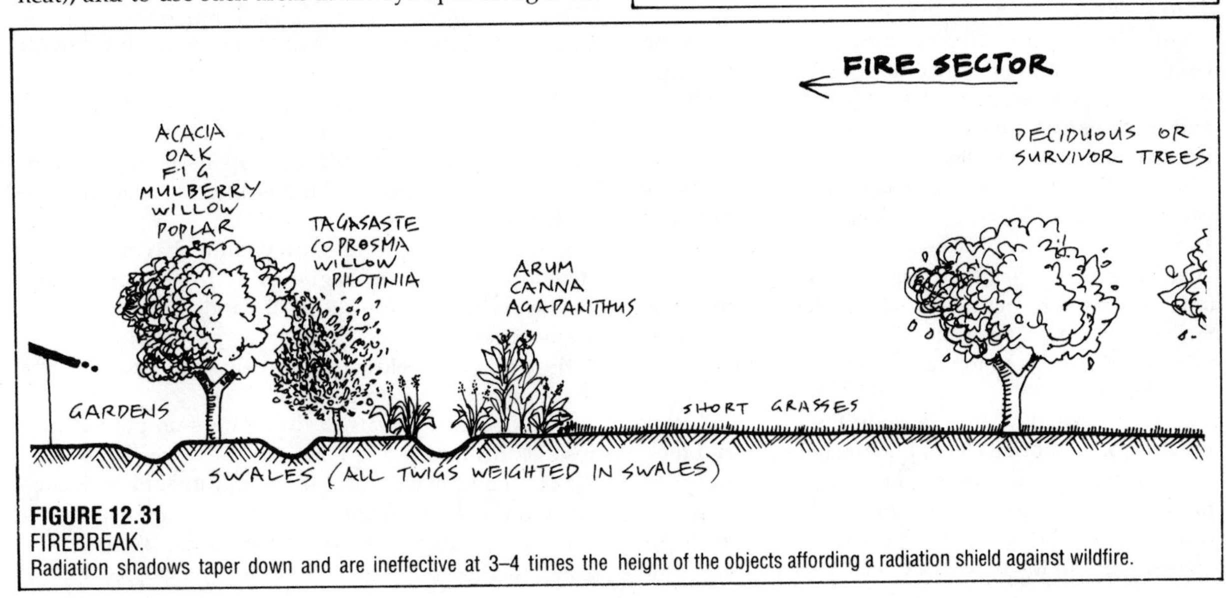

FIGURE 12.31
FIREBREAK.
Radiation shadows taper down and are ineffective at 3–4 times the height of the objects affording a radiation shield against wildfire.

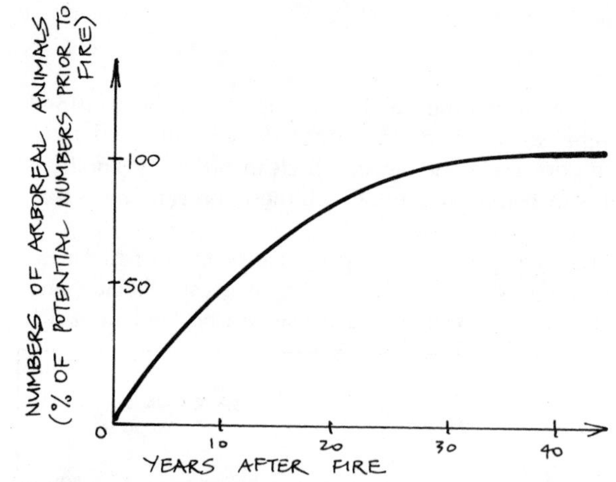

B. This curve for recovery of arboreal animals after fire could as well serve for the slow increase of fire–removed nutrients in soils and leaves.

pollard on such areas, and thus encouraged to create marsupial lawns (or rabbit swards). A forest of young deciduous trees, fig, mulberry, or oak will make a stock refuge if closebrowsed before fires are expected (dry grass period). Short swards are fairly quickly developed by close grazers if first slashed, then limed and fertilised (phosphate). Eventually, such fire refuges become popular places with stock.

With respect to a house or village, close–grazed or mown areas form part of the (upwind or downslope) firebreak system, and in fire danger periods stock can be fed in or mustered, and secured in these areas or in the settlement itself. Chickens survive well in solid housing, as they often produce bare areas around such houses, or chicken sheds can be part–buried in banks to give complete protection.

PERSONAL SURVIVAL IN AND AFTER FIRE

For people, the main survival factor is to cover the body, wearing wool or cotton, and to shelter from fire front radiation behind a tree, car, house, or in a trench; all the better if the whole body can be caped in a wet blanket. Wait until flames have passed, then move cautiously on to burnt ground.

For civil authorities, radio stations must be commanded to keep constant reports going on fire direction, open escape routes, family location centres and refuge areas, and to give constant instructions to householders and travellers.

Just as the fire passes, well–equipped ground teams preceded by a bulldozer should clear roads, and put out spot fires in unoccupied houses. Police need to roster guards, or *well–disciplined* volunteers, to prevent looting until the area is re–occupied and services re–connected. Such services must be on stand–by at the fire periphery as it is suicidal to put forward teams in the path of the wildfire.

In firestorms, oxygen is periodically exhausted, and the fire goes out briefly. People cannot breathe and they faint, so they should never take refuge in small water tanks, dams, or rivers, as they will drown while unconscious. The sea, and large rivers of 100 m or so across, are safe to run to, but beware of fainting from lack of oxygen in the air and water (the skin needs oxygen too). Drink a lot of water to prevent dehydration, and make sure children and stock also get water to replace lost body fluids.

We survive better if we have planned ahead, built fish ponds or bought metal buckets, prepared blankets and important papers to be picked up by the rear door, filled roof gutters, removed doormats, hosed down the garden, tied up or penned the stock, prepared woollens to wear, taken in the clothes, filled baths and sinks with water, and so on.

We *recover* faster if we have expected the noise, confusion, sense of isolation, and are prepared for looters and for "survivor guilt" (that is, "why should *I* have survived when people around me have died?"); for weeks, months, perhaps years, we can feel desolated by the losses we know of. It greatly helps if we have assisted others before, during, and after the fire, as we know we did our best. This applies to all catastrophes, not just fire. Even so, in every large wildfire in settled areas, people will be lost or very badly injured, much property and stock will be destroyed, and the psychological and social effects will persist for months or years.

And if your house burns down, do *not* build another "just the same" as most people do; build one to survive the next disaster. Thus, have realistic expectations, act on them, have some planned moves, and prepare better for next time!

12.17
DESIGNER'S CHECKLIST

Design solid housing to cope with winter cold; shelter the house with earthbanks, walls, or dense crescentic plantings to poleward winds. Drain house and barn areas well for muddy thaw periods.

Choose midslope (thermal belt) locations for house and garden sites, and be aware of frost extent on slopes.

Where summer–dry regimes dominate, design the house and inner zones for fire protection, and provide radiation shields. Develop swales in these areas.

Use plant/animal guilds in orchards, and balance these to reduce or eliminate the need for biocides.

Use minimal or no–tillage systems in crop, maintaining or providing soil humus and soil structure. Use hedgerow for windbreak.

Beware avalanche in snowy climates, and do *not* develop ski runs in fragile slope forests or tundra. Beware acid rain effects on uplands above villages. Hedgerows and swales will harvest snowmelt.

Bury all pipes below frost level in soils, and allow for ice depth on water storages.

Do not destroy peats or dwarf trees on permafrosts; resod bare areas.

Retain and extend hedgerow and upland forests on farms around settlement.

Do very careful fencing and grazing rotations in grasslands, and try to include (or develop) a mix of grazing species in savannahs. Encourage native game species on all farms.

Study the text for data on free-range development for bees, chickens, pigs, and so on. Livestock products are important winter foods.

Plant to create winter stores of grain, roots, fodder for barn animals; integrate the house and barn in severe winter climates.

Beware the ill effects of hyperinsulation, radon, carbon monoxide in buildings, and arrange for solid–mass heaters, good ventilation.

Maximise forestry, social income, free–range animals, and special products.

OUTLINE FOR A PROPERTY DESIGN REPORT

Frequently, people are called on to locate, design, or supervise work on projects and properties, Some of these call for a written report;and a useful format is as follows:

1. GIVE CLEAR ADDRESSES of designer and client, location (with map) of the property, i.e. *identify* place and people.

2. STATE MUTUALLY AGREED–UPON SCOPE (instructions) for this job; this sets the limits of the report. Establish whether any discretionary money is available for (e.g.) maps, travel, soil analyses, aerial photos, water analysis, minor equipment.

3. DESCRIBE THE PROPERTY IN GENERAL TERMS, including climate, rainfall, winds. Such factors as soil erosion, fencing, access, fire risk, water sources or potential for water harvest, slope, and aspect should be clearly marked on maps. Energy supply, market potential, human resources and skills, capital available for development, wildlife and potential pest species can be included as part of the report.

4. PARTICULARISE LAND USE on the basis of soils, slopes, potential species and activities—do not forget *social income* from accommodation, teaching, and so on.

a. Areas: Treat similar areas (marked on a map) alike, e.g. "steep eroding slopes for forestry"; "swampy lowland suited to chinampa", etc.

b. Themes: Themes cover access, roading, fencing, soil treatment, specific crop or forage systems, water sources and control, and potential for farmlink, commonwork, trust management, village development. Point out unique existing resources (gravels, clean water, young forests).

5. ATTACH APPENDICES such as maps, drawings, plans, layouts, details, plant lists, resource lists (addresses of people or services), some photocopies of excellent relevant studies, and any photos taken.

However poorly you think of yourself as a designer, you cannot do a worse job of settlement and agricultural planning than those you see about you! However, to persuade people to accept an externally *imposed* design is a form of insult, an implicit assumption of superiority on the part of the planner. Our best strategy is to work with and educate owner–designers and community groups, who are then part of the design process; or to proceed in stages, working alongside the people who ask for a design. What all good designers come down to is the willingness of the people on the ground to make it work, and they will only try to do this if a great many of their ideas are also incorporated.

For myself, I keep property reports informal, offer free enquiry on details, try to revisit at various stages, and increasingly prefer to train people to self–design, to visit other properties, and to join support networks. It is a good idea to supply a one–year subscription to a permaculture journal.

Work is appreciated if it serves a local or community need, or if it is taught to others; work that nobody wants done is rightly deprecated. For these reasons, and others, it is essential that we have the courtesy and sense:

- Never to knock on closed doors; to go only where we are invited.
- Always stay on the ground. Do not instigate grand projects that nobody really wants, and then expect people to accept the project. Rather, try to define needs as stated or requested and work towards supplying these.
- Always pay your way. Save people more than you cost them.

12.18 REFERENCES

Bishop, John P. *in* G. de las Salas (ed.), *Workshop on Agroforestry Systems in Latin America*, Turrialba, Costa Rica, 1979.

Brown, L., *The Life of the African Plains*, McGrawHill Book Co. New York, 1952.

Farallones Institute, *The Integral Urban House*, Sierra Club Books, 1979.

Pain, Ida and Jean, *Another Kind of Garden*, self–published in France, 1982. Available in USA from Biothermal Energy Center, PO Box 3112, Portland, ME 04101.

Reid, Rowan, and Geoff Wilson, *Agroforestry in Australia and New Zealand*, Goddard & Dobson, Box Hill, Victoria 3128, 1985.

Snook, Laurence C., *Tagasaste (Tree Lucerne) High Production Fodder Crop*, Night Owl Publishers, Shepparton, VIC 3630, 1986.

Stoddart, Laurence A. and Arthur D. Smith, *Range Management*, McGrawHill Book Co. New York, 1955.

Whyte, R. O., T. R. G. Moir, and J. P. Cooper, *Grasses in Agriculture*, FAO Rome, 1975.

Chapter 13

AQUACULTURE

Whoever discovers how to cultivate the eel should get a Nobel Prize, don't you think?
(Daijiro Murata, eel chef, Tokyo)

Catfish farming is living; everything before or after is just waiting.
(Don Carr, entrepreneur, Eagle Pass, Texas. Quote in Huke and Sherwin, 1977)

The highest fish production per hectare can only be obtained by using a combination of species of different feeding habits.
(Swingle, 1966)

Water conservation and irrigation is the lifeblood of agriculture. Grain can be taken as the basic crop, and around it developed industry, animal husbandry, forestry, aquaculture, and other integrated occupations.
(Mao Tse Tung)

13.1 INTRODUCTION

The term WETLANDS covers both natural and artificial impoundments of water: lakes, ponds, bogs, swamps, marshes, and the shallow water or intertidal areas of estuaries and marine marshes. The emphasis here is on those areas used by people for foraging and fish–farming. Because of the complexity of the wetlands environment, and of the species within it, an attempt is made to cover some of the design and planning principles and to restrict descriptions to a relatively few systems. The breeding and rearing of any one species is also omitted, as a comprehensive specialised literature exists on this aspect of aquaculture.

Except in very favoured areas, or on recently glaciated shields, water as ponds is a minor part of the total landscape, although we are constantly increasing the area of impoundments in an attempt to cope with the energy demands and water supply to cities and industry. In impounded areas, on wet terraces, and in canals and raceways, water ceases to be a merely erosive influence and becomes a very productive medium for plant and animal cultures. Associated with water are those specialised plants and animals which are adapted to or seek out water margins, shallows, and the water meadows (plains) which are inundated in floods.

Of all existing systems, tropical rainforest and shallow–water aquatic environments have the greatest natural yields. Mangrove swamps, marshes, and estuaries produce sometimes prodigious biomass of great complexity. Our attitudes to these systems has historically been ambivalent, and since the invention of the bucket dredge and bulldozer, typically destructive. Many marshes, estuaries, saltflats and ponds have been drained or deeply flooded to suit our private purposes, often in disregard of their total yields and values. Only in recent times have we begun to appreciate that a great many of our other activities, such as salmonid and inshore fisheries, depend on the conservation of the wetland habitat.

For millenia, we have occupied the shorelines and islands of marshes, and developed complex civilisations on the floodplains and deltas of major rivers. Swamps have always yielded a variety of foods for people. One of the bases for the concept of bioregion is that of WATERSHED and this too is an ancient natural division of tribes and languages. Cooperative communities and bioregional democracies have been based on water rights, as have hydraulic tyrannies; both exist today.

There is an intimate and indissoluble link between the health of a river, and the health of the catchment; here we can sense the literal truth of the concept of the "upstream and downstream costs" of our activities;

Pollution, in all its forms, most rapidly permeates landscapes and societies via water. Wherever we have used biocides in a catchment, we can reap death in the rivers and the seas offshore as far out as the reef areas, and in the sea itself. The death of coral reef areas is closely correlated with the use of 2,4–D and 2,4,5–T inland. Preservation of the cleanliness and variety of water habitats is as critical to the survival of nations as is the conservation of soils. Both are in peril today.

In specialised cultures, water becomes the main medium for life sustenance, and all island cultures neatly combine the water and earth resources, as does the old Hawaiian *ohana* synthesis (**Figure 10.28**). Elsewhere, as on the Euphrates and on Lake Titicaca in Peru, whole cultures based on reed beds derive forage, boats, housing, bread, and meat from *Typha, Phragmites, Cyperus* or like reeds, while it is said that the Aztecs at Lake Tenochtitlan in Mexico had what is possibly the most productive system of polyculture yet devised, with chinampa crop, waterfowl, reedbed, and fish culture combined.

The cultures in terraced padi (rice and taro) serve much of Asia and Oceania, with minor subsidiary crops. The deltaic mazes of the Fly, Chao Praha, Mekong, Nile, and Ganges rivers support rich cultures of great variety and resource choice, with marine, estuary, freshwater, mangrove, and land organisms to choose from. Many of these advantages can be fairly cheaply created in most humid and some arid environments, even if in miniature.

Before the 1960's, there was a lively and large global aquarium trade, and plants, molluscs, fish and no doubt disease organisms were widely distributed. Many of these were released into local streams and became naturalised. Even this effect was dwarfed by the government–assisted distribution and protection of salmonid fish as a preserve of the idle and the affluent. A few fish species, notably *Tilapia* and carp, were brought in as a basic farmed food for terrace rice and taro culture in Asia or in famine areas. However, it is still true that the great majority of aquacultures have a predominantly local flavour and species composition, and few introduced species have proved to be as adapted and productive, or as uniquely suited, as indigenous species. Only if there are no local species, or if all trials in using local species are fruitless, do people look to exotic forms.

Aquacultures now range from open–water cage and ring–net systems to highly intensive tank or channel–flow cultures. On islands, atolls, and in deserts, totally innovative systems have evolved—and will in future evolve—where local species were absent or of minor productive capacity. As pelagic fisheries are exploited by the oil–rich western nations, island people have been forced to develop alternative systems, and aquaculture is one of these. Unlike terrestrial cultures, many aquaculture developments have been polycultural from their inception.

13.2
THE CASE FOR AQUACULTURE

Until the last few decades, we have been able to harvest sufficient fish, molluscs, and plants from natural water systems. This is no longer the case, and a new impetus is evident in the creation and culture of organisms in the aquatic habitat. Even though a limited production has existed for millenia in all continents, new species are brought into culture annually, and the problems of breeding and rearing a wide range of organisms are being solved.

There are complex reasons for the sudden revival and expansion of aquacultural systems; some of these are undoubtedly connected with over–fishing of the marine resource, but perhaps as important is the change in food habits resulting from global travel and information, with a consequent change from fatty and red meats to fish and shellfish, and a general widening of the demand for variety in foods that can be eaten raw or lightly cooked.

Water cultures had long–tested and undoubted stability, and many have persisted without external imputs for thousands of years. The stability and productivity of aquaculture systems are superior to the terrestrial culture systems so far developed. Given the same inputs in energy or nutrients, we can expect from 4–20 times the yield from water than that from the adjoining land. To summarise why this is the case, we need only to note that:

- Water supply is constant for plant and animal growth in aquatic and semi–aquatic habitats.
- Plant nutrients in particular are available in soluble and easily assimilable form.
- Water and nutrient flow is a factor not represented in fields but is a critical boost to production in water.
- Water organisms (fish, shellfish) need waste little energy in movement; they are largely free of gravitational effects and weight disadvantages.
- Light, nutrient, and plants occupy a three–dimensional medium. There are complex edges, surfaces, and conditions developed as a result, and a variety of species to occupy these.
- The often rigid (and very recently monocultural) inhibitions of farmers are not as yet evident in water cultures, where the advantages of polyculture have been recognised from the beginning.
- Energies lost in cultivation are eliminated or reduced in aquatic systems, although management may need to be as skilled as it is for land crops.
- Impounded water has a great variety of products other than food; it flows on to energy production, recreation, irrigation systems, and transport.

Although all these reasons for greater yields with less inputs have always been there, dirt farmers generally have been slow to convert to aquacultures, and perhaps rightly so where a supporting infrastructure of storage, transport, and sales is lacking. But as these begin to develop, then it becomes worthwhile to abandon the production of the many surplus commodities now clogging world food markets in the developed countries for the more stable and specialised products of aquaculture. Another reason to delay action has been to await the development of earthworking equipment, as it has taken some tribal groups thousands of years to develop terrace areas that we can now create in a few weeks or months of earthworking with machines.

There are even more commonsense reasons to develop aquacultures in areas where people are at risk from famine or flood, as ponding, water harvesting, or diversion cannot help but aid food production generally, and reduce the extremes of drought and flood, even in semi–arid areas. Fish production has long been combined with wet terrace crop, and now in many terraced areas the emphasis is changing from starch to protein production. Fish are replacing rice as a yield over large areas of abandoned padi in Indonesia, for a great gain in protein yield and a reduction in the fuels needed for food preparation (Pullin and Shehadeh, 1982).

Aquaculture, in short, is as much a stable future occupation of responsible societies as are forests, and between these two beneficial systems we will see a great reduction of the areas now given over to pastoralism and monocrop. Both these latter occupations are enterprises less and less favoured by society, and their products are an obvious risk from any point of view one cares to take (fiscal, health, social welfare, energy efficiency, or general landscape stability).

Aquaculture is no more valid as a high–energy–use monoculture than its historical predecessors—the large grain or single–crop farms. It is at its most enjoyable, convivial, and socially valuable when encountered as community taro–terrace culture, and at its most depressing as 100 ha intensive prawn or catfish farms. Thus, my attitude throughout is to stress sensible yield and procedure, but to discourage the "maximum yield of one species" outlook.

To design for greatest energy efficiency, we need to look at the whole pond landscape and configuration to aid aeration, heating, nutrient flow, and the numerous accessory benefits we can get from hydraulic technologies such as water wheels or ram pumps. We as designers need to apply the same methodologies to aquaculture as to any designed system. We can perhaps hold one idea (or species) at centre, and see how many of our designed features connect to and from it, and how many other benefits can be nested within the system, supplying as broad a range of needs for ourselves and other species as we can reasonably achieve. Wild duck do not annoy catfish, and pay their way in phosphatic fertilisers; there is no need to deprive them of islands, shallows, or nest boxes. **Figure 13.1** shows some elements of pond polycultures.

A pond can act as a mirror, a heat store, a run–off area, a cleanser of pollutants, a transport system, a fire barrier, a recreation asset, an energy storage, or an irrigation accessory. All this, and it is intrinsically productive as well. It presents a host of opportunities for aesthetic and functional placements of trees and plants; buildings and pond furniture such as jetties,

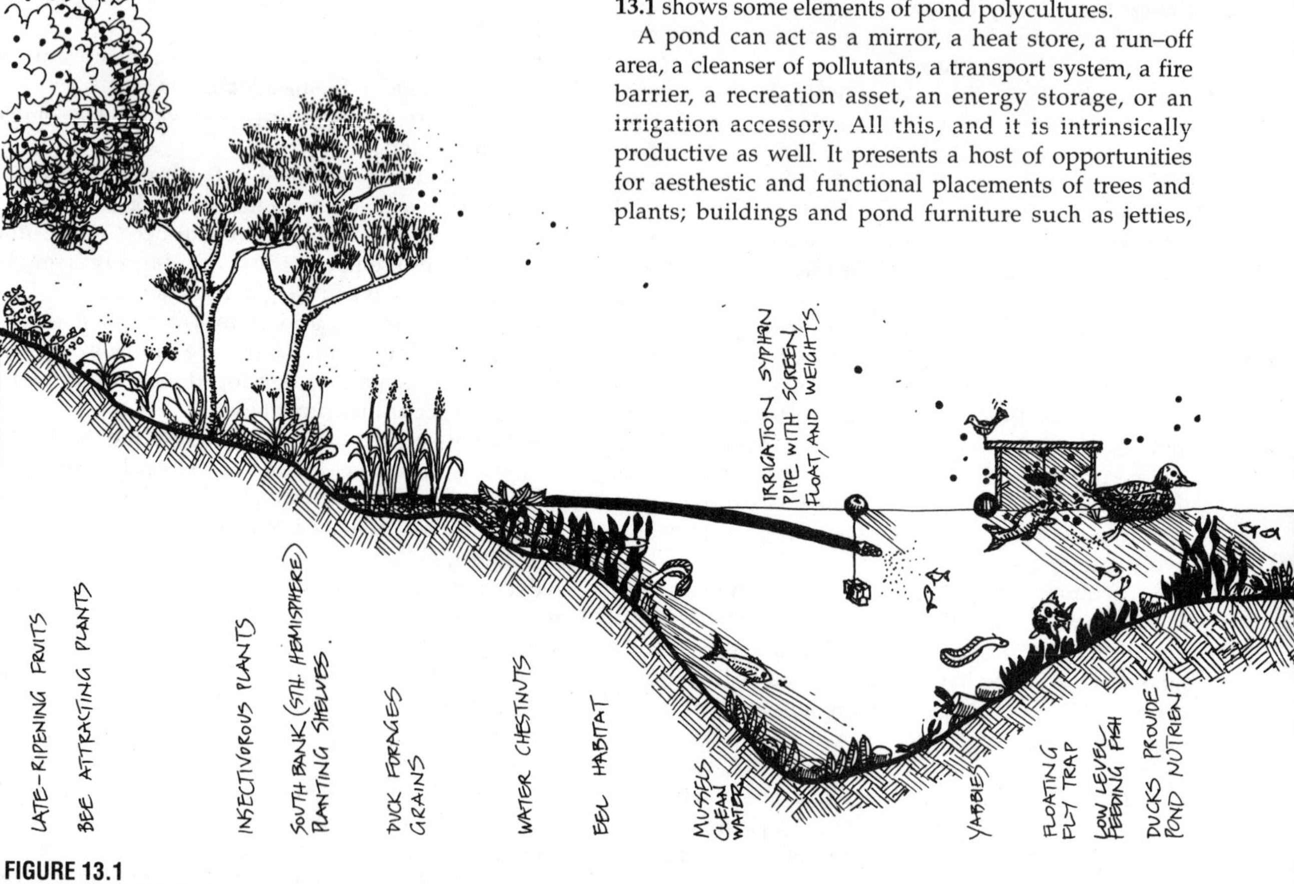

FIGURE 13.1
POND POLYCULTURES.
Some typical pond elements and furnishings.

rafts, boats; and habitat for birds and wildlife, for beavers, water–rats and turtles amongst fringing vegetation, logs, stones, and hummocks.

Thus water (unless treated as a monoculture) has great potential for beneficial design, both in pond configuration and for species mixes. It is an exciting challenge to the innovative farmer, and I will try to give some design parameters with an emphasis on beneficial design for nutrients, plant control, multiple use, and thus higher yields. People rearing specific plant or animal species should spend some time researching the large amount of literature on those cultures.

Skills in pond management, and especially in integrating species and yields, or in judging and regulating water quality are hard–won. To gain in yields and not lose out in costs or disease is a difficult balancing act. We need to start small and build on successes, planning better strategies at each point. For the home–owner a set of tanks or a small pond is pleasurable and probably profitable, as labour is seldom assessed. Nor is such a pond anything but a recreational and relaxing place, but for those who wish to profit from aquaculture, planning and design, and monitoring and research, are essential to success at the intensive level. Those who are able to select and develop an extensive site (20 ha or much larger) can accept lower total yields with less costs and risks. It is a question of options, lifestyle, and preferred approach and aquaculture may well be the accessory enterprise (e.g. medicinal products from algae, or silkworm culture, or cabin fishing licenses) that turns a profit.

13.3
SOME FACTORS AFFECTING TOTAL USEFUL YIELDS

GENERAL CONSIDERATIONS

Figure 13.2.A and **B** illustrate the classical relationship between weight gains in a well–stocked pond, and the reduction in the numbers of fish in one brood or spawning over time.

When we buy an aquarium, fish retailers will tell us to stock 2 cm of fish per gallon, or 4 litres. When we go to a fish nursery for fingerlings, the grower will tell us to stock one 8 cm fish to every square metre of water *or* linear metre of pond edge, as in **Figure 13.3**.

These are approximate measures. What they mean is that a certain VOLUME, AREA, or EDGE of pond will supply so much oxygen and nutrient to the fish. Ponds are said to have a specific "carrying capacity". If we overstock or crowd water, fish cease to grow (or some may die) and the water is then said to be fully stocked. Growth rates of fish approximate **Figure 13.4.A**. We can

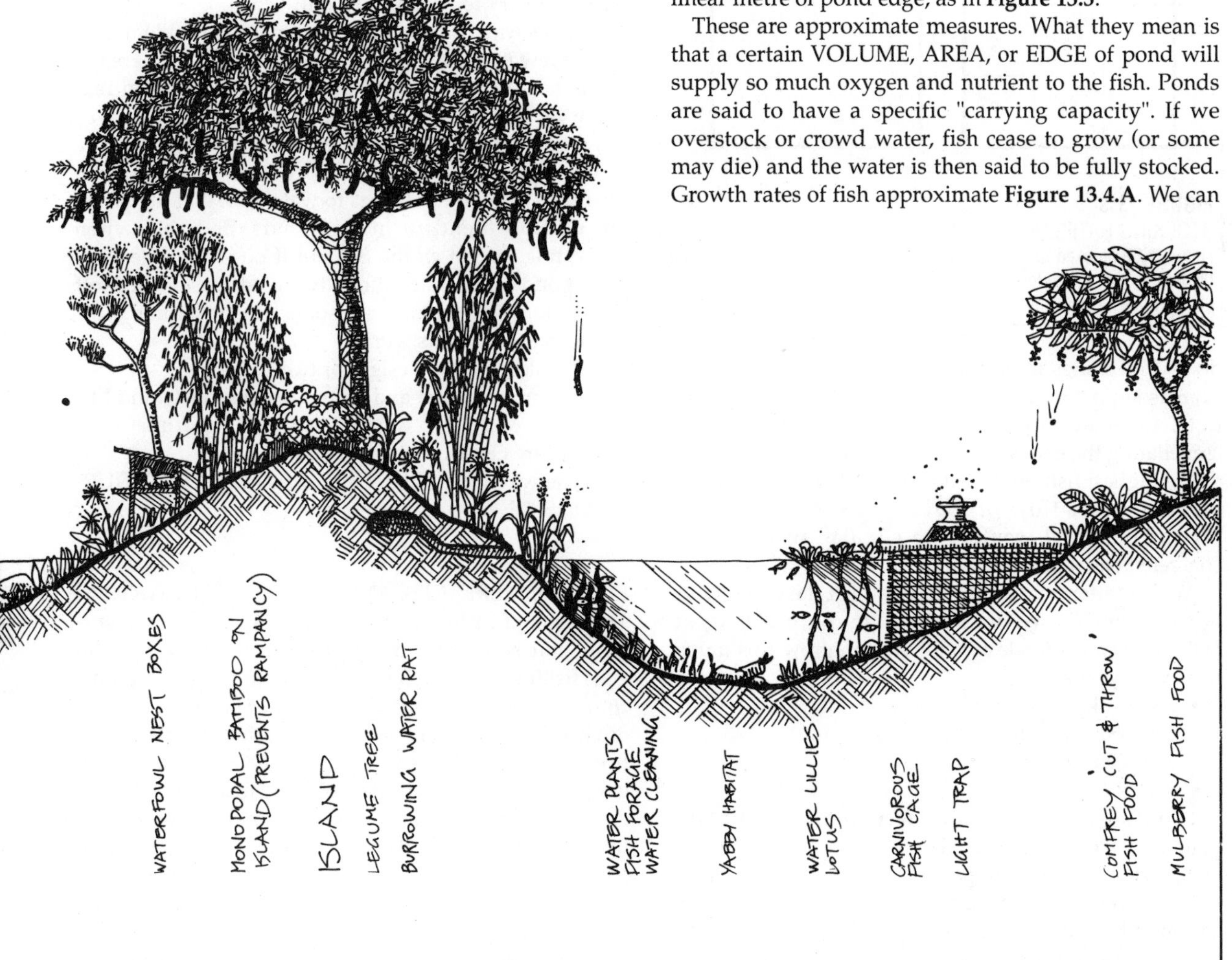

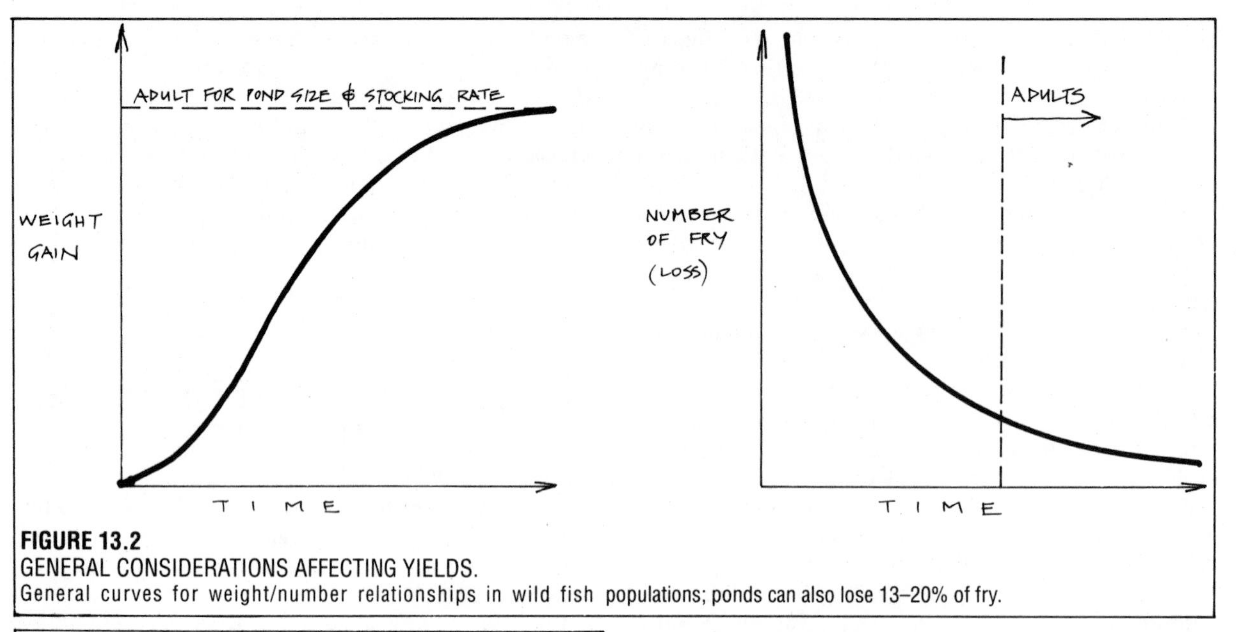

FIGURE 13.2
GENERAL CONSIDERATIONS AFFECTING YIELDS.
General curves for weight/number relationships in wild fish populations; ponds can also lose 13–20% of fry.

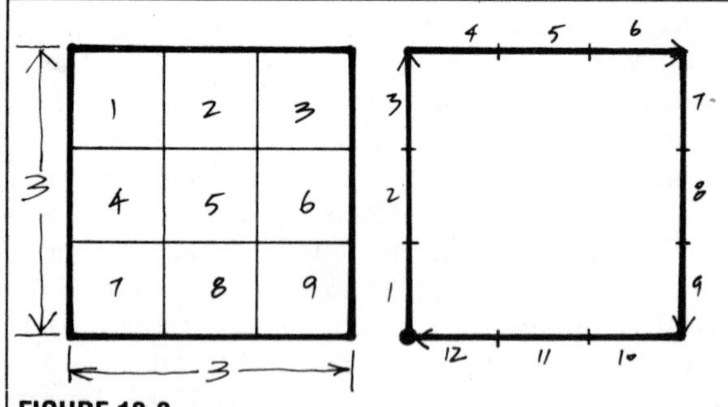

FIGURE 13.3
STOCKING RATES.
Fish can be stocked as a fixed weight per surface area for planktonic feeders, or per metre length of shore line for omnivores; for the latter group it pays to maximise edge.

do several things to lift the carrying capacity of ponds, such as supplying more oxygen by stirring in air, more nutrient by supplying manures, and more edge by crenellating the edges.

The highest fish production recorded was achieved in a rapid canal flow of mainly sewage. The least productive pond would be a circular, warm, clear, concreted basin in a quiet valley.

We can (after a year of trial) establish the capacity of our own ponds. We will take a figure of 200 kg of fish/ha/year—a modest yield. We can grow this much fish every year as:

- 200 x 1 kg fish;
- 400 x 500 gram fish;
- 800 x 250 gram fish; or
- 1000 x 125 gram fish.

About 300 g is generally accepted as minimum pan–fish size. Below this size fish can be utilised by being dried or made into paste.

This gives as a good guide to stocking procedure, although by liming, manuring, providing high protein food, or aerating, we can lift the base productivity. It is the same as pasture management (except we expect higher yields/ha). However, *if* our fish breed in the pond it soon becomes crowded, and we get a lot of little fish. If our fry die off quickly, we get too few fish and harvest a very few larger fish only.

A good way to overcome the breeding problem is to keep a few predatory fish in a cage or netted–off area of the pond. A good way to overcome fry loss is to rear these in small covered ponds, and release them as fingerlings or well–grown fish in order to give them a flying start. All this becomes routine after a while. There are a few other things to watch for: that predators are not too plentiful in the ponds, and that when harvesting we get *all* the fish out if possible, unless we have pond breeders continually culled by screened–off predators, when we can take some fish at all times. This is how natural ponds work.

We could go on to design for two factors:

- The mix of plant and animal species we would like to grow, or know will grow well together—our polyculture GUILD or association.
- The way we will lay out our ponds— CONFIGURATIONS.

EDGES, INTERFACES, AND GRADIENTS IN WATER

The edge effect in water is rather like that of land surfaces, but somewhat more pronounced. Anyone who goes fishing will vouch for the importance of shore-line, channel edge, weed–bed, or reef as productive environments. When stocking rates are estimated for fingerlings, the concepts of surface area or margin length are interchangeable, so that a relatively narrow drain will hold as many fish as a broadwater several times its surface area. Hence the sinuous canal is a rich environment for life forms compared with the circular or square pond, and much cheaper to construct in clay soils.

Maximum edge is assured by either swale, canal, or **chinampa** systems (**Figure 13.33**), which themselves can be sinuous in a flat landscape. Even modest canals make for low–energy transportation, either as barge, flatboat, or float. Harvesting, too, can be simplified by skimming, floating, or trapping products in flow.

Edges occur, or can be produced, in great variety at the land/water interface. Forest, shrubbery, reed–bed, mudflat, gravel, marsh, ice and snow all produce unique habitat beside water surfaces, utilised by very different organisms. Amongst waterfowl, the preferences for a great variety of edge is as marked as it is amongst frogs; some preferring barren spits, and others forest and mulch. Quite small ponds can provide the essential nesting and refuge places for many bird and frog species.

The surface of water itself, because of the molecular tension there, supports striding, floating, and sucking organisms and has caused peculiar adaptions e.g. the fringed legs, buoyant seeds, and suctorial mouthparts of insect, plant, and tadpole respectively. The mud generates and hides a host of rooted, tunnel–making, ciliated, and burrowing or sliding lifeforms from mussels to larval lampreys, tubifex worms to tubers. As the pond surface is critical to gaseous exchange, so the mud surface is to nutrient retention, and it is there that phosphates and nitrogenous products are stored, and humic and faecal products accumulated as mulch. Anaerobic and reducing processes can take place within the close confines of the base mulch, while aeration takes place on the water surface or by the medium of plants, wind, or flow turbulence.

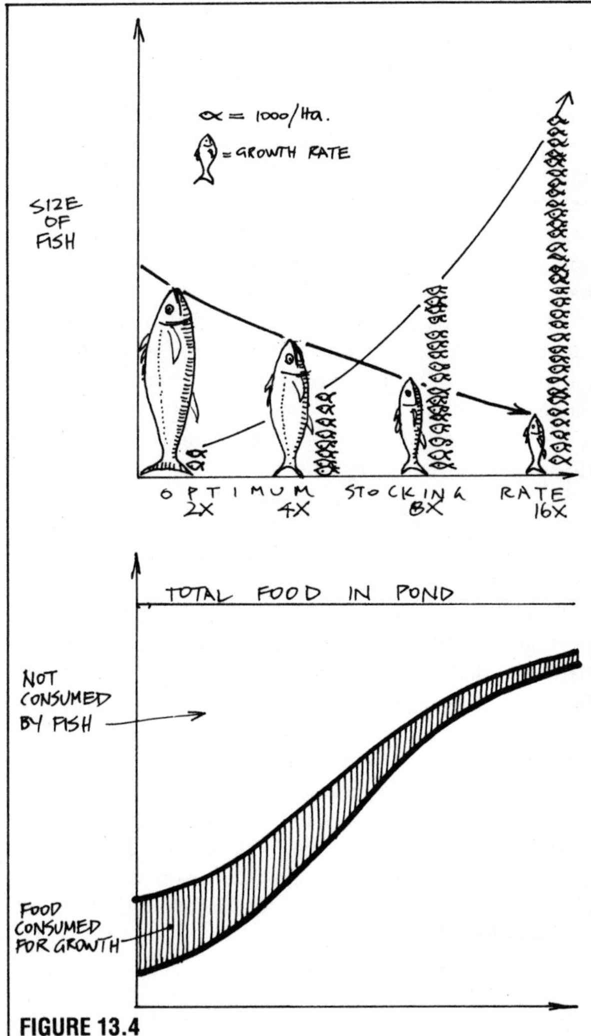

FIGURE 13.4
EFFECTS OF STOCKING RATES
A. As we increase the number of fish per unit area or edge, the size of harvest decreases; too few large fish or too many small fish indicate understocked and overstocked ponds respectively; understocking is the most common error on farms.
B. Only as fish grow to optimum weight are natural food sources fully used; fast–growing minnows or shrimp can hold this food as forage.

ENERGY CONSIDERATIONS

Our design strategies for aquaculture systems must hinge, as ever, on the energy costs of the system: its sustainability in terms of present and future resource costs. The actual excavation of ponds, if well planned and executed, creates a very durable resource, needing little but minor maintenance. Ponds and terraces hundreds (sometimes thousands) of years old are still in production.

Continuing costs, however, are associated with growing fish (or other aquatic organisms) and getting them to market. While both are to some extent site–dependent, the inputs of high quality food or nutrient as an integral part of planning is very much a design–dependent factor.

J. E. Bardach assesses FOOD, FERTILISER, and FUELS for water pumping as the main recurrent fish culture costs, and choice of species as critical to economic production (large carnivores being most expensive to rear). Well over 90% of the energy or monetary production costs are in the three factors mentioned. Labour and fuel energy are to some extent interchangeable, and while the utilisation of waste food is a saving, other pollutants are a cost (sometimes a terminal cost), as the water environment is very susceptible to biocides (*Proceedings of the 14th Pacific Science Congress*, USSR, 1979).

Water supply itself may become the limiting factor in many areas, and recycling of wastes, well–selected polycultures, and careful attention to distances and market are seen as strategies needing attention. Bardach gives some current production costs as:

System	Production Costs MJ/g	kcal/oz
Wild–caught lobster	3.22	21,801
Wild–caught shrimp	2.50	16,953
Intensive silo rearing of perch	2.24	15,167
Pond catfish (USA)	0.58	3,941
Pond polyculture of carp, mullet, *Tilapia* (Israel)	0.027	184
Sewage/stream culture of		

carp in cages (fast stream, no biocidal pollutants)	<0.004	28

Such analyses show why distant–water fishing is collapsing in all sensible (or oil–conserving) countries, and why aquaculture is also likely to eclipse terrestrial production of protein, especially of intensive or feedlot systems. Today, food preferences are also favouring fish and aquatic products generally in health–conscious people, and when I heard recently of an Australian sugar–cane farmer who had converted his riverside canefields to extensive prawn ponds, I was certain that he would be in business long after the sugar refineries had closed down.

As for inherent conversion efficiency, an average figure for catfish, carp, mullet and so on would be 1.5–2.3 kg of high–value invertebrate food to 1 kg of fish flesh (excluding eels and large fish carnivores). While no terrestrial species reaches this sort of efficiency, chickens and some non–domesticated stock come close, at 3.5 or a ratio of 4:1, but beef, mutton, and pork are produced at double or treble the feed cost of fish. We shall see *very* close attention paid to these factors in the near future, or see the bankrupting of western agriculture as fossil fuel energy becomes unusable in terms of its pollutants.

In this section on yields, critical to pond management, we will restrict discussion to those ponds built specifically for fish and aquatic plant culture, and thus presume that they are from 0.5–2.6 m deep, with controlled inflow and outlets, which can be drained or pumped dry if necessary.

WATER QUALITY

All life processes, and decomposition in aerated waters, consume oxygen. A general level of 5 parts per million (ppm) is satisfactory in ponds. At 1 ppm many fish species may die except for a select adapted group of air–breathers used in stagnant and weedy tropical still–ponds. Augusthy (1979) notes that snakeheads (*Ophiocephalus*), Singhi (*Heteropneustes*), the catfish *Clarias*, Mahi (*Notopterus*) and *Anabas* in India all have sacs, labyrinth organs, or special air chambers near the gills from which they can derive oxygen by breathing air. Other fish die in the anaerobic conditions of weedy tropical shallows where the above species, the adapted lungfishes, and many minor species survive. Even derelict swampy ponds can yield 2,400 kg protein/ha/year if a polyculture of air–breathers is stocked, as three or more of the above species, or equally hardy species.

The fish in tropical still ponds combine well with the crop plant *Euryhale ferox*, a spiny floating plant of the water–lily family whose period of yield and harvest coincide with the maturation of the fish. The ponds used are 1–1.5 m deep, and as the fish help control malarial mosquitoes, there are multiple benefits. Seeds of *Euryhale* are marketed as a "popcorn" in India.

The integration of specialised warm–water aquacultures, or the stocking of oxygen–depleted waters is also dealt with by Pullin and Shehadeh (1982). They describe the modification of rice padi to accept specialised species assemblies. Fish refuges are made in padi either as a peripheral canal or as a sump to contain fish between flooding for crop. **Figures 13.5** and **13.6.**

Temperatures in rice pond water may reach 34°C, with optimum growth at 22–28°C. In Indonesia, fish are pan size at 10–12 weeks, while in cooler areas (Japan) it may take 2–3 years of summer seasons to produce pan–size fish in padi. Weedy padi is in part cleared by using (in Thailand and Malaysia) such species as *Trichogaster pectoralis*, *Clarias macrocephalus* or *C. batrachus* for weeds, *Ophiocephalus striatus* as a predator, and *Anabas testudineus*. With such combinations, the extra crop of fish in rice reaches 70–400 kg/ha/year (water 10–34°C). Good fish mixtures, e.g. *Helostoma temmincki* (35%),

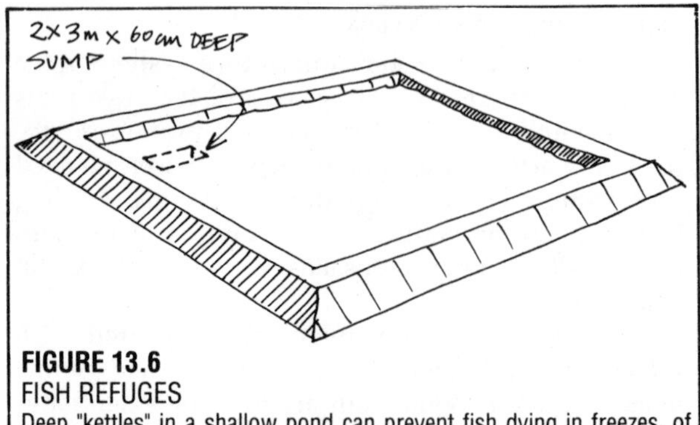

FIGURE 13.6
FISH REFUGES
Deep "kettles" in a shallow pond can prevent fish dying in freezes, of excessive heat, or from accidental pond water loss; they also assist capture in draw–down conditions.

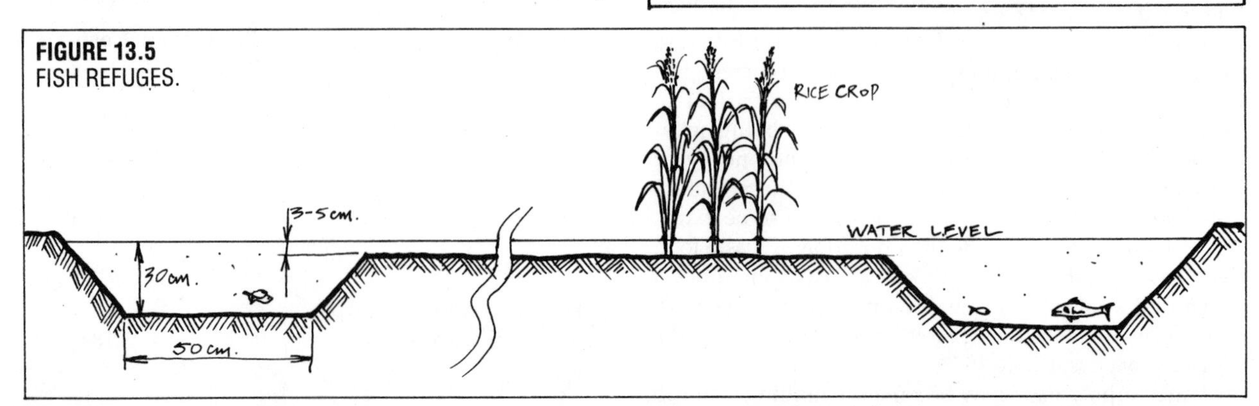

FIGURE 13.5
FISH REFUGES.

Osteochilus hasselti (15%), *Cyprinus carpio* (15%) and *Puntias gonionotus* (35%) give best yields. Alternatively, *Sarotherodon mossambicus* (50%), *Helostoma* (20%), *Osteochilus* (15%), *Puntias* (15%), and *Cyprinus carpio* (10%) are satisfactory. (The percentages are those of fry or fingerlings of these species.) Such polycultures are the result of a series of trials in padi, and all use weed–eating species and predators to control small fish. Crayfish (*Cambarus clarkii*) are also stocked in rice padi, a difficult and specialised pond system, but the fish yields there may add a critical element to nutrition.

We generally classify waters (and in particular sewage lagoons) as:

- AEROBIC: where oxygen is well–supplied by wind overturn, turbulent flow, aerators, or rapids, and where there is a light load of decaying organic material.
- FACULATIVE: (many swamps and weed–clogged ponds, secondary treatment sewage ponds) where the surface may be aerated, but where sediment or sludge collects in cold periods and the pond base becomes anaerobic (sulphur–fixing bacteria then decompose the sediments).
- ANAEROBIC: primary sewage ponds and over–fertilised shallows where there is a low (1 ppm or less) oxygen status.

As less oxygen can be dissolved in warm water than in cold water, and as the plants that produce oxygen by day consume it at night, there is a danger that healthy, weed–filled shallow ponds may develop severe oxygen deficiency on warm summer nights. If valuable fish stocks are held there, care should be taken to both reduce the level of wastes in the pond, and to arrange for automatic aeration at critically low levels. Such dangers are less in clearwater, cold, and open or flowing pond systems where this factor can sometimes be ignored. However, oxygen supply becomes a limiting factor in fish stock density at levels above 5000 kg/ha or thereabouts, and then oxygen as air bubbles needs to be supplied (**Figure 13.7**).

There are several commercial aerators on the market, and many can be solar–powered. But wherever water can be led from a head of 2 m or more, fountains, showers, and "Flowforms" (**Figure 4.34**) will oxygenate ponds. A considerable energy saving is achieved if only part of the pond is aerated at critical periods; fish will gather (kettle) there in times of need. For eels, this aerated section is built with a narrow pass to the main pond to prevent aerated water escaping.

It is often ideal to combine the feeding station with the oxygenated area so that food wastes are also oxidised and waste control simplified; some such arrangement is illustrated in **Figure 13.8**.

ACIDITY–ALKALINITY

Ponds in areas of peats, mangroves, cordgrass flats, samphires, and with water derived from heaths and granites, or siliceous soils, can be very acid in reaction (pH 4.0 or less), the acids being humic acids, tannins, and minor organic acids. In such cases, hydrogen sulphide may also be released by ponded peats or swamps to create sulphuric acid. Peat stripping, mounding, fresh–water or rain leaching, and liming are all used to bring such ponds into production. Below the peats, and especially in basaltic soils, bluish clays produced under reducing conditions may occur (anaerobic or low pH). Fish will thrive in quite acid (natural) waters if calcium is available.

If we drain peats, the organic material dries out, oxidises, and releases carbon dioxide, gradually losing bulk so that some fields in fens slowly sink below drain or even sea level. Fens are derived from alkaline waters, bogs and peats from acidic waters. Marshes, fens, and bogs contain unique plant species and are rich wildlife habitats; hopefully, the era of fen drainage is drawing to a close, as the preservation of wetland habitat is a priority of all enlightened governments and landowners.

As pond pH is ideally 6.0–8.0 or even to 8.5, and only a few fish grow well below these levels (notably salmonids, which are also one of the few satisfactory fish in monocultures), most culture ponds are routinely limed when they are made. They are also regularly checked for pH levels, which are adjusted with unburnt lime after the initial burnt lime dressing. Thus, in the matter of site selection, it is of great advantage to site ponds where run–off from limestone areas can be ponded, or where natural pH levels are already high. Lethal limits for most fish are pH 3.7 (acid) and 10.5 (alkaline).

In practice, we can aim for a pH mosaic in ponds and pond series, as many valuable food organisms prefer soft (acid) water, while fish, molluscs, and freshwater lobsters and prawns prefer hard (alkaline) waters. Crushed dolomite, marble chips, hard limestone gravels, and oyster or mussel shells give a slow release of calcium in tanks, small ponds, intake filter systems, and the upper sections of canal systems.

There is an interaction with pH, respiration, and plant density; pH may rise at night or in clogged coastal waters with *Chara* and other algal forms, or fall as peats form and anaerobic conditions develop.

Not all organisms appreciate hard water (calcium rich), and if accessory ponds are to be developed to breed soft–water fish, forages, or main crop, only organic acids are used, chiefly hydrochloric, humic, or phosphoric acids. The same effect can be sometimes produced by using a peat or dense moss/peat area to filter water, by laying peats on the pond base, and by creating a swampy condition around the culture pond.

MUD, SILT, HUMUS, AND WASTE REMOVAL

Wastes of any nature in ponds reduce water quality if they occur in excess, or if they occur as dead material. We can in some ways distinguish rotting wastes from fertilisers, as the latter need not create excessive growth if herbivorous fish, nutria (a large aquatic rodent), or shrimp are available to eat the detritus or weeds that would otherwise die and create anaerobic conditions.

There is little or no problem in flowing canals or well–aerated ponds, but still ponds need regular monitoring for waste removal. This is most handily achieved by draining and liming the pond and taking off a season of dryland crop (clover, fenugreek) before flooding.

Where ponds cannot be drained, care should be taken to keep at least one bank clear of trees for dredging or bucket removal of muck, which is traditionally layered with green crop for compost in Asia. In padi and terrace, 2–5 cm of mulch is retained over the clay base for cropping. All terrace can be drained to a dryland cycle.

In deeper non–draining resevoirs, recourse must be

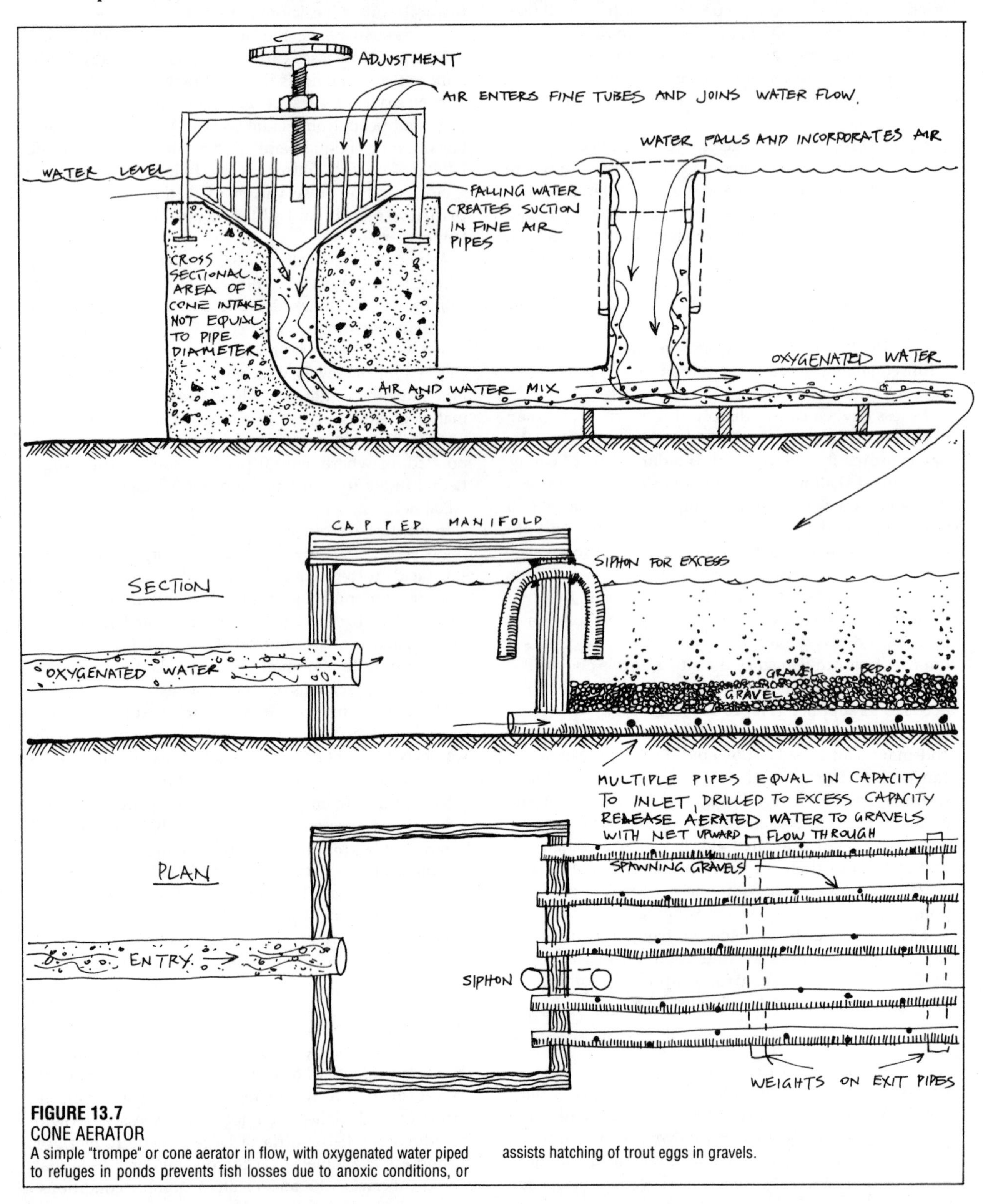

FIGURE 13.7
CONE AERATOR
A simple "trompe" or cone aerator in flow, with oxygenated water piped to refuges in ponds prevents fish losses due to anoxic conditions, or assists hatching of trout eggs in gravels.

made to a jet pump, which can be used as a vacuum cleaner to pump detritus out of the resevoir (and which will remove soils up to gravel size). This can be deposited near the pond as adjacent retaining walls or as low islands and causeways so that silt settles out, and the water flows back to the pond free of detritus. The pumps are typically raft–mounted and liquid fuel powered (**Figure 13.9**).

FERTILISERS

After testing waters, it is likely that minor elements and phosphates may be needed. Caution should be observed in adding nitrates to waters where natural manures are used, or in desert basins where water nitrate levels can be high. Be careful also in using boron–containing detergents, or any biocides not recommended for fish. (Salt, copper sulphate, weak formalin and some antibiotics are used to control diseases or parasites without lasting harm.)

Pig, duck, and second–stage sewage are all used in fish ponds, and any bird or animal manures are useful. We can distinguish between heavily fertilised ponds intended for intensive algal growth to feed milkfish or prawns, ponds where higher vegetation is encouraged to feed grass carp or *Tilapia*, and almost clear–water ponds for bass, perch, and salmonids (trout). That is, we adjust fertiliser to fish food preferences, and in any polyculture a larger proportion of fish can be algal or plant feeders (5:1 herbivore: predator is a usual ratio).

Phosphatic rock and granite dust can be used to supply lime once selected nutrient levels have been achieved. The stocking of waterfowl, the erection of perches for gulls, ducks, and land birds, the siting of pigeon lofts, pig pens, or chicken roosts over water, the erection of martin and swift nest sites in water (or bat roosts in Holland) are all devices to bring in complex plant nutrients to ponds (**Figure 13.10**).

Phosphates, potash, and minor elements are fixed and held in water in a matter of 12–14 hours, or 3–4

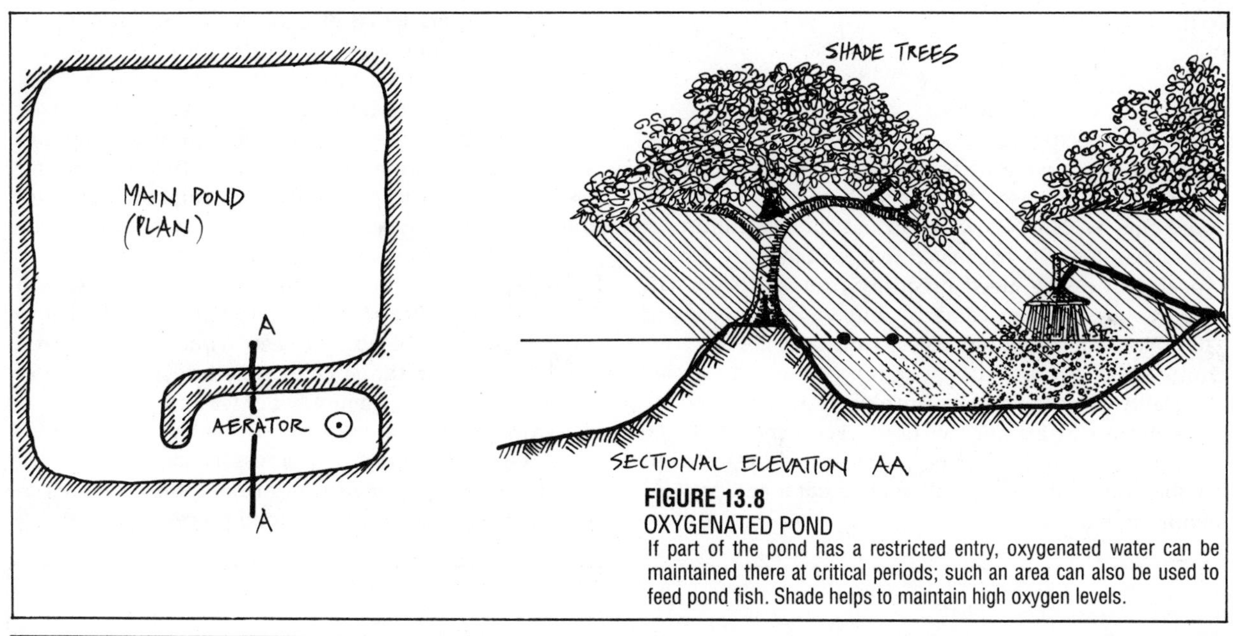

FIGURE 13.8
OXYGENATED POND
If part of the pond has a restricted entry, oxygenated water can be maintained there at critical periods; such an area can also be used to feed pond fish. Shade helps to maintain high oxygen levels.

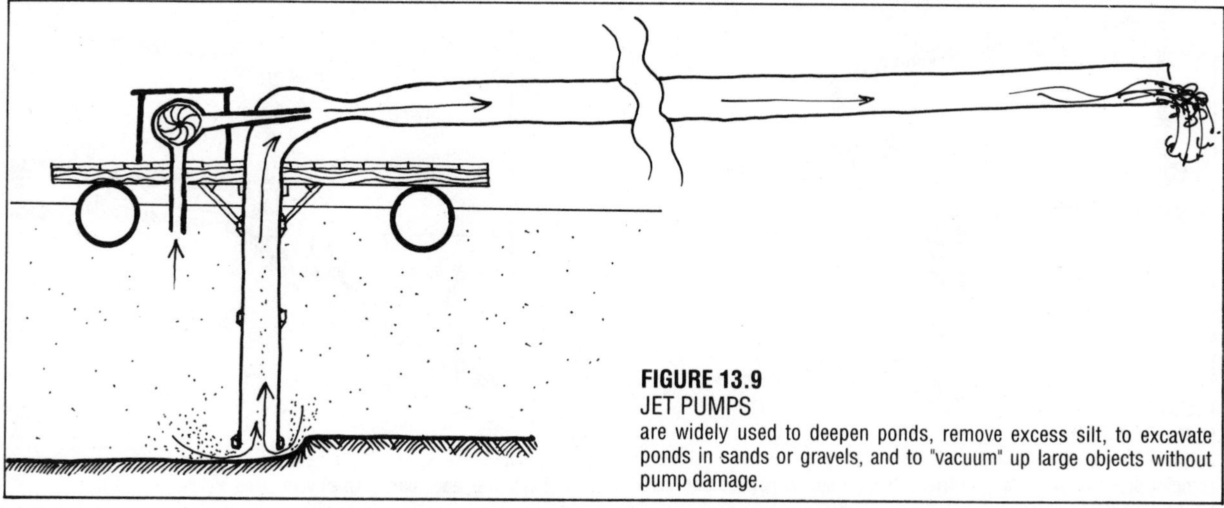

FIGURE 13.9
JET PUMPS
are widely used to deepen ponds, remove excess silt, to excavate ponds in sands or gravels, and to "vacuum" up large objects without pump damage.

days (in oceans), and can be added to quiet marine bays as well as to ponds. Mussels in particular convey phosphates to the mud via their anal siphons, and are the major phosphate resevoir of ponds or rivers. As even modestly fertilised ponds can alter the yield by factors of from 2 to 10 times, attention to nutrient supply is a critical strategy.

As fertiliser uptake is so rapid, we need to stock fertile ponds with shrimp or large numbers of small forage fish while our fingerling pan fish are growing, or we lose yields. Too–high nitrate status can be filtered via a forage food pond of crustaceans, or through reed beds. We rely on land crop downstream to remove excess nutrient from pond overflow. In flow–on of ponds, we can usually achieve a doubling in yield of species such as melons, tomatoes, or fuel forests, so that fish ponds are a key feature in raising land–based yields. Swingle (1966) records a difference in yield for mussels (in the same pond) of from 52.8 kg unfertilised to 1,012 kg fertilised (whole weight; meats are 35% of total) so that the fertiliser factor, well–managed, is a critical yield strategy.

SHELTER AND REFUGES

Stacks of pots, pipes, tyres, and bundles of rope, reeds, or brush can have a decisive yield effect on predation losses of young fish, crayfish, and species subject to predation by their fellows. Crab and crayfish culture, in particular, benefits from such refuges. Similarly, forage species in ponds need breeding refuges to persist. Figures are hard to come by, but yield increases of 10–30% are recorded for crustaceans, and 20–100% for forage fish protected by shallows or weedy areas. Mortality of young and fry are always greater in ponds of simple design and without escapements. Cage culture is not only in itself a high–yield system, but enables the strategic separation of predator and prey (or adults and juveniles).

TEMPERATURES

Within water, surfaces and gradients develop from the effects of heat, solutes, particles, or stream flow. These are utilised differently by different species, so that a river or lagoon opening by a shallow bar to the sea may have saltwater fish at depth and freshwater species on the upper section, and either colder forms at depth or (in icy winters), warm refuges at depth for fish to kettle (crowd in until warmer conditions develop).

The interface between cold and hot water is termed a THERMOCLINE and between salt and freshwater a HALOCLINE; such boundaries can be abrupt transitions if no turbulent flow exists. Where no pronounced surface separation occurs between hot and cold or salt and fresh, gradients develop, especially where wind overturn or current flow is a factor. Fish, molluscs, and plants have specific niches in such gradients, and may move with tide or current to adjust to their specific needs.

At or about 21°C (70°F) we make a categorical difference between *coldwater* fish, to which such temperatures are lethal or debilatory; and *tropical* fish, which suffer cold stress or will not breed below this point. A few common fish (*Gambusia*, the mosquito–eating fish) are temperate tolerant or EURYTHERMAL (1–36°C), but optimum feeding and spawning conditions for coldwater fish are around 15–18°C, and for tropicals 20–25°C. Rainbow trout will die if kept for long periods above 14°C, while *Tilapia* die below 12°C.

Higher temperatures certainly stimulate plant growth and algae production, or general turnover, and it is advantageous to be able to keep winter temperatures at optimum for the species mix selected. We may achieve this in any of a number of ways:

- By including a solar pond (heater) in any other pond (**Figure 13.10**), or as an accessory to a pond.
- Using waste–heated water from industries such as power stations, salt works, or food processing plants.

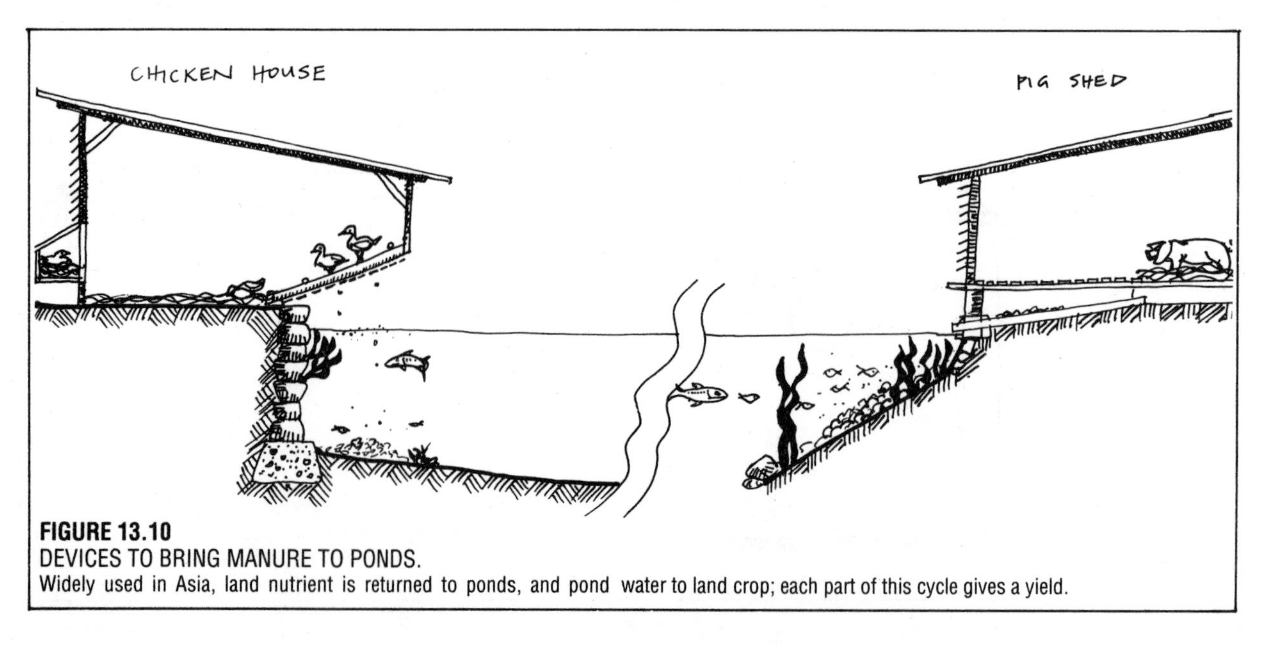

FIGURE 13.10
DEVICES TO BRING MANURE TO PONDS.
Widely used in Asia, land nutrient is returned to ponds, and pond water to land crop; each part of this cycle gives a yield.

• Glazing over canals, refuges, or building raft columns insulated from the earth and colder waters, as heat refuges for cold–tender fish (**Figure 13.27**).

• Providing heat refuges as earth–covered pipes, shaded canals, or kettles of 5 m or so deep in shallow ponds.

Adapted local fish generally need little assistance to survive, but selected stock or exotic fish can require refuges for a week or more of extreme weather. Where no local heat sources are found, pond water pumped over a dark roof, rocks, or cement areas very efficiently extracts solar heat at close to 100% efficiency. Sections of bitumen road can be used in this way, or can have pipes buried under them at 1–5 cm depth as thermal siphoning systems for pond heating in winter. Glasshouses to hold small ponds, compost heat, collectors and trough reflectors are all strategies to maintain heat for higher production. Gains in yields of from 2 to 5 times are common where ponds are regulated to optimise feeding activity and food growth.

Heating devices can raise water levels 3°C or more (glasshouses), and although this doesn't sound spectacular, it can extend the growing periods of fish by 30–40% (*New Scientist*, 30 June, 1977). Collectors, of course, depend on the area of collection relative to the pond. Heat pumps from canals can deliver heat to ponds in the same way, shortening maturation times of fish or plants to about one–third that of cool, open pond areas. Passive collector systems are ideal, and cheap to run by a system of thermosiphon and non–return valves if heat sources are placed below the pond to be heated, as in **Figure 13.11**.

SALINITY

Fresh water contains very little salt. Coastal lagoons usually contain 7–9 parts per thousand (ppt); few frogs can breed above this concentration. Seven ppt is the safe upper limit for human consumption. Brackish water is detectably salty at 11–12 ppt, and oysters may be found from this concentration on. These are the salinities of estuaries and brackish waters. At about 27 ppt many shellfish and marine organisms are found, and the sea itself is at 33–35 ppt.

In areas of high evaporation and restricted tide flow, hyper–saline conditions can develop. Marine organisms seldom spawn above 40–50 ppt, and if fish and crabs do occur there, it is often as one–sex populations, one–age groups, or gigantic forms. Many desert organisms and some desert fish can live to 60 ppt salt.

Fish which live all their lives in freshwater rarely do well above 8 ppt. A great many fish and shrimp, however, migrate fresh to salt and vice versa, some spawning in rivers (trout), some in the sea (eels). Because of their wide range of tolerances, these groups are called *euryhaline*. Many mullet, salmonids, and some oysters and shrimp can be kept in fresh to saline waters. Saltwater fish rarely tolerate salinity below 28 ppt.

Changes in salinity may kill pond fish not adapted to estuaries; salt or freshwater is also used to kill external parasites and disease in fish. Estuaries and brackish waters generally contain more fish species than the river itself, or the sea offshore, although the sea is far richer in molluscs and crustaceans.

In estuaries and lagoons, a halocline (salt–fresh surface) can develop, with the denser salt water at depth, or pushing up under the surface of an estuary as a tidal wedge below the fresh water. Haloclines are common in deep lagoons or rivers with barways. Marine forms live at depth, and freshwater fish at the surface, sometimes for many miles inland. As tides push into and recede from large river estuaries, extreme changes in temperature (especially in winter) and salinity can occur, so that at low tide a mud–living shellfish can be in fresh water at 4°C, and at high tide in seawater at 20°C. Many species in this zone burrow, migrate with the tide, or return to the river at high tide.

Thus, salinity is a species–specific factor rather like temperature, and salinity regulation just as much a matter of concern as temperature regulation. We seldom get sudden increases in salts in ponds, but a hazard of marine pond culture of the inshore tropics is torrential rain. Even with flood diversion drains, the sea itself may become so dilute as to kill lobsters, prawns, or milkfish in marine impoundments.

Very rich brackish–water polycultures of algae, crabs, mullet, sea bass, milkfish, eels, and trout are possible in

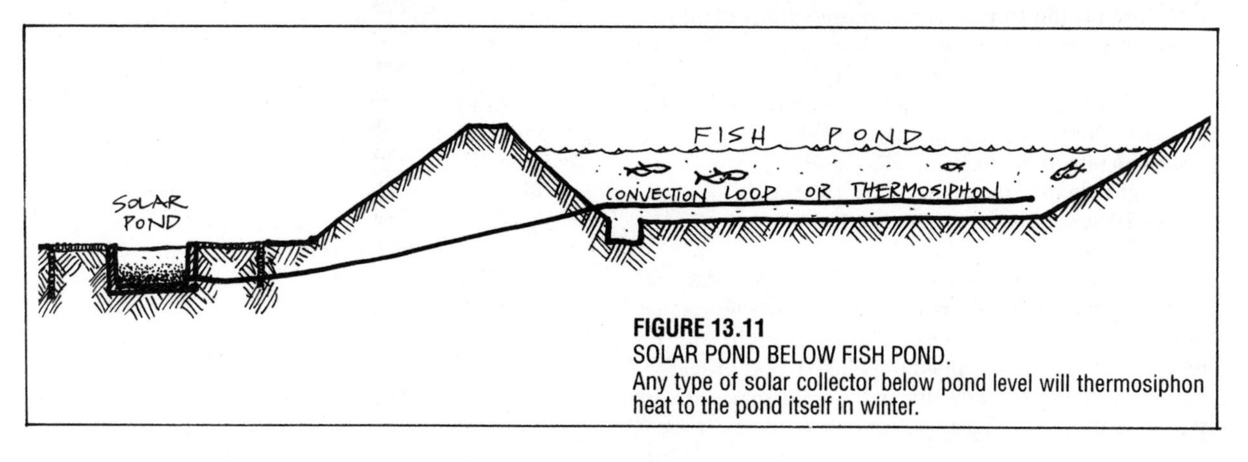

FIGURE 13.11
SOLAR POND BELOW FISH POND.
Any type of solar collector below pond level will thermosiphon heat to the pond itself in winter.

estuarine (and desert or atoll) impoundments. Salinity is controlled by flood by-pass canals and sea weirs or locks. Fish are graded to suit pond conditions, or a salinity mosaic is established to mix species as diverse as carp and mullet. Where tide and streams interact, as in some Hawaiian fish ponds, both freshwater and marine species can be cultivated in one pond series.

Brief immersion in salt water is an old remedy for reducing fungal disease in freshwater fish.

FLOW

Even very modest flow stimulates gaseous exchange, photosynthesis, and therefore growth in plants up to a level where mechanical damage by turbulence can occur. Flow in food-rich waters increases yields far beyond that possible in still ponds, but most streams do not have heavy natural food yields unless they are fertilised or manured, and then a problem of stream clogging with aquatic weeds can occur, necessitating control by nutria, grass carp (White Amur carp) or another efficient herbivore. The other great advantage of flow is in the oxygenation of water, either via a "Flowform", aerator (**Figure 13.7**), or series of rippled falls or weirs in the stream bed.

The balance we would like to achieve in flow is to maintain sufficient flow for aeration and vigorous plant growth, but to restrict flow to a level where we do not suffer rapid leaching of soluble nutrients and calcium. For this reason we often pipe or channel flow to fish and terrace areas rather than working in the stream itself, where floods can affect us in several obviously deleterious ways ranging from fish losses to physical damage to structures, and a wipe-out of balanced fertiliser in waters. In ponds, we may hold back flow for a few days to allow fertiliser uptake, then resume flow for aeration.

13.4

CHOICE OF FISH SPECIES (VARIETIES, FOOD, HEALTH) AND FACTORS IN YIELD

SELECTION FOR EFFICIENT FOOD CONVERSION

Fish low on the trophic scale, those eating plankton, algae, or vegetation, are produced at higher yields. For example (if ponds used are fertilised) we can reasonably expect a maximum production/ha/year of:

Bass (a predator) ....196 kg
Catfish (an omnivore) ....370 kg
Bluegill (an insectivore) ....560 kg
Java *Tilapia* (a plankton feeder)....1,612 kg
(Swingle, 1966)

Obviously, if our choice was for one species, it would be for *Tilapia*. However, real-life tests must be carried out, as some of these neat orders of yield can be reversed if forage is supplied or pelleted food provided. Each trial must be made under the food management schedule proposed, because fish can harvest different foods more or less efficiently. For example, if we add *pelleted food*, the yields are:

Catfish...2,688 kg
Bluegill..896 kg

(Reversing the order found in fertilised ponds only.) Yields in this case depend on the food type.

Having selected a set of *species* in this way, what can we do about variation within species?

If we stock a high density of fingerlings of one size (5000-10,000/ha), and feed these fish, then we can net out the pond after a few weeks or months and select only the larger fish as brood stock, or as an "efficient" variety. After only one year of selection, bass fingerlings from such fast-growing strains converted food at an efficiency of 3.4:1, whereas unselected fish converted at from 7.3:1 average, and the worst cases at 24.2:1—a range of metabolic efficiency greater than the differences *between* species! (Swingle, 1966).

In ponds, and particularly where we raise fish in hatcheries, or where the water supply is free of fish and their diseases, we can both select and ensure stock free of external parasites and internal pathogens, by close attention to cleansing eggs and fry.

The results in ponds can be to double the number of fish growing to maturity or marketable size. Thus, the three factors above give guidelines to raising fish yields by several orders of magnitude. Eels are usually caught at sea (as elvers) for this reason.

Yield increase of from 40-200% are achieved in a pond by the judicious admixture of species. Some examples from Swingle (1966) are given here, and more will be given as examples of traditional systems later. However, it is a sensible initial approach to try to work out what sort of functions any species in a polyculture should serve. Some of these are:

RAPID UPTAKE OF NUTRIENTS

This serves two purposes: the first to fix any fertilisers we may add, and the second to convert wastes produced by dense stocking of other fish. Ideal species are those that breed in ponds, grow fast, and convert decomposers to food. Better still if they form a group which doesn't fly off (as insects do) or migrate out of the pond (like frogs). They then remain as fish food for the young pan fish which are slowly growing up.

Ideal selections are:

• SHRIMP, especially if provided with vegetation or brush bundles for cover; also small crabs, scuds (amphipods), and molluscs.

• MINNOWS or small free-breeding fish; again, provided with cover and screens as the large fish grow, or with shallow water refuges.

Examples (kg/ha/year) Ponds fertilised:

• Bluegill on their own (at 3,900/ha). ......186.4 kg

• Bluegill with *Gambusia* fish ....3,449.4 kg (yield increase of 18 times).

OR

• Bluegill and bass (a polyculture; at 3,750 and 134 fish/ha respectively)..282.6 kg

• Bluegill, bass, *and* fathead minnows.........470.6 kg
(a 66% increase on the above).

Swingle hypothsizes that 50% of a pond devoted to cover or special habitat for small forage species would not reduce the total yield of pan fish, much as 30% tree shelter in fields does not reduce the weight yield of livestock. It is therefore important to plan the shallows of such ponds.

Minnows have increased yields for trout, salmon, *Tilapia*, bass, bluegill, mullet and so on. Our trials should resolve *which* minnows, and how we provide cover or escapement for them. Prawns benefit from brine–shrimp (*Artemia*) mixtures, and most fish benefit from shrimp in ponds. A second reason for adding a pond species is:

WATER QUALITY

This refers mainly to food waste uptake. Species successfully used are fresh or brackish–water mussels, and plants. As plants also add to the daytime oxygenation, and mussels also actively circulate water, both have a dual effect. Their yields are additive to the fish yields, and increase fish yields while they allow higher stocking rates, e.g.

Fish alone..316.6 kg
Fish plus mussels.......................................464.4 kg

(plus 864 kg of mussel meat. We could add 600 kg of water chestnut if plants were also used. Algae activates bacteria and improves the oxygen content of water, as do higher plants).

CULLING OF EXCESS SMALL STOCK (PREDATION)

Small numbers of predators (at any level: fish, turtles, or mammals that eat some of the smaller fish) to prevent over–stocking are essential for a well–maintained growth to market size. Cormorants, otters, or other efficient predators serve this function in large lakes and rivers, but we need better–regulated systems in ponds. Bass, soft–shelled tortoises, snakeheads and pike are just some of the species used to control breeding in carp and bluegill. For example (per ha/year or cycle);

Bluegill alone ....316.6 kg
Bluegill plus bass..484.4 kg.

Increase in yields are modest, of the order of 30–50%, but the predators often fetch higher prices than the prey species, so that the fiscal economics of yields are greater than the figures of weight increase may suggest.

UTILISATION OF DIFFERENT FOODS

This is where a greater potential lies. It is not atypical to find carp ponds with any of the above species mixes, plus 3–7 varieties of carp, all chosen for their distinct food preferences. Even a few additional species help. Examples are (yields per hectare):

Common carp alone (2,500/ha) ....314 kg
Buffalo fish alone (2,500/ha) ....896 kg
Buffalo fish (2,500) plus 250 carp/ha ...925 kg
(300% increase on carp alone)
OR
Channel catfish alone (4,400) ....1,400 kg/ha
Channel catfish (4,400) plus
1,250 *Tilapia* 1,834 kg/ha

Yields are greater for more species added. In the latter case, the food conversion efficiency of catfish (1.7:1) was unaffected by the *Tilapia*; there was no stunting of one of the species by the addition of another. Similarly, the addition of pangas (*Pangasia*) to ponds rich in molluscs does not affect yields of carp or *Tilapia*, nor their feeding efficiency, but adds to yield. It is important to choose fish of different food preferences for maximum yield from polycultures.

THE CONTROL OF BREEDING IN FISH PONDS

We can only talk about optimum stocking rates if we can count on the number of fish in the pond remaining fairly constant. This is not the case where fast- growing pond fish kept beyond 3–7 months can breed in the pond itself. Fish culturalists avoid this sort of overstocking by a variety of methods suited to species, technology, and site. Some of the systems used are:

• by stocking one sex or sterile hybrid fish. Hybridisation in *Tilapia* species give sterile males or one–sex stock fish. "Counts in" are "counts out" less mortality (allow fry losses of 30–40%).

• by crowding fish. Brown bullhead cease to breed above 7,500 fish per hectare (Swingle, 1966). Other species are also inhibited, probably by crowding stress or waste accumulation.

• timing. Fish added and harvested in spring–summer fast–growing conditions can be taken before, or at the point of, breeding, and the pond then restocked.

• predation. Predators with, or screened off from, the breeding fish can reduce their numbers to an optimum level; predator fish yields are a bonus.

• lack of substrate. Ponds lacking the right substrate or niche for breeders will thereby prevent eggs being laid or surviving. Fish will often not spawn if substrate is not provided.

• lack of habitat. Fish that breed in fresh, saline, or seawater will not breed if kept in (or transferred to) a different habitat. This applies to mullet, eels, trout, milkfish, prawns, and shrimp. A lesser effect is that of suitable water temperature. A rise or fall in water level is crucial to carp and mullet species that breed on flooded grasses or reeds.

• hormonal manipulation. Pituitary extract is widely used to induce spawning, and trials of other hormones are being made to inhibit sprawning (*Puntius* is a fish used for pituitary extract).

FISH LOSSES FROM OTHER CAUSES

Losses in ponds can occur from theft, very efficient

predation by numerous or large predators, by diseases or parasites carried in by stream water, as a result of extreme heat or cold, and by accidental draining of the pond. There are obvious precautions to take, involving fences, locks, automatic signalling or pumping systems, and predator control. The fish culturalist needs to live close to or overlooking the pond area for all these reasons.

Losses due to sudden heavy rains in estuarine ponds may be unavoidable in shrimp culture, but flood by-passes are a normal precaution in freshwater pond culture. As fish density increases, the upper limits to yield may be determined by pond water quality and a set of pathogens encouraged by crowding. This latter factor may be the ultimate barrier to yields, as it is in land livestock.

STOCKING RATES

This is a critical factor. Some of the stocking rates are given in the above examples, and they range from 200–300 fish/ha for a predator to 5,000–10,000 fish/ ha for plankton eaters and detritus feeders like *Tilapia* or carp. That is, rates depend on the ability of the fish to stand crowding, find food, and use food efficiently.

Very low stocking rates produce low yields of large fish; thus beware of early trials which show a low yield if fish were stocked (alone) at less than 1000/ ha. Too dense a rate will result in a larger proportion of unmarketable small fish. We can tolerate some 5–10% of stunted fish, but no more than that. Examples of such factors are:

RATE/HA	YIELD (KG)	MARKETABLE (%)
TILAPIA (allowing a 200 g fish as marketable)		
5,000	316.2	97.5
10,000	403.2	50.4
BUFFALOFISH (allowing an average of 250 grams as marketable)		
300	152 averaging 636 g /fish.	
600	273.5 averaging 603 g/fish.	
1,080	656.5 averaging 590 g /fish.	

From the above data, we have *exceeded* the sensible stocking rate for *Tilapia*, and not yet reached the rate that would reduce "marketable size" for buffalofish. Incidentally, *Tilapia* at below market size took 7 months more to reach the limit of 200 g achieved by the fast growers in 6 months! (Swingle, 1966). There is a selection factor involved here, and we can select from the fast growers for future brood fish.

Management helps, in that one heavy stocking can be sold off as they reach market size, or a predator can be added to cull the smaller fish. However, we need a good measure of *optimum* stocking rate for every species used, and this can only be achieved by experiment. Similarly, we need to decide the rates at which polycultures of fish are stocked by assessing the proportion of too-small fish in any one species in the polyculture at harvest time.

<u>In Fertilised Ponds</u>

Fish stocked as fingerlings lose a lot of food unless a fast-breeding forage fish utilises it. Fingerling numbers fall off rapidly as they grow (in predatory fish, by cannibalism), but with supplementary food, much larger numbers can be stocked, saved, and therefore can use the natural foods better. Also, they reach market size much faster, e.g. with brown bullhead (all fish fed while water was above 16°C). See **Table 13.2**.

TABLE 13.2
STOCKING RATES IN FERTILISED PONDS

NO. STOCKED (per hectare)	FISH PRODUCED kilograms
Tilapia alone	
2,500	631.7
5,000	840.1
7,500	1,011.8*
15,000	1,387.0
Tilapia with catfish (all fish fed at 16°C plus)	
560	302
2,500	1,076.7
5,000	1,709.7
7,500	2,646.6

*Above this rate no spawning occurred in thepond

13.5 FISH POND CONFIGURATIONS AND FOOD SUPPLY

POND CONSTRUCTION

Pond or terrace construction on the small scale (to 0.2 ha) can be hand-tuned for drainage, levelling, and spillways, but larger constructions arise either as a result of many years of hand labour, and are then built to suit the landscape, or are built in modern times by survey and machine. Still-ponds have an essential need for a compact clay base, while tidal and diversion ponds can tolerate some losses from seepage.

New ponds need careful assessment and survey for factors of evaporation, seepage, water sources, sealing, and stability. As few fish ponds exceed 2–3 m in depth, earth, clay, and stone-faced earth walls suffice to hold water (**Figure 13.12**). Many trout ponds and fast-flow channels are concrete-sealed in intensive systems, but by far the most productive, common and economical ponds are clay-based. Ponds can evolve from low-walled padi to deeper permanent or seasonal fish systems, or have a dry cycle in summer seasons.

The chapter on water deals with the essentials of construction of dams. The clay core or seal is effective in miniature in fish ponds. A build-up of organic mulch and algae in ponds assists sealing, so that even ponds in sandy loams may gradually evolve to reduce water losses. Long uncompacted canals, however, need good clay bases to prevent water loss in transit, although seepages are not necessarily lost to production if land crop and trees can be planted to utilise seepage water.

It is of great value to elucidate the uses of ponds from very small to commercial sizes, as a proportion of very small ponds have valuable uses in home gardens and for domestic food supply in both urban and rural areas. What we are talking about here, in general pattern terms, is the *order of size*, and therefore appropriate use.

Ponds from 1 – 10 square metres

Very small garden ponds of from 2 to 60 cm deep can be made from old baths, stock watering tanks, plastic–lined holes with protective clay or earth covers, and so on. Pre–cast ponds in plastic, fibreglass, and concrete are sold in most areas.

In shallow ponds, Chinese water chestnut (*Eleocharis dulcis*), kangkong (*Ipomoea aquatica*), watercress (*Rorippa (Nasturtium) aquatica*), taro (*Colocasia esculentus*), frogs, and small fish thrive. Frogs are excellent predators in the garden (as are lizards), and *Hylid* (tree) frogs inhabit the leaves of plants, feeding on insects by day and night.

A square metre of taro gives 20–30 kg of starchy food, while deeper ponds will grow Indian water chestnut (*Trapa natans*), lotus (*Nelumbo*), and arrowhead (*Sagittaria*). Boiled taro, cassava, or plantain can be added to fish food. Stocks of small forage fish for larger ponds can be kept in house ponds, together with mussels and useful molluscs, basic plant stocks, and shrimp.

At 10 square metres, and about 2 m deep, clear fibreglass or Kalwall® ponds, as used at the New Alchemy Institute (NAI—Cape Cod, USA), produce fish and products valued at from $4.50 to $17 per square foot (1984), and amortise costs in 3–5 years, yielding fish, shrimp, and enriched water for semi–hydroponic crops. These ponds provide useful heat storages for night and winter heating, and can be used in greenhouses, or outdoors for much of the year. Outdoor ponds at the NAI are placed in front of reflective (white) walls on white gravels to maximise solar input, increase yields, and lengthen seasons of growth. The Rodale organisation in the USA has also made extensive trials on small tank production of fish.

In glasshouses, such ponds give good yields and moderate temperature extremes, while providing algae–rich waters for terrestrial plant beds, so that fish wastes (and nitrates) are reduced in the water and routed to plants.

It has long been clear that there is a business opening for standard–sized ponds, equipment, and fish stocks for domestic pond culture. This is a field in which there are few suppliers (and most of these have concentrated on decorative plants). I expect that we will see productive pond kits for home owners in the near future. Subsidiary uses are as water stores for fire, tank supplies for gardens in dry periods, productive disposal systems for food wastes fed to omnivorous fish, and sources of recreational interest such as aquariums (fish tanks are very peaceful to contemplate and are recommended for workaholics as a relaxing hobby).

Ponds from 10 – 100 square metres

Ponds of this size are very useful to rear fish fry for sale or stocking, breed forage fish, produce large quantities of vegetable crop, supply aquaculture nurseries and aquariums with plant and animal stocks, and to create significant fire–breaks, water reserves, and heat moderation. One special pond of this size can supply house heating and hot water (the solar pond); it can be built just to heat a larger pond.

For a family, an intensively–managed fish pond of 100 square metres comes close to providing a full protein and vegetable resource if carefully designed and stocked, and if a beneficial polyculture is maintained. Aeration via a photovoltaic cell and air pump may be necessary for high yields, but the modest yields of 300–2,000 kg of protein that can be realistically expected are a significant contribution to food

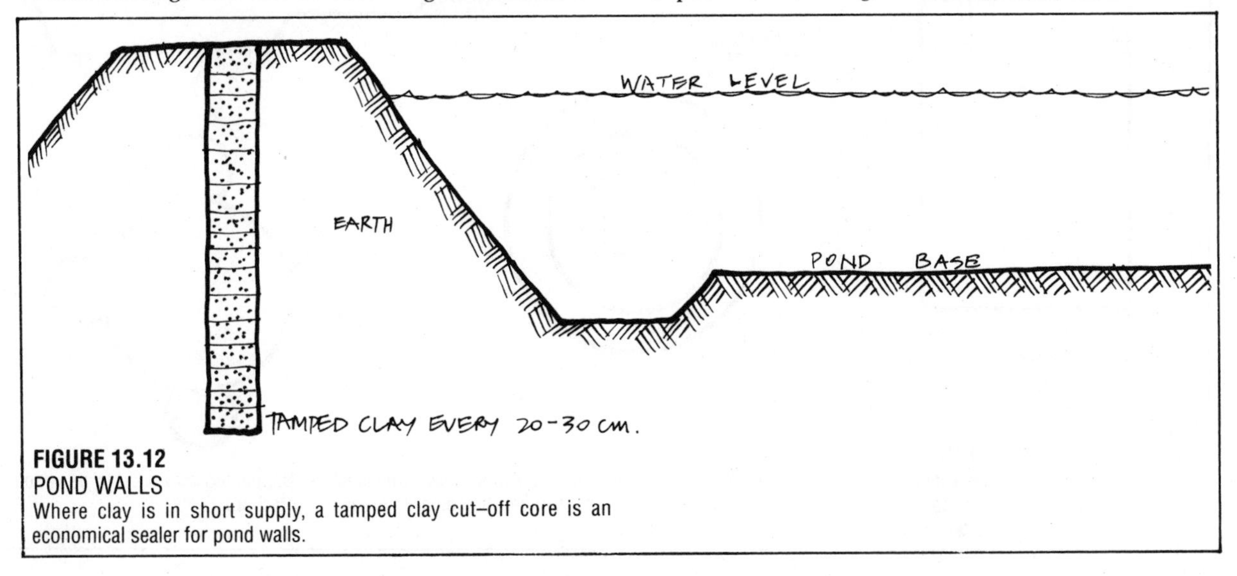

FIGURE 13.12
POND WALLS
Where clay is in short supply, a tamped clay cut–off core is an economical sealer for pond walls.

self–reliance.

In a sense, these are specialty ponds, with a need for very careful planning and subsequent modification (like the home garden, and part of it). Ponds of this size can be an important part of waste food disposal, and can form part of a total wastewater system. In line with a biogas or septic tank unit, they will grow a fairly constant mass of green manures that can be reaped or gathered for garden mulch or biogas recycling.

Ponds 100 – 500 square metres

Chakroff (1982) and others regard ponds of this size (alone or in series) as ideal for fish culture. Easily harvested, netted, or drained, and capable of holding specific age, size, or species assemblies, such ponds give flexibility in management. Devoted to special crop, even one 500 square metres (or five 100 square metres) pond can provide an income from baitfish, aquarium fish, special plant crops, and intensively managed fish (e.g. prawns) of high value. A nursery of any sort, providing eggs, seeds, fry, or fingerlings to growers need not exceed this size. In 1983, incomes of about $30,000 were possible from open–air ponds of this size stocked with prawns (Hawaii).

Ponds 500 square metres – 5 hectares

About 5 ha is estimated to provide a full (affluent) family income in high–value product. Ten to twenty prawn ponds (with ducks, mussels, and edge plants) is a full–time job akin to (but more easily managed than) a

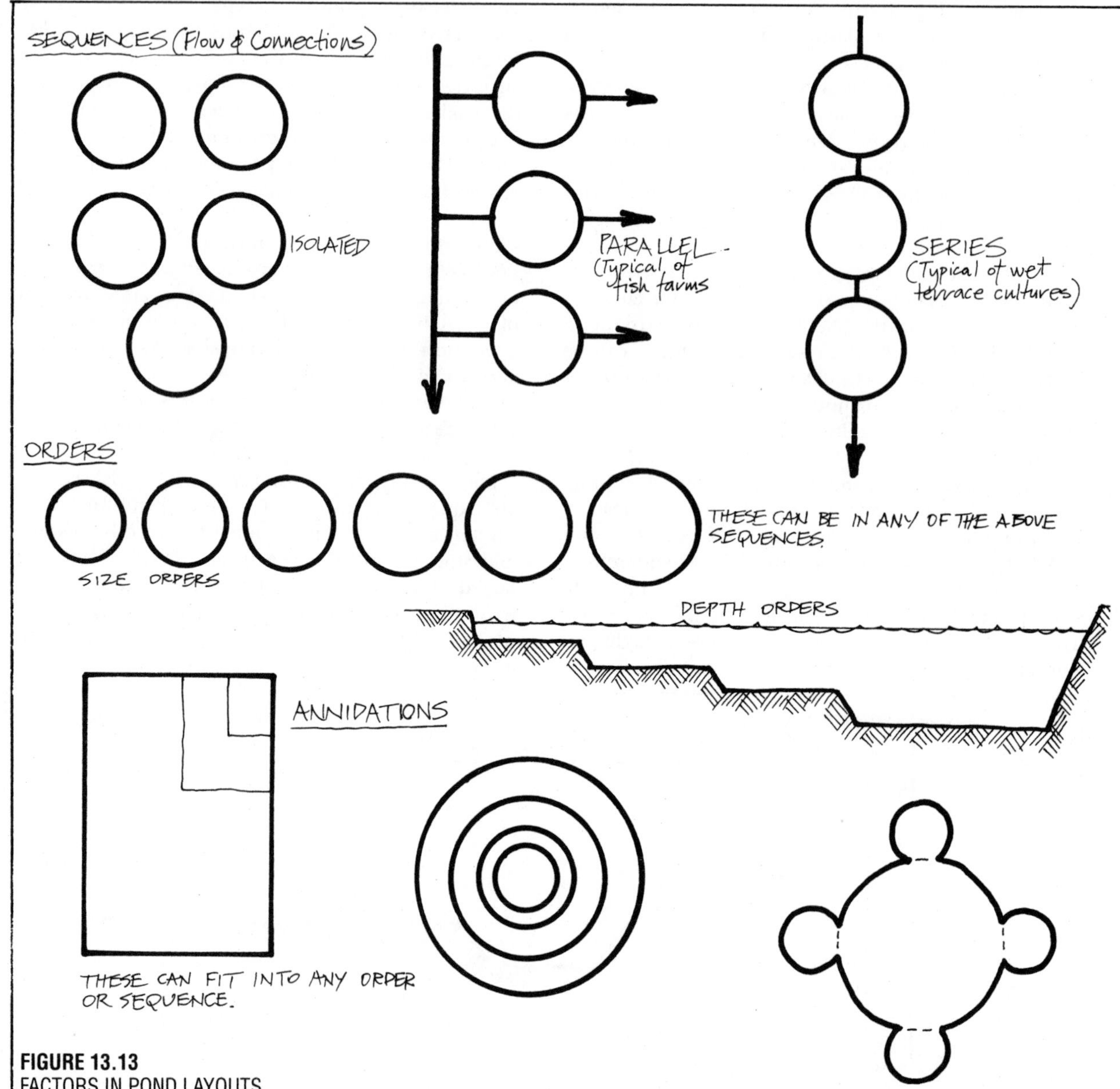

FIGURE 13.13
FACTORS IN POND LAYOUTS.
Ponds can be arranged as isolated, in parallel, in series, or annidated one within another; they can be the same or different sizes, and all these factors must be selected for in particular landscapes, or for a specific fish species or polyculture. Depths too may need to be varied for forages, breeding, or predator protection.

market garden. Many successful enterprises never exceed this size (a family farm). The scale permits multiple use and perhaps leaseholds of other activities, will supply a relatively large area of land crop, and form a complete fire–break for homesteads and fuel forests. We can characterise these as semi– intensive; many eel–rearing establishments operate at about this size. Sewage ponds little in excess of 5 ha will provide safe disposal for small towns.

Ponds greater than 5 hectares

Impoundments of 50–500 hectares are now not uncommon as extensive farms. Even a single impoundment of this size, designed for easy management, gives a living to a family or families. At 10–20 ha, given that the design is fairly adequate, harvest and occasional re–stocking is the main activity. Companies and investors may try to run such systems intensively, and are often limited by cost–benefit, food supply, or disease control ceilings. Innovative leaseholds, recreational fishing, and new scales of planning (including village or shoreline residence development and holiday homes, boating, and sales of water itself) may become feasible.

Every size order can include specialty ponds of the previous order, devoted to specific functions. Some functions may in one or other way serve all orders while themselves remaining small. This is true of an intensive hatchery for trout, carp, or perch.

Given that we have some ideas about the best uses for size or volume of water, what are the best arrangements for series of ponds, and what shapes might ponds take? What order of ponds in total landscape can we define, and what specific constraints or freedoms are permitted or imposed by site selection?

ORDERS OF DEPTH

Depth, as for surface area and volume, has its orders and effects. Mere films of water, or wet mud patches, suffice to grow most productive green salad and root crop, to allow bees to safely drink, revive frogs, and breed a few mosquitoes. Soil at field capacity is effectively a "pond in hiding". From 2 to 6 cm depth, some of the smaller floating plants take root and form mats, and scuds and small crustaceans flourish. From 10–100 cms depth, true floating and anchored plants can separate out their niches, and we can grow Indian water chestnut or almost any fish species covered by the water. Even in rich algal ponds, some light will penetrate.

The maximum commonly used depth to rear fish in culture is 2 m, and at this depth no unstable thermoclines develop to create sudden overturns of oxygen and temperatures; the ponds are easily wind–aerated and yet deep enough to buffer air temperature changes.

From 2–15 m, lakes of clear water are potentially productive. Below this depth, less biological cycling takes place, and less still in the cold, deep, sometimes anaerobic depths of V–shaped lakes with leaf–fall on the margins. Deeps of 4–5 m do have a limited use to fish escaping high surface temperatures, or ice. Caves under water serve the same purpose, and it is common to find trout huddled in drainage pipes under the dam wall on hot clear days (they are then easy to catch if the drainage valve is suddenly opened!)

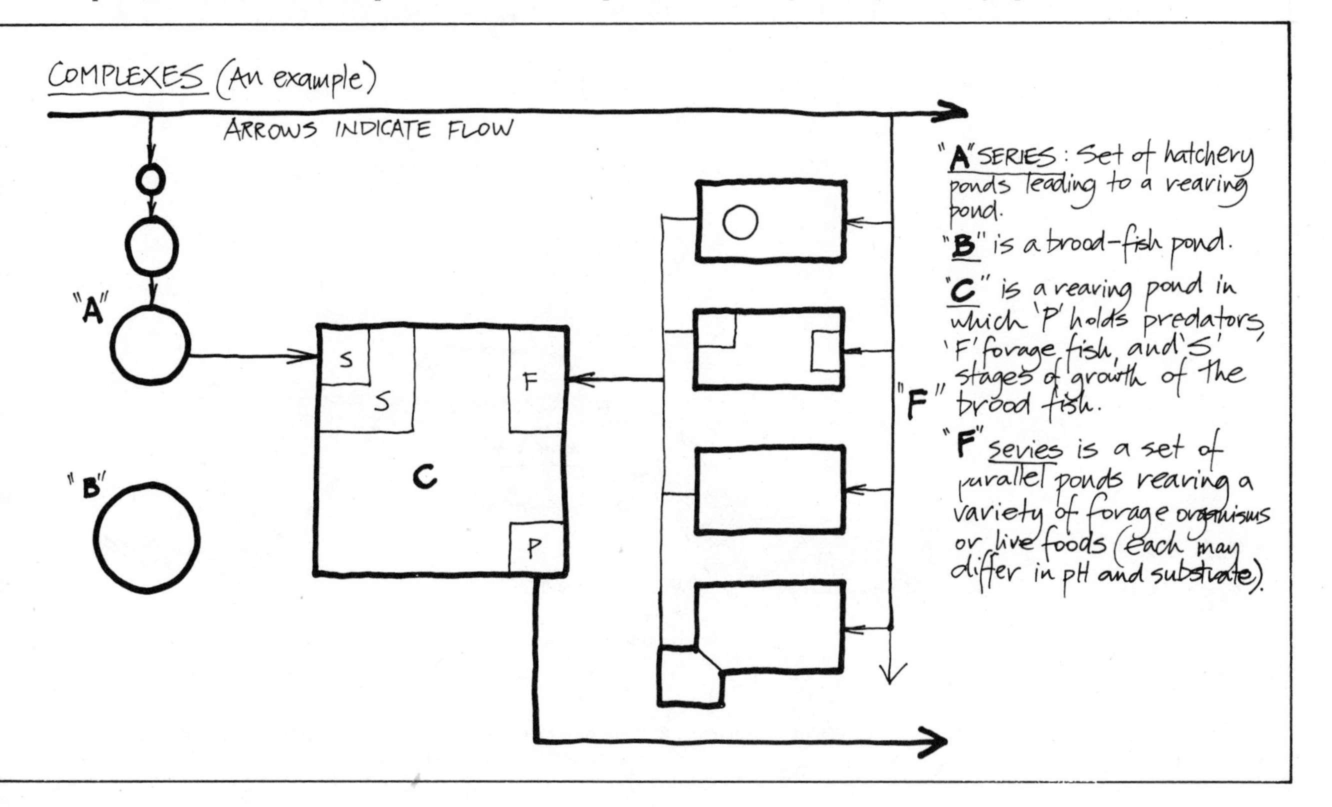

Depth and plants together serve as a set of breeding refuges for the food organisms of omnivores and carnivores. As we will later see, the configuration of levels at different depth may have a profound effect on the total yield or availability of such food organisms.

Thermoclines may develop above 15 m depth, and are fairly typical of lakes subject to sudden seasonal changes in their temperature. Overturns in temperature typically occur in autumn and spring. While in summer the surface waters remain consistently warmer than those at depth, in winter we can get surface ice, then warmer water at depth, and finally ice again deep in the lake, where very cold streams pool up and freeze. Fish may live in a sort of icy sandwich, or tucked away kettled in a modest cave or hollow in shallower water.

PONDS IN SERIES AND FLOW

Ponds can be built isolated from, in parallel with, in consecutive series with, in order sequences with other ponds, or as any combination of these. They can also be annidated (nested) one within each other, and in fact have some features which permit linear arrangements (the pond as canal or waterway).

Isolated Ponds and "Isolation" Ponds are familiar to every aquarium keeper. New stock, diseased stock, trial polycultures, and an insurance against general loss may dictate a set of still–water ponds in quite separate situations (not sharing any flow of water). A set of tanks above grade, or still–ponds filled one at a time can be effectively isolated. Even if treated the same, they will still manage to be different, which is one of the fascinations of aquaculture. Water systems are more connected within themselves, and convey small differences more rapidly than land systems.

Ponds in parallel are perhaps the common fish–culture system, and are analagous to irrigation bays on land in that they possess a head canal (inflow system), an individual flow–through, and a tail canal (drainage system) which also works for surplus water in rains. They are effectively isolated *unless* a disease, pollutant, or qualitative change occurs in the common water supply, when all may fail together, unless (as is also common) flow regimes are staggered so that some flow, others are still.

The parallel series is ideal for valley floor ponds where water at head has been diverted along contour, for single feeder pipes, and for some narrow tidal benches. There are many advantages, and some basic disadvantages in that one cannot "feed the other". That is, we cannot set up a controlled trophic ladder in

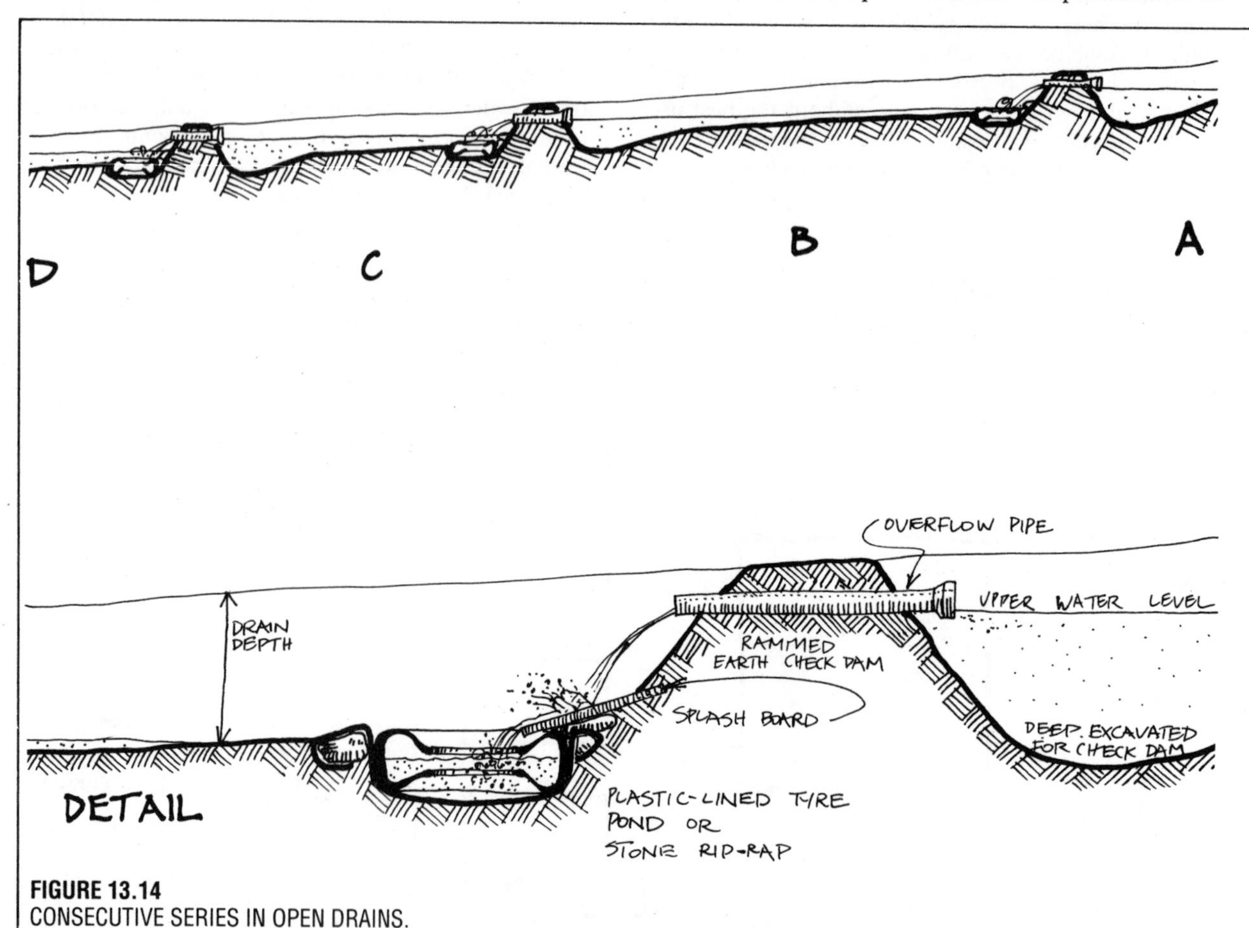

FIGURE 13.14
CONSECUTIVE SERIES IN OPEN DRAINS.
A. Oversized drains, stopped at intervals, give drainage, swale function, and small permanent ponds in landscape.
B. Pond series can be arranged as above to permit one–way travel of forage animals to a rearing pond (see text).

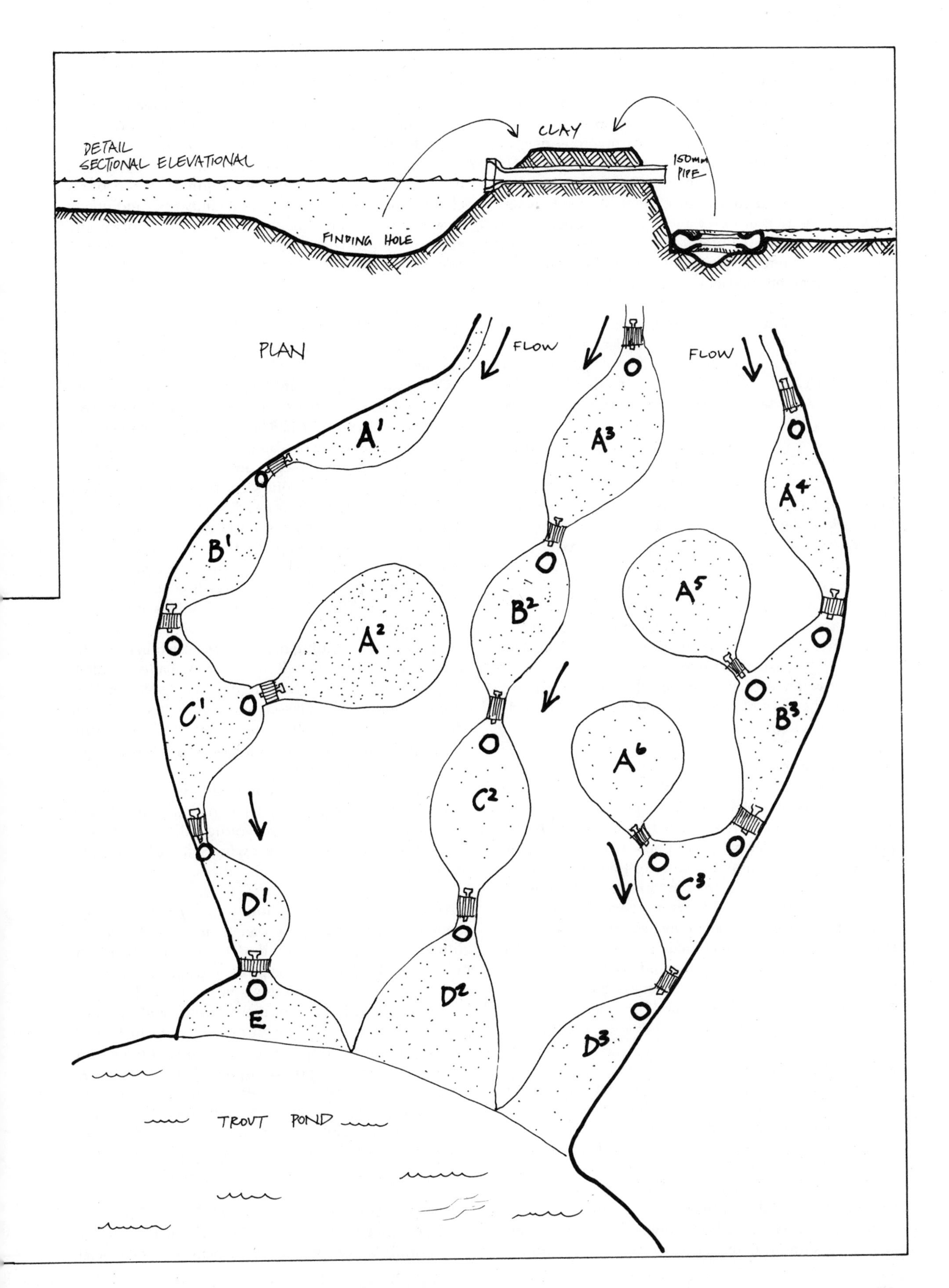
CLAY
DETAIL
SECTIONAL ELEVATIONAL
150mm
PIPE
FINDING HOLE
PLAN
FLOW
FLOW
A^1
A^3
A^4
B^1
A^5
B^2
A^2
C^1
B^3
A^6
C^2
D^1
C^3
D^2
E
D^3
TROUT POND

parallel ponds, where food is cultured and allowed to flow down to higher trophic levels (minnows to trout).

Consecutive Series can be arranged in trophic levels, with orders of ponds breeding forage fish or shrimp spilling over to ponds containing carnivorous fish; or with marshy ponds of tadpoles, scuds, and Daphnia flowing into ponds of omnivores. Even more handily, primary sewage lagoons (anaerobic) can feed secondary lagoons rich in insect and other arthropod fauna and zooplankton (faculative ponds) which in turn cascade into aerobic ponds of useful fish, mussels, shrimp, or green vegetable crop, which in turn...etc. **Figure 13.14**.

Moreover, as quite small ponds (about one–seventh or one–ninth of the size of the next) deal with the anaerobic phase, and as these are best constructed as tunnels or canals from which gas (methane) can be collected, the order of size in this sort of series can be suited to function.

Similarly, a very small hatchery can supply a larger fry pool, which in turn supplies a fingerling tank, which can now provide stock for large ponds. The orders of size here are more like one–tenth or one–twentieth of the next, and their aeration, structure, and therefore construction will differ. Eels are reared in such increasing size order of ponds, as are trout.

Thus, consecutive series has an essential function in terms of orders, but may make little sense in terms of a set of equal–sized ponds of the same function (such as a consecutive series of 500 square metre ponds all stocked with fingerlings). Consecutive series may also be forced on us by valley configuration or other site limitations, and then the risks of change in water quality compels close monitoring.

Evolutionary Pond Systems

Every evolutionary (old) fish culture system or terrace complex has an intricate set of ponds and flows. Elements are added on as needed, as money and time permit, as new information and needs arise, or as new species are incorporated. Although complex, these systems work well and are comfortable to work with. Successful modifications are preserved, mistakes rectified, and catastrophes remedied. However, many such systems are never intended for easy reading, and a novice inheriting one might spend weeks or months working out how to control and manage the system. All are quite unique, and often subtle in operation, with complex water control.

Ruled–up Pond System

In contrast, the flatland rice farmer who converts to catfish may evolve a simple, standardised, all–pervasive and often monocultural rectilinear farm visible from an airliner as a network of precise regularity imposed on the landscape. Anyone can understand it; the system is probably easy to control, but it costs in food, and tends to be a bit boring. The plan may very well have been drawn up by a Euclidean geometry student determined to force an unnatural rigidity on nature. All that is needed to manage is power (energy). But then power, as Mr. Kissinger remarked, is the ultimate aphrodisiac, and there may be some murky compensations hidden in the pattern.

While all of the above evolutions have their admirers, there are potentially a set of quite different approaches, only one of which may be to consider the pond complex as a component of total landscape, and others of which refer more directly to energy (or food) supply and have to do with configurations, or the shaping of ponds themselves.

ANNIDATIONS

We can nest any sort of smaller pond in a larger pond in a variety of ways:

Cages and ring nets in large bodies of water allow control of feeding, harvesting, and disease in caged, netted, or fenced–off fish. They are in (and may benefit from) the larger body of water, but are concentrated for management. Similarly, eggs and fry can be separately reared in partitioned ponds or in aquariums. Their survival is much enhanced, as many fish and crustaceans are cannabalistic (or rather non–selective).

Predators may be kept to thin out populations, separated by a mesh which permits any stunted or young fish to pass. Or, shrimp may live out their lives in shallows which nevertheless adjoin a deep area where their predators lurk.

A solar pond, yielding heat, can be nested in or below a frozen pond, and thaw it in winter. A shaded, chilled, or aerated pond may act as a refuge in hot weather. Part of any pond can be glassed over, even insulated (on shore and in the water) for a heat refuge, while remaining open below or via a base slot for fish to use as a refuge in chilly weather.

Floating basins or mini–nets of live food can be placed in larger ponds, as can a single gravid (pregnant) crayfish, whose young fall free into the larger pond to commence growing without predatory adult competition.

All of the above are operating and valid, although some are energy–storage rather than integrated aquaculture annidations intended to aid fish or plant production. Listing them helps us to decide on how we may use such accessory systems in any pond, but there is no doubt that if we plan them to start with, we can make the original pond much more easy to fit than if we tack them on later, or as afterthoughts. As an after–thought, many "ponds in ponds" are made either to hold wild fish trying to enter a cultured pond, or to hold migrating escapees from a cultured pond. Both are illustrated in **Figure13.15** as one integrated system (upstream and downstream fish traps or sorting cages).

PONDS AS PART OF THE LANDSCAPE MOSAIC

I believe that when creating ponds in barren (or agricultural) landscapes, we must plan for a beneficial

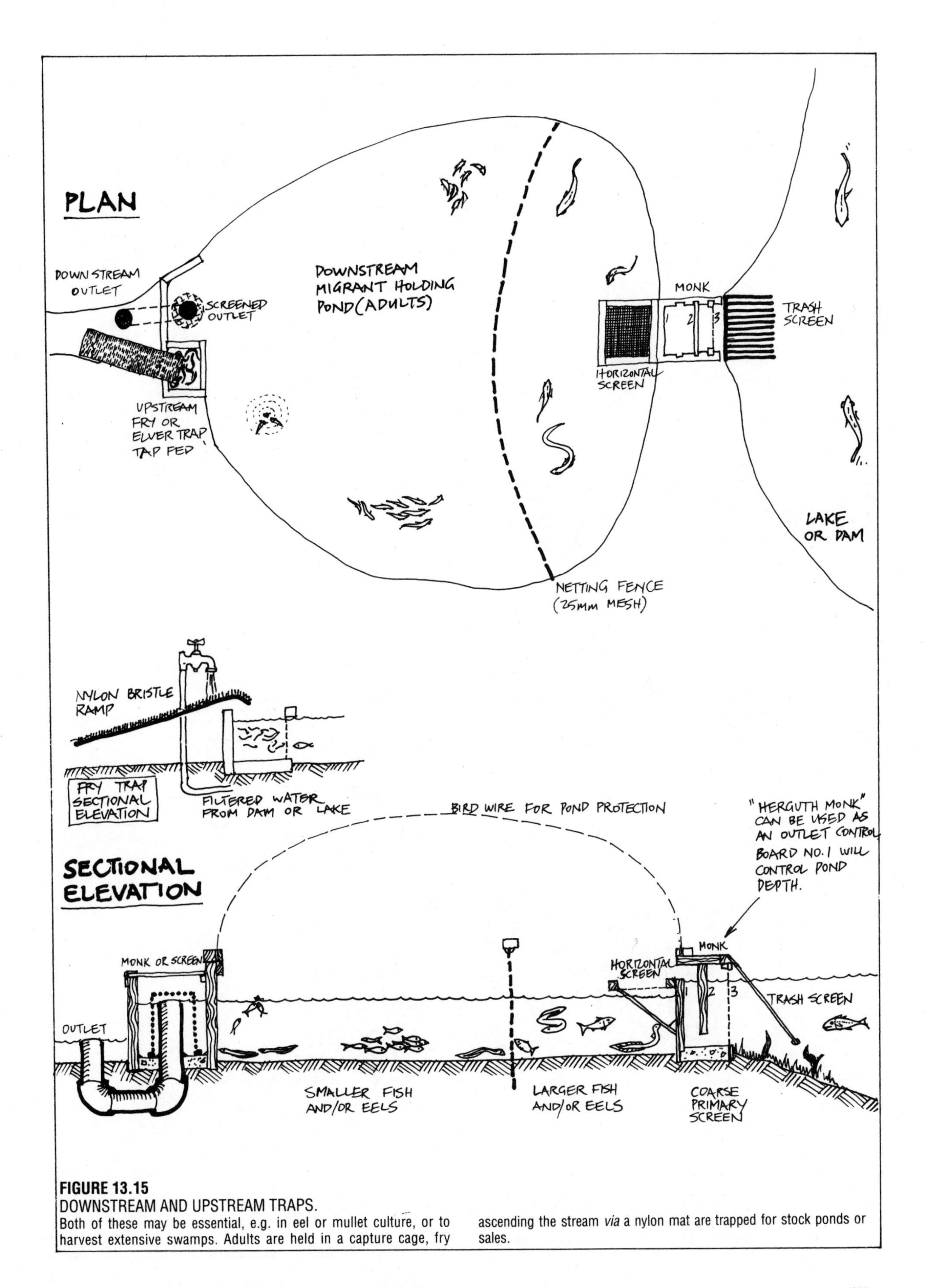

FIGURE 13.15
DOWNSTREAM AND UPSTREAM TRAPS.
Both of these may be essential, e.g. in eel or mullet culture, or to harvest extensive swamps. Adults are held in a capture cage, fry ascending the stream *via* a nylon mat are trapped for stock ponds or sales.

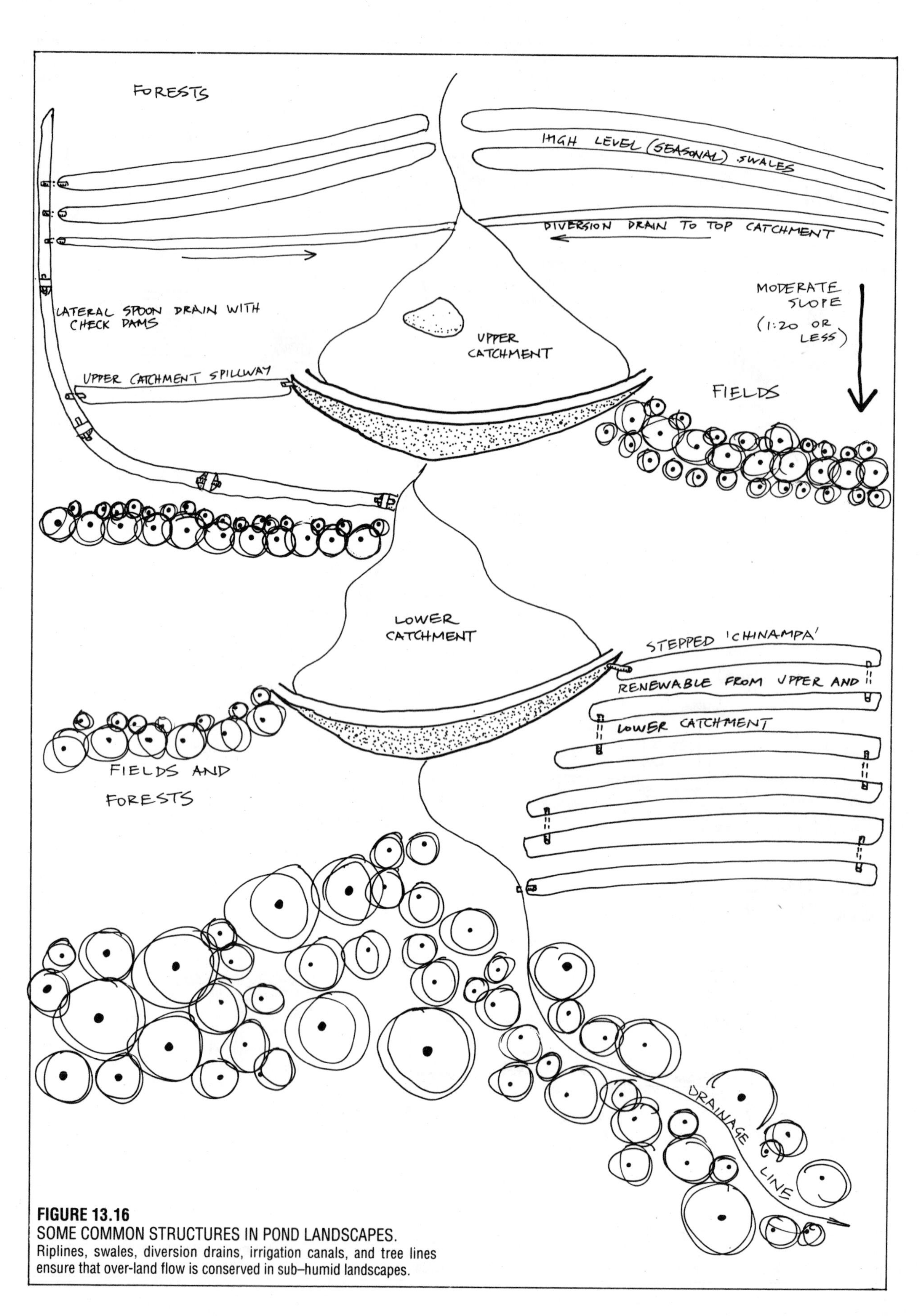

FIGURE 13.16
SOME COMMON STRUCTURES IN POND LANDSCAPES.
Riplines, swales, diversion drains, irrigation canals, and tree lines ensure that over-land flow is conserved in sub–humid landscapes.

mosaic of forest, pond, marsh, and prairie or rangeland. The role of the forest (correctly chosen) is to produce clean water of good nutrient quality, to absorb wastes from fish and their plant associates, and to provide a variety of foods either directly (as fruit) or indirectly (as insect bodies and frass) to the pond in return.

The role of the marsh is to provide a rich habitat for birds and crustacea so that ponds and the forest collect phosphates, and that of the meadow to provide for some mammals and plants that interact with both forest and water. Given that we distinguish four sections or component assemblies (open water, marsh, prairie, forest) in our mosaic, we can have both simple edge effects and other complex edges involving more than two junctions.

As a round figure for sub-humid or humid areas, perhaps we need something like 15% pond, plus 15% marsh (contiguous), plus 30–60% forest, and a remainder in meadow, crop, or pasture (10–40% of the total). Moreover, we need the forest upstream of, downstream from, and between our ponds, the marshes upstream of and in the ponds, and pasture or prairie as downstream and random patches, where trees are difficult to grow. We could perhaps link the whole with a complex of permanent or intermittent drains, streams, canals, and swales (**Figure 13.16**).

We can even define some ideal forests, partly in terms of site and climate, and partly in relation to the pond component. River red gums, some other eucalypts, and some leguminous trees provide an enhanced phosphate drip from rain throughfall. They belong close to the waterways and water edges. Many fibrous-rooted willow and *Casuarina* species either need, or "fix", phosphates. They belong in the downstream forests. So do freshwater mussels; they belong in and are confined to the pond.

We can see many opportunities for sensible local design, encouraged by past successes (mulberries, duck, and silkworms in the Chinese carp-pond complex). Having discussed the series, orders, annidations, and layouts of ponds, we can consider how to shape our ponds.

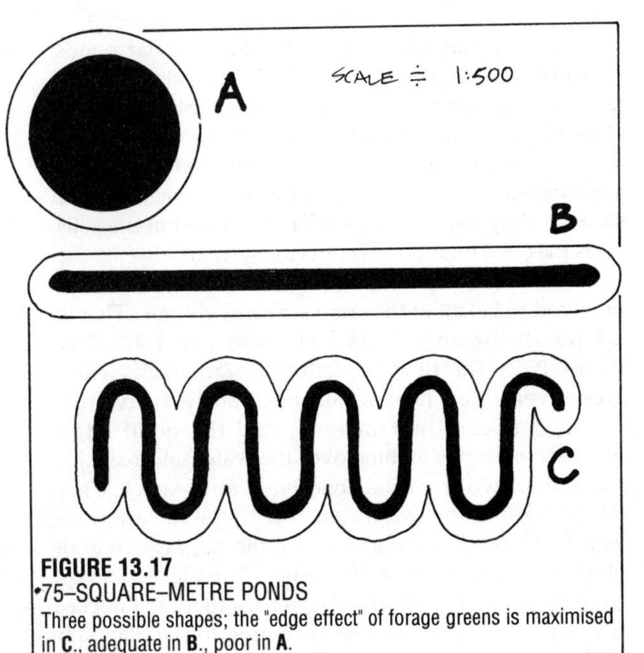

FIGURE 13.17
•75–SQUARE–METRE PONDS
Three possible shapes; the "edge effect" of forage greens is maximised in **C.**, adequate in **B.**, poor in **A.**

POND CONFIGURATION FOR EDGE EFFECTS

Figure 13.17 shows the plan of some ponds of 75 square metres (surface area). They have equal depth, contain the same volume of water (and the same quality of water), and differ mainly in their configuration or ground plan. Pond A (5 m radius) has about 32 m of edge, or margin. Pond B, which is 37.5 m long and 2 m wide has 81 m of margin, and pond C is 1 m wide and 75 m long with 153 m of margin. All are made below grade, or with wide banks, indicated by the thin lines; this "halo" we will call the ZONE OF EDGE EFFECT. Note that B and C differ profoundly in that this zone occupies only some of the field in which B lies, but all of the field enclosed by the folds of C.

Let us consider that a large proportion of the plants around all these ponds can be either eaten by, or will host organisms that can be eaten by, the omnivorous fish we have placed in the ponds. It is at once obvious that pond B provides 2.6 times, and C provides 5 times the food of pond A to our omnivores, and that the land around pond C will do much better at this than the fields around A or B. Moreover, while pond A is rather self-contained and inflexible, pond B is easily partitioned, and pond C easily compartmentalised for parallel flow. Pond C is indeed a very flexible pond in every way.

Further, if our pondside vegetation grows over the pond edge for a modest 0.5 m, then it affects only some 25 square metres of pond A, but reaches half way across pond B and clear across pond C. Thus, if this is of benefit (and we can design it to be so), pond C benefits by a factor of 3 times more than does pond A.

Again, if we wish to partition any or all of the ponds using a set of two 1 metre square sieves, we get only 1 m of bank in pond A in our new enclosure, but any amount of bank in pond B (with 2 partitions) and the same in pond C (with 3 partitions) depending on our desired ends (see **Figure 13.18**).

Given modern machinery, or even pick and shovel, all are equally easy to construct. However, as up to 15 or more times the natural food is available in pond C, our decision is a simple one. There is one last consideration: long narrow ponds can fit easily on slopes, and as troughs they can be stepped and stacked. While circular ponds can be stacked, they become more inaccessible, and as slopes steepen, more expensive to build. So why are most fish ponds we see round or square? Probably because we used a compass and ruler to design them, rather than spend a little time on the consideration of some more basic and life-related implications.

There are serious drawbacks to linear ponds on leaky

sites, as they may lose more water than circular ponds. However, on well–sealed and clay sites, this is not a factor and sealing a linear pond with a plastic liner is also simple. Evaporation from both is equal, or less in shaded narrow ponds. Channel–shaped ponds are appropriate for establishing a modest aquaculture on slopes fed by a spring, where clay is present in the soils, or as part of a total canal connector system.

Note that the ponds in **Figure 13.19** are (or are intended to be) all of the same area and volume. That is, all ponds lie in a 2 ha field and are 1 ha area themselves. The field is planted to tamarack, which grows very well near water, but poorly away from water. Between the tamaracks and the pond edge, blueberries thrive, arching over the water and reaching a metre or two from the pond edge to the trees. They benfit both from the water and the acid tamarack mulch. Through them and above them, grape hybrids climb in the conifers. At the edge of the lake rainbow trout feed, eating both the blueberries that are knocked down by birds and the insects coming to the plants. In addition, the manure of the blackbirds coming to the blueberries encourages a bloom of phytoplankton much appreciated by rainbow trout. At a glance, which field and pond of the four in **Figure 13.19** will produce the best?

There are many potential pond configurations, and many alternative species assemblies to that of trout, grape, blueberry, and tamarack, but wherever we spend a few hours analysing a more efficient or benign configuration before we call in the bucket and dredge, our return may be many times that of the Euclidean or "straight" designer. The yield goes on for years and years, while the digging of the pond is a single event.

Circular ponds and tanks are most appropriate to intensive fish rearing with a water flow at head, or pumped. The usual arrangement is to set water intake jets at an angle, and by this means contrive both to aerate and to induce current flow in a round tank. Most of these intensive rearing tanks, suited also to fry and any active swimmers, have a central water regulating and drainage system to facilitate harvest.

Heat–welded or rivet–silicone tanks of clear plastic or fibreglass (as used at the NAI) have the multiple advantages of growing dense algal food, storing heat, and as well as rearing fish (*Tilapia* is popular), serving as a hydroponic and eventually a terrestrial nutrient source for plants, whose roots are pruned by fish. These indoor tanks are integrated with glasshouse and crop to give multiple benefits; most are still–ponds (aerated but not in continuous flow).

In the earth, the configuration of a round pond has little benefit, giving the least edge for area. Most existing ponds are rectangular (to aid fish–out by netting), and in large series. They are usually built without shallows or bays for forage, often lack drainage, and (as in tank culture) the majority of food has to be supplied from purchase. It is in these essentially simple or factory systems that pumping, food supply, and maintenance of water quality become the major costs of production.

Figure 13.20 suggests that we can set up a separate but interconnected mini–system, rather like a pond with shallows, or a pond with different sizes of boulders and gravels. How could we stock this pond?

We have some 16 possible environments. Ponds A1

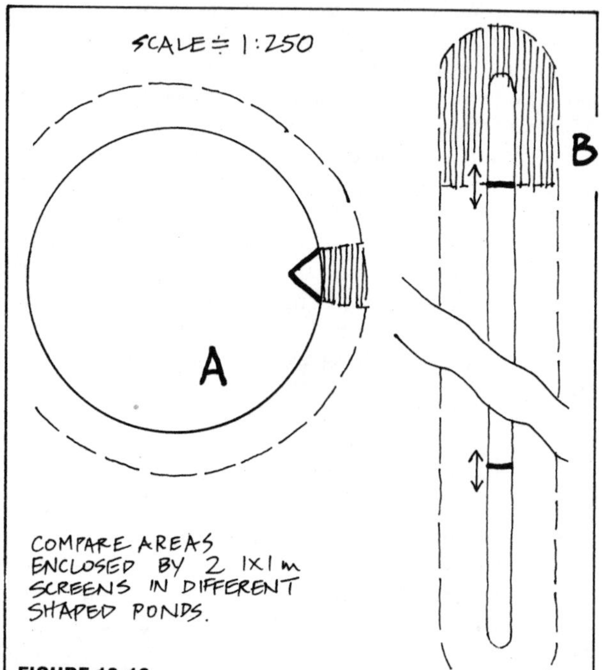

SCALE ≑ 1:250

A

B

COMPARE AREAS ENCLOSED BY 2 1X1 m SCREENS IN DIFFERENT SHAPED PONDS.

FIGURE 13.18
PARTITIONING PONDS.
Small partitions serve to hold year classes, predatory fish, or to isolate manure–rich water to a smaller anaerobic area for culture of fish foods.

FIGURE 13.19
4 PONDSOF SAME AREA,
but differing widely in their capacity to provide for edge plants such as blueberries, to feed fish from edge vegetation, and to irrigate nearby tree roots.

and B1, are for frogs, scuds, marsh plants, and peaty organisms. Ponds A2 and B2 are good bait fish ponds, some needing alkali, some acid, some muddy/humic water. Pond B3 can be our pan fish—or a *predator* species of that pan fish, e.g. A3 can be top minnows, B3 sunfish, and C largemouth bass, so that B3 and C eat A3, and C eats small B3. All these fish species breed in ponds and are carnivores. They (in effect) supply each other with food. All eat tadpoles, frogs, scuds, and water fleas at different life stages. All screens are one way. Even more simply, we could build a single pond and arrange screens as for **Figure 13. 21**.

Thus, we have several choices of configurations, and a pond series of different volumes, areas, aspects, orientations, perimeters, depths, nutrient states, and even handiness for servicing. **Figure 13.21** also gives us good orientation potential, and increases the perimeter of the whole pond. Our "screens" can be as simple as graded gravel or boulder mounds separating the pond into areas. The boulders themselves then become a complex edge and refuge.

When we consider pond margins, we have choices of weedy, woody, mown, or flowering plants. The life forms of woods, flowers, herbage, and lawn can fall in the ponds.

As well, we can *attract* insects in with light, colour, scent, or sound. Some aquatic invertebrate species may take up residence in the boulder screens. I have never seen a pond just like this, although I know of some natural ponds with some of each of these characteristics. But I feel as though we would learn a lot from planning and constructing a pond of this nature.

THE CLIMATIC ORIENTATION OF PONDS: COLD, HEAT, AND WIND

When we have the opportunity to orient ponds, as we do in marshes or flatlands, our concerns are to do so for the benefit of the water environment. The criteria are very like those governing house orientation.

Some conditions are:

- climate basically cold: oxygenation less important than heating (**Figure 13.22**).
- cold winter winds, warm to hot summers: oxygenation in summer, protected in winter (**Figure 13.23**).
- hot at most seasons: oxygenation a primary need, shade necessary for shallow pond (**Figure 13. 24**).
- variable (continental) climate: different needs in any of four seasons (**Figure 13.25**).

POND USES DETERMINED BY SITE

Huet (1975) and some others give a few sensible

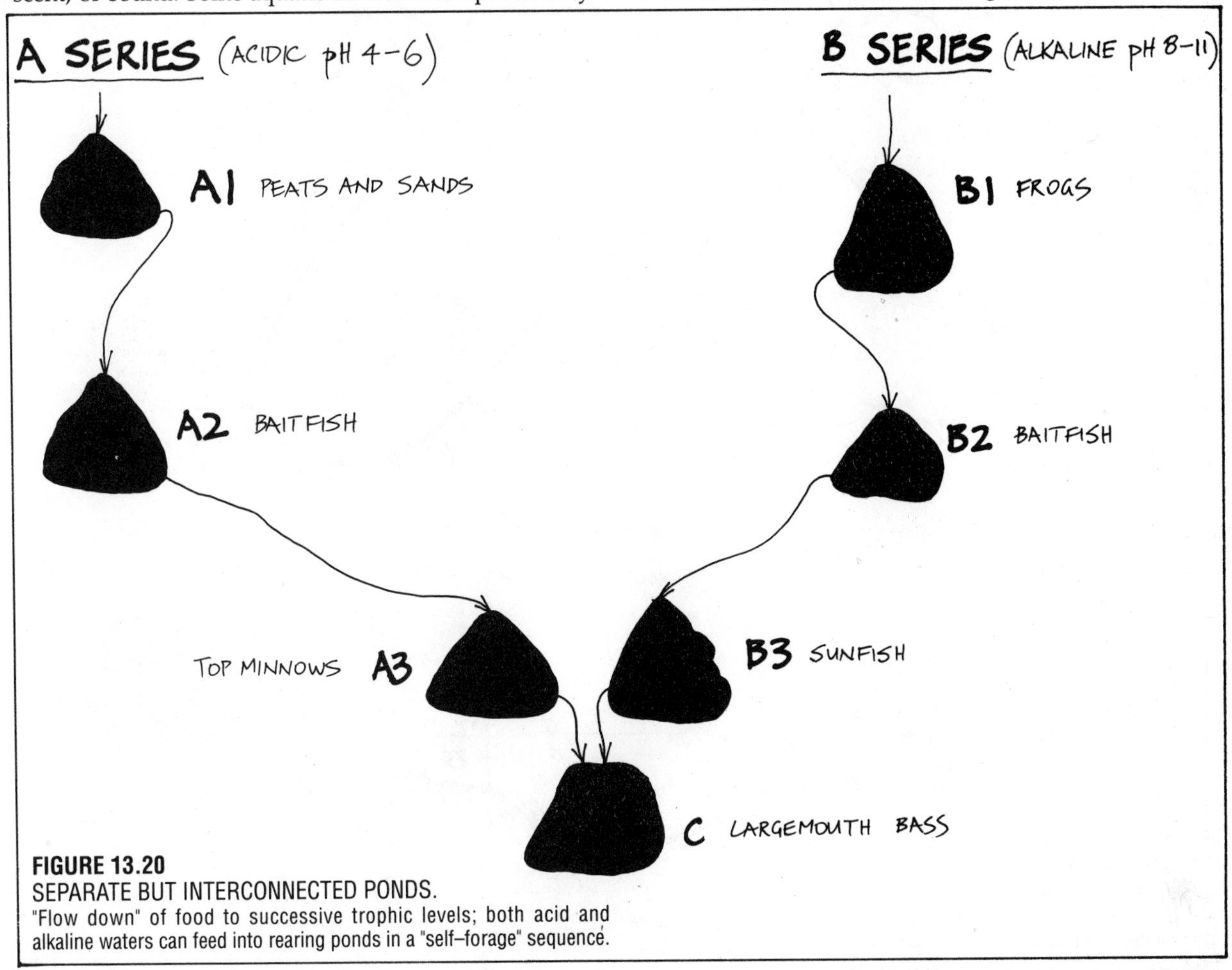

FIGURE 13.20
SEPARATE BUT INTERCONNECTED PONDS.
"Flow down" of food to successive trophic levels; both acid and alkaline waters can feed into rearing ponds in a "self–forage" sequence.

site–related pond arrays for valley sites. While it seems obvious that near–flat sites give greater freedom, these also lack the potential for vigorous through–flow and aeration provided by hill streams directed to valley ponds. It is the width of valleys or estuarine flood plains that may in the end determine a parallel or consecutive flow sequence from river or tide at head.

A second site restriction is that of soil type. Although artificially sealed ponds can be established in any location, stable clay and clay–loam soils are needed for cheap extensive pond systems. Clay is expensive to dig and transport. There are sites obviously subject to soil slump where no ponds at all should be established, as water from even small refuges lubricates the shear planes of soils in slip–off areas, and can trigger earth or mud slides. District inspection, a soil survey, and some simple tests at a soil laboratory will reveal such delicate sites; some are specifically mapped as soil types, or as unstable slopes on land–use plans.

Resevoirs of great size have very similar effects on

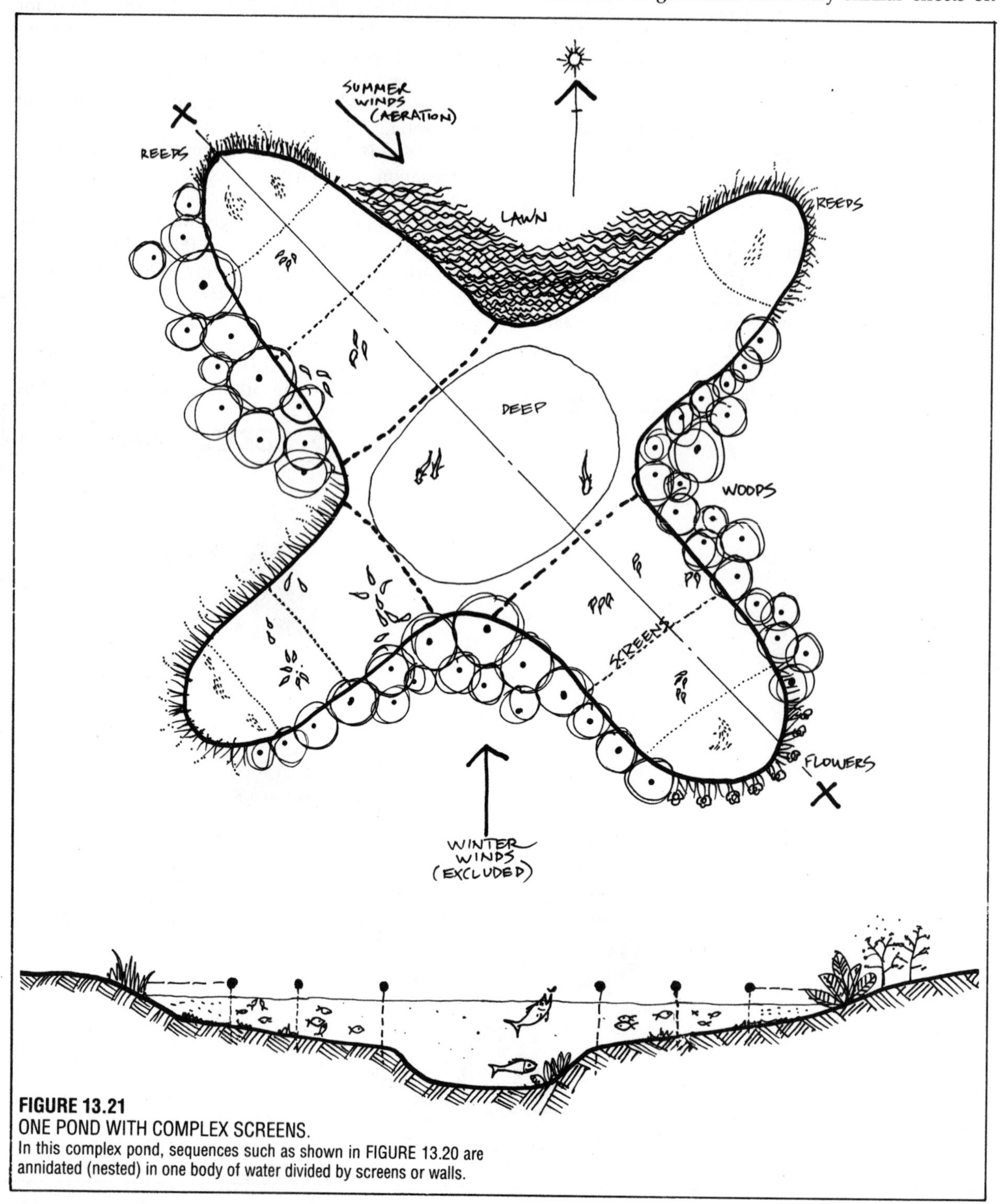

FIGURE 13.21
ONE POND WITH COMPLEX SCREENS.
In this complex pond, sequences such as shown in FIGURE 13.20 are annidated (nested) in one body of water divided by screens or walls.

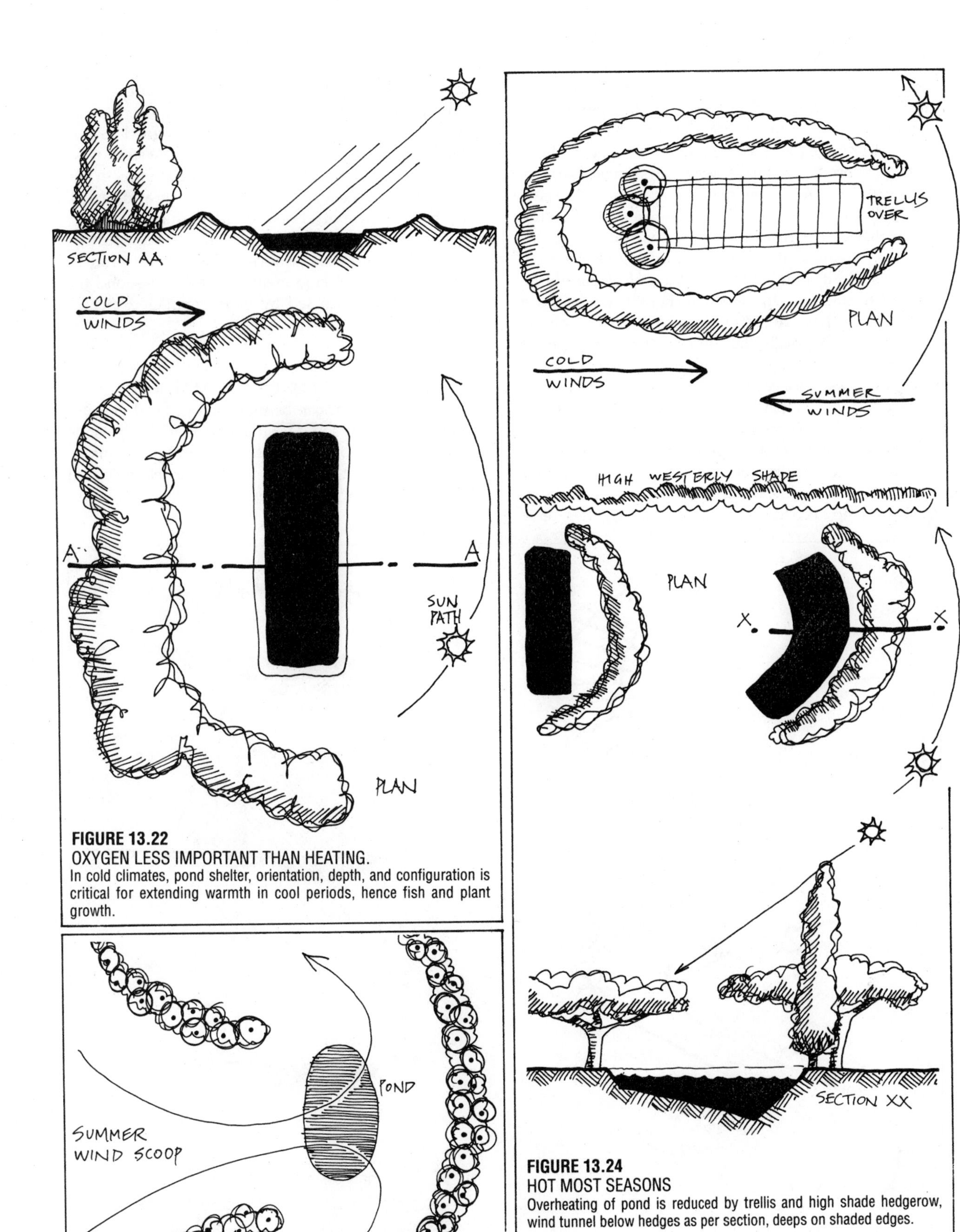

FIGURE 13.22
OXYGEN LESS IMPORTANT THAN HEATING.
In cold climates, pond shelter, orientation, depth, and configuration is critical for extending warmth in cool periods, hence fish and plant growth.

FIGURE 13.24
HOT MOST SEASONS
Overheating of pond is reduced by trellis and high shade hedgerow, wind tunnel below hedges as per section, deeps on shaded edges.

FIGURE 13.23
COLD WINTER WINDS, OXYGEN IN SUMMER.

incipient earthquakes, and water seeping through rock cleavages may either stabilise or destabilise faults, with potentially beneficial or catastrophic effects. For the building of large dams above human settlements, busy roads, or populated valleys, we modest pond–workers need make way for the more heroic aptitude of the civil hydraulics engineer, and even then with a lingering doubt that their structures will, in the end, persist, or that the potential catastrophe will merely be deferred. As a safe limit, ponds in restricted valleys should be limited to one or two metres in height, while those above broad flats can safely disperse a greater volume of water. Fish ponds, however, rarely cause civil catastrophes, and there is almost always good advisory or regulatory services available.

Rarely, we can find a property with a natural constriction between hills that enables us to flood 50–200 ha with one small dam wall. Our capital costs in such a situation can be very small in terms of the total production potential, and a little accessory earth–moving to create peripheral swamps, jetties, or to create islands is all that is needed apart from the small retaining dam and spillway.

Aquaculture must be as seriously designed as any other important production system. If a system is so planned, yields should exceed terrestrial crop in the same region by factors of 10–20 (more in arid areas), and there is no better use of land than as pond–and–forest systems. Of all endeavours, aquaculture (and its polyculture accessories) show greatest promise for the reduction of land areas in present use, and the repair of damage caused by badly–managed pastoralism and monocrop systems.

THE FURNITURE OF PONDS AND MARSHES

Any wetland habitat can be increased in yield and use by the addition of some basic facilities which provide special habitat. Some of these are configurations or earth structures, others are constructs or technological additions. They cover such areas as:

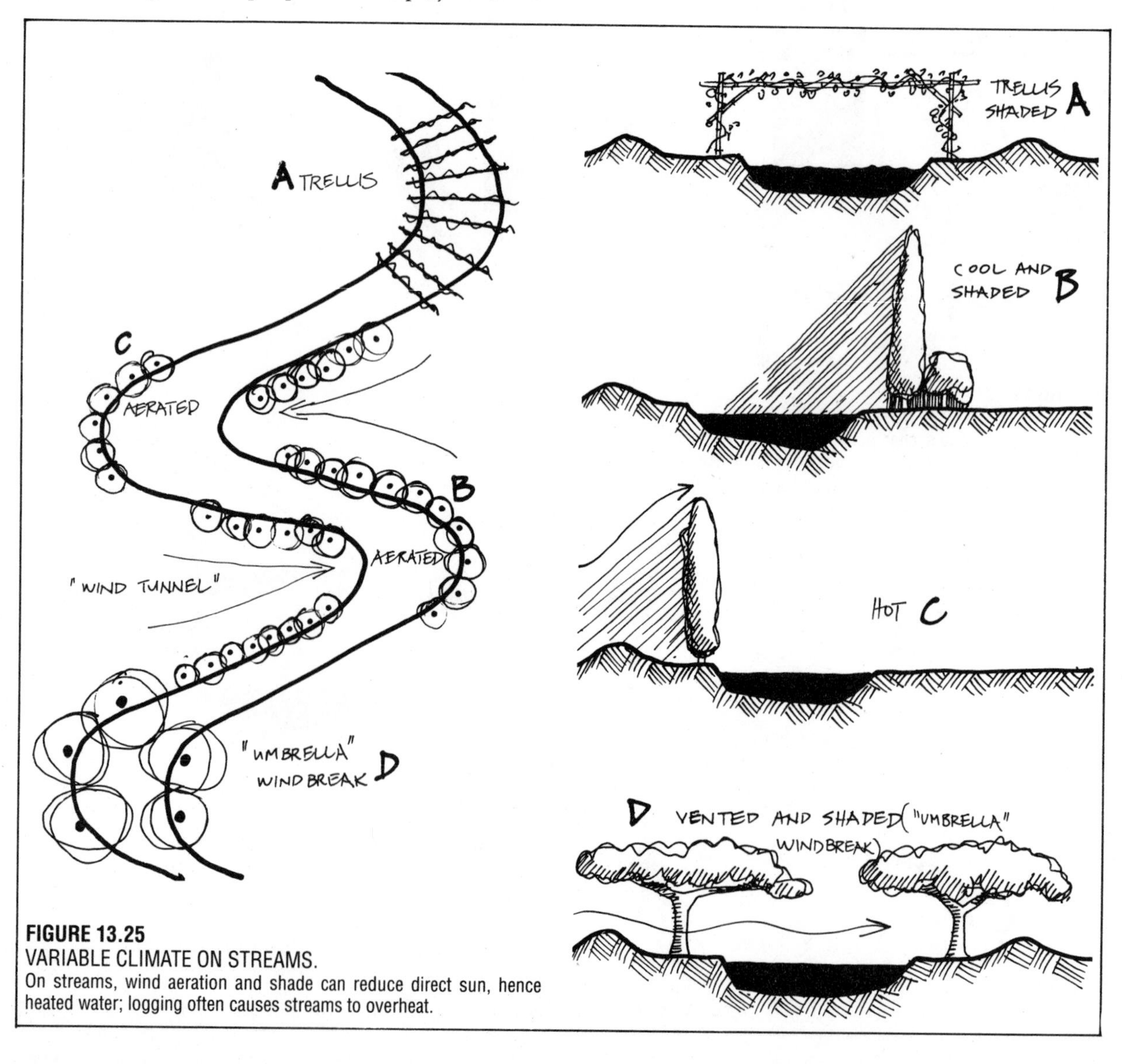

FIGURE 13.25
VARIABLE CLIMATE ON STREAMS.
On streams, wind aeration and shade can reduce direct sun, hence heated water; logging often causes streams to overheat.

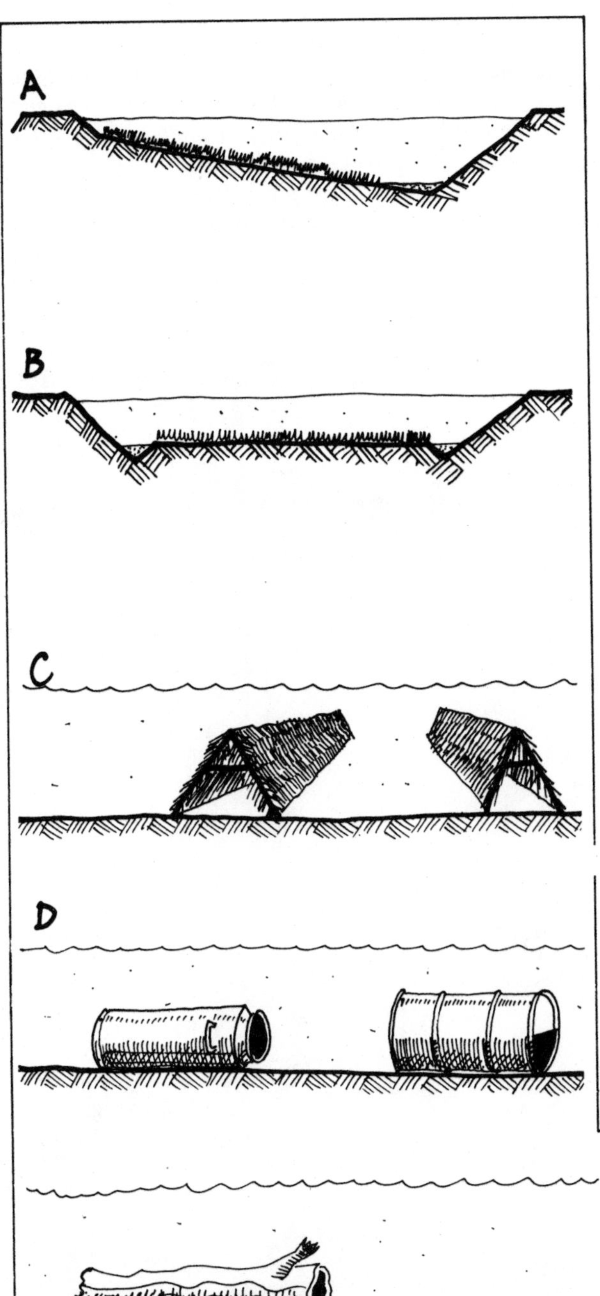

A. Configuration

• *Islands and Hummocks*: Although the construction of quite small islands are excellent wildfowl habitat (quickly occupied by nesting birds), islands have other uses, such as the isolation of useful but invasive plants (runner bamboos), and as a strategy to increase edge for fish. Islands also create sheltered bays in windy areas, or can streamline the winds to better oxygenate water.

Swan and other hummock–nesters may be limited by available (defended) nest sites, and can expand their numbers with small hummocks in shallows. Many territorial waterfowl find these of use for night–roosts, and a good many useful tree species are hummock–dependant. Alligators are the natural hummock architects of Florida swamps.

• *Peninsulas* (islands with narrow causeways) are safe house sites in areas of high fire frequency. They also elaborate the edge effect, increasing the area of shoreline for plants and fishermen.

• *Deeps*. The fish species of shallow waters and marshes may be decided in their composition by the number of deeper refuges in times of extreme cold or heat. Several species occupy such kettles in both tropical and temperate or cold lakes. **Figure 13.33**, in fact, is a series of continuous or extensive deeps in marsh, which is probably the highest–production water of any natural system.

With several carp, galaxiid, catfish, and perch species, deeps flooding out over mud and grass spawning beds are essential to their natural breeding cycle, and we can arrange a "flood" cycle by water regulation, to induce spawning. Many waterfowl also respond to this stimulus.

B. Structures

• *Breeding Substrates*. Depending on species, we can place a series of substrates on which fish will deposit

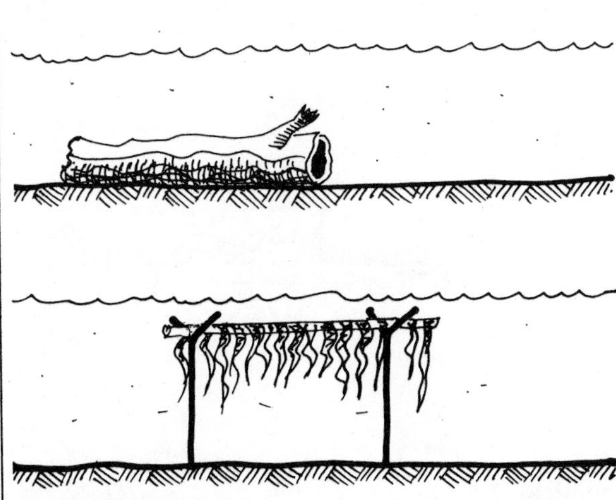

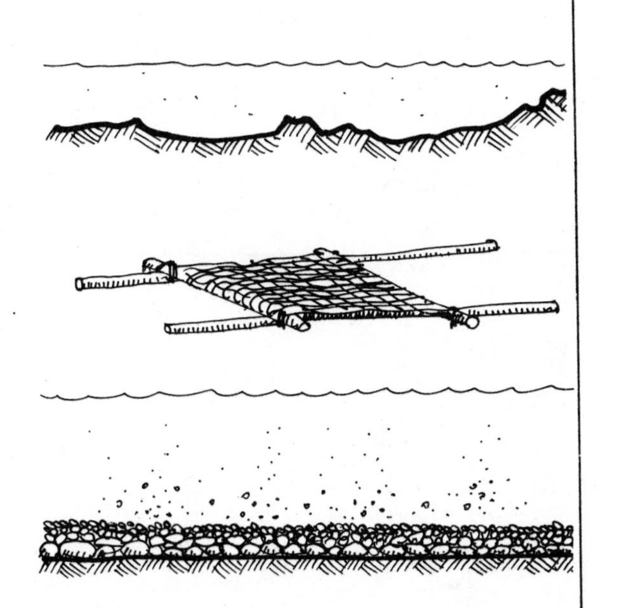

FIGURE 13.26
BREEDING SUBSTRATE.
A. Grassy slopes flooded for carp; **B**. alternative flood system; **C**. thatched shelters for "cave breeders"; **D**. logs or drums for large perch caves (Murray River cod). Gravels, sands, rock piles, mud caves, floating weed, and bundles of reeds or twigs provide other egg sites.

eggs. Some of these are figured (**Figure 13.26**). Others are in the form of earth material, gravels, subsurface aerators, and hollow breeding refuges such as logs, pipes, milk churns or drums and tanks. Many cannabalistic crayfish and territorial fish species defend such homes, and their population density depends on these refuges. Tyre heaps or piles of broken pipe provide condominiums for such species.

For small species such as shrimp, snail, notonectids, and some small fish, bundles of brush perform two functions: that of a breeding substrate, and as a refuge from larger predators. All such refuges can be arranged to be hauled out, and then they operate as "traps" for the species (octopus, eels, and crayfish stay in their holes or in old tyres; shrimp and freshwater crab cling to brush piles), or to collect their eggs and fry.

- *Rafts*: Rafts serve as floating docks in tidal waters, as supports for houseboats or pumps, as walkways to

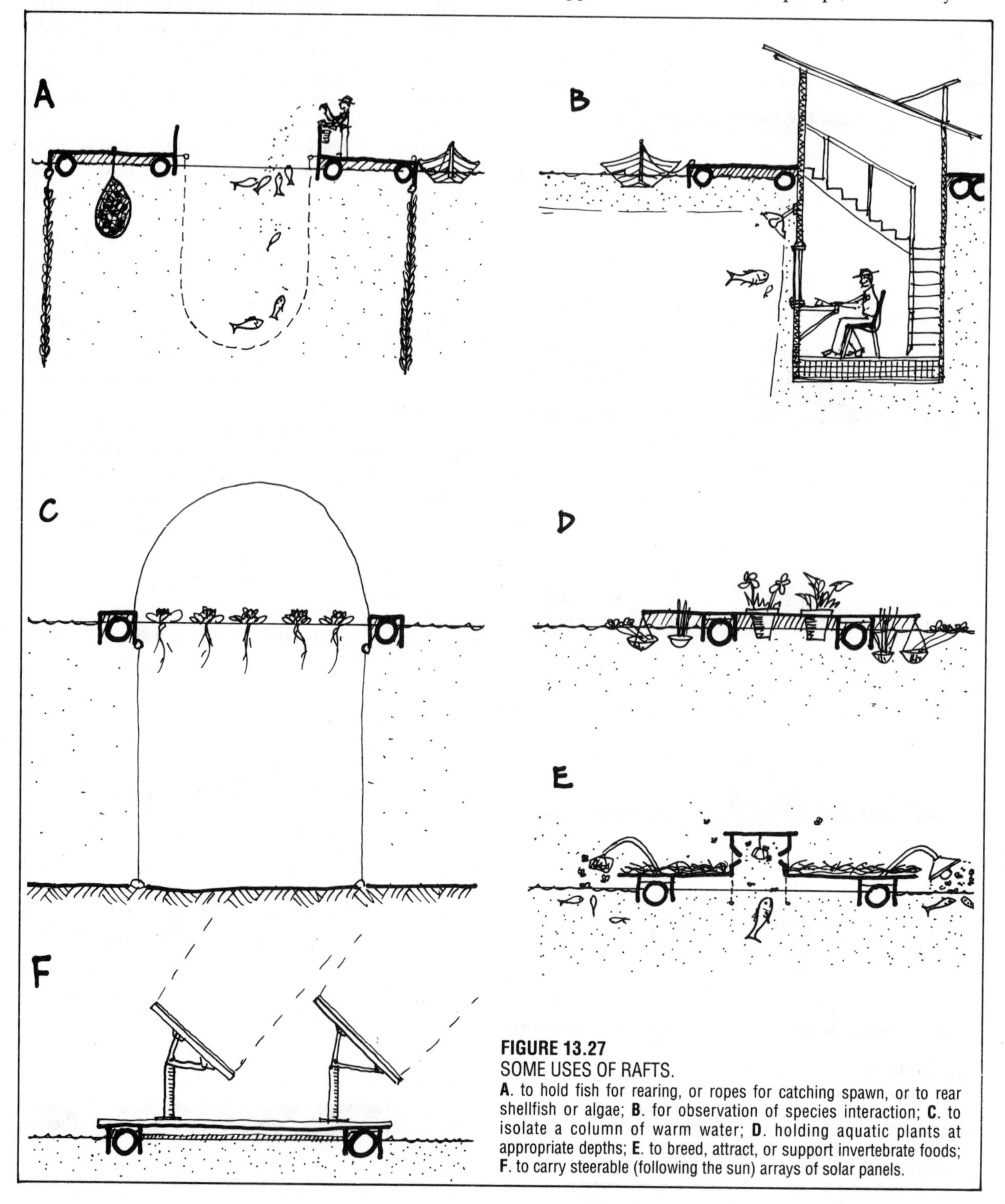

FIGURE 13.27
SOME USES OF RAFTS.
A. to hold fish for rearing, or ropes for catching spawn, or to rear shellfish or algae; **B**. for observation of species interaction; **C**. to isolate a column of warm water; **D**. holding aquatic plants at appropriate depths; **E**. to breed, attract, or support invertebrate foods; **F**. to carry steerable (following the sun) arrays of solar panels.

fish cages and ring–nets, as surface floats for organisms cultured on ropes, brush, and in mesh bags, and as observation platforms. Some of these uses are figured (**Figure 13.27**). With fluctuating surface levels induced by tides or the draw–down of dam storages, only a raft arrangement can cope with the steady level of water needed by certain water plants and nesting birds.

Cultures *on* rafts range from light and lure traps for insect foods to insect incubators of leaf litter or animal wastes. In many cases, multiple uses of rafts are feasible. Rudolf Doernach, a German architect, has actually built raft houses in cistern ponds, and was in this way able to "follow the sun with the house". Heavy arrays of solar cells and solar collectors are most economically oriented to the sun on raft structures of this type. Rafts also hold self–feeders for fish and waterfowl.

- *Screens and fences.* Shallow–water fences and screens are useful in polycultural stocking to keep predators from cultures of forage fish, or to allow stunted fish to be culled by predators. Outlet and inlet screens either prevent or regulate the migration of species (**Figure 13.28**).

Screens (**Figure 13.29**) can be horizontal, sloping, vertical, or as cylinders and cones. The configuration enables, in each case, some degree of self–clearing or deflection of solid particles. Rotating drum screens can be made to be entirely self–cleaning and self–turning, providing a small head of water is available. Drum screens are particularly useful for skimming ponds or collecting floating plants and algae for use as manures or forages.

Fences are a cheap way to screen shallow areas, or to separate the area between deeps. Again, they separate predator–prey or cage populations of carnivorous fish. For a small grower, they enable brood fish to be kept in the same ponds as immatures, and two antagonistic species to be reared in a pond. Screen fences beside deeps in marshes prevent fish escaping from them (extending predation to shallow waters), or permit frogs to breed in shallows without excessive predation

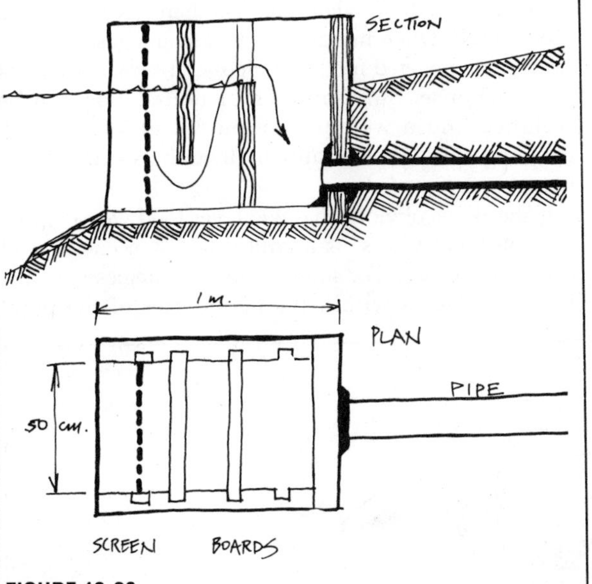

FIGURE 13.28
SCREENS AND BOARDS.
Classical "Herguth monk" outlet incorporates a screen and a level control board.

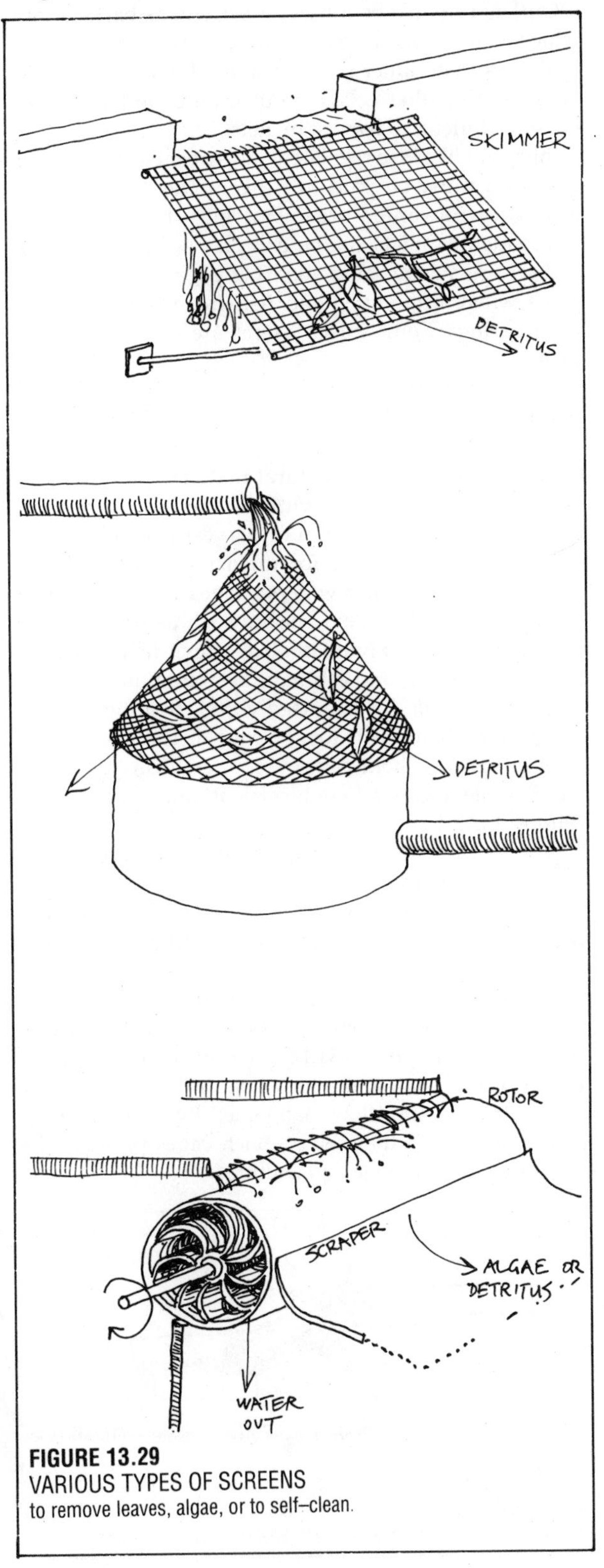

FIGURE 13.29
VARIOUS TYPES OF SCREENS
to remove leaves, algae, or to self–clean.

from trout.

• *Outlets and Inlets*. For ponds, the types of water level controls are a critical factor. Modern production of reliable and flexible pipe has made very simple level control possible, either as an elbow or upturned flexible pipe.

Outlets can also be a harvest system, as baskets into which water falls, or as smaller ponds with screened spillways that gather fish migrating downstream. Both are used to gather eels or trout from complex swamp systems difficult to harvest by nets.

Inlets are likewise regulated and screened, but greater attention must be paid to prevent the entry of silt, weeds, or unwanted organisms into the pond itself. Consequently, inlets can be complex systems of filters and screens where water quality is poor, or very simple pipes where unpolluted and fish-free water is drawn from springs. Some inlet systems are given in **Figure 13.30**.

CAGE CULTURE

Cages of wood, woven natural materials, metal mesh (and nowadays modern synthetic meshes) have been used since antiquity to trap or harvest fish, and are in current widespread use for intensive fish rearing in both still and flowing waters. Cages can be used to protect eggs and fry, and species such as sturgeon are hatched in cages (**Figure 13.31.A**). Live fish, crayfish, prawns, molluscs, and eels have been traditionally held in cages or *caufs* (pronounced corfs), floating barges, and wet wells in boats for the fresh fish market. They are also used to hold eels, oysters, crayfish, carp, and weed or algae eaters to reduce algal taints in fish flesh (**Figure 13.31.C**).

Where flow is rapid, cage mesh large, or wave motion exists, oxygenation in cages is no great problem. There is some advantage, however, in shaping cages subject to rapid tidal flow, or to induce water circulation in the cage, as in **Figure 13.31.B**.

Normally, however, cages are circular, as in salmon ring-net culture, or plain square to suit slatted wood construction (**Figure 13.31.D**). Typical rearing cages in which fish are fed, and which float in larger bodies of water, have a water flow maintained by the swimming action of the fish themselves. Such cages produce the largest yields known to aquaculture.

Mooring cages can be effected by individual anchors, as sets of cages attached to floating docks or walkways, or as gangs of cages fastened to lines, and (allowing a boat-width between cage pairs), the gangs can be stretched across bays, or anchored to float-lines in open water not subject to violent wave action (**Figure 13.31.E**). Old tyres serve well as spacers between cage units. I used caufs for years in sheltered bays to hold shellfish and net fish for market, and rarely suffered losses, but all cages need an annual inspection and watchful maintenance. Predators such as octopus and seals can cause large mortalities in cage fish, whereas pond fish can avoid such losses by evasion. On rare occasions, large shark attack caged fish and destroy cages, a loss not experienced in pond cultures!

NUTRITION AND FOOD SUPPLY

The same care must be paid to the nutrition of fish and shellfish as to that of land species. Vitamin supply from fresh greens, fruits, and algae are essential. High-protein foods, with a modest admixture of starchy foods (and all of these in fresh condition) is a requisite for healthy growth, as fertiliser is for plants. There is no substitute for fresh live food to produce high quality fish. Some strategies for better fish nutrition are given here; they are *the* critical strategies for cost savings (or energy conservation).

Fish maintain growth by eating about 1% of their body-weight daily, and gain most weight (at the least waste of food) when fed at 3% of body-weight daily. Again, trials of feeding are needed for specific species and size groups, but it is a good rule of thumb to feed out as much food as will be completely eaten in 15 minutes, to check on food wasted, and to establish a growth curve from samples taken as the fish grow.

Demand feeding, where the fish themselves trigger a food supply when hungry has its advantages, but may necessitate pelleted foods purchased at some 60–70% of total expense. Sub-samples of netted fish can be weighed and a weighed ration fed at 3% of total, allowing a 10–12% mortality in fingerlings, and 40–60% in fry.

It should be obvious that whatever food we can grow or collect as "wastes" is a critical factor in energy and cost conservation. For some of these strategies we must look to systems within the boundaries of the pond

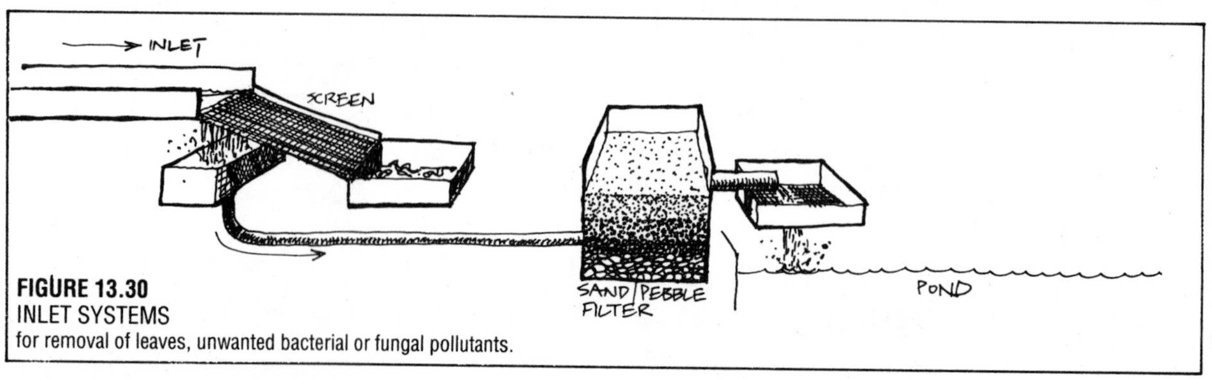

FIGURE 13.30
INLET SYSTEMS
for removal of leaves, unwanted bacterial or fungal pollutants.

(shrimp, minnows, algae) but for others we must closely design the pond margin and create accessory food systems outside the pond itself. While this section completes a summary of the factors that we can manipulate to increase yields, we will follow with a set of food provision strategies, which I believe will have a profound effect on yields, although it is not a factor identified by others unless with respect to breeding or pond management for harvest.

13.6
FARMING INVERTEBRATES FOR FISH FOOD

Dried insects and other invertebrates are high in protein value, thus forming a very important fish food, e.g. if we take the "food quotient" (F.Q.) formula as given in Augusthy (1979):

F.Q. = wt. of food given (e.g. 200 kg of minced fish waste) divided by the wt. of fish gained (e.g. 100 kg of fish gain)

Then: F.Q. = 2 for fish waste in this example.

In these terms, insect larvae are 1.8, compared with guinea grass at 48. Obviously, we do well to encourage

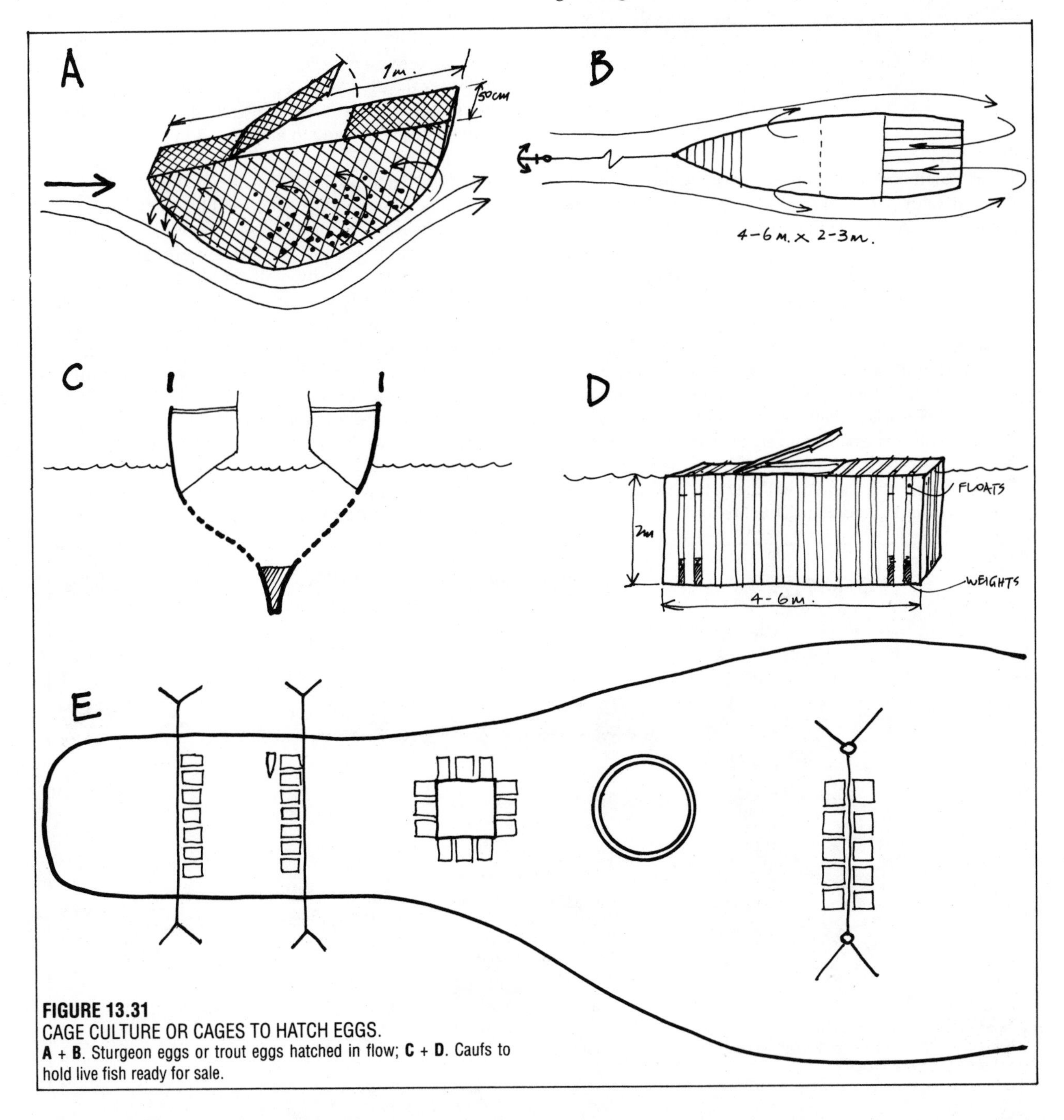

FIGURE 13.31
CAGE CULTURE OR CAGES TO HATCH EGGS.
A + **B**. Sturgeon eggs or trout eggs hatched in flow; **C** + **D**. Caufs to hold live fish ready for sale.

insects. There are several ways to do this, and some are:

Cockroaches, mealworms, and sowbugs:
Scatter food waste or flour, cover with leaves, and "seed" with cockroaches or sowbugs. Add to this pile some leaves and starches from time to time. Millipedes and cockroaches build up in tropical areas, and can be used to feed ducks or fish. Dano Gorsich on Moloka'i Hawaii, has a successful cockroach mulch heap of *Hibiscus* leaves which produces cockroaches for his ducks; if the pile is half–turned every few days, cockroaches are taken by the ducks. The duck manure then stimulates plankton growth in ponds.

Similarly, a "sandwich" mound of boards, paper, leaves, and so on breeds sowbugs (woodlice) and houses earwigs. These can be sieved and shaken out, or the mound demolished and rebuilt with ducks or chickens present. *Zostera* (eel grass) is a good sowbug base.

Termites:
A perforated 200 l drum or loose brick pit, covered can be filled with paper, old wood, cardboard, and straw, and then watered. Termites will invade if they are in the area, or sowbugs can be seeded in cool areas. The pit is periodically dug or sieved out for insects (**Figure 13.32**).

Plague locusts:
Up to the 4th or 5th instar, these insects form ground swarms (flightless), and can be vacuumed, brushed up, or "trawled" in grasslands using a side–towed bar and net. Frozen or dried, they are ideal protein food (people in many cultures eat the singed or dried bodies). They are largely overlooked as a high protein fish food resource. Fermented and dried they can be dry–stored.

Standard light–and–fan floats are available to attract night–flying insects and blow them down to the water surface. Yellow floats attract grasshoppers to ponds, where many fall short of their goal and are consumed.

Pasture grubs:
These are the larvae of *Aphodius, Phyllophaga* and other beetles or moths. They occur at high densities (10 t/ha is not uncommon) in the top 3 cm of soil. Strips of field can be skimmed, sieved, and the grubs "floated" out of the dust with salt solution, dried, or frozen. At an F.Q. of 2–3, every hectare should grow 2–3 t of trout or other high value fish! Also, the adult beetles will cluster around bright lights and can be trapped in funnel–and–drum systems, and also frozen or dried. This is a good way to raise turkeys, when a daily disc–plough line can turn up 100 kg or so of grubs: all this on pasture carrying less than 0.5 t of sheep/ha.

In general, any dense collection of insects can be cultivated, harvested, and converted to fish food; food wastes can be converted to insect food in many cases.

Snails for fish and duck food:
A well–limed area planted to arum lily, *Nasturtium*, root sets of horseradish, *Brassica* seed, broad beans, and cucurbit vines if "seeded" with a few buckets of snails and watered occasionally, will build up a dense snail population over a few years. These can be "fed off" to ducks or gathered and minced for fish food as needed. Every 10 or so square metres, a clump of arum or *Agapanthus* will form permanent snail harbours; *Nasturtium* and horseradish provide food, as do cucurbits and annual *Brassica* species or globe artichokes. Snails can reach high densities under these conditions.

The large tropical snail *Achatina* likes a mixture of papaya, over a ground cover of cucurbits, *Nasturtium*,

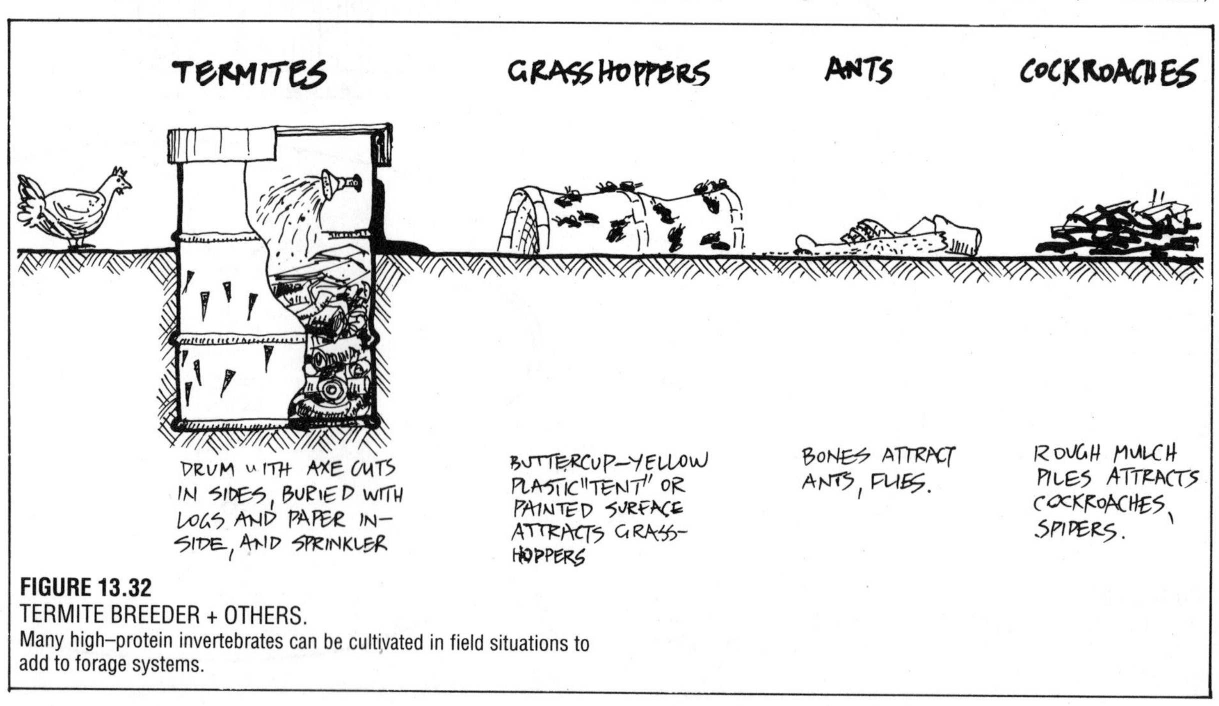

FIGURE 13.32
TERMITE BREEDER + OTHERS.
Many high–protein invertebrates can be cultivated in field situations to add to forage systems.

and fleshy or mucilaginous harbours. Desert snails can be collected from post tops in swales or grasses, where they gather to avoid soil heat.

Zooplankton:
Water fleas, cyclops, ostracods, rotifers, and so on can be cultured in small ponds or tanks supplied with lettuce, potato slices, crushed sugar cane, manioc (cassava) or legume leaves. A shallow bay off the fry ponds can be screened off for this purpose, and the plankton will swim out into the fry ponds. Conditions in the enriched area may not suit fish, but produce ample food.

Midge larvae and tubifex worm cultures in rich, shallow, organic ponds supply essential fry food, as do the brine shrimps (*Artemia*) of salt pans. *Artemia* culture is one of the very few productive uses of saline inland ponds.

Larval flies:
Carrion flies will "blow" waste meats or carcasses suspended over ponds, and near–putrid shallows supplied with kitchen sink water will breed "gentles" (larvae of *Tubifera tenax* flies) in the muddy base (depth of water 1–2 cm).

It is in the development of such high–protein foods as accessory to fish ponds that we save the greatest continuing cost of fish culture—food. In our site planning such areas are as important as the ponds themselves. For herbivorous species, semi–rampageous plants such as *Nasturtium, Tradescantia, Dolichos*, and comfrey supply hardy and palatable foods to fish such as *Tilapia*. In channel cultures, the banks themselves (when planted to such species) are a complete food supply.

Aquatic molluscs:
Species of the genera *Physa, Limnaea, Bythnia, Vivipara, Pisidium, Sphaerium* and so on occupy most alkaline waters. They form a large part of the food of fish, and are easily cultivated on vegetation in organic ooze in shallows and on stream banks, or in pools.

Worms:
Small or large–scale worm beds are invaluable sources of food for fish; worms are collected by flooding the beds at intervals. With a source of hay, food scraps, or manure, worm–growing can be a major fish food producer. The hessian (burlap) cover of worm beds can be immersed in or suspended over ponds after flooding the beds.

FODDER POND SEQUENCES

As each small pond falls through a pipe to the next, upstream migration is prevented while downstream migration is possible or even aided. Ponds I have built were organised as per **Figure 13.33**.

The criteria for upstream species are simple. In plants, it is that they be non–invasive, and in animals, low on the trophic ladder. If such ponds are arranged along spoon or V–drains, several origins and destinations can be achieved, with forage fish or invertebrates migrating always downstream, and even then a perched pond above the trout or predator system can make a trout–free forage–fish polyculture.

From the intake: Ducks add manure; shrimp eat algae produced by the manure. Some shrimp larvae escape to the next pond where a small fish breeds (*Gambusia* for example). These fall again to trout or perch in the last (outlet) pond. Snails can be part of this downflow *if* a separate intake is arranged. Species suitable to each stage are:

A. ORIGIN:
Animals. Manurial species such as ducks, freshwater mussels, amphipods and phreatocids (mud shrimp), small freshwater crabs (*Halicarcinus*), snails, shrimp (*Atya, Macrobrachium*), frog larvae (*Hyla, Rana, Crinea*).

Plants. Non–flowering or non–invasive species such as taro, and manurial species such as *Azolla*. Insects and their larvae will also be represented, like it or not.

Structures. Rotted logs, reed beds, brush and small cover. Taro or other useful crop can also be planted in any pond downstream.

Edges. Comfrey, vining legumes, fruit.

B. NEXT POND DOWN:
Animals. Any of the above plus more predatory invertebrates and very small fish working at planktonic level, e.g. surface–feeding fish such as minnows, *Paragalaxids, Saxilaga* in the mud.

Plants. Useful edible species (kangkong) on mounds.

Structures. Reed bed, small pipes and logs.

Edges. Mulberry, berry fruit, legumes, comfrey.

C and D follow much the same sequences as given in B. Products or yields can be taken off at any level as shrimp, snail, ducks, frogs, taro, *Tilapia*, perch, or trout. Even pH can be altered in some chains to allow different species to enter the chain, and niches arranged for special plant or animal groups, so that high oxygen and low oxygen demand species are accommodated. Such systems can accept water polluted by phosphates and nitrates as part of their intake, providing plants and organisms can be found to cope with that level of pollutant (**Figure 13.33**).

There are at least three strategies to increase the diversity of the waters; all apply to relatively small still–ponds, marshes, or perched ponds. We can:

• Locate ponds at headwater and ridge locations, thus creating small ecological islands;

• Salt small ponds for the development of semi–estuarine species; and

• Manure small ponds and marshes with trace elements, animal manures, and phosphatic or nitrogenous fertilisers in order to produce large quantities of forage fish, algae, or crustaceans which will feed trout.

Swingle (1966) proposes that up to 50% of a catfish pond can be in shallows; these are to provide food for the main fish (as shrimp), not at a cost of reducing fish

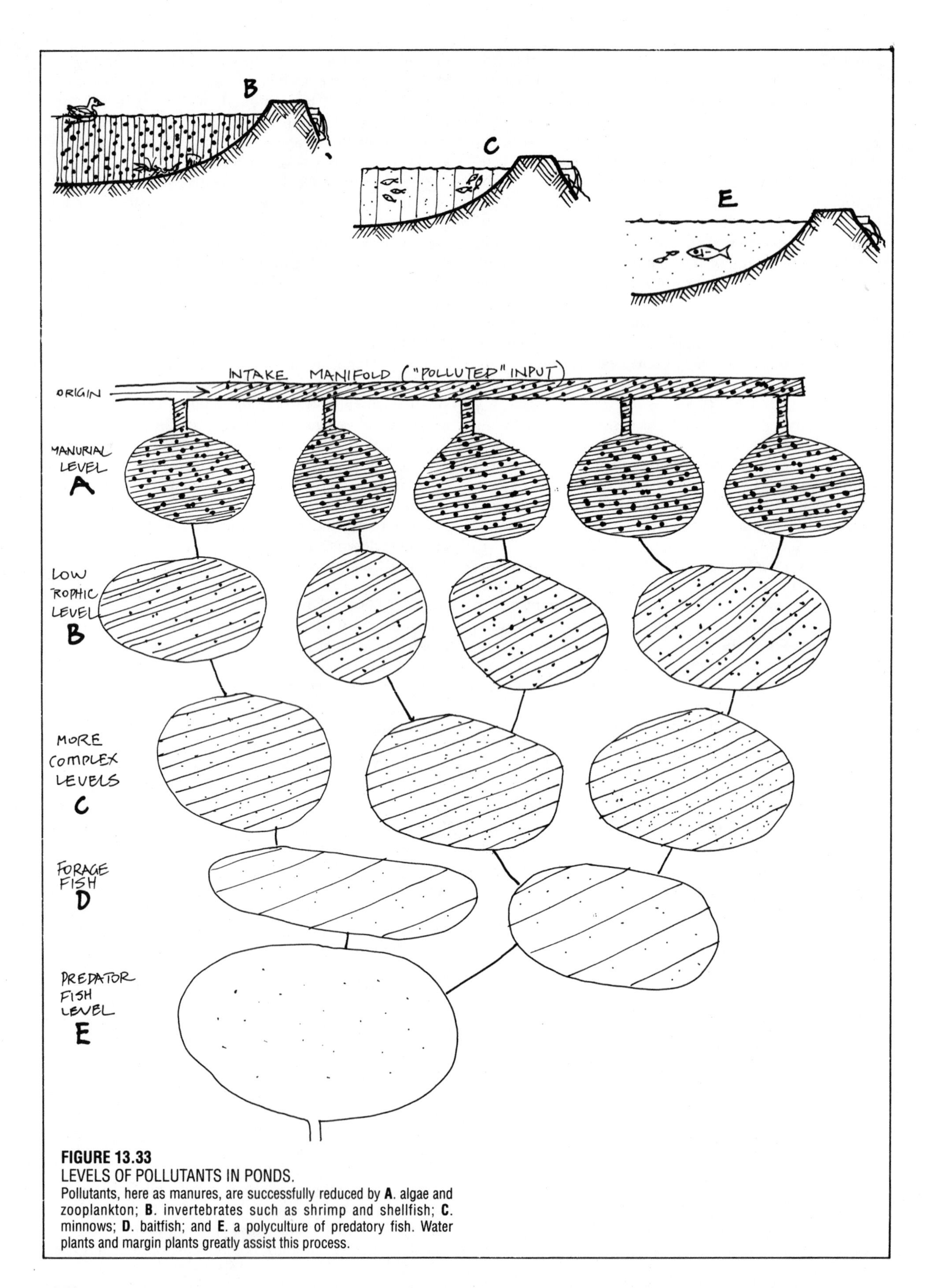

FIGURE 13.33
LEVELS OF POLLUTANTS IN PONDS.
Pollutants, here as manures, are successfully reduced by **A**. algae and zooplankton; **B**. invertebrates such as shrimp and shellfish; **C**. minnows; **D**. baitfish; and **E**. a polyculture of predatory fish. Water plants and margin plants greatly assist this process.

TABLE 13.3
ORDERS OF YIELDS GIVEN SPECIFIED CONDITIONS.

kg/ha/year protein	CONDITIONS
10 – 50	Deep, cold; rocks around lakes with few shallows.
30 – 60	Deep artificial resevoirs
60 – 90	Artificial resevoirs modified for fish and forage culture.
80 – 150	Natural shallow lakes, unfertilized, e.g. glacial outwash areas, natural shrimp yields in coastal lagoons of low pH.
200 – 500	Basic "standing crop" of fish in fertile (but not fed) fish ponds, e.g. bass, trout, bluegill; extensive fertilized waters of one to three species.
500 – 1,000	Unfertilized brackish lagoon cultures of milkfish; modified rice padi crop of carp or *richoseras*; central padi "plateau" cropped, and plants cut for water manure. This is the upper range of totally natural systems.
1,000 – 2,000	Fertilized, extensive carp, milkfish, and mullet cultures, including a dry cycle. About the limit of extensive systems. *Puntius* in padi culture; intensive prawn culture. Fertilizer now 25–50% of costs. Well–chosen and unfed ploycultures can also achieve these yields.
2,000 – 5,000	Intensively fertilized and fed pond polycultures of selected species; water quality monitored. Some feeding or crop residues supplied. Optimum conditions. Food is 60+% of total costs.
5,000 – 20,000	Intensive cage and cauf cultures or small aerated ponds of carp, catfish, tilapia; fertilizer and food now 96% of total costs, and waste products may limit production; disease control is critical. land cycle of wastes essential, to reduce nitrates.
20,000 – 150,000+	Cage culture in oxygenated food–rich streams or with accessory foods. Vigorous flow or waste removal essential and achieved by channel flow. Hardy and disease–free fish stocks essential.

numbers, but at a gain in energy needed to feed them.

Fish such as *Tilapia* and carp are commonly fed on starchy foods from adjacent crop (pumpkin, banana, sweet potato, yam, taro, beans and bean flours, grains and doughs), but also eagerly eat the fallen insects, seed, and fruit from fringing vegetation, along with selected water weeds.

Tilapia, in particular, eat many fruits and edible leaves from garden weeds and vines. For this reason, market gardens and fish ponds belong together, but as fish–pond water is also of good nutrient value to gardens, the relationship is enhanced. Crop production from well–fertilised ponds may be as much as twice that from resevoir irrigation.

Grass carp neatly trim fringing vegetation such as *Dolichos* species, and save encroachment of weeds into taro, while not only mixing quite well with *Macrobrachium* prawns but actually increasing prawn yields without feeding them artificial food. Like *Tilapia,* they appreciate garden waste, plants, and fruits. Barry Costa–Pierce (pers. comm.) has a variety of polycultures under test in Hawaii, and the ponds in which he keeps grass carp with prawns have neatly–trimmed edges (bitten by the fish, not lawn mowers), while the prawns grow as well on grass carp fecal detritus as they do on chicken pellets in nearby monoculture ponds. Even if they grow less, or if we increased pond margins as an edge effect, or planted comfrey and clover, *Dolichos, Tradescantia,* and lucerne along the pond edges for fodder, it is preferable to create the conditions for yield at the pond than to import them from elsewhere at great cost in energy.

13.7
CHANNEL, CANAL, AND CHINAMPA

Next to cages (and sometimes integrated with them) channels of 0.5–2.0 m deep are widely used in fish culture; they are the only economic way to develop "ponds" on slopes of more than 8° unless we develop water terraces. Channels maximise edge effects, and natural foods can be substantially more available in channel culture. The chinampa is probably *the* most efficient culture configuration for natural feeding of fish, and many rice padis have now been modified for this effect.

"The chinampas... of the Valley of Mexico... date back

more than 2000 years and were the main source of foodstuff for the inhabitants of the entire valley, producing as many as seven different crops in a year, two of which were maize." (Tompkins, P. 1976, *Mysteries of the Mexican Pyramids*, Harper and Row.) Properly maintained, chinampas could remain fertile for centuries without having to lie fallow. Rafts of water vegetation were cut from the surface of the canals and towed to mounded banks where they were built up in layers and covered with rich mud scooped up from the canal bottom.

I have extended the use of the word chinampa to include any system in which a sequence of canals and banks in approximate parallels are developed for growing fish and marginal plants. Given a body of standing water such as a lake or swamp, or a humid landscape with a clay base that will retain water all year, or a water table close to the land surface, it is possible to create a cross–section harmonic of land and water whose uses are bounded only by the limits set by climate, the imagination, and the harvest capacity of the designer. Chinampa systems combine the best of both worlds in soil and water culture, and are in use in deltaic regions of Thailand to grow fish and truck crop, ducks and fruit trees (**Figure 13.34**).

There is one other strategic benefit of chinampa systems, in that useful but potentially rampant species such as runner bamboo, vine blackberry, hops, horse-radish and like crops can be water–isolated from other land systems. Small moated islands have the same facility, and waterfowl can nest or rest on these without interference from foxes and feral cats.

We can cheaply create chinampa swamps with a few compacted retaining walls where water levels are regulated to back up over chinampa systems. The ratio of channel to dryland culture is normally about 1:1–3, but if we reverse this ratio, herbivorous fish, plankton eaters, and crayfish are self–foraging.

Crayfish and grass carp in channels 2 m wide, ranging over a swamp strip of 5–10 m wide have a rich forage supply, *plus* land edges. The terrace or swamp can be drained, and in 3–4 weeks harvested or cropped, and the vegetation of terrace and shelf act as fish food and manure. With the right selection of species (ducks, mussels, weed–eating fish, crayfish, or eels) such systems give yields in excess of 1000 kg/ha of water surface, as many food organisms and plants are intimately available to the canal fish.

On slopes, Huet (1964) reports good yields from trout canals in Switzerland (fed or unfed) at 40–60% greater than broad–pond culture. For hill canals, a reliable water intake and clay soils are essential to permanence The vegetation of the edges, shallows, and margins are a critical factor in nutrient supply.

13.8

YIELDS OUTSIDE THE POND

Flow down

We have dealt with "in pond" yields as factors of water quality and fish selection. This does not take into account upstream, incorporated, or downstream yields not directly related to fish flesh. For instance, if we feed industrial fish food pellets to ducks or pigs, and let the manures of these animals fertilise the ponds, we get about the same yield (or even more) of such plankton and detritus feeders as prawns, carp, *tilapia*, and mullet, plus the duck or pig products. This is a question of the correct routing of the food supplied, and involves land yields.

Further, if we use fertiliser on land crop, feed that to pigs and ducks, and use their manures for fish, we get an even greater yield. Beyond that, we can grow permanent low–fertiliser crop for pigs (banana, papaya, acorns), use sparing fertiliser, and get even greater total yields.

Alcohol recovery and the subsequent biogas digestion of green feed, tubers or starchy food, manures, and wastes produces a flow–on slurry not one whit less fertile than the original substances, so that we are now arriving at an integrated flow–down system of tree forage > animal protein > manures > alcohol > biogas > water crop (plants) > forage–fish. Even within this flow, side cycles to worms, notonectids, or *Daphnia* give better utilisation and a yield at every step. The problem that we begin to strike here is that no one family or person can manage a very complex integrated system. We need a higher order of social organisation in order manage these maximum yields (**Figure 13.35**).

Margins

Ponds, depending on their shape, give a new and productive edge in landscape. Trees and vines grow better there, and selected marginal plants drop leaves, insects, insect wastes, fruits, and extend roots into ponds as fish foods. They also shelter ponds or direct winds over them, prevent over–heating and chilling of waters, and provide cover for forage species. Thus, a whole set of yields are available from pond margins.

Ponds

We know that weedy shallows, brush piles, fences, rocks or pipes, and rafts provide additional shelter, escapement, or forage in ponds. There is a rich field here for increasing yields. We could devote 50% of the pond to such cover or forage systems.

Downstream

Water from densely–stocked fish ponds is a rich source of irrigation water for land plants. Yields from sewage or fish–pond water are 2–5 times that from intake water. These land crops, as fuels, food, forage, or structural product, must be integral to pond development if we are to realise the full value of fish ponds, and utilise the

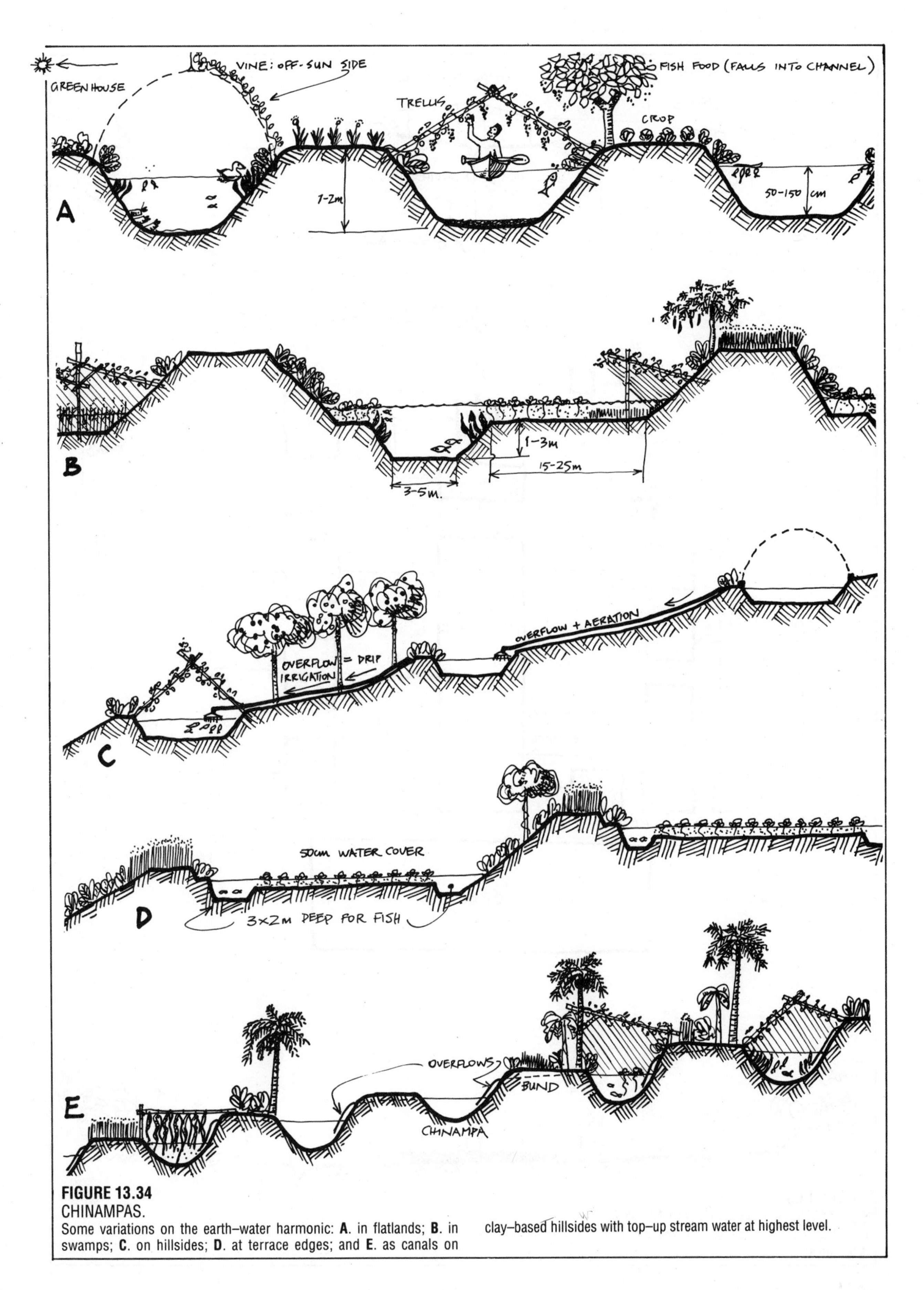

FIGURE 13.34
CHINAMPAS.
Some variations on the earth–water harmonic: **A**. in flatlands; **B**. in swamps; **C**. on hillsides; **D**. at terrace edges; and **E**. as canals on clay–based hillsides with top–up stream water at highest level.

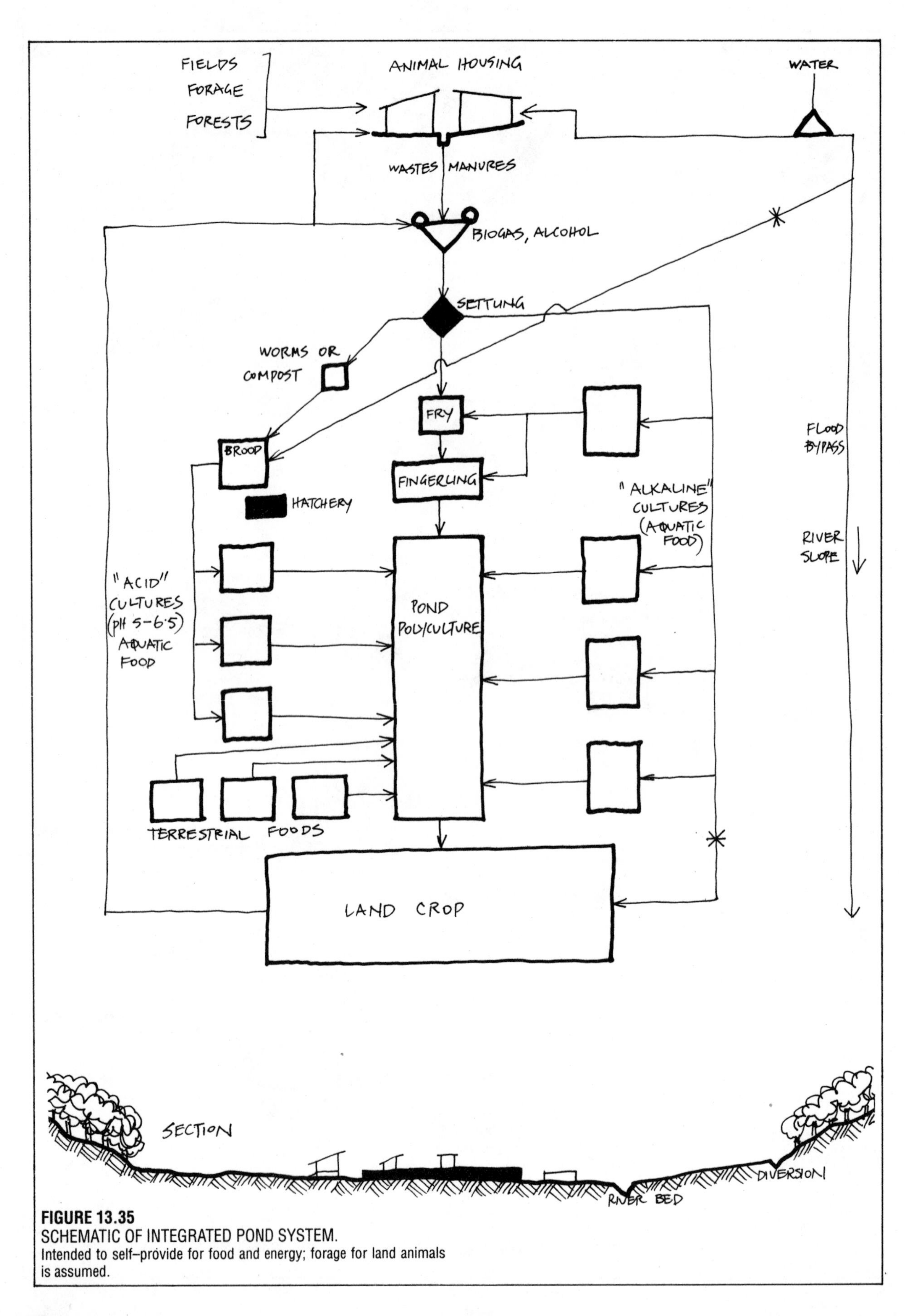

FIGURE 13.35
SCHEMATIC OF INTEGRATED POND SYSTEM.
Intended to self-próvide for food and energy; forage for land animals is assumed.

wastes and algae flowing downstream.

PLANTS OF THE MARGINS

There is a curious lack of data on marginal plants in most literature, even if their value is sometimes noted in passing. What can marginal plants achieve?

- Cover for fish, shrimp, waterfowl, and hence predator protection.
- Spawning and nesting sites for fish and waterfowl.
- Fruits and flowers which manure or directly feed organisms in and on the water.
- Leaves, bark, limbs, and detritus for decomposers in the water such as diatoms, phreatocids, algae, sponges.
- Hence a feeding base for fish and low–trophic feeders or browsers such as shrimp, mullet, and molluscs.
- Insects attracted to blossoms, or falling as larval and pupal forms into the pond.
- pH modification from mulch and leaves, buffering of extreme pH levels.
- Materials to control mosquito larvae and snails, to stupify fish, to make traps and screens, and for conduits and pipes.
- Prevention of bank erosion by mat roots and leaf buffering of wave and flow energy.
- Wind, shelter, shade, and hence evaporation and temperature modification.
- Beautiful reflections.... what more could one ask?

Marginal plants live in a milieu of fairly constant moisture and buffered temperature changes, hence tend to be reliable producers of fruit, nectar, and flowers, tubers and foliage. Many are either very resistant, or very susceptible, to water rot and attack from aquatic organisms, hence making excellent wharf and boat timbers, or rotted mulch in water. Others contain air cells which make them light and buoyant, or conversely are very dense and sink like stones (*lignum vitae*), so that marginal or aquatic timbers have unique values in specific usages.

In crop, those honey–producers of ditches and ponds are very reliable in yield, while tubers are consist-ently produced, and drought is an unknown restraint on yield (**Figure 13.36**).

13.9
BRINGING IN THE HARVEST

There are very few areas of the western world where fresh fish is easily available locally (unless we live near a city fish–market). Modest aquacultural ponds can change that, as fish of known quality and species can be locally supplied. Aquaculture brings shellfish and crayfish or shrimp to areas remote from sea resources. At present, affluent nations eat about 10% (6.1 kg) of fish and 90% red meats (60.1 kg). (Figures for the USA in 1979 from *Science* 206, 21 Dec., 1979.) This is all due to change as aquacultures mature.

It helps a great deal if regional growers can establish a "people's market" on the model of organic growers, when even live fish can be offered. Variety can also be built in if growers allot species to specific members, or if complex polycultures are established.

Processing of fish on farm has many advantages. Drying, smoking, freezing, and salting or reducing to pastes and sauces are all small–scale activities. It is

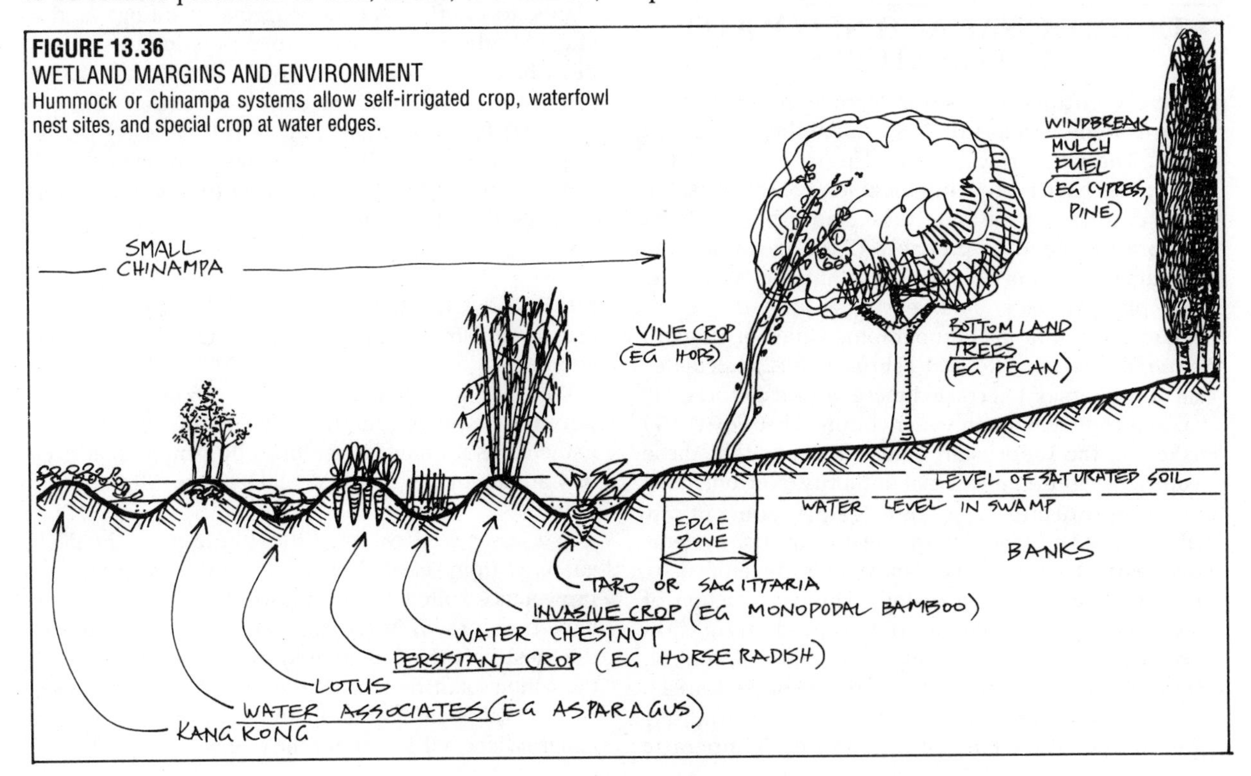

FIGURE 13.36
WETLAND MARGINS AND ENVIRONMENT
Hummock or chinampa systems allow self-irrigated crop, waterfowl nest sites, and special crop at water edges.

critical to protein production in the tropics that rapid processing of fish is achieved in view of the potential for spoilage. More everyday manufacture of clear plastic drying tents or boxes is needed rather than emphasis on increasing fishing efficiency or production, as up to 40% of a catch can be lost to spoilage and damage by rodents and birds.

Ferment of fish wastes (as practiced in Indonesia) produces a protein–rich, salt–stabilised liquid sauce of great value to people on rice diets, and such techniques are an essential part of salvaging fish wastes for market dispersal. Processing also increases on–farm economics and rural employment potential.

For non–carnivorous species, catch–and–restock systems can greatly increase yields over batch or total removal of the species. Many permanent ponds are run on this basis, or that of natural breeding in the case of crayfish (when a *maximum* size limit is used to return large, fast–growing breeders to the pond or breeding area). It is always wise to consider returning or keeping the largest fish, and eating the slightly smaller grades (unfortunately not a strategy well–applied in sea fisheries).

Although a great emphasis is put on fish, true aquacultural planning should always include diverse plant products, with emphasis on local staples. Fish are then accessory to a broad product base, and to tree crops. The disease problems that build up in aquacultures are due to over–emphasis on one species or class of food and high yields, and also to a neglect of shade, wild foods, and the beneficial effects of plants on water quality.

13.10
TRADITIONAL AND NEW WATER POLYCULTURES

The very antiquity of some terrace and riverside cultures demonstrates their sustainability in human terms. The wet terraces of the Ifugao people of the Philippines have been in use for at least 3,000 years; the flood cultures of the Nile (now sadly but perhaps temporarily ended) is probably 17,000 years old; the chinampa system of Mexico is also antique, as are the flood–plain, lakeside, and river cultures of Asia.

Harold Conklin in "Ethnographic Cities of Ifugao", *Scientific American (Review)* February 1982, calculates that 1000 years of toil created these terrace cultures. He makes a commentary on water cultures that Fox (1977) makes on the Indonesian palm cultures—that these peoples recognise no external authority. Not only is it an inappropriate and interfering concept in self–sustained systems, but the creators of such systems have learned a self– governance never developed by any central authority in history. The construction of common canal systems and shared irrigation necessitates community organisation and self–control. Hydraulic despotism, as found in colonised areas, predisposes societies to exploitation and centralised government. Water is, in all sane or democratic societies, regarded as a *public* resource.

Growing 0.5 hectares of padi rice, and 0.25 ha of sweet potato as staples, with 1 ha of woodlot and grove crop, domestic pigs and chickens fed from gleanings and crop, the Ifugao family of 5 spend 400 person days of productive labour per year, or 80 days per able individual. Rainfall is an incredible 3 to 6 m/year, necessitating constant terrace maintenance. The terraces have 2 cm of soil, and half the mass of the bunds is in stone. Canal and ditch maintenance necessitate annual earth and stone repairs; the equivalent value or objective cost is one day of rice for one hour of work.

There are a good many useful plant species of the marine and aquatic environment; only some 20–30 species of water plants are in constant culture or have been subject to varietal selection. Many species are still wild–gathered or (if cultured) are derived from unselected wild stock. Yet there is an obvious potential for wetland forestry, craft product, and specialised forage crop assembles for species such as bees, ducks or waterfowl, and fish. Pigs utilise seasonal swamps and water margins for a variety of fodders. Cattle enter lagoons and shallow waters to browse, while species like the Soay sheep, turtles, and the Galapagos iguana browse the large seaweeds of the seashore, and seagrasses. In samphire pastures geese, swan, and fish species browse at varying tide levels. Anyone interested in specific plant assemblies can evolve a fairly complex array of forbs, shrubs, vines, and trees to suit specific or specialised purposes. There is an obvious role for seed, vegetative, and general aquatic nurseries and suppliers, which are undeveloped in most countries.

Rather than enumerate all the cultured species, I have chosen to deal with a few selected polycultures of plants, to briefly treat the marginal plantings, and to suggest plant assemblies for specific sites. I will describe both a cold area and subtropic plant system to illustrate the aquatic potential, selecting wild rice (*Zizania*) for cool areas, and taro (*Colocasia*) for the tropics. Rice (*Oryza*) itself is grown from temperate to tropical areas, and is not dealt with here, but is very well described by Fukuoka[3,4].

WILD RICE CULTURE

(*Zizania lacustris*) Minnesota, very cold winters and hot summers.

Wild rice is a tall annual grass (to 2.5 m) with several perennial species in the same genus; it has been gathered as a grain for centuries by Amerindian tribes of Canada and the United States, from Minnesota to Florida.

Species or seed collected for culture trials should be gathered from selected plants at the climatic and water provenance suited to the culture site. Some few crops have been grown in the Southern hemisphere, but for the most part the 3 million kg produced are grown in the Minnesotan region. Of the total production, about 50% is still wild–gathered in traditional fashion by Amerindians, who beat out the grains into a canoe, and the remainder in extensive fields of up to 100 ha created by modern growers, with all the inefficiences that moncultures entail.

Wild rice has been brought into culture within the last 30 years, and selection is now taking place. It is therefore one of the most recent grains to be developed, but one peculiarly suited to cold swamps and shallows with intense summer heat. The original culture was for duck flyway (wildlife) seed, and wild ducks harvest perhaps 70% of the total production, as the seed ripens slowly over 3–4 weeks. Mechanical harvest gathers only that seed which is ripe at harvest. Thus, while commercial yields are low (only 50–100 kg/ha), prices are high (at $10 kg in 1984), so that fields of 100 hectares or so give a cash return of up to $100,000 annually in rice alone.

Wildfowl, in dabbling the fallen seed in fields, help thin out seed which will germinate after a winter stratification; the plants crowd if not so thinned. Ideal spacing is 20 x 35 cm, and for small plots, with selected heads, yield can be 200–500 g/head. Obviously, the problem with yield is not in the plant, but in the need to harvest at one time, mechanically. Home plots, gathered over 3–4 weeks, are in fact very productive, and only 200–800 plants should supply a family if well cared for. A strategy used by Indians to save the grain from predation and shattering is to bundle and tie 10–20 heads together, and to untie them periodically for threshing into canoes.

Seed is saved from selected plants and held over winter in bags under water in the icy lakes, or in water in refrigerators at 40°F. It is sown in spring in fields flooded to about 16–20 cm deep, and harvested from mid–autumn. The fields are dried out for 2–3 weeks before machine harvesting, and must be rested or sown to another crop to clean the ground if a new cultivar is to be introduced, as self–seeding otherwise perpetuates the first–sown variety.

In the large fields and deep margin ditches of Minnesota, a self–generated polyculture of crayfish (*Homarus*), mink, beaver, and thousands of waterfowl has evolved with the wild rice culture. Coyote also range the bunds, and packrats store much of the seed. The crayfish live in ditches and browse 5–10 m into the crop, causing some damage, but as they are themselves a potential crop, or a food for mink, this is tolerated. Duck potato (*Sagittaria*) can become a weed of the system, but this is also a waterfowl food and prefers somewhat deeper water than the rice plant, or can be removed by culture or hand weeding if necessary. There is an obvious potential for coldwater fish–rearing in the canals and ditches of the system (species such as trout and bullheads), not as yet developed. Manuring by *Azolla* is not practiced, and the fields are rested dry in winter.

An intriguing potential is the propensity of the packrats to carry and store large quantities of sound, cleaned grain into artificial shelters (like engine manifolds), insulated and plugged by *Typha* (cumbungi, cattail) seed silk. Narrow fields and a set of such artifical storage sites may well be the best way to harvest, and breeding packrats could be supplied with alternative food for the winter, or allowed to keep 15% of their stores full. They may well be the most efficient, as well as the cleanest, harvesters for the crop, and can scarcely do worse than the 50–100 kg (or less than 10%) gathered by machines costing $300,000 or more! Bill Mackently (pers. comm.) uses grey squirrels to collect hickory nuts and acorns in this way, and provides them with buried pipes to store the nuts. He leaves them about 15% of the harvest.

Duck fattening, crayfish harvest, mussel meats, fish production, and vine crop of hops or silverberries (*Actinidia arguta*) are obvious supplementary products to the crop, and could be viable accessory systems, together with restricted fur production, some special forestry, and recreational fishing or tour potential. I have attempted to portray some of this complexity in **Figure 13.37**, and acknowledge my debt to the pioneer growers Hubert and Leonard Jacobson of Aitken, Minnesota, who kindly supplied me with data (and some wild rice to eat). A small terrace or pond of wild rice should be the aim of every home gardener who can obtain seed and who enjoys a nutritious grain, as most Americans do. For wildlife (especially waterfowl) refuges, there is no better autumn fodder.

TARO CULTURE

(*Colocasia esculenta*) Hawaii, subtropics to tropics, and cooler frost–free areas.

I admit to a weak spot for taro fields and terraces, especially those of the traditional Hawaiian culture. Taro is an herbaceous perennial to 1 m high with large arrow–like leaves and a swollen stem base or tuber which can reach 10 kg, but is normally marketed at 1 kg or so, after an 8–15 month growing period (there is no "off season" in the subtropics).

There were some 1,100 Hawaiian varieties, and perhaps 100 or so are still preserved. It is a staple food in Hawaii and some other parts of Polynesia and southeast Asia, where rice or breadfruit are also staples. Grown as a monoculture plant for centuries, the recent accidental or deliberate introduction of *Azolla, Tilapia,* fish, crayfish, and Chinese edible coiled gastropod snails have diversified the wet terrace cultures, although the Hawaiians traditionally rear local fish and prawn species in the ponds.

Taro is propagated by small side tubers or more commonly by a cutting based on a crown disc of about 1 cm thick, and the lower 18 cm of stalk from the old tuber at harvest. Cuttings have the leaves removed by a slant cut just above the first leaf node.

There are special varieties for boiling, baking, ferment to *poi*, or cooking as green leaf spinach (*luau* taro). All taro must be cooked due to the stinging crystals of oxalic acid in the leaves (which can grossly irritate the bare back of non–Hawaiians like myself, incautiously carrying bags of tubers in this way).

There has always been a fish culture tradition with

taro, of local prawns, shellfish, and freshwater fish. *Tilapia* and introduced prawns have merely extended the potential, as have crayfish (Homarus *spp*). What is missing is a good variety of edible freshwater mussels to fix phosphates in the terrace mud. In older times, nutrient supply was of forest leaves (chiefly *Aleurites* or *Hibiscus*). The stems of kukui trees (Aleurites) were stood to rot in the upper terraces where they produced a local edible ear fungus. About 80% of the bulk of a log was in this way converted to fungus, and the remainder broken up as water mulch. Very extensive canal and wet terrace systems were developed for taro culture, and the fermented *poi* used as a staple, as were boiled or baked taro.

Accessory margin crops are ti (Cordyline)—the Polynesian "wax paper" and wrapper for baked foods—papaya, banana, coconut, and sugar cane. No vine crop was grown, but as taro appreciates partial shade, there is a potential for marginal or wide- spaced overhead kiwifruit, passionfruit, pole bean, or cucurbit crop. Bunds planted to *Dolichos hosei* or *Phyla nodosa* (was *Lippia nodosa*) would be useful nitrogenous mulch sources, and would remain short and trimmed at the water level by *Tilapia* or grass carp (which are also on Hawaii).

As pigs and latterly ducks are also traditional livestock, construction of their pens over feeder canals or the upper terraces would add substantially to nutrient supply, although the ducks need confinement to a few ponds as they too appreciate the edible Chinese snails (which are a high-priced local delicacy).

Taro is also grown on dryland areas under irrigation, and responds very well to thick mulch and green crop. In these situations intercrop of *Brassica*, melons, ginger, or lettuce is viable. Beds are normally 1.3 m wide at 30 beds/ha, with taro spaced at 40 x 60 cm in diamond pattern. There are 55,500 bulbs/ha, averaging 1 kg or a little more for each bulb, or a yield of 55 t/ha. Drip irrigation is also used on land crop, at 120 lines/ ha (4 per bed). Spacing in wet terrace is similar, but the "intercrop" there is more profitably prawns, although kangkong and watercress are spot-planted in many taro terraces for greens, and both are planted as low mound crop in special shallow terrace or canal systems with faster flow than is found in taro systems.

In other climates, or mild coasts of the Atlantic and western Pacific (New Zealand), eel rearing would combine very well with taro culture. Taro is set out in wet mud or shallow water, but at maturity can be kept flooded at mid-thigh, or with 0.5 m of free water above the softer muds of the terrace, where prawns and grass carp (which co-exist very well) can be stocked.

It is obvious from the above information, much of which I owe to the hospitality of friends like Chuck and Tina Busby, Imu and Rachel Naki on Moloka'i, and Richard Waller on Hawaii itself (dry taro culture), that taro is a very productive and flexible plant for a basic starchy food, and that a rich polyculture can be developed in the taro which is itself well protected from browsers by its oxalic acid spicules.

Every tropical garden deserves a patch or so of taro; it is often grown as a mulched patch or groundcover below fruit trees in higher rainfall areas, or as a subsidiary crop in banana, guava, avocado, and macadamia nut orchards. There is no more pleasant environment than a rich taro, comfrey, papaya, guava, and banana polyculture, with chilis, ginger, peppers, ducks, pigs, and fish, a visit to the sea for *limu* (seaweed) and crab, and a good earth oven *(imu)* to cook in.

The foregoing detailed accounts of the traditional and possible aquaculture systems may give designers and farmers some different ideas for productive local land-water integrations.

13.11

AQUACULTURE: DESIGNERS' CHECKLIST

Due to the special susceptibility of water life, minimal to zero biocide use is essential near or in waterways.

With fish:

- Stock rates below, at, and above suspected optimum (plus or minus 2,000/ha).
- Fertilise pond and bring pH to 7+.
- Select diseasefree stock.
- Exclude or guard against predators.
- Maximise edge and natural foods.
- Introduce predator fish at about 1:5 ratio.
- Select fish for no pond aeration, or to suit the aeration method proposed.
- Select fish for good local value, but assess food costs for each species.
- Provide shelter from predators and excess light and heat.
- Select fast-growing brood stock for the particular

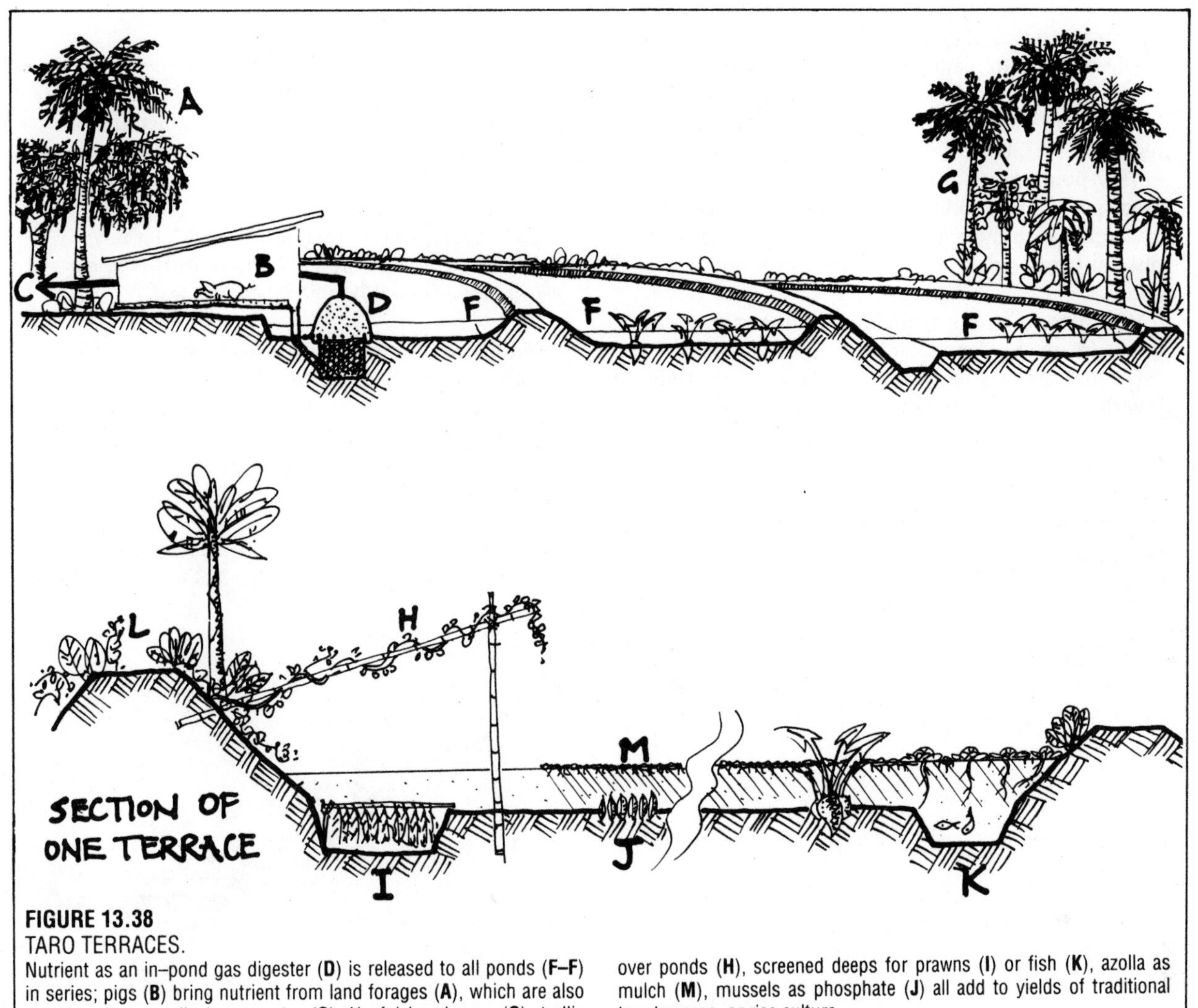

FIGURE 13.38
TARO TERRACES.
Nutrient as an in–pond gas digester (**D**) is released to all ponds (**F–F**) in series; pigs (**B**) bring nutrient from land forages (**A**), which are also used to clean up discharge water (**G**). Useful bund crop (**G**), trellis over ponds (**H**), screened deeps for prawns (**I**) or fish (**K**), azolla as mulch (**M**), mussels as phosphate (**J**) all add to yields of traditional taro terraces, or rice culture.

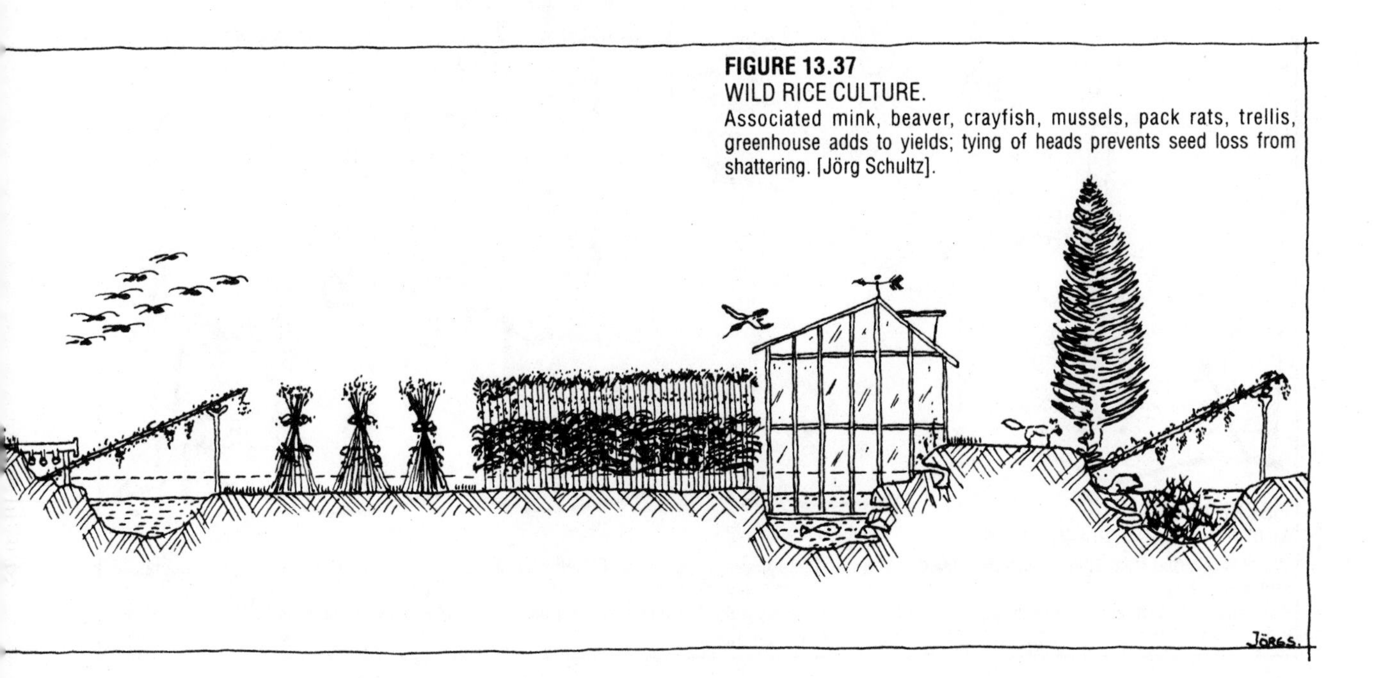

FIGURE 13.37
WILD RICE CULTURE.
Associated mink, beaver, crayfish, mussels, pack rats, trellis, greenhouse adds to yields; tying of heads prevents seed loss from shattering. [Jörg Schultz].

site, from fry to adults.

- Devise a fish/plant polyculture for the ponds.

Analyse landscape for natural or cultivated food resources.

Vary pond depth, pond size, and pH to suit a set of food species and productive fish for the district.

Carefully analyse pond configurations for specific polycultures, easy management, site, and weather effects.

Pay attention to aquatic and marginal crops, downstream crop, and total landscape balance.

Devise accessory food systems for fish or invertebrates, as vegetation, root crop, invertebrates.

In any design, include some appropriate (small or large) wetlands, even if it is from waste water.

13.12
REFERENCES

Augusthy, K. T., *Fish Farming in Nepal*, Institute of Agriculture and Science, Tribhuvan University, Nepal, 1979.

Bardach, J. E., J. H. Ryther and W. O. McLarney, *Aquaculture: the farming and husbandry of freshwater and marine organisms*, Wiley, N.Y. (A basic text.)

Bryant, P., K. Jauncey, and T. Atack, *Backyard Fish Farming*, Prism Press, Dorset, U.K., 1980.

Chakroff, Marilyn, 1982, *Freshwater Fish Pond Culture and Management*, Peace Corps/VITA Publications No. 36E, 1982.

Darlington, P. J. Jr., *Zoogeography: the geographical distribution of animals*, John Wiley & Sons, Inc. London,

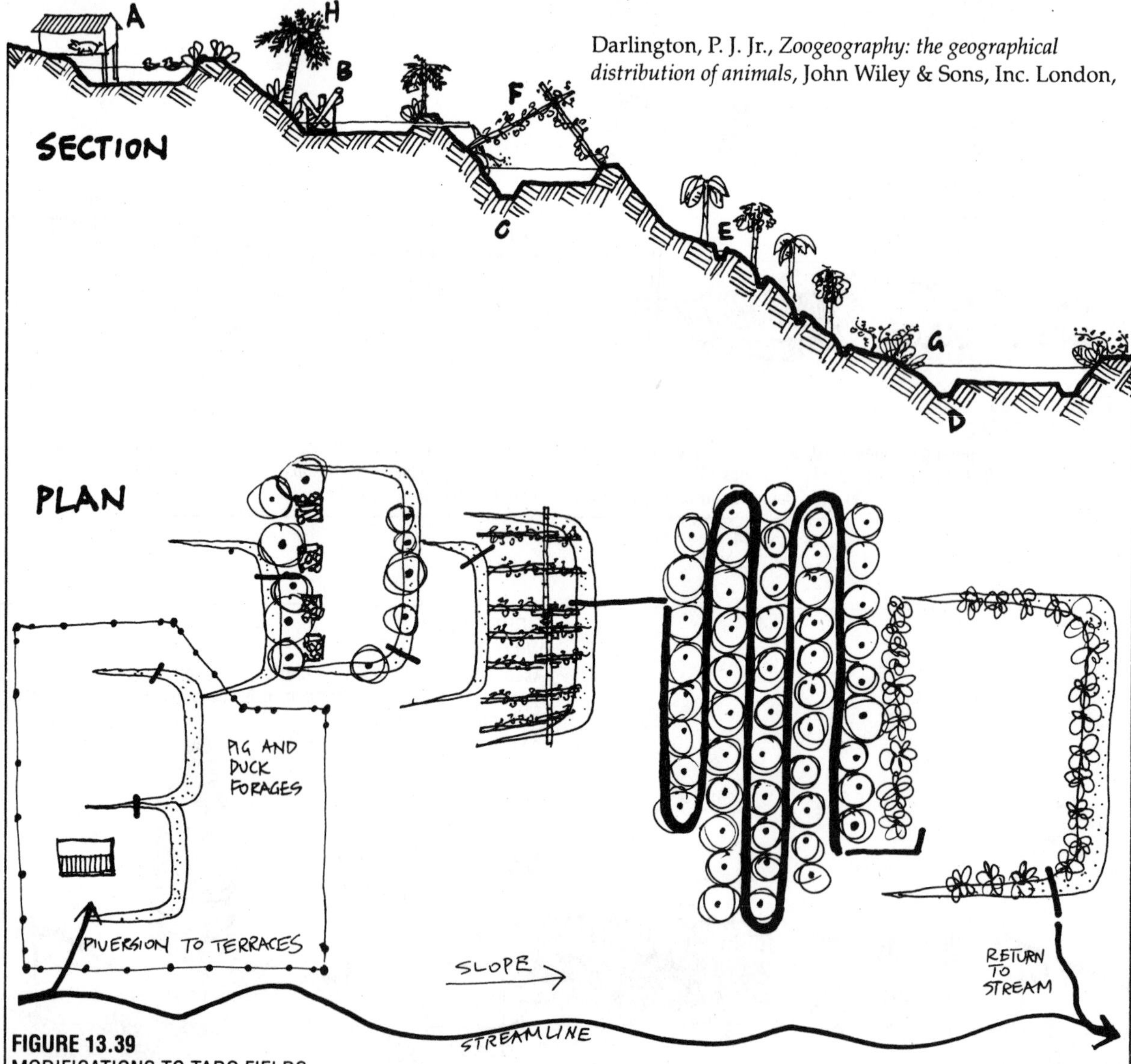

FIGURE 13.39
MODIFICATIONS TO TARO FIELDS.
Water is diverted from stream, fertilised, used in crop, and returned to stream free of pollutants.
A. Animal house; **B**. logs rotting to produce edible fungi (*Aleurites*); **C**. deep channel for shrimp, fish; **D**. fish–taro padi with peripheral channel; **E**. series of irrigation channels in papaya crop, bananas; **F**. cucurbits, beans on trellis; **G**. margin crops e.g. comfrey, taro, Tradescantia for fish food; and **H**. coconut on margins.

1963. (Some good discussions of species groups; for a wider reference.)

FAO Fisheries Technical Paper No. 130, *A Catalogue of Cultivated Aquatic Organisms*,1974.

Hawaii State Center for Science Policy and Technology Assessment, *Aquaculture Development for Hawaii (1978).*

Hills, C. and H. Nakamura, *Food from Sunlight* , University of the Trees Press, California. (Algal culture and aquaculture.)

Huet, Marcel, *Textbook of Fish Culture*, Fishing News Books Ltd. Surrey, U.K., 1975. (Excellent text on breeding and rearing of fish and coldwater species.)

Huke, R. E. and R. W. Sherwin, *A Fish and Vegetable Grower for All Seasons*, Norwich Publications, Vermont, 1977.

Iversen, E. S., *Farming the Edge of the Sea*, Fishing News Books, Surrey, U.K., 1968.

Machado, A. E. de Mafra, *Criacao Pratica de Peixes*, Livraria S/A, Sao Paulo, Brazil, 1980.

Maclean, J. L., 1975, *The Potential of Aquaculture in Australia*, Aust. Govt. Publishing, Canberra, ACT, 1975.

Palmer, E. L. and H. S. Fowler, *Fieldbook of Natural History*, McGraw–Hill, N.Y., 1975. (A useful compendium of species and families.)

Pullin, S. V. and Ziad H. Shehadeh, *Integrated Aquaculture and Aquaculture Farming Systems*, International Centre for Living Aquatic Resources Management (ICLARM), 1982, PO Box 1501, Metro Manila, The Philippines.

Swingle, H. S., *Biological Means of Increasing Productivity in Ponds*,1966. F.A.O. Symposium on warm–water pond fish culture 40–181, Rome, 18–25 May 1966. (A key reference on yield strategies in fertilised and supplemented food ponds.)

Tapiaor, D. D. *et. alia*, *Freshwater Fisheries and Aquaculture in China*, F.A.O. Fisheries Technical Paper No. 68, Rome, 1976.

Toews, D. A. A. and M. J. Brownlee, *A Handbook for Fish Habitat Protection on Forest Lands in British Columbia*, Government of Canada (undated).

United States Academy of Sciences, *Making Aquatic Weeds Useful*, Washington, D.C., 1976.

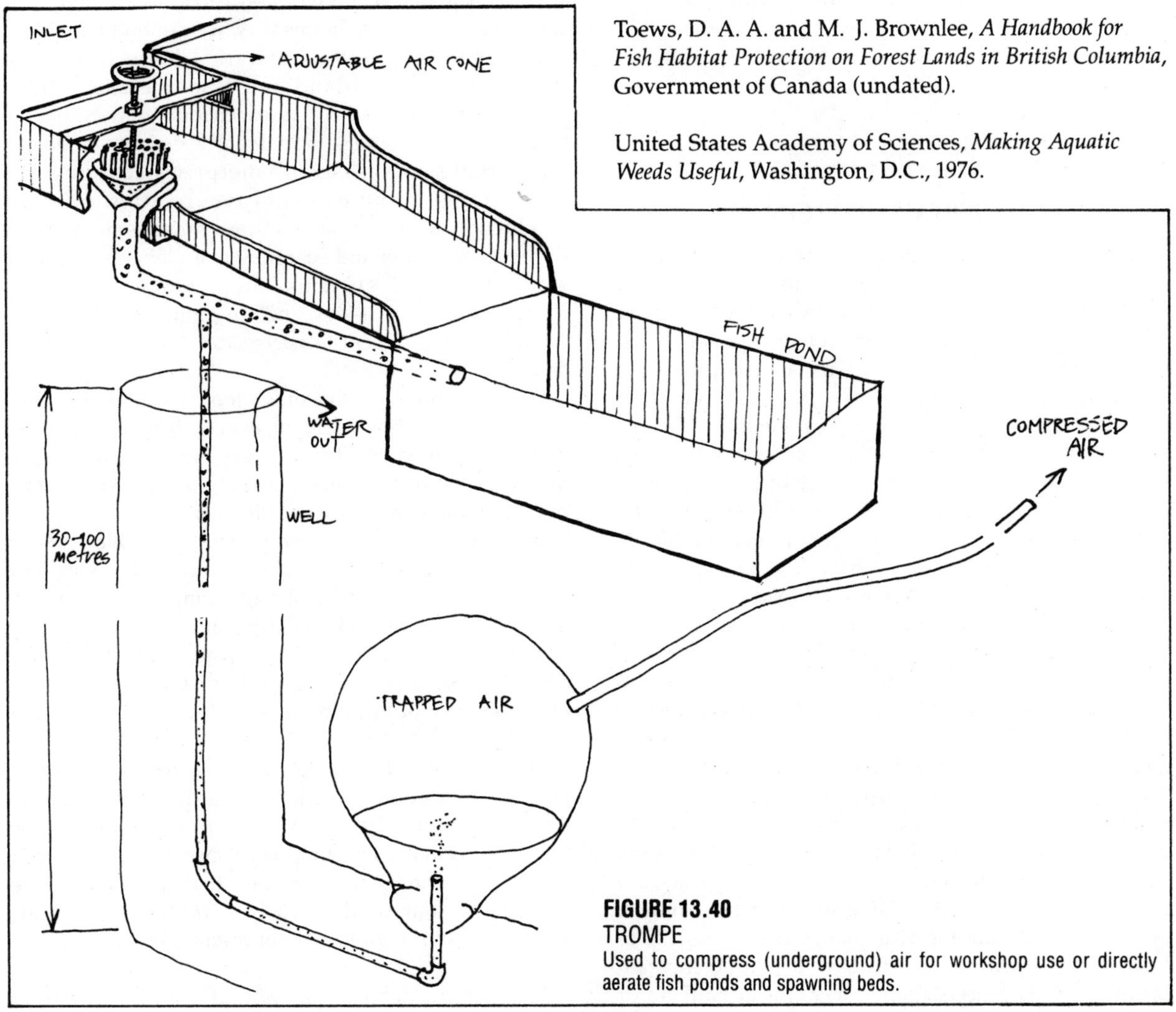

FIGURE 13.40
TROMPE
Used to compress (underground) air for workshop use or directly aerate fish ponds and spawning beds.

Chapter 14

STRATEGIES FOR AN ALTERNATIVE NATION

He who desires but acts not, breeds pestilence.
(William Blake, *Proverbs of Hell*)

The head does not ask for flowers while the belly lacks rice.
(Indian proverb)

14.1 INTRODUCTION

The pragmatic and practical approach to the main body of this work largely omits reference to those visions or beliefs classifiable as spiritual or mystical; not because these are not a normal part of human experience, but because they are arrived at as a result of long contemplation or intense involvement with the mysteries that eternally surround us. We may "dream" understanding, but it is something we cannot demand, define, or teach to others; it is for each of us to develop.

There are things that nobody else can help us with, but in a book written to help people make real–life decisions, to build new landscapes, to regenerate damaged forests, and to lighten our load on earth, the present need is for *clear and practical* approaches.

In the preceding chapters, well–tried and common-sense techniques and strategies of earth restoration have been described and figured. All of this comes to naught if we, as a people, continue to invest in arms and destruction, to permit land abuse, and to fail to tackle the social and political impediments to reclaiming desertified and abused lands, or even to prevent the poisoning of land. Thus, the following sections give strategies for change in the social and economic areas of society. These strategies may, in fact, be of more assistance to real change than the skills of land management, for society has far more competent farmers and engineers than it has ethical bankers or lawyers whose work relates to curing or preventing (not just treating) social and environmental problems.

First we must learn to grow, build, and manage natural systems for human and earth needs, and then teach others to do so. In this way, we can build a global, interdependent, and cooperative body of people involved in ethical land and resource use, whose teaching is founded on research but is also locally available everywhere, and locally demonstrable in many thousands of small enterprises covering the whole range of human endeavours, from primary production to quaternary system management; from domestic nutrition and economy to a global network of small financial systems. Such work is urgent, important, and necessary, and we cannot leave it to the whims of government (always short–term) or industry as we know it today.

We know how to solve every food, clean energy, and sensible shelter problem in every climate; we have already invented and tested every necessary technique and technical device, and have access to all the biological material that we could ever use.

The tragic reality is that very few sustainable systems are designed or applied by those who hold power, and the reason for this is obvious and simple: to let people arrange their own food, energy, and shelter is to lose economic and political control over them. We should cease to look to power structures, hierarchical systems, or governments to help us, and devise ways to help ourselves.

Thus, the very first strategies we need are those that put our own house in order, and at the same time do not give credibility to distant power–centred or unethical systems. In our present fiscal or money–run world, the primary responsibility that we need to take charge of is our wealth, which is the product of our sweat and our region, not representable by valueless currency.

There is no need to stress that we are imperfect

people, living in an imperfect world; "Do not adjust your vision, reality is at fault" (graffiti), so that many strategies given here are starting points rather than endpoints. However, there is so much damage to ecosystems—hence so much rehabilitative work to do—that we will be employed in good works for a few generations to come. In several generations (if we are allowed this time) we may have achieved a truly free world of international affinities, but we always start where we are.

In this chapter, therefore, I will try to set out the currently successful social strategies that enable a small group or a region to define problems and to solve them locally.

14.2 ETHICAL BASIS OF AN ALTERNATIVE NATION

A people without an agreed–upon common basis to their actions is neither a community nor a nation. A people with a common ethic is a nation wherever they live. Thus, the place of habitation is secondary to a shared belief in the establishment of an harmonious world community. Just as we can select a global range of plants for a garden, we can select from all extant ethics and beliefs those elements that we see to be sustainable, useful, and beneficial to life and to our community. It would appear that:

- Sustainable societies emphasise the *duties and responsibilities of people to nature* equal to those of people to people; that any code relates equally to other lifeforms and elements of landscape. To conduct oneself only in terms of response to other people gives a potential to evade responsibility for damage inflicted on the total resource base, and thus ultimately to others.

Beneficial behaviour involves managing natural systems for their own, and our, long–term benefit, not for our immediate and exploitative personal gain. The American Indians (Irequois nation) frame this as a "seventh generation" concept: that our decisions *now* are carried out in terms of their benefit or disadvantage to our descendants in seven generations' time (about 100 years ahead). This helps explain why we always found tribally managed lands to be rich in natural life resources, and why we have managed to ruin much of the resources we inherited.

- As people, we need to adopt an ethic of *right livelihood*, for if we bend our labour and skills to work that is destructive, we are the destroyers. We lay waste to our lives in proportion to the way in which the systems we support lay waste to the environment. Although societies for *social* responsibility are rapidly forming, we need to expand the concept to *social and environmental* responsibility, and to create our own financial and employment strategies in those areas. We should not be passive workers for established destructive systems, but rather we can be investors in life. We cannot profess or teach one ethic, and live another, without damage to ourselves and to common resources.
- We must always study and learn as part of a *total integrated system framework*, conscious of how our knowledge and actions permeate all systems. It is in the fragmenting of knowledge into unrelated disciplines that we can plead ignorance of effects; but we are always responsible for the distant effects of our actions, and in fact should work for foreseen benefits.
- We need to develop *conserver socirties*, with this conservation achieved by close attention to recycling, the avoidance of waste, and to very durable technologies so that their use is prolonged. Similarly, it is unwise to abandon satisfactory older forms of technology even if we install improved forms and processes; just because we develop a new windmill for electricity, we should not allow a dependable older grain mill to sink into disrepair. Because we can use a forage harvester, it doesn't mean we should lose the skill to use a scythe on steep slopes.

Part of being conservative is to concentrate on *developing a mosaic of small, well–managed, and effective systems*. Such modest systems are unlikely to cause widespread upheavals or to be subject to external or unethical control.

- Meaning in life is lost by striving after status and future glory; it is gained and realised by *action* towards a common ideal, in serving the whole according to our physical, mental, educational, and revelationary (understanding) capacities. It is never enough to mean well ("fair words plant no cabbages"),rather, it is necessary to ensure that *it gets done*.
- Security can be found in the renunciation of ownership over people, money, and real assets; insecurity and unhappiness arises as a result of trying to gain, keep or protect that which others need for periods of legitimate access. A lending library enables people to help themselves to information; a locked–up book collection is useful only to the person who owns it.
- If an ethical and responsible community can establish a durable, dependable, and waste–free resource base, then leisure time (time to express our *individual* capacities) becomes a plentiful resource. We will have gained time for life. While leisure is inevitably available for enrichment of cultural life, and to an extent for recreation, emphasis on spectator recreation is just another way to waste the time gained; we then see the professionalisation of arts, sciences, sport, and even education as spectators replace actors.

We should therefore resolve to gain time to evolve ever more effective ways to assist systems or people. It is only when others feel secure that we need not guard our environments, so that the very best preparation for security is to teach others the strategies, ethics, and practices of resource management, and to extend aid and education wherever possible.

I do not, in my lifetime, or that of my children's

children, foresee a world where there are no eroded soils, stripped forests, famine, or poverty, but I do see a way in which we can spend our lives towards earth repair. If and when the whole world is secure, we have won a right to explore space, and the oceans. Until we have demonstrated that we can establish a productive and secure earth society, we do not belong anywhere else, nor (I suspect) would we be welcome elsewhere.

14.3
A NEW UNITED NATIONS

The "United Nations" today is neither united nor represents nations; it is like the oft–quoted "moral majority", which is also neither of those things! Many true nations, such as the Iroquois confederation or any tribal alliance with a common ethic, are not represented by such a body, nor are whole nations such as the Basques, Tartars, Kurds, Palestinians, Hawaiians, Hopi, Tibetans, Pitjatjantjara, Misquito, Aranda, Basarwa, Herrero etc etc etc.

Most nations in the United Nations *repress a majority of peoples on earth.* Talking with Thomas Banyaca, a Hopi messenger of his people, it became clear to me that we need a new concept of "nation", and a new representative body to speak for them. We start by defining a nation as *a people subscribing to a common ethic,* and aspiring to a similar culture. Such nations may not have a common land base, or language, but do have a common ethic, minimally;

- To care for the earth; to repair and conserve;
- To seek peace, and to guard human rights everywhere; and
- To invest all capital, intelligence, goodwill, and labour to these ends.

At present, many thousands of organisations, affinities, tribes, bioregions, and spiritual and non–government organisations aspire to such beneficial ends; in every continent, a majority of people—*the ethical majority*—want peace; a clean and forested earth; a cessation to torture, malnutrition, and oppression; and a right to work towards these ends.

It would take very little additional organisation for these groups to meet together, count their numbers, and recognise each other's rights. There are, for instance, far less paid–up or active members of political parties or oppressive societies now than there are organic gardeners whose life works seek peace and plenty. As groups discuss, and accept, the minimal ethic above, they they can quickly proceed to recognise each other.

Such initiatives have in fact commenced in the Amerindian groups subject to national (i.e. political) oppression in both North and South America. Throughout the world, groups are talking of issuing their own passports, or adopting world citizen status—given a common aim. Perhaps the first move to a new body of *nations united in earth care* are the bioregional and tribal congresses that are occurring today.

Unlike the present United Nations, we do not need a world centre, or paid administrators, but can instead meet as *affinity groups* (e.g. in alternative economic summits, bioregional congresses, tribal conferences, garden and farm design groups) to deal with our specific areas of interest, and to make these affinities global in scope. By avoiding centralised administrations, we avoid power blocs, and by avoiding tax funding, we avoid inefficiency. Fees for a regional secretariat would arise from an annual fee forwarded by participant groups.

Once continental groups and some global groups have allied, these congresses can increasingly bring in less informed or more remote groups to share resources in an humane alliance; after all, global seed exchanges, technology groups, gardening forums, and regional groups already meet and are increasing in cooperation. A concept of a global nation is, in fact, very well developed in such groups, and the idea of war or oppression across race, language, or territory is anthema to those allied in good works. The advantage of such alliances is that even isolated people can find global affinities; this is not necessarily true of regional organisations.

14.4
ALTERNATIVES TO POLITICAL SYSTEMS

Systems of government are currently based on self–interest, economic pragmatism, belief, impractical theory, and power–centred minorities (religious, military, capitalist, communist, familial, or criminal). Almost all such groups set up competitive and "adversary–oriented" systems.

We need to set about, in an orderly, sensible, and cooperative way, a system of replacing power–centred politics and political hierarchies with a far more flexible, practical, and information–centred system responsive to research and feedback, and with long–term goals of stability. And we need to do this in an ethical and non–threatening way, so that the transition to a cooperative (versus conflicting) global society is creative (not destructive).

The world needs a new, non–polarised, and non–contentious politic; one not made possible by those in situations that promote a left–right, black–white, capitalist–communist, believer–infidel thinking. Such systems are, like it or not, promoting antagonism and destroying cooperation and interdependence. Confrontational thinking, operating through political or power systems, has destroyed cultural, intellectual, and material resources that could have been used, in a life–centred ethic, for earth repair.

It is possible to agree with most people, of any race or creed, on the basics of life–centred ethics and commonsense procedures, across all cultural groups; it matters not that one group eats beef, and another regards cows as holy, providing they agree to cooperate in areas which are of concern to them both, and to respect the origins of their differences as a chance of history and evolution, not assessing such differences as due to personal perversity.

It is always possible to use differences creatively, and design to use them, not to eliminate one or other group as infidels. Belief is of itself not so much a difference as a refusal to admit the existence of differences; this easily transposes into the antagonistic attitude of "who is not with me is against me", itself a coercive and illogical attitude and one likely, in the extreme, to classify all others as enemies, when they are merely living according to their own history and needs.

Most human communities function in relation to a long-term sustainability only because they *do* differ from others; what is possible to an Inuit (Eskimo) is not possible to a forest pygmy. Thus, it is not differences in themselves that are important; it is how all groups relate to the basic rules of the local ecology that permit them to function on a long-term basis. Belief, like religion, is a basically private and non-global characteristic, and should not be subject to comparisons. On close examination, we "believe" in those systems that enable us to behave without guilt, with respect to our resources and our own culture.

It has long been apparent that our current political, economic, and landuse systems cannot solve such long-term and worsening problems as soil degradation, ground water pollution, forest decline, the spread of poverty, unemployment, and malnutrition (or its extreme, famine). Despite good scientific prognoses and assessments, effective ground strategies are lacking. The temporary nature of political systems is an impediment to effective action. We could describe all western political systems as those of competing belief elites; whether they are self-described as communist, socialist, capitalist, or democratic, they all function in ways which are essentially short-term.

By their nature, political systems seek to impose a policy control over as wide an area of influence as possible, are power-centred (not life-centred), and are often composed of very few families or (in the case of royalist and feudal societies), one family. Thus, the continuing and long-sustained programmes necessary to reverse forest loss and soil decline are usually sacrificed for the short-term policies of an elite maintaining power. It was said of a recent prime minister of Australia that his national policies all worked to maximise profits from his farm!

"The argument for simplicity is never a political argument...when people practice it in their lives... they don't even need any politics." (*Manas*, 17 Oct 1984). This same statement also refers to the adoption of an ethical basis to action, to the placement of money and resources, and to the determination to act in accordance with one's beliefs. All of these can occur independently of political change, and can be long-term (life-long) personal actions of great effect. That is, people can act independently of political theory (which rarely, if ever, covers the questions of ethics, simplicity, local autonomy, or life-oriented action). Such changes in people come about by education and information, and when enough people change, then political systems (if they are to survive) may follow, or become as irrelevant as they now appear to be in terms of real solutions.

For this reason, the place to start change is first with the individual (oneself), and second in one's region or neighbourhood.

THE RIGHT NOT TO BE IN DEBT.

Some of the most charming and climatically appropriate houses on earth are built without bank loans, architects, metals, concrete, or contractors. However, in every case they are built in areas where trade unions, building surveyors, health officials, and local or state governments do not impede the home builder or the community providing shelter for themselves. While Chile (as an economic system guided by "experts") accumulated a $12 billion foreign debt in 1985, poor people, acting without loans, together built at least $11 billion housing in slum areas by local cooperation without incurring any foreign debt. Why is this the case?

Stone, mud, bamboo, round timbers, rope, thatch, and even baked brick and tiles, are the age-old durable building materials of mankind. All can be locally produced if energy from community forests and people is provided. Even cement and mortar can be made if needed using kilns fired by wood, as can pottery, bricks, and roof tiles. None of this needs money if people work together.

The real cause of a lack of shelter (as with food) in any country is not that of finance, but of restrictive practices by a regulatory bureaucracy. Moreover, state or private ownership (versus *community* ownership) of forests, small mines, and lands is devoted to state or corporate profits to support a largely urban, leisured class of bureaucrats, which denies these basic biological and earth resources to the very people who work to produce or mine them.

We have had "national service" to fight wars, but I cannot recall any but sustainable tribal societies that require every man and every woman to help shelter and feed themselves. Curiously, we are drafted to kill strangers, and denied the right to preserve life; no armies are created to build houses, grow potatoes, or plant forests for the future; unemployment for others is preferred by those who choose power as a method of exploitation.

In very recent societies, our basic "right" is to vote, form unions, protest, or go to law (i.e. to support professional classes). Truly basic rights to grow or protect forests, to build a shelter, grow food, or provide water from our roof areas are commonly denied by local or state regulations. Effective local group action restores the true basic rights, which are those of personal responsibility for our sustenance on earth, and to earth itself. While "natural law" demands a fair return for every gift received, the laws of power demand gifts without thought of return—this is called "economic growth" and means unlimited resource exploitation and the concomitant exploitation of people.

The wealth of any area lies not in banks or cities, but in those basic resources, skills, and natural systems developed by its peoples.

POLITICAL AFFILIATIONS.

There are two ways to ensure the political changes which will bring ecological changes. The first is to mobilise ground support in every electorate where a candidate of any party takes a stand on good ecology, or against nuclear and polluting industry; and the second is to form a local Ecological or Green Party, or a bioregional group.

This would be an easier task if all intentional groups affiliated, and subscribed to a common policy; it is difficult for a small group to evolve a total policy in isolation, and a common policy statement sums up the skills of all groups. Common policy always leaves room for local issues, but gives strong *principles* for guidance in those issues. As well as a guiding ethic, the broad aims of such a party are (as stated in *Planet Drum*, P.O. Box 31251, San Francisco, CA 94131, USA): ecological, socially egalitarian, grassroots democratic, and non-growth. In Germany and other European countries, the Green Party has increasing support and representation, with 23 federal and 48 state seats by 1983. The Green Party's address is Die Grunen, Bundestag, Bonn, West Germany. In the USA, International Green Party, 113 29th St., Newport Beach, California 92663.

EVOLVING A NEW POLICY BASE.

A common global policy can start with a *general ethic* as stated in the beginning of this book; it can then proceed to specific policies, for specific cultures, regions, and landscapes. To structure such policies, we must search out working solutions (e.g. we know that Singapore has solved most housing funding problems, that some towns are energy and food self-reliant, and that many problems have already been solved in other areas or at other times). Thus, the structure under which we should gather common policy is:

- Define what is seen to be the problem or concern; give weight to priorities on a scale from 1–10.
- State *intent* of policy for your region; what it is intended to do (the principles of this policy).
- Collect strategies that have been proven to work; this really means a set of case histories.
- Frame a set of policies based on all successful strategies.

Overall, set policy priorities in rough order, weighted for urgency, public cost or loss of wealth, general or global spread of the problem, long-term effect, and threats to basic resources or life systems. Do not, in the first place, try to frame policy on purely local or trivial matters, unless as a case history applicable to a broad principle.

14.5 BIOREGIONAL ORGANISATION

A bioregional association is an association of the residents of a natural and identifiable region. This region is sometimes defined by a watershed, sometimes by remnant or existing tribal or language boundaries, at times by town boundaries, suburban streets, or districts, and at times by some combination of the above factors. Many people identify with their local region or neighbourhood and know its boundaries.

There is an obvious conflict between the need to live in a region in a responsible way (bioregional centrality) and the need to integrate with other people in other places (global outreach). We need not only to "think globally and act locally", but to "act and think globally and locally".

The region is our home address, the place where we develop our culture, and take part in bioregional networks. Through global associations and "families of common interest" we cross not only the regional but also state and national borders to set up multicultural alliances.

Just as bioregions need a federal congress periodically, so do they occasionally need global congresses; societies or families also need global meetings to break down the idea of defended regional boundaries to humanity. Ethics and principles of self–governance, interdependence, and voluntary simplicity or restriction of human numbers on earth still apply at regional and outreach levels. Intermarriage, visits, mutual trade and aid, skills exchange, and educational exchange between regions of very different cultures enriches both. This is the antithesis of "integration" (bureaucratic genocide) that is promulgated by majority groups who disallow language use and cultural life to minorities. In particular, reciprocal education values *both* sets of knowledge and world concepts, and respects others' lifestyles.

Tribal maps often defined bioregions very well; totems and "skins" (clan groups) of tribes might take, as their totemic mothers, a particular tree or animal, which itself was limited in distribution by the sum of topographic and climatic factors. Other groups occupied ecologies of grasslands, stony deserts, swamps, or mountain ridges. Today, minority language groups (Saamen, Basque, Pitjatjantjara) claim territories that are ancient, and specific to their life mode. Obviously, cities break up into different, often occupational or income, districts, each with its own dialect and ecology, consumption spectrum, and morality. The acid test of a bioregion is that it is recognised as such by its inhabitants.

Ideally, the region so defined can be limited to that occupied by from 7000 to 40,000 people. Of these, perhaps only a hundred will be initially interested in any regional association, and even less will be active in it. The work of the bioregional group is to assess the natural, technical, service, and financial resources of the region, and to identify areas where leakage of resources (water, soil, money, talent) leave the region. This quickly points the way to local self–reliance strategies.

People can be called on to write accounts of their specialities, as they apply to the region, and regional news sheets publish results as they come in. Once areas of action have been defined, regional groups can be formed into associations dealing with specific areas, e.g.:

- Food: Consumer–producer associations and gardening or soil societies
- Shelter: Owner–builder associations
- Energy: Appropriate technology associations
- Finance: An "earthbank" association

And so on...for crafts, music, markets, livestock, and nature study or any other interest. The job of the bioregional office is complex, and it needs 4–6 people to act as consultants and coordinators, with others on call when needed. All other associations can use the office for any necessary registration, address, phone, and newsletter services, and pay a fee for usage.

Critical services and links can be built by any regional office; it can serve as a *land access centre*, operating the strategies outlined later under that section. It can also act as leasehold and title register, or to service agreements for clubs and societies. More importantly, the regional office can offer and house community self–funding schemes, and collect monies for trusts and societies.

The regional office also serves as a contact centre to other regions, and thus as a trade or coordination centre. One regional office makes it very easy for any resident or visitor to contact all services and associations offering in the region, and also greatly reduces costs of communication for *all* groups. An accountant on call can handily contract to service many groups. The regional group can also invite craftspeople or lecturers to address interest groups locally, sharing income from this educational enterprise.

Some of the topics that can be included in the regional directory are as follows. These can be taken topic by topic, sold at first by the page, and finally put together as a looseleaf notebook (volunteers enter local resource centres and addresses under each category; the system is best suited to computer retrieval). The following Resource Index for Bioregions has been compiled by Maxine Cole and myself for the Northern Rivers Bioregional Association of New South Wales, Australia.

The primary categories are as follows:

A. Food and food support systems
B. Shelter and buildings
C. Livelihoods and support services
D. Information, media, communication, and research
E. Community and security
F. Social life
G. Health services
H. Future trends
I. Transport services
M. Appendices (maps, publications of the bioregion)

All of the above sections can contain case histories of successful strategies in that area.

CRITERIA: Practical resources (people, skills, machinery, services, biological products) essential to the functioning of a small region, and assisting the conservation of resources, regional cash flow, the survival of settlement, employment and community security. (Security here means a cooperative neighbourhood and ample, sustainable resources for people.)

CATEGORY A – FOOD AND FOOD SUPPORT SYSTEMS.

Criteria: Native and economic species, organic and biocide free, products of good nutritional value.

A1. Plant resources

1.1 Nurseries and propagation centres, tissue culture, sources of innoculants, mycorrhiza.

1.2 Plant collections and botanical gardens, economic plant assemblies, aquatics.

1.3 Research institutes, horticultural and pastoral agencies.

1.4 Seed sources and seed exchanges.

1.5 Native species reserves and nurseries.

1.6 Demonstration farms and gardens, teaching centres, workshop conveners.

1.7 Government departments and their resources, regulations.

1.8 Voluntary agencies involved in plant protection, planting, and propagation.

1.9 Skilled people, botanists, horticulturists.

1.10 Publications and information leaflets of use in the region, reference books, libraries, posters.

1.11 Contractors and consultancy groups: implementation of plant systems, farm designs.

1.12 Produce: products and producers in region, growers.

1.13 Checklist of vegetables, fruits and nuts which can be grown in the region, and species useful for other than food provision .

A2. Animal resources

2.1 Breeders and stud or propagation centres, artificial insemination, hatcheries.

2.2 Species collections, including worms and like invertebrates.

2.3 Fish breeders and aquatic species.

2.4 Useful native species collections and reserves, potential for cultivation.

2.5 Demonstration farms, e.g. free range, bee culture, workshop conveners, teaching centres.

2.6 Government departments and their resources, regulations.

2.7 Voluntary agencies and animal protection societies.

2.8 Skilled people, farriers, vets, natural historians.

2.9 Contractors (shearers, etc.) and consultancy groups, farm designers.

2.10 Publications, posters, libraries for the region

2.11 Produce: species and suppliers in region.

A3. Integrated pest management (IPM)

3.1 Insectaries and invertebrate predator breeders and suppliers of biological controls.

3.2 Suppliers of safe control chemicals, traps.

3.3 Information sources on IPM.

3.4 Pest management of stored grains and foods.

3.5 References and libraries.

3.6 Checklist of common pests and predators, and safe pest control procedures.

A4. Processing and food preservation

4.1 Suppliers of processing equipment.

4.2 Food Processing Centres (FPCs).

4.3 Information sources on food processing and preservation.

4.4 Sources of yeasts, bacterial and algal ferment materials.

4.5 Processed–product producers in region.

A5. Markets and outlets

5.1 Local markets.

5.2 Delivery services.

5.3 Export markets and wholesalers.

5.4 Urban–rural co–op systems, direct marketing.

5.5 Retail outlets.

5.6 Market advisory skills and groups, contract and legal skills.

5.7 Roadside and self–pick sales.

5.8 Market packaging and package suppliers, ethical packaging systems and designs .

5.9 Annual barter fair.

A6. Support services and products for food production

6.1 Residue testing services for biocides, also nutrient, mineral and vitamin content (food quality control).

6.2 Soil, water and leaf analysis services for micronutrients and soil additives, water analyses, pH levels.

6.3 Hydrological and water supply services (dams, domestic water), design and implementation.

6.4 Fence and trellis suppliers and services, cattle grids and gates.

6.5 Suppliers of natural fertilisers, mulch materials, trace elements, soil amendments.

6.6 Farm machinery, garden and domestic tool suppliers (see also processing), appropriate and tested equipment, fabricators and designers, repair services, hire and contract services.

6.7 Land planning services.

6.8 Glasshouse, shadehouse, food dryers, suppliers, and appropriate materials.

6.9 Lime quarries and sources, stone dusts, local trace mineral sources, regional geological resources.

CATEGORY B – SHELTER, BUILDINGS.

Criteria: Energy–efficient house design and non–toxic materials only

B1. Construction materials

1.1 Timber growers and suppliers, community timber plantations.

1.2 Stone and gravel, earth materials.

1.3 Plumbing and piping, drainage, roofing.

1.4 Bricks and concrete products (tanks, blocks, etc).

1.5 Tiles and surfaces, paints (non–toxic)

1.6 Furniture and fittings.

1.7 Tools and fasteners, tool sharpening services and repairs, glues and tapes.

1.8 Library and research resources.

1.9 Current state of housing in the region (numbers seeking housing, rentals available).

1.10 Sources of toxins and unsafe materials in buildings, appliances, furnishings, paints and glues; high voltage equipment.

B2. Energy systems

2.1 Home appliances for energy conservation and efficiency, energy saving and insulation.

2.2 Hot water systems, solar systems.

2.3 Space heating and house design for the region.

2.4 Power generation systems for region: current and proposed.

2.5 Appropriate technology groups, research centres and demonstrations.

2.6 Designers of low energy home systems and buildings.

2.7 Sources of information, publications, trade literature, library resources.

2.8 Reliable contractors and builders.

B3. Wastes, recycling

3.1 Sewage and greywater disposal (domestic).

3.2 Compost systems and organics.

3.3 Solid wastes disposal and collection (boxes, bottles, plastics).

3.4 Occupations based on waste recycling.

CATEGORY C – LIVELIHOODS & SUPPORT SYSTEMS.

Criteria: Concept of right livelihood or socially useful work. Durable and well–made items.

C1. Community finance and recycling

1.1 Barter and exchange.

1.2 Small business loans.

1.3 Community banking and investment systems.

1.4 Land access systems, commonworks, leases, trusts.

1.5 Legal and information services.

C2. Livelihood support services

2.1 Small business service centres.

2.2 Skills resource bank: business, legal and financial advisory services, volunteer and retired people.

2.3 Self–employment (work from fulfilling regional

needs: job vacancy lists).

2.4 Training courses in region.

C3. Essential trades, and manufacturing services and skills

3.1 Clothing and cloth (spinning, weaving).

3.2 Footwear and accessories, leatherwork.

3.3 Basketry and weaving, mats and screens.

3.4 Functional pottery.

3.5 Steelwork, fitting and turning, smithing and casting, welding.

3.6 Functional woodwork.

3.7 Engines and engine repairs.

3.8 Functional glasswork.

3.9 Paper recycling and manufacture, book trades, printing and binding.

3.10 Catering and cooking (food preparation).

3.11 Draughting and illustrating services.

3.12 Soaps, cleaning materials.

CATEGORY D – INFORMATION SYSTEMS , MEDIA SERVICES , COMMUNICATIONS AND RESEARCH.

Criteria: Essential community information, aids, and research

D1. Communications networks

1.1 Regional radio and C.B., ham radio .

1.2 Regional news and newspapers, newsletters.

1.3 Audio–visual services, photography, television, film

1.4 Business and research communications e.g. fax, telex, modem, card files, computer, journals, libraries, graphics, telephone answering services.

1.5 Computer services and training.

1.6 Libraries and collections of data in region.

1.7 Maps.

1.8 Bioregional groups and contacts—local and overseas.

1.9 Standard documents and data sheets available via the bioregional centre.

CATEGORY E – COMMUNITY AND SECURITY.

E1. House and livestock security,

1.1 House siting.

1.2 Neighbourhood watch.

1.3 Cattle and livestock watch.

E2. Fire volunteers and reports (4 wheel drive clubs).

E3. Flood (cleanup, rubber duckies).

E4. Bush, cliff, beach rescue services.

E5. Communication systems.

6.1 Report centre.

6.2 Emergency communications.

CATEGORY F – SOCIAL LIFE.

Criteria: Assistance for isolated people to meet people of like mind

F1. Introductory services.

F2. Think tanks.

F3. Expeditions.

F4. Work groups.

CATEGORY G – HEALTH SERVICES.

Criteria: Basic preventative and common ailment treatment, necessary hospitalisation, accident treatment, local resources

G1. Medical and pharmaceutical services.

G2. Surgical and hospitalisation services.

G3. Gynaecological and midwifery services, home birth support.

G4. Profile of morbidity in region, life expectancy, infant mortality, causes of death, ailments in order of importance, under:

4.1 Accidents & injuries; infectious diseases; addictions & drugs.

4.2 Genetic and birth defects; nutritional problems.

Note: until the above listing is made, no region can assess health priorities.

CATEGORY H – FUTURE TRENDS & POTENTIAL THREATS TO THE REGION (AS A SERIES OF RESEARCH ESSAYS).

H1. Sea level rises, greenhouse effect.

H2. Ozone depletion.

H3. Water pollution and biocides; radioactives and chemical or waste pollution.

H4. Financial collapse; recession.

H5. Implications for policy making.

CATEGORY I – TRANSPORT (SEE ALSO CATEGORY H).

I1. Barge and sea systems.

I2. Draught animal systems.

I3. Joint or group delivery/cartage.

I4. Innovations; local fuels and new sorts of vehicles.

I5. Transport routes, bikeways.

I6. Air and ultralight craft, blimps.

CATEGORY M – APPENDICES.

Maps – Bioregional map
- Geological
- Plant system
- Soils
- Sources and references to maps, suppliers
- Regions, parishes,
- Land titles
- Access and roads
- Reserves and easements
- Rivers and water supplies

Note that if essential servies are listed, deficiencies noted, and leaks of capital detected, then there is immediately obvious a category of "jobs vacant". If, in addition, there is a modest investment or funding organisation set up (itself a job), then capital to train and equip people to fill these gaps is also available. When *basic* needs are supplied locally, research and skills will reveal work in producing excess for trade—this excess can be as information and education to other regions.

Bioregionalism is an excellent concept, given the irrational land use systems and land divisions developed by the present power structures. However, it is rarely an achievable reality, unless enough people gather in one area and manage to attract a sufficient number of like people to achieve a viable internal economy and trade infrastructure, together with the community common funds that make such enterprises possible.

And that is the secret of success: assembling sufficient commonsense people in one area. If we are one isolated biodynamic gardener in a district of contract vegetable growers or graziers trained in chemical agriculture, we find both the practice and infrastructure support of the isolated system difficult; there may be no one to talk to, let alone share resources with. On the other hand, as land titles in a region are bought out and occupied by any group who share an ethical philosophy, so the shops, markets, processing centres, equipment, and support services for the new economy become worthwhile and available.

As much as "the will to do" indicates health in the individual, so an increasing biological resource indicates health in the community. Every bioregion should monitor tree cover, wildlife, seaweed beds, bird colonies, species counts, and productive cultivated land at regular intervals. If these have increased in yield and maintained in species, the area maintains health. If no increase, or a *decrease*, is evident, something is wrong and should be immediately assessed for correction.

It is only the increase in the variety, quantity, and health of natural systems that indicates the health of any area. Where species disappear, trees or fish die, farmland and forest yields are reduced, and species lists simplify, there is trouble, and a degenerative effect is operating. A "life census" needs to be compiled every 2–5 years, and some data needs continual records, as absences are harder to detect than presences. Modifications to habitat can result in a constantly increasing biological resource, both qualitatively and quantitatively.

Every region needs to act as a curator and refuge for some critical life elements of allied regions, so that *absolute* loss of species is unlikely short of global catastrophe. In some land trusts, it is this biological–environmental accounting which sets the basis for the "economic rent", and (in the event of a degenerative trend) even the basis for continuing in occupation and use of the land.

14.6
EXTENDED FAMILIES

Chiajen (the family): The family is society in embryo; it is the native soil on which the performance of moral duty is made easy through natural affection, so that within the family circle a basis of moral practice is updated; this is later widened to include human relationships and society in general.

(From *The I Ching*)

The concepts of village and bioregion refer to a base or home area, but today many people travel about. Many societies extend as close affinity groups across many nations, thus forming a non–national network. Such groups develop a familial, rather than a competitive or conflicting, inter–relationship. With a common interest and ethical base, cooperative interdependence supplants competition. A "family" of this type, with 1,000 or less members, can ally with like groups to create a tribe, and 20–40 such tribes form a nation. Families, unlike many societies, have child care and the welfare of their members at heart.

Such families already exist in Europe, with small groups living in a scatter of households and locations across many existing national boundaries; some have existed for 18 or more years, and members report individual satisfaction with a larger support group. In practice, any person has 3–5 close friends (who change slowly over time), a support group of 30 or more acquaintances, and resource access to the whole system. A familial system of shared ethics can:

- Keep in touch via a registrar and news service;
- Co–invest in family property development and joint enterprises;
- Rationalise resources for most efficient sharing of space and equipment;
- Develop a series of social contacts via visits;
- Care for the children as one group, and set up a fund for their needs;
- Widen the cultural and support base for households.

Membership in a family of shared ethical values does not conflict with any other membership or duties, and is mainly a matter of organisation of a family registrar and some common funds. Such families need to define each adult as an individual, with a right to the essentials of their own space (bed and work space), garden, and occupation. As nuclear family households are a minor part of modern societies (13–18% of all households), households based on friendship, or work affinity, or designed for students, single, or elderly people, are needed.

Like land ownership, the ownership of people is an illusory aim. Some couples can tolerate years of close work, but many might prefer a slightly more independent existence, close to but not necessarily living with each other.

In particular, children need a wider alliance and support group than just one or two parents. People can find "aunts and uncles" to take part of the responsibility for children in any such extended family, and if the children have a common fund (like their own credit union) for basic needs, then their care at a basic level is assured. They also have more than one household to relate to, or to visit or dwell in when educational needs change.

Families can, for instance, maintain a student dormitory near secondary or tertiary institutes, whereas at present many rural families have no such facility to send children ages 12 or over when they need or request higher education.

People can feel, and sometimes are, trapped in the nuclear family or the "compound" family of blood relatives who may share no ethical or interest base. At times, traditional extended families grossly exploit (in particular) younger women, as household serfs, or are exploited by indolent members. Blood relationship is no guarantee of freedom of choice, or fair dealing. Besides, as people grow and age, they develop differing needs for space and relationships, and other (intellectual or interest) factors call for different personal relationships.

Many of us have been locked in to unsatisfactory work or personal relationships, or too much alone in the context of nuclear family "ideals", which in real societies are for the few. It is good to be able to visit, stay with, and cooperate with a few households and to form new relationships as needed; it is also necessary to have the freedom to choose new work alliances.

In the extended family, problems such as lack of shelter, land access, access to capital and services, deserted or neglected children (or adults), transmission of infective diseases, and population control can largely be dealt with by internal behaviour on some ideal of (dynamic) stability. By selective recruitment, skills and resources can be acquired, or developed by education and group capital investment. Funds can be established as follows:

- COMMON ENTERPRISE FUND: the family fund, held in 2–3 places and convertible to a variety of currencies, and managed by a few individuals as a full–time job. All savings and contributions are accounted to individuals, and available as loans as for any credit union or revolving fund, across all currencies and regions.
- ANNUAL MEMBERSHIP FUND: invested and the interest used. This services the registry, newsletter, and pays part or full–time wages to a collator.
- CHILDRENS' PERMANENT FUND: all adults (age 17 or over) contribute $50 to this fund, with additional gifts encouraged. The fund is managed with the Common Enterprise Fund for essential child–oriented ventures or for education. This is a non–returnable fund. Each mother or mother–to–be would encourage 5–6 support people to contribute donations, and agree to help in other ways. Until age 12, parents can apply for loans, and thereafter (until age 17) the children can themselves apply, after which they are recognised as adults.
- SPECIAL VENTURE FUNDS: to be raised by proposal, open to groups or individuals, and handled by the Common Enterprise Fund administrators for such group ventures as a shared house, boat, overseas programme, or business.

It is probable that some (or all) members could run one or more enterprises to fund a charity or "trust–in–aid" programme for areas of need. Given such basic financial tools, secondary needs are to rationalise resources as a type of real estate service and resource listing internally, so that all members can assist others as producers, consumers, trustees, or by land, appliance, and shelter rationalisation.

As some ideal, groups of 30 or so people could gather in core regions (with some outlier households) and so make travel locally an easier affair. Meta– networking (tribe to tribe) enables such higher–level organisations as travel and accommodation nets to be set up on a global basis, cash to be transferred to areas of need, and larger joint enterprises developed.

As for the touchy subject of population control, taking group responsibility for a very few children sometimes cures the urge to breed, and those who want more than two children are far outnumbered by those who don't, so that where children are not seen as a "future insurance"—as in very poor rural families— the population can soon achieve a steady state. The schematic of **Figure 14.1** gives the basic parameters for both steady–state and out–of–control or declining populations. A registry can therefore inform people of the balances in sex ratio and age structure, and recruitment in the late stages of enrolment can be adjusted to give a fair balance of sex–age distribution. It does not matter, of course, if perfect balances are not achieved, but resources remain plentiful only if people remain relatively few.

Given an extended family, a bioregional network locally, and some form of common work opportunity, any individual is assured of access to resources, capital, cultural exchanges, and good work. We need not only fixed villages and bioregions, but open corridors to other regions, other people, and across nations.

As I see it, conflict arises on "national" boundaries that are fixed or disputed. A web of multi–racial, multi–cultural, and multi–occupational families and global nations obliterates these "defended territories" and suits peaceful lifestyles. The framework for such nations already exists; it remains to give those frameworks the mechanisms that create true interdependence via a new type of extended family.

We all value cultural and environmental diversity—or the world would become one vast Toyota–Coca Cola–McDonald–Hilton monoculture. Thus the concepts of unique bioregions, intact language and culture, and cross–cultural enrichment is central to a permaculture of human resources, and an ecumenical global nation.

14.7
TRUSTS AND LEGAL STRATEGIES

Trusts in the public interest are the legal basis on which churches, universities and many schools, research

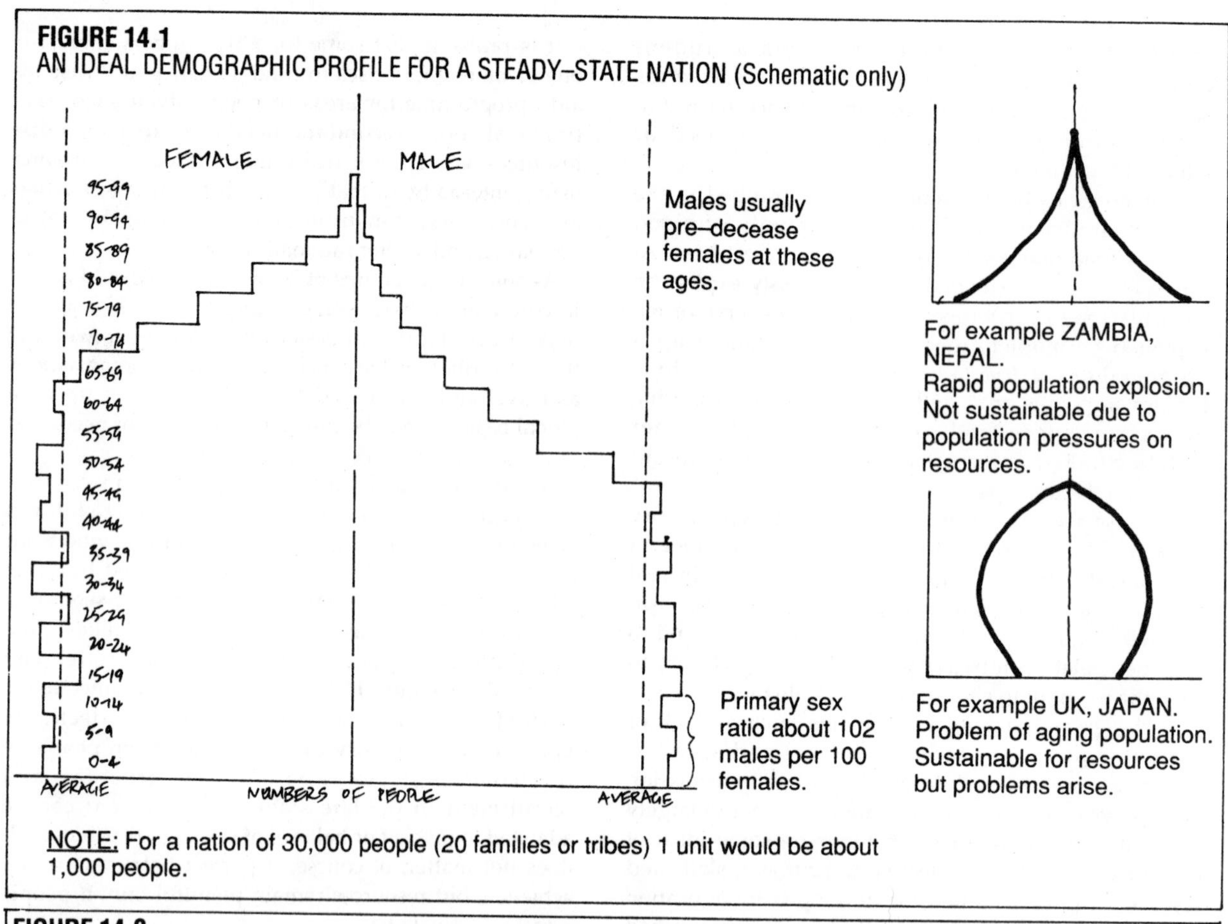
FIGURE 14.1
AN IDEAL DEMOGRAPHIC PROFILE FOR A STEADY–STATE NATION (Schematic only)
FEMALE
MALE
95-99
90-94
85-89
80-84
75-79
70-74
65-69
60-64
55-59
50-54
45-49
40-44
35-39
30-34
25-29
20-24
15-19
10-14
5-9
0-4
AVERAGE
NUMBERS OF PEOPLE
AVERAGE
Males usually pre–decease females at these ages.
Primary sex ratio about 102 males per 100 females.
For example ZAMBIA, NEPAL.
Rapid population explosion.
Not sustainable due to population pressures on resources.
For example UK, JAPAN.
Problem of aging population.
Sustainable for resources but problems arise.
NOTE: For a nation of 30,000 people (20 families or tribes) 1 unit would be about 1,000 people.

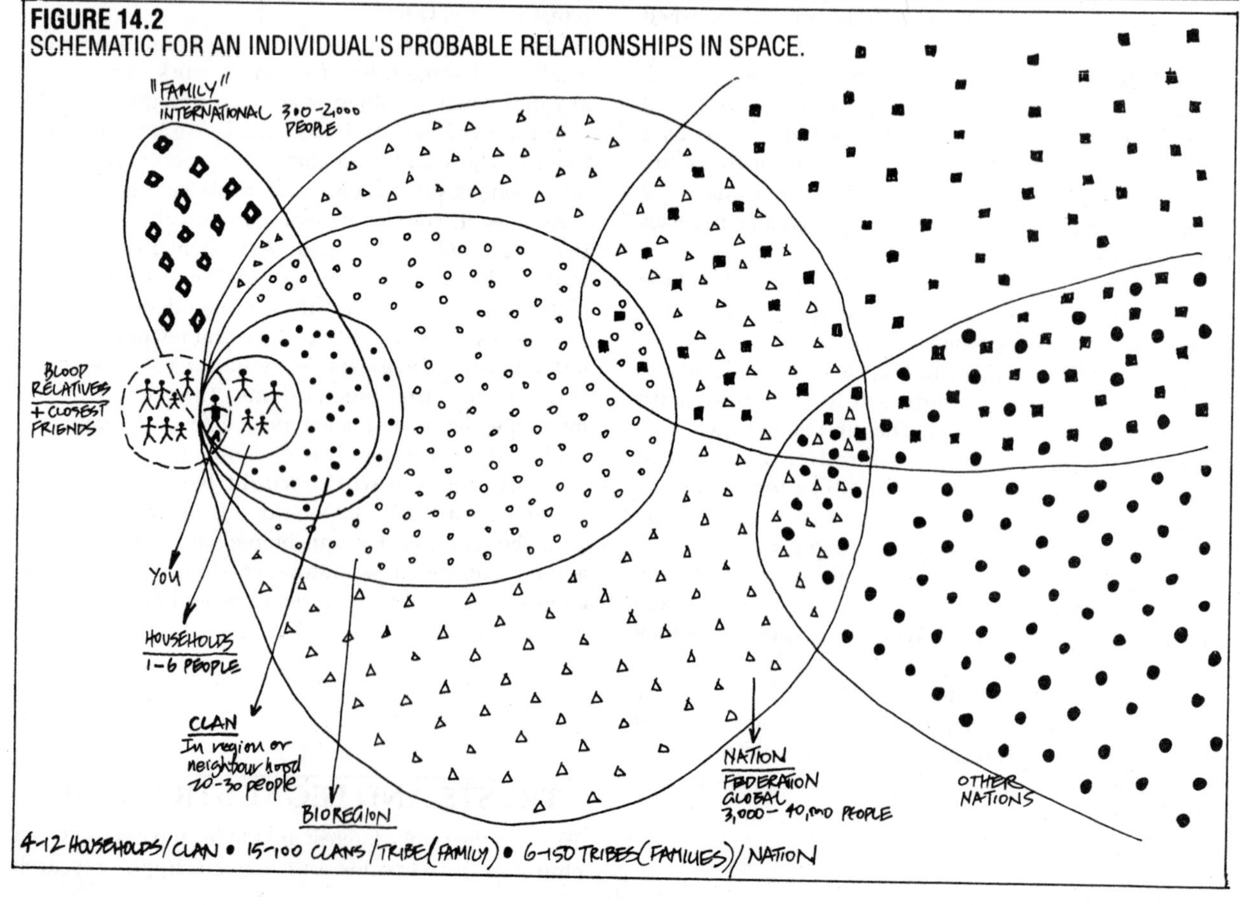
FIGURE 14.2
SCHEMATIC FOR AN INDIVIDUAL'S PROBABLE RELATIONSHIPS IN SPACE.
"FAMILY"
INTERNATIONAL 300-2,000 PEOPLE
BLOOD RELATIVES + CLOSEST FRIENDS
YOU
HOUSEHOLDS
1-6 PEOPLE
CLAN
In region or neighbourhood
20-30 people
BIOREGION
NATION
FEDERATION
GLOBAL
3,000– 40,000 PEOPLE
OTHER NATIONS
4-12 HOUSEHOLDS/CLAN • 15-100 CLANS/TRIBE(FAMILY) • 6-150 TRIBES(FAMILIES)/NATION

establishments, some hospitals, many public services, aid programmes, and charities rest. Few people realise how many, and how varied, are the trusts that serve them in one or other way. About 18–20% of businesses may also be non–profit trusts owned or operated by the charitable trusts that benefit from (are beneficiaries of) them.

It is quite possible, even sensible, to completely replace the bureaucracy of public services with a series of locally administered trusts, and Holland (in particular) largely supplants expensive paid public services (burdened as they are with heavy salary and capital costs, and liable to inaction, self–interest, and executive inefficiency) to publicly formed trusts (called *stichtings*). In the case of any small country, such trusts can run all public operations, and the "government" becomes simply a way of conveying tax capital back to the regions via local trusts. However, trusts can also self–fund via non–profit businesses to become foundations, fully equipped with their own income sources.

Trusts are usually formed, operated, and staffed by people (often initially volunteers) motivated to perform one or other public duty, or who seek to assist a defined or special group in need. Such trusts often have names including the words: church, foundation, institute, communion, school, congregation, charity, bureau, trust, or even company. When the trusts are formed to *trade*, they can take or own any business name that suits their work; such businesses are administered by a trustee.

Trusts are formed just to conduct businesses and trade, giving away their profits annually to named beneficiaries. If the beneficiaries are individuals, such gifts are taxed as private income; if the beneficiaries are charitable trusts or churches, the gift is not only not taxable but can be tax deductible to any giver. Trading, or "unit discretionary", trusts are also known as non–profit corporations (not to be confused with for–profit organisations).

Many large companies set up, and to some extent fund, non–profit organisations or even charitable trusts as a means to reduce taxable income, to carry out educational services, or to obtain public goodwill; some businesses tithe to worthy trusts that they believe in (a tithe is usually a tenth of income, but in practice ranges from 5–15%).

Legally, a trust body consists of a TRUSTEE and a document or TRUST DEED, registered with the public company registrar. There are many good reasons to make the trustee a private company, as directors of such companies need to be few in number (3 or 4 are enough), can appoint others if one dies or resigns, and can be anonymous. A company does not die, unlike its directors, and the small group of trustees can act quickly and decisively without reference to the cumbersome and often uninvolved "board of directors" that some trusts have appointed. It is wise to restrict directorships of a trustee company to those who are very active in trust affairs, and preferably live close to each other in one region. Such a set–up is diagrammed in **Figure 14.3**.

Should any person wish to set up a trust, the very first thing to do is to closely define the purposes of the trust, the group to whom it will apply ("all the citizens of Australia" or "those suffering from spina bifida"), and to instruct a lawyer to draw up the trust deed and to register this and the trustee company.

It is usually possible to buy a copy the trust deeds of other ethical organisations, and to use these as a model for a local trust, so reducing legal costs. Some law societies service ethical trusts at no charge for their time.

Any trust can have (unregistered, no cost) an ASSOCIATION of volunteers, aides, or clients who can publish a newsletter and generally assist the trust in its affairs.

It is also very wise for any charitable trust to establish a non–profit trading (business) trust to help finance its activities, and this trading trust can refund costs to volunteers, pay wages, and gift profits to the charity or to any other charity. Thus, if the charitable trust is TRUST A, and the trading trust is TRUST B, the system as a whole works as per **Figure 14.4.**

The trust deeds state not only the purposes of the trusts but in addition the "will" of the trust is usually included, leaving its assets to an allied trust if this trust completes operations, closes down, or fails from lack of interest or of funds. Also, the trust deed gives an estimate of the *duration* of the trust; if this is intended to be "forever", then legally the statement is likely to be on the lines of "until 21 years after the death of the last descendant of Ming emperors" or some such legally indefinable period.

Trusts are durable, efficient, easy to administer, and of great public service; everybody should be associated with one! There are several small independent but cooperative Permaculture Institutes and allied groups in existence which have associated non–profit trusts operating businesses to fund them; in this way, many trusts are independent of gifts or grants, and become self–reliant for funds. It is estimated that France has 100,000 public interest groups, each with its own areas of interest and subscribers, and that about 10,000 form up annually; one can only suppose that others also fade out, their work redundant or completed.

As so few (dedicated) people can operate a trust effectively, it is far better to set up many such local trusts than to risk the power–centred inefficiences of a monstrous hierarchical system, such as some religious sects and foundations have become. These are essentially fossilised and no longer of relevance to ordinary people. Every dissenter or group of dissenters should therefore set up trusts to promulgate their own views, or form an independent trust in a cooperative network of like trusts.

Unless the formation of trusts is a common practice in a particular country, very few lawyers can set up (or even know about) trusts. They often give bad advice to groups, setting up litigious or cumbersome systems,

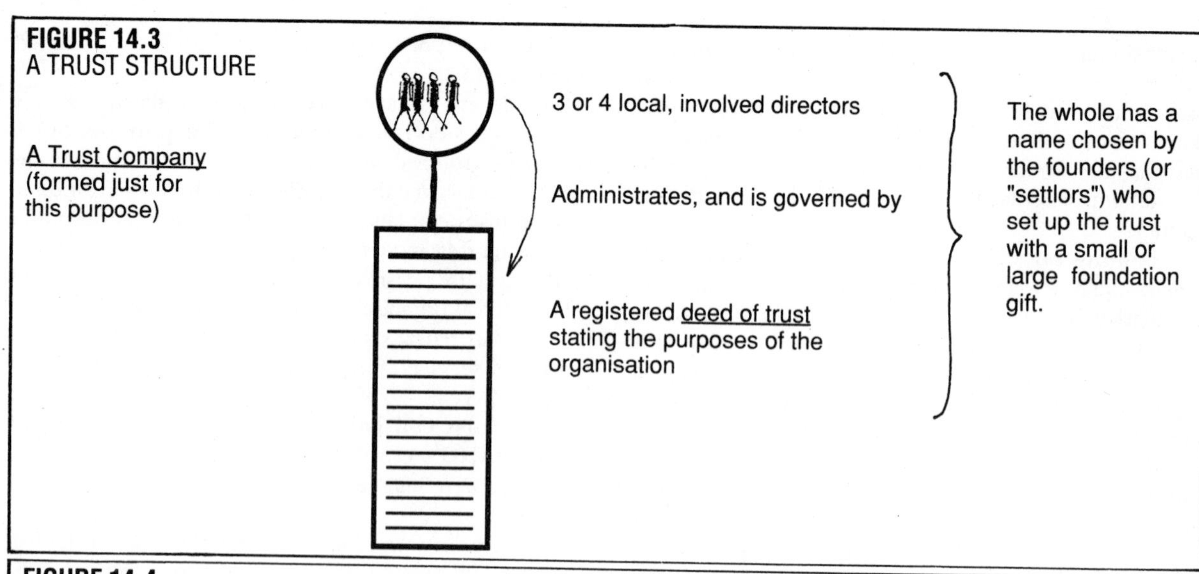

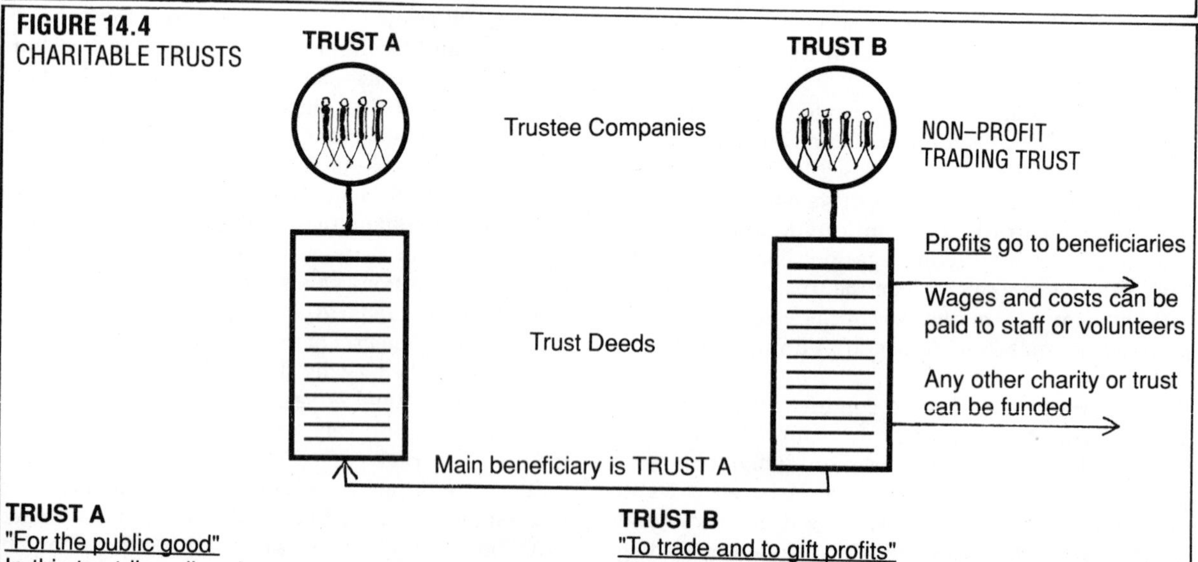

giving endless trouble and necessitating agreement among many people (an end which is, in practice, impossible to honestly attain), and which involves distasteful accomodations and compromises, explicit or hidden. Therefore, a careful search for the right lawyer is essential (corporate lawyers are often knowledgeable about trusts).

Other simple legal structures necessary to companies, cooperatives, credit unions, public investment trusts and so on are all well–outlined in company law, have excellent support services, and are routine arrangements. A good accountant to lay out the bookkeeping and give advice is necessary, as is an efficient office manager to communicate with the trust's target population.

14.8 DEVELOPMENTAL AND PROPERTY TRUSTS

(Appropriate to village development, land rehabilitation)

No investment in glamour stocks (coal, oil, uranium, city properties, paper pulp, agrochemicals, mining) is likely to yield anything to us but more pollution and to hasten global collapse. The evidence on acid rain alone (well documented) will convince any sane person that further "progress and development" will cause social and environmental upheaval.

We need to turn our money resources to truly rehabilitative ends. We accept the need, therefore, for accelerated reafforestation, the preservation of existing forests, sane village development, and the rehabilitation of eroded and misused lands.

In forming a development trust, our aims are not just financial, but also ensure community survival by community involvement. With good management and skillful work, there is no reason why this should not also pay for itself, or show a financial advantage to investors. It is an invaluable experience to model such a property trust, and to teach others how to follow any successes that we achieve.

A property trust purchases real estate for improvement, lease, or rental on behalf of many small investors who cannot afford to individually own or develop such properties. By improving properties so purchased, their value increases, and (under present rulings) taxation is not incurred on that increase in value if the property is held for 10 or more years, nor is the trust itself taxed on its income from investors. A "small business centre" can be a property trust. Many such trusts concentrate on city office properties or rural monocultures; we can concentrate on other aspects of property investment, as outlined herein for village development, or land rehabilitation systems.

The management group obtains backing from investors (via a public prospectus) to float a Property Trust on the investment market. The prime purpose is to give every person a chance to do more than object to or protest inappropriate land sales to overseas investors, land misuse, and poor planning, and to invest in saving critical or endangered national resources (such as wildlife and forests), while actively rehabilitating eroded lands.

In the first trust of this type, the aim can be to stop accepting investors at $2–5 million, which will develop a property or properties as listed in the prospectus. A low unit price ($100) enables even poorer people to invest; a single unit can be held by a partnership, society, or other corporate group so that even less money need be contributed per person in order to assist (e.g.) unemployed people. There need be no limit to the number of units held by any person.

Unlike other property trusts, investors should be given every opportunity to involve themselves in their investments via on-site work, consultancy, leaseholds, tree nursery supply, preference in sales of titles in villages, access to products or services, and (controlled) recreational access to lands and buildings. The trust can inform investors of any opportunity for their involvement at any level from volunteer or recreational use to paid consultancy, building, or in leaseholds available.

Funds can be used for the following:

- To set up the trust deed and management company;
- To pay for the work of the trustee;
- To pay management a retainer for their work; and
- To pay for normal running costs of the trust, including office expenses, printing, bookkeeping, accounting, valuation, and the travel of management to properties.

The precise amount so used should not exceed 4–8% of funds (based on figures from other property trust expenses), and the remainder is devoted to the purchase and development of properties as outlined in the prospectus. Costs reduce as trust income grows.

Any surplus or unused funds accumulating in the trust can be invested in ethical systems, including housing cooperatives, inventory for development projects, and shares in ethical businesses.

The specific project areas in which ethical trusts operate are:

- Purchase of threatened wildlife and forest habitat (usually with a specific plant or animal group in mind). Managed for the aims stated in a specific prospectus.
- Purchase and development of eroded, salted, deforested, or misused land for rehabilitation.
- Development of an energy-efficient, sustainable village on trust lands, for sale as developed, or pre-sale to raise capital for development.
- Special interest/group developments such as trout fisheries, aquaculture, or seed orchards for lease or sale.
- Bioregional development as purchase of community resources in a specific region(s) for use by residents, community groups, trust unit holders, or by residents buying trust properties in that area.
- Purchase of selected properties for assisting developing countries (overseas outposts).
- Bioregional clean energy systems or clean transport methods.

The way such trusts stage development and show a return is as in **Figure 14.5**.

14.9 VILLAGE DEVELOPMENT

As individual designers gain field or applied skills in house, energy system, and property design, and as ethical investment comes of age, the idea of "client work" can be joined to that of earth repair, and to real estate development. In order to do this, finance managers need to join forces with good managerial or design groups. The whole development group thus evolved can then purchase lands, capitalise them, and get them in order as a complex of lake, forest, and village settlement. We need well-designed villages today more than any other enterprise: villages to re-locate those soon-to-be-refugees from sea-level rise, villages to house people from urban slums, and villages where people of like mind can find someone else to talk to and to work with.

Villages can pool their surplus or current financial resources in a developmental credit union, and create land titles to sell in order to develop public service facilities. Nobody need pretend all problems are solved—conspicuous consumption can still ruin the idea of energy self-reliance—but *with good management*, the plan that follows comes very close to a sane village development.

An intentional village should have a group ethic acceptable to all who come there. Ethics, if shared,

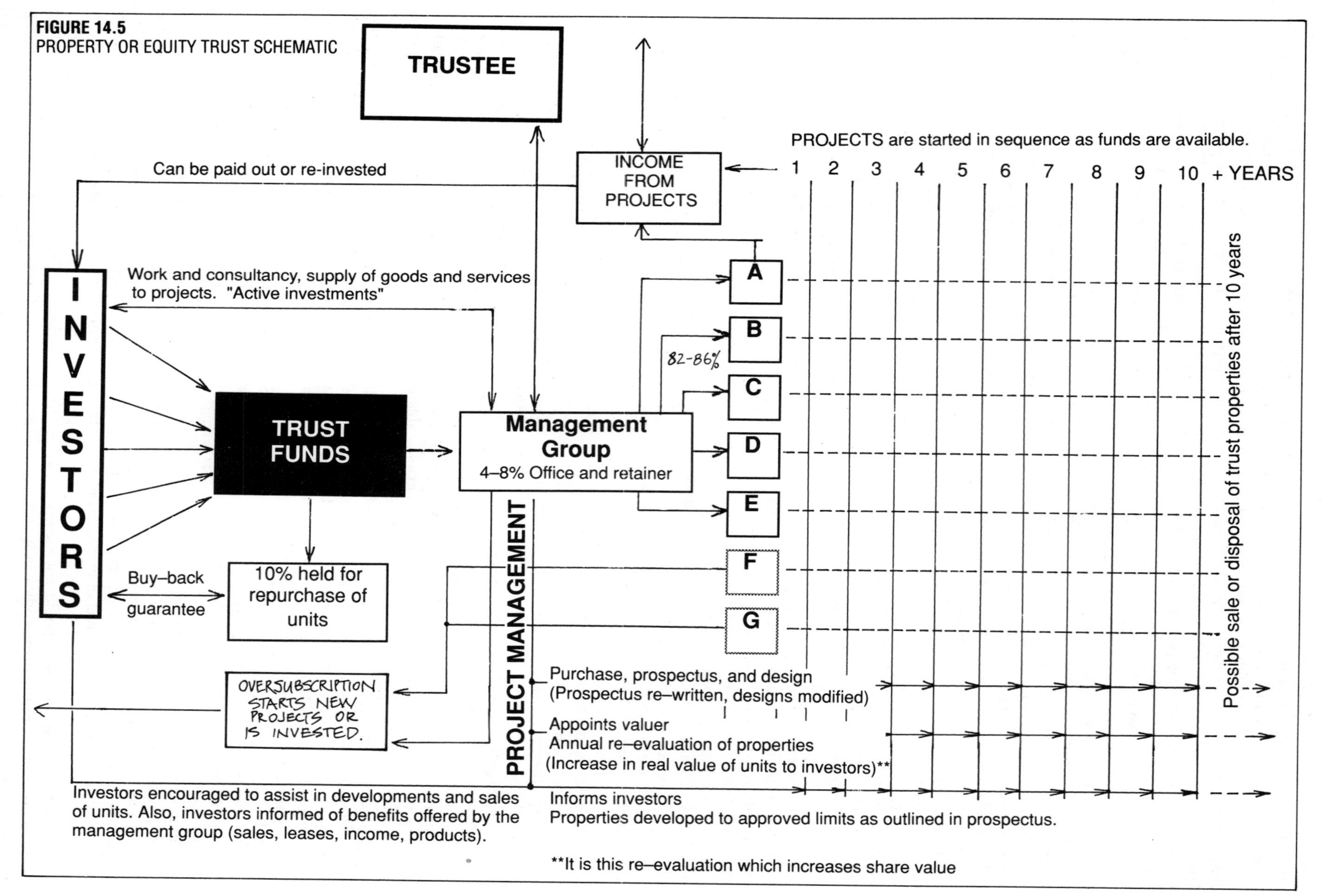

FIGURE 14.5
PROPERTY OR EQUITY TRUST SCHEMATIC

discussed, and acknowledged, give unity to groups, villages, and nations, indicate a way to go, and control our use of earth resources. They can be reflected in our legal, financial, domestic, and public lives.

The aims of a sensible village group might be to:

- REDUCE THE NEED TO EARN, by developing food, energy, and shelter self–reliance;
- EARN WITHIN THE VILLAGE IF POSSIBLE, reducing transport and travel needs; thus to recruit people who could fill most essential village occupations, or who are self–employed;
- PRODUCE A SURPLUS from services to others, thus maintaining a strong economy and outreach potential;
- PROVIDE MANY OF THE NON–MATERIAL NEEDS of people, perhaps of children in particular, by devising meaningful work, relevant education, and a rich natural environment; and
- COOPERATE in various enterprises and small associations.

A village can provide PRIVACY in homes and gardens; ACCESS TO TOOLS as leased, rented, or easily accessed equipment from computers to tractors; ENTERTAINMENT from local folk groups to video cassettes; CONSERVATION as a village wildlife, water, and forest reserve, and RECREATION in the near environment. It can also provide the BASIC LIFE ESSENTIALS of shelter, food, and energy.

No isolated or scattered group of people can self–provide for the above, but it is probable that about 30 to 200 houses can support these services and basic facilities, especially if there is planning for cooperative funding. What is easy for a group may be impossibly stressful for a nuclear family. It is possible for a group to provide many services, and for many people to earn a living in so doing.

It is quite practical to create such new villages without much initial capital in the actual development phase. This can be achieved in these ways:

- By receiving a gift of land to a land trust, intended for village development.
- To find suitable land for sale, and to take out an *option to buy* from the owner, dependent on pre–sale of the idea to buyers.
- To work on tribal or trust lands already communally owned.

To work with an *investor* or finance group to purchase land for village development, and to stage the development.

- To convene a *group* which wants such a village and use their capital to stage a village development.

The first few of these options presume a developer, are faster, and probably easier than the last option. All, however, need careful forward planning. The development *may* give considerable profit (but that is not guaranteed). In fact, fair or normal profits can be used to benefit both people and land, can give young or poorer people titles, and can rehabilitate landscapes otherwise neglected.

As a guide, 30% of titles available should cover (in value) all land and development costs, so that surplus titles are available for community access, profits, gifts, labour equity, and new project development. Land should be priced to local real estate values, and only very poor management would then show a loss on development.

For example, for 100 titles in the village:

Development Stage:

- 60% are sold by the development group to a village group at best prices (30% of that covers all costs, and actually sets the land price);
- 30% are given to the village group for later use;
- 10% are allotted by the developers as labour equity to such people as surveyors, earthmovers, landscapers.

Village Trust Stage (the titles given to the village group by the developer):

- 20% are sold by the village trust to fund village projects, and can be sold at lower (50% or less) value to low–income families;
- 10% are reserved for sale to crucial new recruits, e.g. medical, computer, energy–producing people.

OR

- 10% issued on easy terms to low–income families;
- 20% sold to finance village services.

SITE CRITERIA FOR VILLAGE DEVELOPMENT

There are a variety of locations that can be used for village development:

- TYPE 1. In a city block or suburbs.
- TYPE 2. Adjoining an existing village.
- TYPE 3. Within a part–vacant village.
- TYPE 4. Isolated from any existing integrated settlement.
- TYPE 5. On the site of a pre–existing but now vacant or destroyed village.
- TYPE 6. As a new suburban development.
- TYPE 7. Specialised settlements on coasts or near wilderness.

All need a different real estate and planning approach, so that Type 2 and Types 4 to 6 are probably outright purchase or option systems; Types 1 and 3 are part of a gradual takeover or buy–in system over some years; and Types 2 and 6 may have both factors operating at once, i.e. some land is purchased for development, while older village resources are also purchased for use. Type 4 is the pioneering or kibbutz approach and needs the most intensive planning, especially for water resources, access to market, and specified enterprises.

Type 2 is probably the easiest to plan and administer, and allows a whole graduation of involvement and committment. It also attaches to pre–existing essential services, although these are unlikely to be as useful or appropriate as those indicated here. Purchase of existing homes and lots results in little delay in the pioneering stage.

While the site choice may be very much influenced by opportunity, a criteria that is essential is that any village should be able to catch, store, reticulate, and clean up its own water supply. It is also advantageous that wood, wind, solar, or high-pressure water is available for energy production, and that clear ideas of how clean energy can be obtained, or developed and maintained, is part of the design.

Likewise, road, rail, boat, and not-too-distant air access are also advantageous for trade and travel. Computer and telecommunications will enable most villages to be in a data network, but real-object trade needs transport.

Finally, one cannot stress too much the factor of mixed ecologies. Any village which has access to or can develop forest, aquatic, marine, agricultural, and market areas has many more options open to it than a village marooned in a simple ecology.

Procedural Stages to Follow

1. Formation of a group *or* location of a site.

2. Arrange site option or purchase terms. Options are cheaper, often as little as, e.g. $50 per year for several years *if* the price offered is about 20–30% of the developed site value. The seller may retain a house title on the land if so required, and the price is then discounted.

3. Obtain an "agreement in principle" from the local shire or planning authority, establishing:

- Road type needed, e.g. gravel or sealed;
- Number of allottments allowed in the cluster or per acre;
- What services the developer must provide;
- The stages of development, if the shire requires building to be completed in stages.

4. Do careful sums, establishing prices based on roading, water supply, and sewerage.

5. Prepare a detailed and careful site plan and proposal for the village.

6. Convene, by advertisement, prospective customers and obtain firm commitments. Issue the proposal and site plan.

7. Obtain sealed permission for subdivision from the local authority.

8. Sell to prospective buyers, using a trust fund for road, water, and site preparation.

9. When costs are cleared, decide on future projects from profits and skills gained.

There are various ways to finance the process:

- Developer has all funds for survey and plans.
- Developer raises funds for proposal from investors.
- Potential buyers form a trust account to purchase and develop the land.

Some mix of the first two is possible, with trust funds established by the developers. These funds can be released when the development is fully approved by the shire council, and roading and services can be installed at those stages that may be demanded by the shire or region.

There are appropriate legal structures for a village. The developer needs to set up a trust (Trust A)—a land bank—to hold commons (village land) for the common good and for later development. Here, the developer acts as a foundation director (settlor), and should retire as soon as the site has 10–12 residents, who then assume directorship of the trust lands and cash assets for the village (about 3–4 directors are enough).

Residents should, as early as possible, set up a separate unit discretionary trust (Trust B) for trading operations. Such trusts are currently immune from company tax; they also reduce family tax and enable a wide variety of enterprises to be initiated by one organisation.

The essentials of Trust A are that it holds assets for the public good, does not take risks, and leases or rents to Trust B, which *does* trade and take risks, and has Trust A as one of its beneficiaries. Trust B can duplicate or triplicate itself to accommodate new enterprises and to insulate from risk those successful operations which may later develop. It can also handle financial systems such as leasing and lending units.

Although cluster and strata titles give privacy, negotiability, and autonomy to individuals and families, and titles and houses can be resold, traded, or given to Trust A, it is necessary to set up a land trust if only to administer the common lands, recreation reserves, and sites for future structures such as schools, restaurants, machine shops, or primary production.

In fact, as soon as possible after the developers pay land and developmental costs, they should seek to have village trustees elected. As the current costs of such trusts are minor in the total, and as they are so useful in planning and in income-earning, they should be part of all new village design. It is superfluous and unwieldly to have any more than 3 or 4 trustees for each trust, although small sub-committees can also be allotted part of any specific project work, and other people can contribute their special skills.

The developmental group works best as a small core of 2–4 people, each with special skills such as real estate, design, planning, or law. Surveyors, road-builders, builders, and landscapers are usually locally available for such projects. (Although these may also be developers, some contractors will work for equity in the project either as village occupants, or for resale at a later date.)

The development group should hand over a site design and user's manual to the directors of Trust A, who can display and circulate the initial design, record changes and modifications, and keep clear the essential land areas for productive use. Of course, all initial designs are made to be changed. The challenge is to change the design for the better! A design gives a starting point, not an end point.

Size of Villages

Human settlements vary in their ability to provide resources, to develop a high degree of self-reliance, and in their alienating or (conversely) neighbourly behaviour according to population size and function. At about 100 income-producing people, a significant financial institution can be village-based; at about 500 all people *can* know each other if social affairs are

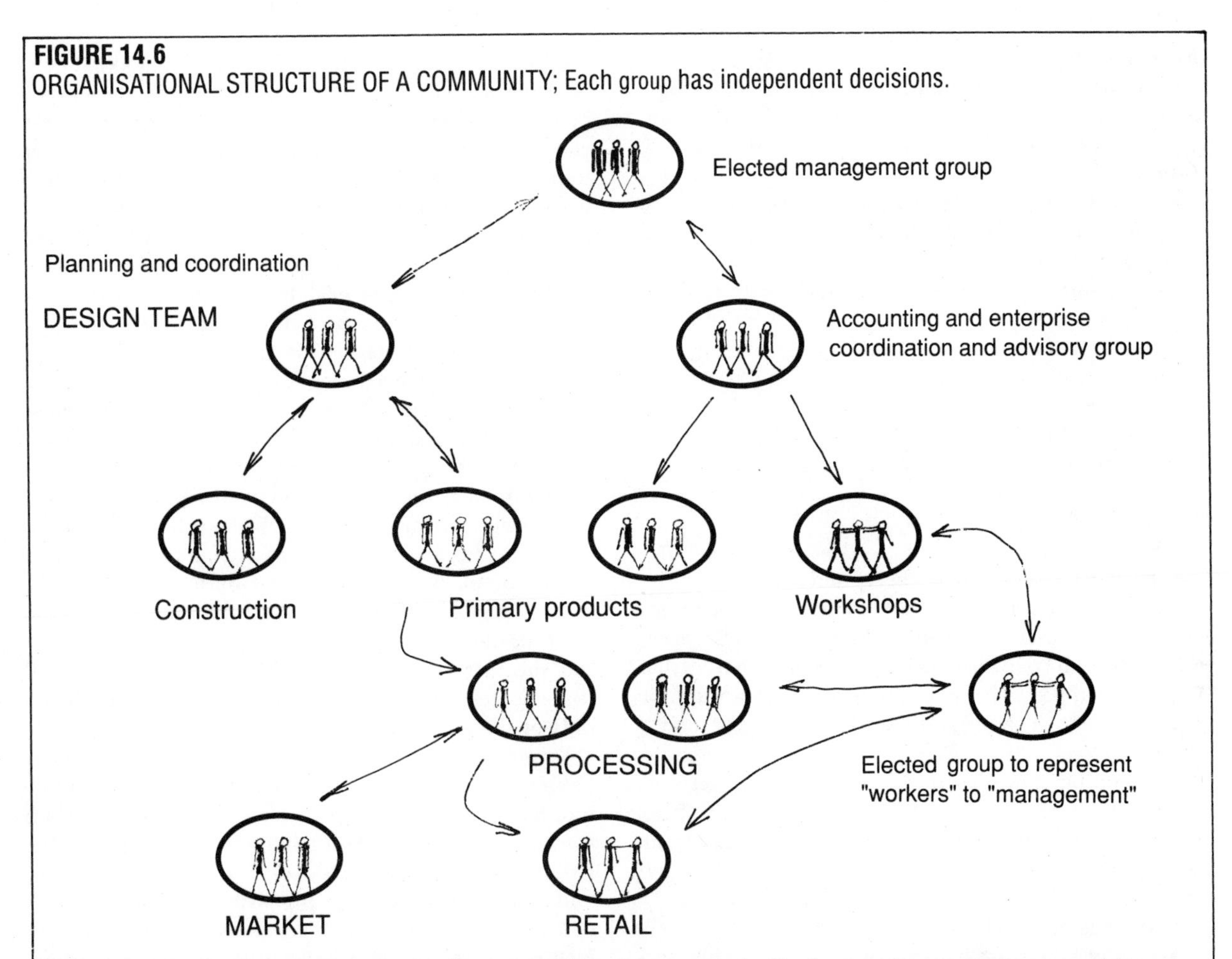

FIGURE 14.6
ORGANISATIONAL STRUCTURE OF A COMMUNITY; Each group has independent decisions.

Feedbacks are obvious; any group can offer external services or goods. Each group is separately accounted and manages their own money. All groups pay basic rent, rates, time–use fees for vehicles and equipment. A community of 30–300 adults. Includes light industrial commercial areas, arable and forest or orchard land.

organised from time to time.

At 2,000 people, theft and competitiveness is more common, and sects set up in opposition—the "ecumenical alliances" are lost. Perhaps we should start small, at about 30 or so adults, build to 200–300 people, and proceed slowly and by choice to 500, then "calve" into new neighbourhoods or new villages.

However, alliances of 200–500 household–size hamlets can make a very viable manufacturing or trading alliance and maintain a safe genetic base. Many tribes of 200 or so confederate to alliances of 4,000–7,000 in this way, share special products by trade, or arrange out–marriages. Thus, pioneer villages can seek alliances with others for the common good.

The Mondragon Cooperatives of Spain at first grew large (3,000–5,000), but later reduced to coopseratives of 300–500 to preserve the identity of every individual. Nevertheless, a group of such small coopseratives can make any vehicle or machine if each produces a part, and this is in fact organised by the smaller cooperatives in the Mondragon system.

In my view, the neighbourhood factor—knowledge of each other's names—is a primary factor, and has proved to be a major factor in survival in disaster, as assessed (e.g.) in the 1967 Hobart fires, where casualties in "anonymous" areas and commuter suburbs were many times higher than in neighbourhoods where people knew and cared for others.

Land Allotment and Village Infrastructure

Infrastructures for energy, commerce, and land allotment are an integral part of a self–reliant village. Few villages own all of these, however, and new villages need to reserve off land and areas for future priority development of both structures and primary product areas. A general plan of these resources must be published for all participants.

The following areas can be reserved for future use:

- School, seminar, and workshop rooms, computers, library, artwork, some crafts areas.
- Food processing centre, cafe, coffee shop, home baked goods, some commercial or surplus sale products.
- Noisy/oily work, woodwork, metal–work, machine shop, repairs, bulk fuels and oils, grease, air pump, vehicle service.
- Retail shops (including plant nursery retail), reception for visitors.
- Dairy and dairy processing centre.
- Domestic livestock housing, chickens, goats, pigs,

sheep, rabbits.

- Methane generator and ferment tanks, sewage, sullage, biomass conversion systems and water clean-up. All sites for energy and water provision are public reserves.
- Glasshouses, commercial crop, and special crop.
- Camp area for casual visitors (toilets and showers); can be a youth or student hostel.

Some community tools are needed for the site at large:

- Tractor – slasher (fire control)
  - posthole borer (fencing)
  - trailer
  - chisel plough/soil conditioner
  - chipper (biomass and mulch provider, fire control)
- Tow truck (mulch and goods)
- Van (goods)
- Back–pack slasher (blackberry and brush)

In workshops, essential tools such as drill press, lathe, radial–arm saw, welder, planer or thicknesser, and router can be available on lease or time share. Trust B can undertake a charge on any tools, accounting for replacement cost, wear and tear, fuel or power, and a service charge.

In planning, first designs should be for water and energy, then access, then dwellings and other structures. Next, landuse can be indicated, and finally legal, social, and financial systems discussed for the place and time.

Dwellings need to be of varying types:

- Family homes (individual or trust ownership; 2–5 bedrooms);
- Singles quarters (flats with strata title; some guest accommodation, 1–2 bedrooms);
- Elderly and hospital quarters (strata and community trust title; single and double bedrooms); and
- Terrace housing where appropriate.

A mix of such housing provides much more for needs and age differences than does the traditional family home. Every village could maintain one empty strata title for emergencies such as family break up, or to stage people in to permanent housing.

Recently some American towns have enacted ordinances to force buildings to comply with a 60% selfsufficient space heating requirement. Every house built today can be close to 100% efficient *by design alone*, at no extra cost in construction. Solar hot water systems are now routine installations, and photo-voltaics almost so. That is, energy needs are solved mainly in the home by a combination of good design and hardware. However, energy can be generated in other ways, and site allowances should be made for this.

The very modern "urban planning" where city or town sectors are designated as industrial, commercial, residential, or recreational are in fact the very antithesis of good planning for transport energy conservation, and bear little or no relationship to the zonation of function and available time around a settlement or house, as outlined in Chapter Three.

Wherever possible, life, work, and recreation should be integrated in a dwelling; not only are households better informed, children less alienated, and adults less isolated from social contact, but the need for complex transport systems is eliminated. We have a great deal to learn from older cities, which evolved in an energy conservative environment; cities such as Florence and Vienna, older parts of Berlin, and almost every village that functioned before 1930. In all such settlements, the cultural, crafts, trades, commercial and domestic functions were *integrated*. Old city blocks in Berlin have housing over street level shops on the sunny side, trades and work in the easily accessible interiors of *hofs* that penetrated and opened up the centre of the block, and a market or supply depot close to this assembly.

Such integrations are conducive to the development of complementary skills in the neighbourhood. In Istanbul and India trades may well be grouped in streets or market areas, so that both new materials, assembly, and sale are facilitated (and branches of each craft allotted, or adjusted to production by demand), but the total market or neighbourhood contains all trades except those based on rare resources or needed only on a regional basis.

Young people growing up in such an environment have a capacity to use many materials, or to make whatever they require, as a result of the informal everyday association with the 'open shops' that are the hallmark of small tradespeople, and where neighbours and family come and go the workplace itself. Davis (California) is one modern town where energy–conserving legislation is in place, and people are encouraged to conduct all non–polluting businesses from their homes; and where bikeways are available throughout the settled area. Elsewhere, "zoned" industries create vast traffic problems, and the separation of people from services.

How is the land not attached to dwellings to be allotted? The following categories are of use in villages everywhere:

- Land for public buildings and services.
- Land for recreation, trails, and walks.
- Water storages and reserves, energy sites.
- Productive land for food: livestock, orchard, nursery, greenhouses.
- Forested areas (fuels and timbers).
- Reserves for wildlife and flora; conservation areas.
- Easements, wayleaves, roads, and public utility allotments.
- Commercial areas for retail, manufacturers, offices, professional services, common markets.

Family dwellings and their 0.2–2 ha lots can accommodate some of these, but in miniature. Many homes are in fact commercial premises for home services and industries; many store their own roof water, provide much of their food from the garden, and may contain recreational assets. However, a larger site plan does allow more convivial access to land, some commercial crop potential, significant forests and ponds, and access and utility easements.

Public access and service centres owned by the village, such as food processing, freezer, and laundrette facilities not only provide a part–time income for a resident but sharply reduce the energy needs for each house to provide and maintain such facilities, and provide a wider district resource.

An even greater saving is realised by a modest tractor–tool–truck hire service, in which Trust B leases these infrequently used assets to residents as needed. In fact, a sensible village would closely investigate the advantages of a total vehicle leasing system, fleet purchase and insurance, local maintenance, and bulk fuel supply.

Village Energy

Coupled with domestic energy conservation, modest power units can supply small villages or regions with their energy, and can certainly be started by the same protest groups who rightly oppose giant coal, nuclear, oil–powered, centralised and polluting energy systems.

Like any other enterprise, a diverse approach is recommended, with energy from wind, tide, river, solar, and methane used where appropriate. **Table 14.7** on energy conversion efficiency has been compiled from several sources. It includes primary conversions (gas to electricity) and secondary conversions (waste heat to high grade gas). However, mechanical efficiency is perhaps the least important concept for people, and is relevant only if:

- The process is non–polluting, or relevant to the place.
- The technology is socially acceptable, and locally benign.
- The cost is affordable, and amortises under 10 years.
- The technology can be locally produced and maintained.

Finally, no matter how efficient a technology may be, if it lays waste to or destroys the basic quality or quantity of soil, water, or clean air, then it must be rejected, as this is the "economics of extinction". For this reason, I have not included fission processes with radioactive by–products, coal as now used, or mineral petroleum beyond initial or transition use.

Financing Public Services

When all titles sell, the monies generated by the sale of the 30% of titles vested in the village trust would ensure a very large interest yield annually for village development. This would, in effect, build fences, terraces, and eventually schools, workshops, and alternative energy systems.

At about year 5 of development, when residents have a clear idea of future needs, the capital itself can be

FIGURE 14.7
THE CONVERSION EFFICIENCY OF SOME DEVICES AND SYSTEMS

DEVICES OR SYSTEMS	CONVERTS (COMMONEST)	CONVERSION EFFICIENCY (%) Lowest	Average	Highest
Amorphous silica solar cells	Sunlight to electricity directly (cheap to produce in quantity)	2	3	5
Cow dung burnt as fuel	Dung to water heat (reduces available fertiliser compared with biogas)		7	
Crystalline photovoltaic solar cells	Sunlight to electricity		13.5	
Open fires	Wood to room heat (very wasteful of wood)		15	
Solar ponds for electricity	Sunlight to electricity generator (110 sq. m/kW)		15	
Humphrey liquid piston engine (pumps)	Gas fuel to water lifted (engine and pump). Very durable and trouble–free engines	10	17	20
Gallium arsenide photovoltaic cells (If these were cheap, all home energy could be electrical from 100 sq. m roof)	Sunlight to electricity		17.5	
Undershot straight–paddle wheels (see poncelot models)	Water flow to shaft horsepower	18	18	20
Modern photovoltaic solar cells	Sunlight to electricity		25	29
Stirling hot air engines	Solar to mechanical energy		20	
Solar flat plate collectors	Sunlight to hot water (Conditions vary widely)	10	30	50
Kenyan bell–bottom metal cookstove (also applies to fairly efficient "pot belly" stoves)	wood to cooking heat		30	
Gas turbines	Gas to electricity (waste heat produced)	30		

Wood burnt to generate power	Wood to electricity		33	
Piston air compressor on water wheels	Shaft horsepower to compressed air	30	35	
Typical diesel engine	Fuel to horsepower		36	
Turbo–charged diesel engine	In production (ceramic turbines)		41	
Adiabatic (cool) diesel	Prototypic in 1984		48	
Gas fuel cells (not widely in use; models are in operation)	Gas to electricity	40	50	60
Solar attic (Compared this with an open fire)	Sunlight to heated air		59	
Well–made slow combustion stove (May pollute air)	Wood to hot water, cooking, and space heat		±60	
Cow dung converted to methane (Useful fertilizer residue)	Gas to cooking heat		60	
Elevated dams or ponds as "batteries"	As energy storage (Water pumped up with off–peak electricity)		60	
Undershot Poncelot (curved vane) water wheels	Stream flow to shaft horsepower		60	
Breast water wheels (curved vanes)	Water power to shaft horsepower		60	
Organic Rankine cycle engines	Thermal efficiency only		65	70
Concentrating solar collectors	Sunlight to water heat		65	
Propellor turbines (hydraulic)	Streamflow to shaft horsepower	65	68	70
Farley Triplate wave machine	wave motion to compressed air or water lifted		70	
Plenum heat exchangers (Combine well with a solar attic)	Air to air, save waste heat		75	
Rotary hydraulic turbines	Water at head to shaft horsepower	75	78	90
Pelton wheel coupled to air compressor	Supplies compressed air		79 at motors	
Solar ponds used for heat	Solar radiation to water heat		80	86
Heat cells	Heat stored to heat given out		81	
Well–designed overshot water wheels	Streamflow to shaft horsepower	75	80	82
Diesel co–generation (providing the heat is used constructively)	Diesel fuel to heat (60%) and electricity (30%)		89	
Biogas conversion	Heat rating of material to gas BTU		90	
Thermal mass heater (compared with most stoves)	Wood to heat	80	90	92
Good pelton wheels or Turgo impulse wheels	Falling water to shaft horsepower		93	
Inverters (electrical)	12V to 110 or 240V	75	90	95
Electrolysis of water	Water to hydrogen fuel at 200°F, at 250 atmospheres		92	93
Unglazed fast–flow collector (Stream flow over black surfaces)	Solar radiation to water heat		±100	

AMPLIFIED EFFICIENCY		**Factor of gain**	**Efficiency**
Heat pumps	Extracting low grade heat to useful space heat or cooling (as compared with using the same fuel more directly with heating)	(x 3.5 to 7)	350%
Compressed air heaters applied to air lines	Expanding cool compressed air just before using in engines (compared with same fuel used in engines)	(x 5.5 to 6)	5-600%

used to install wealth–producing assets for employment in the village (glasshouses, computers, machinery). By allotting 30% of sites to the body corporate (Trust A), the developers ensure that:

- All necessary public services are funded;
- Residents, over a period of 20 years, have effectively paid nothing for land; and
- As soon as sales exceed 60% of total titles, either capital or interest for public services is available.

The Trust can use this asset in many ways, but would be most effective in ensuring either conservation of energy or business development on site, and in using the common wealth for increasing local productive assets. If a credit union has been established, much of the trust capital can be transferred to a loans account, so that residents can draw on it at low interest for local occupational development. Conscientious use of a credit union by village residents would greatly increase capital flow to village enterprises. There may be some capital available as housing loans, but a building cooperative would be much more effective in this instance.

The result of having such capital and interest flowing in to the Trust is that village morale is greatly increased, with every resident seeing long–term plans fairly rapidly achieved. Well– managed, the capital should actually increase, giving a large annual capital for village use.

The Trust will always need income for maintenance of roads, fences, water supply, fire control, and other site factors. This can be raised from small charges for leases and loans to residents, by charging an hourly rate for lease, use, or hire of machines and facilities, and by a 10% levy on net profits of locally–funded cooperatives. This levy is the same as that paid by the Mondragon cooperative system to fund their banks, schools, and research facilities in Spain, and applies only to net profits of trust–funded cooperatives.

A Community Services Council at any new community or village with common lands may be elected to administer policy on publicly owned or common assets, to collect lease monies on utility plots, to administer funds for schools and medical services, and to see that rates are used in maintaining roads, water supplies, and other public services.

"Community Services" in any community can encompass the following:

- Fire control services
- Land leasing of common properties for business use.
- Public reserves and sports grounds
- Ambulance and health services; cemetery administration
- Education and educational facilities
- Community centre: use and hire
- Power and water supply
- Forest management
- Tourism/promotion (if desired)

Some of these are income–producing, some subject to state aid or tax immunity, and some are income–consuming. The Services Council needs to attempt to balance these costs, allot land and assets for income to service groups, and ensure that common property and rights are fairly assigned and well managed.

Council should be comprised of a selection of those *active* in the above areas, not of an uninvolved group. Each area of action can have a basically independent management sub–group, reporting to Council regularly. These sub–groups can appoint one of their number to represent that group on Council.

Council needs to meet monthly, or even more infrequently, with the sub–groups normally handling everyday business within their budget and allotted areas of operation. A Council can call for and act on submissions or reports relating to specific policies and strategies. Sub–groups can raise their own funds (as well as have access to public funds) may have an active business management role in income–producing areas, and supply workers to carry out such businesses.

Income is needed by a community for the continued upgrading of public services. There are several ways in which this may be done:

- LEVY: A yearly levy on each family or property can be collected by the Community Council. This would probably be in addition to the normal Shire Council rates, and may create a hardship on some households.
- TITHE: Individuals, groups, and businesses (including the cooperatives) can tithe 10% of net profit to the public services. All the Mondragon cooperatives do this, and so fund schools, libraries, churches, and recreational grounds. This means "rates" would be on a sliding scale, not a flat rate for everybody regardless of earning power.
- LOCAL CURRENCY: The Community Services Council can print its own currency, backed up by a valuable community resource (firewood, clean bottled water) to fund projects, new buildings, etc. Local currency operations are detailed later in this chapter. A timber forest planted early on and harvested at various stages of maturity can produce most income needed for subsequent years, as it does in some modern towns which have developed social forestry.
- CHARGES: Charges by the school, community centre, and health services can be offset against costs on a "user–pays" basis for tuition, use of facilities, medical needs, etc. This is a method which should be used in conjunction with another model, as it covers only those public facilities which not everybody uses or wants; universally used services are usually pro–rated ("rated").

Potential Enterprises and Occupations

It is of great advantage to analyse just how village occupants can self–employ in service to the village itself and to nearby districts. Let us presume a 50–house (100 adults) village situation. Costs are high in three areas: food, energy, and transport.

We can now speculate how residents can earn their living in the village. Much depends on a village development credit union which is founded by the village under Trust B to serve the village needs.

• *Food*: About 20 adults can support 1 adult providing a food supply; thus, 5 to 7 families can earn a living from food provision, e.g. open air market gardens, glasshouse crop, co–op store (with local trade), domestic livestock and fishery enterprise, and a part–time livelihood from cafe, food processing, and baked goods. All surplus can be sold off to visitors or locals from outside the village proper via a village market.

• *Energy* Energy needs include space heating, cooking, electric, hot water, gas, and appliances. To establish energy–efficient systems, plans need to include prior excellent house design, design for the retrofit of inefficient houses, insulation, and good appliances. Most need capital to start, but almost all amortise in 3–7 years. Two to five families can earn a living from house design and retrofit, with some contract and extension work outside the village, hardware and appliance sales to the village and to the district, and installing, repairing, and tending a village energy system. Over time, some manufacturers in the village would enable a balanced trade in appliances.

• *Vehicles*: These are perhaps the most costly item and need a careful and planned approach, better on a village network level than on a one–village system. Two to four livings are indicated in service, repairs, maintenance of village vehicles, bulk fuel and oil supply; in growing, distilling, or fermenting fuels for engines; and in fleet lease and insurance, special vehicle lease, tractor lease, etc. Village alliances can establish special engine or chassis manufacture.

• *Financial*: Handling the income, loans, accounting of enterprises, and running a credit union is an essential job for a village. A computer is certainly needed, as are accounting and managerial skills. Two people could perhaps handle the financing. A community credit union, holding insurance money as well as incomes, can fund enterprises such as energy systems, glasshouses, dairies, and food processing centres.

• *Medical and pharmaceutical*: One or two people can offer medical and pharmaceutical services to the village or area, including prescription, massage, counselling, and local treatment. Some medications can be made and sold more widely, and specific remedies grown or manufactured. Community health should always be based on prevention of most illnesses.

• *Building*: For some time, a plumber–builder–mechanic can serve village establishment and later maintenance needs locally, and can produce useful furnishings and sale items for extra income, if a woodworking area is made available.

The above occupations cover the essentials of shelter, food, economy, energy, transport, and health. This would initially fund about 20 of the 50 households, and more as manufactures develop.

Other than the essential occupations, there are a range of potential village enterprises. Some can be based on land resources (glasshouse crop, special crop, cut flowers, herbs, pharmaceutical, processed dairy products, fish and aquaculture). Others can develop from local skills: teachers are needed for children, for adult education, and for applied workshops on site. Careful forward planning can yield one or two livings in workshops (craft, medical, or design). A small business service centre may be needed in the mid–term, employing 3–4 people.

Consultancy for other sites in architecture, landscape, and design is possible, as is implementation and provision of plant materials from a nursery on site, which can further develop special crop for site, fire control, bees, orchards, or forages for animals (comfrey, tagasaste, etc.) Some people may like to cooperate in an animal–breeding programme for special poultry, pigeon, sheep, or goat breeds as a small stud.

Computer services to a network, programming, and data bases on special subjects are now in demand, and can be placed in homes. Publishing is greatly assisted by computer word processing and allied computer typesetting services.

Trade, as distribution rights, wholesale potential, import–export trade and village trade networks are yet another probable enterprise, as are craft products from metal or wood workshops, pottery, and art. There is a modest income from guests, visitors, and site tours if the village concurs on that aspect, and from sales to visitors and travellers in the district; educational services and accommodation are much in demand.

All of these, and many others, need little transport. Many can operate on site, and the cooperative store would serve as an outlet, or other leasehold retailers can offer goods and services for the village. Physical therapists and paramedics, especially in massage and stress–related problems, often find plenty of customers in a rural district.

These enterprises depend on two basic factors: capital (enough money to start up and develop), and management (careful accounting, forward planning, market research and development, product development, sensible costing and staffing, correct lease and rental agreements, appropriate legal structures); thus a small business service centre is economical and necessary for 20 or more businesses.

Many small enterprises yielding products or services can pre–sell their wares for initial capital, and many others need no capital to start up, but these may need time to develop. Some small businesses (massage, paramedical) need only skills to start, while others (market garden, orchard, furniture making) need skills, capital, and time.

A careful assessment of skills and available capital will indicate priorities in any village. Income–earning for local development is a priority, while other capital–intensive schemes (e.g. commercial pottery) need to wait. A very reliable early income can be made from information, by way of workshops and classes, although workshop income fades over time as students gain skills themselves. However, workshops yield capital for further developments locally, and a village can, in fact, develop as a special education centre if enough skilled people are attracted to that idea.

It should be the long–term aim of any village to own and operate its own employment enterprises. In past times, it was unusual for a villager to hold just one job; the banker was also a part–time barber and trader and perhaps gardener. Thus it is wise to share even simple occupations, so that individuals have occupational shares in 2–3 enterprises.

In this way, total failure is unlikely, as is unemployment. Holidays can be taken, and wet days spent on indoor work. Thus, in every occupation, job–sharing should be the rule, not the exception. Although the total village structure is complex, the work of any individual is simple, as is the case with a plant in a polyculture.

RECYCLING IN THE COMMUNITY OR NEIGHBOURHOOD

The borough of Devonport, in the city of Auckland, New Zealand, has a total solid waste–recycling system, and from this conservative endeavour manages to return a cash benefit to households (versus the cash payment or rates paid to Councils in non–recycling areas). Data is available from the borough, but there are two or three key features that make the system work:

1. The borough issues an annual calendar, colour coded, and picks up *only* one or other category of waste on any one day, e.g. clean glass, metal, tyres, paper, organic waste, oils, etc. No other, and no mixed loads are collected. Thus, recycling begins with separation of wastes by the consumer.

2. At the waste disposal site, loads must also be in one category. Here, wood is sorted into useful wood (a community woodwork centre is available), firewood (issued to elderly people), and chipping or mulch wood. All organics are composted by a small tractor windrow system, and all oils are collected for re–sale as lubrication oils after filtration. Compost is sold, as is mulch, and any surplus is used to carpet a "ziggurat" (ascending spiral ramp) made of broken pipe, brick, clay, concrete, and clean fill; as the ziggurat ascends, community organic gardens follow the fill.

The waste site is supervised, loads directed, and mixed waste sorted; saleable or recyclable items are grouped in clearly marked areas.

3. People who will not sort their waste (about 4%) must buy a strong plastic bag from the borough (hire cost $7) which itself is recleaned for use. The charge on the bag pays for the bags, into which people separate the waste, and for the calendar; the borough also gets income from the sale of paper, glass, metals, wood, plastics, and compost.

Obviously, the opportunity for local co–ops to enter into recycling industries is there; many small local industries can buy wood, glass, paper, or oils from the borough. Such examples dictate that no Council has any excuse for *not* recycling; not only does waste cost the ratepayers money, but there is also a vast waste disposal problem. It is up to ratepayers to elect officials who *will* recycle sewage and solid wastes, and to vote out waste–promoting councils, who "cost the earth"!

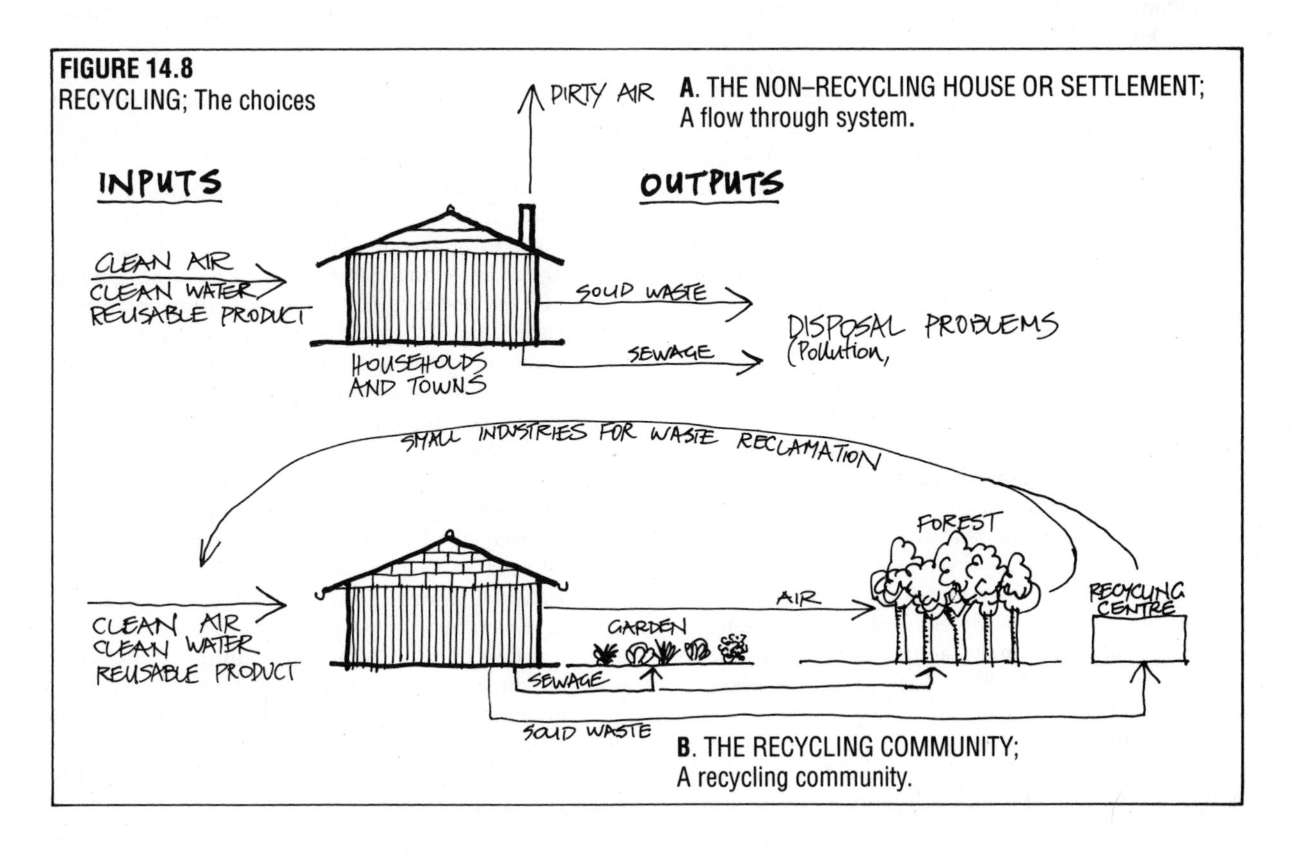

FIGURE 14.8
RECYCLING; The choices

EVOLUTION IN COMMUNITY

No group can achieve financial self–reliance overnight, but within 5–7 years of a determined start, a cooperative group using their creative talents can succeed in making a living for themselves and building up a strong business sector in their community. Any person can feel a sense of social cohesion and group spirit in such a situation. The danger point of "going under" is past, and now it is time to think of diversifying and disinvestment before the group gets too affluent, or too big. At some point, therefore, a decision must be taken to take some positive action (and avoid the fate of affluence):

1. To hand over to other some income–earning but superfluous (to the existing group) enterprise, and so reduce income.

2. To extend aid and services to areas in desperate need, such as those experiencing real poverty, natural catastrophe, or medical insufficiency.

Many groups put off such actions, but it is better to start them (at modest levels) very early in the whole process. There are many ethical groups and individuals who regularly tithe 10% of their gross income to such endeavours. Most of us do not have large appetites; we wish only to have a shelter, enough food, some small luxuries, money to travel, and friends. These are modest needs to achieve; beyond them lies adventure in helping others get on a firm footing. The only real security in life is a secure society of interdependent people, thus the only valid "defence" is aid to others.

Any village group can help others become more self–reliant and give sound management advice to new groups. Fiscal management, like energy management, also needs *social, environmental,* and *ethical* accounting. Money is of no use if its ends are destructive to society, life forms, or values.

Cooperative groups, communities, associations, and shared work groups fail in part to foresee and to plan for evolution. Also, more tragically, to educate their children in the basics of village systems. If any intelligent, hard–working, and ethical group pool some resources and take only a fair living wage, then they *must* amass spare capital in time.

As and when independent villages do achieve an identity, an ethic, and unity, beneficial connections can be made ranging from radio and land links to bulk purchase, trade, and share facilities, so that coastal, urban, arid, tropical, and primarily rural villages can access and share the resources of others on an agreed–upon lease, hire, or exchange basis.

Village coalitions can fund and operate larger systems such as mutual investment funds for special purposes, engage in manufacturing on a reasonable scale, and exchange skills and strategies. At present, few villages have the initial sound legal, financial, and social structures to achieve this.

There is no reason why a village could not own and operate a boat, trucks, or pack animals to facilitate trade, why a mountain or urban village could not purchase and manage a foothill farm for food production, and why an inland village could not finance part of a coast development, as many villages already do in India (along the Ganges) where towns and regions own pilgrim houses for voyagers. All these strategies enrich village or regional life, and give access to a wider world; this is particularly important for children and young people.

14.10

EFFECTIVE WORKING GROUPS AND RIGHT LIVELIHOOD

In any human group endeavour, there are practical and effective, or impractical and ineffective, ways to manage a complex system. Impractical, frustrating, and time–consuming systems are those governed by large boards, assemblies, or groups (seven or more people). These "meetings"; have a chairperson, agendas, proposals, votes, or use concensus, and can go on for hours. Concensus, in particular, is an endless and pointless affair, with coercion of the often silent or incoherent abstainer by a vociferous minority. Thus, decisions reached by boards, parliaments, and concensus groups either oppress some individuals (votes) or are vetoed by dissenters. In either case, we have tyranny of a majority or tyranny of a minority, and a great deal of frustration and wasted time.

The way to abolish such systems is to have one meeting where the sole agenda is to vote to abolish decision meetings—this is usually carried unanimously!—and another where a concensus is reached to abolish concensus—this too shouldn't take long. What do we put in place of such impediments to action?

In every group, there is work to do. This work needs to be set out clearly, as jobs or tasks. Tasks fall into two categories: those which are creative, productive, or constructive, hence pleasing, and those which are basically maintenance (domestic, office, and garden work). Of the first category, we seek volunteers to take up the tasks, and if they come forward, we ALLOT that task to them, agreeing on a timetable and stages of completion. Of the second category, we ROSTER people to do the work, laying out a worksheet and a (usually weekly or monthly) roster.

Wherever no volunteers appear for any task, then the group as a whole contributes a tithe to pay for the task to be carried out by a contractor (as in many trade tasks); thus all work gets done one way or another.

An essential strategy for rapid and flexible action is to limit the number of people responsible for any one area of action or task. Some ideal number is between one and three individuals, who *manage independently*, but who may work to a general plan and schedule to fit in with others. Completion dates are set and notified to all people, and some form of report, diary, or plan is made public or minuted.

Thus, we can form *small* groups of one to three people who are responsible for management of a specific area of activity. It is a fail–safe strategy to attach occasional understudies to this small group, or to stand ready to duplicate the function if it is not being administered. (There is no more time–wasting process than that of believing people will act, and then finding that they will not.)

This "troika" approach (1–3 people per function) ensures that meetings in any one area are few; news can come out as reports, available to anybody. It also means that no one person or group has "rights of decision" over other functions or groups. Unfortunately, despite our most devout wishes, there are very few people who can start up and maintain a function; we are lucky if we can find 6 or 7 of these in any group of 30–40 people. Thus, for all functions needing entrepreneurial skill, we need key people.

However, there are many functions (from crafts and arts to gardening and building) that do not need entrepreneurial skills, but which follow if these resources are available. Thus, many people can be involved in primary production, processing, and building if only a few can manage the essential coordination and funding.

In such a web of function, any one person can be in two or three teams, thus achieving a "portfolio of occupations". Also, each group depends on each other being in function, and this is important for group unity; we presume a shared ethic and values, which are clearly spelt out, but do not assume love, trust, or any particular form of personal diet and behaviour except in line with ethics (we are never perfect, just moving towards improvement).

1. Only in the initial planning do people need to assign or choose functions; once chosen, no group meetings for business are necessary.

2. Each group or sub–group is small enough to reach fast agreements and know of each others' movements and work.

3. No consensus beyond that of an initial ethical and value consensus is necessary; everyday decisions are made by small groups.

Certain behaviours occur at various group sizes; here are some approximate size and function groups:

- 1–3 people: Executive decision, least meeting time, greater pressure to act, fast changes possible, fast replacement of key people.
- 4–6 people: Good volunteer or cooperative group work, or work group for special single projects; good size for work exchange systems.
- 7–20 people: Function well only in social conditions; can be a recreational group or team, but at 7 or so, a *chairperson* is needed and decisions are slow and frustrating, often creating dissent.
- 30–40 people: Acknowledged as the minimal group of people in which most human functions can be covered, and who (if well chosen) can cope with almost any type of problem.
- 40–200 people: Rarely found as a group or settlement, but a good size for a regional organisation.
- 200–300 people: The basic number for genetic variability; such a group can, by careful breeding, maintain their numbers as a tribe and allow for some losses to disease. Probably the minimal human village size (called a hamlet).
- 300–600 people: About the limit at which people know every other person by name; thus, about the limit of "identity". This is the largest satisfactory size for educational or learning systems if personal attention is valued. Acknowledged to be the upper limit for successful cooperatives for real participation.
- 1,000–5,000 people: Usual upper limit of federations of tribes; a good size for a bioregional group or sub–region. Also, a village size limit. Cliques, theft and cheating common and possible; hierarchies are needed.
- 7,000–40,000 people: Towns, large bioregions. Chinese communes start about here. This number is not satisfactory unless broken into small cooperatives and villages. Crowds and very large audiences can reach this size, and can be difficult to control if aroused. It is about the upper limit for any real control by strict hierarchical systems.
- 15,000–10,000,000 people: Cities; mainly disorganised on every level. Effective anarchy and crime, and social isolation in many areas.

As even very small numbers of people (4–6) can be very effective, it is better to set up independent but friendly alliances of small groups than to coalesce into groups of 600 or more. Any alliance of 4–10 villages (12,000–50,000 people) can, by agreement, run a sophisticated trade and travel organisation.

Most groups start with 1–3 people, and recruit slowly. Slow or organic growth is easily coped with, while sudden influxes can be disruptive. As groups pass 30 or so in size, it may be possible to contemplate selective recruitment to make up deficiencies in skills, sex ratio, age differences, or specialties needed.

Where a very large number of people is needed for a job, a calendar is set and conveners let everybody know when and where the personpower is needed. Thus, every larger group needs to delegate responsibility for work to smaller autonomous groups, who are trusted to do the job, and only replaced if they persistently fail to do so. In this way, every person who wants to work controls their work, and non–involved people have no say. This eliminates control by inactive people in tasks they are not familiar with, and nullifies power seekers.

As for dissenters, there never is any impediment to their setting up their own ideal system, and living in it; or setting up a parallel work group to show how it should be done. Above all, there is *no one way* to do anything. "One solution" systems evolve from the concentration of power in one or other form of dictatorship (business, government, or military).

In this way, all group meetings can therefore be social and convivial, and for information exchange. As these meetings are pleasant, we can look forward to them, and so a pleasant and informative occasion replaces a frustrating and stressful "group decision" meeting.

On a wider scale, cells of one to three people can run a very large network; in this way, given occasional (every 2 years) meetings in affinities or work integration groups, attended by small autonomous groups, positive WORK–NETTING (*not* pointless networking) is possible. No person can force another group to cooperate, but must offer reasonable, rewarding, and fair cooperation.

Above all, no group or community need last forever; a group set up to achieve certain ends can disband with a clear conscience if those ends are substantially achieved. Individuals can then take on new and more current tasks, or adopt a different level of effective action based on past experience.

No–one would deny that people are *the* most difficult factor in any design or assembly. It is not that people lack the will to cooperate; it is more often that they have not adapted those sensible legal and administrative, or social mechanisms which allow them to cooperate. At various periods of history, usually coincident with economic downturn, groups of people have left mainstream society to set up intentional communities. This phenomenom occurred in the 1890's, 1930's, and 1960's and at various times between. The most recent (1960–1990) is also the longest period of out–migration, and is still continuing after 20 years; it is a migration of skilled family people towards a smaller society.

Studies of such groups reveal that those who were effective adopted a set of values which ensured their continued internal and external interdependence; of those, perhaps the most important factor was that the group adopted "voluntary simplicity" as an ethic. It is no mere coincidence that there is both an historic and present relationship between community (people assisting each other) and a *poverty* of power due to financial recession.

Thus, the legal and ethical basis for successful community cooperation must stress sharing, trusteeship, and modest consumption; the latter is the more important, as individual power over land, real assets, finance, or group membership leads inevitably to power over others, and we are back where we started. The habit of frugality is perhaps the most important of those assisting other life forms.

Like landscape planning, there are community systems which can cause more time spent in conflict than can be made up later; such errors we can still call *primary errors* as they will lead to constant problems and expense later. Some of them are:

- Group consensus on all decisions (tyranny of the dissenter).
- Group leadership by one or a few (tyranny of the leader).
- Rules about lifestyle (tyranny of proscription).
- Attempts to live in large group households (lack of privacy).
- Overscheduling of meetings.
- Neglect of income–producing activities.
- Poor financial accounting, hence poor maintenance of facilities and equipment.

The individual in the community must recognise the need to subscribe to a group fund for maintaining roads, fences, and infrastructure, or to donate work in lieu of money on a regular basis. There are no "free" machines or free lunches. The essentials to concentrate on are sound land planning, shelter, a capital base, and the development of livelihoods.

Many "communal" systems fail if very few people are legally liable for capital risk. Good ideas and equipment cost money to implement, thus *all those who vote for equipment must be made equally liable to pay for it.* This always keeps the community healthy, as unused tools are expensive, but only for those who buy them! This is a lesson in modesty and responsibility. Moreover, tools on hire should completely repay their cost by charge on a piecework or hourly rate, over a period realistically estimated as half their working life (vehicles, tractors, office equipment and so on).

Every major tool needs to be costed for running costs, repairs, and replacement *plus* any interest on capital. The very powerful principle here is that "everything must pay"; more specifically, in community enterprises "proposers pay capital" and "users pay costs". In natural systems this is the "law of return". We cannot use soils, crops, or forests without costing total upkeep and replacement, or we impoverish the common wealth. Thus, "users pay" should apply at every level of community, except for hardship or welfare services.

There are two unhappy states of human existence; the first and worst is to be defined by your community or nation as unemployed, that is to say, of no use to anyone. In a world where such a great deal of work has to be done just to repair past damage, replace forests, secure soils from loss, house people, or build local self–reliance, unemployment is an obscene concept. Where relief benefits are paid, the state rewards people to accept this role of "no work", and in effect fines them if they work.

Secondly, it is an unhappy state to be employed, but not free to use initiative; any person can go daily to a job, no matter how useless or boring, no matter how destructive, and be paid to be defined in a single role, e.g. as a teacher, clerk, process worker, or labourer; the worker has no say in policy, social value, hours, product quality, or environmental worth.

In most cases, *other people* define the lives of the recipients of relief or salaries; as all such money comes from the pool of public wealth, then all such people are, in effect, on "relief payments", just as a company supported by public subsidy is on public relief (usually our primary production systems in the western world).

The only people who are *self–defined* are those who are self–employed, or who work in community work cooperatives. The consumers pay for their products or services directly, and their houses, products, and choice of work is self–determined; they are only unhappy to the extent they oppress themselves! I could never understand why people struggle to maintain a job down a coalmine, especially when their pooled capital and labour could create a forest, with all the pleasure one gets from working in the open air, and the varied work a forest provides. We can all seek for right livelihood to do work that assists in caring for the earth or other people, work that is congruent with our beliefs.

When we discuss the principle of "commonwork", or study the varied roles of an individual in a village, we can see that no person is just a miner, or clerk, or banker, but that on different days one can be a banker, forester, bee-keeper, writer, printer, or carpenter. It is only the combined pressure of trade unions and monoculture industry that keeps people bound with the invisible shackles of custom to those unguarded slave camps termed industrial suburbs, with all their malnutrition, poor housing, and human suffering.

In boring work, or where people are deprived of intellectual life, emotional life may dominate and so their lives become a drama or series of dramatic events. A balanced life has all three outlets, so that contented people may spend part of their time in: physical exertion (walking, gardening, sport); intellectual pursuits (design, research, education); and emotional–sensual areas (celebration, contemplation, love).

A healthy and balanced life consists of being able to access all such pursuits. In modern life, some time spent in primary production or in manufacture, some in service to a wider group, and some in relaxation–celebration is an ideal; few achieve it. In Central Australian tribes, at every event there may be three "function" groups (independent of totems or "skins"): one group "knows" or records time—*orchestrates* (the intellectual); one carries out dances, increase ceremonies or activity (the physical); and one encourages, applauds, and appreciates (the emotional). Thus, every person fits a matrix of totem and function (**Figure 14.2**).

14.11 MONEY AND FINANCE

In small and unified groups (tribes), what is achieved by financial systems elsewhere is achieved by a set of exchanges, gifts, obligations, and feasts; here social accounting replaces fiscal accounting and to a great extent, everybody "owes" the others. In many smaller villages, barter and exchange occurs as non–formal financial transactions, and a modest financial component is maintained only for travel and trade external to the region; symbolic wealth such as cowries are used in trade.

Only in very mobile societies does money start to replace fair dealing, objective value, and hospitality shared, and the abstract and intrinsically valueless "money" (usually cheap strips of paper or lumps of metal) replaces real goods and services. Even in fiscal societies however, barter and exchange are highly developed (even by multinational firms), and formal barter centres are now also evolving locally to distribute surplus goods for real or imagined needs. Faith in the fiscal system (an essential delusion if money is to maintain any barter value) is fading as nation states and giant corporations fail to meet their debts, and either repudiate debts or go into voluntary liquidation. In every case, the cost falls back on us. Large banks not only lose our money to start with, but make us pay for the loss. Large companies receive public subsidies (often direct cash subsidies, e.g. the sugar industry) that would make millionaires of paupers.

Fiscal (moneybased) societies give a false impression of security, which quickly falls apart every 40 or so years when inflation—which is itself due to greed—makes currency valueless. The final "inflation" is caused by the misuse of money, and is now upon us. It is seen in the collapse of the *environmental* system. No amount of gold or diamonds can avert, reduce, or soften the blows that nature is raining upon us, and in the final accounting, a cabbage can be worth a king's castle (or more) if it saves your life. For the last 40 years or so, money has been made by destruction of real wealth (soils and forests) and the debts are now being called in by nature herself.

Money is in itself not a resource, it represents (or should represent) a resource which lies "somewhere else". Often, however, that resource is a useless object (a diamond) which people rarely find a need for in any lifethreatening crisis, and never in any global crisis such as now threatens us. Money, in a sane society, must therefore be tied or fixed in value relative to a *useful real asset;* this is the very basis of fair trade in large societies.

All money arises from the wealth of the natural world (plants, clean water, clear air, stored energy). The accumulation of unused wealth, or wealth that does not lead to the proliferation of life, is a pollution of the same nature as any unused resource. Manure and money have much in common.

Insecure people can never have enough material resources, or the appearances of security. They tend to spend this money on monuments and protection rather than in assisting nature to produce wealth. Hence, we can find them associated with addictive, ostentatious, and exploitive occupations. Some tend to erect monuments to contain acquisitions (loot) in such places as museums, art galleries, stately homes, castles, libraries, and churches. Curiously, such monuments often display natural things portrayed in paintings and objects, but in so doing use up nature (the cedar table becoming more revered than the cedar tree, the leopard–skin coat more valued than the leopard).

While natural resources fuel such "wealth", artisans and architects develop the monuments, artists decorate them, and bankers, miners, and oil people fund or value them. The erection of monuments itself becomes a reason for existence. The rich are conspicuously represented in societies devoted to monument repair, but not in the area of landscape rehabilitation.

It is but a short step from worshipping inside monuments to worshipping monuments themselves (people often being more proud of their church than they are of the trees and stones which were destroyed to build it). It is an even easier step to confuse oneself with the creator, and all the easier if one adopts a belief system in which god is portrayed as a man! (Some would say this is an insult to god.)

Money, however, is not intrinsically evil; it is the *accumulation* of money and its use to exploit others that is evil. The evil (privilege, power, stupidity, willfulness) lies within people, not within money itself. Nor is the making of money necessarily evil, providing the uses of money are creative and assist the natural world to proliferate. Thus, we can have a clear conscience on money put to earth rehabilitation.

We should develop or create wealth just as we develop landscapes, by concentrating on conservation of energy and natural resources (reducing the need to earn), by developing procreative assets (proliferating forests, prairies, and life systems), by reducing the creation of degenerative assets (roads, monuments, cities), and by constantly divesting ourselves of any surplus wealth to these ends.

Money is to the social fabric as water is to land-scape. It is the agent of transport, the shaper and mover of trade. Like water, it is not the total amount of money entering a community which counts; it is the number of uses or duties to which we can divert money, and the number of cycles of use, that measures the availability of that money. Leakage from the community must therefore be prevented and recycling made the rule.

Money itself is not a resource, and has no intrinsic value or use, but it can create categories of resources or assets, which we can identify as follows (*after* Turnbull, 1975):

- DEGENERATIVE: Those assets that decay, rust, or wear out: the buildings, roads, cars, furnishings, and appliances of society. Too many of these "assets" in any region will impoverish the region in the long term.
- GENERATIVE: The tools of society; those things which manufacture or process raw materials into useful products (huskers, grinders, blenders, lathes, furnaces, and so on). These do wear out, but can be used to *repair each other* in workshops. All groups need some of the tools of processing and repair; a wise farmer hires out or shares such tools.
- PROCREATIVE: The trees, wildlife, fish, invertebrates, mammals, and domestic livestock of a region. People who maximise a procreative asset base can support the use of some tools, and modest degenerative assets. People who maximise the possession of degenerative assets eventually fail in their attempts to organise upkeep and repair—hence so many ruined castles and stately homes.

I would also add to the above categories:

- INFORMATIONAL: Information (education and data), plus applied intelligence makes the best use of all assets, decides balances in the asset base, assesses future trends, and foresees needs and changes. Seeds have a high information content, as do books or data bases.
- CONSERVATIVE: Insulation, dams, money re-cycling systems, good storage areas, and strategic forests to guard against erosion or desertification are all categories of conserver society assets. All these guard resources for future use, and are essential to a sustainable system.

It follows that expenditure on categories 3 to 5 conserve and create wealth in any society. If a great many wealthproducing assets are available, then *some* degenerative assets can be supported, but any society which spends only on categories 1 and 2 will first pollute, and then eventually extinguish, its resource base.

Apart from the asset categories given above, careful consideration must be given by any bioregion to what is locally conserved and used (the basis of regional wealth, such as soil) and what can be exported as a trade item (surplus water or surplus manufacture). It then follows that financial institutions should themselves pay close attention to their function in that region, preventing leakages of essential resources, and expediting the export of local surplus in order to bring scarce resources into the region. Such surplus should not, however, be based on the loss of any irreplaceable resource such as soil or humus.

Above all, any financial institution should pay attention to two necessitous "foundation stones":

- AN ETHIC, expressed as a published, legally binding, and publicly known charter; and
- RESTRICTION TO APPROPRIATE RESOURCE DEVELOPMENT AND TRADE, in its operations (for not all financial institutions suit every objective of community).

Without ethics or restrictions, any financial institution is a danger and a weakness in a community. With sound ethics and resource usage restrictions, any financial institution can prevent leakage of wealth and the erosion of basic resources, so that it is itself an asset to community, and builds wealth for re–investment.

Financial institutions (those which deal in public funds) are of the following nature:

- Credit unions
- Credit cooperatives
- Trusts and foundations
- Savings and loan banks, or associations
- Insurance agencies
- Finance companies and lending organisations
- Commercial or merchant banks
- Investment brokers and stock exchanges
- Limited liability companies (risk capital)
- Trading or public companies
- Cooperatives

There are other and minor systems in use, but each of the above are now worldwide, have specific *appropriate* uses, and can be fairly easily understood or created by any community. The essentials of an ethical banking system is not only that it has an ethical charter and is used for appropriate assets, but that it belongs to and is governed by the community it serves, and therefore is not open to distant or centralised control.

One of the more extraordinary features of many of the strategies outlined in this chapter is that they have arisen in (and been developed and applied by) poor, depressed, minority, often unskilled, and frequently "powerless" groups. Good people everywhere *can* take financial and developmental control of their regions, give equal service to all people, and rise from an ethical but outcast sum of minorities to be a driving force in world stability. So go to it, as the sum of minorities is always the majority!.

In this section, we are apparently talking about money, but keep it clear in your mind that we are actually talking about a philosophy of true democracy, peace, and "lifetime". Lifetime is that little space we are given to experience this world, which shapes up to what we can imagine to be heaven, but where the achievement of paradise is constantly set back by the "serpents" of greed, power, stupid exploitation, and war.

Time and money are often interchangeable. To control the cash flow of our society is to control our lives. No price is too high to pay for the right to work at "right livelihood", to consume what we can help produce, to feel secure, and not only to avoid harming, but to actively assist, other people and life forms. People who steal our independence steal our lifetimes; our personal independence relies on a cooperative human society.

By changing ourselves, and living in closer harmony with life processes, we reduce the conflicts brought into our lives by the opposing demands of a truly sound economy and that of "unlimited growth" in the capitalistic sense; between a false assertion of human dominance over nature, and the certainty that we depend on all of nature; between the injunction to treat all people as equal, and the status given to those who consume and prosper at the expense of others; between the tyranny of need created by gadgets and luxury, and the satisfaction of working with others to achieve our basic needs; between our natural drive to accumulate possessions, and the realisation that it is only what we share that gives us access to all necessary possessions.

THE INFORMAL ECONOMY

Barter is a common economy practiced particularly in rural or neighbourhood areas where people are more likely to know one another. At the household level, people exchange garden products and plants, share labour, and exchange goods and services. Occasionally people may form 35 person *work groups* to build houses, create gardens, or clean up housework; these work groups may be episodic, forming the pattern of a roundrobin until all present needs are met.

On a community level, or with more than 6–8 people involved, labour exchange may need to be coordinated or regulated. The Bendigo Home Builder's Club in Victoria, Australia is a group of 35 people building individual homes. They pay $5 a year per family, mainly to cover the printing and distribution costs of the Club's newsletter. Each member can either be a recipient or donor of labour. The units of exchange are hours of labour, and all labour is considered equal. Using a standard labour exchange form (which is legally binding), the recipient is debited and the donor credited for every hour's work he or she performs. There is a Labour Organiser in the group to sort out the balance of payments, and to despatch labourers to a recipient (who must have at least 60 hours in credit).

A Community Barter Club also works on a system of debits and credits, where residents offer goods, services, and skills, from landscaping to massage, from mowing to printing. Even the Club secretary or organiser is paid in credits. A credit is calculated at one hour, and the donor and recipient agree among themselves what they consider the job is worth. People are not limited to a one–to–one exchange; as the Club organiser keeps records of the debits and credits of each individual, transactions occur as long as services are desired. The Community Barter Club can be an asset for people in the community who are unemployed or underemployed, and for those who need services but cannot afford to pay cash for them.

Internal economics are greatly aided by exchange newsletters, computer services, and advertisement. These represent a good medium to swap *goods* in particular, with the Barter Centre charging only on a proportion of successful swaps. Several newsletters, like "Exchange and Mart" in the U.K. cope with this service. Brokerage houses now deal in large surplus barter systems for industry, using a Trade Unit (T.U.) valued at about $1, for pricing and exchange value. These can then be placed in smaller blocks for a variety of exchanges in goods and services, and are a good way to turn a large surplus of one commodity into a range of services and goods needed.

L.E.T. System

Conventional money derives from many agencies external to a community, and circulates throughout all communities, tending to be accumulated in cities, multinational coffers, and banks supporting large investors. Community money (or credit), however, is not usable or necessarily wanted outside that community, hence circulates indefinitely in the community, providing a constantly available resource.

The LET System (Local Employment Trading System) centres in a community: every joining member must be willing to consider trading in local "green" dollars. The LET dollars carry no interest, and administration costs are charged on a "cost of service" basis. Any taxes applicable are the responsibility of members, and any member can know the turnover or balance of any other member. Every member gets periodic statements of accounts. The currency, although equivalent to legal tender, is not issued and cannot be cashed in. Green dollars are "earned" by goods or services to others, and "lost" by using services or goods. All trade, or credit standing, is a public act, and refers to the community as a whole. However, unlike simple barter, a member in credit can spend over the whole range of services or goods offered.

Production, as time spent by members in service to others, is thus never limited by the lack of money. Businesses can charge federal currency for spare parts, and green dollars for labour. Price is agreed upon by the individuals, and reported in to the LET centre by the consumer. "Foreign" goods are thus more expensive, and local components increase; local businesses thrive. Charities and local farms benefit greatly, as charity donors can see their funds as likely to return to them. Anyone who wants work can offer services; they need not wait for "jobs". As only members can trade with each other, the community account is at all times balanced. In effect, any member (by working or selling) issues their own currency, and could return any community to full employment. An ideal member has many transactions, but accumulates modest debits and credits. See under Resources at the end of this chapter for addresses.

Finally, the informal economy includes purely volunteer labour, exchanges of gifts, and taking responsibility for a certain community project or area. For example, convivial treeplanting on community common areas should be a part of every household's responsibilities. This may well be achieved by the "adoption" of a few acres of community forest by a household. Other community projects can be helped along by volunteer efforts, gifts of materials, and gifts of time as advisors or entertainers.

THE FORMAL ECONOMY

"Formal" means that goods or services are conducted under a legal umbrella, and are regulated by accounting procedures. Exchange can still take place, but it is *accounted for* in terms of stocks or services. Such formal economics are necessary where people (managers) act for a group of members or investors, and not just for themselves or their households. Legal procedures must also be followed by selfemployed people or family businesses, where cash is received for goods or services rendered or offered publicly.

A community may have at least these formal structures:

- Cooperatives
- Community savings and loans
- Small businesses service office
- Investment funds for special projects
- Leasing company or system

Cooperatives

A cooperative is a group of people acting together for the benefit of members. It is a legal entity with limited liability, and perpetual succession (no dissolution for individual gain). It has several principles:

- Open membership (open to all who can make use of the services and will accept the responsibilities of membership).
- Democratic organisation (members participate in decisions affecting the cooperative, with affairs to be administered by people elected by the members).
- Strictly limited interest on share capital (fair but limited award for capital and restricted influence of people holding share capital).
- Surplus or savings out of the operation of the cooperative belong to members (with members to decide on the use of the surplus, whether to develop the business, to provide common services for members, and/or to distribute among members according to the degree of involvement in the cooperative).
- Education (cooperatives should provide education for members, officers, employees, and the public on principles and techniques of cooperation).
- Cooperation between cooperatives (encourages the development of more cooperatives).

The worker–cooperative centred around Mondragon, in the Basque region of Spain, are worthy of note. In less than 30 years, 96 workercooperatives, employing 17,000 worker–members, have emerged. Each person is required to invest about $5,000 when joining. This can be borrowed from the bank or obtained by installments deducted from wages over a two–year period. Of this investment, 20% is a contribution to collectively owned funds, and 80% is for the purchase of an individual shareholding or capital account (which is normally not drawn upon by the worker except on retirement, death, or in cases of extreme hardship). In this way, a co–op can partially fund itself, with generous help from the cooperative bank.

The Mondragon cooperatives have several features:

1. 10% of the profits must be returned to the community for public services. 20% of the profits are held as capital reserves, and 70% are distributed to workers, although not all of this is available for withdrawal until a worker leaves the cooperative, at which point all of their financial interest in the business must be withdrawn. The worker is, in effect, "loaning" the cooperative the money, and so receives interest.
2. A cooperatively run bank oversees the functioning of all new cooperatives in the group, finances new cooperatives (up to 90%), and offers expert management skills.
3. No redundancies in the cooperatives—workers are retrained and new jobs found in other expanding co-operative groups.
4. The ratio of the lowest to the highest paid person is never greater than1:5.
5. An annual meeting of all workers in a particular enterprise elects both directors (managers) to run the business and a social council (union) to negotiate with directors on work conditions, pay, education, etc. The meeting observes the principle of one worker, one vote.
6. Each cooperative averages about 200–300 worker–owners; large numbers become too impersonal, and large cooperatives are divided into smaller independent units.
7. The community has cooperative schools, hospitals,

a university, housing, health and welfare services, a technical research laboratory, super-markets, banks, and computer centres; all of these are cooperatives, and schools earn part of their costs by contract to manufacturing cooperatives.

Unlike the Mondragon cooperatives, which are usually appliance manufacturing factories, a small community cooperative might have three categories of membership, as below:

1. *Worker–owners:* These manage the cooperative, and are split into "management" and "union" groups. They contribute a set amount of capital into the capital fund, of which a percentage can be withdrawn should the worker depart. Only the worker–owners have a vote. Managers are elected, and are responsible to the rest of the group; they can also be sacked!

2. *Corporate members*: These are the primary producer, manufacturing, or public service associations. They are the *users* of the store cooperative in that their products are sold there, and/or they receive bulk supplies through the store for their business. They also pay a joining fee to capital funds. They may cooperate together for group insurance purposes. Voting powers can be allotted on the basis of involvement.

3. *Households:* Basically consumers, each household pays a nominal joining fee, goods are bought at a discount, and an annual dividend is received for the bulk purchases over the year. There is no vote.

Cooperatives also involve *sharing*, achieved by spreading the skills needed for any one job over more than one person (rotating jobs); by having near equality in shareholding; and by being able therefore to assess how others are coping with a job. Cooperatives have a greater demand on the energy and time of their workers—there are often planning or assessment sessions after working hours. However, *productivity* in such cooperatives is very high, and incomes or profits correspondingly high.

Even in cooperatives, the functions of management (supervision, administration, accounting, and assessing) and worker representation (unions) are necessary, but unlike privately owned businesses, the whole workforce are shareholders, and all vote for people to fill these positions. Thus, the work force has total control over the composition of representatives, rather like a bioregion.

In fact, a bioregion is a sort of multi–cooperative, where smaller groups take on specific services, thus specific responsibilities. Nobody "represents" a bioregion or cooperative in the ungovernable sense that elected politicians "represent" their electorate (i.e. every "representation" or policy decision of a coopera-tive comes from the ground up; whereas almost every politician makes purely *personal* decisions over a vast range of policy and expenditure—and that is "manage-ment out of control"). In fact, today's governments are not only in themselves irresponsible, but they often fund secret and far more irresponsible agencies, responsible to nobody!

Community Savings and Loans

A worthwhile goal of any community would be to keep the money saved and earned in the community *cycling within* itself. The only way to do this is to establish financial and economic systems onsite, such as a credit union, revolving loan fund, or local currency.

Credit Union

Anyone who belongs to an identifiable group of 30 or more people can start a credit union. The purpose or charter of this credit union can be to fund local or neighbourhood self–reliance. A community credit union can pay all routine accounts of a household; some credit unions even have a cheque account service. Credit unions or friendly societies can set aside 10% of income to satisfy instant requests for money from depositors; larger sums can be withdrawn at short notice—often within a week. Friendly societies handle health and insurance.

The credit union can carefully assess loan appli-cations. Money borrowed in order to save money is soon repaid, and so is safe to lend. Thus, money advanced for gardens, fuel conservation, energy generation, or for appropriate vehicles and appliances is soon returned. The savings (in time) exceeds the cash borrowed; from then on, the borrower has some spare capital. Usually, money borrowed to save energy is amortised over periods of from 2–7 years.

The Revolving Loan Fund

The basic principle of a revolving loans fund is that people put in $500–5,000 capital at a *nominated interest* (from 010%) into an established financial institution, and this is then loaned out to new businesses within the community. The group in charge of administering the loan checks references, offers advice, and acts on the recommendation of people who will service the loan (usually a skills or research group of volunteers, some of whom may take part in, or service, the business).

This can be called a loans trust, credit union, finance cooperative or enterprise fund pool; it can include barter, a labour exchange, a regular fair or market, and it needs an open register of local skills and resources, well displayed. Such a modest fund can operate out of a house or old shop front, or from a counter in an established food co–op or cooperative business.

On average, informed and concerned people will initially contribute a few hundred dollars, just "to see a good thing go". This is enough; others will have good ideas about small essential services and businesses, and the research group can be very busy researching and publicising "leaks" of money from the area, so that under or unemployed people can start up services and supplies to stop these leaks, e.g. Does the area make its own bread, yoghurt, sausages, shoes, clothes, pots and paper? Does it reuse its waste wood, glass, metal, paper, or organic wastes? Does it provide a wide range of services from haircutting to legal advice? If not, jobs are open and funds to start them are available! Loans, at low local interest (6–11% is fair) are made, and every borrower must be a contributor (active investment). The skills group help to select equipment, test markets (presale of products is ideal), train young entrepreneurs in bookkeeping, and find resources and materials. Very few of such publicly needed, publicly funded and publicly open businesses fail. Everybody is self–interested in their success!

As confidence in the local fund grows, loans can start to cover energy–saving house additions, insulation, or new well–designed housing, small vehicles, small fuel supply technology, and land purchase for approved projects. Even so, funds subscribed may always exceed demand (businesses are slow to develop), so the fund managers should always be ready to *fund the start–up of more advanced money systems* such as investment advisors in ethical trusts, local insurance and banking, and a local "mint" to print a district currency of non–inflatable money, which in the end is also non–interest bearing.

Every place where this has started (and since 1980, there are dozens or hundreds of funds, currencies, barter fairs, and investment trusts established to build a sustainable future) has benefited. Imports are greatly decreased, local employment rises rapidly, good products (and security) are available, and community morale is enhanced.

Thus, community savings and loans associations are appropriate for reducing community and household costs, and freeing more capital into the community, which leads us to the S.H.A.R.E. and C.E.L.T. systems of "wealth–producing" loans. These are revolving loan funds that provide capital to community–based groups, enriching the community and forming a strong support base for the businesses established.

S.H.A.R.E. stands for SelfHelp Association for a Regional Economy. It is a local nonprofit corporation formed to help encourage small businesses that are producing necessary goods and services for the community (in this case, the Berkshire area in Massachusetts, U.S.A.). It works in conjunction with a local bank in the area. Members of the community can become S.H.A.R.E. members, which means they open a S.H.A.R.E. joint account with the bank. They receive only 6% interest (but this means small loans can be given out at 10% interest). The person receiving the loan must first collect references from people who know them as responsible and conscientious. They must show that the proposed business will attract customers from the community or even from outside the community. By doing this preliminary work, the borrower gets to know many people, and the community has a keen interest in seeing that the business succeeds.

C.E.LT. .stands for Community Enterprise Loans Trust, a New Zealandwide charitable trust to promote and support small businesses and cooperatives. C.E.L.T. helps people form and run cooperatives and other enterprises by providing advice, running training sessions so that people can learn cooperative business skills, and by providing loans.

C.E.L.T. services are funded by subscriptions from the public ($5), by donations, and by government special schemes. Education and other work is funded by the interest from deposits and loans. C.E.L.T. accepts cash deposits, and lends out to enterprises working closely with them until they are on their feet.

Depositors receive from 0–12% interest per year depending on the amount of time the money is in the account, and whether the depositor wants interest paid. The borrowing criteria is that the entrepreneur must be willing to work closely and regularly with CELT during the loan so that a business has the greatest chance to succeed. CELT has now achieved the status of a bank, and can offer services such as a bank offers.

Southern Cross Capital Exchange Ltd, operating out of Wentworth Falls, NSW, Australia, is a nonprofit organisation that brings together those who want to borrow from specific (socially conscious) projects, and those who have money to loan to such projects. The role of the S.C.C.E .is to review applications for loans and to recommend individuals and businesses to receive these loans. It is not a bank or finance company, and so loans through the S.C.C.E. are not secured. However, they are *guaranteed* by the Exchange (though only if loans are made through S.C.C.E., not *directly* to the project). The borrowers make personal guarantees, and "guarantee circles" are set up to spread the risk (e.g. parents who want to build a school will all guarantee to pay back the loan). This sort of capital exchange format may be one way in which community schools and other socially conscious projects can become financially viable.

<u>Local Currency</u>

As the community gains skills in financial management, there is no reason why an internal and district economy should not be bolstered by a noninflating currency *printed by the community*. Already this is done by individuals and businesses who have a product or skill to sell (*real* value), and who print up vouchers or coupons to pay for setting up their business. For example, a publishing company sends out pre–publication order slips to people *before* a particular book is published. People buy or prepurchase the book at a slightly reduced price, which enables the publisher to print the book (this is how the book you are reading was printed).

In another example, a restaurant (Zoo Zoos) in Washington State, USA, in transferring from single ownership to a work cooperative, needed to raise funds to buy out the owner. They printed meal vouchers, redeemable up to one year, and sold them to future customers and friends. Most people came in to eat their promised meal, but some vouchers were traded to other people for some other service in the community, and thus the "currency" starts circulating as vouchers relating to a real commodity.

There are many currencies in every society, such as

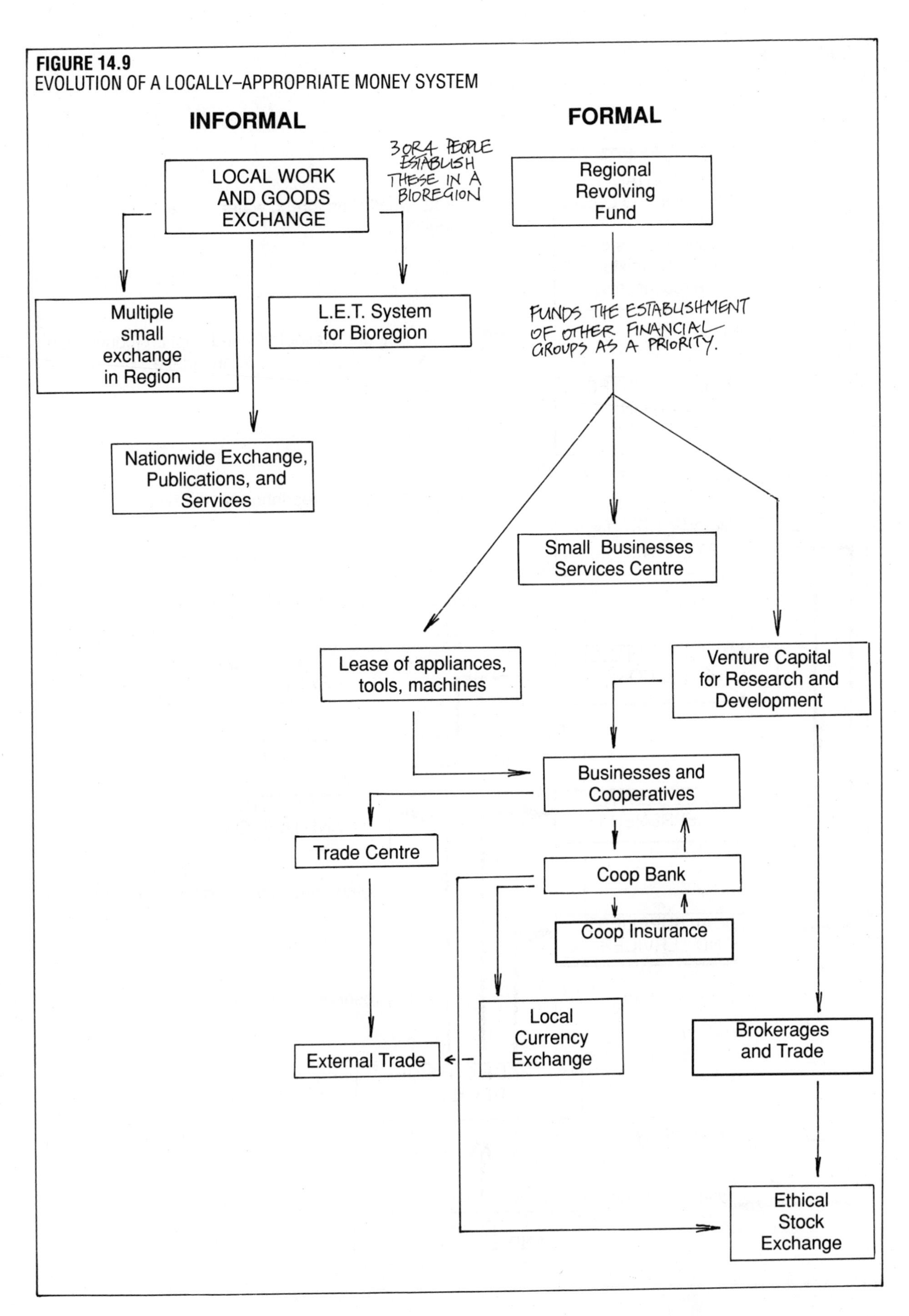

FIGURE 14.9
EVOLUTION OF A LOCALLY–APPROPRIATE MONEY SYSTEM

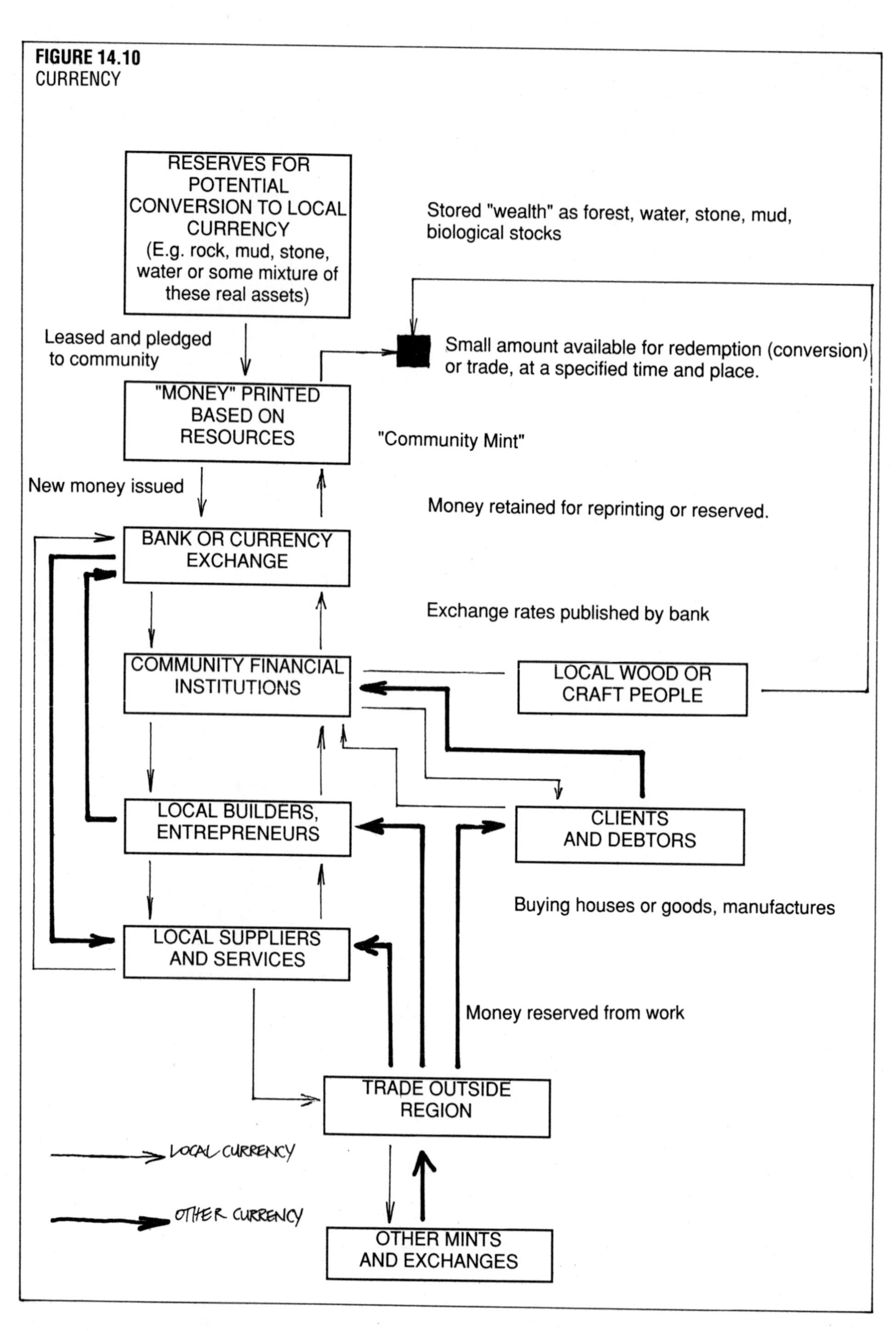
FIGURE 14.10
CURRENCY
RESERVES FOR POTENTIAL CONVERSION TO LOCAL CURRENCY (E.g. rock, mud, stone, water or some mixture of these real assets)
Stored "wealth" as forest, water, stone, mud, biological stocks
Leased and pledged to community
Small amount available for redemption (conversion) or trade, at a specified time and place.
"MONEY" PRINTED BASED ON RESOURCES
"Community Mint"
New money issued
Money retained for reprinting or reserved.
BANK OR CURRENCY EXCHANGE
Exchange rates published by bank
COMMUNITY FINANCIAL INSTITUTIONS
LOCAL WOOD OR CRAFT PEOPLE
LOCAL BUILDERS, ENTREPRENEURS
CLIENTS AND DEBTORS
Buying houses or goods, manufactures
LOCAL SUPPLIERS AND SERVICES
Money reserved from work
TRADE OUTSIDE REGION
LOCAL CURRENCY
OTHER CURRENCY
OTHER MINTS AND EXCHANGES

promises, exchanges, stamps, coupons, vouchers, cigarettes, hugs and kisses. All are freely and legally exchanged for goods or services, which are themselves interchangeable (a song for a lettuce is a good bargain).

However, we need a redeemable, solid, real and objective currency for trade and exchange. For currency to be valid and usable, some preconditions are necessary. First, it must be backed up by a real, objective resource. Secondly, other people must have confidence in it, which is *why* it is backed up by a local resource and can therefore be traded. Lastly, there must be a demand for its use, and a place to exchange it for other currency so that it works as well for other people as it does locally.

Demand means that there must be a real need for some item or service lacking in the local society. Now, any community printing its own currency has these things to do:

- Find and lease a resource that can be pledged to the exchange institution as a redeemable asset. If this is timber, the currency is based on "a cord of timber, cut and stacked in the woodyard at such–and–such a place", or if clean water "a gallon of water bottled and redeemable at a certain place." This is *the reserve* and gives confidence in the currency.
- Print the currency itself, preferably in a solid local material difficult to duplicate or forge elsewhere, and numbered, dated and counted. The value so printed should not exceed the value of the reserve by a factor of more than 3 times. Reserve pledges need to be publicly available and assessed. Notes can exceed reserves if used for procreative assets only (e.g. forests). Forests represent firewood, but reserves are never cut except in dire emergencies.
- Educate local businesses and banks (exchanges) in the use of the currency, which can be changed for the national currency (while this lasts) at the bank, or for other regional currencies. Post exchange rates publicly at the money exchange or bank and issue the currency to banks.
- Start satisfying the demand (e.g. for homes) by lending the currency to builders, who can exchange it at the bank for other currency needed in trade, or locally for products and services.

The bank should not itself decide uses; the community (via a set of financial, advisory or monitoring institutions owned by them) should do that. Currency should be used to satisfy real needs of the community for food, shelter, trade. As we are talking about community money, its uses need to be decided by that community.

Producers of *local* goods, accept *local* currency only; this creates a demand, and other regions must "buy" such currency to obtain local goods. Most local businesses will accept a regional currency.

Note: That if reserves are living things such as trees, wealth increases and can be created. Even if a currency was originally based on bricks, it can be used to create such biological reserves as forests over time.

To prevent hoarding, notes can be dated and a new issue us made every 4–5 years; this is also a check on unwarranted accumulation, and possible forgeries.

Wherever there is a need (for housing, roads, small businesses, farms), the dollars needed are supplied by the exchange, but the borrower or user *must repay in local currency*, thus creating a demand for it. Most small businesses accept the currency, and much of the local trade can be carried on in this currency. When most needs of the region are met, the currency can be collapsed. Many small towns funded their public works this way in the 1930's. The E. F. Schumacher Society, Great Barrington, Mass. U.S.A. has data on these systems, and runs one such currency (see Resource listing at the end of this chapter).

Critical personnel to attach to a revolving loans office are in the following categories:

1. *Asssessor–designers:* People skilled in good house, factory, and farm design, energy budgeting, and appropriate technology—the "permaculture" team, used to help assess proposals to be funded.
2. *Accountants:* People skilled in setting up appropriate accounting systems to monitor progress and profit in ventures, working closely with team (a).
3. *Broker–bankers:* People actually handling cash flow, assessing reserves, and operating the banking and insurance functions of the office.
4. *Lawyer–trustees:* People able to package a set of legal strategies for community or family groups, and to advise (with **2**) on taxation, export–import, trust and leasing documents, labour exchange agreements, and company, commonwork, or cooperative law.
5. *Real estate or realty*: The office can handle the bookkeeping and serving of local industry and services, supply goods, advise, provide labour and goods exchange, and arrange legal forms. Through the designers, users can obtain help in building and land design, nursery and livestock services, and appropriate tools.

For a developed region, services such as a travel club, credit union, food cooperative (or rather, group purchase cooperative), farm club, and educational and medical services can also be provided. It must be stressed that the "revolving loans office" is not a *place* but a *group* in cooperative function. Each may operate in their own home, but all services are listed in an educational or informational newsletter and can combine where needed.

<u>Small Business Services Centre</u>

There will no doubt be many businesses run in any community which are not run as cooperatives. However, that should not preclude their sharing in certain commonlyused services, such as accounting, telephone services, secretaries, telex, insurance, distribution, cocataloguing of goods, group advertising, and export assistance. A small business services centre is itself a small business, now very popular and effective in India. It may be a key organisation in a bioregion.

A great benefit to having many business offices located in one place is the increased number of consumers and the ability to concentrate many products in a single *product catalogue*. Direct marketing is a fastgrowing selling technique; it cuts out the retailer and so enables a product to be sold at a lower price than can be offered by a conventional retailer. A product catalogue is often a valuable product in itself, offering product information and advice. Many people want to support small businesses and cooperatives, and often get into the spirit of the venture when they read about individuals and businesses in the catalogue. The real saving to small businesses is shared facilities such as premises, accounting, and office services.

In addition, the possibility of a group label for products exists in a community. Although each business may be a separate entity, the label can be of a similar design, with the words "Another Product from Boon Dock" (or some such) printed at the bottom. Otherwise, each label has its own business name and information. This generates interest in the community products, which should gain a valuable reputation for quality, durability, or taste as standards for the group label are established.

Small business service centres can offer the following facilities to businesses:

- *Secretarial*: Letters, prospectus, submissions, writing and presentation, filing, mail order.
- *Bookkeeping*. Banking, cost accounting, summary and position analyses, presentation to accountants at years' end. Billing and collecting accounts, paying bills.
- *Accounting*: Tax assessment, broad strategies, trust advice, plan of networking monies, common funds, trust administration.
- *Legal*. Legal forms, leases, representation in litigation, land and propertyconveyancy, new legal structure design, seminar services, trust structure.
- *Communication*: Phone, telex, fax services, photocopy, answering services, travel arrangements and despatch.
- *Gift, tithe, and public service deductions*. Setting up funds for aid, tree planting, and accepting gifts to aid trusts; servicing land access and investment systems.
- *Education*. Holding seminars, inviting speakers, and instructing new members in services and procedures; assisting neighbouring regions to set up parallel and allied services.
- *Skills register*. Keeping a file on key people available for special advice and assessment.
- *Research services*. Retrieval and basic research for region and projects.
- *Lease* of seminar rooms, small warehousing, equipment, office space.

In business, there is no substitute for good management, budgeting, accounting, and marketing skills. Most of what makes a successful business is the combination of human–centred values with good management, which we have listed below (paraphrasing from the findings of the book *In Search of Excellence* by T. Peters and R. Waterman, 1984, Harper & Row).

1. *Shared and stated values:* All concerned with the company believe strongly in a set of values, often restated. Such values need to be carefully framed, realistic, and simple to remember. They are also inherent in the following:

2. *Respect and encouragement:* Management should give staff control of their own areas, encourage them to develop new ideas, and to follow guidelines and values rather than a rigid set of rules.

3. *Reputation:* The company maintains a reputation for high–quality products, service, and reliability. This cannot be stressed often enough. Most customers will deal with a firm over and over again if it proves to be reliable, rather than take a chance with a company which may be cheap, but which maintains such sloppy standards as slow service and shoddy products.

4. *Lean management:* Successful businesses use a simple organisational structure, a minimum of staff directing operations, and no "corporate planners" or analysts. Management is often in close contact with both producers (staff) and customers, and involved in production.

5. *Action:* Once a decision is made, effort is made to get it done with all possible speed. Customer reaction is then gauged in a matter of weeks or months, not years.

6. *Familiarity:* In expansion, or in new products, good businesses stick to what they know best (and don't expand into or acquire a business in an unfamiliar field).

In summary then, the successful business:

- states group values and maintains a sense of loyalty to those values,
- personalises services, good quality, prompt service, and reliability,
- innovates in its area of expertise and looks ahead, and
- has a lean structure, and no uninvolved staff.

Attention to current and future trends (social, climatic, economic, political) is essential for any business group, and small business centres can research on such trends; societies change according to new information, products, and materials, and businesses must change or expand with these trends.

In addition to normal business principles, the more intense and more democratic operation of cooperatives demands that co–op staff must participate in planning, seriously contributing to policy, procedures, and innovations. The sharing of "power" is really a sharing of *responsibility*; part of that responsibility is the capital risk of any enterprise.

Leasing Systems

Any cooperative or village could run a leasing service for seldom–used items of capital equipment (photocopiers to trucks) which individuals or businesses do need on occasion.

User Pays Principle: From privately to publicly owned

FIGURE 14.11

SCEMATIC OF CAPITAL FLOW WITHIN AND WITHOUT A VILLAGE

The aim is to keep as much money as can be usefully and productively used within the system; thus to reduce taxation, consumption of outside resources, energy costs, and outside investment, and to return resources to income and consumer products within the village.
Consumption is reduced by lifestyle changes, vehicle sharing, etc.
Taxation is reduced by legal strategies
Energy, by conservation and technological strategies.
Trade, by developing local resources.
Little of this is possible unless investment, <u>via</u> the village bank, is controlled by the village.

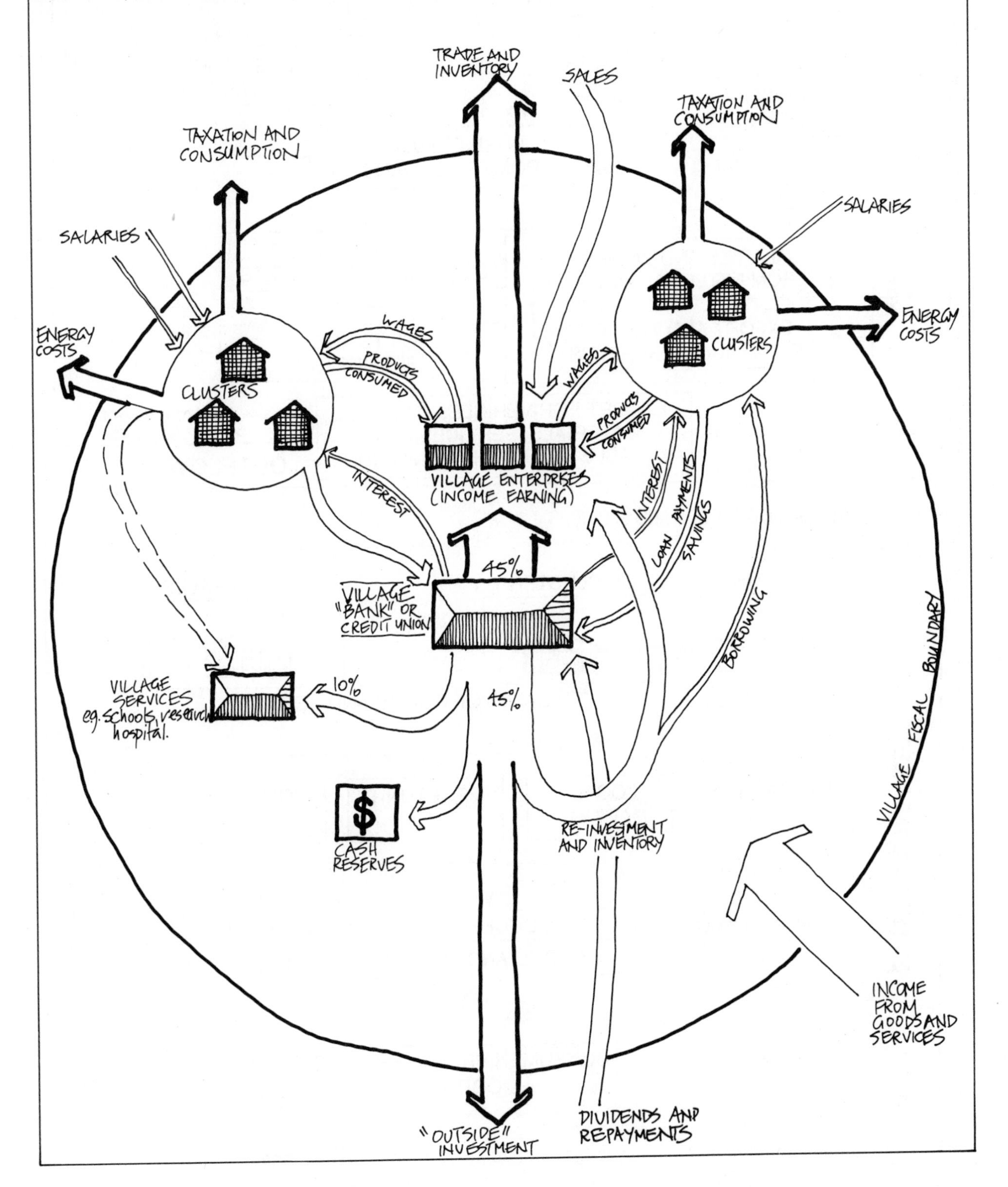

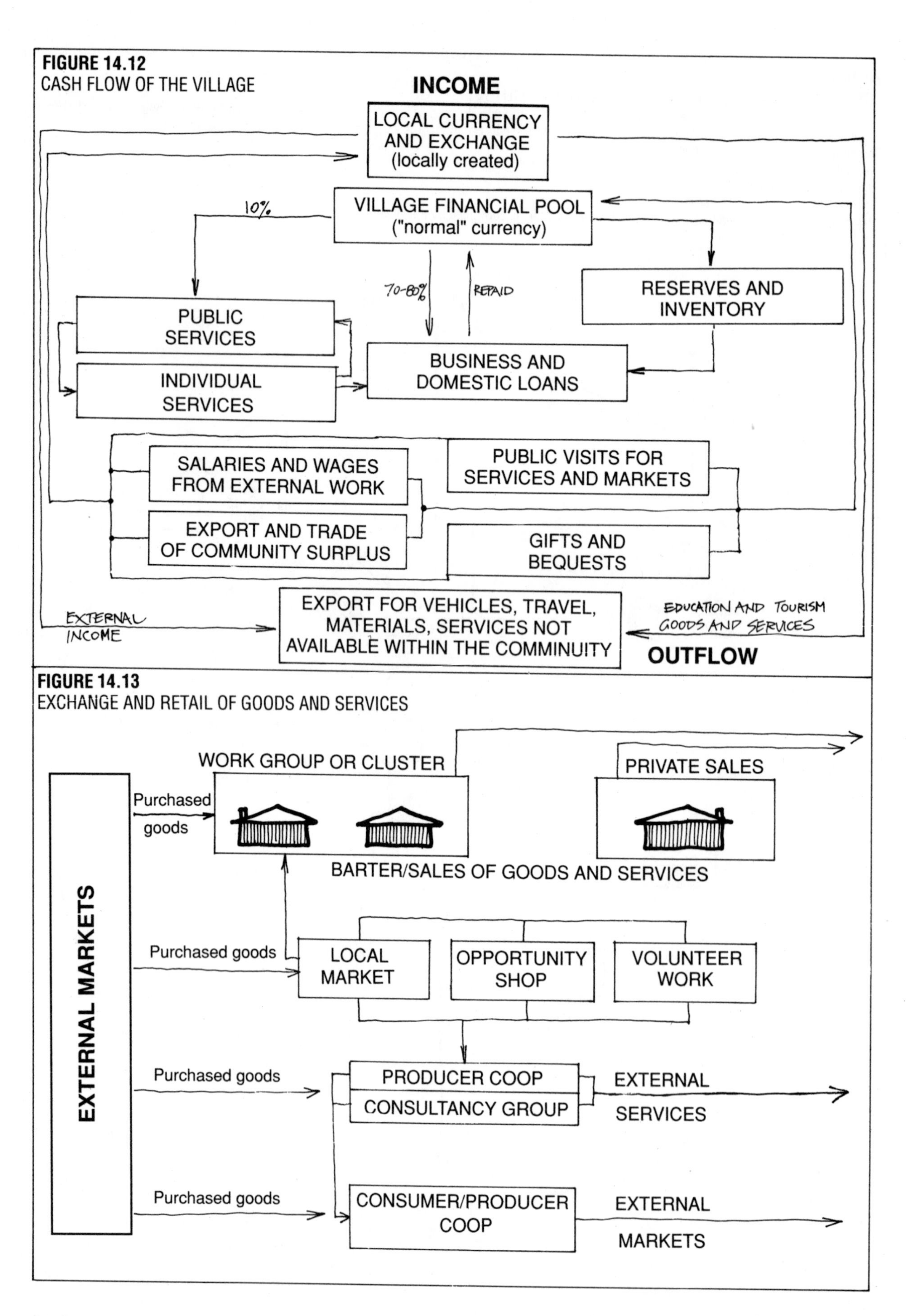

FIGURE 14.12
CASH FLOW OF THE VILLAGE

FIGURE 14.13
EXCHANGE AND RETAIL OF GOODS AND SERVICES

assets (sewing machines to reference libraries), a charge sufficient to cover running costs, repairs, and replacements *must* be placed on that item. This may be subscription (library) or piece work (photocopier), hourly rate (computer) or miles travelled (vehicle). These charges apply equally to businesses, administration, trusts, and private groups or individuals. Persistent misuse (e.g. of a vehicle) results in a withholding period or permanent withdrawal of permission to use.

Personal Accountability Principle: This applies to any group purchases, whether public or private. The *group purchasing* is held totally and individually responsible for payment for any item. This is rigorously applied and holds *even if a member leaves a group* or the community. This principle stops "I'vegotabright idea" and a "let'sgetit" approach—ideas must be paid for. If the idea is a good one, it will pay itself off in time through lease. An example of this sort is if a group of five wanted to purchase a large brush chipper to create compost for themselves and to hire out to others in the community. A chipper would be purchased, and an hourly charge put on it to cover purchase, maintenance, and replacement. Eventually, if the chipper is used by enough people, it may even be possible for the original five to get a return on their money, although this was not necessarily an aim.

Special coinvestment on projects can be initiated by advertisement in the community. Examples are: group water storages or energy systems, group refrigeration facilities, coownership of a fishing vessel or coastal holiday home, etc. These are not working cooperatives or businesses, but rather projects that save money or give the opportunity for a wider range of resources than if each person had to fund them individually. The investing group decides expense, location, and use payments.

14.12

LAND ACCESS

TRUSTEESHIP OF LAND

Our own lifetimes are, in terms of soils, trees, or climate, as ephemeral as snowflakes. For a little while, we have the use of the earth, and our time here is bounded by birth and death. Thus the very concept of land ownership is ludicrous, and we need only to use what is needed for the brief time that we are here; even birth and death are small events in a total life pool continuum.

The law clearly distinguishes between *ownership* or entitlement to a resource, and the *rights of the use of it.* Laws of ownership are relatively modern, and are foreign to tribal or clan law. Laws of trusteeship are ancient, philosophical and realistic. Ownership, in effect, gives the titleholder (person or state) a "right to exploit in the short term". Trusteeship governs any resource for the very long term, with no right to exploit resources beyond essential needs, or replacement time.

The way that land passed from clan management to personal ownership is well documented; since the year 1400 or thereabouts, the methods used were as in **Figure 14.14**.

Most of us live on lands once tribal, now "owned". Very few of us have any rights to share the resources of such land, which is either state (army), church, or corporation–controlled.

However, with the benefit of scepticism gained from hindsight, many people are working to reverse this historical trend; tribes are still forwarding their claims to common ground after 200–400 years of occupation, and thousands of people are forming trusteeship organisations to remove land from private ownerships, church, and state control, and to return it to use by those people who live on and near the land. In fact, educated people of good will, and traditional people, have seen where ownership has ruined common resources, and are returning to the concept of taking local stewardship of the land itself. Thus, in the evolution of land concepts, we have **Figure 14.15.**

Gifts or deeds of land can be vested in a tax–deductible trust for use by a specific group or the public generally, under certain reasonable conditions; many community gardens run this way, usually at small rental. Many people with large incomes actually benefit from tax–deductible land gifts. They can purchase and improve land, and gift it at the improved value at a paper profit.

Essential land for local food, fuel and structural forestry, recreation, and conservation can be planned, and secured *under a set of public trusts* by public investment, gift, bequest, taxdeductible donation, transfer from other authorities or trusts, or outright purchase. Thus, the district *secures its initial land resource.* Each and every parcel of land needs 3–4 involved, active, and interested *trustees,* and under a legal limit, its plans and purposes should be set for the long–term for 10–50 years ahead). Some areas will be under sports centres, some in trust to conservationists, some under lease to organise gardens and farms, and some reserved for educational and public bodies for public services. Community forestry on steep and rocky lands will provide fuels, food, and buildings for the future. Even here, every *household* can plant and tend an area, and profit from or manage it; it is also an improving asset that can be sold or transferred. This works; "public" forestry does not. Industry should grow every stick that they use by a *charge on product,* and investment in the community forestry owned by local households. Good models of village forestry are operating in Indian and Taiwan; poor models of public forestry are all about us.

Land trusts need be few, close to settlements, and cover all essential uses; the rest of the land can go back to natural forests or prairie. Every scrap of land *in* settlement should first be planned and used. Many, if not most, small towns need no other land assets. Any society that develops lawns beyond those used for

FIGURE 14.14
TRUSTEESHIP OF LAND

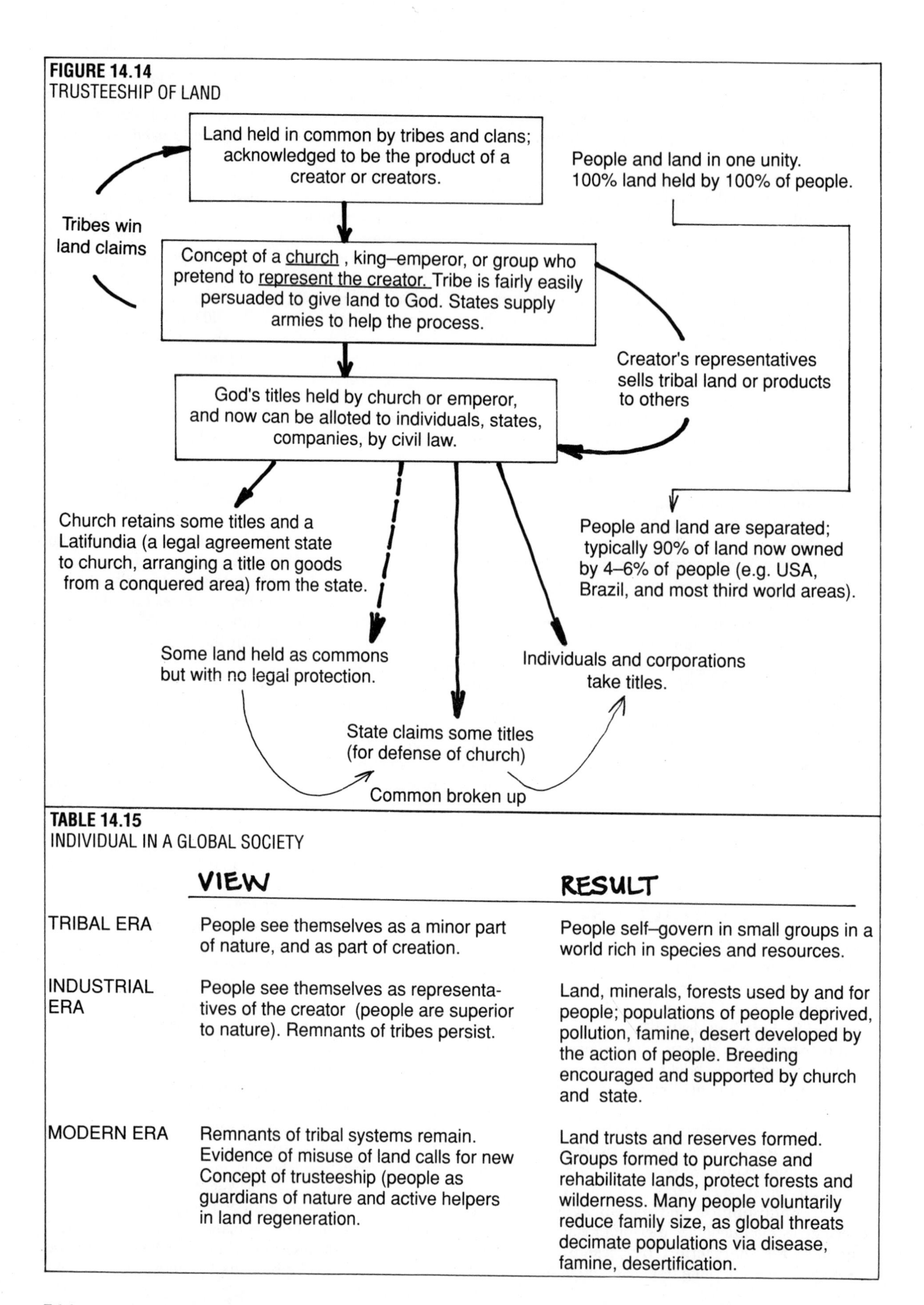

TABLE 14.15
INDIVIDUAL IN A GLOBAL SOCIETY

	VIEW	RESULT
TRIBAL ERA	People see themselves as a minor part of nature, and as part of creation.	People self–govern in small groups in a world rich in species and resources.
INDUSTRIAL ERA	People see themselves as representatives of the creator (people are superior to nature). Remnants of tribes persist.	Land, minerals, forests used by and for people; populations of people deprived, pollution, famine, desert developed by the action of people. Breeding encouraged and supported by church and state.
MODERN ERA	Remnants of tribal systems remain. Evidence of misuse of land calls for new Concept of trusteeship (people as guardians of nature and active helpers in land regeneration.	Land trusts and reserves formed. Groups formed to purchase and rehabilitate lands, protect forests and wilderness. Many people voluntarily reduce family size, as global threats decimate populations via disease, famine, desertification.

recreation can support itself for wood, fuel, and building materials by the conversion of lawns to use.

If people who *really* want land set up a determined research group on "ways and means", or open an advisory centre on such methods, they would achieve their needs much faster and with far less expense than if they rely on undertaking a political "revolution" (a transfer of power), or saved their pennies, and with much greater longterm benefit to society.

Just keep in mind that there is more than enough land already cleared for all people, and that it has long ago been paid for in labour or cash; there is really no need to buy it again, just the get the *right of use*. Every country has some, or many, methods to obtain usage rights.

There is usually only one title, or a few titles, but there are probably thousands of possible rights upon any land. Think of the "ownership" as a blank canvas or empty wardrobe. We can paint on or hang up an array of rights, and (unless we are very thorough) never fill the land space available.

Land in trust can be developed in a number of ways, and can include garden clubs, commonworks, and as leased land for specific purposes. Many have water conservation, developing forest, wildlife corridors, wetlands, and special species reserves as primary aims.

Practical warnings are *not* to accept land gifts that have many restrictive conditions attached; in fact, it is wise to perhaps limit acceptance to unconditional gifts. These can then be sold to purchase more suitable land, or to capitalise land elsewhere.

Secondly, unless a large cash reserve is available, each trust can only manage one area of land, and only slowly expand. Every parcel of trust land is a unique and longterm development, and if too much land is accepted, simple maintenance and land tax costs can bankrupt the trust.

Above all, a land trust should have a very clear idea of what it wants to achieve, and to set a practical time limit to do the job. All trusts need an income, and may thus need a business or trading arm. The trust can gift land to other regional trusts or associations of whom it approves, so that local gifts can be routed through a tax-deductible trust.

Why should people give land away? Some of the reasons are:

- To continue land in its use: as an organic farm, wildlife refuge, economic botanical garden, or as an example of good land use.
- Because people believe in trusteeship: Many of us do *not* believe in private land ownership, but in trust ownerships for public use under sound environmental control.
- Land is surplus to their needs, and a cost: A few people own too much land (are what is termed "landpoor"), thus achieve low productivity and incur land maintenance costs. By giving some land away (especially to a tax-deductible trust) they can concentrate on a smaller productive area. Good accounting advice will often dictate that land should be creatively gifted to such a trust (i.e. it benefits the giver; this is especially true of donors who have a large income from other sources).
- As a bequest: Some goodhearted older people will bequeath land, yielding up successive rights as they age, to younger landless people or to a land trust; this is form of public bequest, and is a fairly common occurrence to establish parks, wildlife areas, or demonstration and teaching farms. Trusts can also give to other trusts under a bequest basis.
- Because they want a village, or more people on the land: In modern times, nuclear families or individuals are socially very isolated on large farms. By establishing a land trust, they can legally encourage others to develop parts of the property, and so set up a socially rich area with multiple potential.

There are many more reasons why people gift land to trusts, so that (generally speaking) there are more lands available than there are reliable stewards to occupy them.

As with money, land ownership and thus land usage in society is unbalanced, except where tribal land councils still exist; even in tribes, cattle or resource ownership can become unbalanced (as in Botswana, where 9% of the people own 80% of the cattle) if crops and herds cease to be tribally owned and are privately owned on common lands.

LAND ACCESS OFFICE

People often complain that they lack access to land resources; at the same time, we live in a delinquent or devastated landscape. How do we marry needs and land resources? The establishment of a regional office (a land access office—LAO) opens up the potential for offering a set of strategies enabling better land use, and suited to the finances and involvement of people using the service. A selection of strategies follow, and can be modified for local conditions:

- Landlease system within urban areas;
- Garden or farm club;
- City farms;
- Towns and cities as farms;
- Farm link system; and
- Commonwork.

Land lease system within urban areas (Oxfam Model): This is particularly suited to young families in rental accommodation. The regional office posts paired lists: List A is for those who want 200–1,000 sq. feet of garden to grow food. List B comprises those people (usually elderly or absentee landlords) who will lease either vacant land or the land around their houses on an annual, renewable basis. People list themselves and, as local land comes up, introduce themselves. The LAO prepares a standard lease specifying rental (if any), goods exchange, length and type of lease, access, and the names of the parties.

Thus, many young families get legal access to garden land, on an "allotment" basis. The regional office may need to map and actively seek land, and should make a small service charge for registration of leases.

Garden or Farm Club: These suit families with some capital to invest as shares, with annual membership (shares can be sold). A farm is purchased by the club or society on a public access route 12 hours from the city. This property is designed by the club or society to serve the interests of members, whether for garden, main crop, fuelwood, fishing, recreation, camping, commercial growing, or all of these. Depending on the aims and share capital, people can lease small areas, or appoint a manager. Rich clubs develop motel–style accommodation and recreational fisheries. Worker–based clubs usually develop private plots with overnight (caravan–style) accommodation for weekends. A management committee plans for the whole area (access, water, fences, rates, etc) and can be selected by the club.

Many such clubs exist in Europe, and some in Australia; they offer multiple use of one lot of land by many people. Membership in such clubs can be made saleable or transferable, and may increase in value over time.

City Farms: A local group of 100 or more families forms a *city farm association*, and invites local, state, or federal authorities (via their local representatives) to allot from 1–80 ha (preferably with a building) to a city farm. Such invitations are irresistable to those who hold office by virtue of local goodwill or votes.

On this land, the following activities are promoted:

- Demonstration gardens;
- Garden allotments (where space permits);
- Domestic animals (rabbit, pigeon, poultry, sheep, goats, cows, horses) kept and used as demonstration and breeding stock;
- Reycling centre for equipment, building materials (income–producing).
- Tool rental and access;
- Gleaning operations;
- Plant nursery;
- Seminars, demonstrations, training programmes, educational outreach; and
- Seed, book, plant, and general retail sales.

In New York alone, the "Open Space Coalition" counts 1,100 parcels of land as one or other form of city farm. The Federation of City Farms in the UK numbers some 46 active farms, and many more groups forming at any one time.

The essentials of a successful city farm is that it lies in an area of real need (poor neighbourhoods), that it has a large local membership, and that it offers a wide range of social services to the area. Many city farms become totally or mainly self-supporting from sales of goods or services, plus modest membership fees. The one essential is a long–term legally binding lease. Coalitions of such farms represent a large lobby or vote group in society, and are therefore politically respectable!

Each city farm has a small management group, and most have numerous volunteers, or a few paid staff.

Towns and Cities as Farms: A twist on the above, which can be operated by a city farm group. There are several ways to use cities as farms—many German towns carry on an active city forestry along roads and on reserves. From 60–80% of total city income is thus derived from city forest products.

Surplus city garden or food product is collected, sorted, packaged, and retailed. Some groups collect, grade, and sell citrus or nut crop, and many provide young trees to gardeners on contract for later product off the trees. Others range sheep, duck, or geese flocks for fire or pest control. All seem to make a very good income by treating the city as a specialist farm. A processing, shearing, or like facility may be needed by the group.

Nonprofit groups often collect unwanted food from orchards, canneries, etc. and distribute them to the poor, or sell at a small profit to keep running costs down. This is known as a "gleaning" system; many thousands of tons of unwanted food is so redistributed in the USA. Givers take a tax reduction on gift to a gleaning trust (any church or public trust).

Farm Link (Producer–consumer cooperative): These are appropriate to highrise or rental accommodation in an urban area. From 20–50 families link to one or more farms in the nearby countryside. Although they can purchase and manage a property, they usually come to an arrangement with an already established market gardener. Quarterly meetings are held between both parties to work out what products can be trucked direct from the farm to the families, who use the product and can retail any surplus to others.

The farmer adjusts production to suit family needs, and as the "link" grows, the system can also accommodate holidays on the farm, educational workshops, and city help on the farm at peak work periods (planting and harvesting).

This strategy enables us to build compact urban areas while retaining farmlands that are uncluttered by settlement. The alternative (as in Australia, the USA, and increasingly in Europe) is for cities to become sprawling monstrosities of suburbs that reduce whole production areas to lawns and rotary clotheslines, and forests and trees to chopsticks and newspapers, while over–extending public utilies and creating insolvable transport and waste problems.

The system is best developed in Japan, although scattered examples operate elsewhere for products as diverse as fish and game, wheat and firewood. Benefits are numerous:

- Producers have an assured market at a mutually agreed price set a year ahead.
- Consumers can reduce costs to below wholesale by assisting with harvest.
- Townspeople can access farm facilities; this is valuable education for children.
- The system is convivial; in fact its main aims are to

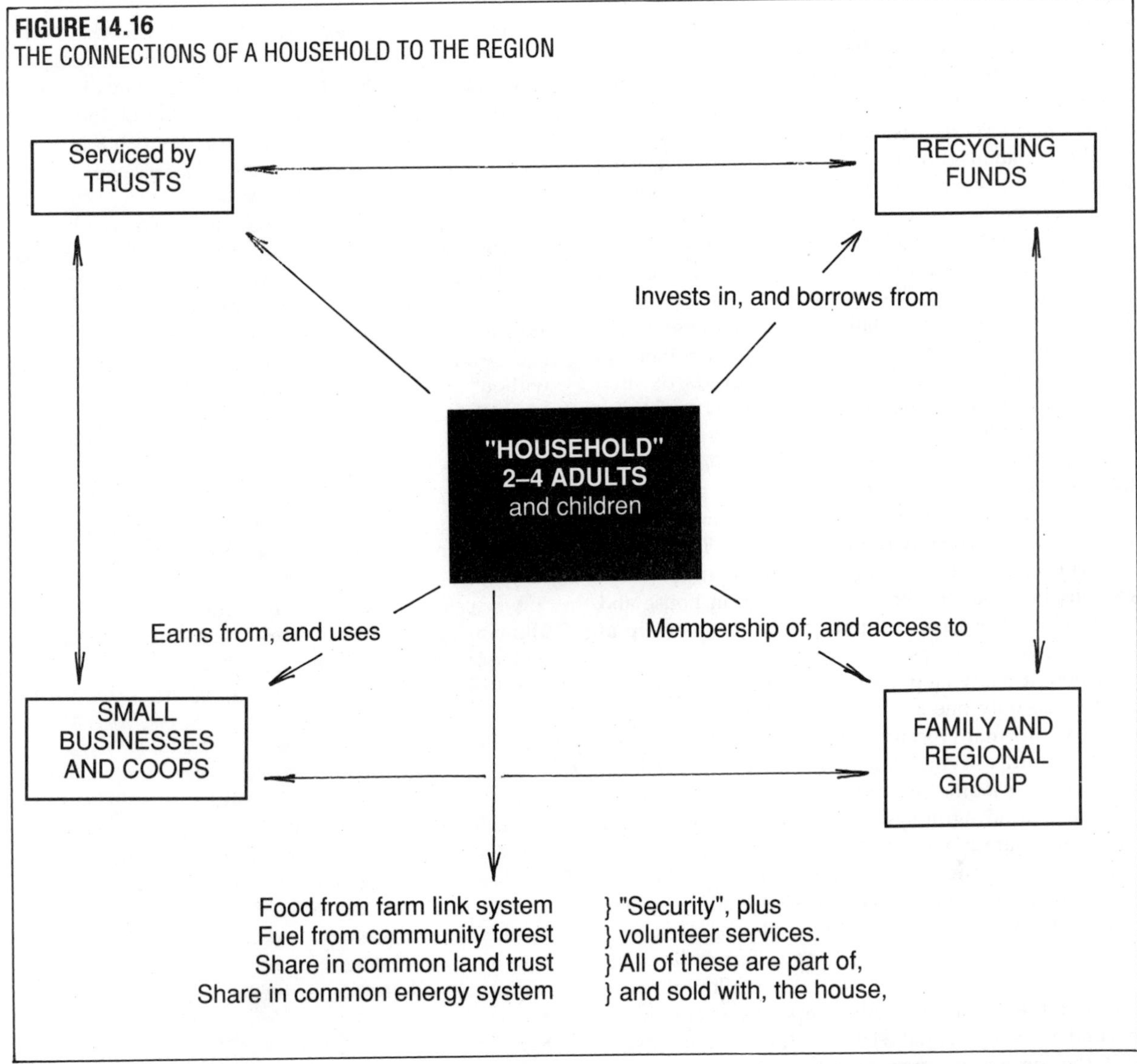

develop *friendships*.

- Consumers control quality and quantity by pre–order and stated preferences.

Commonwork systems: Any trust, tribe or even an individual farmer can allow multiple land use by setting up a commonwork system, akin to the African *mahisa* or livestock loan system of Botswana.

Briefly, the land area is closely assessed for earth resources, wildlife, forest and aquaculture potential, small and large livestock (bees to bullocks), arable land and mixed orchard, and socio–educational potential. The local needs for primary processing, building, and consultancy or implementation services can also be researched.

This basic design work completed, the trust or owners can advertise locally for people to run any one of these enterprises with levels ranging from that of a hobby (developing a butterfly forage system) to a fulltime occupation income (a trout farm). Proper legal safeguards (lease documents) setting out rights of use, length of lease, responsibilities of the trust and lessee, and terms of payment (adjustable, but usually 10% of gross income) are drawn up to safeguard users. This 10%, credited to the donor, forms a capital fund together with any capital raised by the trust or the lessees (some as grants or gifts, some as business loans). The trust also needs income for maintaining services such as roads, fences, power, and to pay land taxes.

All lessees are selfemployed; land costs are minimal (their share of services) and land access is secure. As enterprises develop, the capital fund enables further research and development. It is ideal to plan so that any one enterprise (energy supply, bees, tree nursery) helps to supply others, as well as regional needs. More than one farm or trust can join in a commonwork system; the trust or land owners set broad conditions of sustainable use, and allot space or resources to enterprises, but the contributors to the capital fund vote in their own cash management group. Such a landuse system promises full and beneficial use of lands, and can take up much of the unemployment in a district.

However, commonworks need to be close to towns or on a public transport route, so that townspeople can participate; remote areas do not suit this system unless it is paired with a village development. Commonwork members are free to leave, sell out to new lessees, and eventually be refunded their 10%, less base maintenance and service costs, but plus any interest paid.

All occupations can be "open", available to all, but a proportion of occupations (adjusted to that proportion of the society that needs such work to be available) can be assigned to specifically disadvantaged groups at the time of life or state of health where such work is perhaps the only possible useful work one can do. In Turkey, for instance, the totally blind have the sole right to sell pigeon food in public places; this gives them an independent income. There are hundreds of such essentially minor incomes available in every society for otherwise–neglected groups or ages.

"Livings" are occupations which return a living wage to a family. The fair assessment of these rest in family size, especially the number of dependents, base costs in the society, and frugality of the family unit. The need to earn is *most* reduced by a set of strategies ranging from gardens for food, efficient use of energy in house and work, sharing of basic equipment, and membership of bulk purchasing groups.

Although many employees (unfortunate people) are paid to do only one job, members of a commonwork can take up many occupations, the net return from which afford a living plus some net profit for local investment as a tithe on earnings.

In the Mondragon system, actual cash or income differentials are limited in the ratio 1:4 or 5; that is to say, a sweeper cannot receive less than 25% of the total wage of a doctor; this is a good basis for adjusting any ethical sharing in any system. If people, via education, retraining, or selfhelp can improve their skills, their political mobility (*not* fixed castes) can allow them to improve their relative earning capacity, although not beyond a fair differential relative to their community.

In the concept of commonwork we have arrived at a new synthesis, a future model not only for farms but for complex small communities. Although tribal peoples had (and have) a clear idea of total ecologies—the sum of fire, regrowth, wallaby, and pioneer fire species—and although they had a word for this, there is no word–concept for a "total human family ecology" beyond "a living".

In my meaning of "a living", it is the beekeeper, the bees, their water, flowers, pollen, propolis, and the means to make the beehives. It is a clear legal access to forage, and a registered and secure bee site; and it is the right to sell or market product to pay for other life needs. In fact, it is a human ecological totality, provided with abstract and real self–reliance, and sufficient to pay for any tuition, travel, health service, and insurance needed by the family accessing a living.

This right, and many similar rights, overlay all landscapes, all societies, but can only be designed to be beneficial if that society (or that segment of society who want self–reliance) entrust their lands and lives to a public–interest deed of trust, and also trust each other to carry out a function. In fact, commonwork systems based on a social edifice of responsibility to and for others. For a nation–state, taxes were perhaps intended to supply social needs, whereas in fact they have always been used to raise armies and enrich a minority far beyond the needs of a living; to create wealth for a few, and privelege for bureaucrats.

If taxes were not in fact so used, we would all live in a society where the need to work would be negligible, and both employment and unemployment (thus, workers and owners) absent. Employment, like suburbs and institutions such as gaols, are as modern as lawns and politics; all ancient societies of people arranged life without any of these impediments, but only by seeing life as livings, and living things as basic to life.

A short list of developmental design programs for a commonwork land trust is as follows:

- *Priority 1*:: Maximise water storage by constructions of Keyline, swale, soil conditioning, ridge forestry, and broad canals or dams for diversion and irrigation.
- *Priority 2*: Specify forest sites and forest types, allowing specific forage forests for locally acceptable or endemic animal species.
- *Priority 3:* Specify crop areas for orchard and perennial crop, based on best soils, low slopes, windbreaks, and water access.
- *Priority 4*:: Specify complex access and trackways, critical edges for flowers, bee fodders, trellis systems, wayside crop.
- *Other*: Select a few caretaker house sites if necessary, and designate and stockpile earth materials if and as these are discovered in earth works. Check history and archaelogy, and keep a journal and open plan of all development.

Probable livelihoods for 100 ha, 40 or more families:

- Bees
- Poultry
- Ducks
- Fish species
- Forage systems
- Fuel forestry
- Structural forestry
- Cut flowers
- Nursery
- Methane and wind energy systems, solar systems (manufacture and power sales)
- Accommodation and tours
- Workshops and field days
- Training programmes
- Cooperative store
- Credit union
- Insurance
- Leasing
- Processing of raw materials to pottery, bricks, dried, pickled and smoked food products.

All of this asserts that we are not "just clerks" or "just housewives" but that we have many roles in any free society. Our freedoms are, in fact, a choice of those chains of responsibility or social duties with which we feel comfortable—not the freedom to do nothing, or to do what we like (to be self-serving)—but the freedom to choose among occupations—the portfolio of self–expression, work or duties that we in fact do evolve in non–hierarchical societies, villages, and tribes.

Thus, more formal cooperatives need to include retraining, education, and work mobility for their membership, or risk frustration and boredom with work. The work factor itself can include some proportion of time or output devoted to social services generally, so that everyone feels that they are contributing to their society as a whole. The very concepts of employment/unemployment deny this potential, and again frustrate people so that rebellion as strikes or riots follow. Guilds and unions (in the sense of trades) may actually reinforce this sense of irrevocable fate (no choice), as do caste systems in India.

Thus, in bioregional networks, commonworks, and intentional villages, the individual can choose a set of duties and occupations that fit skills, choice, and age. A few also develop special skills and become teachers in trades or disciplines.

14.13

AN ETHICAL INVESTMENT MOVEMENT

Prior to 1980, very few innovative or consciously ethical (legally structured to be so) financial systems existed; today there are hundreds of such organisations, holding their own summits and handling, via their stockbrokers, in excess of 160 billion dollars annually (in the USA alone). Many other such organisations exist in Europe, Australia, New Zealand, India and southeast Asia.

In 1983, the Permaculture Institute, following seminars with the E. F. Schumacher group in the USA, started teaching in local community funding and ethical investment: "banking on the earth". Any local group is able to set up a resource list of data from existing models, to invite fund managers to visit their region to give seminars, and to adopt or devise local financial recycling systems, ethical brokerages, or non–monetary community exchange.

A local "earthbank" group is at first a research, teaching, or seminar convening organisation, but as local money systems are established, some members obtain employment by running financial, exchange, or barter systems. In the UK, USA, Australia, and Canada, annual seminars (in the UK called "The Other Economic Summit"—TOES) are now convened to hear from advisors and fund managers, and to supply education or materials to new groups. If no such summits exist in your area, convene one; we started with only 12 people in Australia, but 60–100 interested people and organisations now attend these seminars, and banks, insurance companies, cooperatives, and credit unions send representatives to assist new groups.

The rise of a large, popular, efficient set of services to divert public money to good ends (and get it back) is a reaction to (or revulsion with) the current misuse of money by governments, large aid agencies, and rapacious investors whose sole motive is profit, power, or greed. This movement is one of the truly new phenomena of this century, and its growth is exponential.

The large amount of investment capital redirected through ethical brokerages is the tip of an iceberg which involves many thousands of ordinary people who are members of guarantee circles, ethical credit unions, community loans trusts, common fund agencies for bioregions, or nonformal systems of labour and workday exchanges, barter systems, direct market systems, or no-interest, pre–purchase, "green dollar" systems.

Moreover, existing banks, credit unions, cooperatives, businesses, and allied groups are discussing the rewriting of their charters to include the values of earth care, people care, and the production of socially useful (or socially sensitive) products. Some credit unions and banks already have such a charter, and keep corporate watchdogs on the staff whose sole job it is to monitor companies for unethical behaviour. Not only corporations but volunteer groups and consumer groups produce monitoring publications on multinationals on a global scale, and publish "nonbuyers" guides of the products of unethical organisations.

The negative ("non–buy") emphasis of the early years involved disinvestment in companies which:

DO NOT CARE FOR THE EARTH, producing:

- Pollutants and dangerous wastes, waste product as excess packaging, and nonreturnable or not recyclable containers.
- Shoddy and quickly superseded products, unrepairable items, or those lacking good servicing and spares.
- Poisons, biocides, armaments, and dangerous materials (radioactive waste, mercury, asbestos, leaded petrol, chlorinated fluorocarbons in insulation or spray cans, radioactive paints and so on.

DO NOT CARE FOR PEOPLE, as assessed by:

- Dangerous foods or medicines.
- Have unsafe or polluted workplaces; this includes noise pollution.
- Deal in addictive substances or provide addictive services (alcohol, tobacco, gambling).
- Do not permit organised labour, do not deal with employees on a fair basis, nor pay fair wages.
- Exploit people directly via slavery, bonded labour, excessive profit margins, by forms of prostitution, racial and sexual discrimination, or harassment.
- Support or cooperate with regimes using torture or imprisonment without charge, dictatorships, corrupt regimes restricting voting, disenfranchising people by gerrymander, or by allowing votes only to certain groups.

DO NOT SHOW A PUBLIC CONSCIENCE

- Use bribery and price–fixing.
- Operate on excessive profits.
- Monopolise resources or markets.
- Do not themselves invest in ethical groups.

As the ethical investment movement matures, however, this negative approach is evolving into a very *positive* search for, and willingness to fund and support (or establish), enterprises which:

- Assist conservation and reduce waste (not *treat* waste and so grow to depend on more wastes!) or energy use.
- Grow clean food free of biocides or dangerous levels of contaminants.
- Are involved in community reafforestation.
- Build energy conserving houses or villages.
- Produce clean transport or energy systems.
- Assist people's self–reliance.
- Found cooperatives, self–employed ventures, or profit–sharing systems.
- Produce durable, sound, useful and necessary products.

Thus, local or bioregional funds can establish small or large enterprises necessary to that region, using money raised by residents. Brokers or enterprise trusts can direct surplus investment to socially and environmentally responsible industries and developments such as new, well–designed villages. All such ethical organisations state their criteria in their legal or informal charters.

RECOMMENDED TYPES OF INVESTMENTS

Investments need to be staggered in terms of ultimate return, so that some money is always on call. To these ends, a set of loans or investments can be scattered over short to longterm enterprises, e.g.

- Short Term: Loans for draught–proofing, insulation, attached glasshouse, clean water tanks, trickle irrigation, and dam building. All amortise in from 1–2 years, some in much shorter time (one winter). Also, loans can be given to good local industries with careful management and market assessment.
- Medium Term: Bee, chicken, and pig forage systems plus stock, large water catchments for aquaculture, irrigation; nurseries, subdivisions, tours and tourism, buying and selling farms after environmental rehabilitation; larger local industries or their expansion; clean power systems.
- Long Term: Town and city reconstruction or development; fuel crop and processing; small farm development; large–scale property retrofit; orchard establishment; research into new energy forms.
- Permanent: Forestry, and the purchase of natural remnant forest (shares can be traded as values increase over the long term); wildlife reserves and rainforest in good condition (as revalued property trusts dedicated to species preservation).

We can order investment value under some such simple system as follows:

1. Active e.g. a group of people investing in reafforestation, and working in that area.
2. Passive e.g. buying the products of an ethical company
3. Neutral e.g. funding a film which may have no message, ethical or otherwise.
4. Unethical e.g. retailing dangerous and persistent pesticides or herbicides.

This gives us a set of priorities based on "greatest effect and involvement". Category 4 above is, of course, not permitted by any ethical charter, and Category 3 need be funded only when other needs are satisfied.

We are acting at our best level, and have the greatest chance of success (or least chance of failure) when we are active workers in, and consumers of, the products or services that we fund. Within the "active" category (1 above), we can set priorities based on local or current problems in the biosocial context of the times. Today these would be:

Biological:

- The prevention of soil deterioration, i.e. *soil creation*
- The prevention of deforestation, i.e. *afforestation*
- Conservation of species, i.e. *creation of species reserves*

Environmental:

- Preserving the quality of air, i.e. *pollution control*
- Reduction of waste, i.e. *recycling and elimination of waste products*
- Cleaning up water i.e. *water storages and disinvestment in polluting industries.*

There is no implication in the above list that any investment fund should fund industries based on (*relying* on) the production of wastes.

Wherever a body of laws has been formed on the basis of the responsibility of people to their environment, a dynamic, long–maintained, and relatively harmless occupancy of the earth has resulted. I cannot think of any better examples than the long–term tribal occupancies of deserts, rainforests, and prairies.

But wherever a body of laws has been formed based on our "rights" to property, to protect material resources and accumulations, and to permit destruction of the public resource, we will not only destroy whole environments and species, but in the end ourselves.

It is already unlawful to clear forests in South Australia, to light fires in many areas, or to destroy protected wildlife, trees, or reserves in many countries; but we can murder with impunity by using biocides, destroy whole forests with acid rain, destroy the ozone layer, and risk sea level rise without penalty. Many organisations are demanding that this too changes, and that those responsible are charged with the damage, as people bereft of social conscience; we may yet live to see a class of corporate criminals brought to book for their conscious crimes.

Auditing is the periodic assessment of the validity of any financial enterprise or investment strategy. Whereas conventional financial systems propose a single economic criteria to such audits, I propose that we of the alternative nation apply three criteria:

1. The economic audit: "Where did the money go? Was it honestly used? Is the system economically viable?" (*The European audit*)

2. The ethical audit: "Was the enterprise concerned with its ethical (people care) accounting? Did the enterprise benefit people in the long run? " (*The Iroquois audit*)

3. The environmental audit: "Were the activities life–enhancing? Is the earth therefore more productive in terms of life forms? "(*The Pitjatjantjara or "life increase" audit*)

Active and Passive Investment Involvement

Many investors never see or experience the systems they fund via brokerages. As the ethical investment process evolves, many more projects involving investors as residents, builders, primary producers, or suppliers of goods and services can be developed. Bioregional funds do, in fact, offer their investors a chance to at least define the sort of goods they want produced in their region, at the quality level they would prefer. As an example, the development of a permaculture village does just that; the shares (not identical units) actually fund the whole development, including a common development fund, a local revolving loan fund, commercial and light industrial leases; areas are also set aside for primary producers, regional markets, recreation, conservation, and energy reserves. The process of village development has been outlined herein; in such developments every resident can be a participant at most levels.

We should, I feel, discourage passive investment; all brokers can introduce investors and producers in a mutually supportive web. There is no inducement greater than self–interest, and self–interest dictates that every investor should use, assist in, and consume the products and services they invest in. Investment centres should be active in person–to–person introduction—even investment parties!

Analysis of those sections of society and managed funds that prop up the whole investment system means identifying the source of such funds. Retirement funds (superannuation), union funds, insurance funds, and common trust funds are all large sources of investment monies; it is wise, therefore, to include representatives of or contributors to such funds in earthbank societies, and to invite them to ethical investment conferences.

After all, why should coal miners' union funds pay for the takeover that closes down their colliery, rather than the forest development funds that offers them retraining as foresters (and forests will be needed forever!) Only the corruption of fund managers would prevent such sensible provision for future work, and corruption cannot be exposed without investor pressure. Why should an insurance company have money in motor vehicle manufacturers producing unsafe or faulty vehicles? Why,should we fund our own destruction when the alternative is wide open for profitable, ethical development?

Proportional Investment

If one has $100, how should this be spread about to do the most good with the least risk? This is a matter of personal choice or good advice, but some sensible propositions area:

- 10% to risk ventures (new ideas, new ventures).
- 10%20% to a local S.H.A.R.E. programme or credit union as community development funds.
- 10%20% in any existing clean public power utility.
- 20%40% in a social investment fund.
- 10% to a public interest investment (school, hospital, research centre).

Thus the risk is spread widely, home and regional assets funded, and public services supported. This can also alter as new opportunities arise, but like self–employment, money should cover a wide portfolio of ventures.

Investment Sources

A good many people inherit, earn, or win sums of money from $1,000-100,000 surplus to their present needs. They do not want to invest in their own destruction by supporting polluting or addictive industries, and instead seek socially responsible investment. Another class of investors are members of churches and organisations which profess an ethic of peace and goodwill. These groups also need to place surplus funds in organisations which work towards their aims. Various lay bodies such as the Sierra Club, Friends of the Earth, and organic growers have funds to place in investment for at least short periods, and cannot always trust the local banks, which invest in adversary systems or which do not reveal their investment policy publicly.

In fact, everybody who uses a bank to store money is an investor by default, and if unaware of the bank's investment strategy, may most probably be investing (via the bank) in systems which are creating local problems or global disorder. Thus, it is necessary to locate or found ethical investment groups and put our money with them.

About 70% of the total "free world" investment is American, and of the 70%, the majority (or about 40% of the total) is in the hands of women, who tend to inherit as well as save. It is obviously important for womens' groups to direct this money to life-enhancing enterprises. Thus, at least one of our "minority" groups can invest in their own salvation, and also change the world, not only without losing money but in fact getting more return from regenerative investment than from the death system. For example, as stated in *CoEvolution Quarterly*, Summer '83, p. 91, investment returns for public utilities supplying power were as follows:

Nuclear based stocks appreciated 24%
Mixed groups (some nuclear) 52%
Non–nuclear power utilities 82%

Dividends for those groups were 30%, 59% and 81% respectively. It is now certain that socially responsible investment pays. We can also *make it cost to be socially irresponsible* by withdrawing investments from unethical groups. Socially sensitive investments not only pay better, but are basically an insurance against (rather than susceptible to) stock market crashes, as local facilities are always important in hard times, and local investments do not crash.

As money begins to flow to regional and socially responsible funds, even those people not personally persuaded of the need to invest in the future will have cause to think about investing in unethical systems from which public and financial support is being withdrawn. Pursued vigorously as a strategy, reinvestment in ethical systems can change the total direction of capital flow towards beneficial systems.

Some Strategies for Investment in the Environment

Recently third–world debt (heavily discounted by the creditor bank as a poor investment, so that one can buy $1,000,000 of debt for $120,000 of cash) has been purchased by conservationists (via a tax–deductible trust such as the World Wildlife Fund, or one of many conservationist societies). For this debt, the trust asks not for repayment of capital and interest in hard currency, but in forest or wetland assets in the debtor country.

In this way, everybody benefits: the banks, the debtor nation, the environment, and the purchasers; moreover the wildlife and forest reserves so purchased can be sensitively developed for nature, tourism and research rather than being cut down for debt repayment. The whole world benefits from forests, and such benign strategies need to be operated on the widest possible scale. This is not so difficult, as almost every world nation except perhaps Japan, Botswana (diamonds), and Nauru (rock phosphate) is today a debtor nation and few have any hope of repaying their debts—such is the stupidity of governments and banks that invest in corruption and exploitation (note that ethical funds don't share in the stock market crashes if they hold real assets that support *self–reliance*).

It is also possible to deposit funds with a tax–deductible trust, which purchases critical or species–rich areas discounted by farmer debt or by misuse or overcutting (salted and eroded lands). By putting aside a sum for management, income can be made from wildlife reserves, seed, or new forests. In the case of capable farmers, they can themselves be appointed as co–managers, and many would gladly accept this role in restoration and earth care. Many good farmers are made bankrupt by trying to restore land to health!

Investors can do the same for a public (for–profit) trust that reassesses its share prices annually. As a property is developed with lakes, forests, and wildlife, so its value increases, and the increased share value can be traded. It is in such restoration work that management teams (some of them co–investors, by an issue of shares for labour) can test their rehabilitation skills, as outlined in this book and elsewhere in the land restoration literature.

Company takeovers or raids are often used to enrich ruthless individuals by "assetstripping"—selling off a public company's assets and keeping the profits, regardless of the effects on the work force or national economy. However, the same methodology can be used by conservationminded takeover teams, who "strip" polluting companies, and develop land and urban assets to serve the needs of the society and of nature. Many failing logging companies have vast areas of degraded lands suited to small forest farm operations, and would–be forest farmers would love to manage a small area of forest properly, as would many theoretical botanists and academics who have long known how to develop a forest for eternal yield, but never had the land or capital to do so.

By these and other methods, the public can start to go to work via the normal financial and market procedures of the capitalist world to set their environment in order, to preserve species, and to educate and train an effective work force in assuming control for good purposes.

TOTAL APPROACHES TO FINANCE IN SOCIETY

Margrit Kennedy (in a manuscript *Toward an Ecological Economy: money, land and tax reforms*, Oct. 1987) is convinced that an interest-free financial system is not only the sole sustainable medium for exchange, but that such a "no growth" fiscal system encourages and preserves all natural resources. It would at one stroke abolish the condition where the third world and the poor in affluent countries pay out more in interest than it has received in loans, prevent the growth of a minority wealthy elite, and stabilise resource use.

In everything we use there are hidden interest costs: about 12% of garbage collection charges, 38% of drinking water charges, and 77% of social housing charges are accountable as interest. The gains go to the rich or lenders, and the losses to the poor or borrowers even though the earth may provide the wealth of their labour (the *production sphere*). This wealth is removed by interest charges in the fiscal or *circulation sphere*. Thus, wealth used by unethical investment strategies is rapidly transferred via global stock or money markets to most efficiently exploit the poor; a wry comment on the people devising or running such systems.

The gross imbalance of wealth promotes "big" spending in capital–consuming but publicly paid investments (big dams, big power stations, big housing corporations) where governments refer the costs to the people. In the end, only military (waste) funding can use all this misbegotten wealth. So, sensible smallscale and cost–effective solutions are prevented or actively discouraged by governments and fiscal managers alike.

With no or very low interest rates, people buy goods at a *steady* rate, and industries do not need to be geared to cope with the fluctuations in market caused by the swings inherent in a global money supply which starves some regions and floods others with only a profit motive in mind. These goods or services need about a 5% *maintenance cost*, just to pay for repairs or people to run the system. It only needs any town, region, or nation to set up such a constant system to put right many social and ecological ills. This was, in fact, the system tried successfully by some small Austrian towns in the 1930's depression; as such systems strengthen and grow, so regions can stabilise and pay for all their essential longterm resources.

At any rate, the present system is in the process of collapse, and the new barter systems are expanding; the only question we have is if the life support systems of earth will still be intact, or whether sanity in fiscal affairs will be delayed until no human survival is possible on a polluted earth.

14.14

FUTURES

I have borrowed, in part, from the publications of the infant world regional and familial alliances to detail (as a thematic structure) those global problems and local disturbances that will concern all of us over the next few decades. These are:

• ENVIRONMENTAL DETERIORATION.

A. Desertification, under the topics of:

1. Deforestation.
2. Water balance disturbance.
3. Soil salting and collapse.
4. Overgrazing.

• POLLUTION.

A. Of the atmosphere, leading to acid rain and climatic change.

B. Of soils via chemical waste.

C. By radioactives in the soil and food chains.

D. Of inland and fresh waters.

E. Of the estuaries and marine systems.

F. Of food by biocides, radiation.

• THE EXTINCTION OF NATURAL SYSTEMS AND SPECIES.

A. By rainforest destruction.

B. By desertification of arid area borders.

C. By clearing for agriculture.

D. By draining wetlands.

• CLIMATIC CHANGES.

A. Heating of earth by carbon dioxide and gaseous pollutants in the global atmosphere.

1. Rising sea levels
2. Reduction of stratospheric ozone.
3. Intensification of local ozone at ground level.
4. Acidic particles leading to acid rain.

• SOCIO–POLITICAL AND ECONOMIC CONCERNS.

A. The use of torture and imprisonment for repression of people; arrest or detention without charge or public trial.

B. The continuous oppression of minority ethnic, language, cultural, or tribal peoples.

C. Corruption, and the misuse of public monies by selfinterest groups.

D. Replacement of crafts and skills with machines and mass production.

E. Intolerable employment; unsafe, unhealthy, waste–productive .

F. Essentially short–term solutions to long–term, chronic problems.

G. Cash resources sequestered via addiction and crime.

• DIRECTLY HUMAN CONCERNS

A. Meaningful work (employment in right livelihood).

B. Adequate nutrition.

C. Adequate and easily maintained (low–energy shelter.

D. Access to a land base for sustenance.

E. Access to finance for development.

• RESEARCH AND SCIENTIFIC CONCERNS

A. The perversion of science to assist war, torture, oppression.

B. The lack of common, practical translation of scientific findings to those who can use the information.

C. The ineffectiveness of researchers in applying findings and obtaining feedback.

D. Setting priorities for research via morbidity and global analyses, and funding such priorities.

E. The monopolisation of socially useful inventions by patents, especially in seed and technology.

In every one of the above categories, *effective solutions to the stated problems exist*, have been applied, and have solved that problem locally or even nationally. Some are imperfect and need adjustment, others work in the context of specific cultures or landscapes, while a lesser number are effective across the whole range of specific phenomena. But in almost every case, "case histories" of solutions are not locally available over the range of current problems. Any such library of solutions needs an educational outreach. Educational programmes themselves need orientation to practical problem-solving using successful models.

Modifying all climatic and plant data given herein is the global warming effect and stratospheric ozone loss now expected to continue for the foreseeable future. This will mean increased air and seawater temperatures, the extension of typhoons and monsoon rains away from the equator towards Latitudes 20–25°, drier winters in western coastal and southwest Mediterranean climates, and a general increase in carbon dioxide, hence increased plant growth in the semiarid areas of tropical deserts.

As a background to any trends such as global warming, the basic 11year (22 year) sunspot cycle, and more importantly the 18.6 year lunar cycle (the latter affecting the shift north or south of cyclonic systems, the former affecting incoming radiation) will continue to determine drought and wet years. Such factors are now firmly tied to food crises and drought in most continents (*New Scientist* 8 Oct 87, p. 28).

The lunar atmospheric tide is the overriding effect, and the chief collator of such data (Robert Currie, State University of New York, Stonybrook, N.Y. USA) warns of agricultural shortages in the northern hemisphere in 1990–1992. These years will be wet in the southern hemisphere, and we can expect a repeat in the year 2009 or thereabouts. Thus, governments and farmers can plan to reduce herds, store grains, increase tree crop, enlarge or increase water storages and swales, and select plant crop species for such regular or cyclic variations. At present, Africa is in drought, and Europe in a flood cycle; every 9+ years this reverses.

It is now time to diversify bioregional resources to afford a flexible response to these changes. Such preparations may mean that several strategies need to be applied, including;

- Assessment of the local and national relief, and sea level rise effects in every region. For example, most low islands may be at some risk, while lowlying cities and coasts can be either evacuated or protected (at great cost).
- Extension of hurricane–proof housing design to Latitudes 30° north and south.
- Preparation for much hotter climactic factors both in gardens and homes (from 8–12 °C warmer in meso–thermal areas, from 1–5°C warmer in the tropics), thus far greater attention to high tree and palm canopy, insulation, building ventilation, and shadehouses.
- Increases of 16–25%in ultraviolet radiation, hence a change in human activity, and clothing. Animal breeds, resistant to heat effects, need to be selected now.
- A generally more mixed ecology with many more forest components in both urban and rural areas, including species from warmer regions.

As for future occupational changes and migrations, there will need to be far more emphasis on home gardens or city farms, specialist crop, forestry, and local processing for regional needs, as no one can ensure continued agricultural food supply or industrial crop sustainability.

Just to halt or modify the worst aspects of the current atmospheric trends, there is no doubt that any nation wishing to survive will redirect public and private investment monies into sustainable practices, and ban many substances and industries. For example:

- Biocide residues will be disallowed in food, water, and soils, and thus only sophisticated biological controls will apply on farms. Many chlorine products will be discontinued, as well as plastics such as polystyrene.
- Padi rice and feed–lot beef will be disallowed or eliminated as major pollutants, and supplemented with less polluting alternatives (e.g. potatoes, free range beef, chicken flocks).
- Coal burning for power and industry, and fossil fuel use generally will be greatly curtailed, thus solar, tide, hydroelectric, wave, wind, and biogas systems will come into general use. "Clean" fuels will be in global demand especially solar-electric vehicles and wind–assisted freight systems (ships and balloons, or dirigibles).
- Community forestry will be of critical importance in bioregions, cities, and in arid or salted areas.
- Farm tenure could be tied to good husbandry, not "yield", and disallowed if soil losses exceed soil creation on site, or water sources are polluted. Thus, annual cropping or plough culture will greatly decrease, and with it a host of problems.
- Wise government will ban cattle, goats, and sheep from delicate arid or highland areas, and accept the high yields of adapted wildlife under careful mangement.
- Any activity producing acid rain or persistent radioactives will be eliminated or minimised.

There will need to be a global response to environmental refugees, especially from atolls, low coasts, estuaries, and coastal cities (due to sea level rises). These will need to be housed in new, well–designed, low–energy inland villages or self-reliant settlements, minimising transport and fossil fuel uses. Moreover, many of these strategies will need to start now, and many sensible people and regions are already moving in these directions. Books such as this one will be needed for planners and designers, and general media services (radio, television, educational services) will have a key role to play in avoiding chaos and directing a well–orchestrated series of changes via positive advice and documentary education. New approaches to total bioregional policy planning by coalitions of scientists, planners, conservationists, and cooperating essential industries can replace short–term, nationalistic political processes.

Factors such as coastal seawater contamination of groundwaters, increased flood damage, coastal erosion, profound changes in fisheries and agriculture, and social disruption are foreseen as inevitable for a century to come, but if all the causative factors continue unabated, worse and more rapid changes can occur over the next 30 years.

Essential industries (small steel, cement, pipe, glass, and workshop-based enterprises, plus energy industries) will need to be relocated inland, and extensive road rerouting plus barge or water transport services will be needed; areas that will become islands will need efficient water or air transport.

As we cannot predict effects on fisheries or crops, diverse planning will be needed to establish inland aquacultures, forests, and gardens; economic species should be collected and preserved for future changes.

Above all, people need now to be well–informed so that they can act for themselves, or in concert, and we should all prepare for selfreliance and regional interdependence. As the problems are truly global, global concern and action will be needed.

I believe that only group or community (bioregional) survival is meaningful and possible; individual survival is meaningless, as is survival in fortresses. Thus, we must plan for total regions, and include all the skills of a global society.

The profound change we must all make is internal; everybody needs to realise that there is no group coming to their rescue, that it is only what each of us does that counts; thus, those who cooperate with others, and take on a task relevant to all people, will be valued above those who seek personal survival.

14.15
AID AND ASSISTANCE IN AREAS OF NEED

In 1946, the ecologist Aldo Leopold (in *A Sand Country Almanac*, Oxford University Press) foresaw two seemingly inevitable trends: one is the exhaustion of the wilderness as a resource in itself, although a remnant may be preserved in museums or as a genetic resource, and "the other is the world–wide hybridisation of cultures through modern transport and communication...the question arises whether certain values can be preserved that would otherwise be lost." Thus, in developing permaculture, we have the following factors in mind:

- We need to cultivate or construct the resources that we use, not plunder a failing wilderness.
- All remaining genetic resources are to be preserved, as far as possible on their native sites (this includes cultivated plants).
- We need to accept that the hybridisation of culture does occur, but at the same time preserve the values in all cultures that assist human happiness, responsibility, sharing, conservation, and good management. That is, we need to put an ethical or value base to our actions.

Many of us offer aid to areas of need, and there are numerous non–governnment organsiations (NGO's) at work in the area of aid; some are supported by churches, others by civil institutions, and many are groups of people organised locally for self–help. Aid is a necessary but delicate affair; some forms of aid can produce dependency, facilitate further inequities in a society, destroy or impair cultural values, decrease the yields of the environment, upset delicately balanced nutritional habits, or actually destroy sustainable local ecologies or agricultural systems.

Perhaps we can approach the matter of successful aid by defining what such a success would entail; by setting criteria for judging "success". Thus, successful aid should:

- Address real and basic problems of the region (nutrition, drought relief, land and resource inequality) and to recognise that political and financial action may be necessary.
- Devise strategies to offset the effects of such problems, and to educate a group of local people to apply trials of those strategies.
- Assess trials for side effects and sustainability.
- Leave a local group able to further extend or educate others in such strategies.
- Provide modest support services and monitoring; above all, to record and circulate case histories to other non–government organisations.

Problem areas are no place for fools, amateurs, or people who will not listen to others or assess results. For example, many alkaline desert soils lack available zinc; whole grains and seed legumes may exacerbate metabolic zinc loss. Thus, traditional diets need to be examined and supported if they provide sources of zinc from meats, bone, ashes, or animal testicles. A new stove or cooker may prevent the incorporation of ashes in the diet, or a new diet may create a severe deficiency. There is no substitute for thorough analysis of soils and foods, the use of trace elements or soil additives, respect for traditional methods of food preparation, and an excellent education to accompany the project. Some of the factors that greatly assist effective aid are therefore:

- Excellent research, and excellent teaching locally; local teachers to be encouraged for the long term.
- Courtesy and respect for traditional diets and methods, cultures and languages.
- Honest, modest, and practical (achievable, affordable) advice as to *trials* of new systems.
- Feedback assessed, and flexibility of approach maintained (no one solution).
- Congruence in lifestyle and advice of advisers; advisers or educators should adopt their own advice! It is also important to teach models based on successful trials, not theoretical models.
- Effort to reach all sections of society.
- A positive, cheerful, enthusiastic approach to projects, inspiring by example.

The core of successful aid lies in modest trials, careful extension, and provision for widespread education, so that after aid has ceased (or ended a phase) local people can continue the education process, maintain any system (financial, technological, or agricultural), and can call for additional modest resources if necessary.

Many problems are very long–term, and short–term aid (typical of emergency programmes) is not able to address these; drought has an 18–20 year periodicity, and needs to be coped with by food storages on good years, emergency food and forage from tree crop, pre–drought reduction of herds, widespread rainfall harvesting systems, and a well–informed public assisted by appropriate policy such as equable adjustment of livestock herd size, and government aid to establish drought refuges *locally* for essential livestock and for people.

All these strategies need careful long–term planning, and firm policy implementation; these need to be in place over several decades before fine–tuning is possible. As political rule can change so rapidly, and is often repealed by opposing rulers, planning for the very long–term is possible only as a resident regional involvement. "Advisors" are short-term, and if they do not leave a corps of well-informed people, are of ephemeral effect; even such a basic technology as a water tap needs a trades-person capable of descaling vents and reseating valves, or replacing washers over the long term.

In catastrophes, only residents are effective over the short term; it is they who need, and can effectively use, relief housing and supplies. Outside aid is far less effective except in the matter of supply of *requested* resources; in areas of India where drought was offset by storage of hardy crops such as ragi (a sorghum), the introduction of exotic wheat varieties has meant that ragi is often unavailable for storage. Eucalypt monoculture for rayon fibre (textiles) has obliterated many ragi fields, and in total this may add up to a deferred catastrophe. Aid-financed deep wells and pumps in the same region have enabled large livestock herds and more annual cropping at the cost of a rapidly falling water table. So "improvements" in short-term finances (to large landowners or industrialists) add up to a greatly impoverished population and environment; in short, desertification due to "improvements".

Aid as Joint Enterprises

What is a joint enterprise? It is a mutual agreement, written and legal, that two groups, one third world (TW) and one western world (WW), work out for a mutual ethical enterprise. Accounting is:

- *Financial*: Most of the cash is from the WW group to establish a small manufacturing industry (seed production, craft, publishing, modest technological). The TW group supplies mainly skills and labour equivalents based on *local* average incomes. Jointly, costs are accounted up to product sale level, then net profits are *equably split* (profit minus costs and agreed-on cash reserves for future materials and expansion).
- *Ethical*: The product is life-enhancing and benefits people and the global ecology.
- *Social*: The product does not impoverish a local resource, or benefit an already rich group. The social effects are consistently assessed and accounted.

Note: All forms of accounting are assessed annually, and the results circulated to all investors or co-owners.

It is probable that the WW group sets up sales, ads, investment in the first place, and acts for the enterprise in *their* country, thus generating capital. The TW group sells locally and supplies mainly labour and skills, but also teaches skills to the WW people. Both groups set aside 10-15% of nett profit as research and development funds, or fund socially needed health and education. Trade is always reciprocal.

The *long-term aim* is to:

- Satisfy needs of both groups in a specific area.
- Set up reciprocal beneficial trade and travel.
- Teach others how to do this.
- See that both ends of the arrangement have their essential needs supplied.
- Reduce the need for trade goods, and increase the information flow.

The *main aim* is to make friends with each other; to draw closer together socially. This is the primary written rule "To become friends for mutual enrichment".

For example, a dryland group in the WW sends a convener to a host group in the dryland TW, and assesses local needs, both ways. A mutual decision is reached on an enterprise, e.g. seed growing. *Both* grow and exchange seed, set up a single seed catalogue and packaging system, agree to split profits, make arrangements for reciprocal travel, and devise ways to be closer friends.

A FINAL LIMIT TO DEVELOPMENT.

Few economic systems, including those outlined in this section, give thought to some ultimate end. Even if we do achieve the goals of global community self-management, we are as much in danger of destroying the world by producing goods endlessly in Mondragon cooperatives, communist, or capitalist factories, or as individuals. There are certain rules for earth care which lie beyond the economic realm. I believe we should always tend towards minimising the spread of people and their works on the face of the land.

When we replace agriculture with gardens, then we should close down, as a priority, the most distant or most damaging agricultures. We can retain as land stewards the very few broadscale graziers and managers who now use vast tracts of land or who crop huge monocultural acreages. Better still, we can make foresters of our farmers. Some of them are already on this path.

If we close down farms and wasteproduct factories, we need to greatly enlarge true wilderness, for it is the ultimate grace to give room on earth to all living things, and the ultimate in modesty to regard ourselves as stewards, not gods.

14.16

REFERENCES AND RESOURCES

Kennedy, Margrit, 1987 *Toward an Ecological Economy: Money, Land, and Tax Reforms*. Ginsterweg 45, D3074, Steyerberg, West Germany.

Max Neef, Manfred, *From the Outside Looking In: experiences in barefoot economics*, Dag Hammarskjold Foundation, Sweden, 1982.

Mollner, Terry 1982, *Mondragon Cooperatives and Trusteeship* (the design of a non-formal education process to establish a community development program based on Mahatma Ghandhi's theory of trusteeship and the Mondragon cooperatives), Doctoral thesis, University of Massachusetts USA. Copies available for $20 + postage from The Trusteeship Institute, Inc, Baker Road, Shutesbury MA 01072 USA. (Highly recommended reading for community groups and organisers.)

Morehouse, Ward 1983, *Handbook of Tools for Community Economic Change*, Intermediate Technology Group of North America Inc, PO Box 337, CrotononHudson, New York 10520 USA. (Available from the publisher. A basic explanation of land trusts, self–management, community banking, self–financing social investment, and S.H.A.R.E . programs. Highly recommended.)

Peters, T., and R. Waterman, *In Search of Excellence*, Harper & Row, 1983. (Has principles of good enterprise based on existing companies.)

Sale, Kirkpatrick, "Bioregionalism: a new way to treat land", *The Ecologist 14* (4), pp 167–173, 1984. (partly on cultural, but mainly on landscape, factors: that is, an *ecoregion*.)

Tukel, George 1982,*Toward a Bioregional Model*, Planet Drum Foundation 1982, PO Box 31251, San Francisco, California, USA. (This short treatment sees "bioregion" as watershed; while valid, many tribal and urban bioregions may not fit this model.)

Turnbull, Shann, *New money sources and profit motives for democratising the wealth of nations*, The Company Directors Association of Australia Ltd., 27 Macquarie Place, Sydney 2000, Australia. (A basic reference for a reformist economy. Shann Turnbull is involved with planning new systems. *And* he is an old caving companion of mine.)

Turnbull, Shann, *OPTIONS: Selecting A Local Currency*, The Australian Adam Smith Club, June 1983.

For some contacts in the US (or elsewhere) I have listed below public service organisation with good advice. They also need *your* input as new ideas and services, or new investment opportunities arise in your area.

To start your own money handling, write to and get a publications list from: The E.F. Schumacher Society, Box 76A, RD 3, Great Barrington, Massachusetts 01230. This group accumulates and publishes on successful community financial strategies for no–capital enterprises (pre–selling), local SHARE programmes, and how to print your own currency, as well as other strategies. If you have worked one of these strategies, notify them, and give a clear account of your system. This will reach alternative people via their conferences and publications.

C.E.L.T. (Cooperative Enterprise Loan Trust): people's banking and seminars advisory services; includes S.C.O.R.E. Service Corps of Retired Executives. P.O. Box 6855, Auckland, New Zealand.

Directory of Socially Responsible Investments, 1984 et sequ. Was $5 from The Funding Exchange, Room A, 135E 15th St. New York, NY 10003.

L.E.T.S. (Local Employment Trading System): organised credit/debit non–currency systems. Kits, games, software, information from: Micahel Linton. Landsman Community Services Ltd., 375 Johnston Ave., Courtenay, B.C. CANADA, V9N 2Y2, or the Maleny and District Community Credit Union, 28 Maple St., Maleny QLD 4552, Australia.

S.H.A.R.E (Self–Help Association for a Regional Economy) PO Box 125, Gt. Barrington, MA 01230, USA).

To report on dirty business locally or regionally, and to find out who has dirty work afoot elsewhere, contact:
1. The Interfaith Center for Corporate Responsibility (I.C.C.R.), a coalition of churches, issuing a newsletter *The Corporate Examiner*, which reports on local topics and their follow–ups. They also offer a phone advice service from their New York office (Phone: 2128702295). The newsletter is $35 per year, 11 issues, from ICCR, Room 556, 475 Riverside Drive, New York, N.Y. 10115.

2. Council on Economic Priorities (non–government research). Newsletter $25 per year, 10 issues, 84 Fifth Avenue, New York, N.Y. 10011.

Good news should be remitted to:
Good Money ($36 per year, 6 issues), from the Center for Economic Revitilization, Inc., Box 363, Worcester, Vermont 95682. This is probably the best source for investors who want to make their money work well.

APPENDIX

LIST OF PLANTS MENTIONED IN TEXT BY COMMON NAME

Acacia, *Acacia spp.*
African boxthorn, *Lycium ferrocissimum + spp.*
African marigold, *Tagetes erecta, T. minuta*
Agapanthus, *Agapanthus africanus*
Agave, *Agave spp.*
Aggie's pants, *Agapanthus africanus*
Albizia, *Albizia spp.*
Albizia, coast, *Albizia lophantha*
Alder, *Alnus spp.*
Alfalfa, *Medicago sativa*
Allspice, *Pimentia dioica*
Aloe, *Aloe spp.*
Alyssum, *Alyssum spp.*
Amaranth, *Amaranthus spp.*
Apple, *Malus pumila*
Apricot, *Armeniaca vulgaris*
Arrowhead, *Sagittaria spp.*
Arrowroot (Queensland), *Canna edulis*
Arrowroot (West Indian), *Maranta arundinaceae*
Arum lily, *Arum; Zantedeschia spp.*
Aspen, *Populus tremuloides +spp.*
Asparagus, *Asparagus officinalis*
Aster, *Aster spp.*
Aubergine, *Solanum melongena*
Autumn olive, *Elaeagnus umbellata*
Avocado, *Persea americana*
Azolla, *Azolla spp., A. filicoides*
Balsa, *Ochroma pyramidae*
Bamboo, *Bambusa, Phyllostachys, Arundinaria, Dendrocalamus*, and allied genera.
Banana, *Musa paradisiaca + spp.*
Banana passionfruit, *Passiflora mollissima*
Banksia, *Banksia spp.*
Baobab, *Adansonia digitata*
Barley, *Hordeum vulgare*
Basil, *Ocimum basilicum*
Basswood, *Tilia spp.*
Beach convulvulus, *Ipomoea pes–caprae*
Beach pea, *Lathyrus littoralis*
Bean, broad, *Vicia faba*
  common, *Phaseolus vulgaris*
  Dolichos, *Lab–lab purpureus*
  fava, *Vicia faba*
  four–winged, *Psophocarpus tetragonolobus*
  Goa, *Psophocarpus tetragonolobus*
  lima, *Phaseolus lunatus*
  mung, *Vigna radiata*
  phasey, *Macroptilum lathyroides*
  soya, *Glycine max*
  tepary, *Phaseolus acutifolius*
  velvet, *Lab–lab purpureus*
  yam, *Pachyrrhizos tuberosus*
Beet, *Beta vulgaris*
Bell pepper, *Capsicum annuum*
Birch, *Betula spp.*
Black nightshade, *Solanum nigrum*
Black walnut, *Juglans nigra*
Blackboy, *Xanthorrhea australis + spp.*
Blackwood, *Acacia melanoxylon*
Blue bush, *Mairana spp.*
Blue gum, *Eucalyptus globulus*
Blueberry, *Vaccinium spp.*
Boobialla (Dune), *Acacia sophorae*
Bower vine, *Tetragonia implexicona*
Bracken fern, *Pteridium aquilinum*
Brambles, *Rubus fruticosus + spp.*
Breadfruit, *Artocarpus altilis*
Breadroot, *Psoralea esculenta*
Broccoli, *Brassica oleracea*
Bromeliad, *Fam. Bromeliaceae*
Buckthorn plantain, *Plantago cornopus*
Buckwheat, *Fagopyrum esculentum*
Buddleia, *Buddleia spp.*
Busy Lizzie, *Impatiens mallerana*
Cabbage, *Brassica spp.*
Calapo, *Calopogonium mucunoides*
Calliandra, *Calliandra spp.*
Camass lily, *Camassia quamash*
Canary Island date, *Phoenix canariensis*
Candle nut, *Aleurites spp.*
Canistel, *Lucuma rivicoa*
Cape gooseberry, *Physalis peruviana*
Cape weed, *Arctotheca calendula*
Capsicum, *Capsicum annuum*
Carob, *Ceratonia siliqua*
Carrot, *Daucus carota*
Cashew, *Anacardium occidentale*
Cassava, *Manihot esculenta*
Cassia, *Cassia spp.*
Casuarina, *Casuarina spp.*
Casuarina, coastal, *Casuarina equisetifolia*
Cat tails, *Typha spp.*
Catchfly, *Lychnis, Silene*
Catmint, *Nepetea cataria*
Catnip, *Nepetea cataria*
Cauliflower, *Brassica oloeracea*
Cedar, *Cedrus spp.*
Celery, *Apium graveolens*
Centro, *Centrosema pubescens*
Chayote, *Sechium edule*
Cherry, *Prunus cerasus, P. avium*
Chestnut, *Castanea spp.*
Chicory, *Cichorium intybus*
Chilli pepper, *Solanum frutescens*
Chinese gooseberry, *Actinidia chinensis*
Chinese water chestnut, *Eleocharis dulcis*
Chinquapin, *Castanea pumila*
Chives, *Allium schoenoprasum*
Choko, *Sechium edule*
Cilantro, *Coriandrum sativum*
Cinnamon, *Cinnamomum zeylandicum*
Citrus, *Citrus spp.*
Cleavers, *Galium aparine*
Clover, *Trifolium spp.*
Cocksfoot, *Dactylis glomerata*
Cocoa, *Theobroma cacao*
Coconut, *Cocos nucifera*
Coffee, *Coffea spp.* incl: *C. robusta, C. arabica*
Comfrey, *Symphytum spp., S. officinale*
Common reed, *Phragmites communis*
Convulvulus, *Convulvulus spp.*
Copperburrs, *Bassia spp.*
Cordgrass, *Spartina spp.*
Coriander, *Coriandrum sativum*
Corn, *Zea mays*
Cotton, *Gossypium spp.*
Cowpea, *Vigna sinensis*
Cranberry, *Vaccinium macrocarpon + spp.*
Crocus, *Crocus sativus*
Crotolaria, *Crotolaria spp.*
Cucumber, *Cucumis sativus*
Cucurbit Fam., *Cucurbitaceae*
Cumbungi, *Typha spp., T. angustifolia*
Currants, Black, *Ribes nigrum*
  Gold, *Ribes aureum*
  Red, *Ribes rubrum*
Custard apple, *Annona spp.*
Cypress, *Cupressus spp.*
Daffodil, *Narcissus spp.*
Daikon radish, *Raphanus sativus*
Dun pea, *Pisum sativum arvense*
Dandelion, *Tarascum officinale*
Derris, *Derris spp., D. elliptica*
Desmodium, *Desmodium spp.*
Dill, *Anethum graveolens*
Dock, *Rumex spp.*
Douglas fir, *Pseudotsuga spp.*
Drumstick tree, *Moringa oleifera*
Duck potato, *Sagittaria spp.*
Durian, *Durio zibethinus*
Eggplant, *Solanum melongena*
Elderberry, *Sambucus spp.*
Elephant garlic, *Allium schoenoprasum var.*
Endive, *Cichorium endivia*
Erythrina, *Erythrina spp.*
Fennel, *Foeniculum vulgare*
Fenugreek, *Trigonella foenum–graecum*
Field daisy, *Bellis perennis, Aster spp.*
Fig, *Ficus carica + spp.*
Filbert (nut), *Corylus avellana + spp.*
Firs, *Pseudotsuga* and *Abies spp.*
Fuchsia, *Fuchsia spp.*
Ginger, *Zingiber officinale*
Gingseng, *Aralia quinquefolia*
Gladioli, *Gladiolus spp.*
Globe artichoke, *Cynara scolymus*
Glycine, *Neonotonia wightii*
Glyricidia, *Gliricidia sepium*
Gooseberry, *Ribes uva–crispa + spp.*
Goosefoot, *Chenopodium spp.*
Gorse (Furze, Whin), *Ulex europaeus*
Gourds, *Lagenaria spp.*
Granadilla, *Passiflora quadrangularis*
Grape, *Vitis vinifera*
Grape hyacinth, *Muscari neglectum*
Grass, banna, *Pennisetum purpureum*
  barley, *Hordeum maritimum*
  beach, *Ammophila spp., A. arenaria*
  bent, *Agrostis spp.*
  blady, *Imperata cylindrica*
  Buffalo, *Buchloe dactyloides*
  Cocksfoot, *Dactylis glomerata*
  elephant, *Pennisetum spp.*
  guinea, *Panicum maximum*
  kangaroo, *Themeda australis*
  kikuyu, *Pennisetum clandestinum*
  lemon, *Cymbopogon citratus*
  makarikari, *Panicum coloratum*
  napier, *Pennisetum purpureum*
  pampas, *Cortaderia sellowiana*
  panic, *Panicum spp.*
  para, *Brachyaria mutica*
  perennial rye grass, *Lolium perenne*
  purple pigeon, *Setaria porphyrantha*
  rhodes, *Chloris gayana*
  rice, *Oryzoides hymenopsis*
  sea barley, *Hordeum maritimum*
  signal, *Brachyaria spp.*
  sudan, *Sorghum halapense*
  tall wheat, *Agropyrum elongatum*
  vetiver, *Vetivaria zizanoides*
  Wimmera rye, *Lolium rigidum*
Grasstree, *Xanthorrhea australis + spp.*

Greenleaf desmodium, *Desmodium intortum*
Guava, *Pisidium guavaja*
Haifa white clover, *Trifolium spp.*
Haole koa, *Leucaena leucocephala*
Hau (Hawaiian hibiscus), Hibiscus *rosa-sinensis*
Hawthorn, *Crataegus oxycanthus + spp.*
Hazelnut, *Corylus avellana + spp.*
Heather, *Calluna, Erica, + spp.*
Hickory, *Carya ovata*
Honey locust, *Gleditsia triacanthos*
Honeysuckle, *Lonicera spp.*
Horseradish, *Armoracia rusticana*
Horseradish tree, *Moringa oleifera*
Huckleberry, *Gaylussacia, Vaccinium*
Huskberry, *Physalis peruviana*
Hyacinth, *Muscari neglectum*
Ice plant, *Carpobrotus + allied genera*
Impatiens, *Impatiens mallerana*
Indian water chestnut, *Trapa spp., T. natans*
Inga, *Inga spp., I. edulis*
Iris, *Iris spp.*
Jacaranda, *Jacaranda spp.*
Jak fruit, *Artocarpus spp.*
Jerusalem artichoke, *Helianthus tuberosus*
Jicama, *Exogonium brachteatum*
Jujube, *Ziziphus jujuba*
Juniper, *Juniperus spp.*
Kale, *Brassica oleracea*
Kang kong, *Ipomoea aquatica*
Kenya white clover, *Trifolium semipilosum*
Kiawe (Hawaii), *Prosopis pallida*
Kiwi fruit, *Actinidia chinensis*
Kiwi, hardy, *Actinidia arguta*
Kniphofia, *Kniphofia spp.*
Knot weed, *Polygonum spp.*
Lab–lab, *Lab–lab purpureus*
Lantana, *Lantana camara*
Laurel, *Laurus nobilis*
Lavendar, *Lavendula spp.*
Leatherwood, *Eucryphia billardierii*
Leeks, *Allium Ampeloprasum*
Legumes, Fams: *Fagaceae, , Papilionaceae*
Lentils, *Lens culinaris*
Leucaena, *Leucaena leucocephala*
Levant garlic, *Allium schoenoprasum var.*
Lettuce, *Latuca sativa*
Lignum vitae, *Guaiacum sanctum*
Lime (basswood), *Tilia spp.*
Lime (West Indian), *Citrus aurantiifolia*
Lippia, *Phyla (Lippia) nodiflora*
Loquat, *Eriobotrya japonica*
Lotus, *Nelumbo nucifera*
Lucerne, *Medicago sativa*
Lupin, *Lupinus alba + spp.*
Lychee, *Litchi chinensis*
Macadamia nut, *Macadamia spp.* esp: *M. integrifolia*
Macro, *Macrotyloma spp.*
Mahogany, *Swietenia mahogani*
Maize, *Zea mays*
Mango, *Mangifera indica*
Mangroves, *Several genera of trees;, Rhizopora spp.*
Mariposa lily, *Calochortus nuttallii + spp.*
Matsudana willow, *Salix matsudana*
Medlar, *Mespilus germanica*
Melia, *Melia azedarach*
Melons, usually *Cucumis melo*
Mesquite, *Prosopis spp.*
Mile–a–minute vine, *Mikania spp.*
Milk vetch, *Astralagus*
Millet, *Various genera.*, includes *Pennisetum, Panicum, Setaria, Paspalum*
Mirror plant, *Coprosma repens*
Monkeypod, *Samanea saman*
Mulberry, *Morus spp.*
Muscari, *Muscari neglectum*
Mustard, *Brassica nigra, B. hirta*
Nasturtium, *Nasturtium var.*
Neem tree, *Azedarachta indica*
Nettle, *Urtica dioica*
New Zealand hemp, *Phormium tenax*
Nutgrass, *Cyperus rotundus, Eleocharis , sphacelata + spp.*
Oak, *Quercus spp.*
Oak, Holm, *Quercus ilex*
Oak, silky, *Grevillea robusta*
Oats, *Avena sativa*
Oca, *Oxalis crenata*
Okra, *Abelmoschus esculentus*
Olives, *Olea europea*
Oncoba, *Oncoba spinosa*
Onion weed, *Allium triqetrum*
Onions, *Allium spp. especially A. cepa*
Onions, multiplier , *Allium aggregatum group*
Oxalis, *Oxalis spp.*
Paeony, *Paeonia spp.*
Palm, *Mauritia spp.*
betelnut, *Areca catechu*
borassus, *Borassus flabellifer + spp.*
Butia, *Butia capitata*
date, *Phoenix dactylifera*
doum, *Hyphaene thebaicus*
jelly, *Butia capitata, B. yatay*
nypa, *Nypa fruticans*
oil, *Elaeis guineaensis*
rattan, *Dendrocalamus, Calamus*, and related genera
salak, *Salacca spp.*
sugar, *Arenga pinnata*
Pangola, *Digitaria decumbens*
Papaya, *Carica spp., C. papaya*
Parsley, *Petroselinum crispum*
Parsnip, *Pastinaca sativum*
Passion fruit, *Passiflora spp.*
Peach, *Amygdalus persicae*
Peanut, *Arachis hypogaea*
Pear , *Pyrus communis + spp.*
Peas, *Pisum spp., P. sativum*
Pennisetum, *Pennisetum spp.*
Pepino, *Solanum muricatum*
Pepper, black (vine), *Piper nigrum*
chilli, *Solanum frutescens*
sweet, *Solanum annuum*
Perennial rye grass, *Lolium perenne*
Persimmon, *Diospyrus kaki*
Philodendron, *Philodendron selluum*
Photinia (red–tipped), *Photinia spp.*
Pigeon pea, *Cajanus cajan*
Pigface, *Mesembryanthemum spp.*
Pine, Araucaria, *Araucaria spp.*
Australian, *Callitris spp.*
Canary Island, *Pinus canariensis*
Norfolk Island , *Araucaria heterophylla*
Oregon, *Pseudotsuga spp.*
Stone, *Pinus edulis*
Pineapple, *Ananus comosus*
Pineapple guava, *Feijoa sellowiana*
Pistachio, *Pistachia vera + spp.*
Plantain, *Plantago spp.*
Plantain (Cooking banana), *Musa spp.*
Plum, *Prunus domestica + spp.*
Pomegranate, *Punica granatum*
Pongamia, *Pongamia pinnata*
Poplar, *Populus spp.*
Potato, *Solanum tuberosum + spp.*
Pride of India, *Melia azedarach*
Pride of Madeira, *Echium fastuosum*
Prosopis, *Prosopis spp.*
Puccinela, *Puccinela spp.*
Puero, *Pueraria phaseoloides*
Pultenea, *Pultenea spp.*
Pumpkin, *Cucurbita maxima*
Pussy willow, *Salix caprea*
Pyrethrum daisy, *Pyrethrum spp., , P. cinerariifolium*
Quillaja, *Quillaja saponaria*
Quince, *Cydonia oblonga*
Quinoa, *Chenopodium quinoa*
Raintree (Canary Is.), *Ocotea foetens*
Raintree (Indian), *Samanea saman*
Rape, *Brassica napus*
Raspberry, *Rubus idaeus + spp.*
Raspberry jam acacia, *Acacia acuminata*
Rattle pod, *Crotolaria spp.*
Red ink vine, *Rhagodia baccata*
Redwood, *Sequoia sempivirens*
Red–hot poker, *Kniphofia spp.*
Reed grass, *Phragmites spp.*
Reedmace, *Typha spp., T. angustifolia*
Rhubarb, *Rheum rhaponticum*
Rice, *Oryza sativa*
River red gum, *Eucalyptus camaldulensis*
Roly–poly, *Bassia Quinquenerva*
Rosella, *Hibiscus sabdariffa*
Rose (hedgerow), *Rosa multiflora*
Rosemary, *Rosmarinus officinalis*
Rosewood, *Tipuana tipu, Pterocarpus spp.*
Rosewood (Burmese), *Pterocarpus indicus, P. erinaceus*
Round rushes, *Juncus effusos + spp.*
Rubber hedge (Africa), *Euphorbia tirucalli*
Rubber tree, *Hevea braziliensis*
Rue, *Ruta graveolens*
Russian olive, *Elaeagnus angustifolia*
Rye, *Secale cereale*
Safflower, *Carthamus tinctorius*
Sage, *Salvia officinalis*
Salt bushes, *Atriplex spp.*
Salvia, *Salvia spp.*
Samphire, *Crithmum, Salicornia + spp.*
Sand–leek, *Allium scorodoprasum var.*
Sapote, *Diospyros. Casimiroa, , Calocarpum, Lucuma spp.*
Sea grape, *Coccolobus uvifera*
Sea grasses, *Posidonia, Zostera spp.*
Sedge, *Scirpus spp., S. validus, Cyperus spp.*
Service berry, *Amelanchier canadensis*
Sesbania, *Sesbania spp.*
Shallot, *Allium aggregatum* group
Shiitake (fungus), *Lentinus edodes*
Shungiku, *Chrysanthemum spp.*
Siberian pea shrub, *Caragana aborescens*
Silk sorghum, *Sorghum almum*
Silky oak, *Grevillea robusta*
Silver beet, *Beta vulgaris (a spinach)*
Silverberry, *Actinidia arguta*
Silverleaf Desmodium, *Desmodium uncinatum*
Siratro, *Macroptilium atropurpureum*
Sisal, *Aloe sisalana*
Sissoo, *Dalbergia sissoo*
Skunk cabbage, *Symplocarpus foetidus*
Snow pea, *Pisum sativum + spp.*
Sodom apple, *Solanum spp.*
Sorrel, *Rumex acetosa*

Spinach, *Spinacia oleracea*
St. Johns wort, *Hypericum*
Stinking Roger, *Tagetes erecta, T. minuta*
Strawberry, *Fragaria vesca + spp.*
Stylo, Townsville, *Stylosanthes guianensis*
Sugar beet, *Beta vulgaris*
Sugar cane, *Saccharum officinarum*
Sunflower, *Helianthus annuus*
Sunroot, *Helianthus tuberosus*
Sweet potato, *Ipomoea batatus*
Swiss chard, *Beta vulgaris* (a spinach)
Tagasaste, *Chaemocytisus palmensis*, (was *Cytisus proliferus)*
Tamarack, *Larix americana*
Tamarind, *Tamarindus indicus*
Tamarisk, *Tamarix apetala and spp.*
Tapioca, *Manihot esculenta*
Tares, *Vicia spp.*
Taro, *Colocasia esculenta*
Taupata, *Coprosma repens*
Tea, *Camellia sinensis*
Teak, *Tectona grandis*
Tephrosia, *Tephrosia candida*
Thyme, *Thymus spp., T. vulgaris*
Til, *Ocotea foetens*
Tipuana tipu, *Tipuana tipu*
Tobacco, *Nicotiana tabacum + spp.*
Tobacco bush, *Nicotiana spp.*
Tomato, *Lycopersicon lycopersicum*
Tropical acacia groups, *Acacia auriculiformis,*
A. mangium + spp.
Tumbleweed, *Amaranthus spp., Sisymbrium spp.*
Tumeric, *Curcuma domestica*
Tung and oil nuts, *Aleurites moluccensis*
Turnip, *Brassica rapa*
Vanilla, *Vanilla planifolia*
Verbena, *Verbena spp.*
Vetch, *Vicia spp.*
Violet, *Viola odorata*
Wallflowers, *Cheiranthus cheiri*
Walnut, *Juglans regia*
Wandering jew, *Tradescantia albiflora*
Water chestnuts, *Eleocharis, Trapa spp.*
Water hyacinth, *Eichornia crassipes*
Water lettuce, *Pista spp.*
Water lily, *Nymphaea spp.*
Water mint, *Mentha aquatica*
Water plantain, *Alisma plantago–aquatica*
Watercress, *Rorippa amphibia + spp.*
Watermelon, *Citrullus vulgaris*
Wattle, green, *Acacia mearnsii*
silver, *Acacia dealbata*
Waxberry, *Myrica cordifolia + spp.*
Wheat, *Triticum spp., T. aestivum*
White acacia, *Acacia albida*
White cedar, *Melia azedarach*
White clover, *Trifolium repens*
Wild rice, *Zizania lacustris*
Willow, *Salix spp.*
Wood fungi, *Pleurotus spp.*
Wormwood, *Artemesia absynthium*
Yam, *Dioscorus spp.*
Yatay, *Butia capitata, B. yatay*
Yucca, *Yucca spp.*

LIST OF PLANTS MENTIONED IN TEXT BY SPECIES NAME

Abelmoschus esculentus, Okra
Abies spp., Firs
Acacia spp., Acacias
A. acuminata, Raspberry jam acacia
A. albida, White acacia
A. auriculiformis,
A. mangium + spp., Tropical acacia groups
A. dealbata, Silver wattle
A. mearnsii, Green wattle,
A. melanoxylon, Blackwood
A. sophorae, Boobialla (dunes)
Actinidia arguta, Silverberry, hardy Kiwi
A. chinensis, Kiwi fruit, Chinese gooseberry
Adansonia digitata, Baobab
Agapanthus africanus, Agapanthus, Aggie's pants
Agave spp., Agave
Agropyrum elongatum, Tall wheat grass
Agrostis spp., Bent grass
Albizia spp., Albizia
A. lophantha, Coast Albizia
Aleurites moluccensis, Tung and oil nuts
spp., Candle nut
Alisma plantago–aquatica, Water plantain
Allium aggregatum group, Multiplier onions
A. Ampeloprasum, Leeks
A. ascolonicum,
A. aggregatum group, Shallot
A. schoenoprasum, Chives
A. schoenoprasum var., Elephant,garlic levant garlic
A. scorodoprasum var., Sand–leek
A. spp. especially *A. cepa*, Onions
A. triquetrum, Onion weed
Aloe spp., Aloe
A. Sisalana, Sisal
Alyssum spp., Alyssum
Amaranthus spp., Amaranth, tumbleweed
Amelanchier canadensis, Service berry
Ammophila spp.,
A. arenaria, Beach grass
Amygdalus persicae, Peach
Anacardium occidentale, Cashew
Ananus comosus, Pineapple
Anethum graveclens, Dill
Annona spp., Custard apple
Apium graveolens, Celery
Arachis hypogaea, Peanut
Aralia quinquefolia, Gingseng
Araucaria spp., Araucaria pine
A. heterophylla, Norfolk Island pine
Arctotheca calendula, Cape weed
Areca catechu, Betelnut palm
Arenga pinnata, Sugar palm
Armeniaca vulgaris, Apricot
Armoracia rusticana, Horseradish
Artemesia absynthium, Wormwood
Artocarpus spp., Jak fruit
A. altilis, Breadfruit
Arundinaria spp., Bamboo
Asparagus officinalis, Asparagus
Aster spp., Aster, daisies
Astralagus spp., Milk vetch
Atriplex spp., Salt bushes
Avena sativa, Oats
Azedarachta indica, Neem tree
Azolla spp., A. filicoides, Azolla
Bambusa spp., Bamboo
Banksia spp., Banksia
Bassia spp., Copperburrs
B. Quinquenerva, Roly–poly
Bellis perennis, Field daisy
Beta vulgaris, Beets, sugar beet, silver beet, swiss chard
Betula spp., Birch
Borassus flabellifer + spp., Borassus palm
Brachyaria spp., Signal grass
B. mutica, Para grass
Brassica napus, Rape
B. nigra, B. hirta, Mustard
B. oleracea, Broccoli, cauliflower, kale,
B. rapa, Turnip
B. spp., Cabbage
Fam. *Bromeliaceae*, Bromeliad
Buchloe dactyloides, Buffalo grass
Buddleia spp., Buddleia
Butia capitata, Butia palm, jelly palm
B. yatay, Yatay
Cajanus cajan, Pigeon pea
Calamus spp., Rattan palm
Calocarpum spp., Sapote
Calliandra spp., Calliandra
Callitris spp., Australian pine
Calluna spp., Heather
Calochortus nuttallii + spp., Mariposa lily
Calopogonium mucunoides, Calapo
Camassia quamash, Camass lily
Camellia sinensis, Tea
Canna edulis, Arrowroot (Queensland)
Capsicum annuum, Bell pepper, capsicum
Caragana aborescens, Siberian pea shrub
Carica spp., C. papaya, Papaya
Carpobrotus + allied genera, Ice plant
Carthamus tinctorius, Safflower
Carya ovata, Hickory
Casimiroa spp., Sapote
Cassia spp., Cassia
Castanea spp., Chestnut
C. pumila, Chinquapin
Casuarina spp., Casuarina
C. equisetifolia, Coastal casuarina
Cedrus spp., Cedar
Centrosema pubescens, Centro
Ceratonia siliqua, Carob
Chaemocytisus palmensis, Tagasaste (was *Cytisus proliferus*)
Cheiranthus cheiri, Wallflowers
Chenopodium spp., Goosefoot, fat hen
C. quinoa, Quinoa
Chloris gayana, Rhodes grass
Chrysanthemum spp., Shungiku
Cichorium endivia, Endive
C. intybus, Chicory
Cinnamomum zeylandicum, Cinnamon
Citrullus vulgaris, Watermelon
Citrus spp., Citrus
C. aurantiifolia, Lime (West Indian)
Coccolobus uvifera, Sea grape
Cocos nucifera, Coconut
Coffea spp., Coffee
Colocasia esculenta, Taro
Convolvulus spp., Convolvulus
Coprosma repens, Taupata, mirror plant
Coriandrum sativum, Coriander, cilantro
Cortaderia sellowiana, Pampas grass
Corylus avellana + spp., Hazelnut, filbert
Crataegus oxycanthus + spp., Hawthorn
Crithmum spp., Samphire

Quillaja saponaria, Quillaja
Raphanus sativus, Daikon radish
Rhagodia baccata, Red ink vine
Rheum rhaponticum, Rhubarb
Rhizopora spp., Mangroves
Ribes uva–crispi, Gooseberry
  R. aureum, Gold currant
  R. nigrum, Black currant
  R. rubrum, Red currant
Rorippa amphibia + spp., Watercress
Rosa multiflora, Rose (hedgerow)
Rosmarinus officinalis, Rosemary
Rubus fruticosus + spp., Brambles
  R. idaeus + spp., Raspberry
Rumex spp., Dock
  R.acetosa, Sorrel
Ruta graveolens, Rue
Saccharum officinarum, Sugar cane
Sagittaria spp., Arrowhead, duck potato
Salacca spp., Salak palm
Salicornia + spp., Samphire
Salix spp., Willow
  S. caprea, Pussy willow
  S.matsudana, Matsudana willow
Salvia spp., Salvia
  S. officinalis, Sage
Samanea saman, Monkeypod, raintree (Indian)
Sambucus spp., Elderberry
Scirpus spp., S. validus, Sedge
Secale cereale, Rye
Sechium edule, Chayote, choko
Sesbania spp., Sesbania
Sequoia sempivirens, Redwood
Setaria spp., Millets
  S. porphyrantha, Purple pigeon grass
Sisymbrium spp., Tumbleweed,
Solanum spp., Sodom apple
Solanum annuum, Sweet pepper
  S. frutescens, Chilli pepper
  S. melongena, Eggplant, aubergine
  S. muricatum, Pepino
  S. nigrum, Black nightshade
  S. tuberosum + spp., Potato
Sorghum almum, Silk sorghum
  S. halapense, Sudan grass
Spartina spp., Cordgrass
Spinacia oleracea, Spinach
Stylosanthes guianensis, Stylo, Townsville
Swietenia mahogani, Mahogany
Symphytum spp.,
  S. officinale, Comfrey
Symplocarpus foetidus, Skunk cabbage
Tagetes erecta, T. minuta, African marigold, stinking roger
Tamarindus indicus, Tamarind
Tamarix apetala + spp., Tamarisk
Tarascum officinale, Dandelion
Tectona grandis, Teak
Tephrosia candida, Tephrosia
Tetragonia implexicona, Bower vine
Themeda australis, Kangaroo grass
Theobroma cacao, Cocoa
Thymus spp., T. vulgaris, Thyme
Tilia spp., Basswood, lime
Tipuana tipu, Tipuana tipu, rosewood
Tradescantia albiflora, Wandering jew
Trapa spp., T. natans, Indian water chestnut
Trifolium spp., Clover
  T. semipilosum, Kenya white clover
  T. repens, White clover
  T. spp., Haifa white clover
Trigonella foenum–graecum, Fenugreek
Triticum spp., T. aestivum, Wheat
Typha spp., T. angustifolia, Reedmace, cat tail, cumbungi
Ulex europaeus, Gorse, furze, whin
Urtica dioica, Nettle
Vaccinium spp., Blueberry, huckleberry
  V. macrocarpon + spp., Cranberry
Vanilla planifolia, Vanilla
Verbena spp., Verbena
Vetiveria zizanoides, Vetiver grass
Vicia faba, Fava bean, broad bean
Vicia spp., Vetch, tares
Vigna radiata, Mung bean
  V. sinensis, Cowpea
Viola odorata, Violet
Vitis vinifera, Grape
Xanthorrhea australis + spp., Blackboy, grasstree
Yucca spp., Yucca
Zantedeschia spp., Arum lily
Zea mays, Corn, maize
Zingiber officinale, Ginger
Zizania lacustris, Wild rice
Ziziphus jujuba, Jujube
Zostera spp., Seagrass

GLOSSARY

a 'a lava: A type of lava characterised by its jagged, angular and sharply pointed blocks. Its black, clinker–like appearance results from the explosively escaping bursts of gas through the rapidly congealing surface as the molten lava beneath tugs at the hardening crust, causing it to break into irregular masses (Hawaiian term).

Albedo: The ratio of light reflected to that received.

Allelopathy: The study of plant or plant-animal interactions.

Annidation: From *nidus* (a nest). Nesting within; a strategy of stacking a like form (as bowls within bowls) or unlike forms (vine in a tree) one within the other.

Anti-feedants: Usually bitter or unpalatable plants or chemical substances used to discourage browsers (wormwood or quassia extract for insects, or fat and powdered egg for hares).

Bioregion: A watershed, neighbourhood, or area of like natural phenomena and climate, linked by water, culture, ridges and valleys, or local recognition—"the perceived neighbourhood".

Complexing: Here used to denote the continual addition of elements or connections to a system (a process).

Consociations: A mix of different species allied for mutual benefit (e.g. a mixed flocks of birds).

Coppice: Resprouted limbs on trees which can be harvested; many tree species coppice from budwood in stumps or lignotubers.

Copse: A small wooded area.

Crenellated: Wavy-edged, or with regular gaps in the profile.

Edge: The junction/zone of mixing or effect that lies between two media or landscape forms, as between forest and grass, hill and plain, fresh and salt water. Edges are normally richer in species than the contributing media themselves.

Entropy: Unavailable energy; state of disorder or chaos in a system.

Forb: A broadleaf herb, e.g., nettle.

Fractal: A term coined by B. Mandelbrot to describe natural or contrived phenomena of indefinable length (e.g. the length of a coastline) or constant repetition of a "generator" or basic form. The fractal index describes the density of such forms at a given scale.

Guild: A species assembly of plants and animals of benefit to each other or to a selected crop species. Guilds need to be placed in sensible patterns for management and to effect the benefits of interaction.

Harmonics: as in "edge harmonics". Used to describe sinuous edges or surfaces, or a more or less regular variation along an edge or surface.

Hedgerow: An edge or row of trees bordering or partitioning a field.

Interstitial: Space between cells or soil crumbs "in the gaps between".

Latticing: Faced or criss-crossed with lines or matrix forms.

Lobulate: Possessing lobes, as a coral reef edge or wave-washed strand line.

Matrix: The media surrounding a component; a tabulation of data as in a 4 x 4 column table.

Microclimate: The localised climate around landscape features, buildings, and in forests or edges of forests; important for selecting sites for specific crops or species.

Mid-sky: In valleys, the central place of the sun's traverse.

Mosaic of species or cultivation: a composite of different elements, each of one or more species, in the landscape (e.g. a pattern of diverse crops and farms); tesselations of different forms or colours or components.

No tillage; low tillage: No tillage means no cultivation of the soil, usually as a combination of tree crop, mulch,

green manure, and seeding out in mulch. Weeds are controlled by slashing, mulching, browsing, flooding. **Low tillage** is allied, as shallow and wide-spaced seed furrows drawn through mulch, or crops grown on mounds in a field of green manure.

Nurse plants (also nurse crop): Pioneer species used to provide green manure, nutrient, or shade for suceeding or main crop.

Ogives: Curvilinear (often semi-linear) lines across the axis of flow or growth; the pulse lines in a glacier or the outline of a tree crown in suceeding years.

Predator species: Insects or vertebrates eating plants or pest species, e.g. ladybird larvae controlling aphids, lions culling game.

Recruitment, period of: the period when new generations of plants in forests or shrublands receive favourable conditions for growth due to many variable mosaics like fire and rain. Can be as rarely as every 7–20 years.

Retrofit: Re–designing and re–building a house or landscape system to reduce energy costs or to achieve sustainability.

Schedule: The time/place cycle or time-use of an area by the same or different species or age group of a species; the usual way space or resources are shared by mammals.

Solar chimney: A black, thin metal chimney which acts as a heat engine to exhaust air from a room or enclosed place, thus drawing in fresh or cool air.

Tessellation: From *tesserae* (tiles). Fitting forms together to make a surface or enclosed space; a mosaic of like or unlike forms.

Thermal belt: Sun-facing mid-slope site defined by early blossom and leaf emergence, thus a good home site for climate control (above frost, below rime).

Thermodynamics: The study of the relationship between heat and mechanical energy, or of the fate of energy in systems.

Thermosiphon effect: A heat-activated closed loop containing a liquid or a gas, so that the fluid within the loop rises (on the heated side) and falls (on the cooler side). Used to convey heat from a low source to a higher storage area.

Wildlife corridors: Belts of trees, marsh, or river forest (usually 300-600 feet wide) connecting two or more larger habitat areas; also **migration corridors** and **flyways**.

Wind-flagging: The degree to which a tree or its branches is adpressed away from a prevailing wind, as a flag is blown out from its pole in wind.

RESOURCES

More information about permaculture and permaculture training programmes can be obtained from the following addresses (correct at the time of printing). The publications mentioned list national and regional centres.

International Permaculture Journal, 113 Enmore Rd, Enmore, NSW 2042, Australia. The journal is a must for all permaculture enthusiasts; it contains informative articles and sample design work, resource listings, a directory of permaculture associations, book reviews, news of design course trainings, and much more. Back issues are highly recommended.

North American Permaculture, PO Box 573, Colville, WA 99114, USA. Provides networking and facilitation for North American and international permaculture organizations.

Permaculture Communications, P.O. Box 101, Davis, CA 95617. Publishes *The Permaculture Designer's Directory* coordinates permaculture mailing lists and gives data for North America.

REFERENCES

1. Mollison, Bill, and David Holmgren, *Permaculture One: a perennial agriculture for human settlements*, 1978, Tagari Publications, Tyalgum, NSW 2484, Australia.

2. Mollison, Bill, *Permaculture Two: practical design for town and country in permanent agriculture*, 1979, Tagari Publications, Tyalgum, NSW 2484, Australia.

3. Fukuoka, Masanobu, *The One-Straw Revolution*, 1978, Rodale Press, Emmaus, PA.

4. Fukuoka, Masanobu, *The Natural Way of Farming*, 1985, Japan Publications, Inc., Tokyo & New York.

5. Yeomans, P. A., *Water for Every Farm/Using the Keyline Plan*, 1981, Second Back Row Press, PO Box 43, Leura, NSW, Australia.

6. King, F. H., *Farmers of Forty Centuries: permanent agriculture in China, Korea, and Japan*, 1911, Rodale Press, Emmaus, PA.

7. Howard, Sir Albert, *An Agricultural Testament*, 1943, Oxford University Press.

8. Smith, J. Russell, *Tree Crops: a permanent agriculture*, 1977, Devine-Adair, Old Greenwich.

9. Kern, Ken and Barbara, *The Owner-Built Homestead*, 1977, Charles Scribner's Sons, N.Y.

10. Turner, Newman, *Fertility Pastures and Cover Crops*, 1974, Bargyla and Gylver Rateaver, Pauma Valley, CA.

11. Douglas, J. Sholto, and Robert de Hart, *Forest Farming*, 1976, Watkins, London.

12. Geiger, R. *The Climate Near the Ground*, 1950, Harvard University Press, N.Y.

13. Watt, Kenneth, *Principles of Environmental Science*, 1973, McGraw-Hill.

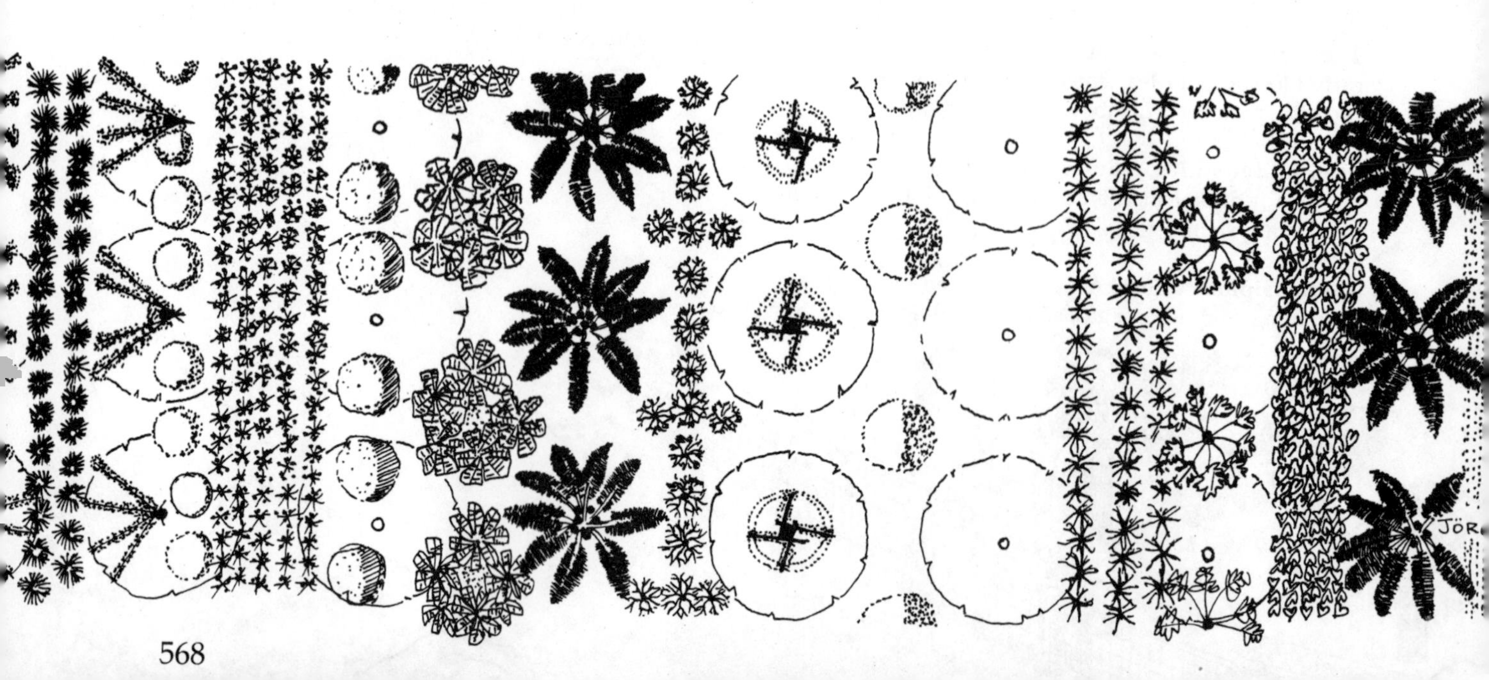

INTERNATIONAL PERMACULTURE ORGANIZATIONS

AFRICA

B.B.I.N. Project
P.O. Box 162
Maun, Botswana
Offering: Community gardens, horticulture projects.

Botswana Permaculture Institute (B.I.P.)
Private Bag 47
Serowe, Botswana
Phone (Serowe): 430550/430930
Contact: Garry Sawdon, Russell Clark

Lesotho
P.O. Box 682
Masero, Lesotho
Contact: Ron Savage, Leopa Rafutho
Offering: Horticulture and community forestry.

Permaculture Club
No. 11865 Zengeza 4, P.O. Zengeza
Harare, Zimbabwe
Contact: Robert Guzha

Permaculture Design Committee
P.O. Box 121
Serowe, Botswana
Offering: Permaculture design and education.

Permaculture Seed Exchange
P.O. Box 328
Lobatse, Botswana
Contact: Mr. M. I. Rampou

Zimbabwe Institute of Permaculture (Z.I.P.) and Permaculture Association of Zimbabwe
P.O. Box 8515
Causeway
Harare, Zimbabwe
Contact: Bridget O'Conner, John Wilson
Offering: Newsletter, 150-acre demonstration site, rainwater harvest, small livestock, market garden, co-ops, education.

AUSTRALIA

Bamboo Network
P.O. Box 500
Maleny 4552, Queensland
Phone: (071) 94 3177
Contact: Hans Erken

Crystal Waters Permaculture Village
MS 16 Ahern's Road
Connondale via Maleny 4552, Queensland

Epicentre Sydney
113 Enmore Road
Enmore, New South Wales 2042
Phone: (02) 51 2175
Offering: Publishes *International Permaculture Journal*, information, book service, education.

Jurnkurakurr Aboriginal Resource Centre
P.O. Box 909
Tennant Creek 0860, Northern Territory
Phone: (089) 623076 or 623020
Offering: Nursery and land management center, tree and garden programs for outstations, Woofers.

Mid North Coast Bioregional Association
494 Wingham Road
Taree, New South Wales 2430
Phone: (065) 510533

Permaculture Association of South Australia (P.A.S.A.)
Hindmarsh City Farm
17 Green Street
Brompton 5007, South Australia
Offering: Newsletter, $10/year.

Permaculture Association of Western Australia
P.O. Box 430
Subiaco 6008, Western Australia
Offering: Memberships, $15/year includes quarterly newsletter.

Permaculture Canberra
P.O. Box 12
Bungendore 2621, Australia Capital Territory
Contact: David Watson
Offering: Permaculture education, millpost nursery (SAE for catalogue).

Permaculture Chiltern
RMB 1130
Chiltern 3683, Victoria
Phone: (057) 261596
Contact: V. R. Gravestein

Permaculture Consultancy
56 Isabella Avenue
Nambour Heights 4560, Queensland
Offering: Design, consulting, publishing, information, education. Subtropical fruits compendium and *The Best of Permaculture* wholesale/mail order.

Permaculture Design
16 Fourteenth Street
Hepburn 3461, Victoria
Phone: (053) 48 3636
Contact: David Holgrem

Permaculture Gympie
MSF 626 Middle Creek Road
Pomona 4568, Queensland
Phone: (071) 85 1511
Contact: Julie Cortenbach

Permaculture Institute
P.O. Box 1
Tyalgum, New South Wales 2484
Phone: (066) 793442
Offering: Information, publishing, education, books.

Permaculture Institute of Western Australia (P.I.W.A.)
The Environment Centre
P.O. Box 7375, Cloister Square
Perth 6001, Western Australia
Phone: (09) 3215942
Contact: Miles Durand
Offering: Publishes *The Permaculturalist*, $6/year.

Permaculture Melbourne
24 Lucerne Crescent
Alphington 3078, Victoria
Phone: (03) 49 4232
Contact: Christine Pinniger, Ian Batchelor 560 4174 (E mail peg: batchelor)
Offering: Workshops—Bridget Seidlich 387 7196, PC site register for Victoria.

Permaculture Nambour
P.O. Box 650
Nambour 4560, Queensland
Offering: Quarterly *Permaculture Edge* magazine, videos, books.

Permaculture Narangba
Callaghan Road
Narangba, Queensland
Phone: (07) 888 1223

Permaculture North Eastern Victoria
Spring Valley Road
Flowerdale 3717, Victoria
Contact: Ric Stubbings
Phone: (057) 80 1301 or Kerry Butler (057) 62 2210 or Janet Mckenzie (057) 78 9581

Permaculture Northern Rivers
Room 11 104A Molesworth Street
Lismore, New South Wales 2480
Phone: (066) 220020

Permaculture Resource Group of the Soil Association of SA
GPO
Box 2497
Adelaide 5001, South Australia
Offering: Newsletter, *The Living Soil*, meetings, field days.

Permaculture Service Ltd.
Stoddards Road
Tyalgum, New South Wales 2484
Phone: (066) 793242

Permaculture Service Ltd.
59 Isabella Avenue
Nambour 4560, Queensland
Phone: (71) 412749

Permaculture Southern Highlands
18 Gilbraltar Road
Mittagong 2575, New South Wales
Contact: Ralph Long

Peninsula Permaculture and Organic Growers Group
P.O. Box 39
Dromana 3936, Victoria
Phone: (059) 2292
Contact: Graham Cooke
Offering: Soil testing and consulting.

Tangentyere Council Land Management Department
P.O. Box 8070
Alice Springs 5750, Northern Territory
Phone: (089) 525855
Offering: Nursery, permaculture development for urban and rural settlements.

CANADA

Maritime Permaculture Institute
c/o Fraser Common Farm
1374 256th Street, Aldergrove, B.C. VOX 1A0

DENMARK

Eco-Village
Produktionshoejskolen i Floeng
Hedelykken 14, Hedehusene 2640
Contact: Jesper Saxgren

FC—Danish Centre for Renewable Energy
SDR. TDBY.
DK-7760 Hurup Thy
Phone: 45 7 95 65 65

Urban Ecology
Baggesensgade 54
Copenhagen N. 2200
Contact: Floyd Stein or Toni Anderson Sonder Bouley 53 Copenhagen V. 1720

FEDERAL REPUBLIC OF GERMANY

Permaculture Institute Europe
Ginsterweg 4–5, Steyerberg 3074
Phone: (05764) 2158
Contact: Margit and Declan Kennedy

Earthcare: Institute fur angewandte Okologie und Permaculture
Podbielskistr 32, 300 Hanover
Phone: (0511) 660592
Contact: Wulf Blume

Permaculture Regensburg
c/o Willi Schmidt
Phone: (0941) 51639

FRANCE

Fruit and Nut Farm Pyrennes
11190 Couize, Sougraigne
Phone: 68-69-84-52
Contact: A&J Darlington
Offering: Ten-Hectare Permaculture, no-till cereals, forage systems. Accommodations by prior arrangement.

Permaculture Pyrennes
Las Encantades, 11300 Bouriege
Phone: 68-31-51-11
Offering: Educational programs and projects.

GREAT BRITAIN

Permaculture Association and English Permaculture Institute
8 Hunters Moon, Dartington
Totnes, South Devon
Phone: (0803) 865115 or 867546
Contact: Andy Langford
Offering: Newsletter, *Permaculture News*, book service, courses.

Scottish Permaculture
The Lee Stables, Coldstream
Berwickshire, TD12 4NN
Phone: (0890) 2709
Offering: Courses

GUATEMALA

Alter Tech
A.P. 2
Momostenango, Totonicapa'n
Demonstration farm of indigenous, sustainable, and permaculture agriculture and appropriate technology.

INDIA

C.E.D. Center for Education and Documentation
3 Sulaman Chambers
4 Battery Street, Bombay 400 039
Phone: 202 0019
Contact: John D'Souza E mail Geo2: indialink

Permaculture Association of India
117 East Marredpalli
Secunderabad 500 26, Andhra Pradesh
Contact: Dr Venkat

Permaculture Consultancy
Kirehully Estate
Saklaspur P.O., Hassan District
Karnataka 573 134, South India
Contact: J. P. Pereira
Offering: Consulting/teaching, researching/writing, *Permaculture Design and Natural Pest Control*.

S.E.W.A. Self Employed Women's Association
S.E.W.A. Reception Centre
Ahmedabad 380 001
Contact: Manju S. Raju, Coordinator
Offering: Urban/community forestry, women's banking co-op.

JAPAN

Permaculture Japan
c/o Arthur Getz
5-12-6 Kyodo
Setagaya-Ku, Tokyo, T156
Offering: Educational project to disseminate permaculture principles and to integrate with traditional Japanese agricultural approaches.

MEXICO

Northern Mexico Permaculture Institute
Rio Conchos
150 Col. Fuentes del Centenario
Hermosillo, Sonora
An institute for education and research. Offering: Design courses, workshops, and lectures in Spanish.

Mariposa Foundation—Fundacion Mariposa, A.C.
Venta Grande, Pueblo
or
A.P. 21140
Mexico, D.F. 04000
Offering: Design courses in Spanish and networking in Mexico.

NEPAL

Institute for Sustainable Agriculture Nepal (INSAN)
Baneshwore-10
GPO Box 3033, Kathmandu
Phone: 220448
Offering: Educational programs, publications, demonstrations, consulting.

NEW ZEALAND

Coromandel Peninsula, Thames Valley Permaculture Institute
32 Waiotahi Creek Road, Thames
Phone: (0843) 89 955
Offering: Establishing a rural permaculture radio station (AM), develops bee forage systems.

Koanga Permaculture Consultancy
RD 2
Kaiwaka, Northland
Contact: Bob Corker

Permaculture Canterbury
15 Macauley Street, Christchurch
Contact: Ofelia Chambers
Offering: Education and projects.

Permaculture Hokianga
Taheke
Hokianga, Northland
Contact: Miriam and Jim Tyler

Permaculture Institute of New Zealand
P.O. Box 37030, Parnell 29
Offering: *International Permaculture Journal* subscription plus local newsletter insert.

Permaculture Nelson
RD 1 Todd Valley, Nelson Bay
Contact: Richard and Bella Walker
Offering: Hill reforestation, permaculture orchard gardens. Conveners of local courses.

Permaculture Poverty Bay
Tairawhiti Polytechnic, Gisborne
Contact: Ron Taiapa

Permaculture Waikato
P.O. Box 57, Raglan
Contact: Halley Paine

Permaculture Wanganui
Wanganui Regional Community College
Private Bag, Wanganui
Contact: Dave Aslabie

Permaculture Whangarei
14 Charles Street
Kamo, Whangarei
Contact: Jamie Hancox

Tasman Permaculture Association
Waiora Mara Farms
RD 1 Graham Valley, Motueka
Contact: Earl Convoy

Tasman Permaculture Seed Exchange
PO Box 211
Motueka Point, Motueka 88718
Contact: Dieter Proebst

Tui Landtrust Permaculture
Wainui Inlet RD 1, Takaka Bay
Contact: Robyn McCurdy
Offering: Compilation of educational material aids, curriculum outline for schools.

NORWAY

Eco-Valley
Gronta's Torggt 8
Byrne 4340
Phone: (04) 481919

PACIFIC ISLANDS

Permaculture Western Samoa
Box 4228, Apia
Western Samoa 21417
Contact: Fay and Vai Ala'ilima

PORTUGAL

Agrobio
Rua D. Dinis 2 R/C
1200 Lisboa
Contact: Angelo Rocha

SOUTH AMERICA

Centro de la Permaculture Chilena
Correo 3
Casilla, 9091
Vina del Mar, Chile
Contact: Julio Perez Diaz

Permaculture Brazil
Av. Bage 1110/02 cep 90000
Porto Alegre, Rio Grande de Sul, Brazil
Phone: (0512) 332-167 or 417-949

SPAIN

Permacultura Iberica
Apdo. 391, Castellon 12080

UNITED STATES

Alter Project
Permaculture Institute of the Allegheny Valley
Slippery Rock University
Slippery Rock, PA 16057
Offering: Design courses, workshops, consultancy, and demonstration site. Masters of science degree in sustainable systems with permaculture design as a core course. Publishes *The Alternator*.

Central Rocky Mountain Permaculture
P.O. Box 631
Basalt, CO 81621
Offering: Courses, workshops, and internship programs centered around integrated market garden and greenhouse operation.

Drylands Village Development
535 Cordova Road, Suite 285
Sante Fe, NM 87501
A land-based community development group for the establishment of sustainable systems of energy, food production, economics, and governance.

Earth Advocates
Nobody's Mountain
Route 3, Box 223
Livingston, TN 38570-9547
Offering: Design courses, workshops, and internships. Demonstration site researching lesser-known food crops, bamboo, etc.

Elfin Permaculture
P.O. Box 16683
Wichita, KS 67216
Offering: Design courses, workshops, and lectures. Publishes *The International Permaculture Species Yearbook, Robin*, and other educational literature.

Epicenter Hawaii
P.O. Box 1612
Kealakekua, HI 96745
Offering: Design courses, workshops, lectures, and networking within the Hawaiian bioregion. Publishes the *Earthbank Guide to Sustainable Economics*.

Flowering Tree Permaculture
P.O. Box 4302
Fairview, NM 87533
Offering: Consulting, design courses, and internships. Research and education on high desert permaculture species and techniques.

Friends of the Trees
P.O. Box 185
Port Townsend, WA 98368
An international networking center for permaculture, reforestation, and the Travellers Earth Repair Network. Publishes *International Green Front Report*.

Gap Mountain Permaculture
9 Old County Road
Jaffrey, NH 03452
Center for research, demonstration, and education for the Northeast focusing on cold weather design, composting toilets, and constructed wetlands systems.

Gray Wolf Environmental Education Center
1812 South W. White Road
San Antonio, TX 78220
An experimental, self-sustaining farm/ranch/community using permaculture for the establishment and preservation of native plant and animal species, and conservation of water and energy preservation of Native American skills.

Great Northwest Permaculture Institute
2073 Marble Valley Basin Road
Addy, WA 99101
Institute organized to assist in global efforts of earth repair and environmentally responsible community and agricultural development.

Heartland Permaculture
8801 Scarlet Circle
Austin, TX 78737
Dedicated to community organizing and putting permaculture in cities by providing education, designers, and speakers' bureau.

Maui Epicenter
P.O. Box 400
Kihei, HI 96753
Offering: Courses, workshops, and lectures.

North American Permaculture
P.O. Box 573
Colville, WA 99114
Providing networking and facilitation for international permaculture organizations.

Panhandle Permaculture Association
Box 148
Bushland, TX 79012
Offering: Design courses, workshops, and practical demonstration of sustainable agriculture.

Permaculture Activist
P.O. Box 101
Davis, CA 95617
A quarterly journal for the permaculture movement of North and Central America.

Permaculture Communications
P.O. Box 101
Davis, CA 95617
Mail-order sales of permaculture-related books, publications, and mailing lists. Offering: Consulting in permaculture and key line design.

Permaculture Drylands Institute
P.O. Box 1812
Sante Fe, NM 87504-1812
Offering: Drylands design courses, workshops, lecturers, and demonstration of drylands strategies. Publishes *Sustainable Living in Drylands*.

Permaculture Institute of Southern California
Sprout Acres
1027 Summit Way
Laguna Beach, Ca 93651
A multifaceted organization responsible for education, research, demonstrations, workshops, and urban/rural design strategies for sustainable living.

Running Rain Society
P.O. Box 74
Datil, NM 87821
Offering: Design courses and educational programs for sustainable lifestyle in drylands.

SEED (Sustaining the Earth through Environmental Design)
861 East 25th Street
Houston, TX 77009
Promoting harmonious interface of human community and the environment.

Sonoran Permaculture Services
738 North 5th Avenue, Suite 212
Tucson, AZ 85705
Offering: Public relations, consulting, and other professional services.

Tucson Permaculture Guild
6570 West Illinois
Tucson, AZ 85746
A network of friends in the Tucson/Phoenix area meeting monthly to discuss permaculture projects.

INDEX

The capitalised listings refer to headings in the text.

ISLAND PRESS BOARD OF DIRECTORS